D0858207

TUNNEL ENGINEERING HANDBOOK

TUNNEL ENGINEERING HANDBOOK

Edited by

John O. Bickel

Civil & Structural Engineer Graduate
Swiss Federal Institute of Technology
Former Partner now Principal Associate
Consultant
Parsons, Brinckerhoff, Quade & Douglas
Engineers, Architects, Planners.

T. R. Kuesel

Senior Vice President
Parsons, Brinckerhoff, Quade & Douglas
Engineers, Architects, Planners.
Member National Academy of Engineering

VNR VAN NOSTRAND REINHOLD COMPANY
NEW YORK CINCINNATI TORONTO LONDON MELBOURNE

Van Nostrand Reinhold Company Regional Offices:
New York Cincinnati

Van Nostrand Reinhold Company International Offices:
London Toronto Melbourne

Library of Congress Catalog Card Number: 80-24536
ISBN: 0-442-28127-7

Manufactured in the United States of America

Published by Van Nostrand Reinhold Company Inc.
135 West 50th Street, New York, N.Y. 10020

Published simultaneously in Canada by Van Nostrand Reinhold Ltd.

15 14 13 12 11 10 9 8 7 6 5 4 3 2 1

Library of Congress Cataloging in Publication Data

Main entry under title:

Tunnel engineering handbook.

Includes index.
1. Tunneling. I. Bickel, John O. II. Kuesel
T. R.
TA805.T82 624.1'93 80-24536
ISBN 0-442-28127-7

Preface

There has been hardly a period in recorded history during which tunnels in some form were not used. When people started living in cities in the Eastern Mediterranean countries, one of the most urgent problems was bringing water to those places, often from considerable distances. For centuries, starting with the Assyrians in about 1200 B.C., to the declining years of the Roman Empire, tunnels were constructed as part of often elaborate aqueduct systems. In order to reduce evaporation losses, the ancient Persians built miles of quanaats, tunneling between closely spaced shafts, many of which are still in use today. The modern tunnel age started with the construction of the railroads, in particular with the long Alpine tunnels and in the western part of the U.S. The automobile age, with its rapid acceleration of major highway development, brought a new surge in tunnel construction, including the first subaqueous highway tunnels, with ventilation posing a new problem. Parallel with this was the growth of cities in water-deficient regions, which called for increased activity in the construction of aqueducts with long tunnels, tapping areas of abundant precipitation. The trend to place major traffic arteries through and around cities, underground, and the need for more rapid transit systems in urban areas will involve a steady growth in tunnel construction in the future, accompanied by the search for new and more rapid methods for building them.

This handbook presents to the engineer, constructor and student a comprehensive outline of the state-of-the-art in the design and construction of tunnels in all their ramifications. To do this in one volume of manageable size limits the scope to an exposition of the many problems involved and a guide to their basic solutions. The material herein, thus, ranges over the entire spectrum of engineering–including geology; civil, structural, mechanical and electrical disciplines; plus a measure of ecological and environmental considerations. The direction of research into new methods and tools applied to tunnel construction is discussed, as well.

The authors of the various sections contribute from wide experience in their respective fields. For particular problems, consult the bibliography, which lists sources of detailed information available to the engineer who is making this work his specialty and who may have to take into account many special considerations.

Tunnel engineering, and especially construction, requires a great deal of know-how, acquired (often the hard way) by many years of experience and work in this field.

Contents

TUNNEL ENGINEERING HANDBOOK

1
Tunnel Characteristics

JOHN O. BICKEL

Associated Consultant
Parsons, Brinckerhoff, Quade & Douglas

TUNNEL GLOSSARY

Adit: A short transverse tunnel connecting two parallel tunnels or an entry from the face of the slope to a side-hill tunnel.

Air lock: A compartment in which air pressure can be equalized to the compressed air inside a shield-driven tunnel as well as the outside air to permit passage of men and material.

Bench: Part of a tunnel section with the approximately horizontal upper surface temporarily left unexcavated.

Blow-out: A sudden loss of a large amount of compressed air at the top of a shield.

Breastboards: Timber planks to support the face of tunnel excavation in soft ground.

Bulkhead: A water- and airtight partition, usually of steel, to temporarily close off the working portion of a compressed air shield tunnel or the ends of a floating tube of a trench type tunnel.

Drift: One or more small tunnels, excavated within and ahead of the full cross-section.

Dry packing: Filling a void with stiff mortar, placed in small increments, or gravel packed into the space between rock excavation, and poured in place tunnel lining to permit drainage of seepage water.

Evasé stack: An air exhaust stack increasing in cross-section in the direction of airflow to regain air pressure.

Face: The vertical surface at the head of a tunnel excavation.

Grommet: Rings of compressible material inserted under heads and nuts of bolts connecting sections of tunnel liners to seal the bolt hole.

Heading: Same as **drift**.

Jumbo: A frame with platforms to support men and drills for rock excavation, rolling on steel rails or rubber wheels.

Lagging: Timber planks, steel plates or other materials inserted above tunnel-supporting ribs to hold soil or rock.

Laser: A concentrated, monochromatic light beam to control the direction or tunnel excavation, particularly of tunnel-boring machines.

Liner plate: A steel plate segment, generally preformed to support a tunnel excavation.

Lining: A temporary or permanent concrete structure to secure and finish the tunnel interior.

Mixed face: The portion of a tunnel where both rock and soft ground occur in the same cross-section.

Mucking: Removal of material excavated or blasted from the tunnel face.

Pilot tunnel: A small tunnel excavated over the entire length or over part of a tunnel, to explore ground conditions and assist in final excavation.

Pioneer bore: Same as **pilot tunnel**.

Poling boards: Timber or steel planks driven into the soft ground at the tunnel face over supporting steel or timber sets to hold back soil during excavation.

Rock bolts: Steel bolts inserted and anchored in bore holes around the periphery of a tunnel excavation to hold rock in place.

Scaling: Removing loose pieces of rock from the tunnel surface after blasting.

Sets: Steel ribs or timber framing to support the tunnel excavation temporarily.

Shaft: A vertical excavation to gain access to tunnels or mines from the surface.

Shield: A steel cylinder with open or closed face equal to the tunnel diameter for tunnel excavation in soft ground.

Shotcrete: Concrete mixture sprayed onto the rock surface.

Slick line: Pipe or hose inserted between the rock surface and forms to place concrete lining.

Spiling: Same as **poling boards**.

Wall plate: Footing placed on the rock shelf to temporarily support the arch ribs of tunnel sets.

TUNNEL SERVICE CLASSIFICATIONS

Tunnels can be classified according to their service as follows.

Highway Tunnels. These accommodate all types of vehicles permitted on public highways, except that their use by bicycles and horse-drawn vehicles may be limited or prohibited.

Railroad Tunnels. These tunnels serve standard railroad trains and may need special clearances for electric traction from catenaries.

Rapid Transit Tunnels. These tunnels serve urban and metropolitan rapid transit trains, to meet standards of particular systems.

Aqueducts and Sewers. Used to convey fresh water or sanitary wastes and storm water, the sizes and construction of these tunnels vary according to local conditions.

Underground Caverns. These are tunnels built to house underground hydroelectric power plants, hardened defense facilities and special waste storage. The tunnels vary widely according to service requirements and local conditions.

Shafts. These are vertical or inclined excavations that serve as access to mines or tunnel construction, or for tunnel ventilation. They are built to suit requirements.

Special Tunnels. Tunnels are also used to carry water pipes, electric cables or other utilities.

TUNNEL LOCATIONS

Tunnels may be classified according to location as follows.

Underwater Tunnels. Built by various methods under rivers, harbors or other waterways to serve any one of the purposes listed above, these tunnels are used when clearance requirements or land use prevent construction of bridges.

Mountain Tunnels. Tunnels through mountains are used to carry transportation facilities or water.

Tunnels at Shallow Depth and Under City Streets. These are primarily used for rapid transit or other transportation in urban areas.

Bored Versus Cut-and-Cover Tunnels. Bored by whatever means, these tunnels require a minimum of overburden, depending upon soil conditions. Shallow tunnels are most economical by cut-and-cover construction unless other conditions (such as interference with city street traffic) preclude this method.

TUNNEL MEDIA

Geological conditions greatly affect the cost of tunnel construction, as indicated below.

Rock Tunnels. Rock tunnels are excavated in a firm, cohesive medium which may vary from relatively soft marl and sandstone to the very hard igneous rocks such as granite. Bedding and fissuring of rock layers, and the presence of water, control construction methods and difficulties.

Soft Ground Tunnels. This category includes all tunnels built in soft, plastic or non-cohesive soils where water may or may not be a problem.

Mixed Face Tunnels. These tunnels have part of their cross-sections in rock and part in soft ground, with rock interface often weathered and frequently difficult to construct.

2
Clearances and Alignment

JOHN O. BICKEL

Associated Consultant
Parsons, Brinckerhoff, Quade & Douglas

CLEARANCES FOR HIGHWAY TUNNELS

Standards for lane and shoulder width and vertical clearance for highways have been established by the Federal Highway Administration and by the American Association of State Highway Officials (AASHO) according to classification. Figure 2-1 shows AASHO tunnel clearances for a two-lane primary highway. For an additional lane, width shall be increased by at least 10 feet, preferably by 12 feet. For truck highways and interstate systems, the vertical clearance shall be at least 16 feet over the entire width of roadway plus an allowance for resurfacing. Figure 2-2 shows clearances for a three-lane subaqueous tunnel on interstate highway 64 across Hampton Roads. Figure 2-3 shows clearances for a three-lane rock tunnel on an interstate highway. For curved alignments with superelevations, clearances may be increased to provide for overhang.

ALIGNMENT AND GRADES FOR HIGHWAY TUNNELS

Alignment. Alignment should be straight, if possible. If curves are required, the minimum radius is determined by stopping sight distances and acceptable superelevation in relation to design speed.

Sight Distances for Stopping

Design speed (mph)	30	40	50	60	70
Sight distance (ft)	200	275	350	475	600

Passing distances do not apply, since passing in tunnels is not permitted.

Tunnels on Interstate Highways. Design speed cannot be less than 60 mph unless otherwise restricted in urban areas; the minimum radius of curvature preferably should not be less than 1500 feet. Other tunnels should be designed for the speeds governing on the approach highways according to state or local regulations.

Curvature and superelevation should be correlated in accordance with the AASHO "Policy on Geometric Design of Rural Highways." Superelevations should not exceed 0.10 to 0.12 foot/foot.

Grades. Upgrades in tunnels carrying heavy traffic are generally limited to 3.5% to reduce ventilation requirements. For long two-lane tunnels with two-way traffic, a maximum grade of 3% is desirable to maintain reasonable truck

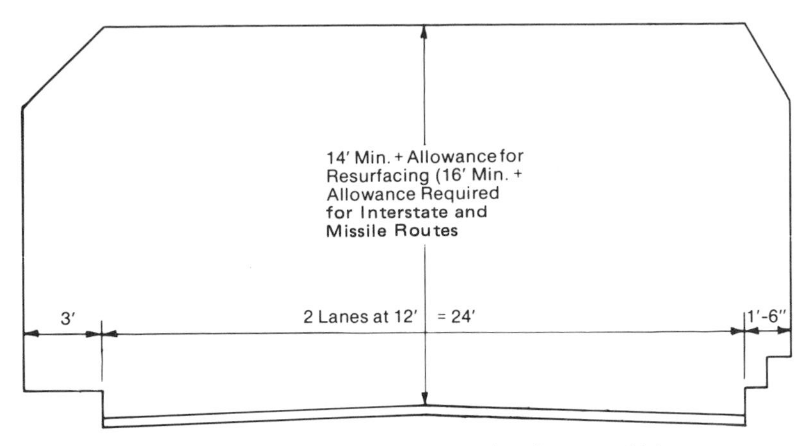

Fig. 2-1. Clearance diagram for two-lane interstate highways.

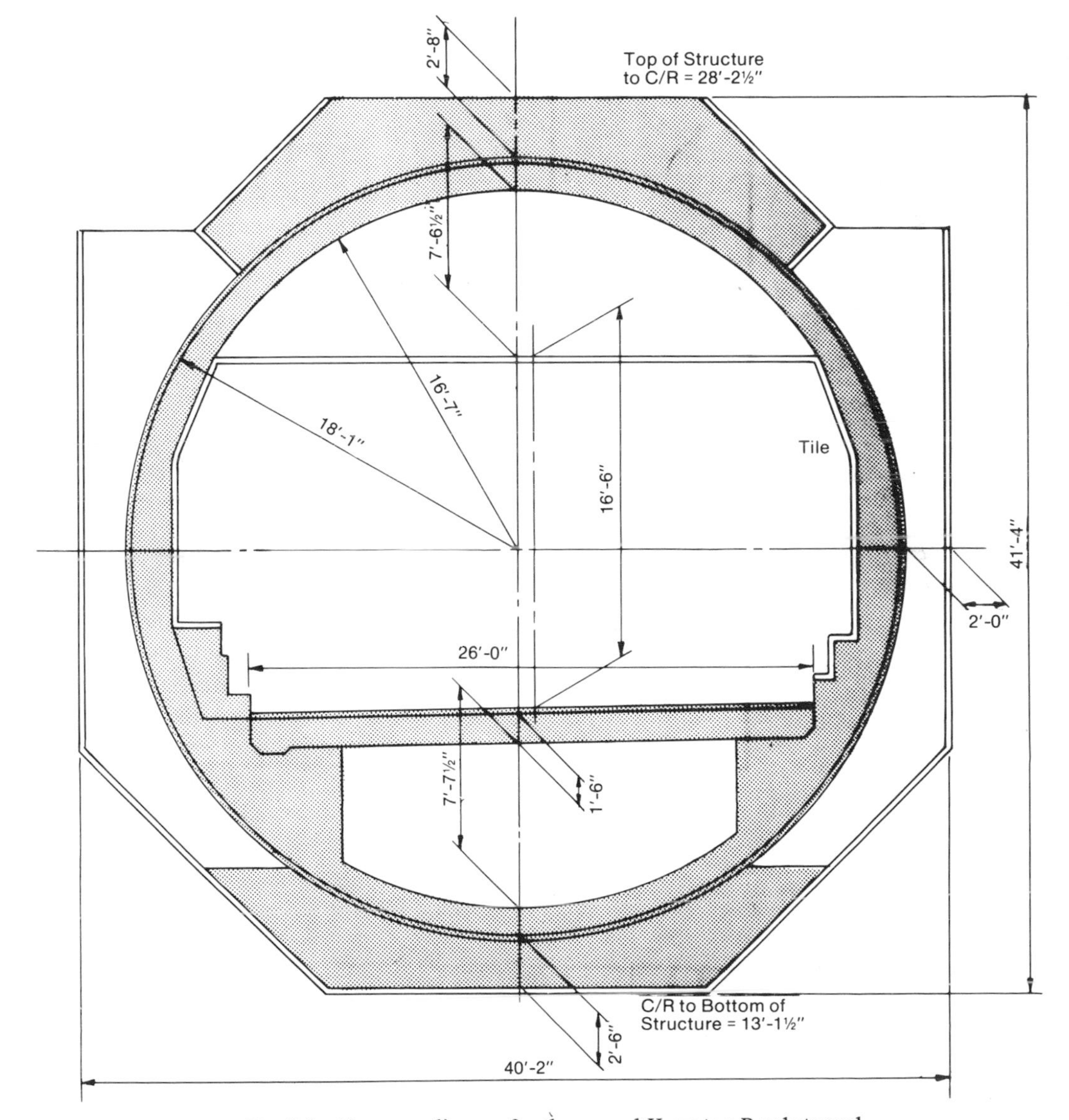

Fig. 2-2. Clearance diagram for the second Hampton Roads tunnel.

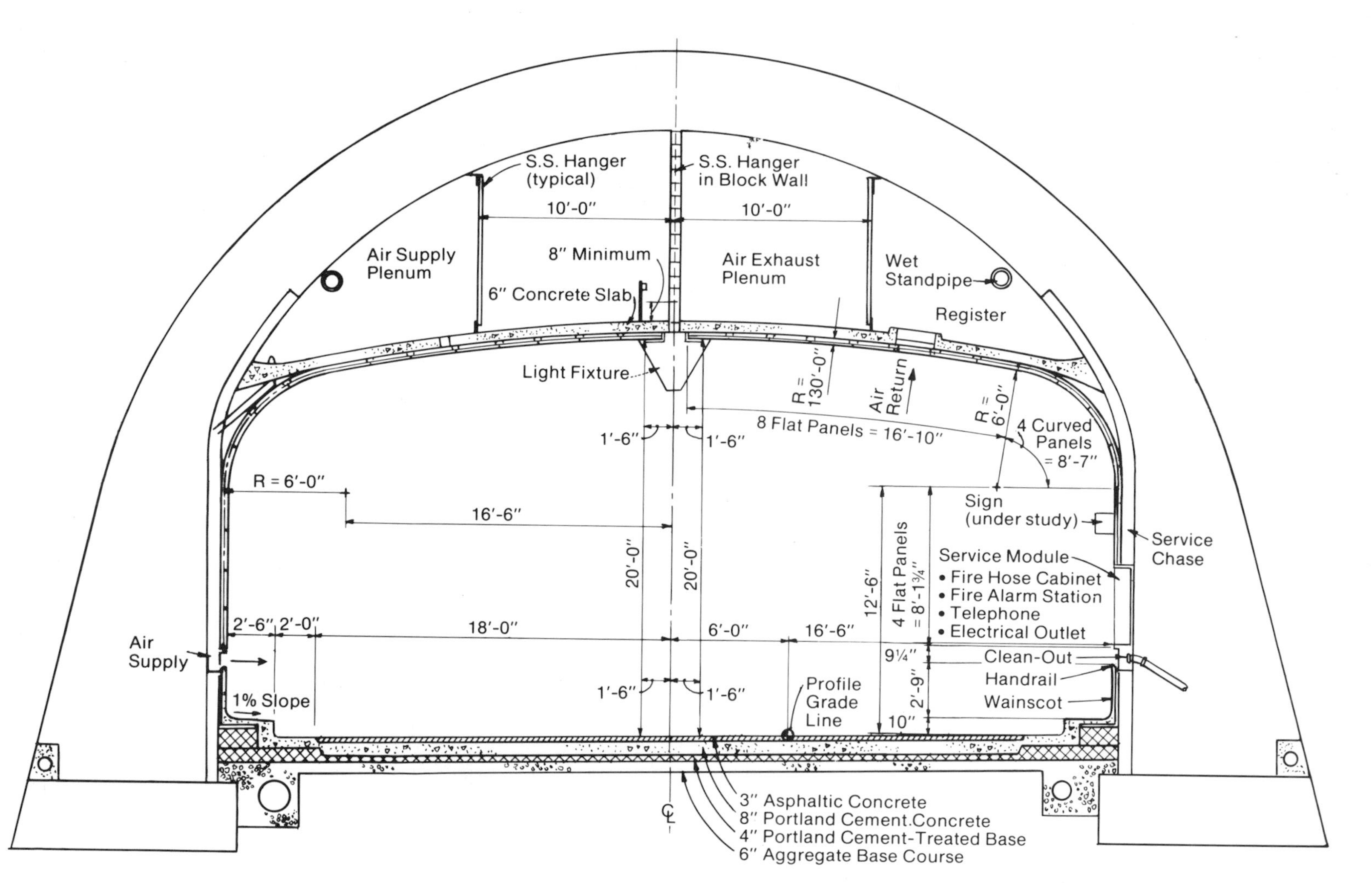

Fig. 2-3. Clearance diagram for a three-lane rock tunnel on an interstate highway.

speed. For downgrade traffic, 4% or more is desirable. For lighter traffic volumes, grades up to 5% or even 6% have been used in subaqueous tunnels for economy's sake. In these tunnels, grades between channel lines controlling navigation depth are at a minimum adequate for drainage to a low point, preferably not less than 1%. Lengths of vertical curves at grade changes are governed by design speed and stopping sight distance.

CLEARANCES FOR RAILROAD TUNNELS

Individual railroads have clearance standards to suit their equipment. Minimum clearances recommended by the American Railway Engineering Association (AREA) for bridges on tangent track are shown in Fig. 2-4. These same standards apply to tunnels.

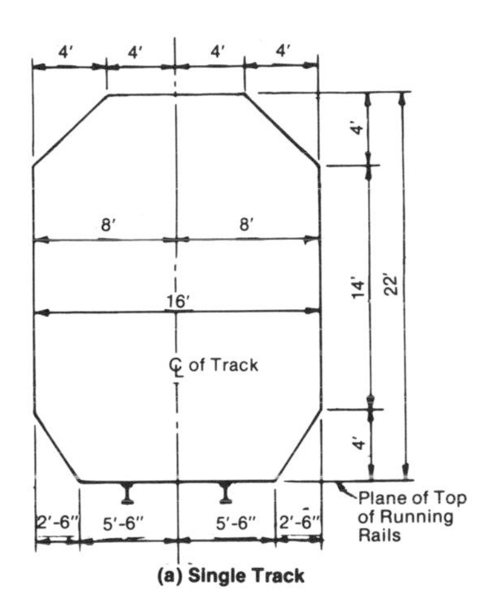

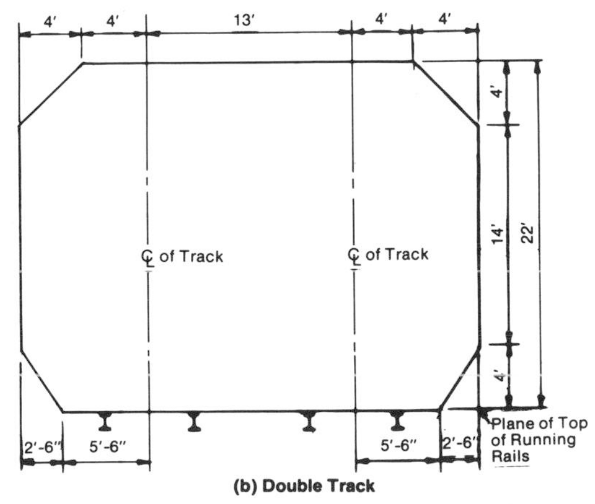

Fig. 2-4. Clearance diagram for railroad tunnels—area.

On curved track, the lateral clearances shall be increased for the midordinate and overhang of a car 88 feet long and 62 feet between the centers of trucks, equivalent to 1 inch per degree of curvature. Clearances for superelevation are governed by AREA recommendations.

Additional clearance requirements for third rail, catenary or pantographs should conform to the diagrams issued by the Electrical Section, Engineering Division, of the Association of American Railroads.

In all cases, the latest standards of AREA should be used for new construction, and special legal requirements and standards of using railroads should govern if exceeding the above clearances.

Circular tunnels are fitted to the clearance diagrams with such modifications as may be acceptable.

ALIGNMENT AND GRADES FOR RAILWAY TUNNELS

Alignment. Where possible, from the standpoint of general alignment and cost, tunnel alignment should be straight to facilitate construction. Curved and even spiral tunnels have been used on many railroads, particularly in mountainous areas.

Radii of curvature and superelevation of track are governed by maximum train speeds of the particular railroad.

Grades. If possible, maximum grades in straight tunnels should not exceed 75% of the ruling grade of the line. This grade should extend about 3000 feet below and 1000 feet above the tunnel.

Grades in curved tunnels should be compensated for curvature in the same manner as for open lines.

CLEARANCES FOR RAPID TRANSIT

Tunnels. Each rapid transit system establishes its own rolling stock, power supply system and signal space.

Figure 2-5 shows the normal clearance diagram of the New York City Independent Subway System.

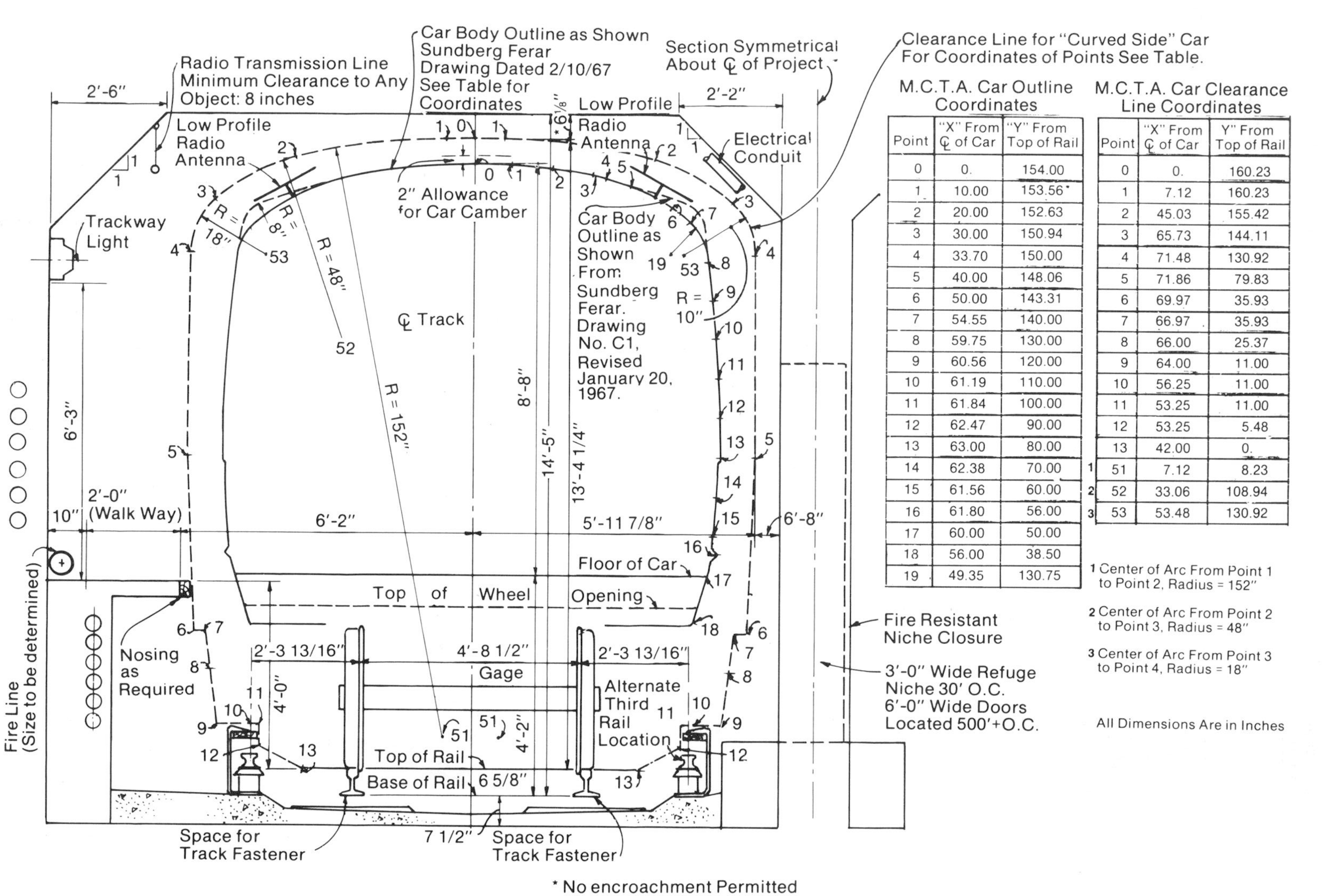

M.C.T.A. Car Outline Coordinates

Point	"X" From ℄ of Car	"Y" From Top of Rail
0	0.	154.00
1	10.00	153.56*
2	20.00	152.63
3	30.00	150.94
4	33.70	150.00
5	40.00	148.06
6	50.00	143.31
7	54.55	140.00
8	59.75	130.00
9	60.56	120.00
10	61.19	110.00
11	61.84	100.00
12	62.47	90.00
13	63.00	80.00
14	62.38	70.00
15	61.56	60.00
16	61.80	56.00
17	60.00	50.00
18	56.00	38.50
19	49.35	130.75

M.C.T.A. Car Clearance Line Coordinates

Point	"X" From ℄ of Car	"Y" From Top of Rail
0	0.	160.23
1	7.12	160.23
2	45.03	155.42
3	65.73	144.11
4	71.48	130.92
5	71.86	79.83
6	69.97	35.93
7	66.97	35.93
8	66.00	25.37
9	64.00	11.00
10	56.25	11.00
11	53.25	11.00
12	53.25	5.48
13	42.00	0.
[1] 51	7.12	8.23
[2] 52	33.06	108.94
[3] 53	53.48	130.92

1 Center of Arc From Point 1 to Point 2, Radius = 152"

2 Center of Arc From Point 2 to Point 3, Radius = 48"

3 Center of Arc From Point 3 to Point 4, Radius = 18"

All Dimensions Are in Inches

* No encroachment Permitted

Fig. 2-5. Clearance diagram, MTA–New York City.

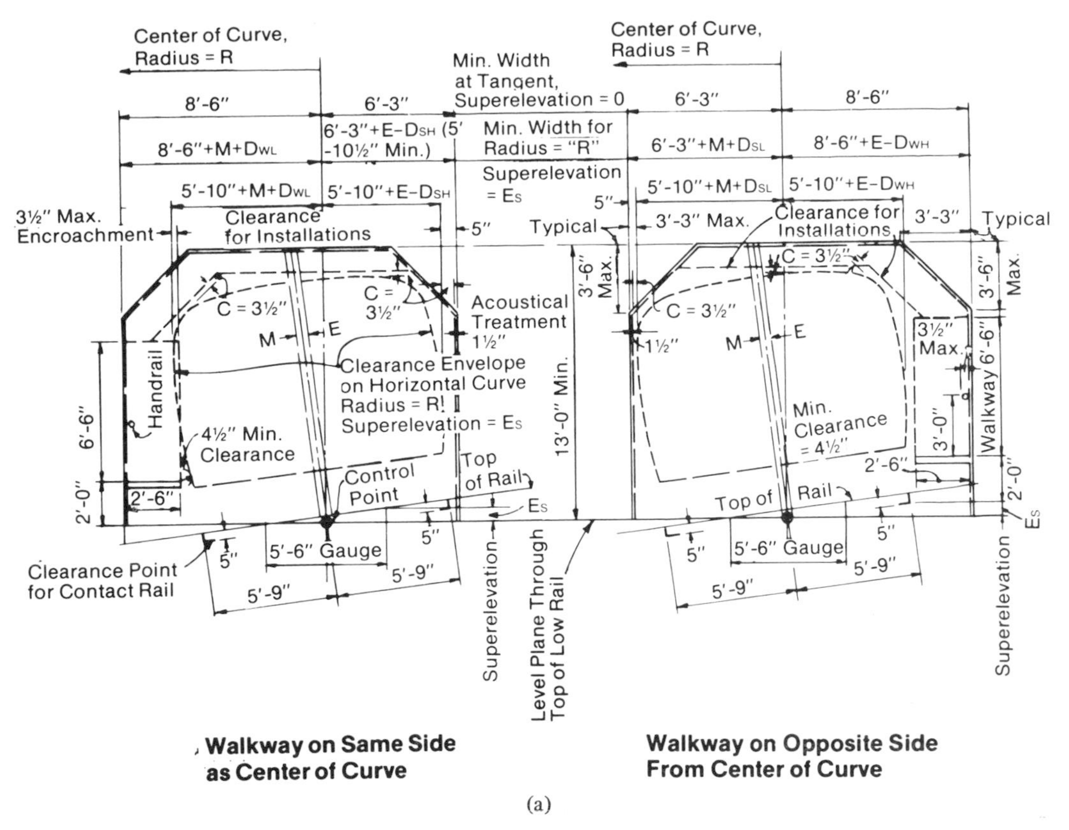

(a)

Fig. 2-6. Clearance diagram–BART.

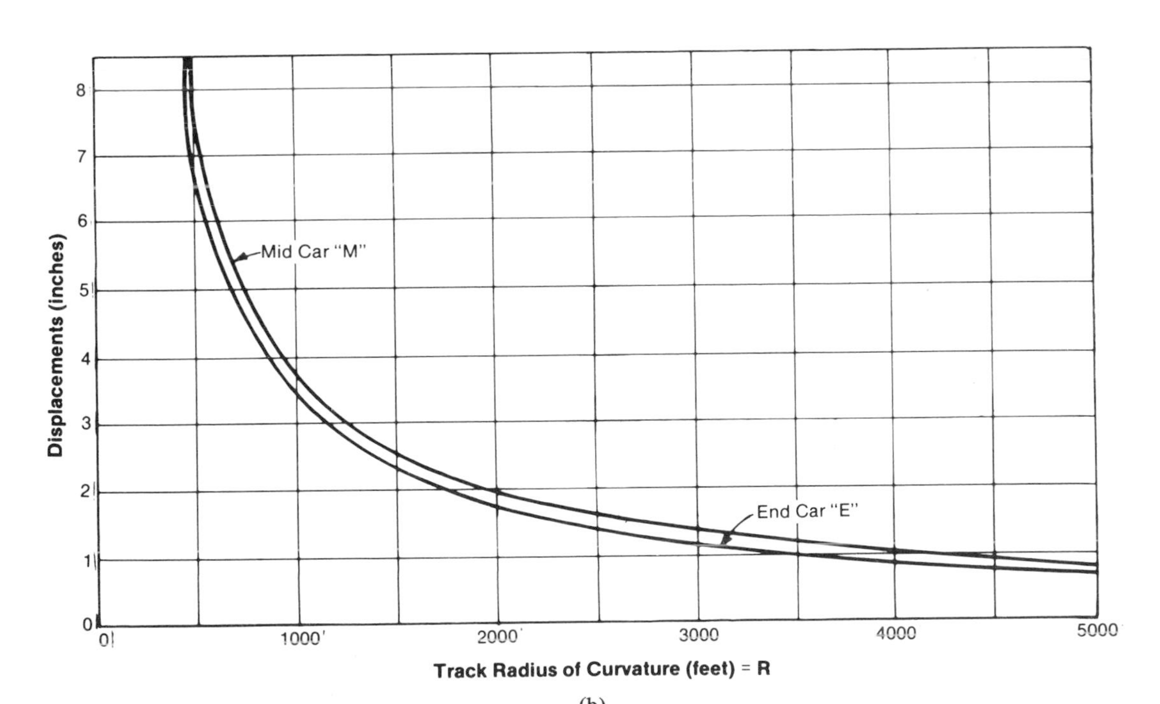

(b)

Fig. 2-6. *(continued)*

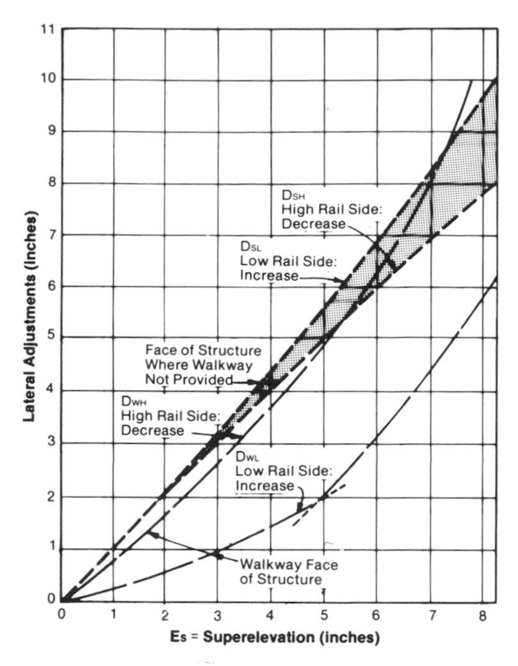

Note
Curves for D_{WH} & D_{WL} are for walkway tread 2'-0" above adjacent rail. Where walkway tread is at the level of adjacent rail or below, the curves for D_{SH} & D_{SL} shall apply.

(c)

Fig. 2-6. *(continued)*

Figure 2-6 shows the clearances for the San Francisco Bay Area Rapid Transit System in the shield-driven circular tunnels for tangent line and with superelevation in curves. The cars are 10 feet wide and 75 feet long on a 5-foot, 6-inch gage track. Clearances allow for overhang of cars, tilting due to superelevation and sway and also for a broken spring or defective suspension.

ALIGNMENT AND GRADES OF RAPID TRANSIT

Tunnels. Operating requirements of the system govern the curvature and limiting grades of its lines. The New York City Independent Subway has a minimum radius of 350 feet with transition curves for radii below 2300 feet. The maximum grades for this system are 3% between stations and 1.5% for crossovers and turnouts.

The San Francisco Bay Area Rapid Transit, designed for train speeds of 80 mph, determines the relations among speed, radius and superelevation of horizontal curves by the following formula:

$$E = \frac{4.65V^2}{R} - U$$

where

E = superelevation (in.)
R = radius (ft)
V = train speed (mph)
U = unbalanced superelevation, which should not exceed $2\frac{3}{4}$ inches optimum or 4 inches as an absolute maximum.

At 80 mph design speed, the radius with an optimum superelevation would be 5000 feet. For a maximum permissible superelevation of $8\frac{1}{4}$ inches, the minimum radius required would be 3600 feet.

For yards and turnouts, the absolute minimum radius is 500 feet. Maximum grade is 3.0% in line and 1.0% in stations. For good drainage, grades should not be less than 0.5%.

3
Preliminary Investigations

R. KENNETH DODDS
President
Foundation Sciences, Inc.

For the tunnel designer and builder, the rock or soil through which a tunnel is excavated is a construction material. It is just as essential to understand its engineering characteristics as it is to understand the concrete or steel used in other phases of the work. A thorough job of exploration is always made on every tunnel completed; however, on some projects, it is accomplished by the actual driving of the tunnel when it can no longer be an economic benefit to the owner. Difficult ground conditions do not in themselves make difficult construction problems, provided they have been properly defined and taken into account in design. The unanticipated problems are those that create costly delays and redesign.

Underground, it is what is not known that causes problems. It is no coincidence that the more thoroughly investigated tunnels are usually completed with fewer cost overruns and on schedule. Explorations for tunnels are made to help determine the feasibility, safety, design and economics of a project. More specifically, some of the purposes of explorations are:

1. To define the physical characteristics of the general material through which a tunnel is to be driven;
2. To provide specific rock or soil design parameters;
3. To help define the limit of certainty for the project, and to alert the engineer to possible conditions that may arise during construction so that he may prepare contingency plans;
4. To remove uncertainties of material conditions for the bidder;
5. To establish design conditions so a "change condition" can be fairly determined during construction;
6. To improve the safety of the work; and
7. To provide experience working with the material, which, in turn, will improve the quality of field decisions made during construction.

DEVELOPING THE PROGRAM

Exploration techniques, procedures and interpretations are not normally part of a design engineer's training; therefore, it is advisable to seek the assistance of an experienced engineering geologist or exploration engineer with this work.

The first step in an exploration program is to establish its objectives. This is done by the designer and it guides the exploration specialist in the development of a program that will fulfill the requirements of the project. A sensible exploration program starts with a statement of what is already known of the physical conditions of the project area and develops logically from there. Each phase of the program is built on what is learned from the preceding work, and the program is subject to continuing review and improvement. An exploration program begun and completed without modification is often poorly conceived, poorly conducted and wasteful.

A few general points should be kept in mind as the exploration program is being developed.

1. Exploration is done to develop knowledge of the physical conditions at the project site. The locations of borings, drifts, shafts, etc. should be based only on this need; they should not be justified as part of the completed project. Experience over the years has established that few projects are constructed precisely on or along their original alignment.
2. The amount of exploration required to provide the answers needed by the designer is in no way related to the funds available for the work. The funds available may control the work accomplished, but they do not delineate the work actually required.
3. There are certain levels of exploration effort below which it is dangerous to go. In fact, in some cases, it may be better to have no data and recognize the lack of knowledge than to have inconclusive, misleading data.
4. Explorations are made to provide reliable, useful knowledge, and this is realized by good, experienced field technical supervision. It is no exaggeration to say that a well-conceived and controlled boring program can develop more useful information with 10 borings than a poorly conceived and inspected one can in 30 or 40.
5. It is poor economics for the owner to perform exploration ahead of design and expect the designer to do the best he can with what he is given.

Geologic Factors to be Developed by Exploration. A tunnel designer must either know or "estimate" geologic factors before he can proceed with his design. Some of the factors to be evaluated are the type and distribution of materials through which the tunnel will pass; the deformational and strength properties of each material; the intensity and direction of the in situ stress field and its relationship to the rock strength properties, such as the potential for stress relief failure or "popping" rock, the potential for bending stresses on sets, etc.; groundwater conditions; geologic structural features of the rock and their influence on its physical properties; the presence and influence of shear zones; and seismic activity potential.

ORDER OF THE INVESTIGATION

An investigation should proceed from the general to the specific. Each of the methods or steps is a tool to be used in the analysis of the project. Each improves the accuracy of the work, but none is a solution in itself. All exploratory methods require interpretation and judgment, for it is only by the application of interpretation and judgment that they become useful. The steps of a complete exploration program are as follows:

1. Search of available literature and records;
2. Aerial photograph study;
3. Surface geologic reconnaissance;
4. Geophysical suurvey;
5. Exploratory borings;
6. Test pits, drifts and shafts;
7. In situ testing;
8. Laboratory studies;
9. Full-scale model testing;
10. Actual construction;
11. Post-construction monitoring and performance.

GEOLOGIC RECONNAISSANCE

It is seldom possible to thoroughly explore a tunnel alignment directly; therefore, the final geologic conditions incorporated into the design assumptions are based on detailed knowledge of selected local areas connected by the interpretive skills of the engineering geologist. These essential interpretations are based on a knowledge of the rock types; their macro- and microscopic, mineralogic and structural features; and the geomorphic situation and agents at work.

Geologic reconnaissance includes a search of available literature, the study of aerial photographs and surface geologic mapping. In developing an investigation program, it should be remembered that the answers to some geologic questions are not always found at the work site but that relationships between rock units or structural features obscured at the work site may be obvious some distance away.

Literature Review. Before any time is spent in the field, a review should be made of published information about the geology, soils, groundwater, seismic history and performance of structures in the project area. In urban areas, some knowledge of the history of the site is also important, as it can identify old landfills or alterations to drainage patterns which may affect the project. Geologic and soils maps are especially useful, as are aerial photographs. General sources of geologic and soils information are state geologic surveys, the U.S. Geological Survey, the U.S. Department of Agriculture Soil Conservation Service and the U.S. Forest Service (aerial photographs).

Aerial Photographs. On the ground, only a limited area may be viewed, and large-scale features that would be obvious from a distance are not recognized. The use of aerial photographs allows the viewer a broad look at the area under study. Aerial photographs are especially useful in geomorphic analysis, and much insight into the engineering properties of rock is available from the skilled evaluation of rock's response to the natural testing laboratory of its environment.

A variety of techniques are now used in aerial reconnaissance. Included are vertical, oblique, color and infrared photography and side-looking radar.

Detailed interpretation of aerial photographs requires the services of a specialist, but they can be helpful to the designer with a small amount of experience. Some of the features that can be readily recognized are topography, drainage patterns, vegetation, land use and potential construction material sources.

To use aerial photographs, the scale, shadow orientation, compass orientation and date of the photography must be known; the photographs are best studied as stereoscopic pairs viewed in reflected light with one of a variety of stereoscopes. A description of the types of instruments available is found in Ray[1] and in *Agriculture Handbook 294.*[2]

The Significance of Slopes. Topographic slopes are of two types: erosional and depositional (see Fig. 3-1). Erosional slopes are the direct result of the forces acting on them and of the physical properties of the materials composing the slope which resist these forces. The forces

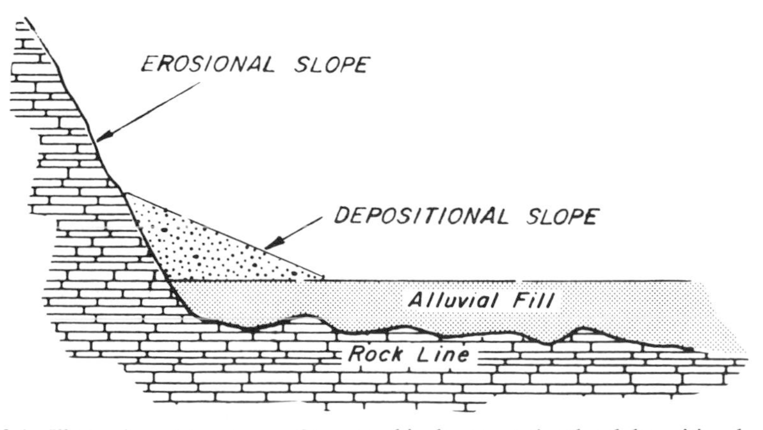

Fig. 3-1. Illustration of two types of topographic slopes: erosional and depositional.

working on slopes include gravity, in situ rock stresses from present and past overburden loads and tectonic forces, chemical changes and the abrasive agents of running water, wind and ice.

The shape of a slope can provide knowledge of the physical characteristics of a rock, especially in areas of young topography; that is, where drainage channels nearly fill the valley floors. Slopes can be classified as linear, composite, concave or convex.

Linear slopes result from rock of nearly uniform physical properties. Composite slopes result at significant changes in the strength properties of the rock. Concave slopes form when the slope is delivering material to its base faster than it can be removed by natural processes. Convex slopes result when the erosional agents at the foot of the slope are removing material faster than the slope can deliver it.

Other features that may be easily recognized on aerial photographs are landslides, major faults and geologic structures such as anticlines, synclines and domes.

Other Remote Sensing Methods

Infrared Sensing. Infrared photography and infrared sensing are not identical methods. Infrared sensing employs an optical scanning system that projects the thermal radiation which is between the visible spectrum and radar wavelength, to an infrared detector. It is then converted to an electrical signal, amplified and modulated, passed through a cathode ray tube and recorded on a photographic film.

Infrared photography, as the name implies, uses film sensitive to the radiation in the infrared wavelengths.

Features which have marked differences in their heat radiation characteristics, such as water courses and springs, can often be recognized more easily by the use of this technique. Additional information in infrared imagery can be found in Colwell.[3]

Side-looking Radar. Radar scanning methods are now available to provide good quality imagery of terrain below cloud cover. The technique is of value where conventional methods of photography are not applicable because of weather interferences. Interpretation and use of this type of imagery is similar to that used in aerial photographs.

GEOPHYSICS

Geophysics is a tool which can be used in geologic studies for tunnels. It is a tool that depends heavily on interpretation and, therefore, on the skill and experience of the person evaluating the data collected. Geophysical methods have the obvious advantage of being nondestructive; they are relatively fast, and their unit cost is low. However, their precision is usually also low and restraint must be used to avoid assigning more accuracy to their results than the method justifies.

The most important use of geophysical methods in exploration programs for tunnels is the location of anomalous conditions that require detailed identification by other, more direct methods. Ideally, geophysical work is done before boring programs are begun so the results can be used to help locate borings where they will provide the most information.

Some geophysical methods include seismic refraction and reflection surveys, electrical resistivity soundings, gravity surveys and magnetic surveys.

Geophysical techniques which can be used down-hole include neutron density, sonic and seismic velocity determinations, electroresistivity and gamma-gamma techniques. Others are used very rarely in tunnel design evaluations, although up-hole and cross-hole seismic surveys are used to develop shear wave velocity data (see Stokoe and Woods[4]). A discussion of down-hole geophysical methods can be found in Keys *et al.*[5]

Seismic Refraction Method. The velocity of an elastic wave passing through a material is a function of the material's structure, composition and in situ stress condition. Velocities increase with density, water content and compactness. Seismic waves follow principles of propagation, reflection and refraction similar to those of light waves (see White[6] for a complete treatment of this subject). Equipment used for these measurements includes a source

Table 3-1. Relative seismic velocities (P-waves), feet/second.*[7]

Material	Velocity
Dry, loose topsoils and silts.	600–1200
Dry sands, loams; slightly sandy or gravelly soft clays.	1000–1600
Dry gravels, moist sandy and gravelly soils; dry heavy silts and clays; moist silty and clayey soils.	1500–3000
Dry, heavy, gravelly clay; moist, heavy clays; cobbly materials with considerable sands and fines; soft shales; soft or weak sandstones.	3000–4800
Water, saturated silts or clays, wet gravels.	4800–5000
Compacted, moist clays; saturated sands and gravels; soils below water table; dry medium shales, moderately soft sandstones, weathered, moist shales and schists.	4800–6000
Hardpan; cemented gravels; hard clay; boulder till; compact, cobbly and bouldery materials; medium to moderately hard shales and sandstones; partially decomposed granites; jointed and fractured hard rocks.	5500–8000
Hard shales and sandstones, interbedded shales and sandstones, slightly fractured hard rocks.	8000–12,000
Unweathered limestones, granites, gneiss, other dense rocks.	12,000–20,000

*Note that the velocity of sound in air at sea level and 32°F is 1087 feet/second.

of seismic energy, a detector (geophone) and a timing device that records the wave travel time. A good discussion of the seismic method is contained in Henbest *et al.*[7] (see Table 3-1).

Application of Seismic Studies. Some uses of seismic studies are:

1. To identify general material types;
2. To locate anomalous geologic conditions, such as deeply weathered zones, buried valleys and shear zones;
3. To determine approximate elevation of the top of the hard rock;
4. To locate construction material sources; and
5. To assist in the efficient location and use of borings.

Considerations and Limitations of Seismic Investigations. The degree of reliability of seismic velocity measurements improves above the groundwater table and where the difference in wave propagation velocity between two adjacent materials increases. Seismic lines must be reversed (shot points at both ends) to identify sloping features; vertical features narrower than three times the geophone spacing probably will not be isolated and identified. Low velocity features below a higher velocity material will not be detected. Water velocity is about 5000 feet/second, which is the same as some compact gravels, mudstones and weathered rocks; this leads to difficulties when the groundwater table is at or near the contacts with these types of rock. Because of traffic vibrations and other interference, seismic surveys are difficult in populated areas, near heavily traveled roads or near operating heavy machinery. The use of explosives to generate a shock wave is often prohibited in these areas.

To overcome some of the problems of using explosives in populated areas, a truck-mounted unit has been developed that generates a constant vibration frequency in the ground. The reflection of these elastic waves is recorded and interpreted similarly to conventional methods (see Roman[8] and Mossman *et al.*[9]).

Resistivity Surveys. The resistance a material has to the passage of an electrical current depends on its chemical components and its degree of saturation. The resistivity method applies an electrical current to the ground through two electrodes. The changes in potential across the known distances between these electrodes is then used to help evaluate material types. A thorough discussion of this method is found in Van Nostrand and Cook.[10] Wet clays and silts, mineralized water and some metal ores are good conductors; dry sands and gravels and crystalline rocks without metal ore are poor conductors.

Resistivity surveys usually provide the most useful information when they are made in conjunction with a seismic study, as shown by Table 3-2, where the characteristics of different materials are compared. Resistivity values for

Table 3-2. Comparison of material characteristics.

MATERIAL	RESISTIVITY	SEISMIC (P-WAVE) VELOCITY
Dry gravel	High	Low
Dense rock	High	High
Pure water	High	Medium
Saline water	Very low	Medium
Dry compact boulders and cobbles	Very high	Moderately high
Saturated boulders and cobbles	Moderate	Moderately high

Table 3-3. Resistivity values of typical materials.

Brine	Less than 1,500
Moist clay and saturated silts	1000–10,000
Dry silts, clayey, sand or gravelly till	10,000–20,000
Sandy clays, saturated sands, well-fractured rocks filled with moist soils	15,000–50,000
Moist sands, moist sand-silt-gravel mixtures, slightly fractured bedrock (moist)	30,000–100,000
Moist to saturated gravels, sandy gravel with some silt, slightly fractured bedrock with dry soil filling	100,000–300,000
Dry gravels and coarse sands, massive hard bedrock	300,000 plus

some materials are given in Table 3-3. Moisture content has a significant effect on the results.

In both the resistivity and seismic methods, good vertical control and carefully logged borings are essential for reliable interpretation.

BORINGS

Borings are another tool used in the evaluation of the engineering geology of a site; they require skillful interpretation and application. The boring program is developed on the facts learned during site reconnaissance and during geophysical investigations.

Borings are the most common method for detailed exploration of civil works. As a result, the mere fact that they are made can provide a false sense of security to the designer. Even a good boring program will not provide all the answers about materials and their properties, but a good boring program will provide enough answers so that the designer can be prepared for the significant variations in geological conditions to be encountered. The usefulness of borings per dollar spent becomes less as the tunnel is placed farther from the surface.

Borings are usually made to provide more specific knowledge of rock units, the variations in material and their physical properties. They provide small, local samples for detailed study. Only where there are insufficient surface rock exposures should they be relied upon to investigate general rock types, as this knowledge should be available from the geologic reconnaissance.

Because tunnels are linear features, boring programs for them should concentrate in the areas of greatest potential difficulty. Except in special cases, borings should not be arbitrarily spaced at some given distance from each other along the alignment.

Some of the areas requiring more detailed exploration are:

1. Portals;
2. Topographic lows above the tunnel, as these often represent structurally weak rock;
3. Rock types with deep weathering potential;
4. Water-bearing horizons; and
5. Shear zones.

In deep rock tunnels, borings are made to obtain specific knowledge of the rock along the tunnel alignment. This usually requires sampling the rock above the tunnel for the proper preparation of geologic cross-sections. Figure 3-2 shows a group of borings along a tunnel alignment which are poorly located because they leave large gaps in the rock units involved. The layout in Fig. 3-3, on the other hand, produces a much more complete picture of the geologic situation, and thus is to be preferred, even though the borings do not reach the tunnel grade.

Alignments of drill holes often deviate

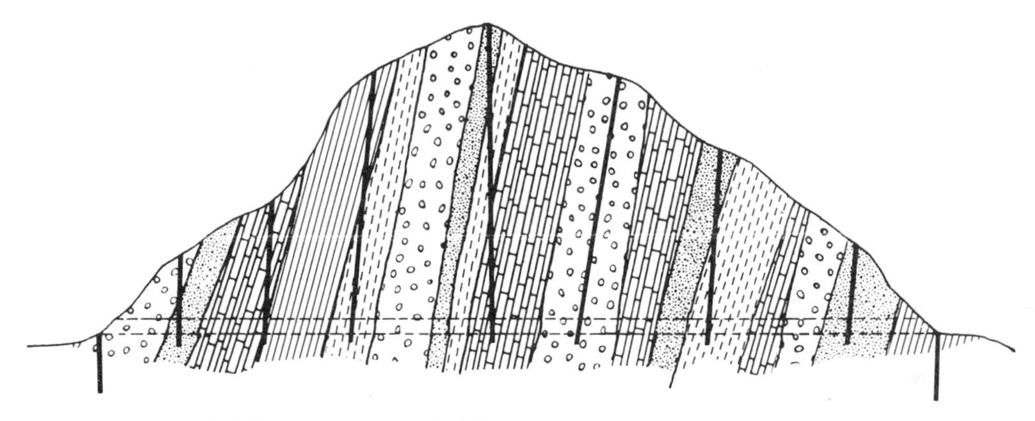

Fig. 3-2. A group of poorly laid out borings along a tunnel alignment.

appreciably with depth. This deviation is usually between 1 and 3 feet per 100 feet, but, in special cases, it can be much more. Borehole deviation surveys are required if the precise locations of samples from deep holes are necessary.

The number of borings required is a function of the geologic complexity of the area. Projects through competent, nearly horizontal rock may require very few borings, while those in geologically complex areas may require many more. The success of even a carefully planned boring program is dependent, to a very large degree, on the quality of on-site inspection during the drilling operation. It should be the concern of both the designer and the owner to be sure that trained and experienced persons are in charge of the technical aspects of a boring program.

Soft Ground Boring Programs. Soft ground can be encountered at any depth of cover in a tunnel due to large shear zones, hydrothermal alteration, poor cementation in sedimentary beds, etc. However, exploratory techniques especially adapted to soft ground conditions are primarily used for tunnels with shallow soil cover; these are, principally, drive sampling and cable tool churn drilling.

Drive Sampling. Drive sampling is the most satisfactory method available for recovery of representative soil samples from shallow depths, and, depending on the scope and aims of the work, it can be accomplished with light, portable rigs capable of drilling to a depth of approximately 75 feet, or a heavy churn drill capable

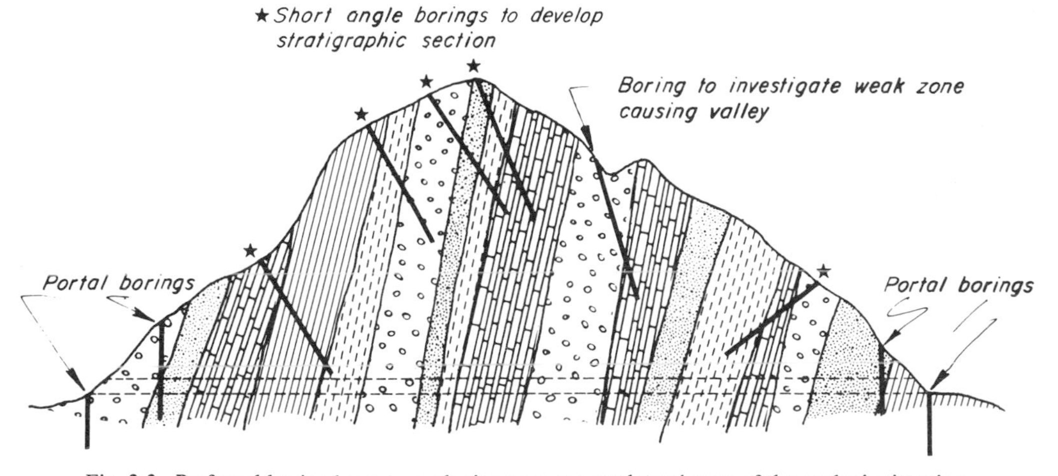

Fig. 3-3. Preferred boring layout, producing a more complete picture of the geologic situation.

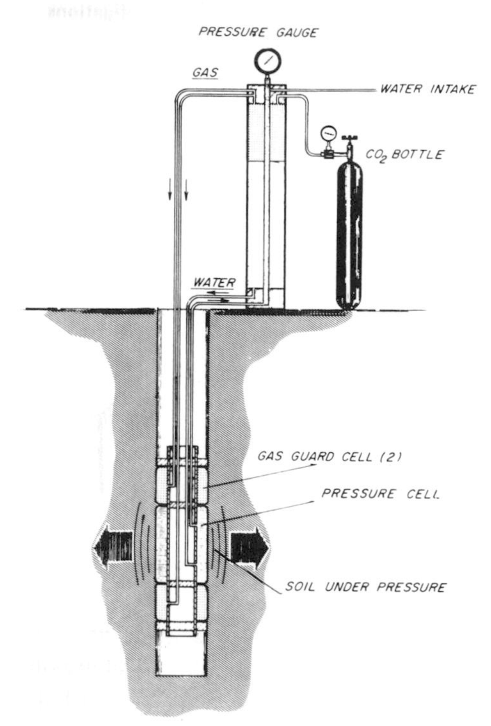

Fig. 3-4. Soil Dilatometer (Menard system).

of drilling 300 feet or more. In coarse-grained material, the churn drill must be used for the large drive barrel, 6 to 8 inches in diameter, necessary to properly sample gravels.

The simplest equipment necessary for drive sampling includes a drive weight (usually between 140 and 300 pounds) and guide; a power-drive cathead, a tripod and rope to lift the weight; a drive casing and shoe to hold the hole open for sampling; a set of drill rods; and a sample barrel. This method is described in ASTM Standard D-158611.

Soft Ground Down-hole Testing. The deformational properties of soils can be evaluated using a dilatometer, as shown in Fig. 3-4. In this type of instrument, the deformation of the soil is measured by the change in volume of the pressure cell as gas pressure is varied. The accuracy to which the volume change can be measured limits the effective range of the method to soils with a modulus of deformation of less than 5.0×10^5 psi.

Hard Rock Boring Program. Once the layout for the borings has been established, the specifications for the work must be prepared. The specifications should be written to suit the purposes of the work, some of which may be as follows:

1. To define the geologic stratigraphy and structure through which the tunnel is to be driven.
2. To determine physical properties of the materials.
3. To study the fracture pattern in various rock units.
4. To measure permeability and groundwater conditions.
5. To collect data on in situ stress levels.
6. To evaluate general blasting or mechanical excavating characteristics.
7. To evaluate probable support requirements.
8. To provide access for down-hole testing and logging equipment.

Each of these purposes requires some adjustment of the techniques used, and it is the job of the exploration engineer to get the best balance in the program. For detailed information about drilling technology, see Campbell and Lehr.[12]

Specific conditions the bidder must know if he is to submit a responsive bid on an exploration contract are given below:

1. The maximum depth of the hole.
2. The maximum angle of the hole.
3. Limits of permitted hole deviation.
4. Limits on types of drilling fluids permitted.
5. The type of sampling required.
6. Specific special equipment required.
7. Types of testing to be accomplished, and duties during testing.
8. Time available to complete work.
9. Access to work sites.
10. Controls of the engineer over his operations.
11. Limits of his liability for the quality of the works.

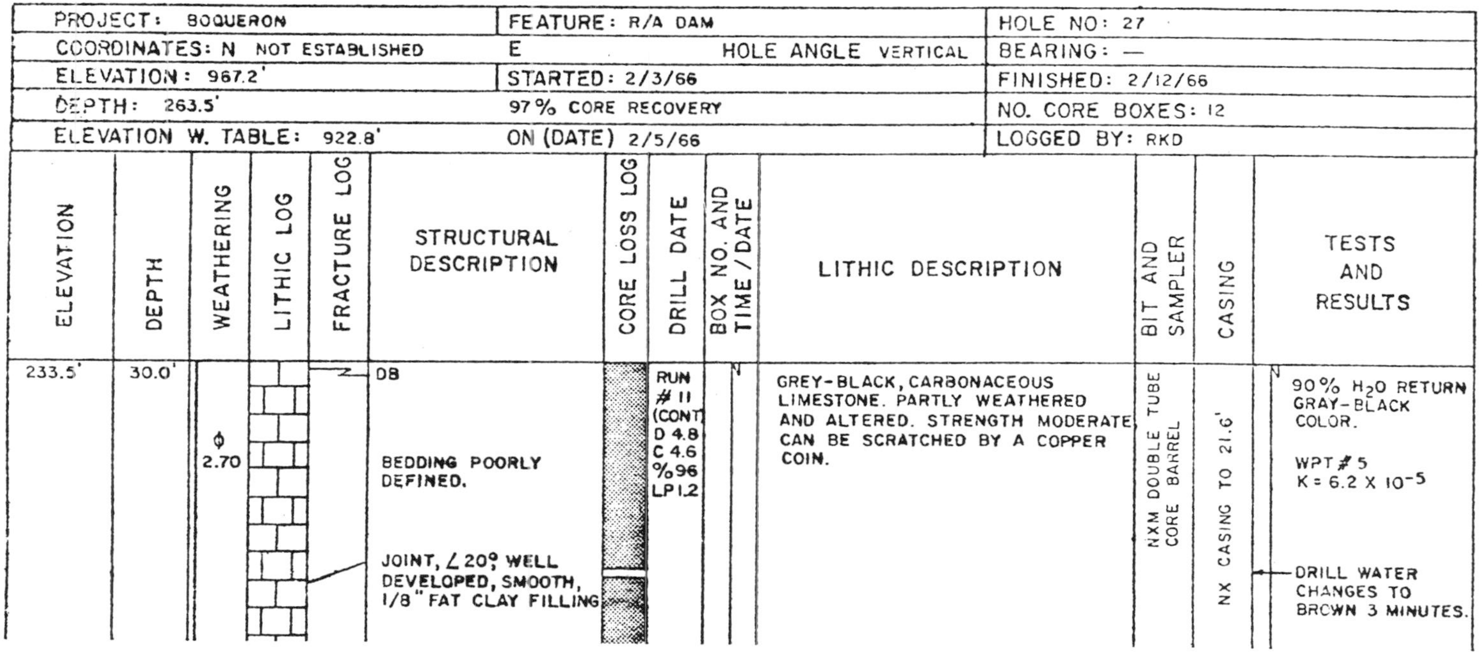

PROJECT: BOQUERON	FEATURE: R/A DAM	HOLE NO: 27
COORDINATES: N NOT ESTABLISHED	E HOLE ANGLE VERTICAL	BEARING: —
ELEVATION: 967.2'	STARTED: 2/3/66	FINISHED: 2/12/66
DEPTH: 263.5'	97% CORE RECOVERY	NO. CORE BOXES: 12
ELEVATION W. TABLE: 922.8'	ON (DATE) 2/5/66	LOGGED BY: RKD

ELEVATION	DEPTH	WEATHERING	LITHIC LOG	FRACTURE LOG	STRUCTURAL DESCRIPTION	CORE LOSS LOG	DRILL DATE	BOX NO. AND TIME / DATE	LITHIC DESCRIPTION	BIT AND SAMPLER	CASING	TESTS AND RESULTS
233.5'	30.0'	ϕ 2.70			DB BEDDING POORLY DEFINED. JOINT, ∠20°, WELL DEVELOPED, SMOOTH, 1/8" FAT CLAY FILLING		RUN #11 (CONT) D 4.8 C 4.6 %96 LP1.2		GREY-BLACK, CARBONACEOUS LIMESTONE. PARTLY WEATHERED AND ALTERED. STRENGTH MODERATE. CAN BE SCRATCHED BY A COPPER COIN.	NXM DOUBLE TUBE CORE BARREL	NX CASING TO 21.6'	90% H_2O RETURN GRAY-BLACK COLOR. WPT #5 K = 6.2 X 10^{-5} DRILL WATER CHANGES TO BROWN 3 MINUTES.

EXPLANATION OF SYMBOLS USED ON CORE LOGS

DB = Drill Break (Not a Natural Fracture)
CL = Core Loss
DCR = Core Loss Distributed Over Run (When exact location for core loss can not be determined)
WPT = Water Pressure Test
RB# = Rebound Number (Schmidt Hammer)
SG = Specific Gravity
Degree of Weathering _ _ _
(Decreasing ——→)

ϕ = Core diameter
D = Length Drilled
C = Core Recovered
% = Percent Core Recovery
LP = Longest Piece of Core in Run
DR = Drill Rate
HB = Hammer Break (Core broken by Inspector to fit in box)
H_2O = Water

Fig. 3-5. Sample boring log with combined graphic and narrative.

With all of the variables in exploratory boring, in most cases it has proven more in the owner's interest to let exploratory boring contracts on a per-hour rig time basis rather than on a per-foot-drilled basis. This gives the engineer full control over the work without creating areas of potential dispute with the contractor.

Field Inspection. Field logs should include the most precise description possible of the material, and they should be prepared as the hole advances. The field log is the most important record of the work. It, and not the hole in the ground, is the real product of the boring program that is purchased by the project owner.

A good field drill log is prepared with the idea that the information will be used for a variety of reasons, only some of which are known at the time of its preparation. It, therefore, must record all available information. The scale of the field log should be no less than 1 inch to 1 foot. The log must also be designed so that it can be used with a minimum of searching for needed data. This requires a compromise between a graphic and a narrative log. Figure 3-5 gives an example of a combined graphic-narrative log, and shows some of the abbreviations used.

Boring logs are most efficiently prepared by persons experienced in the interpretation of engineering geology and in evaluating the behavior of drilling equipment in relation to subsurface conditions. The boring log should contain the following:

1. Geologic descriptions of materials encountered.
2. Descriptions of rock fabric elements.
3. Records of field tests.
   a. Water permeability tests.
   b. Rebound hammer tests.
   c. Core caliper tests.
   d. Specific gravity tests.
   e. Dynamic velocity tests.
4. Graphic logs of lithology, structure and core recovery for rapid scanning of data.
5. Basic data on each core run.
   a. Length of run.
   b. Percent core recovery.
   c. Longest piece of core.
   d. Location of core loss zones.
6. Drilling rate.
7. Return drill water condition and amount.
8. Depth in hole where drill behaved abnormally.
9. Opinion of logger on location and reason for core losses.
10. Other conditions which could affect results of the work, such as the type and condition of equipment, bit and water pressure and RPM of bit.
11. Groundwater conditions.
12. Casing or cementation requirements.
13. Location of core in box, for easy future reference.
14. General administrative details.

Water Pressure Tests in Borings. Water pressure tests are part of most rock exploration programs. Carefully done, they produce a great deal of useful information about subsurface conditions.

Before water pressure test data can be effectively used, the field results must be converted to a coefficient which gives a common base to the variables in the test. One used is:

$$k = Q\mu / 2\,\pi L P_c \log_e L/r$$

where

Q = the absorption of the test section, in cc/second
μ = the viscosity of the test fluid, in centipoises
L = the length of the test section, in cm
P_c = the corrected average pressure in the test section, in atmospheres (atm)
r = the radius of the hole, in cm.

This formula gives a result in:

k = cc/second centipoise/cm^2/second atm/cm or darcy units.

Water Test Evaluation. In few rock types is the permeability more or less uniform throughout a rock mass. Generally, the value of k measured is an average of a wide range of values over the test section. Taken by itself, the test water loss often can give good approximations of the amount of water which will pass through a given stratum, providing, of course, that the test hole intersects rock and fractures which are typical of the rock mass overall. For other applications, such as grouting, the test water loss can be very misleading. For example, a test may give a permeability of 50 md (5×10^{-5} cm/second), suggesting a fairly tight rock, while, in fact, most of the water may have been lost through one fracture having a permeability of 3 or 4 thousand md (3×10^{-3} cm/second). Consequently, when establishing grouting criteria, one must be careful about arbitrarily picking a value for k below which no grouting will be done.

Interpreting Core Loss. It has often been said that, in borings, it is not so much the core that is important, but the places where no core is recovered. Some of the value of good inspection is found in interpreting core loss. When core loss occurs, the geologist must refer to his drilling notes to locate the actual point of the loss. Such things as a change in drill water color, a short period of rapid penetration, or a drill water loss may help to indicate the core loss zone and its extent. Where a weak clay filling of a joint has been washed out by drill water, some clay staining will usually remain on the joint surface, although it must be looked for carefully. Of course, such obvious causes as grinding of the core in the barrel, or dropping core because of a weak core spring, must not be overlooked, for these are core losses which do not necessarily indicate a weakness in the rock.

Some apparent core losses are really not actual losses at all, and must be identified if errors in interpretation are to be avoided. The most important examples of these apparent losses are:

1. Voids in the rock due to solution, open fractures, lava tubes, etc.;
2. Core left in the hole because the core did not break at the bottom of the hole when the core barrel was removed; and
3. Core not placed in the box in the proper order.

Control of Core Loss

The following procedures will help to minimize core loss.

1. The proper core barrel must be chosen. In most cases, this will be a double-tube M-series barrel with a bottom discharge bit.
2. Drill water pressure should not exceed 150 psi.
3. In weak rock, core runs should be kept short: 2 feet or less.
4. Bit pressure should not exceed 100 psi.
5. Drillers should never be allowed to drill through core blocks.
6. Before each run, the condition of the core barrel, bit and core lifter must be checked.

Supplemental Bore-hole Logging. Once a boring is completed and the core is in the box, the largest expense of the exploration is past, and other uses for the hole should be considered. Depending on the project requirements, some of the uses can be:

1. Groundwater head or flow measurements, permeability tests;
2. Groundwater temperature and salinity measurements;
3. Rock structure studies using bore-hole or television cameras;
4. Rock physical property measurements using down-hole seismic velocity probes or bore-hole deformation pressure cells;
5. Geophysical logging to locate permeable strata, changes of material in core loss zones, voids, etc., using electrical resistivity, or gamma or neutron density logs;
6. Hole diameter calipering used in conjunction with core calipering;
7. Bore-hole directional surveys.

Each of the above methods is designed to help produce additional insight into the composition and/or physical condition of the rock. *None* produces data that should be used without interpretation and reference to other exploration methods in the solution of an engineering problem.

HORIZONTAL BORINGS

Before discussing exploratory adits, the developing technique of long horizontal borings, which may be partially substituted for adits, will be discussed.

Emphasis here is on the word "partially,"

because horizontal borings have definite limitations as to the type and detail of data produced. Harding *et al.* discussed the state-of-the-art for this technique, and listed 17 horizontal borings between 5300 and 1034 feet in length completed to the time of their report. Of these 17 borings, only nine were diamond core borings, and most were drilled for mining projects.

Majtenyi and Rubin listed four advantages of long horizontal borings:

1. Safety and environmental problems minimized;
2. High rate of advance at relatively low cost;
3. Minimum rock disturbance; and
4. Discrete decision points at which process can stop at a lower cost level.

Four methods have been used to drill long horizontal borings; diamond wireline core drilling, rotary drilling (tricone type bit), down-hole motor drilling and down-hole percussion drilling. For the purpose of this section, only diamond wireline core drilling usually produces enough useful facts to be worthwhile. Before such a system can be used for many applications, the problems of hole deviation and hole stabilization must be addressed.

When used in conjunction with down-hole testing techniques (such as bore-hole camera, oriented core or geophysical), and when the core logging procedure described in this section is followed, long horizontal borings can be a valuable part of a tunnel exploration program. However, there are questions that only direct access to the rock by means of exploratory tunnels can answer. These questions relate to the rock mass deformational properties, the in situ stress, the strain response to openings and the extent, and properties, of soft material not recovered by core drilling methods.

EXPLORATORY ADITS

Exploratory adits and shafts provide an opportunity for the designer and the builder to have direct access to the material through which a tunnel is to be excavated. In many parts of the world, because of low labor rates, the cost is not much more than borings and the return in information much greater. For large underground rock chambers, exploratory adits are a necessity.

Exploratory adits should not be less in cross-section than 6.5 × 6.5 feet (2 × 2 meters), and, in areas of high labor costs, larger cross-sections, such as 10 × 12 feet (3 × 3.7 meters), are often less expensive because of the ability to increase the use of mechanical equipment.

Beside providing direct access to the in-place rock for detailed engineering geologic analysis, exploratory adits and shafts provide an opportunity to measure the physical properties of the rock in place, to measure the response of the rock to the opening, to test the effectiveness of various types of support systems, to evaluate the effectiveness of excavation methods, to monitor groundwater or gas conditions and to presupport sections of difficult ground. It also allows bidders to observe rock conditions directly, making possible more responsive bids and easier evaluations of actual "changed conditions" during construction.

One of the most common errors made in planning exploratory adits is to let the excavation contract and to treat the contractor in the same manner as on a regular construction project. This approach, based on maximum advance of the heading and minimum interference with the contractor's "production," defeats the true purpose of the work, which is to produce a maximum of useful information for the designer and builder. Exploratory adit contracts should be written to allow the engineer the greatest possible amount of freedom in directing the work.

Records. Detailed records should be kept of the excavation work for exploratory adits and shafts because it is never known when this information will be important in the analysis of some phase of the testing or stability analysis programs.

IN SITU ROCK MECHANICS STUDIES

The science of measuring the physical properties of rock in situ is advancing rapidly. The art of evaluating and interpreting these measurements for application to engineering problems, while proceeding at a more moderate pace, has also improved greatly in the past few years. When the variables affecting in situ measurements are recognized and applied, rock mechanics knowledge can make a significant contribution to the economics and safety of a tunnel.

In situ rock mechanics studies are made in order to:

1. Evaluate temporary and permanent rock reinforcement requirements and design;
2. Provide loading parameters for tunnel lining design, including rock strength utilization in pressure tunnel linings;
3. Evaluate stability of rock columns between two adjacent openings;
4. Determine suitable construction excavation practices for inclusion in specifications;
5. Provide background knowledge on rock physical properties for bidders; and
6. Develop a sufficient general knowledge of the rock and its physical situation to answer design and construction questions as they arise.

The work completed to date in rock mechanics has established some basic facts which every tunnel designer and builder should be familiar with.

1. Rock strengths are time-dependent and are controlled more by a limiting strain than by a limiting stress.
2. Rock strength (strain limit) is very sensitive to the in situ stress field (confining pressure).
3. Variations in rock strength due to the effects of fabric (bedding, joints, schistosity, etc.) can be extreme.
4. The in situ stresses in rock are seldom due only to the present vertical height of overlying material, but usually contain elements of past tectonic stresses and the previous loading history of the rock.
5. The direction of maximum principal stress is seldom vertical and very often is nearly horizontal.
6. The individual elements of bedded or layered rock units often carry very different in situ stresses. This is more true as the direction of the maximum principal stress approaches the direction of the layering; in that case, the stress carried by each rock unit may be related to the ratios of their moduli of deformation.
7. Poisson's ratio for rock is not a constant, and is sensitive to stress level.

Tools of Rock Mechanics

Various test methods are employed to establish the physical parameters of a rock. A description of a selected few of these "tools" follows.

Plate Loading Tests. Plate loading tests measure the in situ deformation characteristics of a material and help evaluate its strength. The data collected are used to help evaluate such things as:

1. The probable load and the pattern of load build-up on support systems and lining;
2. The percentage of load taken by rock in a pressure tunnel;
3. The required thickness for stable rock pillars between adjacent openings; and
4. The design of foot blocks for steel supports.

Plate loading equipment comes in a variety of sizes and configurations, but is usually one of two types: 1) a rigid die that is forced into the rock (see Fig. 3-6, or 2) a flexible diaphragm that adjusts its shape to apply a "uniform" pressure across the test surface. Rigid die tests are explained in detail by Dodds[13] and by Deere[14]; and diaphragm tests are described by Wallace[15] and by Clark.[16]

Variables affecting plate loading tests must be taken into consideration during analysis of test results. They include:

1. Variations in material behind the loaded plate.
2. Damage to the rock during the excavation of the test area or improper preparation of the rock surface to be tested.
3. Stress concentration around the test area caused by the shape of the opening in the rock.
4. The in situ stress field, either as a confining pressure or as a cause of stress relief fractures behind the test area.

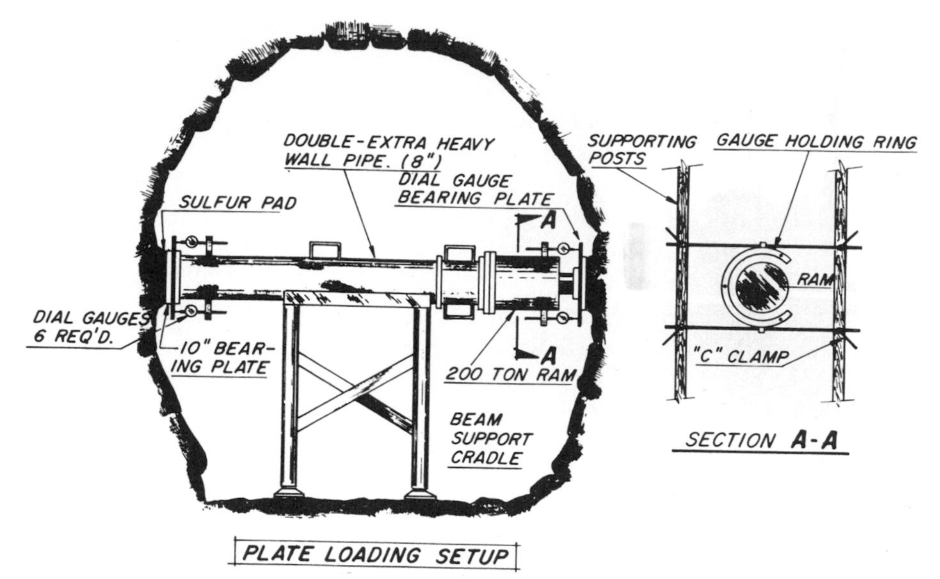

Fig. 3-6. Rigid die plate loading test.

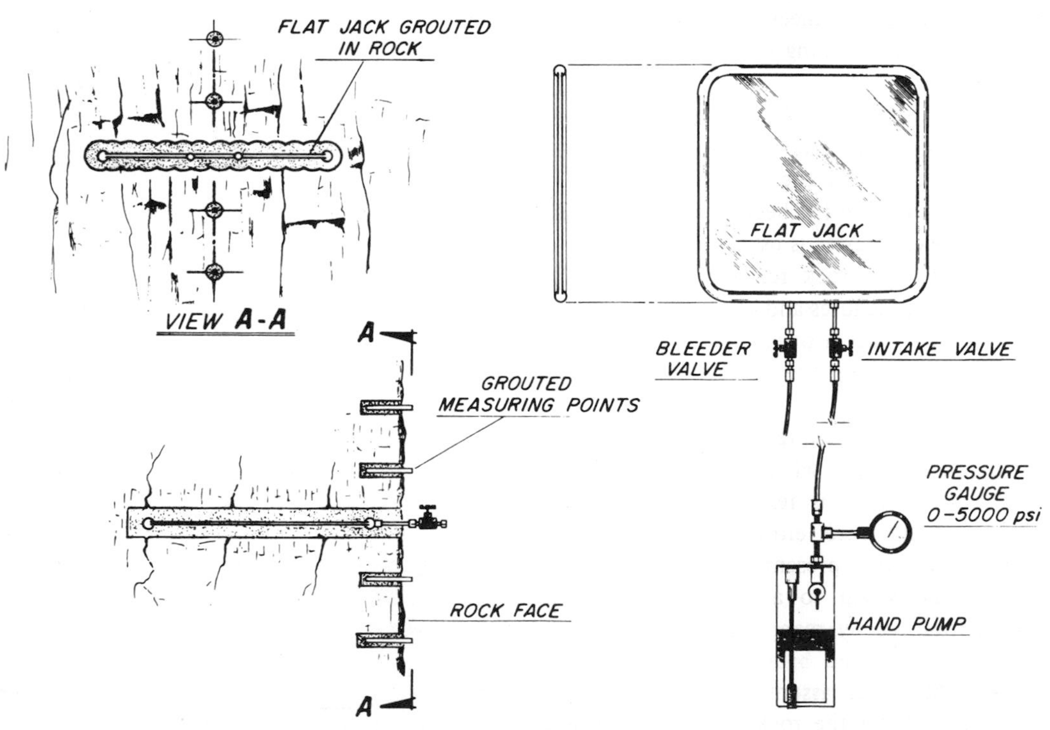

Fig. 3-7. Plate loading equipment.

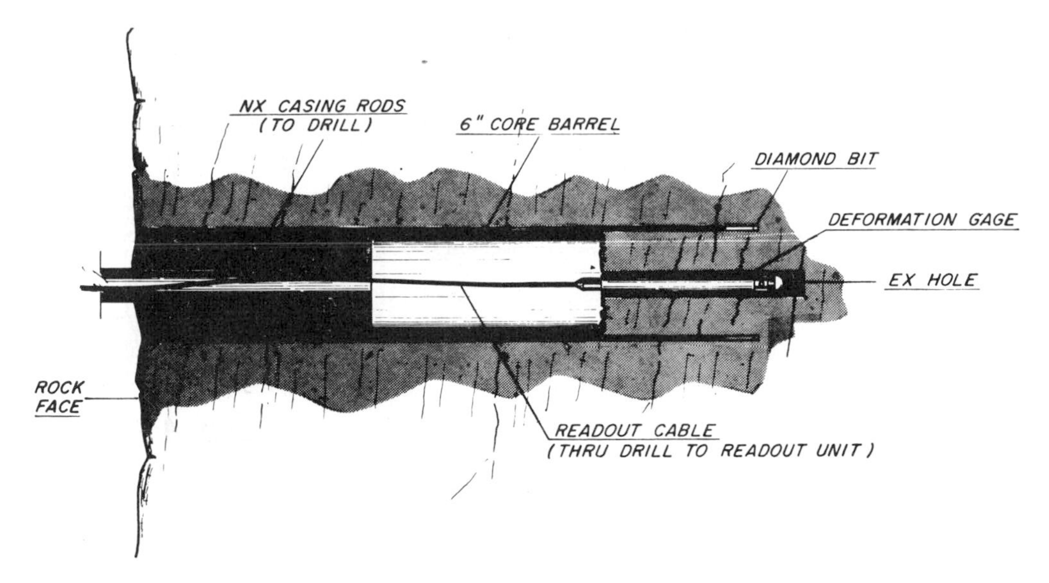

Fig. 3-8. Overcoring technique used to measure strain recovery in the rock.

5. Possible mechanical difficulties in the test equipment.

In Situ Stress Measurements. The in situ stresses present in a rock mass have an important effect on the physical properties and behavior of a rock as an engineering material. The in situ stress is a confining pressure which affects the viscoelastic and strength properties of the rock. The intensity and direction of these stresses control the strain patterns and redistribution of stress in the rock mass remaining after an excavation has been made. Knowledge of the in situ stresses is essential to the sound, logical design of rock reinforcement systems, excavation procedures and opening layouts, and to the interpretation of expected rock strength and deformational properties.

Flat Jack Tests. These tests measure the biaxial relaxation of a rock when the influence of the in situ stress field is removed by isolating a section of the rock either by drilling or by sawing. The stress present is then calculated by relating the strain to a known modulus of deformation, or by reapplying a pressure until the rock is strained back to its approximate original shape and assuming that the pressure required to force the rock back to its original shape is equivalent to the stress released when the confining medium was removed. For a discussion of the flat jack method, see Alexander,[17] Dodds[18] and Clark.[19] Figure 3-7 shows the layout and equipment for a typical test.

Flat jack tests are affected by the following variables, which must be considered when using test results.

1. When the test is made in an exploratory adit, the stress level measured includes the stress concentrations caused by the adit.
2. Stress levels in the rock, which are high in relationship to the rock's strength, can cause stress relief failure, which results in misleadingly low measurements of stress.
3. Unless special provision is made in the test measuring system, shear stresses occurring in the closing of the slot are not determinable. The result is that measured stresses are lower than actual stresses.
4. Rock is seldom fully elastic, although the method assumes this to be so.
5. Even isolated blocks of rock retain some locked-in stress. This means that measured stress is some modest percent lower than that actually present.

Overcoring Strain Meters. All commonly used methods of measuring in situ stress in a rock mass assume that the release of confining stress results in rock strain that is elastically reversible. The overcoring technique measures strain recovery in the rock by means of an electrical or photoelastic instrument, when the strain meter, which is set in a small-diameter borehole, is isolated from the existing rock stress field by drilling a large-diameter hole around it (see Fig. 3-8).

In the U.S., the instrument usually used for this work is a six-element, three-axis unit modeled after one developed by the U.S. Bureau of Mines (Merrill[20]). The strain measured during overcoring is converted to an equivalent stress by reloading the large-diameter cores in a biaxial chamber, as shown in Fig. 3-9, and calculating the measured moduli. Calculations are based on the work of Panek.[21] The overcoring technique can be used to evaluate stress patterns, not only at the surface of an opening, as does a flat jack, but also at some depth behind a free surface.

Beside applications similar to those listed for flat jacks, overcoring tests can help evaluate the stress pattern behind an opening and how it may change with time. This information is essential if minimum size rock columns between openings are to be safely established. Figure 3-10 shows a generalized pattern of stresses carried in a rock section at various stages in the history of an opening.

The significance of the behavior of a rock, as shown in Fig. 3-10 at Time[3], is that the effective span of the opening is increasing by the stress relief near the surface, with resulting changes in the zone of influence of the opening in the surrounding rock; as an example, the height of the tension zone around the axis of minimum principal stress.

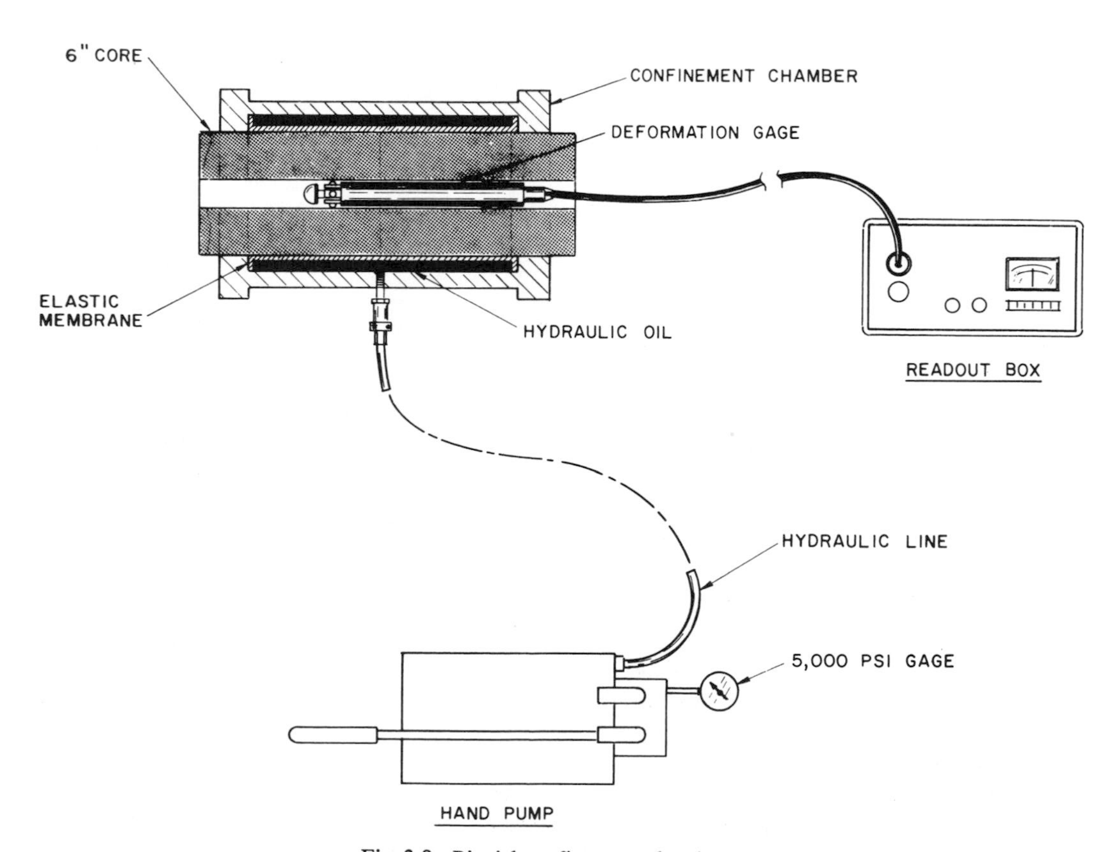

Fig. 3-9. Biaxial confinement chamber test.

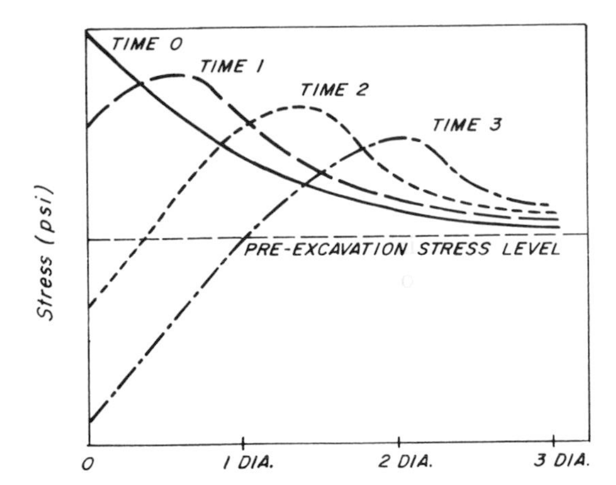

Fig. 3-10. Generalized pattern of changes in the stresses carried in a rock section at various stages in the history of an opening.

Factors affecting the accuracy of overcoring strain meter tests are:

1. The precision with which the modulus of deformation used in the calculation is measured and the degree of anisotropy in the rock;
2. The care with which the actual overcoring is completed; and
3. The degree to which the rock fulfills the assumption of elasticity.

Strain Around Openings. A knowledge of the pattern of rock strain, its magnitude and rate in response to an opening, is used to evaluate:

1. The need for and types of reinforcement or support required;
2. The allowable time for installing temporary supports or rock reinforcement systems;
3. The in situ strength and failure patterns for the rock mass;
4. The long-term stability of the opening;
5. The stability of rock sections between openings; and
6. The effectiveness of various support systems and opening shapes.

Creep. The stability of an opening in rock is complex and difficult to measure, short of actually causing a failure, and no single method can be used. Rock masses usually fail because some limit of strain has been exceeded; that is, a modest stress level held for a sufficient time will eventually cause most rocks to fail. Rocks creep.

In order to judge long-term stability, the rate at which a rock creeps under a given load intensity and the relationship of this creep rate to the material's strain limit should be known. Figure 3-11 shows a simplified strain pattern followed by a rock mass with time.

When an opening is made in a rock, the resulting strain history is divided into four parts:

1. Instantaneous elastic strain (*AB* in Fig. 3-11);
2. Primary creep (*BC* in Fig. 3-11);
3. Secondary creep (*CD* in Fig. 3-11);
4. Tertiary creep (*DE* in Fig. 3-11).

Elastic strain occurs as soon as the opening is made; in primary creep, also called elastic flow, the semilogarithmic decrease in creep rate between the elastic strain and the establishment of a steady state secondary creep rate occurs. Secondary creep occurs at an essentially linear rate, which, depending on the stress level, is an approximate constant for a material. In tertiary creep, an accelerating rate of strain occurs when the bonds in the rock are actually being ruptured. Not including tertiary creep, these phenomena can be expressed by:

$$\epsilon_t = A + B \log (t + 1) + C{:}t$$

where

ϵ_t = creep at time t
A = instantaneous elastic creep

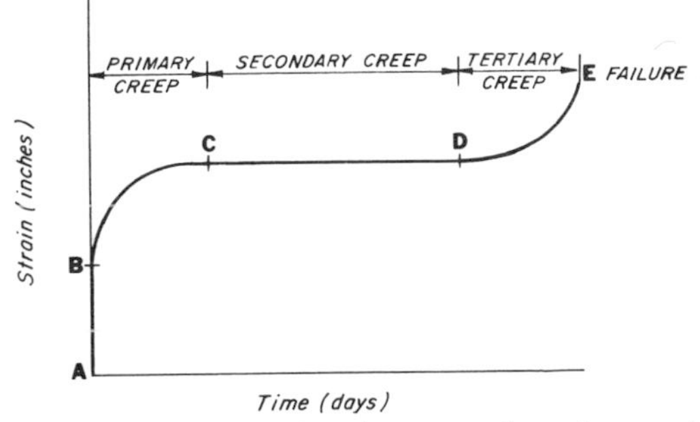

Fig. 3-11. A simplified strain pattern shown by a rock mass with time.

$B \log (t + 1)$ = primary creep
$C{:}t$ = secondary creep.

If the data from various tests are combined and some interpreteation is used, a general strain curve for the material can be approximated, and ϵ_t can be computed. The value for A comes from the jacking and laboratory tests; B is derived from the creep phase of the flat jack tests and laboratory sustained load tests; and C is evaluated from the extensometer tests.

Extensometers. Extensometers measure strain differentials between two points, one of which is fixed in space. Figure 3-12 shows one type of extensometer in which the anchor point at A is set deeply within the rock mass, out of the zone of influence of the opening, and is the reference point against which the other anchor locations are measured.

Many different types of extensometers are commercially available, so their physical characteristics can be matched to the project situation.

Some of the factors which complicate the interpretation of extensometer data are the following:

1. Delays in installing the instrument after the rock is exposed;
2. Instability of electrical instruments over the extended periods of time required to establish secondary creep rates;
3. Effects of nearby blasting on the anchor points.

Support or Reinforcement Performance Monitoring Instrumentation. Rock support is vastly different from rock reinforcement. Rock support holds in place rock which either has failed already or is expected to fail eventually. Rock support can be accomplished by steel ribs, timber sets, concrete lining or piers. Where rock support is required, the loads almost always increase with time, and one of the necessary factors in designing the support is evaluating the maximum load build-up within the span of the project life. This evaluation is improved by using knowledge gained from extensometer and load cell tests.

A sound, economical support design requires an understanding of the direction and intensity of the ground stresses.

Rock bolts can be used for temporary support, but their most efficient use is as permanent reinforcement. Rock bolting binds the rock into a continuously-acting member or beam and thereby mobilizes some of the strength of the rock itself. It reinforces the rock and resists failure by reducing the effects of fracture systems on the mass strength. Where applicable, rock bolts generally provide a safer opening at less cost than other available methods. In a typical underground opening, rock bolts are installed so as to form a continuous stress beam or arch in the rock. This beam or arch can be analyzed by conventional arch theories if one knows the internal rock load, the amount of translation and rotation of the end of the beam, the beam thickness and the applicable strength characteristics of the rock. These parameters are developed from testing previously described.

It is advantageous to monitor actual loads being developed on supports as a safety measure and to improve design knowledge.

Load Cells. Solid load cells are used to measure the loads being developed by steel sets. These

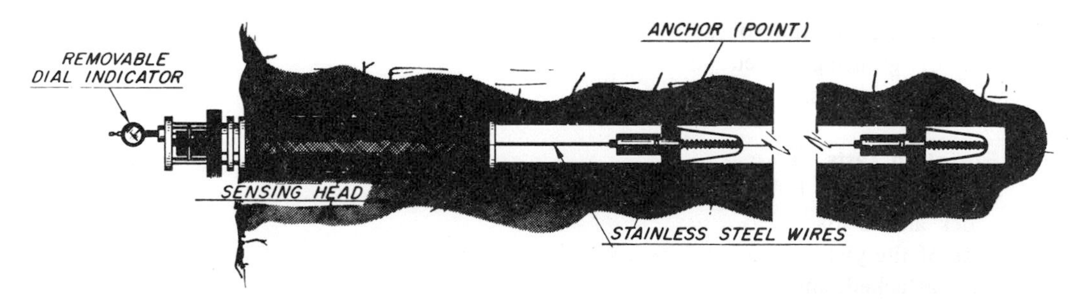

Fig. 3-12. Multiple point extensometer equipment.

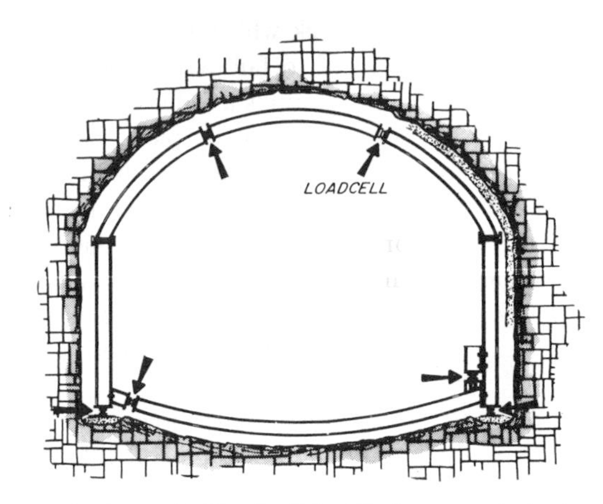

Fig. 3-13. Typical load cell installation on a steel set.

instruments use bonded electrical resistance strain gauges or hydraulic pressure cells as transducers. It is also possible to use the steel set as a load cell by mounting electrical, mechanical or photoelastic strain measuring devices directly on the set. A typical load cell installation on a steel set is shown in Fig. 3-13. Hollow load cells are also available to monitor the loads being carried by rock bolts.

Rock Bolt Tests. The economic design of rock bolt systems requires that the engineer specify an anchorage system both down the hole and at the surface that will maintain the load placed on the bolt. This requires:

1. Unit rock loads at the anchor and beneath the surface bearing plate less than the creep range of the rock;
2. Anchorage systems balanced to the size of the bolt, because it is a waste of steel to use a large-diameter rod when the anchor will slip before its strength is utilized; and
3. Specifications for time of installation and time of grouting suited to the rock's physical properties.

Load Retention Tests. To make the rock bolt load retention test, the bolt is anchored solidly in the rock and is loaded to a predetermined percentage of the yield strength of the rod. A load cell is attached, and the tensioning device is removed. The load carried by the bolt is measured for a specified number of days, dependent on the time the bolt must retain tension prior to grouting.

Pull Tests. A satisfactory rock bolt anchor must retain, without slippage, a load at least equal to the yield point of the rod to be used. The pull-out test checks this property by loading a properly installed rock bolt until the rod or the anchor fails. During the test, careful records are kept of the bolt load and deformation. Figure 3-14 shows the results of a typical rock bolt pull-out test.

Examples of Rock Mechanics Instrumentation Programs. Recent rock mechanics instrumentation projects are discussed by a number of authors in Lane and Garfield.[22] To make this work effective, it is essential to have some built-in redundancy, as shown by Dodds and King,[23] from which Tables 3-4 and 3-5 have been taken.

Testing in Hard Rock. A typical hard rock testing program, as depicted in Fig. 3-15, will consist of these test methods:

1. Plate loading tests;
2. Flat jack tests;
3. Overcoring strain meter test;
4. Extensometer installations; and
5. Loads on support systems.

The data from these tests are combined with detailed geologic mapping and adit excavation

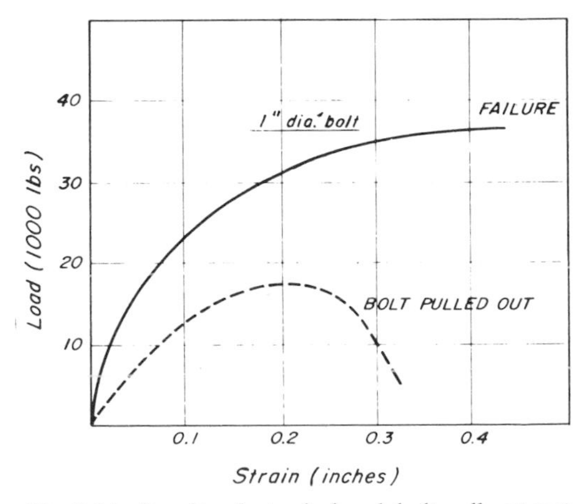

Fig. 3-14. Results of a typical rock bolt pull-out test.

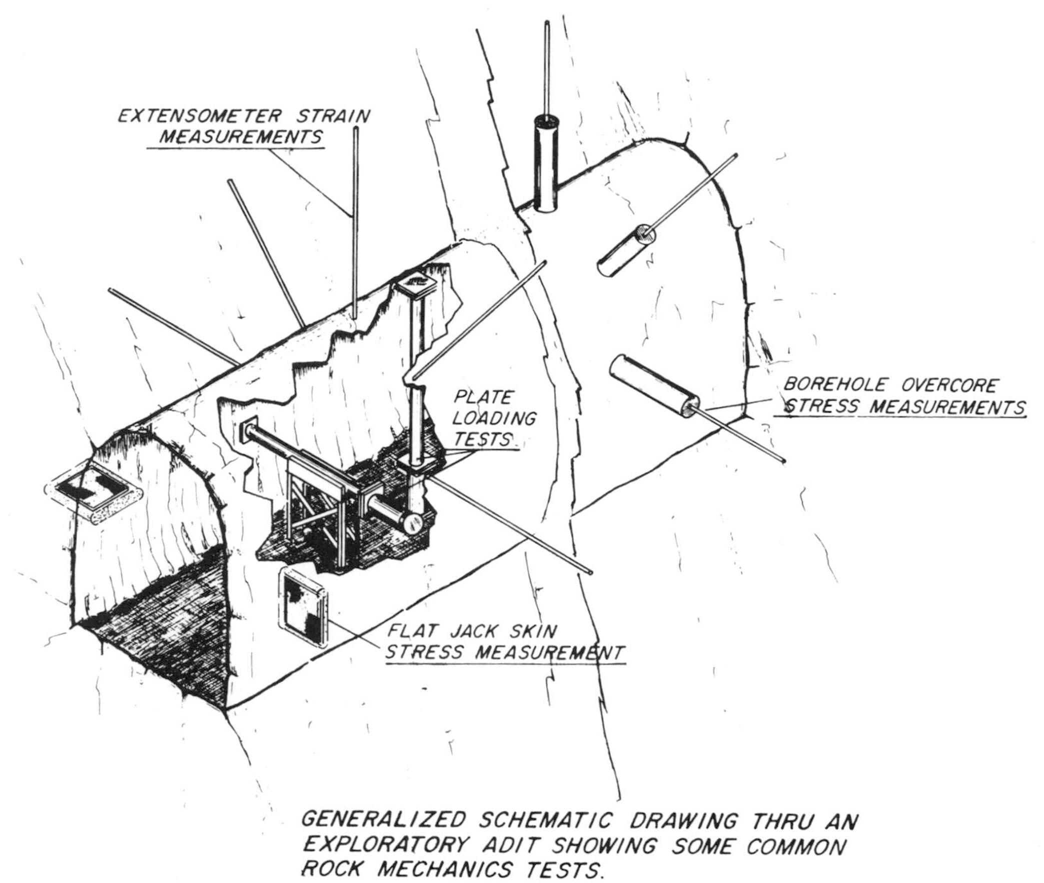

Fig. 3-15. Generalized schematic drawing through an exploratory adit showing some common rock mechanics tests.

records to produce a final analysis of the physical situation at the site, as shown on Tables 3-4 and 3-5.

Soft Ground Physical Property Tests. Soft ground conditions can include not only low-strength soil and sedimentary deposits, but also weathered zones, shear zones or poorly cemented strata in hard rock tunnels. In many cases, soft ground areas which occur as a small percentage of a hard rock tunnel are more difficult to handle, because their extent is difficult to determine before construction begins, and because work procedures are not set up to handle them efficiently. However, most soft ground tunnels are located near or in urban centers at shallow depth, so exploration by shallow borings is possible. Specific items to be investigated are as follows:

1. Strength and deformational properties, which are often measured in place with dilatometers as previously discussed;
2. Grain size of soil and, specifically, the presence of any bouldery ground;
3. Groundwater conditions;
4. Depth to hard rock and the possibility of "mixed face" conditions;
5. Presence and condition of buried, abandoned, man-made structures;
6. Most applicable excavation method: e.g., full face, multiple drift, compressed air shield, atmospheric pressure shield;
7. Method of ground support: e.g., steel sets and lagging, liner plate, shotcrete or concrete.

Much of the information on the physical characteristics of soft ground is interpreted from laboratory tests on representative samples.

Table 3-4. Information to be obtained from in situ tests.[23]

INFO	MODULUS DEFORMA.	CREEP COEF.	IN SITU STRESS	STRENGTH	RELAX PATTERN	EMPIRICAL DATA
Flat Jack	•	•	•			
In situ Shear	•			•		
MPBX		•			•	
SPBX		•			•	
Rock Bolt L.C.			•			•
Prop Load Cell			•			•
Gloetzl Cell			•	•		
Strain G Rebar			•	•		
Photo Cell			•		•	
Prop Extension			•		•	•
Plate Loading	•	•	•	•		
Anchor Pull						•
Anchor Retention						•

Soft Ground Instrumentation. On large projects, where the physical properties of the ground remain reasonably uniform, significant savings can be realized when the designer has positive knowledge of the stresses that will develop on the lining and the rate of their build-up. A short exploratory drift, instrumented as shown in Fig. 3-16, will produce this knowledge.

ROCK MECHANICS LABORATORY STUDIES

Measurements of rock physical properties in the laboratory are an important part of a rock mechanics study. These tests allow the investigation of the influence of a changing environment, such as confining stress and temperature, on the performance of a rock. Laboratory tests provide the first knowledge of a rock's physical properties in the initial development stage of a project, before in situ measurements can be made. During final design, the tests provide economic means of expanding in situ test results over the project area.

The usefulness of laboratory test results in the development of design parameters is controlled by how representative the test specimens are of the rock in place, by the influence of rock discontinuities on mass physical strength and by the changes to the rock's environment imposed by the project. Sound laboratory testing programs provide sufficient numbers of tests to allow some statistical control over the measurements.

Table 3-5. Use made of data obtained from in situ tests.[23]

USE	TEMP SUPPORT	ROCK BOLT DESIGN	PAC	STEEL SET DESIGN	ALIGN-MENT	LINING DESIGN	SPEC.	CONSTRUC. DECISIONS
Flat Jack	•	•	•		•	•		•
In situ Shear	•				•	•		•
MPBX	•	•	•	•	•		•	•
SPBX	•	•	•	•	•		•	•
Rock Bolt L.C.	•	•						•
Prop Load Cell	•			•		•	•	•
Gloetzl Cell	•		•					•
Strain G Rebar	•		•					•
Photo Cell	•		•			•		•
Prop Extension	•			•		•	•	•
Plate Loading				•		•		•
Anchor Pull	•	•					•	•
Anchor Retention	•	•					•	•

Rock mechanics laboratory tests include the following.

1. Strength measurements in compression and tension.
2. Confined compressive strength (triaxial strength).
3. Modulus of deformation and Poisson's ratio.
4. Confined modulus of deformation and Poisson's ratio.
5. Creep under sustained load.
6. Creep, when confined under sustained load.
7. Creep, when heated, under sustained load.
8. Dynamic modulus of elasticity and Poisson's ratio.
9. Recovery modulus.
10. Sliding friction.
11. Shear strength.
12. Permeability.
13. Specific gravity.
14. Porosity.

Confined strength tests are especially important for underground projects, as they help establish the coefficient of internal friction for the rock and define the effect of confinement on strength. Confinement increases the strength of rock to 1.5 to 15 times the actual confining stress level. It is essential to know the effects of confinement in order to evaluate the performance of the rock during and after excavation and the stability of rock pillars.

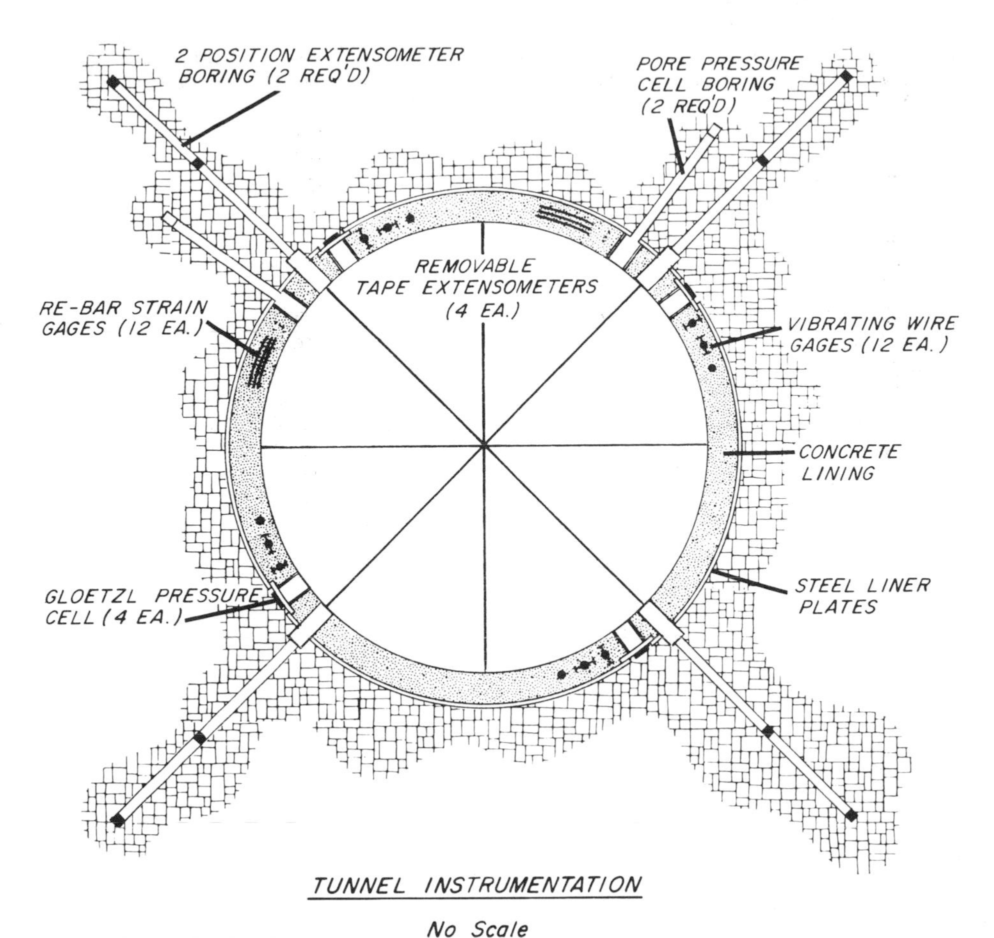

Fig. 3-16. Soft ground instrumentation scheme.

REFERENCES

1. Ray, Richard G., "Aerial Photography in Geologic Interpretation and Mapping," U.S. Geological Survey Professional Paper 373, 1960.
2. Soil Conservation Service, *Agriculture Handbook 294,* U.S. Department of Agriculture, 1966.
3. Colwell, R.N., "Photo Interpretation in Geology," in Tator, B.A. (Ed.), *Manual of Photographic Interpretation,* Americal Society of Photogrammetry, 1960.
4. Stokoe, K.H. and Woods, R.D., "In Situ Shear Wave Velocity by Cross-hole Method," *ASCE Journal of the Soil Mechanics and Foundation Division Proceedings,* p. 443, 1972.
5. Keys, W. Scott and MacCary, L.M., *Application of Borehole Geophysics to Water Resources Investigations,* U.S. Geological Survey Water Resources Investigations, 1971.
6. White, J.E., *Seismic Waves: Radiation Transmission and Attenuation,* McGraw-Hill, New York, 1965.
7. Henbest, O.J., Erinakes, D.C. and Hixson, D.H., "Seismic and Resistivity Methods of Geophysical Exploration, Soil Conservation Service," Technical Release #44, U.S. Department of Agriculture, 1969.
8. Roman, Irwin, "Apparent Resistivity of a Single Uniform Overburden," U.S. Geological Survey Professional Paper 365, 1960.
9. Mossman, R.W., Heim, G.E.. and Dalton, Frank E. (VIBROSEIS), "Applications to Engineering Work in an Urban Area," *Forty-first Annual International Meeting of the Society of Exploration Geophysicists.*
10. Van Nostrand, R.G. and Cook, K.L., "Interpretation of Resistivity Data," U.S. Geological Survey Professional Paper 499, 1966.
11. ASTM, *Annual Book of ASTM Standards, Part II,* ASTM, Philadelphia, 1971.
12. Campbell, M.D. and Lehr, J.H., *Water Well Technology,* McGraw-Hill, New York, 1973.
13. Dodds, R.K., *Measurement and Analysis of Rock*

Physical Properties on the Dez Project, Iran, Testing Techniques for Rock Mechanics, ASTM STP 402, p. 52, 1966.
14. Deere, Don U. (Ed.), *Determination of the In Situ Modulus of Deformation in Rock,* ASTM STP 477, 1970.
15. Wallace, G.B. and Olsen, O.J., *Foundation Testing Techniques for Arch Dams and Underground Powerplants, Testing Techniques for Rock Mechanics,* ASTM STP 402, p. 272, 1965.
16. Clark, G.B., *Deformation Moduli of Rocks, Testing Techniques for Rock Mechanics,* ASTM STP 402, p. 133, 1965.
17. Alexander, L.G., "Field and Laboratory Tests in Rock Mechanics," *Proceedings Third Australian-New Zealand Conference on Soil Mechanics and Foundations,* England, p. 1161, 1960.
18. Dodds, D.J., *Flat Jack Tests, Report to Missouri River District Laboratory,* U.S. Army Corps of Engineers, Omaha, Nebraska, 1969.
19. Clark, *ibid.*
20. Merrill, R.H., "In Situ Determination of Stress by Relief Techniques," *International Conference on State of Stress in the Earth's Crust,* Rand Corp. Memo RM 3583, 1963.
21. Panek, L.A., "Measurement of Rock Pressure with a Hydraulic Cell," Trans. *American Mining Engineers,* No. 220, 1961.
22. Lane, K.S. and Garfield, L.A. (Eds.), *Proceedings of North American Rapid Excavation and Tunneling Conference,* Society of Mining Engineers, 1972.
23. Dodds, D.J. and King, E., "Rock Mechanics Instrumentation, Trans Koolau Pilot Tunnel," *Proceedings of North American Rapid Excavation and Tunneling Conference,* Society of Mining Engineers, 1972.

4
Tunnel Surveys and Alignment Control

Edward W. Peterson

Construction Engineering Consultant

and

Peter K. Frobenius

Project Engineer,
Hydro and Community Facilities Division,
Bechtel Incorporated

GENERAL SURVEY PROCEDURES

Scope of Survey Work for Tunnels and Underwater Tubes

Preliminary Survey. A preliminary horizontal and vertical control survey is required to obtain general site data for route selection and for structure design. The preliminary survey relies on existing survey records and monuments. Additional temporary monuments and benchmarks are placed where required. In addition, photogrammetric mapping, recording of seismic activity and geophysical profiling are performed. Where construction of an underwater tube is planned, hydrographic mapping and current velocity surveys are included in the preliminary work. The width of the corridor to be surveyed varies from 200 feet to half a mile, depending on terrain and right-of-way conditions. A topographic map and a geologic profile of the surveyed corridor are prepared to locate the horizontal and vertical projection of tunnel centerline.

Primary Control During Final Design and Construction. After completion of route selection, a horizontal and vertical survey of a high order of accuracy is conducted. Triangulation stations are established. Primary surface traverses incorporating permanent monuments are tied to the triangulation stations. Benchmarks are placed to establish project vertical control and the new system is tied into the existing control network of the area (see Fig. 4-1 and Fig. 4-2.

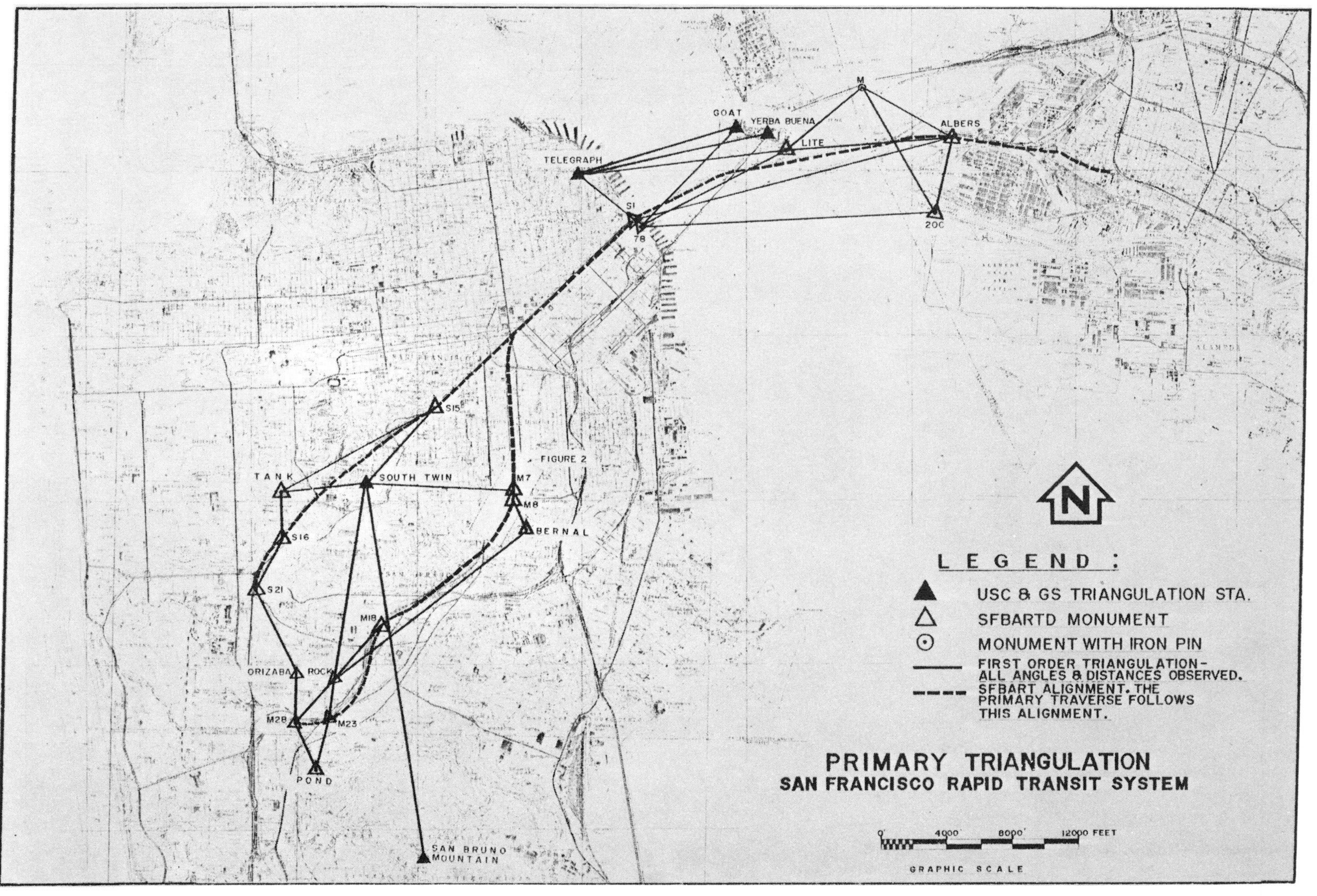

Fig. 4-1. Primary triangulation–Trans-Bay Tube.

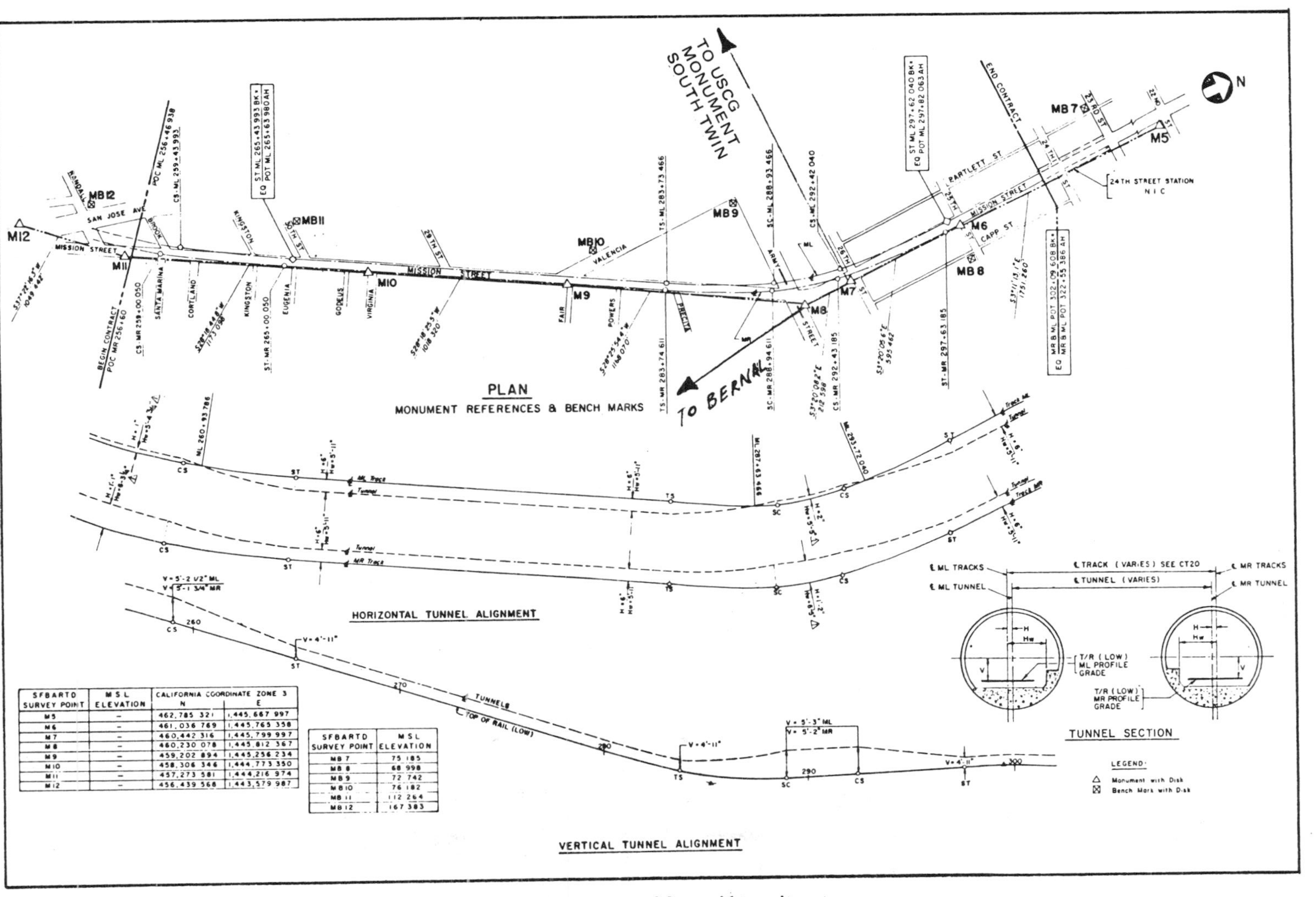

SFBARTD SURVEY POINT	M S L ELEVATION	CALIFORNIA COORDINATE ZONE 3 N	CALIFORNIA COORDINATE ZONE 3 E
M5	–	462,785 321	1,445,667 997
M6	–	461,036 769	1,445,765 358
M7	–	460,442 316	1,445,799 997
M8	–	460,230 078	1,445,812 367
M9	–	459,202 894	1,445,256 234
M10	–	458,306 346	1,444,773 350
M11	–	457,273 581	1,444,216 974
M12	–	456,439 568	1,443,579 987

SFBARTD SURVEY POINT	M S L ELEVATION
MB 7	75 185
MB 8	68 998
MB 9	72 742
MB 10	76 182
MB 11	112 264
MB 12	167 383

Fig. 4-2. Primary control for rapid transit system.

Survey Work During Construction

During construction, the following survey work is performed.

Transfer of Line and Grade. Tunnel centerline, stationing and grade are transferred from primary control surface monuments and benchmarks to the tunnel and carried forward by the traverse methods as the tunnel is constructed.

Tunnel Construction Control. A tunnel construction control system is developed that will assure driving of tunnels or placement of underwater tubes within the allowable tolerance.

Observation Wells. Observation wells to monitor the groundwater level adjacent to tunnels and underground structures are installed and read periodically.

Settlement Points. Measurement of surface movements is performed by level readings on settlement points installed over the tunnel and on adjacent buildings.

Special Recording Devices. Installation and observation of special recording devices to record vertical and lateral soil movement or stresses adjacent to tunnels or underground structures is sometimes required.

As-built Monumentation. After completion of the tunnels, permanent centerline monuments and benchmarks are placed in the tunnel at intervals of ±1000 feet and at TS (tangent spiral) and SC (spiral–circular curve point). From these monuments, measurements are taken radially to critical clearance points to ensure that the clearance envelope is in accordance with design requirements for future installations.

Accuracy Requirements

The following survey specifications are recommended for major projects.

Triangulation: Second order Class I[1,2], closing error not to exceed 1:50,000.

Vertical control: Establish permanent benchmarks to the requirements of second order Class I[1,2].

Primary traverse: Second order Class I[1,2].

The survey methods used to transfer working line and elevation underground and to set the laser beam of the tunnel construction control to line and grade should provide for this precision:

1. Angular measurements to the nearest one second of arc;
2. Stationing to the nearest thousandth of a foot; and
3. Benchmark elevation to the nearest thousandth of a foot.

The precision of the target readings of the laser control system and tunnel ring measurements as performed after every shove should be in the range of one to two hundredths of a foot. The short time available for the performance of these measurements explains the lesser precision required.

Obtained Accuracy and Correction of Deviation. Survey specifications on the BART (Bay Area Rapid Transit) project were essentially as outlined above, and the obtained accuracy is typical for present tunneling conditions. Tunnel driving specifications on the BART project required that the as-built centerline of the tunnel should be located within the limits of a 3-inch bull's eye centered on the theoretical tunnel centerline. Thus, for a tunnel heading of about 5000-foot length, the ratio of precision was:

$$\frac{0.125 \text{ ft}}{5000 \text{ ft}} = 1:40{,}000.$$

These alignment errors had to be absorbed by the permissible tolerance:

1. In consistency of the primary surface survey;
2. Errors encountered during transfer of line and grade from the surface to the heading; and
3. Inability of the construction forces to keep the tunneling equipment on the indicated alignment.

Experience shows that a deviation of the tunnel alignment (item 1 and 2 above) in the magnitude of one-tenth of a foot has to be expected for a heading of 5000-foot length. Proportionally larger or smaller deviations have to be anticipated for longer or shorter headings.

The deviation caused by the inability of the construction forces to keep the tunneling equipment on line and grade (item 3 above) depends on the equipment operation and ground conditions. Under normal ground conditions, deviation is kept to less than one-tenth of a foot. In soft soil, however, the deviation may reach half a foot. Therefore, it is apparent that the required accuracy may not be obtained in areas where difficult tunneling conditions exist.

In rock tunnels, it is possible to reset steel sets in order to meet the specified tolerance requirements. Since it is virtually impossible to reset lining in soft ground tunnels after the tunnel is driven, it is recommended that some additional vehicle clearance beyond the specified tolerance is incorporated in the tunnel design to absorb deviations of the magnitude outlined above. If clearance is not provided, realignment of the track or roadway centerline may be necessary to fit the as-built conditions.

Division of Responsibility Between Engineer and Contractor

Survey costs are small in comparison to the expenditures involved in tunnel driving. Nevertheless, if tunnel driving is held up because of faulty survey work or because of interference of the survey crew with driving operations, resulting losses are large.

The cost analysis of heading costs shown below illustrates this point.

Unit bid price of tunnel—$3000 per foot
Average advance per day—40 feet per day
Average income per hour—$5000 per hour

Even if only one-third of the average income is related directly to heading time, prevention of delay of tunneling operations is important to the contractor. For this reason, specifications relating to tunnel driving accuracy should be written as a performance specification and the contractor should get full responsibility for transferring line and grade from the primary surface control into the tunnel and for development of tunnel construction control procedures.

The engineer performs all survey work before the start of construction, such as preliminary surveys and primary control surveys on the surface. During construction, the Engineer's responsibility should be confined to checking surveys, which can be conducted on weekends when no tunneling work is performed. The engineer should also be concerned with coordination of survey work at contract interfaces. He makes sure that underground survey control of two adjacent contractors agrees at the interface.

Information concerning groundwater levels as obtained from observation well readings is of vital importance for the contractor's tunneling operation. It is, therefore, reasonable to include installation of observation wells, maintenance of the wells and periodic reading of water levels in the contractor's contractual obligation. Water level records should be made available to the engineer at the time of recording.

Level readings of surface settlement points, which serve as indication of construction problems at the tunnel heading, are not in the immediate practical interest of the contractor. As a matter of fact, the chance of inaccurate level readings during times where heading problems are encountered is greater than during normal operation. The contractor's best people are preoccupied with the construction problems at the heading during times of trouble and, therefore, spend a minimum of time on required surface level readings. For this reason, surface levels over the tunnels should be run and evaluated by the engineer and the results should be made available to the contractor. The same goes for installation and monitoring of special recording devices, such as subsurface settlement points, inclinometers and strain gauges.

PRIMARY SURVEY NETWORK

Survey Control

To establish survey control for a tunnel or underwater tube, control has to be tied into the basic survey network of the area. Any major project in the U.S. should be tied to horizontal and vertical control networks established by the National Geodetic Survey.[1] Data sheets, published by the Geodetic Survey, show location diagrams and coordinates of the triangulation stations, based on state plane coordinate systems, such as the Lambert conformed conic projection in California.[3]

The U.S. Coast and Geodetic Survey also publishes location diagrams and elevations of established benchmarks, based on the 1929 mean sea level datum. It should be noted, however, that benchmarks located on soft ground may settle, and their elevation should be checked periodically during construction of the project from reference benchmarks located on firm ground.

Electronic Distance Measurement

The spacing of triangulation stations is so wide that the expense of connecting project control to the existing stations by triangulation and baseline measurement by tape has been quite expensive in the past. Development of electronic distance measuring instruments, such as the "Geodimeter" and the "Tellurometer," makes it economically feasible to connect project control to existing survey networks. In addition, it is now common practice to check not only the baseline but almost all calculated distances between triangulation stations using electronic distance measurement instruments. Electronic distance measurement instruments are also used to run primary surface traverse along the tunnel alignment. The basic principle of electronic distance measurement is the determination of time required for a light beam or radio waves to travel between the two stations.

For distance measurement, a clear line of sight is required. The range of measurements is one-half mile to 50 miles. Instruments have an inherent instrumental error of about 0.04 foot plus or minus 2 or 3 parts per million of the distance measured. In general, the accuracy of measurement will be greater for long distances.

Primary Control for Tunnels Under Flat Land—Rapid Transit System

Monuments. Triangulation stations are established and tied into the existing National Geodetic Survey triangulation network by angular and distance measurement. Primary traverses are run between triangulation stations before the contract drawings are completed. Each traverse is several miles long and approximates the future tunnel alignment (see Figs. 4-1 and 4-2).

Benchmarks. Permanent benchmarks are placed as required to carry elevation into the tunnel. If settlement over the tunnel is a concern, as, for example, in urban areas, additional benchmarks are placed along the tunnel alignment at at a spacing not exceeding 600 feet. The benchmarks should be placed about 200 feet away from the centerline of the tunnel to avoid settlement caused by tunnel excavation (see Fig. 4-2). During construction of the tunnel, surface and subsurface settlement points are set over the centerline of the tunnel and on adjacent buildings. Level loops for settlement surveys are run through the settlement points and closed on the permanent benchmarks.

Primary Control for Tunnels Through Mountains

It is usually more difficult to establish primary control for tunnels through mountains than for tunnels in flat land. In most cases, the rugged terrain makes it necessary to establish a triangulation system connecting existing survey control on each side of the mountain. Triangulation stations are established at the top of the mountain, from where other triangulation stations in the vicinity of both portals can be observed.

Figure 4-3 shows such a triangulation network. Targets and reference points are set on the working line, and backsights are established at least 500 feet away from the tunnel portal.

Monument Construction and Records

Permanent monuments and benchmarks are constructed of brass discs secured in concrete. They are described by narrative and sketch, showing coordinates, elevation, bearing and distance to adjacent monuments. Diagrams and descriptions are included in the contract drawings. Reference marks are placed around the monument to make it possible to reestablish the monuments in case they are destroyed during construction.

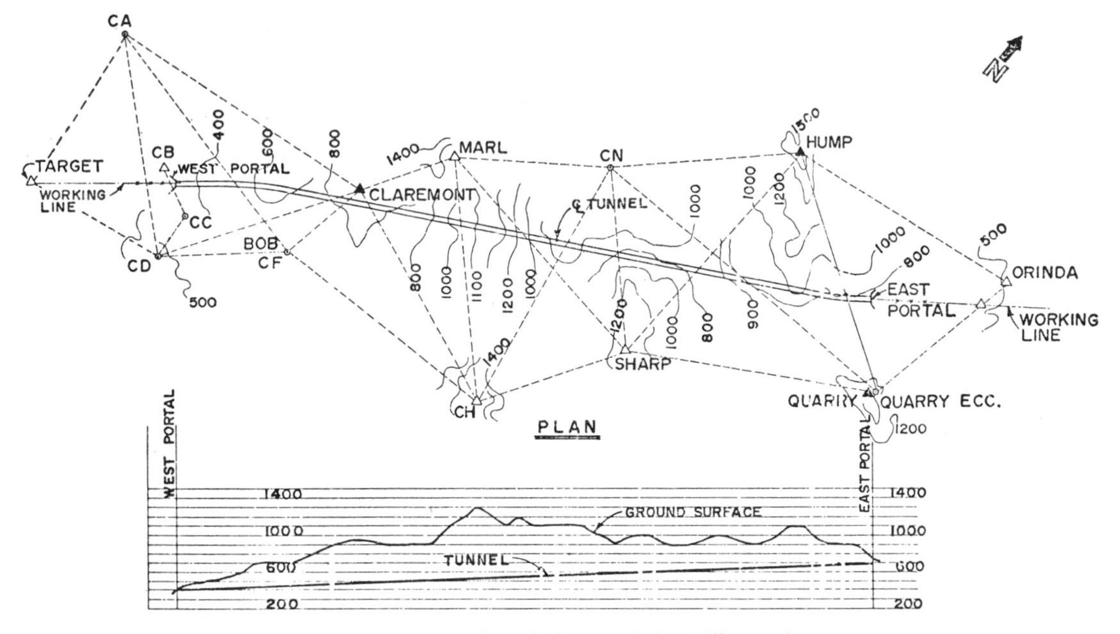

Fig. 4-3. Primary triangulation–Berkeley Hill tunnel.

TUNNEL GEOMETRY

Relationship of Centerline Track and Centerline Tunnel on a Rapid Transit System

On a rapid transit system, centerline of track and centerline of tunnel are normally not identical because of clearance requirements. Centerline of track is the basic control during layout of the system. During construction of the tunnel, however, it is desirable from a practical standpoint that the contractor's and the engineer's field personnel use centerline of tunnel rather than centerline of track as the basis of tunnel control.

The vertical and horizontal offset from centerline of track to centerline of tunnel varies with the superelevation of track (see Fig. 4-2). The resulting tunnel centerline is a curve of complex mathematical definition and cannot be produced in the field using standard survey procedures. Therefore, a tunnel centerline should be developed which is composed of tangent, circular and transition spiral sections and approximates the complex theoretical tunnel centerline within a specified tolerance (0.25 inch). This centerline should be incorporated into the contract drawings of the tunnel contract and all tunnel control should be based on this curve.

A computer printout listing coordinates of points; tangent bearing; and elevation of points and slope at 5-foot intervals on the tunnel centerline should also be incorporated into the contract documents. Since stationing of centerline tunnel and centerline track will not agree because of different curve radii, station equations between centerline tunnel and centerline track should be incorporated at the beginning and the end of each construction contract, at TS (tangent spiral) points, at SC (spiral–curve) points and at such points as vent shafts or cross passages. Stationing for tunnel centerline should start at station 0 + 00 for each tunnel contract. Stationing of track proceeds through the entire system, which, generally, is made up of several tunnel contracts.

If the rapid transit system has a natural center point from which several lines branch out in different directions, the station 0 + 00 should be assigned to this point. Stationing then proceeds to the outlying areas, and future extensions of the system can be added without upsetting the sequence in stationing.

Working Line. The working line is the survey

line used by the contractor's field personnel to establish shield or tunneling machine guidance in the tunnel. The working line may coincide with the tunnel centerline or may run through the laser position points for the laser set-up. The selection of the working line must be left to the contractor to suit his tunnel equipment and methods. The working line is usually established by traverse survey methods.

Tunnel Lining Geometry and Taper Rings

Steel, cast iron or precast lining rings are installed in tunnels to support the ground. The lining used for the construction of the soft ground tunnels of the San Francisco BART system is typical. The lining rings were composed of six welded steel segments and a small key segment (see Fig. 4-4). The segments were bolted together and to the flange of the last-erected ring in the tail of the shield to form a standard lining ring of 2.5-foot width and 17.5-foot outer diameter. The standard rings were intended for use on tunnel sections with curve radii over 1500 feet. Special rings of 18.0-foot outer diameter were installed in tunnel sections where curve radii are shorter than 1500 feet. On these curves, the overhang of subway cars in curves required additional clearance.

If straight lining rings are erected in sequence, they will form a straight tunnel section, except for deflections of rings to one side or the other caused by ring distortion. Therefore, where the tunnel was constructed on a vertical or horizontal curve, taper rings were required to change the horizontal or vertical direction of the tunnel. The width of these rings tapered from the standard 30-inch width to a reduced width of 29.25 inches, 28.75 inches and 28 inches, as shown in Fig. 4-4 and Table 4-1. The radius of

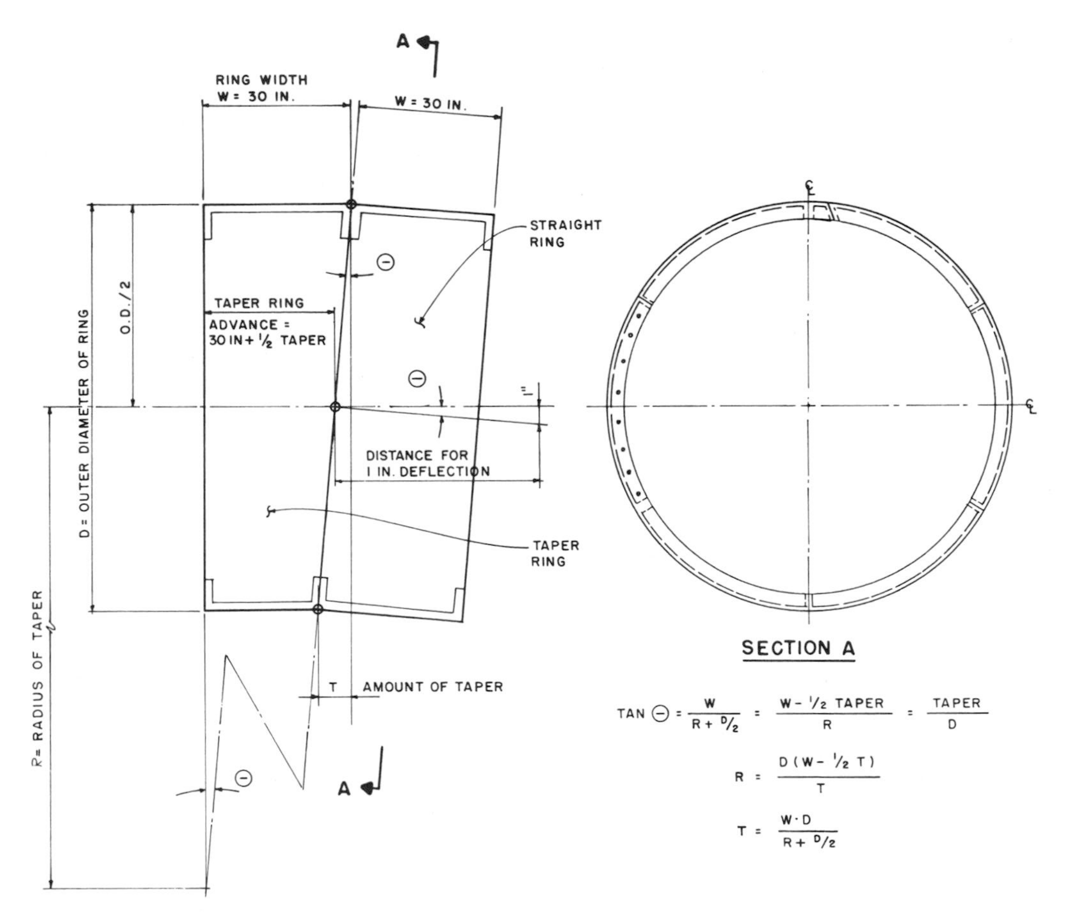

Fig. 4-4. Taper ring geometry.

Table 4-1. Taper Ring Data.

ITEM	RING O. D.		
	17 FT, 6 IN.	18 FT, 0 IN.	18 FT, 0 IN.
Ring O. D. (inches)	210.0	216.0	216.0
Amount of taper (inches)	0.75	1.25	2.00
Angle θ (degrees)	0°, 12′, 16.65″	0°, 19′, 53.65″	0°, 31′, 49.81″
Deflection in 30 inches (inches)	0.107	0.174	0.278
Distance to Deflect 1 inch (feet)	23.33	14.44	9.0
Radius of Continuous taper rings (feet)	691.25	423.0	261.0

the tunnel curve determines the type and number of taper rings required. Table 4-1 shows that taper rings of 17.5-foot outer diameter which taper 0.75 inch form a tunnel curve of 691.25-foot radius. For a longer curve radius, alternating straight and taper rings are installed.

Usually, the contractor has to make an estimate of the number of straight and taper rings required before the tunnel is started. Based on the radius and length of curves involved and the geometry of the taper ring, he makes an estimate of the theoretical number of taper and straight rings. The actual number of taper rings required will always be greater than the theoretical number because some of the taper rings are used up to correct tunnel driving deviations. To develop a feeling for the actual number of taper rings required, the number of installed rings is compared to the theoretical number of taper rings for tangent and curved tunnel sections in Table 4-2. Extra taper rings can always be installed back to back to form a double straight ring. A comparison of Lines 3 and 4 of Table 4-2 shows that the theoretical tunnel footage as computed by multiplying the number of rings installed by the width of tunnel rings has to be reduced by a fraction of the taper for each taper ring installed.

When a taper ring is erected so that its widest point lies exactly on a vertical or horizontal diameter, the change of direction is entirely vertical or horizontal. Quite often, however, the widest point of the taper ring must be erected at an intermediate point because of the necessity to break ring joints. When this is done, changes of direction take place in both vertical and horizontal planes. Taper clocks, as shown in Fig. 4-5, are convenient to visualize the relationship between vertical and horizontal change of direction for various positions of the taper ring. In order to use the taper clock, a reproduction of this clock is mounted on cardboard and the inner ring is separated from the outside ring. The outside ring is rotated in position to resemble the ring installed before the taper ring which is under consideration. The taper ring in the position shown in Fig. 4-4 will produce an overhang of 0.15 foot and a lead of 0.057 foot on the left, which will throw the tunnel to the right. If it is not desired to elevate the tunnel, the taper ring could be rotated two bolt-holes clockwise, resulting in zero overhang and a lead of 0.06 foot on the left. This would, however, cause the ring joints to line up with the joints of the previously erected ring, which is not desirable from a standpoint of lining strength.

SURVEY WORK DURING CONSTRUCTION OF THE TUNNELS

Tunnels Driven from Portals

Where tunnels are driven from portals, one work point at the portal and a backsight, both on working line, are sufficient to extend working line into the tunnels. The theodolite is set up over the work point at the portal. A backsight is taken to the previously established backsight on the working line. Then the telescope is plunged and work points are set on the working line in the tunnel (see Fig. 4-3).

Table 4-2. Straight and Taper Rings Required for Varying Curve Radii.

				TUNNEL LINED WITH 2.5-FOOT WIDE STEEL LINING RINGS		
				17.5-FOOT DIAMETER		18.0-FOOT DIAMETER
			LINE	STRAIGHT TUNNEL	1785-FOOT RADIUS TUNNEL	655-FOOT RADIUS TUNNEL
Theoretical number of rings required	Straight rings		1	660	234	111
	Taper rings	0.75-inch taper	2	–	148	–
		1.25-inch taper	3	–	–	119
		2.0-inch taper	4	–	–	37
Theoretical footage			5	1650.00	955.00	667.50
Actual Footage*			6	1649.10	950.25	656.85
Actual number of Rings installed	Straight rings		7	621	204	85
	Taper rings	0.75-inch taper	8	39	178	–
		1.25-inch taper	9	–	–	138
		2.0-inch taper	10	–	–	44
Overrun as a percentage of all rings	Taper rings	0.75-inch taper	11	6%	8%	–
		1.25-inch taper	12	–	–	7%
		2.0-inch taper	13	–	–	3%

*The difference between theoretical and actual footage is caused by the fact that each taper ring is about one-half a taper shorter at the centerline tunnel than the straight ring.

Tunnels Driven from Work Shafts

Transfer of Working Lines and Levels from the Surface to the Work Shaft. In many cases, tunnel work is conducted from work shafts. Two methods of transferring line and levels from the surface to the bottom of the shafts are described. The first method is by transit sights. Two work points located on the working line are set at opposite edges of the shaft (see Fig. 4-6).

The theodolite is set up over one of the points (WP1) and a backsight is taken to a target on the extension of the working line (WP2). Then the scope is plunged and the point on the opposite edge of the shaft (WP3) is sighted to ensure the tangent line. After the theodolite is thus aligned, the line is extended down and across to the bottom of the shaft where a work point is established (WP4). Then the theodolite is set up over the work point at the opposite edge of the shaft (WP3) and the same procedure is repeated, resulting in two work points on the working line at the bottom of the shaft.

Another method of transferring the line into a shaft is by means of steel wires supporting heavy weights hung in pails filled with oil. The working line on the surface is marked by two work points at opposite edges of the shaft. A theodolite is set over one work point and sighted on the work point at the opposite edge of the

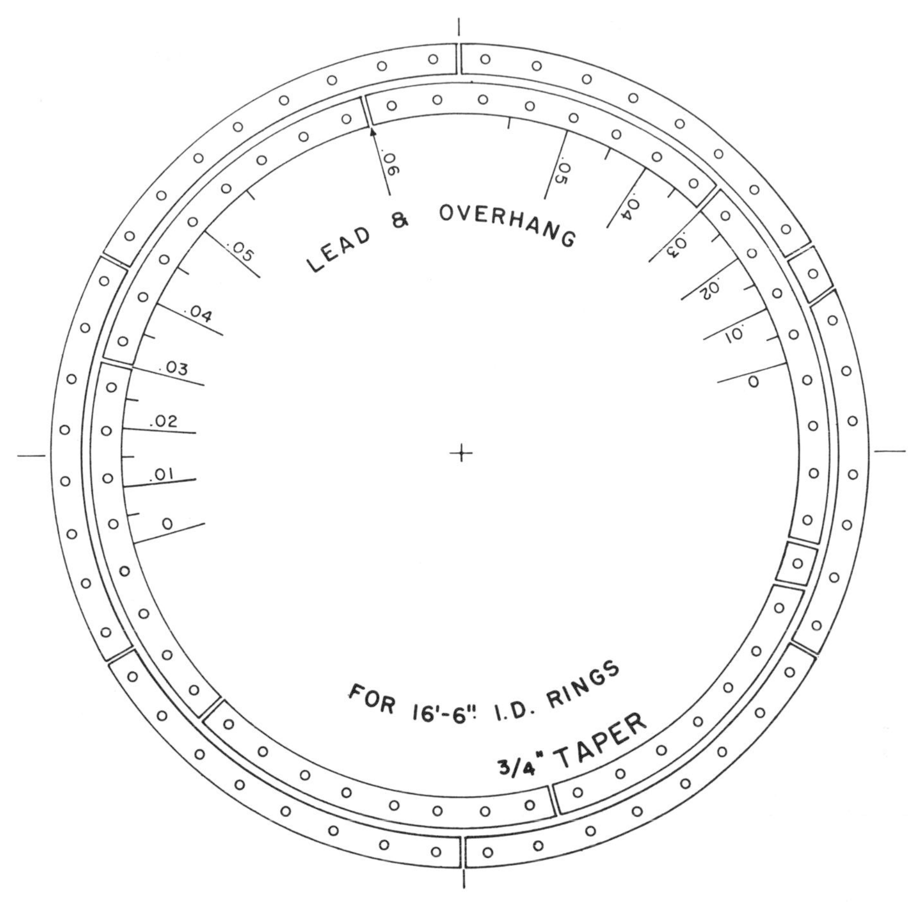

Fig. 4-5. Taper clock.

shaft. Two steel wires, each supporting heavy weights hanging down the shaft, are brought in line with the theodolite sight. An instrument set up in the shaft can be moved (wiggled in) to be in line with both wires. This instrument is then on working line and can be used to establish work points at the bottom of the shaft. In lieu of the wire weights suspended from the surface, an optical plummet (optical plumbing device) may be set up in the shaft under a surface target on the working line to establish a work point at the bottom of the shaft. Stationing is transferred into the shaft by the same methods as described above.

Levels are transferred down by suspending a standard steel tape into the shaft, supporting a weight producing standard tension. On the surface, a tape reading is made from a known height of the instrument. Then the level is set up on the bottom of the shaft and a reading is made on the suspended tape. The height of the instrument at the bottom of the shaft is obtained by subtracting the difference in tape readings from the height of the instrument at the upper level and correcting for temperature.

Carrying Working Lines Through a Compressed Air Lock

When the working line is extended through a pressurized air lock, care must be taken to position the instrument and work points at locations in the lock where distortion which takes place during pressure changes inside the lock is at a minimum. The instrument or work points should be located as near to the bulkhead as possible. The lock, which has the best foundation, should be used for locking through. A muck lock supported on concrete foundation is

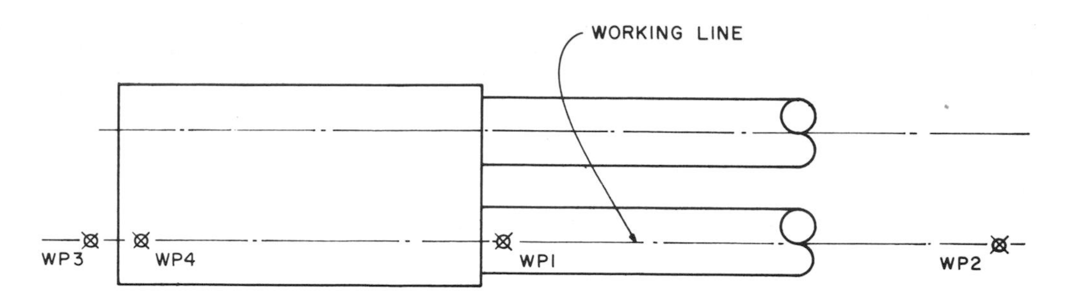

PLAN OF SHAFT AND TUNNELS

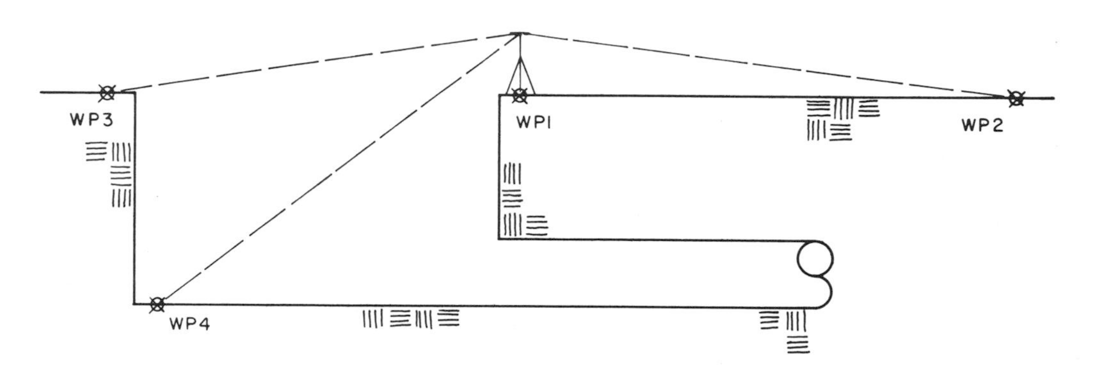

SECTION THROUGH SHAFT AND TUNNELS

NOT TO SCALE

Fig. 4-6. Transfer of working lines in shaft.

better than a man lock supported by steel framing.

Two methods of transferring line through the air lock are described. In the first method, the theodolite is set over the work point (WP5) in the shaft adjacent to the air lock (see Fig. 4-7). A backsight is taken to the work point at the far end of the shaft (WP4). Then the scope of the theodolite is plunged and three points are established in the bottom of the air lock in line with the working line in the shaft. During this operation, the free air door of the air lock is open and the air lock is in free air. Then the air lock is pressurized and the compressed air side door of the air lock is opened. The theodolite is set up in the tunnel and is moved (wiggled in) until it is in line with the three points in the air lock. Then the scope of the theodolite is plunged and working points are established in the tunnel.

Another method of transferring line through the air lock involves setting up the theodolite in the air lock over a working point (WP8) which was established in the air lock from the free air side (see Fig. 4-8).

After the theodolite is set up over the point in the air lock, a backsight is taken to the work points previously installed in the shaft on the working line (WP4 and WP5). Then the telescope of the theodolite is plunged and the lock is pressurized. The door on the compressed air side is opened and a foresight established in the tunnel and work points set on the working line. Level and stationing are carried through the air locks and by similar procedures.

Survey Pipes from the Surface

In some compressed air tunnel installations, compressors are located close to the access

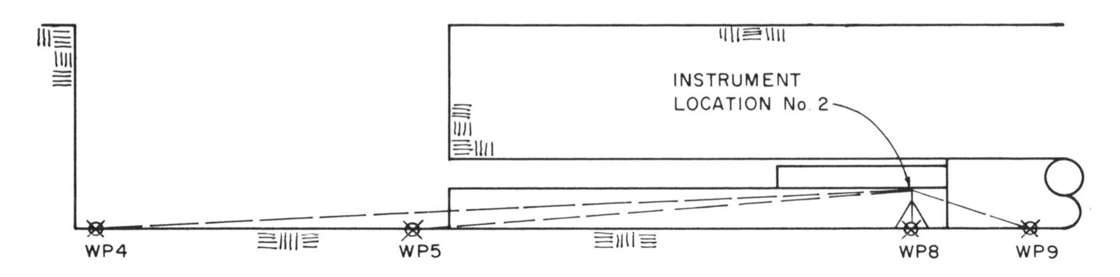

SECTION OF SHAFT AND AIRLOCKS

NOT TO SCALE

Fig. 4-7. Extending horizontal control through airlocks–Method 1.

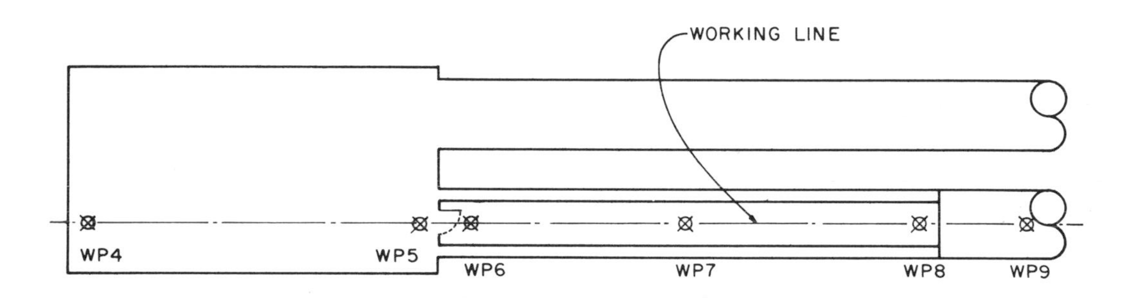

PLAN OF SHAFT AND AIRLOCKS

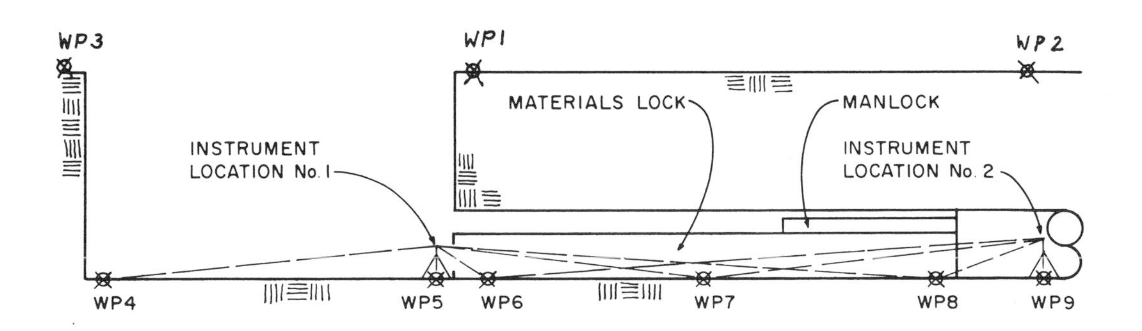

SECTION OF SHAFT AND AIRLOCKS

NOT TO SCALE

Fig. 4-8. Extending horizontal control through airlocks–Method 2.

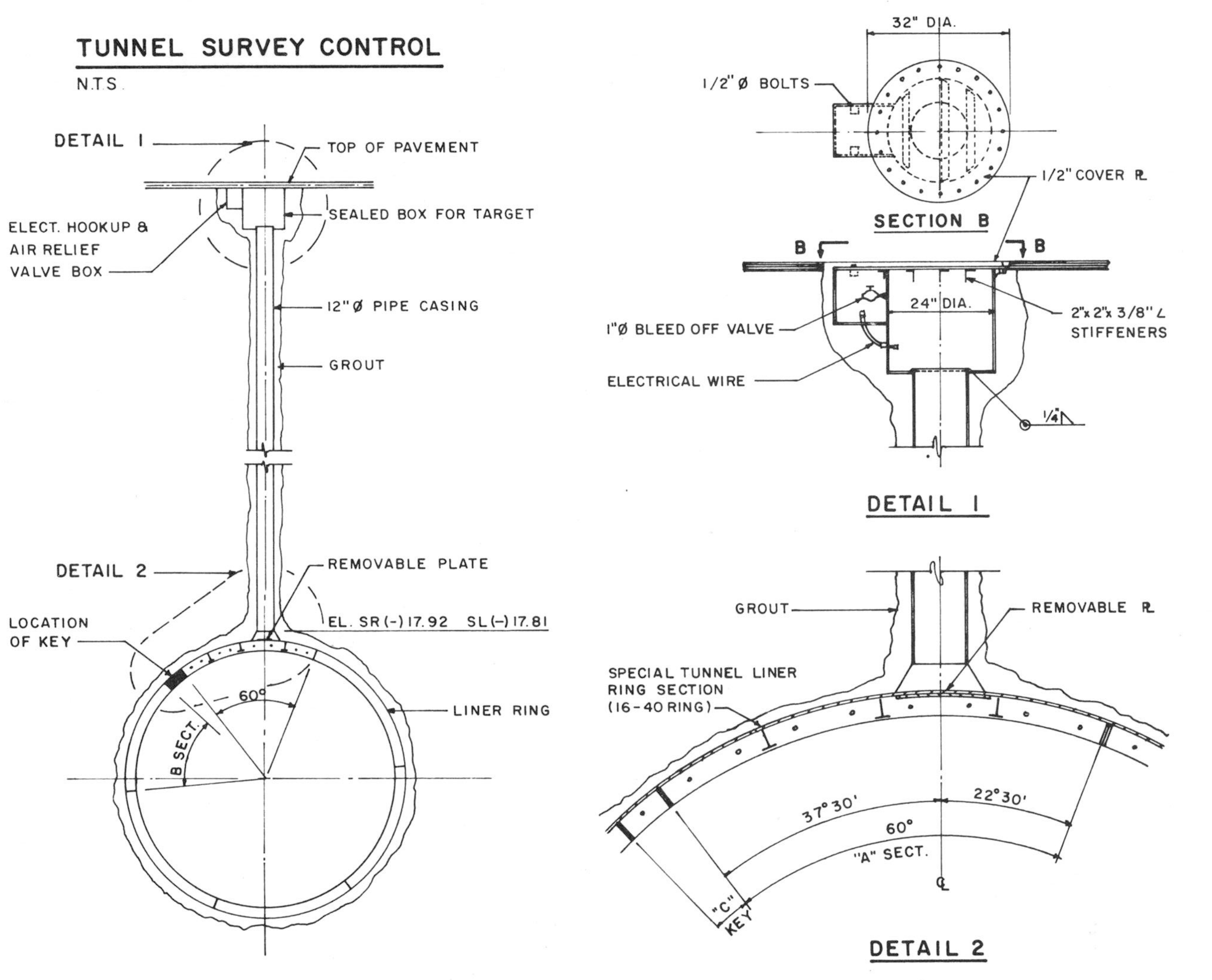

Fig. 4-9. Survey hole from surface.

shafts, and compressor vibrations, combined with the short backsight available in the access shaft and obstruction by construction equipment, make the repeated transfer of working lines through the shaft into the tunnel extremely difficult. In this case, it is advisable to bring control into the tunnel through survey holes sunk from the surface (see Fig. 4-9). Two holes at a distance of about 200 feet on the working line are sufficient to transfer two work points from the surface into the tunnel and thereby establish working line and stationing in the tunnel.

A target on the working line is set in the upper part of the survey pipe. Then the point is transferred into the tunnel either by wire weights or by the use of an optical plummet.

Survey holes can also be sunk before the tunnel reaches the receiving chamber to allow a final check in time to make necessary corrections in the direction of driving before holing through. Survey pipes shown in Fig. 4-9 are for compressed air tunnels. Free air tunnels simply require a drill hole from the surface of adequate diameter to drop a line from the surface.

Maintenance of Line and Grade in the Tunnel

From the work shafts or tunnel portals, the working line is carried ahead through work points; stationing is established by tension chaining. If the tunnel is in stable ground, pins are driven into the crown of the tunnel to serve as work points. The theodolite is centered under a plumb bob suspended from the pin. To establish a work point in soft ground where tunnel movement is anticipated, a plumb line and target are attached to movable slides bolted to ring flanges or tunnel ribs in the crown of the tunnel (Figs. 4-10 and 4-12). A survey platform is mounted below the slide and plumb bob positions for each survey run are recorded on waterproof graph paper attached to the sur-

Fig. 4-10. Slide and plumbline bolted to tunnel rings.

face of the platform. The transit is set up on the survey platform under the plumb bob suspended from the slide marking the work point. With this arrangement, surveyors can work in the crown of the tunnel without interfering with the passage of muck trains and other equipment (see Fig. 4-11). Although normal sight conditions usually permit sights around 600 feet, it is advisable to set work points at a distance of not more than 300 feet to allow for bad sight conditions due to smoke or fog. Levels are carried ahead by similar methods and benchmarks are established on tunnel lining at distances not to exceed 300 feet.

It is important to keep clear, concise records of survey work. Every time the line is rerun, a record of scale settings should be entered on a summary sheet on which the scale settings of previous survey runs are recorded. Chaining differences for distances between scales are shown on the same summary sheet (see Fig. 4-12). This survey record makes it possible to analyze the results of survey runs and to differentiate between survey errors and a pattern of tunnel movement. In soft ground, the lines are rerun at least once every week until good agreement of scale readings indicates that the location of a given control point or a section of the line is stable (see Table 4-3 for typical survey equipment and Table 4-4 for manpower requirements to transfer and maintain line and grade in tunnels).

TUNNEL CONSTRUCTION CONTROL

Construction Control For Drill and Blast Method

Where the tunnel is excavated by drill and blast methods, centerline is extended to the face before drilling for the next round starts. The centerline location is marked on the face and the drill pattern is centered on that mark. Surveyors also give centerline location for the setting of steel sets.

Fig. 4-11. Surveyors in tunnel.

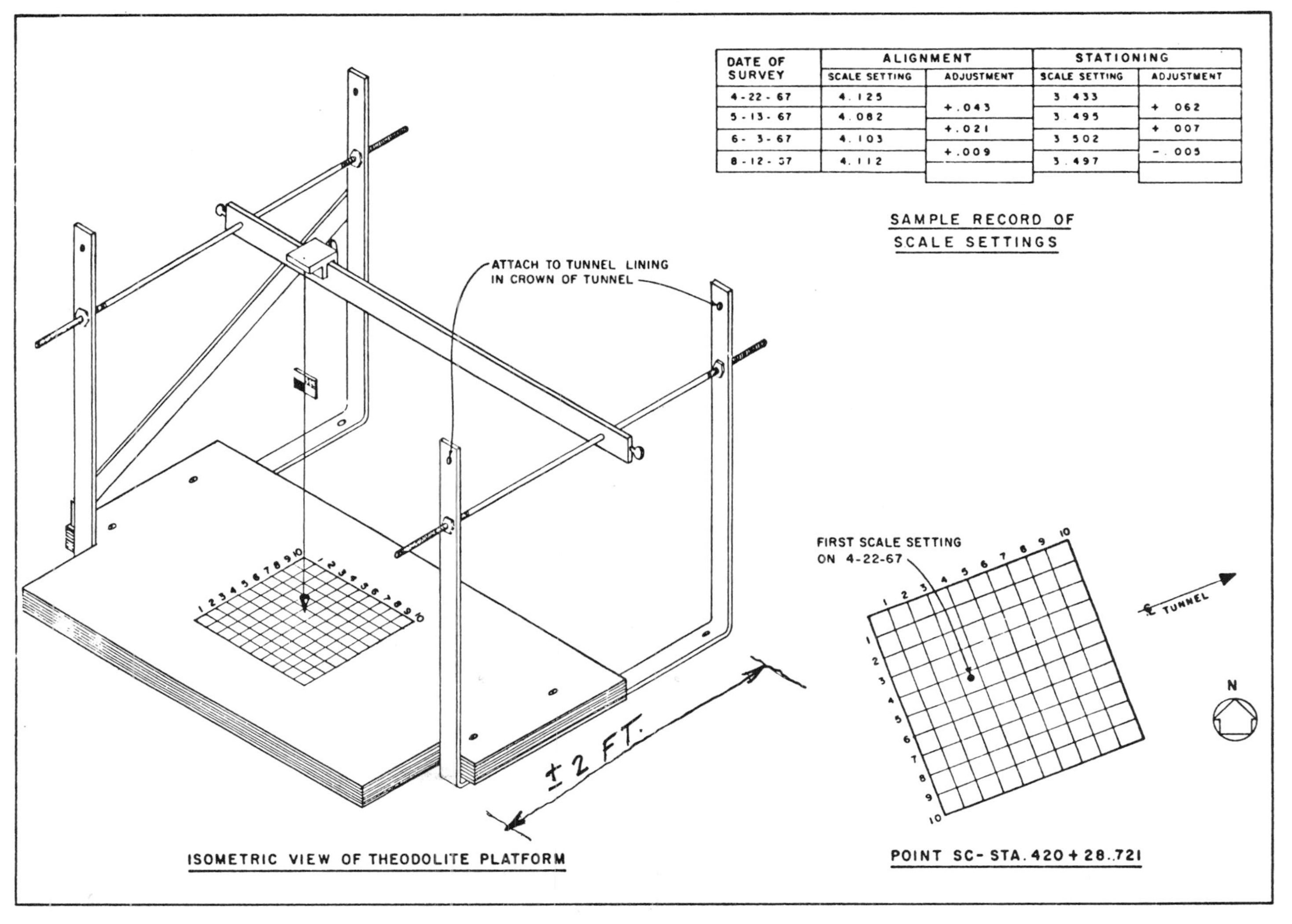

DATE OF SURVEY	ALIGNMENT		STATIONING	
	SCALE SETTING	ADJUSTMENT	SCALE SETTING	ADJUSTMENT
4-22-67	4.125	+.043	3.433	+.062
5-13-67	4.082	+.021	3.495	+.007
6-3-67	4.103	+.009	3.502	-.005
8-12-37	4.112		3.497	

Fig. 4-12. Survey platform.

Table 4-3. Surveying Instruments and Accessories. (The following survey instruments and accessories have been found to provide the degree of precision for tunnel work necessary to comply with the section on "Accuracy Requirements,")

NUMBER REQUIRED	
2	Theodolites, Wild T-2 complete with container, tripod, lateral adjusting trivet, illuminating accessories.
1	Optical plumbing device to be used with T-2 theodolite.
1	Set of traverse equipment consisting of three targets with tribrachs interchangeable with T-2 theodolite, battery boxes for illumination of targets (for survey work on the surface).
	Scales and survey platforms as required to use as work points in the tunnel.
2	Levels, Zeiss Ni2, complete with tripod and trivet.
1	Laser for shield control system (several models available).
1	Constant voltage transformer to ensure constant voltage to laser from fluctuating power supply due to varying demands of tunneling operations.
4	Level rods, Philadelphia type, 7 feet/13 feet, with micrometer target.
2	Rod levels.
1	Steel tape, 300 feet, standardized by the National Bureau of Standards.
4	Steel tapes, 100 feet, standardized, graduated in hundredths of a foot.
6	Steel tapes, 50 feet.
4	Clamp handles for steel tapes.
4	Tension handles for steel tapes.
2	Tape thermometers and cases. Folding rules graduated in hundredths of a foot. Plumb bobs.

Table 4-4. Manpower Requirements. (This is a breakdown of manpower used for surveying and shield or tunneling machine control on typical projects.)

Contract 1S0022: San Francisco Bay Area Rapid Transit (BART) Project

Contractor: Morrison–Knudsen, Brown & Root, and Perini Corporation, a joint venture.

Length and Diameter of tunnel	Two 17.5-foot diameter tunnels of 5,200-foot length each.
Tunneling equipment:	Two tunneling machines driving both tunnels simultaneously.
Average footage per 24-hour day:	40 feet per day per machine.
Maximum footage per 24-hour day:	107 feet per day per machine.

Manpower for transfer of line and grade to heading, maintenance of line in tunnel and surface settlement readings:

One party chief One instrument man Two rod men	One-day shift every day and many weekends
Tunnel machine control:	One surveyor (heading engineer) on each machine each shift.

Contract 1S0024: San Francisco Bay Area Rapid Transit (BART) Project

Contractor: Peter Kiewit Sons' Company.

Length and Diameter of Tunnels:	Two 17.5-foot diameter tunnels of 1640-foot length each.
Tunneling equipment:	Two tunneling shields driving both tunnels simultaneously.
Average footage per 24-hour day:	20 feet per day per shield.
Maximum footage per 24-hour day:	32.5 feet per day per shield.

Manpower for transfer of line and grade to heading, maintenance of line in tunnel and surface settlement readings:

One party chief One instrument man	One-day shift every day and some weekends
Tunnel machine control	One surveyor (heading engineer) for both shields for each shift (one-half surveyor per shield.

Construction Control for Shields or Tunneling Machines

A shield or tunneling machine progresses in a sequence of shoves. After each shove, the shield or machine is stopped and its location and attitude are determined.

If the shield or machine is found to be off-line, adjustments of the steering mechanism are made to guide it back to its desired location.

Where tunnel lining is erected in the tail of the shield, its location and attitude are determined and recorded. The decision of whether to install standard or tapered lining sections after the next shove is based on this record.

The most practical method of shield or machine control is by laser beam and double target (for a description of the laser principle, see the *Encyclopedia of Science and Technology*[4]). A laser tube (see Fig. 4-14) is set up at a distance behind the shield or tunneling machine to emit a laser beam from a predetermined point of origin along a predetermined line to the targets mounted on the shield or tunneling machine (see Fig. 4-13 and Fig. 4.20). In the horizontal plane, the laser line is a chord line or a tangent to the tunnel centerline (see Fig. 4-15). In the vertical plane, the laser line approximates the

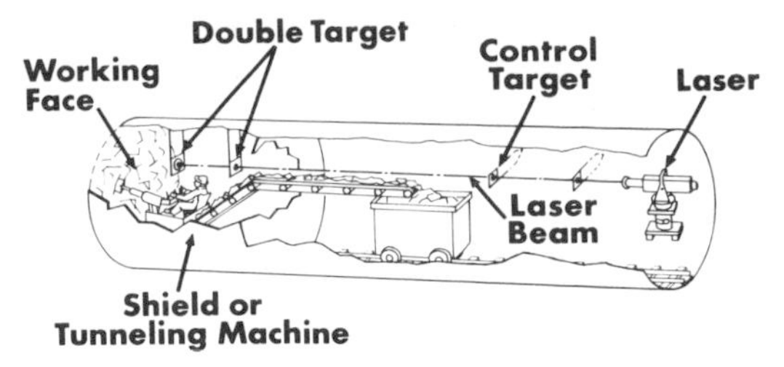

Fig. 4-13. Tunneling control by laser and double target.

slope of the tunnel centerline (see Fig. 4-16).

After the tunnel is driven to the end of one laser beam line, the laser is moved to the next laser position point and the laser tube is set to emit the laser beam along the next predetermined laser beam line.

Two targets, called the front target and the rear target, are mounted on the shield or tunneling machine, centered on a line parallel to the longitudinal axis and from 4 to 10 feet apart. The rear target is transparent and the leading target opague. The targets are intersected by the laser beam, which produces a bright red spot on the target. Theoretical points of intersection between laser beam line and targets are calculated in advance for each

Fig. 4-14. Laser setup in tunnel.

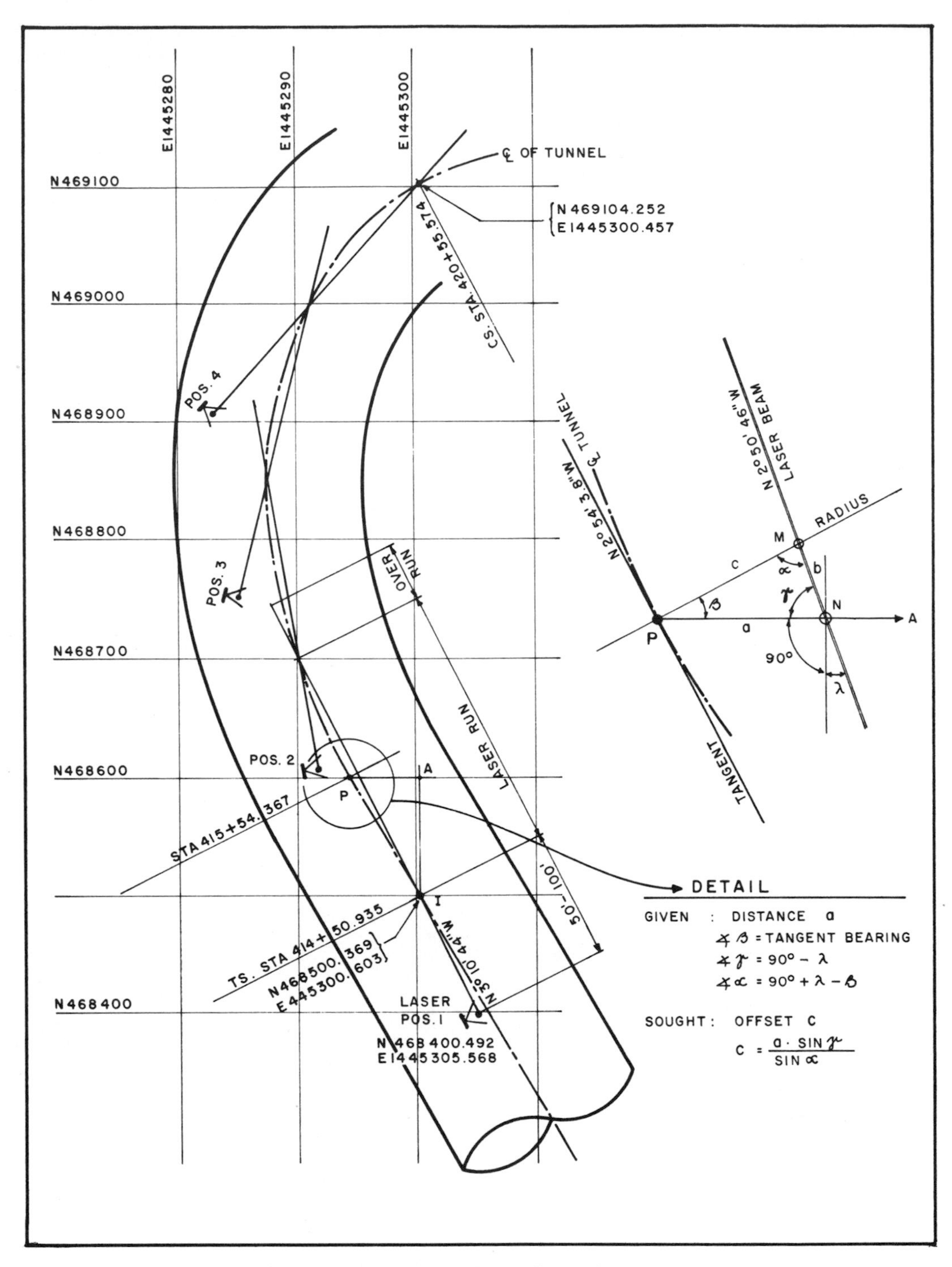

Fig. 4-15. Plan of tunnel centerline and laser line.

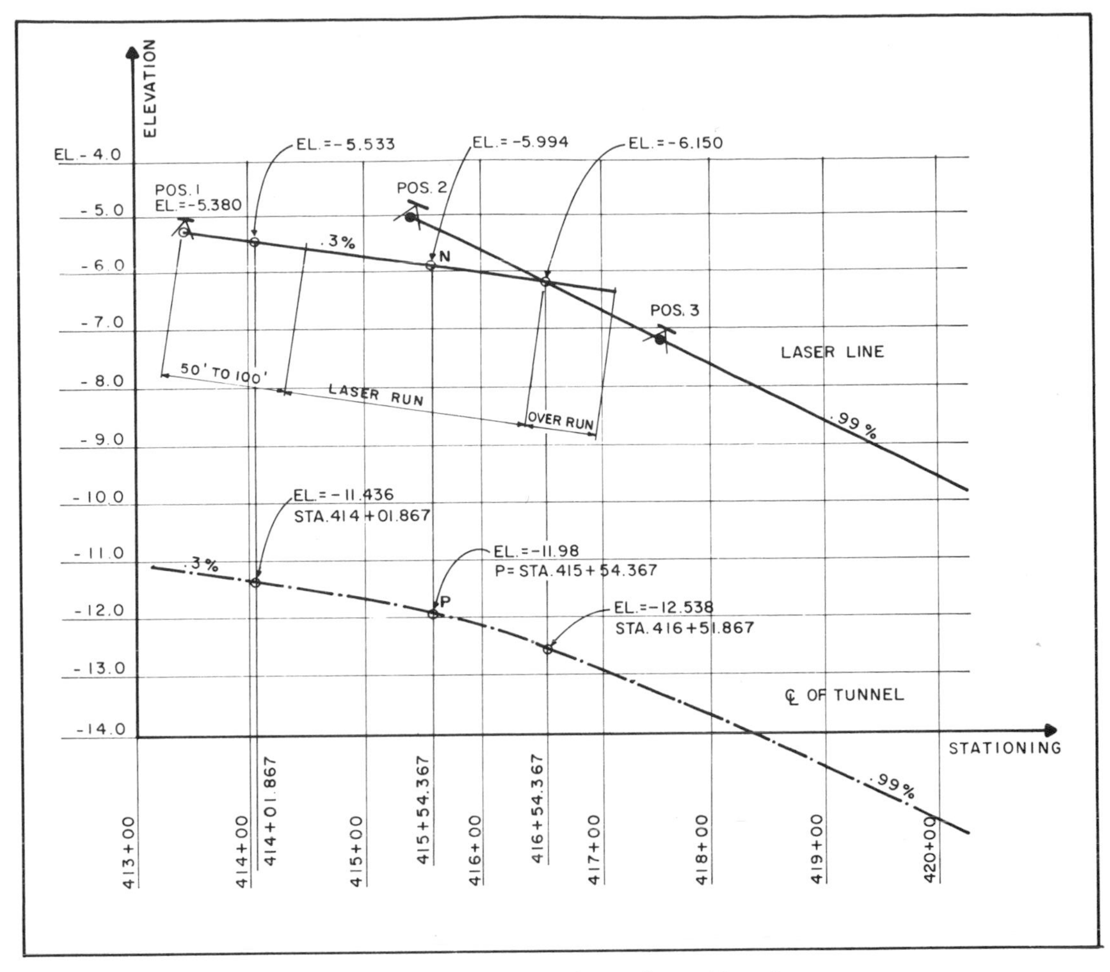

Fig. 4-16. Profile of tunnel centerline and laser line.

shield location. The theoretical points of intersection are plotted on the targets and connected by a curved line (see Fig. 4-17).

The shield or tunneling machine is guided by attempting to maintain coincidence of the actual laser line intersection points with the predetermined intersection points on the target.

Calculation of Offsets to Laser Line from Tunnel Centerline

The laser line and centerline of the tunnel are plotted in plan and elevation (see Figs. 4-15 and 4-16). Several trials may be necessary to find the best location of the laser line, and the following guidelines are observed in locating the laser line.

1. Find the longest unobstructed line of sight to reduce the required number of changes of laser positions.
2. Select a laser position which is out of the way of passing tunneling equipment.
3. On tangent and flat curves, the length of laser line is limited by diffusion of the laser beam (practical limit of present models about 1000 feet). The beam may be focused at any point from 50 feet to infinity. At 1000 feet, the spot can be concentrated to one-inch diameter.
4. The target offsets for the projected laser line cannot exceed the size of the laser targets, which is often restricted due to space requirements for other equipment.
5. At the end of each laser line, an overrun is

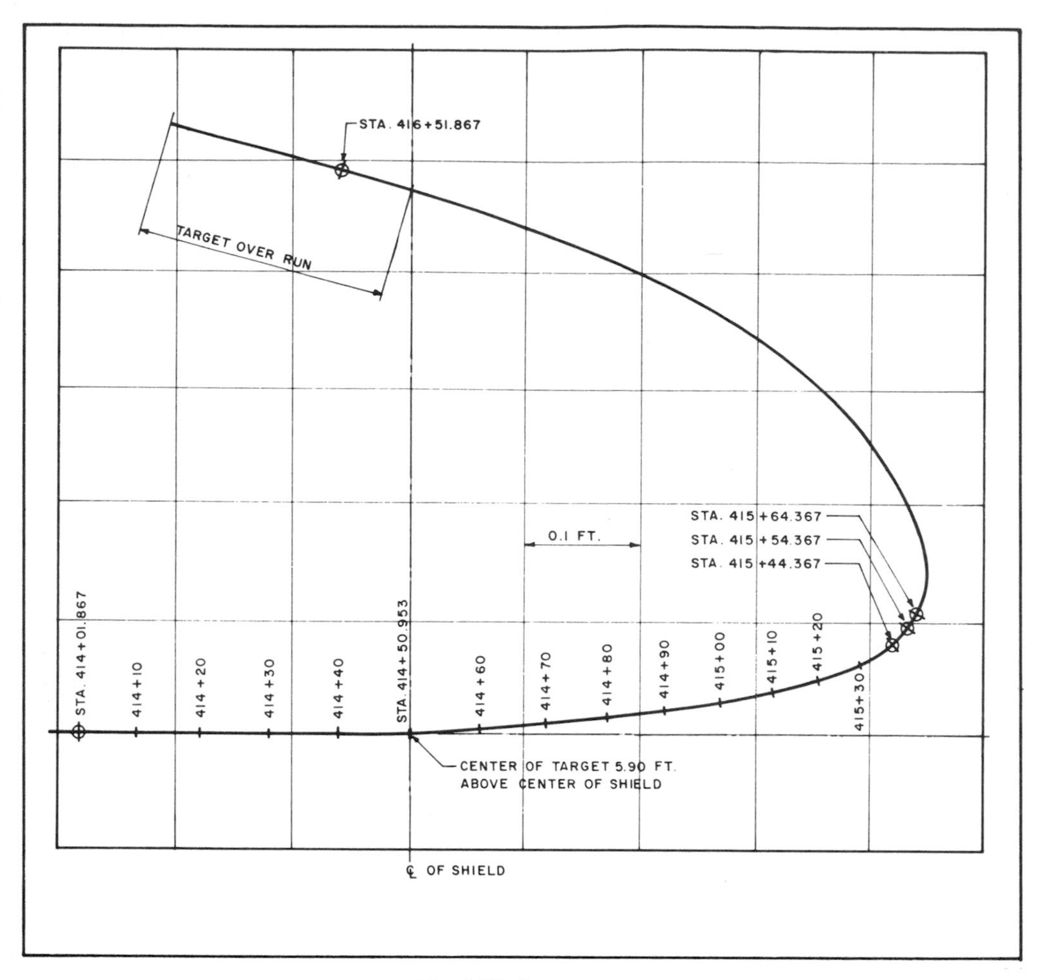

Fig. 4-17. Laser target.

provided. This gives the heading engineer the opportunity to make the change of laser positions and targets at any convenient time while the shield or tunneling machine operates in the tunnel section covered by the overrun.

Once the coordinates and the elevation of the laser position, as well as the lateral bearing and slope of the laser line, have been determined, horizontal and vertical target offsets are calculated as shown in Table 4-5 and Figs. 4-15 and 4-16. Then the target offsets are plotted on the target, as shown in Fig. 4-17. The calculations are performed under the assumption that coordinates and elevation of points on the tunnel centerline are given at 5-foot intervals. Also given are the tangent bearing and the slope of the tunnel centerline at each point. If they are not given, they must be determined by standard survey methods.[3,4]

The calculations for target offsets are of a repetitive nature and the use of a computer is helpful to reduce the time required for the calculations.

Positioning of the Laser

The laser has to be positioned in its X, Y and Z coordinate positions at the laser position point and its laser beam has to be set to beam along the predetermined laser line. Several ways of mounting the laser tube are in use. Recently developed laser models are equipped with a standard hub for tripod or bracket mounting (see Fig. 4-14). The base is leveled by leveling

Table 4-5. Horizontal and vertical offsets from centerline tunnel to laser line.*

		Column 1	Column 2	Column 3	Column 4	Column 5
Horizontal Offset	1. Point P at station	414 + 01.867	415 + 44.367	415 + 54.367	415 + 64.367	416 + 51.867
	2. Tangent bearing	N3° 10′ 44″W	N2° 57′ 7.9″W	N2° 54′ 3.8″W	N2° 50′ 41″W	N2° 7′ 48.7″W
	3. Coordinates of point P	N 468 451.359 E1 445 303.325	N 468 593.646 E1 445 295.546	N 468 603.633 E1 445 295.035	N 468 613.620 E1 445 294.534	N 468 701.035 E1 445 290.686
	4. Coordinates of intersection laser line and curve	N 468 500.369 E1 445 300.603	N 468 500.369 E1 445 300.603	N 468 500.369 E1 445 300.603	N 468 500.369 E1 445 300.603	N 468 500.369 E1 445 300.603
	5. Distance $P - A$ = Line 4 – Line 3	2.722	5.057	5.568	6.069	9.917
	6. Distance $I - A$ = Line 3 – Line 4	49.010	93.277	103.264	113.251	200.666
	7. Distance $A - N = I - A \times \tan \lambda$	2.437	4.637	5.134	5.630	9.976
	8. Distance $P - N$ = Line 5 – Line 7	0.285	0.420	0.434	0.439	0.059
	9. Horizontal laser line offset: Distance $P - M = \dfrac{P - N \sin \gamma}{\sin \alpha}$	0.285	0.420	0.434	0.439	0.059
Vertical Offset	10. Elevation point P	−11.436		−11.987		−12.538
	11. Slope of centerline tunnel at point P	0.30%		0.48%		0.64%
	12. Elevation of laser line at point P	−5.533		−5.994		−6.150
	13. Slope of laser line	0.30%		0.30%		0.30%
	14. Distance $P - N$ = Line 10 – Line 12	5.903		5.993		6.388
	15. Vertical laser line offset: Distance $P - M = \dfrac{P - N \sin \gamma}{\sin \alpha}$	5.900		5.991		6.388

*All dimensions are in feet.

Fig. 4-18. Laser mounted inside pipe with adjustable wingbolts.

screws and circular levels. In addition, coarse and fine adjustment of azimuth and elevation of the laser beam are provided.

Another way of mounting the laser tube is to secure it inside a pipe with adjustable wingbolts (see Fig. 4-18). The pipe, in turn, is mounted by adjustable brackets to the tunnel lining.

Regardless of the method of mounting the laser, safety checks have to be installed to alert the surveyor if the laser drifts off alignment or is hit by construction equipment. A good method of checking utilizes a control target made of a piece of metal with a hole just large enough for the laser beam to pass through (see Fig. 4-19). The control target is set on the laser beam line between the laser and the shield and the laser beam passes through the hole of the control target. Should the laser move, the disturbance is quickly noticed.

Instruments and Manpower Requirements

One-second theodolites and other survey equipment, as shown in Table 4-3, are used to set the laser on the laser position point and to set the control target. Typical manpower requirements for tunnel survey are shown in Table 4-4.

Shield or Tunneling Machine Driving Records

The target observations and measurements are recorded by the heading engineer after each shove. Following is the description of a typical progress report, illustrating the required records and the use of the shove jacks to maintain the shield on alignment (see Tables 4-6 and 4-7). The report was made on Contract No. 1S0024 of the San Francisco BART project, Peter Kiewit Sons' Company, contractor. The shield is of 18-foot outer diameter. The steel tunnel rings installed behind the shield are of 17 feet, 6 inches outer and 16 feet, 6 inches inner diameter. Each ring was made up of a key segment and six segments of 2.5-foot width.

From the data recorded on the progress report, the location and attitude of the shield are determined. In this specific case (after shove for Ring No. 311), the shield is high ($1\frac{1}{4}$-inch front target, $1\frac{3}{4}$-inch rear target). It is to the left of the tunnel centerline ($\frac{1}{4}$-inch front target, $\frac{3}{8}$-inch rear target). In the vertical and horizontal planes, the shield axis is not parallel to the tunnel axis.

Fig. 4-19. Control target mounted in crown of tunnel.

Fig. 4-20. Laser target mounted on tunneling machine.

The deflection of the shield axis from the theoretical grade of the tunnel centerline is called "inclination" or "pitch." The algebraic difference of vertical front and rear target deviations serves as a measurement of shield inclination ($1\frac{1}{4}$ inches - $1\frac{3}{8}$ inches = $\frac{1}{8}$ inch). (See Fig. 4-21.)

The deflection of the shield axis from the tunnel centerline in the horizontal plane is called "Yaw." It is measured as the algebraic difference of the horizontal front and rear target deviations ($\frac{1}{4}$ inch - $\frac{3}{8}$ inch = $\frac{1}{8}$ inch). (See Fig. 4-22.)

The preceding progress report was developed for a shield-driven tunnel. However, the above described principle of alignment control also applies to soft ground and rock tunneling machines.

SURVEY OF CONSTRUCTION OF UNDERWATER TUBES

All tubes and bridges, when under construction, have different conditions for survey and construction control due to terrain, climatic conditions, reach of water crossing, dense development, restricted military reservations for both shore and water encroachments, merchant vessel anchorage, etc. There are but few locations where conditions are ideal. Hence, one might be reduced to using a less than desirable control system but obtain excellent control by careful surveying procedures.

Short tubes, a mile in length more or less from shore to shore, usually need but one strong quadrilateral approximately square or any other four-sided figure without small angles. One side of the quadrilateral should, if possible, determine the centerline of the tube and the other two monuments or station (one on each shore or close proximity) for angular control. At least two of the monuments, when possible, should be accurately tied in to National Geodetic Survey control stations of first order so that the distance between them can be used as a baseline. Otherwise, a baseline of sufficient length, (3000 to 4000 feet) would have to be measured precisely either by tape or by some type of distance meter.

Table 4-6. Ring 311
PROGRESS REPORT

LINE 1 — SHIFT G.Y. WALKER TOMLIN DATE 11-17-71

LINE 2 — PUSH DATA 311 TIME REMARKS T.L.

START 11:30
STOP 11:33
START 11:44
STOP 11:46
START 11:50
STOP 11:53

LINE 3 — TOTAL TIME 23 DOWN TIME 15 PUSH TIME 8

SHIELD POSITION AT END OF PUSH

LINE 4 — PUSH PRESSURE (3,706) JACKS USED (1/2 ALL BUT 7, 8) ROTATION ($1\frac{1}{2}$ C.W.)

LINE 5 — FRONT ($1\frac{1}{4}$) (1/2 " " 8) REAR ($1\frac{3}{8}$)

(1/4) ( ) (3/8) ( )

LINE 6 — RING POSITION AT END OF PUSH

RING NO. 309 ( ) SEE SWING SHIFT REPORT — PITCH (1") OVER-HANG ( )

RING NO. 310 (3/4) LEAD (1/8) 9/16 $8\text{-}3\frac{3}{4}$ $8\text{-}2\frac{5}{8}$ $16\text{-}6\frac{3}{8}$ ( )

LINE 7 — TIE ROD YES STA. 440+00 TIE ROD NO TYPE ST. 440+02.5

For tubes longer than one mile, more elaborate survey procedures are necessary. The survey work for the construction of the Trans-Bay Tube between San Francisco and Oakland is typical for the construction of long underwater tubes and is described below. For tubes of shorter length, survey requirements may be relaxed as suggested in the preceding paragraphs.

The twin track transit tube connects the rapid transit rail lines from downtown San Francisco with Oakland, Berkeley and Richmond. The length of the tube is 3.6 miles, and alignment of the tube changes both horizontally and vertically, dipping to about 135 feet below sea level (see Fig. 4-23).

Preliminary Surveys

Prior to design and construction, preliminary surveys were conducted to obtain general site

Table 4-7. Description of Progress Report.

Line 4. *Rotation or Roll* ($1\frac{1}{2}$ inches clockwise) is defined as the rotation of the shield about its longitudinal axis. If roll is indicated, the laser targets are adjusted for roll by moving them along the slotted holes until the error in target location caused by roll is eliminated (see Fig. 4-20 for slotted holes).

Line 5. *Front target deviation* (vertical $1\frac{1}{4}$ inches high, horizontal $\frac{1}{4}$ inch to the left). Vertical or horizontal displacement of the shield at the location of the target from its desired location.

Rear target deviation (vertical $1\frac{3}{8}$ inches high, horizontal $\frac{3}{8}$ inch to the left).

Line 6. *Ring pitch* (1 inch). Ring pitch is determined by hanging a plumb bob from the lead flange at the crown of the ring. The distance from the lead flange in the invert of the ring to the plumb line is measured as ring pitch.

Ring lead (Ring No. 310–$\frac{1}{8}$ inch). Lead boards are maintained on the tunnel lining at springline within 50 feet of the heading. On each board, a mark is so placed that a line joining the two marks is at right angles to the tunnel centerline. The ring lead is determined by measuring the distance from the marks on the lead boards along both springlines of the tunnel to the lead flange of the ring. The difference of both measurements is the lead.

Horizontal ring deviation (Ring No. 310–$\frac{9}{16}$ inch). A plumb line is set on the laser line. Measurements are made to the left and right of the plumb line to the ring springline. The offset from laser line to tunnel centerline(0.434 inch for Ring No. 310 at Station 415 + 54 in Table 4-5) is subtracted from the measurement to the right springline and added to the measurement to the left springline. The result represents the distance from left or right springline to tunnel centerline (left, 8 feet, $3\frac{3}{4}$ inches; right, 8 feet, $2\frac{5}{8}$ inches). The two distances are added to arrive at the actual ring diameter (16 feet, $6\frac{3}{8}$ inches). The difference of the two distances is divided in half and represents the ring deviation ($\frac{9}{16}$).

Vertical ring deviation (Ring No. 310–$\frac{3}{4}$ inch). A measurement is taken from the laser line to the top flange of the ring. The theoretical distance for this measurement is calculated in the office (2.259 feet). The theoretical distance is subtracted from the actual measurement (2.319 feet) to arrive at the vertical ring deviation (0.06 foot = $\frac{3}{4}$ inch). Measurement to the invert of the ring are not included in the ring check because the presence of muck and mucking equipment in the invert makes such measurements impractical.

Line 7. (Ring No. 309, tie rod installed; Ring No. 310, tie rod not installed). Indicates whether the measurements were made after or before the tie rod, pulling both springlines together, was installed.

data for route selection and for structure and foundation design. These surveys included photogrammetric mapping of the proposed terminal sites, hydrographic mapping of a 1500-foot wide corridor from San Francisco to Oakland, current velocity, salinity and temperature surveys, and geophysical sub-bottom profiling to determine geological data of the Bay bottom.

An interim triangulation network, based on National Geodetic Survey positions in the area, was established to serve as control for the preliminary surveys, with temporary monuments set near the San Francisco and Oakland terminal sites and on the footings of the San Francisco-Oakland Bay Bridge.

Primary Control During Final Design and Construction

After completion of route selection, a high order control survey was conducted to serve as primary control for construction. This survey was based on National Geodetic Survey positions which were tied into the Bay Area Rapid Transit primary survey system controlling the entire project. Survey monuments were established near the terminal areas and on footings of the Bay Bridge in locations overlooking the entire construction area (see Fig. 4-24). The horizontal survey was conducted using one-second theodolites and Geodimeter. The network consisted of a Geoimeter traverse combined with simple triangulation figures, using the traverse courses as bases. A minimum accuracy of 1:25,000 was specified for the primary survey and, to attain this, ten sets of angles were observed with a rejection limit of four seconds from the mean. Distances were measured on all traverse courses and on sufficient triangle sides to assure that the specified accuracy of survey was attained. Azimuth was determined by observing across the Bay from the National Geodetic Survey monuments in San Francisco and Oakland, and checks on azimuth were made into previously established azimuths in the Bay Area Rapid Transit primary control survey system.

During the survey, it became apparent that survey accuracy could have been increased with-

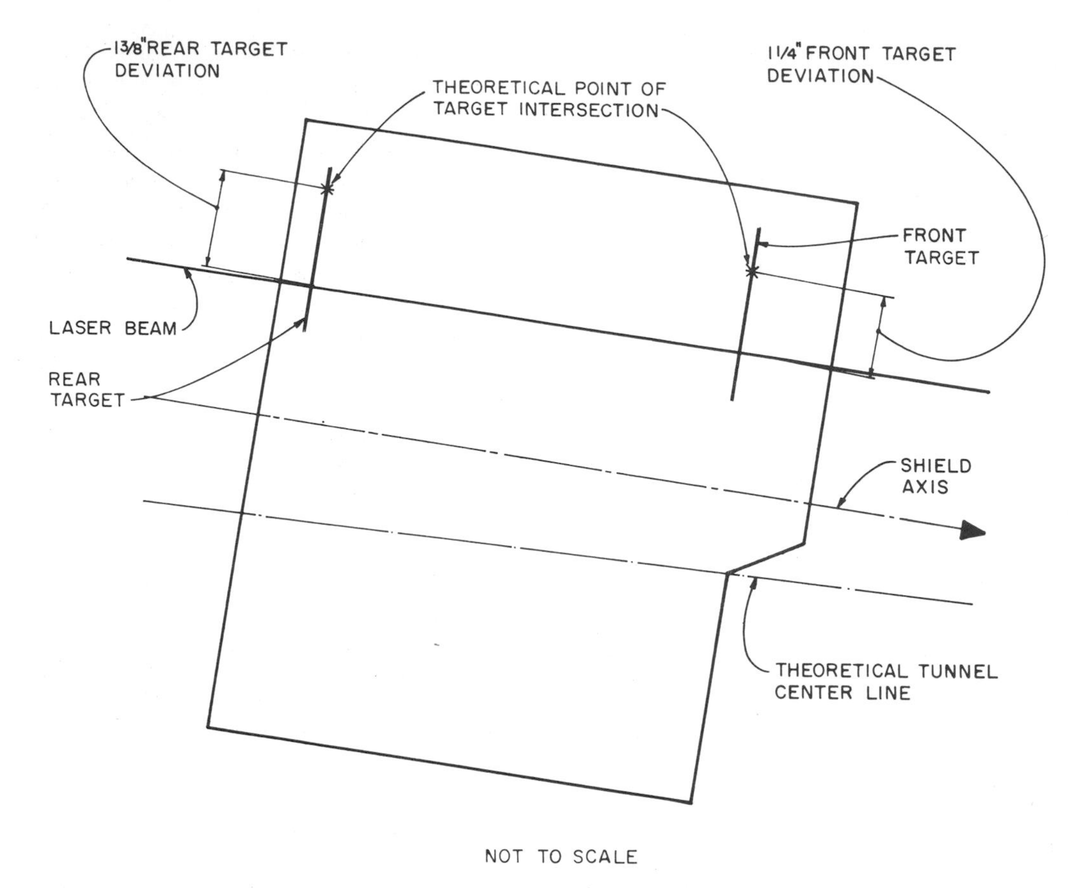

Fig. 4-21. Side view of shield.

out large additional expenditure to 1:50.000, for triangulation and primary traverse. This increase in accuracy avoids control discrepancies during construction and is recommended for future work.

Computations were based on the National Geodetic Survey published coordinates (California Coordinate System, Zone 3; Lambert Conformal Projection). Adjustments in the survey computations were made by simple triangle observing strength of figure, and by compass rule adjustment of the Geodimeter traverse.

After completion of final alignment design, additional control monuments were established on the San Francisco and Oakland shores to precisely define the two tangents in the final Trans-Bay Tube alignment. These monuments defined the N 61°51′54″ E tangent and the N 79°06′18″ E tangent of the final alignment, as shown on Fig. 4-23.

Vertical control was based on mean sea level datum, as defined by recently established Bay Area Rapid Transit System benchmarks, set near the tube terminals. The terminals are located in a landfill area, and the published elevations of National Geodetic Survey benchmarks in this area were outdated because of subsidence.

Levels were carried across the Bay by leveling the benchmarks on the footings of the Bay Bridge. The "Valley Crossing" method of leveling was used, with two level instruments sighting simultaneously in opposite direction at each other. This method tends to minimize (but does not eliminate entirely) the effects of vertical refraction anomalies in the line of sight between the two instruments. At least six separate readings were made at each instrument, with a rejection limit of 0.04 feet from the Mean. During the level survey, it was found necessary to reobserve several lines to attain a suitable series of observations. This was because of variations in the light path due to dynamic refractive conditions over the long (up to 3400 feet) lines being observed. Elevation differences on each line were computed by averaging all observed differences, extending those that fell outside of the 0.04-foot rejection limit. Computation of mean sea level elevations was by distribution of the total closure error, proportional to the square of the length of the observed line.

The vertical error of closure between previously fixed elevations in San Francisco and Oakland was 0.02 foot. A tie to National

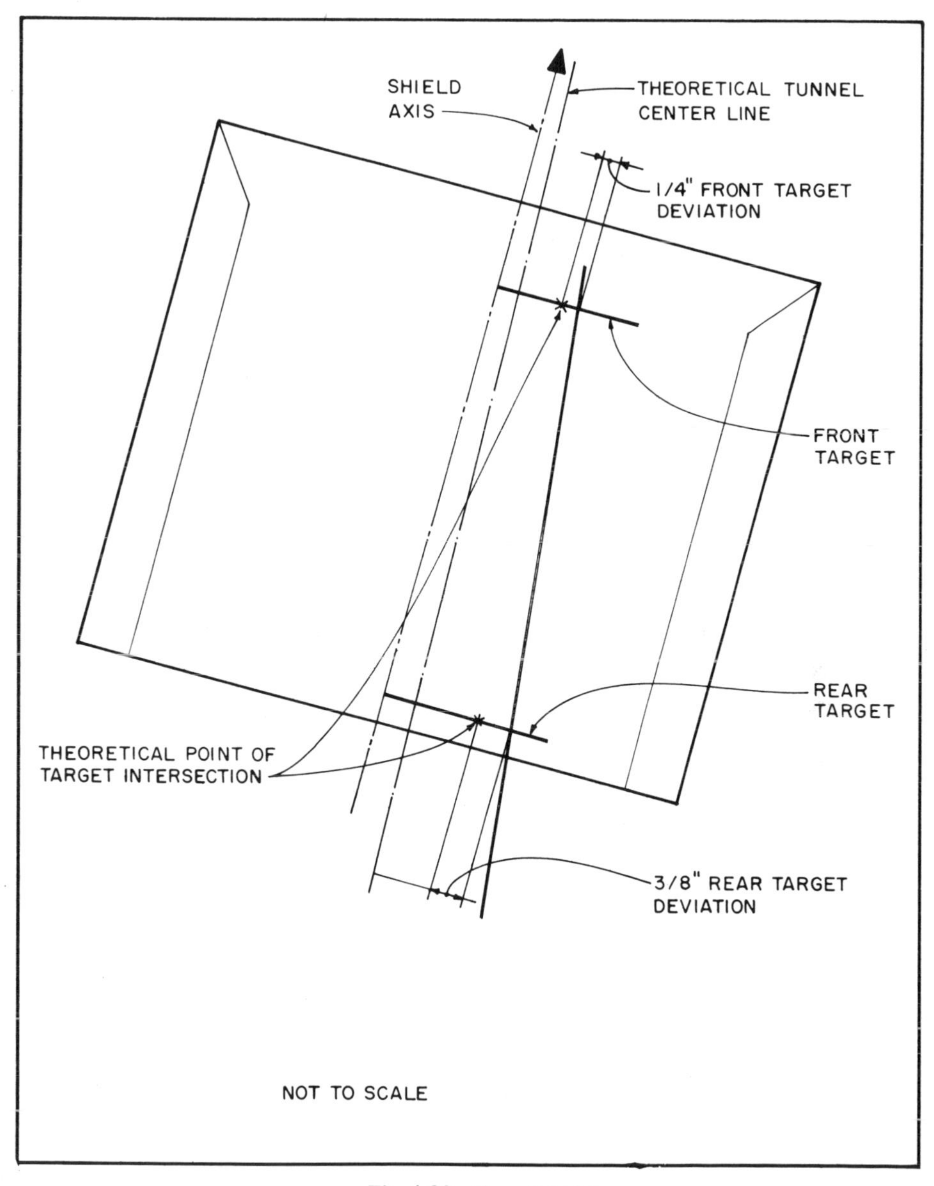

Fig. 4-22. Plan of shield.

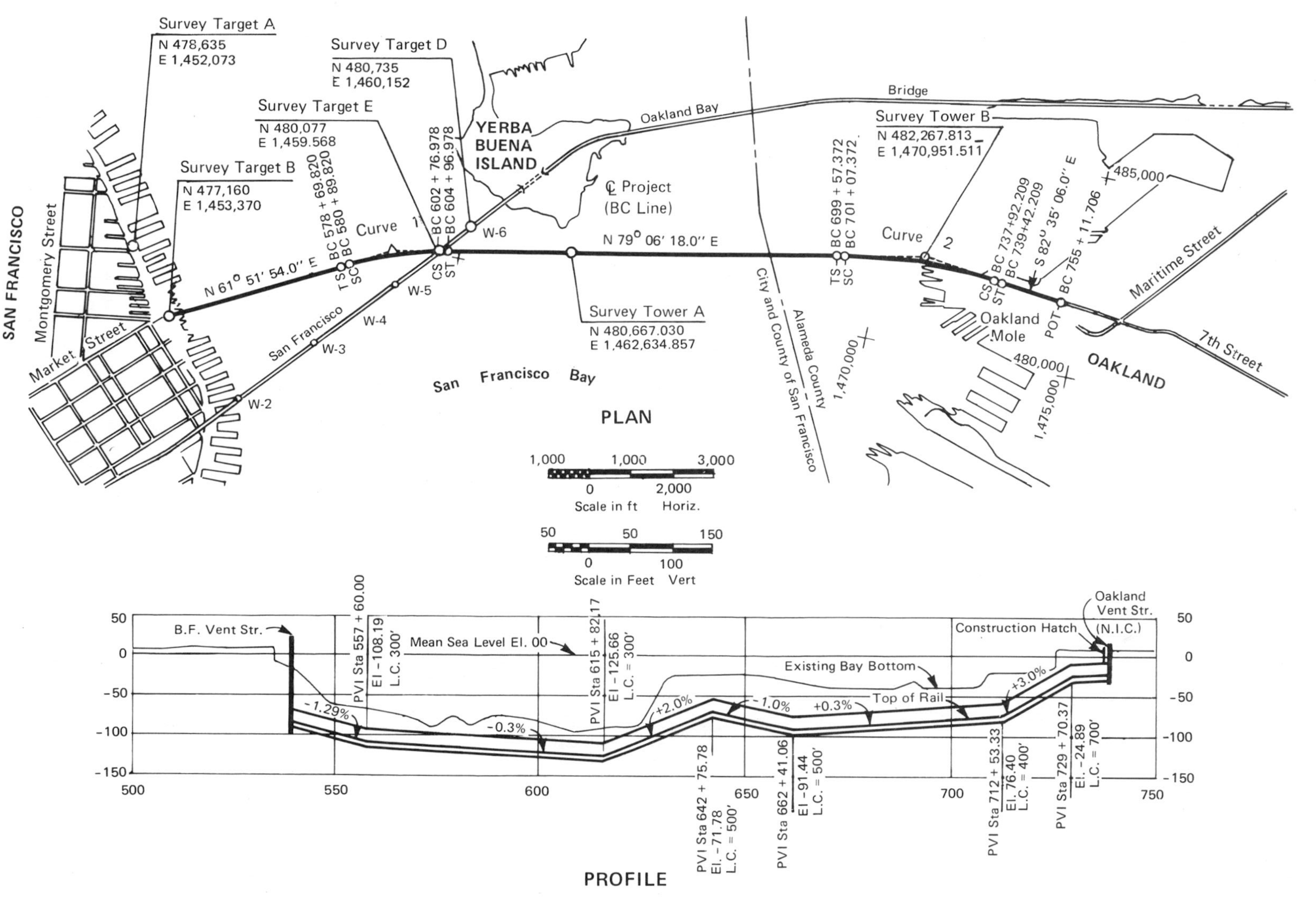

Fig. 4-23. Plan and profile of Trans-Bay Tube.

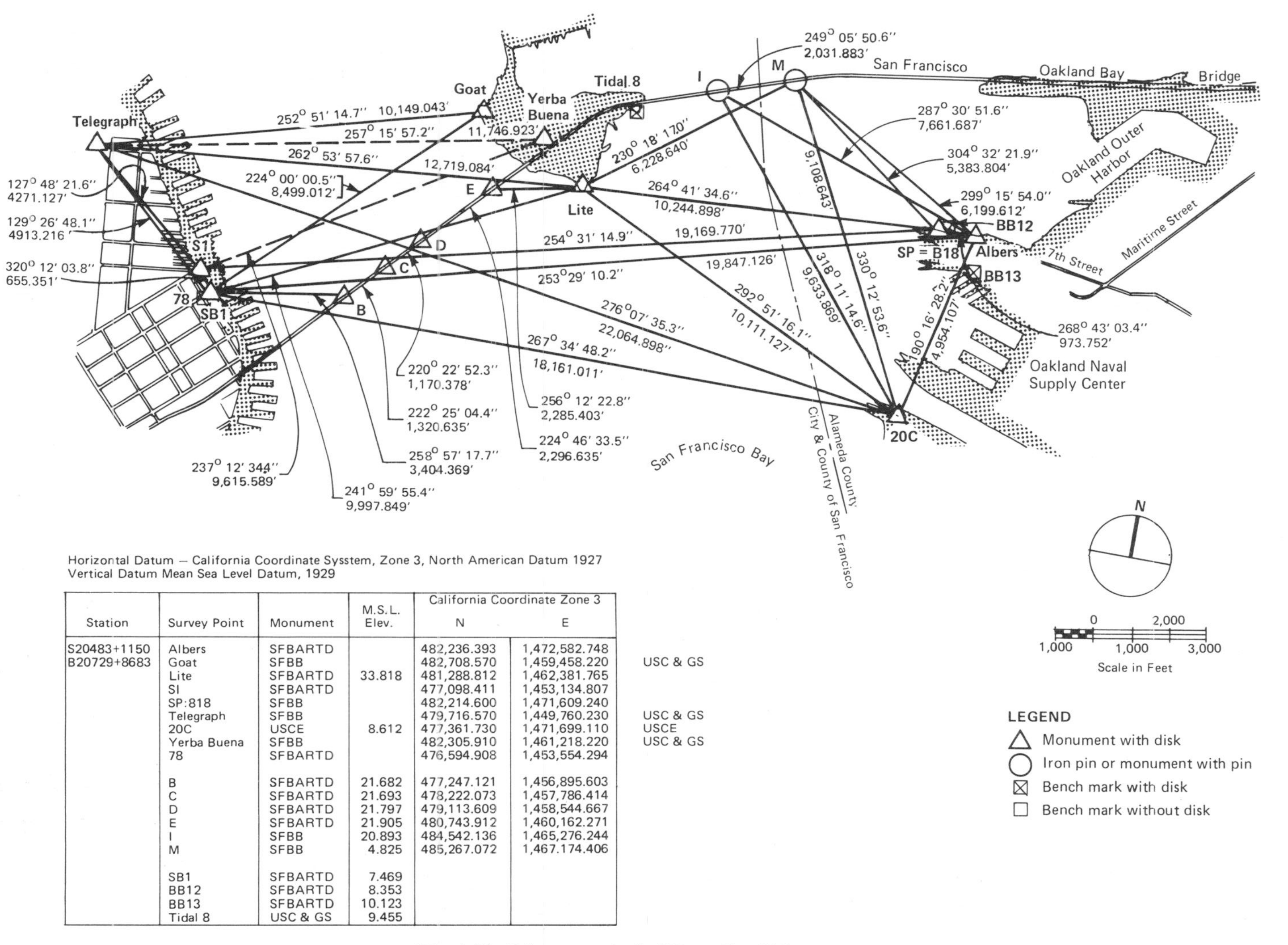

Horizontal Datum – California Coordinate Sysstem, Zone 3, North American Datum 1927
Vertical Datum Mean Sea Level Datum, 1929

Station	Survey Point	Monument	M.S.L. Elev.	California Coordinate Zone 3 N	California Coordinate Zone 3 E	
S20483+1150	Albers	SFBARTD		482,236.393	1,472,582.748	
B20729+8683	Goat	SFBB		482,708.570	1,459,458.220	USC & GS
	Lite	SFBARTD	33.818	481,288.812	1,462,381.765	
	SI	SFBARTD		477,098.411	1,453,134.807	
	SP:818	SFBB		482,214.600	1,471,609.240	
	Telegraph	SFBB		479,716.570	1,449,760.230	USC & GS
	20C	USCE	8.612	477,361.730	1,471,699.110	USCE
	Yerba Buena	SFBB		482,305.910	1,461,218.220	USC & GS
	78	SFBARTD		476,594.908	1,453,554.294	
	B	SFBARTD	21.682	477,247.121	1,456,895.603	
	C	SFBARTD	21.693	478,222.073	1,457,786.414	
	D	SFBARTD	21.797	479,113.609	1,458,544.667	
	E	SFBARTD	21.905	480,743.912	1,460,162.271	
	I	SFBB	20.893	484,542.136	1,465,276.244	
	M	SFBB	4.825	485,267.072	1,467.174.406	
	SB1	SFBARTD	7.469			
	BB12	SFBARTD	8.353			
	BB13	SFBARTD	10.123			
	Tidal 8	USC & GS	9.455			

Fig. 4-24. Primary control of Trans-Bay Tube.

Geodetic Survey benchmark Tidal 8 on Yerba Buena Island indicated a discrepancy of 0.06 foot. Investigation into National Geodetic Survey records revealed that the published elevation of Tidal 8 was determined by "Valley Crossing" methods from Oakland and San Francisco prior to construction of the Bay Bridge. As these lines are approximately 2 miles long, a 0.06-foot discrepancy in elevations can be expected; therefore, no adjustment of the Trans-Bay survey elevations was made.

Use of Laser Beams to Align Dredges and Screeding Barges

A trench of 60-foot width was dredged in the Bay mud and a 2-foot gravel foundation course was placed in the bottom of the trench to serve as the foundation of the tube sections (see Fig. 4-23). Both the dredge excavating the trench and the screeding barge which placed and screeded the foundation course were guided by laser beams. Dredge heading accuracy of 3 inches was accomplished and the foundation course was screeded to an accuracy of 1.8 inches. The principle of dredge control is illustrated in Fig. 4-25. The circular beam spot of the alignment laser was converted into a line beam shape by a fan beam accessory lense. Dimensions of the beam were:

4 to 6 inches per mile–narrow dimension
850 feet per mile–wide dimension.

The dredge or barge operator sights back on the narrow laser line to align the dredge or barge. He assures that the laser line has not shifted from its predetermined bearing by turning around to see the beam reflected by a retroreflector mounted on the other shore.

Construction Surveys During Fabrication of Tube Sections

Between ventilation buildings, the 19,113-foot length of tube was composed of 57 individually constructed reinforced concrete sections, each with a continuous exterior steel shell. The steel shells were fabricated in a shipyard and launched prior to installation of the reinforced concrete. The sections ranged in length from 273 to 336 feet and were 24 feet high and 47 feet wide. The tube contained both horizontal and vertical curves requiring design of a number of curved tube sections. Of the total of 57 sections, 15 were curved horizontally, 4 were curved vertically and 2 sections had vertical and horizontal curves built in (see Fig. 4-26). The remaining 36 sections were straight.

Precise fabrication procedures of the steel shell and continuous dimensional checks by survey methods were essential for final alignment accuracy. A deviation in the deflection angle of the tube end section results in horizontal or vertical alignment deviations of the tube centerline during tube placement at the bottom of the Bay.

The tube sections were fabricated on an inclined plane to facilitate launching. Conventional surveying techniques employing leveled instruments were impractical and specialized shipyard techniques were employed.

To minimize effects of structure deformation due to temperature, check surveys of the fabricated sections were conducted at night.

Surveys During Placement

Tubes were fitted with "gland" or "snorkel" type pipes and survey towers extending from the tube at the bottom of the Bay to the water surface (see Fig. 4-27). Control of the tube position during placement was accomplished by using primary control monuments as reference and triangulation to the survey towers. On some tubes, the survey pipes penetrated into the tube so that the survey tower location mark in the gallery floor inside the tube could be observed directly from the top of the survey tower above the water surface. Other tubes were fitted with survey marks in the top skin of the tube, and this also could be sighted from the top of the survey tower. Prior to placement, the survey tower and sight pipe were installed on the floating tube with sufficient inclination of the sight pipe so that, when the tube was placed on its prepared foundation, the 18-inch sight pipe would be approximately vertical. A vertical collimator (optical plummet) was mounted on a stand on top of the sight pipe, with lateral motion provided to ensure vertical centering over the survey mark at the

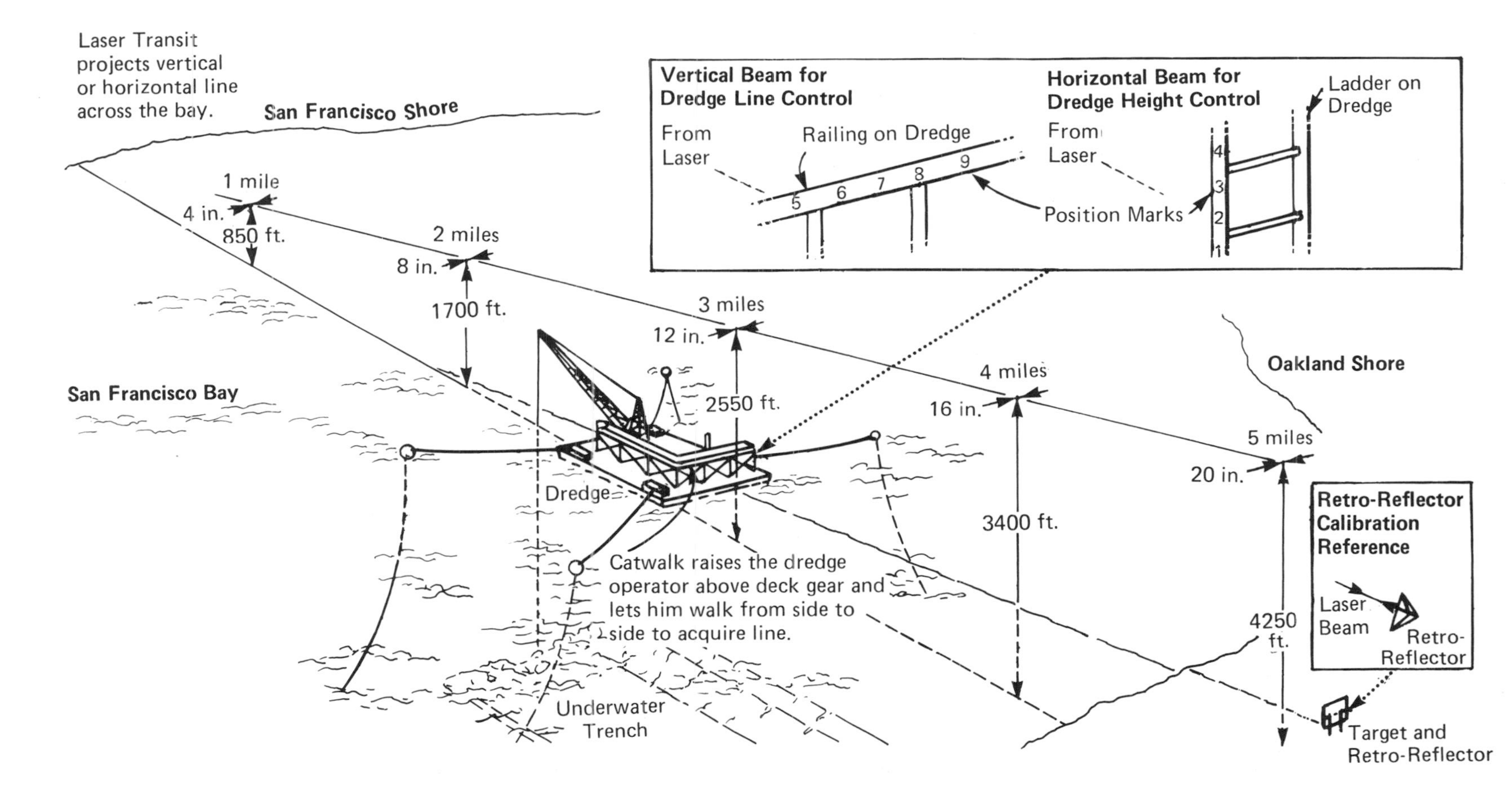

Fig. 4-25. Principle of dredge and screeding barge control.

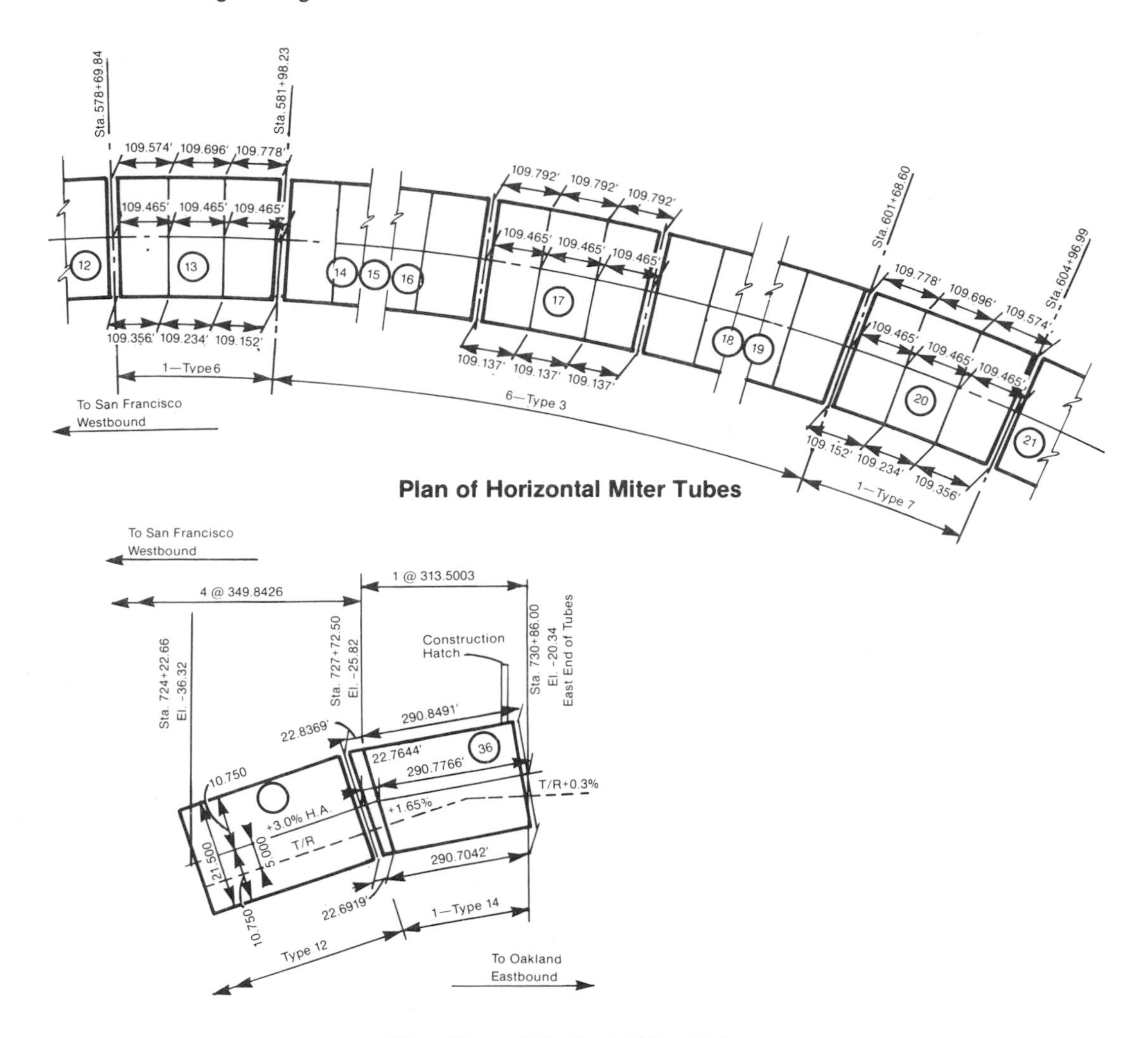

Fig. 4-26. Plan and elevation of tube sections with built-in horizontal and vertical curves.

bottom of the pipe, or over the gallery floor mark, in the case of "gland" type pipes. A light was lowered into the pipe for illumination. During placement, it was necessary to relevel and recenter the collimator frequently, so that the alignment surveyors could accurately observe the placement line. After the joint was made and dewatered, the collimator was again leveled and centered, and alignment surveyors made their final observation on as-placed line.

As-placed grade of the tube was observed by taking rod readings to the top of the sight pipe from level instruments set up on shore or bridge footing control monuments.

REFERENCES

1. Federal Geodetic Control Committee, 1974 *Classification, Standards of Accuracy, and General Specifications of Geodetic Control Surveys.* US Dep. of Commerce, NOAA, Rockville, MD 20852.
2. Federal Geodetic Control Committee, 1975 *Specifications to Support Clarification, Standards of Accuracy, and General Specifications of Geodetic Control Surveys.* US Department of Commerce NOAA, Rockville, MD, 20852.
3. Davis and Foote, *Surveying, Theory and Practice,* McGraw-Hill, New York, 1966.
4. *Encyclopedia of Science and Technology,* McGraw-Hill, New York, 1966.
5. Talbot, Arthur N., *The Railway Transition Spiral,* McGraw-Hill, New York, 1927.

Fig. 4-27. Tube placement: survey tower and sight pipe.

5
Soft Ground Tunneling

T. G. McCUSKER

Construction Engineering Consultant

In tunneling, soft ground is material which can be removed with reasonable facility using hand tools, even though such tools may not, in fact, be the ones employed. The degree of difficulty and the consequent cost of performing tunnel excavation and the methods to be employed are chiefly determined by the stand-up time of the tunneling medium. The utility of the completed tunnel depends upon its ability to sustain both construction and permanent loading. Construction shall result in minimal disturbance of or damage to utilities and structures potentially affected by the construction. All of these matters are interdependent and require proper knowledge of subsurface conditions to enable the designer and the constructor to make correct choices and informed decisions.

This chapter discusses the variety of soil conditions which may be encountered, their properties, methods of improvement of these properties, appropriate construction methods, the design and construction of tunnel linings and related matters.

SOIL CHARACTERISTICS

Ground Classification

Terzaghi[1] prepared a Tunnelman's ground classification in six principal categories, later modified by Heuer[2] into the table shown below (Table 5-1).

Stability of Cohesive Soils

It is apparent that firm ground requires no special treatment and that squeezing ground may present problems. Soil does not appear in distinct categories but as a continuum. It is therefore desirable to be able to distinguish those properties which will enable quantification of behavior.

Investigation by Broms and Bennermark[3] shows that failure in a vertical clay face will occur when

$$\frac{P_z}{C} \geqslant \frac{2\pi}{1 + \frac{1B}{6Z}}$$

where

P_z = the total overburden pressure at depth Z,
C = the undrained shear strength of the clay
B = the width of the opening.

For values of Z/B of 2 or greater, the critical value of P_z/C is about 6. However, if the value

Table 5-1. Tunnelman's ground classification.*

CLASSIFICATION		BEHAVIOR	TYPICAL SOIL TYPES
Firm		Heading can be advanced without initial support , and final lining can be constructed before ground starts to move.	Loess above water table, hard clay, marl, cemented sand and gravel when not highly overstressed.
Raveling	Slow raveling Fast raveling	Chunks or flakes of material begin to drop out of the arch or walls sometime after the ground has been exposed, due to loosening or to overstress and "brittle" fracture (ground separates or breaks along distinct surfaces, opposed to squeezing ground). In fast raveling ground, the process starts within a few minutes, otherwise the ground is slow raveling.	Residual soils or sand with small amounts of binder may be fast raveling below the water table, slow raveling above. Stiff fissured clays may be slow or fast raveling depending upon degree of overstress.
Squeezing		Ground squeezes or extrudes plastically into tunnel, without visible fracturing or loss of continuity, and without perceptible increase in water content. Ductile, plastic yield and flow due to overstress.	Ground with low frictional strength. Rate of squeeze depends on degree of overstress. Occurs at shallow to medium depth in clay of very soft to medium consistency. Stiff to hard clay under high cover may move in combination of raveling at execution surface and squeezing at depth behind face.
Running	Cohesive running Running	Granular materials without cohesion are unstable at a slope greater than their angle of repose (± 30°–35°). When exposed at steeper slopes they run like granulated sugar or dune sand until the slope flattens to the angle of repose.	Clean, dry granular materials, Apparent cohesion in moist sand, or weak cementation in any granular soil, may allow the material to stand for a brief period of raveling before it breaks down and runs. Such behavior is cohesive running.
Flowing		A mixture of soil and water flows into the tunnel like a viscous fluid. The material can enter the tunnel from the invert as well as from the face, crown, and wall, and can flow for great distances, completely filling the tunnel in some cases.	Below the water table in silt, sand, or gravel without enough clay content to give significant cohesion and plasticity. May also occur in highly sensitive clay when such material is distrubed.
Swelling		Ground absorbs water, increases in volume, and expands slowly into the tunnel.	Highly preconsolidated clay with plasticity index in excess of about 30, generally containing significant percentages of mont morillonite.

*Modified from Terzaghi[1] by Heuer.[2]

of Z/B is less than 2, the critical value should be computed from the following.

$$\frac{P_z}{C} = \frac{2\frac{Z}{B} + - 1}{1 + \frac{1B}{6Z}}$$

Soft Clays. These commonly have a shear strength which increases with depth, and strict application of the above formulae will result in an unconservative critical value of P_z/C. In such cases, Deere *et al.*[4] suggest that as an approximation, one may, under normal circumstances, use a value of $(P_z/C) = 6$, with a value of C representing the average shear strength to a distance of one diameter from the opening.

While the available case histories indicate that 6 is the correct critical value for P_z/C, it must be noted that time of exposure and rate of excavation are important because undrained shear strength is dependent on pore water pressure and it is desirable to take advantage of the negative pore water pressure which is induced at the time of initial exposure of the freshly cut surface. As this pressure dissipates, there is a possibility of reduction of strength and consequent failure.

The Stability of Cohesive Soils. With P_z up to six times the value of C, the stability of cohesive soils is dependent on the plastic behavior of these materials. Brittle or fissured materials such as some hard clays or clay-shales may become unstable at lower ratios because of inability to accommodate to local overstressing by plastic deformation. This is unlikely to be a problem except at considerable depth, because of the high shear strength of this type of material. Peck[5] has adapted this analysis to include the support effect of compressed air within a tunnel and presents the formula in the form shown below.

$$N_t = \frac{P_z - P_a}{S_u}$$

where N_t is the stability factor, P_z is the overburden pressure at depth Z to tunnel centerline, P_a is the air pressure above atmospheric pressure and S_u is the undrained shear strength of the clay.

Peck finds, from case studies, that for $N_t > 5$, the clay may squeeze sufficiently fast to fill the tail void behind a shield before the void can be filled, indicating that hand mining would be difficult under such conditions. For $N_t > 6$, ground will move into the face regardless of normal breasting procedures; and for $N_t > 7$, it becomes extremely difficult, if not impossible, to control the shield. In all the above cases, the effect of ground movement will be reflected at the surface.

For $N_t < 4$, no unusual difficulty should be anticipated in tunneling, this range of values between 1 and 4 representing raveling to firm soil. As noted above, however, the soil strength is time-dependent both because of local failures related to stress concentrations around flaws and dissipation of negative pore pressures.

Table 5-2 summarizes the relationship between stability factor and the behavior of cohesive soils.

Overconsolidated Clays. These may exhibit a different kind of instability, known as swelling. The construction of a tunnel through such material results in relief of confining pressure at the boundaries of the excavation. In saturated clays, the resultant pressure difference will result in a flow of pore water within the clay to the relaxed zone. Depending on the degree of overconsolidation, the resultant swelling can continue until the opening is completely closed. The most efficient method of dealing with such material is to install a lining of circular cross-section as soon as practical after excavation. The lining may have to be designed for pressures considerably in excess of overburden pressure.

Sensitive Clays. These clays have an unconfined compression strength after molding less than about one-quarter of the unconfined compressive strength of undisturbed material. Such clays have a high water content and their loss of strength on remolding is associated with a change in permeability. The Leda clays of the St. Lawrence valley typify a clay so sensitive that it is instead called a quick clay. Similar clays are found in the Anchorage area in Alaska and in Scandinavia.

The process of tunneling in sensitive clays always causes sufficient remolding to result in loss of strength of the material close to the boundary of the excavation and the clay will flow quickly to fill the tail void behind a shield. As drainage of pore water proceeds, such clays regain their strength. Surface settlement is to be expected, both as a result of physical movement of the material and volume change associated with drainage of pore water from the remolded material.

Stability of Non-cohesive and Mixed Soils

While non-cohesive sands are found in many locations, there is more commonly a gradation

Table 5-2. Criterion for stability in plastic clays at depths greater than two diameters.[2,5]

$$N_t = \frac{P_z - P_a}{S_u}$$

N_t = stability factor
P_z = total vertical pressure, depth z
P_a = air pressure above atmosphere
S_u = undrained shear strength of clay

<table>
<tr><th>VALUE OF N_t</th><th colspan="3">EFFECT ON TUNNELING*</th></tr>
<tr><td></td><td>General shear failures and ground movement around tunnel heading cause shield control to become difficult; shield tends to dive.</td><td rowspan="3"></td><td rowspan="6">Squeeze loads on tunnel supports must be considered.</td></tr>
<tr><td>7</td><td>Shear failure ahead of tunnel causes ground movements into the face even in shield-tunneling.</td></tr>
<tr><td>6</td><td>Clay may squeeze rapidly into shield void.</td></tr>
<tr><td>5</td><td></td><td rowspan="3">Tunneling without unusual difficulties.</td></tr>
<tr><td>4</td><td>Rate of squeeze does not present a problem.</td></tr>
<tr><td>1</td><td></td></tr>
</table>

*The analysis may be applied to silts only if their properties are adequately defined by their undrained shear strengths.

of grain sizes, including silts and clays in some proportion. For ease of reference, they will be treated together as granular soils.

Granular soils differ from cohesive soils in that their properties are much more rapidly affected by the presence or absence of water. Sand with some clay binder above the water table may have excellent stand-up time, whereas below the water table the same soil may be classified as fast raveling. The stability of granular soils can be adequate for tunneling if there is some clay binder present or if the soil is moist and the negative pore pressure developed at the freshly cut surface does not dissipate too rapidly.

Compact unsorted soils with angular grains are the most stable, while well-sorted soils or dune sands with rounded grains are the least stable. With high permeabilities, no granular soil is stable below the water table. As well as physical support, control of groundwater is the most important factor in lending stability to granular soils.

It will sometimes happen that a granular soil is cemented to a greater or lesser extent by chemicals leached from above by seepage of rainwater down to the water table or deposited as part of the weathering process in residual soils. Improvement of stand-up time results from such cementation, but if a layer of hardpan has developed from accumulation of leached salts at the water table, difficult mixed-face conditions can result.

Other Conditions Affecting Stability

A quite common and more difficult type of mixed soil is found in till or boulder clay. These soils, associated with glacial retreat, will often contain very large boulders in a poorly sorted and very variable soil. The difficulties in tunneling through such material usually arise from the time taken to dispose of awkwardly placed boulders. The ground around the boulders must be left unsupported to provide working room so that greater stand-up time is required. In some areas, such as Milwaukee, the till can be very permeable, thus compounding the problem, especially where construction is at great depth below the water table.

In urban areas, man-made obstructions are likely to be encountered in earth tunnels. It is not uncommon to find abandoned pile foundations, particularly near the waterfront. Since the necessity for deep pile foundations is associated with weak soils, the problem of tunneling is once again compounded. However, most such piling is of wood and can be dealt with with reasonable facility.

In some places, abandoned mine-workings have been encountered. This is particularly a problem where a community has grown since the time when mineshafts were closed off and abandoned in unrecorded locations. No ready solution is available to deal with this problem.

GROUND WATER CONTROL

Terzaghi[1] observed that "all the serious difficulties that may be encountered during the construction of an earth tunnel are directly or indirectly due to the percolation of water toward the tunnel."

The control of this water is of the utmost importance in tunneling. The presence of water above the water table may be beneficial in providing an increase in stand-up time because of capillary forces until they dissipate, but below the water table, the presence of water serves to reduce the angle of friction drastically, and seepage pressures cause rapid and complete failure in non-cohesive soils.

The presence of water in clays is of primary importance in determining the strength, sensitivity and swelling properties of the material. The type of control to be exercised, whether in construction or in design of the final lining, is directly dependent on these properties.

The methods available for groundwater control are dewatering, grouting, compressed air, freezing and special construction methods.

Figure 5-1 was prepared by Schmidt[6] to illus-

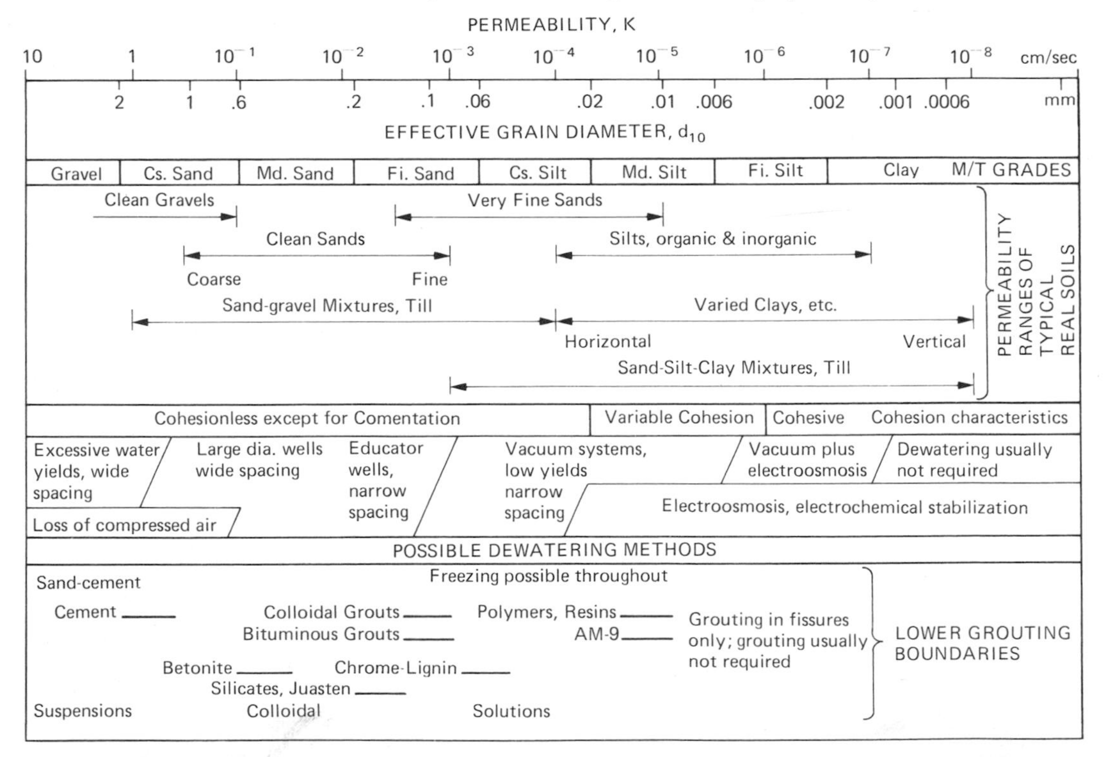

Fig. 5-1. Correlation of permeability, soil character and grouting response from various sources.

trate the relationships between permeability of soil types and applicability of various groundwater control methods.

DEWATERING

The simplest and cheapest method of controlling groundwater is to remove it by pumping from wells of the kinds described below. Unfortunately, there may be undesirable side effects resulting from consolidation of the soil under its increased effective weight following removal of the water. The resulting settlement can damage structures near the dewatered area. As the water is removed from the ground by pumping from a single well, a cone of depression is formed in the water table. The shape of this cone and its areal extent depend upon ground characteristics, but it is usually depicted as a reflected hyperbola in cross-section with the apex at the bottom of the well and the water table as a horizontal asymptote. The radius at the surface may well be of the order of 1000 feet, in relatively permeable soils, but the amount of groundwater depression at large distances from the well is small. If calculations based on the expected amount of groundwater depression and the consolidation characteristics of the soil indicate the probability of damage caused by differential settlement of nearby structures, the shape of the cone of depression can be modified locally by reintroducing part of the extracted water into the ground through recharge wells to maintain the watertable at or near its original level in the vicinity of the affected structure.

Figure 5-2 shows the realtionship between grain size and dewatering method, compressed air use and grout type. The coefficient of permeability is related closely to the D10 grain size (10% of the soil is finer than the D10 grain size).

Soil having a coefficient of permeability greater than approximately 10^{-3} cm/sec will permit water to drain freely into the tunnel excavation. It will be troublesome because the flow will carry in fines from the soil and may produce fast raveling to running conditions as well as resulting in surface settlement from loss of ground.

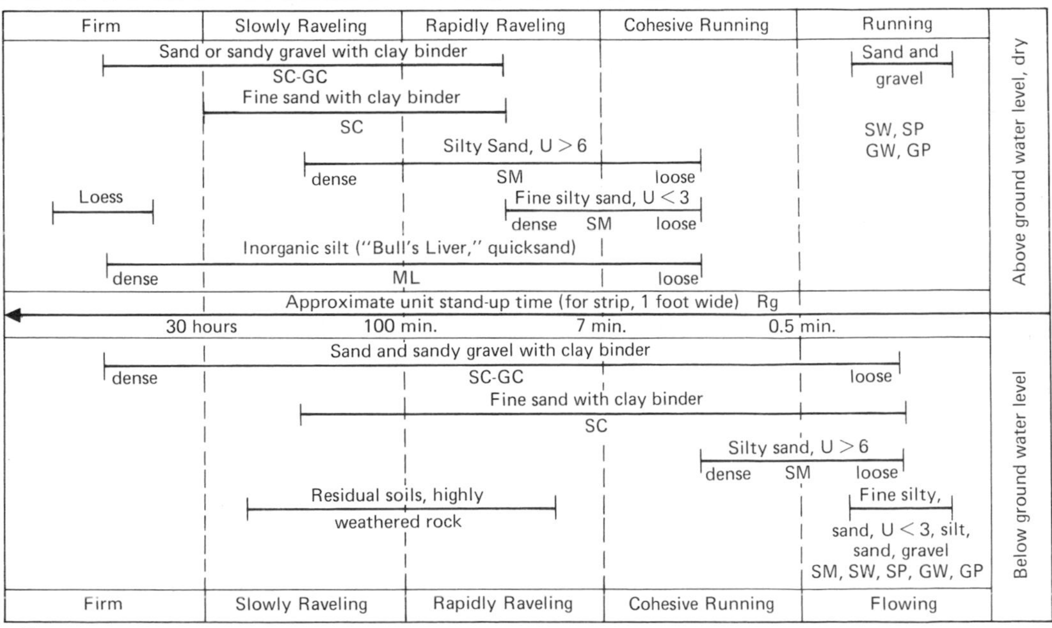

Note:

1) Air loss (in tunneling under compressed air) and water inflow is governed by the permeability, largely a function of D_{10}.
2) Behavior below ground water table under suitable air pressure is approximately the same as above ground water level.
3) Loose is here defined by N < 10 (standard penetration test), dense by N > 30.
4) Descriptive terms of materials according to the Unified Soil Classification.
5) Behavior may be somewhat better than shown above ground water level, if material is moist and fine or silty.

Fig. 5-2. Behavioristic classification of various soils.

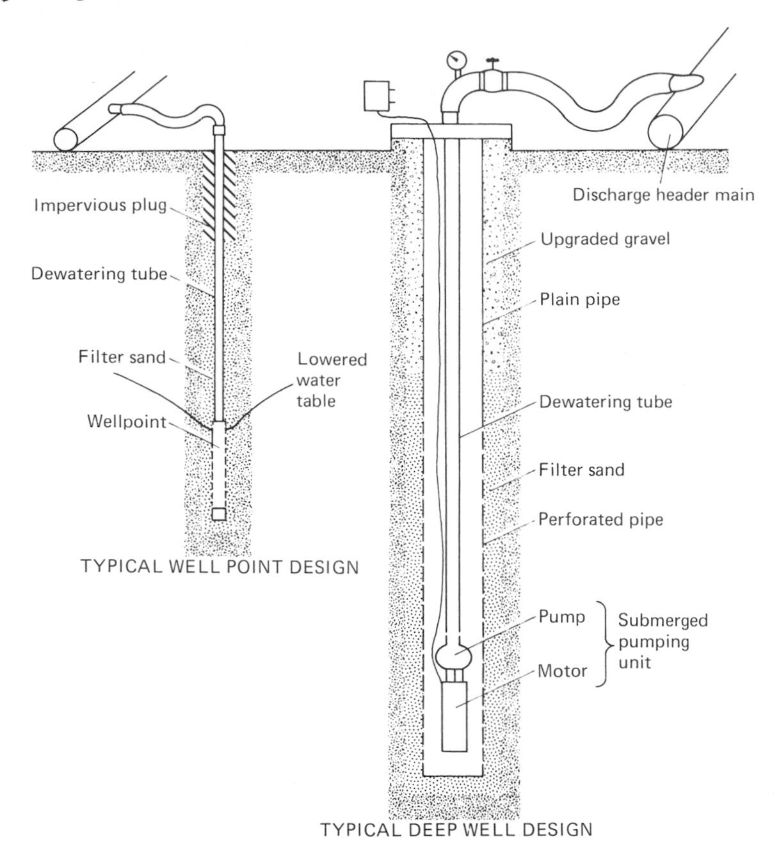

Fig. 5-3. Schematic design of wellpoint and deep well.

Stability may be a problem in saturated soils even where the coefficient of permeability is as low as 10^{-5} cm/sec, if adequate support measures are not provided.

Although flows will be low, the clay binder may be removed and, in any case, the angle of repose will be reduced and seepage pressures will cause steady local failure of the working face. Even with these relatively impermeable soils, an unsupported working face can be subject to progressive and serious failure. Various methods of dewatering are available and are discussed below.

Wellpoints

A wellpoint consists of a perforated casing (typically 3 feet x $1\frac{1}{2}$ inches), surrounded by filter sand, installed below the water table and connected to a header pipe at the surface. A series of wellpoints is connected to a single header, which, in turn, is connected to a suction pump. The layout is illustrated in Fig. 5-3, although without the impervious plug.

Figure 5-4 is a pair of nomograms from which the spacing of wellpoints can be determined for different soils types and different amounts of groundwater lowering.

The theoretical limitation for suction pumping is one atmosphere of pressure, or about 34 feet. However, leakage and inefficiencies make the practical limit only about 15 to 18 feet. The method is therefore very little used in tunnel construction except in combination with compressed air when reduction of piezometric head will enable a more economically low working pressure to be used in the tunnel.

Vacuum Wellpoints

When the coefficient of permeability is less than about 10^{-4} cm/sec, capillary forces prevent the flow of water from the soil into conventional wellpoints. However, by applying vacuum

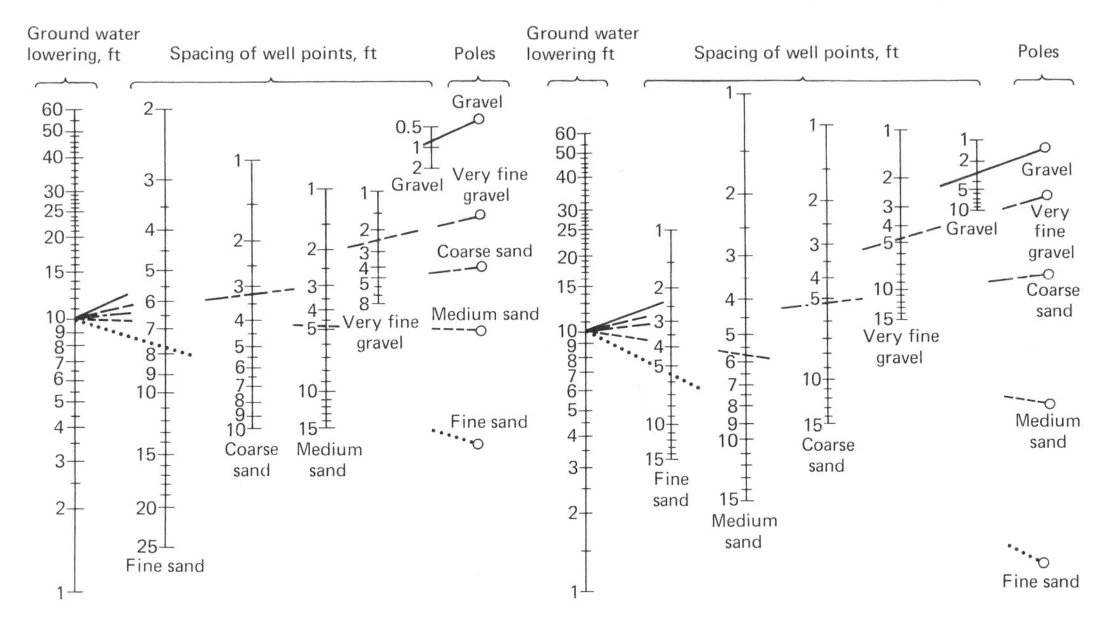

Fig. 5-4. Wellpoint spacing diagrams.

to the well, the ground surface is effectively loaded by the difference between atmospheric pressure and the reduced pressure in the well. The resulting consolidation forces water out of the soil. Figure 5-4 illustrates the layout.

The process, though effective, may take several weeks to reduce the water table to its new stable level.

Eductor Wells

Since it is not practical to use staged wellpoints for tunnel work, a means is required for single-stage deep level pumping. This can be accomplished, in soils of fairly low permeability, by the jet-eductor system. The wellpoint is attached to a jet-eductor pump. The pump is supplied with high-pressure water from the surface and a return pipe which carries not only the operating water but also the water extracted from the wellpoint. The efficiency of the system is low, but it can be used to lower the water table as much as 100 feet.

Deep Wells

A typical deep well installation is illustrated in Fig. 5-3. The deep well is most useful in relatively permeable soils, where adjacent wells can be well-separated, typically 20 to 200 feet as against 2 to 5 feet for wellpoints. The installation is capable of pumping large volumes of water against high head using submersible pumps rated at several tens of horsepower.

Deep wells perform most efficiently if they are drilled down to an impervious formation. However, the water table between parallel rows of wells cannot be lowered more than about three-quarters of the original piezometric head.

Drainage Within the Tunnel

Dewatering has sometimes been accomplished by installing horizontal wellpoints in the working face. This method affords control of the degree of saturation of the soil at its best, but at its worst, particularly in stratified material, it is very inefficient and ineffective.

The excavation efficiency is in all cases drastically reduced because of the presence of the wellpoint and header system and the care which must be taken to avoid damage to them.

System Design

The design of a dewatering system can be accomplished quite readily in soils which are uniform over a wide area and considerable

Table 5-3. Characteristics of Major Grouts

		TYPE OF GROUT	UNCONFINED STRENGTH (1)	MATERIAL (2) COST (RATIO)	FIELD OF (3) APPLICATION	GROUTING PROCEDURE
Suspensions	Unstable Grouts	Suspension of cement in water (water/cement ratio ~10:1)	Similar to concrete	4.2	Fissures or cracks in rock or masonry	No limited quantities. Pumping until refusal.
Suspensions	Stable Grouts	Cement and activated mortars: Prepakt, Termocol, Colcrete	Similar to concrete		Filling of large voids	Limited quantities.
		Cements + clay (+ sand)	1 to 700 psi	1	Wide rock fissures and sands and gravels with high permeability	
			<2 psf	1.1		
Liquid Grouts (Chemicals)	Hard Gels	Sodium silicate +$CaCl_2$	150–300 psi mortar up to 600 psi	10.7	$k > 5 \times 10^{-3}$ cm/sec	Grouting in 2 phases Grout of one type
		+ Ethylene Acetate		11		
		Lignosulforate + bichromate k	5 psi (mortar 50 to 70 psi)	6.5 to 8	$k > 2 \times 10^{-4}$ cm/sec	
	Plastic Gels	Sodium Silicate + reagent	7 psi	2 to 4	$k > 5 \times 10^{-3}$ cm/sec	Limited quantities
		Deflocculated bentonite	1.5 to 3 psi	1.8	$k > 5 \times 10^{-2}$ cm/sec	
	Organic Resins	AM9	<14 psi	50 to 130	$k > 10^{-5}$ cm/sec	
		Resorcinol formaldehyde	15 to 1400 psi	10 to 40		
		Urea formaldehyde (acid grout)	300 to 1400 psi			
		Precondensed polymers (Epoxy)	up to 14,000 psi comp. up to 4300 psi tensite	150 to 500	Concrete cracks	
	Hydro-Carbon Based Grouts	Bituminous emulsion with silicate or resorcinol	1.5 psi mortar 150 psi	6 12	$k > 10^{-3}$ cm/sec	
		Hot bitumen	very viscous fluid		abundant water circulation	

(1) Strength given for pure grout (for very low values taken as twice the rigidity)
(2) Base 1 for material cost comparison
(3) Permeabilities shown correspond to granular soil susceptible to impregnation. Excessively fluid viscous grouts should not be introduced into too pervious soils. (inefficient technically and waste of money.)

Modified from Tallard, G. (1975), "Dewatering and Grouting as Supplementary Ground Engineering Techniques," Proceedings of Seminar on Underground Problems, Techniques & Solutions, Chicago, Oct. 20–22.

depth. Such conditions are rare and the manipulation of the available mathematical formulae to achieve the required solutions is best left to those expert in the field. With the best information, pumping tests should be carried out initially to test the validity of the design before the total investment installation is made.

GROUTING

Grouting as a means of stabilizing soils has more often been used in the U.S. in shaft-sinking and to repair collapses than as a routine method. It is normally an expensive and time-consuming process and is not perfectly reliable even when very great care is exercised.

Table 5-3 compares the characteristics and material costs for the major grout materials. Figure 5-5 illustrates a range of suitability for various types of grout in terms of permeability.

Grouting is much more commonly used in Europe even though the same cost factors apply. The main reasons for such use, apart from stabilization of the tunnel zone in water-bearing soils, are to reduce surface settlements and to provide or supplement conventional underpinning. Two examples of appropriate uses of grouting are reporduced below as Figs. 5-6 and 5-7, taken from a report by Clough[7]. These figures illustrate grouting from the surface to stabilize a zone involving mixed face tunneling and highly pervious material (Fig. 5-5) and grouting from the surface to provide underpinning support directly above a tunnel (Fig. 5-7).

Grouting has also been used for ground stabilization in subaqueous tunnels at considerable depth to reduce the pressure of compressed air required and where shallow cover would have made the use of high-pressure compressed air dangerous. Such grouting is performed from within small-diameter pilot tunnels which can be driven more safely in difficult conditions. The main tunnel can then be constructed to full size within the grouted zone.

It should be noted that grout will not be effective in the presence of moving groundwater. For this reason, attempts to stabilize a working face by grouting from within the tunnel or from the surface into the working face area will

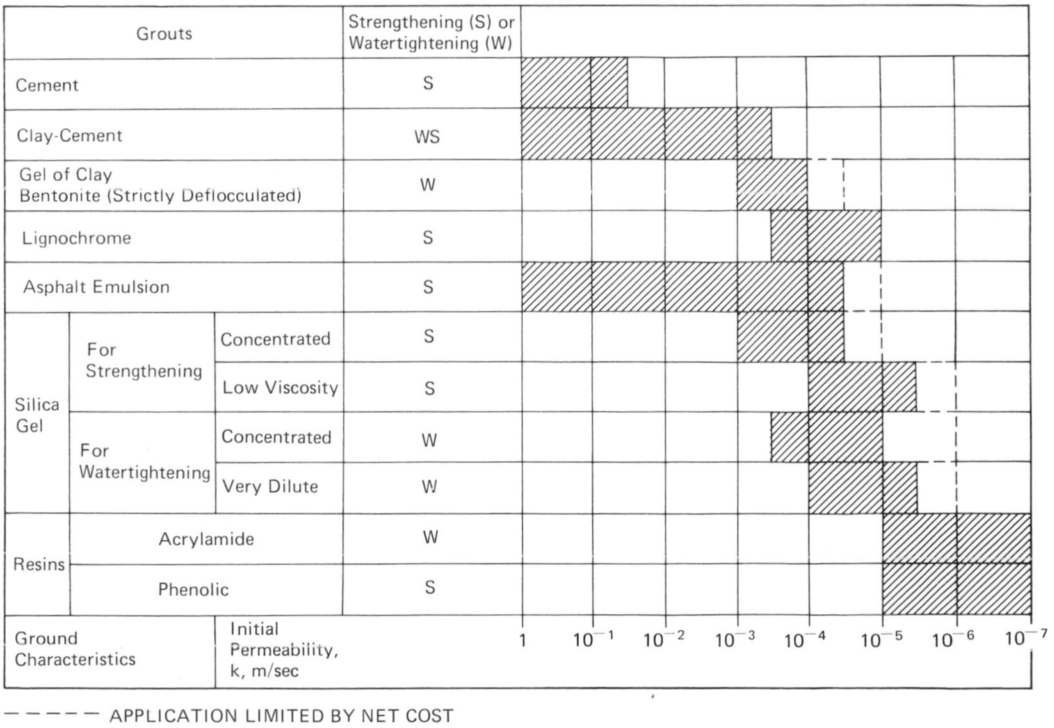

Grouts			Strengthening (S) or Watertightening (W)
Cement			S
Clay-Cement			WS
Gel of Clay Bentonite (Strictly Deflocculated)			W
Lignochrome			S
Asphalt Emulsion			S
Silica Gel	For Strengthening	Concentrated	S
		Low Viscosity	S
	For Watertightening	Concentrated	W
		Very Dilute	W
Resins	Acrylamide		W
	Phenolic		S
Ground Characteristics		Initial Permeability, k, m/sec	1 10^{-1} 10^{-2} 10^{-3} 10^{-4} 10^{-5} 10^{-6} 10^{-7}

– – – – APPLICATION LIMITED BY NET COST

Fig. 5-5. Grout applications in loose soil.

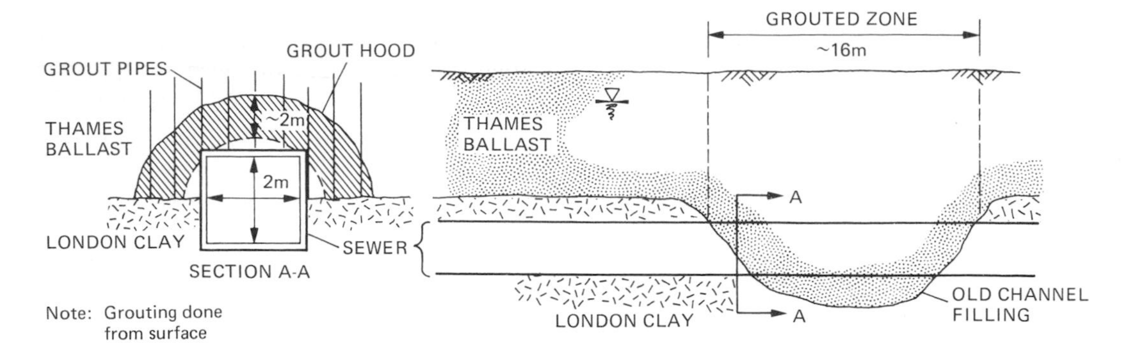

Fig. 5-6. Use of grouting for sewer relocation.

rarely be successful. In stratified soils, each pervious stratum must be individually treated. It is very important to prevent grout from escaping up the outside of the injection pipe.

Grouting Methods

The methods in common use are stage grouting, series grouting, circuit grouting and packer grouting. The spacing of injection holes may be predetermined, especially if a high degree of certainty is required as to continuity of the grouting volume. In such cases, hole spacing will be 3 to 5 feet, depending upon permeability of the soil. It is more common to grout to refusal, using primary holes at wide spacing, then intermediate secondary holes and so on until grout take reaches a low value, indicating that all voids have been filled.

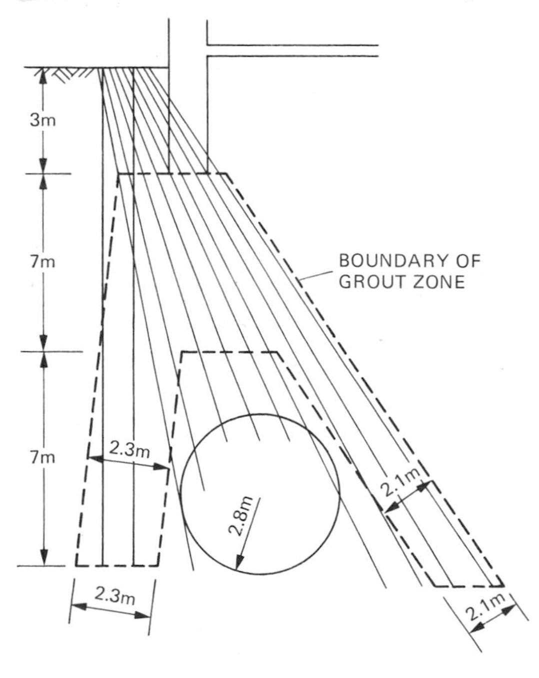

Fig. 5-7. Grout hole patterns to create underpinning grout arch.

In *stage grouting*, the holes are drilled and grouted successively deeper, the hole being washed out between stages before the grout hardens.

In *series grouting*, new holes are drilled from the surface for each successive deeper zone, the holes remaining full of grout after each step.

In *circuit grouting*, a double pipe is used. The injection pipe is at the bottom of the hole, the top being closed by a packer. Grout is forced in under pressure and any not flowing into the formation returns via the outer annulus to the holding tank. The advantage of this method is that segregation of the grout is avoided and premature plugging of the formation is minimized.

In *packer grouting*, part of the hole is isolated for grouting by expanding packers above and below the zone of interest. Grouting is performed from the bottom up when packers are used. The advantage of this method is that different treatment can be given to different zones of a stratified material. However, it is difficult to get reliable seating of the packers against the soil, and, to overcome this, Ischy invented the "Tube á Manchette" or sleeved pipe. Figure 5-8 shows the basic design.

The device consists of a steel pipe perforated at 12-inch intervals with rings of grout orifices. Each of these rings is covered by a rubber sleeve. Inside the pipe rides a grouting pipe with expandable packers to isolate the injection

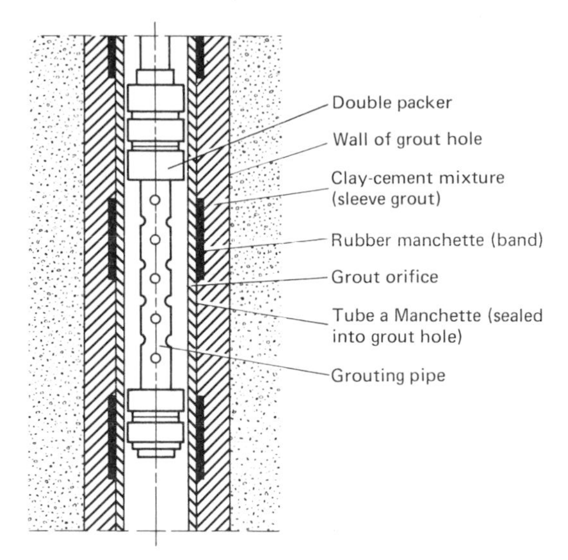

Fig. 5-8. Tube á Manchette.

area from the rest of the sleeved pipe. In use, the Tube á Manchette is inserted into the drilled hole and a weak clay-cement mixture is pumped in to fill the annular void between the Tube á Manchette and the ground. The grouting pipe is then lowered into position, the packers are expanded and grout is pumped in. The pressure allows the grout to force its way past the rubber sleeve, fracture the clay-cement filling and enter the ground. The sleeved pipe necessarily remains in place after grouting is completed, and regrouting through the same hole is not possible.

Grout Types

Colloidal Grouts. These are viscous and their viscosities increase from the time of injection to the time they gel. Due to the high viscosity, such grouts can only be used in soils coarser than fine sand with permeabilities greater than 10^{-3} cm/sec.[8] Collodial grouts have been used in tunneling to stop water inflows, to harden running sand, to prevent settlement of surface structures, to prevent blowouts in compressed air work and to consolidate ground.[9] There are both silicate and non-silicate processes.

The basic ingredient in silicate processes is a solution of sodium silicate in water. The solution contains free sodium hydroxide and collodial silicic acid. Addition of a solution containing salts or acids causes the formation of a solid gel, the set time of which can be varied from 20 minutes to 3 hours by varying the concentrations of the reagents.

Both single-shot and multiple-shot methods are in use. Although the single-shot methods assure more complete mixing of the reagents, and therefore decrease the occurrence of "windows" in the grouted area, it becomes necessary to dilute the grout to avoid gelling in the mixing tank with a consequent loss of strength of the grout in most cases. Use of an organic ester, aldehyde or amide avoids this difficulty and results in gels with compressive strengths of 300 to 450 psi.[10]

Multiple-shot grouts have the disadvantage of producing ungrouted windows, as noted above, and in addition, the solutions react instantaneously upon mixing and therefore lose penetration ability; the necessary closely spaced drill holes increase cost; and solutions must be injected under high pressure, which may not always be possible where overburden is limited or jacking might damage structures.

Non-silicate colloidal grouts commonly have lignosulfate or lignosulfite oxidized by bichromate to form a salt. Setting time can be varied from 10 minutes to 10 hours and compressive strengths of 75 psi have been recorded.

The liquid bichromate solution is toxic and it is important to avoid contamination of groundwater where this is a source of drinking water.

Resin Grouts. These grouts rely on the in-situ polymerization of liquid organic solutions to form a solid product. The viscosity of this class of grouts is very low and soils with permeabilities as low as 10^{-5} cm/sec can be treated. The viscosity remains constant until the grout has set, so that very high penetrations can be achieved.

With all grouts, it is necessary to perform tests on the grout-soil interaction before use. Tests with groundwater alone may give misleading results.

COMPRESSED AIR

Compressed air is most often used to stabilize the ground in tunnels constructed in permeable soils below the water table, where dewatering is impractical, especially in subaqueous tunnels. It is also used to provide a measure of ground support in tunnels in soft, squeezing clays to increase the stability factor above the critical level. Compressed air is introduced into the working chamber of plenum to balance the pressures to the required extent.

The hydrostatic pressure cannot be balanced over the whole face, since it increases at 0.43 psi/foot of depth. The imbalance of hydrostatic pressure in an 18-foot O.D. tunnel between crown and invert of the excavation is almost 8 psi. In practice, it is usually found that balancing the hydrostatic pressure at springline or tunnel axis level will afford a reasonable compromise. The objective is to stabilize the lower part of the working face without causing too great a loss of air from overpressure in the upper part. In very permeable ground, such as gravel, this compromise will not work. It then becomes necessary to render the exposed soil impervious. This is traditionally effected by covering the working face with reworked clay and limiting the area of active excavation. However, such soils are candidates for grouting and, if the surface is inaccessible, the grout can be applied from a previously driven small-diameter pilot tunnel, which can be constructed with less difficulty.

The action of the air overpressure in the upper part of the face in permeable soils will result first in driving water out of the pore spaces while leaving a capillary film on the soil particles with resultant apparent cohesion. Prolonged exposure or rapid escape of air will dry this film and the support effect of the capillary water will be lost. In the lower part of the face, the air pressure will reduce seepage pressures significantly and thereby enhance stability by moving the failure characteristics to a higher level (e.g., running to raveling). Since the characteristics of the ground are being modified continuously and dynamically by the use of compressed air, it is important to guard against failure by providing support to those areas of the face of the excavation not being actively worked.

In cohesive soils, the effects of imbalance between air pressure and overburden pressure are less critical, since the requirement is only to reduce the stability factor to a safe level. However, in silts and silty clays, there is still the possibility of drying out part of the face and, by inducing shrinkage resulting from loss of pore water, to develop fissures which propagate away from the face and provide preferred escape paths for the compressed air. This possibility should also be guarded against where extensive beds of pervious material are included in the otherwise relatively impermeable soil. The danger in these cases resides in the possibility of opening up escape paths to the surface or to deep sewers or other conduits which can permit the possibility of sudden catastrophic loss of air pressure from the working chamber.

It is conventional to depict the action of compressed air in the ground ahead of the tunnel in keeping the water out of the heading by expanding into the ground like a balloon with the air pressure and hydrostatic pressure balanced at all points on the surface. This idealized representation is rarely, if ever, correct. At least some of the air always escapes to the ground surface and may be observed as bubbles breaking in puddles and ponds as much as half a mile from the working face, although the greatest concentration of escaping air is usually vertically above the face. This escaping air has found paths to a free air surface, and so long as the separate paths remain narrow and distinct, no problems will develop. However, and particularly for the vertically escaping air, there exists a significant risk that the various paths will expand and join with the possibility of catastrophic loss of air from the tunnel. In subaqueous tunneling, it is important to observe the water surface and monitor the characteristic "boils" on the water surface. It is better to accept the more difficult working conditions resulting from reduced plenum pressure than to run the risk of a "blow." Prudence will sometimes dictate that work should continue through weekends and holidays, when the risk of a blow is present.

In all cases where a compressed air tunnel is stopped for a substantial period, such as a holiday or a weekend, it is essential that the working face be fully covered by breastboards and all remaining exposed material covered

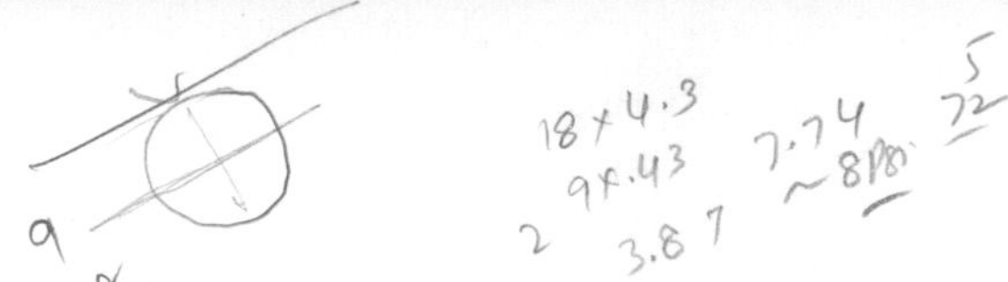

with mud or cement grout. It is wise to seal the joints between breastboards in the same way.

Where an impervious stratum lies above the permeable material in which the tunnel is being driven, the compressed air flowing through the face and expanding into the soil may remain at a pressure greater than the hydrostatic head at the underside of the impervious stratum. In such cases, the difference is additive to the hydrostatic head and may result in increased pressure and water flow at the working face. The air pressure must then be reduced as much as possible for safe working and it may be necessary to remove excess water from the invert by wellpoints. (For further details refer to Chapter 6.)

FREEZING

Basic Conditions

As with grouting, freezing is more often used in shaft-sinking than in tunneling, but the method is useful where nothing else will serve, provided that surface access over the tunnel alignment is available.

It should be noted at the outset that for the method to be successful, sufficient water must be present. Freezing will not change the stability characteristics of dry soils. The groundwater is frozen by extraction of the latent heat and this is done most efficiently by a steep temperature gradient.

Refrigerated Brine

The typical installation consists of a refrigeration plant that cools a brine solution, which is then pumped down the center of an annular freeze pipe to the bottom of the hole, returning via the outer annulus in contact with the soil. The warmed brine is returned to the refrigeration plant and the cycle continues. In practice, a number of freeze pipes are connected to a pair of headers for the flow and return lines. The circulation is maintained continuously until the expanding frozen zones outside each pipe overlap. Thereafter, during construction, the circulation may be sustained only to the extent necessary to maintain the freeze wall.

For tunnel construction, it is only necessary to stabilize the soil, and strength is not important. There is, therefore, no need to maintain continuous freezing to achieve lower temperatures at which the ice is stronger. In fact, it will be a disadvantage in tunneling to have ground any stronger than necessary.

Difficulty in closing the freeze wall will be experienced in the presence of flowing groundwater, and closure may be impossible.

It should also be noted that the presence of organic material (common in silts and clays) or salt water will result in greater difficulty in freezing, since the groundwater freezing point will be depressed.

Instead of calcium chloride brine (the most commonly used), other brines with lower freezing points are available but more expensive. In addition, liquid nitrogen or liquid carbon dioxide can be used, but their cost is high and special precautions must be taken when they are used.

Liquid Nitrogen

For special purposes, especially for projects of limited extent and duration, boiling of liquid nitrogen in the freezing elements may be appropriate. However, it is better to provide a degree of control, since there is a risk that the liquid nitrogen will be forcefully ejected from the open pipe by bubbles formed during the boiling process. In addition, considerable irregularity may be expected in the frozen zone. If the liquid nitrogen is supplied to a freezing element in the same manner as the brine in a normal system, but with the supply pipe nearly as large as the freezing element, the flow can be regulated to match the rate of boiling.[12]

Positive exhaust must be provided for the vapor, which is heavier than air and will cause suffocation if it accumulates in sufficient quantity. Also, steel pipes will be embrittled by the low temperature, and non-ferrous metals or stainless steel are required.

Design

It is necessary to determine the thermal characteristics of the soils to be frozen and the freezing point of the groundwater. The effect

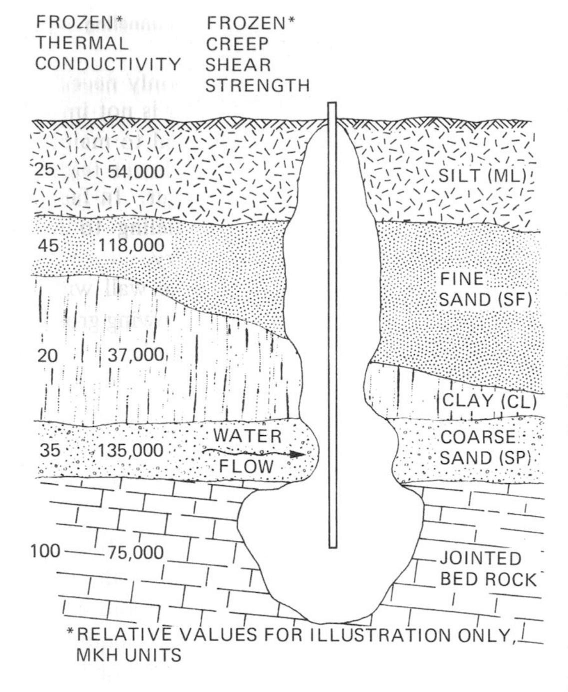

Fig. 5-9. Typical effect of thermal properties on frozen ground.

of variations in thermal conductivity is illustrated in Fig. 5-9 from Shuster[12]). The analysis will be governed by the conditions giving rise to the thinner frozen zones.

The computation of thermal energy and time to complete freezing is usually based on simplified assumptions as to uniformity of heat transfer within the soil. Computer analyses can be made for more complex or critical projects. Figure 5-10 shows the relationship between freezing time and spacing of freezing elements for brines and for forced conversion and boiling of liquid nitrogen.

Operation

It is essential to include instrumentation as part of any freezing installation. Thermocouple strings can be used to monitor the ground temperatures between freezing elements and to monitor refrigerant temperature. A simple method for determining closure of a frozen ring in free-draining soils is to monitor a centrally

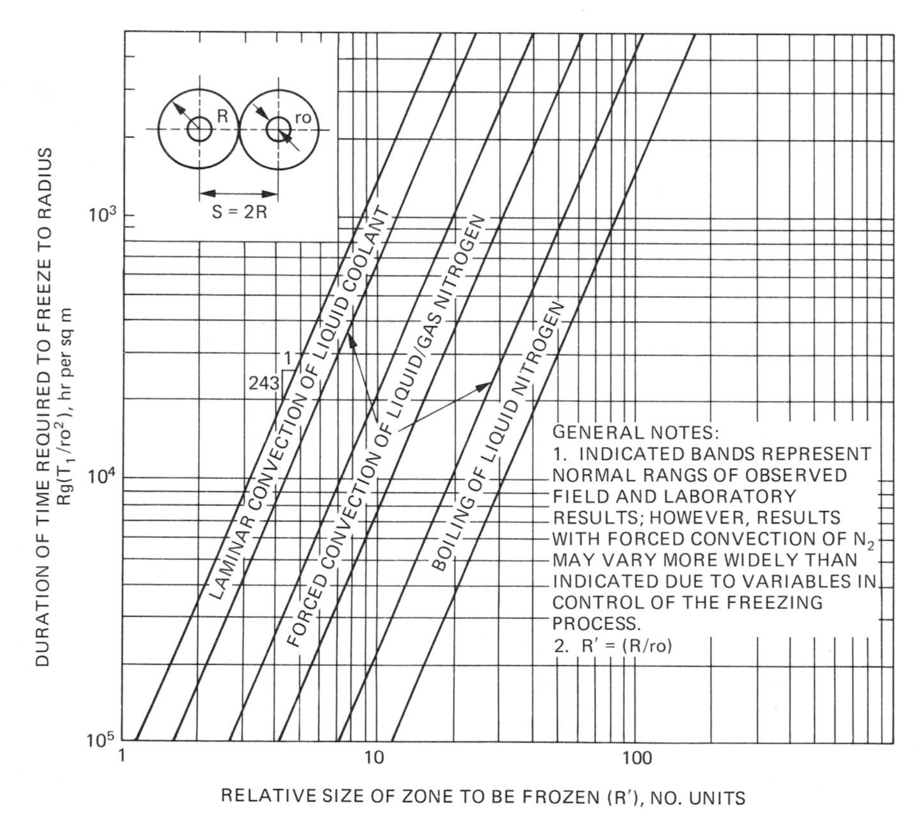

Fig. 5-10. Determination of desired freezing time.

located piezometer. The advancing freeze front pushes water out of the pores as it expands on cooling and, once the wall is closed, there is no means of exit from the area and water rises in the piezometer.

During installation, it is important to ensure that the outer casing of the freezing element retains its integrity. Loss of brine into the ground will render completion of the freeze wall difficult and it may be necessary to resort to a more expensive circulating fluid.

As previously stated, provided that sufficient information is available as to the location of the freezing isotherm, the refrigeration plant can be shut down part of the time.

Any utilities in the area, such as water, sewer or stream lines, subject to be affected by freezing must be protected by insulation if they are within the freeze zone. They should be monitored to ensure that this protection is effective.

The freezing method is likely to be most appropriate in silty soils, not subject to stabilization by dewatering or grouting. These soils are the most likely to be affected by frost heave, caused by water migrating towards the freezing zone under capillary action and freezing into lenses of ice. The possibility of damage from this phenomenon must be kept in mind.

Excavation of the frozen ground may be by drilling and blasting, although it is sometimes difficult to keep the drilled holes open. Conventional mining equipment has been used successfully in permafrost and roadheaders appear to be effective in the conditions created by freezing.

Short tunnels may be frozen by driving the freezing elements into the ground horizontally around the perimeter of the excavation. If the surface is inaccessible, it may be practical to drive a higher level tunnel than that which required frozen ground stabilization and install the freeze system from the higher tunnel.

CONSTRUCTION METHODS

Due to high labor costs and stricter safety requirements, and to shorten construction time, most soft ground tunneling today is done by means of shields and increased use of mechanization. This applies particularly to tunnels in pervious ground below water level and subaqueous tunnels. If the size of the project warrants the high initial investment, tunnel-boring machines may be used if the ground is reasonably uniform and there are no serious obstacles encountered which would interfere with its operation. (See Chapter 6 on shield tunnels and Chapter 10 on tunnel-boring machines.) When investment in this equipment is not economical in short tunnels or there are too many obstacles, such as large boulders in glacial fill, which interfere with the advance of a shield or tunneling machine, manual excavation may have to be used. In non-cohesive or running soils which cannot be stabilized by one of the methods described above, a shield may be used even in a short tunnel.

Excavation in cohesive soil or ground which has been stabilized, not subject ot hydrostatic pressure of free water, may be mined by one of the following methods.

Drifts or small headings are the basis of manual excavation. They should be as large as possible but not less than 5 feet wide and 7 feet high. The ground is supported by temporary timber or steel framing. The roof is held by driving of poling boards held by cross-timbers and posts. Excavation at the face is advanced as steep as the soil will stand and the boards are driven forward, with the rear end supported on the frame and the front by the soil. A new set of supports is placed under the front of the poling boards and the process is repeated. Part of the face may need breasting. In larger head-

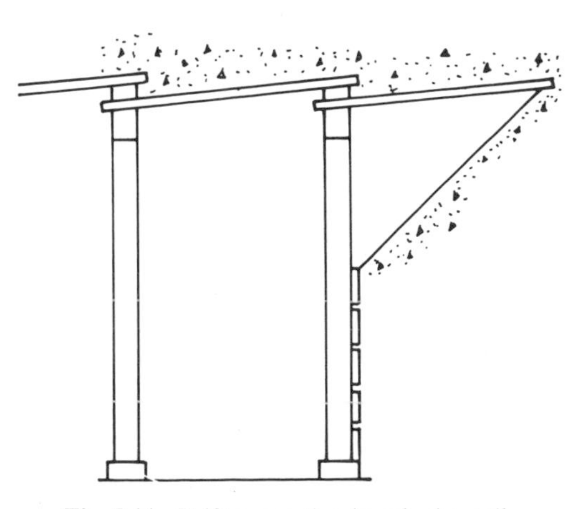

Fig. 5-11. Drift excavation in cohesive soil.

ing steel spiling consisting of small wide flange, beams may be used, in conjunction with steel supports, instead of timber. They are driven into the face by jacks or air hammers. The sides of the excavation are held by boards or lagging as required. Figure 5-11 illustrates this process.

Steel liner plates may be used instead of poling boards if the roof, or part of it, will stand long enough to permit their placing. Erection of the plates starts at the top of the arch and works down. They are available in various shapes and sizes. For small headings, corrugated plates may suffice. In larger openings, the plates are backed up by steel ribs, against which they are blocked. Voids behind the plates should be filled with pea gravel or grout to prevent uneven loading.

Small tunnels may consist of a single heading, and for larger tunnels a series of drifts is used, their number depending upon the total cross-section and soil characteristics. Different methods have evolved of which the following are now used most often.

The American method starts with a top heading in crown, supported by posts, caps and poling. The excavation is widened between posts to the spring line, and the arch segments adjoining the crown are set, supported by extra posts or struts, and the ground held by lagging. The excavation is benched down along the sides and ribs are set, temporarily supported on sills. This is repeated down to the invert; then the bench is fully excavated. The soil is held by lagging as needed (see Fig. 5-12).

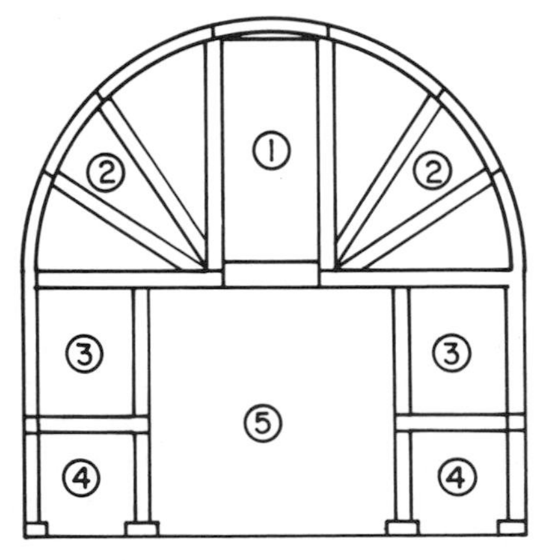

Fig. 5-12. American method of excavation.

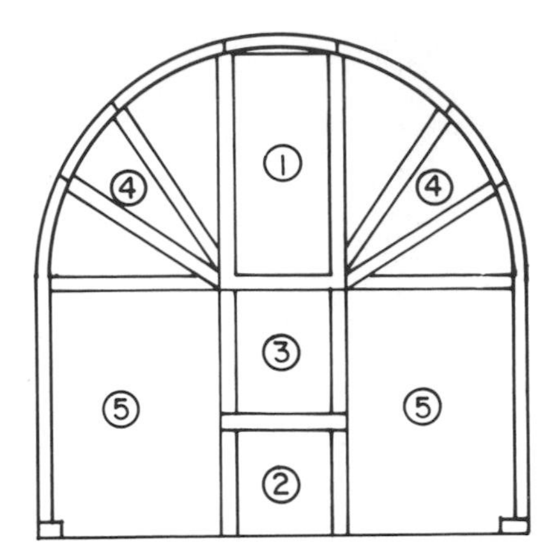

Fig. 5-13. Austrian method.

The Austrian system uses a full height center heading, which may start with a top heading which is cut down to the invert in short length, or as separate bottom and top headings with the core between the headings excavated in short length and the short posts replaced by long ones. The arch section is widened in short length similar to the American method. The remaining benches are excavated to full face in short increments. Posts are set to support the sills under the arch ribs. The soil is held by poling boards and lagging. This method is suitable if soil pressure is generally uniform on all sides (see Fig. 5-13).

The Belgian method places the permanent arch lining as soon as possible if loads are primarily vertical and side pressures are minimal. Excavation starts with a center heading from the crown to the spring line. This is widened on both sides, the ground being held by segmental arch ribs and transverse poling supported on longitudinal timbers. The ribs are supported by struts extending from sills in the center heading, and sills on the benches. The arch lining may then be placed for permanent support of the vertical load. The center heading is then excavated to the invert, leaving benches to support the arch of the tunnel lining. Slots are cut in the benches to underpin the arches. The benches are removed and the invert and the rest of the lining is placed (see Fig. 5-14).

For excavation in non-cohesive soils, every effort should be made to stabilize loose or running soil to use methods described above, or a shield should be used. If water is no problem,

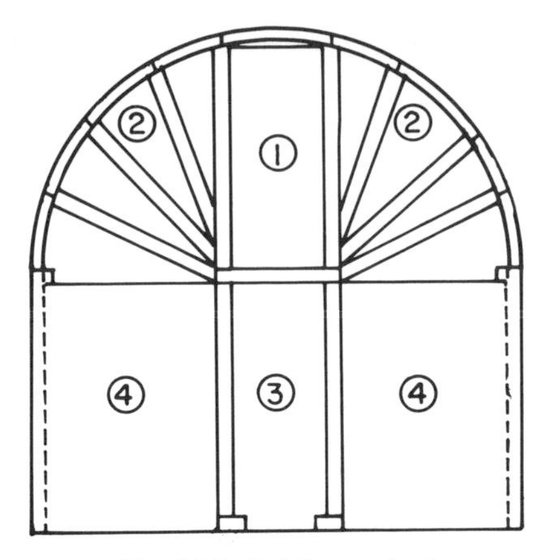

Fig. 5-14. Belgian method.

drifts may be excavated and sections enlarged by multiple headings, a slow and costly procedure.

Drifts may be excavated as previously described, but with complete ground support. A shallow slot, about two feet deep and one or two poling boards wide, is excavated in the top of the face and a short vertical breastboard is held by a strut placed immediately to hold the ground. After the vertical breastboards have been set across the width of the heading, a cap, supported on short posts, is placed. The rest of the face is then excavated downwards in small increments and held by breastboards (see Fig. 5-15).

SURFACE EFFECTS OF TUNNEL CONSTRUCTION

The unfavorable effects of tunnel construction are usually of consequence only in urban areas,

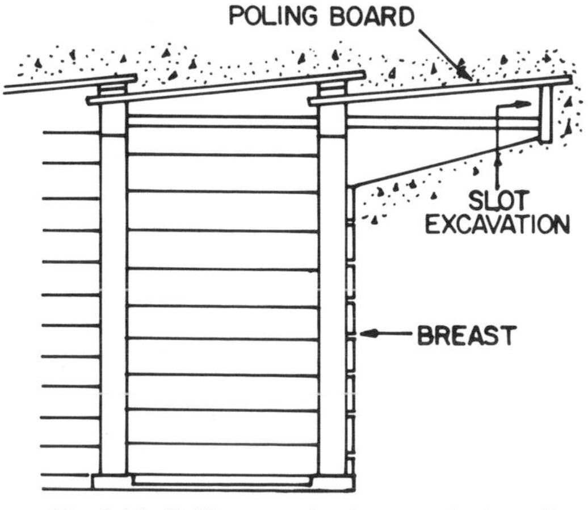

Fig. 5-15. Drift excavation in non-cohesive soil.

where there are structures and people to be affected. The majority of problems result from subsidence due to excavation or dewatering and permanent lowering of water table. In addition, the construction of a tunnel may impose limitations on future construction in its vicinity.

Water Table Depression

When the water table is lowered, the effective load on the subsoil is increased by the difference between the drained and submerged weight of the soil between the original and depressed water tables. The settlement due to this additional load depends upon the compressibility of the soil. The amount of settlement in sands and gravels is usually very small, unless the sand is very loose. Repeated fluctuations of the water table will cause more settlement because the settlement after each event is permanent and the effects of repetitive loading by dewatering are additive.

If the affected zone contains strata or lenses of soft clay, silt or peat, much larger settlements are likely to be encountered. The amount of settlement to be anticipated can be calculated from the known consolidation characteristics of the affected soil.

Permanent depression of the water table commonly occurs in the vicinity of tunnels because they act as drains which collect infiltrating groundwater. To the extent that the filtration rate exceeds the recharge rate, the water table will be lowered. In addition, the loosened soil along the outside of the tunnel may act as a drain and, if water can escape from the soil at the low point (such as a portal), permanent water table depression will result from this drainage.

Piles embedded in the drained soils will be subject to drag-down forces related to the settlement, and this may be particularly serious for buildings founded on friction piles. Wood piles newly exposed above the water table may be subject to rapid deterioration.

Subsidence Due to Lost Ground

The loss of ground in soft ground tunnel construction is normally reflected as a settlement trough at the ground surface above the tunnel alignment. The potential for such loss of ground is controlled by soil properties, the construction method and the quality of workmanship. More settlement is experienced from two tunnels side by side, but usually less than double that for a single tunnel. If the separation of twin tunnels is greater than one or two diameters, depending on the soil, they will act essentially independently.

For tunnels in clay, the stability factor referred to in the discussion of stability of cohesive soils will provide an indication of what may be expected. Cording and Hansmire[13] report the work of other investigators to the effect that for stability ratios of 1 to 2, only small elastic movements will be expected, while the volume loss will increase rapidly for stability ratios much greater than 4. The rate at which the soil moves will be affected by the stresses and disturbance accompanying advance of the shield, during which time the rate would be expected to be a maximum.

The presence of saturated soils overlying impervious soils is conducive to loss of ground as the water flows towards the tunnel and induces running or flowing conditions. Groundwater control by closely spaced wellpoints is the best solution to this problem.

Sources of lost ground in shield tunneling are described in Chapter 6. In manually mined tunnels, which would rarely be used in soft ground under urban areas, ground is lost during mining operations. The amount depends on the nature of the soil and the care exercised in supporting the ground during excavation, and is unpredictable. Every effort must be made to limit the loss.

Settlement Trough

The volume of the settlement trough is typically equal to the volume of ground lost in the tunnel. However, exceptions may occur in granular soils because of volume increase above the tunnel, and in clays because of consolidation close to the tunnel. Cording and Hansmire,[13] reporting the work of Saenz and Utesa, state that in soft clays, maximum settlements increase with increasing depth of cover, both because higher stresses squeeze more soil into the tunnel and, more importantly, because of increased consolidation of the clay.

The shape of the settlement trough in the normal range is taken to approximate the normal probability curve, the displacement (S) below the horizontal at any point being given by

$$S = S_{\max} e - \frac{x^2}{2i^2}$$

where $S_{\max}$ is the settlement at the center of the trough, e is the distance from the center and i is the distance from the center to the point of inflexion. The volume of the settlement trough thus defined is given by

$$V_s = 2.5 i \, S_{\max}.$$

For practical purposes, the trough cross-section can be approximated by a triangle whose base is $5i$ and height $S_{\max}$. In cases where $S_{\max}$ exceeds $Z/200$ (where Z is the depth to tunnel springline), according to Cording and Hansmire,[13] the settlement trough no longer approximates a probability curve. Further settlements are concentrated at the center of the trough.

Cording and Hansmire[13] use Fig. 5-16 to show the interrelationship of the various elements significant in predicting or determining the settlement trough parameters.

Damage Criteria

The available information indicates that buildings partially within and partially outside the trough will be subject to damage if the slope of the trough exceeds 1:1000. Damage to load-bearing walls occurs at lower values than for frame structures, especially when subjected to the hogging type of deformation at the edge. Horizontal strains of approximately the same magnitude as the vertical strains are present near the edge of the settlement trough, according to Cording and Hansmire,[13] and these can result in damage even to underpinned structures. The fact that a structure is subject to damage from the estimated slopes, strains or settlements is not necessarily an indication that underpinning is required, since underpinning procedures generally cause differential settlements of the same order of magnitude as

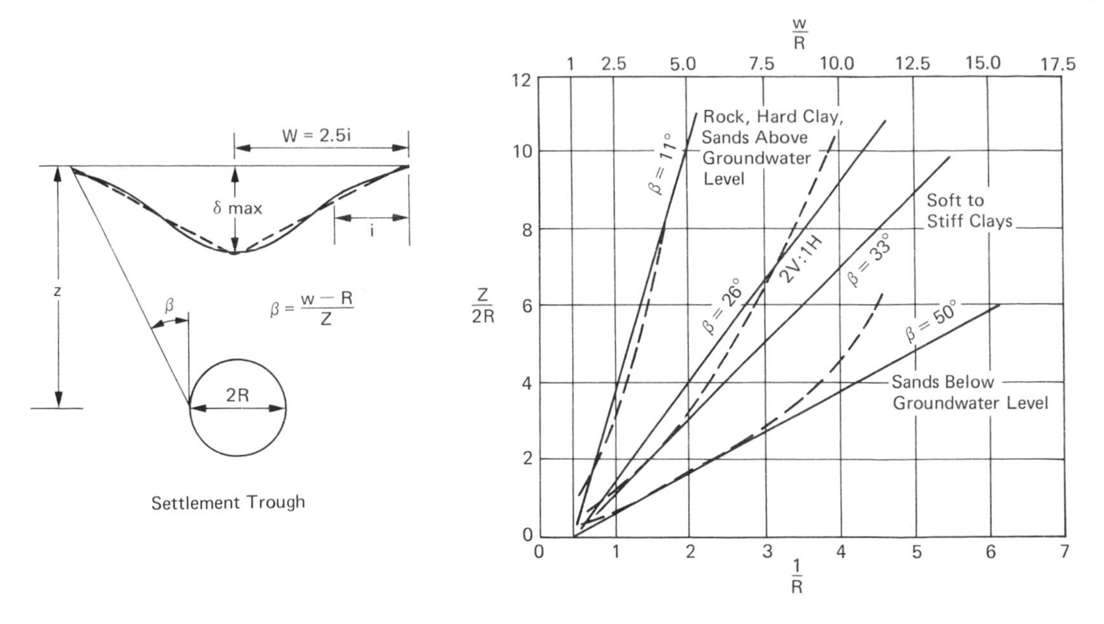

Fig. 5-16. Settlement diagrams.

those observed at the edge of the settlement trough. Moreover, underpinning alone will not provide protection against horizontal strains.

TUNNEL SUPPORTS

Primary Lining

All tunnels in soft ground need preliminary supports as excavation proceeds unless the soil has been solidified by grouting or freezing, in which case the supports may follow the excavation of some distance without interfering with work on the face.

For lining of circular, shield-driven tunnels, see Chapter 6.

Sets and Lagging. These are the most common preliminary supports. Timber supports used formerly have generally been replaced by structural steel arches and posts. Erection starts at the crown and proceeds towards the spring line in segments, depending upon the excavation method used. Poling boards, either steel or wood, packed tightly, support the ground. In loose soil, care is needed to prevent open joints, through which soil may be lost. Posts supporting the arches are set on foot blocks. If the ground shows any tendency to squeeze, horizontal struts are installed at the invert. Posts and ribs consist of wide-flange H sections, their size depending upon the cross-section of the tunnel and the soil pressure expected. Lagging has to be blocked tightly against the surface. Spacing of the ribs may vary from 2 to 4 feet, according to soil characteristics.

Liner Plates. Liner plates of steel are available in various sizes and shapes. If stand-up time of the soil permits, they may be used instead of lagging. In all but small tunnels, where ribbed plates may suffice, they are backed up by steel ribs and posts. The plates are bolted together to form a tight support. Voids behind the plates must be filled by grouting or with pea gravel.

Perforated steel plates, reinforced with steel ribs, have been used in Europe and are available in the U.S. They may be used in conjunction with shotcrete, the latter filling the voids behind the plates and reinforcing the liners.

Shotcrete. This has occasionally been used as preliminary lining in firm soil, particularly in Europe, where shotcrete technique has been developed earlier than in the U. S.

Permanent Lining. With the exception of shield-driven railroad or rapid transit tunnels, which have tight circular segmental linings, all tunnels in soft ground require a permanent secondary concrete lining. Depending on soil condition, this may be placed after full face excavation, or parts may be poured as soon as possible to relieve the primary lining from exposure to full loads for longer periods. The arch may be poured as soon as excavation has reached the springlines and can be supported on benches. As excavation proceeds, posts are erected in slots cut into the bench to underpin the arch. Walls and inverts are placed after completion of excavation. Thickness of the lining is governed by the loads and the space required to pass the slickline by the steel ribs for pouring. Pouring procedure is similar to that used for rock tunnels. The amount of reinforcing steel is determined by the soil pressure, but in general a minimum of steel is placed in the inner face to resist shrinkage and temperature changes.

DESIGN OF LINING

Loads on linings depend on the characteristics of the soils. Circular liners installed in shield-driven tunnels in soft ground below water level are designed as flexible rings which deform to the readjustment of the soil pressure so that they carry only compressive stresses and no bending moments (see Chapter 6).

A concrete secondary lining in highway tunnels is usually placed after the primary lining has adjusted. Its thickness is determined by construction and service requirements, and, with temperature and shrinkage reinforcement, is adequate to supplement the capacity of the primary lining.

For loads on horseshoe sections excavated by any of the previously described methods, Terzaghi[16] has developed values, based on tests on field experience.

Running ground consists of clean gravel or coarse sand above water level, with zero stand-up time, requiring continuous support as excavation proceeds. By arch action due to side friction of the ground, the actual loads are considerably smaller than the depth of the overburden, more than 1.5 times (width + height) of excavation (see Fig. 5-17).

Primary Lining is designed for the following loads, given in feet.

Dense ground–$H_r = 0.30\ (B + H_t)$.
Medium ground–$H_r = 0.40\ (B + H_t)$.
Loose ground–$H_r = 0.50\ (B + H_t)$.

The load per lineal foot of tunnel is

$$Q = WH_rB$$

Where W is the unit weight of soil. In all but arid regions, the saturated weight is used, since ground may be saturated after heavy rains.

Secondary lining of concrete is usually required for interior finish. Its thickness is determined by construction requirements and it has temperature and shrinkage reinforcements. It is generally installed some distance behind the face, in which case it will support long-term additional loads which may occur, in conjunction with the primary lining. In very loose soil, it may be necessary to install the secondary lining, particularly the arch as soon as possible

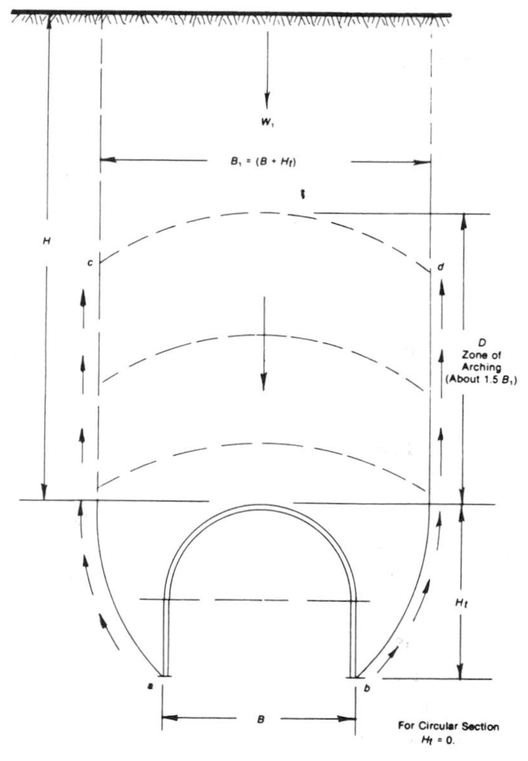

Fig. 5-17. Roof load in running ground.

after the primary lining has been placed, and it has to support the initial soil pressure as well. The load may reach $H_r = 0.60\ (B + H_t)$.

Raveling ground has a stand-up time at least long enough to permit the installation of one support, either ribs and lagging, or liner plates.

Primary lining is designed for the following loads, given in feet.

Above water table–$Hp = \dfrac{T - t}{T} Hr$ (running).

Below water table–$Hp = \dfrac{T - t}{T} 2Hr$ (running),

in which case a bottom slab or a circular lining may be needed.

T equals time elapsed in minutes between excavation and erection of primary supports, and t equals stand-up time.

Secondary lining of concrete is required similar to that mentioned for running ground. Maximum load may reach $H_p = 0.60\ (B + H_t)$.

Squeezing ground consists usually of soft or medium clay. In mose cases, excavation would be by the shield method, possibily combined with a tunned-boring machine, with circular liners (see Chapter 6).

Where a horseshoe shape is indicated, the load on the liner is determined by the arching action of the ground (see Fig. 5-18) and by the homogeneousness of the ground. The reduction in pressure is determined by the shearing strength, s, of the soil, with $s = \dfrac{Q_u}{2}$, where Q_u is the unconfined compressive strength of the clay.

Primary lining in free air tunneling has to support the following height of soil, in feet, above the arch

Homogeneous soil–$H_p = H - \dfrac{SHQu}{2W(B + 2H_t)}$.

Soft roof, stiff sides–$H_p = H - \dfrac{HQ_u}{2WB}$.

Stiff roof, soft sides–$H_p = H - \dfrac{HQ_u}{2W(B + 6H_t)}$.

Unit pressure on roof is

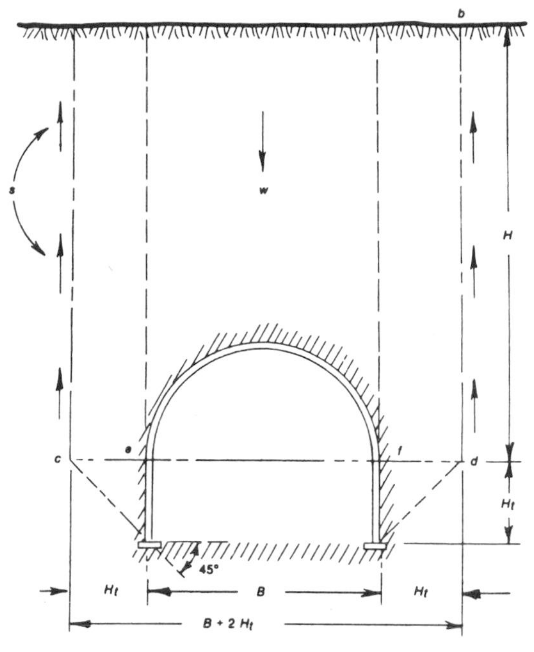

Fig. 5-18. Loading in squeezing ground.

$$P_s = WH - \frac{2\,SH}{B + 2\,H_t}$$

If P_s is positive, supports should be placed as close to the face as possible; if it is negative, excavation may proceed 10 to 12 feet ahead of the last support.

Average horizontal pressure on the side of the lining is

$$P_h = P_s + \frac{WH}{2} - Q_u$$

Secondary lining of concrete is placed as interior finish and to carry additional loads which may develop. In very soft clay, it should be placed as soon after primary lining as is practicable and designed to carry the above loads in conjunction with the primary lining. In firm clay, the lining may be placed after a longer time interval and reinforced for temperature and shrinkage only.

REFERENCES

1. Terzaghi, Karl, "Geologic Aspects of Soft Ground Tunneling," Chapter 11 in *Applied Sedimentation*, Trask, Parker D. (Ed.), John Wiley & Sons, New York, 1950.

2. Heuer, Ronald E., "Important Ground Parameters in Soft Ground Tunneling," *Proceedings of Specialty Conference on Subsurface Exploration for Underground Excavation and Heavy Construction,* ASCE, 1974.
3. Broms, B.B. and Bennermark, H., "Stability of Clay at Vertical Openings," *Journal ASCE* 3: 71-94, 1967.
4. Deere, D.U., Peck R.B., Monsees, J.E. and Schmidt, B., "Design of Tunnels, Liners and Support Systems," report prepared for Office of High Speed Transportation, U.S.D.O.T. Contract No. 3-0152, 1969.
5. Peck, R.B., "Deep Excavation and Tunneling in Soft Ground," *State-of-the-Art Volume*, Seventh International Conference on Soil Mechanics & Foundations, pp. 225-290. Mexico City, 1969.
6. Schmidt, B., "Exploration for Soft Ground Tunnels—A New Approach," *Proceedings of Specialty Conference on Subsurface Exploration for Underground Excavation and Heavy Construction,* ASCE, 1974.
7. Clough, G.W., "Development of Design Procedures for Stabilized Soil Support Systems for Soft Ground Tunneling," Volume 1, Report prepared for U.S.D.O.T. under Contract DOT-05-50123, 1977.
8. Janin, Jean J. and LeSciellour, Guy F., "Chemical Grouting for Paris Rapid Transit Tunnels," *Proceedings, Construction Division, ASCE* 96, *No. COI:* 71-74, 1970.
9. "Chemical Grouting," *Proceedings Soil Mechanics and Foundation Division, ASCE* 83, *No. SMI, Pt.1, Paper 1426,* 1957.
10. Caron, C., "The Development of Grout for the injection of Fine Sands," *Grouts and Drilling Muds in Engineering Practice,* Butterworths, London, 1963.
11. Mayo, Robert S. and Richardson, Harold W., "Practical Tunnel Driving," McGraw-Hill, New York, 1941.
12. Schuster, J.A., "Controlled Freezing for Temporary Ground Support," *Proceedings, First North American Rapid Excavation and Tunneling Conference,* 2, ASCE/AIME, 1972.
13. Cording, Edward J. and Hansmire, W.H., "Displacements Around Soft Ground Tunnels," *Proceedings, Fifth Panamerican Conference on Soil Mechanics and Foundation Engineering,* Buenos Aires, 1975.
14. Terzaghi, Karl, "An Introduction to Tunnel Geology in Rock Tunneling with Steel Supports," by R.V. Proctor and T.L. White, Commercial Shearing & Stamping Co., 1946.
15. Terzaghi, Karl, "Earth Tunneling with Steel Supports," by Robert V. Proctor and Thomas L. White, Commercial Shearing & Stamping Co. 1977.

6
Shield Tunnels

ROBERT S. MAYO

Robert S. Mayo & Associates

HISTORY

Tunnels in loose, non-cohesive or soft ground are generally driven by means of a steel shield, most often circular in shape, either in free air or, if necessary, under compressed air. The shield permits the excavation of soil and the erection of primary lining under safe conditions; allows better control or ground settlement; and is a must in subaqueous tunnels.

The first shield in free air was invented by Marc Brunel, who, together with his son Isambart Brunel, built a tunnel under the Thames in London. The shield was 38 feet wide and 22 feet high. Under great difficulties and with several floodings, the 1200-foot tunnel took from 1825 until 1845 to build. In 1896, it was taken over by the East London Subway and still is part of that system.

Compressed air in a shield was first proposed by Lord Cochran but was never used by him.

The first circular shield, very similar to those used today, was designed by J.J. Greathead in 1896 in building the second tunnel under the Thames in the stiff London clay, without compressed air. It had a diameter of 7 feet, 3 inches, and a length of 1350 feet for pedestrian use.

The St. Clair Tunnel of the Grand Trunk Railway, between Port Huron, Michigan and Sarnia, Canada was built in 1888 using a Beach hydraulic shield with hydraulic jacks, erector arm and cast iron segments. It was 21 feet O.D. and 6500 feet long. It used air pressure as high as 28 psi but was completed uneventfully and is still in daily use now.

Once it was demonstrated that shields with cast iron lining could safely construct subaqueous tunnels, such shields have come into extensive use, in free or in compressed air. These shields permitted the building of large subway systems in cities like London and New York, large sewers and, later, highway tunnels. Contractors have even selected them for relatively short tunnels in difficult ground.

Some notable shield-driven tunnels are described below.

The Hudson River tunnel, which became a part of the Hudson-Manhattan Railroad, was started in 1879 by D.C. Haskin with compressed air without a shield, but was abandoned after some costly disasters in 1882 (after 1600 feet). It was resumed in 1889 with a Greathead shield but closed down again in 1891 due to financial difficulties. The whole project was taken over in 1902 by William McAdoo and completed.

Shield-driven highway tunnels began in 1922 with the Holland Tunnel under the Hudson River, followed by the three Lincoln Tunnels and the Midtown Tunnel under the East River. Shields were used extensively on the BART system in San Francisco and on the Washington Metropolitan Transit System.

Meanwhile, new and improved techniques for soft ground tunneling have been developed in many countries. In some fields, they are more advanced than American practice. England, France, Russia and Spain have built large subway systems. The Japanese, in the last decade, have been building subways using innovative methods. Mexico City, Sao Paolo and other cities south of the U.S. are building subways and large interceptor sewers using shields with and without compressed air. Many more rapid transit tunnels will be built in the years to come using shields and new and cheaper linings.

SOIL CONDITIONS REQUIRING SHIELD CONSTRUCTION

Weak Non-cohesive Soils. Running sand and loose gravel or silt, which are practically cohesionless and which cannot be safely and efficiently excavated by ordinary soft-ground tunneling techniques, can be handled by the use of a shield, usually working in free air. This will also provide better control of ground settlement over the tunnel.

Weak Plastic Soils. Soft plastic clays, particularly those with a critical water content and which become easily liquefied when worked, call for shield excavation, either in free air or, in certain cases, with compressed air.

Soils Under Water Pressure. For tunneling below water level, a shield is required unless the groundwater can be lowered or the soil solidified by injection. For tunnel construction below bodies of free water, a shield, usually with compressed air, is a must, except in solid rock at an adequate depth.

CONSTRUCTION OF SHIELD

Open Shield. This shield has an open working face, divided only by the bracing, and is used in soil which is firm enough to stand or which can be held by breasting. It may be used in free or compressed air. It usually consists of a circular steel cylinder with stiffening ribs and bracing to withstand the external pressure. The circular shape permits erection of the lining segments even if the shield has a tendency to roll during advance, which is usually the case. Horseshoe-shaped shields have also been used. They are statically less efficient but require less excavation and do not have a tendency to roll. Although the majority of shields have been circular, there is no reason why under proper conditions a horseshoe shape may not be appropriate. Figure 6-1 shows a shield used for a single-track tunnel of the Washington Metropolitan Transit Authority (WMTA) with a 17-foot, 8-inch O.D. fabricated steel primary lining with 2-foot, 6-inch wide rings.

The *body of the shield* is a steel cylindrical plate, stiffened by at least two circular ribs and vertical and horizontal bracing members. These divide the face into a number of pockets which are used for mining and mucking. The size of the pockets is adapted to the methods and equipment used for excavation and removal of soil. The length of the body is made as short as possible, usually between 6 and 8 feet, depending on the diameter. For machine mining, the shield may have to be longer in order to house the mechanical equipment.

The *tail of the shield* extends backwards from the body and provides space to erect the liner rings. Its length is at least one and one-half times the width of the liner so that, at the end of the shove, not less than half of the last erected ring remains within the tail. There is also room for the jack shoes and the portion of the jacks which extends through the rear diaphragm. As a rule of thumb, the tail should have a thickness of 1 inch for every 10 feet of diameter. There must be clearance of $\frac{3}{4}$ to 1 inch all around between the inner face of the tail and the lining, to prevent the shield from becoming "iron bound" during corrections of alignment. Figure 6-2 shows a small shield for a crossing under a railroad.

The *cutting edge* extends around the forward end of the shield and is generally $2\frac{1}{2}$ or 3 inches thick. It is braced to the forward diaphragm by

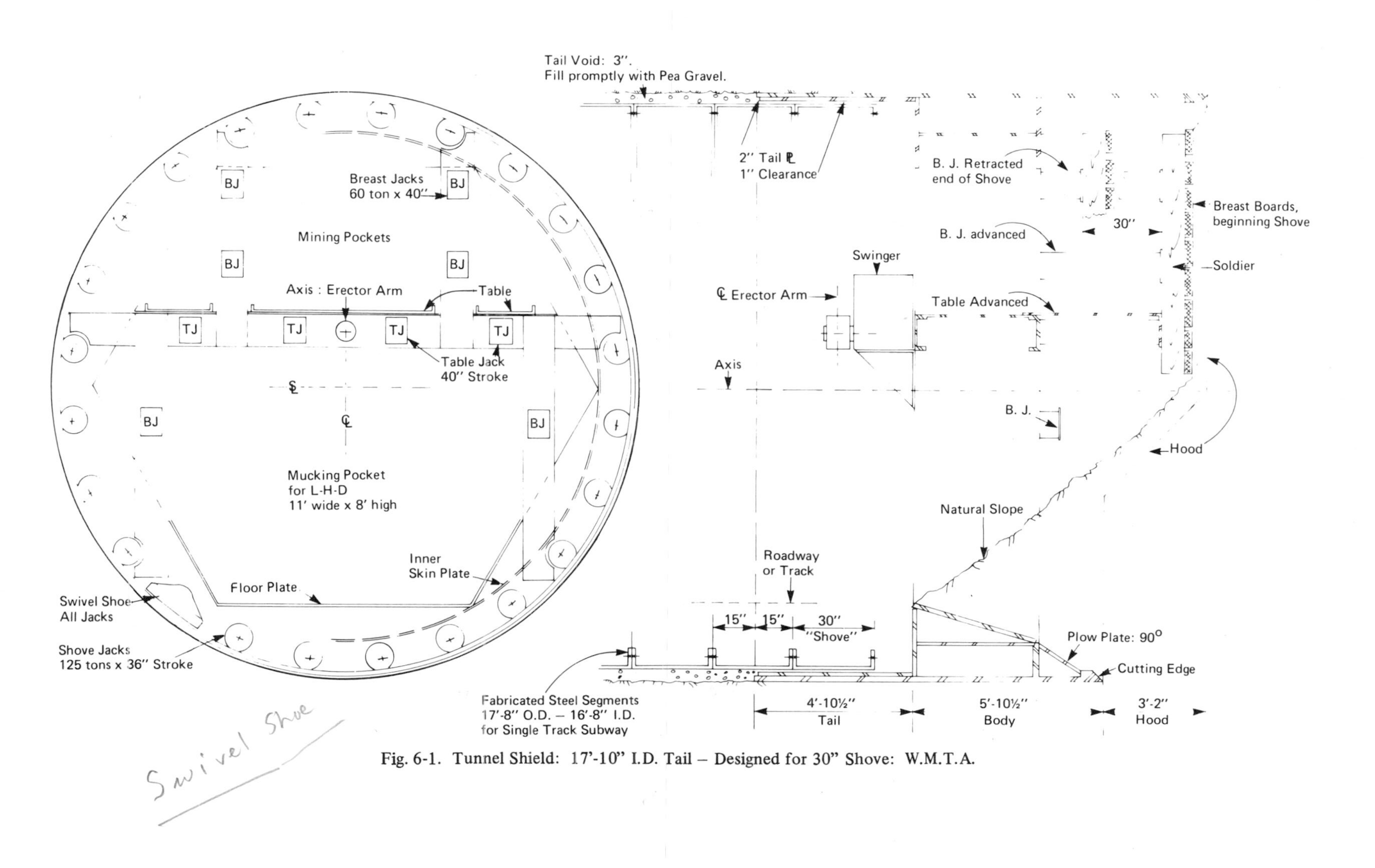

Fig. 6-1. Tunnel Shield: 17'-10" I.D. Tail – Designed for 30" Shove: W.M.T.A.

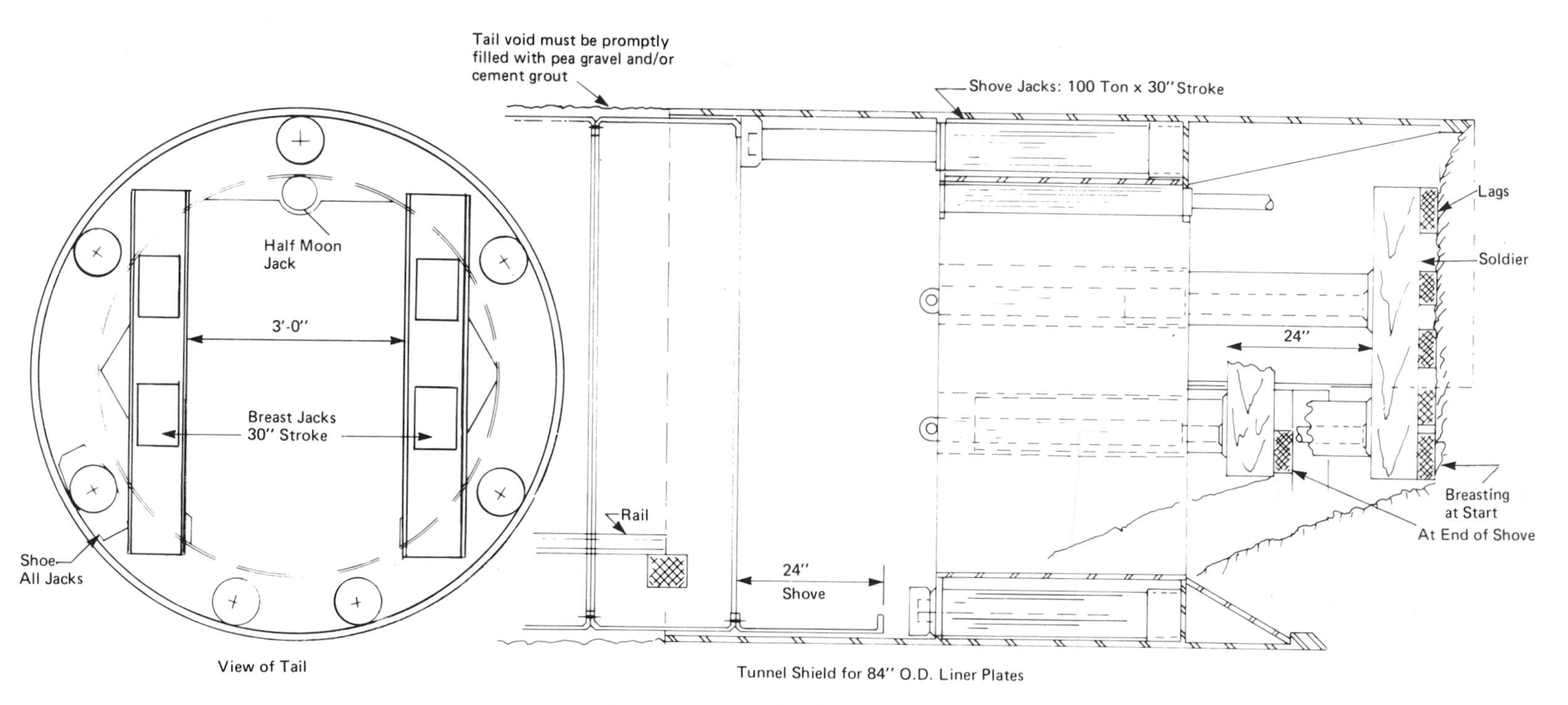

Fig. 6-2. Small tunnel shield for railroad crossing.

Fig. 6-3. Cutting edge of tunnel Shield.

frequent gusset plates. To aid in cutting harder ground, it is coated with abrasion-resistant welding wire (see Fig. 6-3).

The *hood* projects at least one shove ahead of the shield generally for the upper 180 degrees. Under the protection of this hood, the miners can advance the breastboards, one at a time, beginning at the top, to hold the face prior to the next shove. A hood is not required on closed shields passing through soft silts like those under the Hudson River or in the soft clays found around Detroit.

Closed Shields. In very soft clay or silt or fine running sand, the front face of the shield is closed by a solid steel bulkhead. This is equipped with ports through which the soil is excavated and removed. The ports can be closed with steel doors. The bulkhead is braced with steel ribs, welded to the body to resist the soil pressure. The construction of the shield otherwise is the same as for the open shield.

Half Shields. Semicircular or semielliptical shields are sometimes used to support the roof of a tunnel during excavation in dry or dewatered ground. This may apply to a tunnel at shallow depth where open excavation is not permitted or practical. The shield is supported on temporary steel beams set on steel posts or concrete sidewalls. Its construction and operation are similar to the full shield.

SHIELD ACCESSORIES

Shoving jacks. The shield is advanced by a series of hydraulic jacks mounted around the periphery. They should have a total capacity of at least $6\frac{1}{2}$ tons/foot^2 of face. Their stroke should be 6 inches larger than the shove to allow for rebound of the primary lining. The bodies of the jacks are set in the stiffening rings of the body of the shield and the pistons bear against the primary lining through jacking shoes which distribute the pressure over the liner face.

Breasting Jacks. A number of breasting jacks are mounted in the upper face to hold the breastboards which bear against the soil (Fig. 6-1).

Fig. 6-4. Breasting jacks.

Fig. 6-5. Erector arm.

The jacks are allowed to retract during the shield advance but maintain pressure on the face. Figure 6-4 shows the breasting jack arrangement and valves for the shove jacks in a 10-foot, 6-inch diameter shield.

Table Jacks. Steel plates at the level of the working deck can be advanced by jacks to give the miners a working platform or "table." These are also used for holding the breastboards.

Erector Arm. An erector arm is mounted at the rear diaphragm just above the horizontal axis. It picks up the liner segments as they are delivered in the bottom and raises them into position where it holds them until they are bolted in place. Figure 6-5 shows an erector arm.

Hydraulics. All jacks and the erector arm are operated from a hydraulic system. The shove jacks work off banks of valves in quadrants: upper, lower, right and left. The breast jacks and table jacks have individual controls, as has the erector arm. The system uses a fire-resistant hydraulic fluid. The electically-driven pump is generally located in the power house near the shaft. It has remote controls for starting, stopping and changing pressure. The maximum available pressure of 10,000 psi is seldom required. If the shield does not move under 4000 psi, a check should be made for obstruction of the cutting edge by a boulder or rock ledge.

Tail Seal. Rubber seals may be bolted to the inside of the tail at the rear end to prevent soil, pea gravel, grout or water flowing into the back of the shield. Unfortunately, the seals are subject to damage by projections on the outside of the liner rings. They may also become "frozen" into the grout. If they are torn, replacement is practically impossible.

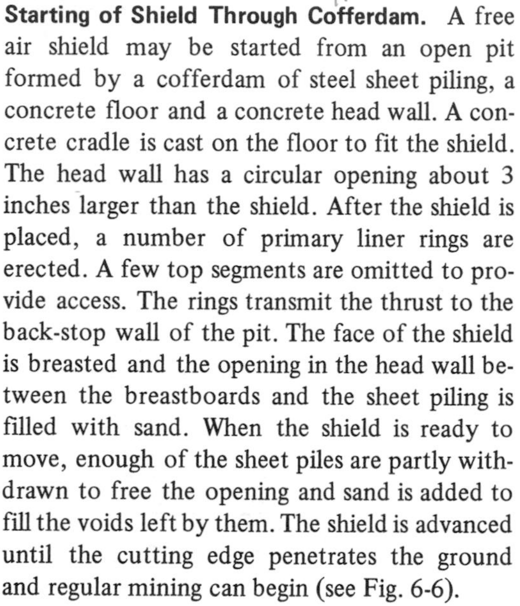

HANDLING OF SHIELD

Starting of Shield Through Cofferdam. A free air shield may be started from an open pit formed by a cofferdam of steel sheet piling, a concrete floor and a concrete head wall. A concrete cradle is cast on the floor to fit the shield. The head wall has a circular opening about 3 inches larger than the shield. After the shield is placed, a number of primary liner rings are erected. A few top segments are omitted to provide access. The rings transmit the thrust to the back-stop wall of the pit. The face of the shield is breasted and the opening in the head wall between the breastboards and the sheet piling is filled with sand. When the shield is ready to move, enough of the sheet piles are partly withdrawn to free the opening and sand is added to fill the voids left by them. The shield is advanced until the cutting edge penetrates the ground and regular mining can begin (see Fig. 6-6).

Starting Shield from Shaft. Subaqueous tunnels often have to be started from a shaft at the shore, particularly when compressed air is required. The shaft may be a temporary pit or may later become a permanent ventilation shaft. A bulkheaded circular opening is provided in the shaft wall for advancing the shield which is erected on a concrete cradle on the shaft floor. If groundwater cannot be lowered to permit the initial advance of the shield under free air, an air deck designed to withstand the air pressure has to be installed in the shaft or pit, in which air locks for men and material are placed. Pri-

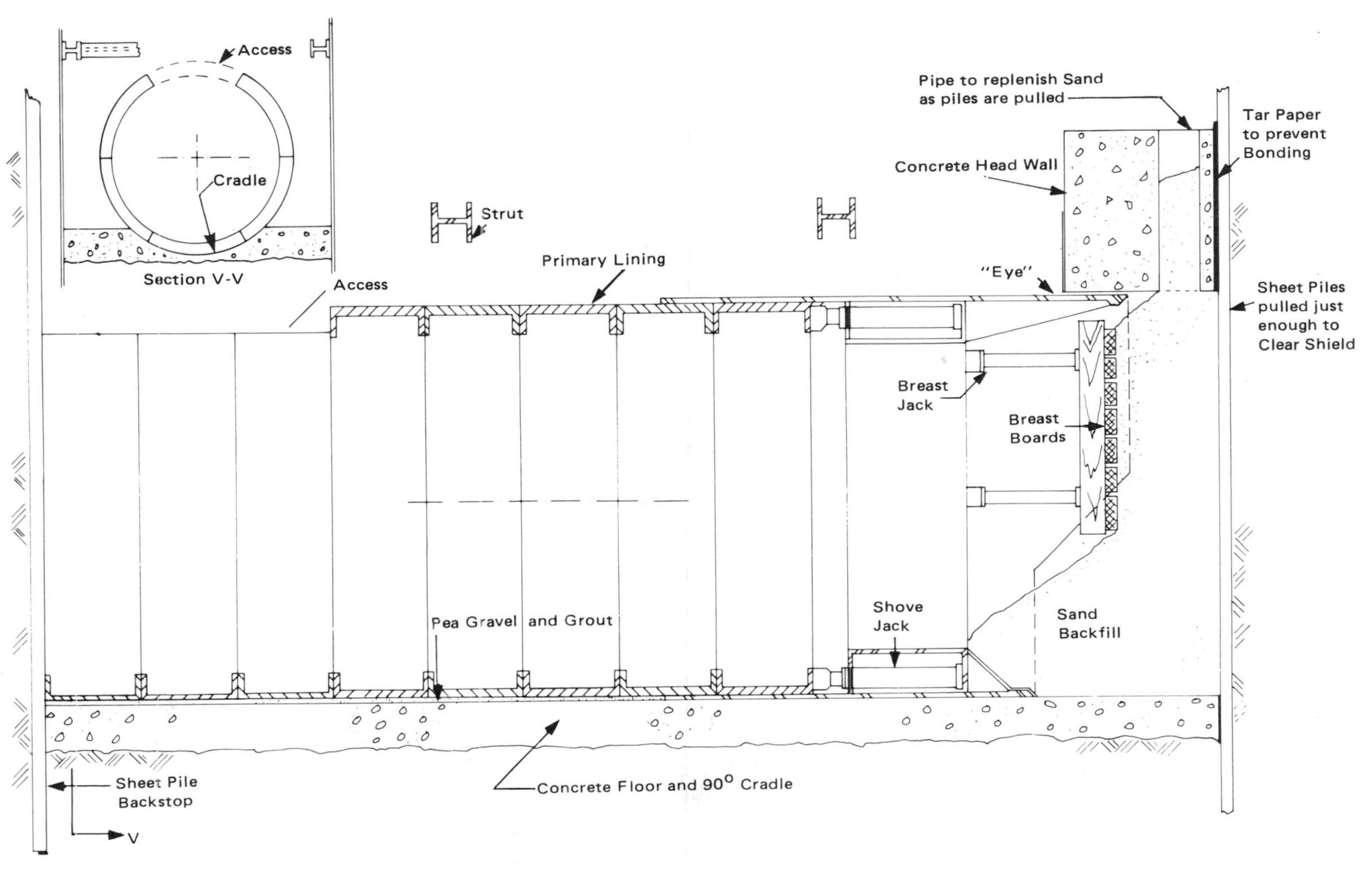

Fig. 6-6. Starting the shield in free air.

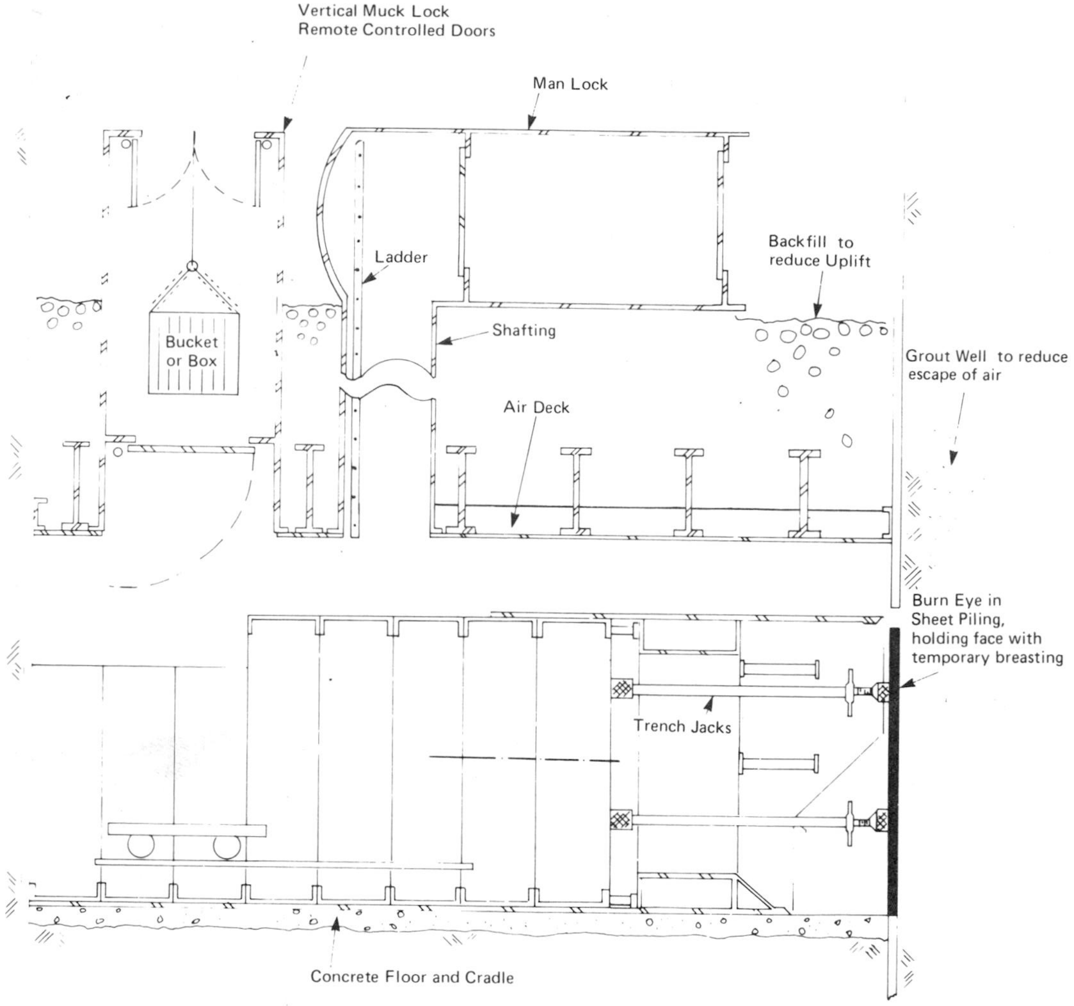

Fig. 6-7. Starting the shield under air.

mary liner rings are erected behind the shield to transmit the thrust to the opposite wall. When the shield is ready to advance, the bulkhead in the wall opening is removed, or, in case of a pit, a circular opening is burned in the cofferdam, and the soil is held by temporary breasting (see Fig. 6-7). After the shield has advanced sufficiently to erect an air bulkhead and locks in the tunnel, the air deck in the shaft may be removed to provide free access for material handling.

The shield may also have to be driven from a rock shaft into soft ground, as is commonly the case in the New York area. The shield is assembled in a rock tunnel adjacent to the river where the bulkhead and locks are assembled for driving the subaqueous part of the tunnel.

Advance and Steering of Shield. In each move, the shield is advanced by the shoving jacks a distance equal to the width of a primary lining ring. Its alignment must be continuously checked, because it tends to drift sideways or up or down. As shown in Fig. 6-8, the standard method of steering the shield horizontally is to establish the lead before beginning the shove. Leadboards

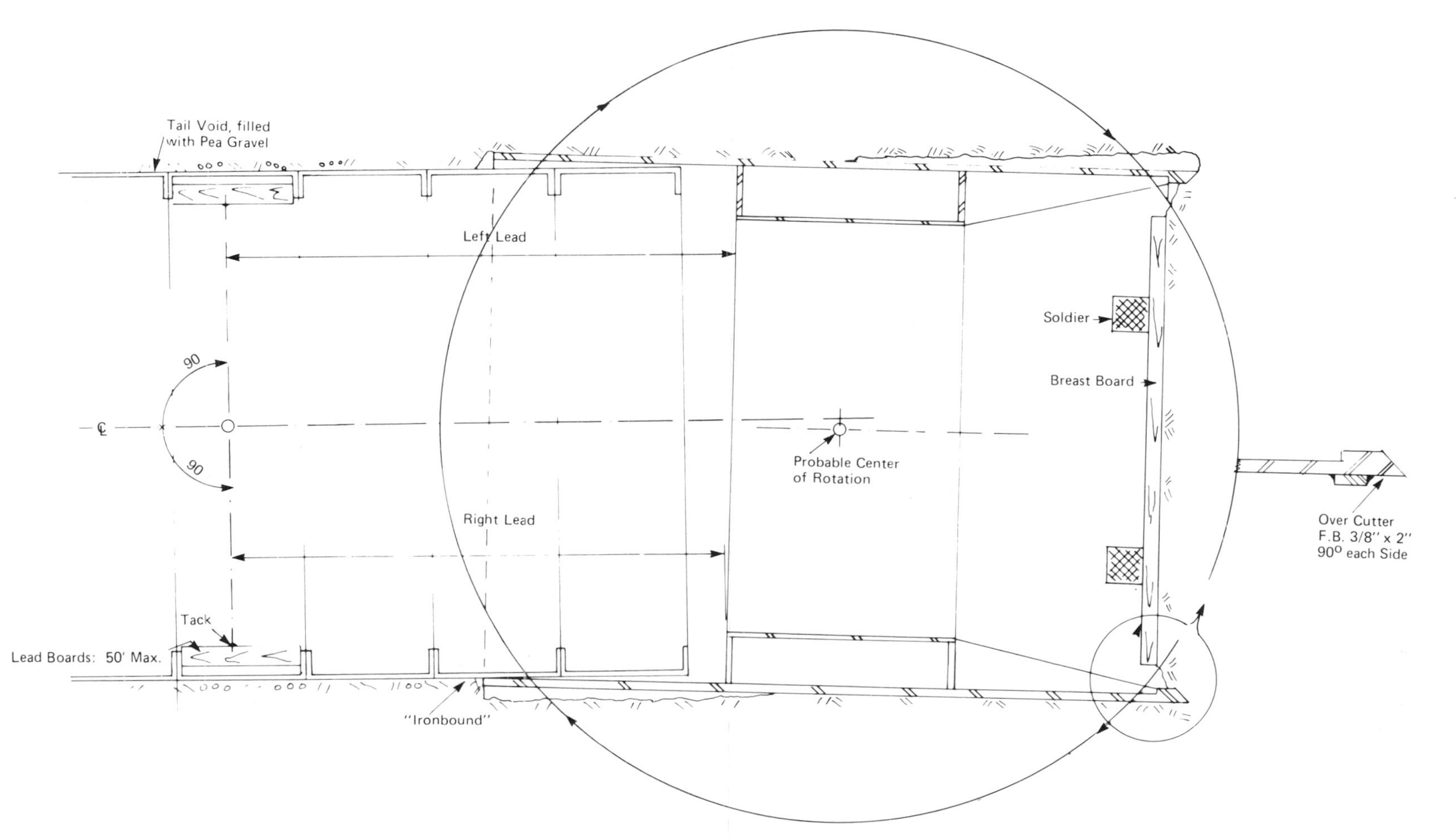

Fig. 6-8. Steering the shield.

are wedged into the segments at distances not greater than 50 feet. At the beginning of the shove, one side or the other is advanced a small amount by operating all the jacks on that side. This tends to slightly turn the shield, crowding the ground or leaving a void, as indicated by the arrows in Fig. 6-8. Once the desired lead is established, the shield is advanced by using all shove jacks. A shield is said to be iron bound when the tail is hitting the primary lining. There should be a surveyor on each side of the tunnel to constantly observe the lead distances as the shield advances to see that it is moving uniformly in that direction.

Most shields are nose-heavy and have a tendency to "dive." This must be carefully watched and, if necessary, only the lower group of jacks used to make the shove. Often an extra jack is added at the bottom to correct this tendency.

In firm ground, it is difficult to establish a lead because it is impossible to crowd the ground sideways. Therefore, an overcutter is attached at the nose of the shield to get a slightly larger hole. This generally is a flat bar, $\frac{3}{8}$ inch thick, applied to the right and left hand quadrants, or occasionally around 270 degrees of the circumference, but never at the bottom. This gives additional space for correction of the alignment and greatly reduces skin friction, but it also increases the tail void and may increase settlement, particularly if the shield is shut down for the weekend.

Curves are very difficult to drive with a shield. The minimum radius should be 575 feet and the segments are supplied with tapered rings. Sometimes every ring is tapered but more often only every other ring.

Rolling of Shield. All circular shields have a tendency to roll. Some shields have been designed with removable fins on the sides. They have retractable wedges to cut a spiral groove in the ground. Others have used two adjustable platforms at the front end, which may be raised or lowered a small amount to twist the shield as it is advanced into the ground. Probably the simplest way is to install part of the shove jacks in slotted holes, as shown in Fig. 6-9. The cover plates can be removed and the jack forced over to the end of the slot by wood wedges. This will give a spiral effect to the shove. A hooded shield is top heavy and rolling should be corrected before it gets more than 5 degrees.

A horseshoe or straight-sided shield should not roll, but if it does, it is a good deal more serious and more difficult to correct.

EXCAVATION AND MUCKING

Running Sand and Gravel. When driving through running sand and gravel, the shield must have a hood and the face is breasted at all times. At the beginning of the mining shaft, the upper breastboard is carefully removed and the muck raked into the tunnel (see Fig. 6-10). This board is advanced one shove and held temporarily by two trench jacks. In fine, running sand, it will be necessary to place a handful of hay, straw or excelsior outside the breastboard, particularly at the corners. The remaining breastboards are advanced, one at a time, until the soldier beams are free and can be reset and held by the breast jacks. The trench jacks are then removed, and the shield is ready to be shoved ahead.

Advancing breasting in fine, water-bearing sand is a slow and costly operation to prevent loss of ground. Perhaps the job should be put under compressed air, unless the water table can be lowered by external pumping or the ground solidified by chemical gouting, either from the surface or from the face.

Stand-up Ground. This may be hard clay, cemented gravel, dry silt or loess. As shown in Fig. 6-11, the face is broken down by miners

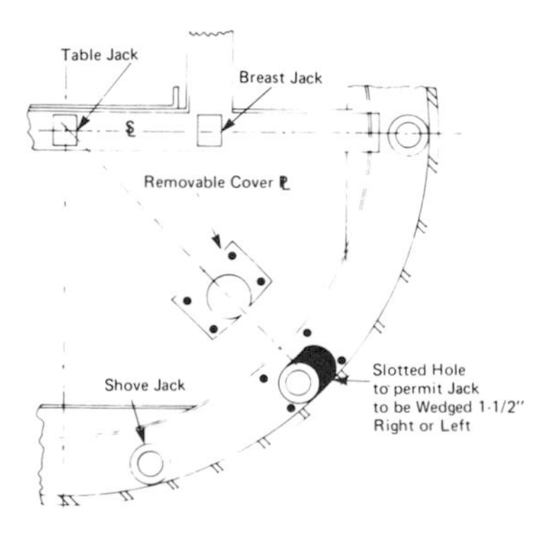

Fig. 6-9. Slotted shove jack mounting to correct roll.

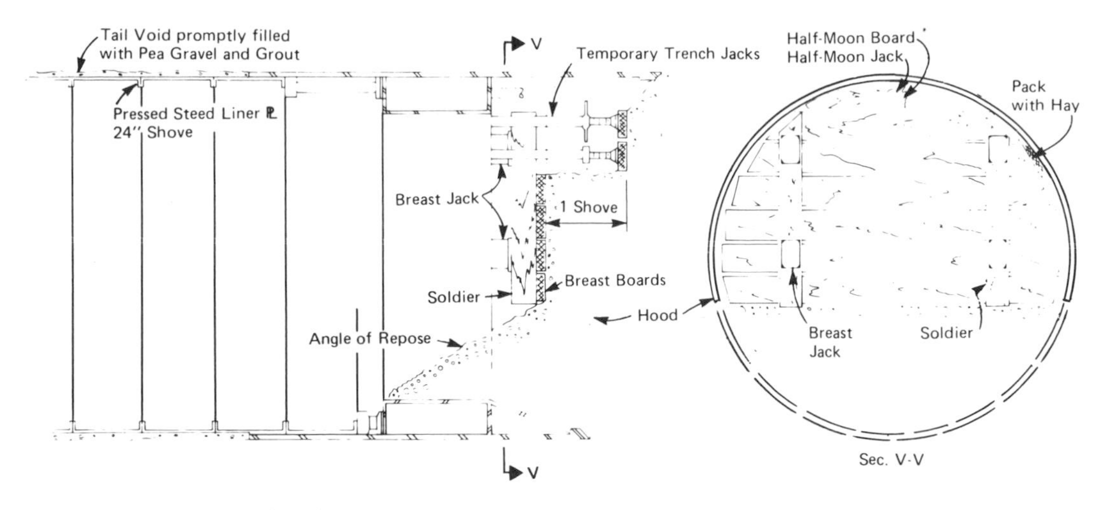

Fig. 6-10. Breasting a small shield in running sand or gravel.

with pneumatic clay spades before the shove. About 6 inches of ground is left to be trimmed by the cutting edge. Modern practice is to mount a small hydraulic backhoe to one side of the lower pocket, which not only can break down the face but also rakes the muck back to where the mucking machine can reach it, or it may deposit it directly on the belt conveyor. The face is seldom breasted except over the weekend. If this type of ground is anticipated for the whole length of the job, the shield may not need a hood. This type of ground is ideal for a mechanical mole if the length warrants the initial investment.

Plastic Ground. When tunneling through plastic blue clay or silt, it may be possible to extrude the muck into the tunnel. As shown in Fig. 6-12, the shield is closed by stop logs, leaving two openings about 12 X 12 inches to allow the muck to enter. As the shield advances, the muck will flow into the tunnel where it is cut by wire loops into chunks that one man can handle. Using this method, there is enough pressure on the ground to cause it to flow into the tail void and perhaps heave the street. It is better to raise the surface $\frac{3}{4}$ inch than to let it settle the same amount. If the shoving pressure becomes too great, it may be necessary to enlarge the muck openings. The shield foreman must keep a close watch on pressure of the hydraulic fluid when shoving. If it suddenly rises, it may mean a boulder or other obstruction has been encountered (and should be removed before it damages the cutting edge).

Hudson River Silt. This is a very fine silt with a high water content, making it nearly semi-liquid. Special methods have been developed in the course of about 60 years for advancing through this material. The shield is of the closed type with four ports for access to the face or to admit muck. The shield does not require a hood unless mixed face is anticipated at the beginning or the end of driving. In the earliest tunnels, it was attempted to "shove blind," displacing all the silt, but this made it difficult to maintain line and grade, and the tunnel had a tendency to float, since the primary lining alone weighed less than the ground displaced. Present practice is to admit 20 to 25% of the silt into the tunnel through the upper openings (see Fig. 6-13). This remains in the tunnel as ballast until it can be removed in free air. There is a traveling dam or bulkhead attached to the shield and towed along behind it. This keeps the muck out of the pit behind the main bulkhead where work is done. Segments are delivered on the working deck and must be lowered into the pit to be picked up by the erector arm. The upper bulkhead doors are hydraulically controlled to regulate the inflow and to close the openings between shoves.

Mixed face. On occasion, a shield must be driven from rock to soft ground, resulting in

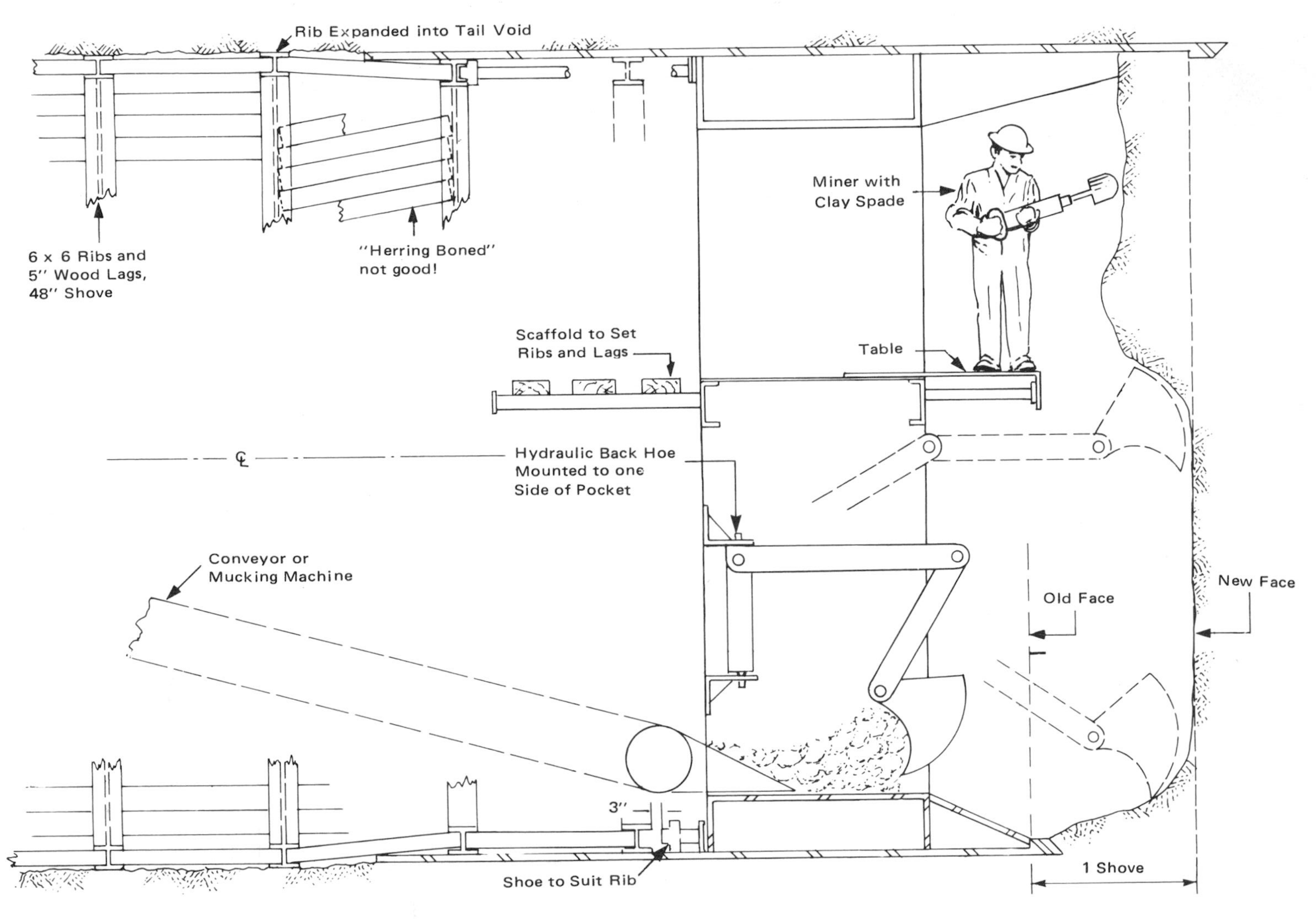

Fig. 6-11. Stand-up ground.

Fig. 6-12. Plastic clay excavation.

a mixed face. In the New York area, it is common to assemble the shield in a rock tunnel adjacent to the river, where the bulkhead and locks are installed ready for driving the subaqueous part. As the shield breaks away from the rock into the soft ground, mining becomes very complicated and costly. As shown in Fig. 6-14, a bottom drift is advanced as far forward as is considered safe. In this is poured a concrete cradle to support the shield and maintain alignment. The shield is then advanced cautiously, breaking out rock in short sections and breasting the face when the soft ground is exposed. Blasting should be avoided and the rock broken with "busters."

On occasion, the shield may encounter high points of the bedrock, sometimes only 2 or 3 feet above the bottom of the shield. Full breasting must be carried down to the bottom of the shield so that the ledge can be broken out by chipping, busting or (if it is really hard) blasting. On some shields, the lower cutting edge has been made of short sections of cast steel, bolted to the shell plate. If they are damaged by blasting, they can be replaced.

Mixed face tunneling is slow and tedious and will increase the cost three to five times.

Filling the Tail Void. The void left by the tail of the shield around the primary lining has to be filled as soon as possible during the shove. For this, pea gravel is the standard material. This is washed gravel with pebbles $\frac{1}{4}$ to $\frac{3}{8}$ inch in diameter. Washed, crushed stone is used occasionally. The pea gravel can be purchased in bulk or in paper bags. The latter is more expensive but more convenient in small tunnels. As soon as the shield begins to move and the grout-holes have passed beyond the end of the tail, the gravel is injected, beginning at the bottom.

Low-pressure cement grout is injected two or three rings behind in the pea gravel. This fills the interstices of the gravel and any voids not filled by the gravel and may stop groundwater seeping into the tunnel. Sometimes, high-pressure grout is injected through the same holes about a week later. This operation must be done cautiously in order to prevent grout from seeping into sewers and basements of buildings.

Some contractors use a slow-setting grout of cement-sand-lime to fill the tail void instead of pea gravel. If the mix is too thin, it will flow into the tail and it may "freeze" the tail, making it difficult to establish the lead when beginning the next shove. In theory, a tail seal would prevent this leakage, but tail seals are often torn off or damaged and cannot be replaced.

Settlement. Settlement is inherent with shield-driven tunnels except in the rare cases when

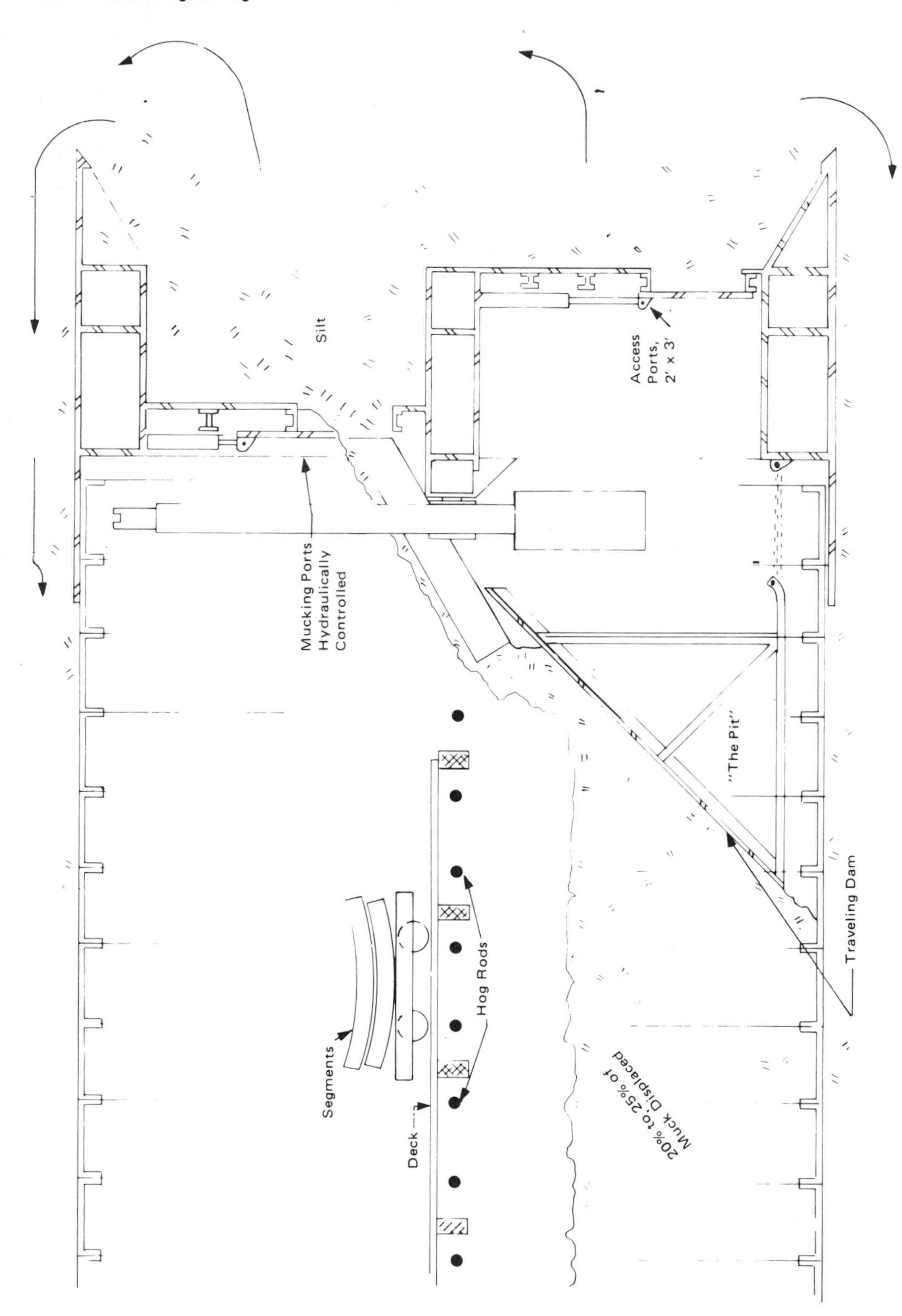

Fig. 6-13. Subaqueous tunneling in Hudson River silt.

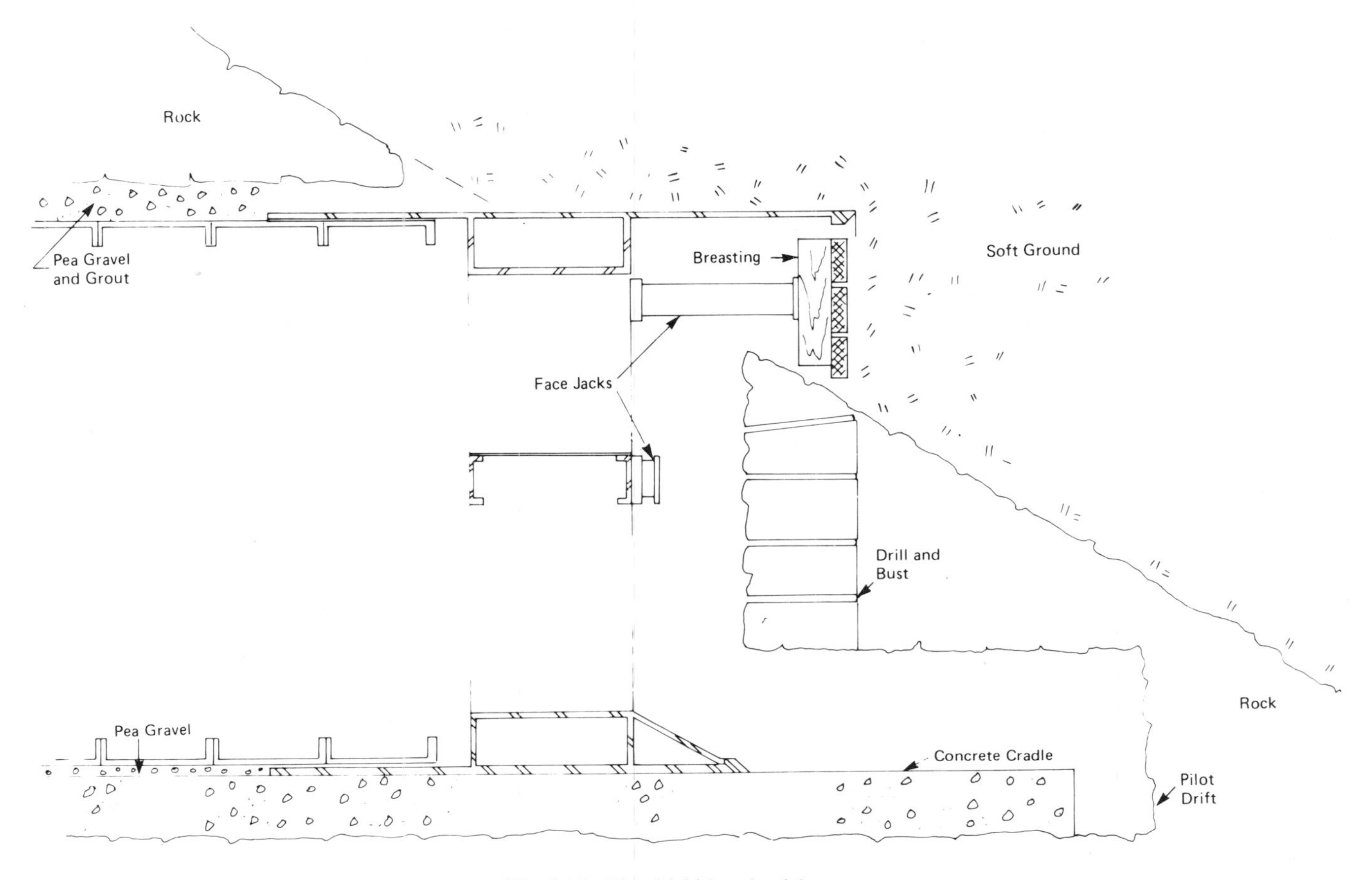

Fig. 6-14. The shield in mixed face.

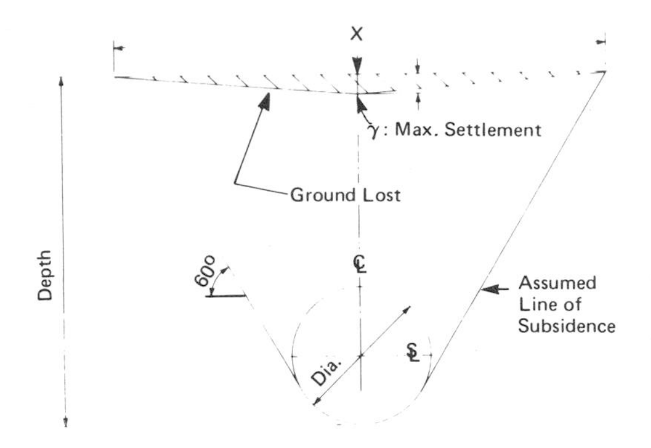

Fig. 6-15. Diagram for estimating ground lost per linear foot.

enough pressure can be maintained against the face to actually heave the surface. This settlement may be due to the following causes.

1. Failure to fill the tail void completely. Since the void can be calculated, the heading foreman should keep a careful record of the gravel placed.
2. Loss of face probably accounts for most of the settlement. This may be due to careless mining or improper breasting. Even in stand-up ground, there may be an imperceptible drift of the face, particularly over the weekend. It takes very little movement of the face to cause a lot of settlement.
3. Other causes may be crushing of old sewers or ducts, broken water mains which wash out ground or collapsing vaults under the sidewalks.

Figure 6-15 shows a diagram for estimating the volume of lost ground based on the total settlement on the center line of the tunnel, as suggested by Professor E.J. Cording of the University of Illinois. On a tunnel for WMTA, the initial settlement was 8 inches. By placing as much as 80% pea gravel in the tail void and breasting the face except when actually mining, the settlement was reduced to $2\frac{1}{2}$ inches.

Machine Excavation. Tunnel-boring machines are used more and more in conjunction with a shield for excavating firm and hard ground (for details, Chapter 10).

Removal of Material. An efficient system of removal of excavated material (mucking) and bringing in liners and supplies is of great importance. This applies particularly to machine excavation. At the face, the material is placed on a short conveyor, either by shovel, in case of hand mining, or by an excavating machine. It is then transferred to a loading conveyor which dumps it into mine cars. The muck trains are shuttled back to the shaft where the cars are hoisted to the top for dumping. The liner segments are carried to the face on empty mine cars (for details of material handling, see Chapter 11).

PRIMARY LINING

All shield tunnels must have a primary lining to support the ground and take the thrust of the shove jacks. Traditionally, segmental cast iron liners have been used; other methods are ribs and lagging, fabricated steel liners, pressed steel liner plates and precast concrete segmental liners.

Cast Iron Liners. The liners are cast in segments, generally 30 to 32 inches wide. Some foundries can cast $\frac{7}{8}$-inch width, 6-inch deep flanges. All four flanges have machined surfaces to ensure a tight fit, and all are provided with caulking grooves all around. Figure 6-16 shows liners for a 16-foot, 6-inch I.D. rapid transit tunnel. There are 5 standard 60-degree "A" segments in each ring and a "B" segment with one end tapered to fit a 12-inch wide closing key. The segments are placed by the erector arm and bolted to the previous ring by steel bolts. The bolt-holes are sealed by plastic grommets inserted with the bolts.

There usually is a tendency for the crown of the tunnel to settle, and most specification require that a "hog rod" be set horizontally across the tunnel in every other ring. These rods are generally $1\frac{1}{4}$-inch turnbuckles, which remain in place about two weeks, until the outside pressures have equalized. In certain materials, when the shield is shoved blind, the lateral pressures

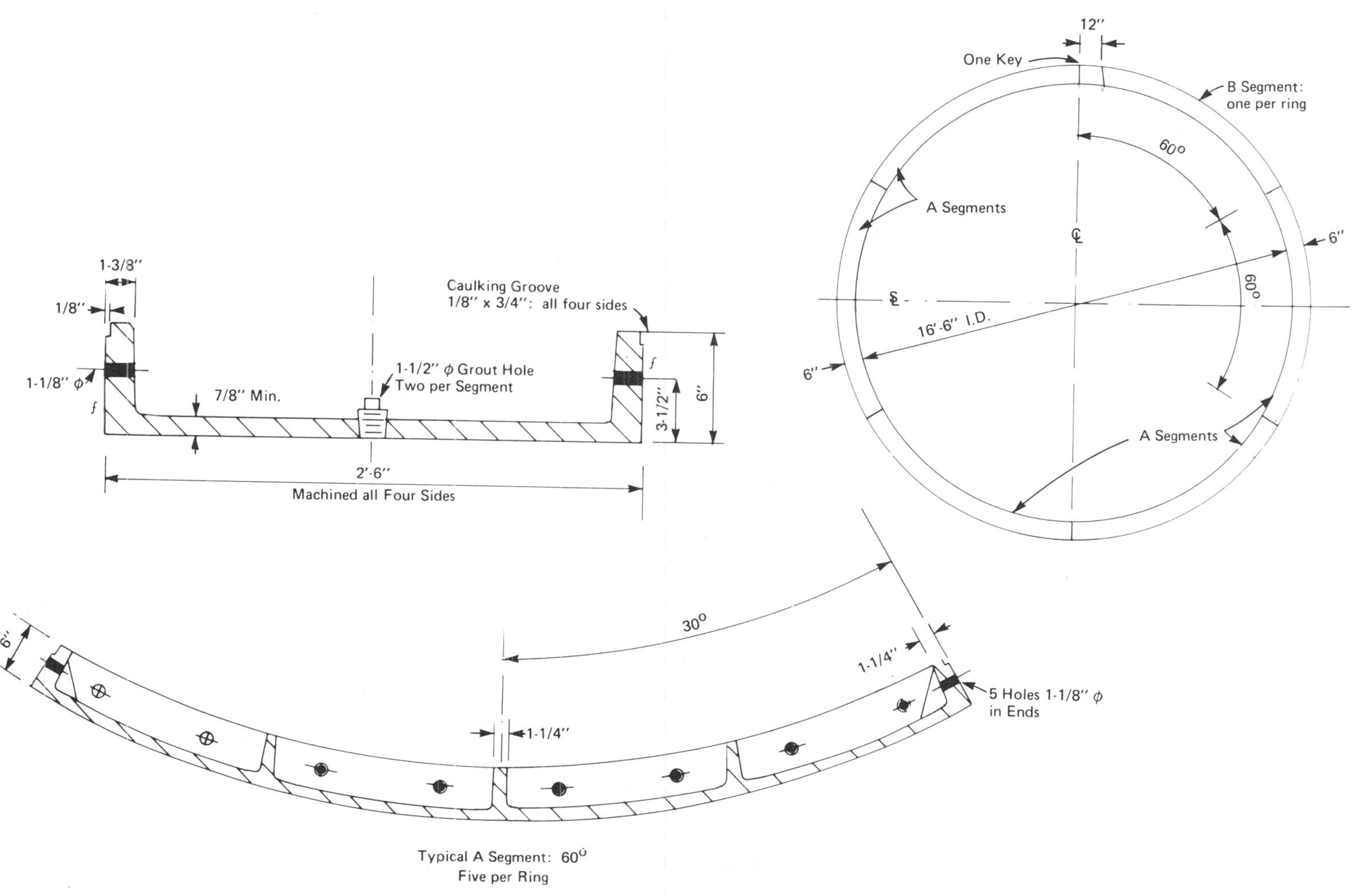

Fig. 6-16. Cast-Iron: 16′-6″ in I.D.

may initially be greater, so that the vertical diameter may be elongated temporarily until pressures are equalized.

There is no evidence, based on 100 years experience, of corrosion of cast iron liners. Unless there is an interior finish required, as in highway tunnels, cast iron primary liners do not have a secondary lining.

Fabricated Steel Liners. Recently, fabricated steel segments have been used in subway and highway tunnels. Figure 6-17 shows details of liners for a 16-foot, 6-inch rapid transit tunnel of the BART system. It has a $\frac{5}{8}$-inch web and $\frac{13}{16}$-inch flanges 6 inches wide provided with caulking grooves. By fabricating the 30-inch wide segments to extremely close tolerances in width and squareness in special jigs, a tight fit was accomplished without machining the flanges. The segments were sand blasted in the shop and sprayed with a coal tar silicate or coal tar epoxy. After completion of the tunnels, a cathodic protection system was installed to guard against corrosion of the plates by stray electric currents. There is no secondary lining.

Pressed Steel Liner Plates. A typical pressed steel liner plate is shown in Fig. 6-18. They may be obtained in $\frac{1}{4}$-inch or $\frac{3}{8}$-inch plates flanged on all four sides. They are fabricated in lengths of π feet, so that there is one plate for every foot diameter, and are either 16 inches or 24 inches wide. In order to take the jacking pressure, they are reinforced with steel angles. Liner plates without ribs are seldom used in tunnels larger than 12 feet in diameter, and in all cases require a secondary lining of concrete for strength and protection against corrosion. Plates can be obtained with neoprene gaskets glued to all four edges to improve water tightness.

Ribs and Wood Lagging. In tunnels which require a secondary lining for interior finish or smooth surface, steel ribs and wood lagging have been used in firm and relatively dry ground. Figure 6-19 shows such a lining used in some WMTA subway tunnels, consisting of 6 X 6 WF ribs with 6-inch wood lagging. The ribs are always spaced 4 feet apart and the finished lagging is inserted with a tight fit. For smaller tunnels, 4 X 4 WF ribs and 4-inch lagging are often used. After the shove, the rings are expanded by hydraulic jacks to fill the tail void, eliminating pea gravel.

Precast Concrete Lining. Precast concrete liners have not yet found great use in the U.S. except for the O'Rourke block, used in a number of tunnels in Detroit. The blocks are heavy, generally 2 feet long and 30 inches wide. They are not watertight and require a secondary lining.

Precast concrete segments were designed for a test section of the Baltimore subway tunnels, as shown in Fig. 6-20, but were not competitive with steel liners.

Similar liners have been used extensively for subway construction in Europe, Mexico and Japan. They are cast in accurate molds and made watertight with neoprene gaskets. They are erected and bolted like cast iron liners.

Precast concrete liners with articulated joints were used in the stiff London clay in the Victoria Line of the subway. The rings were expanded by hydraulic jacks to fill the tail void. Individual rings were not connected together but depended upon the thrust of the shove jacks to press them into close contact and were held in place by the stiff clay.

In subway construction, concrete segmental liners generally do not receive a secondary lining, since slight seepage, not exceeding 1 gallon/minute/1000 feet is tolerated.

Extruded Concrete Lining. For a 16.4-foot O.D. tunnel under the Riachuelo River in Buenos Aires, the author of this chapter designed a shield, air locks and steel forms for use with an extruded concrete lining. The jacking pressure was applied directly to the wet concrete, forcing it into the tail void, thus preventing settlement. This type of construction may offer a promising development for most grounds requiring a shield.

Curves. Curved tunnels require tapered liner rings. They are usually 3 inches narrower on the inside than on the outside. Sometimes every ring has to be tapered, other times only

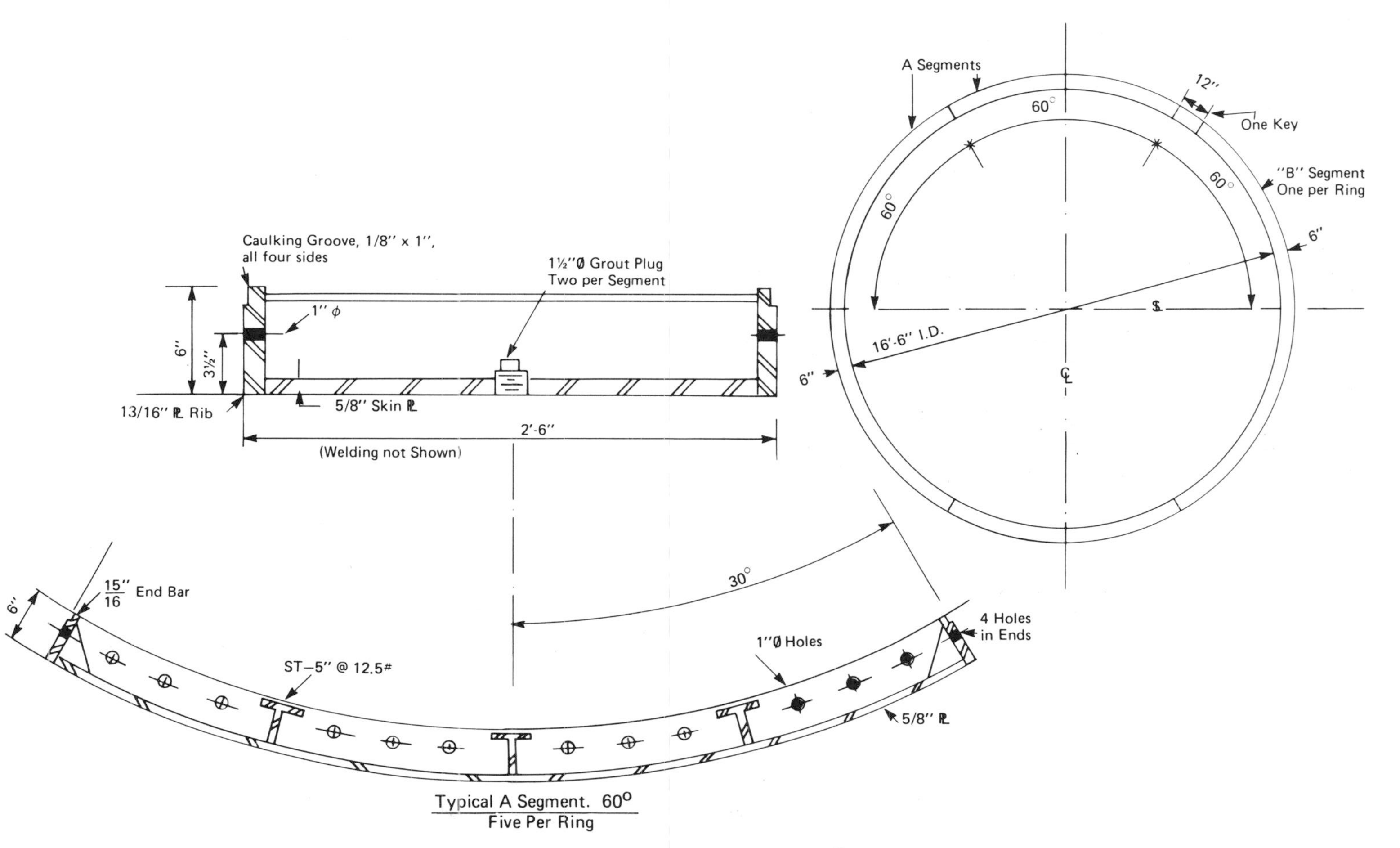

Fig. 6-17. Fabricated steel segments: 16'-6" in I.D.

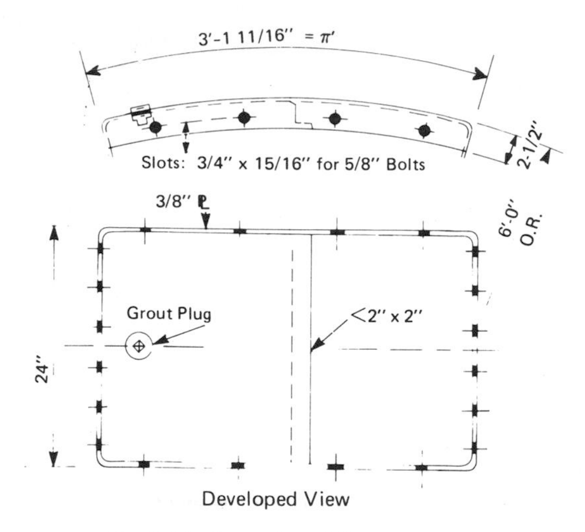

Fig. 6-18. Pressed steel-linear plates.

every other. The radius of the curve should be at least 600 feet, preferably more.

Sealing and Caulking. Sealing segmental liners subject to water pressure always presents problems. Cast iron and steel liners have caulking grooves in the flanges into which a sealing medium is inserted. In most cases, lead wire is pounded into the grooves. New developments in synthetic caulking compounds which can be applied against water seepage should be investigated and tested on account of the high labor cost of lead caulking.

Compressive gaskets of neoprene or similar material, attached to the segments outside the bolt line, have been used. They require careful handling of the segments to prevent damage prior to installation. Also, the corners may be troublesome.

Heavy precast concrete segments were used on an extension of the Paris Metro. They were

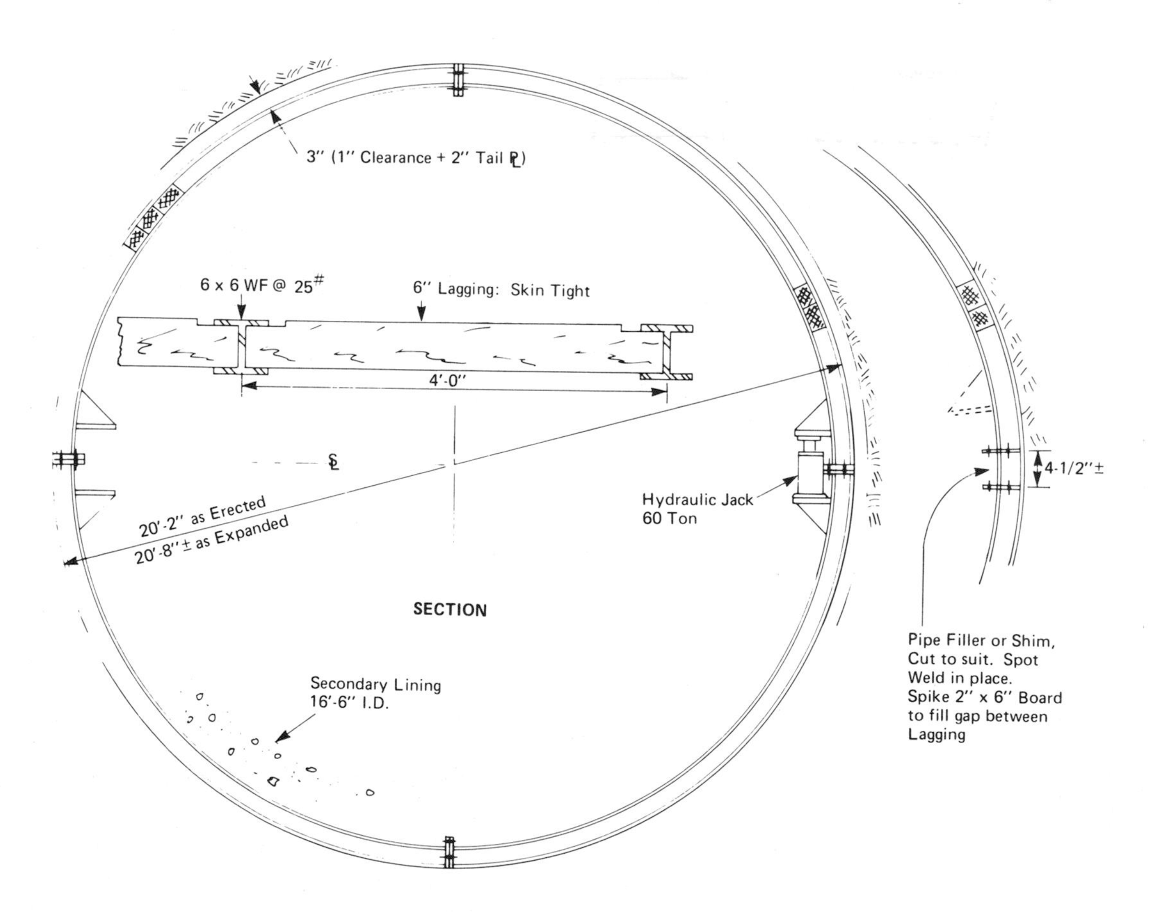

Fig. 6-19. Ribs and wood lagging using 6 × 6 W.F.

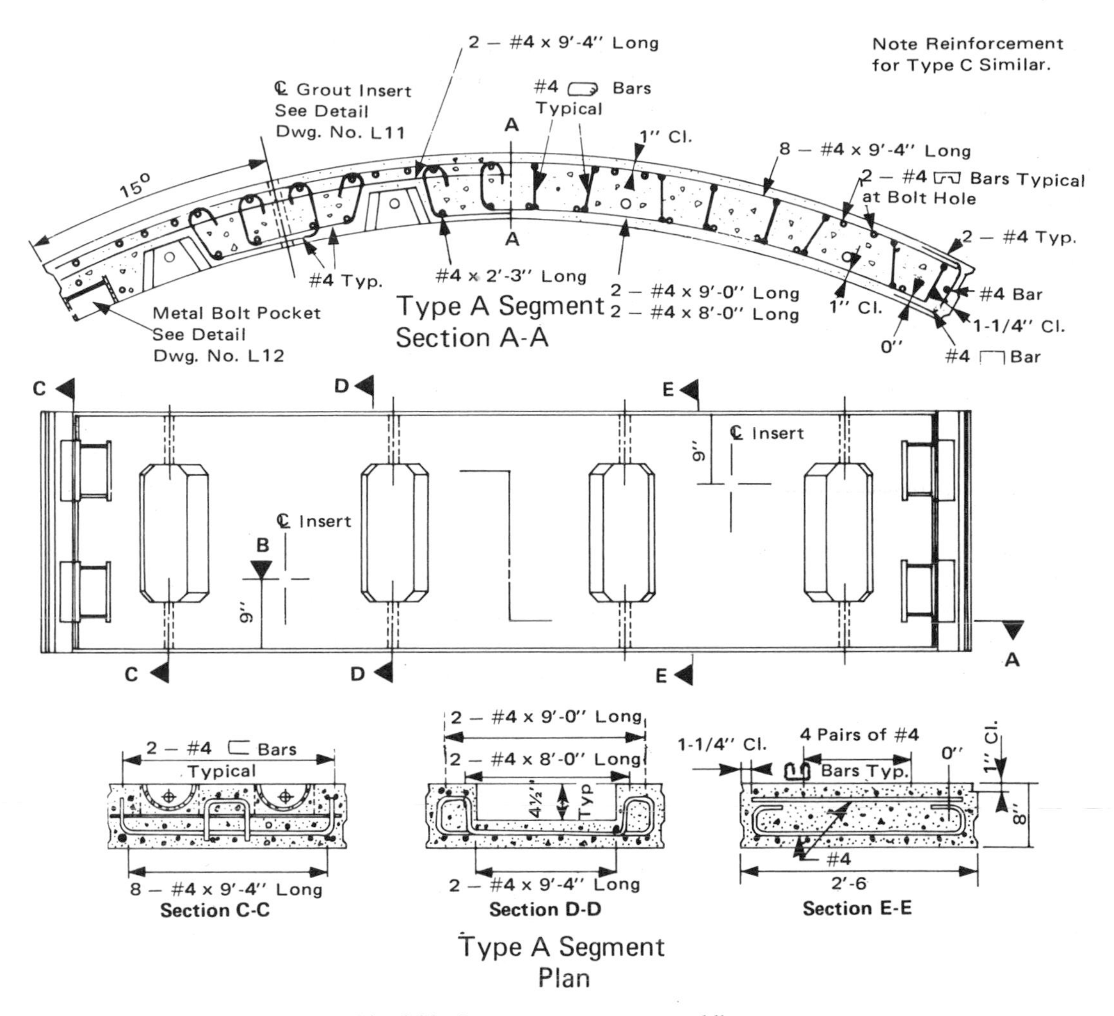

Fig. 6-20. Precast concrete segmental liner.

coated with a heavy mastice compound at the site. Although under about 40 feet of water head, the process was quite successful (but rather messy). Occasional leaks at corners were sealed by drilling holes and injecting compound.

It is prudent to provide caulking grooves regardless of what other method of sealing is used.

Transportation and Erection. Where the segments are not longer than about 6 feet, they can be carried into the heading on muck cars, two segments per car. They are lifted out of the car by a trolley hoist and laid in the pit where the erector arm can reach them. Where segments are longer, they are carried on a segment car which has a turntable to swing the segments around so they can be picked up by the erector arm. Often the car may have a depressed center to pass the stack of segments through the muck lock.

The erector arm picks up the segments and holds them in place for bolting to the previous ring, starting at the bottom. A special key segment is inserted last in the top. Longitudinal joints in adjacent rings are offset. Tightening of the bolts is done with torque wrenches after all segments of a ring are in place.

Design of Flexible Lining. All primary linings described have only limited bending resistance and therefore adapt themselves to the soil pressure by deformation until equilibrium is established, whereby they become essentially compression rings. Usually the vertical pressure tends to flatten the rings, tops and bottoms, pressing them against the sides. This deformation is generally used as a design criterion and varies according to the soil. A deformation of $1\frac{1}{4}$ inch/10 feet of diameter is admissible. If this is exceeded in very soft soil, temporary horizontal tie rods are installed and left in place until the ground has adjusted.

SECONDARY LINING

A secondary lining is required if the primary lining consists of ribs and lagging or O'Rourke blocks or if the tunnel interior needs a finish (as in highway tunnels). Rail tunnels with segmental metal or precast concrete lining usually do not receive a secondary lining.

Poured Concrete Lining. Generally used in highway tunnels, which receive a tile finish or other light-reflecting surface, poured concrete lining will be of a minimum practicable thickness for pouring and enclosure of conduits, fixtures for lighting and other recessed equipment. The concrete is poured behind steel forms, starting at the invert. The length of the pours depends on space limitations to bring in equipment, capacity of the concrete plant, and the work cycle selected by the contractor. In tunnels with non-watertight primary lining, the concrete secondary lining is placed as close to the face as possible. With segmental primary lining, the placing of the secondary lining may be deferred until the shield is well advanced or the tunnel holed through in order to set up an efficient construction procedure, or, in the case of compressed air operation, the lining can be placed in free air.

Sprayed Linings. Sprayed linings of cement mortar may be used to protect the interior of steel primary linings if necessary. Sprayed concrete lining may be applied to a sufficient thickness to provide a fairly smooth interior surface, adequate for an epoxy type of finish. Concrete is applied in successive layers, each about 3 to 4 inches thick (see Chapter 12). Acoustical lining may be sprayed directly to the primary lining or on a sprayed mortar lining for the reduction of train noise in rapid transit tunnels, particularly adjacent to the stations (see Chapter 17).

Design of Secondary Lining. Secondary linings are usually applied after primary linings have been in place for a period during which they have adjusted to the soil pressure and deformation, so that they have to absorb only long-time secondary settlements. The stresses will be primarily compressive so that a system of temperature reinforcing in the concrete lining will be adequate. In subaqueous tunnels, the thickness of the secondary lining may be governed by the requirement that its weight, together with that of the rest of the structure, be sufficient to counterbalance the buoyancy forces.

COMPRESSED AIR TUNNELING

Compressed air is a valuable, sometimes essential, adjunct to shield-driven tunnels. It is necessary in tunnels exposed to positive hydrostatic pressure under rivers or waterways, or in tunnels in porous soil below groundwater level if the latter cannot be lowered by pumping. A small amount of air pressure may also be useful in non-cohesive wet soil to dry up the face and facilitate breasting.

Air Pressure. In theory, 0.43 psi of air pressure is required to balance 1 foot of hydrostatic pressure. To account for irregular behavior of groundwater, and as a safety margin, the accepted relation is

2 feet water pressure = 1 psi air pressure.

Figure 6-21 shows the unbalanced pressures at the top and bottom of the shield relative to the nominal pressure at the center. At the top of a 20-foot diameter shield, the air pressure is 4.3 psi too high and air tries to escape to the surface. At the bottom, the pressure is 4.3 psi too low and water or the ground itself will tend to

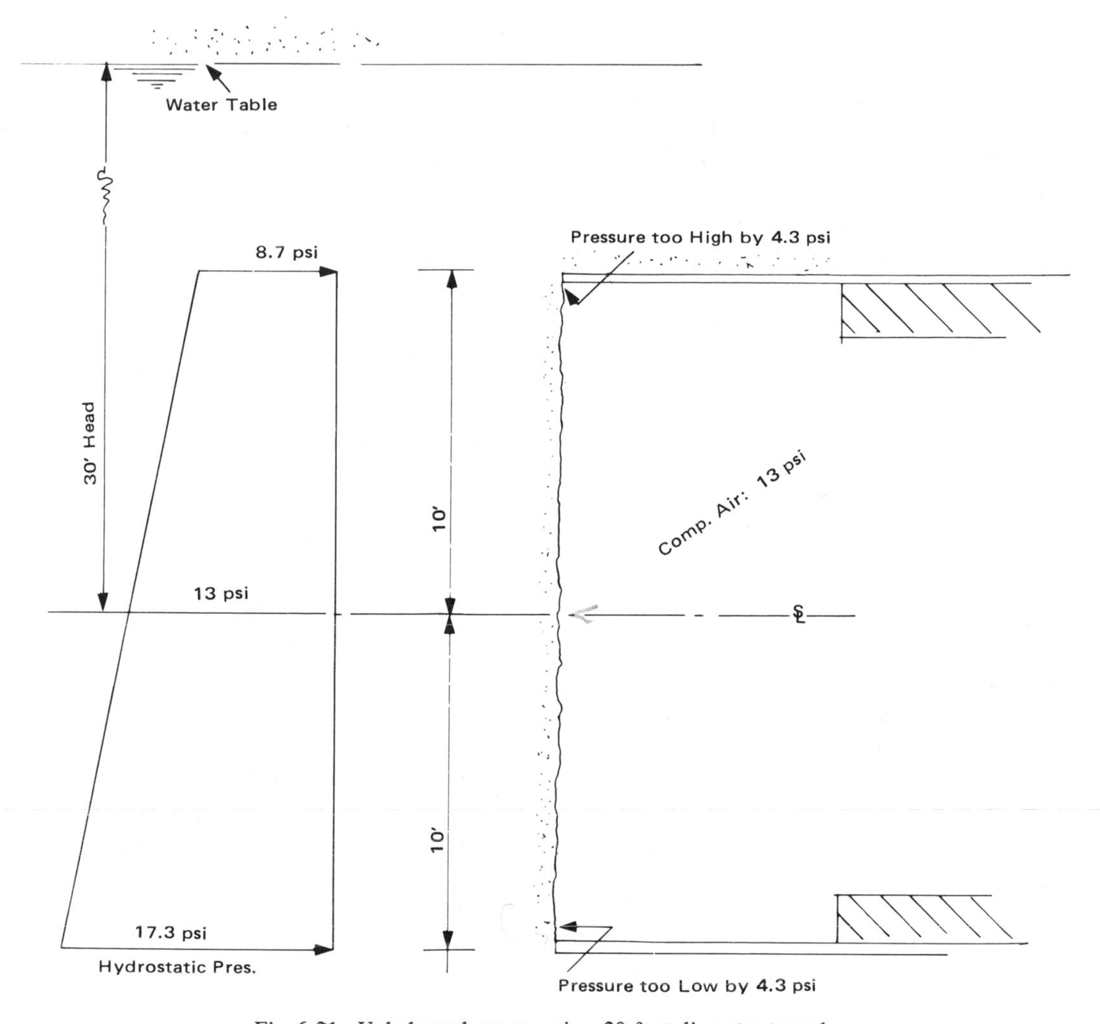

Fig. 6-21. Unbalanced pressures in a 20-foot diameter tunnel.

flow into the tunnel. This is a typical condition in all tunnels but in most cases is not serious. In a subaqueous tunnel, it is very important to check the amount of cover. As a rule of thumb, 10 feet of cover are required for 3 feet of unbalanced pressure. If the cover is insufficient, it may be necessary to place a temporary clay blanket in the river bed.

In very pervious ground, the air pressure may have to be varied several times during a shift. It is reduced when working the upper face until the breastboards have been moved ahead. The space between the boards may have to be "mudded" to preserve air. The pressure is raised while the lower breastboards are advanced.

Volume of Air. The volume of air required is generally estimated on the basis of 20 cfm/foot^2 of face. This volume must be checked to make sure that it includes 200 cfm ventilation for each man in the heading. The volume also depends on the type of ground and the tightness of the primary lining. A small tunnel through sand in Philadelphia required 32 cfm/foot^2. A 12-foot tunnel in Brooklyn needed 83 cfm/foot^2 in one heading and only 28 cfm in the other. Much of the loss may be due to leaks in the joints of the liner plate, not only through the face. If necessary, the losses may be reduced by keeping the concrete lining as close to the face as practical.

The low-pressure air is supplied by low-pressure compressors located in a fireproof building near the head of the shaft or portal. The intake of fresh air should be located high enough to prevent sucking in smoke or dust. OSHA specifications require that the compressors obtain their power from dual sources. This may mean two sets of compressors, one electrically driven, the other by diesel engines. The air should be passed through after coolers to keep the ambient temperature in the tunnel below 85°F. This will also remove oil and odors from the air (see Chapter 11).

AIR LOCKS AND SAFETY DEVICES

Figure 6-22 shows a longitudinal section through a complete compressed air heading with the shield, material handling, man-walks, safety screen and locks.

Bulkhead. The part of the tunnel under air pressure is closed off by an airtight bulkhead if tunneling has to start under compressed air from a shaft. In order to keep the total volume of compressed air as small as possible, new bulkheads are advanced from time to time. Some safety regulations limit the distance between bulkhead and face to a maximum of 1000 feet. The bulkhead may be concrete between 5 and 10 feet thick or structural steel. The former is lower in first cost but has to be chipped out for removal. Steel bulkheads have come into more frequent use recently. The bulkhead must be securely anchored in the lining and sealed airtight against it.

Pipe sleeves for the following services are required in the bulkhead. It is well to add a few spare sleeves in case additional services are needed, since their later installation is practically impossible.

Two low-pressure supply
High-pressure air for pneumatic tools
Blow line
Foul air exhaust
Pump discharge
Hyraulic fluid supply and return
Emergency fire line, for city firemen
Safety valve exhaust
Low- and high-pressure alarm
Pressure gauges
Telephone conduit
Electric light conduit—110-volt
Power conduit—220-volt or 440-volt
Conduit for recording instruments
Extra pipes, capped

The low-pressure air pipes generally are 8 or 10 inches in diameter. One can terminate just inside the bulkhead but the other should be carried to within 100 feet of the face. Both should have flap valves so that in the event of a break in the pipe no air will escape. The blow line, 4 inches in diameter, is peculiar to compressed air tunnels. Its outside end should point downwards into the sump. The heading end should have a valve and 20 feet of suction hose. When water, including small amounts of mud and sand, are blown out into the sump, it can also be opened for ventilation (or to remove smoke after blasting).

Air Locks. In order to pass men and material between free air and compressed air, locks are installed in the bulkhead, as shown in Fig. 6-23.

In very small, short tunnels the bulkhead may serve as a lock being equipped with an airtight steel door. In most tunnels, regular locking chambers are needed. They are equipped with airtight doors at each end, one of which must always be closed. The doors open towards the high-pressure side so they cannot be opened until the pressures on both sides are equalized. This is done by valves in the ends. There should be two of these at each end, one for operation from the inside, the other from the outside.

Muck locks are steel cylinders 30 to 60 feet long with doors at both ends and the necessary valves. The lock should be as large as possible because its doors will govern the size of cars that can be handled. A 6-foot diameter lock will have doors 37 inches × 4 feet; an 8-foot diameter lock door will be 48 inches × 6 feet; and a 9-foot diameter lock door will be 54 inches × 6 feet. In small tunnels, the muck lock must also be used for passing men through, for which there should be folding benches on one or both sides of the lock. This combined muck-man lock is not desirable for pressures greater than 20 psi, since there is too much time lost in decompress-

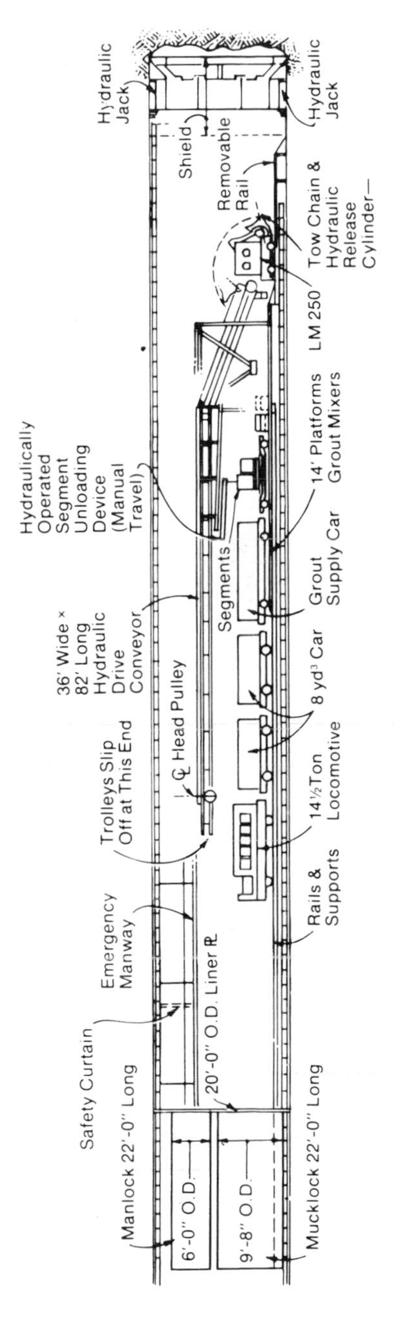

Fig. 6-22. Section through compressed air heading.

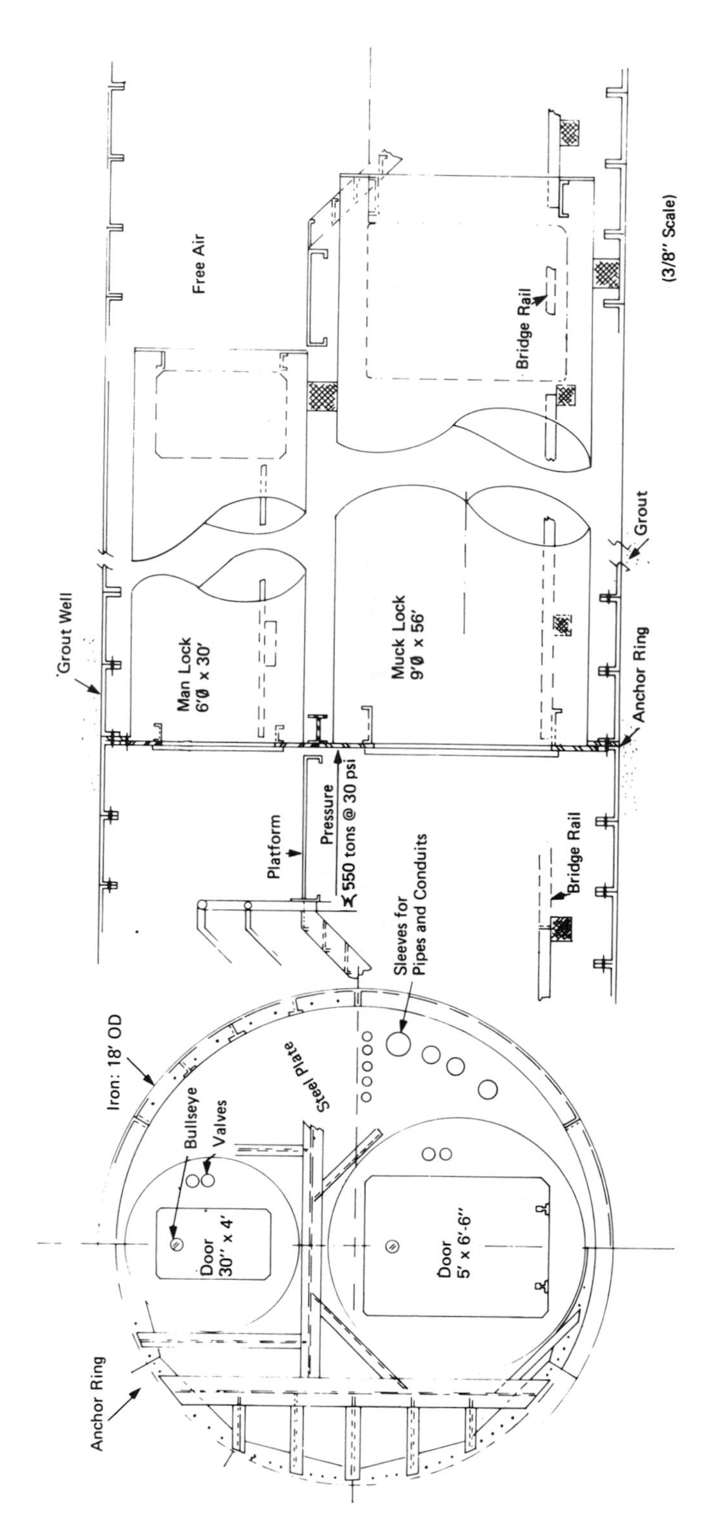

Fig. 6-23. Steel bulkhead and locks, San Francisco Subway.

ing men, tying up the lock for material and muck handling.

Man locks are usually 6 feet in dia. and 20 to 30 feet long, with benches along both sides. Electric heaters should be mounted under the seats to dispell chill and fog during long decompression. Valves on the inside and outside serve to admit and release air. When working air is pressured greater than 12 psi, a clock-operated decompression valve is required to meet the OSHA decompression cycle.

On tunnels 14 feet or larger, regulations require two locks, a man lock separate from the muck lock. For a 24-man gang, the lock should be 30 feet long. The inspectors and mechanics and any other men working in the heading must be included in the gang.

In large tunnels two man-locks may be provided to speed operations.

Emergency locks are required in subaqueous tunnels or other tunnels in danger of rapid flooding. This lock is set as high in the bulkhead as possible and is normally also 6 feet in diameter and fitted with benches. On occasion, a 5-foot diameter lock is all that can be fitted in. This lock should always stand with the inside door wide open and ready for emergencies. With permission of the saftey engineer, the survey gang may use this lock for prolonging the line.

Special decompression chamber, also called "luxury lock," is required when the decompression time exceeds 75 minutes. This is located in the tunnel so that men may walk directly from the man lock into the decompression chamber without lowering the pressure (see Fig. 6-24). This lock must be 7 feet in diameter and at least 1.33 feet long for every occupant. It must be electrically heated and lighted and equipped with toilet, urinal and drinking fountain. It may be well to install shower baths so the men can clean up and change clothes at the end of their compression time. On a large job, such a lock might well be installed even for compression times less than 75 minutes in order not to tie up the man lock for long periods.

Pressure gauges are required in all air locks. In the man locks, gauges indicate the pressure inside the lock and in the tunnel. A gauge mounted outside the lock shows the pressure inside of it to be read by the lock tender. Gauges mounted outside the muck locks show the pressure inside the lock.

Flying Gangways. In tunnels 16 feet or larger, it is well to install a flying gangway, or hanging walkway, of steel, 3 feet wide with at least 6 feet of headroom. For passing under the safety screen, the minimum headroom is 42 inches. There are railings on both sides and an occasional ladder to the invert so that men walking back in the tunnel can climb to the walkway. This keeps men out of the way of muck trains and avoids flooring over the ties. A flying gangway is essential in tunnels subject to rapid flooding, where it gives direct access to the emergency lock.

Safety Screens. In any tunnel 16 feet or larger, subject to rapid flooding, there should be a safety screen in the upper half of the tunnel, located as close to the heading as practical. In case of flooding, this will trap air above the walkway, permitting men to escape, and facilitate the task of regaining the heading. Two screens may be required which can be leapfrogged ahead in order to keep the foreward one no more than 300 feet from the shield. The screens are of steel attached to an anchor ring placed between two liner rings.

Hazards. There are several specific hazards to working under compressed air. The most common is fire. Because of the large amount of oxygen, wet wood and rubber hose will burn fiercely. No smoking is allowed in the tunnel. When it is necessary to cut or burn, extreme caution must be taken. A number of fire extinguishers should be located along the tunnel, and the amount of flammable material taken into the tunnel should be strictly controlled.

The second major hazard is "blows." With the unbalanced air pressure at the crown, air may escape through an opening in the river bottom above. Unless this is quickly checked, it may open a large hole, allowing pressure in the tunnel to fall and water to rush in. The only way to fight a blow is from within the tunnel. Once the

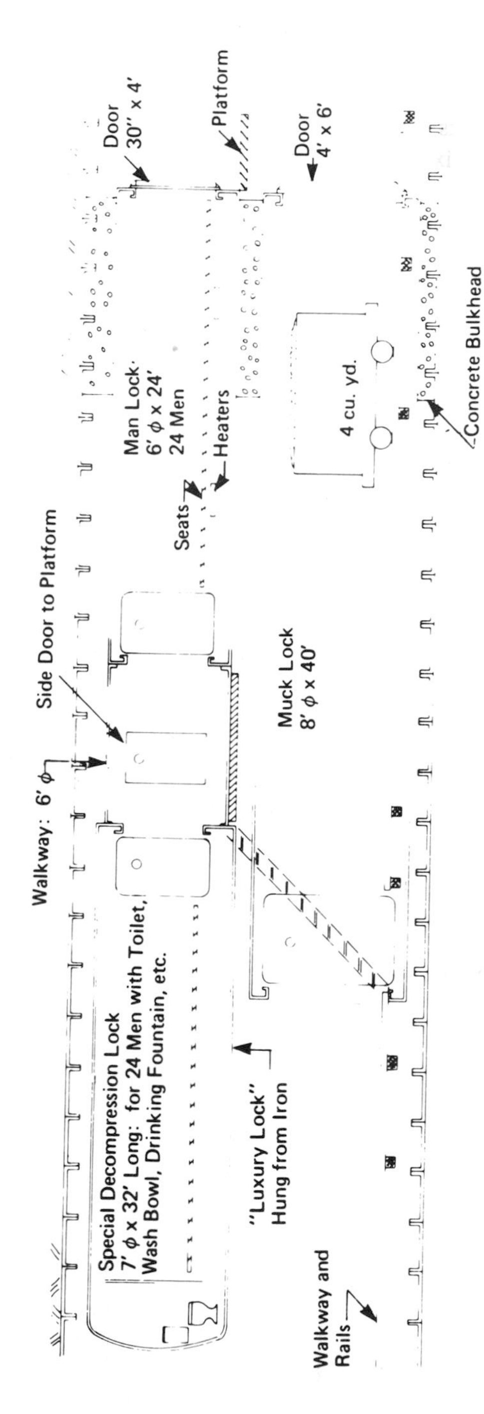

Fig. 6-24. Special decompression lock when decompression time exceeds 75 minutes.

hissing sound of escaping air is heard, the foreman must put all his efforts into checking this incipient blow. Hay, lumber, tools or anything loose should be thrown into the opening to stop the flow. Timbered breasting should be liberally plugged with soft clay to seal the joints between the boards. Fine leaks in certain soils can be checked by a 1:1 mixture of cement and dry sawdust.

WORK AND SAFETY RULES IN COMPRESSED AIR

Traditionally, work in compressed air was governed by the New York State Code, which called for split shifts, with the sandhogs eating lunch in free air. In 1963, the State of Washington formulated new codes for hours of work and decompression times, which eliminated the split shift. This code was adopted by the state of California and used in the San Francisco BART system. The Occupational Safety and Health Administration of the U.S. Department of Labor (OSHA) has prepared new regulations on hours of labor and decompression times. They are much stricter and should be studied carefully before preparing an estimate on compressed air work. Local union contracts should also be checked, because they may be more restrictive than the government regulations.

Table 6-1. Occupational Safety and Health Administration's safety and health regulations for construction, June 24, 1974.

PRESSURE (PSIG)	HOURS OF LABOR	DECOMPOSITION TIME (MINUTES)
0–12	8	3
16	6	33
20	4	7
20	4	43
	3	15
24	4	92*
	3	52
28	3	98*
	2	41
32	2	85*
	$1\frac{1}{2}$	43

*A special decompression chamber, 7 feet in diameter, must be installed when decompression time exceeds 75 minutes. Only one work period allowed per 24 hours.

Hours of Work and Decompression Times. Table 6-1 shows selected hours of work and decompression times under OSHA.

The decompression rate is not a straight line but requires an automatic decompression valve to reduce the pressure in accordance with the OSHA regulations.

Medical Locks. Recompression is the only cure for Caisson disease, known as "the bends." The affected sandhog is placed into the medical lock and the pressure is quickly raised slightly higher than that in which he was working. He is then decompressed at twice the normal time. In the past, a medical lock was only required when the pressure exceeded 15 psi. Present OSHA rules call for a medical lock on every job where the pressure is above normal.

The medical lock has a 6-foot, 6-inch diameter and is about 20 feet long, designed for 75 psi pressure. OSHA regulations specify that it must be equipped with an automatic decompression valve, oxygen breathing apparatus, a sprinkler system, benches, lights and heaters. There must be a trained lock attendant who shall remain on duty five hours after the last man has left the tunnel. A qualified physician must be on call at all times that compressed air work is under way.

"The bends" manifests itself, from the author's experience at 23 psi, in about two hours as rheumatic pains in the elbow and knee, which disappear in an hour or so in a light case. More serious is the "blind staggers," when nitrogen bubbles form in the spine, which causes excruciating pain and may result in permanent injury or even death. This may be taken for drunkeness and every sandhog should wear, at all times, night and day, a badge stating: "This is a compressed air worker. If found unconscious, DO NOT TAKE TO HOSPITAL. Rush immediately to MEDICAL LOCK located at . . ."

Medical Checks. All personnel working under compressed air must undergo a thorough medical check before being employed for this duty. Periodic checks should be made thereafter as may be required by the safety rules or as determined by the medical staff.

CONSTRUCTION PLANT

The construction plant, including the compressor building, the change house with sanitary facilities, maintenance and repair shops, offices and emergency facilities, are located at the head of the shaft or near the portal. For details, see Chapter 11.

7
Rock Tunnels

LYMAN D. WILBUR

Vice President, Morrison-Kundsen Company, Inc. (*Retired*)
Chairman of the Board and President, International Engineering Company, Inc. (*Retired*)
Consulting Engineer
Member National Academy of Engineering
Honarary Member American Society of Civil Engineers

GOVERNING ROCK CHARACTERISTICS

Definition and Classification

From a geological standpoint, any one of the materials composing the earth's outer solid shell is rock. However, in tunneling parlance, rock has been considered to be any of the materials of the earth's outer crust that can be excavated only by drilling and blasting, as contrasted to softer materials that can be excavated readily by spade or pick and shovel. More recently, this definition would have to include hard materials that can be bored without blasting by large boring machines (that really are, in effect, large drills).

Rocks may be classified into three main categories; igneous, sedimentary and metamorphic. Igneous rocks that have emerged from the interior through fissures or vents, such as basalt, are classified as extrusive, whereas molten rocks that may have pushed up without reaching the surface, such as granites, are classified as intrusive. Sedimentary rocks consist principally of recemented or consolidated fragments of previously existing rocks, such as conglomerates, sandstones and shales, or the cemented particles of the hard parts of marine organisms such as clams and corals (limestones and dolomites). Metamorphic rocks, such as slätes, schists and gneiss, are the result of recrystallization that took place under high temperatures and pressures.

Fractures and Faults

Fracture of rocks is due to volume contraction while cooling or to movement of the earth's crust (among other causes). Sedimentary rocks, such as sandstones, although originally laid down in level or nearly level layers, can be found tilted at every angle from the horizontal as a result of folding from up-thrusts of the earth's crust. A crack or fracture where no visible displacement has occurred is known as a joint. There is no resistance to separation at surfaces of joints even though they may be closed and almost invisible. The joints near the surface are likely to be more open that those at greater depth. Joints divide the rock into blocks and provide access for air, water and dissolved chemicals to infiltrate the rock. Freezing, thawing and chemical action cause further changes in the rock that we call weathering. Joints in igneous rocks may be closely spaced a few inches apart, such as in columnar basalt, or widely spaced (many feet) as in some granites. In many rocks, the joints are so irregular that the blocks between them are tightly interlocked and cannot move relative to one another without fracturing the rock. Joints in sedimentary rocks usually contain three sets of joints, one parallel to the bedding planes and the others intersecting at right angles. In limestones and sandstones, these joints may be several feet apart, whereas, in shales, the jointing may be less than an inch apart. Joints in metamorphic rocks are usually normal to the direction of cleavage. The location and arrangement of joints bears an important relationship to overbreak, required tunnel support and water problems.

Fractures of great length with relative displacement of adjoining rock masses are known as faults. These are caused by major movements of the earth's crust. A fault may extend over a distance of several hundred miles. The slippage along the fault may not be confined to one plane or surface but along several adjacent surfaces over a considerable width. This results in a zone with the rock badly fractured and with some of the rock being ground into powder known as gouge. Great pressures often exist along the fault. As a result of rubbing under pressure, the rock faces may be smoothed and polished into a condition known as slickensides. Fault zones can be the most difficult ground through which to tunnel, particularly if water is encountered. A normal fault resulting in elongation of the region is shown in Fig. 7-1.

The vertical displacement or "throw" may vary from an infinitesimal amount to many feet. The horizontal displacement at right angles to the fault is known as the heave. Reverse faults resulting in shortening are less common. Sometimes horizontal forces develop in the rock crust much greater than forces due to the weight of the rock. These can cause an upthrust that results in folds and overthrust faults with horizontal displacement of many miles.[1]

Dip and Strike

The angle at which various strata intersect a tunnel has an effect on the difficulties of excavating, the amount of overbreak and the amount of support required. The angles are known as *dip* and *strike*. *Dip* is the angle of inclination of the plane of stratification with the horizontal plane. *Strike* is the direction of the line of intersection of the plane of stratification with a horizontal plane with reference to the true north-south line.

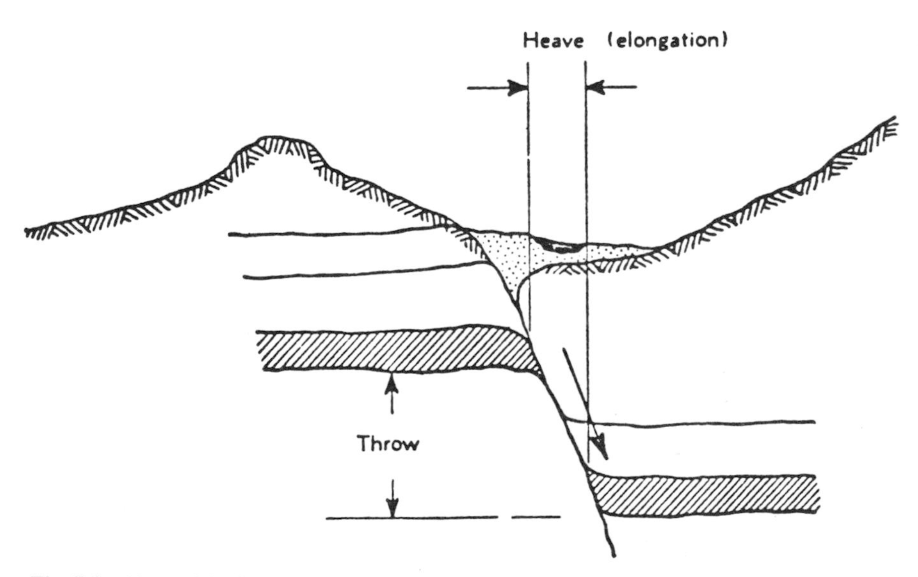

Fig. 7-1. Normal fault. (Commercial Shearing, Inc., Rock Tunneling with Steel Supports).

Water

Water is found in almost all rocks in the upper limits of the earth's crust where tunneling takes place. Joints in the rock, even if tight, permit water to penetrate by capillary attraction. The larger the openings of the joints, the more water that can enter a cavity (such as a tunnel). Since joints in rocks are usually more open the closer they are to the surface, less water is likely to be found in deeper tunnels. Below 100 or 150 feet, joints tend to be tight and produce little water. This is not always true, however. The San Jacinto tunnel in California at depths of over 800 feet produced flows of 40,000 gpm, 15,800 gpm near a single heading and pressures up to 600 psi.[2] In limestone formations, water channels are sometimes formed by the solution of the limestone, creating large caverns such as the Mammoth Cave of Kentucky. Large streams can flow in such caverns. Sometimes caverns or faults in rock formations are filled with sand saturated with water. These cause some of the most difficult tunneling conditions. If an otherwise hard rock tunnel intersects such a sand formation, it may stall progress for many months. Many feet of the tunnel previously driven may be filled with sand, as in the Litani tunnel in Lebanon.[3]

When a tunnel is excavated, the friction in the joints previously noted helps to hold the individual blocks in place. When the joints are dry, the rock may stand, whereas if they later become wet during a rainy season, the roof and sides of the tunnel might fail. Except in well keyed formations, water, regardless of quantity, causes extra problems. In certain areas, the water may be hot, such as in the Tecolote tunnel in California[4] and the Graton tunnel in Peru.[5]

Chemical Defects

Chemical defects are those defects due to removal or alteration of rocks due to chemical reaction with water. Even the densest rocks have porosity of 0.5 to 2.0%. Water reaches the innermost particles of rock through the joints and pores, decomposing such rocks as feldspar and dissolving such constituents as calcium carbonates in limestones. This weathering occurs more rapidly in warm, moist climates than in cool, dry climates. In the southeastern U.S., weathering extends to 100 feet in regions of moderate rainfall. In the semi-arid regions of the southwestern U.S., the depth of weathering is small. Chemical rock weathering may be absent in areas that have been covered with ice where glaciers have scoured the rock formation.

Where there is no surface runoff in a limestone region, it is likely that there will be solution channels underground.

Some metamorphic rocks can be altered by weathering into clay-like material that squeezes or swells into tunnel openings. The squeeze may be produced by viscous flow of the rock. If the squeeze is due to an increase of water content, it is referred to as swelling rock.

Some of the dark basic igneous rocks can be altered into a greenish material called serpentine. Such rock can produce heavy pressures on tunnel support.

Drillability

Drillability of rock is affected by many of its properties: mineral composition and hardness, texture and friability, density and the general structure of the formation. Abrasive wear on tungsten carbide bits is directly related to the hardness of the minerals in the rock. The penetration rate of a drill is affected by all of the above noted properties. Hardness of minerals is measured on the Mohs scale shown in Table 7-1. This scale is limited in its application to rocks, as they are usually composed of several minerals.

Field Inspection. A rule of thumb method for determining drillability of rock has been developed,[6] as described in the following paragraphs.

The Mohs scale of hardness is not a mathematical relationship, the higher number meaning only that it is harder than the next lower number. Minerals with a higher number can scratch any one with the same or lower number. If a

Table 7-1. Mohs hardness scale.[6]

1. Talc	6. Feldspar
2. Gypsum	7. Quartz
3. Calcite	8. Topaz
4. Fluorite	9. Corundum
5. Apatite	10. Diamond

Table 7-2. Method of determining Mohs number.

MATERIAL	WILL SCRATCH	
	UP TO	NOT OVER
Fingernail	$2\frac{1}{4}$	
Copper penny	3	
Knife blade	5.5	
Glass		6
Tungsten carbide bits	8.5	

NOTE: Quartz can be found almost anywhere and can be used to check the harder rocks.

material can be scratched with quartz but not with feldspar and it will scratch feldspar but not quartz, it has a hardness of $6\frac{1}{2}$ Mohs. When testing one rock with another to determine its hardness, care must be taken to determine whether or not it has a groove. The powder from the softer rock left on a harder rock can be misleading if not rubbed off. Rocks may contain more than one mineral, so tests should be made at several places on a piece of rock in order to determine the average hardness.

Each of the four properties affecting the drillability of rock, hardness, texture, fracture and formation are considered separately. Experienced drillers can tell how a rock will drill. The important thing to know is how fast it will drill. Rock drillability may be classed in five conditions: fast, fast average, average, slow average and slow. Kits for testing of minerals are on the market. If one is not available, a simple test with objects at hand, although not as accurate, can be used. Table 7-2 gives approximate Mohs numbers found by this means.

Table 7-3. Texture (grain structure of rock).

DRILLING CONDITION	TYPE OF ROCK
Fast	Porous (cellular or filled with cavities)
Fast average	Fragmental (fragments, loose or semi-consolidated)
Average	Granitoid (grains large enough to be readily recognized–average grained granite)
Slow average	Porphyritic (large crystals in fine-grained mass)
Slow	Dense (grain structure too small to identify with naked eye)

Table 7-4. Fracture.

DRILLING CONDITION	TYPE OF ROCK
Fast	Crumbles into small pieces when struck lightly
Fast average	Brittle (rock breaks with ease when struck lightly)
Average	Sectile (when slices can be shaved or split off and crumble when hammered)
Slow average	Tough (rock resists breaking when struck with heavy blow)
Slow	Malleable (rock that tends to flatten under blow of hammer)

Texture can be determined by visual inspection of the grain structure of the rock and then classified for the drilling condition in accordance with Table 7-3.

Fracture as used herein in connection with drillability refers to how a rock breaks apart

Table 7-5. Formation.

DRILLING CONDITION	TYPE OF ROCK
Fast	Massive (solid or dense, practically no seams)
Fast average	Sheets (layers or beds 4 to 8 feet thick with thin horizontal seams)
Average	Laminated (thin layers 1 to 3 feet thick with horizontal seams with little or no earth)
Slow average	Seamy (many open seams in horizontal and vertical positions)
Slow	Blocky (wide open seams in all directions and filled with earth, or shattered or fissured)

ROCK CHART

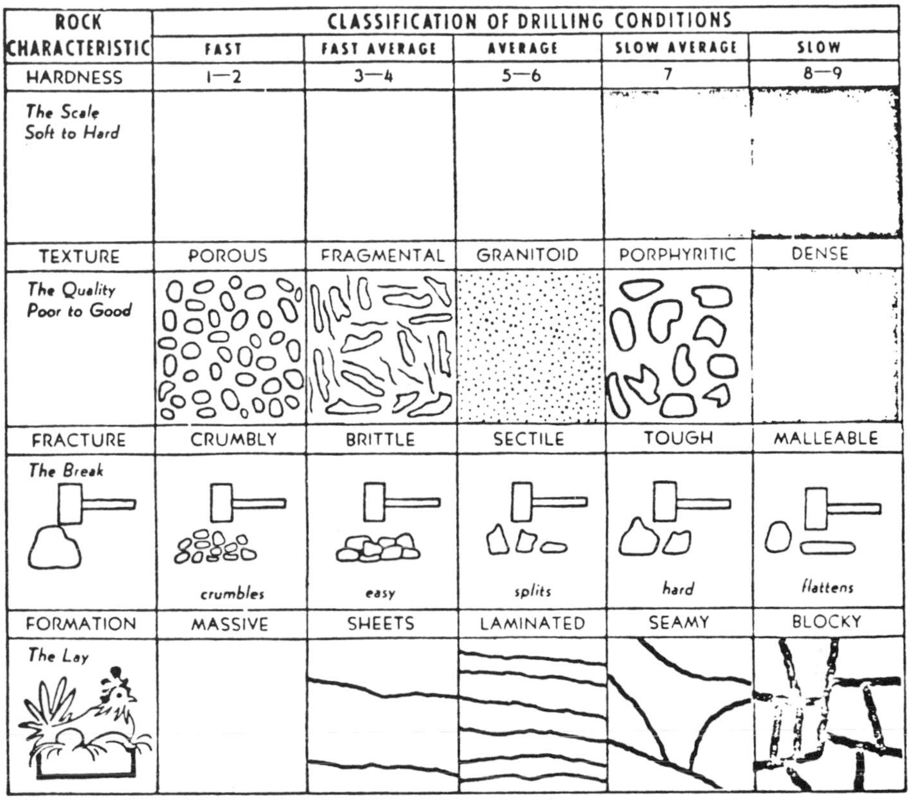

ROCK CHARACTERISTIC	CLASSIFICATION OF DRILLING CONDITIONS				
	FAST	FAST AVERAGE	AVERAGE	SLOW AVERAGE	SLOW
HARDNESS	1—2	3—4	5—6	7	8—9
The Scale Soft to Hard					
TEXTURE	POROUS	FRAGMENTAL	GRANITOID	PORPHYRITIC	DENSE
The Quality Poor to Good					
FRACTURE	CRUMBLY	BRITTLE	SECTILE	TOUGH	MALLEABLE
The Break	*crumbles*	*easy*	*splits*	*hard*	*flattens*
FORMATION	MASSIVE	SHEETS	LAMINATED	SEAMY	BLOCKY
The Lay					

Fig. 7-2. Rock Drilling Characteristics. (Davey Compressure Company, Drillers Handbook on Rock).

when struck by a blow with a hammer. The five drilling conditions are shown in Table 7-4.

Formation describes the condition or structure of the rock mass. The five drilling conditions are shown in Table 7-5.

The rock chart in Fig. 7-2 shows the drilling characteristics for the five drilling conditions.

When the characteristics of a rock fall into different drilling conditions, which is usually the case, it is necessary to compute the final drilling condition. This may be done by the point system shown in Table 7-6.

To determine the final drilling condition, the points for each characteristic must be added together to get the total points. Few rocks would have either extreme of 32 points (fast) or 4 points (slow), most rocks falling within the three inner conditions. A porous, brittle, laminated rock with hardness of Mohs 8 to 9 would have a "fast average" drilling condition. This rock has fast, fast average, average and slow drilling conditions to combine into a total of 16 points. When both extremes of drilling conditions are encountered, judgment must be

Table 7-6. Drilling condition point system chart.[6]

NATURE OF ROCK	FAST	FAST AVERAGE	AVERAGE	SLOW AVERAGE	SLOW
Hardness	8	4	3	2	1
Texture	8	4	3	2	1
Fracture	8	4	3	2	1
Formation	8	4	3	2	1
Total	32	16	12	8	4

Table 7-7. Drillability versus drilling speed.[6]

DRILLING CONDITION	FAST	FAST AVERAGE	AVERAGE	SLOW AVERAGE	SLOW
Drillability factor	2.67	1.33	1.0	0.67	0.33
Drilling speed (inches/minute)	20	10	known 7.5	5	2.5

used. If three characteristics are fast and one (formation) slow, the three fast ones would be revised to average, or to a total of 10 points, correcting a fast condition to average. If three characteristics are slow and one (formation) is fast, the fast one would be revised to average, or, to a total of 6 points, correcting average to slow.

Drillability may be measured by the drilling speed (inches per minute) at which a drill may penetrate different rocks. A drillability factor has been determined from drill performance on rock drilling jobs, on field tests and in the laboratory. This factor for the various drilling conditions, together with its application to drilling speeds at all drilling conditions for a given drill and bit when the drilling speed for one condition is known, is shown in Table 7-7.[6]

In Table 7-7, the first line shows the relative speed that a given drill will penetrate rocks having the five different drilling conditions. The second line gives an example in which a drill that we shall call Model X has previously been used on a rock classified as having average drilling conditions, determined from Table 7-6, and found to drill at the rate of 7.5 inches per minute. For another rock, with a fast drilling condition, this drill should penetrate at 20 inches per minute. In yet another rock, with a slow drilling condition, this drill should penetrate at the rate of 2.5 inches per minute. Similarly, if a particular drill had been used on a slow rock and found to penetrate at a certain rate, is should penetrate a rock with average drilling conditions three times as fast.

The drilling speed of all conditions can be determined by multiplying the known average drilling speed by the drillability factor.

Laboratory Tests for Drillability. The best way to determine drillability, other than by actually drilling in place, is to run laboratory tests on rock samples. Tests for hardness, density, abrasiveness and crushing strength, together with microscopic analysis, miniature drilling, crushing the rock, core drilling and grinding, may be performed in the laboratory.

Hardness Tests. Scleroscope hardness readings,[7] as used by Joy Manufacturing Company in its laboratory, give more definitive results in determining drillability of rocks. A small diamond-pointed hammer is permitted to fall free from a height of 10 inches through a thin glass tube and strike the rock sample. The height of rebound shows against a scale, and the harder the sample, the higher the rebound. Typical readings are as shown in Table 7-8.[7] The soft rocks are crushed to powder by the hammer, while the hard rocks are partly shattered, with most of the energy being returned in the rebound. This action is analogous to the percussion drill.

Petrographic Analysis. Super-thin segments of rock cemented to microscopic slides, when viewed through a petrographic microscope under polarized light, aids in identifying the minerals contained in the sample, permitting

Table 7-8. Rock hardness.[7]

MINERALS		IGNEOUS ROCKS	
Gypsum	12	Basalt	90
Calcite	45	Diorite	90
Feldspar	90	Rhyolite	100
Quartz	115	Granite	100–110
SEDIMENTARY ROCKS		**METAMORPHIC ROCKS**	
Shale	30–50	Marble	40–50
Limestone	40–60	Slate	50–60
Sandstone	50–60	Schist	60–65
Taconite	90–115	Quartzite	100–115

the laboratory technician to determine the texture and approximate percentage of the various minerals therein.

Laboratory Drilling. Mini-bits $\frac{3}{8}$ inch wide, ground to extreme accuracy, are used to drill small samples in the laboratory under rigidly controlled speed and pressure. The wear on the bits is measured under magnification and the depth drilled is carefully measured to give an indication of expected prototype bit wear and a penetration factor.

Deleterious Materials

A deleterious material may be described as any material that will slow progress of excavation below the average that might be otherwise attained. Water in any appreciable quantity in a tunnel being driven downgrade can cause trouble in drilling the bottom holes. Large quantities of water in any tunnel can be troublesome, particularly if it must be pumped. In chemically altered formations, the rock may be disintegrated to the point where earth tunneling methods must be used. Sand, when found in cavities, is particularly troublesome when wet. Gouge found in fault zones, slickensided material and bentonite seams may cause great delays. Soft ground in a tunnel set up for hard rock driving methods will greatly slow progress and perhaps bring it to a halt until revised procedures are developed and additional equipment is provided.

ROCK LOADS

When a tunnel or cavern is excavated in intact rock, an adjustment takes place in the stresses that previously existed. Prior to excavation, the vertical load is equal to the weight of the rock mass above (see Table 7-9). Horizontal unit pressures may vary between wide limits. Circumferential stress at the wall of a tunnel in intact rock is approximately twice that existing prior to excavating, whereas radial stress equals zero. The effect of the excavation rapidly decreases with increasing distance from the tunnel walls. At a distance from the wall in intact rock, equal to about twice the diameter of the tunnel, the stress approaches the condition prior to excavating. Because of the high crushing strength of rocks, the walls will not fail in compression except at great depths, over 2000 feet for softer sandstones and more than 19000 feet for the strongest rocks. However, slabbing of the wall may occur at lesser depths, particularly if high horizontal in-situ stresses occur. Because of arch action, a tunnel or cavern excavation may require no support or very heavy support, depending on the strength of the rock and its condition. Rocks with few fractures require little or no support, while badly fractured rocks, regardless of how hard they are, require support. Drilling and blasting tends to fracture the rock and open existing joints for several feet from the tunnel periphery. For this reason, tunnels excavated with the use of explosives, except in the most intact rocks, will develop heavier loads on supports than those excavated without blasting.

Table 7-9. Weights of various rocks (lb/foot^3).

	RANGE	AVERAGE
Sandstone	132–170	154
Limestone	145–172	162.5
Granite and gneiss	163–171	167.5
Slate	174–175	175
Trap rock	178–187	185

Deere *et al.*[8] refers to many theoretical analyses of rock loads but concludes that none of them can be applied to the usual hard rock situation without also considering the effect of rock fractures and faults that are usually the governing factors. Because of the complex nature of fractures, joints and faulting, it has been necessary for tunnel support designers to rely on empirical formulas.

The loads that must be supported in hard rock tunnels are due to the tendency of rock to drop out of the roof, largely due to fractures. In some hard and brittle rocks, thin slabs of rock are suddenly detached from the walls after the rock has been exposed. This is known as *popping*. In some regions popping appears to be due to horizontal pressures leading to the formation of geologically young mountain chains and, in other cases, due to unknown causes. Pressures required to prevent popping are small. Tunnel lining in areas of popping rock should be able to withstand 400 lb/sq. ft. and footings of ribs should be able to resist horizontal displacement. Support should be wedged tightly against the tunnel wall.

Terzaghi[9] was the first to develop a rational approach to loading for design for tunnel support. This method was developed for use in tunnels excavated by conventional drilling and blasting techniques. His approach is still used by many designers and the loads he developed are given in the following paragraphs.

Tunnels in Unweathered Stratified Rocks and Schists

The load on supports depends on the dip and strike of the strata, the spacing between the joints, the shattering effect of blasting, the distance between the working face and the roof support and the length of time between removal of natural support and installation of artificial support. The vertical load should not exceed the weight of rock in an area with a height of 0.5**B** in horizontally stratified rocks and 0.25**B** in vertically stratified rocks, where **B** represents the width of the tunnel. In tunnels in steeply inclined stratified rocks, the lateral force on tunnel walls can be estimated as shown in Fig. 7-3.

If the bedding or cleavage planes rise at a steep angle to the horizontal, a wedge-shaped body of rock, **aed** in Fig. 7-3, tends to slide into the tunnel and subjects the post at **ac** to bending. The lateral force, **P**, per unit of length of the tunnel, which acts on the post can be estimated as indicated in Fig. 7-3. The estimate is based on the assumption that the rock indicated by the shaded area to the right of **ce** in Fig. 7-3(a) has dropped out of the roof and that there is no adhesion between rock and rock along **de**. On these assumptions, the wedge-shaped body of rock **ade** is acted upon by its weight, **W**, and the reaction, **Q**, on the surface of sliding **ad**. In order to prevent a downward movement of the wedge, the vertical post **ac** must be able to resist a horizontal force, **P**. The reaction, **Q**, acts at an angle ϕ to the normal on the surface of sliding **ad**. The angle ϕ is the angle of friction between the wedge and its base. The weight, **W**, is known. The intensity of the forces **Q** and **P** can be determined by means of the polygon of forces shown in Fig. 7-3(b).

The angle of friction, ϕ, depends not only on the nature of the surfaces of contact at **ad** but

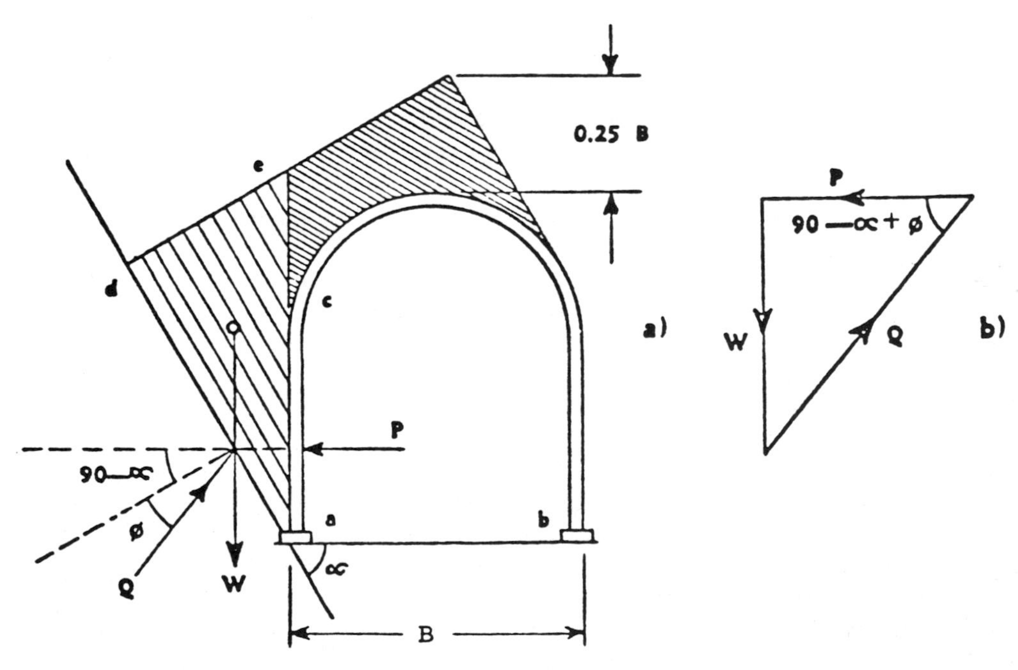

Fig. 7-3. Forces acting on tunnel. Support in inclined strata. (Terzaghi, in Commercial Shearing, Inc.).

also on the hydrostatic pressure in the water which percolates into the space between the two surfaces. Experience with slides in open cuts in stratified rocks indicates that the value of ϕ for stratified rocks with clay or shale partings may be as low as 15°. If no such partings are present, 25° seems to be a safe value.

The highest value for the unit pressure on the roof depends on the slope of the strata. For steep strata, it will hardly exceed 0.25**B**, whereas for gently inclined strata it may approach the value of 0.5**B**.

Tunnels Through Moderately Jointed Massive Rocks

The ultimate load on the roof should not exceed 0.25**B**. Horizontal loads are negligible except where the rock is in a state of elastic deformation, due to tectonic stresses or other causes, in which case support may be required as stated for "popping" rock.

The hard rock tunneler must be prepared to handle crushed zones near faults that behave similarly to tunnels in cohesionless sand. Above the water table, the loads, as determined by model tests with dry sand, are affected by arch action. After the tunnel support is in place and back-packed, the load will increase about 15%. Referring to Fig. 7-4, they are as shown in Table 7-10.

The load $\mathbf{H}_{p\ min}$ is produced by a very small downward movement of the arch which satisfies the deformation condition for arching. If the arch is allowed to settle more, the load increases to $\mathbf{H}_{p\ max}$. It is important, therefore, to install and back-pack the support as quickly as possible after a section of the tunnel is excavated in order to keep the loads as low as possible. Experience shows that the loadings on most tunnel supports are nearer the minimum than the maximum.

The average unit pressure in crushed rock zones on the sides $\mathbf{P}_h$ may be determined by the equation

$$\mathbf{P}_h = 0.30\mathbf{W}\,(0.5\ \mathbf{H}_t + \mathbf{H}_p)$$

in which **W** is the weight per cubic foot of the rock. In crushed rock below the water table, tests on sand with percolating water indicate that $\mathbf{H}_p$ is about double that in the dry.

Blocky and Seamy Rock

When a new round opens up new ground, the rock temporarily supports itself as a half-dome. Immediately after blasting a round, some rock will fall from the roof between the face and the last rib of tunnel support, leaving a gap in the half-dome. If the newly opened ground is left unsupported, additional blocks will fall and the rock in the roof will loosen at the joints with more and more rock falling out until a general breakdown of the roof occurs. The time interval between blasting and the general breakdown of the rock half-dome is called the bridge-action period, or stand-up time. Support must be installed before the bridge-action period expires. The earlier it is installed, the less will be the ultimate load on the support.

Lauffer[10] has emphasized the importance of the relationship of active span (width of tunnel or distance between face and supported tunnel, whichever is less) to "stand-up time," Fig. 7-5.[11]

According to Terzaghi,[9] the load on the roof of a tunnel in blocky and seamy rock will increase with time and distance from the working face. Later tests,[12] made in the pilot bore of the Straight Creek Tunnel at Loveland Pass, Colorado, showed that loads build up sharply near the working face and then taper off to lower stable loads about 200 feet from the heading.

As determined by tests on various railroad tunnels in the Alps, loads on the roof will be in the range shown in Table 7-11.

In squeezing ground at depths of about 300 feet, the load to be carried by roof supports can be estimated as $\mathbf{H}_p = 2.1\mathbf{C}$, and, at depths of more than a thousand feet, $\mathbf{H}_p = 3.5\mathbf{C}$, where $\mathbf{C}$ = width + height of tunnel in feet.

In the squeezing ground of Straight Creek Tunnel, Zone III loads were estimated using an elastoplastic analysis that was developed in soil mechanics.[13] This section of the tunnel was 56 feet, 2 inches wide by 64 feet, 8 inches high. Rock cover was 700 to 1000 feet. The analysis indicated loads of 47 kips/foot^2. The final lining was designed for 75 ksf. The maximum load measured at the time work was nearing completion was 40 ksf. This compares with Terzaghi's estimate of 42 ksf at 300-foot depth and 70 ksf at depths of more than a thousand

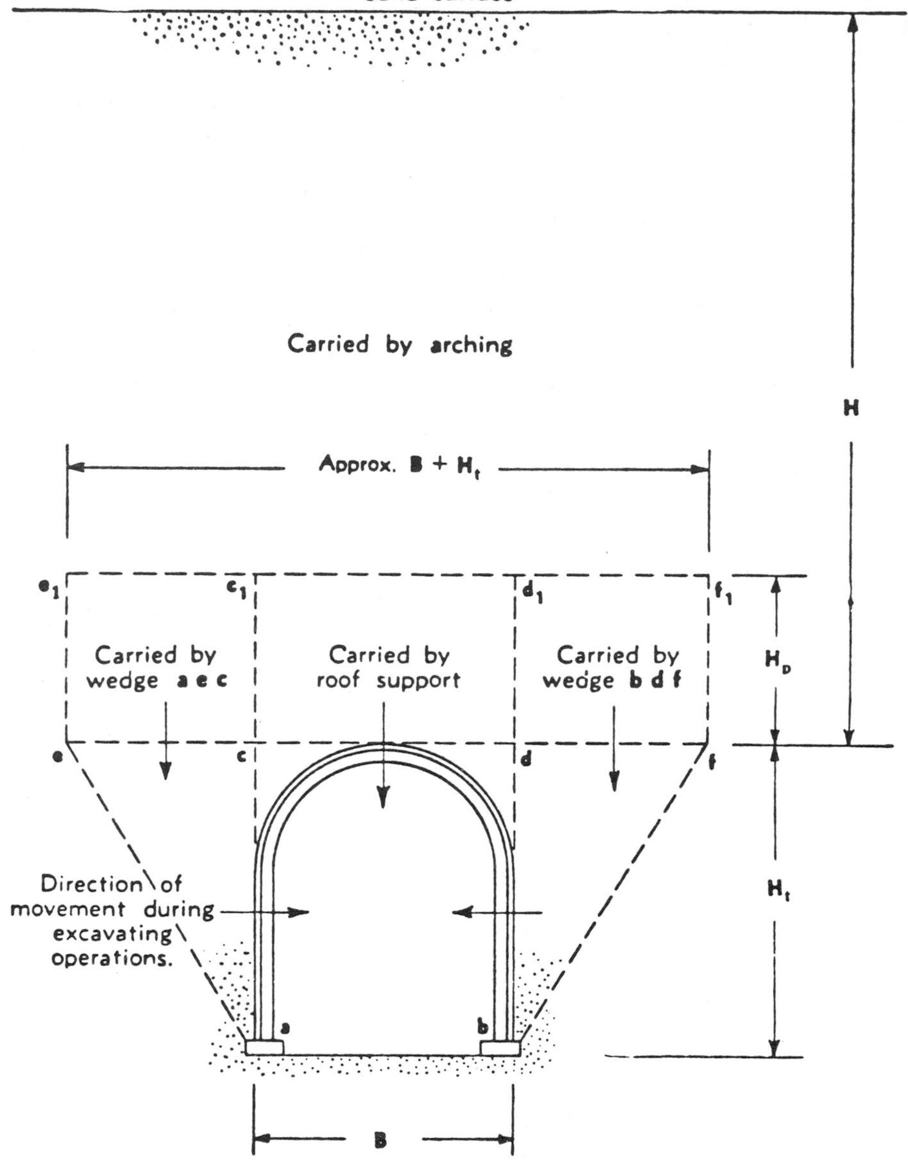

Fig. 7-4. Loading of tunnel support in sand (Terzaghi, Commercial Shearing, Inc.).

Table 7-10. Rock loads in dry crushed rock.

Dense sand	$H_{p\,min} = 0.27C$	for yield of 0.01C
	$H_{p\,max} = 0.60C$	for yield of 0.15C or more
Loose sand	$H_{p\,min} = 0.47C$	for yield of 0.02C
	$H_{p\,max} = 0.60C$	for yield of 0.15C or more

where C = width of tunnel plus height of tunnel, ($B + H_t$ Fig. 7-4)

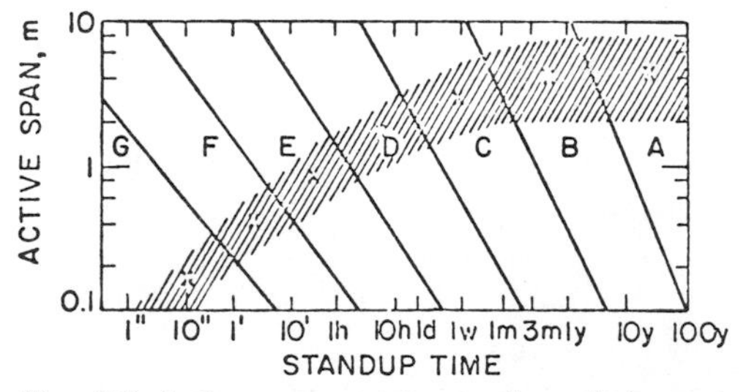

Fig. 7-5 Active span versus stand-up time. A- best rock mass, G- worst rock mass. Shaded area indicates practical relations. (AIME)

Table 7-11. Rock loads (in feet) in blocky and seamy rock.

	Initial Value	Ultimate Value
Moderately blocky rock	$H_p = 0$	$H_{p\ ult} = 0.25B$ to $0.35C$
Very blocky and shattered rock	$H_p = 0$ to $0.60\ C$	$H_{p\ ult} = 0.35C$ to $1.10C$

H_p = height of rock to be supported (feet)
B = width of tunnel (feet)
C = width + height of tunnel (feet)

feet. The unit pressure on the sides and floor of the tunnel are, respectively, about one-third and one-half that on the roof.

In swelling rock, pressures as high as 10 tons/foot^2 are not uncommon in deep tunnels and, in some cases, have been as high as 20 tons/foot^2.

A summary of Terzaghi's recommendations is shown in Table 7-12.

Deere *et al.*[8] have shown that in many cases Terzaghi's rock loads are ultraconservative. With more advanced methods of smooth wall drilling and blasting to limit the shattering of the rock

Table 7-12. Rock load Hp in feet of rock on roof of support in tunnel with width B (ft) and height H_t (ft) at depth of more than 1.5C, where $C = B + H_t$

Rock Condition	Rock Load H_p in feet	Remarks
1. Hard and intact	zero	Light lining, required only if spalling or popping occurs.
2. Hard stratified or schistose**	0 to 0.5**B**	Light support.
3. Massive, moderately jointed	0 to 0.25**B**	Load may change erratically from point to point.
4. Moderately blocky and seamy.	0.25**B** to 0.35**C**	No side pressure.
5. Very blocky and seamy	(0.35 to 1.10)**C**	Little or no side pressure.
6. Completely crushed but chemically intact	1.10**C**	Considerable side pressure. Softening effect of seepage towards bottom of tunnel requires either continuous support for lower ends of ribs or circular ribs
7. Squeezing rock, moderate depth	(1.10 to 2.10)**C**	Heavy side pressure, invert struts required.
8. Squeezing rock, great depth	(2.10 to 4.50)**C**	Circular ribs are recommended.
9. Swelling rock	Up to 250 feet irrespective of value of **C**	Circular ribs required. In extreme cases, use yielding support.

*The roof of the tunnel is assumed to be located below the water table. If it is located permanently above the water table, the values given for types 4 to 6 can be reduced by 50%.

**Some of the most common rock formations contain layers of shale. In an unweathered state, real shales are no worse than other stratified rocks. However, the term shale is often applied to firmly compacted clay sediments which have not yet acquired the properties of rock. Such so-called shale may behave in the tunnel like squeezing or even swelling rock. If a rock formation consists of a sequence of horizontal layers of sandstone or limestone and of immature shale, the excavation of the tunnel is commonly associated with a gradual compression of the rock on both sides of the tunnel, involving a downward movement of the roof. Furthermore, the relatively low resistance against slippage at the boundaries between the so-called shale and rock is likely to reduce very considerably the capacity of the rock located above the roof to bridge. Hence, in such rock formations, the roof pressure may be as heavy as in a very blocky and seamy rock.

beyond the tunnel periphery, and the complete elimination of blasting by the use of tunneling machines for boring tunnels, the loads on supports are appreciably reduced.

A more recent engineering classification of rock for use in determining the behavior of rock in tunnels is known as *rock quality designation*, or, more briefly, RQD. It is based upon a modified core recovery procedure that takes into account the number of fractures and the amount of softening of the rock as observed from cores of drill holes. Included in the core recovery are only pieces 4 inches (10 centimeters) in length or longer; all other pieces are discarded. RQD is expressed as the percentage of modified core recovery, as shown in the example of Fig. 7-6(a).

This method is not as accurate in sedimentary and foliated metamorphic rocks as for igneous rocks and thick bedded limestones and sandstones. In shales, the cores must be logged immediately after removing from the core barrel.

(a)

CORE RECOVERY	MODIFIED CORE RECOVERY
10″	10″
2″	
2″	
3″	
4″	4″
5″	5″
3″	
4″	4″
6″	6″
4″	
2″	
5″	5″
50″	34″

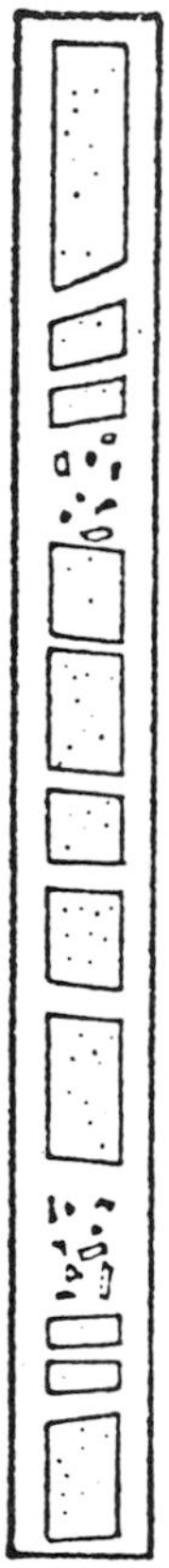

Core Run = 60″

Core Recov. = 50/60 = 83%

RQD = 34/60 = 57%

(b)

RQD (ROCK QUALITY DESIGNATION)	DESCRIPTION OF ROCK QUALITY
0–25	very poor
25–50	poor
50–75	fair
75–90	good
90–100	excellent

Fig. 7-6. Modified core recovery as an index of rock quality. (After Deere et al., 1967).

Poor drilling and equipment can give erroneous results. Proper supervision is a requisite. Double tube core barrels of at least NX size ($2\frac{1}{8}$ inches in diameter) are required. Breaks caused by handling are ignored.

Deere *et al.*[8] have made an extensive report on the use of the RQD method for assisting in determining the supports required in tunnels. This method relies less on judgment and gives promise of being more accurate than Terzaghi's method. Deere *et al.* have classified the rock quality in terms of RQD as shown in Fig. 7-6(b).

Rock with an RQD of 25 or less is in a class with soft ground tunnels insofar as tunneling methods are concerned.

It has been shown that a qualitative relationship exists between the RQD and the support required for tunnels in rock. Tentative recommendations[8] for load to be carried in tunnels 20 to 40 feet in diameter are as shown in Table 7-13.

These recommendations for tunnels excavated by blasting, which are about 20% below the average estimated by Terzaghi's method, will probably be satisfactory in most cases. Rock load factors for machine-driven tunnels are about 25% below those for tunnels driven by drilling and blasting. Additional experience is needed before final values can be adopted.

A comparison of Terzaghi and Deere loads are shown in Table 7-14.

ROCK EXCAVATION IN GENERAL BY DRILLING AND BLASTING

Tunnels have been driven of almost any size and shape conceivable, from the smallest size that a man can crawl through to mammoth 50-foot diameter bores and underground powerhouse caverns over 100 feet wide, 175 feet high and over 1200 feet long. The small crawl tunnels are called "coyote holes" and their use is generally limited to quarry blasting. The smaller the bore, the less will be the drilling and explosive cost. However, progress is limited in small bores and the cost per unit of volume is much greater than in larger bores. Tunnels are frequently built larger than necessary for their ultimate use requirements in order to permit more rapid progress with equipment that can be used in the larger sections. The minimum practical size is 4 feet wide by 6 feet high inside support so as to permit a man to stand upright. A more usual minimum is 5 feet by 7 feet. With the use of tunneling machines, the size of the tunnel has less effect on progress than when using the drilling and blasting method. Smaller bored tunnels, under 10 feet in diameter, become so crowded with equipment that they are difficult to operate efficiently. Smaller tunnels of considerable length require widening for passing tracks. This requires excavation outside pay lines and usually backfill concrete.

Tunnels smaller than 10 feet square are limited to relatively small equipment that restricts progress. Tunnels 12 feet wide permit the use of 3-foot gauge track, a commonly used gauge. Tunnels 12 feet or more in width will permit the use of equipment large enough to permit rapid progress. With railroad transportation, tunnels up to 16 or 18 feet in diameter are ideal in size. The larger tunnels permit a greater selection of equipment to suit the particular conditions of the project.

In good sound rock having a high RQD, a tunnel or cavern of very large size may be excavated with little difficulty and may require little or no support. However, in "bad" ground (low RQD), the smaller the bore of the tunnel the less will be the difficulties in supporting the tunnel. An exploratory tunnel through the Continental Divide in Colorado at Loveland Pass was excavated without great difficulty. However, when the Straight Creek, two-lane highway tunnel was excavated parallel to it, the rock could not be held with any conventional support. This illustrates the need for caution in extrapolating the behavior of a small tunnel to a proposed larger tunnel.

There is no standard for tunnel size. Engineers frequently design to the last theoretical inch of diameter, whereas the cost of the tunnel would not change appreciably for a section a few inches larger. The cost of forms for concrete is large. Frequently, tunnel forms and equipment are used on only one project. Cost savings could be realized by designing tunnels to standard sizes or to the size of a recently completed tunnel.

Types of Headings

Full Face Advance. This is universal in small size tunnels. In excellent rock, large tunnels

Table 7-13. Support recommendations for tunnels in rock (20 to 40 feet in diameter).[8]

Rock Quality	Tunneling Method	Alternative Support Systems: Steel Sets**	Rock Bolts+	Shotcrete
Excellent* RQD > 90	A. Boring machine	None to occasional light set. Rock load 0.0 to 0.2**B**.***	None to occasional.	None to occasional local application.
	B. Conventional	None to occasional light set. Rock load 0.0 to 0.3**B**.	None to occasional.	None to occasional local application 2 to 3 inches.
Good* 75 < RQD < 90	A. Boring machine	Occasional light sets to pattern on 5 to 6 foot center. Rock load 0.0 to 0.4**B**.	Occasional to pattern on 5 to 6 foot center.	None to occasional local application 2 to 3 inches.
	B. Conventional	Light sets, 5 to 6 foot center. Rock load (0.3 to 0.6)**B**.	Pattern, 5 to 6 foot center.	Occasional local application 2 to 3 inches.
Fair 50 < RQD < 75	A. Boring machine	Light to medium sets, 5 to 6 foot center. Rock load (0.4 to 1.0)**B**.	Pattern, 4 to 6 foot center.	2 to 4 inches on crown.
	B. Conventional	Light to medium sets, 4 to 5 foot center. Rock load (0.6 to 1.3)**B**.	Pattern, 3 to 5 foot center.	4 inches or more on crown and sides.
Poor** 25 < RQD < 50	A. Boring machine	Medium circular sets on 3 to 4 foot center. Rock load (1.0 to 1.6)**B**.	Pattern, 3 to 5 foot center.	4 to 6 inches on crown and sides. Combine with bolts.
	B. Conventional	Medium to heavy sets on 2 to 4 foot center. Rock load (1.3 to 2.0)**B**.	Pattern, 2 to 4 foot center.	6 inches or more on crown and sides. Combine with bolts.
Very poor+ RQD < 25 (Excluding squeezing or swelling ground.)	A. Boring machine	Medium to heavy circular sets on 2 foot center. Rock load 1.6 to 2.2**B**.	Pattern, 2 to 4 foot center.	6 inches or more on whole section. Combine with medium sets.
	B. Conventional	Heavy circular sets on 2 foot center. Rock load (2.0 to 2.8)**B**.	Pattern, 3 foot center.	6 inches or more on whole section. Combine with medium to heavy sets.
Very poor (Squeezing or swelling)	A. Boring machine	Very heavy circular sets on 2 foot center. Rock load up to 250 feet.	Pattern, 2 to 3 foot center.	6 inches or more on whole section. Combine with heavy sets.
	B. Conventional	Very heavy circular sets on 2 foot center. Rock load up to 250 feet.	Pattern, 2 to 3 foot center.	6 inches or more on whole section. Combine with heavy sets.

*In good and excellent quality rock, the support requirement will in general be minimal but will be dependent upon joint geometry, tunnel diameter, and relative orientations of joints and tunnel.
**Lagging requirements will usually be zero in excellent rock and will range from up to 25% in good rock to 100% in very poor rock.
+Mesh requirements usually will be zero in excellent rock and will range from occasional mesh (or straps) in good rock to 100% mesh in very poor rock.
***B = tunnel width

Table 7-14. Comparison of Terzaghi and Deere Rock loads for tunnels supported with steel sets.

RQD (%)	JOINT SPACING cm.	JOINT SPACING in.	DESCRIPTION	TERZAGHI CLASSIFICATION: DESCRIPTION	ROCK LOAD: Initial	ROCK LOAD: Final	DEERE ROCK LOAD FOR TUNNELS UP TO 40' DIAMETER: CONVENTIONAL	DEERE ROCK LOAD FOR TUNNELS UP TO 40' DIAMETER: BORING MACHINE
	50	24	Excellent	Hard & Intact	0	0	0 to 0.3 B	0 to 0.2 B
				② Hard, stratified or schistose ③	0	0.25B		
90		12	Good	Massive, moderately jointed	0	0.5 B	0.3 B to 0.6 B	0.0 B to 0.4 B
80	20			④ Moderately blocky & seamy	0	0.25B to 0.35C		
70, 60		6	Fair	⑤ Very blocky, seamy and shattered	0 to 0.6 C	0.35C to 1.1 C	0.6 B to 1.3 B	0.4 B to 1.0 B
50, 40, 30	10	4	Poor				1.3 B to 2.0 B	1.0 B to 1.6 B
20, 10	5	2, 1	Very Poor	⑥ ⑨ Completely crushed, gravel and sand and squeezing and swelling rocks	0.54C to 1.2 C	0.62C up to 250'	Excluding squeezing or swelling ground: 2.0 B to 2.8 B; In squeezing and swelling ground: Rock load up to 250 ft.	Excluding squeezing or swelling ground: 1.6 B to 2.2 B; In squeezing and swelling ground: Rock load up to 250 ft.

Where B = width of tunnel
C = width plus height of tunnel

can also be excavated full face by the drilling and blasting method. Where drilling and blasting is used to break the rock, the work is accomplished in cycles, known as "the round," in the following order: drill, blast, smoketime, muck (excavate) and place support. During the drilling period, the drilling jumbo or drills are moved in and set up. While drilling is in progress, water lines, air lines, ventilation and power lines are extended. Usually, more men are required for the drilling cycle than for the mucking cycle, so it would be advantageous for the extra men not needed for mucking and transportation to install the utilities during the muck cycle. Most union agreements do not permit this use of drillers and helpers. During the blast period, the drill holes are loaded with explosives and primed, the drill jumbo and men are moved back to a safe location and the charge detonated. Smoketime is the period required to reduce the smoke from the blast to a level safe for men to work. In earlier years, the ventilation line was connected up to a blower capable of only pushing air into the tunnel heading so that the smoke traveled out through the tunnel. Now reversible blowers that can exhaust the air through the ventilation line, as well as blow air through the line to the heading, are available. Exhausting rather than blowing reduces smoketime to a matter of a few minutes (as few as ten) and eliminates the health hazard of blowing the smoke past the miners. The excavation period covers moving the mucking machine to the face, scaling down loose rock in the roof or upper side walls, loading the muck into cars, trucks or on conveyor belts or into a slurry pipe, extending the track if rail transport is used, and moving the mucker back to a safe place. Extension of utilities not completed during the drilling period are continued during excavation. The final period of the cycle is for placing the required ground support.

During the mucking period, all transport is usually busy taking out the muck. During the drilling period, transportation is available to move ground support materials, drill steel, explosives and utility materials to the heading. On at least one shift per day, a bull gang maintains track or roadway and other services.

Alternate Headings. Because the size of the drilling crew is usually larger than the mucking crew and frequently union rules do not permit interchange of men from one type of work to another, it is possible to make better use of crews if two headings are worked at the same time with the same crews. While the drill crew works in one heading, the muck crew works in the other, and vice versa. This operation is economical until the time spent in transporting men from one heading to the other offsets the man hours of idle time in a single heading operation, or unless varying rock conditions in the two headings do not permit synchronizing operations. This type of operation also permits greater overall progress with the use of one set of equipment where it is possible to transfer the equipment from one heading to another. This method can be used to good advantage in the construction of two parallel tunnels, particularly where rubber tired drill jumbos and hauling equipment permit easy transfer from one heading to the other. When the heading is too far advanced for economical alternate operations, work can be continued on one to completion with one set of equipment, after which the second heading can be completed or additional equipment and personnel can be brought to the project to operate both headings simultaneously without alternating. Alternate headings may also be used on a single short tunnel, such as those used for diversion, if the men and equipment can be easily transported from heading to heading. Alternate headings are frequently operated from shafts, adits or daylighted sections between two tunnels in a single line. In laying out a tunnel, the engineer should consider the advantages of a location that will permit alternating between headings.

Pilot Headings. If the bridge-action period of a tunnel in blocky and seamy rock does not permit the rock to stand long enough to install the required support, it is necessary to excavate only a portion of the cross-section at first and support it.

If ground conditions are uncertain and "heavy" ground (i.e., rock with low RQD) is suspected, a pilot heading within the design section, of a size that can be readily supported, can be driven. The condition of the rock at the roof of a tunnel in poor ground is the most critical situation. Therefore, if the principal purpose of a pilot heading is to explore the ground ahead, it is desirable to locate it at the roof of the full section. The minimum size for efficient operation is 10 × 10 feet. As many headings as are necessary, within the design section, to control the ground may be driven consecutively based on knowledge gained from the pilot heading. In poor rock, wall plate drifts at the spring line may be driven first, followed by a top heading with the bench being taken out last. In earlier years, before the advent of modern equipment, when miners were accustomed to hard labor, it was customary in larger tunnels to drive a top center heading, between the spring line of the tunnel and roof, and support it with segments of ribs fitting into the support system of the full tunnel section.

This is rarely done now except in poor rock. Following this operation, the remaining rock above the spring line was excavated by winging out and completing the rib segments to wall plates located at the spring line. Excavation of the bench was the final operation.

Center Drift. Another method frequently used in earlier years for large tunnels in good rock was to drive a center drift and ring drill to the periphery of the tunnel, blasting each ring consecutively (see Fig. 7-7). Because larger and better equipment is now available, this method is rarely used for excavating tunnels but is frequently used for shafts, vertical or inclined.

Tunnel Cross-section

The *choice of cross-section* should be made to accomplish the cheapest overall cost of the tunnel. If a boring machine can excavate the rock, so that the overall cost of excavating, supporting and lining will be cheaper than for a tunnel excavated by drilling and blasting, a circular section will generally be used, as this is the only cross-section that most machines can accomplish. Machines have been developed that can excavate other cross-sections (see Chapter 10). In larger tunnels excavated by drilling and blasting, a circular section is usually more expensive to excavate because it requires two excavation operations. During the excavation of the major part of the cross-section, a portion of the rock in the bottom must be left in place to provide a flat surface for truck haul road and is usually left for rail track. The excavation of the small remaining invert section is costly and requires additional time. As an alternate, smaller circular tunnels may be widened out at the invert to provide for haul roads or tracks. This requires a high percentage of excavation outside of theoretical lines. In difficult ground with low RQD's requiring heavy support, particularly in swelling or squeezing ground, a circular section will be likely to be the cheapest section because the temporary and permanent support required will be less expensive.

Tunnels for conveying water under pressure require reinforcement at least where the rock cover is too small to provide the necessary resistance. A circular section provides for the least costly reinforcement. Even where a finished circular section is required, the excavated section may be horseshoe in shape. If it is necessary to keep groundwater out of a concrete-lined tunnel, large external pressures may develop that will be more easily resisted by a circular section.

Tunnels in rock with little or no side pressure can be most cheaply excavated if the sides are nearly vertical and the bottom is flat. A flat or arched roof makes little difference in cost of excavation, but except in the best of rock or in very small tunnels, an arched roof will be cheaper because of support requirements. Railroad tunnels, except in difficult ground, are usually constructed with a semicircular top and vertical sides, because of clearance requirements.

The horseshoe section is a compromise between the straight side-walled tunnel with flat bottom and a circular section. It is usually designed with a semi-circle on top and with curved sides and bottom, the radius of the sides below springline and bottom being twice the radius of the arch (Fig. 7-8). The advantage of the horseshoe section is that it permits use of ribs utilizing arch action to support the sides of the tunnel and the invert in case of swelling or squeezing ground, thus giving much of the advantage of a circular section while at the same time providing a nearly level bottom. A level bottom is most important for tunnels excavated by drilling and blasting. The slight decrease in width at invert level does not appreciably affect the size of equipment compared to what could be used in a vertical walled tunnel of same maximum width. It is usually possible to drill and blast to the bottom of the invert, in the initial driving operation, leaving loose muck in the bottom to provide a level surface for hauling. Prior to concreting, the loose muck can be removed quickly as part of the cleanup for the lining operation. In tunnels conveying water, the curved bottom permits drying up the tunnel with fewer and smaller "bird baths."

Portalling In

Except in the best of rock, the first rounds near the portal can result in large overbreak or slides of the surface rock at the portal. If the rock

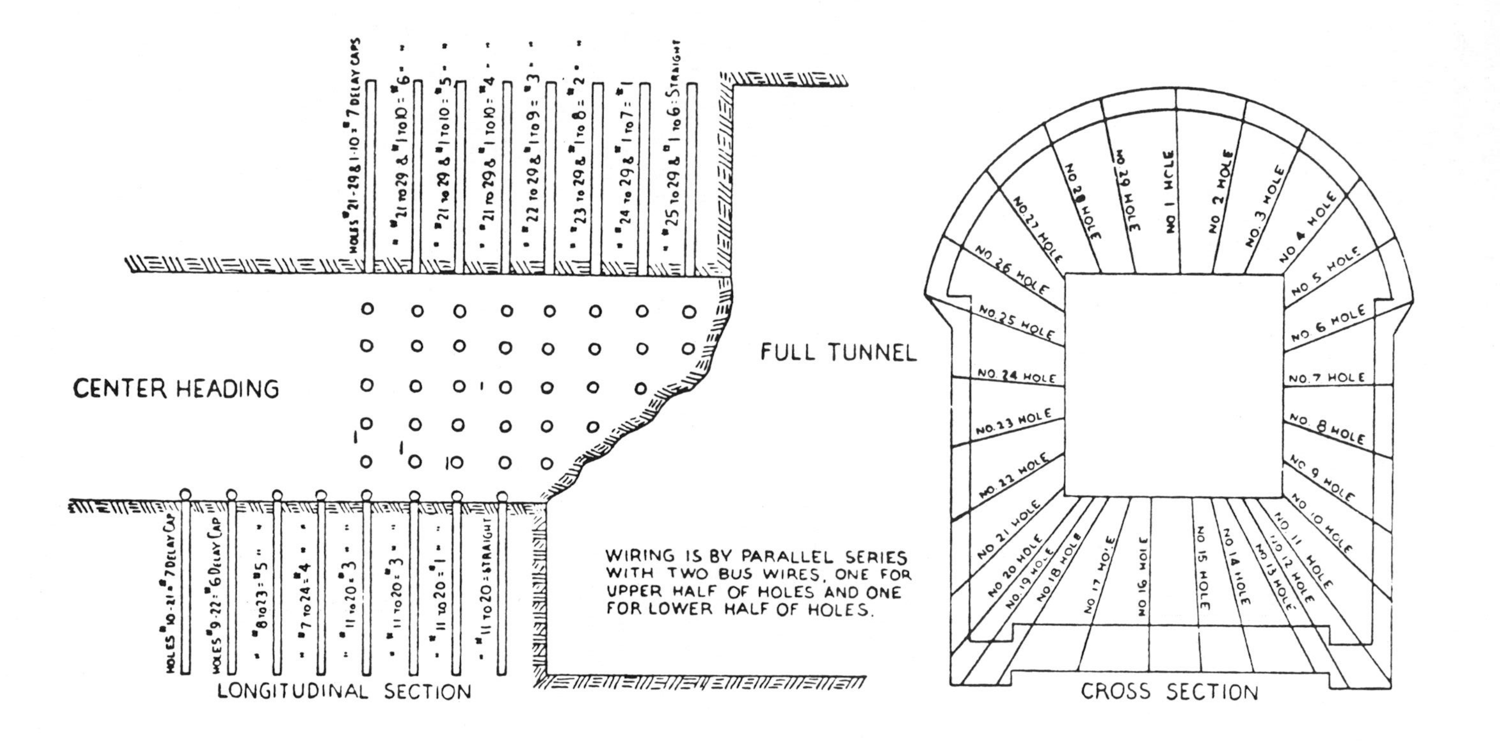

There are 29 holes in each ring of an enlargement round. Angles are spotted by clinometers and distances are measured from the center the arm. Sub-grade holes extend 2 ft. below that line. Eight rings (32 ft.) are fired in one shot, but the lower half are four rings ahead their uppers thereby giving the lower holes a four ring "lead." Upper half holes contain 40% gelatin, while lower ones are charged w 60%. Four cartridges of 80% gelatin are in each ditch hole (Nos. 11, 12, 13, 18, 19, and 20). Connections were in parallel series.

Fig. 7-7. Cascade tunnel. (Rock Tunnel Methods, Hercules, Inc., 1931).

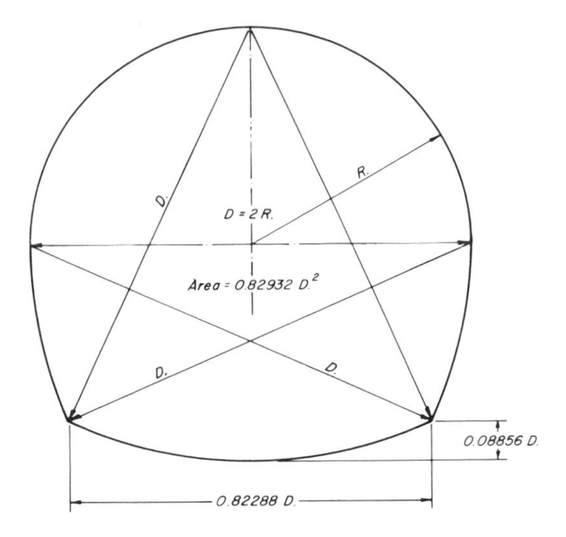

Fig. 7-8. Typical horseshoe tunnel section.

will not stand long enough to permit installation of supports, after blasting and mucking, without large rock relaxation and falls, it may be necessary to start the excavation by driving pilot headings until sounder rock is reached. A good method is to drive wall plate drifts at the springline, install wall plates, enlarge to the full section down to the springline in sound rock and then back out with short rounds to the portal. The bench may then be excavated. The portal is thus secured by adequate support and further advance of the tunnel may proceed normally.

Tunnel Grade

It is desirable to drive a tunnel upgrade for several reasons: 1) Most tunnels are wet naturally and all except bored tunnels have some drill water at the face (drainage can be by gravity rather than by pumping and water at the face will cause less difficulty. It is always a nuisance to have to mop up water at the heading, and pools of water interfere with drilling and blasting of the bottom holes. With gravity drainage, it is easier to keep the bottom of the heading clear of standing water. When driving down steep grades, water tends to pack the muck, requiring more excavating time); and 2) Less power is required for hauling muck (the same power required for downgrade haul of loaded cars or trucks will take the empty cars or trucks upgrade on moderate grades).

The optimum grade for rail haul is something not over 0.4 of 1%. For steeper grades cars, when standing, must be blocked. For grades over 1%, the length of trains must be sharply curtailed or extra power supplied, with resultant rapid run-down of batteries and increase of power or fuel consumption. Rail haul with grades over 3% is very difficult to accomplish. The steeper the grade above optimum, the greater will be the difficulty and the slower the advance. Steeper grades downhill can be used satisfactorily for short tunnels, where equipment and muck cars can be lowered or raised with a hoist. Where truck haul and wheel or tractor mounted excavating equipment is used in wet tunnels, grades up to 10%, such as at Manapouri powerhouse access tunnel,[14] can be driven on a downgrade. A haul up this 10% grade for a distance of 7000 feet required 25 minutes. Steeper grades are possible in dry ground to the extent equipment can get traction. The access tunnel to the top of Dez Dam in Iran was constructed heading downward in dry conglomerate on a 10% grade without difficulty, whereas the wet conditions in the Manapouri Tunnel are estimated to have slowed progress by 40% over a comparable horizontal tunnel.

Tunnels driven on steep slopes from the bottom, such as for penstocks, are usually more cheaply constructed if the slope is sufficient to permit the excavated material to slide or fall to the bottom without hanging up on the slope. Experience has shown that the minimum practical slope is 42 degrees to the horizontal and that a slope of 48 degrees is about the minimum that should be used unless water is used to wash the fines down. The Kemano penstock tunnels with drops of 1400 feet were constructed with an inclination of 48 degrees to the horizontal. If the horizontal distance is greater than occurs with the minimum slope, a horizontal section may be constructed either at the top or bottom of the penstock. Even if the overall length is greater than for a straight alignment, the total cost is usually less. At Brownlee Powerhouse in Idaho, where the 24-foot diameter penstock tunnels were about 500 feet long, they were constructed on a 30-degree grade. Here it was possible to use bulldozers to push the muck economically to the downstream portal. The spillway incline tunnel at Yellowtail Dam, at an angle of 55 degrees to the horizontal, was excavated by first raising the pilot tunnel by use of the Alimak Raise Climber, a device developed in Sweden that operates on overhead track to carry drills and men to the face.

Problems With Tunnels Driven on Steep Slopes

The grade of Cubatao penstock tunnel, 12 feet in diameter, was 42 degrees from the horizontal. It was excavated full face from the bottom. Ten-inch I-beams spaced at two-meter centers with solid lagging carried track and utilities. Muck was allowed to fall freely in the bottom half to the bottom. It was necessary to wash the fines down twice a week to allow the muck to roll free. The Kemano penstock tunnels, 15 feet in diameter, were driven by raising a pilot tunnel 6 feet wide by 10 feet high. A horizontal timber partition located at mid-section followed the driving upward and supported track, steps and utilities with the muck falling in the bottom half. Most of the time, the chutes were kept full, with muck being drawn off at the bottom by pneumatically controlled gates discharging into trucks, but excess fines and excess water could build up pressure on the timber deck. Then it would be necessary to empty the chutes. Occasionally, blockages would occur in the chute. On breaking loose, they caused damage to the deck on reaching the lower stream of muck, and, on one occasion, the muckway deck was pushed to the roof of the raise. If the chute was kept free, there were no problems in getting the muck to the bottom. On completion of the pilot shaft, the deck was removed with enlargement being carried on from the top down.

DRILLING AND BLASTING

Drilling patterns depend on the size and shape of the tunnel and the quality of the rock to be excavated. Rock that is badly fractured will require a lesser number of holes than massive formations with few seams. There must not only be enough volume in drill-holes to permit placing sufficient explosives to break out the rock, but the holes must be placed sufficiently close in blocky rock so that the resulting muck will be in small enough pieces to be handled efficiently by the mucking machine. The square feet of tunnel face per hole will be greater for a large tunnel than for a smaller one in the same rock and greater for softer or highly fractured rock than for hard massive formations.

It is possible to estimate the number of drill holes for a given rock condition based on physical characteristics of the rock and theoretical design of the explosive charge required. The required pattern must be determined by trial and error. It is unusual that a given condition will be as estimated or continuous for any great length of tunnel. It is more usual that the conditions will change from round to round. It is necessary for an experienced tunnelman to be at the heading during each round where changing conditions are being experienced so that the drill pattern can be adjusted as required.

The square feet of tunnel face per hole will usually vary between the limits shown in Table 7-15 or Fig. 7-9.

In order for the rock to break out of its confined condition ahead of the tunnel face, it is necessary to eject some of the rock back into the tunnel to make room for the expansion of the main body of rock. Formerly, this was accomplished by using a number of cut holes, usually four to six drilled in a wedge or pyramid shape near the center of the face, with the apex in the direction of advance. These holes should be drilled a little beyond the depth of the other holes and are blasted first so that the rock within the cut holes would be thrown into the open tunnel. Sometimes this throw is for a considerable distance, causing damage to timber or steel sets and more time for mucking the scattered rock. This method has been largely replaced by

Table 7-15. Blast hole requirements

AREA OF TUNNEL FACE (FEET^2)	TUNNEL FACE PER HOLE (FEET^2)	
	SOFT OR HIGHLY FRACTURED	HARD OR MASSIVE
100	4.5	2.25
200	5.3	3.25
250	6.0	4.0
400	6.4	4.5
500	7.0	4.9

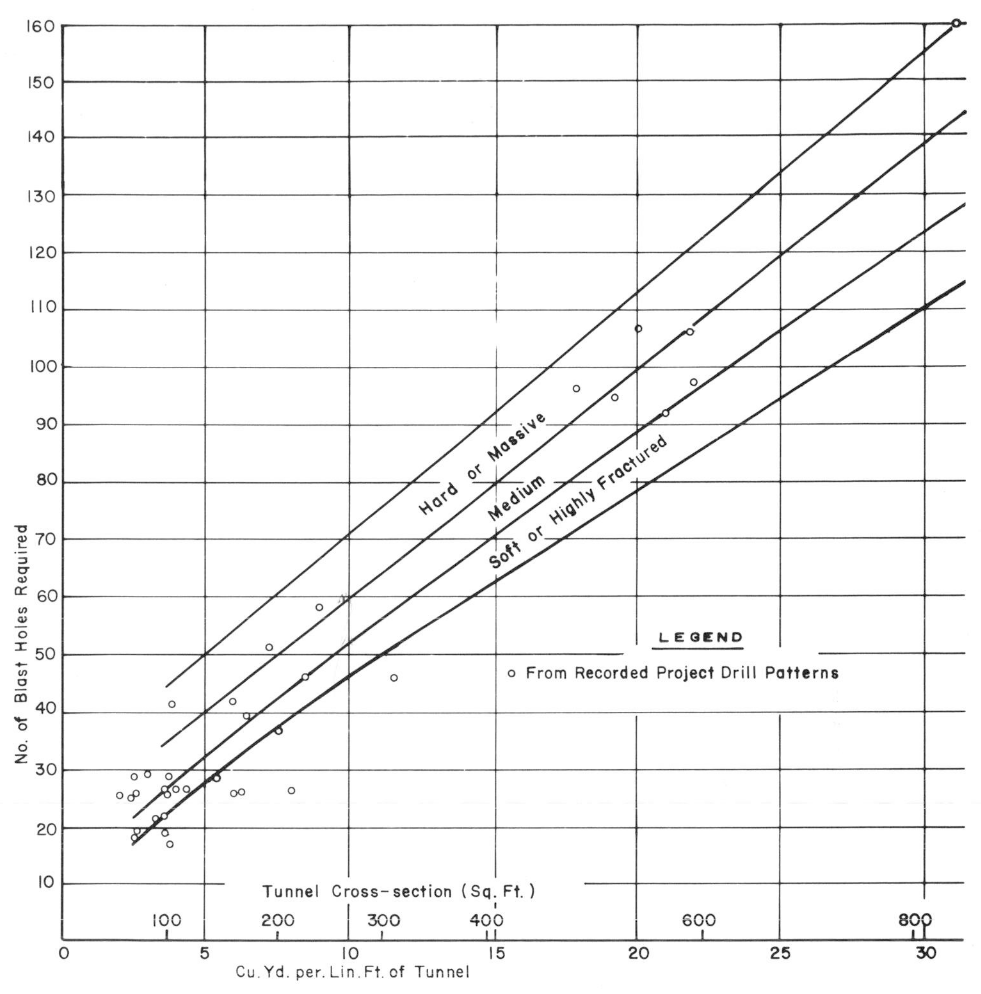

Fig. 7-9. Drill Hole Requirements.

the "burn" cut, which provides one or more large holes near the center of the face, not loaded with explosives, into which the rock inside the first ring of closely spaced loaded holes expands when blasted. These large holes are up to 5 inches in diameter. The advantage of the "burn" cut is that all drill holes are approximately parallel to the tunnel, rather than at an angle, and the rock is not thrown so far from the face as with wedge cuts. The burn cut hole may be enlarged by loading and exploding closely spaced surrounding holes so that all shot rock is ejected from the hole, thus making more room for the expansion of rock from following explosions. This is called a cylinder cut. Figures 7-10 and 7-11 show typical drill patterns. The holes outside of the cut holes and inside the perimeter are called relief holes. One set of relief holes may be sufficient in small headings. For large headings, two, three or more sets may be necessary to prevent an undue burden on the trim holes. The outside holes along the perimeter are called trim holes. The bottom holes are called lifters.

Drills and Mountings. The most economical drill is a jackhammer, which should be supplied with a wet head when used in a tunnel. Because horizontal drilling with a hand-held drill is very tiring, jack legs or air legs have been developed to support the drill on extendable pneumatic legs, which, in turn, are supported by the tunnel floor or drill jumbo platform (Fig. 7-12). A refinement is ladder drilling, in which the air

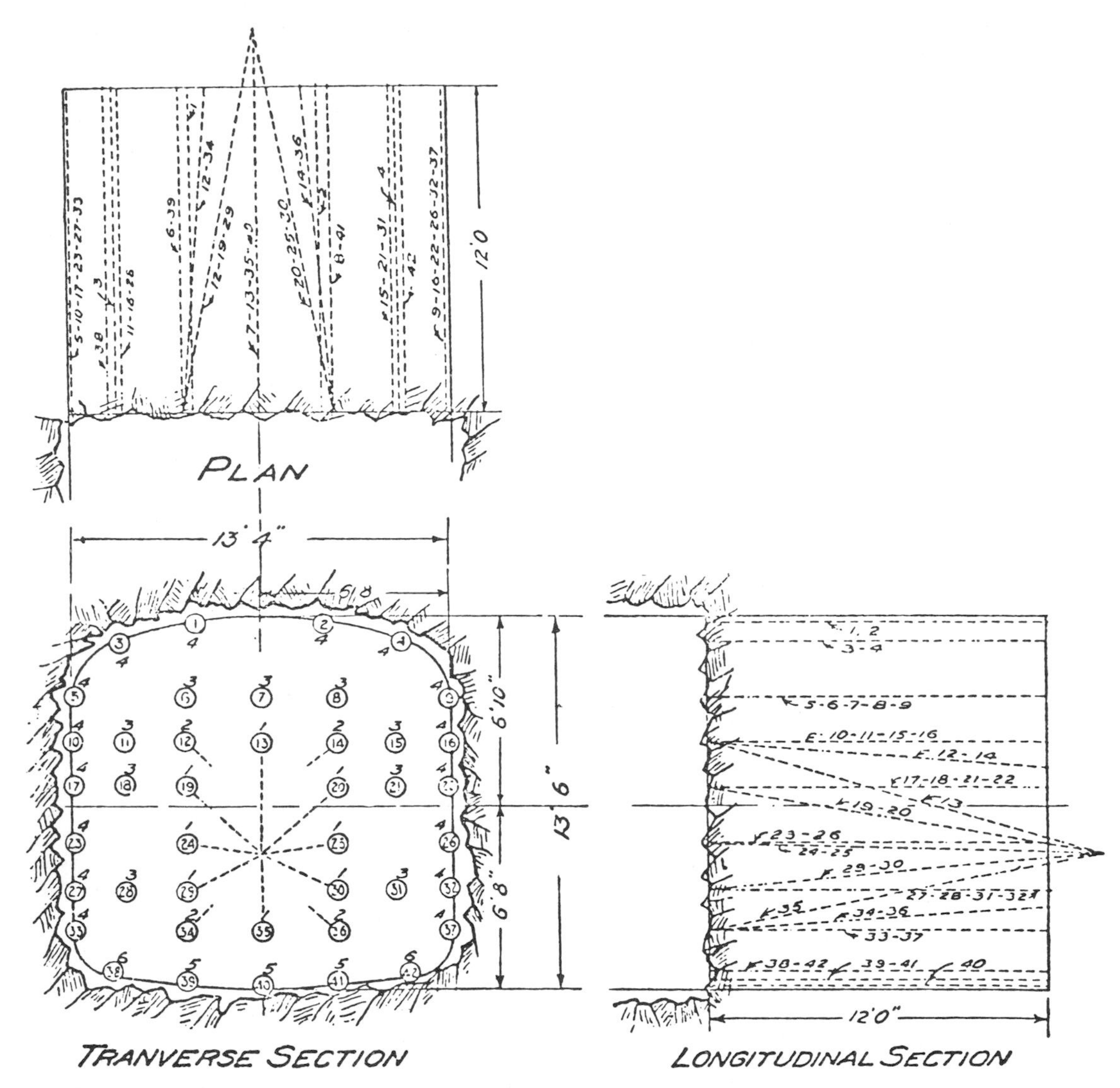

Fig. 7-10. Mountain division tunnel. When drilling South Fork headings, a 42-hole round was adopted as standard. In it, holes 12, 14, 34, and 36 are shown as first relievers but were given additional angle in drilling and made into cut holes when required in hard rock. When necessary, an additional hole was placed between holes 1 and 2 to shape roof. On the transverse section, the figures not circled indicate the firing order of the detonators. The numbers of the holes are shown by the figures inside the circles. The South Fork headings are in hard, close-grained granodiorite. The dimensions of the transverse section are of one of the finished unlined sections. (Rock Tunnel Methods, Hercules Incorporated.)

leg drill propels itself along a steel ladder frame. Although one man can operate an air leg drill, a helper per one to three drills is advisable to get the best production. For drilling in high labor cost countries, $3\frac{1}{2}$- to 4-inch bore drifter drills with automatic feed devices, mounted on hydraulically positioned booms supported on drill jumbos, have been found to be most economical (Fig. 7-13). One miner can supervise several automatic drills (and, on some of the Swiss tunnels, as many as seven). Union regulations may limit this to fewer and even require helpers on each drill, greatly increasing the cost. Prior to the advent of hydraulic booms, drills were mounted on pipe column bars spanning the width or height of the tunnel. Later the column bars were supported on jumbos.

Drill jumbos have made it possible to drill any

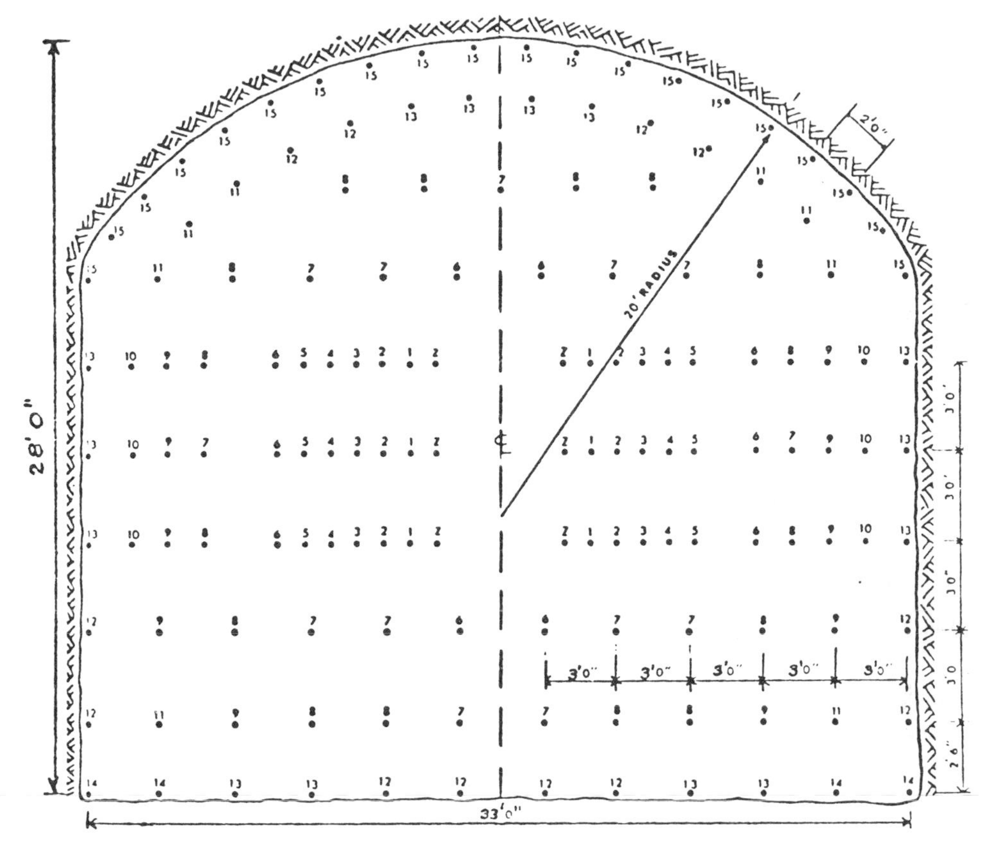

Fig. 7-11A. Drilling and delay pattern for tunnel using "perimeter blasting" technique. 1st North American Rapid Excavation and Tunneling Conference, AIME, June 5–7, 1972.

size tunnel full face. With decks at 6 to 8 feet spacing, the hydraulic boom-mounted drills can cover the entire face, drilling holes at any angle. Jumbos are of three general types. Main line track-mounted jumbos ride on the haulage track. They are pulled back to a siding near the face when not in service (see Fig. 7-14). This type is suitable for small tunnels. Straddle track jumbos are designed to ride on wide gauge track to permit haulage on the main line through them. They straddle the haulage track, so when not in service they need to be hauled back only far enough to avoid damage from the blast at the tunnel face. This type is suitable for large tunnels. Rubber-tired or crawler-mounted jumbos are suitable for use in tunnels using truck haulage. Jumbos are designed so that deck platforms cover the entire width of the tunnel. The sections outside the frames or interfering with haulage can be dropped on hinges to provide clearance for moving the jumbo away from the face and to permit trains to pass through the straddle track type. A minimum of 25 square feet of face should be allowed for each drill in small tunnels. In larger tunnels, enough drills should be provided to limit the number of holes per drill to a maximum of 8 to 10 per round.

Tungsten carbide bits are now almost universally used for rock drilling. Bits integral with the $\frac{7}{8}$-inch or 1-inch drill rods used with air leg drills are usually $1\frac{1}{2}$ to $1\frac{5}{8}$ inches in diameter, and thrown away after their useful life of 400 to 900 feet of drilling. The bits used with $1\frac{1}{4}$-inch diameter drifter drill steel are detachable and in sizes from $1\frac{1}{2}$ to $1\frac{7}{8}$ inches. They have a life of 300 to 400 feet of drilling in hard rock such as granite. Bits require sharpening on a grinding wheel after 40 to 50 lineal feet of drilling.

Perimeter Drilling. If perimeter holes are drilled far apart, the resulting perimeter of the tunnel

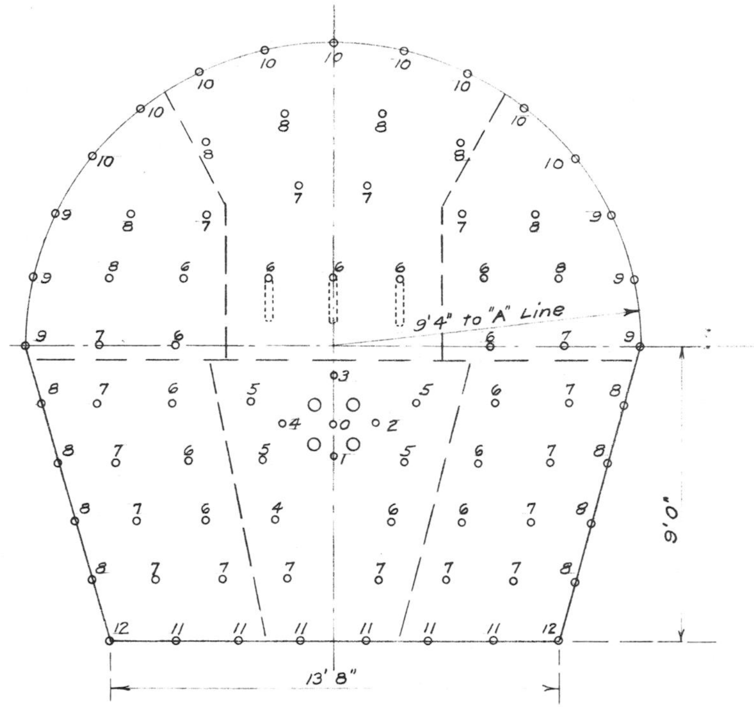

Fig. 7-11B. Drill pattern with burn cut holes Teton dam River Outlet works. Figures indicate firing order of the detonators. 6 Drills used. Diagram marked off for each drill. Holes $1\frac{3}{4}$ in. diam. except burn cuts 4" dia. Holes drilled 10 ft to pull 8.3 ft $1\frac{1}{4}$" × 8" dynamite used.

will be very jagged, with much resulting overbreak or expensive trimming, or both. If the rock is of sufficient quality so that falls do not occur at open seams before it can be supported, it is possible to reduce the amount of overbreak by reducing the spacing of holes. Line drilling with holes spaced as close as 6 inches, combined with controlled blasting, will reduce overbreak to a minimum. It is more usual to space perimeter holes at about 2-foot centers. As an alternate to line drilling, presplitting may be used to reduce overbreak. This is accomplished with holes spaced at 1-foot centers and alternate holes only loaded with a light charge and fired before the rock breaking holes inside the presplit lines. This method has been used in powerhouse caverns.

Fig. 7-12. Jackleg drill.

Depth of Rounds. It is desirable to make the depth of the round as long as conditions will permit, as this will reduce the lost time between each operation of the round. For small tunnels, the depth of the round is limited by the depth that can be efficiently removed in a single blast.

Fig. 7-13. Drifter drill with hydraulic booms.

This depth is usually the minimum dimension of the tunnel face. The actual depth of the round is usually less than this minimum dimension (not over three-quarters). If the rock requires support, the length of the round will also be limited by the stand-up time, sometimes called the bridge-action period, of the rock. The stand-up time is the period during which the rock, except for initial spalling, will stay in place after blasting, and maintain the integrity of the roof prior to the general breakdown of the arch or half-dome action. A few blocks will drop immediately after blasting. If support is not installed in badly fractured rock, more and more rock will drop out until the tunnel is filled with broken rock. The time during which this will occur depends on the shape and size of the blocks between joints and the nature of the material in the joints. In ground supported by ribs, the depth of the round should be a multiple of the rib spacing. If the stand-up time is short and support is by means of rib steel, the length of the round may be limited to the spacing of the ribs. The depth of the round will also be governed by the available travel or feed distance of the drill on its frame. Drill feeds are generally 8 or 10 feet. Changing steel takes time so the equipment and round should be planned for as few changes as possible. In tunnels up to medium size, it is possible to produce as many as six rounds per day.

Fig. 7-14. Track drill jumbo.

In larger tunnels (over 20 feet in width), the depth of the round depends more on the quantity of muck to be removed. Since mucking will require considerable time, the depth of the round is frequently set so that one round can be completed in an 8- or 10-hour shift.

Because the rock will not break, on an average, to the bottom of the hole, it is necessary to drill holes from 6 to 30 inches deeper than the depth of the round desired. The over-depth drilling depends on the quality of the rock and the spacing of the holes and will usually average about 12 inches. In the Big Creek No. 4 tunnel in California, a 24-foot diameter horseshoe section, in granite, requiring practically no support, the length of rounds drilled were 8 to 15 feet, with an average of 12 feet.

Drilling Time. Rock drills will penetrate up to 4 feet/minute, depending on the hardness and quality of the rock. The drills are usually mounted on a drill jumbo that must be parked several hundred feet from the face. Time for moving the jumbo to and from the face and for changing steel and moving drills from hole to hole must be added to the actual drilling time. Actual production will usually be from 1 to 2 feet/minute, allowing for lost time. At Big Creek No. 4 tunnel in granite, 1416 feet of hole was drilled, on the average, with 13 drills, and loaded with explosives in $1\frac{1}{2}$ hours, or at the rate of $1\frac{1}{4}$ feet/drill/minute, including moving the jumbo to and from the face.

Explosives: Type and Quantity. There are two types of explosives (called powder, in tunnel parlance) in general use for tunnel excavation. They are nitroglycerine dynamites and ammonium nitrates. Dynamites are mixtures of nitroglycerine and other substances. They are rated in strength according to the percentage of nitroglycerine. They are manufactured in ratings of 15 to 90%. The energy produced per pound is not in proportion to the percent of nitroglycerine, as other ingredients add to the energy produced: 60% dynamite is only $1\frac{1}{2}$ times as strong as 20% dynamite. The blast effect is determined not only by strength, but also by density and velocity. Density can be expressed by the number of cartridges of a particular size contained in a 50-lb case. There may be only one-third as many of the denser cartridges as the least dense. In hard rock, the denser dynamites are required to break the rock, so these are generally used in tunneling. Cartridges are manufactured in various sizes. The most generally used size in tunneling is $1\frac{1}{4} \times 8$ inches. Gelatine dynamites are dense and most water-resistant, and up to a 60% rating they have better fume characteristics. Thus, they are preferred for tunneling (40 and 60% gelatines are best adapted to tunnel work). The range of dynamites available is shown in Table 7-16.

Ammonium nitrate, in the form of fertilizer prills, when mixed with from 1 to 12% carbon black, powdered coal or fuel oil, has become a favorite explosive in tunnel work because of its lower cost (5 to 7% by weight of fuel oil gives the best performance). It requires a dynamite cartridge in each hole primed with an electric blasting cap to cause it to explode. Ammonium nitrate is not water-resistant so it can only be used in relatively dry tunnels or when encased in plastic bags. Powder companies now market ammonium nitrate prills ready mixed with the necessary admixtures, although mixing can easily be done on the job. This powder may be pneumatically placed in the drill holes.[15]

In order to detonate dynamite, it is necessary to insert a blasting cap in one cartridge in each hole. Blasting caps are manufactured for instantaneous explosion or with 15 or more delays and with various wire lengths up to 100 feet. Ordinary delays provide for time intervals of from 1 to $2\frac{1}{2}$ seconds or more between delays. Millisecond delays have time intervals in the order of hundredths of a second.

The amount of powder required to break rock out from a tunnel heading is much greater than in an open cut because of its confined location. Smaller tunnels require more explosives per unit of excavation than do larger tunnels. Rocks with low RQD's require less powder than those with high RQD's. Harder massive formations require more powder than softer formations. Powder consumption can vary from over 9 lb/yard^3 in a massive hard granite formation in an 11-foot diameter tunnel to less than 2 lb/yard^3 in a softer or less massive formation and large-diameter tunnel. At Mecca Pass Tunnels in California, in a badly fractured

Table 7-16. Properties of gelatin and semi-gelatin dynamites.*

TYPE	GRADE	BULK STRENGTH	DENSITY (CTGS. PER 50 LB, $1\frac{1}{4} \times 8$ INCHES)**	VELOCITY (FEET/SECOND)	WATER-RESISTANCE	FUMES†
du Pont gelatin	20 to 60% 75 to 90%	32 to 66% 70 to 79%	85–96 101–107	10,500–19,700†† 20,600–22,300††	Excellent Excellent	Very good Very poor
"Hi-velocity" gelatin	40 to 60% 70 to 90%	38 to 47% 53 to 71%	94–107 113–120	16,700–19,700 20,300–22,000	Excellent Excellent	Very good Very poor
Special gelatin	25 to 80%	38 to 75%	88–107	13,100–17,100††	Excellent	Very good
"Gelex"	1 2 5	60% 45% 30%	110 122 150	13,100 12,600 11,300	Very good Very good Good	Very good Very good Very good

*From *Blasters Handbook, Fifteenth Edition*, published by E. I. du Pont de Nemours & Co., Inc., Explosives Department.
**Subject to a variation of plus or minus 3% from standard.
†Grades rated with "very poor" fumes are not recommended for underground use.
††The velocities shown for these gelatin dynamites are the high values.

granite formation called Alaskite, requiring support throughout its length, only $1\frac{1}{2}$ lb/yard^3 were required. In an 11-foot diameter tunnel in California, driven through siltstone and sandstone with some clay, only $1\frac{1}{2}$ lb/yard^3 were required. The chart in Fig. 7-15 shows the range of powder consumption that can be expected for various sizes of tunnel and rock quality.

Blasting Procedures. It is important that the holes in the drill pattern be blasted in the proper sequence to obtain the best breakage and muck pile formation. To accomplish this, electric detonators with built-in delays of fixed periods are used. These have largely replaced the caps and different lengths of fuse used in earlier days. The cut holes are fired first, followed by the inner set of relief holes and progressively toward

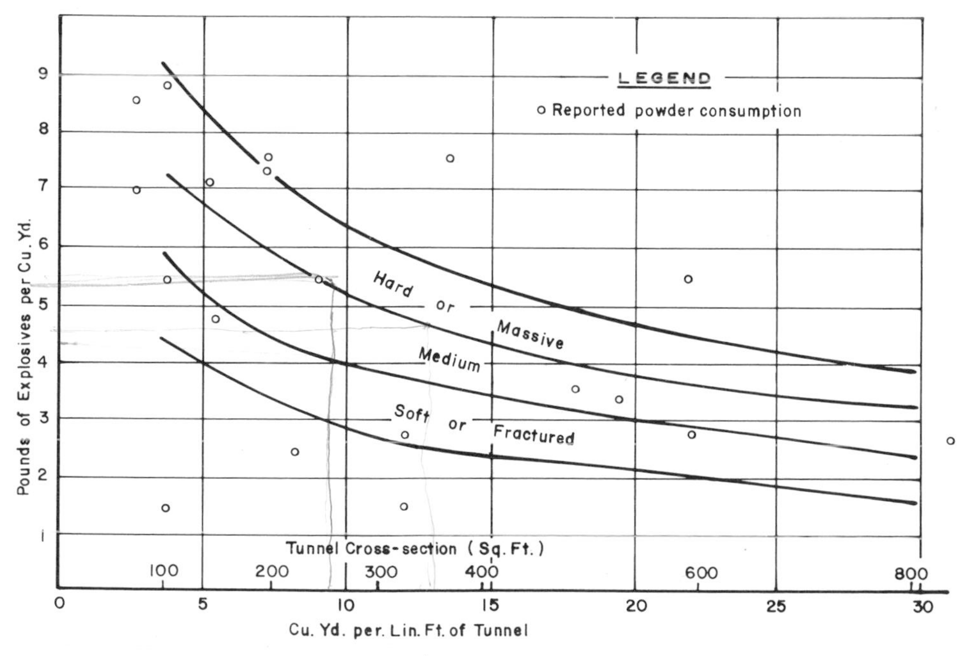

Fig. 7-15 Explosive requirements.

the outer relief holes. The corner relief holes perform best if fired after the adjacent holes have fired. Trim holes are usually fired in rotation with the bottom holes heavily loaded and fired last for hand mucking. For machine mucking, top holes are usually fired last. The number of delays historically used varies between wide limits (see Fig. 7-16).

Dynamite cartridges, if not perforated, should be slit and pressed into place one at a time by means of a wooden tamping stick. The detonator known as a blasting cap should be near the bottom of the hole, to prevent cut-offs, and is usually placed in the first cartridge with the cap pointing to the collar of the hole. One detonator is used in each loaded hole. It is important that the right delay is used in the proper hole in order to ensure the proper sequence of explosions. Stemming is not generally practiced but it is recommended, as it prevents ejection of explosives by earlier delays in the same round, saves explosives and improves fume conditions.

Caps may be connected in series or parallel circuits and detonated with a blasting machine or a power circuit. Two hundred and twenty volts is preferable to lower or higher voltages. Sixty cycles is preferable to 25 cycles. Electric blasting caps are detonated by the heating up of their resistance bridge wires by the passage of electric current through them. The voltage required is based on Ohm's law:

$$E = IR$$

where

E = voltage of power supply
I = current (amperes)
R = resistance of circuit (ohms).

The current requirement for a series of instantaneous or delay caps is 1.5 amperes. If a mixture of instantaneous and delay caps is in a series, a minimum of 2.0 amperes should be provided. The current requirement for parallel circuits is usually 0.5 to one ampere per cap, but may vary widely, depending on the circuit arrangement. Du Pont[16] recommends a com-

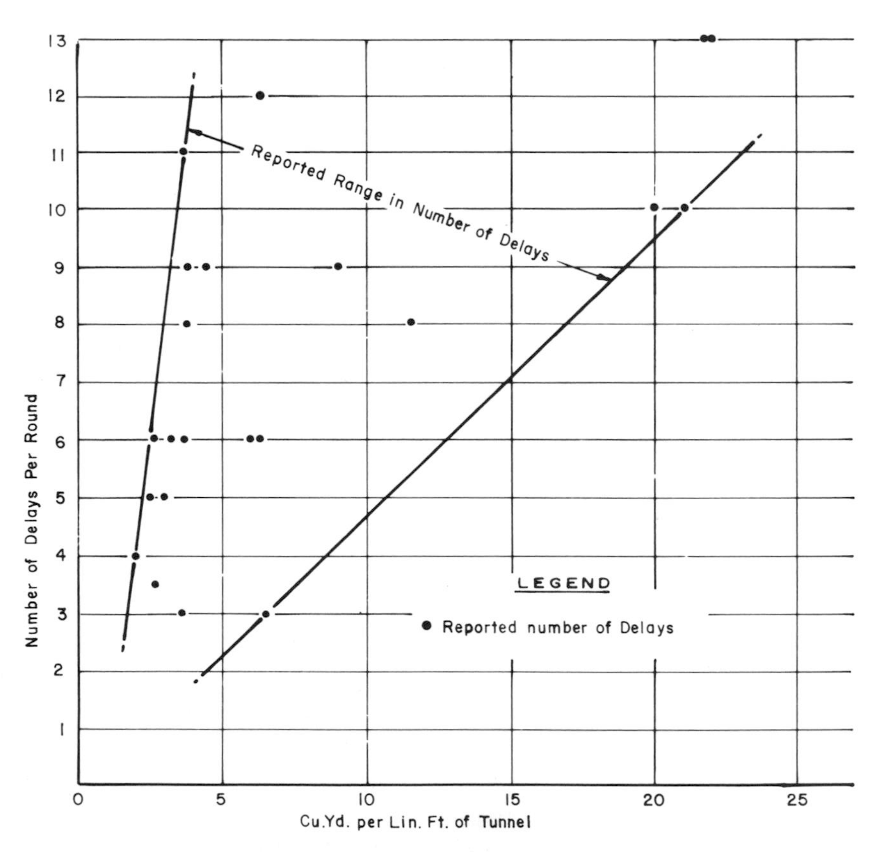

Fig. 7-16 Number of delays per round.

Table 7-17. Nominal resistance* of Du Pont electric blasting caps in ohms per cap.**

	COPPER WIRE		IRON WIRE		ALUMINUM WIRE		
LENGTH OF WIRE IN FEET	INSTANTANEOUS CAPS	DELAY CAPS	INSTANTANEOUS CAPS	DELAY CAPS	INSTANTANEOUS CAPS	DELAY CAPS	LENGTH OF WIRE IN FEET
4	1.23	1.13	1.89	1.79	–	–	4
6	1.29	1.19	2.29	2.19	1.30	1.20	6
7	–	–	2.49	2.39	–	–	7
8	1.36	1.26	2.68	2.58	1.37	1.27	8
9	–	–	2.88	2.78	–	–	9
10	1.42	1.32	3.08	2.98	1.43	1.33	10
12	1.49	1.39	3.48	3.38	1.50	1.40	12
14	1.55	1.45	3.87	3.77	1.57	1.47	14
16	1.62	1.52	4.27	4.17	1.63	1.53	16
20	1.75	1.65	5.06	4.96	1.77	1.67	20
24	1.87	1.77	–	–	1.90	1.80	24
30	1.71	1.61	–	–	–	–	30
40	1.91	1.81	–	–	–	–	40
50	2.12	2.02	–	–	–	–	50
60	2.32	2.22	–	–	–	–	60
80	2.72	2.62	–	–	–	–	80
100	3.13	3.03	–	–	–	–	100
120	3.54	3.44	–	–	–	–	120
150	4.14	4.04	–	–	–	–	150
200	5.16	5.06	–	–	–	–	200
250	6.19	6.09	–	–	–	–	250
300	7.19	7.09	–	–	–	–	300
400	9.22	9.12	–	–	–	–	400

*At 68°F.
**From *Blasters Handbook, Fifteenth Edition,* published by E.I. Du Pont de Nemours & Co., Inc., Explosives Department.

puter analysis be made of parallel circuit requirements. Caps from different manufacturers may have different characteristics and will likely lead to misfires if used in the same blast. In straight parallel hookups, the hookups fire progressively, beginning nearest the source. If instantaneous caps are used for the cut holes, they should be attached at the far end of the bus wires so that their detonation will not disrupt the circuit before the delays have been activated.

The resistance of blasting caps and copper wire are shown in Tables 7-17 and 7-18.

The total resistance of a series circuit is equal to the combined resistance of the number of caps in the series plus the resistance of the wires leading from the blasting machine or power line switch. The total resistance of parallel circuits is equal to the resistance of a single cap divided by the number of caps in the circuit plus the lead wire resistance. The calculated voltage

Table 7-18. Resistance* of copper and aluminum wire.**

AWG GAUGE NO.	OHMS PER 1000 FEET	
	COPPER	ALUMINUM
8	0.628	1.03
10	0.999	1.64
12	1.588	2.61
14	2.525	4.14
16	4.02	6.59
18	6.39	10.50
20	10.15	16.70
22	16.14	26.50

*At 68°F.
**From *Blasters Handbook, Fifteenth Edition,* published by E.I. Du Pont de Nemours & Co., Inc., Explosives Department.

drop should not exceed 90% of the available voltage. The resistance of a circuit can be tested after everything is ready for the blast by touching the leading wires to the terminals of a blasting galvanometer. This test will also indicate short circuits or breaks in the line.[16]

It is general practice to have the blasting switch locked in open position with one man, usually the shifter, having the only key. He is responsible for seeing that all men leave the face and get to a safe place before the switch is unlocked and the blast fired. For blasting purposes, power may be drawn from transformers or generators up to four or five times the rated capacity. Firing lines should be well away from other power and light lines and service pipes and, if possible, on the opposite side of the tunnel. The portable leading lines should be fully insulated and extend only to a point reasonably safe from the blast and should not be smaller than No. 14 gauge. Connecting wires beyond this point should be at least No. 20 gauge and used only once.

While drill holes are being loaded, there should be no lighting circuits near the face except for a floodlight served with well-insulated cord.

Power circuit firing has been the practice because of former limitations in the capacity of blasting machines. Larger machines are now available. Du Pont, one of the leading manufacturers of blasting equipment, has the following to say about their new blasting machines, placed on the market in 1971.

"Parallel wired Delay E.B. cap circuits have been used for many years by tunnel contractors because of the speed and simplicity with which the caps can be connected to bus wires. Power lines were used to fire the parallel rounds because power was readily available and the rounds exceeded the parallel firing capability of conventional blasting machines.

"The probability for arcing caused by excessive energy from power lines was significantly reduced by the use of timing switches (e.g., Du Pont Arcontroller), but the hazard of misfires and hangfires was not eliminated, particularly for small rounds and pop-shooting. Another drawback of power firing is the high cost of large transformers, switch gear and the need to have electricians on hand to operate and maintain the high voltage system.

"Because tunnel diameters and length of bore are ever increasing, it was evident to Du Pont that there was a need for developing high energy condenser discharge blasting machines capable of reliably firing from one to 200 delay electric blasting caps in parallel with no danger of arcing. In January, 1971 the manually operated XSS-2003 model condenser discharge blasting machine having a rechargeable battery feature and the remote controlled XSS-2002 model condenser discharge blasting machine were made available to the tunnel construction industry on a monthly fee lease agreement. These machines are not for sale because of the specialized nature of their use. The manufacture of the circuit breakers used in Atlas Arcmasters and Du Pont Arcontrollers has been discontinued, leaving the industry without an alternate source of timers for power firing.

"All internal high voltage components are potted in a waterproof plastic case for maintenance-free service in moist atmospheres. Special recessed firing terminals and recessed operating controls as well as plastic case construction protect the machine operator against electric shock hazards. A minimum firing voltage interlock unit incorporated in the design of the new machine, in combination with highly reliable capacitors, assures full voltage output.

"A Du Pont computer analysis of the parallel bus wire circuit is mandatory before installation of a Du Pont XSS-2000 model CD blasting machine in order to establish the most reliable and efficient blasting systems at the start of the project.

"With Du Pont XSS-2002 and XSS 2003 CD blasting machines, firing stations can be readily moved at 1000 or 2000 foot increments in conformance with federal, state and local regulations without additional lead line costs at extended distances from the shaft or entry. There is also no need to provide a large primary transformer, blasting switches, Arcontrollers, extensive power distribution systems along with step-up step-down transformers, or employ an electrical crew to primarily maintain, test and install fixed distribution cables for the blasting circuit."

BLASTING HAZARDS AND LIMITATIONS

Damage to surrounding structures may be caused by air blast or acceleration of the rock beneath a structure. The time of a shock, and its acceleration, frequency and amplitude all have an effect on damage. It has been found that acceleration is not well adapted as a criterion for blast damage from explosive shocks. Accelerations in excess of $0.1g$ (where g is acceleration due to gravity) caused by earthquakes can result in severe damage because such shocks may last for several minutes. Shock caused by an explosion is over almost instantaneously, with the result that structures can stand a much higher acceleration without damage. It has been determined from the results of many experimental blasts that no damage is to be expected unless the acceleration exceeds g. However, caution should be exercised if the acceleration is between $0.1g$ and $1g$.

Crandall[17] has proposed the use of a factor called energy ratio to measure the limits for safe blasting without damage to structures. It is defined as

$$ER = a^2/f^2$$
$$= 16\pi^4 f^2 A^2$$

where

ER = energy ratio
a = acceleration (feet/second)^2
f = frequency (cycles/second)
A = ground displacement or amplitude of the shock wave (feet).

It has been found that energy ratios below 3 are safe for structures, and that those in excess of 6 are hazardous. Massachusetts and New Jersey limit ER to 1 down to frequencies of 10 cycles/second. These limits are illustrated in Fig. 7-17.

Frequencies vary from 3 to 30 cycles/second, but are seldom greater than 10 to 15 cycles/second and travel at velocities ranging from 1000 feet/second in dry sand to 10,000 feet/second in granite. Lower frequencies occur at lower velocities and vice versa. Damage is caused by differential motion. Wavelength varies but is roughly 300 feet. Smaller structures will have less differential motion and will be damaged less than larger structures. Ground amplitude can be approximated by using a frequency of 15 cycles/second. The U.S. Bureau of Mines[18] has developed the relationship between ground amplitude, distance and weight of explosive shown in Table 7-19.

ER has been empirically related to distance and amount of explosives[19] by the equation

$$ER = \left(\frac{50}{D}\right)^2 C^2 K$$

where

C = explosive charge in (lb)
D = distance from blast (feet)
K = constant (see Table 7-20).

From this equation, the safe quantity of explosives may be computed. Each delay in a round should be considered a separate blast. The different delays all produce their own individual shock. They are not compounded. The amount of explosive in each delay of each round is limited by the size of the tunnel. It will usually be found that it is impossible to load sufficient explosives fired in sequence, as required for tunnel construction, to damage structures on the surface a hundred feet away. In a 13-foot diameter sewer tunnel in Pittsburgh, the vibrations from blasting were generally less than those from traffic and nearby industry.

Instrumentation

Where damage might occur, measurements of acceleration should be made. This can be accomplished with seismographs but it is more usual to use calibrated pins. The following instruments for testing for air quality are available.

- Carbon monoxide tester, to determine presence of carbon monoxide (CO).
- Carbon monoxide detector, to determine concentration of CO.
- Carbon monoxide indicator, to measure the concentration of CO from 0 to 500 ppm.
- Explosimeters and combustible gas indicators.

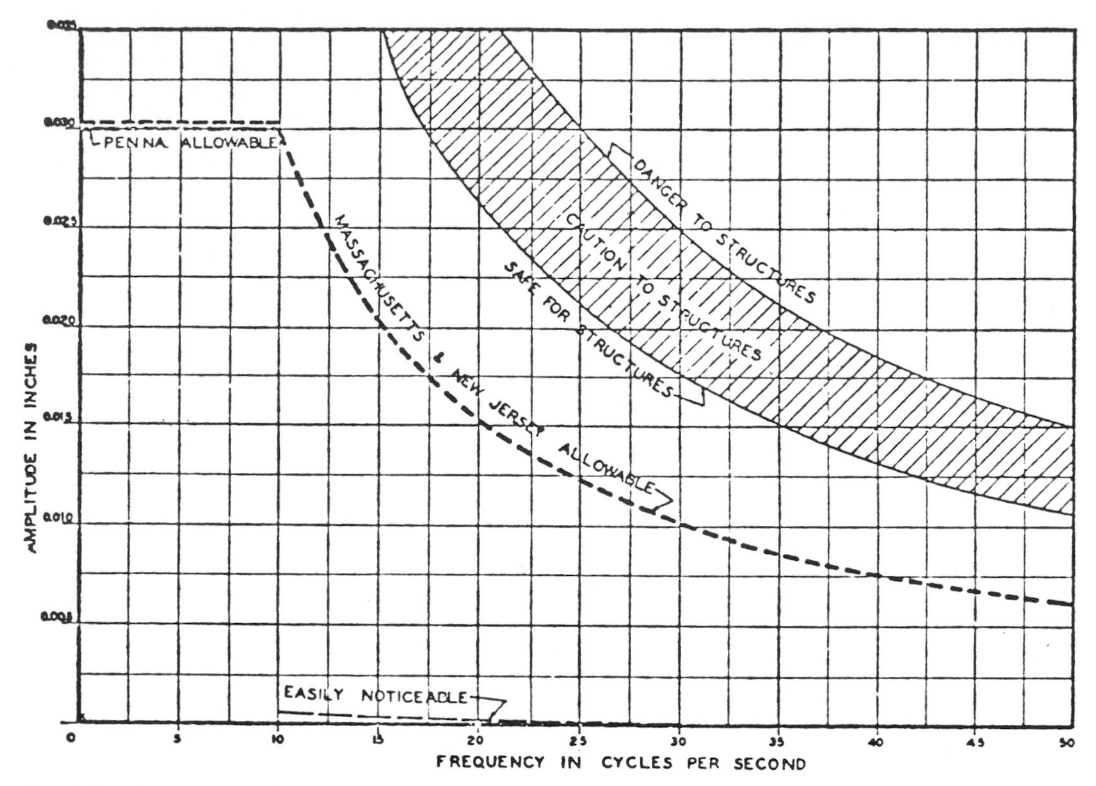

Fig. 7-17. "Energy ratios" related to frequency and amplitude. The line marked "Danger to structures" represents an energy ratio of 6; that marked "Safe for structures," and energy ratio of 3; the dotted line marked "Massachusetts and New Jersey allowable" represents an energy ratio of 1 down to a frequency of 10 c/sec. (L. Don Leet, Vibrations from Blasting Rock, Harvard University Press, 1960).

Table 7-19. Ground amplitude-distance-weight of explosive (amplitude in inches) for normal overburden.*

WEIGHT OF EXPLOSIVE (LB)	DISTANCE (FEET)										
	500	600	700	800	900	1000	2000	3000	4000	5000	6000
1,000	.035	.031	.027	.023	.020	.018	.0050	.0020	.0012	.0010	.0010
2,000	.056	.049	.044	.037	.032	.028	.0080	.0032	.0019	.0016	.0016
3,000	.073	.064	.057	.048	.042	.037	.010	.0042	.0025	.0021	.0021
4,000	.089	.078	.069	.059	.051	.045	.013	.0051	.0030	.0025	.0025
5,000	.100	.090	.080	.068	.059	.052	.015	.0058	.0035	.0029	.0029
6,000	.120	.100	.090	.076	.067	.058	.016	.0066	.0040	.0033	.0033
7,000	.130	.110	.100	.085	.074	.065	.018	.0073	.0044	.0036	.0036
8,000	.140	.120	.110	.093	.081	.071	.020	.0080	.0048	.0040	.0040
9,000	.150	.130	.120	.100	.088	.076	.022	.0086	.0052	.0043	.0043
10,000	.160	.140	.130	.110	.094	.082	.023	.0093	.0056	.0046	.0046

For rock or very thin overburden, divide the above amplitudes by 10.
For abnormal overburden (thicker than a half wave length), multiply by 3.

*From *U.S. Bureau of Mines Bulletin* 442, p. 66.

Table 7-20. Table of derived values* of the soil constant S from seismic velocities in various subsurface materials. (K = ground transmission coefficient.)**

	V (FEET/SECOND)		S (LB/INCH^2)		K	
SOIL TYPE	MIN	MAX	MIN	MAX	MIN	MAX
Top soil:						
light dry	600	900	262	590	.00382	.0017
moist, loamy silty	1,000	1,300	812	1,370	.00123	.00073
clayey	1,300	2,000	1,420	3,370	.000707	.000297
semiconsolidated sandy clay	1,250	2,150	1,510	4,150	.000662	.000241
Wet loam		2,500		5,600		.000179
Clay, dense wet, depending on depth	3,000	5,900	8,850	34,100	.000113	.0000294
Rubble or gravel	1,970	2,600	6,400	11,100	.000156	.00009
Cemented sand	2,800	3,200	9,700	12,600	.000103	.0000793
Sand clay	3,200	3,800	10,000	13,900	.0001	.000072
Cemented sand clay	3,800	4,200	17,800	21,700	.000056	.000046
Water-saturated sand		4,600		22,500		.0000444
Sand	4,600	8,400	26,200	87,000	.0000382	.0000115
Clay, clayey sandstone		5,900		45,000		.0000222
Loose rock talus	1,250	2,500	1,750	7,000	.00057	.000143
Weather fractured:						
rock	1,500	10,000	3,100	140,000	.000323	.00000715
shale	7,000	11,000	63,000	156,000	.0000159	.0000064
sandstone	4,250	9,000	23,500	116,000	.0000435	.0000086
Granite slightly seamed		10,500		160,000		.00000625
Limestone, massive	16,400	20,200	390,000	590,000	.00000257	.0000017

*Values derived from relationship $S = \frac{1}{2} eV^2$.
**From F.J. Crandall, "Transmission Coefficient for Ground Vibrations Due to Blasting," *Journal of the Boston Society of Civil Engineers* XLVII, *No. 2*, April 1960.

- Combustible gas alarm, with audiovisual alarm signals.
- Combustible gas and oxygen indicator, to measure percent oxygen and explosive gases.
- Oxygen indicators, to determine percent of oxygen.
- Toxic gas detectors.
- Dust instruments.
- Smoke instruments.

Automatic gas monitors, continuously recording dangerous gas, are available and are sometimes connected to audible and visual warning devices that will give an alarm when the concentration of gas reaches a predetermined percentage of the lower explosive limit. They can also be arranged to automatically shut off power to all equipment if the percentage of gas reaches another predetermined limit.

Protection Against Damage Claims

Because residents in the vicinity of tunneling operations can sometimes hear a dull rumble or feel a slight vibration when a round is blasted, they frequently start looking for damage and, upon finding cracks in the plaster, make a claim against the tunnel constructor. In most cases, it will be found that the cracks were already in existence before the blast. It is good practice to make a survey, with plenty of photographs and written records, of all property along the proposed tunnel line before tunneling operations start. This will discourage claims and in most cases give incontrovertible proof that the claimed damage existed prior to the tunnel construction. The Broadway Tunnel in San Francisco was constructed underneath a street with houses overhead from portal to portal. This tunnel was in an altered serpentine formation. The pilot tunnels were excavated by drilling and blasting and caused practically no damage to the structures above.

Hazards of Premature Explosions

Electric blasting caps are used almost exclusively in modern tunneling. To prevent a premature explosion, it is important that no electric current from any source be permitted to enter the wires leading to the explosives prior to the planned time of the blast. Extraneous electricity can come from lightning storms, static radio inductive fields, stray currents, galvanic action, high tension power lines and electric train propulsion.

Du Pont warns users of their electric blasting caps of the possibility that dangerous accidental ignition and injury can occur if they are handled or exposed "in the vicinity of lightning storms, electrified clouds, sand or snow storms [or] near sources or conductors of stray or galvanic voltages capable of delivering 0.05 ampere or more through a one-ohm resistance (test with Du Pont Blaster's Multimeter)." Du Pont suggests: "Keep the ends of the electric blasting cap circuit insulated from ground and from sources of stray currents until they are connected to the power source. Never connect the cap circuit to the power source until after loading is complete and personnel are safely cleared from all areas of danger and the circuit is ready to be fired."

Du Pont also warns of possible accidental ignition if the caps are handled "in the vicinity of electrostatic or magnetic fields such as high voltage transmission lines, capable of inducing 0.05 ampere or more in one-ohm cap leg wire circuits;

"inside minimum safe distances from radio frequency transmitters (radio, television, radar, etc.) listed by the Institute of Makers of Explosives in Pamphlet No. 20;

"near generated sources of static electricity such as improperly grounded pneumatic loading machines (see Du Pont pamphlet "Static Electricity Hazards and Pneumatic Loading Systems"), automotive vehicles, etc.;

"to physical abuse or abnormal shock or vibration or to temperatures in excess of 150°F."

Lightning strikes will detonate a blasting circuit. A near miss can also cause detonation. The lightning charge can be carried several miles by induced currents. No satisfactory method has been developed to render blasting caps safe against lightning. In the open, blasting operations should be suspended if a lightning storm approaches within 5 miles. This can be estimated by observing the time interval between the lightning and thunder. This precaution is applicable to the first rounds near the portal and, in some cases, underground if differences in potential in the ground result from lightning strikes.

Static electricity is generated by dust storms. Electric blasting caps have exploded from this source when the leg wires of the caps were suspended in the air. Static electricity generated by compressor belts have caused premature explosion of ordinary caps and escaping steam has generated static and caused the detonation of electric blasting caps. Du Pont[16] makes the following recommendations.

"Recommended precautions to be taken where static electricity can be generated mechanically include: (1) all parts of moving equipment in the vicinity of the blasting operations should be electrically connected together at a common point, and this point should be connected to a good earth ground rod, (2) all conductors and metal parts of such a system should be kept away from blasting caps and blasting circuit wires or otherwise electrically insulated, (3) the ground wires and earth ground rod for such a system should be kept away from rails, other wiring and piping which might conduct stray electric currents from these sources to the blasting site, (4) all moving equipment should be shut down while blasting circuits are being connected and until the blast has been fired."

The Nonel shock tube, developed recently in Sweden, to transmit the impulse to explode blasting caps, is a non-electrical system that eliminates all electrical hazards of a premature explosion except, possibly, direct lightning strikes. This product consists of a plastic tube $\frac{1}{8}$ inch in diameter with a thin coating of reactive material on the inside surface which detonates at 6000 feet/second. The detonation is mild and the outer surface of the tube remains intact during and after functioning.

Pneumatic loading systems are sometimes used to load ammonium nitrate-fuel mixtures. Static electricity can be generated when the explosive is blown in over electric blasting cap leg wires.

Radio transmitters may cause induced currents to flow in blasting circuits, causing electric blasting caps to detonate. Such induced currents may be caused by radio broadcasting stations, automobile radio telephones, television or radar. Electric blasting within one mile of a broadcasting or high power short wave transmitting station or within one-quarter mile of other radio transmitters is a potential hazard. It is recommended[20] that within these distances the blasting circuit should be laid out with a radio pilot lamp inserted in place of the cap to check for induced currents. "Primacord" and regular caps and fuse should be used instead of electric blasting caps if any glow appears in the lamp.

Stray currents may be found in electric cables, conductors such as rails, ground returns, pipes and ventilating ducts. These have caused premature explosions. Electric haulage, either in the tunnel or above, is a principal source of stray currents. The resulting hazard can be reduced by connecting all electrical equipment to low-resistance grounds, preferably not over one ohm. Cross-bonding of utilities, including rails, pipes, ventilating ducts and armored cables should be done at frequent intervals. The rails should be bonded and grounded. Power lines should be kept in good repair and back from the face during loading operations. Lead wires should be insulated from the ground and blasting circuits separated from possible conductors. Occasional tests should be made to show whether or not stray currents are present at the face.

Misfires

Occasionally, one or more holes in a blasting circuit will not detonate when the circuit is fired. This is called a misfire or missed hole and working near it is the most hazardous operation associated with blasting. Before returning to the heading, the lead wires should be disconnected from the source of power. If possible, the leg wires should be checked with a galvanometer and, if found usable, the misfire should be fired in the usual manner. If this fails to detonate the charge, a new primer should be used. It is frequently effective if exploded on top of a foot or so of stemming. Metal tools should not be used for any digging necessary to place the new primer close enough to explode the charge. Stemming may be blown or worked out of the hole using a stiff rubber hose. If other means fail, it may be necessary to drill holes around the missed hole to be fired to expose or detonate the unexploded powder. If this operation is required, great care must be exercised in drilling to avoid intersecting the original hole.

Gas

The explosion or burning of dynamite or ammonium nitrate produces noxious gases, including oxides of nitrogen (NO, NO_2) that must be reduced to harmless levels before men can return to the heading. Dynamites are rated according to their fume characteristics, as shown in Tables 7-16. Only those with the better characteristics should be used underground in order to reduce the time and ventilating air necessary to remove or dilute the gases to safe levels. Explosive gases or gas-air mixtures, particularly methane, are frequently encountered underground.[21] Poisonous gases, such as carbon monoxide, are encountered particularly after a fire. The best way to handle such situations is to provide ample ventilation.

Time Limits for Blasting

Explosives have varying resistance to water before deteriorating. In wet tunnels, the blast should be fired as soon after loading as practicable. Because of the hazard of premature explosion, no hole should be left loaded any longer than necessary. Blasting may be limited by public authorities to specific hours.

MUCKING

Excavated material in a tunnel is known as muck and the loading of it into vehicles for transportation is known as mucking.

Scaling of Loose Rocks

After a round has been blasted, some of the rock that has been broken loose will hang in a precarious position, endangering the safety of the crew. On returning to the face after a blast, the walls and roof should be washed down to permit careful inspection of the rock surface. The first operation is to scale down any loose rock in the tunnel roof or sides that might cause injury to personnel. This usually takes only a few minutes and is accomplished while equipment and services are being brought forward and prepared for mucking.

Mechanical Loaders

The largest capacity mucking machine that can operate in the tunnel without restricting motion will usually produce the best results. In tunnels using rail for haulage, mucking machines are usually of two types, rail-mounted with a bucket arranged 1) to scoop up the muck and cast is back over the top of the machine directly into cars, such as the Eimco (Fig. 7-18), or 2) to scoop up the muck, tilt the bucket, and slide it onto a conveyor belt, such as the Conway (Fig. 7-19), which not only discharges into the cars but also acts as a minor storage reservoir while cleaning up. In very small tunnels, it is usually more economical to use an air motor for power. If all other equipment, such as pumps, is also operated by air, it is unnecessary to extend a power cable into the tunnel, thereby saving its cost. For larger tunnels, the cost of air for power becomes too great and is usually replaced by electrically driven equipment. In tunnels using rubber-tired hauling equipment, the mucker can be a caterpillar tractor-mounted or rubber-tired loader. In larger tunnels, revolving shovels are sometimes used, either powered electrically or with diesel engines where the tunnel is wide enough to permit the truck to stand beside the shovel.

Fig. 7-18. Eimco rocker shovel.

Fig. 7-19. Convey mucker.

In larger tunnels, rubber-tired front end loaders are sometimes used either as conventionally built or (preferably) with the bucket arranged so it can discharge into a truck standing alongside.

Haulage of Tunnel Muck

For haulage of muck by rail or rubber-tired vehicles, see Chapter 11.

VENTILATION

Tunnels must be constantly supplied with fresh air to provide workers an adequate amount of quality air, to remove smoke, dust and fumes from the heading, to cool hot tunnels, to dilute natural gases and to remove the exhaust of diesel engines. A tunnel may not be safe to enter even when no operations are in progress because of carbon dioxide from decaying timber or methane gas that may exude from the tunnel walls.

The ventilation system is designed to bring fresh air to the tunnel face through a pipe called the fan line, or to draw it through the tunnel by exhausting through the fan line. A blower can be made reversible so that air can be either blown into the tunnel through the line or sucked out through the line. The most common system in use at present consists of light guage (20 to 26) spirally-crimped pipe with axivane type of in-line electric fans. These have better efficiency when blowing in one direction, so they are usually installed to exhaust air from the face through the vent pipe. It is undesirable to blow the air into the heading because it forces the smoke and noxious gases from the blast to travel the full length of the tunnel, creating a hazard to the men working away from the heading or passing through the contaminated air. For short lines or blowing air from the end of the main ventilation duct to the face, "Ventube," a flexible collapsible tube, made by impregnating selected fabrics with a high grade rubber, is sometimes used. The end of the ventilation line should be kept generally within 40 feet of the face.

No one should be allowed to return to the face until the smoke, dust and fumes from the explosion have been expelled. Even explosives with the best fume characteristics produce toxic gases that may be dangerous to breathe. Spraying the muck pile will reduce dust, dissolve soluble toxic gases and can displace nonsoluble gases in the muck pile to permit their earlier removal by the ventilation system.

The most commonly encountered dangerous natural gas found in tunnels is methane (CH_4). It is odorless. Air mixtures with 5 to 15% CH_4 by volume are flammable and will explode in confined areas. Methane, being lighter than air, tends to gather in pockets in the roof. The gas is most commonly found in coal seams, shales and in regions where oil and natural gas is found. Detection can be made by several methods.[22] Flame safety lamps, designed to indicate the percentage of gas, have been commonly used. Regulations now require the use of electrically operated methane detectors. A spark from equipment or from striking a steel tool on rock can cause an explosion. When such conditions are encountered, it is necessary to use "permissible" equipment designed to prevent sparking. Permissible dynamite is also available where required. In order to prevent concentration of methane above a safe level, it is necessary to provide ample ventilation with small blow pipes or exhaust ducts, directed into pockets where the lighter methane gas tends to concentrate. Concentration of methane gas should be kept to not over 1%. If it exceeds 1.5%, the men should be withdrawn from the areas of excess concentration.

Other mine gases and their characteristics are shown in Table 7-21, (from the U.S. Bureau of Mines).[22]

Fans must be operated continuously while men are working underground. If they are not operated when the tunnel is idle, they must be turned on sufficiently in advance of men entering the tunnel to clear it of contaminated air. Atmospheres in all active areas of tunnels should contain at least 19.5% oxygen, and not more than 0.005% carbon monoxide, 0.5% carbon dioxide, 5 ppm nitrogen dioxide. There should be no harmful quantities of other toxic gases, fumes, mists or dusts. Instruments should be available for testing the quality of air and quantity of pollutants.

Design of Ventilation System

For design and construction of a ventilation system, see Chapter 11.

POWER SUPPLY AND LIGHTING

Power Supply

Where electrically powered equipment is used inside the tunnel, it usually operates at 440 volts. The equipment is served by a power line running the length of the tunnel. This may operate at 440 volts for a short tunnel. In longer tunnels, it is economical to take the power into the tunnel at voltages of 2300 or more. These higher voltages are usually carried in metal clad insulated cables strung on hangers dowelled into the walls. Dry-type transformers reduce voltage to that required by the equipment. The

Table 7-21. Summary of important characteristics of mine gases.

GAS	CHEMICAL FORMULA	MOLECULAR WEIGHT	WEIGHT LB/FOOT^3 AT 70° F AND 29.92" MERCURY PRESSURE	SPECIFIC GRAVITY (AIR = 1)	EXPLOSIVE OR FLAMMABLE RANGE IN AIR (PERCENT BY VOLUME)	OXYGEN PERCENTAGE BELOW WHICH NO MIXTURE IS FLAMMABLE		COLOR	ODOR	TASTE	DANGEROUS TO BREATHE
						NITROGEN AS DILUENT OF AIR	CARBON DIOXIDE AS DILUENT OF AIR				
Air			0.075	1.0000	Nonexplosive[a]			None	None	None	No
Oxygen	O_2	32.00	0.083	1.1054	do[a]			do	do	do	No[f]
Carbon dioxide	CO_2	44.00	0.115	1.5291	do			do	do	[e]	Yes[g]
Methane	CH_4	16.03	0.042	0.5545	5 to 15	12.1	14.6	do	do	None	No
Carbon monoxide	CO	28.00	0.073	0.9672	12.5 to 74	5.	5.9	do	do	do	Yes[h]
Hydrogen sulfide	H_2S	34.09	0.089	1.1906	4.3 to 45			do	Yes	Yes	Yes[h]
Ethane	C_2H_6	30.05	0.079	1.0493	3.0 to 12.5	11.0	13.4	do	None	None	No
Propane	C_3H_8	44.09	0.117	1.5625	2.2 to 9.5	11.4	14.3	do	Slight[c]	do	No[i]
Butane	C_4H_{10}	58.12	0.151	2.0100	1.9 to 8.5	12.1	14.5	do	do[d]	do	No[i]
Nitrogen dioxide	NO_2	46.01	0.119	1.5894	Nonexplosive			Red[b]	Yes	Yes	Yes[h]
Sulfur dioxide	SO_2	64.07	0.170	2.2638	do			None	Yes	Yes	Yes[h]
Hydrogen	H_2	2.016	0.0052	0.0695	4.0 to 75	5.0	5.9	do	None	None	No
Nitrogen	N_2	28.02	0.073	0.9674	Nonexplosive			do	do	do	No

[a] Supports combustion.
[b] In fairly high concentration; lower concentrations, which are not visible, may be dangerous.
[c] Above 2 percent.
[d] Above 0.5 percent.
[e] Acid or biting taste in high concentration.
[f] Not dangerous at ordinary pressures; at pressures above 1 atmosphere oxygen poisioning may occur.
[g] In fairly high concentrations (5 percent or more).
[h] Extremely dangerous even in very low concentration.
[i] May produce symptoms of drowsiness or dizziness.

transformers are moved ahead as the face advances, being kept far enough back to be safe from damage by fly rock from the blasting and close enough to avoid an excessive drop in voltage; 440-volt power may be supplied through insulated wires attached to insulators dowelled into the tunnel walls. Power is supplied either from commercial sources or from job-site generating plants. General practice is to use alternating current.

Emergency Power

Tunnels driven from shafts or down grade, where flooding could be serious to life or property, require emergency power from an alternate source, another utility company and transmission line, or on-site standby generators. The alternate source need be sufficient to supply lighting, pumps and hoists only. Elsewhere, batteries can be used to supply a minimum of lighting to permit evacuation of the tunnel. Otherwise, an adequate supply of flashlights should be provided.

Construction Lighting

Circuits for lighting are required in addition to power circuits. Adequate lighting to permit walking the entire length of the tunnel in safety should be provided. At the heading, flood lights should be sufficient to provide safe and efficient conditions.

Shooting Lines

Separate circuits, preferably at 220 volts, should be used for detonating explosives. These should be well separated from other power lines.

Safety Requirements

Electric circuits should be installed and hooked up to equipment by properly trained personnel, with proper safety installations to avoid stray currents or other conditions that might cause premature blasts of explosives. Safety rules have been published.[23]

For further details of design and construction of power supply, see Chapter 11.

EMERGENCY PROVISIONS

Tunneling is one of the most hazardous occupations that man undertakes. It is necessary to have equipment available to meet any contingency.

Personnel

Government regulations require that a first aid attendant, qualified according to the U.S. Bureau of Mines or American Red Cross Standards, must be available at a first aid station on the project during all work hours.

First Aid Equipment

In addition to the usual first aid kits, stretchers, etc., it is necessary to provide resuscitation equipment where oxygen deficient air or carbon monoxide causes asphyxiation or poisoning. The U.S. Bureau of Mines recommends the use of an oxygen-inhalator in conjunction with manual artificial respiration treatment. The simplest device for mouth-to-mouth rescue breathing without actual physical contact consists of a tube to fit in the rescuer's mouth connected to a flexible mask which fits over the victim's mouth. The simplest oxygen inhalator consists of an oxygen cylinder with valves connected through a tube to a mouth mask. Portable resuscitators delivering preset alternating positive and negative pressures to victims of respiratory failure are convertible to constant flow inhalators. Carbon monoxide poisoning test kits will determine the presence and percent of carbon monoxide in a patient's blood.

Fire Protection

Fire underground is less likely today in tunneling because less timber is used for support. However, fire extinguishers should be available near the heading, particularly when flammable gases or combustible substances are present. Water is always available from the line supplying drills. Outlets and fire hoses should be located near possible fire hazards.

Rescue Equipment

After fires or explosions, the presence of poisonous or asphyxiating gases or lack of oxygen necessitates use by individuals or rescue crews of equipment providing protection against carbon monoxide and other poison gases, or to supply oxygen. Three types are available:[24]

1. The carbon monoxide self-rescue respirator;
2. Gas masks for protection against gases, smokes and vapors in air that contains enough oxygen to support life, with canisters; and
3. Self-contained breathing apparatus.

Self-rescuers can be carried on a man's belt but are only suitable for self-rescue in atmospheres that are not deficient in oxygen. They are tested to give protection in air containing 1% CO for a minimum of 30 minutes. Gas masks are designed to filter out gases through cannisters (which can be replaced), giving protection for about two hours. Timers on the cannisters indicate the time of use. There are four types of self-contained breathing apparatus:

1. Compressed oxygen, using a closed circuit (this is the most satisfactory type);
2. Liquid oxygen (not widely used);
3. Self-generating oxygen (cannister provides for 45 minutes of hard work); and
4. Compressed air (air tank lasts only about one-half hour; this type is not recommended for tunnels except as auxiliary equipment).

TEMPORARY TUNNEL SUPPORTS

General

Because most rocks have joints that are either natural or induced by blasting, there are few tunnels that do not require some support in at least part of their length. It is only in the best of rocks, with an RQD of 90% or more, that no support is required. Supports are needed to protect the tunnel workers from rock falls and popping rock, to limit overbreak and to prevent the blocking of the tunnel by fallouts or in squeezing or swelling ground.

Because of the time lag between the placing of temporary supports and the placing of permanent tunnel lining, it is usually necessary to design the temporary supports to take the entire rock load. It has been customary to also design the permanent lining to take the entire rock load. This results in a duplication of support. The tunnel designer should consider the use of temporary tunnel support that can also be used either as permanent support or as an integral part of the permanent supporting system.

Timber Supports

Timber was used exclusively where supports were required in earlier tunnels. At the present time, it is rarely used except for lagging and blocking or in emergencies where there is not time to obtain steel ribs or where the length of tunnel to be supported is minimal. The simplest form of timber support is the square set, as shown in Fig. 7-20(a). For larger tunnels,

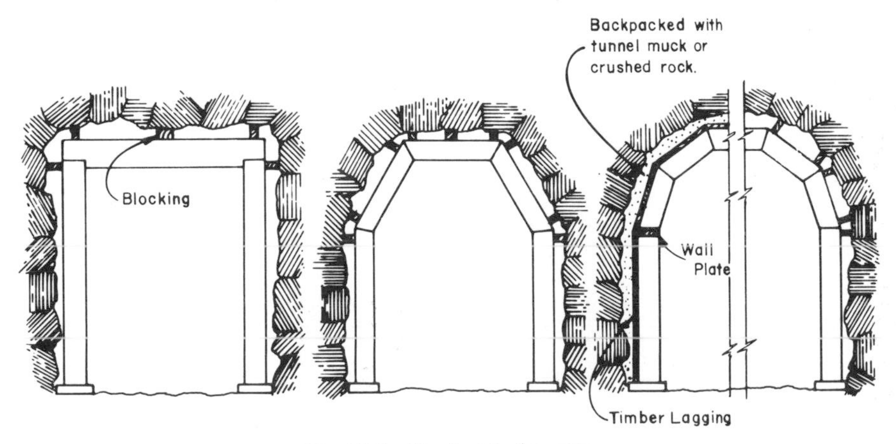

Fig. 7-20. Typical timber sets.

segmental sets are required. These may contain three or more segments, as shown in Figs. 7-20(b), (c) and (d), or may be segmented around the entire periphery in bad ground.

Timber sets must be blocked tightly at all corners with wooden blocks and wedges and at other places where blocks of rock could become loose and drop. In badly fractured rock, the sets must be lagged and back-packed with crushed rock, sand or tunnel muck. Collar braces between sets, consisting of timbers parallel to the tunnel, must be installed to resist forces longitudinal to the tunnel and to prevent sets from leaning. Timber sets are rarely made of less than 8 X 8 timbers in the smallest tunnels. In the larger tunnels or in very bad ground, the sets would be of 12 X 12 or larger timbers. In the 10-foot, 6-inch diameter Coast Range tunnels of the Hetch Hetchy Project in California, 14 X 16 timbers were crushed like matchwood in squeezing ground. In some of the older large tunnels, such as those that later became part of the Pennsylvania Turnpike, oak timbers were used. The usual timber now used is pine or fir.

Tunnels requiring support, and too large or in too difficult ground to be excavated full face, will require wall plates at the springline on which to install the arch segments before the lower part of the tunnel is excavated and posts installed, as shown in Fig. 7-20(c). In many railroad tunnels, the timber sets that have been used for permanent support are still in place (Fig. 7-21). However, the danger of fire is always present, and many tunnels have been lost or seriously damaged by fire. Most new railroad tunnels requiring support are being lined with concrete. Where concrete lining must be used, as in aqueduct tunnels, it must be designed to take the full rock load, because in time the timber will rot away.

Posts must be set on footing blocks to spread the load on loose tunnel muck. Allowance must be made for some settlement of the timber sets by placing them slightly higher than their ultimate required position.

Steel Ribs

With the increasing scarcity and cost of large timbers, their use as tunnel support has been replaced largely by steel ribs made of structural beams. H-beams or wide flange beams rolled to fit the shape of the tunnel are preferable to I-beams, as they are easier to block and lag. Steel is ideally suited for this purpose, as it can be bent to any shape, and the fewer pieces to be handled take less time to install. It has been the practice in the past to design the steel

Fig. 7-21. Timber sets being replaced by steel sets.

support to take the entire rock load and, where concrete lining is later installed, to also design it to take the full load. The steel has been permitted to encroach within limits into the concrete section but without taking any credit for its strength. Because the steel can be adequately protected by the concrete lining and grout, to last the useful life of the tunnel, its strength should be incorporated into the design of the permanent lining, with the thickness of concrete being only sufficient to protect the steel from corrosion.

Steel rib support can be fabricated in many different ways. Rock tunnel steel support systems are roughly of six types: continuous rib with or without invert strut, rib and post, rib and wall plate, rib wall plate and post, full circle rib and liner plate. Sometimes, the full circle rib is used together with liner plate, but this is usually in soft ground tunnels so it will not be discussed here.

In making the choice of a support system, the method of attacking the excavation, the rock behavior and the size and shape of the tunnel cross-section must be considered. Since most tunnels in the better rock qualities are now driven full face, the wall plate systems would not be used. The continuous rib type should be used wherever possible. In tunnels with roofs joining the side walls at an angle, the rib and post type would be indicated. This type is commonly used in large, double-track railroad and two-lane highway tunnels. In tunnels with light loading and high sides or circular sections, the rib and wall plate type with the wall plates supported on hitches (Fig. 7-22) or on pins drilled into the side wall might be used to save the expense of the posts. This type is also adapted to areas where spalling rock occurs only in the roof. However, this type of support for the wall plate is often difficult to accomplish and it may be necessary to carry posts to the floor.

Sometimes the designer will specify this type of support with non-uniform posts going down to wherever the overbreak will permit a footing. This is impractical and will result in excessive costs. Therefore, this type of support should rarely be used and only in very special cases. On large tunnels with bad rock conditions, or where the tunnel is not excavated full face, the rib, wall plate and post type would ordinarily be used. In squeezing or swelling rock or where heavy side pressures are encountered, the full circle rib type would usually be used. The continuous rib type may have vertical, sloping or curved sides, as in a horseshoe shape. The rib should be formed of only two pieces where they can be so handled in order to achieve the cheapest erected cost. Two section ribs have been installed in horseshoe-shaped tunnels of 15 X 15 feet excavated size where light support was required. Three section ribs have been installed in an 18-foot, 10-inch high by 15-foot, 4-inch horseshoe-shaped tunnel. The more usual types are shown in Figs. 7-23 to 7-26.

Rib posts must be set on footing blocks, usually 3 X 12 or 4 X 12 square blocks, with allowance made for some settlement. Collar braces or spreaders are usually used at 4- to 6-foot spacing. These may be of wood (Fig. 7-27) placed next to tie rods to tie the two adjacent ribs together. The collar braces are removed before placing concrete lining. An alternate construction is to use tie rods inside of $1\frac{1}{2}$-inch pipe spreaders. Where the rock must be supported between ribs, lagging must be used. This may be timber planks or steel such as channels or liner plates in very heavy ground.

Rib Design

Ribs must be designed to support the anticipated rock loads. The greater the spacing, the cheaper will be the ribs, but the greater will be the cost of the lagging. The cheapest cost of the ribs and lagging has been found by experience to be with ribs spaced about 4 feet for moderate rock loads, 2 to 3 feet for heavy loading and 5 feet for light loading. This spacing must be adapted to the special requirements of the job. Where the support must be placed close to the face, the spacing should be an even fraction of the round pulled. Where safety niches or embedded items are to be placed in the lining, the spacing should be so the ribs will not interfere with such items. If prefabricated lagging is to be used, the spacing of ribs must be compatible. Rock loads in a single tunnel can cover a great

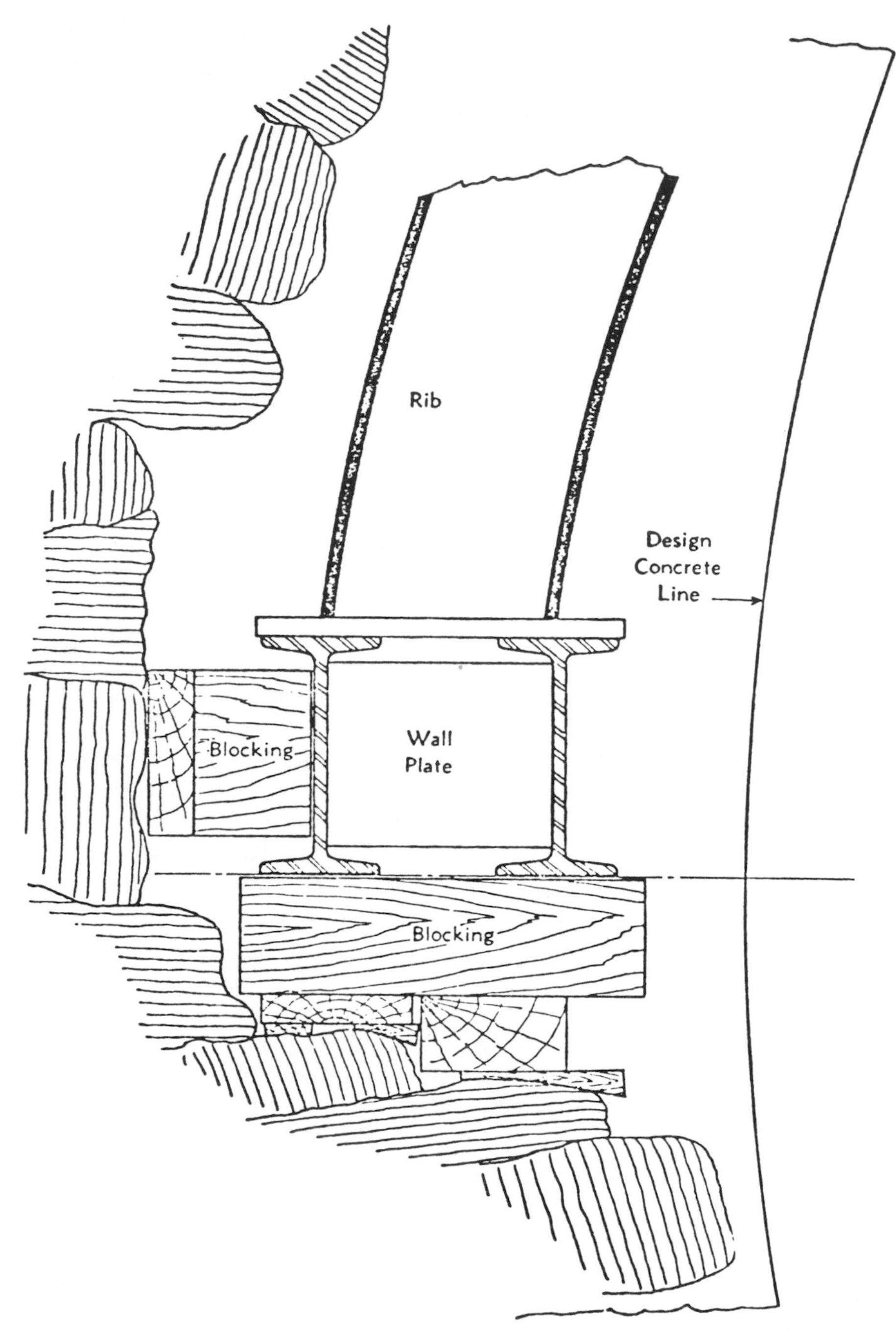

Fig. 7-22. Wall plate in hitch (Commercial Shearing Inc.).

range of values, so it is necessary to use considerable judgment in arriving at the best design for the tunnel ribs. The extremes of roof loading can be tabulated, with a selection being made of what is expected to be the normal condition. The design of a rib to meet this condition may be considered the standard rib. Lighter and heavier ribs should also be selected to meet the different loading conditions, as it is cheaper to change the weight than to change the spacing. The different ribs should be of the same depth. If fault zones or squeezing rock is

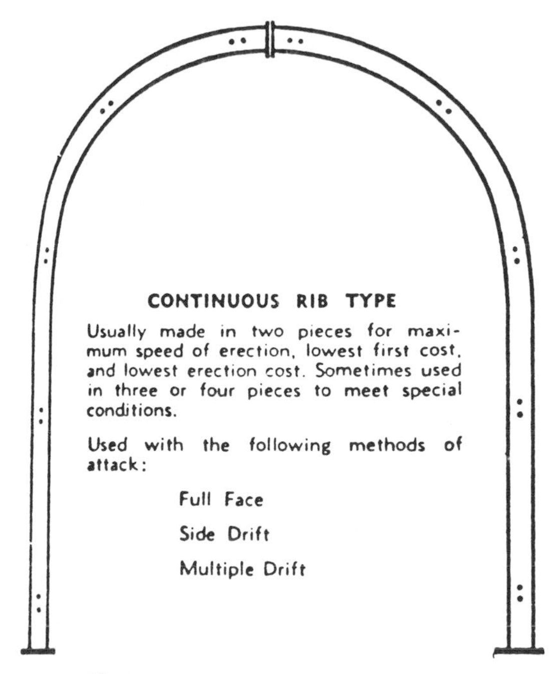

Fig. 7-23. Continuous rib type. From rock tunneling with steel supports. (Commercial Shearing, Inc.).

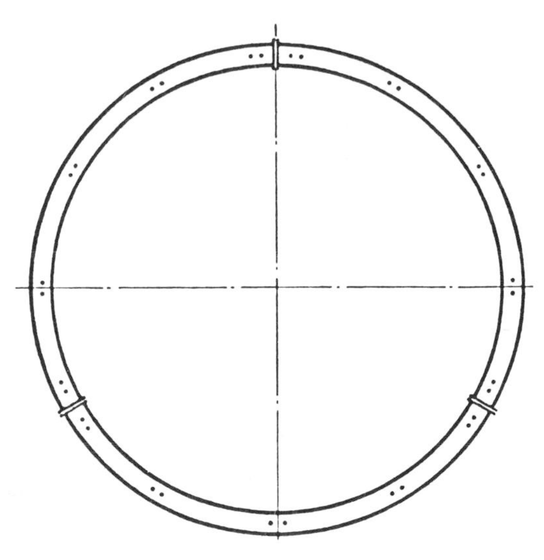

Fig. 7-25. Full circle rib type used with the following methods of attack:

Full Face: In tunnels in squeezing, swelling and crushed rock, or any rock that imposes considerable side pressure. Also where bottom conditions make it impossible to carry roof loads on foot blocks. In tunnel conditions sometimes encountered in rock tunnels.

Heading and Bench: Under earth tunnel conditions with joints at spring line.

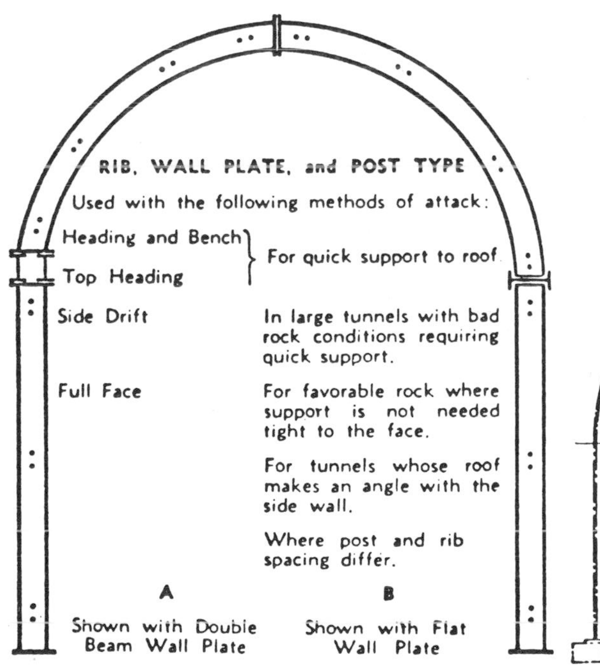

Fig. 7-24. Rib, wall plate and post. From Rock tunneling with steel supports. (Commercial Shearing, Inc.).

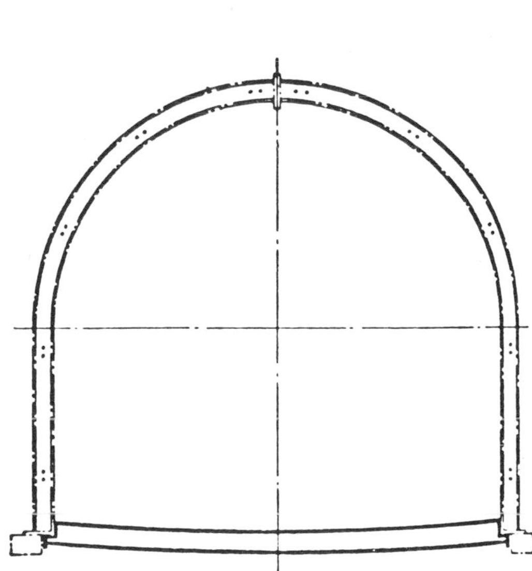

Fig. 7-26. Invert strut. Used where mild side pressures are encountered, and to prevent bottom from heaving.

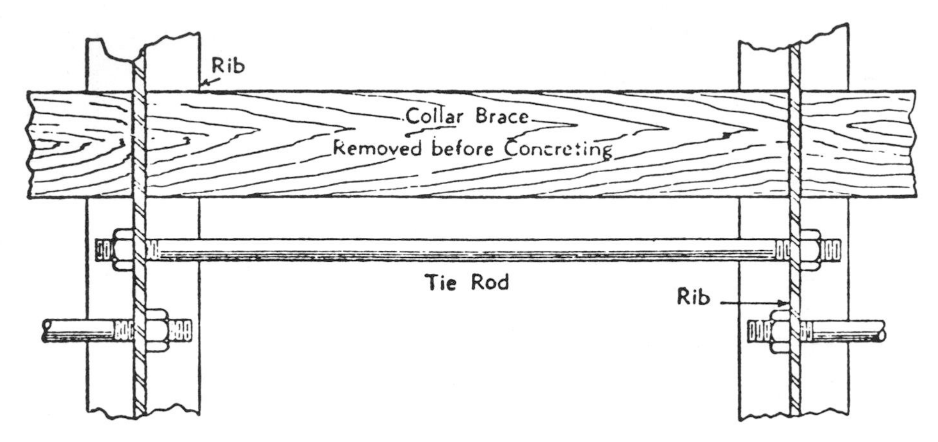

Fig. 7-27. Collar brace, (Commercial Shearing; Inc.).

encountered, it may be necessary to go to a circular section. It is better to use a little more strength of rib than actually required than a little less, as the cost of putting jump sets between ribs already placed is much more expensive than using the extra weight in the first place. The initial supply of ribs to the job should include all three weights of ribs. The supply on hand should only be sufficient to keep the job running until another order, based on the experience with the ribs in place on the job, can be delivered. Figure 7-28 shows one condition that might be encountered.

The cheapest steel sets are two-piece ribs. Posts as a rule are made of H-beams. Where side pressure exists, there is often an upward movement of the bottom, frequently accompanied by a soft bottom that will not support the load on the post footing. Invert struts may remedy this situation.

Support Determination Based on Geologic Prediction

A study of data from 53 tunnel projects, together with various practical and empirical applications relating to tunnel construction, has resulted in a correlation of rock structure rating (RSR) and rib ratio (RR) to determine support requirements.[25] RSR is a number based on geologic conditions usually known during the design stage and prior to construction. Six conditions are combined into three parameters and given maximum values as follows:

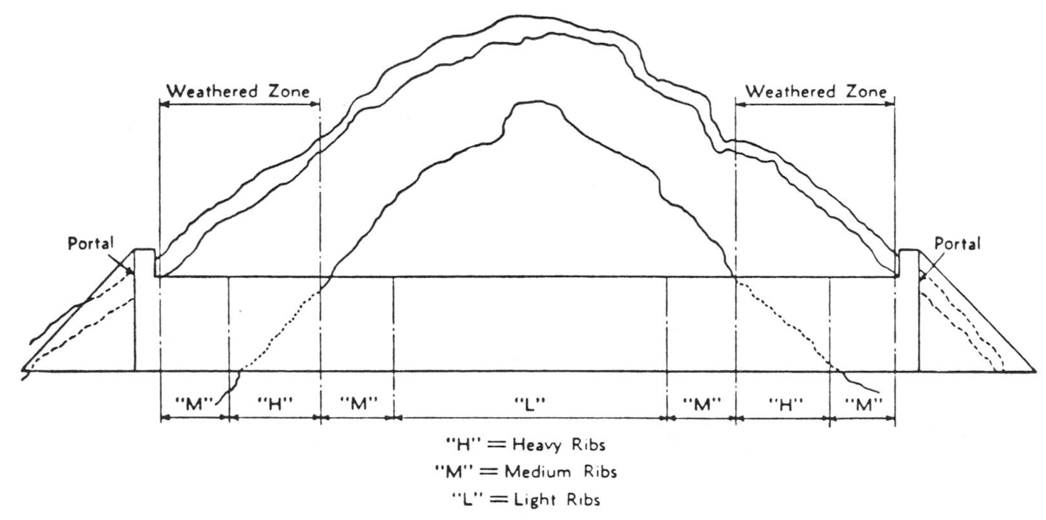

Fig. 7-28. Proper use of various weight ribs in tunnel through a ridge. (Commercial Shearing, Inc.).

Parameter	*Conditions*	
A	Rock type and folding or discontinuities	30
B	Joint pattern and joint orientation	45
C	Water inflow and joint condition	25
	RSR value, maximum	100

Parameter *A* (Table 7-22) is a general appraisal of rock structure through which the tunnel is to be driven.

Parameter *B* (Table 7-23) relates the joint pattern and direction of drive.

Parameter *C* (Table 7-24) takes into account 1) the overall quality of rock as indicated by the numerical sum of values assigned to parameters *A* and *B*; 2) the condition of the joint surfaces; and 3) the anticipated amount of water inflow.

The RSR value is the numerical sum of parameters *A*, *B* and *C* and will range from a minimum of 19 to a maximum of 100.

The rib ratio (RR) is the ratio of actual support required to that developed from Terzaghi's formula for determining roof loads for loose sand below the water table, expressed in percent.

Terzaghi's empirical formula for maximum roof load for loose, cohesionless sand below the water table (datum condition) is

$$\mathbf{P} = 1.38\,(\mathbf{B} + \mathbf{H})\gamma$$

where

$\mathbf{P}$ = load per square foot (lb)
$\mathbf{B}$ = width of tunnel (feet)
$\mathbf{H}$ = height of tunnel (feet)
γ = unit weight of sand (assumed 120 lb/foot^3).

The rib ratio[26] developed from the studied tunnels is shown in Fig. 7-29.

Spacing of ribs for circular tunnels or tunnels with approximately the same height as width, using the RSR method, would be determined by dividing the spacing shown in Table 7-25[26] by the rib ratio taken from the RSR-RR curve in Fig. 7-29 and multiplying by 100.

Example: If RSR equals 50, RR equals 30. A tunnel excavated to 20 feet in diameter would require 8WF31 ribs spaced at $2.14 \times 100 \div 30 =$ 7.1-foot centers or the equivalent.

The rib ratio shown in Fig. 7-29 reflects the condition for tunnels excavated by drilling and blasting. Information available for tunnels excavated by boring machines was insufficient to determine a correlation of support requirements between the two methods. It is well known that support requirements for a bored tunnel are less than for one excavated by drilling and blasting. From available data, it is suggested

Table 7-22. Rock structure rating, parameter *A*, general area geology.*

MAX. VALUE 30

BASIC ROCK TYPE	HARD	MED.	SOFT	DECOMP.
Igneous	1	2	3	4
Metamorphic	1	2	3	4
Sedimentary	2	3	4	4

	GEOLOGICAL STRUCTURE			
	MASSIVE	SLIGHTLY FAULTED OR FOLDED	MODERATELY FAULTED OR FOLDED	INTENSELY FAULTED OR FOLDED
Type 1	30	22	15	9
Type 2	27	20	13	8
Type 3	24	18	12	7
Type 4	19	15	10	6

*From "Ground Support Prediction Model: RSR Concept," *Second North American Rapid Excavation and Tunneling Conference* 1:691–707, AIME, 1974.

Table 7-23. Rock structure rating, parameter *B*, joint pattern, direction of drive.*

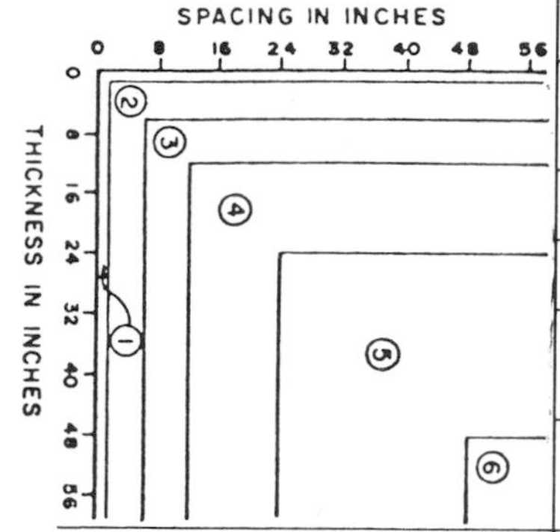
SPACING IN INCHES
0 8 16 24 32 40 48 56
THICKNESS IN INCHES
0 8 16 24 32 40 48 56
1
2
3
4
5
6

	MAX. VALUE 45							
	STRIKE PERPENDICULAR TO AXIS					STRIKE PARALLEL TO AXIS		
	DIRECTION OF DRIVE					DIRECTION OF DRIVE		
	BOTH	WITH DIP		AGAINST DIP		BOTH		
	DIP OF PROMINENT JOINTS					DIP OF PROMINENT JOINTS		
	FLAT	DIPPING	VERTICAL	DIPPING	VERTICAL	FLAT	DIPPING	VERTICAL
1 Very closely jointed	9	11	13	10	12	9	9	7
2 Closely jointed	13	16	19	15	17	14	14	11
3 Moderately jointed	23	24	28	19	22	23	23	19
4 Moderate to blocky	30	32	36	25	28	30	28	24
5 Blocky to massive	36	38	40	33	35	36	34	28
6 Massive	40	43	45	37	40	40	38	34

NOTES: Flat 0–20°; dipping 20–50°; vertical 50–90°.
*From "Ground Support Prediction Model: RSR Concept," *Second North American Rapid Excavation and Tunneling Conference* 1:691–707, AIME, 1974.

Table 7-24. Rock structure rating, parameter *C*, groundwater, joint condition.*

MAX. VALUE 25

ANTICIPATED WATER INFLOW (GPM/1000′)	SUM OF PARAMETERS *A* + *B*					
	13–44			45–75		
	JOINT CONDITION					
	GOOD	FAIR	POOR	GOOD	FAIR	POOR
None	22	18	12	25	22	18
Slight (< 200 gpm)	19	15	9	23	19	14
Moderate (200–1000 gpm)	15	11	7	21	16	12
Heavy (> 1000 gpm)	10	8	6	18	14	10

NOTE: Joint condition—good = tight or cemented; fair = slightly weathered or altered; poor = severely weathered, altered or open.
*From "Ground Support Prediction Model: RSR Concept," *Second North American Rapid Excavation and Tunneling Conference* 1: 691–707, AIME, 1974.

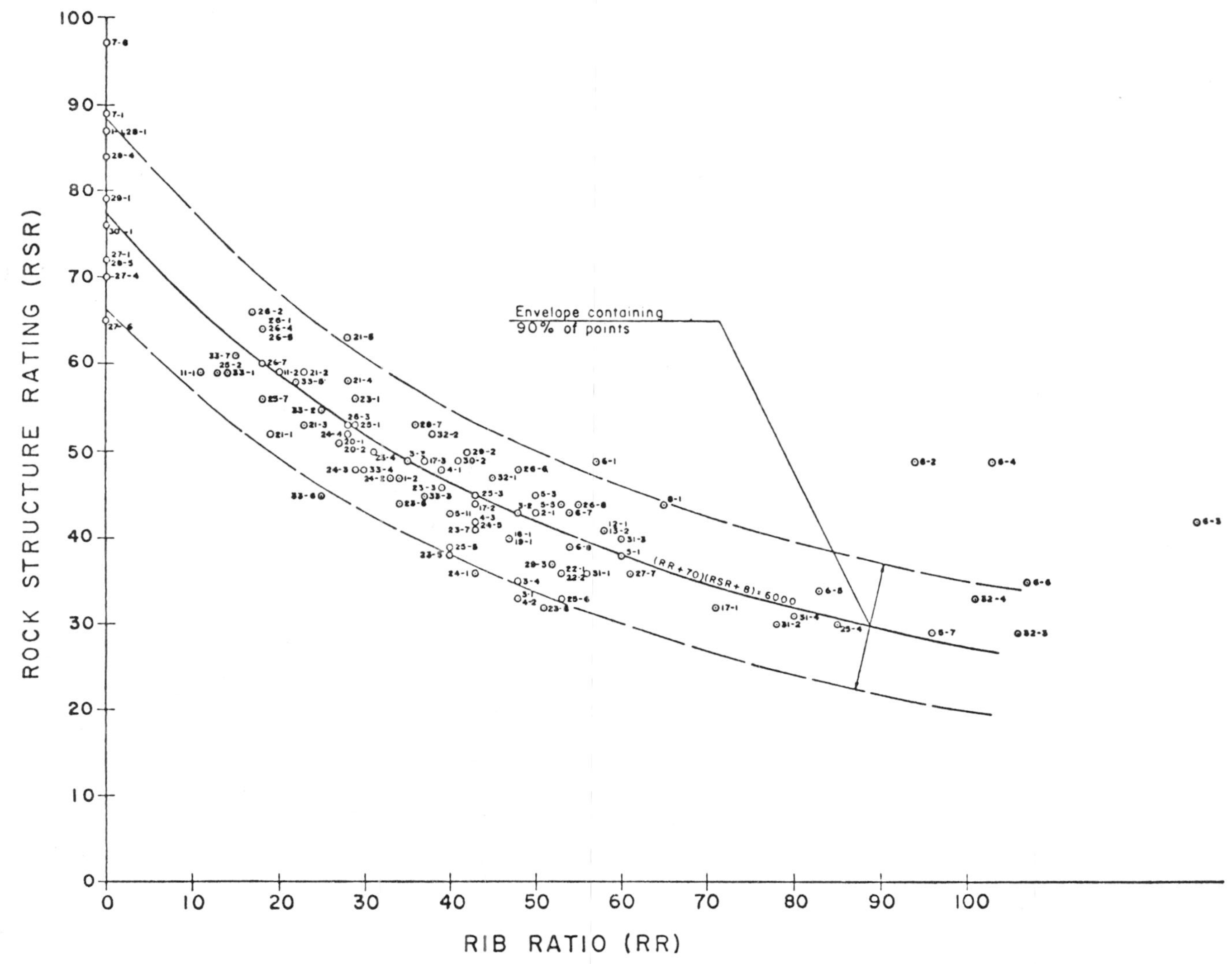

Fig. 7-29. Rock structure rating. Correction of RSR and RR.

Table 7-25. Theoretical spacing of typical rib sizes for datum condition (spacing in feet).

RIB SIZE	TUNNEL DIAMETER 10	12	14	16	18	20	22	24	26	28	30
4I7.7	1.16										
4H13.0	2.01	1.51	1.16	0.92							
6H15.5	3.19	2.37	1.81	1.42	1.14						
6H20		3.02	2.32	1.82	1.46	1.20					
6H25			2.86	2.25	1.81	1.48	1.23	1.04			
8WF31				3.24	2.61	2.14	1.78	1.51	1.29	1.11	
8WF40					3.37	2.76	2.30	1.95	1.67	1.44	1.25
8WF48						3.34	2.78	2.35	2.01	1.74	1.51
10WF49								2.59	2.22	1.91	1.67
12WF53										2.19	1.91
12WF65											2.35

Table 7-26. Correlation of rock structure rating to rock load and tunnel diameter

TUNNEL DIAMETER (*D*) (FEET)	(Wr) ROCK LOAD ON TUNNEL ARCH (K/FOOT^2) 0.5	1.0	1.5	2.0	3.0	4.0	5.0	6.0	7.0	8.0	9.0	10.0
	CORRESPONDING VALUES OF ROCK STRUCTURE RATINGS (RSR)											
10	62.5	49.9	40.2	32.7	21.6	13.8						
12	65.0	53.7	44.7	37.5	26.6	18.7						
14	66.9	56.6	48.3	41.4	30.8	22.9	16.8					
16	68.3	59.0	51.2	44.7	34.4	26.6	20.4	15.5				
18	69.5	61.0	53.7	47.6	37.6	29.9	23.8	18.8				
20	70.4	62.5	55.7	49.9	40.2	32.7	26.6	21.6	17.4			
22	71.3	63.9	57.5	51.9	42.7	35.3	29.3	24.3	20.1	16.4		
24	72.0	65.0	59.0	53.7	44.7	37.5	31.5	26.6	22.3	18.7		
26	72.6	66.1	60.3	55.3	46.7	39.6	33.8	28.8	24.6	20.9	17.7	
28	73.0	66.9	61.5	56.6	48.3	41.4	35.7	30.8	26.6	22.9	19.7	16.8
30	73.4	67.7	62.4	57.8	49.8	43.1	37.4	32.6	28.4	24.7	21.5	18.6

that, for bored tunnels, the RSR value should be adjusted upward by multiplying by a factor of 1.2 for a 10-foot tunnel, 1.15 for a 25-foot tunnel and 1.05 for a 30-foot tunnel. (If the RSR value has been determined to be 50, it would be increased for a 10-foot diameter tunnel to 50 X 1.20, or 60 when using Fig. 7-29 to determine the rib ratio.)

RSR values can be expressed in terms of unit loads for various sized tunnels, as shown in Table 7-26.

The correlation between RSR and rock loads can be extended to show the general relationship between ribs, rock bolt and shotcrete types of support (as shown in Fig. 7-29).

Wall Plates

Where it is necessary to install arch ribs before posts, as in top heading and bench construction, or where posts are not required, wall plates are used. They transmit the load from arch ribs to posts, hitches or pins, as the case may be. Posts are usually installed under each arch rib, but this may not be required where loading is light. The three types of wall plates generally used are shown in Fig. 7-30.

The double beam wall plate, made with two I-beams, is more commonly used because ribs and posts are clamped by toggle plates and bolts, thus avoiding the time required for matching bolt holes as required with the single wall beam plate made with an H-beam reinforced at each rib seat with vertical T-shaped diaphragm plates, and also permitting variable spacing of ribs. Flat wall plates require support under each arch rib and function only as a cap for posts and sill for erecting roof ribs. An alternate wall plate may be formed by setting the arch ribs on wood blocks, followed by construction of a shotcrete beam reinforced with $\frac{5}{8}$-inch round bars. Wall plates may be temporarily supported on pins before posts are installed to meet excavation and loading conditions.

Bracing is used between ribs to increase their resistance to buckling about their minor axes. Lagging firmly attached to ribs will serve this purpose. Otherwise, collar bracing and tie rods, as shown in Fig. 7-27, or spreaders are required. Tie rods are commonly $\frac{5}{8}$ inch or $\frac{3}{4}$ inch and collar braces may be any convenient size (3 X 4 inches, 4 X 6 inches or 6 X 6 inches). Tie rods are inserted through holes in the rib web. Spreaders may be pipe around tie rods, or structural shapes with a clip angle or plate at each end to permit bolting to the rib sets.

Blocking, generally with wood, at sufficient intervals to support the rock load and prevent undue bending in the ribs is required. Lagging and back-packing to resist loads between the ribs are as required for timber sets. Lagging may be of steel or wood. Steel lagging is usually attached to ribs as shown in Fig. 7-31. All lagging must be securely blocked or back-packed. It may be tight or open, as required for rock conditions.

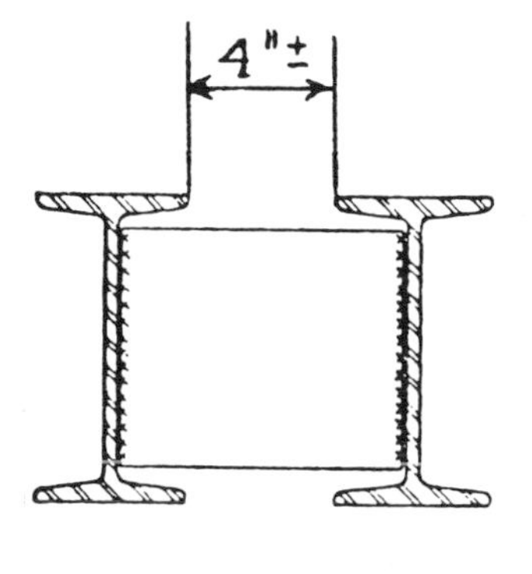

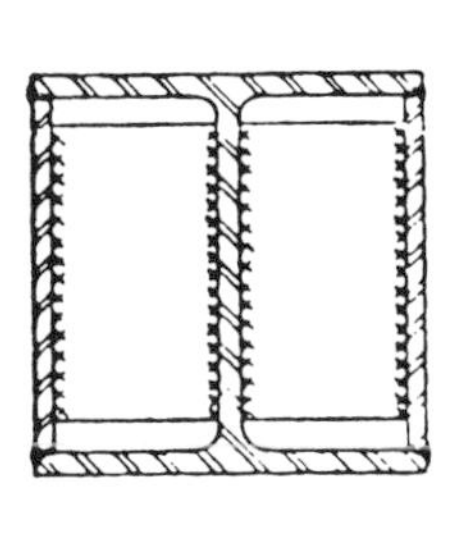

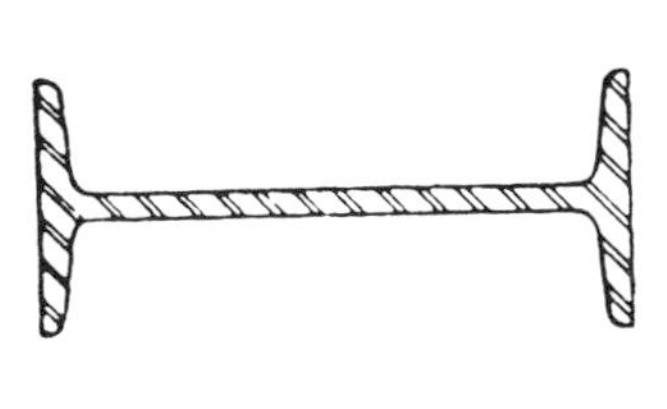

Fig. 7-30. Wall plates. (Commercial Shearing, Inc.).
(a) Double Beam Wall Plate. Diaphram welded under each rib seat.
(b) Single Beam Wall Plate.
(c) Flat Wall Plate.

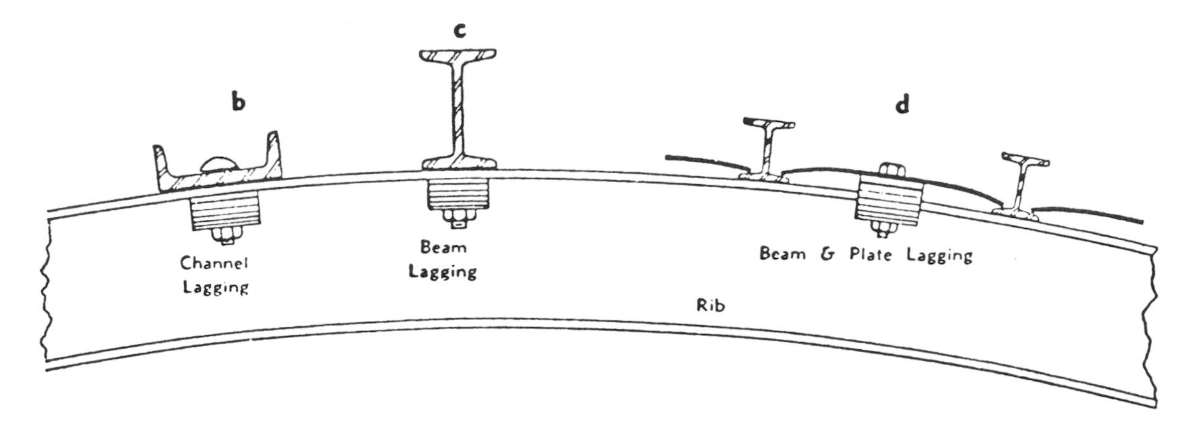

Fig. 7-31. Types of lagging. (Commercial Shearing, Inc.).

Back-packing may be used to fill the voids caused by overbreak between the lagging and the rock surface. It usually consists of rock from tunnel muck but can be of other materials if more economical to use. It may be required to support the rock surface prior to placing permanent lining, but is usually not as economical as concrete placed with the permanent lining unless it can be mechanically placed. Where dry back-packing is used rather than concrete, more grout will be required.

Liner Plates

Where tight lagging is required, steel liner plates are frequently used either with or without steel ribs. They are almost universally used for shield-driven tunnels. Until recent years, liner plates were made mostly of cast iron. Now liner plates of steel have largely supplanted cast plates. Where used with ribs, the construction procedure is much the same as with steel ribs and lagging. However, where the plates must take arch action, it is usually necessary to provide a "key" plate in the arch for closing a ring of plates. In circular sections, erection of the rings starts at the bottom with closure at the top. Liner plates used for lagging may be made of pressed steel, as in Fig. 7-32. Liner plates used for ring pressure made of plate steel must be carefully fabricated. Typical plates used in the San Francisco BART subway tunnels (mostly in soft ground) are as shown in Fig. 7-33. Concrete liner plates have been designed to be competitive with steel.

If the steel support is adequately protected by concrete and grout, it will do the job of supporting the rock so that the concrete lining

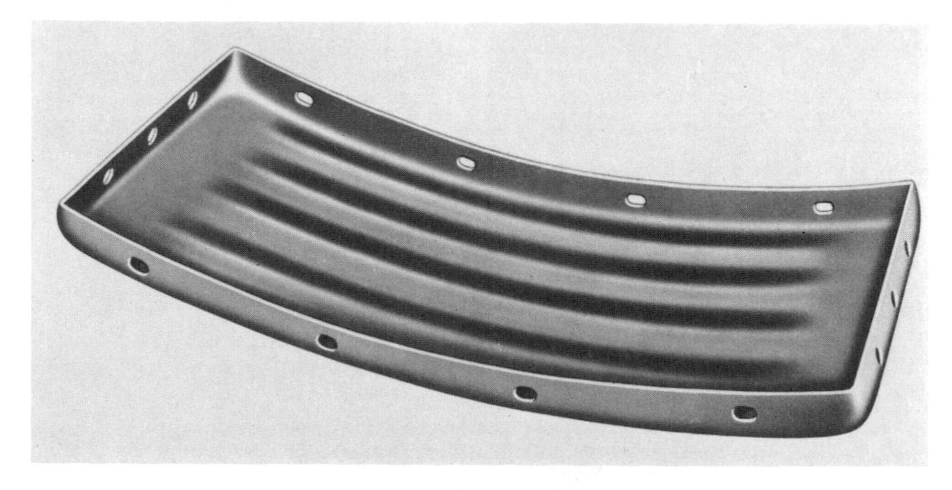

Fig. 7-32. Liner plate.

Fig. 7-33. Steel liner plates.

need be only the thickness required to protect the steel.

Rock Bolts

As previously stated, almost any intact rock tunnel needs no support. It is the breaking away at joints and loosening of the joints that causes rock falls. If the rock can be held together, it will then act as intact rock. To accomplish this, the use of rock bolts (Fig. 7-34) has replaced steel rib sets in many cases where the rock is not too badly fractured and jointed. Where the rock will stand up for a

Fig. 7-34. Cubato underground powerhouse—rock anchors.

sufficient length of time before requiring support, holes for rock bolts and their installation can be made from the drill jumbo while the next round is being drilled. The amount of steel and cost installed of rock bolts is much less than for steel ribs so they have largely replaced steel ribs where the rock is suitable for their installation. Rock bolts serve their purpose best when installed close behind the face. However, the optimum design must be compromised with the practical considerations of progress in the tunnel. Rock bolts are usually used in relatively good rock but can be used in some rock that would be classified as poor. They have the advantage of extending only a few inches outside of the rock, which reduces the size of the tunnel that would be necessary when using wood or steel sets for supports. In relatively good rock, they are installed where inspection shows they are required, whereas in the poorer rock, a regular pattern would be used.

Rock bolt systems are usually selected, rather than designed, by one of several methods that can be grouped into three categories: 1) experience, 2) rock support and 3) rock reinforcement. The selection of rock bolts, based on experience (category 1), would be to use bolts with a length of one-third to one-half of tunnel width (but not less than 8 feet or the width of the tunnel if it is less) and spacing of 4 to 6 feet. Patterns should be regular and specified by the designer to meet the most critical conditions expected. Designs based on other methods should be compared with the above guidelines. For rock support (category 2), bolt spacing may be taken from Fig. 7-35. Where a joint block or bed might fall out of the roof or a side wall may spall, the selection of rock bolts may be based on the support such rock will require (category 2) to keep it in position, and will be based on engineering judgment, geological information and experience. The rock bolts in horizontally stratified rock are designed to carry the weight tending to fall. To hold blocky rock, it is often necessary to select rock bolts on a trial and error basis. Selection of rock bolts based on rock reinforcement (category 3) is made on the assumption that the rock can be confined so that it will become a part of the structure supporting the opening. This is accomplished by confining the rock, increasing the friction of the joints so as to increase horizontal shear-resistance in the joints, thereby increasing beam strength or arch action at the roof of the tunnel. If the bolts are installed early enough, a stable arch or beam will be formed in the rock as originally excavated.

Guidelines for anchorage selection indicate cement or epoxy in very hard or in weak rock and shale and expansion anchors in hard or medium hard rock. Rock bolts are usually one inch in diameter. It is important that they be tightly installed and prestressed as close to the face as practicable. Prestressing is usually done with a torque wrench. The type of nut, bearing plate, condition of threads and amount of lubrication affect the torque-load relationship so the torque on the wrench must be calibrated on each project to ensure obtaining the desired amount of stress in the bolt. Ductile steel in bolts is preferable to high strength steel as it will yield if overstressed, permitting the rock to adjust and redistribute forces with less likelihood of sudden failure, giving warning so that additional support may be added if required.

Rock bolts may be used as permanent support if they are protected by completely filling the drill hole with grout. Grouting may be accomplished through a plastic tube inserted into the drill hole with a return through a hole in the center of the bolt (Fig. 7-36).

Deere[8] has given his recommendations for rock bolt support in Table 7-13.

Shotcrete (refer to Chapter 12) is a sprayed concrete with small aggregate (maximum, one inch) that has been successfully used in recent years for both temporary and permanent support. It is most effective if placed immediately behind the face and can be used to stabilize almost any kind of rock. In a Vancouver, B.C. railroad tunnel[27,28] constructed in shales, sandstones and conglomerates, the only temporary support required was shotcrete applied to the roof and walls immediately after each round up to the working face. The next round was blasted within two hours without deleterious effect (Fig. 7-37). The thickness of shotcrete required may be no more than enough to prevent air slacking or small blocks falling out (2 to 3 inches) in good rock or 6 inches or more

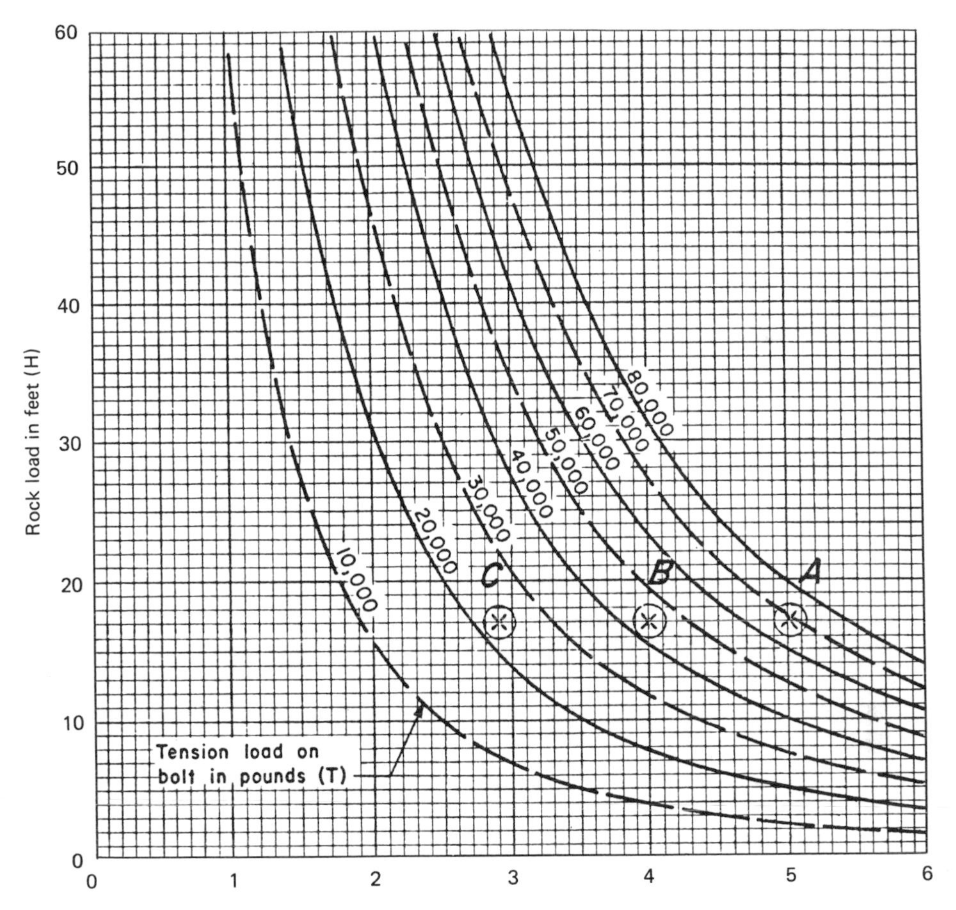

Fig. 7-35. Rock bolt spacing diagram for suspended rock loads.

$$\text{Rock load in feet} = H = \frac{T}{WS^2}$$

where T = tension load on bolt in pounds
S = spacing of bolts in feet, square pattern
W = weight of rock in pounds per cubic foot (taken as 160 lb/cu ft)

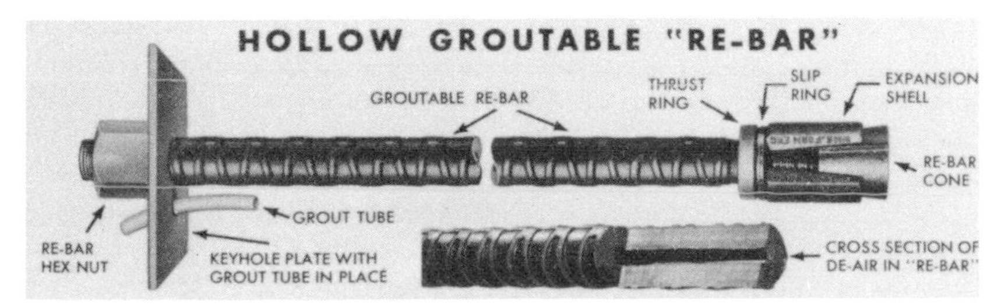

Fig. 7-36. Hollow groutable rock bolt.

Fig. 7-37. Shotcrete application.

in poor rock. Before the advent of shotcrete, gunite with quick-setting cement, essentially the same as shotcrete except composed with sand aggregate only, was successfully used to support very bad squeezing ground in the Hetch Hetchy Coast Range tunnels in California, where 16 X 24 inch timbers were broken like matchsticks in a 10-foot, 6-inch diameter tunnel in squeezing ground.[29] Sublining of gunite varied from 15 to 24 inches. A 9-inch finished concrete lining was placed inside the gunite. In the worst ground, the excavated diameter was 16 feet. Most shotcrete designs evolve with experience in a given tunnel on a trial and error basis. The thickness can be readily changed to meet the conditions encountered. The loads on shotcrete lining are a function of rock condition and time of installation. In a machine-driven or smoothly blasted tunnel, a relatively thin lining will suffice. Deere[8] has suggested thicknesses for shotcrete as shown in Table 7-13, but warns they are tentative and subject to change as greater experience is obtained.

Table 7-27. Shotcrete mix and strength, Vancouver Tunnel.[27,28]

MIX	TIME	COMPRESSIVE STRENGHT (PSI)
Cement, 650 lb	2 hours	Minimum 200–250
$\frac{3}{4}$-inch stone, 900 lb	12 hours	Minimum 800
$\frac{1}{4}$-inch stone, 850 lb	24 hours	Minimum 1500
Sand, 1520 lb	28 days	4000
Water-cement ratio 0.35		
Accelerator 25 lb		

The concrete mixes used on the Vancouver railroad tunnel and their strengths are shown in Table 7-27. Aggregate recommendations are shown in Fig. 7-38.

Where smoothness of surface is unimportant, shotcrete may also serve the purpose of final lining. Shotcrete is also useful together with rock bolts in supporting rock from falling out between bolts and as a protective cover for wire mesh supported by rock bolts.

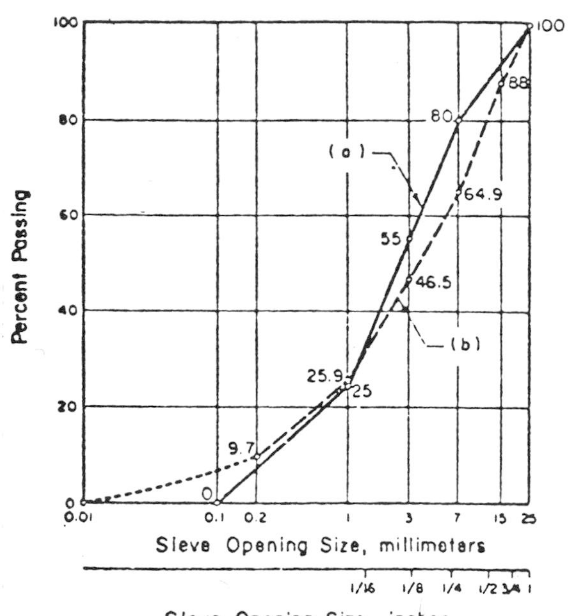

Fig. 7-38. Aggregate grading for coarse-aggregate shotcrete. (After Kobler, 1966).[30]
a. Recommended by Aliva Company, Switzerland
b. Recommended by Rothfuchs
Note: ACI Standard 506-66 contains grading requirements for fine and coarse aggregate individually.

The Bernold System

In 1968, Bernold of Switzerland introduced a new system of tunnel support to act as temporary support as well as all or part of the permanent support. The basic feature consists of patented expanded metal sheets somewhat similar to a very heavy duty metal lath, installed as close to the face as required by the rock conditions to provide temporary support. The Bernold sheets are available in thicknesses varying from 1.25 to 3 millimeters and in two sizes (3 feet, $6\frac{1}{2}$ inches $\times$ 3 feet, $11\frac{1}{4}$ inches, and 4 feet, 4 inches $\times$ 3 feet, $11\frac{1}{4}$ inches) and weigh from 25 to 96 lb each. The sheets are used both as forms and reinforcing steel for the concrete placed behind them. The sheets are lapped, clamped together and temporarily supported on hinged ribs that can be removed after the concrete has attained sufficient strength (approximately 36 hours) and moved ahead through the ribs remaining in place to support the sheets being placed at the heading. Low slump plastic concrete is pumped behind the Bernold sheets and may be placed within two feet of the face. Bulkhead forms composed of Bernold sheets are placed so as to allow lapping of sheets after concreting.

In machine-excavated tunnels, the Bernold sheets may be placed against the rock and gunite shot through the perforations. If a smooth finish of the tunnel lining is not required, the Bernold sheets can be protected with a thin layer of sand-cement shotcrete as the final finish. If a smooth surface is required, a lining of concrete can be placed inside the Bernold support. In bad ground, poling plates supported on movable ribs have been used in conjunction with the Bernold system, wherein concrete is placed against the tail end of the poling plates.

The Bernold system has been used extensively in Europe and in some mines in the U.S.

TECHNIQUES FOR SUPPORT OF DIFFICULT GROUND

The previous paragraphs have considered methods of support where the rock will stand after drilling, blasting and excavation, without support and damaging relaxation of the roof rock, for a period sufficient to permit the installation of the support required for a complete round. Most of the footage of rock tunnels has been constructed under these conditions. However, many rock tunnels encounter "bad" rock that can add materially to the total cost of the project even if the difficult conditions are for a relatively short distance. Delays or slow progress for weeks or months are not uncommon. A typical example is the Straight Creek Tunnel in Colorado.[31] In formations of chemically altered or badly fractured rocks or where sand pockets may be encountered in hard rock tunnels, support of the tunnel roof, walls and face may be required as the face is advanced. In some cases, the rock loads that develop as time goes on are in excess of that for which the original support was designed. Where these conditions occur, the previously described support must be augmented by additional measures to hold the tunnel, and the length of rounds may have to be reduced.

Crown Bars

Where wood or steel rib sets are being used for support and rock tends to fall out of the roof in relatively large blocks, quick support can be obtained by the use of crown bars. These may consist of timbers, H-beams, channels fabricated to form rectangular steel members or other similar bars capable of supporting the rock ahead of the last set by cantilever action. Crown bars may be installed on the outside of the sets (Fig. 7-39) or set in hangers inside the sets (Fig. 7-40). They may be used with any type of rib sets, with or without wall plates. As soon as a round is blasted or a short advance of the face excavated by other means, the crown bars are installed to support the newly opened ground. The rock must be blocked against the crown bars as soon as possible to keep the rock movement to a minimum. The underslung crown bars may be moved ahead after the blasting of each round. At each advance of the bars, the blocking must be removed and the rock reblocked. This is done one crown bar at a time so that the area subject to rock falls is kept to a minimum. The crown bars are intended to support the rock only until the next set is in place, at which time the load is transferred by blocking to the set. Where the top

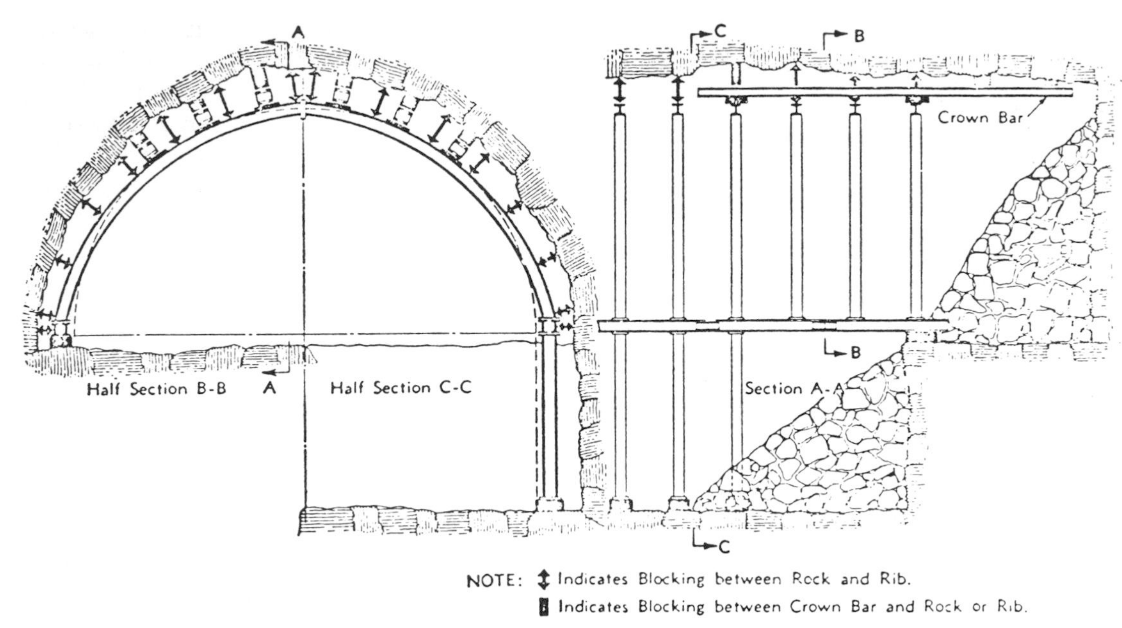

Fig. 7-39. Crown bars mounted on the ribs support the new roof by cantilever action (Commercial Shearing, Inc.).

heading and bench method of excavation is used and the wall plates will not support the steel arch ribs, underslung crown bars may be helpful in bridging the load to the bench. Another type of crown bar is shown in Fig. 7-41.

Poling Plates

In rocks requiring tight lagging for support, poling plates are sometimes used. These are used in somewhat the same manner as crown bars except they consist of flat sections such as

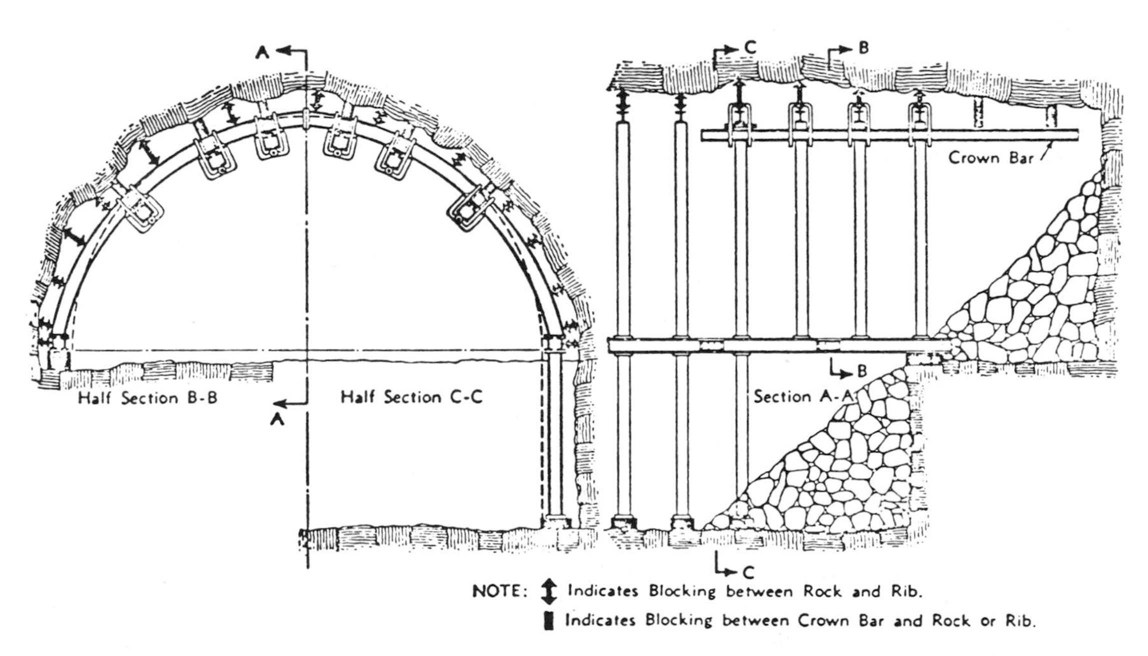

Fig. 7-40. Crown bars, hung from the ribs, support the new roof by cantilever action. (Commercial Shearing, Inc.).

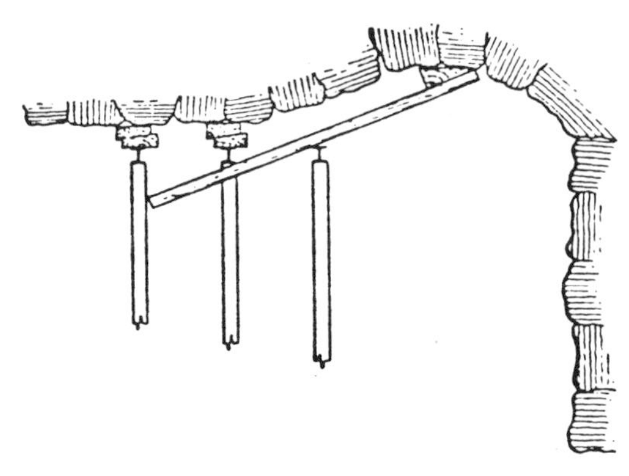

Fig. 7-41. Emergency crown bars (Commercial Shearing, Inc.).

steel channels placed flat to the rock load. They are located on the top of the rib sets and are moved ahead by jacking or leverage with permanent lagging being installed in back of the poling plates as they advance, one at a time. This is a difficult operation and is used only in the worst ground.

Spiling

Where the amount of difficult ground is not too great, spiling is often used as support to advance the tunnel heading. Spiling consists of timber or steel members that may be driven on a slant from the inside of the second rib set from the heading over the first set. The spiling supports the rock by cantilever action until the next set can be installed. This method of support can be continued indefinitely, but it is a slow, tedious, expensive process for advancing the face. This type of support can be used to solidly lag soft ground sections, as shown in Fig. 7-42, with spiles driven ahead of the excavation. On the Mono Craters Tunnel in California, spiles were 3 X 6, 4 X 6 or 6 X 6 timber and, in some cases, 6-inch steel channels driven by a long timber used as a battering ram.

Breastboards

Where the rock at the heading face will not stand long enough to advance the heading to the next set, it is necessary to support the face. Otherwise, the face will collapse and, with no support for the roof, it will also fall in. To avoid this, timber or steel beams called breastboards, placed against the face and supported by stulls to previously placed tunnel support to the rear, may be used. The face may require only a few breastboards, but in the case of wet sand could require tight breastboards with excelsior or other packing to caulk the joints. On the Broadway Tunnel in San Francisco, which consisted of twin vehicular bores, each 44 feet wide, a breastboard jumbo (Fig. 7-43) was devised to hold the face in badly fractured serpentine rock. Because of the size of the tunnel section, it was originally planned to drive the tunnel by constructing two wall plate drifts and a top center heading. The top center heading was to be widened to the wall plate drifts and finally the core was to be removed. After the drifts were completed, it was found that progress of only 6 linear feet of tunnel per day in widening the top heading was possible. With

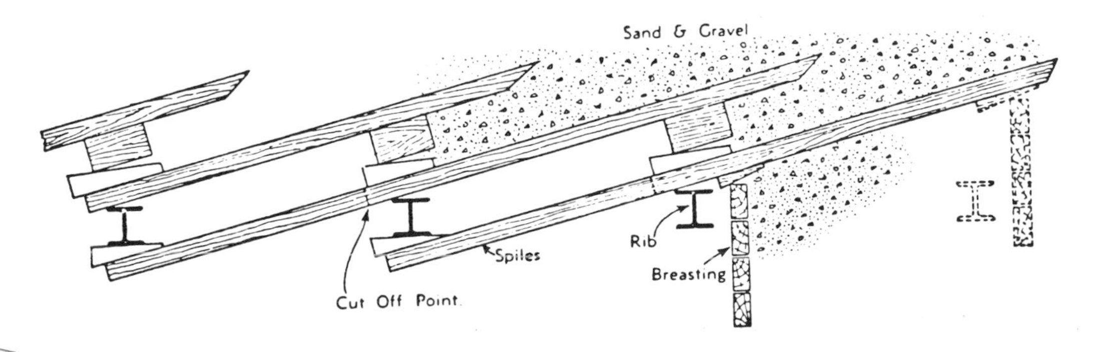

Fig. 7-42. Diagram of forepoling method of supporting loose or running ground.
The spiles are driven ahead before mining out. The breast boards are removed, one by one, the space ahead excavated and the breasting installed in its new forward position. Thus only a small portion of the face is open at any one time. The tails of the spiles are cut off prior to concreting. (Commercial Shearing, Inc.).

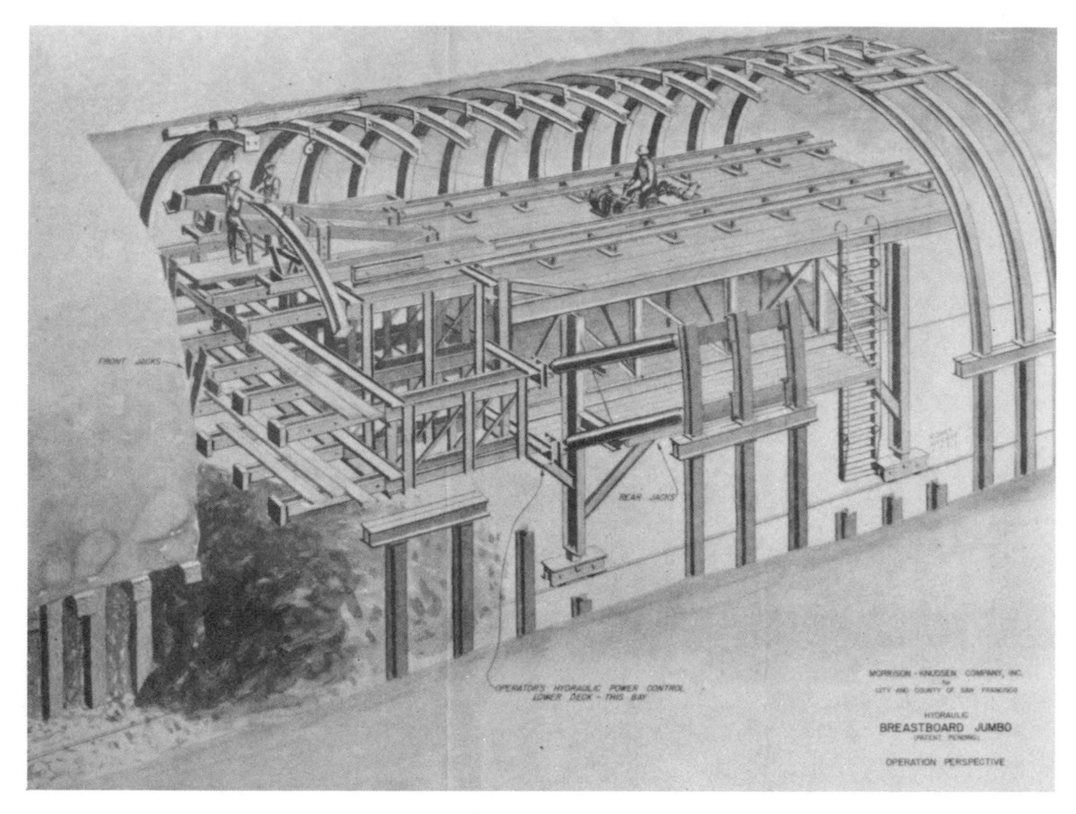

Fig. 7-43. Breastboard jumbo.

the breastboard jumbo, an average progress of 18 feet per 24-hour day was accomplished. Furthermore, the top center heading was not needed. Eighteen hydraulic jacks, not all of which were needed, held breastboards tightly against the face. The boards could be removed one at a time and the rock excavated by spading or, where necessary, by light blasting. Time for placing arch ribs was reduced from one hour to five minutes.

Chemical Grout

Chemical grout has been used not only to stop or reduce water flow, but also to add strength to rock surrounding the tunnel.[32]

Treatment in Squeezing Ground

In squeezing ground, due to rock deformation, it is necessary to provide support for side pressures as well as for the roof and bottom. The circular section, such as shown in Fig. 7-44, is best adapted to this type of ground so as to take advantage of the arch action utilizing the strength of the ring in compression rather than bending. The space outside of the lagging must be packed thoroughly, so as to produce uniform loading on the ribs. Rock that squeezes because of volume expansion is known as swelling rock and here again the circular section for support is the best, although other sections can be used. Because swelling will proceed only so far, the loads on the support in stiff swelling ground can be reduced by overexcavating, as shown in Fig. 7-45, and by providing crush lattices in the ribs, as shown in Fig. 7-46. In the Mono Craters Tunnel in California, excavated as a horseshoe section 11 feet, 9 inches wide by 12 feet $3\frac{1}{2}$ inches high, the 6-inch, 20-lb H-beam supports were spaced in some places as close together as 9 and 12-inch centers. The loads were so great that reinforced concrete blocks $12 \times 18 \times 24$ inches and $9 \times 12 \times 18$ inches were used to support the ribs. Similar foot blocks in wet sand sections may be necessary.

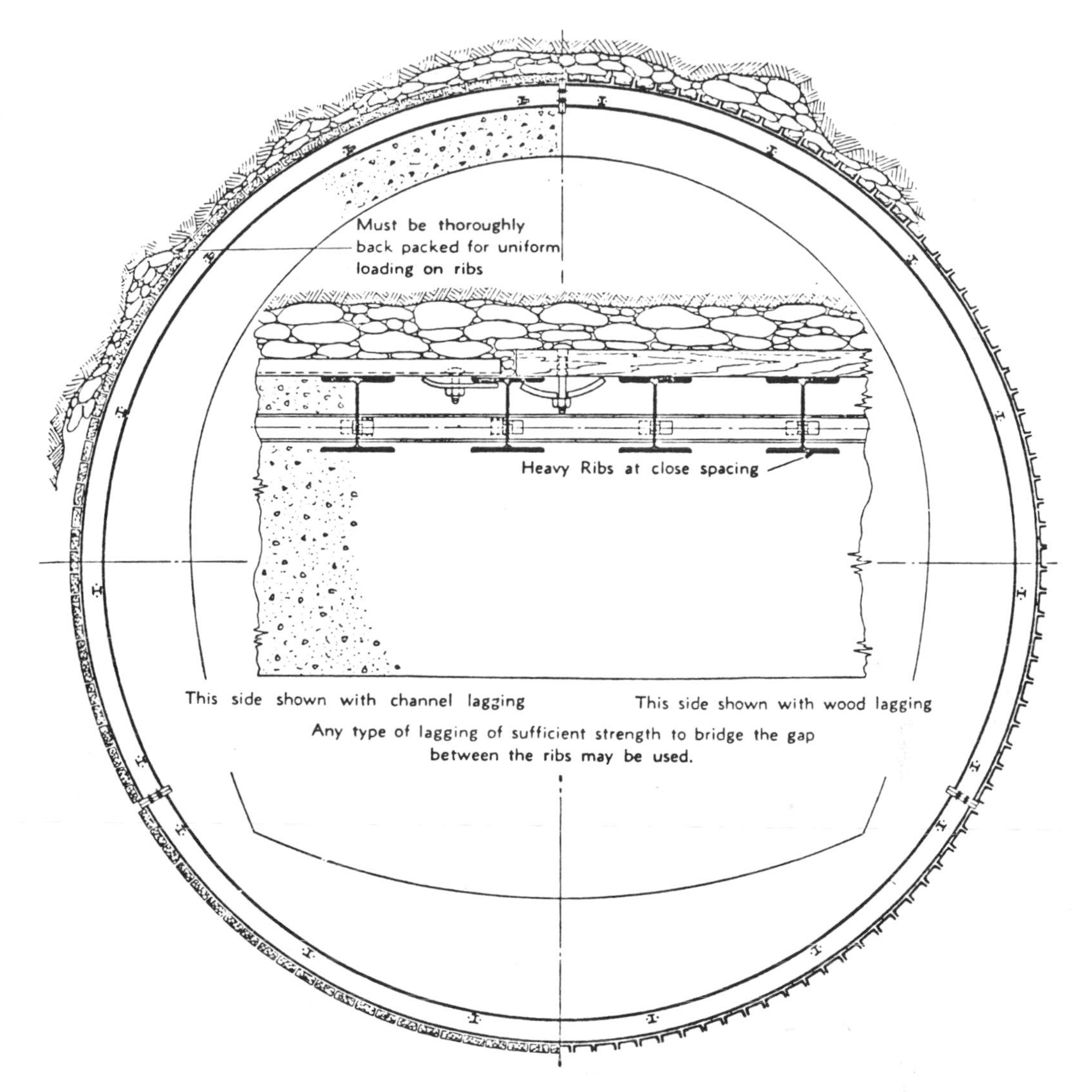

Fig. 7-44. Full circle ribs closely spaced and heavily lagged for heavy loads associated with squeezing conditions.

Relieving Sets

To strengthen overloaded sets, relieving sets may be installed inside the originally placed support with squeeze blocks between the two sets to permit better distribution of the load on the inside set.

Jump Sets

Where loads develop after installation of supports in excess of their capacity, it may become necessary to install intermediate supports between those initially installed. These are called jump sets. Their cost, combined with the initial support, is much greater than would be that of support adequately designed in the first place.

Wall Plate Drifts

To permit installation of wall plates of reasonable length, it is frequently found desirable to excavate a wall plate drift of minimum size permitted by the selected excavation method. The invert of the drift is usually at the bottom of the wall plate, but it may be at the bottom of the tunnel, as at the Broadway Tunnel in San

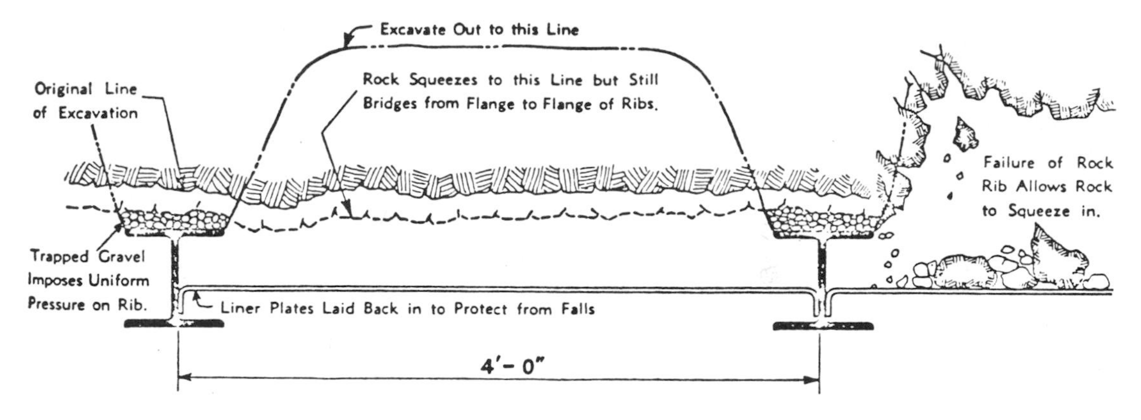

Fig. 7-45. Over-mining in stiff swelling ground to induce softening.
In stiff swelling ground it is necessary to allow the ground to squeeze some undetermined amount to soften it. If the ground is not sufficiently soft to extrude between the ribs when the shrinkage provided in the crush lattices (Fig. 113) is used up, slots are excavated beyond the ribs, as shown here. This is repeated until extrusion between the ribs is established. The squeeze is then allowed to run its course before concreting.

Francisco, where a concrete foot block, steel posts and wall plates were installed in the drift before proceeding with the excavation of the core and arch (Fig. 7-47).

Liner Plates

In soft ground or in running sand sections, tight lagging with rib sets or metal liner plates must be used.

Portalling In

Frequently, the most difficult ground is at the portals. In tunnels of large cross-section, the rock above the tunnel may come down with the first rounds. This may be largely avoided by starting with a pilot bore until sounder rock is reached. Enlargement may then be made to the full or top heading section, as the case may be, at the inside end of the pilot bore. After installing support, the full section may then be excavated to the portal by backing out.

Auxiliary Headings and Galleries

Where difficult ground is anticipated or encountered, an auxiliary heading of smaller dimensions may be driven paralleling the desired tunnel. This tunnel may be advanced more rapidly than the main bore, permitting, by means of cross-drifts, access to additional headings in the main bore. A good example of this method of attack is the Moffat Railroad Tunnel through the Rocky Mountains in Colorado and the adjacent access tunnel, later used for transporting water across the divide.[33]

In large tunnels, where the rock will not support itself long enough to permit installation of the required support, it is necessary to perform the excavation in steps by advancing smaller tunnels or drifts within the larger section. This may consist of a top center drift, wall plate drifts and any number of intermediate drifts around the periphery of the tunnel. A typical operation is shown in Fig. 7-48. In its worst sections, the Straight Creek Tunnel required 15 separate drifts outside the periphery of the tunnel. These were each filled with concrete to form a continuous support, after which the main tunnel itself was excavated.

Shields

For soft ground sections of rock tunnels, shields have been used (see Chapter 6). These have not always been successful.[34]

WATER FLOW

One of the most troublesome obstacles to tunnel advance is water. Even a small amount in a

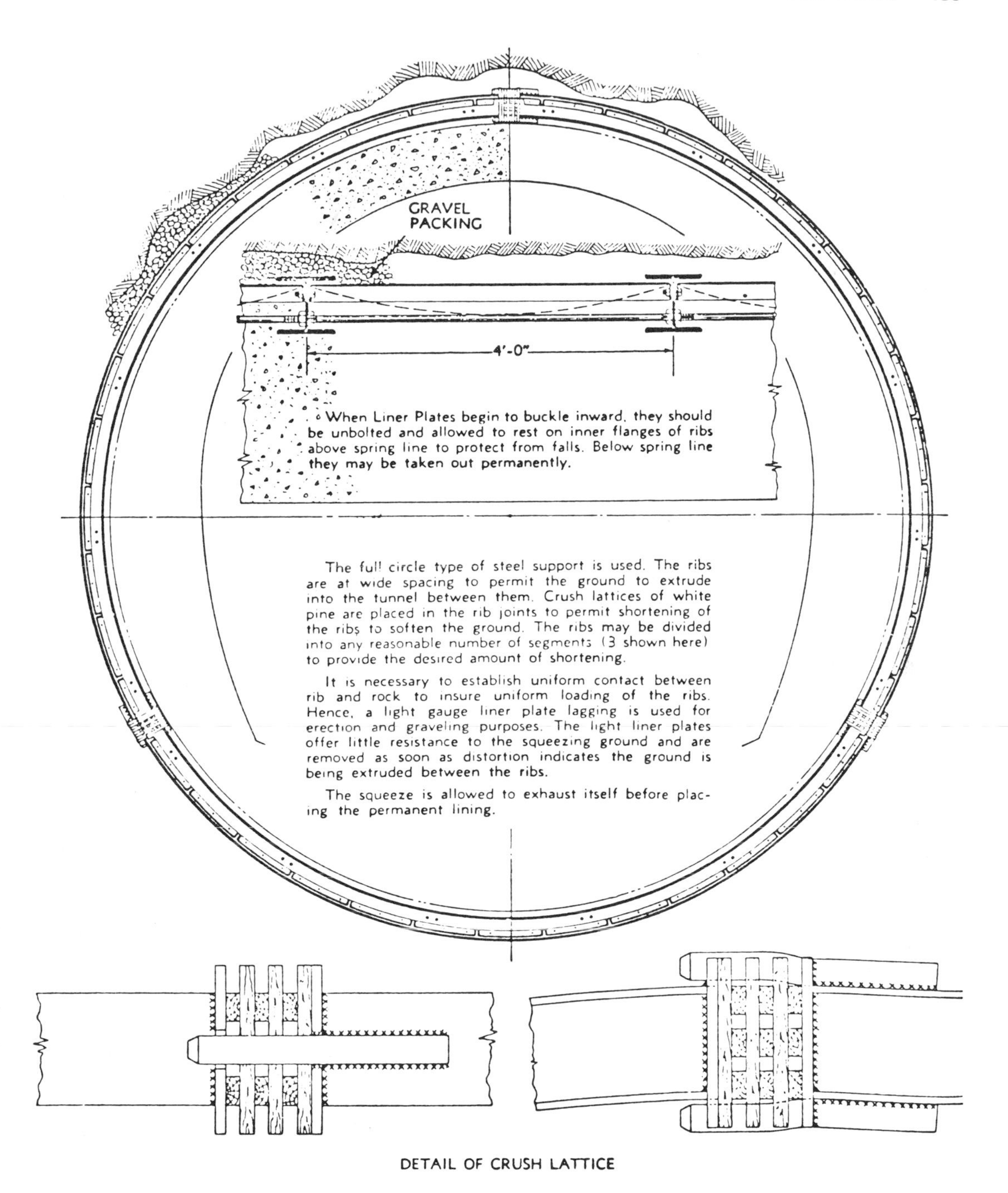

Fig. 7-46. Yielding lining for swelling rock.

tunnel being driven downgrade collects at the face and is a nuisance. Any appreciable amount of water at the face will interfere with progress regardless of the grade of the tunnel. Water tends to lubricate the joints in the rock, particularly the softer rocks, and weaken its ability to stand without support. Most tunnels have some water inflow. When the Litani Tunnel in Lebanon advanced into a loosely cemented sandstone, a flow up to 95,000 gpm filled 9800 feet of the tunnel with 130,000 cubic yards of sand and finally receded to 3,000 gpm.[35,36] The

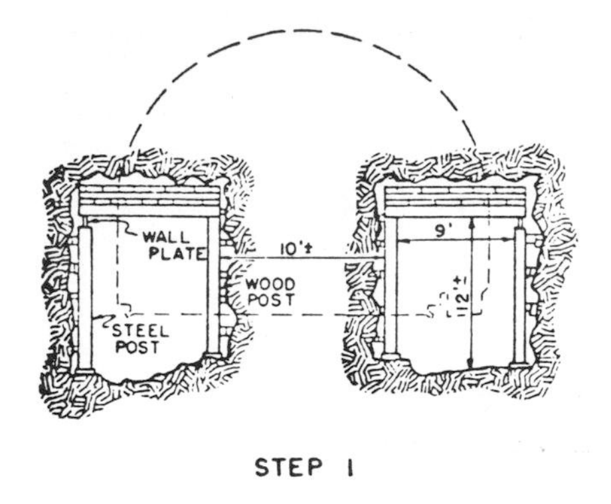

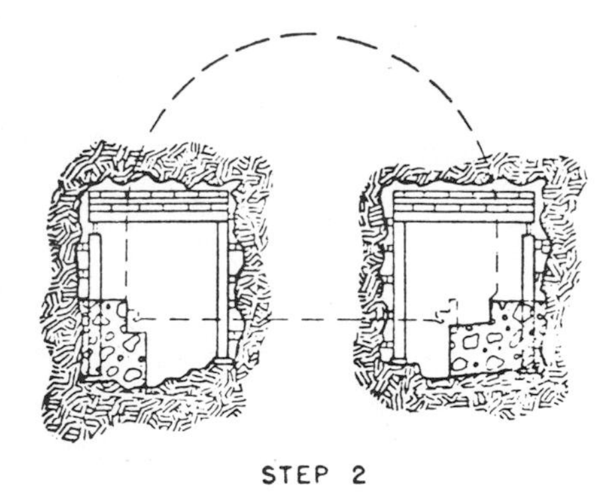

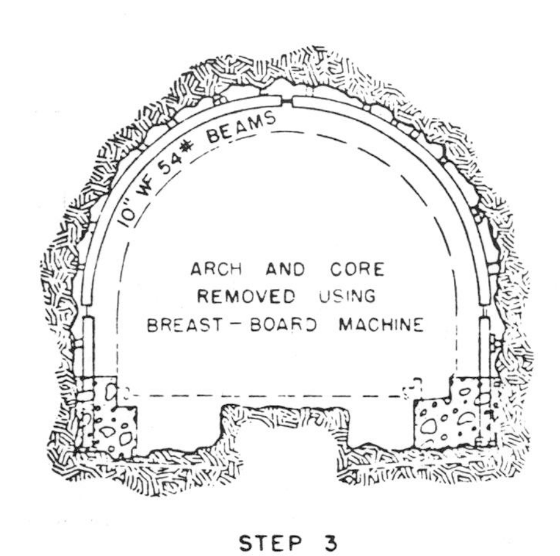

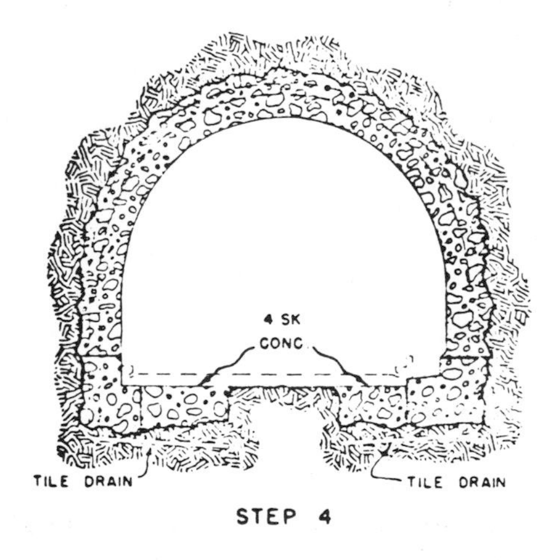

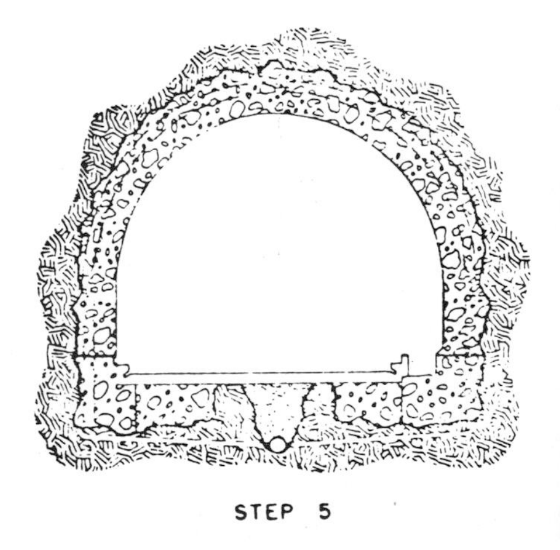

Fig. 7-47. Broadway tunnel. Type B Section.

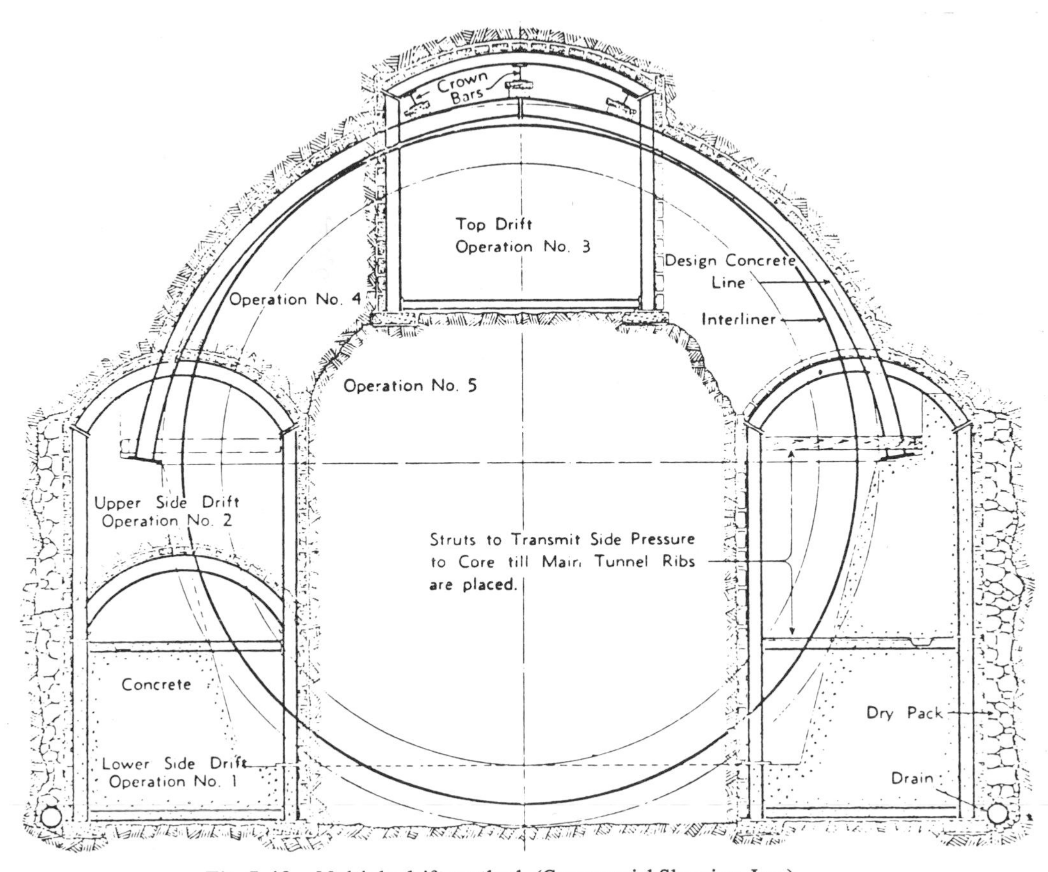

Fig. 7-48. Multiple drift method. (Commercial Shearing, Inc.).

work was stopped for two years. In the Graton Tunnels in Peru, where two tunnels 4 meters high by 3 meters wide were driven side by side, 18 meters on centers, one used for drainage, 2 meters lower than the other which was used for haulage, water flow reached 134,000 gpm. It was only by transferring the water from one bore to another that progress was possible.[37,38] Both were used for ventilation, one bringing in air and the other exhausting it. Excessive water flows in the San Jacinto Tunnel in Southern California were stopped by building a concrete bulkhead and grouting behind the bulkhead. This method has been used successfully in a number of cases, but is costly because of delays to tunnel progress.

To control the water, improve working conditions and prevent washing of wet concrete lining, sheet metal "pans," sometimes of thin plain or corrugated metal or as solid lagging, are used to divert the water to the sides and floor of the tunnel. Gravel or pipe drains may be used to carry the water to the tunnel portal. If the water is not carried in drains to the portal, weep holes in the concrete lining may be necessary to prevent excessive external pressures.

The worst kind of water in a tunnel is hot water. The Tecolote Tunnel in California had water temperatures of 112°F. Men were transported through the hot water area in muck cars filled with cold water. The Graton Tunnel in Peru had water at 156°F and high rock temperatures so that special dynamite capable of withstanding temperatures of 175°F was used.[37] An elaborate refrigerated cooling system was installed in a station near the hot water area to make it possible for men to work. As soon as possible after excavation, the hot water and hot rock areas were lined with concrete with arrangements to drain the water out of the haulage tunnel and into the parallel drainage tunnel.

Long pilot exploratory holes are sometimes drilled ahead of the face in tunnels where heavy inflows exist or are expected. In the Graton

Tunnel, such holes were drilled from recesses in the side of the tunnel several hundred feet ahead of the face, so that the normal excavating cycles could continue without interruption.

Water containing chemicals, such as sulfides, can react with temporary steel support or permanent concrete lining, causing deterioration and reduction in strength and possibly ultimate failure. If corrosive water is present, suitable steps must be taken, such as waterproofing or use of special cements.

PERMANENT LINING OF TUNNELS

Most tunnels that require temporary support during construction also require support of a permanent nature. It is the usual practice to place permanent lining after excavation of the entire tunnel is complete, even in large tunnels, where interference between excavation and concrete haul would be minimal. However, where muck trains can pass through tunnel lining forms, and equipment and ground conditions are such that early lining is desirable, concrete lining can be successfully accomplished in a 15 X 17 foot diameter horseshoe section prior to completion of excavation, as was successfully accomplished at Rimutaka Tunnel in New Zealand.

Vehicular Tunnels

If there is any possibility of blocks of rock dropping out of the roof or sides of a vehicular tunnel after construction is completed, it is imperative that they be permanently supported even though not supported during construction. This is the case except in formations of the highest rock quality (RQD).

Water and Sewer Tunnels

Tunnels conveying water can be dewatered if rock falls restrict capacity. Therefore, it will usually be more economical not to line such tunnels if the rock is of such quality that it will stand without support during construction and the expected rock falls will be minimal and scattered. If serious falls occur, the tunnel can be cleared and lined in spots, if necessary, at much less cost than lining the entire tunnel.[39] It will usually be found that a tunnel excavated as required for a lined section and left unlined will carry the design flow. It is desirable that the invert of all water tunnels be paved. Otherwise, when inspections are made, the uneven bottom will contain deep puddles of water, making access difficult. It is not necessary to clean up the invert except for pressure tunnels. It is costly to remove loose muck from the tunnel floor and this should be avoided where possible, thereby saving time and money. If the velocity of flow exceeds 12 feet/second, tunnels for power and diversion are usually lined.[40] Because it is difficult to unwater sewer tunnels, and because smooth flow is desired, all sewer tunnels should be lined.

Penstock Tunnels

When the rock cover of pressure tunnels is inadequate to resist the internal pressure, such as at the lower end of a penstock tunnel, it is customary to line with steel, backfilled with concrete. Penstock tunnels should be steel-lined in zone *A* to take full internal pressure for some distance from the powerhouse. It has been customary, in connection with underground powerhouses, to allow for the steel to take the full load for a distance up to 100 feet. At Churchhill Falls, in excellent rock (RQD 94) and with close rock bolting, only 20 feet were so lined, with the total length of steel lining only 130 feet from the upstream wall of the powerhouse. The major portion of steel lining is in zone *C*, where the steel liner shares the resistance to internal pressures with the rock and concrete backfill. Zone *B*, between zones *A* and *C*, is the transition zone in which the thickness of the steel liner is stepped down to the requirements of zone *C*. Concrete-lined portions of penstocks will crack. If water pressures are higher than in-situ stresses acting normal to joints in the rock, they may open and result in damaging seepage.[41] A failure in the penstock caused flooding at Bryte Underground powerhouse in Norway. In pumped storage plants, some authorities suggest that in preliminary design stages the steel lining should be carried to a point where rock cover is about 50% of maximum static pressure.[39]

Types of Lining

Ashlar masonry was used for lining early highway tunnels. Until recently, railroad tunnels relied on timber supports with more or less solid lagging for support. Many early sewer tunnels were lined with brick. Concrete lining, usually cast in place, has largely replaced all other materials in all types of tunnels. Concrete lining is frequently placed without reinforcing, but where bending stresses are serious, reinforcement may be required. Low pressure grout is usually used to fill the voids that would otherwise always occur in the tunnel roof. If the tunnel design relies on the rock to resist internal pressures of a water tunnel, high pressure grout is usually used to fill shrinkage cracks between the concrete and rock. Shotcrete or gunite, as used for temporary support, is coming more into favor as permanent lining where smooth surfaces are not required. Rock bolts are sometimes used for permanent support. In this case, they must be grouted for protection against corrosion and are usually used with wire mesh, covered with a thin layer of shotcrete. Highway tunnels sometimes have a tile lining on the walls, attached to the main concrete lining, to permit easier cleaning of the deposits from emissions of automotive exhausts. Brick or concrete lining of sewer tunnels is subject to corrosion from gases, acids and other deleterious substances. Under these conditions, special coatings must be applied to the surface of the linings.

Waterproofing

Some tunnels must be kept dry to serve their purpose. Dry tunnels are desirable in other cases, particularly those constructed for highways, especially in cold climates. Nothing causes more annoyance and adverse comments than a leaky, messy-looking tunnel. Expecting the concrete to be watertight in wet rock is unrealistic. Control of the water can be done in several ways. In moderately wet rock, the few leaks that will show up can be controlled by cutting chases into the lining and installing drains. Provisions must be made for cleaning out these drains. In really wet rock, more effective methods are required, such as those discussed below.

Dry pack can be used behind the tunnel lining, and through this, water can pass to drains discharging into the tunnel drainage system. This requires considerable labor for placing and there is danger that the dry pack may be plugged by grouting.

Membranes to shut out water, consisting of polyvinyl chloride (PVC)[42] or synthetic rubber sheets (neoprene) or multi-ply coal tar layers attached to a preliminary lining of shotcrete, have been used by Swiss engineers. The final lining is placed against the membrane. This method requires much labor, is expensive and is not completely reliable.

Perhaps the best solution is to provide a furred-out interior finish that hides the water and lets it drain off into the tunnel drainage. This also provides the bright, smooth interior finish that is needed in all long highway tunnels.

METHODS OF PLACING CONCRETE LINING

Types of Forms

Concrete forms are constructed as follows.

Steel Telescoping Forms. These are designed to collapse so that they can be transported through other sections of forms that are in place ready to receive concrete. This type is used for continuous placing of concrete and permits maximum production per shift. Bulkheads are not required except when concrete placing is stopped (such as at week-ends). Sufficient forms must be provided, usually several hundred feet, to permit concreting at a rapid rate and so as to allow time for the concrete to set before stripping and moving the forms from the back end through the remaining forms to the front end of the operation. This type of form will be found most economical for long tunnels. It is suitable for use in small and intermediate size tunnels where the concrete surface can be kept alive. If the thickness of concrete or reinforcing steel does not allow room for a concrete placing pipe, called slick line or shooting pipe, on top of the arch form, it is necessary to provide a curved recess in the form to contain the slick line. If tunnel ribs are used for support,

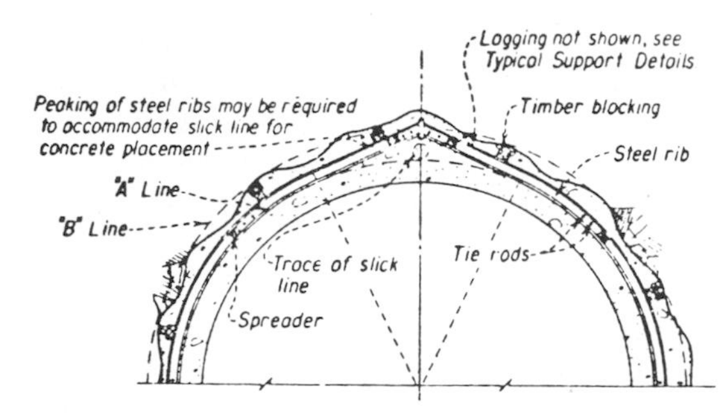

Fig. 7-49. Typical peaked structural steel rib.

they are sometimes designed with a peak at the top to make room for the slick line (Fig. 7-49).

Steel Collapsible Forms. These are designed to collapse sufficiently to permit stripping of the forms. This type (Fig. 7-50) is used for short tunnels not justifying the capital expenditure for continuous pouring; where the mass of reinforcing steel does not permit continuous placing; where the slope distance of a continuous pour is too great to keep it alive; or where telescoping is not required. In large tunnels, it is sometimes possible to get a good operation with two or three sets of forms. They can be started at the end and middle or third points of the tunnel, with pours being made in each form on alternate days. The disadvantage of this form is the requirement of placing a bulkhead at the end of each pour, the necessity of making long moves of placing equipment between pours and the limitation of the pour to a fixed length of tunnel. Depending on the amount of overbreak or placing troubles, the time for placing will involve overtime, idle time or inefficiency of moving men from one operation to another. Bulkhead forms are costly in both labor and materials.

Hinged windows in the forms, of manhole size, are provided except where the thickness of concrete permits men to move freely behind the forms. Vibration of concrete is accomplished through these windows. They are closed as the concrete reaches them. Supplemental vibration is sometimes accomplished by vibrating the forms.

Wood Forms. These are not much in use except for very short tunnels and transitions.

Fig. 7-50. Collapsible steel form.

Full circle forms may require strutting to the rock at the top to prevent floating. The inverts usually have removable panels that are placed after the concrete has been screeded to approximate line and grade. (Fig. 7-51)

Form Jumbos

Frames, called travelers or jumbos, riding on mainline track, wide gauge track or rubber-tired trucks, are used to strip the forms and move them to the next pour. They are equipped with jacks to raise or lower and to collapse or extend the forms. In the case of full circle forms for continuous pours (Fig. 7-52), the jumbos ride on track made integral with the forms. The outer end of the form is supported on spuds or jacks that are removed as the fresh concrete approaches them. The invert form may be integral or separate from the arch.

Form Stripping

Concrete lining in tunnels does not need much strength to support itself. The bond to the rock, the rough surface of the rock and the confined situation affect the ability of tunnel concrete to stand by itself. It has been found in practice that arch forms can be stripped safely 12 hours after placing concrete.

Grouting

It is impossible to completely fill the voids at the top of the tunnel with concrete. Low-pressure grout injected through holes in the concrete is used to fill the major voids and is required in most concrete lined tunnels. If the rock is relied on to resist internal pressures, high-pressure grouting is used to fill the shrinkage cracks between the concrete and rock. Holes may be made by drilling through the concrete. Grouting connections may be made to a nipple cast in the concrete or by inserting into the drill hole a tapered nipple threaded on each end into a soft lead sleeve at the collar of the hole.

Placing Equipment

The basic problem in placing concrete in tunnel lining is to elevate the concrete to the top of the arch and pack it as tightly against the roof as possible. Two methods are in general use, as described below.

Pneumatically Placed Concrete. This utilizes a tank with a quick-acting airtight door through which the concrete is loaded at the top. Leading from the bottom is a 6-inch or 8-inch pipe called a slick line. After loading, and with the door closed, compressed air is supplied to the placer tank, forcing the concrete out through the slick line. Air jets at various angles and places help the concrete get started and prevent blockages. The placer is usually located near the invert of the tunnel, necessitating two bends to get the discharge pipe to the top of the forms. The sloping pipe between the bends rises at an angle between 30 and 45 degrees. The flatter slope is preferable. The concrete traveling at high velocity causes heavy wear of the pipe line. The pipe is made of high carbon steel. If rotated 90 degrees at a time before it is worn through, the pipe will give better service. The greatest wear occurs in the bottom of the lower pipe elbow and the top of the top elbow. Former slick lines included special elbows with replaceable manganese liners. These have been largely replaced by rubber hose elbows. The discharge end of the slick line is kept buried in the concrete to force the concrete to as high an elevation as practicable. In a large tunnel, in order to keep the concrete alive, it may be necessary to start placing the concrete near the invert. This can be done by utilizing two slick lines, one on each side of the tunnel, alternately discharging into one and then the other so as to bring the concrete up on both sides at the same rate. An alternate plan is to run the slick line along the top of the arch, discharging into a box that feeds two radial pipes leading down the sides of the forms, as was done in the 30-foot diameter San Gabriel Dam No. 1 Tunnel in California. The entire assembly can be moved back and forth to bring the concrete up uniformly on both sides of the form.

Placing by Concrete Pump. A concrete pump is hooked up to a slick line similar to that used for pneumatic placing. The concrete travels at a slower speed so there is less wear on the pipe.

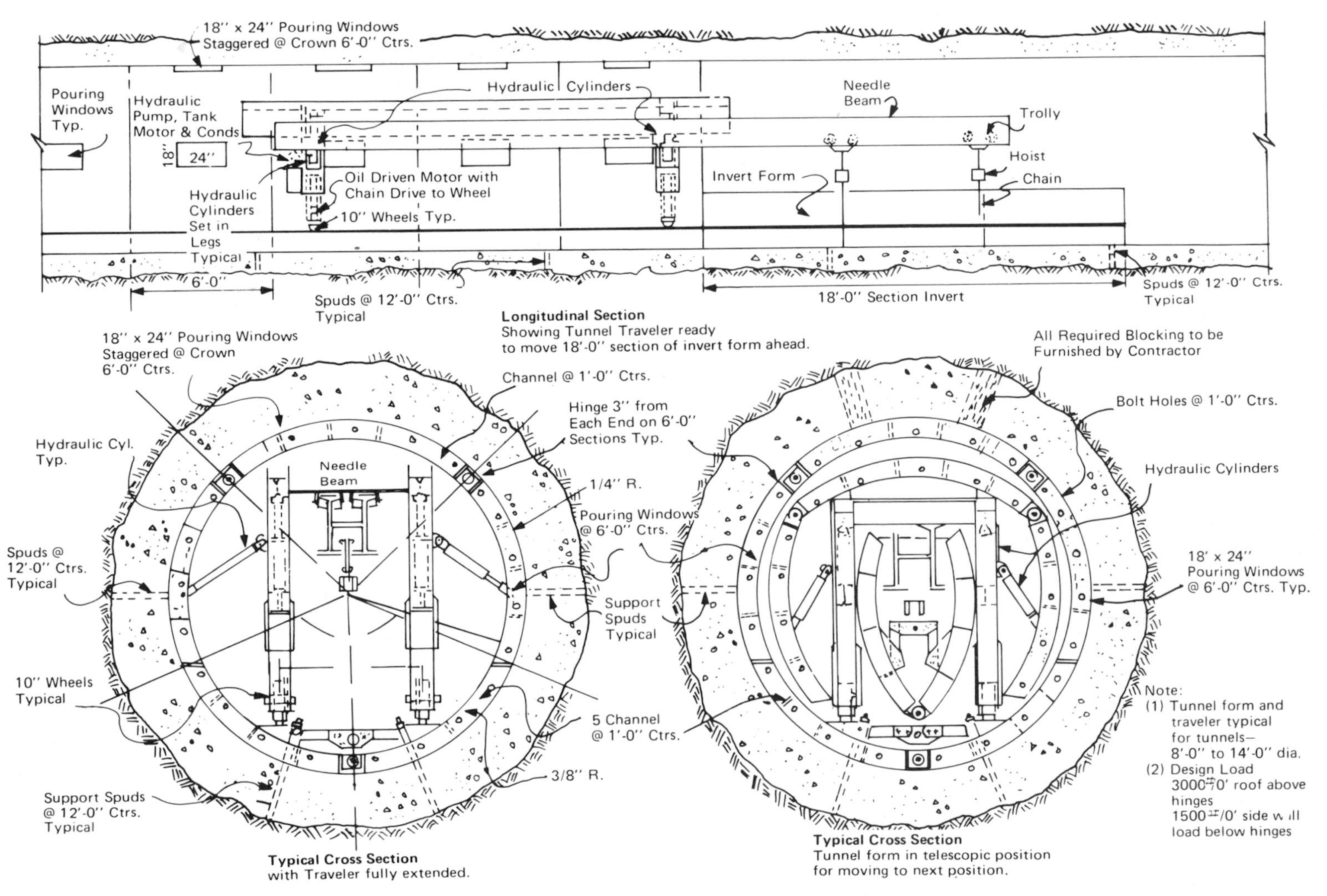

Fig. 7-51. Form jumbo and telescoping steel forms. (Economy Forms Corporation).

Fig. 7-52. Telescoping form in place.

Sometimes, compressed air jets are used to assist the concrete along its way. The end of the placing pipe is kept buried in the fresh concrete as for pneumatic placing.

Placing equipment is always mounted on a mobile jumbo so that it can be moved as the concreting proceeds.

Invert Concrete

Since invert concrete need not be elevated, it can be placed by the following methods.

1. Mixer or agitator cars discharging onto a belt extending the length of the section to be concreted. As the concrete is placed, the belt is withdrawn. The outer end of the belt may be supported on rubber tires if cleanup to rock is not required; otherwise on track supported on invert forms or screed which is removed as the concrete advances.
2. Discharging from cars traveling on an overhead track.
3. Pumpcrete.
4. Mixer trucks dumping directly into place in large tunnels where track is not used for excavation.

A heavy screed riding on invert forms or guides placed with the arch pour is pulled along the invert as concrete is placed. The screed may have vibrators incorporated into it to speed the operation. Good practice is to embed nuts in the invert for anchoring the arch forms. The nuts are held in place by tapered dummy bolts set in a template. These dummy bolts are withdrawn after the concrete has set and is replaced by others when the arch forms are set.

Batching, Mixing and Transport

Concrete may be delivered to the placing equipment by the following systems.

1. Batching outside the tunnel, hauling in batch cars to mixers near placing equipment and discharging from the mixer into the placer. In this case, a rather elaborate jumbo must carry the necessary equipment and move along the entire length of the tunnel as the work proceeds (Fig. 7-53). In large tunnels, transportation of batches can be by trucks using a turntable to turn around.
2. Batching and dry mixing outside the tunnel, and hauling in mixer or agitator cars that discharge onto a conveyor belt that delivers the concrete to the placing equipment. Water is added near the placing equipment.
3. Batching and mixing outside the tunnel, and hauling mixed concrete to the placing equipment.

All of these systems require car switching near the placing equipment similar to that used during excavation except in the case of systems 2 and 3 and where agitator equipment is used that permits the transfer of concrete from one car to another along the entire length of the train. Systems 1 and 2 are suitable for any length of tunnel if the aggregate is dry. However, the cost of agitator cars tend to rule out system 2 for longer tunnels. System 3 is suitable only for shorter tunnels. Many things can go

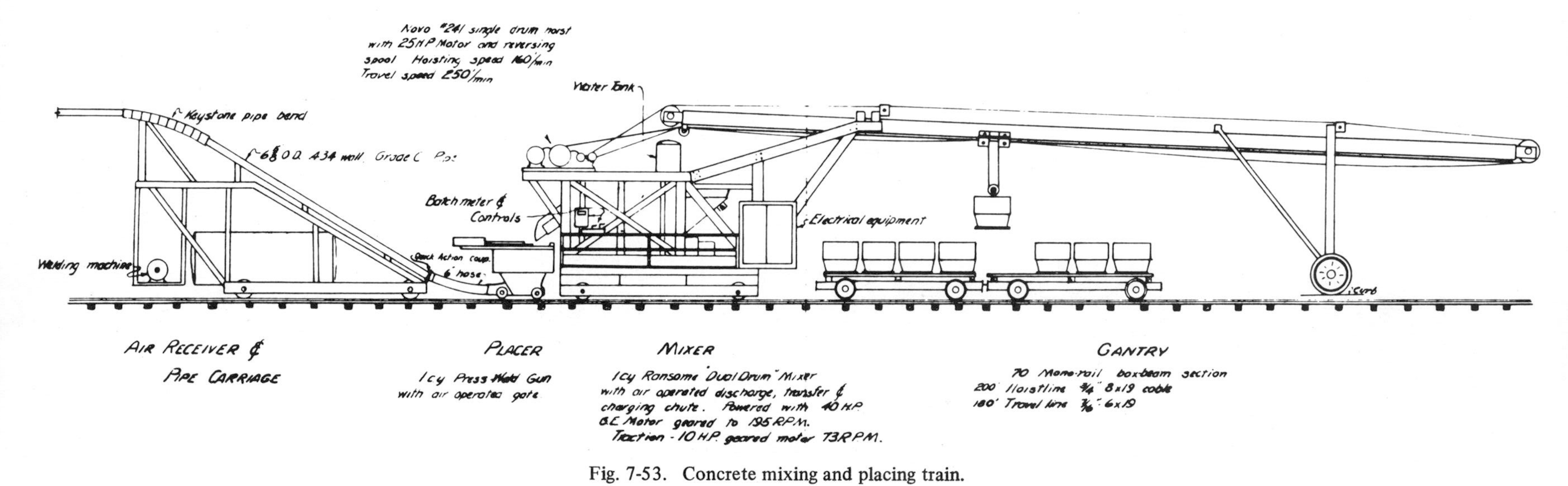

Fig. 7-53. Concrete mixing and placing train.

wrong with the concrete placing system so it is undesirable to have mixed concrete traveling a long distance in the tunnel. If placing operations shut down for any length of time, which is all too frequent in tunnel concreting, much concrete will have to be dumped and it is difficult to get it outside of the tunnel before it sets up in the cars.

Sequence of Operations

The lining of a small circular tunnel is usually placed in one operation. Larger circular tunnels may be placed in the same manner but it is more usual to place the invert first. Longitudinal forms on radial planes between 30 and 45 degrees from the vertical centerline are usually placed first. The concrete is placed within the forms and a heavy screed, shaped to the desired radius and riding on the forms, is dragged along them. In some cases, it may be desirable to leave the track in place on the rock bottom until the arch is placed. In this case, longitudinal forms may be set to form curbs on both sides of the track with their tops shaped to the tunnel radius. The curbs will contain holes and nuts into which tapered bolts to support the arch forms can be screwed. After the arch concrete is placed and the tunnel invert prepared for concrete, the curbs act as a guide for heavy invert screed that is dragged along them as the concrete is placed. After the screed has passed, the green concrete may be trowelled if a smoother surface is required.

For tunnels of horseshoe section, it will usually be found desirable to first place concrete curbs at the base of the walls on each side of the tunnel to support the arch forms. It is good practice to include a few inches of the invert with the arch pour. This provides a shelf for the invert screed to ride on. The advantage of this sequence is that haulage track or road from excavation operations can remain in place to serve the concrete arch placing and may not need to be replaced for the invert. After the concrete arch is in place, the track can be removed in sections, followed by placing of the invert concrete in that section before continuing with track removal.

DESIGN OF CONCRETE LINING

Concrete lining of tunnels may be needed to permanently support the rock loads, to protect temporary support from deterioration, to withstand external pressures and internal pressures, to prevent water losses or simply to provide a smooth surface. In a water conveyance tunnel, even if a concrete lining is not required for structural reasons, an economic study should be made to compare the cost of the concrete-lined tunnel with a hydraulically equivalent unlined tunnel. Depending on rock conditions, location of the construction site, and availability of aggregates, the concrete-lined smaller tunnel, with equal capacity, might prove to be more economical.

Lining for a Smooth Surface

In this case, it is necessary to have only a minimum workable thickness of concrete. This will depend on the concrete mix and should not be less than twice the maximum aggregate size. The average thickness will be much greater than the minimum because of the uneven rock surface.

Lining for Support of Rock Loads

It has been, and still is, customary practice to design concrete lining to resist the entire external load to which it is subjected, neglecting the strength of temporary support. This was proper when timber was used almost exclusively for temporary support because of its short life. Where steel ribs or rock bolts provide adequate support during construction, their capacity should be used in the permanent design, the concrete serving only as corrosion protection and a bridge between ribs if the temporary lagging has been wood or other material with a short life. The design load is the same as given for temporary support in Table 7-14. Low-pressure grout injected into the arch and sides will provide full corrosion protection and fill voids that might lead to future rock falls above the lining.

Lining for External Hydrostatic Pressures

The simplest solution where external hydrostatic pressures are present is to provide a drainage system of weep holes through the concrete to relieve the pressure. Such holes are usually 1 to 2 inches in diameter with a penetration of 6 to 15 feet, and are preferably located at the tunnel invert and lower part of the side walls. Transversal construction joints in the lining also help to relieve the pressure. In tunnels that must be kept dry, in pressure tunnels subject to unwatering (where leakage through the rock seams cannot be tolerated) or in other tunnels where leakage into them is undesirable, it is necessary to design the lining to withstand possible hydrostatic loads. A particularly difficult situation is at the entrance to large diversion tunnels or outlet works from reservoirs, where the gates are located at the upstream portal of the tunnel in seamy rock. Transition structures frequently include large, flat areas that must be designed to withstand full hydrostatic head. Circular sections require less concrete than other sections to be able to withstand hydrostatic loads. Grouting helps to reduce the hydrostatic loads, but since its effectiveness is never sure, the concrete must be designed for the theoretical load.

Lining for Internal Pressures

A pressure tunnel or penstock may be subject to great heads. From the structural standpoint, concrete lining is usually required at each end of the tunnel, extending inward to a point where the rock cover above is not less than the maximum internal head. Beyond this point, in a sound rock formation, the weight of the rock will counteract the internal pressure with a safety factor of approximately 2.0. Near the portals, the concrete must be lined with steel or other watertight lining or otherwise treated to prevent tension cracks and resultant leakage therefrom. The internal pressure in a tunnel may be resisted by reinforcement in the concrete. Where rock cover is sufficient, grouting of the voids around the lining will bring the concrete into contact with the rock so that the rock can take the load. High-pressure grouting will put the concrete in compression and eliminate or reduce longitudinal cracks and reduce or eliminate the amount of reinforcing steel required. Reinforcing is not required where vertical rock cover (C_V) is equal to or more than the hydrostatic head, and where the horizontal distance (C_H) to the rock surface is about twice that amount (Fig. 7-54).[43] Some authorities state that the minimum requirement for cover of an unlined tunnel varies from 1.5 to 0.75 of the static head.[40]

Design Criteria for Internal Water Pressure

The design of a tunnel lining is generally based on theories that equate the deflections of the lining and the rock. A scientifically correct solution because of uncertainties inherent to this problem is not possible. One of the approximations in such theories is treating the three-dimensional stress problem as two-dimensional by considering only the stresses in the two directions perpendicular to the axis, ignoring the influence of the rock stresses in the direction parallel to the tunnel axis. Other approximations result from the fact that the properties of the rock and the magnitude of the stresses in the tunnel walls are based on in-situ tests performed at several locations in the tunnel or in lieu thereof, estimating such stresses.

The lining of a tunnel subjected to internal water pressure has to be designed for the following three loading conditions.

1. During construction–external load due to maximum grouting pressure.
2. During operation–full internal water pressure and external rock load.
3. Under inspection or repair–no internal water pressure; maximum external load due to rock or groundwater.

In determining the maximum grouting pressure for the loading condition during construction, the following criteria should be established.

1. The external grout pressure should counteract internal hydrostatic pressure so that the lining does not have to be reinforced heavily for tensile stresses. Since the grout pressure on the outside of the lining is uneven and tends to relax after setting,

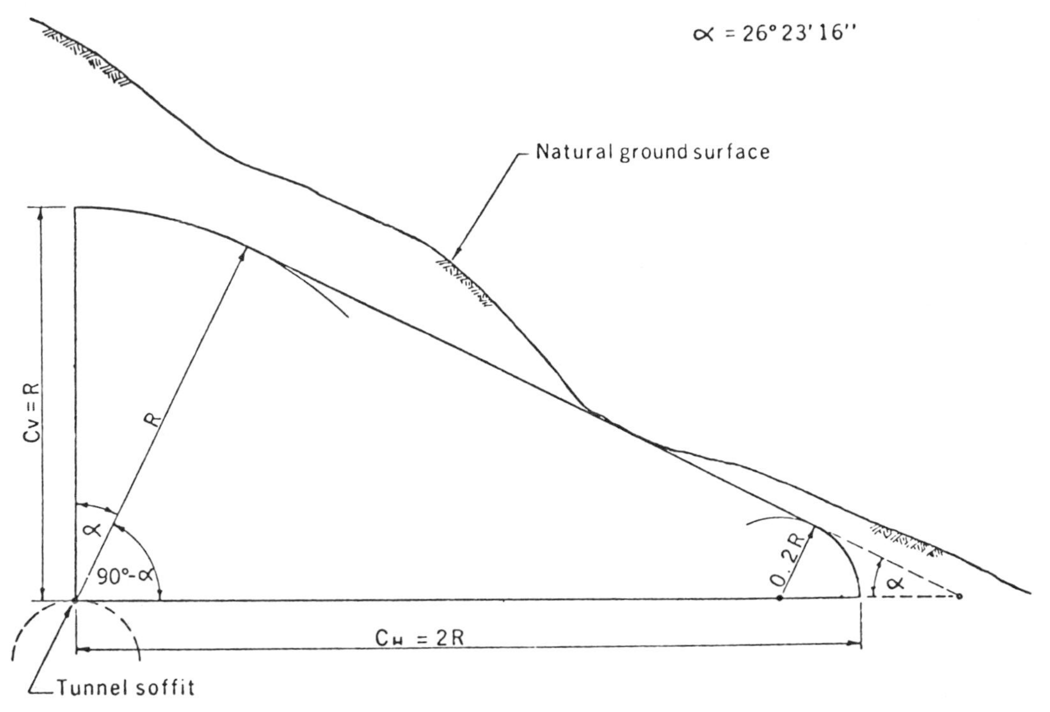

Fig. 7-54. Current practice, equivalent cover. From Unlined Tunnels of the Snowy Mountains, Hydroelectric Authority, ASCE Conference, Oct. 1963.
Curve C is undeformed tunnel lining.
Curve D is deformed lining due to all active forces.
Curve E is final lining deformation after passive pressures were brought into action.

it is considered necessary to double the pressure of injection. Jaeger[44] even recommends $P' = 4P$, where P' is the grout pressure at the pump and P is the internal water pressure.

2. The maximum grout pressure, on the other hand, should not exceed half the rock pressure. This will prevent lifting or weakening of the rock.
3. The resultant compressive stress in the lining, caused by the outside grout pressure, should not exceed the allowable value.
4. The buckling of the lining under assumed uniform radial external load (grout pressure) should be checked. The critical external buckling pressure is given by Roark:[45]

$$P = \frac{t}{R}\left(\frac{S_y}{1 + 4\dfrac{S_y}{E}\left(\dfrac{R}{t}\right)^2}\right)$$

where

P = grout pressure (psi)
t = lining thickness (inches)
E = modulus of elasticity of concrete (psi)
S_y = compressive yield point of concrete (psi)
R = radius of the tunnel (inches).

An economical lining design in poor rock, considering the second loading condition (during operation), is accomplished when the passive rock resistance due to the lining pressure against the rock is included in the calculations, as suggested by M.A. Drucker[46] in his analysis of passive soil pressures. The more flexible (thinner) the lining is, the less pressure required to expand it. Consequently, more pressure is transmitted to the rock surface, causing deformation of the rock and at the same time more deformation of the lining. On the other hand, the stiffer (thicker) the lining is, the greater the pressure needed to expand it (and the less pressure, therefore, transferred to the rock). If the rock is very sound, its deformation is small and thus the stresses in the lining are small. In such a case, a small amount of reinforcement is required to prevent possible leakage. If, however, the rock is weak (low values of K = the constant of elastic compression), the rock and lining deflections will be excessive and heavy reinforcement will be necessary to carry the high tensile stresses in the lining. For very weak rock, the lining must be designed for full water pressure. It is sufficiently accurate to assume that the lateral tunnel lining deflection is due to a uniform vertical loading acting downward on top of the lining over the entire horizontal diameter and resisted by an equal uniform pressure acting upward below this diameter, and that the intensity of the passive lateral pressures are proportional to the lateral deflections. M.A. Drucker has developed the following simple formulas for moments and thrusts due to external loading on circular tunnels (Fig. 7-55).

For bending moments at top and bottom centerline in foot-pounds:

$$M_T = M_B = -0.182\,Kr^2 \times \frac{Pd_1}{Kd_1 + P}.$$

For bending moments at horizontal centerline in foot-pounds:

$$M_H = +0.208\,Kr^2 \times \frac{Pd_1}{Kd_1 + P}.$$

For thrust at top and bottom centerline:

$$H_T = H_B = 0.59\,Kr \times \frac{Pd_1}{Kd_1 + P}.$$

where

$$P = \frac{0.0089\,EId_1}{r^4}$$

$$P_2 = K \times \frac{Pd_1}{Kd_1 + P}$$

d_1 = unrestrained lateral deflection at the horizontal diameter due to all the active forces (inches)

d_2 = deflection at the same point as d_1, after the passive pressures were brought into play (inches)

d_3 = deflection due to the passive resistance (inches)

K = constant of elastic rock compression representing the pressure in lb/foot^2 that would cause the rock to compress 1 inch (lb/foot^2)

E = modulus of elasticity of the concrete (psi)

I = moment of inertia of the lining section (inches^4)

r = radius (feet)

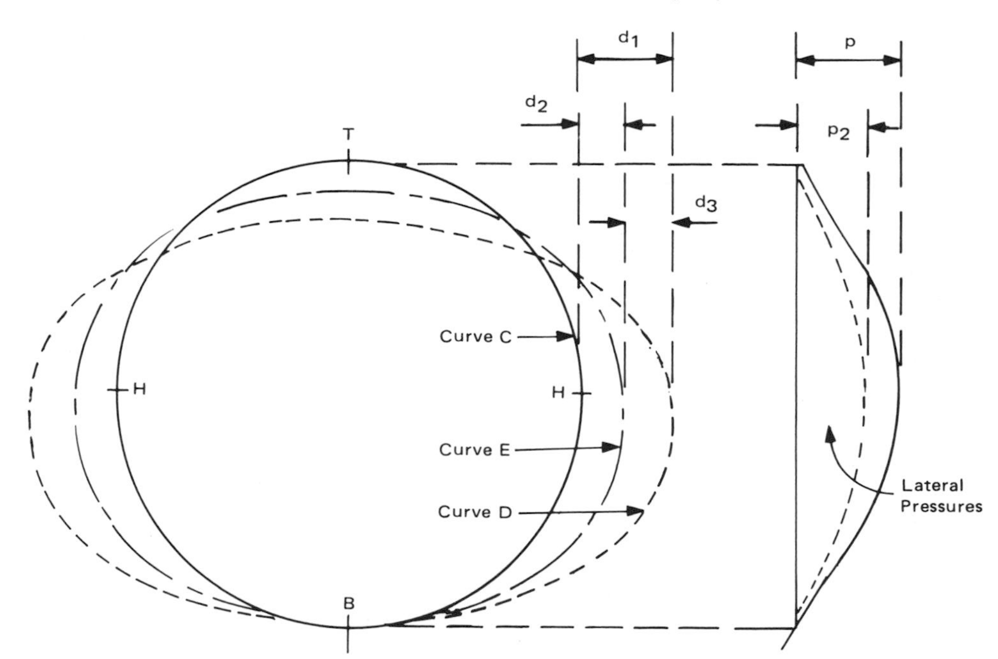

Curve C is undeformed tunnel lining.
Curve D is deformed lining due to all active forces.
Curve E is final lining deformation after passive pressures were brought into action.
Fig. 7-55. Tunnel Lining Conditions According to Drucker.

P = outside lateral pressure, with no restraint ($lb/foot^2$)
P_2 = passive resistance ($lb/foot^2$)

Concrete Mix

The maximum practicable size of aggregate for pneumatic placing is 2 inches. Larger aggregate will tend to plug the line. It is preferable not to use any larger aggregate in concrete pumps. High-slump concrete (5 inches) is necessary for pneumatic placing and desirable for concrete pumps. In the case of tunnel concrete, it is really "poured" rather than "placed." Restricted access inside the forms makes it necessary to have a concrete that requires little vibrating to fill the voids around tunnel ribs, lagging and packing. Because of the high cement content, tunnel concrete gives off more heat during curing than more readily accessible structures. This results in high temperatures in the tunnel.

Use of *plasticizing admixtures,* such as Sika Plastiment, are useful for placing lining, as they produce a higher slump with less water and reduce shrinkage. Certain rock formations may contain minerals requiring special cements. Groundwater with deleterious chemical content may react with concrete made with ordinary cements. In a highway tunnel in Switzerland, the normal cement was leached out of the roadway slab in a matter of months because of an underlying layer of gypsum that had been ignored in the design of the tunnel.

SPECIAL PROVISIONS FOR SEISMIC CONDITIONS

Tunnels should not be located across known active faults if another solution is possible. Even faults thought to be inactive may come alive. Earthquakes may cause displacement, either horizontal or vertical, at the fault, with the result that the use of the tunnel may be prevented or impaired. Except immediately at the fault, tunnels in competent rock (with high RQD) and with strong lining should not be seriously damaged by earthquake shaking. In poor rock formations, close to a fault, shaking may loosen the rock so that loads increase to those shown for crushed zones in Table 7-10.

There is little record of earthquake damage to tunnels. Four concrete-lined 16 X 22 foot Southern Pacific Railroad tunnels were severely damaged or destroyed by an earthquake in 1952 by movement along a fault intersected by the railroad.[47] Two of the tunnels were within the zone of faulting. The most distant damaged tunnel was within 1000 feet of the faulting. This line in the Tehachapi Mountains of California was out of service for 21 days. Repairs were made by daylighting two tunnels and part of a third with cuts up to 150 feet deep. In the 1170-foot long fourth tunnel, a 372-foot section was relined with a 2-foot thickness of concrete and a 600-foot section with 8 inches of gunite. A 12-inch concrete invert was placed throughout the tunnel.

The 18-foot diameter San Fernando tunnel, 29,000 feet long, under construction by the Metropolitan Water District of Southern California, was two-thirds excavated at the time of the February 9, 1971 earthquake. The primary support system consists of four 8-inch thick by 4-foot wide precast concrete segments placed in continuous abutting rings. Surveys indicated that the east portal of the tunnel is now 7.2 feet higher relative to the Magazine Canyon terminus than it was before the earthquake. There was no evidence of shear offsets resulting from the earthquake. The support system adjusted itself to the irregularities which developed throughout the excavated length so that damage to the tunnel was very minor, consisting principally of cracking and spalling of a few of the concrete tunnel supports. Because it will operate as a pressure conduit, the change in grade due to the earthquake is not expected to have any serious effect on the operation of the tunnel.

The Berkeley Hills railroad tunnel in California was designed 6 inches larger than normally required to allow for possible movement at a fault crossing the alignment.

UNDERGROUND CAVERNS

Many *hydroelectric power stations* have been built underground. Heights of the underground caverns for existing stations are up to 175 feet, widths to 95 feet and lengths to 1233 feet.

Exploration and Instrumentation

Because of the great widths and heights involved, it is necessary to do much more exploratory work, testing of the rock quality and instrumentation than is necessary for the usual size tunnel. If support fails or rock falls from the roof or sides of a tunnel 20 to 30 feet high, there is no serious problem in reaching the point of trouble and correcting it. The falling rock will choke itself off before irreparable damage will be done. However, in a powerhouse cavern, a failure of the roof or wall support could make it necessary to abandon the site. It is necessary to determine the dip and strike of the rock, its strength, joint patterns and, particularly, the location of weak joints. A study of geological records and surface geology will determine whether further consideration of a site is warranted. The second step *may* be seismic surveys, but it is usually taking cores from a number of drilled holes. The drill-holes may be used for water absorption tests, testing strength of the rock and bore-hole photography. If the results appear favorable, the following steps may be necessary to prove the adequacy of the site.

1. *An exploratory drift to and along the top of the proposed cavern*, from which:
   a. Observations of the geology can be made;
   b. In-situ stress measurements can be taken;
   c. Plate load tests can be accomplished;
   d. Flat jack tests can be accomplished; and
   e. checking can be done of orientation of joints and their strength.
2. *Checking of in-situ stresses by the bore-hole deformation method developed by the USBM.* This consists of drilling a 1.5-inch diameter pilot-hole and measuring the change in diameter when the pilot-hole is stress-relieved by over-coring with a 6-inch diameter core barrel.[48]

During construction, it is important to instrument the rock to determine how it may behave. Extensiometer readings and seismitron measuring of noise will help to locate any trouble spots in time to reinforce them before serious damage is done. (See Section 3)

Arch Excavation

Because of the great heights and widths, it is necessary to excavate the arch first in such a manner that the openings are always under control. Of the great variety of excavation sequences possible, only a limited number are adaptable to any given site where rock conditions will limit the choices. In poorer rock, crown and side drifts, as shown in Fig. 7-56(a), is a good method of attack. After the drifts (areas 1) are secured, the rock between them (areas 2 and 3) is removed either at one time or sequentially, depending on the quality of the rock. In poor rock, a center heading with multiple side slashes, as shown in Fig. 7-56(b), can be used. After the center heading is removed and secured, side slashes (area 2) may be advanced simultaneously or consecutively, followed by outer slashes (area 3). The relative widths and sequencing of side slashes may be varied to suit conditions as they are encountered. In smaller caverns or in better rock, a center heading with two side slashes, as shown in Fig. 7-56(c), can be used.

In strong competent rock, the full face of the powerhouse arch above the springline may be blasted at one time, providing the "stand-up" time is sufficient to permit reinforcement or support of the roof before damaging relaxing occurs in the roof rock. At Kemano in British Columbia, in granite formation, the arch was taken out by excavating first a center drift, followed by successive rings of rock around the drift designed so that the broken rock from successive rings would approximately fill the space left by the previous excavations. To excavate the final ring, at the perimeter of the arch, slots were excavated at 200-foot centers, from which holes were drilled 100 feet in each direction along the excavation line. These provided for blasting with little overbreak.[49,50]

Arch Support

As the arch is excavated, it must be supported to the extent required before proceeding with the balance of the excavation. Except in the most exceptional rock, some support will be required. Because spalling rock can damage equipment installations below, most underground power stations require the rock surface to be covered. Most stations built to date have a solid concrete lining over the full length of the powerhouse, even though rock bolts may be designed to support the load. Of 40 stations re-

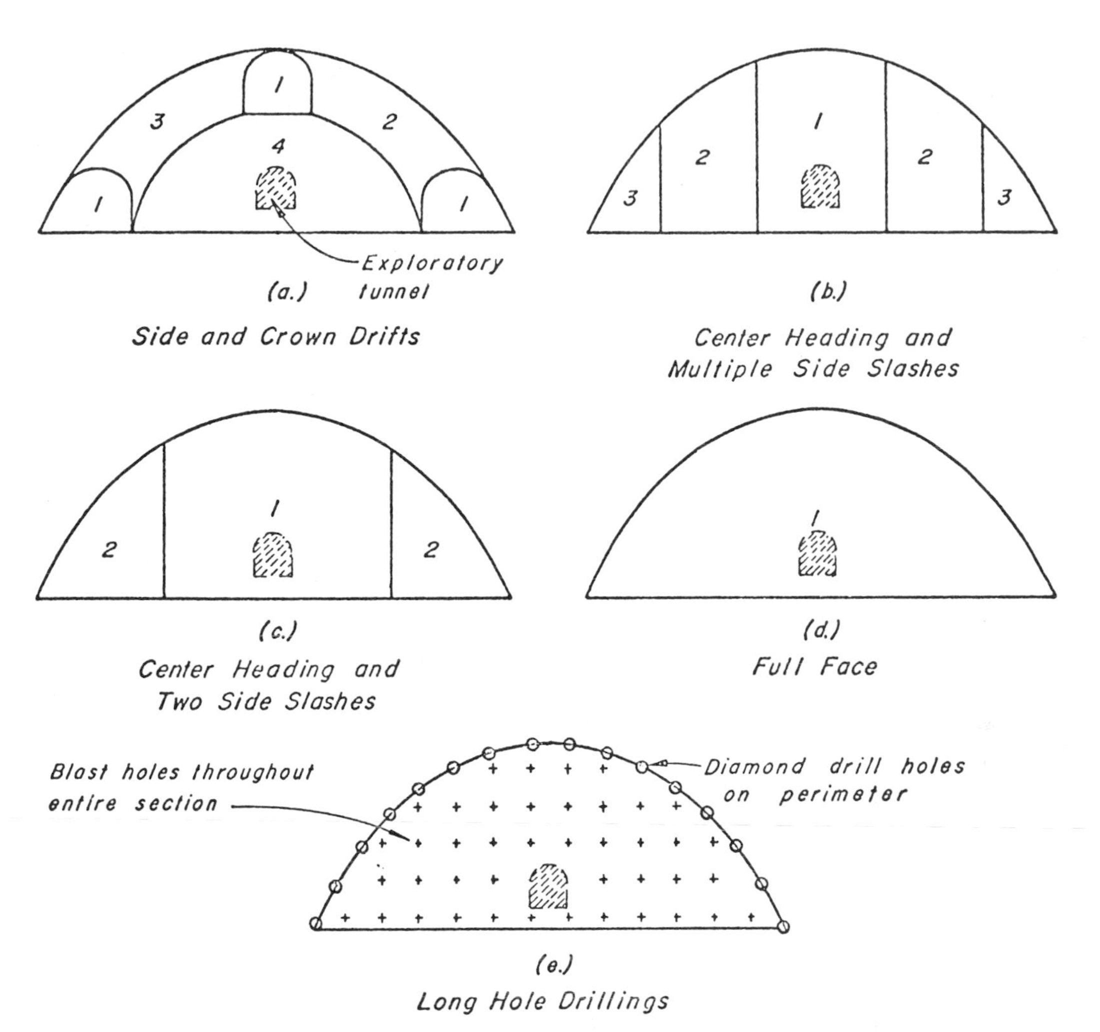

Fig. 7-56. Excavation sequence schemes.

viewed, 5% had no support, 12.5% had rock bolts without concrete, 10% had concrete ribs and 72.5% had solid concrete lining, mostly designed to take the rock load. More of the latest stations are being supported by rock bolts with wire mesh and shotcrete covering the rock surface. Loads may be estimated as for tunnels.

Rock Bolts in Caverns

The loads capable of being supported by rock bolt systems installed in large caverns, with RQD's in excess of 50 based on their yield strength, have usually been the equivalent of columns of rock between 0.10 and 0.25**B** in height. This is less than for steel ribs with relatively soft wood blocking that permits more adjustment of the rock and therefore greater loads on the support. Bolts should be installed as quickly as practicable after opening a section of the crown so as to limit rock adjustment to a minimum and thereby reduce loading accordingly. Joint orientation and quality of joints (clay-filled, slickensided, smooth surfaces or tight rough surfaces) will have considerable effect on crown stability. Rock with a high RQD of 90 to 100 might not behave as well (be considered "good-excellent") if the joints were smooth planes filled with a thin film of clay. The wedges of rock that must be supported by bolts depends on the angle of friction along joint surfaces and their orientation. It has been shown[51] that for caverns at large depth, the wedge that must be supported cannot be forced into the opening if the angle of friction is greater than half the angle of the wedge. This is illustrated in Fig. 7-57. As the wedge displaces, the following relation applies.

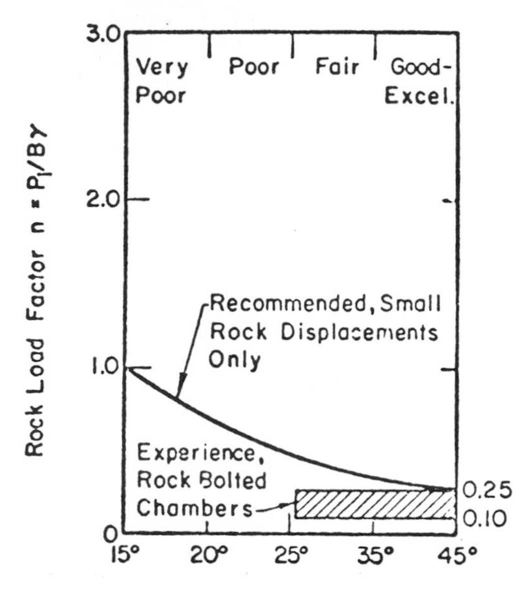

Fig. 7-57. Relation of rock load factor to angle of friction, ϕ, along wedge boundary. (1st North American Rapid Excavation and Tunneling Conference, 1972, AIME)

$$P_i = P_n\left(1 - \frac{\tan\phi}{\tan\theta}\right) + \frac{B_r}{4\tan\theta}$$

where

θ = one-half of included angle of wedge
P_i = internal pressure required to hold the wedge in place
P_n = average normal pressure acting on the side of wedge
ϕ = angle of friction along the joint surface.

The loads occurring from the largest wedge that could move into the flat crown of a deep cavern is represented by Fig. 7-58, providing the rock joints above are not permitted to open up. Joint orientations may reduce the possible load. It will be noted that recommended loads are conservative for fair to excellent rock. For shallow depth of cover, loads can be greater.

At Oroville Dam, the arch was supported by 20-foot rock bolts spaced at 4-foot centers. The walls were reinforced with rock bolts at 6-foot centers and the time allowed for installation was increased from 3 hours to 48 hours after each blast.[52] At Churchill Falls, the arch was supported by bolts spaced at 5-foot centers installed in two patterns, labelled I and II, at

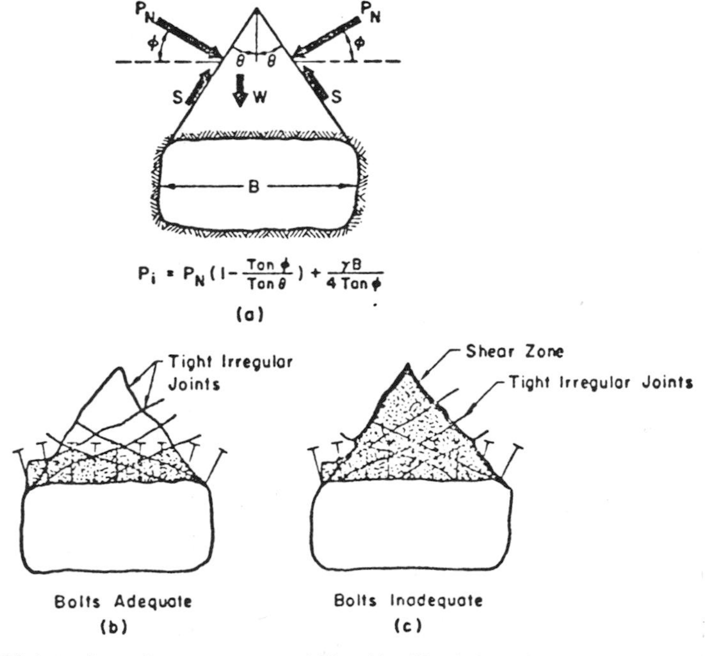

Fig. 7-58. Effect of joint orientation on crown stability (1st North American Rapid Excavation and Tunneling Conference, 1972, AIME).

7.1-foot centers each.[41] Design load of bolts with an ultimate strength of 68 kips was 45 kips. Pattern I was installed within 10 feet ot the working face, with lengths as shown on the plans within 8 hours after blasting. Lengths of bolts in pattern II were varied from 15 to 25 feet and were required to be installed to within 60 feet of the working face within 3 days. In poor quality rock, installation of both patterns was required to within 5 feet of the face. Wall rock bolts were installed on a 7-foot grid and had a design load of 30 kips with an ultimate strength of 45 kips. They were placed at 20 degrees to the horizontal so as to reinforce the most joints and were designed to hold gravity wedges inclined at 50 degrees to the horizontal. Recommendations and trends in the length of rock bolts are shown in Fig. 7-59.

Excavation below the Arch

Excavation below the arch is usually taken out in benches by down drilling, blasting and loading with shovels or front-end loaders into trucks. While this excavation is less costly than arch excavation, many estimators frequently overlook the extra cost of line drilling and pre-splitting, as well as the dental excavation at the bottom of the cavern.

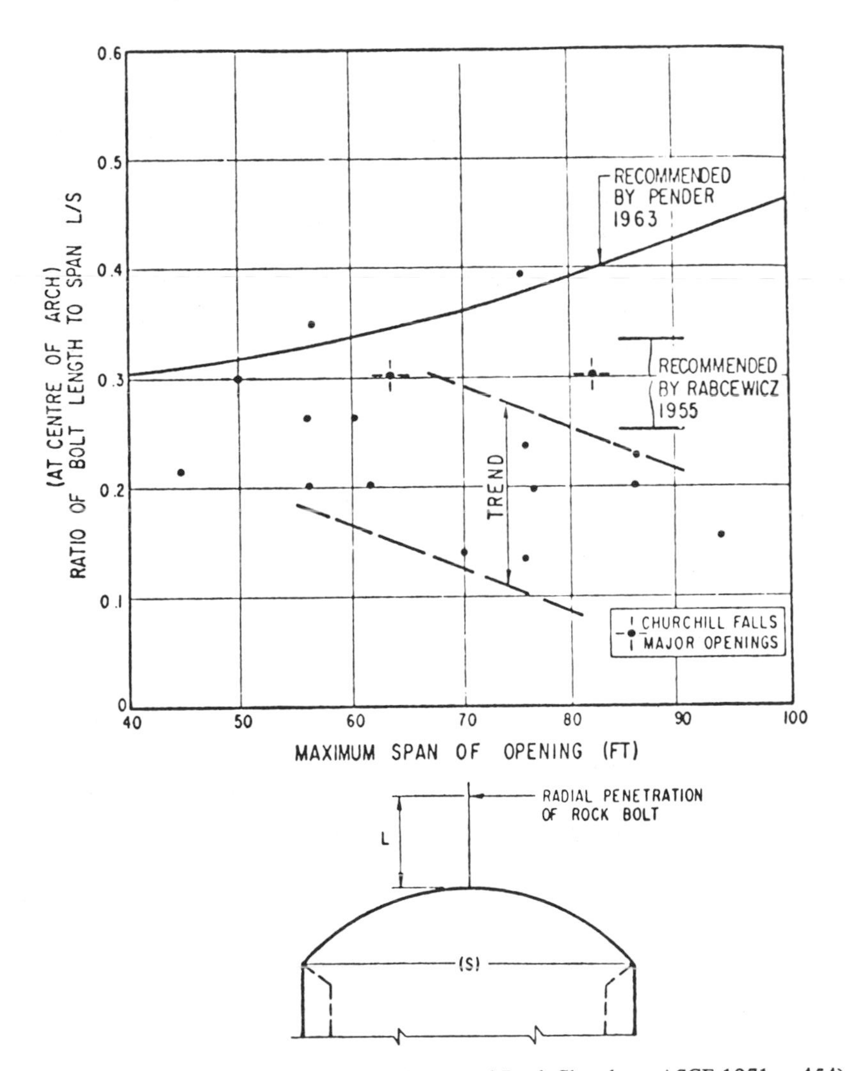

Fig. 7-59. Arch bolting for precedent openings. (Underground Rock Chambers, ASCE 1971 p. 454).

Wall support in underground powerhouses is frequently limited to rock bolting.

Access

Because the excavation must be carried on from many different levels, access is always a problem. The first work in the powerhouse must be done near the arch, whereas the normal access tunnel comes in at the powerhouse floor and the draft tubes come in at the extreme bottom. Anything that the designer can do to make possible the utilization of design excavations for access to the powerhouse at the springline of the arch would tend to reduce costs of construction. Because of the congested working space and interconnections between penstock shafts, access tunnels, tailrace tunnels and other openings into the powerhouse, it is necessary to coordinate all operations that may be going on at one time. No operation can be carried on by itself once a connection is made to the main cavern. If a "shot" is to be made, it may be necessary to clear all personnel from the cavern, thus interrupting all work, as well as that where the blast is to take place. Proper scheduling and sequencing of the operations is nowhere more important than in an underground powerhouse.

Underground Industrial and Storage Facilities

Because of danger from land or snow slides, war damage or other causes, it may be desirable to place storage tanks, manufacturing plants and warehouses underground. In Chicago, large caverns are being excavated to provide temporary storage for rainwater run-off where sanitary sewers are not in a separate system. These caverns are built as very large tunnels.[53] In Chile, because of snow avalanches, a copper ore deposit at the Rio Blanco mine site was not developed in the area until it was decided to place the mill underground. The large caverns required were excavated in a manner similar to that used for underground powerhouses.

Concrete Work

Because of limited access, concrete in underground power stations and facilities must be carefully scheduled. The saving in form cost by forming on one side only instead of two, as for outside sections, is more than offset by the cost of anchoring forms and overbreak backfill. The lack of storage and equipment maneuverability space within the cavern requires materials to be brought into the station only as they are needed and to be removed and returned between uses. It is usually necessary to install the temporary powerhouse gantry cranes before placing much of the concrete. This more difficult, and expensive handling of materials results in costs much higher than for outside construction.

PAYMENT QUANTITIES

Excavation

Because it is impossible to excavate to neat lines in rock tunnels, payment in unsupported tunnels is usually specified to a theoretical line "B" outside of the required excavation line "A," Fig. 7-60. The "B" line is usually a fixed distance beyond the "A" line. It should be established to encompass the average expected overbreak using the best excavation practice. Payment for excavation is usually made to the "B" line whether or not all excavation between the "A" and "B" lines is removed. In supported tunnels, the "A" line is sometimes specified as the line within which no unexcavated material of any kind, no timbering, and no metallic or other supports shall be permitted to remain. In this case, and with rib support, the "B" line should be established to encompass the average overbreak beyond a theoretical excavation line at the outside of the support.

In machine-bored tunnels in high-quality rock, overbreak will be negligible. In poorer rock, lining will normally be placed within the tail of a shield so rock falls are eliminated. For these reasons, the "A" and "B" lines are not as important as with tunnels excavated by drilling and blasting. The distance between the "A" and "B" line should be about 2 inches for machine-bored tunnels.

Payment for steel support is made on the basis of weight of steel support used. Because it is impossible to determine in advance the actual amount of support that will be required,

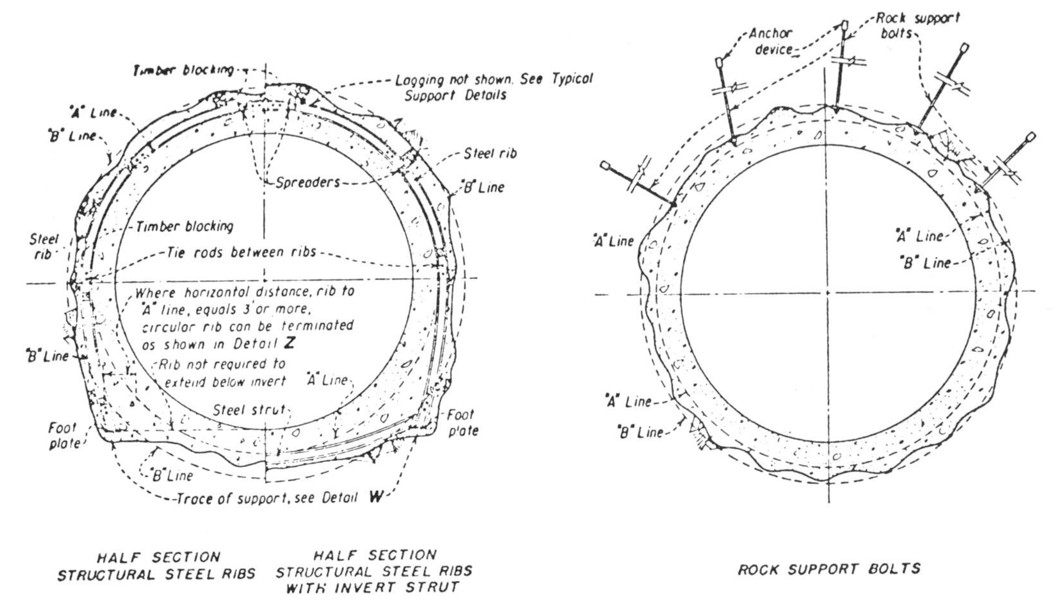

Fig. 7-60 Pay lines.

this item frequently varies widely from the estimated amount in the bid schedule. Engineers hesitate to restrict the amount of support a contractor may use, because of safety considerations and responsibility in case of failure. For these reasons, advantage is sometimes taken, if there is a good price on support items, to install more than necessary. Efforts have been made to avoid this problem by the engineer fixing the price before the bids are taken. A better solution is to fix the price only on quantities in excess of those in the bid schedule. The best solution is to make a thorough study of the site of a proposed tunnel and as realistic an estimate as possible of probable support requirements. Because a contractor must have several weeks supply of support on hand, the engineer usually agrees to buy back, at cost, any support left over at the end of the job.

Rock bolts are usually paid for by the linear foot for the actual length of bolts installed. The unit price may include the cost of drilling the holes; of furnishing and installing the rock bolts complete with all accessories except bearing plates and structural steel shapes; and of retightening the bolts. Bearing plates and other structural shapes are measured and paid for separately.

Concrete lining is measured for payment to the "B" line regardless of the quantity actually placed. It is common practice to pay for cement used outside the "B" line.

Unpredictable Factors

Excessive water is sometimes covered by a price for pumping or handling water. If "bad" ground is anticipated, but of unknown characteristics or extent, a provision is somtimes made that if progress is less than a given footage per day, the work will be paid for on a cost plus basis. The cost plus arrangement should start after some days of reduced production, to encourage the contractor to make all possible progress.

REFERENCES

1. Longwell, Knopf and Flint, *Outlines of Physical Geology,* John Wiley & Sons, New York,
2. Leadbetter, B.C. (General Superintendent, San Jacinto Tunnel), "Driving an Extremely Difficult Tunnel," *Engineering News-Record* **121**: 669–673, November 24, 1938.
3. "Sand-filled Water Tunnel Holed Through," *Engineering News-Record* **174**: 20, May 6, 1965.
4. "Heat Being Conquered at Tecolote Tunnel," *Engineering News-Record* **151**: 25, November 26, 1953; *Engineering News-Record* **152**: 45–48, June 17, 1954.

5. "Twin Tunnels Will Drain and Ventilate Flooded Peruvian Mine," *Engineering News-Record* **173**: 24, September 3, 1964.
6. Nast, Paul H., *Drillers Handbook on Rock*, Davey Compressor Company, Kent, Ohio, 1955.
7. Bateman, W.M., *Rock Analysis*, Joy/Air Power, Joy Manufacturing Company, March–April 1967.
8. Deere *et al.*, *Design of Tunnel Liners and Support Systems*, PB 183799, Customer Services Clearinghouse, U.S. Department of Commerce, Springfield, Virginia, 1967.
9. Terzaghi, Karl, "Rock Defects and Loads on Tunnel Supports," in Proctor and White (Eds.), *Rock Tunneling with Steel Supports*, pp. 17–99, Commercial Shearing, 1964.
10. Lauffer, H., "Gebergsklassefezierung fur den Stollenbau," *Geologie und Bauwesen* **24**, *No. 1*: 46–51, 1958.
11. Brekke, T.L. and Howard, T.R., "Stability Problems Caused by Seams and Faults," *Proceedings of First North American Rapid Excavation and Tunneling Conference*, AIME, Chicago, Illinois, June 5–7, 1972.
12. "Pilot Bore is Laboratory for Twin Road Tunnels," *Engineering News-Record* **173**: 38–39, August 13, 1964.
13. Hopper, R.C. *et al.*, "Construction of Straight Creek Tunnel, Colorado," *Proceedings of First North American Rapid Excavation and Tunneling Conference*, AIME, Chicago, Illinois, June 5–7, 1972.
14. "Manapouri: Where New Zealand is Mining for 700,000 KW," *Engineering News-Record* **178**: 26–37, June 8, 1967.
15. "Air-placed Explosives in Carolina Tunnels," *Engineering News-Record* **171**: 28–30, September 5, 1963.
16. E.I. Du Pont de Nemours & Co., *Blasters Handbook, Fifteenth Edition*, pp. 157, 139, 170–171.
17. "Ground Vibration Due to Blasting and Its Effect Upon Structures," *Journal of Boston Society of Civil Engineers*, pp. 222–245, 1949.
18. U.S. Bureau of Mines Bulletin 442, p. 66.
19. Crandall, F.J., "Transmission Coefficient for Ground Vibrations Due to Blasting," *Journal of Boston Society of Civil Engineers* **XLVII**, *No. 2*, April 1960 (Reprint).
20. Du Pont Technical Bulletin: *Static Electricity Hazards and Their Control in Pneumatic Loading Systems.*
21. Lyra, Flavio H. and MacGregor, W., "Furnas Hydroelectric Scheme, Brazil: Closure of Diversion Tunnel," *Proceedings of the Institution of Civil Engineers* **36**, *Paper No. 6993*: 21–46, January 1967.
22. *Mine Gases and Methods for Detecting Them*, U.S. Bureau of Mines Circular 33 (Revised March 1954).
23. *Tunneling, Recommended Safety Rules*, U.S. Bureau of Mines Bulletin 644, 1968.
24. *Protection Against Mine Gases*, U.S. Bureau of Mines Circular 35, 1954.
25. Wickam, George E., Tiedman, Henry R. and Skinner, Eugene H., "Support Determinations Based on Geologic Predictions," *Proceedings of First North American Rapid Excavation and Tunneling Conference* **1**: 43–64, AIME, Chicago, Illinois, June 5–7, 1972; *Proceedings of Second Conference* **1**: 691–707, 1974.
26. Jacobs Associates, *Research in Ground Support and Its Evaluation for Coordination with System Analysis in Rapid Excavation*, Bureau of Mines, Contract No. HO210038, Technical Report No. 115.
27. Mason, E.E., "The Function of Shotcrete in Support and Lining the Vancouver Railroad Tunnel," a paper presented at the Tunneling and Shaft Symposium, Minneapolis, Minnesota, April, 1968.
28. Mason, R.E., *Instrumentation of the Shotcrete Lining in the Canadian National Railway Tunnel*, Vancouver, B.C., M.S. Thesis in the Department of Mineral Engineering, University of British Columbia.
29. "Hazardous Tunneling at Hetch Hetchy," *Engineering News-Record* **110**: 701–704, June 1, 1933.
30. Kobler, H.G. *Dry-Mix Coarse Aggregate Shotcrete as Underground Support, Shotcreting*, ACI Special Publication 14, pp. 33–58, 1966.
31. "Miners Winning Battle at Straight Creek Tunnel," *Engineering News-Record* **186**: 20–22, April 22, 1971.
32. "Tough Tunnel Bows to Chemical Grouting," *Engineering News-Record* **168**: 32–34, March 29, 1962.
33. "Construction Methods on Six Mile Moffat Tunnel," *Engineering News-Record* **94**: 966–971, June 11, 1925.
34. Hopper, R.C., Lang, T.A. and Mathews, A.A., "Construction of Straight Creek Tunnel," *Proceedings of First North American Rapid Excavation and Tunneling Conference*, pp. 501–508, AIME, Chicago, Illinois, June 5–7, 1972.
35. "Sand-Filled Water Tunnel Hole Through," *Engineering News-Record* **174**: 20, May 6, 1965.
36. "Flowing Sand Plugs Two Miles of Tunnel," *Engineering News-Record* **165**: 38–40, August 25, 1960.
37. Kincaid, Charles G., "The Graton Tunnel Project in Peru," *Tunnels and Tunneling* **2**, *No. 5*: 281–285, September 1970.
38. "Graton Tunnelers Nightmare Complete with Scalding Deluge," *Engineering News-Record* **185**: 22–25, November 26, 1970.
39. "Kemano Repair Bill $2 Million," *Engineering News-Record* **167**: 23, July 20, 1961.
40. Vasilescu, M.S., Benziger, C.P. and Kwiatkowski, R.W., "Design of Rock Caverns for Hydraulic Projects," *Symposium on Underground Rock Chambers*, pp. 21–50, ASCE, January 13–14, 1971.

41. Benson, Raymond P. *et al.*, "Rock Mechanics at Churchill Falls," *Symposium on Underground Rock Chambers,* pp. 407-486, ASCE, January 13-14, 1971.
42. "Two-layer Plastic Liner System Will Keep Austrian Alpine Tunnel Dry," *Engineering News-Record* **190:** 68, April 12, 1973.
43. Dann, H.E., Hartwig, W.P. and Hunter, J.R., "Unlined Tunnels of the Snowy Mountains Hydro-electric Authority in Symposium on Unlined Power Tunnels," ASCE October 1963 Conference, San Francisco, California.
44. Jaeger, Charles, "Present Trends in the Design of Pressure Tunnels and Shafts," *Water Power,* p. 127, April 1955.
45. Roark, R.J., *Formulas for Stress and Strain,* McGraw-Hill, New York, 1954.
46. Drucker, M.A., "Determination of Lateral Passive Soil Pressure and its Effect on Tunnel Stresses," *Journal of the Franklin Institute,* May 1943.
47. *Earthquakes in Kern County, California, During 1952,* Bulletin 171, Division of Mines, Department of Natural Resources, November 1955.
48. Wild, Philip A. *et al.*, "Northfield Mountain Underground Station," *Symposium on Underground Rock Chambers,* pp. 287-331, ASCE, January 13-14, 1971.
49. Libby, James W., "Planning and Construction of Underground Facilities for Kemano, Kitimat Project," *Proceedings of Second Protective Construction Symposium* **2:** 711, Rand Corporation, Santa Monica, California, 1959.
50. Wise, L.L., "Methods Spur Underground Powerhouse Tunnels, *Construction Methods and Equipment,* p. 72, December 1952.
51. Cording, Edward J. and Deere, Don U., "Rock Tunnel Supports and Field Measurements," *Proceedings of First North American Rapid Excavation and Tunneling Conference* **1:** 601-622, AIME, Chicago, Illinois, June 5-7, 1972.
52. Kruse, George H., "Power Plant Chamber Under Oroville Dam," *Symposium on Underground Rock Caverns,* pp. 333-379, ASCE, January 13-14, 1971.
53. "Tunnel Will Store Storm Run-off," *Engineering News-Record* **179:** 32, November 30, 1967.

Bibliography

Tunneling Technology, an appraisal of the state of the art for application to transit systems, by Golden Associates, Consulting Geotechnical Engineers and James F. MacLaren, Ltd., published by The Ontario Ministry of Transportation and Communications.

8

Mixed Face Tunneling

LYMAN D. WILBUR

Vice-President, Morrison-Knudsen Company, Inc. (Retired)
Chairman of the Board and President, International Engineering Company, Inc. (Retired)
Consulting Engineer Member National Academy of Engineering
Honorary Member American Society of Civil Engineers

SOIL CONDITIONS

When one part of the tunnel face is in soft ground, usually the top, and the other in hard rock, it is called a "mixed face." If the top of the tunnel is in soft ground and the bottom is in rock, it can produce one of the most difficult tunneling conditions, much more so than if the tunnel were completely in soft ground. Hard rock miners are not familiar with soft ground and thus are not as well equipped to handle the soft ground (just as soft ground workers are not familiar with hard rock). Futhermore, except when boring machines are used, the hard rock must be drilled and blasted without damaging the support of the soft ground.

Mixed faces occur as a result of 1) intersecting different strata, 2) a seam of decomposed rock, 3) ancient river beds that have been filled with sand, gravel or mud and, 4) fault zones. Frequently, such formations contain water and will cause serious inflows of sand and water into the tunnel.

The preliminary geological investigations and exploratory borings for a tunnel should endeavor to locate possible mixed face conditions. The heading must be continually watched for the unexpected condition. When approaching suspected mixed face conditions, an exploratory hole should be drilled 8 or 10 feet beyond the normal depth so as to obtain advance warning. If a mixed face condition is encountered, exploratory drilling should be made to determine its extent. This will assist in a determination of the best method of attack.

PROCEDURES FOR TRANSITION

When a tunnel advances from full face rock to a mixed face, methods of operation must usually be radically revised. Care must be taken to prevent the soft ground, if water-bearing, from being washed into the tunnel. If exploratory holes indicate water-bearing sand and gravel, grouting in advance of the breakthrough into the soft ground may prevent a serious run. Provisions must be made for tight breastboarding and tight lagging of the soft ground area, or other measures must be taken—such as freezing, grouting or chemical treatment—that will prevent runs and provide the necessary support. Soft ground appearing first in the roof of the tunnel is more difficult to handle, but care must be exercised to prevent runs even if the soft ground appears first in the bottom of the tunnel. If the tunnel is being excavated by drilling and blasting, the length of the rounds should be reduced so as to open up the soft ground area in small increments until the revised methods for handling the changed conditions are fully developed. The earth portion, if at the top, should be excavated first. After the soft ground is thoroughly supported, the rock must be carefully blasted so as not to destroy the support of the earth section.

If the soft ground is in the bottom, with rock on top, it will be necessary to support the entire perimeter of the tunnel above the invert as soon as the bottom of the rock stratum reaches a plane that permits sloughing of the earth to undermine the rock walls sufficiently to weaken the rock arch of the tunnel.

As a tunnel advances from a mixed face with soft ground at the top to full face earth, temporary supports must be extended downward to support the earth portions of the perimeter of the tunnel until the entire perimeter is supported. If the transition zone is short, it may be best to use supports reaching to the invert for its full length. Then the methods for excavating soft ground tunnels (see Chapter 5) will prevail. If the tunnel is being excavated with a mole, full ring support will be required with liner plates or lagging covering the earth portions of the tunnel surface.

Transition from full face earth to mixed face with rock in the bottom will require drilling and blasting of the rock after the earth in a round has been excavated and supported. If the rock appears in the top of the tunnel, it should be excavated before taking out the earth portion of the round. The face may have to be supported by breastboards. Blasting must be done with care to avoid disturbing the support.

EXCAVATION IN DRY GROUND

Steel Poling Plates. A combination of soft ground and hard rock techniques must be used. If the top of the hard rock is nearly horizontal, the soft ground above may be taken out as a top heading (Fig. 8-1). The soft ground must be supported by liner plates, ribs and solid lagging, or similar tight support. Steel poling plates jacked into the face of the heading one at a time can give temporary support until the primary support is placed.

Spiling. Another method is to use wood or steel spiling with ribs (Fig. 8-2). Wood breastboards are used, if necessary, to support the face. After the roof is securely supported, the rock bench must be excavated by careful drilling and blasting. This method of construction requires much hand labor and is tedious and slow.

Shield. For a long mixed face tunnel, consideration should be given to a half-shield supporting the soft ground on top, or a full shield (see also Chapter 6). The shield should be equipped with jacks to hold breastboards against the face, as well as jacks to move the shield ahead (Fig. 8-3). After the soft ground is excavated a few feet ahead of the rock and supported with breastboards, the rock is blasted carefully to avoid damage to the shield and to

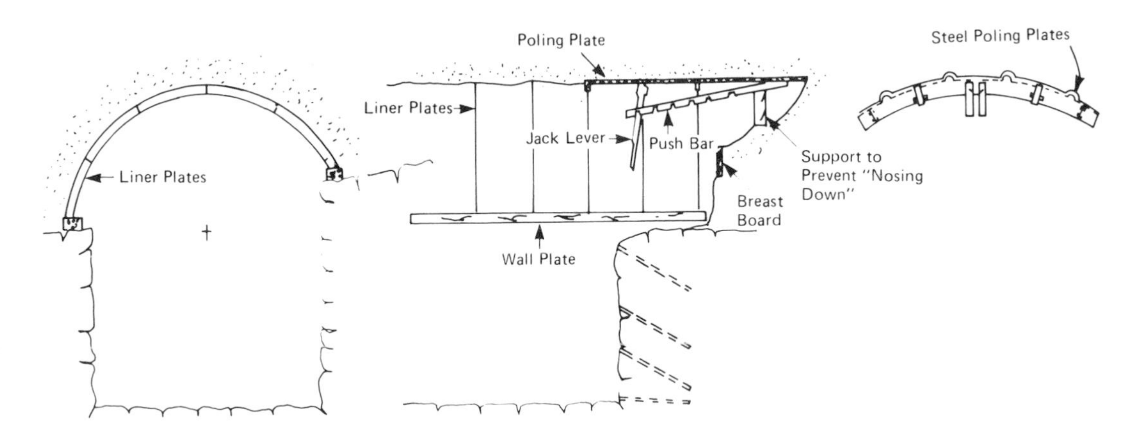

Fig. 8-1. Rock tunnel in mixed face using steel poling plates.

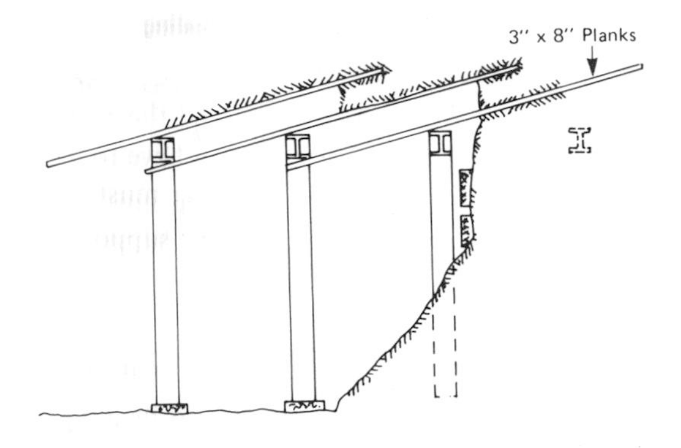

Fig. 8-2. Fore poling with wood spiles.

the breasting. The shield is then moved ahead to its next position.

Machine Excavation. Most tunnel-boring machines are not adaptable to working in mixed face. If a machine designed for rock excavation encounters a mixed face, it may not work in soft ground. Tunneling must stop and methods changed when mixed face is reached, so it is important to determine in advance the presence of such conditions. It may be necessary to excavate a bypass around the machine and excavate by other methods until a full rock face is reached, when the machine can be moved ahead to continue the excavation.

EXCAVATION IN GROUND WATER

For excavating the soft ground in mixed face tunneling, one of several methods (described in Chapter 5) may have to be applied.

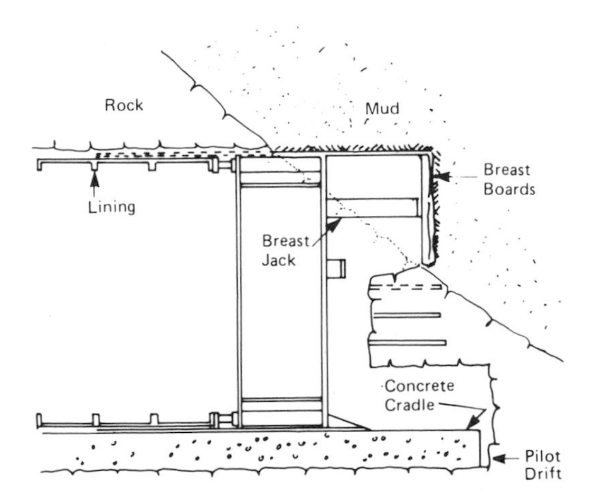

Fig. 8-3. Tunnel shield in mixed face.

If the ground is not too fine, the same methods described for dry ground may be used. This also is the case if the soil can be consolidated by dewatering or grouting.

Dewatering. This may be done by a series of wellpoints, extended from the surface if the depth does not exceed about 15 feet, or by deep wells for greater depth. Test wells with pumping tests should be made to determine the permeability of the soil.

Grouting. Various methods of grouting may be used, depending on the soil conditions (see Chapter 5). Chemical grouting of silty fine sands, silts and course sand and cobbles at mixed faces below the water table was accomplished in a 10-foot, 6-inch and 14-foot, 6-inch sewer tunnel constructed in New York.[1] Permeabilities of soils were in the 10^{-3} to 10^{-6} centimeters/second range with those at the lower rate predominating. Two chemical grouts were used, a polymer grout, AM-9, which is a mixture of two monomers (acrylamide and N, N'-methylenebisacrylamide), and Terranier, consisting of low molecular weight polyphenolic polymers catalyzed by solutions containing a formaldehyde and metal salt. The latter grout is available in mixtures to give higher strengths or for penetration into very fine-grained soils. The manufacturer claims AM-9 will penetrate soils with an effective diameter as small as 0.013 millimeters and will result in compressive strength of stabilized soils up to 280 psi. On approach to the mixed face, a shield was installed and AM-9 injected ahead of the shield and to a radial distance of 5 feet beyond the excavation line. The initial plan of consolidating 5 feet ahead of the face and 5 feet beyond the excavation line was found to be inadequate. It was concluded that a minimum of 15 feet ahead of the working face needed to be consolidated. Most of the grouting was done from two rows of holes 8 feet from the centerline and 10 feet on centers. Better results were obtained by grouting additional holes along the tunnel centerline. One and one-quarter-inch pipes were used for injecting pipe directly into the soil. Grout was injected from 10 feet below invert grade to 8 to 10 feet above the crown. The grout take was 175 to 200 gallons of 35% gel per vertical foot of hole treated. A shield was used in conjunction with the grouting, but it was not necessary to use compressed air. The consolidated ground attains more strength with age. Where strength is required, grouting of this type should be performed at least 30 days in advance.*

Freezing of Ground. In some cases, freezing can provide a temporary ice dam to hold back the soft ground. In Milwaukee, a sewer tunnel 90 feet deep had soft ground on the sides. Three-inch pipes were driven into the ground and capped. Two $\frac{3}{4}$-inch copper tubes were inserted through the cap. Freezing was done with liquid nitrogen, delivered to the site in a tank truck. Liquid nitrogen can produce a temperature of $-320°F$ when expanded. In two days, the ground was frozen solid. In New York, where the top third of a tunnel was in soft ground and the bottom third in rock, in order to avoid the noise of a compressor plant at night, frozen dikes were put on each side of a sewer tunnel with cross-dikes at 100-foot intervals.[2] In this case, an elaborate refrigeration plant was installed to freeze the 100-foot stretchers, one after the other.

Shield Tunneling. Shield tunneling with compressed air may have to be resorted to in fine running soil under water pressure for longer stretches of mixed face if consolidation of the ground is not practical or is too costly at greater depth. The soft ground at the face is tightly breastboarded (see Chapter 6).

TEMPORARY SUPPORTS

Ribs and Lagging. In the soft ground portions of a mixed face tunnel, the type of support will be similar to that required for soft ground tunnels (see Chapter 5) or in bad ground conditions of hard rock tunnels. Timber sets or partial sets founded on the top of the rock surface with solid timber lagging can be used for short stretches. Where *steel ribs* are used, wood or steel lagging is required along the perimeter of the soft ground section. The arch ribs must be set on wall plates above or at the top of rock. Posts extending below the wall plates are required if the top of rock is below the arch or if the rock overbreaks too much or is too weak to provide support for the arch.

Liner Plates. Used along the perimeter of the soft ground section, as used for soft ground tunnels (either with or without steel ribs), these will serve the purpose of solid lagging as well as arch support. When used under such conditions, the liner plates are supported on wall plates. The wall plates are supported on the rock surface or on posts, as rock conditions require.

Shotcrete. This is one of the best support systems if the ground will stand for a sufficient period to permit its application. It would only be suitable where the soft ground has sufficient cohesion to remain in place, after excavation of a complete ring, for a period of a few hours. The shotcrete can be placed in small amounts at any one time so that advance of the soft ground section may be made in small increments (see Chapter 12).

SETTING CRADLE FOR SHIELDS

Where it is necessary to install a shield in a rock tunnel preparatory to excavating through a mixed face transition zone and on into soft ground, a concrete cradle is constructed having a radius conforming to the outside radius of the shield. This cradle must have a length adequate to support the shield during erection and to provide a track long enough ahead of the shield for about two shoves. This will permit the shield to bite into the ground ahead sufficiently to keep it on line and grade before it leaves the cradle. The arc of the cradle should be about 60 degrees. Steel rails can be imbedded in the concrete for accurate alignment and to reduce friction.

*AM-9 is now prohibited by OSHA in the U.S., as it has been determined to be carcinogenic.

REFERENCES

1. Anderson, E. Roy and McCusker, Terence G., "Chemical Consolidation in a Mixed Face Tunnel," *Proceedings of First North American Rapid Excavation and Tunneling Conference* 1: 315-329, AIME, Chicago, Illinois, June 5-7, 1972.
2. *Constructioneer,* p. 20, September 22, 1969.

9
Shafts

ROBERT J. JENNY

President
Jenny Engineering Corporation
South Orange, New Jersey

GENERAL

There are generally two modes of access to tunnel construction: through a portal providing direct access from the surface, or through a shaft, providing vertical access to the level of tunnel operations. Since urban land is valuable, and interference with existing services must be minimized, most tunnels built through urban areas require shafts to reach the working area, and to provide egress for tunnel muck.

Tunnel shafts can be temporary or permanent: temporary shafts are for the contractor's use only; permanent shafts will become part of the tunnel structure. Permanent shafts can be used for ventilation, pumping, utility lines or manholes, or they may be enlarged to house stations.

Locating a shaft at the midpoint of a tunnel will permit tunnel driving in two directions; also, one compressor plant, one hoist, one shop and one office can serve both headings. Locating a shaft near vacant land will facilitate the erection of temporary buildings. The proximity of muck disposal locations and routes should also be considered.

Once the shaft has been sunk to grade, a pump chamber and a sump is excavated if required. The pump should provide sufficient capacity to handle the maximum anticipated flow. An estimate of groundwater seepage can be based on previous experience in the same soil or rock medium. Unexpectedly large inflows may occur if water-bearing strata or seams are encountered during excavation.

SHAFT EXCAVATION IN SOFT GROUND

Shafts in soft ground are normally excavated with a crane using a clamshell bucket to hoist the muck from the shaft and drop it into a hopper or truck on the surface.

Temporary shafts in soft ground are often circular in shape. A concrete collar, a ring of concrete usually 2 feet wide and 4 feet deep, with the top surface at least 12 inches above ground level, should be placed around the top of the shaft. The collar prevents distortion of the shaft's primary lining and prevents surface water and debris from falling into the shaft. (see fig. 9-1). Handrails must also be provided.

Primary shaft linings are normally installed at every 4 or 5 feet of advance. However, shafts have been sunk up to 30 feet without supports. The rate of installation depends on the type of lining and the nature of the soil medium.

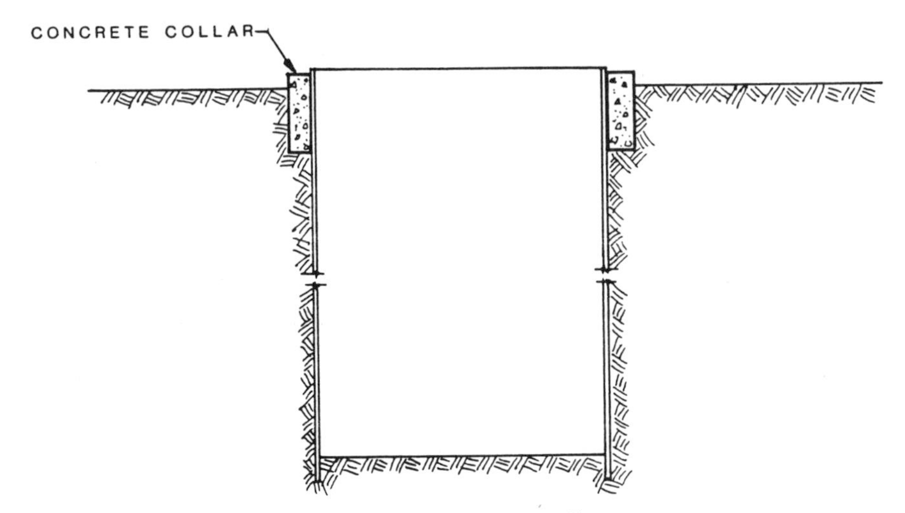

Fig. 9-1. Concrete shaft collar.

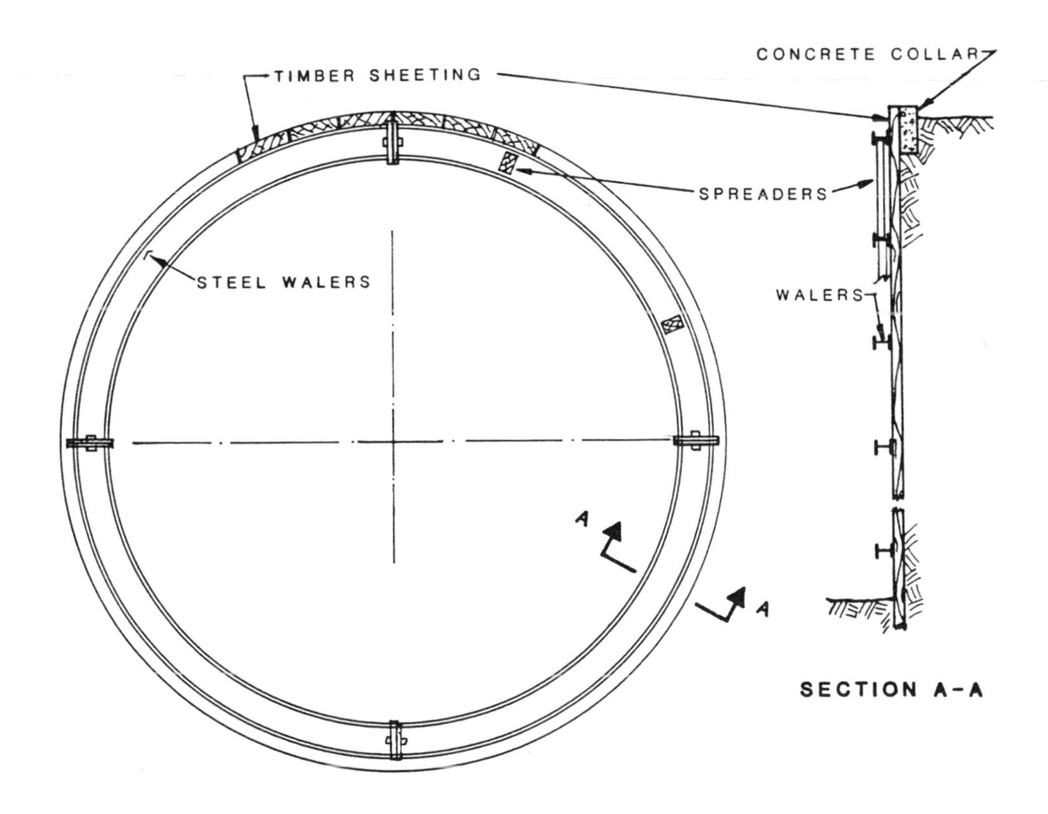

Fig. 9-2. Timber sheet piling.

Permanent shafts are usually concrete lined. They may be round or rectangular in shape; concrete for shaft lining may be cast either with forms on both sides or forms on the inside only with the ground support system on the outside.

Soft ground shaft sinking may disturb or damage neighboring buildings, utilities, pipelines or streeets. The problem is especially acute with soft plastic soils; when plastic soil is excavated, the load over the excavated area is reduced, and plastic yielding may result, causing ground yielding at the surface. Plastic yielding can result below stresses which are associated with shear failure. A properly designed sheeting system can prevent plastic yielding.

Soil characteristics, shaft depth and economic factors will dictate the choice among the many available sheeting and bracing systems.

Timber Sheet Piling

Timber sheet piling is normally used only in shallow shafts, since driving the thick sheeting is difficult. The method can be economical to start excavating in soft material, not deeper than about 20 feet, overlying rock.

Three-or four-inch thick timber is driven into the ground and excavation is usually performed simultaneously, thereby reducing friction and also preventing the soft timber tips from catching on obstructions and splintering. Horizontal

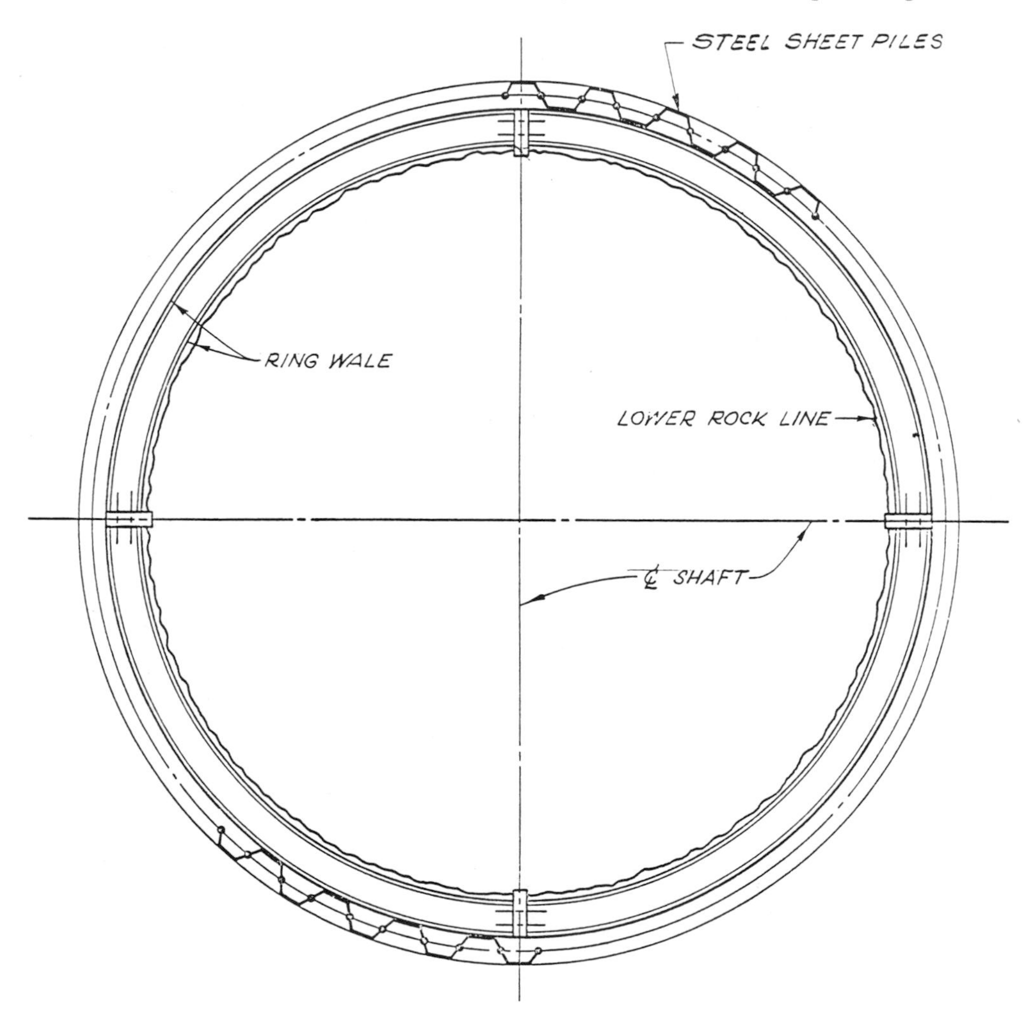

Fig. 9-3. Steel sheet piling. a) Cross-section. b) Elevation.

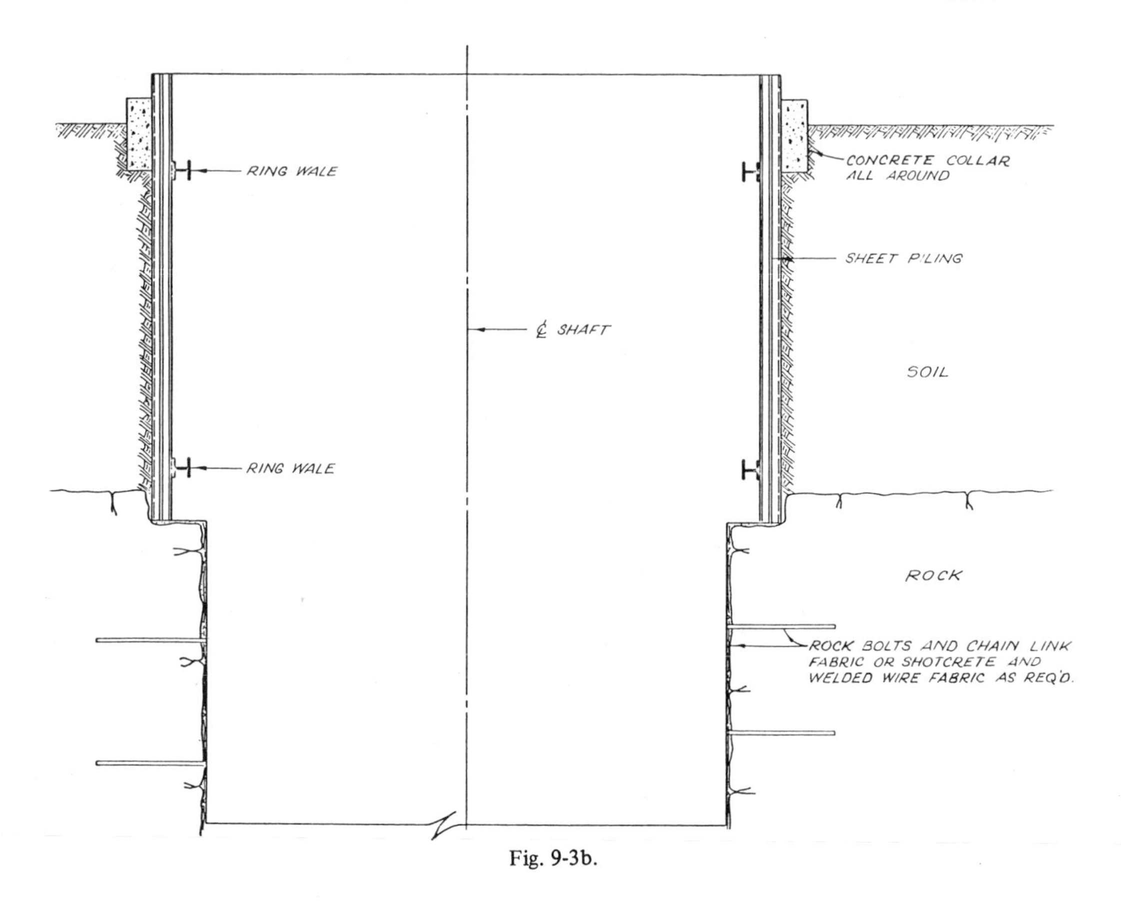

Fig. 9-3b.

circular or rectangular steel rib sets are installed against the interior of the sheeting. All the lateral earth pressure acting on the sheeting is transmitted to the steel ribs, which carry the stress in ring compression or bending, depending on whether the shaft is circular or rectangular.

The timber piling is inexpensive, and, once installed, is very easy to work with (see fig. 9-2).

Steel Sheet Piling

Interlocking steel sheet piles are usually used to brace soft, water-bearing ground if the excavation exceeds about 20 feet.

Steel sheet piles must be driven carefully to ensure proper interlocking of the joints to cut off water seepage. Excavation usually occurs

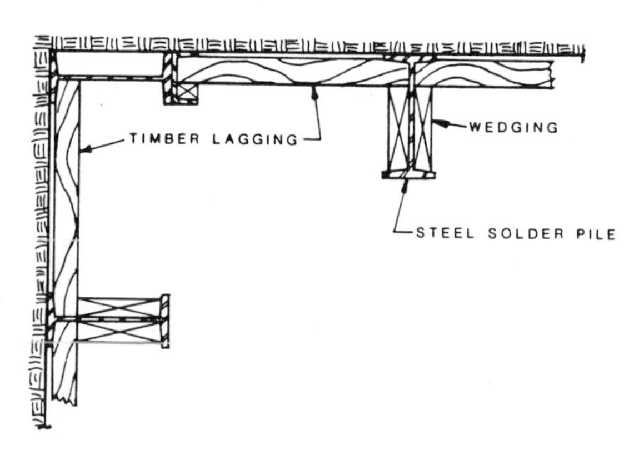

Fig. 9-4. Soldier piles and lagging.

after driving operations are completed, unless the shaft is unusually deep.

Horizontal steel rib sets are then installed progressively with the depth of excavation. (see fig. 9-3).

In deeper excavations, where larger earth pressures are encountered, horizontal steel rib sets are installed. The steel rib sets must be designed for either ring compression or bending, depending on whether the shaft is circular or rectangular (see fig. 9-4).

Soldier Piles And Lagging

Steel H piles called "soldier" piles usually spaced from 6 to 10 feet apart, are driven to the required depth. Excavation begins after all the piles are driven. As the excavation proceeds, horizontal timber lagging is placed against the face of the excavation and wedged between the flanges of the H piles. Usually, a small space is reserved between the lagging to allow for drainage. In very wet soil or in running sand, hay is forced into the spaces to prevent the ground from flowing into the excavated area.

Liner Plates

Most liner plates are corrugated pressed steel pans with bolt-holes in the flanges on the sides and ends to permit bolted erection of the ring.

Shaft excavation begins by precisely erecting the first ring on the ground surface, and placing a concrete or earth collar around it. The soil is then excavated and liner plates are added progressively to the lowest ring.

The soil pressure is carried by the liner plates in ring compression. In large-diameter shafts, more than about 35 feet circular hori-

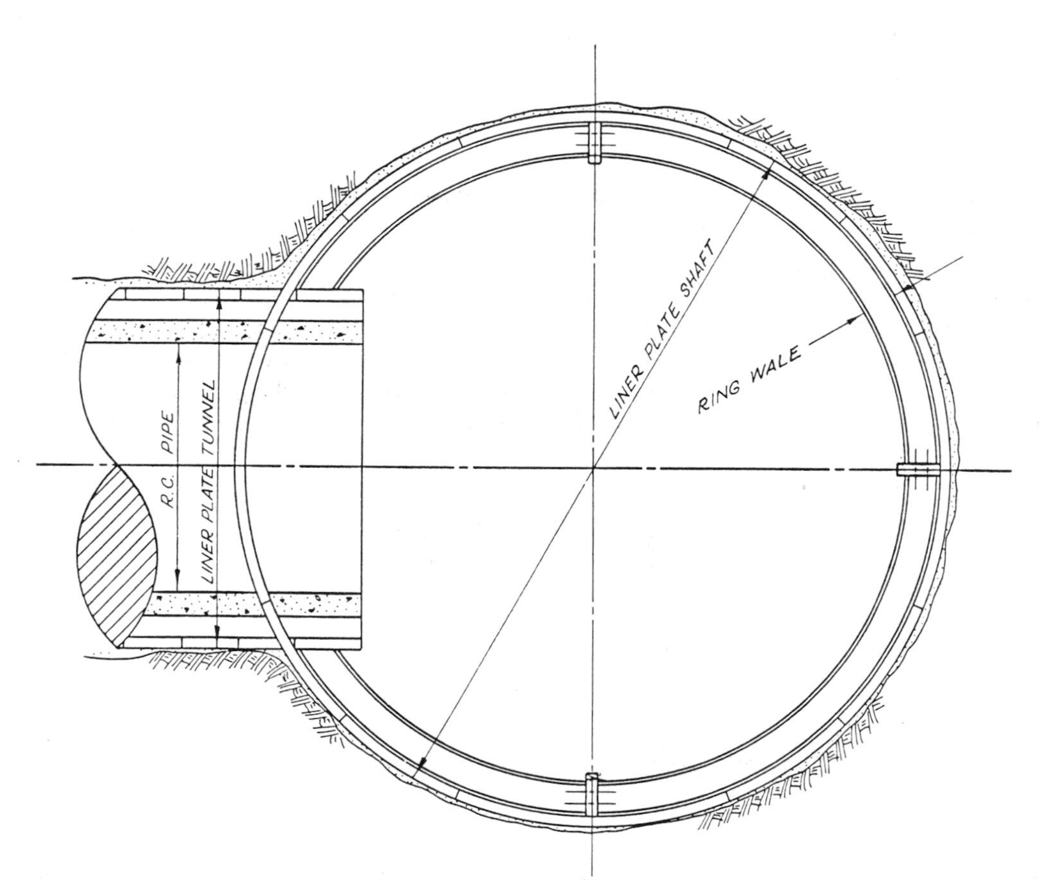

Fig. 9-5. Liner plate shaft. a) Cross-section. b) Elevation.

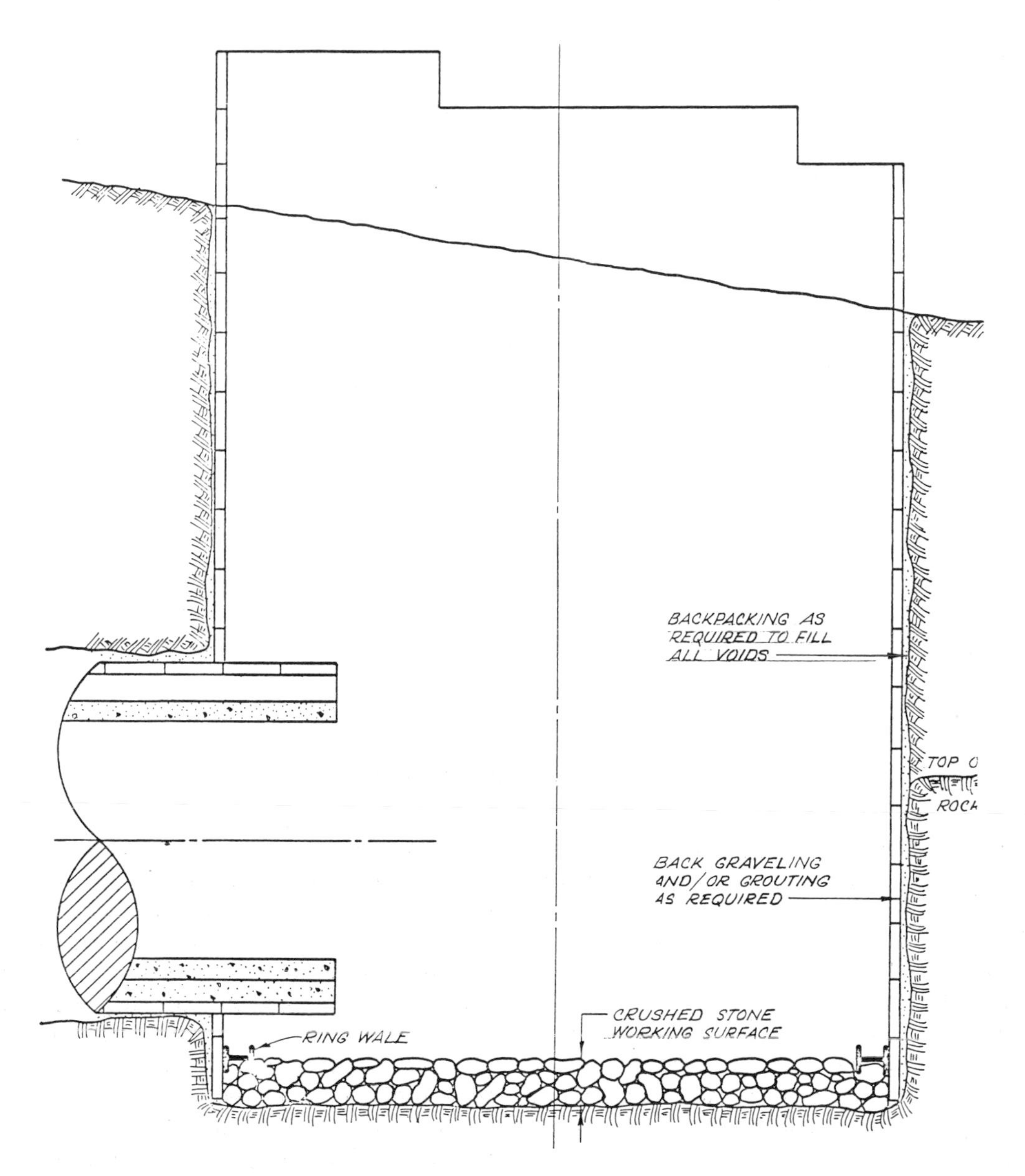

Fig. 9-5b.

zontal steel rib sets can be set at a predetermined spacing in between liner plate rings for additional strength.

The main advantage of liner plates is that their small size permits ease of operation in limited working spaces, such as small shafts and tunnels (see fig. 9-5).

Horizontal Ribs And Vertical Lagging

The ribs and lagging method is somewhat similar to liner plate construction. Rings are made of structural steel members, cold-formed to required curvature, the sections butted at each end. Butt plates welded to the ends of the segments are provided with bolt-holes. Six- to eight-foot lengths of timber are usually used for lagging (see fig. 9-6).

This method requires excavation of the soil to a distance equal to the length of the lagging. Curved ring segments are bolted together and held in place by tie rods and spacers. The tie rods and spacers are placed between the webs of the rings. Placement of the vertical lagging

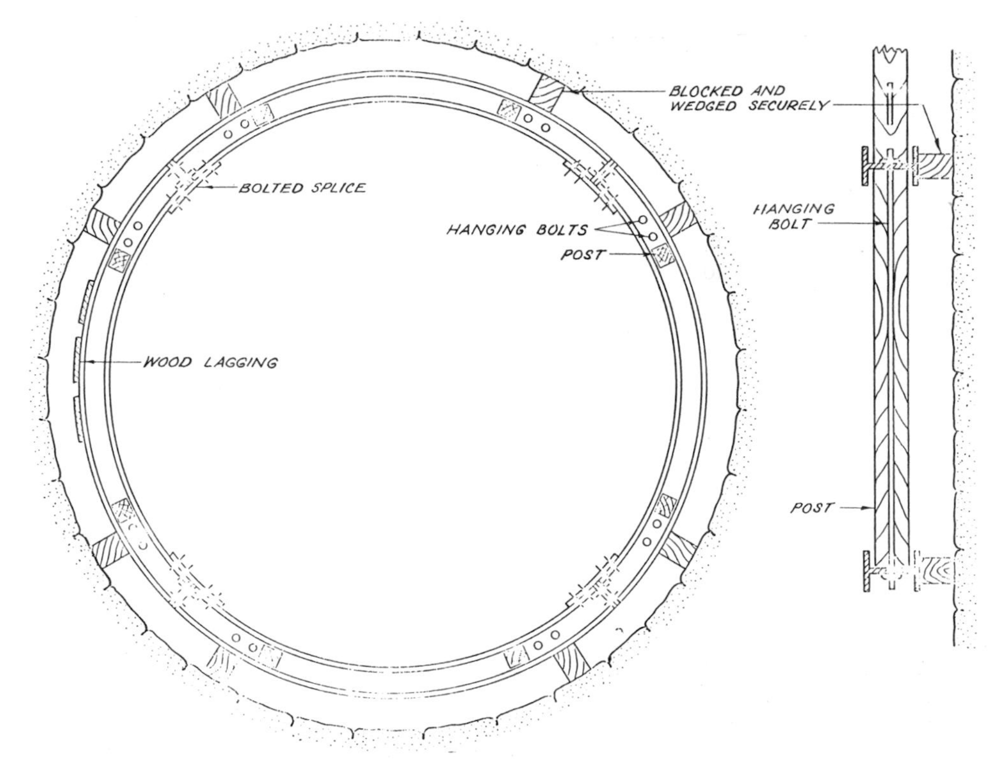

Fig. 9-6. Ring wales and lagging.

follows. The rings can be placed on varied centers to resist such lateral earth loads as may be encountered.

Since the soil must be somewhat initally self-supporting for the height of the lagging, this method is usually employed in cohesive soils, although it can be used when the ground has moderate "stand-up" time. During construction, the interval between excavation and lagging placement should be minimal to prevent ground loss.

Slurry Walls

Bentonite, a naturally occurring clay, has a large capacity for absorbing water. In suspension, it is a liquid when agitated, and a gel if left to stand. The liquid suspension can be injected into a permeable soil mass and allowed to congeal in the voids, thereby decreasing permeability. There are generally two methods of excavation with slurry.

The first method uses excavating equipment suitable for ground formations consisting of soft to medium hard, loose or cohesive soils, without boulders or other obstructions. The excavation is performed by clamshell buckets, which open the trench in a panel sequence with the trench kept filled with slurry. Clamshell rigs are available in a wide variety of types, depending on the site conditions. Some are hydraulically or mechanically operated and are cable suspended, while others are mechanical or hydraulic clamshells controlled with a kelly bar. Concrete is placed in each panel by tremie with the bentonite being recirculated to the next panel excavated. A cofferdam can then be created by bracing the concrete with steel or reinforced concrete wales and struts.

The second method, known as the "interlocked" element type, is used when the soil is very hard and bouldery, or when the excavation must reach considerable depths. The technique used under these conditions is shown diagrammatically in Fig. 9-7. Since clamshells are no longer effective in these formations, the excavation must be performed solely with the aid of percussion tools. Primary holes are first drilled using percussion rigs with the assistance of a bentonite slurry. In most cases, the spacing of the primary holes (center to center) is twice their diameter, which means that the space to be filled by the secondary elements is equal to the diameter of the primary holes. This spacing is also convenient with respect to the permeability of alluvial deposits, since the penetration of the soil between primary holes by the bentonite slurry is almost complete, and, therefore, the excavation of the secondary elements involves a zone of soil stabilized through gelation in its pores.

Concrete is placed in the primary holes with the aid of tremie pipes. An hydraulically expandable chisel is then used to excavate the panels between the primary holes.

Structures built by these methods can be reinforced by placing the steel in the slurry-filled holes or trench before concrete placement begins. After the concrete walls are in place, the interior excavation for the shaft can begin.

EXCAVATION IN SOFT, WET GROUND

Excavation in soft, wet ground can be accomplished in a number of ways. The most common method is to lower the groundwater table in the working area. Other methods include freezing

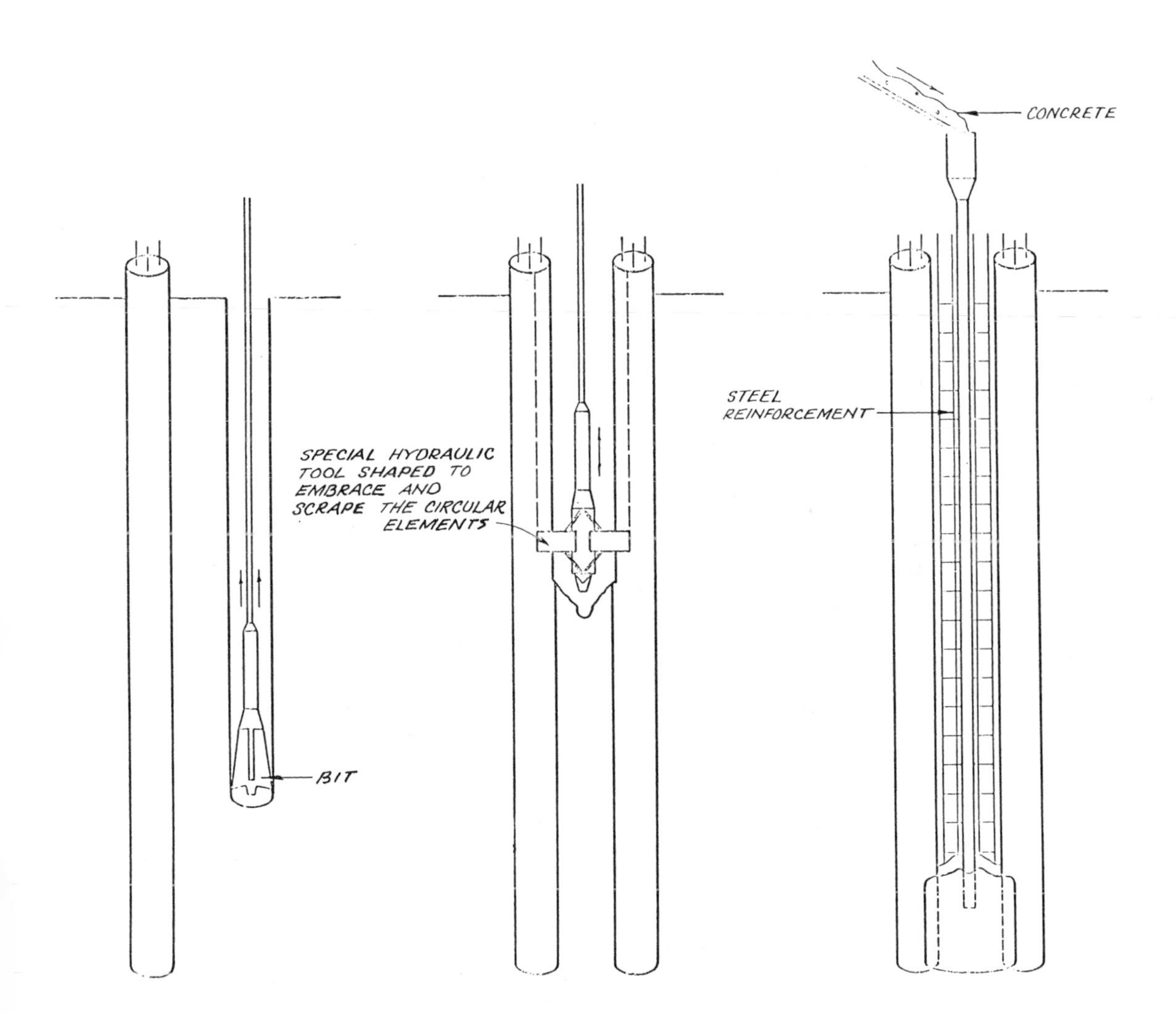

Fig. 9-7. Slurring wall construction by ICOS interlocked element method. a) Constructing primary elements. b) Drilling secondary elements. c) Concreting secondary elements. d) Completed primary elements-section. e) Completed wall-section. (*Adapted from Xanthakos, Underground Construction in Fluid Trenches*).

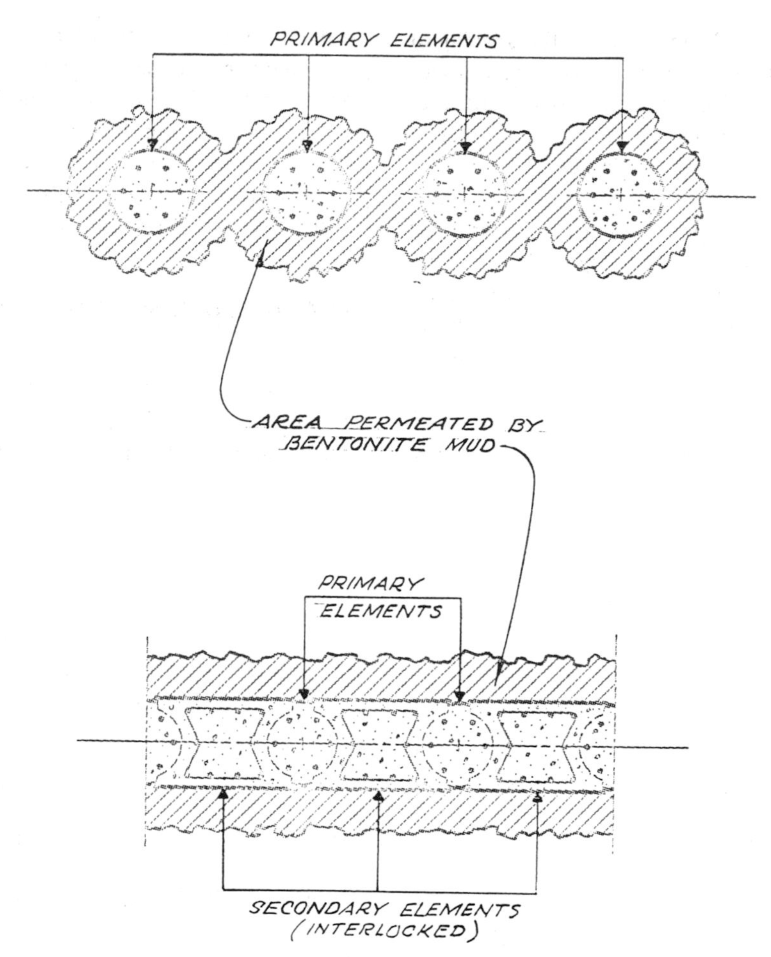

Fig. 9-7b., e.

of the soil, the use of slurry, grouting, sinking a pnuematic caisson and sinking a dredged drop caisson with a tremie concrete seal.

The methods of open pumping, wellpoint systems, deepwell and freezing will be discussed. Typical pneumatic and dredged caissons are shown in Figs. 9-8 and 9-9.

Lowering Of Groundwater

Lowering the water table for shaft excavation will ensure dry, safe and firm working conditions. Dewatering operations in shafts are time-consuming.

To determine the type of groundwater control system in advance, a pumping test should be performed. Geologic and soils information should first be evaluated. The test should yield a distance-drawdown versus time curve and the test should continue until the slope of the curve in relation to the pumping rate becomes constant. The pumping test should yield results such as water volume pumped, well yield and time required to reach equilibrium.

It is advisable to perform a chemical analysis of the ground water to check for dissolved salts or gases. Calcium salts and iron oxides in the water could corrode metal portions of the dewatering system. If well screens are used, the screens could become plugged by precipitating salts.

Dewatering operations can result in the lowering of the water table under adjacent areas, in some cases as far as 2000 feet from the well.

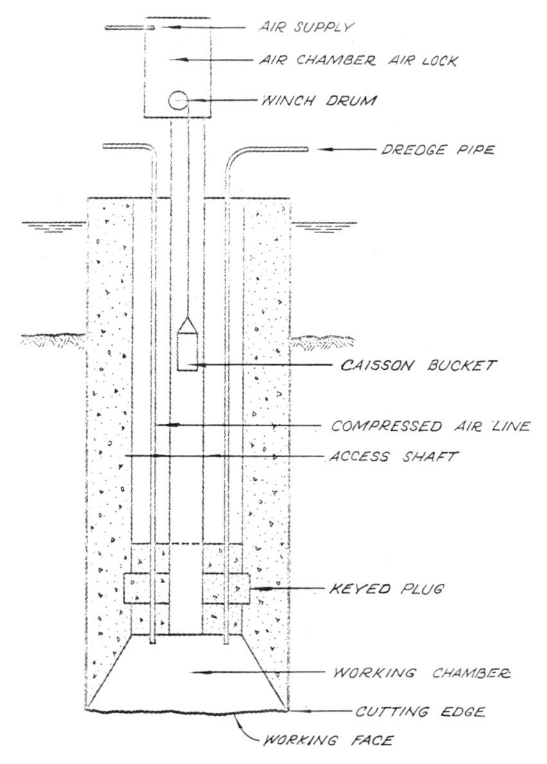

Fig. 9-8. Schematic representation of pneumatic caisson excavation.

Therefore, extreme caution must be observed if large structures are in the vicinity of dewatering operations. Recharging of the groundwater may be required to minimize settlements of adjoining structures.

Dewatering increases the effective stress in the soil, which, in turn, causes settlements. Also, in cases where adjacent structures are supported on piles, enough dragdown can be developed on the pile foundation to cause settlements.

Open Pumping

The most simple shaft excavating method through water-bearing pervious soil is the sheeting and open pumping method. This consists of driving steel sheet piling, excavating and pumping water from the bottom of the excavation.

However, the pumping operation causes seepage of water (and loss or fines) around the toe of the sheeting. Furthermore, if the pressure due to the upward seepage of water becomes greater than the soil pressure at the bottom of the excavation, a quick or "boiling" condition in the soil results. Also, if the seepage of water around the toe of the sheeting significantly dislodges soil particles, the sheeting can be undermined (see fig. 9-10).

Wellpoint System

The wellpoint system is generally used for dewatering to a depth of about 15 feet. This is the most common method of dewatering in the U. S. for shallow open excavations. The technique is best suited for use in medium to fine sand for work of short duration.

The method consists of placing wellpoints on 3- to 12-foot centers around the area to be excavated. The wellpoints are attached to a common header pipe, which is connected to a pump. Wellpoints are well screens which require suitable filter material around the screen to prevent the collection of soil particles with the water. Once the dewatering is accomplished, shaft excavation can begin. However, the wellpoint system must operate during excavation lest the water table return to its original level.

The main disadvantages of a wellpoint system for shaft excavation are that the system must be within the area of excavation, and that the depth of dewatering is limited to 15 feet (see fig. 9-11).

For depths over 15 feet, a vacuum wellpoint system can be used.

Deepwells

Deepwells can be used to dewater pervious materials to whatever depth the excavation requires, and they can be installed outside the zone of excavation.

The deepwell system consists of spacing 6- to 18-inch diameter wells on 20- to 200-foot centers, depending on perviousness, depth of dewatering required, etc. The wells have a commercial type of water well screen surrounded with a properly graded sand-gravel filter. Each well is equipped with its own submersible pump. The excavation for the shaft can begin after water drawdown to the required elevation has been accomplished.

For shaft excavation, deepwells can provide dewatering to the depth desired; relatively few

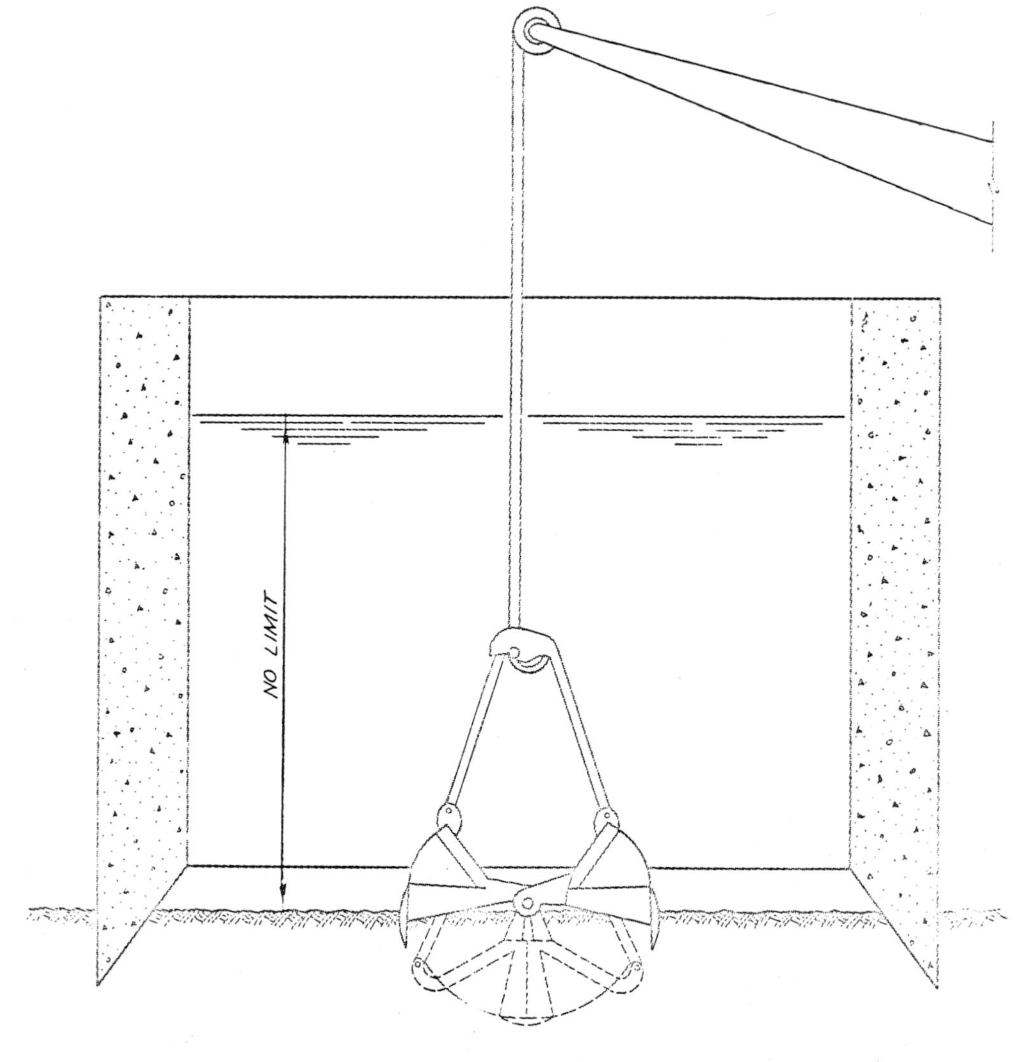

Fig. 9-9. Schematic representation of dredged caisson. a) Excavation by clamshell. b) Excavation by airlift.

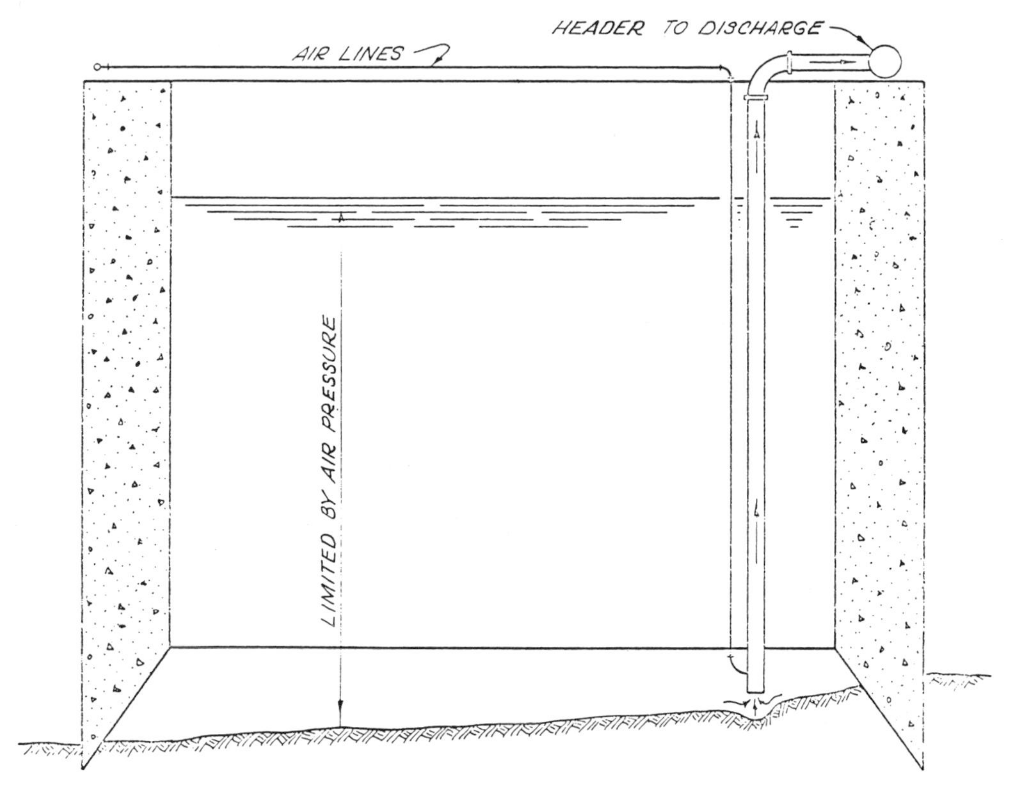

Fig. 9-9b.

units have to be installed; and, once installed properly, only maintenance is required. However, only pervious strata can be dewatered by this method (see fig. 9-12).

Freezing

In water-bearing ground where surface subsidence is not allowed, such as adjacent to large buildings, the most reliable method of excavation is freezing the soil, then excavating. The freezing eliminates both seepage and plastic flow. There is no limitation on the depth to which freezing may be used.

The procedure consists of sinking pipes around the area to be excavated and circulating a cold brine solution through the pipes, thereby freezing a wall of soil. Excavation can then begin. If a concrete lining is used, the concrete can be placed against the frozen soil.

The refrigerating circuit in each bore-hole consists of a tube closed at one end, containing another smaller-diameter open-ended pipe. Brine is cooled at a surface refrigerating plant, then circulated down the inner pipe and up through the annular space between the pipes back into the refrigerating plant. This process is continued until sufficient soil is frozen. The time interval usually varies between two and five months.

The only disadvantages to the freezing technique are the time required to freeze the soil and the cost of the equipment. It can be noted however, that after use, the refrigerating plant can be salvaged (see fig. 9-13).

SHAFT EXCAVATION IN ROCK

Shaft excavation in rock is usually performed by the drill and blast method. The limit of shaft excavations for tunnels is often less than 120 feet; therefore, the use of sophisticated shaft excavating equipment is often unwarranted. When sinking shafts deeper than 120 feet, other methods within the realm of the mining engineer are used.

Prior to the start of rock excavation, the shaft

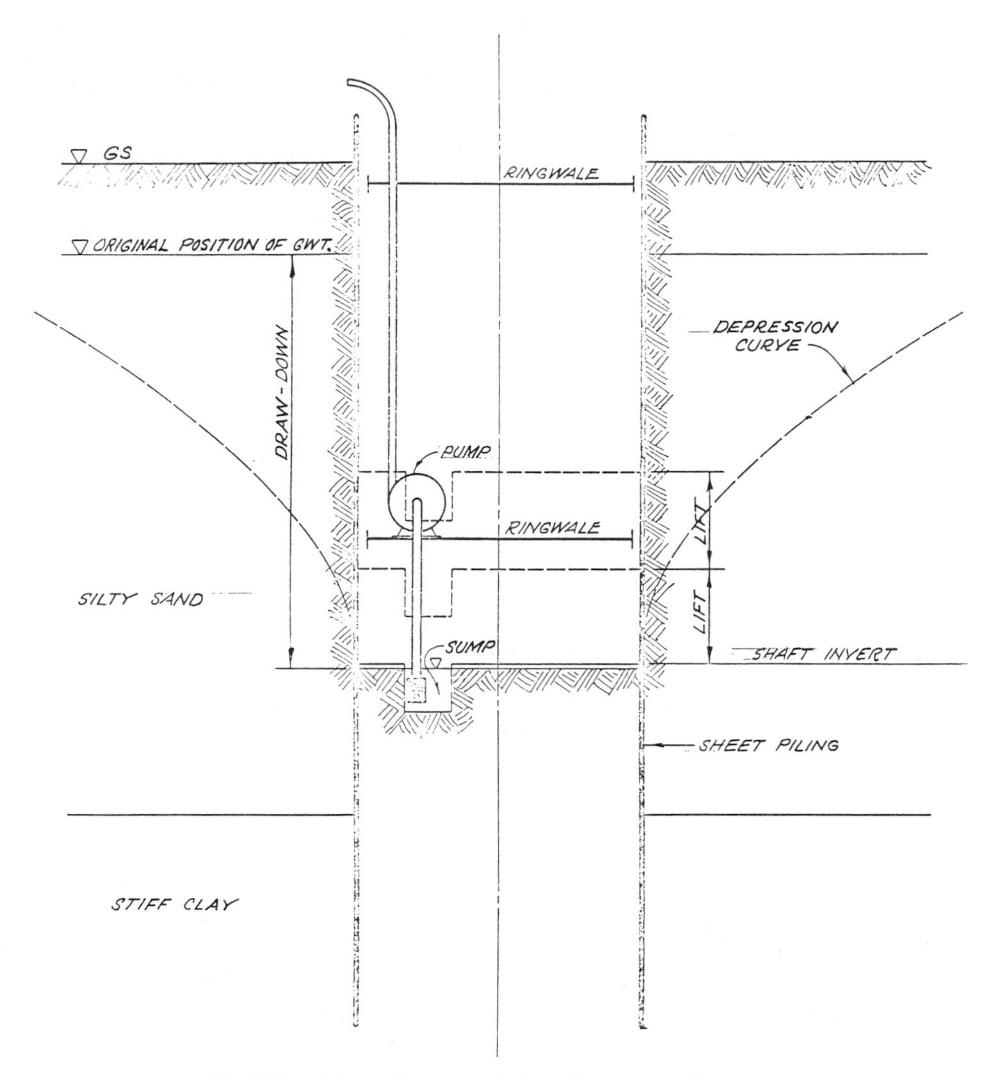

Fig. 9-10. Schematic representation of open pumping.

lining through the overburden should be grouted and sealed where necessary to avoid groundwater infiltration.

If water infiltration is expected when excavating through poor rock, a grout curtain can be installed to achieve a water barrier. A ring of holes about 10 feet from the shaft is drilled for the depth of the shaft, then grouted to refusal.

Drilling

Shaft drilling can be done by either hand-held air drills or by collapsible shaft jumbos. The shaft jumbos can be set and levelled in the hole; the bull hoses can be attached to the boom-mounted drills; and drilling can commence. Usually, only the deeper shafts can justify the use of a jumbo.

Blasting

The major considerations in designing a shaft round are ease of drilling and minimizing overbreak. Special arrangements can be made for creating sumps if there is a water problem.

The most suitable blasting arrangement for

circular shafts is the pyramid cut. The pyramid cut consists of several holes drilled angularly to meet in a common apex near the center of the face. Peripheral relief holes and trim holes may also be incorporated (see fig. 9-14).

Rectangular shafts are usually sunk with a V cut. Each V consists of two holes drilled from points as far apart as possible on the face to meet at the bottom of each hole (see fig. 9-15).

A series of V cuts parallel to one another will control the width of the shaft.

Mucking

After blasting, mucking can usually be carried out by cranes with a clamshell bucket. Hand mucking is used in shafts up to 10 feet in diameter. In shafts over 100 feet in depth, the methods and equipment used in the mining industry must be employed. A sinking frame is set up over the shaft and fitted with buckets called "kibbles" that travel on cable or rail guides to prevent swinging. At the top of the shaft, a kibble dumper tilts the bucket to dump muck onto the ground or into a hopper. At

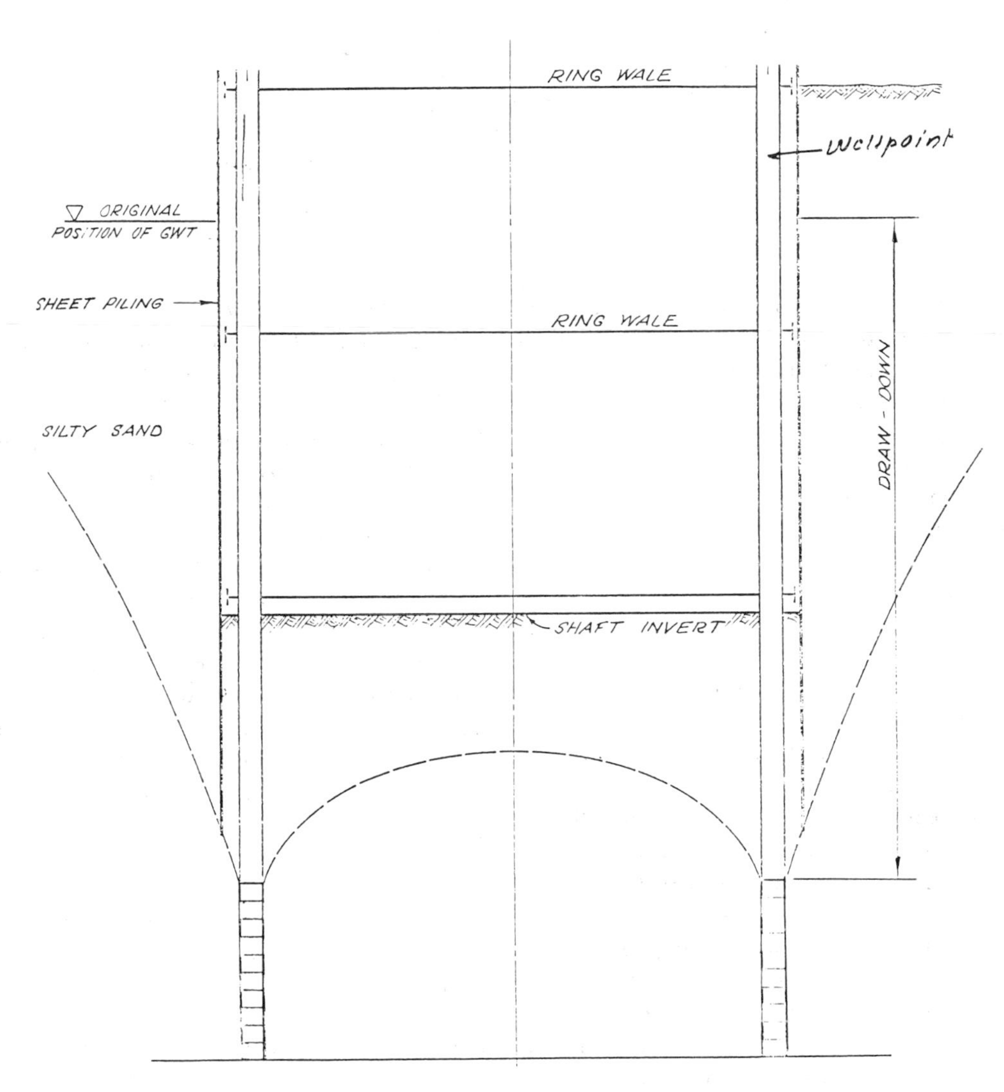

Fig. 9-11. Schematic representation of wellpoint system. a) Elevation. b) Section.

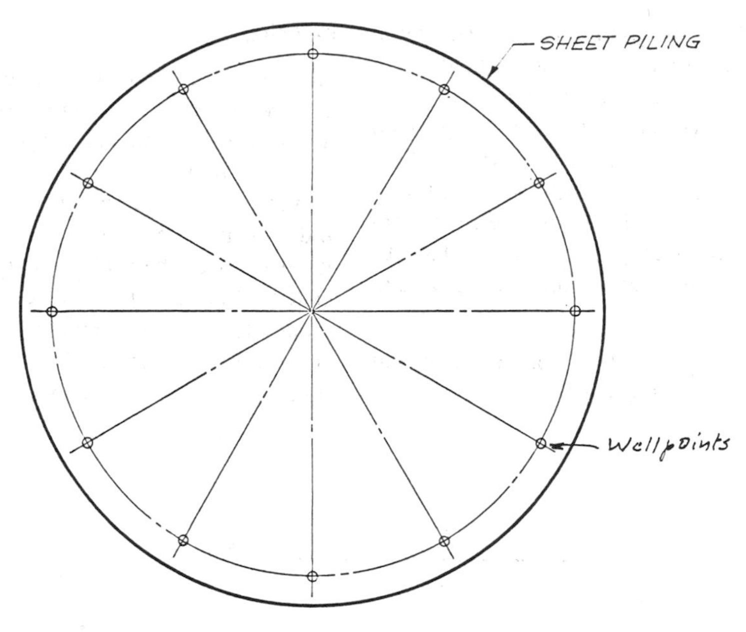

Fig. 9-11b.

the bottom of the shaft, the buckets are generally loaded by a small air-operated clamshell grapple, by an "orange peel" bucket or by an air-operated crawler-mounted rocker-type mucking machine.

When the shaft is less than about 60 feet deep, the muck may be hoisted in muck boxes that are lifted by crane.

Raises

Raised excavation is excavation toward a higher elevation, often from one tunnel to another. The raised shafts may be used for manways, rock passes or ventilation. The raise can be used to intersect another tunnel above an existing one. Shaft raising is sometimes used in urban tunnel construction to minimize surface disruption.

Raises are excavated by drilling pilot holes, usually about 9 to 12 inches in diameter, then reaming the hole to the proper diameter. The deviation of the pilot hole can be about 1% of the length, using modern techniques. Good crews have been known to achieve less than 0.5% deviation. The most common and successful system to date has been drilling the pilot hole down and reaming up the required raise. Sometimes the pilot hole is drilled up and reamed down. Only a few types of raise drills are widely used, and the nature of the rock through which the raise will pass should be studied carefully to assure the use of proper cutters for efficiency and economy.

The conventional method of raise mining by blasting is quite hazardous, especially the scaling down of loose rock after the blast. A comparison of drill-blast raising versus raise boring in the Western Cordilleran Region of the U. S. revealed a much higher daily advance rate by the boring method.

Temporary Supports

In sedimentary, fractured or blocky rock, the walls can be quite treacherous. When rock support is required, it should be placed quickly after excavation. Usually, the support can

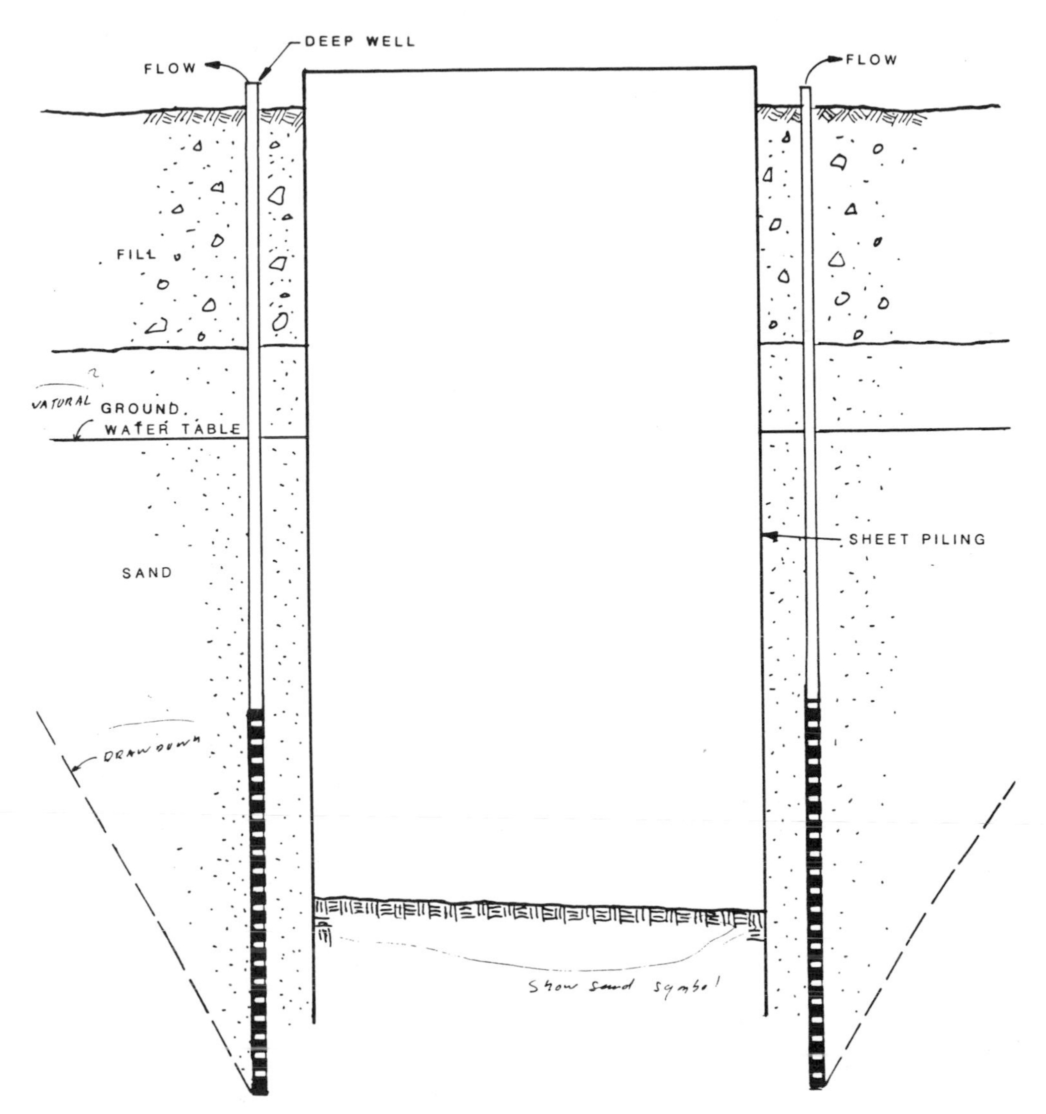

Fig. 9-12. Schematic representation of deepwell dewatering.

consist of steel ribs and liner plates, or steel ribs with lagging, roof bolts and wire mesh, or pneumatically-placed concrete (shotcrete). The temporary supports are sometimes known as a "primary lining." Prior to the placement of the support system, the loose rock should be scaled down.

Steel ribs and liner plates and ribs with lagging have already been covered previously in this chapter. These methods are usually used in loose material.

If the rock is of reasonably good quality, it may be advantagous to merely install rock bolts into the wall and fasten wire mesh to the bolts to keep rock spalls from dropping on the workers.

It is sometimes good practice to apply shotcrete to the walls. This method has been applied to a number of shafts and tunnels during the past few years. Shotcrete linings are also becoming more popular as permanent linings (see Chapter 12).

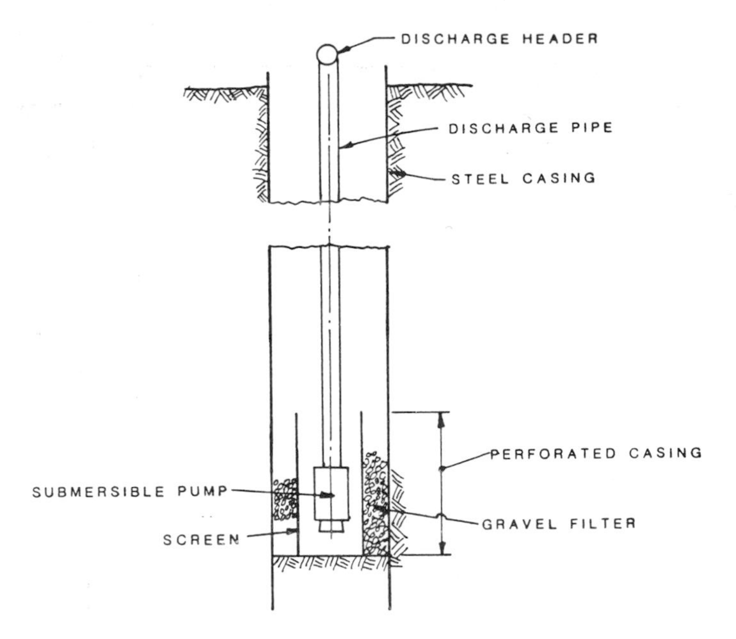

Fig. 9-12b.

Typical installations of rock bolts, wire mesh and shotcrete are shown in Figs. 9-16 and 9-17.

LINING OF SHAFTS

Secondary linings are required if the shafts are of a permanent nature. The shafts are usually concrete lined. Typical shaft forms are shown in Fig. 9-18.

Concrete Lining In Soft Ground

Any shaft sunk through soft ground requires a primary lining for initial support. As outlined previously, this temporary support could consist of various sheeting and bracing systems, steel ribs with lagging or liner plates, or shotcrete if in rock. If a long-term functional shaft is required, a secondary lining will be required.

Prior to the construction of the concrete lining, the primary lining should be true in shape, and direct continuous contact between the ground and the primary lining should be ensured.

The secondary lining can either be poured against the primary lining, or it can be formed from both the outside and inside. If the lining is formed on the outside, the annular space between the primary and secondary lining should be tightly backfilled or packed with pea gravel, well-graded sand or other suitable material.

Rocks Bolts And Wire Mesh

Rock bolts and wire mesh are usually used in relatively sound rock to maintain stability and prevent spalling. Sometimes shotcrete is applied in addition to the bolts as a primary, or final, lining. Since the rock is of fairly good to good quality, the construction of the secondary lining is rather straightforward, and the concrete can be poured directly against the rock. If water infiltration through seams in the rock is a problem, the seams should be well grouted before the construction of the secondary lining. Final contact grouting will also be required to fill any voids between the concrete secondary lining and the rock.

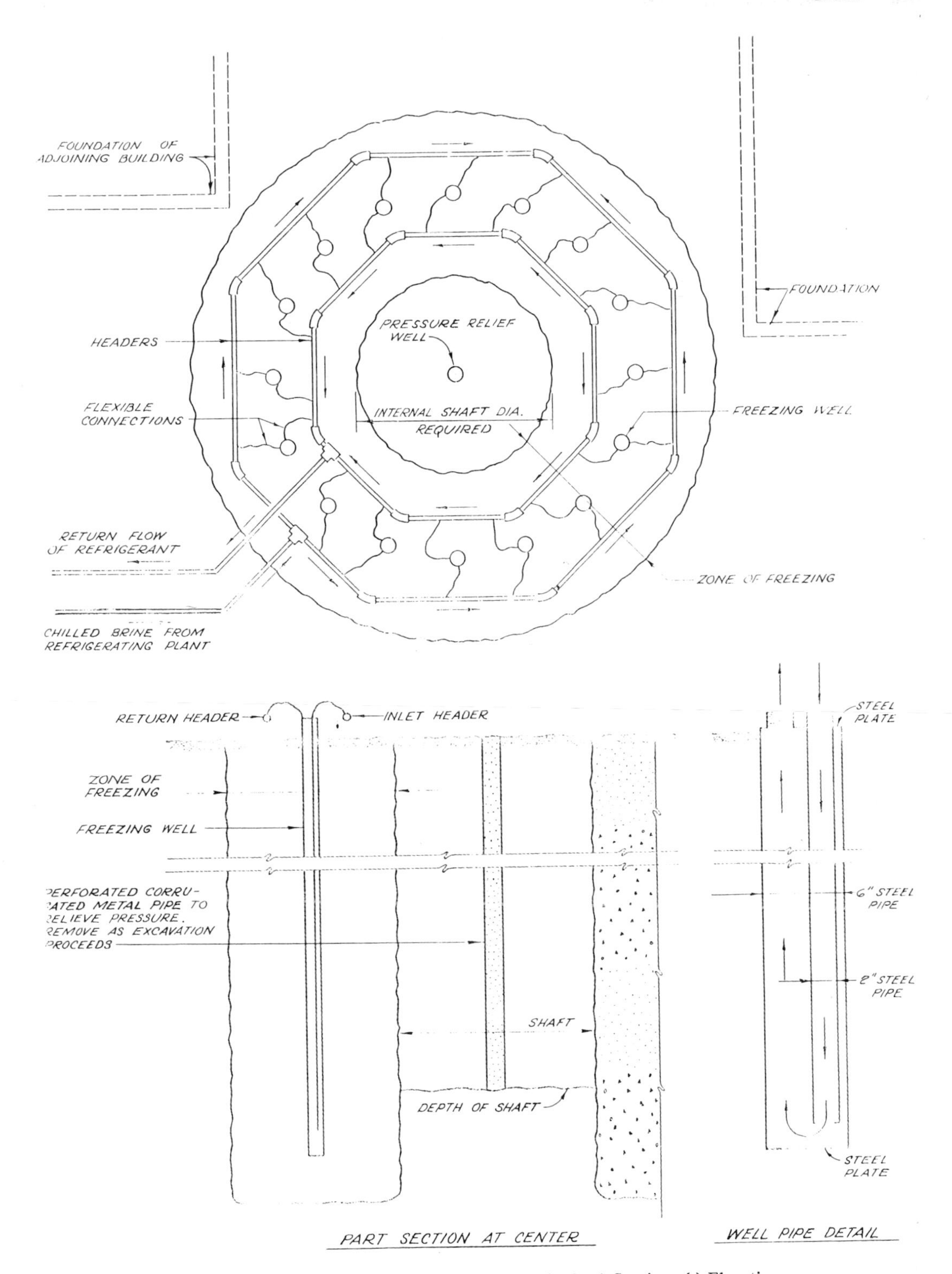

Fig. 9-13. Ground-freezing method. a) Section. b) Elevation.

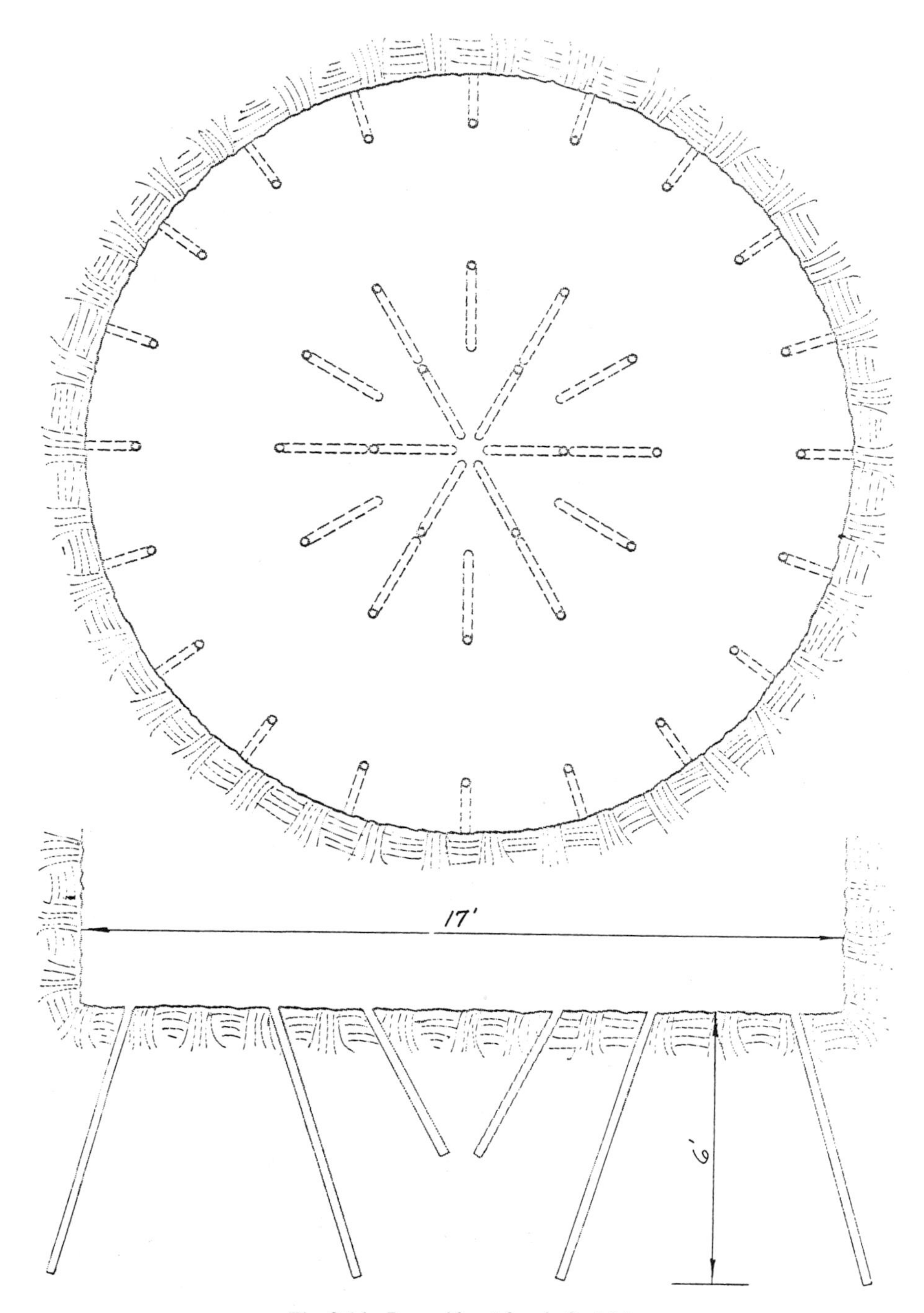

Fig. 9-14. Pyramid cut for shaft-sinking.

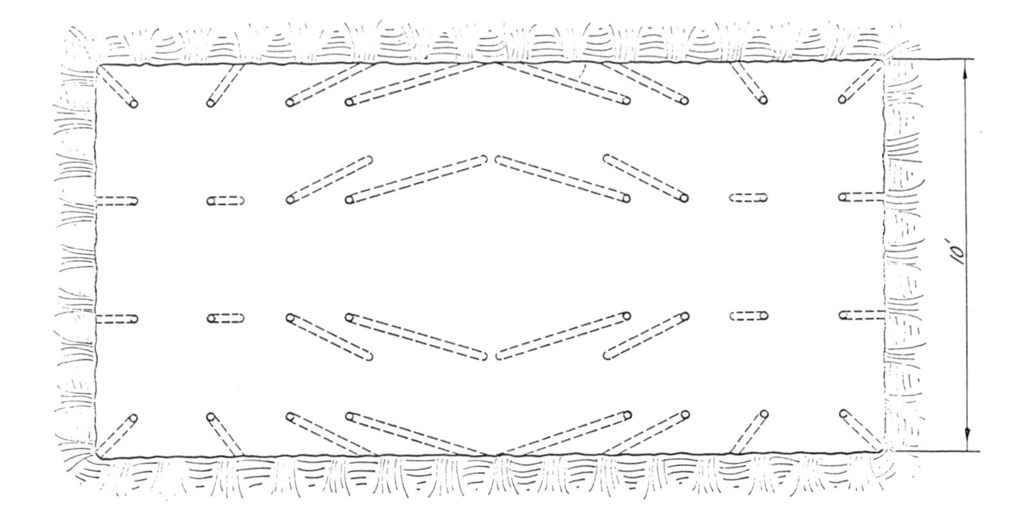

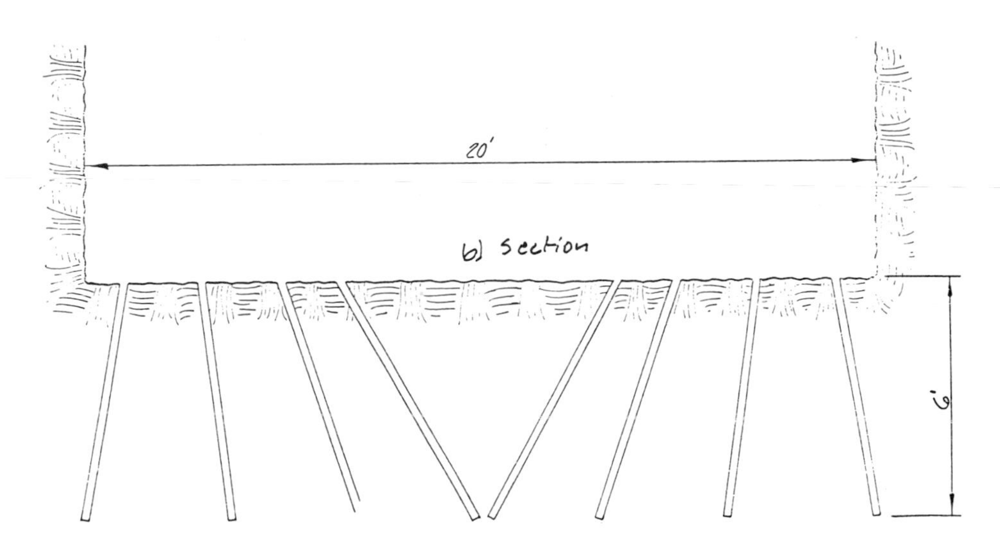

Fig. 9-15. V-cut for shaft-sinking.

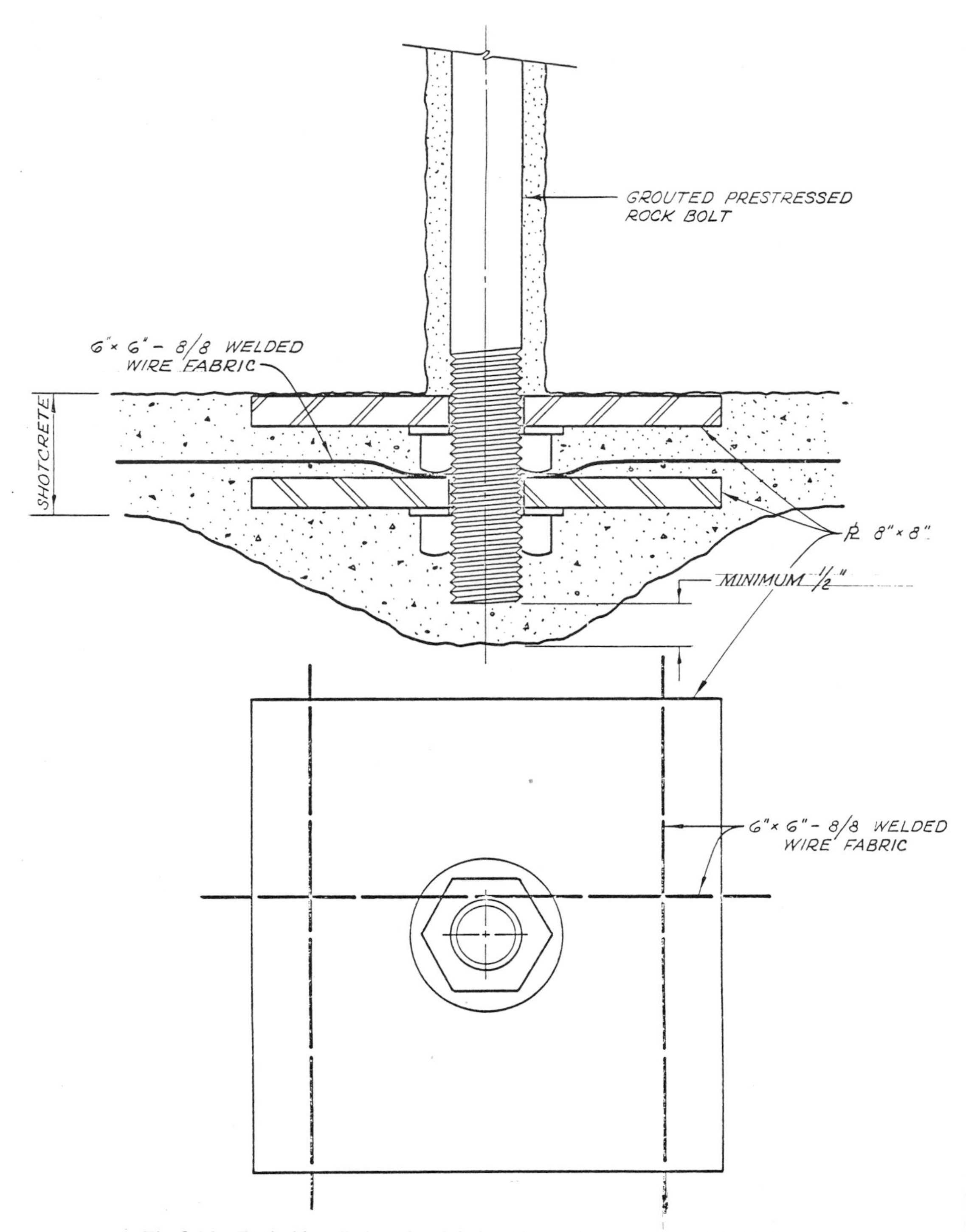

Fig. 9-16. Typical installation of rock bolts, wire mesh and shotcrete, Washington Metro.

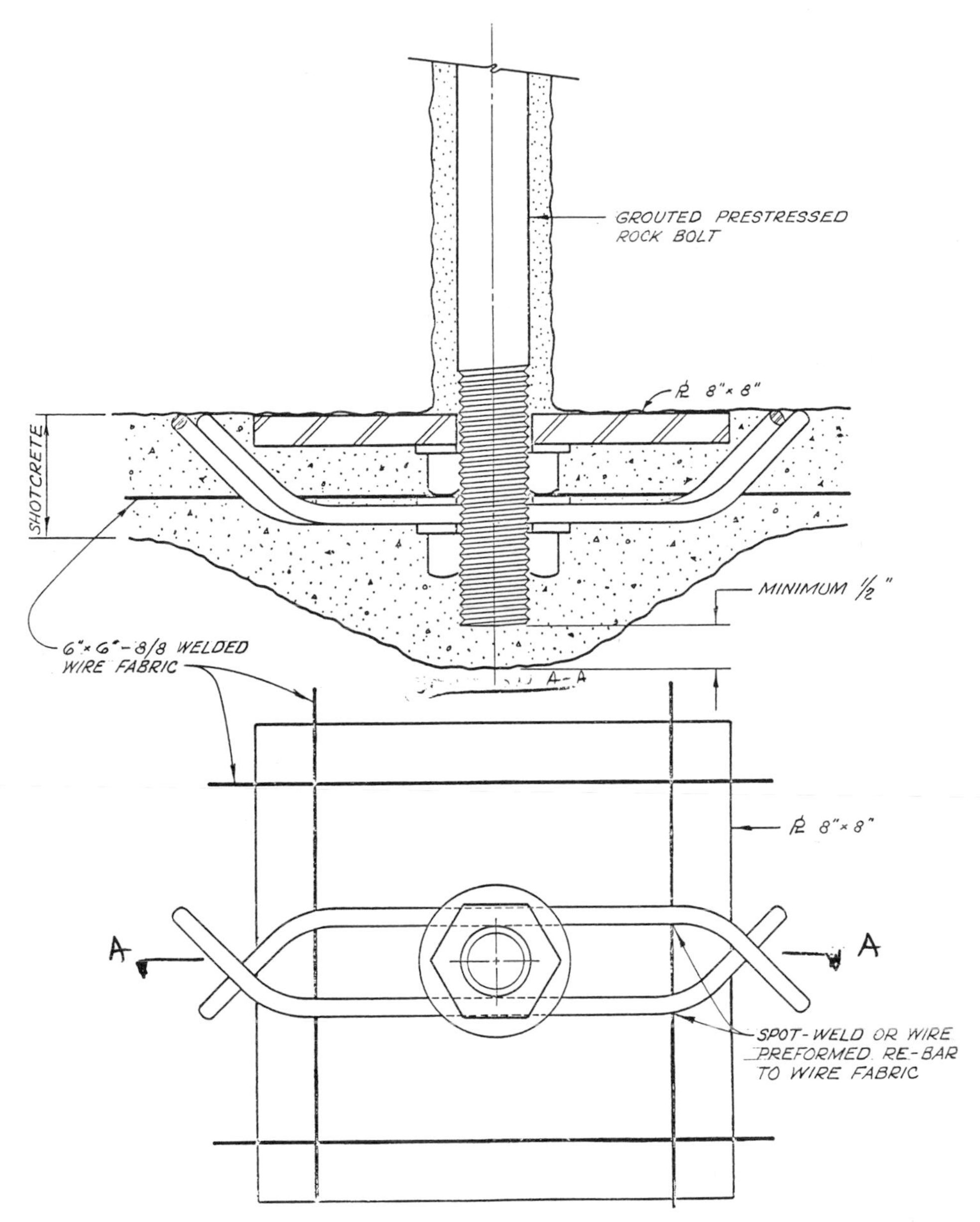

Fig. 9-17. Alternate installation of rock bolts, wire mesh and shotcrete, Sweden.

Fig. 9-18. Forms for concrete shaft lining. (*Courtesy of R. Mayo.*)

Shotcrete

Shotcrete is used for temporary support to prevent rock spalling. Sometimes the shotcrete is used with wire mesh for added strength. Currently, some work is underway using shotcrete as a permanent lining, since the quality and strength of the mixture has been greatly improved. However, many designers still require the placement of a concrete secondary lining.

REFERENCES

1. *Blaster's Handbook,* I.E. DuPont de Nemours & Co., Wilmington, Delaware, 1958.
2. Hammond, R., *Tunnel Engineering*, Macmillan, New York, 1959.
3. Havers, J.A. and Stubbs, F.W. Jr., *Handbook of Heavy Construction,* McGraw-Hill, New York, 1971.
4. Jumikis, A.R. *Foundation Engineering*, Intext, Scranton, 1971.
5. Lane, Kenneth and Garfield, L.A., *Proceedings of first North American Rapid Excavation and Tunneling Conference, Chicago, Illinois, June 5-7, 1972,* Port City Press, Baltimore, 1972.
6. Leonards, G., *Foundation Engineering*, McGraw-Hill, New York, 1962.
7. McWilliams, J.R. and Erickson, E.G. *Methods and Cost of Shaft Sinking in the Coeur D 'Alene District, Shoshone County, Idaho*, U.S. Department of Interior, 1960.
8. Mayo, R.S., Adair, T. and Jenny, R.J. *Tunneling: The State of the Art,* U.S. Department of Housing and Urban Development, 1968.
9. Merritt, F.S., *Standard Handbook for Civil Engineers,* McGraw-Hill, New York, 1968.
10. Powers, J.P., "Groundwater Control in Tunnel Construction," *Proceedings of North American Rapid Excavation and Tunneling Conference,* ASCE-AIME, New York, 1972.
11. Széchy, K., *The Art of Tunneling*, Akadémiai Kiadó, Budapest, 1966.
12. *Use of Shotcrete for Underground Structural Support*, ACI-ASCE, Detroit, 1974.
13. Xanthakos, S.E., *Underground Construction in Fluid Trenches*, University of Illinois, Chicago, 1974.

10

Tunnel-Boring Machines

J. GEORGE THON

Executive Consultant
Bechtel Incorporated
San Francisco, California

BASIC COMPONENTS OF A TUNNEL-BORING MACHINE

A tunnel-boring machine generally is a device for excavating a tunnel in such a way that the material to be removed is disintegrated by the continuous rotation of a group of cutting tools thrust against the surface of the material at the working face. The following paragraphs describe the essential components.

All tunneling machines incorporate tools for breaking up the material to be excavated into sizes that can easily be removed. In most, but not all, machines, the full circular cutting head is employed. The cutting tools are mounted in an arrangement suitable to excavate a tunnel of the required diameter when the head is rotated under thrust against the working face.

The machine body is mounted immediately behind the cutting head and remains stationary while the cutting head excavates. It incorporates a mechanism to maintain its stationary position during excavation and to move itself and the cutting head forward to continue the excavation. The machine body also contains the mechanical equipment to provide the necessary thrust and torque transmitted through the cutting head to the cutters.

In most cases, muck is removed from the excavated face by a number of buckets on the cutting head and is dropped onto a conveyor belt system. It then passes on a series of conveyors to the back of the tunneling machine, where it discharges into another transportation system for removal from the tunnel, which is usually entirely independent of the tunneling machine.

All tunneling machines contain these components in some form or another; however, their general arrangement depends primarily on whether the machine is intended for use in soft ground or in rock. Detailed design varies with manufacturers. The main design difference arising from the type of ground to be encountered is that soft ground requires some kind of support as soon as the excavation is made, a limitation not imposed by rocky ground. In soft ground machines, the main function of the forward thrust generally is to support the vertical face of the excavated ground against collapse, whereas, in rock tunneling machines, the thrust is to supply the energy to disintegrate the material.

HISTORICAL DEVELOPMENT OF TUNNELING MACHINES

Although the development and use of tunneling machines have undergone rapid growth only in recent years, the idea is far from new. The first tunneling machine on record was made in the

U.S. by John Wilson in 1856. It was tested on a part of the Hoosac railway tunnel in Massachusetts, a structure 4.7 miles long, which took a total of 21 years to complete. Wilson's machine bored only 10 feet before it was abandoned. Two other machines were used on the same tunnel without notable success.

The first successful tunneling machines were developed by Colonel Beaumont in England in the 1880's. They had a diameter of 7 feet, and one used on a tunnel under the Mersey River reached a rate of advance of about 115 feet/week. One of Beaumont's machines worked part of a pilot tunnel for the proposed English Channel Tunnel in 1882. It excavated pilot headings each a mile long under the sea from both shores. It was able to maintain a rate of 49 feet/day in chalk, over a period of 53 working days. The Beaumont machine excavated by means of kerfing tools attached to a rotating head and was powered by compressed air. The project was eventually stopped for political reasons, not because of shortcomings in the tunneling machine itself.

During the next 70 years, only about 15 machines were designed and constructed, but none operated with the success necessary to initiate any rapid growth and improvements.

A new era opened in the mid-1950's, when James S. Robbins of Seattle entered the field of tunneling machine design and manufacture. The first success was a machine that worked on Oahe Dam, South Dakota, in 1954. It drove a 25-foot, 9-inch diameter tunnel through faulted and jointed shale with a compressive strength of 200 to 400 psi. It had a cutterhead horsepower of 400 and a thrust of only 100,000 lb. This machine and a similar one supplied in 1955 drove a total of 22,500 feet of tunnel, reaching maximum rates of advance of 140 feet in a day and 635 feet in a week.

Robbins achieved notable success and recognition when one of his machines drove a sewer tunnel in Toronto through limestone, sandstone and shale with compressive strengths between 8000 and 27,000 psi. This machine was the first one to be completely equipped with rotary disc cutters (24); it had a diameter of 10 feet, 9 inches, a cutterhead horsepower of 340 and a thrust of 314,000 lb. Rates of advance of up to 10 feet/hour were achieved.

The performance of the tunneling machine on this project attracted worldwide attention and initiated a period of intense development as other manufacturers sought to enter the field. This period continues, and tunneling machines are still being improved as field experience grows.

CUTTERS FOR TUNNELING MACHINES

The characteristics of tunneling machines are determined by the kind of formation in which they are designed to operate—namely, soft ground, soft rock, medium hard rock or hard rock. There is no universally recognized system of making the classification, but Table 10-1 shows a grouping according to the unconfined

Table 10.1. Classification of formations on the basis of their characteristics for tunneling.

Classification	Unconfined Compressive Strength (psi)	Typical Formations
Soft Ground		Uncemented deposits of clay, silt, sand and gravel, possibly saturated; marl
Soft Rock	Less than 8000	Shale, tuff, claystone, siltstone, sandstone.
Medium Hard Rock	8000 to 25,000	Some basalt, granite, and andesite; average sandstone and limestone; dolomite, chalk, rhyolite, gneiss, schist.
Hard Rock	Over 25,000	Some basalt, granite and andesite; well-cemented sandstone and limestone; marble, chert, diorite, quartzite, argillite.

compressive strength of the rock, which is based on the classifications used by several manufacturers of tunneling machines. Table 10-1 also shows some different types of rock that commonly fall into the different groupings. The classification is for general guidance only, since the compressive strength and the difficulty of tunneling in a particular rock type can vary considerably, depending on the crystalline structure, particle shape, amount of cementation and degree of weathering.

Several different types of cutters have been developed for working most efficiently in various formations.

Cutters for Soft Ground Tunneling Machines

Some type of drag cutter is almost always used in a soft ground tunneling machine. Drag cutters are all basically simple picks or teeth and are made in a wide range of shapes and sizes. They are fixed rigidly to the cutting head. When forced into the ground and moved, they tear material loose from the working face. Because they are of much simpler construction than the cutters used in rock, they are much cheaper and much easier to change. Drag cutters are generally not used in harder rock because they wear too quickly to be economical.

Cutters for Rock Tunneling Machines

Different types of cutters are used in rock tunneling machines, depending on the rock hardness. Some of these cutters are shown in Fig. 10-1. To break up the rock, the cutters are thrust against the surface and are rolled across it under pressure. The cutters are mounted on roller bearings, which are frequently critical in determining the life of the cutter. Present rock tunneling machines usually incorporate a system of automatic bearing lubrication to prolong the cutter life.

Disc Cutters. These have been used in many of the most successful tunneling machines operating in soft and medium hard rock. A single cutter consists of a disc-shaped base mounted on a roller bearing with a replaceable cutting edge of hardened steel around the disc. They may consist of single, double or triple cutting discs mounted together. A triple disc cutter is shown in Fig. 10-1(a).

The disc cutter operates on what is referred to as the "Kerf" principle, by which the cutter attacks the rock by cutting a groove into it while exerting a shearing force that causes the remaining ridges of rock to break away.

Disc cutters are the best means of cutting rock with compressive strengths up to 25,000 psi, but they may be limited by a highly abrasive rock with a much lower compressive strength. The range of disc cutters is now being extended to much harder rock by the use of discs with tungsten-carbide inserts around the edge. A disc cutter of this type is shown in partial section in Fig. 10-1(b).

Roller Cutters. This type of cutter rolls over the rock and fractures it by applying a direct mechanical force to all parts of the rock surface under attack.

Rolling cutters are of two types: milled-tooth cutters and tungsten-carbide-insert cutters. A milled-tooth cutter is shown in Fig. 10-1(c). This type is produced from heat-treated alloy steels in which rows of teeth have been cut on the outside of a truncated, cone-shaped base. The teeth may be self-sharpening by single hard-facing on the tooth flank, or both faces of the teeth may be hardened to give maximum abrasion-resistance.

When the abrasiveness of the rock puts it beyond the capabilities of the milled-tooth cutter, tungsten-carbide-insert cutters are used. A cutter of this type is shown in Fig. 10-1(d). The cutter has a truncated cone shape, with the cutting surface studded with small tungsten-carbide inserts.

The shape of the individual inserts varies. They may be chisel-crested, with a shape similar to a tooth; they may have a conical nose with a rounded apex where the angle of the cone may be 60 to 120 degrees; or they may be hemispherical. The choice of the shape depends on the hardness of the rock to be shattered.

The tungsten-carbide roller cutter disintegrates the rock more by a process of pulverizing it than by chipping out pieces. This type of cutter has been used to excavate the highest-strength rock that has been economically worked to

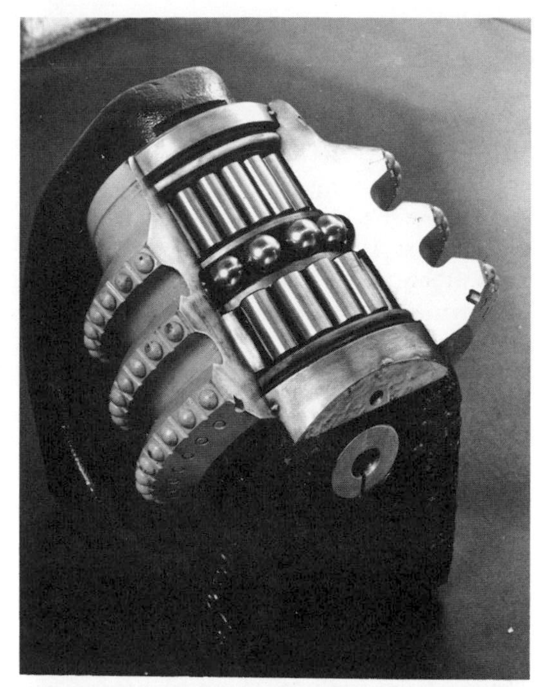

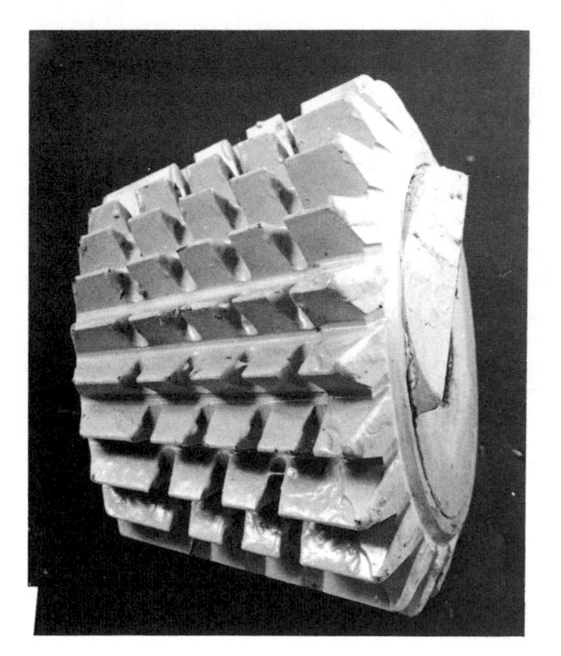

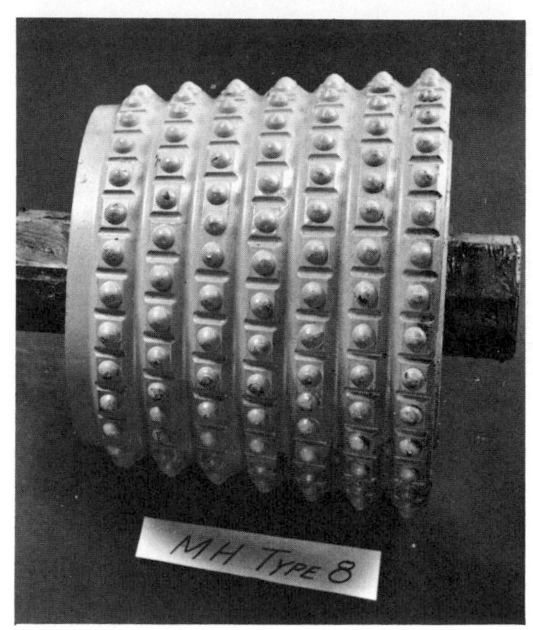

Fig. 10-1. Type of rock cutters.

date, but it has the disadvantages of high cost, slow penetration rate and the production of a high percentage of fines.

Layout of Cutters on the Cutting Head

In tunneling machines for use in soft ground, one type of drag cutter is generally used over the entire face of the cutting head. However, rock tunneling machines may employ different types of cutters to advantage on different parts of the cutting head. In the choice and arrangement of cutters over the face, there are three distinct zones with different requirements: the center, the face and the outside edge.

Center Cutters. Drilling the center of the hole in harder rock is usually accomplished by one or more roller cutters, since the fracturing action of the disc cutters cannot take place effectively until the hole is started. The cutters may be arranged as a tricone bit, or a single cutter may be cantilever mounted.

Face Cutters. These are the cutters that attack the main area of the face. They are generally disc cutters or roller cutters, depending on the hardness of the rock, but drag cutters are sometimes used in soft rock.

Gauge Cutters. Gauge cutters are mounted at the extreme outside edge of the cutterhead, and their function is to drill and maintain the hole diameter at the required size. They are usually disc cutters or roller cutters. To provide clearance for the outer end of the cutter bearing, it is necessary to slope the face of the cutter so it is no longer parallel to the working face. Because this positioning would lead to uneven and faster wear than the face cutters, the gauge cutters are usually treated with special hardfacing or tungsten-carbide inserts.

Theoretical Background of Cutters

Until recently, there had been few theoretical analyses of the requirements for a well designed and efficient tunneling machine. Improvements in design were based largely on empirical data. Recent papers have tried to remedy this.[1]

A rock tunneling machine operates by attacking the rock with a wedge-shaped tool in such a way as to set up high stress concentrations, which cause the brittle rock to shatter and disintegrate around the cutting tip. Hence, in an efficient rock tunneling machine, the volume of rock excavated is much greater than the volume of rock swept and penetrated by the cutting tools. The ultimate interest is the volume of rock removed, not the nature of the incision. The cutting tools may be relatively efficient in themselves and produce a high penetration into the rock for a low energy absorption. However, their performance in a tunneling machine may

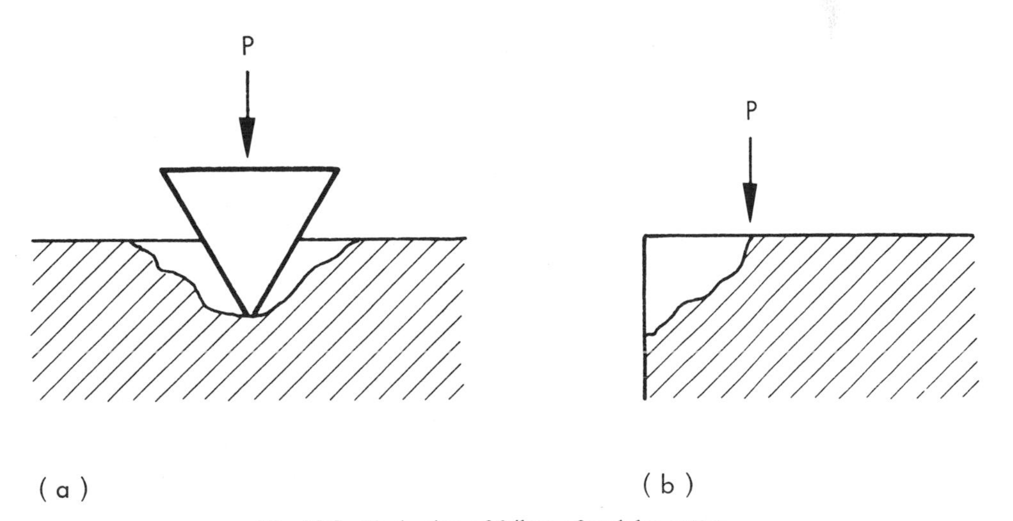

Fig. 10-2. Mechanism of failure of rock by cutter.

fall below expectations and produce a comparatively low rate of excavation, because there may be too few or too many cutters incorrectly spaced on the cutting head.

The mechanism of failure in rock is shown in Fig. 10-2. Under a normal thrust, the action of the cutting edge may be compared with the sharp point of a right-circular cone attacking a flat surface of hard rock. The practical effect is that the cone produces a small crater surrounding the tip by spalling off a small amount of rock, until the stress at the point of the tip is counterbalanced by the rock strength, as shown in Fig. 10-2(a). The subsequent spalling of the rock depends on the shape of the rock being attacked. When an open face exists, as shown in Fig. 10-2(b), much higher shearing stresses result from a given thrust P, so that the rock can break off more easily.

The above process is followed by the majority of rock tunneling machines. They first indent the rock surface with a sharp cutting tool and subsequently burst off pieces of the adjacent rock. In terms of the energy input per unit volume of rock excavated, the indentation process is a highly inefficient one and the bursting process is highly efficient. The aim of the design of the machine head is to arrange the layout of the cutters so as to enhance the latter process by the continuous creation of free faces with the optimum orientation with respect to the size, shape, thrust and rigidity of the cutting tools.

Experience with the disc cutter has shown it to be the best tool for meeting both requirements. An angle on the cutting edge of 55 or 60 degrees has been found to give the best compromise between effective cutting and rate of wear. Since discs are rotated by friction, slipping may occur with some types of rock, resulting in a translational motion of the disc edge over the rock surface. This slipping causes a high production of fines, a high rate of energy absorption, and a rapid wearing of the cutting edge. At the present time, it seems that this problem will be reduced by increased use of disc cutters with tungsten-carbide inserts.

For extremely hard rock, toothed roller cut-

Fig. 10-3. Atlas Copco hard rock tunneling machine.

ters penetrate the rock more effectively than disc cutters because of the shape of the teeth and the support provided to the teeth by this type of cutter. In the operation of the cutter, the teeth shatter the rock in the same direction as the direction of motion of the cutter, whereas a disc cutter shatters the rock in the direction perpendicular to the direction of motion of the cutter. The difficulty of arranging roller cutters with such an advantageous orientation towards a series of free faces makes the second phase of the excavation process, the shattering of the rock, less efficient than an arrangement of disc cutters. Recently developed large-diameter disc cutters with tungsten-carbide inserts are beginning to displace toothed cutters in hard rock tunneling.

Machines using the Undercutting Principle

The majority of tunneling machines manufactured are based on the empirical and theoretical considerations of cutterhead design described above. Some European manufacturers provide important exceptions to these methods. Atlas Copco machines use an entirely different way of disintegrating the rock and have an entirely different arrangement of the cutterhead. Drag cutters are used successfully for boring hard rock. The machine shown in Fig. 10-3 cuts rock by separately driven cutter units equipped with tungsten-carbide tips. Four cutter units are used on the driving face of the machine, mounted inclined to the machine axis. The whole head is propelled forward at a speed with an exact relationship to the rotational speed of the individual cutter heads. This causes the cutters to penetrate the rock in concentric helical paths, as shown in Fig. 10-4. The rock is undercut by the rotating drag cutters, and the remaining uncut portions are broken away by a light backward pressure exerted by a wedge-shaped protrusion behind the cutting tips and rotating with them. In this manner, only about a third of the whole volume of excavated rock is worked by the tips. Consequently, although drag cutters are considered by most manufacturers to be uneconomical for use in hard rock because of the high rate of wear, the arrangement of the cutters in this machine counteracts such a disadvantage by cutting at a slow speed and great cutting depth. Although this principal has been patented since 1951, its efficient application in a successful tunneling machine had to wait for development of strong and wear-resistant grades of tungsten carbide.

ROCK TUNNELING MACHINES

Basic Components of a Rock Tunneling Machine

A typical rock tunneling machine is shown in Fig. 10-5, together with the ancillary equipment necessary in an actual tunneling operation. Many variations are possible on the system and

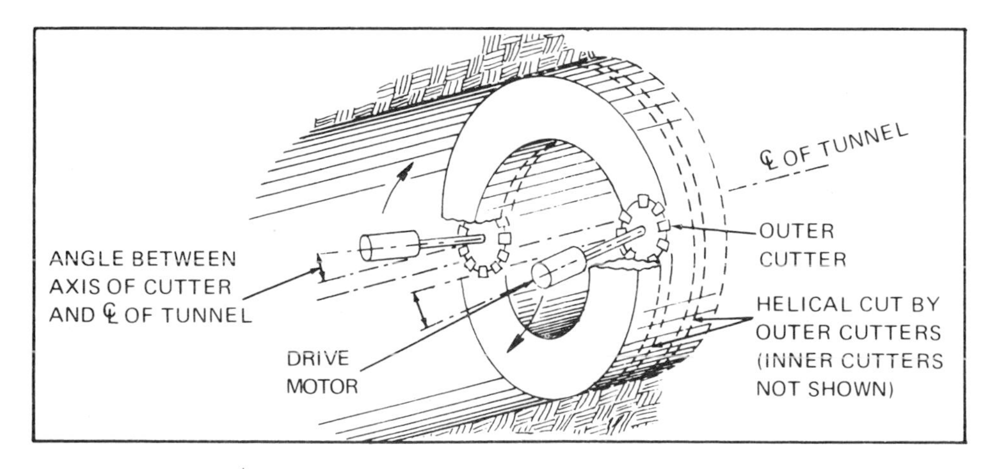

Fig. 10-4. Principle of the undercutting method as used in Atlas Copco hard rock tunneling machines.

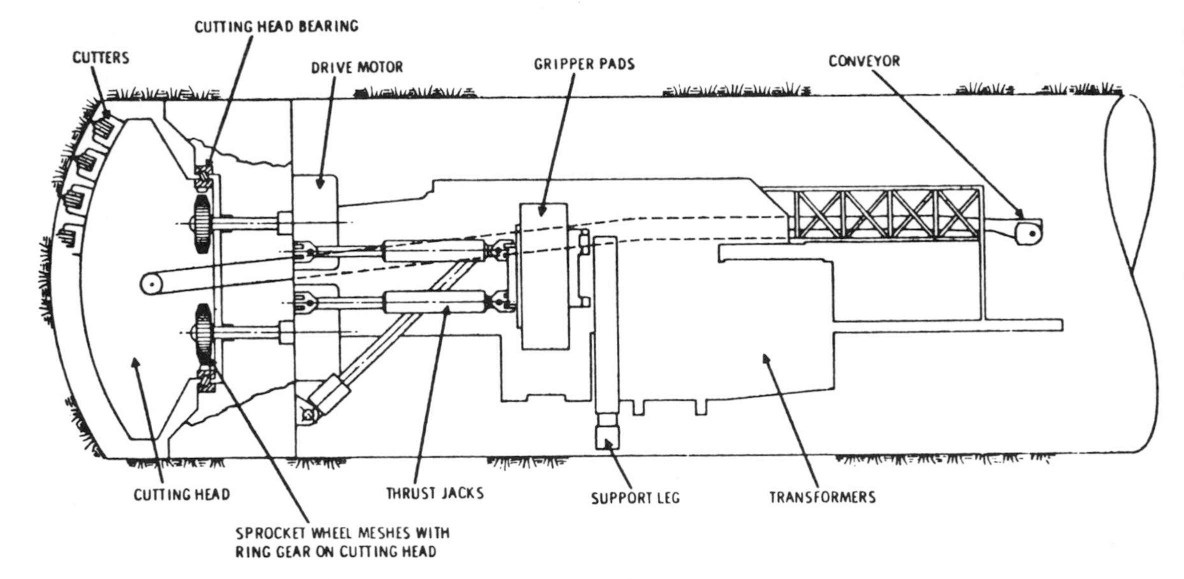

Fig. 10-5. Rock tunneling machine and ancillary equipment.

equipment shown, but Fig. 10-5 illustrates the basic essential components.

The Excavation System. The excavation is performed by a cutting head of welded-steel construction, usually convex in the direction of advance, with the cutting tools arranged on it for optimum cutting effect. The cutting head must be structurally capable of withstanding the thrust and torque that it transmits to the cutters, but in rock it does not have to support the material at the working face. The cutting head incorporates a means of removing the excavated rock from the face; in this instance, a series of buckets around the edge of the cutting head is used to raise the material and drop it on the start of the conveyor system.

The Machine Body. This supports the cutting head, connects the principal components of the machine, and accommodates the driving motors, other electrical and hydraulic equipment, and ancillary equipment. The drive motors supply the torque for rotating the cutting head. In Fig. 10-5, the motors are arranged peripherally and supply the torque directly through sprocket wheels to a ring gear at the back of the cutting head. In some machines, the drive motors are mounted at the back of the machine and transmit the torque through a main drive shaft passing along the centerline of the machine to the cutting head.

The method by which a tunneling machine is advanced and steered is similar in principle in a large number of machines. Figure 10-6 shows the propulsion method used by Jarva Incorporated. Immediately behind the cutting head are eight hydraulically actuated legs with gripper pads on the ends. The gripper pads consist of curved shoes with conical button inserts that are thrust against the tunnel wall during excavation to hold the machine in position. In addition, there are smaller support legs at the rear of the machine that are used only during the advance cycle.

Step 1 shows the position of the legs at the start of the boring cycle, with the eight main legs clamping the machine in position against the tunnel walls and the rear legs retracted. During boring, the rotating cutting head is pushed forward by means of four hydraulic thrust jacks, and excavation proceeds until the

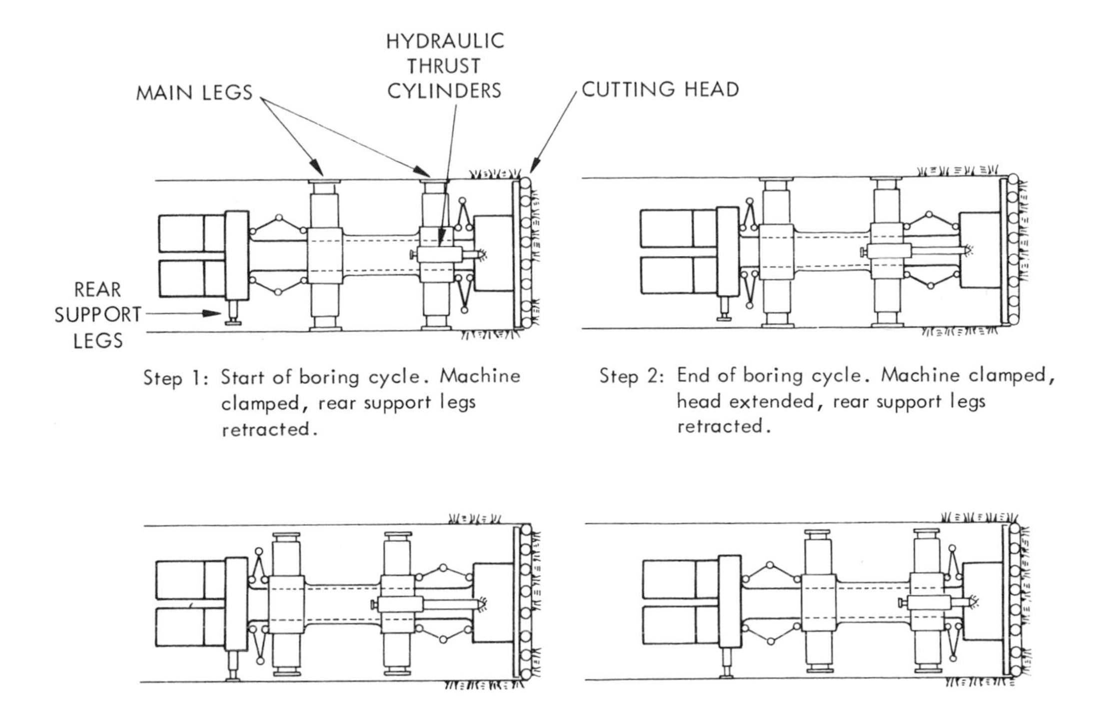

Fig. 10-6. Method of advance of a rock tunneling machine.

cutterhead reaches the end of its stroke, as shown in step 2. The length of this advance stroke is 1.5 to 2.0 feet on Jarva machines, but varies with other manufacturers. The position at the start of the reset cycle is shown in step 3. The rear support legs have been extended, and the cutting head rotated so that one of its arms rests vertically on the tunnel floor. The main legs are then retracted, and the machine is supported by the rear support legs and the cutting head. The eight main legs are advanced as a unit as the thrust jacks are retracted, until the machine is in the position shown in step 4 at the end of the reset cycle. The main legs again clamp the machine into position, the rear support legs are retracted, and the machine is once more in the position shown in step 1, ready to begin the boring cycle.

Other manufacturers generally use similar systems. In the machine shown in Fig. 10-5, one pair of side gripper pads is used, and continuous steering adjustments can be made during the excavation by means of the thrust jacks and steering jacks.

The Mucking System. Figure 10-5 shows a system that is similar to those often used on projects employing rock tunneling machines. In the muck removal system, the excavated material is scooped up by buckets around the cutting head, dropped onto a conveyor system and taken to the back end of the tunneling machine. There, it moves onto a separate conveyor system and is loaded into a train of mucking cars. The conveyor system is long enough to allow a complete train to back in underneath it so that the train can be filled without the need for a switching system. Included is equipment for controlling dust and removing it from the tunnel spoil and for transporting and handling the ventilation lines. The conveyor system must be long enough to allow sufficient space for track laying immediately behind the tunneling machine.

If a concrete lining is to be placed in a hard rock tunnel, there is seldom any urgent reason for it to be poured immediately behind the excavation, and the concreting operations can proceed independently. The same is usually also true of rock bolt and steel rib installation, although provision is sometimes made for placing rock bolts immediately behind the tunneling machine.

Manufacturers of Hardrock Tunneling Machines

The largest manufacturers of tunneling machines and their products are described. With time, more of these will enter the field.

The Robbins Company of Seattle, Washington has manufactured more than 100 machines, the majority designed for medium or hard rock, using disc cutters for the face and either tricones or tridiscs for the center cutters. A 12-foot diameter machine with this arrangement is shown in Fig. 10-7.

Robbins built one of the largest machines yet used with a diameter of 36 feet 8 inches for driving the combined diversion and power tunnels for Mangla Dam in India. Table 10-2 shows its basic characteristics. During operation, the original disc cutters, which could only be replaced by backing up the mole, were replaced by drag tungsten-carbide bits and a door was provided in the cutting head to given access for replacement. The machine drove five tunnels, each 1650 feet long, in 1963 and 1964, through sandstone and clay with some hard sandstone and limestone seams.

This machine was further modified for driving the twin road tunnels under the Mersey River in England through wet sandstone crossed by several faults. The diameter was reduced to 33 feet, 11 inches, and double disc cutters were installed. During driving the first tunnel, the 16-foot diameter main bearing was replaced. The motors and hydraulic equipment were relocated in waterproof housings for protection against the large inflow of water.

The record of speed for boring was claimed for a Robbins machine of 10 feet, 2 inches in diameter with 22 disc cutters and a tricone center cutter. It achieved a maximum advance of 156 feet in an 8-hour shift, 419 feet in a 3-shift day, and 1905 feet in a week in soft shale with a maximum compressive strength of up to

Fig. 10-7. Robbins tunneling machine for medium-hard rock.

Table 10-2. Typical characteristics of rotary full-face rock tunneling machines.

Manufacturer	Project Location	Machine Diameter (ft-in)	Thrust (lb)	Cutting Head Torque (lb-ft)	Cutting Head Horsepower
Robbins	Mangla Dam, West Pakistan. Mersey Road Tunnels, England	36-8 then 33-11	600,000	5,250,000	1,000
Robbins	Oso Tunnel, Colorado	10-2	372,000	175,000	300
Jarva	Bay Area Rapid Transit, Calif.	20-0	2,200,000	660,000	825
Ingersoll-Rand	Port Huron Tunnel, Michigan	18-4	1,500,000	500,000	750
Hughes	Navajo Irrigation Project, New Mexico	19-10	900,000	---	1,000
Dresser	Navajo Irrigation Project, New Mexico	20-6	1,080,000	703,000	720

5000 psi in the Oso Tunnel in Colorado. The average rate of advance excluding time lost in bad ground, was 133 feet per calendar day (see Table 10-2).

Jarva Incorporated, has manufactured a number of machines. These have been designed for the whole range of ground conditions from soft ground to hard rock, and diameters from 8 to 21 feet. One of the larger machines built by Jarva was a 20-foot diameter mole used in San Francisco on a section of the Bay Area Rapid Transit (BART) project. The cutters were triple-disc type with tungsten-carbide inserts, and the rock types included serpentine, greenstone, chert, and breccia. In this ground, penetration rates of up to 5 feet/hour were obtained, with maximum advance of 108 feet in a day and 364 feet in a week. More data on this machine are given in Table 10-2.

The Lawrence Division of the Ingersoll-Rand Company has manufactured a number of machines that have been used on projects in the U.S. and in South Africa for boring hard rock. A photograph of an Ingersoll-Rand hard rock tunnel machine is shown in Fig. 10-8. Its unique feature is that it drills a central pilot-hole ahead of the main working face, anchors itself into the pilot-hole, and then pulls on the anchor to obtain the necessary thrust at the cutting head instead of applying the thrust from the back of the machine. The machine supports itself on pads jacked against the tunnel walls while the pilot drill bores a hole ahead of the cutting head. The pilot drill is 11 feet in front of the cutterhead but has to be reset every 2 feet for boring. The auxiliary jacks extend to support the machine while the pads are retracted. The pads and pilot anchor, which are on a common shaft, advance simultaneously, then the pilot anchor expands to grip the inside of the pilot hole, while the pads are reset and the auxiliary jacks retract. The main boring operation starts next, with the cutting head and the pilot bit drill advancing together. During the excavation, the cutting head receives support from the side pads and from the pilot anchor. It is claimed that this system allows for very precise positioning of the machine because it can be pivoted about the pilot anchor to assume a new line or grade.

Two Ingersoll-Rand machines drove the Cookhouse tunnel in South Africa and a sewer tunnel in Rochester, New York. One of their largest machines was used on the Port Huron

Fig. 10-8. Ingersoll-Rand hard-rock tunneler.

tunnel in Michigan, where it bored an 18-foot, 4-inch diameter hole through shale and limestone boulders during 1969–1970. The maximum rates of advance were 112 feet in one shift, 210 feet in 1 day, and 3500 feet in 1 month (see Table 10-2). In the Dorchester tunnel in Boston, another machine worked successfully in hard argillite and andesite with a compressive strength of 24,000 to 28,000 psi at the upper limits of what has been economically excavated by tunneling machines.

The Hughes Tool Company has produced several tunneling machines, two of which were experimental in nature. Some of these machines were produced in association with foreign companies, particularly Mitsubishi Heavy Industries, Japan. Hughes' largest machine was used on the Navajo Irrigation Project, New Mexico when it drove a 19-foot, 10-inch diameter tunnel, 2 miles long, through sandstone with a compressive strength of 4000 to 6000 psi. The cutting head had 44 cutters; most of these were toothed roller cutters, and some were disc cutters. The machine can drill a five foot length of tunnel in a cycle before it has to be reset. The average rate of advance was 1160 feet/month over the total elapsed time of the project. The maximum rate of advance was 171 feet in a day and 662 feet in a week.

Dresser Industries has built tunneling machines, one of which was used on the Navajo Irrigation Project, where it drilled a 20-foot, 6-inch-diameter tunnel through sandstone and shales with compressive strengths of between 2000 and 9000 psi, using double disc cutters on the face. Maximum rates of advance have been 178 feet in a day and 1690 feet in a month (see Table 10-2).

Wirth Corporation in Germany is a major European manufacturer of hard rock tunneling machines. Since 1967, Wirth has manufactured more than 14 full-face rotary tunneling machines for 20 different projects. The machines use disc cutters or tungsten-carbide roller cutters, depending on the rock hardness. A Wirth machine worked on 34.3-foot-bored diameter road tunnel in Switzerland, through sandstone and marl, which is close to the largest-diameter hard rock tunneling machine yet produced. The tunnel was bored in three steps: first, an 11-foot diameter pilot was driven by a separate machine for the entire length of the tunnel; then, another machine, with two cutting heads in the form of concave annular rings, enlarged it to 22 feet in diameter; finally, a similar cutter head increased the excavation to its fullsize all in one one operation. A Wirth machine excavated the 3-mile long Nast Tunnel in Colorado, driving a

9-foot, 10-inch-bored diameter hole through blocky abrasive granite. Progress averaged 400 feet/month.

Atlas Copco produces hard rock tunneling machines that operate on a principle that is basically different from that of other full-size boring machines; this approach, called the undercutting method, was fully described earlier. The cutting head of an Atlas Copco machine is shown in Fig. 10-3. They drive circular bores between 10 and 15 feet in diameter and have operated in rock types from lava ash to hard rhyolite with compressive strengths in the range of 10,000 to 26,000 psi.

Atlas Copco machines worked on the Seikan tunnel in Japan, an undersea railway line under the 13.5-mile wide strait between the islands of Honshu and Hokkaido. One end passed through consolidated volcanic ash with a compressive strength of 4400 psi. At the other end, the rock was andesite containing water-laden faults. Three versions of the machine have been in use: the first two were 11.9 feet in diameter and the third was 13.2 feet, with four cutting wheels, each 3.6 feet in diameter, mounted on a 11-foot diameter rotating head. The difficult geologic conditions have limited the rate of advance to 5 feet/hour, with the best performance 300 feet in one month.

Another Atlas Copco machine can drive smaller tunnels of noncircular cross-section, to meet the demand for small noncircular service tunnels in metropolitan areas. The machine has one cutting head of the same design as used in the larger full-face tunneling machines and with a diameter equal to the required width of tunnel. The single cutting head is mounted so that it can swing vertically. It moves slowly from the tunnel floor to the roof, cutting and removing the rock by the undercutting principle, with a $1\frac{1}{2}$-inch penetration. Muck is removed by a hydraulically driven chain-drive conveyor and is pushed onto the end of the conveyor by the downward swing of the cutterhead. This machine can drive a tunnel with a width of about 4 by 7.5 feet.

The McAlpine/Greenside group has produced a tunneling machine that has the capability of driving tunnels of different diameters without major modifications. The rotating cutting head is about 3 feet in diameter and moves along an arm that is also rotated about a horizontal shaft. A basic machine was developed that cuts a range of sizes of tunnel with 85% of the machine remaining unchanged. Two have been used successfully to drive the Severn-Wye cable tunnel in England. The machines are both fitted with two rotary cutting heads, with tungsten-carbide drag cutters, and they are mounted on a single rotating arm to drive a bored diameter of 11.5 feet. The two machines were used on two sections of the tunnel, through quite different rock types and with widely different performance rates. One machine operated in hard dolomite and limestone with compressive strengths ranging over 20,000 psi. The average rate of advance was 27 feet/week, with a maximum rate of 56 feet/week. The other section was through siltstone with a compressive strength of 400 to 8000 psi, where the rates of advance doubled to an average of 58 feet per week and a maximum of 115 feet/week.

Experience With TBM's On WMATA. Washington Metropolitan Area Transity Authority is constructing a transit system which will serve Washington, D.C. and adjacent areas of Northern Virginia and Maryland. The Metro comprises 86 stations and over 102 miles of rapid transit lines, of which approximately 16 miles are in hard rock tunnels. Approximately 8.5 miles of twin tunnels were driven using tunnel boring machines (TBM).

Results obtained on five contracts are shown on Table 10-3.

Use of Rock Tunneling Machines in Inclined Shafts

The use of hard rock tunneling machines has been extended to steeply sloping tunnels. This was first accomplished in the Emosson underground hydroelectric station in Switzerland in 1968-1969 and, more recently in 1975 in the Hydro Penstock Grimsel Oberaar. At Emosson, a hard rock machine drove a 9.85-foot-bored diameter, 3760-foot long penstock inclined at 33 degrees through hard granite rock, by driving upwards from the lower end of the shaft, The machine was equipped with 26 carbide-insert roller cutters, which produced spoil con-

Table 10-3. WAMATA Hard rock tunnel data.

CONTRACT	ROCKVILLE ROUTE	WISCONSIN AVENUE	NORTH WISCONSIN AVENUE	ROCKVILLE PIKE	N. 16TH STREET
TUNNEL LENGTH	19,012'	15,239'	15,411'	19,026'	4,860'
STARTED FINISHED	3/1/74 9/19/75	9/19/75 11/8/76	12/4/75 10/4/78	3/25/77 10/10/78	4/9/76 5/5/77
ROCK TYPE	GRANITIC GNEISS, SCHIST	DIORITE GNEISS, SCHISTOSE GNEISS, AND SCHIST	DIORITE GNEISS, SCHISTOSE GNEISS, AND SCHIST	GRANITIC GNEISS, SCHISTOSE HORNBLENDE GNEISS & SCHIST	SCHISTOSE GNEISS
ROCK STRENGTH, PSI	2000-14000, AVE. 8000	3000 - 17,500, 90%: 4000-10,000	3000 - 17,500, 90% 4000 - 10,000	2000 - 18,000 AVE. 8000	8000 - 10,000
MOLE TYPE	ROBBINS MODEL 191 - 161	ROBBINS 191 - 161	DRESSER MODEL 190	ROBBINS, MODEL 191 - 161	JARVA MARK 22-2008
MOLE O.D.	19' - 1"	19' - 1"	19' - 2"	19' - 1"	20' - 8"
THRUST: AVAILIBLE USED	1,850,000 LBS 1,380,000 LBS	1,850,000 LBS 1,380,000 LBS	2,262,000 LBS 1,080,000 LBS	1,850,000 LBS 1,380,000 LBS	2,000,000 LBS
ADVANCE LENGTH (SHOVE)	5'	5'	4'	5'	4'
RPM/THRUST	MAX: 41,000 LBS NORM: 30,000 LBS	MAX: 41,000 LBS NORM: 30,000 LBS	–	MAX: 41,000 LBS NORM: 30,000 LBS	–
HORSEPOWER	900	900	720 - 1080	900	1000
CREW SIZE/SHIFT	20 - 22	20 - 22	10 - 13	17	6-8 IN HEADING
SHIFTS/DAY	3 SHIFTS, 5 DAYS	3 SHIFTS	3 SHIFTS	3 SHIFTS	2 SHIFTS
ADVANCE: MAX. AVERAGE	634'/WK, 168'/DAY 82'/DAY	482'/WK, 125'/DAY 65'/DAY - 56'/DAY	457'/WK 230'/WK - 65'/WK	502'/WK 246'/WK - 228'/WK	228'/WK 95'/WK - 76'/WK
MOLE AVALABILITY	51% WORKING DAY	38.5%	54%	50%	43%

taining pieces up to $1\frac{1}{2}$ inches, with a high percentage of fines washed out as drilling mud. The sloping tunnel was found to have a distinct advantage over a horizontal tunnel for the removal of spoil from the working face. The cuttings discharged through an opening in the cutting head and were washed out through a closed duct by the drilling mud used to cool the cutters, thus obviating the need for a chain conveyor or conveyor belt. The inclined shaft was completed in 9 months, with an average rate of progress of 17 feet per 2-shift working day. The maximum rates of advance were 52 feet in a day and 469 feet in a month.

The machine, which weighed 90 tons, was maintained in its inclined position by a specially designed automatic locking device. The proper functioning of this feature was important in calming the fears of personnel that the machine would slide back on top of them.

There was an estimated saving in the concrete lining of about one cubic yard per linear foot, partly due to the reduction in overbreak that always occurs with conventional tunneling methods and partly due to a redesign of the concrete lining to take account of the smooth and undisturbed nature of the rock surrounding the excavation. There was a further considerable saving because of the much reduced amount of support required.

Following this success, a 7.38-foot-bored diameter shaft with a 42-degree slope was driven through similar rock, also on the Emosson project.

SOFT GROUND TUNNEL-BORING MACHINES

Types of Soft Ground Conditions

The characteristics of a soft ground tunneling machine are determined to a large extent by the type of soft ground in which it will be working. The following descriptions of soil conditions encountered in tunneling are in general use.

Firm Ground. In firm ground, a tunneling machine may be advanced several feet or more without immediate support. No problem occurs with uncontrolled sloughing if the excavated face is unsupported. Stiff clays and cemented or cohesive granular materials are generally of this type.

Raveling Ground. If raveling ground in the roof of the tunnel and in the upper part of the working face is unsupported for some period of time, it flakes off and falls into the excavation. The action is progressive and may lead to open cavities growing above the excavation. Slightly adhesive sands, silts and fine sands with apparent cohesion, and residual soils with relict structures often form raveling ground.

Running Ground. Running ground slumps from any unsupported lateral face until a stable slope is formed at the angle of repose. Cohesionless soils, such as dry sand or clean, loose gravel, fall into this category.

Flowing Ground. If seepage pressure develops towards the working face of what would otherwise be raveling or running ground, the soil may be transformed into flowing ground, which can advance like a thick, viscous liquid into the tunnel through any available opening.

Squeezing Ground. If squeezing ground is unsupported, it gradually moves inward, but without necessarily breaking and raveling. This behavior is a characteristic of soft to medium clays, and the time dependency of the movement is related to the plasticity of the soil.

Basic Components of a Soft Ground Tunneling Machine

Because of the nature of the ground in which it is designed to operate, a soft ground tunneling machine differs from a rock tunneling machine in several aspects. Firstly, a complete circular shield must support the top and sides of the excavation as soon as it is formed; frequently, the working face must also be supported. Secondly, some kind of lining system must be placed to provide support behind the shield as soon as the shield moves forward.

The Excavation Equipment. If the tunneling operation is in firm ground, the most effective method of working the face is generally by an articulated backhoe or a similar excavating tool. This method works most efficiently when no breasting of the face is required and the backhoe can work with a minimum of restriction.

This type of machine will be discussed in this section, although it does not fit the description of a tunneling machine given earlier.

More commonly, the excavation is accomplished by a circular rotating cutting head using drag-type cutters. The cutting head may be of spoked construction with the cutters distributed along the spokes, leaving open spaces between them. The spoked cutting head provides easy access to the working face and to the cutters for maintenance or replacement, but it does not provide a great amount of support at the working face. This design works well in firm ground and usually provides sufficient support in raveling ground.

A commonly used alternative type of cutting head has a completely closed working face, in which case it is usually known as a drum digger. This design provides the maximum degree of support at the face and is the best machine to use in running ground. The cutters break up the material at the face, and the spoil is removed through slots in the face of the drum. The size of opening can usually be adjusted to prevent uncontrolled flow of running material from the face. Doors provide access to the working face, but access is usually limited, which must be weighed against the need for full support of the face. A machine of this type produced by the Mining Equipment Manufacturing Corporation (MEMCO) of Wisconsin is shown in Figs. 10-9 and 10-10.

Both the spoked and the drum cutting heads can be made to operate with a full rotation or with an oscillating motion over a suitable arc.

Machines that rotate fully can frequently be operated in both directions of rotation. This feature will prevent the whole machine rotating

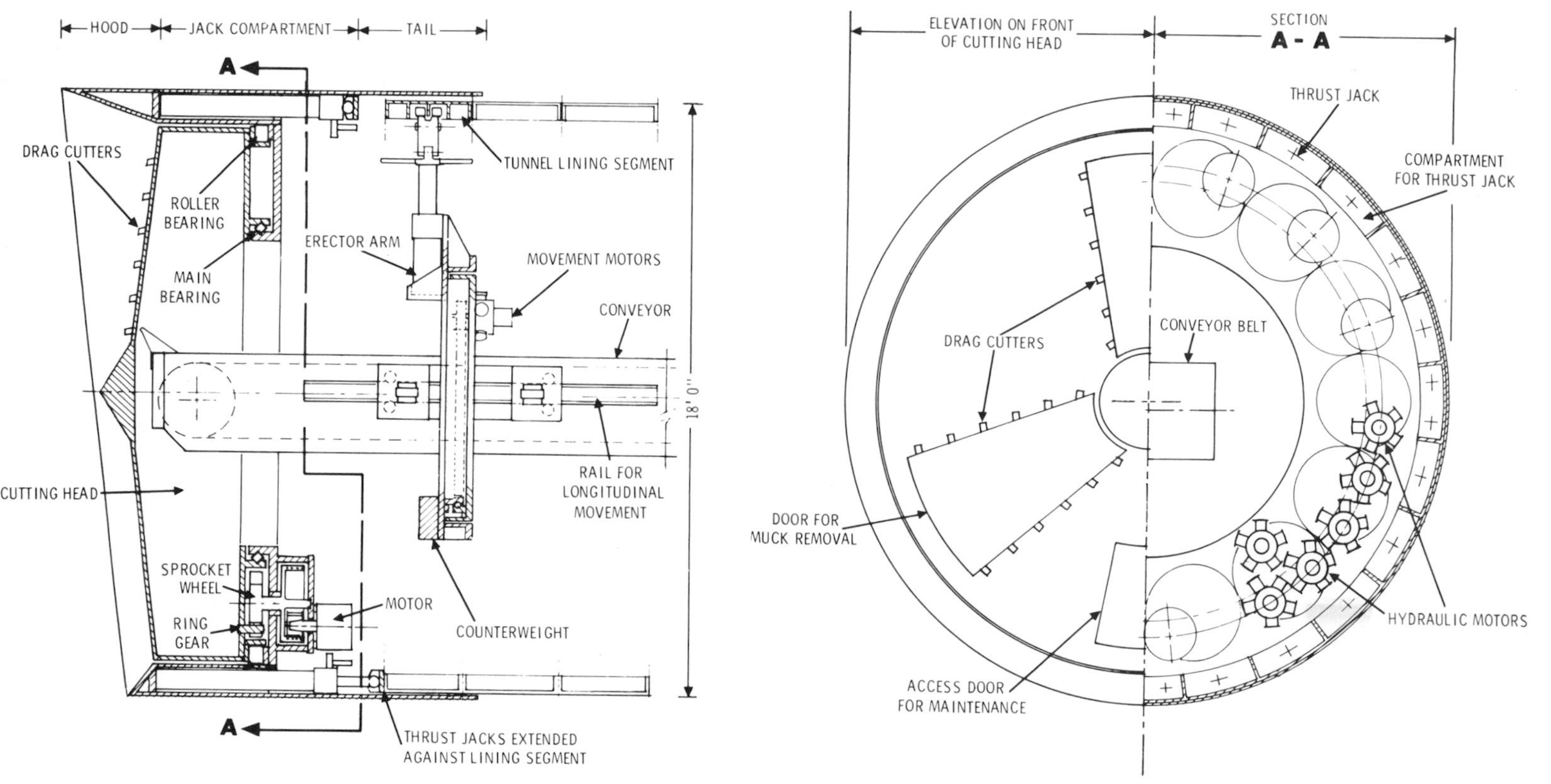

Fig. 10-9. Soft ground tunneling machine and lining erector arm.

Fig. 10-10. MEMCO soft ground tunneling machine.

off the plumb position or allow it to be brought back on line if this occurs. Spoked cutting heads can also be made so that separate halves or quadrants can oscillate independently of each other.

The Machine Body. The shield is the predominant feature of the main body of a soft ground tunneling machine. It is located immediately against the working face and protects the men and equipment from falling ground and supports the excavation until a permanent lining can be placed. The shield is quite similar to shields used in conventional hand-tunneling operations. If a full circular cutting head is used, it is usually mounted within the shield together with the drive motors.

The forward movement of the machine is effected by a number of hydraulically operated jacks arranged around the rear periphery of the shield, usually reacting against the last erected lining ring. These are shown in Fig. 10-10. Each jack can be operated individually, or several jacks can be operated together as a unit. The jacks are made capable of shoving the tunneling machine forward a distance equal to the width of the lining rings. When this part of the working cycle has been completed, the shield has been moved forward a sufficient distance to allow the next lining ring to be erected in the tail. If the machine has been designed to shove against the previously placed lining ring, excavation and advance have to stop while the next lining ring is placed and the jacks are reset against it.

The jacks also provide the means of steering the shield and adjusting it to its required alignment between shoves. By engaging more jacks on one part of the perimeter than on the remainder, the advance of the shield may be directed up and down or right and left.

The Mucking Equipment. During the forward movement of the machine, the muck is pushed and deflected onto a conveyor or dragchain, which removes it to the back of the shield. There it is transferred to a conveyor train, which is pulled behind the shield. This train transfers the muck to a loading point of the muck removal system. This system may be similar to those used in rock tunneling machines, and the mucking train shown in Fig. 10-5 is also suitable for soft ground tunneling.

Compressed Air Systems. The tunneling machine in soft ground may need to be operated under compressed air behind an airlock. The condition of the ground and the position of the water table determine whether a compressed air system is needed. A tunneling machine cannot perform in flowing ground, and measures must be taken to change the flowing nature before tunneling operations commence, either by lowering the water table or by the use of compressed air. These steps may not always suffice, as one experience during construction of the BART project showed. Attempts had been made to dewater a section of the tunnel line, but these were not entirely successful because of the presence of clay layers in the granular materials. The tunnel was being driven under compressed air when a deposit of flowing fine sand was penetrated. The machine was a drum digger type with a cutting head completely closed, and the flowing ground was controlled with only minor intrusions. If an open-spoked cutting head had been used, much greater problems could have developed.

In squeezing ground, the degree of support that must be provided by the tunneling machine depends on the flow characteristics of the ground and the planned rate of advance. During machine close-down, support may be required even where it is not needed during a moving operation. In very plastic ground, the machine may become unmanageable due to a strong tendency to tilt as it advances. This condition can be greatly improved by the use of compressed air.

Manufacturers of Soft Ground Tunneling Machines

Backhoe-type Machines. Five backhoe-type machines were used in San Francisco for the BART system, where they performed well. The backhoe was used to rip the face and load the spoil either into muck cars or into the tunnel invert, where it was removed by a loader. These machines had no means of retaining the excavated face and could only be used in ground where the face did not need support. Hence, they were used basically in cemented sands that had been dewatered, although in one case the shield drove through an area where soft and weathered rock occurred together.

Exceptionally good performance was achieved with a backhoe tunneling machine on four tunnels totaling five miles, constructed for the Metropolitan Water District of Southern California. This machine was manufactured by MEMCO. The tunnels have a driven diameter of 26 feet and pass through partly consolidated sandstone and siltstone of the Saugus formation. The machine achieved an advance rate of 202 feet in one 24-hour day on the Castaic No. 2 tunnel. Average progress over one 2.5-mile length of tunnel was 113 feet/day.

An interesting variation on the backhoe-type machine was used in the construction of some sections of the Hamburg underground railway system. In one section, the tunnel passed through hard clay on the lower half of the excavated face and a mixture of fine sand with boulders on the upper half. A partly mechanized device was designed to meet these conditions. In the lower part of the shield, two small hydraulic backhoes were incorporated to excavate the free-standing clay and load it onto conveyor belts. The upper part retained the features of a hand shield, with a series of three platforms onto which the loose soil could slope freely and be removed by hand. If necessary, the working face could be blocked up by conventional methods during breaks in the work. The machine excavated a 25-foot diameter hole, and was advanced by 24 jacks with a total thrust of 7.8 million pounds. The rate of advance averaged 10 feet/day with a maximum rate of 30 feet/day.

An improved version of the backhoe-type excavator machine has been introduced by the Robbins Company. The shield is equipped with hydraulically operated breasting doors by which the upper half of the face can be closed off immediately if running ground is encountered. A highly maneuverable, hydraulically operated digging arm is equipped with a scraping tool with a ripping point on the reverse side, and the operator selects the appropriate tool by rotating the boom head. He first works over the complete excavated face with the ripper, and then uses the scraper to move muck onto a conveyor system for spoil removal. The excavating head may be easily replaced with a rotary auger where ground conditions are suitable for that method of excavation. A machine of this type was used to drive the 22-foot diameter San Fernando Tunnel in Southern California, and another drove a 14-foot, 6-inch diameter tunnel in Detroit.

Machines with Circular Cutting Heads. A Calweld machine used to excavate part of the BART system tunnels in San Francisco is shown in Fig. 10-11. The cutting head is a spoked-wheel-type, with each of the four quadrants capable of oscillating independently through a 30-degree arc. The length of the shield is 12 feet $1\frac{1}{2}$ inches at the bottom, and projects forward 4 feet, 2 inches at the top. Each quadrant consists of three spokes with pick-type drag cutters mounted along them. The two lower quadrants were mounted vertically, and the two upper quadrants were in a slanting position, parallel to the edge of the hood. Each cutting quadrant is mounted on a rotating arm. The oscillating motion is caused by two hydraulic jacks that engage in the arm and alternately rotate it in opposite directions. Twenty shove jacks were spaced around the perimeter of the shield to give it forward motion and directional control.

Behind the cutting arms was a bulkhead, but this was partially removed soon after construction began in order to improve access to the face. As the machine advanced, the bulkhead channeled the muck through the bottom of the shield, where it fell onto a dragchain conveyor that carried it onto a more extensive conveyor system for the removal from the tunnel. (More data on this machine are shown in Table 10-4). The Calweld machine reached maximum rates of advance of 105 feet in a 3-shift day, 367 feet a week, and 1190 feet in a month (4 weeks). It was able to maintain an average rate of advance of 49 feet/working day for the final 3500-foot length of tunnel.

A machine with a circular cutting head had been used successfully in soft ground in sections of the Hamburg underground railway system. A Bade-Holzman machine was used to drive three sections of the system with a total length of 14,000 feet of single-line tunnel of 18.3-foot diameter. Further development of this machine produced the BADE MDS 610 GS tunneling machine, which drove tunnels for subway projects in Vienna, Austria; Sao Paulo, Brazil; and Hong Kong. The machine is generally used in soft ground, such as sand and clay.

Eight rams rotate the wheel of the machines at $\frac{1}{2}$ rpm by the rack and pinion principle. The front of the wheel is composed of star-shaped steel plates, which are held against the face by

Fig. 10-11. Calweld soft ground tunneling machine.

Table 10-4. Typical characteristics of soft ground rotary full-face tunneling machines.

Manufacturer	Project Location	Machine Diameter (ft)	Thrust (lb)	Cutting Head Torque (lb-ft)	Cutting Head Horsepower
Calweld	Bay Area Rapid Transit System, San Francisco	18.0	5,040,000	3,888,000	1,000
Bade	Sao Paulo System	20.30	10,000,000	5,000,000	1,100
K.M. Tunneling Machines	Various in S.E. England	7.5 20.0	672,000 3,300,000	45,600 630,000	100 800

independently operable jacks. The jacks allow tilting of alternating steel plates slightly to admit soil during the rotation of the wheel (Figure 10-12. The wheel is advanced independently of the shield by four push jacks, which are supported by the rear diaphragm. The maximum stroke of the jack is 2 feet, advancing the wheel 1 foot ahead of the hood. The four push jacks can be operated at different pressures to incline the wheel with respect to the shield axis. The forward motion of the shield is provided by 25 shove jacks, pushing against the erected tunnel lining. Shield progress for each tunnel cycle is 3.28 feet, corresponding to the length of one tunnel ring. During the tunneling cycle, the shield and wheel advance in several alternating steps. Steering of the 23-foot long and 20.30 foot diameter shield can be facilitated by mounting overcutting teeth on the cutting wheel and overcutting ahead of the hood (see Fig. 10-13). The basic characteristics of the Bade machine are shown in Table 10-4. The maximum rate of progress achieved was 40 feet in a 24-hour period of tunneling.

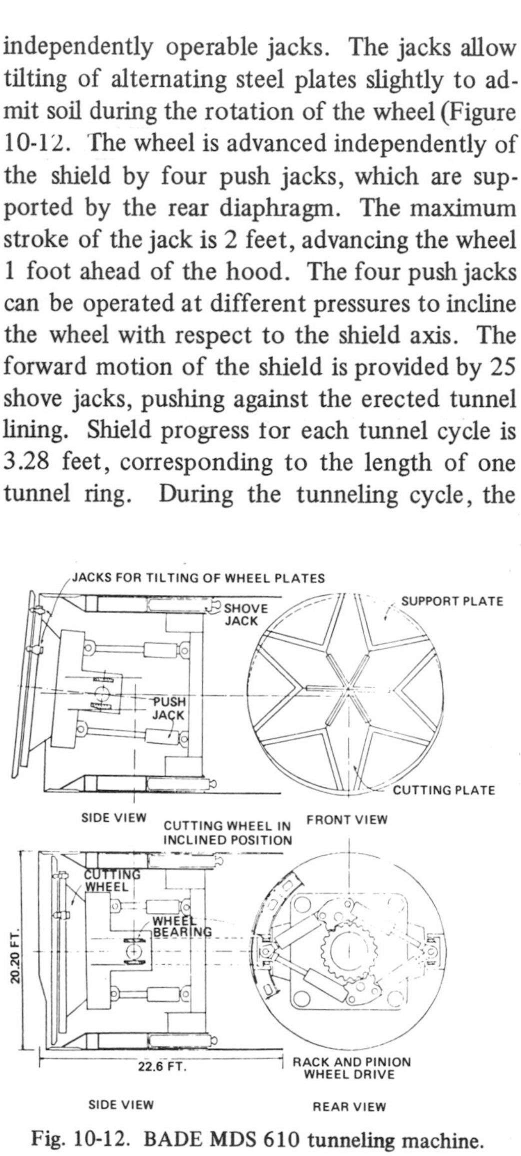

Fig. 10-12. BADE MDS 610 tunneling machine.

Three drum diggers, fully closed at the cutting face, were used on the BART system tunnels. These were manufactured by MEMCO, and one of them is shown in Fig. 10-11. The machine consisted of a cylindrical shield of 1-inch thick steel, with an outside diameter of 17 feet, 11 inches and an overall length of 12 feet, 5 inches. The steel cutting wheel, 15 feet, 11 inches in diameter, was centered in the front of the shield, forming a completely closed drum-shaped surface. Three radial doors formed part of the face of the cutting head. Pick-shaped drag cutters were mounted along the radial edges of each of the doors. The doors could be opened and closed by hydraulic rams. In order to advance, the doors were opened, the machine was rotated, and the teeth scraped away the ground from the face. The spoil was forced through the openings in the doors, where it fell onto a conveyor belt. In suitable ground, the doors were removed completely during tunnel-driving and were only reinstalled for weekend shutdown.

The firm of K.M. Tunnelling Machines Limited has gained considerable experience in full-face tunneling in soft ground during the last decade. Their machine was designed for operation in soft, cohesive ground such as clay, marl, soft shale, chalk and sandstone. It consists of a conventional tunneling shield with the cutting head attached to a drum that revolves inside the outer shield. Usually, pick-type drag cutters are attached to the head, but other types of cutter may be used, depending on the ground conditions. Rotary drive of the cutterhead is provided by high-torque, low-speed hydraulic motors. The rotary drive system is placed between the inner drum and the outer shield, which leaves free access to the working face in the center of the machine and allows easier removal of the spoil, inspection of the working face and replacement of the cutters. Forward motion is provided by a ring of heavy-duty hydraulic rams at the rear of the machine. These can thrust against the last tunnel ring installed, if a precast lining is being placed immediately following excavation. An alternative system is provided by which the rams can react against an independent jacking ring that is expanded against excavated tunnel walls so that excavation can proceed independently of the lining operation. This can only be done where the ground is strong enough to take the thrust from the ring and also stand unsupported until the lining is placed.

The machines are designed with a certain amount of flexibility in the cutting diameter. Within a certain range of diameter and power, the outside diameter of the machine may be altered by changing the outer shield and modifying the cutterhead.

This machine has been used extensively and with great success in the London clay for aqueducts, gas storage chambers and underground railway tunnels. Some of the characteristics of the range of machines are shown in Table 10-4. Rates of advance of up to 514 feet/week of bored and lined tunnel have been achieved with the 7.5-foot diameter version of the machine.

SOFT GROUND TUNNELING MACHINES IN FLOWING GROUND

When tunneling operations have to be carried out in soft ground that would turn into flowing ground in the presence of water, the usual practice is to work under compressed air. Drum digger tunneling machines have been used successfully under such conditions behind an airlock. However, this type of operation has the problem that all men and materials brought to or from the excavated face have to pass through the airlock. Also, there is a limit to the amount of useful time that men can spend in the working area; and, as the air pressure necessary to maintain the face increases at greater depths, this kind of operation becomes more and more difficult and uneconomical.

Attempts have been made to overcome this problem by limiting the pressurized space to a bulkheaded chamber at the cutterhead, using air, slurry or water as the pressurizing medium to support the excavated face. These methods have been tried on construction projects with differing degrees of success, and show promise of leading to development of a safe and reliable device at some time in the future.

One of the first such machines was designed by the Robbins Company and was used on the construction of part of a new line on the Paris Metro between 1964 and 1967. The machine drove a 9,430 foot length of tunnel, 33.7 feet in diameter. The soils were limestone overlaying sands, and the head of water reached a maximum of 65 feet. The cutting head was a drum digger with 56 fixed cutters, the thrust was 16 million pounds and the maximum torque was 5.2 million pound-feet. Compressed air was used at the working face to support the soils, while the working chamber was in free air. Problems were experienced in operation, and it proved difficult to maintain the air pressure at the face, partly because of leakage of air through the face and partly because of the small size of the air reservoir formed by the bulkheaded chamber. Despite the problems, the maximum rates of advance achieved were 41 feet in a day and 5600 feet over a 12 month period. The average rate of advance for the operation was 10,8 feet/day.

The principle of using a pressurized bulkhead chamber at the working face was further developed by using pressurized slurry rather than compressed air as the pressurized medium. A combination of rotary breasting and fluid pressure within the bulkhead maintains pressure on the face.

In 1972, the British National Research

Development Corporation (NRDC) and the London Transport Executive (LTE) jointly financed a development contract to drive a test tunnel at New Cross as part of the Fleetline of the London Transport. The contractor was Edmund Nutall Ltd. and the tunneling machine was build by Robert L. Priestley Ltd. A total length of 472 feet was driven at New Cross in sandy ground below the ground water table. Subsequently, the same machine was used to drive a sewer tunnel at Warrington U.K. in loose sand. Excavation rates of 18 feet/10-hour shifts and 130 feet/week were achieved. One of the problems in the loose sand was vibration. The bottom of the tunnel was in rock and there were boulders present. Both of these conditions caused an increase of the normal vibration associated with tunneling machines. For this reason, ground treatment by chemical grouting was necessary in stretches of tunnel to reduce surface settlement.

The German firm of Wayss and Fraytag has also developed a slurry face machine, called Hydroshield. This development uses both bentonite slurry and compressed air. The compressed air chamber is used as a pressure regulator.

As more experience is gained, it becomes apparent that slurry pressure regulation at the tunnel face and measurement and regulation of rate of muck removal from the cutterhead are of prime importance in practical application. Perhaps the greatest experience gained and progress in this area of development was made by Japanese companies. The Japanese list more than 60 tunnel projects, dating back to 1969, in which slurry face machines have been used. The Tekken Construction Company developed a machine that completed a 3440-foot, 16.7-foot diameter tunnel in interbedded loose sand, silt and clay below groundwater level in downtown Tokyo in 1975 (see Fig. 10-13). Progress rates of 40 feet/day and 900 feet/month were achieved. The main technical improvements on the machine were in the development of an automatic slurry control system for control of:

- Optimum slurry pressure.
- Slurry composition.
- Measurement of volume of excavated muck.

The system uses gamma ray density meters, electromagnetic flow meters and automatic control valves to regulate slurry flow (see Fig. 10-14).

Earth Pressure Balanced Shield Tunneling. A new development in soft ground tunneling designed to control ground water and prevent the collapse of the tunnel face has been devel-

Fig. 10-13. TEKKEN slurry face tunneling shield.

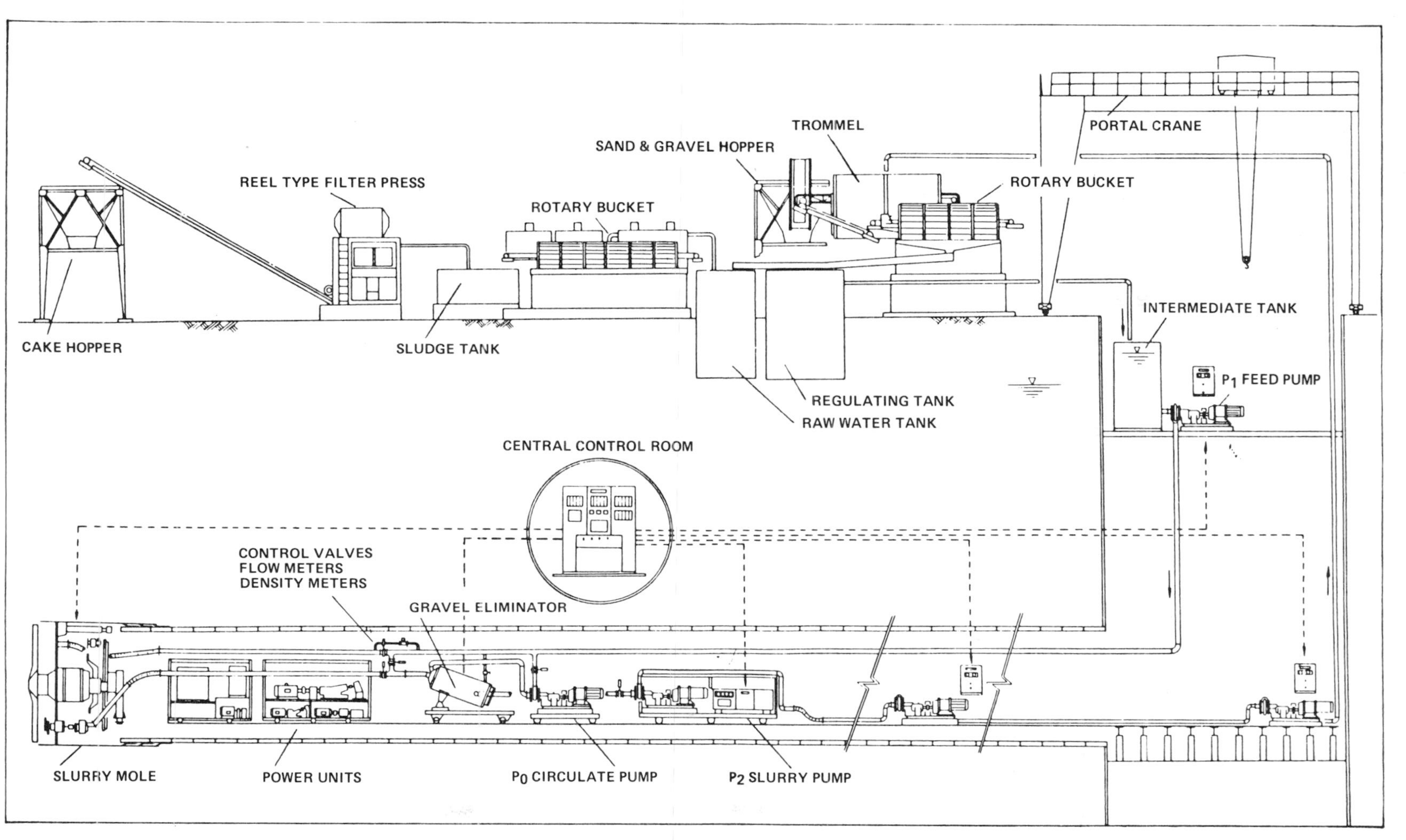

Fig. 10-14. TEKKEN slurry circulation system for tunnel machine.

oped in Japan. The principle of the earth pressure balanced method is that for protection of the tunnel face, excavated material fills the pressure chamber of the shield machine and balances the earth pressure at the face. A bulkhead is provided between the cutterhead and the shield chamber and as the shield advances the muck is squeezed through an opening in the cutterhead. The material is excavated by a rotating cutter and discharged by a screw conveyor.

The excavated volume and discharge volume must be the same. To achieve this condition, the advancement of the shield is controlled either by monitoring the discharged muck volume or by monitoring the earth pressure of the face. The discharge volume can be measured by scale reading of muck hoist or by reading of a scale conveyor underneath the gate of the screw conveyor.

Earth pressure within the pressure chamber is controlled by observation of the thrust of the shield jacks, torque and revolving speed of the cutter frame and torque and revolving speed of the screw conveyor.

The advancement of the shield and the rotation of the screw conveyor are synchronized to equalize the existing pressure at the tunnel face with that of the muck in the pressure chamber.

The Earth Pressure Balanced (EPB) tunneling method has been successfully used on several jobs in Japan. In the United States, this method has been adopted on the North Point section of the sewer tunnel, part of the San Francisco Clean Water Program. The 3100-foot long, 12-foot diameter tunnel will be driven through recent bay deposits consisting of sandy and silty clays. The depth of ground above tunnel crown is about 30 feet and the ground water depth is about 15 feet. Figure 10-15 shown diagrammatically the EPB shield tunneling machine to be used in San Francisco by the contractor Ohbayashi-Gumi, Ltd. The shield is scheduled to start operation in January 1981.

EXPERIENCE IN VARIABLE GROUND

Problems of Tunneling in Variable Ground

One of the major complaints by tunneling machine users has always been the lack of ability to operate successfully and continuously in variable ground. Usually, if a tunneling ma-

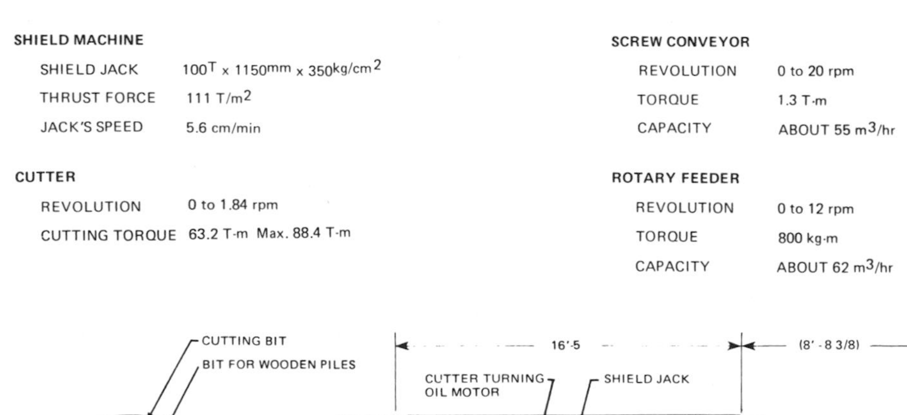

Fig. 10-15. EPB shield tunneling machine.

chine suddenly encountered ground other than that in which it was designed to operate, it would not function, and would cause considerable delay and a return to hand-mining methods until the difficult ground was tunneled through.

With increased use of tunneling machines, more and more data have become available describing this kind of problem and the solutions developed in the field to overcome them. The following case histories describe this kind of experience.

Fairmont Hills Tunnel. The Fairmont Hills Tunnel forms part of the BART system in San Francisco. The tunnel consists of twin 20-foot diameter tubes about $\frac{3}{4}$-mile long. Bore-holes were put down at about 350-foot intervals, rock samples were taken and tested, and a geological profile of the tunnel was prepared. As planned, the tunnel line would pass through chert, sandstone, greenstone and serpentine, with sheared zones in places. However, a buried valley at the north end of the tunnel was not discovered.

The contractor elected to drive the tunnel by machine. A shaft was sunk at the north end of the tunnel, and the machine was assembled at the shaft bottom. The boring machine had only advanced a short distance before it hit stiff clay at the top of the tunnel. Within a few feet, the cutters had clogged up and the machine was inoperable. The machine had to be pulled out and the first 220 feet of tunnel excavated by hand until a full face of rock was exposed. Crews for the conventional mining methods had to be mobilized and larger steel sets had to be ordered that would permit the tunneling machine to pass through. As a result, this first length of 220 feet of tunnel took more than 10 weeks to drive.

Futher problems occurred about halfway along the tunnel line, when the machine had to pass through a zone of meta-sandstone that had deteriorated into serpentine in places. The machine arrived at this area on New Year's Eve and the tunnel was shut down for the holiday. Three days later, when work recommenced, the machine could not be started because the bad ground had squeezed in around it. After considerable effort, the machine was pulled back; then the roof collapsed. It became necessary to back-pack the roof, shore the tunnel walls and hand-mine through the squeezing ground. It took 5 weeks to drive the 20-foot length of tunnel under these conditions.

When the second of the twin tunnels passed through the same zone, progress was maintained continuously, and no problem occured.

Southeastern Trunk Sewer, Melbourne. Field experience enabled the Melbourne Metropolitan Board of Works to modify and considerably improve the performance of a Robbins hard-rock tunneling machine working in variable ground during construction of part of the Southeastern Trunk Sewer Tunnel, Melbourne.[2] In this case, the user's modifications have since been incorporated by the manufacturer into a new line of machines, specifically intended to handle similar variable ground conditions.

The Board's requirements for the sewerage system for Melbourne called for a main trunk sewer with a 55,000-foot length of concrete-lined rock tunnel, varying in lined diameter from 8 feet, 6 inches to 11 feet, 6 inches. The tunnel would be driven through the Silurian strata, part of which is capped by a thick layer of water-bearing tertiary sandy sediments. The Silurian strata are intensively folded with extensive faulting and shearing, and weathering can penetrate to tunnel level. The rock generally consists of weak to soft mudstone, intensively jointed, with a compressive strength of 2000 to 3000 psi and interbedded with thin beds of stronger but more highly jointed sandstones. Ferruginization of some of the sandstones has occurred, producing bands of hard and abrasive rock in places. There are frequent dikes fully weathered to clay along the length of the tunnel.

A Robbins rock runneling machine with a thrust of 356,000 lb was purchased with a cutting head incorporating 27 disc cutters and a central tricone cutter. During the first months of use, the machine made poor progress, averaging just over 4 feet per shift. The major problems encountered were as follows

- The horizontal distance from the front edge of the shield to the central cutter was

3 feet, leaving the rock virtually unsupported in this region.

- The cutters dragged blocky material across the working face, disturbing unstable ground ahead of the machine.
- The buckets became jammed either with clay or with large blocks of rock, which then gouged out more rock and caused extensive overbreak.
- Instead of cutting the rock in loose ground, cutters and buckets frequently stuck and stalled the machine.
- Poor access to the front of the machine made it impossible to inspect or change the cutters in bad ground. During cutter maintenance in unstable ground, the rock ahead of the machine frequently caved in and then had to be stabilized by hand-mining operations around the head of the machine.
- The front side gripper pads tended to push into soft ground as the machine moved forward, putting it off-line and damaging the ground.
- The disc cutters required a high thrust for proper operation. In faulted ground, they tended to skid without rolling, causing flat edges to develop.
- Ground failure tended to occur along the top of the shield, partly because the shields subsided when adjustments were made to the hydraulically actuated support system and partly because of contact between the shields and the surface of the bored tunnel.

After a half-year of operation, the machine was taken out of service and major modifications were carried out. The modified machine is shown in Fig. 10-16. The redesign included the following.

- A new shield was placed over the cutting head. The shield segments were mounted on spring-loaded arms, giving them greater flexibility. The radius of the shield segments was only $\frac{1}{2}$ inch less than the bored radius of the tunnel to support the roof effectively. In addition, the shield was fitted with a series of trailing fingers made from steel plate that would allow forepoling operations to take place ahead of the machine in bad ground.
- The cutting head was modified. The tricone center cutters were replaced with a group of eight drag cutters with tungsten-carbide tips, clamped together as a unit. Drag cutters were used as face cutters, mounted in front of the disc cutters, so that the drag cutters would operate in soft material; if the drag cutters wore too rapidly, the disc cutters would operate. Finally, drag cutters were used as the gauge cutters.

Fig. 10-16. Robbins rock tunneling machine modified for operation in variable ground conditions.

- The rotational speed of the cutting head was reduced, which lessened the problem of boulders being dragged across the working face and also helped considerably to increase the life of the drag cutters.
- Various modifications were made to the buckets to obtain an improved flow of material into and through them and so to reduce the number of blockages.
- The surface area of the side gripper pads was increased. This reduced the bearing pressure against the tunnel walls and eliminated the tendency to dig into the softer ground.
- Other modifications included an overload cutout and absorption brake to reduce damage when the cutting head jammed in soft ground. The main beam was extended 10 feet and included a 5-foot long hole to provide access for men and materials to the working face.

After these modifications, the machine was returned to service and the rate of advance increased by a factor of four or five.

New Machine for Variable Ground Conditions

Since the modified machine performed satisfactorily the manufacturer has incorporated most of the new features into a new line of tunneling machines especially intended for use in variable ground and capable of boring in both hard rock and soft ground.

One such machine was used for a hydroelectric power tunnel in Southern Italy in the Province of Calabria. This job, referred to as the SILA Project, was bored by a 14.5-foot diameter machine with a spoked type cutterhead and a double telescoping shield completely surrounding the machine, as shown in Fig. 10-17. This shield permits articulation for steering as well as protection from rockfalls while the machine advances. During the boring, the rear portion of the shield remains stationary, with its grippers providing the thrust reaction for the forward shield and cutterhead. Precast concrete segments are installed in the tail of the rear shield while the forward shield advances.

The rock at this site is a highly weathered granite, very little of which is stable or self-supporting. The cutterhead of the machine was designed to permit easy access to the face for cutter changing or hand mining in very bad ground, and some circumferential grill bars were added in the openings between the bucket scoops to limit the size of boulders that could pass through into the cutterhead openings and conveyor in case of face instability.

Fig. 10-17. Design for boring rock with bad ground.

This machine encountered ground that was so unconsolidated that the face ran in against the machine, carrying large boulders more than 2 cubic yards in size with such force that they moved the machine backward. For these conditions, the machine's cutterhead was not adequate as designed, since it would become plugged and stalled by the large, hard boulders. The contractor, S.E.L.I. Corporation of Rome, modified the machine's cutterhead face to the configuration illustrated in Fig. 10-18. A rotating face support was added which, in normal cutting, is only a few inches from the rock face. In caving conditions, the rock moves only a small amount before being supported by the face plates. Spaces between these rotating face plates are not more than 8 inches, which determines the size of rock that can pass from the face through the machine cutterhead to the conveyor.

The effect of this on-the-job modification was dramatic. The machine successfully progressed through bad ground that would have been impossible previously. These conditions would have required driven or drilled-in spilling over closely spaced steel support ribs and shotcrete or wood breasting to support the face with traditional methods.

GEOLOGICAL DATA FOR TUNNELING MACHINES

Geological Exploration

By their very nature, tunneling machines are less adaptable to changes in ground conditions than the equipment used in conventional tunnel excavations. The performance of a machine is therefore more sensitive to changes in rock type, the reaction of the rock to the excavation,

Fig. 10-18. Modified cutterhead for unstable face.

groundwater inflow, and the presence of faults, shears, zones of alteration and other structural defects. Definition of the geologic conditions is therefore more important for a machine-bored tunnel than for a conventionally driven tunnel.

In areas of complex geology, the present exploratory techniques for determining ground conditions are relatively imprecise, and this fact alone may be the basis for a contractor deciding against using a tunneling machine. Present exploratory methods involve geologic mapping of exposures and exploratory holes drilled at specific locations along the alignment and at the portals, with laboratory testing of selected rock samples, geophysical surveys, and in some cases a pilot bore. This exploration is limited to specific areas of the alignment, and it is usually difficult to project the observed features along the tunnel line.

The amount of investigation necessary to define the tunneling conditions depends upon the complexity of the geology and the amount of previous underground experience in the area. Obviously, the most informative geologic investigation is the excavation of an exploratory tunnel within the limits of, or immediately adjacent to, the entire length of the proposed bore. This degree of exploration is usually not practical, but it has been done in conjunction with a few large-diameter transportation tunnels where the pilot bore is used in the design of the finished structure. An example of a major tunnel where a pilot tunnel was driven prior to the main drive is the Seikan Tunnel, an undersea railway link beneath the 13.5-mile wide strait between the islands of Honshu and Hokkaido, Japan. The ground is consolidated volcanic ash containing some water at one heading and numerous water-laden faults at the other. The main concern was that sea water could enter the tunnel and completely flood it. During the driving of the horizontal section of the pilot tunnel under the strait, engineers were probing ahead of the driving face to study geological conditions. To do this, they drove L-shaped tunnels from the sides of the pilot bore and used special drills to explore 1600 to 2000 feet ahead. Although the rigs were designed to drill much greater horizontal distances, the holes curve downwards too far to be meaningful beyond 2000 feet. The drilling was conducted with $3\frac{1}{4}$-inch rods and diamond-studded tricone cutters, which were advanced in 10-foot sections. Water was pumped through the stem to wash out the cuttings. The amount of return water was used to estimate whether water-bearing faults or fractured rock zones were penetrated. When an adverse condition was detected, additional holes were drilled to provide more information. When necessary, grouting crews moved in and grouted the zone ahead to reduce inflow before the pilot tunnel finally reached and passed through the zone.

Rock Strengths

During geological exploration drilling, rock cores are obtained, and attempts are made to determine the suitability of the rock for excavation by tunneling machine. The most common property used as an index of the drillability of a rock is its compressive strength, although toughness, brittleness and abrasiveness also influence the rate of penetration and cutter costs. To date, no indices have been developed that accurately predict the performance of a tunneling machine, but manufacturers and contractors are accumulating data and experience that will provide a basis for making reasonable forecasts.

The compressive strength of rocks that have been successfully bored with a tunneling machine should always be evaluated in the light of the structural character of the rock and other geological features of the particular occurrence. A tunneling machine that has successfully penetrated hard but closely fractured rock may not necessarily perform as well in the same rock in a massive occurrence, even though the laboratory compressive strength tests, performed on uncracked samples, may give equal results.

A tunneling machine on a particular project is sometimes used to bore a short segment of rock that is much too hard to be driven economically by the machine in order to gain access to a longer section of softer rock. Here the gains from using the machine in the softer rock more than offset its disadvantages in the short segment of hard rock. However, it frequently happens that when such a job is reported, the higher rock strengths are quoted as having been successfully bored, even though it would not have been economical to do so for more than a short distance.

ALIGNMENT CONTROL OF TUNNELING MACHINES

A tunneling machine progresses in a sequence of shoves. After each shove, the location and attitude of the machine are determined. If the machine is found to have moved off its theoretical line, adjustments are made to guide it back to its desired location during the next shove.

The use of conventional optical instruments as a means of control has now been almost entirely replaced by methods based on laser instruments. The most practical method of machine control is by laser beam and double target. A laser tube is set up at a distance behind the tunneling machine to emit a laser beam along a predetermined line to the targets mounted on the tunneling machine. The targets are centered on a line parallel to the longitudinal axis of the machine and from 4 to 10 feet apart. The laser beam produces a bright red spot on the target. Theoretical points of intersection are calculated in advance for each machine location. The machine is guided by attempting to keep the laser spots on the predetermined intersection points of the target.

SPOIL REMOVAL

The means for removing the excavated material from the working face is a critical link in any excavation system. Any inadequacy or interruption will cause complete disruption of the tunneling cycle. Within the tunnel, there is generally no space available for spoil storage, and to prevent the tunneling machine from being unnecessarily idle, the transportation system must be capable of handling the peak output from the machine.

Rail haulage using a locomotive and dump cars is the most common form of tunnel transportation. Figure 10-5 shows a typical arrangement at the tunnel face of a rock tunneling machine, auxiliary equipment and the associated rail-haul unit. Figure 10-7 shows a photograph of the conveyor system used during construction of the River Mountains tunnel, Nevada. The trucks are fed by a conveyor system moving the muck from the machine to the haulage train. It is essential that the conveyor system be long enough to cover the complete length of the train in use. Alternatives to conveyor systems for loading the trains are front-end and overshot loaders, when some switching system for the muck cars must be used.

Another type of train that has been used consists of four or five bunker cars operated together without intermediate bulkheads. Spoil is fed from one end to the other so that a train can be loaded and unloaded from either end. Because there are no gaps between cars, this type of train cannot take bends as sharp as can conventional trains.

On a group of four tunnels totalling 5 miles for the Metropolitan Water District of Southern California, passing through sandstones of the Saugus formation, the contractor used a single-rail car with a capacity of 80 cubic yards for spoil removal. It was capable of removing all the muck produced during a cycle of one 5-foot advance of the tunneling machine used on the project.

Slurry pipelines may be used more widely in tunnels for spoil removal in the future. In this system, solids of suitable size are suspended and transported in a liquid medium. In other fields, the slurry pipeline system is an established method of transporting solid materials. Recent developments in mining have been impressive. An 18-inch diameter pipeline, 273 miles long, is used to transport coal at Black Mesa, Arizona, and a 9 $\frac{1}{2}$-inch diameter pipeline, 53 miles long, carries ore concentrate at a project in Tasmania. A critical item for this method is an adequate water supply.

The Azotea tunnel, a 13-mile water tunnel between the states of Colorado and New Mexico, was driven through shale and sandstone. The bored diameter varied between 12 feet, 8 inches and 13 feet, 4 inches. Excavation rates averaged 55 feet/calendar day, 153 feet/working day in the shale and 72 feet/working day in the sandstone. At the start of the project, the contractor made an attempt to remove the spoil by a slurry pipeline. The muck from the mole conveyor went through a finger crusher into a mixing tank and then entered the pipeline. The greatest problem was making a proper slurry. The material came out of suspension in the pipeline, plugging it frequently. As a result, this spoil removal system was discarded early in the project and a conventional rail car system was used.

Slurry pipelines have been used successfully on some soft ground tunnel projects where the spoil can more readily be maintained as slurry. An example is provided by the twin undersea railway tunnels recently driven under Tokyo Bay through soft silt and sand by a Mitsubishi soft ground tunneling machine. The excavated material fell into a pressure tank just behind the cutting head, fresh seawater was pumped from the surface, and a pair of agitators mixed the soil and water into a slurry. Pumps forced the slurry through an 8-inch pipeline to a separation plant at the surface. The system required a pump every 1300 feet. At the separation plant, the material passed through a sand classifier screw, which removed granular materials. The remaining slurry flowed through both primary and secondary settlement systems before the seawater was finally discharged.

A continuous belt conveyor system was successfully used for muck removal during construction of the Balboa Outlet tunnel, part of the distribution system of the Metropolitan Water District of Southern California. This tunnel was 3,760 feet long, had a bored diameter of 16 feet and passed through sandstone, siltstone and conglomerate. The conveyor system consisted of 30-inch wide PVC conveyor belts located on the side of the tunnel just below the springline and supported on brackets at 5-foot centers anchored to the wall of the tunnel. The system allowed for additional lengths of conveyor belt to be added in 75-foot increments as the tunneling machine advanced.

TUNNEL LININGS, AS RELATED TO TUNNELING MACHINES

When a tunnel is driven through soft ground, a primary lining has to be placed to support the ground as the excavation proceeds. In some cases, this is the only lining provided; in others, a secondary lining is later poured within the primary. Past practice has been to place as the primary lining preformed circular segments of suitable size, made of cast iron, steel or concrete. In firm ground, the lining can be erected behind the shield as it moves forward. If the ground tends to be running, the lining must be erected within the tail of the shield in position to provide support as soon as the shield moves forward.

The length of the tunnel ring used and the number of segments into which it is divided vary from one tunnel to another. The design of the precast lining units must take account of the weight and size of individual segments to be handled in the limited working space within the tunnel and allow for soil pressures and deflections as well.

On the recently completed Balboa Outlet tunnel built for the Metropolitan Water District of Southern California, the usual precast units for the primary lining were replaced with a shotcrete lining. The mix contained $\frac{3}{4}$-inch maximum aggregate; high-early-strength-cement was used plus a quick-set additive. The shotcrete was applied in a 3-inch thick layer, and was discharged pneumatically by hose. This lining provided temporary support of the tunnel until the permanent lining was placed.

Shotcrete was also used to provide the lining on the Heitersberg railroad tunnel, Zurich, a hardrock tunnel passing through sandstone, shale and marl. On this project, the contractor tried to create an integrated tunneling operation, with the installation of rockbolts and the placement of the concrete lining following immediately behind the excavation. The tunneling machine laid a precast concrete invert section immediately behind the cutting head, so that all of the machine and auxiliary equipment, except for the cutting head, rides on finished invert. The system was equipped to install rock bolts within 15 feet of the excavated face. The first layer of shotcrete was applied within 30 feet of the face, and a final layer of shotcrete was applied by a separate machine.

MACHINE TUNNELING COSTS

Factors influencing the relative costs of constructing a tunnel by use of a tunneling machine or by conventional hand methods are illustrated by relatively detailed cost estimates for tunnels driven by each method in both rock and soft ground.

Costs are derived on the basis of cost per linear foot of tunnel and are considered as being made up of a fixed cost and a variable cost. Fixed costs consist primarily of the net cost of plant and equipment investments necessary for a particular job, which will be written off at a certain sum per linear foot of tunnel. This cost will be constant throughout the construction of a project of given layout and ground conditions, and will not be affected by the efficiency of the performance of the machine and the tunnel personnel. In addition to major construction equipment, tunnel services such as ventilation and temporary tracks are also fixed costs.

Variable costs consist primarily of labor costs, together with power costs and operation and maintenance costs of the equipment. These costs are generally constant on a daily basis and therefore vary per foot of tunnel as the daily rate of advance changes.

Cutter costs should depend only on the nature of the rock being worked, in which case they would be part of the fixed costs. In practice, these costs also depend greatly on the efficiency of the tunneling crew in properly maintaining the cutters. In this analysis, the effectiveness of the cutter maintenance procedures will not be included, and cutter costs will be assumed to be part of the fixed costs.

Because of the high purchase price of tunneling and supporting equipment, the fixed costs for a machine-driven tunnel are much higher than for a hand-driven operation under the same tunneling conditions. Thus, for machine tunneling to be economically successful, the relatively high fixed equipment cost per foot of tunnel must be offset by a much lower variable cost (principally labor) per foot of tunnel than the handtunneling operation. Variable costs can be kept lower by using a smaller crew on the tunneling machine and reducing labor costs, or by achieving a higher rate of progress than is possible by handtunneling, or by a combination of both.

The following factors are critical in determining the cost of any tunnel:

- Type of equipment and amount of labor used.
- Type of ground or rock.
- Length, diameter and grade of the tunnel.
- Tunnel access.

In the following paragraphs, the use of a tunneling machine is compared with conventional handtunneling that uses drilling and blasting methods in rock or manual shield methods in soft ground, for tunnels in rock, both medium-hard and hard, and in soft ground. The effect of changes in the length of the tunnel on the relative costs is considered. Finally, the relative rates of escalation of labor and equipment are used to show the future trend of machine and conventional tunneling costs. Throughout, the tunnel diameter is kept constant, access conditions are assumed to be the same for both types of equipment, and the grade is assumed to be less than 1%.

Cost Comparison of Tunneling by Machine and by Conventional Methods in Rock

A comparison of the cost of machine tunneling and conventional tunneling will be given for rock of two different strength ranges:

- Medium-hard rock (compressive strength 15,000 psi).
- Hard rock (compressive strength 25,000 psi).

In both cases, it is assumed that the rock can stand up after excavation without support, and that a permanent lining, if required for other reasons, could be installed at a rate independent of the excavation rate. Ground conditions are assumed to be uniform throughout the length of the tunnel. The tunnel costs given include only excavation and mucking, since these are the only factors changed by the two methods of attack. Other costs, such as portal construction, dewatering, tunnel clean-up, tunnel lining, contingency, overhead and profit, are not included in the estimate.

The tunnel is assumed to be 20,000 feet long, with a driven diameter of 12 feet. Access is through a portal, and muck removal is by train. Work is assumed to proceed at a rate of three shifts a day, five days a week. Crews, equipment and progress rates are based on actual job experience. To illustrate the principles involved, costs for the year 1979 in California are used. To obtain current costs, all rates for labor and materials must be adjusted to account for escalation. Any development that would increase the productivity of tunnel machines will also influence the results of cost comparison.

The following rates of progress are assumed.

	Average Progress Rates (feet/day)	
	Machine Tunneling	Hand Tunneling
Medium-hard rock	100	60
Hard rock	60	50

Figure 10-19 shows the average and maximum rates of advance achieved in the excavation of tunnels of different diameter in medium-hard rock. The average rate of 100 feet/day used in this comparison is well below the actual experience curve.

In addition, as shown in Fig. 10-20,[5] machine tunneling rates are expected to increase in the future. It should be pointed out that the curves illustrated in Fig. 10-20 are approximate and indicative only of trends, since the points of actual experience required to plot such curves are sometimes available only in rock of an average strength or a diameter somewhat different from that plotted.

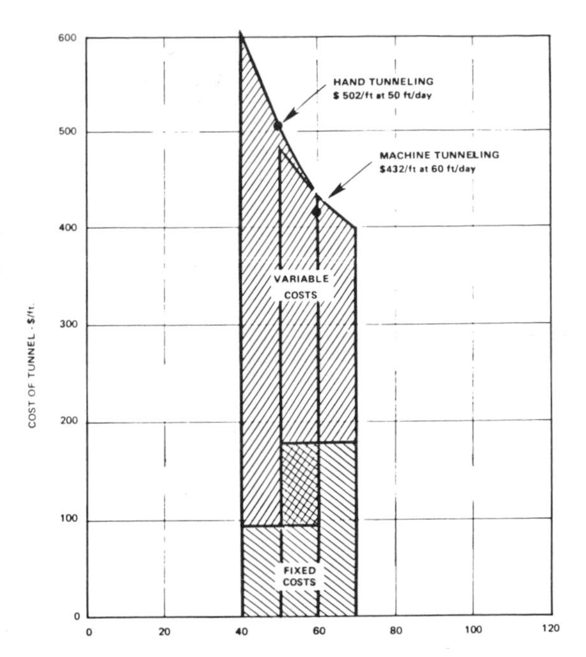

Fig. 10-20. Penetration rate versus time showing effect of rock strength at 12-foot ϕ tunnel.

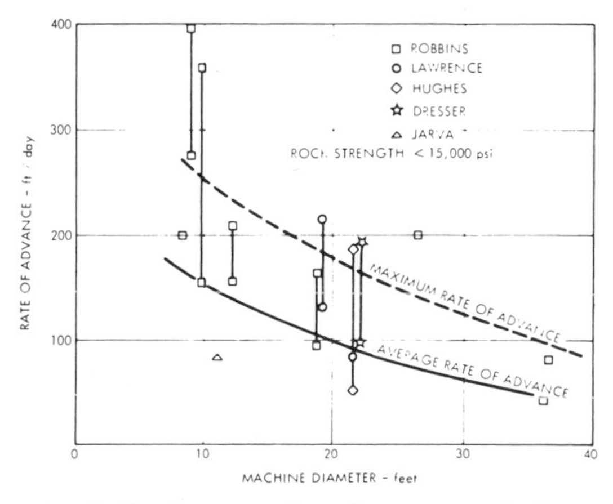

Fig. 10-19. Average and maximum rates of advance for different tunnel diameters in medium-hard rock.

The costs per foot of tunnel for the cases considered are developed in Tables 10-5 and 10-6. Figures 10-21 and 10-22 show the cost curves when the assumed rates of progress are varied with the same equipment and labor inputs.

Figure 10-21 shows that in medium-hard rock the tunneling machine can drive a tunnel at a lower cost than hand tunneling methods for the range of rates of advance that can reasonably be expected with the two types of equipment. The cost for machine tunneling is $302/foot and for hand tunneling $417/foot. The curves are for rock with a compressive strength of 15,000 psi, but the cost curves would have a similar relationship over a range of about 5000 to 20,000 psi.

As stated previously, the cost figures were developed for the assumption that the rock stands up without any kind of temporary support. If either rock bolts or steel sets have to be installed to support the rock immediately behind the cutting head, much of the advantage of machine tunneling is lost, because present methods of steel-set erection and rockbolt installation require considerable manpower and time. Since the machine would not be

Table 10-5. Equipment and plant and labor costs for machine and hand tunneling in medium-hard and hard rock.

Item	Machine Tunneling			Hand Tunneling		
Equipment and Plant Costs	Cost ($1,000's)			(Cost ($1,000's)		
	Purchase	Salvage	Cost	Purchase	Salvage	Cost
Tunneling machine, conveyor, dust collection	1250	175	1075			
Mainline jumbo, drills, mucker				375	150	225
Other equipment for mucking	700	175	525	625	250	375
Temporary track, ventilation and power	1250	500	750	1250	500	750
75% of shop and yard costs	250	125	125	375	200	175
Total	3450	975	2475	2625	1100	1525
Cost per foot for 20,000-ft tunnel		$124/ft			$76/ft	
Labor Costs	Day Shift	Swing Shift	Night Shift	Day Shift	Swing Shift	Night Shift
Excavation crew	6	6	6	17	13	13
Mucking crew	4	4	4	4	4	4
Track maintenance and supply	3	3	3	4	3	3
Portal crew	4	3	3	5	3	3
Manshifts/shift	17	16	16	30	23	23
Total manshifts/day		49			76	
Total manhours/day		392			608	
Cost per day @ $28/hour		$10976/day			$17024/day	
(Average wage in 1979, including payroll burden and overtime)						

tunneling during support erection, advance rates would be reduced and the cost per foot of tunnel increased to the rates for handtunneling methods.

Figure 10-22 shows a similar cost comparison for machine tunneling in hard rock of 25,000 psi compressive strength. The curves are very close together, showing that at the present time tunneling machines reach their economic limit at about this level of rock strength. The main reason for this limit lies in the reduced performance of the cutters as the rock strength increases. The rate of penetration of the cutting head is reduced approximately in inverse proportion to the compressive strength of the rock, and the lower penetration rate causes increased cutter costs per foot of tunnel. Hence, there are two effects tending to increase costs in hard rock: 1) increased cutter costs, and 2) higher labor costs per foot of tunneling, resulting from the slower rate of advance.

Present-day technology places the economic break-even point of tunneling machines at a rock compressive strength of 25,000 to 30,000 psi. However, cutter manufacturers are continuously improving the bearings and lubrication systems of the rock cutters, which have been the weak points in the past. Harder steel alloys are being developed to increase the life and effectiveness of the cutting edge in actual contact with the rock, and disc cutters with tungsten carbide inserts are being used for the harder rocks. These improvements will lead to longer cutter life at increasing penetration rates. As the efficiency and wear characteristics of cutters improve, tunneling machines will be able to work economically in harder and more abrasive rock formations.

Table 10-6. Summary of cost comparison

Item	Machine Tunneling	Hand Tunneling
MEDIUM HARD ROCK		
Fixed Costs ($/ft)		
Equipment and plant	$124/ft	$76/ft
Cutter costs	$36/ft	---
Powder costs	---	$14/ft
Total fixed costs	$160/ft	$90/ft
Variable costs		
Labor	$10,976/day	$17024/day
Equipment operation and maintenance	$2400/day	1560/day
Power, supplies, etc.	828/day	1056/day
Total variable cost/day	$14204/day	$19640/day
Total variable cost/ft		
By machine, at 100 ft/day	$142/ft	---
By hand, at 60 ft/day	---	$327/ft
Total cost of tunnel (Excavation only)	$302/ft	$417/ft
HARD ROCK		
Fixed costs ($/ft)		
Equipment and plant	$124/ft	$76/ft
Cutter costs	$55	---
Powder costs	---	$20/ft
Total fixed costs	$1.79/ft	$96/ft
Variable costs		
Labor	$10976/day	$17024/day
Equipment operation and maintenance	3000/day	1800/day
Power, supplies, etc.	1200/day	1500/day
Total variable costs/day	$15176/day	$20324/day
Total variable cost/ft		
By machine, at 60 ft/day	$253/ft	---
By hand, at 50 ft/day	---	$406/ft
Total cost of tunnel (Excavation only)	$422/ft	$502/ft

Cost Comparison of Tunneling by Machine and by Conventional Methods in Soft Ground

In the case of softground tunneling, one always present factor has a major influence on the relative costs of excavating by tunneling machine or by conventional hand methods. This is the need in soft ground to place some type of ground-support system immediately behind the shield of the tunneling machine as soon as it moves forward. Thus, a lining-erection system must be provided that is able to keep up with the maximum rate of advance of the tunneling machine. Present systems usually include a mechanical erection arm, but such systems also have a heavy labor requirement, which constitutes a high unproductive cost during tunneling machine downtime.

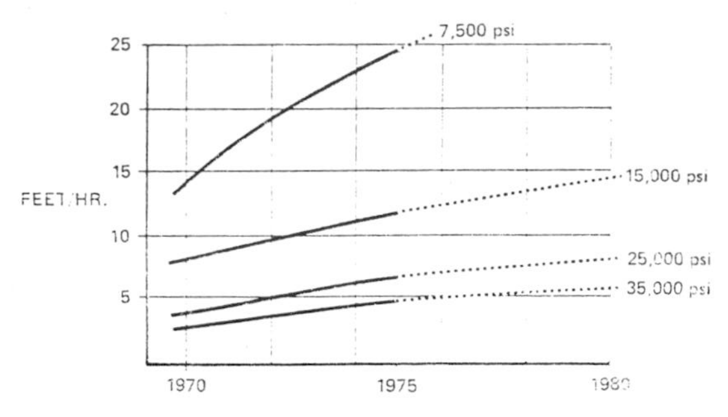

Fig. 10-21. Cost comparison between machine and hand tunneling in medium-hard rock.

Detailed field construction data from an actual project illustrate the influence on the costs of ground-support requirements. During the building of the San Francisco BART project, tunnels were driven both by tunneling machines and by hand methods. Two particular tunnels lend themselves to a comparative cost study. Both were 7000 feet long, 18-foot driven diameter, and lined with prefabricated welded steel lining segments, with seven pieces making up a 2.5-foot long ring. Both tunnels were in similar ground, consisting of sands and clays. The natural water table was above the roof of the tunnels, and water was controlled during construction by pumping or by compressed air.

The soft ground tunneling machine had a

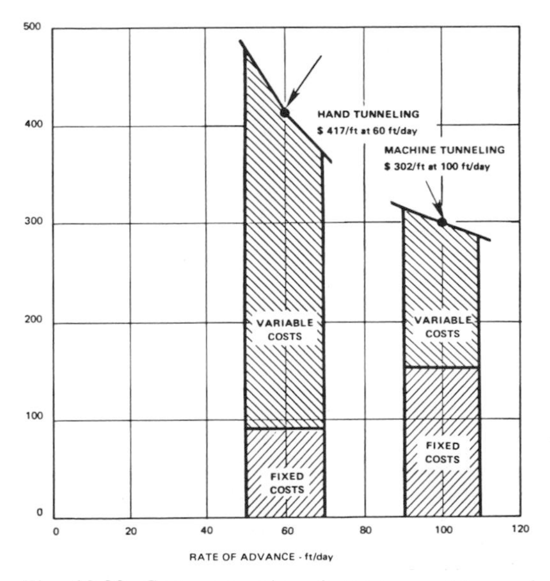

Fig. 10-22. Cost comparison between machine and hand tunneling in hard rock.

shield and an oscillating cutting head, and forward movement was provided by jacks acting against the previous lining ring, and mucking and breasting were performed by hand. Placing of the lining rings was again assisted by a mechanical erector arm. Advance of the shield had to wait until the ring was in position and the jacks could thrust against it.

Average times for the component operation of the working cycles, downtime, and resulting rates of advance are given for both cases in Table 10-7. A breakdown of the cost per foot of each tunnel is given in Table 10-8, which shows that the final costs per foot of tunnel were substantially the same in each case at the rate of advance achieved. A number of items, such as construction of access shafts and tunnel clean-up, which were equal in both methods, are not included because they do not affect the relative costs of the two methods.

The average working cycle for the machine tunneling operation took 65 minutes, corresponding to a maximum possible rate of advance of 55 feet/day of lined tunnel. The actual average rate achieved was 40 feet/day because of a 27% downtime. The principal cause of the lost time was erection of lining. Table 10-7 shows that the erection of the lining rings took 70% of the working cycle with the tunneling machine, and Table 10-8 shows that lining-ing machine, and Table 10-7 shows that lining-erection laborers were 73% of the total crew at the working face. This number was considered necessary to place the lining in a reasonable time and prevent excessive slowdown of the tunneling machine during advance, even though these laborers were idle during the remainder of the tunneling cycle.

With design improvements, downtime of the machine might be reduced to 5%; Fig. 10-22 shows the variation in cost per foot of tunnel at various advance rates that would result from such improved performance. Since downtime for hand tunneling was only 10%, there is little possibility of improvement beyond another 5%; on the other hand, downtime is unlikely to exceed 30%. These are the probable limiting rates for the hand-excavated tunnel.

Table 10-8 shows that the costs for machine and hand tunneling were $407/foot and $403/

Table 10-7. Time cycles for machine tunneling

Operation	Machine Tunneling	Hand Tunneling
Remove material from face, breasting, mucking	Included in shove time	50 min
Shove shield or machine forward	20 min	35 min
Erect lining ring and inject grout behind lining	45 min	45 min
Time required for one cycle (an advance of 2.5 ft)	65 min	130 min
Projected advance per 24-hour day	55 ft/day	28 ft/day
Average advance achieved per 24-hour day	40 ft/day	25 ft/day
Downtime of equipment	27%	10%

foot, respectively. However, Fig. 10-23 shows that the tunneling machine has a much greater potential for an improved advance rate. If downtime could be reduced to about 5% for both types of equipment, the relative costs would be \$340/foot for machine tunneling and \$383/foot for hand tunneling, at the corresponding rates of advance for the two methods.

Further significant cost reductions would be difficult to achieve with the equipment used on the BART project. Savings could be achieved by an improvement in the lining-erection system to cut down on time and labor. There would also be savings if the lining could be placed without stopping excavation, which would require a change in the present method of providing thrust reaction during advance of the tunneling machine.

Effect of Tunnel Length on the Cost Comparison

The previous cost comparison was based on two tunnels, both 7000 feet long in soft ground needing continuous support, and for this tunnel length the costs per foot were virtually equal. Figure 10-24 extends the cost comparison for machine and hand tunneling to different lengths of tunnel. The curves are based on the assumption that the same equipment would be suitable for longer tunnels, which is valid until the tunnel reaches a length that would require additional mucking trains to keep up with the excavation or that would make excavation at a second heading economical. In a longer tunnel, as the equipment is written off over a greater footage, the cost reduction for machine tunneling is relatively greater than for conventional tunneling. Thus, with the equipment used on the BART tunnels, the cost curves suggest that machine tunneling would have shown a distinct cost advantage if the tunnels had been longer. This comparison did not include allowance for the greater rates of advance that can be expected in a longer tunnel as the working crews develop and maintain tunneling skills, which would have made machine tunneling even more advantageous.

Changes of Relative Cost Comparison with Time

Figure 10-25 shows changes in cost indices for construction, labor and equipment over the last 13 years. Not only is the labor index higher than the equipment index, but it is increasing at a much steeper rate. The data presented previously showed that the cost of a machine-driven tunnel had a greater component of equipment cost, and the conventional driven tunnel had a greater component of labor cost.

Table 10-8. Cost comparison of machine and hand tunneling is soft ground (San Francisco BART Project)

Item	Machine Tunneling			Hand Tunneling		
Equipment and Plant Costs	Cost ($1000's)			Cost ($1000's)		
	Purchase	Salvage	Cost	Purchase	Salvage	Cost
Tunneling machine, erector, conveyor	500	50	450			
Tunneling shield, conveyor				135	13	122
Other equipment for mucking and lining	464	214	250	372	174	198
Temporary track, ventilation and power lines	95	35	60	95	35	60
75% of shop and yard costs	143	51	92	143	51	92
Total	1202	350	852	745	273	472
Cost per foot for 7000-ft tunnel		$122/ft			$67/ft	
Labor Costs	Day Shift	Swing Shift	Night Shift	Day Shift	Swing Shift	Night Shift
Excavation crew	5	5	5	7	7	7
Ring erection and grouting crew	14	14	14	8	8	8
Muck train and ventilation crew	8	8	8	5	5	5
Shaft crew	19	11	11	15	9	9
Manshifts/shift	46	38	38	35	29	29
Total manshifts/day		122			93	
Total manhours/day		976			744	
(Cost per day at $10/hour		$9760			$7440	
(Average California wage in 1968, including payroll burden and overtime)						
Cost Summary						
Fixed costs						
Equipment and plant		$122/ft			$67/ft	
Variable costs						
Labor		$9760/day			$7440/day	
Equipment operation & maintenance		1020/day			610/day	
Power, supplies, etc.		610/day			360/day	
Total variable cost/day		$11,390/day			$8410/day	
Total variable cost/ft						
By machine, at 40 ft/day		$285/ft				
By hand, at 25 ft/day					$336/ft	
Total cost of tunnel		$407/ft			$403/ft	

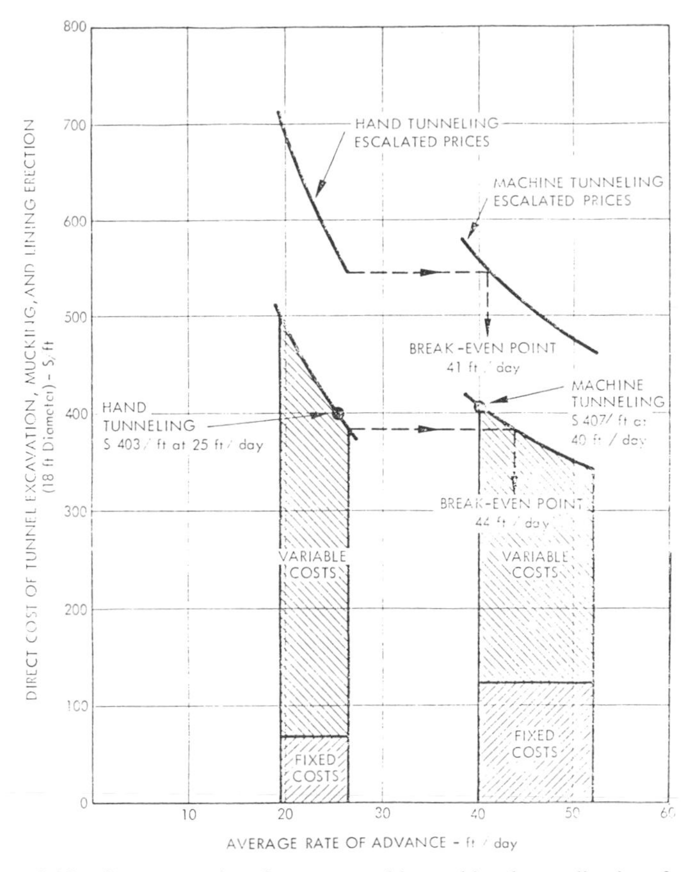

Fig. 10-23. Cost comparison between machine and hand tunneling in soft ground.

Thus, over a period of time, the machine tunneling costs will become smaller relative to the conventional tunneling costs. In Fig. 10-23, the lower curves show the relative costs at different rates of advance at the actual labor and equipment costs pervailing in the period from 1967 to 1969, when the BART tunnels were constructed. At that time, and for the type of equipment used, the break-even rate for a tunneling machine was about 44 feet/day. As the rate of advance of the machine increases above that amount, the cost per foot of tunnel becomes less than could be obtained by conventional methods.

Other Cost Differences Between Machine Tunneling and Conventional Tunneling

There are a number of other factors that affect the relative costs of machine tunneling and hand tunneling that were not considered in the above comparisons. These include the following.

- Use of a machine in a rock tunnel provides greater stability of the material immediately surrounding the excavation because the rock has not been disturbed by the dynamic shock of blasting. A saving can be realized in the thickness of the concrete lining required, since it can be designed for lower pressures. Furthermore, fewer temporary roof supports will be needed at the construction stage, although it is difficult to estimate this saving quantitatively.
- A further saving effected by a tunneling machine in rock results from the greatly reduced overbreak and the consequent reduction in concrete required for the lining.

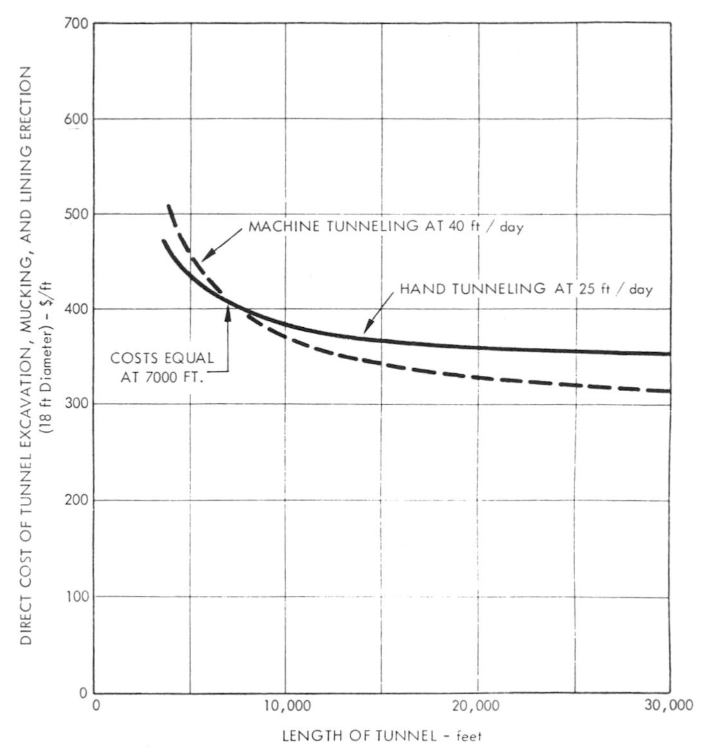

Fig. 10-24. Cost comparison between machine and hand tunneling for varying tunnel length in soft ground.

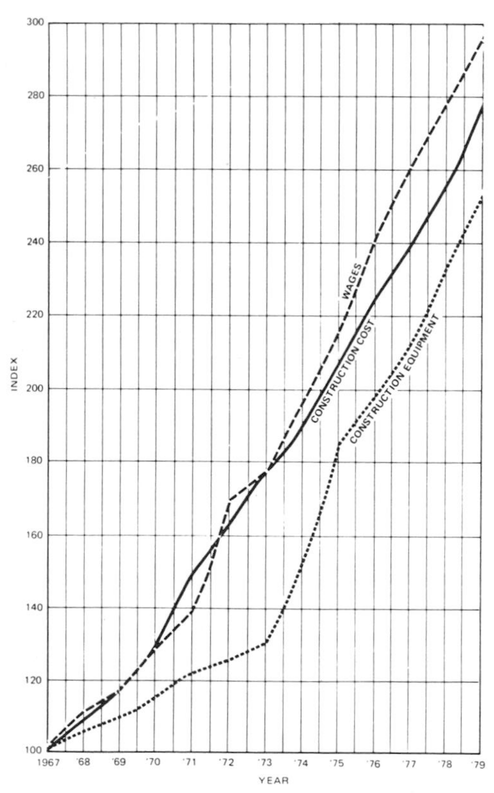

Fig. 10-25. Construction, labor and equipment costs in U.S. from 1967 to 1979.

In addition, a number of other savings occur with the use of a tunneling machine that will seldom show in the contract prices but which should nevertheless be considered in an overall evaluation of any project.

- The faster average progress rates for machine tunneling allow shorter overall construction times than conventional methods, resulting in lower amounts of interest on money borrowed for construction and an earlier beginning of the income stream from the project.
- In urban areas, drilling and blasting in hard rock under a residential area may cause considerable damage and result in substantial third-party claims. Likewise, severe restrictions may be placed on working hours and acceptable noise level of the blasting operatings, causing increased costs in this form. Both costly alternatives can be avoided by the use of a tunneling machine.
- When tunnels are driven in soft ground, tunneling machines generally result in lower

surface settlements than comparable hand-tunneling methods. Considerable savings result from lower damage claims from owners of adjacent property. Savings also result from the fact that many buildings need not be underpinned if a tunneling machine is used, whereas conventional methods would require underpinning.

DEVELOPMENTS IN METHODS OF ROCK DISINTEGRATION

A vast amount of research is being carried out on new ways of disintegrating rock, with the hope that these methods could lead to more efficient means of tunnel excavation. Much of the work is still at the level of laboratory experimentation and is a long way from providing a cheap and effective excavation tool. The methods might eventually be developed as aids to tunneling machines already in use, or they could develop into entirely new types of tunneling machines.

The variety of methods now being investigated for rock disintegration may be considered under the following headings.

Mechanical	Thermal	Chemical
Water cannon	High-velocity flame	Softeners
Vibration	Flame jet cutting	Dissolvers
Abrasion	Electric arc	
Cavitation	Electron beam	
Pellet impact	Plasma	
	Freezing	
	Laser	
	Atomic fusion	

Out of this range of exotic techniques, two show particular promise for development into practical excavation methods.

Water Cannon

The first method is by high-pressure, low-volume impulse water jets, produced by a device known as a water cannon. In this method, a high-pressure slug of water is fired at the rock and impacts with pressures of five to ten times the compressive strength of the rock, producing tensile and shear fractures. The potential advantages of water cannons over current mechanical means are as follows.

- Larger rock fragments are produced.
- The major problem of cutter wear is eliminated.
- Their ability to be powered electrically, pneumatically or hydraulically minimizes environmental problems in the tunnel.
- Energy may be highly concentrated on any part of the rock.
- There seems to be no limit to the type of rock that can be broken, simply by using higher jet pressures.
- The use of water helps keep down dust.
- A low volume of water is needed.

One of the problems associated with this method is control of the extent of the fracturing, so as to produce an excavated profile without excessive breakage of the surrounding rock. To fracture medium-hard to hard rock, jet pressures of up to 200,000 psi are necessary, and, for the hardest rocks, pressures of more than 600,000 psi are needed. The U.S. Department of Transportation has sponsored considerable study and experimentation on the use of high-velocity water pulses to fracture rock and has completed the design of a water cannon to discharge pulses at pressures up to one million psi. They plan to fabricate the unit soon and conduct laboratory and field tests. If successful, this device could be employed as a primary means of fracture or as a component of the present type of boring machines.

The firm of Calweld has recently joined with Exotech, Incorporated, with the purpose of developing the use of high-pressure water jets in a tunneling machine.

Flame Jets

The other system of rock disintegration that shows high promise is by flame jet cutting. This has been amply demonstrated as a practical means of cutting rock for a number of years. It is currently used in quarrying all kinds of rock for architectural purposes throughout the U.S., and produces smooth-faced blocks with relative ease. Flame jet cutting is in use throughout the mining industry and has economically driven thousands of feet of 6- to 9-inch diameter blasting holes during mining operations for taconite, an ore so hard it is almost undrillable.

United Aircraft Laboratories have conducted research in flame jet cutting and have compared the rates of cutting on different kinds of rock. Granites, gneisses, and sandstones were cut rapidly, whereas gabbros and mica schists responded more slowly.

The technique of flame jet cutting has been demonstrated. What has not been solved are the environmental problems of using this method in a tunneling operation: control of fumes, heat dissipation, ventilation, etc.

In another approach, the U.S. Department of Transportation is initiating a research program into the technical and cost feasibility of using heat as an assist only at the outer periphery of a tunnel excavation. Since the major portion of cutter replacement costs are for the gauge cutters, this is an obvious area in which to obtain maximum benefits from an improved method of cutting.

Other Methods

Some research has been directed to applying the phenomenon of cavitation to rock disintegration. When a flowing liquid containing dissolved air hits a solid barrier such as a rock face, the air comes out of solution as bubbles, generating intense explosive pressures. In an experiment, cavitation bubbles were generated in water moving at 500 feet/second. The peak intensity of cavitation erosion was found to vary as the sixth power of the jet velocity. This idea is still in the laboratory research stage.

The use of a gun to fire pellets at a rock face has been explored. In tests, a lightweight gas gun fired plastic projectiles at a rock target. Both solid and water-filled missiles have been used, with impact velocities of more than 20,000 feet/second. Another firm has developed a pneumatic gun capable of firing concrete missiles at a high velocity, but this method has only been tested on a laboratory scale.

Tests have been made at cutting rock with an electron beam, generated by a non-vacuum electron gun. Using a 25-kW gun, slots were cut in sandstone, and the rock could be made to spall with the jet at an appropriate distance.

Tests were made with a laser for penetrating rock, but results have not been encouraging because of high costs, power problems and present inefficiency of the laser. Some testing has been performed to investigate the laser's ability to weaken the rock, and it is claimed that this method causes a substantial reduction in the flexural strength of the rocks tested.

The use of inexpensive chemicals to decrease rock hardness and strength has been investigated for more than 20 years. The Massachusetts Institute of Technology has conducted numerous experiments in using surfactants to weaken rock as a tunneling machine aid. It is claimed that field experiments in the Chicago Sewer System have indicated increased advance rates of up to 12% after the use of chemicals.

RESEARCH IN RAPID TUNNELING

The potential economies of the more rapid rates of excavation of tunneling machines cannot be fully realized until the other activities in the tunneling operation are coordinated with the working of the machine to produce a fully integrated tunneling system. Generally, new methods of spoil removal, roof support systems and placement of permanent lining have not kept up with improvements in the design of tunneling machines. The growing demand for tunnels for more purposes throughout the country has created a strong economic incentive to develop new systems and reduce tunneling costs.

The private sector of industry does not have sufficient resources to conduct the necessary research because of the large investments required and the long delays before payoff. To fill this need, publicly funded research might develop better technology.

Tunneling Machines

Most tunneling machines perform well as long as they penetrate uniform ground conditions. Interruption of unsuitable ground causes severe operating problems, and future research should be directed to improve machine versatility in variable ground conditions. In the same context, it might be well to research the feasibility of rearrangement of machinery for easier access to the working face. Most tunneling machines effectively block access to the face when emergency excavating methods must be developed in variable ground.

Tunnel Waste Removal Systems

Present train and rail systems are not expected to be able to keep pace with accelerated tunnel machine advance, and alternative methods, including slurry pipeline, conveyor belt and unitized service systems should be evaluated.

Hydraulic transportation of solids by slurry pipeline has developed in the last decade, the slurry pipelines now in operation are moving materials with solid specific gravity ranging from 1.05 to 5.0. The technology developed on these projects can be applied to the problem of the long-distance movement of excavated material produced by a tunneling machine. The question is not one of technical feasibility, but one of equipment layout, size selection, and optimum system efficiency.

The particle-size distribution and the maximum particle size are key parameters in any slurry pumping system. The optimum top size is determined by balancing the cost of grinding down the maximum sized material against the cost of pumping the larger sizes. Data gained on existing slurry pipelines can be used as a basis for computing pressure losses and designing the power requirements for new systems. Pumping would be done by metal-lined centrifugal slurry pumps in series. Recycling of the water would be necessary; investigations may cover the use of bentonite slurry as the transporting medium.

As the tunneling machine advances, skid-mounted pump units will advance with it. Flexible, coiled rubber hoses on the suction and delivery ends of these units can be used to increase the period between shutdowns for extensions of the system.

Conveyor belts have been used in mining operations more than in tunneling operations, because conventional drill-and blast tunneling did not produce muck at a rate to justify a high-speed continuous removal system. Present-day tunneling machines have changed this. For use in a tunnel, the conveyor belts must have mobility, flexibility and minimum size and weight. They may be supported by a wire rope hanging from rock bolts in the tunnel roof to conserve space.

A unitized service system for tunneling has been designed and includes provision for all the tunneling operations in prefabricated units about 50 feet long. The components include a conveyor belt, ventilation pipe, utility conduit, walkway and a light rail system for men and materials. The units are easily installed behind the tunneling machine as it advances. All the services are concentrated in a small area of the tunnel invert, making more room available for placing the supports and the lining.

Supports and Lining

In order to improve the overall efficiency of tunnel-boring machines in ground needing primary supports, further developments are required in the following areas.

- Speed of erection of segmental liners and reduction of their weight, for which the use of polymer-impregnated concrete and reinforced plastics should be investigated.
- Greater use of sprayed concrete, with or without reinforcing, and improved methods of application by automatically controlled equipment.
- Use of perforated steel sheets with shotcrete.
- Lightweight, small-section steel ribs.
- Cast-in-place continuous concrete lining.

Other Topics for Investigation

Other fields of study should include new techniques of geologic and geophysical investigations. Geologic mapping includes statistical methods of interpreting surface geological data. Techniques should be developed in long-distance horizontal boring as a means of investigation. This should be done both from the tunnel portal and ahead of the tunneling machine during construction. Holes up to 3000 feet in length should be considered, and the drilling equipment may operate through the cutting head as an integral part of the machine.

Geophysical methods as presently used in mineral exploration should be modified and tested as investigative tools in tunnel line exploration. An extensive program of laboratory testing of materials should be undertaken and results correlated with the behavior of the materials in the field.

Health and safety aspects of tunneling should be studied and environmental factors investigated.

REFERENCES

1. Gaye, Felix, "Efficient Excavation, With Particular Reference to Cutting Head Design of Hard Rock Tunneling Machines," *Tunnels and Tunneling*, January–February 1972.
2. BECHTEL Associates, WMATA–Hard Rock Tunnel Data, Internal Report.
3. Neyland, A.J., Murrell, R.F., Watson, F.G. and Cusworth, A.J., "Tunnel-boring in Fractured and Weathered Sedimentary Rock" (Melbourne Metropolitan Board of Works, Southeastern Trunk Sewer, Sections 3 and 4), delivered to the Australian Geomechanics Society, Raise and Tunnel Boring Symposium, Melbourne, August 1970.
4. Bartlett, J.V., Biggart, A.R. and Triggs, R.L., "The Bentonite Tunneling Machine," *ICE Proceedings*, November 1973.
5. Miki, Gosaburo, Saito, Takao and Yamazaki, Hironobu, "The Principle and Field Experiences of a Slurry Mole Method for Tunneling in the Soft Ground," *Ninth International Conference on Soil Mechanics and Foundation Engineering*, Tokyo, Japan, 1977.
6. Robbins, R.J., "Development Trends in Tunnelboring Machines for Hard Rock Applications," private communication.
7. "Tunnel Construction, State-of-the-Art and Research Needs," Special Report 171 Transportation Research Board National Academy of Sciences.
8. Robbins, R.I., "Mechanized Tunneling–Progress and Expectations" Tunneling '76 Symposium.
9. Matsushita, H., "Earth Pressure Balanced Shield Method" Proceedings RETC 1979.
10. Kurosawa, S., "Earth Pressure Balanced Shield Tunneling."

11
Materials Handling and Construction Plant

A.A. MATHEWS

Construction Engineering Consultant

Most of the operations involved in tunneling are concerned with materials handling. Excavated material must be removed from the heading; elements of the primary support system must be brought to the heading; water inflows must be removed; and materials for the secondary lining must be transported into the tunnel. Planning and engineering of the materials handling system is most important to tunnel construction because tunneling activities are concentrated at the working face, and access to that point is highly restricted.

Because most of the construction plant required for tunneling is closely related to the materials-handling functions, materials handling and the construction plant are considered together in this chapter. As used here, the term "construction plant" is any physical facility used in the construction of the tunnel. Contractors frequently distinguish between the terms "plant" and "equipment," using the term "plant" for items which are generally assembled on the site to form a fixed facility. For the purpose of this handbook, such a distinction is considered unnecessary.

This section deals with the selection and operation of the physical facilities used for building the tunnel.

BASIC TRANSPORTATION SYSTEMS

Several basic types of transportation systems are available for use in tunnel construction. This section outlines the advantages and disadvantages of each. Later sections will describe the design and operation of each system in greater detail.

Rail

From the standpoint of energy consumption, the rail system is by far the most efficient materials-handling system presently available for tunnel construction.

Advantages:

1. Provides an easily maintained traffic way.
2. Because of high efficiency, minimizes ventilation requirements.
3. Compatible with most excavating and loading methods.

4. Adaptable to nearly all sizes of tunnels.
5. Several power sources available.
6. Satisfies entire materials-handling need.
7. Fixed guidance system permits relatively small clearance limits.

Disadvantages:

1. Requires constant extension at the heading.
2. Passing locations are either fixed or only semi-movable.
3. In case of derailment or other accident, the entire system is generally shut down.
4. Unloading points are relatively fixed.
5. When constructing concrete invert, track must be taken up and then usually relaid.

Limitations:

1. Three percent grade is a good practical limit, with special cases handling up to six percent. Greater than six percent requires special devices, such as cables and hoists, for propulsion.
2. With current construction and maintenance standards, speed is limited to about 15 mph. Much higher speeds are feasible when greater attention is given to construction and maintenance of the track and roadbed.

Rubber-Tired Vehicles

Transportation with rubber-tired vehicles offers one of the most flexible systems because its operation is not restricted to locations having fixed facilities such as railroad track.

Advantages:

1. Roadway extension not rigidly tied to heading progress.
2. In a wide tunnel, passing locations can usually be selected at will.
3. Disruption of entire system due to accident is minimized.
4. Maximum flexibility in operation.
5. Combination load-haul units are available.
6. Work on tunnel invert is usually simplified.
7. Satisfies entire materials handling need.

Disadvantages:

1. Roadbed is difficult to maintain, particularly with soft invert or wet conditions.
2. Requires extensive ventilation system.
3. Not compatible with all excavating and loading equipment.
4. In narrow tunnels, passing points are fixed.
5. Commercial-size vehicles are not usable in small tunnels. Productive capacity, even with special equipment, is reduced in small tunnels.
6. Diesel is the only readily available source of power.
7. Except in very large tunnels, clearances are generally a problem.

Limitations:

1. Ten percent grades are no problem; grades to 25% are possible.
2. With a good roadbed, 25 mph speeds are not impractical (50 mph would be a likely practical limit for a superior type of roadway).
3. Because of ventilation and roadbed maintenance problems, truck transport is usually not economical for hauls in the tunnel of more than about one mile.

Belt Conveyor

Belt conveyors have been used extensively in mining operations, but have received only limited recognition as a prime transport system for tunneling. Technically, they are capable of moving excavated material out of the tunnel and bulk supplies in.

Advantages:

1. Capacity to handle excavated material for any conceivable rate of heading advance.
2. Compatible with most excavating and loading methods.
3. Adaptable to nearly all sizes of tunnels.
4. Relatively small clearance requirements.
5. Excellent reliability and low maintenance.
6. Continuous operation.

Disadvantages:

1. Does not satisfy entire materials-handling need. Auxiliary system for transporting personnel and supplies is required.
2. Particle size of materials to be handled is limited.
3. Requires extensive structural support system.
4. Requires complicated system for extension in the heading.
5. Not compatible with extremes in curvature of tunnel alignment.
6. Breakdown of one part shuts down entire system.

Limitations:

1. An 18 to 20 degree slope, either up or down, is the practical limit.
2. Depending on conveyor width, maximum particle size is about 12 to 18 inches.

Pipeline

Bulk materials can be successfully transported through pipelines, using either air or water as the transporting medium.

Advantages:

1. High capacities available.
2. Occupies minimum space in tunnel.
3. Adaptable to discharge to surface through relatively inexpensive risers.
4. Continuous operation.

Disadvantages:

1. Does not satisfy entire materials-handling need. Auxiliary system for transporting personnel and supplies is required.
2. Excavated material must be reduced to fine size at heading.
3. Requires complicated system for extension in the heading.
4. Reliability and maintenance are questionable.
5. Low efficiency, resulting in high power requirement.
6. Requires large volumes of air or water at the heading.

Limitations:

1. Virtually no limitation as to slope, alignment or capacity.
2. Particle size presently limited to maximum of about $\frac{1}{2}$ inch or less.

Slusher

A slusher is a hoe or box-type scraper pulled by means of a cable and a hoist. Only for special cases could it be considered a basic transportation system. However, it has many other uses in the tunnel, including use to fill a train of muck cars without car changing.

Advantages:

1. Requires no special roadbed.
2. Performs both mucking and haulage functions.

Disadvantages:

1. Low capacity, which decreases directly with travel distance.
2. Restricts access to the heading by other means.

Limitations:

1. With proper scraper design, slopes up to about 30 degrees could be handled. No limit for downslopes.
2. Practical distance limitation is about 500 feet.
3. Tunnel should be straight.

Overhead Rail

This transportation method is generally used only for moving cumbersome items (such as support elements, track panels and boring machine cutters) up to the vicinity of the heading. Except in special circumstances, it cannot compete with other systems for general transportation throughout the tunnel.

RAIL TRANSPORT

Rolling Stock

As a general rule, the muck cars should be as large as possible, considering the capability of the mucking system, clearance requirements in the tunnel, and the method of dumping. While the length of car which can be efficiently loaded with any given mucker is limited, this can be increased by means of accessory equipment such as belt conveyors or slushers.

In many tunnels, the dimensions of the car are limited by clearance requirements. The length of the muck car is also limited by structural design considerations and by the dumping method. The composition and gradation of muck may have some bearing on the choice of muck car. Table 11-1 lists the various types of muck cars which are in general use.

With the end dump and side dump cars, the body is tipped up and the discharge side or end is automatically pivoted away, permitting the muck to slide out. Meanwhile, the chassis is held onto the rail so that the entire car does not tip. The following methods are used to tip the body.

Gravity. Here the body is pivoted in such a manner that, when released by hand, it automatically dumps and swings back into position.

Power Cylinder. A portable compressed air or hydraulic cylinder is used to raise the car body. With this type, the car body can be mounted lower on the chassis and has less mechanism than the gravity type.

Traveling Hoist for Dumping Side Dump Cars. A hoist, traveling on tracks independent from the muck car tracks, raises the car body to the dumping position.

Table 11-1. Generally used muck cars.

CAR TYPE	DUMP METHOD
Side dump	Gravity or power
End dump	Gravity or power
Bottom dump	Trap doors
Rigid body	Rotary dump

Camel Back. The Granby type of side dump car has a roller attached to the side of the car body. At the dump, this roller travels up a ramp, thus causing the body to pivot and dump.

With bottom dump cars, trap doors in the car bottom are unlatched, allowing the muck to drop out. For dumping rigid-body cars, the entire car is turned over in a rotating structure so that the muck is spilled out. The dumper may be designed to handle cars while they are still coupled in the train, using rotary couplers. Otherwise, single cars must be uncoupled from the train. Special cars or dumping systems have been devised for particular tunneling systems or for special conditions.

In addition to muck cars, various special cars must be provided to facilitate the work. These include man-haul, explosives, vent pipe, flatcars and concrete cars.

For any important tunnel job, all rolling stock should be equipped with roller bearings and springs.

Two types of couplers may be used; automatic and manual. The selection of a coupler depends partly on the mucking system, the method of dumping and the size of the car.

Propulsion

The selection of locomotive type is influenced by the length and size of tunnel as well as by the availability of electric power, the grade of the tunnel, labor working rules, etc.

Power is supplied by either internal combustion engines or electric motors. For use in tunnels, the internal combustion engine must be of a type which minimizes the emission of harmful gases and avoids the use of highly flammable fuels. These considerations dictate the use of diesel engines. The advantages are lower capital cost; relief from some union working rules requiring an excessive number of electricians; flexibility; and a virtually unlimited range. The disadvantages are greater ventilation requirements and higher maintenance cost. Table 11-2 lists the general dimensions of a number of typical diesel locomotives.

Harmful exhaust ingredients are reduced by the use of a scrubber. Two basic types are available: washer and catalytic.

Table 11-2. General dimensions of Plymouth diesel locomotives. (*Courtesy Plymouth Locomotive Works*)

LOCO-MOTIVE	TONS	L	W	H	B	T.G.
TMDR	4 to 6	8 feet, 6 inches to 10 feet, 6 inches	41 to 65 inches	52 to 56 inches	36 inches	18 to 56.5 inches
MMD	3 to 15	11 feet, 0 inches to 15 feet, 0 inches	45 to 72 inches	56 to 90 inches	45 to 60 inches	18 to 56.5 inches
FMD	5 to 8	9 feet, 7 inches to 11 feet, 0 inches	41 to 73 inches	50 to 54 inches	39 to 50 inches	21.5 to 56.5 inches
DMD	8 to 16	12 feet, 2 inches to 14 feet, 6 inches	45 to 76 inches	60 to 70 inches	45 to 60 inches	23.6 to 66 inches
JMD	15 to 25	14 feet, 11 inches to 16 feet, 6 inches	66 to 76 inches	70 to 82 inches	70.4 inches	23.6 to 66 inches
MMD	25 to 45	19 feet, 8 inches to 20 feet, 2 inches	78 to 100 inches	96 to 100 inches	84 inches	30 to 66 inches

NOTE: The weights and dimensions above can be altered within certain limitations if necessary. New models and special features are built to order.

The U.S. Bureau of Mines has tested and approved the use of certain diesel locomotives for underground service, the main criterion for approval being the exhaust emission composition.

Electrically powered locomotives are of two types: battery and trolley. A combination trolley-battery type is also available.

Trolley-type Electric Locomotives. These are by far the most efficient, but have the disadvantage of requiring an overhead trolley cable and bonded rail. Since this disadvantage is particularly difficult in the heading, on long tunnels where the capacity and efficiency of the trolley locomotive is required, combination battery-trolley locomotives can be used or the heading may be serviced by battery or diesel powered units. It is also possible to use a cable reel for operation beyond the end of the trolley.

Battery Locomotives. These locomotives use either lead-acid or nickel-alkaline (Edison) batteries. Data regarding battery sizes and capacities should be obtained from manufacturers.

The critical performance factors for a locomotive are its weight and horsepower. Since the locomotive's drawbar pull is developed by means of the adhesion of its wheels to the rails, the maximum drawbar pull which can be developed is directly related to the weight on the driving wheels. Because of slack in the couplers, the drawbar pull required to start a train of cars moving does not depend on the static rolling resistance of the entire train. Also, sand can be used to improve the adhesion of the driving wheels during starting. Consequently, any difference between static and dynamic rolling resistance is not considered when determining locomotive size and type. The usual determining factors in selecting a locomotive are its abilities to move the loaded train and to maintain a satisfactory speed. If long, steep grades are involved, the determining factor may be the locomotive's braking ability or its ability to return upgrade with a train of empty cars.

The maximum available drawbar pull in pounds, D, is found from the formula $D = L[(2{,}000)(A)\text{-}G]$, where L = weight of locomotive on drivers in tons, A = coefficient of adhesion (see Table 11-3) and G = grade resistance in lb/ton (20 lb for each percent of grade).

The drawbar pull required to move the train is $D = TR$, where T = the total train weight (excluding the locomotive) in tons and R = the train's resistance in lb/ton.

Table 11-3. Coefficient of adhesion, *A*.

RAIL CONDITION	CHILLED CAST IRON WHEELS UNSANDED	CHILLED CAST IRON WHEELS SANDED	ROLLED STEEL WHEELS UNSANDED	ROLLED STEEL WHEELS SANDED
Clean, dry rail	0.20	0.28	0.25	0.34
Clean, wet rail	0.20	0.25	0.25	0.31
Slippery, moist rail	0.15	0.25	0.15	0.25
Dry, snow-covered rail	0.10	0.15	0.10	0.15

$R = F + C \pm G + P$, where F = train rolling resistance. It is as low as 4 to 5 lb/ton under ideal conditions. For tunnel work using anti-friction bearings, it will vary between 20 lb/ton for the average tunnel track and 10 lb for especially good track.

C = the resistance created by curves in the tunnel alignment. It can usually be neglected if there is not substantial curvature in the tunnel alignment. If desired, it can be computed by allowing 0.8 lb of drawbar pull per ton per degree of curvature. One degree of curvature = $\frac{5730}{\text{curve radius (ft)}}$.

G = the resistance created by gravity as the train travels an incline. Within the grades which can be traveled by conventional locomotives (0 to 6%), the grade resistance is 1% of the train weight per percent of grade, or 20 lb/ton per percent of grade.

P = acceleration resistance. It is approximately 100 lb/ton for an acceleration of 1 mph/second. For most tunneling applications, an acceleration of 0.1 to 0.2 mph/second is satisfactory, corresponding to a resistance of 10 to 20 lb/ton.

For stopping a train on a grade, the velocity, the weight and the rolling resistance of the train, and the weight and adhesion of the locomotive must be considered.

From the foregoing formulae, the minimum weight of locomotive for any given application can be determined. Most manufacturers list the drawbar pull of their locomotives for various speeds and track conditions, making the selection of the proper locomotive quite simple.

The required horsepower can also be calculated from the following formula:

$$\text{hp} = \frac{(T + L)(R)(V)}{(375)(M_e)(100\% - F_a)} + L_a$$

where

$(T + L)$ = weight of train plus locomotive (tons)
R = train's resistance (lb/ton)
V = velocity (mph)
M_e = mechanical efficiency of locomotive driveline (usually about 0.90 for direct drive and 0.80 for torque converter drive)
F_a = correction factor for altitude (usually 3% for each 1000 feet of elevation over 1000 feet above sea level for naturally aspirated diesel engines; for supercharged engines, there is usually no performance loss up to 5000 feet with 1%/1000 feet from 5000 to 10,000 feet and 3%/1000 feet thereafter—see manufacturers' specifications)
L_a = accessory losses (generator, fan, scrubber, compressor, etc., in hp).

Track

Common track gages are 24, 30, 36, 42 and the railroad standard, $56\frac{1}{2}$ inches. In selecting a track gage, principal considerations are the tunnel width, requirements for rail-operated mucking equipment, gage and size of rolling stock to be used, availability of equipment and possible resale values. To provide passing clearances, the track gage generally should be less than one-fourth the tunnel width, but for stability, should be at least one-half the maximum width of the rolling stock. Most rail-mounted mucking machines require a certain minimum track gage for efficient operation.

The required weight of rail depends upon the maximum individual wheel loading of the equipment to be used and upon the spacing of the ties or other rail supports.

Figure 11-1 shows the minimum weight and maximum single wheel load recommended for $1\frac{1}{2}$- and 3-foot tie spacings. Heavier rails than indicated for given loads will reduce track resistance and maintenance costs, while lighter rails should be used only for short-term jobs.

Track accessories include splice bars, bolts, ties, spikes, tie plates and gage bars, the last two items being used only for high-speed track or special situations. The tie length should be at least $1\frac{1}{2}$ times the track gage, or 24 inches greater than the track gage, whichever is greatest. Ties may generally be of any good quality, locally available wood. Steel ties have the advantages of strength and durability, and they require less headroom (but are more expensive) than wood; also, they are not adaptable to a circular invert. Table 11-4 gives the approximate weights and dimensions of track and accessories per thousand feet of single track, all based on a 2-foot tie spacing.

Track ballast is usually derived from tunnel muck; however, good quality ballast made from sound gravel or crushed rock, uniformly graded from $\frac{3}{4}$-inch to No. 4 mesh, will pay for itself on many jobs.

In machine-bored tunnels in rock, or where a segmented precast concrete tunnel lining is installed, ties may be set directly on the smooth, hard invert surface without ballast.

Careful construction and adequate maintenance of the roadbed are essential. Their cost will be repaid many times over in increased traffic speeds, more efficient locomotive operation, fewer derailments and longer life of rolling stock. The significance of this fact is not always appreciated.

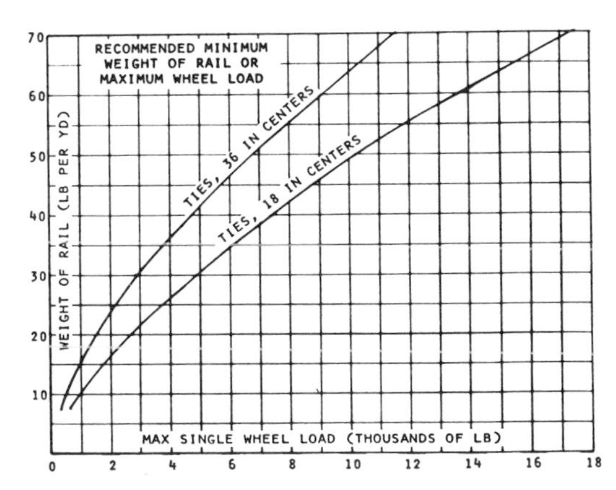

Fig. 11-1. Weight of rail and tie spacing.

A turnout is an arrangement by which rolling stock may be diverted from one track to another. Figure 11-2 illustrates the principle elements of a turnout. "Turnout" and "switch" are not synonomous terms. A switch is that part of a turnout consisting of the pair of switch points and appurtenant moving parts. Switches in general use in tunneling are the standard railroad split switches, such as illustrated in Fig. 11-2. The general configuration of a turnout is determined by the frog number N, which is a measure of the frog angle. A larger frog number means a smaller frog angle and a more gradual turnout. Detailed dimensions of frogs, switches and turnouts have been standardized among all major rail manufacturers.

Except for long car bodies binding one another, or not clearing the tunnel ribs, the minimum possible radius of curve in the tunnel is limited by the wheelbase and wheel diameter on the locomotive and rolling stock, as given in Table 11-5.

The maximum wheelbase refers to the center distance between fixed axles as on locomotives and four-wheeled muck cars.

Track gage on curves must be increased to prevent binding of wheel flanges. Gage is generally increased $\frac{1}{16}$ inch for each $2\frac{1}{2}$ degrees of curvature, to a maximum limited by the width of wheel tread.

Track Layout and Car Handling

Passing tracks in the tunnel and at shaft stations are provided by means of standard turnouts with sufficient passing track or tail track to satisfy the particular operational requirements. Switching cars at the heading, however, requires special consideration.

The Portable or California Switch. This comprises a section of double track with turnouts and ramps at each end, all of which slides on the main track. A typical example is shown in Fig. 11-3.

Other car changing methods include the lat-

Table 11-4. Quantities per 1000 feet of single track.

RAIL WEIGHT (LB/YARD)	TONS	WEIGHT OF SPLICE PLATE, INCLUDING BOLTS* (LB)	SIZE OF SPIKES	WEIGHT OF SPIKES (LB)	NUMBER OF TIES (2 FEET CENTER-CENTER)	SIZE OF TIES	MBM TIES				
							3 FEET	4 FEET	5 FEET	6 FEET	7 FEET
12	4.00	245	$3 \times \frac{3}{8}$ inches	315	500	$3\frac{1}{2} \times 4\frac{1}{2}$ inches	2.1	2.8			
16	5.333	292	$3\frac{1}{2} \times \frac{7}{16}$ inches	494	500	4×5 inches	2.6	3.4	4.3		
20	6.667	334	$3\frac{1}{2} \times \frac{1}{2}$ inches	627	500	4×5 inches	2.6	3.4	4.3	5.1	
25	8.333	441	$3\frac{1}{2} \times \frac{1}{2}$ inches	627	500	4×5 inches	2.6	3.4	4.3	5.1	6.0
30	10.0	579/829	$4 \times \frac{1}{2}$ inches	727	500	$4\frac{1}{2} \times 5\frac{1}{2}$ inches	3.1	4.1	5.2	6.2	7.2
35	11.667	977	$4\frac{1}{2} \times \frac{1}{2}$ inches	792	500	$5 \times 6\frac{1}{2}$ inches	4.0	5.3	6.7	8.0	9.3
40	13.333	977/1,190	$5 \times \frac{1}{2}$ inches	870	500	$5\frac{1}{2} \times 7$ inches	4.7	6.2	7.8	9.3	10.9
45	15.00	1,284	$5 \times \frac{9}{16}$ inches	1095	500	$5\frac{1}{2} \times 7$ inches	4.7	6.2	7.8	9.3	10.9
50	16.667	1,880	$5 \times \frac{9}{16}$ inches	1095	500	$5\frac{1}{2} \times 7$ inches	4.7	6.2	7.8	9.3	10.9
55	18.333	2,187	$5 \times \frac{9}{16}$ inches	1095	500	$5\frac{1}{2} \times 7$ inches	4.7	6.2	7.8	9.3	10.9
60	20.00	2,357/2,143	$5\frac{1}{2} \times \frac{9}{16}$ inches	1194	500	6×8 inches	5.8	7.7	9.6	11.5	13.4
65	21.667	2,420	$5\frac{1}{2} \times \frac{9}{16}$ inches	1194	500	6×8 inches	5.8	7.7	9.6	11.5	13.4
70	23.333	2,624	$5\frac{1}{2} \times \frac{9}{16}$ inches	1194	500	6×8 inches	5.8	7.7	9.6	11.5	13.4
75	25.00	2,780	$5\frac{1}{2} \times \frac{5}{8}$ inches	1509	500	6×8 inches	5.8	7.7	9.6	11.5	13.4
80	26.667	3,060	$5\frac{1}{2} \times \frac{5}{8}$ inches	1509	500	6×8 inches	5.8	7.7	9.6	11.5	13.4
85	28.333	3,293	$5\frac{1}{2} \times \frac{5}{8}$ inches	1509	500	6×8 inches	5.8	7.7	9.6	11.5	13.4
90	30.00	3,669	$5\frac{1}{2} \times \frac{5}{8}$ inches	1509	500	6×8 inches	5.8	7.7	9.6	11.5	13.4
100	33.333	4,259	$5\frac{1}{2} \times \frac{5}{8}$ inches	1509	500	6×8 inches	5.8	7.7	9.6	11.5	13.4

*30-foot rail up to 60#; 33-foot rail above 60#.

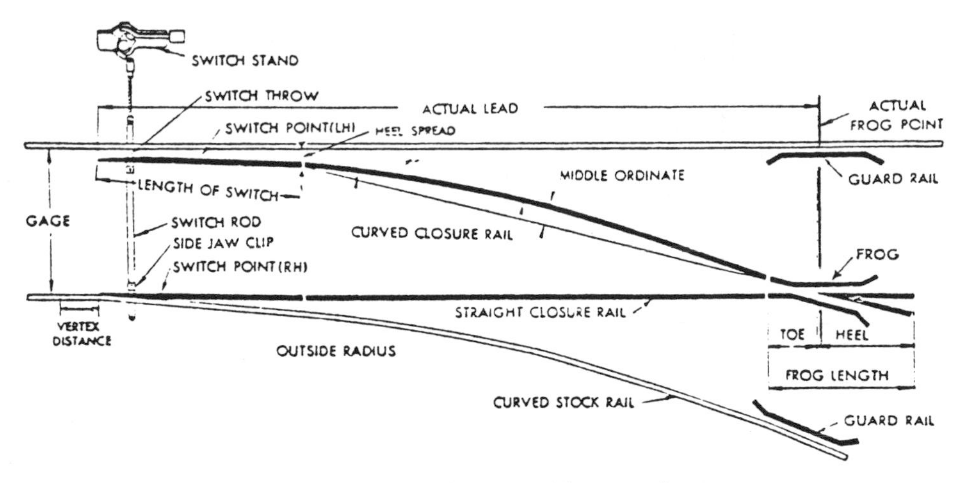

Fig. 11-2. Typical turnout with names of parts.

eral car passer and the overhead car passer or cherry picker.

The Jacobs Sliding Floor. Shown in Fig. 11-4, this is a recent development for track layout in the heading. This comprises a structural steel floor some 200 to 250 feet long and occupying most of the invert width. It is built in three or more sections so that it can be moved along by means of integral hydraulic rams as the heading advances. The sliding floor has tracks for the jumbo, the mucker, and the loaded and empty trains.

Table 11-5. Minimum radius of curve, in feet, over which rolling stock will pass.

MAXIMUM WHEELBASE (INCHES)	DIAMETER OF WHEEL (INCHES) 14	16	18	20	22	24	26	28	30	33	36
36	12	12	12	13	13	13	14				
38	12	12	12	14	14	14	15	15			
40	13	13	13	14	14	14	16	16	16		
42	14	14	14	15	15	15	16	16	16		
44	15	15	15	16	16	16	17	17	17	20	
48	16	16	16	17	17	17	19	19	19	22	22
54		18	18	19	19	19	21	21	21	25	25
60		18	20	21	21	21	23	23	23	28	28
66			22	23	23	23	26	26	26	31	31
72			25	26	26	26	28	28	28	34	34
84			29	30	30	30	33	33	33	39	39
96				34	34	34	37	37	37	45	45
108				39	39	39	43	43	43	51	51
144				52	52	52	56	56	56	68	68

The "Navajo Blanket." While not having all of the sliding floor features, this provides for extending the track in the heading in standard rail length increments.

RUBBER-TIRED VEHICLES

General

Rubber-tired vehicles, typically diesel-powered, are available for most transport needs in tunnel work. This type of transport has advantages over others in being able to deliver muck directly to one or several disposal areas and at some distance from the tunnel. In multiple heading operations, rubber-tired vehicles can move from heading to heading, in sequence with excavation cycles or variances in production rates.

Types of Vehicles

Load-haul Units. Standard front end loaders such as the Caterpillar 980 and the low profile Wagner MS-2 can be used for transport. Although primarily intended for loading, such units may be economically used for short haul distances.

Shuttle or load-haul-dump units are designed in two configurations. In the first type, muck is carried in the bucket that does the digging. Bucket capacities range between 1 and 17 cubic yards. In the second type, the loading bucket

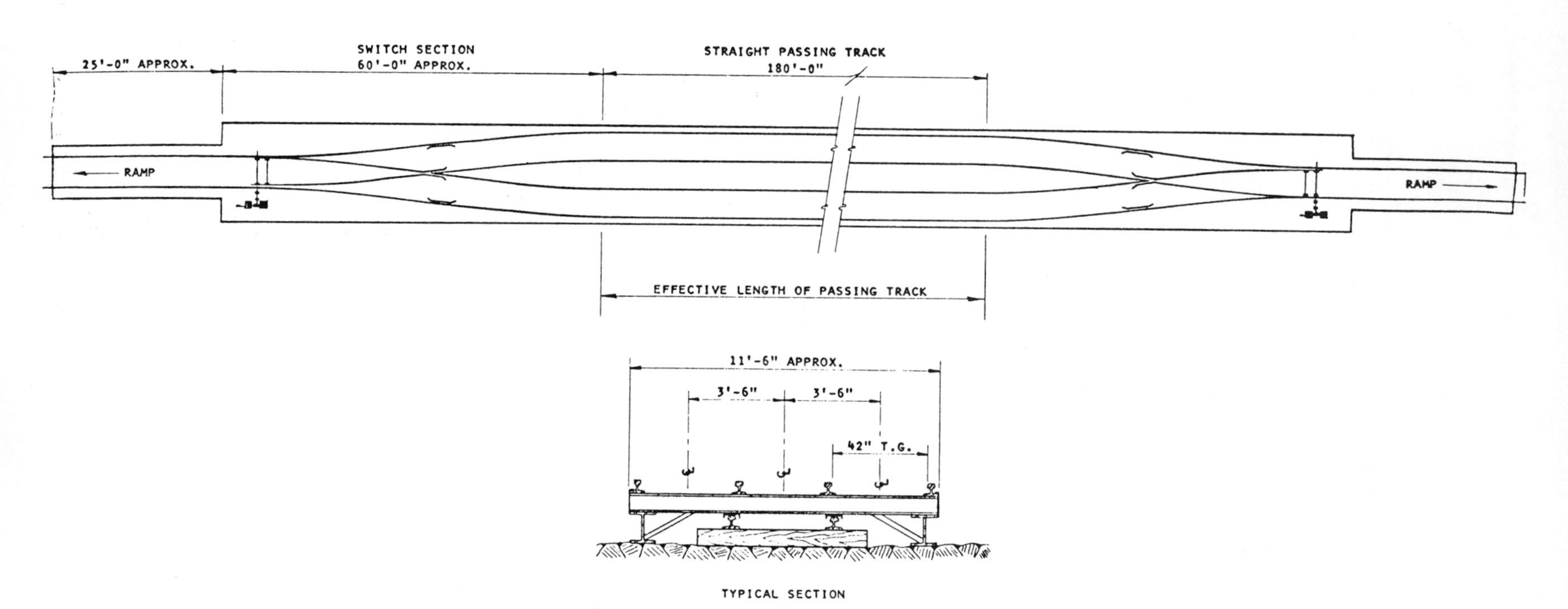

Fig. 11-3. California switch.

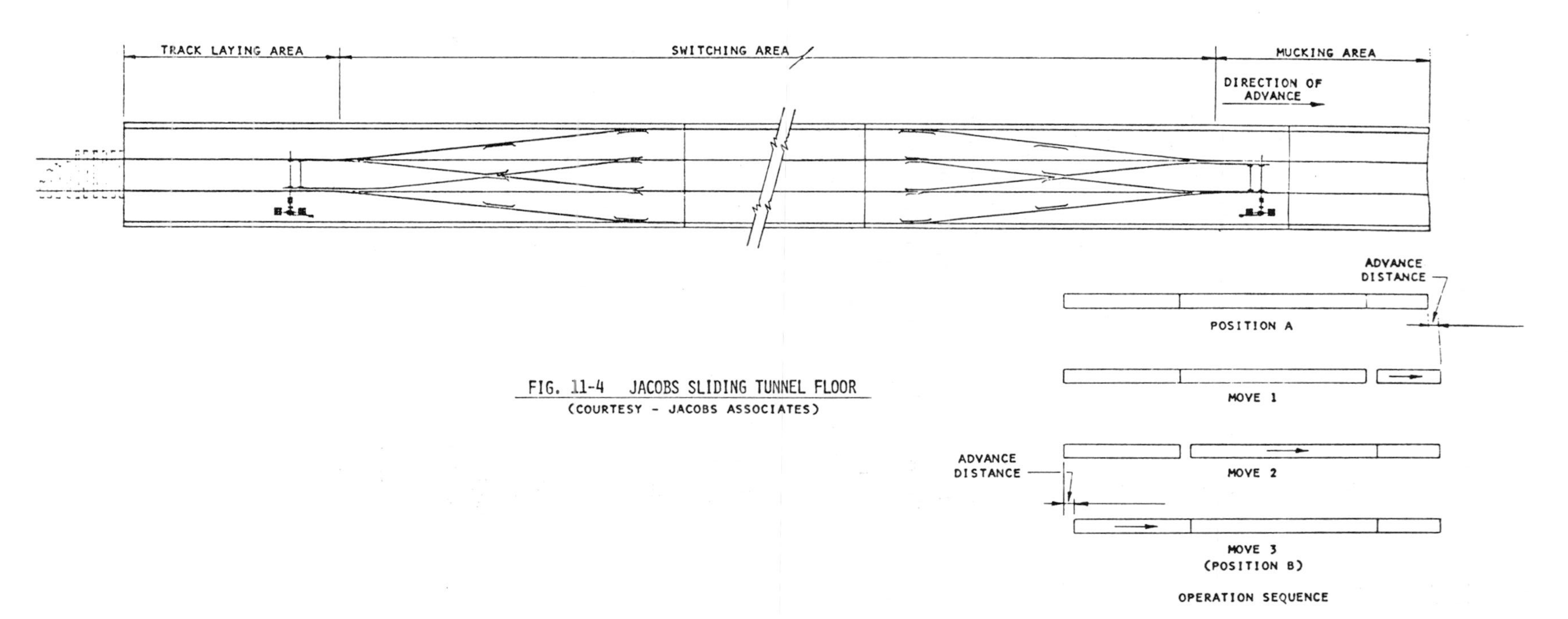

Fig. 11-4. Jacobs sliding tunnel floor.

charges a hopper or bowl for hauling, and capacities range between 2 and 12.5 cubic yards. The economic limit of haul depends upon the size of the load carried and the operating speed. The larger sizes have been used on hauls up to 3000 feet at speeds up to 15 mph. Usually, the speed is in the range of 6 to 8 mph.

Dump Trucks. Trucks designed especially for underground work are available in many different sizes. Vehicles of this class do not require turning in the tunnel; they simply reverse direction, the gearing being the same for each direction. The driver either sits sideways or changes positions using dual controls. The Koehring 1860 is representative of this type, although much smaller units are available. Conventional rear dump trucks such as the Euclid R22 are often used in large tunnels. The rocker type rear dump, such as the Caterpillar-Athey PR 621, 631 and 651, is particularly suitable for muck hauling. For their capacity, they have a short turning radius which is further shortened when the body is in the raised position.

Special Vehicles. For man-haul, vent pipe erection, explosives and supply services, special vehicles are available or may be constructed on truck chassis to suit the needs of the job.

Roadbed

The tunnel muck remaining on the invert or produced at the face is sometimes suitable for the roadbed. If not, sound, well-graded aggregate from outside sources should be provided.

Production

The production of the haul unit depends upon its capacity; the time required to spot, load, pass and dump; and the average travel speed of the vehicle. Speed of travel depends upon several factors. One is psychological, speed in the confined space of a tunnel being magnified so that 15 mph inside seems more like 30 mph outside. Another is safety, and arbitrary speed limits may be required due to other work in progress, turns or intersections. The other factors affecting speed are the condition and gradient of the roadbed.

For these reasons, sophisticated methods for calculating travel times are unnecessary for tunnel construction.

Grades and rolling resistance, as shown in Table 11-6, determine the maximum possible speed for any given unit. Actual average speed is determined according to judgment as to the influence of the other above-described factors.

Performance specifications published by vehicle manufacturers give the maximum attainable speed for all grades, both loaded and empty.

Allowance for loading time depends upon the mucking system being used.

Typical time allowances for other operations are given in Table 11-7. Unusually difficult spotting or turnaround problems will produce times in excess of those quoted.

In narrow tunnels, an additional allowance must be made for passing.

To calculate the number of trucks required for a given situation, the total estimated travel time (travel loaded, turn and dump, travel

Table 11-6. Rolling resistance.

TYPE OF HAUL ROAD SURFACE	POUNDS PER TON OF GROSS VEHICLE	EQUIVALENT GRADE (%)
Concrete and asphalt	30	1.5
Smooth, hard, dry surface; well maintained; free of loose material	40	2
Dry surface but not firmly packed; some loose material	60	3
Wet, muddy surface on firm base	80	4
Soft or unpacked dirt fills	160	8
Loose sand or gravel	200	10
Deeply rutted, or soft spongy base	320	16

Table 11-7(a). Turning and dumping time (minutes).

OPERATING CONDITIONS	REAR-DUMP	SHUTTLE UNIT
Favorable	1.3	0.6
Average	1.5	1.0
Unfavorable	2.5	1.5

Table 11-7(b). Spot at loading machine (minutes).

OPERATING CONDITIONS	REAR-DUMP	SHUTTLE UNIT
Favorable	0.6	0.3
Average	1.0	0.5
Unfavorable	1.5	1.0

empty, passing and spot in heading) is divided by the estimated loading time. The result is the number of trucks required.

The number of standby units which should be provided depends upon the condition of the vehicles, the severity of service and the extent of the job. Practice varies from one spare for each two units in service for unfavorable conditions to one for each eight units for favorable conditions. For average conditions, one spare vehicle for every four units should be provided.

BELT CONVEYORS

General

Most mechanical excavators utilize short belt conveyors to discharge muck into other modes of transport. Complete, high-capacity material handling systems composed entirely of conveyors are used in the mining industry. Their application to tunnel work is technically practical but the economics must be considered separately for each case.

Detailed design criteria can be found in the catalogs published by belt conveyor manufacturers. It should be kept in mind, however, that most tunnel service is extremely severe, requiring conservatism in the selection of equipment, spacing of carriers and selection of power.

Loading Conveyors

Belt conveyors can frequently be used in the heading to facilitate the loading of the muck transport equipment. The Dixon conveyor, which was developed on the Colorado River Aqueduct, permits the loading of an entire train without changing cars. The conveyor is installed in a gantry-type frame, or, in a steel-supported tunnel, it can be supported on the steel ribs. The loading end can be raised to permit movement of the mucker and drill jumbo to and from the face. The discharge end is provided with a swinging chute to prevent spillage between cars.

When there is insufficient space in a tunnel to position a loader alongside a truck for discharge, a belt conveyor can be used to transfer the muck from a much narrower hopper into the wide truck body.

Long Conveyors

Belt conveyors have occasionally been used as the prime system for transporting muck out of the tunnel. Notable examples are the Oahe Dam Outlet Tunnels in South Dakota, the Sepulveda Tunnel in California and some of the Chicago Sewer Tunnels.

For underground service, wire rope-supported conveyors are frequently used in place of rigid frame-supported conveyors. The wire rope conveyor may either be hung from the roof by chains or supported from the invert on light pipe standards, as shown in Fig. 11-5.

Extension of the conveyor to keep pace with the heading advance may be accomplished through a variety of belt storage and take-up devices. The practical limit to which extendable devices can be used depends largely upon the nature of the excavation systems, the speed of advance, the quantity of material being moved and the production cycle.

When it becomes necessary to add belt length, a new section with separate drive and material transfer station may be added or a new length may be spliced into the belt. Both schemes involve potential delays such as moving extendable devices and drives ahead, completing transfer stations and splicing.

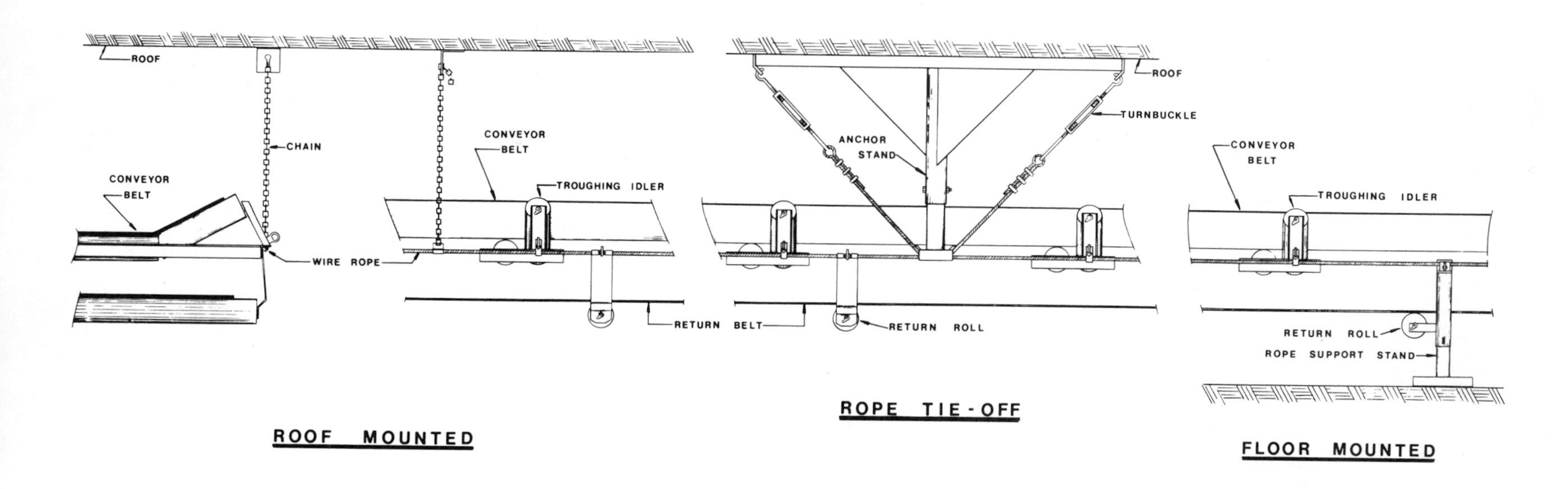

Fig. 11-5. Typical wire rope conveyor installation.

Table 11-8. Maximum lump size and idler spacing for typical idlers.* *(Courtesy Continental Conveyor and Equipment Co.)*

	LUMP SIZE		IDLER SPACING			
			BULK WEIGHT OF MATERIAL (PER CUBIC FOOT)			
BELT WIDTH	UNIFORMLY SIZED	MIXED WITH 90% FINES	75 LB	100 LB	125 LB	RETURN IDLERS
18	3	6	5 feet, 0 inches	5 feet, 0 inches	4 feet, 6 inches	10 feet, 0 inches
24	5	8	4 feet, 6 inches	4 feet, 0 inches	4 feet, 0 inches	10 feet, 0 inches
36	7	12	4 feet, 0 inches	4 feet, 0 inches	3 feet, 6 inches	10 feet, 0 inches
48	10	16	4 feet, 0 inches	4 feet, 0 inches	3 feet, 6 inches	10 feet, 0 inches

*Additional adlers at approximately 2′-0″ are recommended at loading points.

The higher allowable tension in steel cord conveyor belting permits very long single flights. Surface installations over four miles in length have been made. When the steel cords are prestressed, the requirement for take-up travel is greatly reduced. The use of long flights of steel cord wire rope conveyors in very long tunnels could replace a series of shorter conveyors and their transfer stations as excavation progresses, thus minimizing the problems associated with many drives and potential spillage at transfer points.

The return side of a single conveyor flight can be used to transport dry shotcrete or concrete material into the tunnel.

Flexible Conveyors. Conveyors having the flexibility for making turns, climbing at steep angles, dumping to the side and providing full load-carrying return capability are available. In the Serpentix non-linear system, a chain-driven device transmits the tension, thus allowing differential movement in the folds of the belt during twists or turns. Sectionalized construction permits fast extensions of the conveyor and intermediate drive stations reduce chain tension.

Belt Conveyor Design. Belt width, speed and idler spacing is dictated by the lump size, tonnage and unit weight of the material (see Tables 11-8, 11-9 and 11-10).

Generally, inclined conveyors should not exceed a slope of 18 to 20 degrees, even flatter slopes being necessary with material having a great tendency to roll. Skirt boards are necessary at belt transfer points and at loading points.

The power required to drive a conveyor can be obtained from Figs. 11-6 and 11-7 and Table 11-11. The combined total of the required horespower shown gives the horsepower necessary for the system.

Belt selection depends upon the material handled and the tension required (see Tables 11-12 and 11-13). Fabric belts are available for tension requirements not exceeding 1000 lb/inch of belt width. Steel cord belts may be used for

Table 11-9. Normal and maximum belt speeds recommended (maximum speeds in feet/minute). *(Courtesy Continental Conveyor and Equipment Co.)*

BELT WIDTH (INCHES)	LUMP OR MODERATELY ABRASIVE MATERIALS (SUCH AS WELL-SHOT GRAVEL SAND)	HEAVY-SHARP OR VERY ABRASIVE MATERIALS (SUCH AS POORLY SHOT ROCK)
18	300–400	250–350
24	500–600	400–500
36	500–600	400–500
48	700	600

Table 11-10. Speeds and capacities for typical conveyor loading on 35-degree troughing idlers in tons/hour for various weights of materials and widths of belts. (Capacities given are for horizontal conveyors having a uniform feed and load. If there are peak loads, belts of sufficient capacity to handle material at maximum rate should be used.) *(Courtesy Continental Conveyor and Equipment Co.)*

BW (INCHES)	WEIGHT PER CUBIC FOOT OF MATERIAL	CAPACITY (TONS/HOUR)													CUBIC YARDS PER HOUR (AT 100 FEET/ MINUTE)
		BELT SPEED, F.P.M.													
		200	250	300	350	400	450	500	550	600	700	800	900	1000	
18	75	101	128	153	179	204	230	225							51
	100	136	170	204	238	272	306	340							51
	125	170	213	255	298	340	383	425							51
24	75	202	252	303	354	404	455	505	556	606					100
	100	269	337	404	472	539	606	674	741	809					100
	125	337	421	505	590	674	758	843	927	1011					100
36	75	498	623	748	872	997	1122	1246	1371	1496	1745	1994	2244	2493	246
	100	664	831	997	1163	1329	1496	1662	1828	1994	2327	2659	2992	3324	246
	125	831	1038	1246	1454	1662	1870	2077	2285	2493	2909	3324	3740	4155	246
48	75	925	1156	1388	1619	1851	2082	2313	2545	2776	3239	3702	4164	4627	457
	100	1234	1542	1851	2159	2468	2776	3085	3393	3702	4319	4936	5553	6170	457
	125	1542	1928	2313	2699	3085	3470	3856	4242	4627	5398	6170	6941	7712	457

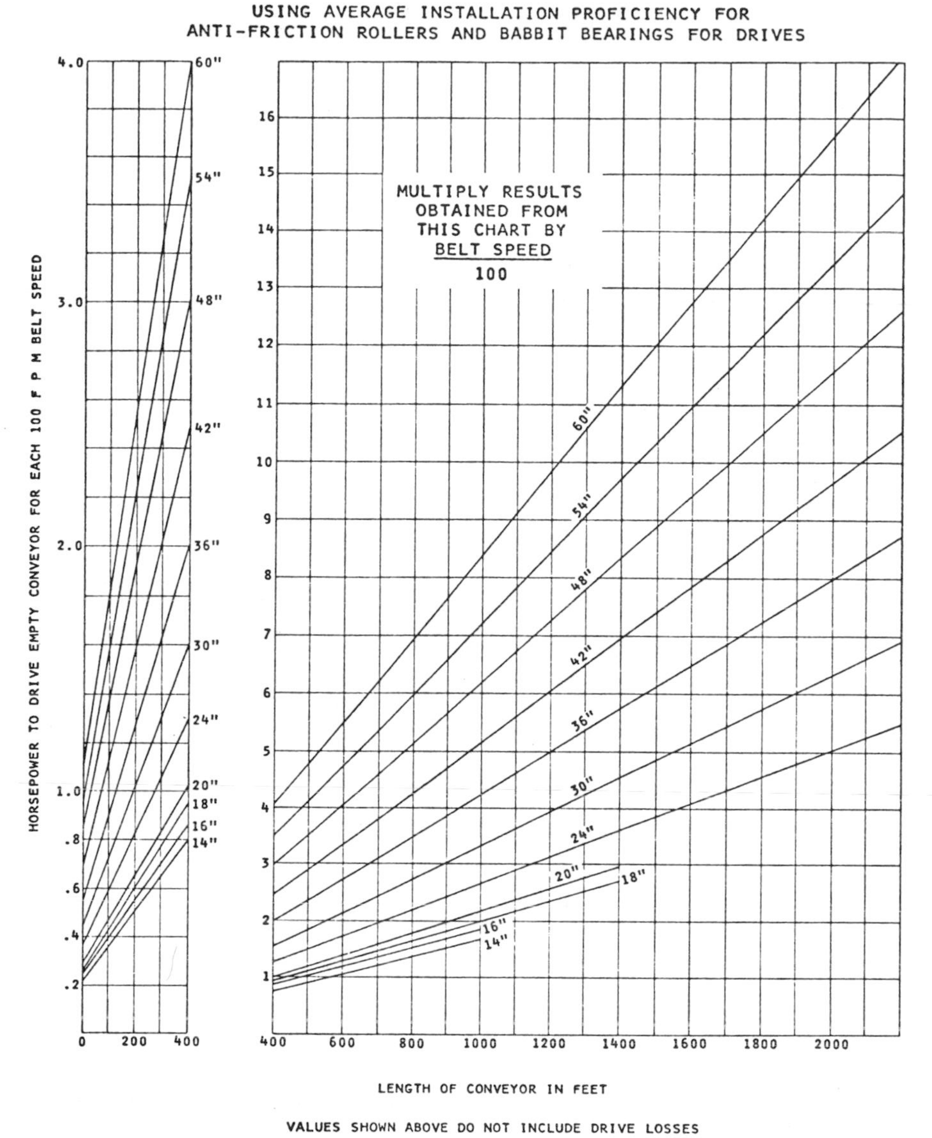

Fig. 11-6. Horsepower required to drive empty conveyor.

required tensions up to 8000 lb/inch of belt width.

Typical pulley sizes and shaft diameter requirements are shown in Tables 11-14 and 11-15. Drive componenets of lesser size than shown may be used for installations with lower belt tensions.

At charging or interchange points, the troughing idler spacing is reduced to protect the belt from the impact of loading. In severe conditions, rubber or pneumatically cushioned troughing idlers may be used to absorb the shock of loading. Each conveyor section must be equipped with a take-up device to allow for stretch and shrinkage of the belt and still maintain the proper tension to support the load and permit the drive pulleys to move the load. A great variety of drive arrangements are available to meet space, power and reduction requirements of each particular installation. Direct coupled, fully enclosed, sealed and lubricated drive systems are preferable for underground service.

Multi-section conveyor systems should be

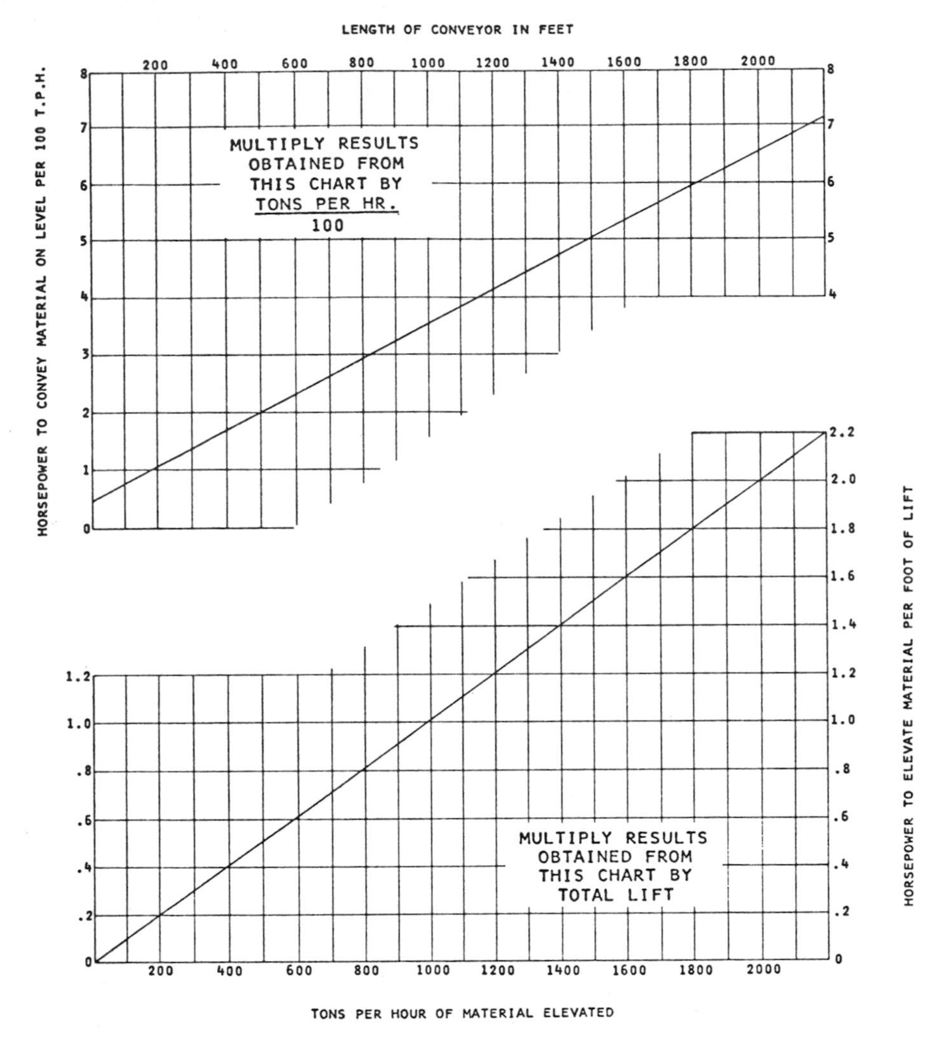

Fig. 11-7. Horsepower requirements to convey material horizontally and to elevate material.

Table 11-11. Additional horsepower for drive losses. *(Courtesy Continental Conveyor and Equipment Co.)*

CAST TOOTH GEARS	ROLLER CHAIN OR "V" BELT DRIVES	ENCLOSED SPEED REDUCERS HELICAL TYPE*
Add 6% for each Reduction	Add 5% for each Reduction	Add 2% for each Reduction

*If worm gear reducer is used, be sure to use manufacturer's efficiency ratings.

Table 11-12. Belt tensions. *(Courtesy Continental Conveyor and Equipment Co.)*

		AUTOMATIC TAKE-UP				SCREW TAKE-UP			
		TENSION (LB)				TENSION (LB)			
		TIGHT SIDE		SLACK SIDE		TIGHT SIDE		SLACK SIDE	
ARC OF CONTACT	TYPE OF DRIVE	BARE PULLEY	LAGGED PULLEY	BARE PULLEY	LAGGED PULLEY	BARE PULLEY	LAGGED PULLEY	BARE PULLEY	LAGGED PULLEY
180°	Plain	1.64 E	1.50 E	.64 E	.50 E	1.97 E	1.80 E	.97 E	.80 E
210°	Snubbed	1.50 E	1.38 E	.50 E	.38 E	1.80 E	1.66 E	.80 E	.66 E
240°	Snubbed	1.40 E	1.30 E	.40 E	.30 E	1.68 E	1.56 E	.68 E	.56 E
270°	Snubbed	1.32 E	1.24 E	.32 E	.24 E	1.58 E	1.49 E	.58 E	.49 E
300°	Tandem	1.26 E	1.19 E	.26 E	.19 E	1.51 E	1.43 E	.51 E	.43 E
330°	Tandem	1.22 E	1.16 E	.22 E	.16 E	1.46 E	1.40 E	.46 E	.40 E
360°	Tandem	1.18 E	1.13 E	.18 E	.13 E	1.42 E	1.36 E	.42 E	.36 E
390°	Tandem	1.15 E	1.11 E	.15 E	.11 E	1.39 E	1.33 E	.39 E	.33 E
420°	Tandem	1.13 E	1.09 E	.13 E	.09 E	1.36 E	1.31 E	.36 E	.31 E
450°	Tandem	1.11 E	1.07 E	.11 E	.07 E	1.33 E	1.29 E	.33 E	.29 E
480°	Tandem	1.09 E	1.06 E	.09 E	.06 E	1.31 E	1.27 E	.31 E	.27 E

$$E = \frac{(\text{hp})(33{,}000)}{\text{Belt Speed, fpm}}$$

Maximum allowable belt tension, lb/ply/inch width rayon fabric belts; nylon or rayon filling.

Splice	MP 35	MP 43	MP 50	MP 60	MP 70	MP 90
Vulcanized	35	43	50	60	70	90
Fastener*	27	33	40	45	55	

*Best available metal fasteners.

interlocked electrically so that the last section to receive material is the first to start. Centrifugal switches provide one convenient interlocking device and also serve to protect the drive system in case of failure of some component. Emergency shutdown arrangements are necessary to prevent excessive damage to belts and other conveyor parts in case of accident. Braking devices may be employed to bring the system to a standstill quickly. Holdback devices are required to prevent rollback of a loaded inclined conveyor that has been stopped. The belt speed of declining conveyors may be controlled by means of regenerative power systems.

PIPELINE

Pipelines can be used to transport many types of solids. Both hydraulic and pneumatic systems are employed and either system could be used in tunneling. The hydraulic system employs water, or a slurry of water and fines, as the transporting agent, while the pneumatic system uses air. Pipelines offer the possibility of fast and continuous muck removal without adding appreciably to the space congestion in the tunnel.

The Hydraulic System

Capabilities. When considering a hydraulic pipeline for the transport of tunnel muck, the following requirements should be kept in mind.

- An adequate supply of water will be needed.
- The design of the facility to extend the pipeline must be resolved so that an uninterrupted operation can be expected.

Table 11-13. Conveyor belt construction.* *(Courtesy Continental Conveyor and Equipment Co.)*

Minimum Number of Plies to Support the Load:

		BELT SPECIFICATION									
		RMA CLASSIFICATION									
	BELT WIDTH	MP35	MP43	MP50	MP60	MP70	MP90	MP120	MP155	MP195	MP240
Material weighing not more than 75 lb/foot^3	18	3	3	3	3	3	3	3	3		
	24	4	4	4	4	4	3	3	3		
	36	5	5	4	4	4	4	4	4	4	4
	48	6	5	5	5	5	5	5	5	5	4
Material weighing not more than 100 lb/foot^3	18	4	4	3	3	3	3	3	3		
	24	5	5	4	4	4	4	4	3		
	36	6	6	5	5	5	5	5	5	4	4
	48	7	7	6	6	6	6	6	6	5	5
Material weighing not more than 125 lb/foot^3	18	4	4	4	4	4	4	4	3		
	24	6	6	5	5	5	5	5	4		
	36	7	6	6	6	6	6	6	5	5	5
	48	8	8	7	7	7	7	7	6	6	6
Maximum Number of Plies for Troughing:											
Maximum number of plies which will trough on 35 degree idlers	18	4									
	24	5	4	4	4	4	4	4			
	36	8	8	7	7	6	6	5	5	5	4
	48	10	9	9	8	8	7	7	7	6	6

*Data contained in this table are based on present practices and ratings of equipment and belting; all of these data are subject to change with new developments.

- The flow of muck from the tunnel face excavation should be fairly continuous, without frequent shutdowns.
- The raw muck from the excavation must be properly sized and this may require crushing.
- Where crushing is necessary, the effect of the water content of the muck on the crushing efficiency should be investigated.
- The final muck characteristics should permit transport under heterogeneous or saltation flow conditions.
- The pipe diameter should be at least three times maximum particle size.

Table 11-14. Recommended minimum pulley diameters. *(Courtesy Continental Conveyor and Equipment Co.)*

	RAYON-NYLON BELT AT OVER 80% OF RATED TENSION									
PLIES	MP35	MP43	MP50	MP60	MP70	MP90	MP120	MP155	MP195	MP240
4	20	20	24	24	24	24	24	30	36	36
5	24	24	24	30	30	30	30	36	42	42
6	30	30	30	36	36	36	36	42	48	48
7	30	30	36	42	42	42	42	48	54	60
8	36	36	42	48	48	48	48	54	60	66
9	42	42	48	54	54	54	54	60	66	78
10	48	48	54	60	60	60	60	66	66	84

Table 11-15. Typical head pulley and shaft diameter requirements. *(Courtesy Continental Conveyor and Equipment Co.)*

BELT WIDTH	HEAD PULLEY DIAMETER (INCHES)	MAXIMUM BEARING CENTERS (INCHES)	SHAFT DIAMETER (INCHES)	MAXIMUM TENSION (LB)
18	20	34	$3\frac{7}{16}$	4,185
	24	36	$4\frac{7}{16}$	7,709
24	20	40	$3\frac{15}{16}$	6,166
	24	42	$4\frac{15}{16}$	10,440
	30	44	$5\frac{7}{16}$	14,481
36	24	56	$5\frac{7}{16}$	12,072
	30	58	7	22,645
	36	60	$7\frac{1}{2}$	25,096
	42	64	9	36,749
48	36	73	8	30,457
	42	77	9	36,749
	48	79	10	46,879
	54	79	10	45,219

- Dewatering of the slurry plus storage and treatment of the solids must be provided.
- The possibility of serious corrosion problems should be considered.

Layout. There are two methods of introducing the solids into the pipeline. One is to feed the mixture of liquid and solids directly into the pump. The second is to inject the solids into the pipeline on the discharge side of the pump. The first method is simple and has a high capacity but is limited to low heads and has an inherently high pump maintenance cost. The second method is capable of handling higher heads, permitting long-distance movement. Pump maintenance costs are lower, but the design is more complex, with a complicated injection system.

Water is the usual liquid medium. If the percentage of fines in the excavated material is low, some means must be employed to increase it.

The muck may be transported horizontally through a pipeline running the entire length of the tunnel, or the pipeline may be diverted vertically through shafts or drill-holes.

Component specifications will be determined by operating parameters, such as hourly capacity, pipe length and static head, as well as by design criteria. Figure 11-8 presents a schematic layout of a hypothetical hydraulic transport system. It features a loop system where the head of the return water is used to partially balance the power demand on the transport line.

The Pneumatic System

Pneumatic transport has been used to backfill open areas of underground mines. Three-inch minus rock has been transported pneumatically at the rate of 300 tons/hour for a maximum horizontal distance of 1000 feet. Transportation of 6-inch material for distances of 4000 feet is the goal of current research.

Capabilities. A pneumatic system will operate most effectively with free-flowing granular material. Crushing may be required to limit maximum particle size. The system is not suitable for wet, sticky material. Vertical lifts should be limited to 1000 feet and horizontal runs to 2500 feet. Greater distances can be handled by installing systems in series.

One drawback to the pneumatic system is the rather large amount of power required. One 1000-foot horizontal system conveying material at the rate of 300 tons/hour was reported to require 800 hp. This indicates a consumption of 14 hp/ton-mile/hour.

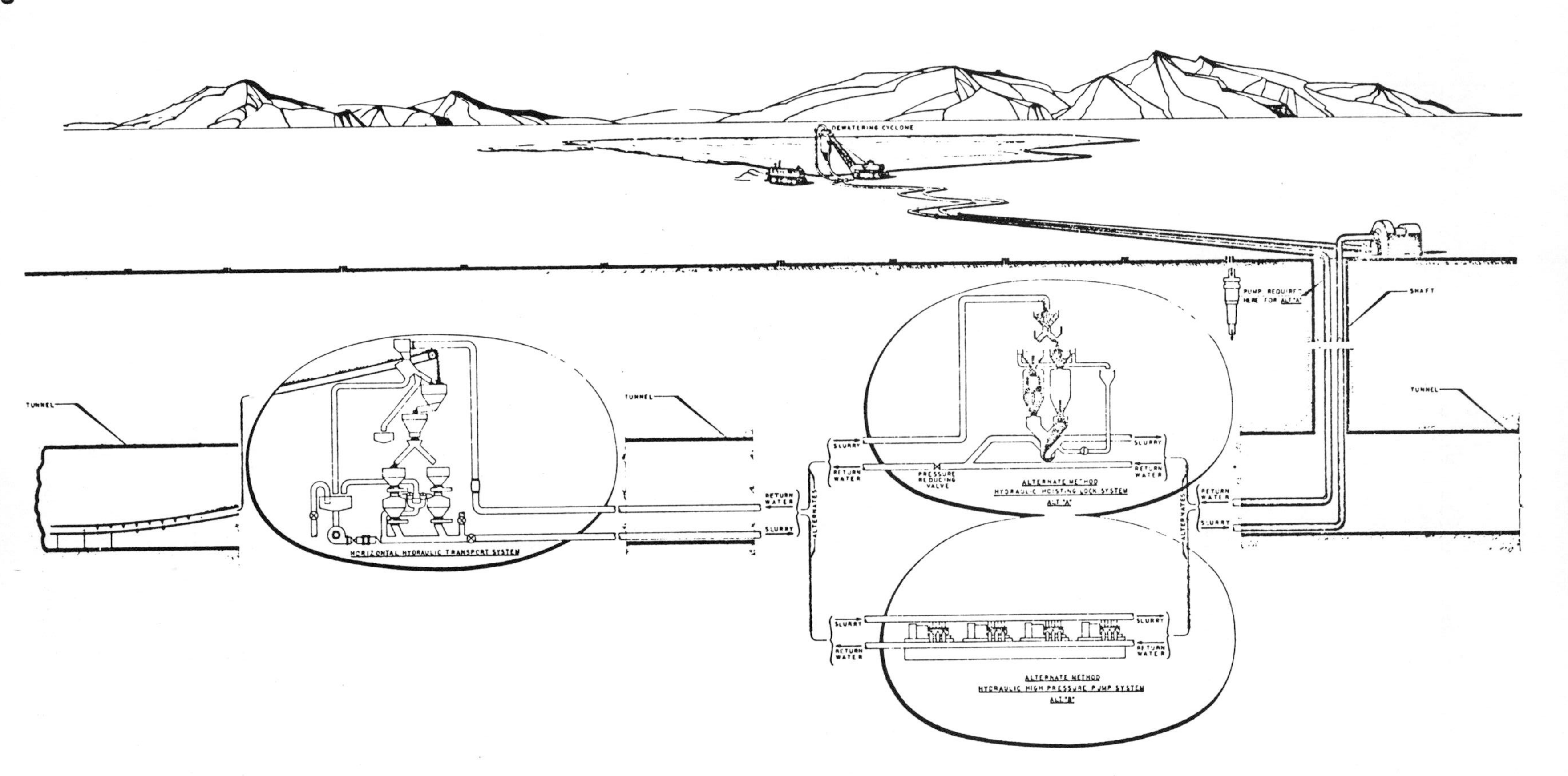

Fig. 11-8. Schematic layout of a hypothetical hydraulic pipe transport system.

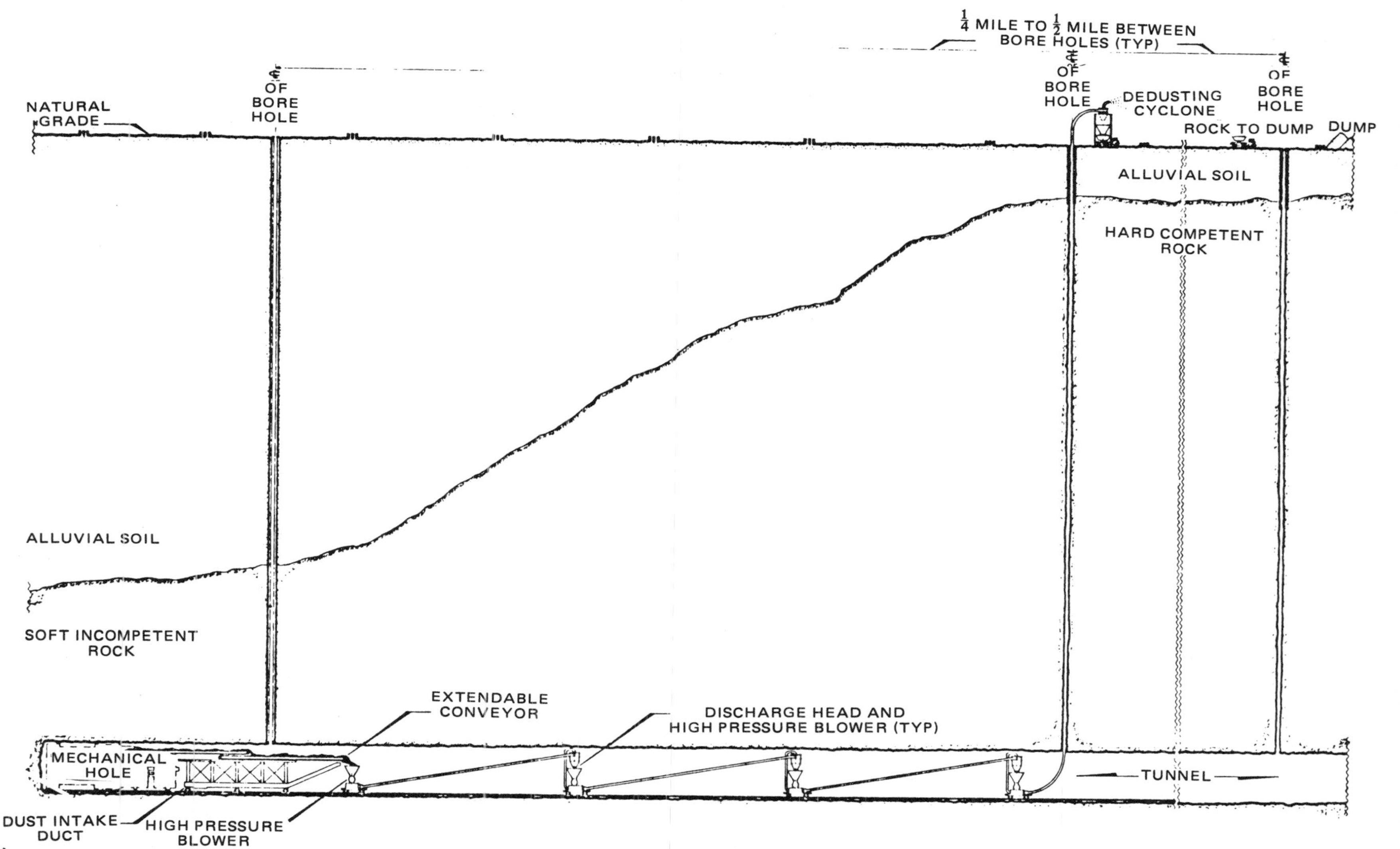

Fig. 11-9. Schematic layout of a hypothetical pneumatic pipe transportation system.

A high noise level and dust present serious problems. A high degree of wear can be expected, being much greater at bends. The wear problem can be reduced by rotating the pipe from time to time and by designing the bends for frequent replacement of wearing surfaces.

Figure 11-9 is a schematic layout of a pneumatic system.

SLUSHERS

A slusher installation moves broken rock or material from a point of origin, such as a muck pile, to a dumping point by means of a scraper pulled by a multiple drum hoist. A pull rope and one or two tail ropes are attached to the scraper, providing forward and backward movement.

Slushers are commonly used for short haulage of muck. They are compact and lightweight and can be conveniently moved through tight quarters.

System Layout and Components

Normally, a slusher is used for special applications where the working area is restricted or congested, where the face is on a steep slope or where the prime objective is to move the material only a limited distance.

Digging, pulling and returning are the three power requirements of a cycle. During the digging period, there is a peak which may considerably exceed the rated output of the motor. However, it is normally of short duration and within the overload capacity of the motor. Following this peak comes the pulling power which should not demand more than the rated capacity of the motor. After the scraper is emptied, the return pull requires one-third to one-half of the pulling power.

Following are brief descriptions of the major components.

Slusher Hoists. Compressed air hoist sizes are available up to 35 hp and electric hoists up to 150 hp in either double or triple drum styles. The movement of the scraper is controlled by the operator or it may be automated. The range is limited only by drum capacity, although normally 100 to 150 feet are practical limits. Slushing over longer distances is possible and two scrapers may also be used in tandem. Rope speeds normally do not exceed 450 fpm.

Scrapers. Scrapers are available in many sizes and types. They are usually made of an abrasion-resistant cast steel alloy, and are built for rugged service. Wearing parts, such as teeth and blades, are replaceable.

Sheaves and Anchor Pins. A tail sheave must be mounted somewhat behind and above the material to be moved. Use of an anchor pin is one convenient means of anchoring the tail sheave. Diameter of sheaves depends on rope diameter, a ratio of 15 : 1 being acceptable.

System Design

The capacity of a slusher system depends on the volume of material transported during one cycle and the time required for the cycle. The former is influenced by the dimensions and shape of the scraper as well as the condition of the path of travel. Maximum capacity for a given scraper is attained in a trough wherein side-spill is prevented. Capacity is also increased when scraping down a slope and decreased when scraping up a slope.

Production cycles should allow for the fixed time of loading and dumping as well as any other lost time. Table 11-16 gives production rates for various slusher combinations, assuming a 50-minute hour and fixed time of 6 seconds/cycle. Table 11-17 lists a number of typical air and electric slusher models.

Wire Rope Selection

The wire rope for a slusher application is subjected to very severe treatment. The effect of abrasion can be minimized by selecting wire rope with large outside wires, such as the 6 × 19 Seale. Uneven and cross-winding on the drum, along with high pressures, tends to crush and distort the rope. This condition can be alleviated by using a 3 × 19 Seale class rope, as it has no core (although flexibility will be sacrificed).

Due to high impact loads, the pull rope for a slusher installation should have a safety factor of seven. Though the tail-rope diameter can be smaller, it is normal to size both ropes the same.

Table 11-16. Mucking rates for slushers.

SCRAPER SIZE (INCHES)	LOOSE CUBIC FOOT CAPACITY	SCRAPER WEIGHT (LB)	MUCK WEIGHT (LB)	REQUIRED LINE PULL (LB)	PRODUCTION–LOOSE CUBIC YARDS/HOUR (50-MINUTE HOUR, 6 SECOND FIXED TIME) HAUL DISTANCE/FEET					MAXIMUM REACH (FEET) AND CABLE SIZE (INCHES)
					50	100	150	200	500	
36	11	550	1100	1650						
	Air–130	fpm–2000-lb	pull		23	12	8	6		300 $\frac{3}{8}$
	Air–215	fpm–2000-lb	pull		36	20	14	10	4	500 $\frac{3}{8}$
	25 hp–400	fpm–2050-lb	pull		58	34	24	18	8	750 $\frac{3}{8}$
42	14	780	1400	2180						
	Air–170	fpm–2500-lb	pull		37	20	14	11		230 $\frac{7}{16}$
	Air–250	fpm–2620-lb	pull		52	29	20	15		250 $\frac{7}{16}$
	25 hp–350	fpm–2350-lb	pull		67	39	27	21	9	950 $\frac{7}{16}$
48	19	1290	1900	3190						
	Air–250	fpm–3,460-lb	pull		70	39	27	21		220 $\frac{1}{2}$
	25 hp–250	fpm–3,300-lb	pull		70	39	27	21	9	950 $\frac{1}{2}$
	35 hp–350	fpm–3,300-lb	pull		91	53	37	28	12	950 $\frac{1}{2}$
54	27	2100	2700	4800						
	Air–200	fpm–4950-lb	pull		83	45	31	24	10	635 $\frac{5}{8}$
	30 hp–200	fpm–4950-lb	pull		83	45	31	24	10	635 $\frac{5}{8}$
	50 hp–300	fpm–5500-lb	pull		116	65	45	35	15	635 $\frac{5}{8}$
60	39	3280	3900	7180						
	75 hp–320	fpm–7810-lb	pull		175	100	71	53		375 $\frac{3}{4}$
	100 hp–300	fpm–11,000-lb	pull		168	94	66	51		260 $\frac{7}{8}$
66	55	4830	5500	10,330						
	100 hp–300	fpm–11,000-lb	pull		235	133	93	72		260 $\frac{7}{8}$
	125 hp–300	fpm–13,750-lb	pull		241	134	94	72		202 1
72	75	6800	7500	14,300						
	100 hp–200	fpm–15,000-lb	pull		232	127	87	66		202 1
	125 hp–250	fpm–15,000-lb	pull		279	155	107	82		202 1

Factors for operating on a slope: Level = 1.0
$-10^\circ = 1.2$ $+10^\circ = 0.8$
$-20^\circ = 1.5$ $+20^\circ = 0.7$
$-30^\circ = 1.9$ $+30^\circ = 0.5$
$-40^\circ = 3.4$ $+40^\circ = 0.4$
Allowance should be made for partially loaded trips and other adverse conditions.

MUCK REMOVAL FOR TUNNEL-BORING MACHINES

Capacity

The muck removal system for a tunnel-boring machine must have the capability of removing the excavated material from the machine at its maximum penetration rate, and the capacity to remove muck from the tunnel at such a rate as to minimize delays to the heading progress.

All tunnel-boring machines operate on a cyclical basis wherein the penetration phase is interrupted by repositioning of grippers or thrust jacks, by support installation, and by other requirements. Unless the muck removal system can handle the maximum production

Table 11-17. Medium capacity heavy duty slushers. *(Courtesy Joy Manufacturing)*

AIR SLUSHERS

NUMBER OF DRUMS	MOTOR (HP)	DRUM HALF FULL 90 LB AIR PRESSURE ROPE PULL (LB)	DRUM HALF FULL 90 LB AIR PRESSURE ROPE PULL (FMP)
2 or 3	20	3200	210
	21	3600	195
	23	4300	180

ELECTRIC SLUSHERS

NUMBER OF DRUMS	MOTOR (HP)	DRUM HALF FULL 60 CYCLES, AC ROPE PULL (LB)	DRUM HALF FULL 60 CYCLES, AC ROPE SPEED (FPM)	DRUM HALF FULL 25 AND 50 CYCLES, AC ROPE PULL (LB)	DRUM HALF FULL 25 AND 50 CYCLES, AC ROPE SPEED (FMP)	DRUM CAPACITY AND SPEED RELATIONSHIP	
2 or 3	15	2475	200	3000	165	Drum Capacity each Drum 640 feet of $\frac{3}{8}$ inch 450 feet of $\frac{7}{16}$ inch 360 feet of $\frac{1}{2}$ inch	Standard tail rope speed is 20% faster than haul drum speed. If so specified, tail drum speed can be furnished the same as haul drum rope speed.
	15	1980	250	2360	210		
	15	1650	300	1980	250		
	20	3300	200	4000	165		
	20	2640	250	3140	210		
	20	2200	300	2640	250		
	25	4125	200				
	25	3300	250	3930	210		
	25	2750	300	3000	250		
	30	4950	200			Drum Capacity, each Drum 900 feet of $\frac{3}{8}$ inch 650 feet of $\frac{7}{16}$ inch 525 feet of $\frac{1}{2}$ inch 350 feet of $\frac{5}{8}$ inch 225 feet of $\frac{3}{4}$ inch	Standard tail rope speed is 31.6% faster than haul drum speed. If so specified, tail drum rope speed can be furnished the same as haul drum rope speed.
	30	3950	250	4700	210		
	30	3300	300	3960	250		
	40	4400	300	5280	250		
	40	3770	350	4550	290		
	40	6600	200	8000	165	Drum Capacity, each Drum 1070 feet of $\frac{1}{2}$ inch 685 feet of $\frac{5}{8}$ inch 475 feet of $\frac{3}{4}$ inch 410 feet of $\frac{7}{8}$ inch 270 feet of 1 inch 215 feet of $1\frac{1}{8}$ inch	Standard tail rope speed is 25% faster than haul drum speed. If so specified, rail drum rope speed can be furnished the same as haul drum rope speed.
	40	5275	250	6280	210		
	50	8250	200	10,000	165		
	50	6600	250	7850	210		
	50	5500	300	6600	250		
	50	4720	350	5700	290		
	60	9900	200	12,000	165		
	60	7925	250	9420	210		
	60	6600	300	7925	250		
	75	12,375	200				
	75	9900	250	11,780	210		
	75	8250	300	9900	250		
	75	7060	350	8525	290		

of a complete cycle, there will be times when tunnel advance will be delayed by this lack of capacity.

Rail haulage has been mainly used in long machine excavated tunnels, although belt conveyors have found application in a few special cases.

There are three basic systems for loading a train of cars behind the machine. In every case, trains designed to hold the muck from one or more complete cycles are loaded by a separate belt conveyor which transports the muck from the boring machine. Train or locomotive switching takes place during the non-productive portion of the cycle, or in large tunnels, can be handled without hampering production. The following systems are common.

Gantry Conveyor with Single Track. Used in tunnels less than about 17 feet in diameter. The empty train is spotted under the conveyor and is pulled away by either a car mover or a locomotive as the cars are loaded. Mainline rail is installed between the boring machine and the end of the string of empty muck cars. The conveyor frame skids on the tunnel ribs or rides on the mainline track. A California Switch is needed behind the conveyor and can be pulled by the boring machine or moved periodically to keep it within 500 feet of the conveyor. Trains are switched during the non-productive part of the cycle.

Deadend California Switch with Transverse Discharge Conveyor. Used in tunnels greater than 17 feet in diameter and incorporates a floor that rides on mainline track. Two parallel tracks are mounted on the floor with a switch at the portal end. A conveyor is mounted on a frame above the muck cars and centered between the two tracks. A traveling transverse discharge conveyor diverts the muck from the main conveyor to the empty muck cars. While one train of muck cars is being loaded on one track, a locomotive is spotting an empty train on the other.

Deadend California Switch with Skewed Conveyor and Car Passer. Used in tunnels greater than 17 feet in diameter. Floor, parallel tracks and California Switch are the same as before; however, empty muck cars are placed on only one track while the conveyor loads cars on the other track. Empty cars are switched from one track to the other by means of a lateral car passer. The empty cars are fed to the car passer by means of an endless chain running between the rails. The loaded cars are moved past the conveyor discharge with a similar chain. The conveyor runs from the boring machine to the discharge point and is on a skew to the tunnel centerline. This eliminates the need for a transverse conveyor or a side discharge.

The above systems are pulled along behind the boring machine and are moved either on the regrip cycle or during the boring stroke. If the pull required for this movement affects the action of the machine, it can be supplied by separate means. Mainline track laying must take place between the boring machine and the trailing floor, or, as in the case of the Gantry Conveyor, track must be laid ahead of the string of empty muck cars.

To negotiate curves, the trailing equipment must be articulated as necessary.

Selection of System

Any system should consider the character of the muck and the possibility that it may change. Means to transport support elements, track supplies, repair parts and personnel should be incorporated into the overall system.

HOISTING

Capabilities

A hoisting system must be capable of hoisting all of the muck which the tunnel produces and still have available time for handling supplies, equipment and personnel. The shaft should be large enough to implement this capability as well as provide space for utilities, emergency ladderways, and ventilation ducts.

The capacity to handle tunnel muck is a function of load, hoisting speed and shaft depth. Since high speeds require greater acceleration and deceleration phases, low speeds are more appropriate for shafts of shallow depth. The shaft compartment size will frequently be determined by the size of equipment to be lowered.

System Design

The tunnel engineer is concerned with the development of the hoisting layout and selection of equipment, either new or used. It is easier to evaluate new equipment than old, because of assistance obtained from the manufacturer.

All applicable safety regulations should be observed. Codes are quite stringent for hoisting operations. For final system design, manufacturers should be consulted.

Hoist Types and Capacities

There are two main types of hoists: drum and friction. With the first, hoisting is done by wrapping the wire rope on one or more drums; with the second, hoisting is accomplished by friction between the wire rope and a powered friction wheel.

A drum hoist can operate as an unbalanced, balanced or counterweighted system. The unbalanced system is primarily used for shaft sinking operations, very shallow shafts, or small jobs. It is not as economical as a balanced system, but the layout is simple and only one compartment is required.

There are several drum configurations, but the most common is the single drum. It can operate as an unbalanced system or can be counterweighted, the counterweight being equal to the empty weight of the skip and cage, plus one-half the payload. This greatly reduces the unbalanced line pull and the required horsepower.

The friction or Koepe hoist has no drum and develops rope movement through friction between a friction wheel and the wire rope. The rope passes from the conveyance in one shaft compartment over the friction wheel and to the conveyance or counterweight in a second compartment. These hoists are commonly equipped with a tail rope to maintain the proper tension ratios of the hoist ropes. One end of the tail rope is fastened to the bottom of one conveyance, passes to the bottom of the shaft and up the other compartment to the bottom of the other conveyance. Slippage between the rope and the friction wheel is prevented if the ratio of rope tension on the tight side to that on the slack side is not too high.

Friction hoists require a synchronizer which can be adjusted manually or automatically to correct the indicating device for creep and slip on the friction wheel. It is common practice to mount friction hoists in the headframe directly above the shaft.

Manufacturers should be consulted for details concerning motors, controls, brakes, emergency brakes and clutches.

For a given productive capacity, the required size of the payload varies inversely with the hoisting speed. Larger payloads require heavier cable, and in deep shafts the weight of the cable becomes an important factor in the hoisting requirement. Consequently, high hoisting speeds are necessary in order to minimize the size of cable. On the other hand, with high hoisting speeds, additional power is required for acceleration.

Generally, speeds from 500 to 1000 fpm are common for shafts of moderate depth, up to 1000 feet. For deep shafts, speeds as high as 5000 fpm have been used. Accelerations of 2 to 4 feet/second/second are common, with extremes of 10 to 12 feet/second/second being recorded.

In selecting a hoisting system, ample allowance must be made for non-productive time. When the muck hoist must also be used for handling men and materials, productive time is not likely to be more than 75% of total time and it may be as low as 50%.

Hoist Ropes

Many variations of hoist ropes can be found in manufacturers' literature and each has its own advantages and disadvantages, depending on the type of application. The final selection of a hoist rope will be based mainly on the strength and service requirements. Normally, a 6 × 19, regular lay, improved plow steel rope with independent wire rope center is used for hoisting purposes.

The safety factor of the hoist rope should vary with the depth of the shaft. Table 11-18

Table 11-18. Static factors of safety for hoisting ropes for various depths of shafts when men are hoisted.

LENGTH OF ROPE (FEET)	MINIMUM SAFETY FACTOR FOR NEW ROPE	MINIMUM SAFETY FACTOR WHEN ROPE MUST BE DISCARDED	PERCENTAGE REDUCTION
500 or less	8	6.4	20
500 to 1000	7	5.8	17
1000 to 2000	6	5.0	16.5
2000 to 3000	5	4.3	14
3000 and over	4	3.6	10

lists the factors of safety recommended by the U.S. Bureau of Mines, depending on the length of rope. The minimum factor of safety decreases with rope length because the elasticity of a long rope in part compensates for the stresses due to loading and starting.

Table 11-19 lists three types of 6 X 19 wire rope, $\frac{3}{4}$-inch and larger (IWRC–Independent Wire Rope Center).

Headframes

Hoisting in shafts less than 100 feet deep will frequently be done with a crane and no headframe is necessary. However, safety regulations in different locations may require a headframe for shallower depths. A headframe straddles the shaft and supports the muck bins. It also must provide working clearances at the shaft

Table 11-19. Breaking strength, tons of 2000 lb.

	APPROXIMATE WEIGHT (LB/FOOT)		EXTRA IMPROVED P/S	IMPROVED P/S	
ROPE DIAMETER	FIBER CORE	IWRC	IWRC	FIBER CORE	IWRC
$\frac{3}{4}$	0.95	1.04	29.4	23.8	25.6
$\frac{7}{8}$	1.29	1.42	39.8	32.2	34.6
1	1.68	1.85	51.7	41.8	44.9
$1\frac{1}{8}$	2.13	2.34	65.0	52.5	55.5
$1\frac{1}{4}$	2.63	2.89	79.9	64.6	69.4
$1\frac{3}{8}$	3.18	3.50	96.0	77.7	83.5
$1\frac{1}{2}$	3.78	4.16	114.0	92.0	98.9
$1\frac{5}{8}$	4.44	4.88	132.0	107.0	115.0
$1\frac{3}{4}$	5.15	5.67	153.0	124.0	133.0
$1\frac{7}{8}$	5.91	6.50	174.0	141.0	152.0
2	6.72	7.39	198.0	160.0	172.0
$2\frac{1}{8}$	7.59	8.35	221.0	179.0	192.0
$2\frac{1}{4}$	8.51	9.36	247.0	200.0	215.0
$2\frac{3}{8}$	9.48	10.4	274.0	220.0	239.0
$2\frac{1}{2}$	10.5	11.6	302.0	244.0	262.0
$2\frac{5}{8}$	11.6	12.8	331.0	268.0	288.0
$2\frac{3}{4}$	12.7	14.0	361.0	292.0	314.0

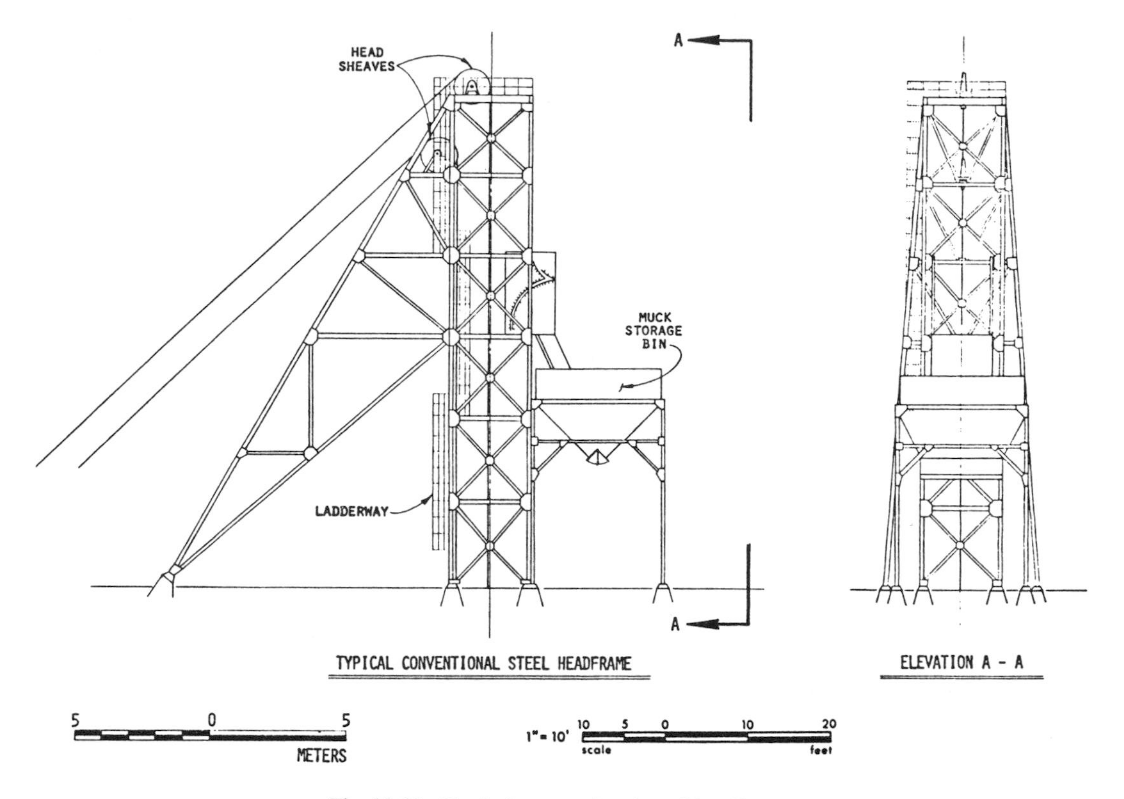

Fig. 11-10. Typical conventional steel headframe.

collar and must be compatible with the skip and cage design. Headframes are fabricated from steel, reinforced concrete or timber. For tunnel work, steel is generally used.

The conventional type of headframe has backlegs and the hoist is mounted on the ground as illustrated in Fig. 11-10. With a drum hoist, the distance from head sheave to drum must be great enough to limit the fleet angle to a maximum of 1 degree, 30 seconds. The resultant of the rope forces on both sides of the head sheave should fall between the backpost and the headframe. Otherwise, the headframe would have to be anchored down.

Friction hoists are usually mounted in the headframe, eliminating need for backlegs (see Fig. 11-11).

The height of the headframe is determined by clearance requirements for equipment entering the cage, height of muck bins, overall height of skip and cage combinations, height required for dumping and allowance for overwinding. For 6 X 19 rope, the head sheave (and any other sheaves) should have a diameter of at least 48 times the rope diameter.

The headframe must be designed for stresses due to dead load, wind load and live load. In some areas, seismic forces should also be considered and allowance must be made for shock and vibration. Rope forces should be assumed at one-half the rope breaking strength.

Cages

Cages are used to convey personnel, material and equipment. Loaded muck cars can also be hoisted in a cage. The cage can be attached to or be part of the skip, or it can act as the counterweight in a separate compartment.

Cages and skips are usually of welded steel construction, although aluminum has been used. Replaceable wearing parts, such as guide shoes, are usually attached with bolts.

For shallow shafts, some contractors prefer to use a light-duty, self-service, passenger elevator for hoisting workmen. This relieves the muck hoist from that duty.

Safety catches are required on cages used for hoisting or lowering men. Details can be obtained from manufacturers.

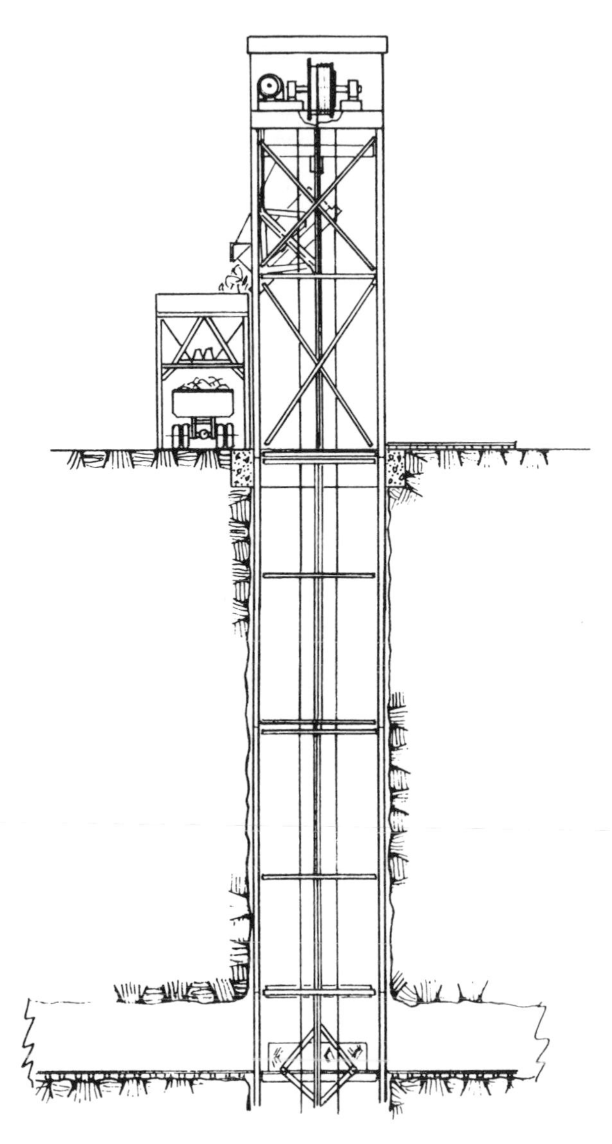

Fig. 11-11. Tower mounted friction hoist.

Hoisting Muck Cars

In addition to being hoisted in a conventional cage, muck cars can be hoisted in a special guide cage which provides for automatically dumping the car.

The guide cage consists of two frames. The outer frame runs in the shaft guides and is attached to the inner frame by a pivot at the bottom. A safety latch prevents the inner frame from accidentally pivoting at locations other than the dumping point. A guide wheel is attached to the inner frame so that the muck car will be emptied. The amount of rotation required of the inner frame will depend on the configuration of the muck car and the ability of the muck to flow. The cage can also be used to transport men if built with the proper safety features, such as cage enclosures, saftey catches and hinged bonnet.

It is also possible to hoist only the muck car body in a guide cage (see Fig. 11-12). In this case, the frames are designed to be lowered over the muck car to engage the body.

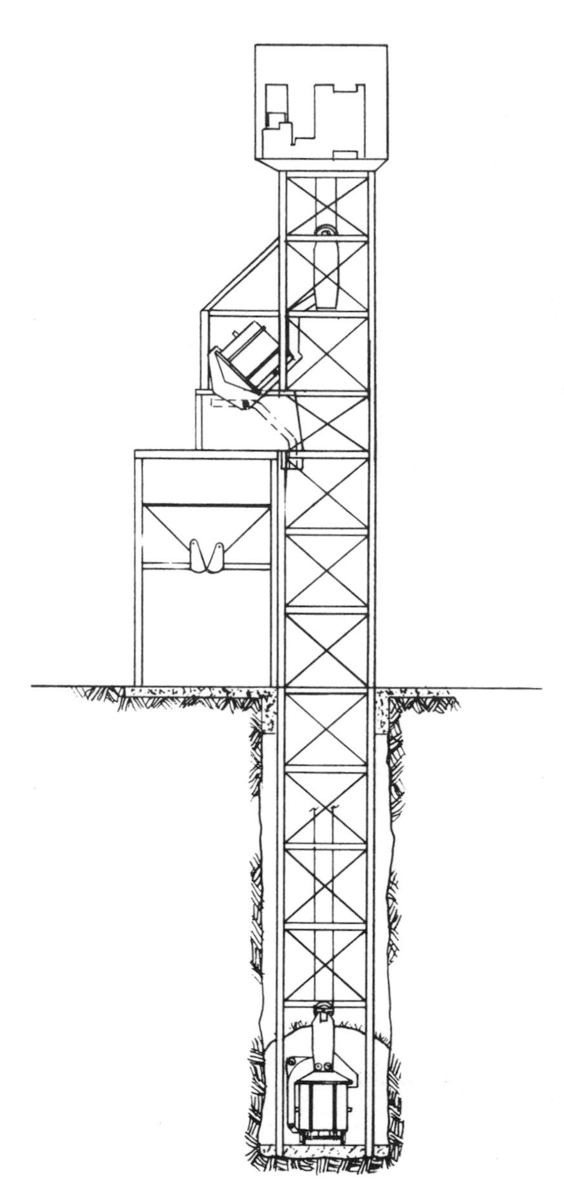

Fig. 11-12. Muck car body in guide cage.

Skips

Compared to cage hoisting, hoisting in skips offers the advantages of less time loading and dumping, large capacity in shafts of small cross-section, less labor for dumping at the top and less dead weight. Empty weight is generally 40 to 60% of the payload weight. Skips are generally one of two basic designs.

Bottom Dump Skip. When a long slender skip is required, the bottom dump type is preferable. Instead of rotating the entire skip for dumping, a guide roller engaging the dump scrolls opens a pivoting door on the side of the skip. A safety latch prevents the door from opening prematurely.

Kimberly-type Skip. In this type, the skip body is pivoted on the bottom of the guide frame. It is dumped by means of guide rollers engaging the dump scrolls. The body should be relatively short. Sloping sides and a round bottom will facilitate dumping.

Shaft Works

Shaft cross-sections may be either rectangular or circular. Rectangular shafts are timbered with wood or steel members or lined with concrete. Circular shafts are either concrete lined or lined with steel ribs and lagging (see Fig. 11-13). A circular configuration offers an advantage from the structural standpoint, but the percentage of usable area is lower than that for rectuangular shafts.

Shaft guides are needed for the skip, cage and counter-weight. Guides are made of either wood, structural steel or wire rope. The first two are fastened to either the shaft lining, to the shaft supports or to buntons. Steel guides are popular because they wear less than wood, maintain alignment better, are non-combustible and are not affected by moisture. Wire rope guides give a very smooth travel and are simple to install and maintain. Locked-coil rope is

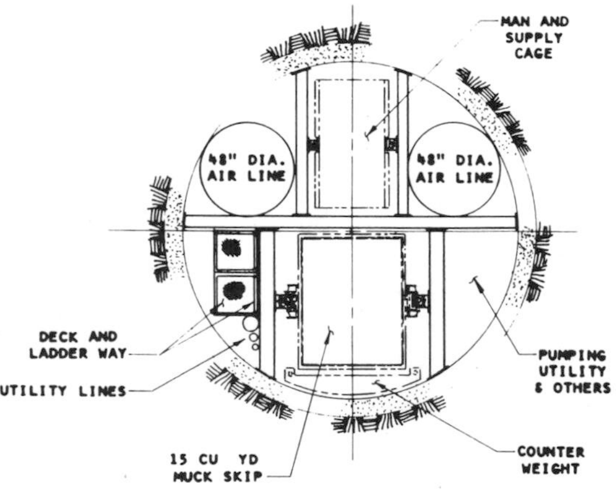

Fig. 11-13. Typical circular concrete lined shaft.

used and tensioned by calibrated railroad car springs at the rate of 7 to 12 lb/foot of length. To prevent skip sway due to the ropes having the same frequency of vibration, no two ropes should have exactly the same tension.

All shafts should be provided with ladderways to be used in case of hoist failure and as a means of inspection and maintenance. In addition, provision must be made for compressed air, water supply, pump discharge, ventilation and electrical lines.

The configuration at the shaft-tunnel intersection will depend mainly on the muck hoisting method. Shaft sumps are generally required for the collection and pumping of water. In the case of skip hoisting, the shaft depth below the tunnel will depend on the dimensions of the skip loading system.

Material Transfer

Several typical shaft bottom transfer arrangements are described below.

Measuring Pocket Fed by a Storage Bin (Fig. 11-14). The amount of muck in the measuring pocket is controlled on either a weight or volume basis. When the skip is in the proper location, an air cylinder moves the retractable chute into position and opens the gate of the measuring pocket. The proper amount of muck is discharged into the skip and is hoisted to the surface. While the skip is gone, the measuring pocket is automatically refilled. This system lends itself quite well to automation.

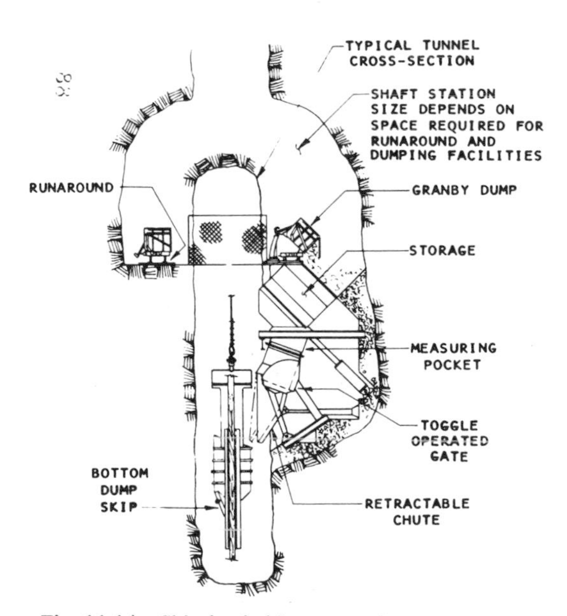

Fig. 11-14. Skip loaded by measuring pocket and retrackable chute.

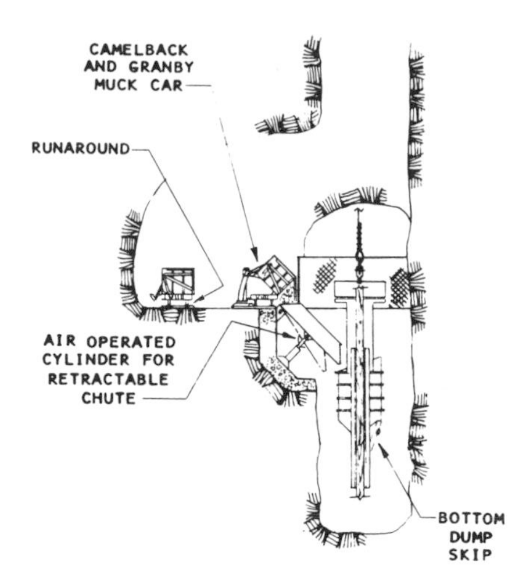

Fig. 11-15. Muck car dumping directly into skip.

Muck Cars Dumping Directly into Skips (Fig. 11-15). The muck car and skip size are equal so that the skip is filled without spilling. This system can also be automated, but it ties up a train of cars for a longer period of time.

Intermediate Conveyors. When the shaft is offset from the tunnel, a belt conveyor can be used to transfer the muck from the car dump to the skip-loading area. Other features can be similar to those described above.

Typical Calculations

The design of a high-capacity hoisting system is rather complicated and should not be attempted by the inexperienced. However, for estimating purposes and for developing a conceptual system, the following formulae will be helpful to the tunnel engineer.

Rope Stress. S (estimated) = W $(2.0 + 0.1a)$, where S = estimated rope stress, W = load (skip and/or cage plus payload, lb) and a = acceleration (feet/second/second). For example, assume live load = 7 tons, skip weight = 3 tons and cage weight = 1.5 tons. W = 11.5 tons or 23,000 lb. Assume acceleration = 4 feet/second/second. S (estimated) = 23,000 (2.0 + 0.4) = 55,200. For a 1000-foot deep shaft, the safety factor should be 7, requiring a rope with a breaking strength of 386,400 lb or 194 tons. A 2-inch rope of Extra Improved Plow Steel with IWRC is indicated (see Table 11-22).

Total rope stresses can be expressed as

$$P = S_r + \left[W + w1 + (a)\left(\frac{W + w1}{g}\right)\right]$$

where

S_r = bending stresses (lb).

$S_r = (E_r)\ (dA \div D)$, where E_r can be taken as 12,000,000; d = diameter largest wire, A = total area of wires and D = diameter of bend over sheave to centerline of rope in inches. For 6 X 19 rope, $d = d_r$ (rope diameter) ÷ 15.52. For a 2-inch rope, d = 0.1289, A = 1.4900 and dA = 0.192. D should be at least 96 inches and S_r = 24,000 lb; $w1$ = 7.39 X 1000 = 7390 lb = weight of rope. Thus,

$$P = 24{,}000 + \left[23{,}000 + 7{,}390 + \frac{4 \times 30{,}390}{32.2}\right]$$

$$= 58{,}165 \text{ lb.}$$

For the factor of safety of 7, the rope should have a breaking strength of 407,000 lb or 204 tons, indicating that the assumed 2-inch rope would be unsatisfactory. However, if the sheave diameter were increased to 120 inches, S_r would be reduced to 19,200 lb and P to 53,365 lb, and the 2-inch rope would be acceptable.

Horsepower. Solving for the exact motor horsepower is a complicated process. When the acceleration period is short, the horsepower in excess of that required for full load speed can probably be handled by the overload capacity of the motor. The approximate required horsepower can be calculated from the formula below.

$$\text{hp} = \frac{(L)(V)}{(33{,}000)(\text{eff})}$$

where

L = unbalanced load (lb)
V = velocity (feet/minute)
eff = hoist efficiency (say 0.85).

In the previous example, L = load (W or 23,000 lb) plus weight of rope ($w1$ or 7,390 lb) minus weight of counterweight (16,000 lb) = 14,390 lb. If V = 800 fpm, hp would be 410.

UTILITIES

Power Supply

An electric system for a tunnel project will have some or all of these major components:

- Power sources
  - Utility company
  - Special generating plant
- Main substations
  - Transformers
  - Switch gear (including circuit breaker)
- Main feeder lines
- Secondary substations
  - Transformers
  - Switch gear (including circuit breakers)
- Secondary feeder lines
- Junction boxes
- Service lines
  - Light lines
  - Temporary power cable
  - Building service
  - Trailing cables
- DC conversion units
  - Rectifier
  - Motor generator

The most economical source of electricity is usually the local power company. This will require a high voltage feeder from a trunk line to the job site. However, if the site is remote and the feeder line long, generators installed at the site may be more economical.

A standby source of power must be available if the prime source is unreliable or if the tunnel is subject to flooding, and for the low-air compressor plant of all tunnels driven under compressed air.

The main substation reduces the primary voltage to some usable secondary transmission voltage, usually 4160, or 2300 volts. Transformers for this are generally provided and installed by the local power company. The voltage of the secondary side of the main transformer must be compatible with equipment or meet the primary voltage requirements of the secondary transformers.

Transformers used on the surface may be of the wet or oil type, whereas those used underground must be of the dry or non-flammable type.

Power for tunnel requirements is normally carried through a plastic-coated, armored, multi-conductor cable (three conductors with one or more grounds).

Within the tunnel, common practice is to install the main power cable in 1000-foot lengths, with a junction box and lighting transformers at the end of each length. Transformers for fans and pumps and for other requirements are installed as needed.

Most power company rate schedules penalize a customer with a low power factor. The overall job power factor can be improved by using compressors with synchronous motors with unity or leading power factors. Capacitors can also be used to improve the power factor.

A qualified electrical engineer should design the distribution system; select voltages, wire sizes and transformers, and choose transformer locations, controls and protective devices. It is the responsibility of the electrical engineer to design the most economical and practical system that will be compatible with the primary power source and comply with all applicable codes and standards.

The tunnel engineer, while not needing to be familiar with all of the details of a power supply system, must have a working knowledge of the fundamentals of electrical engineering so that power demands and equipment requirements can be anticipated, job layouts prepared, and adequate information made available to the electrical engineer.

Power demand can be estimated by listing all

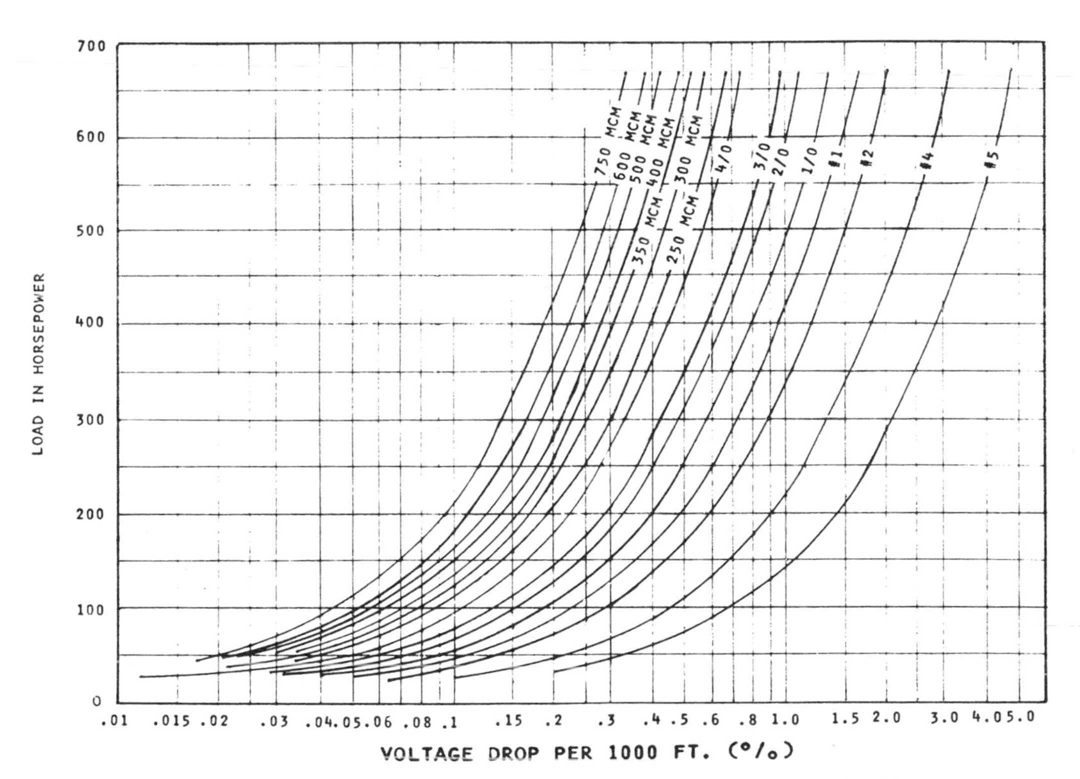

FOR 3 CONDUCTOR 2300-VOLT PARKWAY CABLE WITH VARNISHED-CAMBRIC INSULATION AT 85°C. VOLTAGE DROP IN PERCENT FOR 1000 FT. OF CABLE TO USE GRAPH, DIVIDE DESIRED VOLTAGE DROP (SUCH AS 5%) BY LENGTH OF LINE IN THOUSANDS OF FEET TO FIND VOLTAGE DROP PER 1000 FT. SELECT PROPER SIZE OF CABLE OPPOSITE TOTAL LOAD IN HORSEPOWER

Fig. 11-16. Parkway cable selection graph.

of the anticipated load centers on the job, with a tabulation of the horsepower and lighting requirements at each. The combination of items, each modified by its respective duty factor (if any), which is likely to be consuming the most power at the same time determines the maximum overall demand. Energy consumption can be estimated from this same tabulation, using estimated operating times and duty factors. Kilowatt-hours can be derived from hp-hours by multiplying by 0.746.

It is important to select the proper size of power cable for the tunnel. Ordinarily, the voltage drop should be limited to 5%. Figure 11-16 gives a graph for selecting 2300 v Parkway Cable for a total load up to 700 hp. For other voltages, suitable correction factors can be applied and similar graphs can be made for other types of cable or higher horsepower ratings. In long tunnels, individual reaches can be calculated separately.

Ventilation and Environmental Control

A tunnel may be ventilated either by blowing air in or exhausting it out through a duct. In blowing, fresh air from the surface is forced through the vent line to a point near the face. The exhaust air travels back through the full length of the tunnel to the portal or shaft. This method has the advantage of constantly supplying fresh air to the face where most of the work is done. It has the disadvantage of exposing the remainder of the tunnel to any contaminated air generated at the face.

In the exhaust method, the foul air is exhausted through the vent line, and the fresh air enters at the portal or shaft and travels through the tunnel to the face. While this method creates a better environment along the tunnel, any heat, moisture, dust and smoke generated along the tunnel is delivered to the vicinity of the face.

When the heading is advanced by means of drilling and blasting, the ventilation system may be operated on the exhaust configuration for 15 to 30 minutes after the blast, to remove much of the smoke and dust, and then changed to the blow configuration.

Auxiliary blowers are often used in any type of tunnel to properly distribute fresh air around the face.

Tunnel ventilation requirements are usually specified in the health and safety regulations applicable to the particular project. Usually, a minimum velocity of 50 feet/minute in the tunnel is required together with a minimum volume of 200 cubic feet/minute for each workman in the tunnel. When diesel equipment is used, 75 cubic feet/minute of fresh air for each brake horsepower is often required. Normally, when more than one unit is being operated, they will not all be developing maximum rated horsepower at the same time, and some allowance can be made for that. Some regulations are based on the measured content of contaminants in the tunnel atmosphere.

Properties of various gases and diesel exhaust constituents and their physiological effect on workmen are listed in Table 11-20.

The use of tunnel-boring machines at great depth can produce a real problem. The naturally high rock temperatures, augmented by the heat generated by the boring machine, can produce very high temperatures in the heading. And since there is usually moisture present in the tunnel, the resulting high humidity further aggravates the problem. Then, as this hot, humid, dust-laden air is exhausted toward the portal, it becomes cooler, causing the moisture to condense and precipitate the dust on the bottom of the vent pipe.

Current health and safety regulations require that dust levels in the tunnel atmosphere be kept low. Wet drilling and wetting down the muck pile after blasting is required. With tunnel-boring machines, the dust is confined to the face by a dust shield. From there it is conducted by vacuum to a dust-collector and the discharge air is then released into the exhaust fan line. Other dust-producing sources can be treated by water sprays.

The possibility of toxic or explosive gas in the tunnel should always be kept in mind. Where this is a likely possibility, special precautions should be taken when planning the work and equipping the job. Permissible electrical equipment may be necessary, along with additional ventilation capacity. In conventional drill and blast tunnels, periodic testing should be done, while with tunnel-boring machines, continuous

Table 11-20. Properties of various cases which may be present in tunnel.

GAS	DENSITY	COLOR	ODOR	SOURCE	PHYSIOLOGICAL EFFECT ON WORKMEN
Oxygen (O_2)	1.11	None	None	Air is normally 20.93% O_2.	At least 20% is required to sustain normal health. Workmen become dizzy if concentration drops to 15%. Some workmen may die at 12.5%; most will faint at a concentration of 9%; and death will occur at 6% or less.
Nitrogen (N_2)	0.97	Yellow	None	Air is normally 78.10% N_2.	Nitrogen has no ill effect on men except to dilute air and decrease O_2%.
Carbon dioxide (CO_2)	1.50	None	None	Air is normally 0.03% CO_2. CO_2 is produced by decaying timber and fires, and is present in diesel exhaust.	CO_2 acts as a respriatory stimulant and may increase effects of other harmful contaminants. At 5% CO_2, breathing is laborious. A concentration of 10% can be endured for only a few minutes.
Carbon monoxide (CO)	0.97	None	None	Present in diesel exhaust and blast fumes.	CO is absorbed into the blood rather than O_2. In time, very small concentrations will produce symptons of poisoning. A concentration slightly greater than 0.01% will cause a headache or possibly nausea. A concentration of 0.2% is fatal.
Methane (CH_4)	0.55	None	None	Present in certain rock formations containing carbonaceous materials.	Has no ill effect on men except to dilute air and decrease O_2%. It is dangerous because of its explosive properties. Methane is explosive in the concentration range of 5.5 to 14.8%, being most explosive at a concentration of 9.5%.
Hydrogen sulfide (H_2S)	1.19	None	Rotten Eggs	Present in certain rock formations and sometimes in blast fumes.	Extremely poisonous–0.06% will cause serious problems in a few minutes.
Sulfur dioxide (SO_2)	2.26	None	Burning Sulphur	Present in diesel exhaust and blast fumes.	Strongly irritating to mucous membranes at low concentrations. Can be kept below objectionable levels by limiting fuel sulphur content to 0.5 precent.
Oxides of nitrogen	Approximately 1.5	Yellow-brown	Stings Nose	Present in diesel exhaust and blast fumes.	NO_2 is most toxic. All oxides of nitrogen cause severe irritation of the respiratory tract at high concentrations. Acute effects may be followed by death in a few days to several weeks owing to permanent lung damage.

monitoring is necessary. Suitable instruments are available from safety equipment manufacturers. Attention should be given to the physical properties of the gases, since some tend to collect either in high or low spots in the tunnel.

The ventilation equipment can be located outside the tunnel, using either an axial flow fan, a centrifugal fan or a positive pressure blower. The latter two can be reversed by means of a piping and valving arrangement and the former with a reversing switch. Greater pressure differential can be obtained by placing two or more fans in series and greater capacity by placing them in parallel.

For very long tunnels, current practice is to install axial fans in the vent line as the tunnel progresses. With this system, the pressure extremes are reduced and capacity is added only when it is needed. In all installations, the vent line should be designed to resist external as well as internal pressure, the amount depending on the characteristics and the operating plan for the system.

Portable or collapsible vent tubing is suitable for short jobs, for temporary use and for auxiliary use, such as supplemental ventilation to the working face. It is not satisfactory for long tunnels where a heavy duty system is indicated.

Vent pipe is fabricated from sheet plastic, fiber, paper and steel. Of these, steel pipe is the strongest and most durable. For installations having high external pressures, spiral-weld steel pipe is most economical.

To reduce shipping costs, current practice for most jobs of any magnitude is to fabricate the pipe on the job from steel strip.

Couplings should be strong enough for the particular job and provision must be made for sealing the joints against leakage, a strip of soft rubber being one method employed.

System Design. The two most important variables affecting the design of a system of given capacity are the vent pipe diameter and the power required. The greater the pipe diameter, the smaller will be the power requirement. But larger diameter pipe costs much more than that of smaller diameter, being both larger and stronger, and in any tunnel there is a limit as to the size that can be accommodated. Usually several alternative combinations must be evaluated, considering both first cost and operating costs for the life of the job.

The required volume of air can be ascertained as previously noted. Lacking any specified criteria, a minimum velocity in the tunnel of 50 feet/minute and a minimum of 200 cfm per workman plus 75 cfm per diesel brake horsepower, whichever is greater, should be provided.

From Fig. 11-17, the total friction loss for various pipe diameters can be calculated. Fan manufacturers' catalogs can then be consulted for selecting a suitable fan for each likely pipe diameter. Most axial flow fans develop static pressures of 5 to 15 inches W.G. and, generally, a fan spacing of 1000 to 3000 feet is appropriate.

For example, consider a 20-foot diameter tunnel requiring 20,000 cfm of ventilation air. The chart shows a range of pipe diameters from 18 to 66 inches. However, a 3000-foot fan spacing with a 15-inch W.G. fan would indicate a friction loss of 0.5 inches W.G. per 100 feet, requiring a 31-inch diameter pipe. Similarly, a 1000-foot fan spacing would indicate a 24-inch diameter pipe. For a loss of 5 inches W.G., corresponding figures are 39-inch and 30-inch diameters. From this, it is apparent that vent pipe sizes in the 24-inch to 42-inch diameter range could be considered.

When making final calculations, the velocity head must be added to the total friction loss in order to find the total dynamic head required. The velocity head in inches W.G. is

$$H = \left(\frac{V}{4008}\right)^2,$$

where V = air velocity in the vent pipe in feet/minute.

Additional friction losses, filters, elbows, etc., if significant, can also be added.

Tables of friction loss and manufacturers' fan performance data are based on dry air at 70°F at sea level, having a density of 0.075 lb/foot^3. Since both the friction loss in the vent pipe and the pressure developed by the fan are proportional to density, the altitude of the installation can be ignored except for the determination of

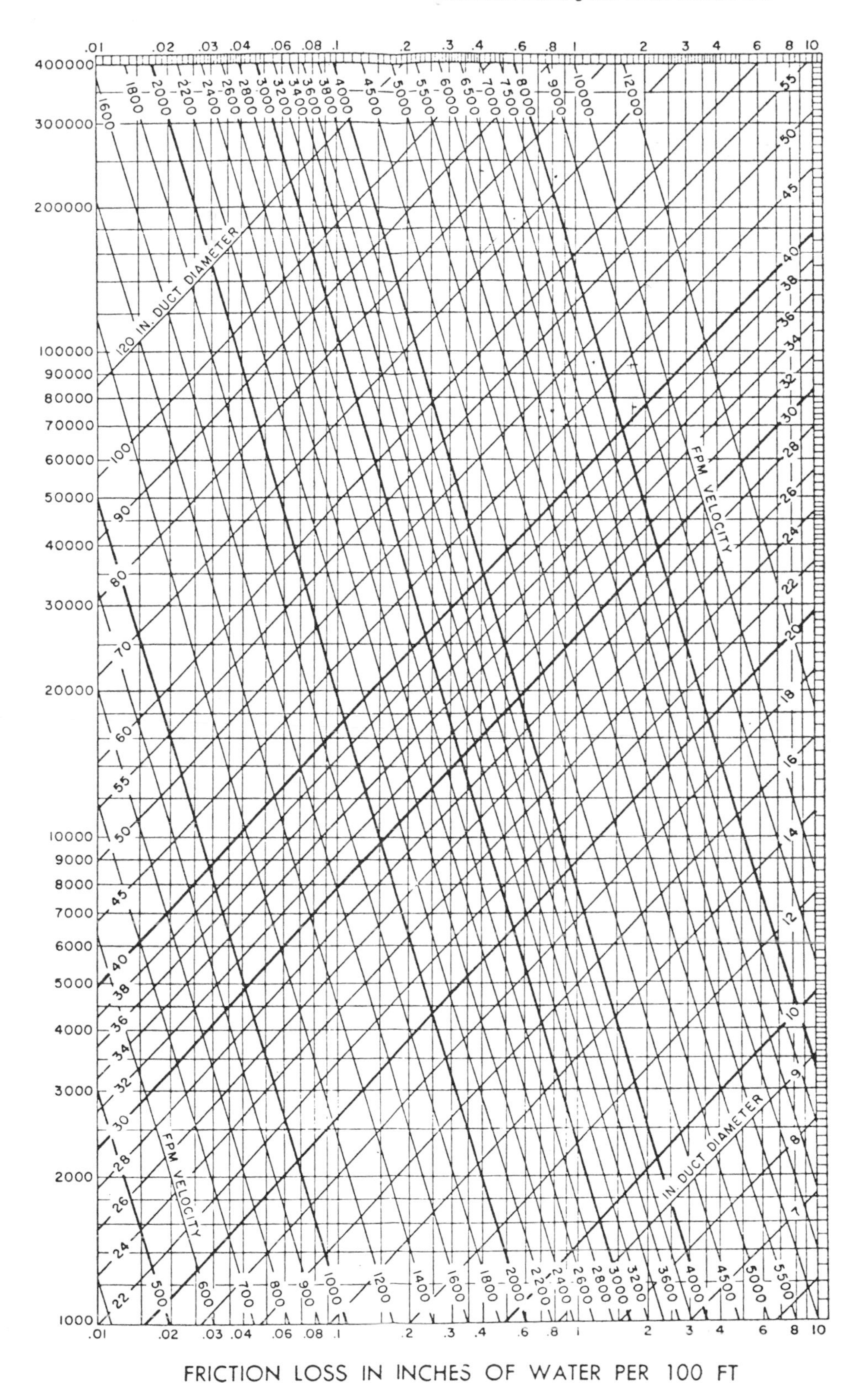

Fig. 11-17. Friction of air in straight ducts.

the required fan horsepower. The actual horsepower required is proportional to the density of the air and the manufacturers' specified horsepower can be modified:

Required hp = density ratio
X manufacturer's specified hp.

The density ratio for various altitudes and temperatures is given in Table 11-21.

Compressed Air

Most tunnel projects utilize tools and equipment powered by compressed air.

For projects of very short duration or in areas remote from electrical power sources, portable diesel-driven compressors may be indicated. These are available in various sizes up to 1200-cfm capacity and can be quickly mobilized and put into operation.

Where electric power is available, electric-driven compressors are most economical for all but jobs of very short duration.

Slow-speed, double-acting compressors, water-cooled and having separate intercoolers and aftercoolers, are the most efficient and offer the lowest ultimate maintenance cost. They are usually powered by synchronous motors providing a favorable power factor. They are more expensive, and require a more costly foundation and more extensive ancillary plant, but are the least expensive ultimately for projects of long duration. Figure 11-18 shows a typical, heavy-duty installation.

The compressed air system usually includes most of these items:

Power supply and driving motors
Two-stage compressors
Intercoolers
Aftercoolers
Cooling water supply
Collective manifolding
Air receiver (about one-tenth of volume of free air produced per minute)
Antifreeze injector
Pipeline with outlet valves
Reheaters
Supplementary receivers
Connection hoses.

At the tunnel heading, the moisture in the exhaust from the drills may cause fogging conditions. This can be alleviated by reheating the air just before it is used. In addition, this reheating increases the pressure and thus may increase the efficiency of the drills by as much as 10%.

Although a long pipeline becomes a receiver

Table 11-21. Density ratios (based on air being dry and weighing 0.075 lb/foot^3 at sea level at 70°F).

ALTITUDE IN FEET ABOVE SEA LEVEL	STANDARD ATMOSPHERIC PRESSURE		AIR DENSITY RATIOS AT VARIOUS ALTITUDES AND TEMPERATURES — AIR TEMPERATURE (°F)				
	psi	INCHES MERCURY	50	60	70	80	90
0	14.69	29.92	1.039	1.019	1.00	0.981	0.964
1000	14.16	28.86	1.003	0.984	0.965	0.947	0.913
2000	13.66	27.82	0.966	0.948	0.930	0.913	0.880
3000	13.17	26.82	0.931	0.913	0.896	0.879	0.847
4000	12.69	25.84	0.898	0.880	0.864	0.848	0.817
5000	12.23	24.90	0.865	0.848	0.832	0.817	0.787
6000	11.78	23.99	0.833	0.817	0.802	0.787	0.758
7000	11.34	23.10	0.802	0.787	0.772	0.758	0.730
8000	10.91	22.23	0.772	0.757	0.743	0.729	0.702
9000	10.50	21.39	0.743	0.729	0.715	0.702	0.676
10,000	10.10	20.58	0.715	0.701	0.688	0.675	0.650

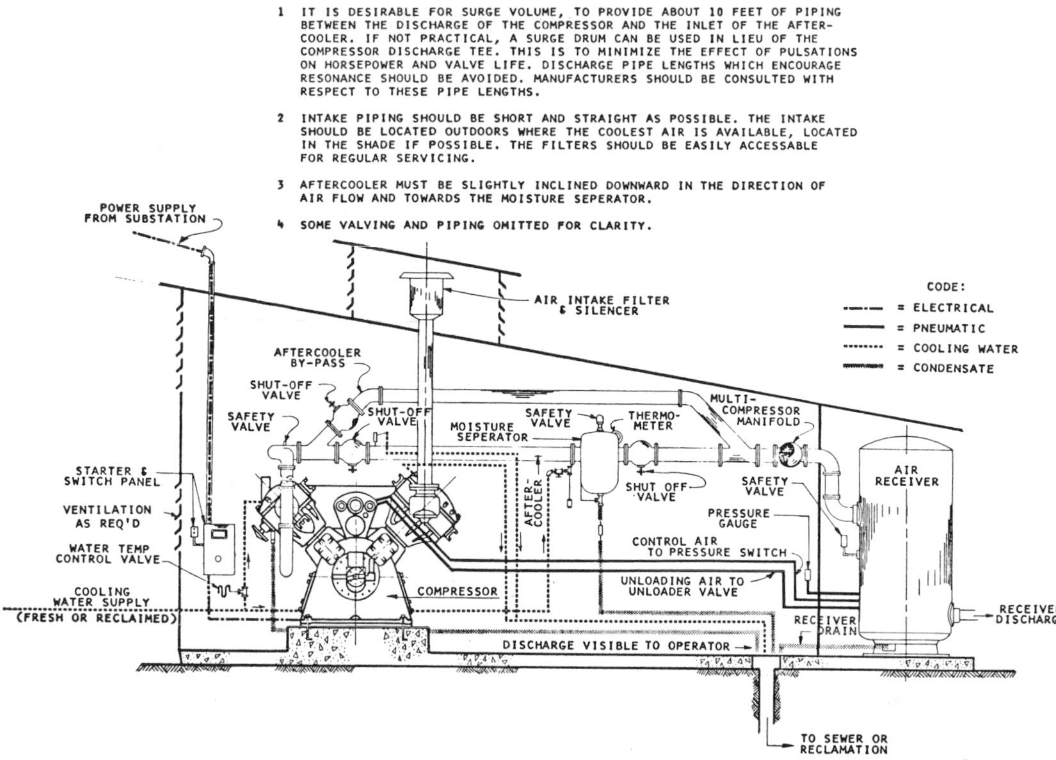
1 IT IS DESIRABLE FOR SURGE VOLUME, TO PROVIDE ABOUT 10 FEET OF PIPING BETWEEN THE DISCHARGE OF THE COMPRESSOR AND THE INLET OF THE AFTER-COOLER. IF NOT PRACTICAL, A SURGE DRUM CAN BE USED IN LIEU OF THE COMPRESSOR DISCHARGE TEE. THIS IS TO MINIMIZE THE EFFECT OF PULSATIONS ON HORSEPOWER AND VALVE LIFE. DISCHARGE PIPE LENGTHS WHICH ENCOURAGE RESONANCE SHOULD BE AVOIDED. MANUFACTURERS SHOULD BE CONSULTED WITH RESPECT TO THESE PIPE LENGTHS.
2 INTAKE PIPING SHOULD BE SHORT AND STRAIGHT AS POSSIBLE. THE INTAKE SHOULD BE LOCATED OUTDOORS WHERE THE COOLEST AIR IS AVAILABLE, LOCATED IN THE SHADE IF POSSIBLE. THE FILTERS SHOULD BE EASILY ACCESSABLE FOR REGULAR SERVICING.
3 AFTERCOOLER MUST BE SLIGHTLY INCLINED DOWNWARD IN THE DIRECTION OF AIR FLOW AND TOWARDS THE MOISTURE SEPERATOR.
4 SOME VALVING AND PIPING OMITTED FOR CLARITY.
POWER SUPPLY FROM SUBSTATION
AIR INTAKE FILTER & SILENCER
CODE:
= ELECTRICAL
= PNEUMATIC
= COOLING WATER
= CONDENSATE
AFTERCOOLER BY-PASS
SHUT-OFF VALVE
SAFETY VALVE
SHUT-OFF VALVE
MOISTURE SEPERATOR
SAFETY VALVE
THERMO-METER
MULTI-COMPRESSOR MANIFOLD
AIR RECEIVER
STARTER & SWITCH PANEL
VENTILATION AS REQ'D
WATER TEMP CONTROL VALVE
COOLING WATER SUPPLY (FRESH OR RECLAIMED)
AFTER-COOLER
SHUT OFF VALVE
SAFETY VALVE
PRESSURE GAUGE
CONTROL AIR TO PRESSURE SWITCH
UNLOADING AIR TO UNLOADER VALVE
COMPRESSOR
RECEIVER DRAIN
RECEIVER DISCHARGE
DISCHARGE VISIBLE TO OPERATOR
TO SEWER OR RECLAMATION

Fig. 11-18. Typical compressor installation.

in itself, supplementary receivers should be installed at points of use where large intermittent demands occur. One example is at a concrete placing setup using pneumatic placers.

Most pneumatic tools are designed to operate at a gauge pressure of 85 psi. However, greater efficiency, particularly of rock drills, is attained when the pressure is from 90 to 100 psi gauge. At higher pressures, maintenance costs increase rapidly. To maintain 100 psi at the drills in the heading, the compressors may have to be set at 120 psi. To conform to varying demands during the life of the job, and to provide partial capacity during repair work, the compressor plant should include several individual compressors. For many projects, an extra unit for standby should be provided.

The volume of compressed air consumed by various items of pneumatic equipment is affected by:

Atmospheric pressure
Compressed air temperature
Gauge pressure
Equipment design
Equipment condition
Work conditions such as the operator's ability and the rock hardness.

Table 11-22 lists the approximate consumption rates, at sea level, of various rock drills and other items of pneumatic equipment, all rated at sea level conditions. To obtain the rated output at altitudes above sea level, the free air volume to be supplied must be increased above the volume required at sea level. Table 11-23 lists the multipliers to be used in calculating the minimum compressor plant capacity when a compressor plant is to be operated at various altitudes above sea level.

Compressors are usually designed for use over a specific range of altitudes. Two-stage reciprocating compressors are not significantly affected by changes in altitude below 3300 feet. At greater elevations, the first-stage piston diameter must be increased to maintain the sea level production rating. If a compressor is to be

Table 11-22. Air consumption data (based on 90-100 psi pressure at sea level).

Drills	CYLINDER DIAMETER OF DRILL (INCHES)										
	$2\frac{1}{2}$	$2\frac{5}{8}$	$2\frac{11}{16}$	$2\frac{3}{4}$	3	$3\frac{1}{2}$	4	$4\frac{1}{2}$	5	$5\frac{1}{2}$	6
AIR TRACK							400	575	650	750	900
DRIFTER				150	200	275	325	400	475	550	
JACKLEG	125	150	175		210						

MISCELLANEOUS EQUIPMENT

		cfm/UNIT
Stopers	Light	155
	Heavy	180
Sinkers	Under 30 lb	70
	30 to 50 lb	100
	51 to 65 lb	125
Paving breakers	30 to 40 lb	30
	41 to 65 lb	60
	66 to 85 lb	75
Spaders		40
Jumbo hydraulic pumps (1 required per 3 jibs)		125
Jumbo air lights		15
Muckers	20 hp	250
	37 hp	325
	50 hp	530
Diaphragm sump pumps: outlet hose	2 inches	50
	$2\frac{1}{2}$ inches	85
	3 inches	100
Shotcrete machines: per CY per hour		50 to 100
Air saws		50

Table 11-23. Air consumption multipliers for altitude operation. (*Courtesy Ingersoll-Rand Co.*)

ALTITUDE ABOVE SEA LEVEL (FEET)	Multiplier
0	1.00
1000	1.02
2000	1.05
3000	1.08
4000	1.11
5000	1.14
6000	1.18
7000	1.22
8000	1.26
9000	1.30
10,000	1.34
12,500	1.46
15,000	1.58

used at an altitude other than that for which it was designed, the manufacturer should be consulted.

Compressor plant capacity should be increased 20% over theoretical to allow for miscellaneous losses such as leaks at couplings, valves, etc.

Friction losses in air lines create a loss of pressure. Table 11-24 lists the approximate losses due to friction in various diameter pipes and hoses over a range of flow rates. Pressure losses due to valves and fittings are tabulated in most compressed air handbooks. The compressor plant discharge pressure should be adjusted so that, as the air line is extended, the pressure at the heading is kept at 90 to 100 psig.

Light-gauge, spiral-welded steel pipe is commonly used for the air line. Couplings should be light and simple, such as the "Victaulic" type. Outlet tees, with valves, should be installed periodically. The diameter of the main

Table 11-24. Pressure losses due to line friction. *(Courtesy Ingersoll-Rand Co.)*

cfm FREE AIR AT 100 psig LINE PRESSURE	PRESSURE LOSS IN psi PER 100 FEET OF STRAIGHT PIPE — NOMINAL PIPE DIAMETER–SCHEDULE 40							
	2 inches	3 inches	4 inches	5 inches	6 inches	8 inches	10 inches	12 inches
150	0.15							
200	0.26							
300	0.57	0.07						
500	1.51	0.20	0.05					
750	3.36	0.44	0.11					
1000	5.90	0.76	0.19	0.06				
1500		1.68	0.43	0.13	0.05			
2000		2.99	0.75	0.24	0.09			
2500		4.67	1.16	0.36	0.14			
3000		6.71	1.64	0.51	0.20	0.05		
4000		11.90	2.90	0.90	0.35	0.09		
5000			4.50	1.40	0.55	0.12		
6000			6.45	2.00	0.73	0.19	0.06	
8000			11.50	3.53	1.38	0.33	0.10	
10,000				5.47	2.13	0.52	0.16	0.06
12,500					3.30	0.80	0.25	0.08
15,000					4.75	1.14	0.36	0.12

cfm FREE AIR AT 100 psig LINE PRESSURE	PRESSURE LOSS IN HOSE (psi)* — HOSE LENGTH AND INSIDE DIAMETER						
	50 feet	50 feet	50 feet	50 feet	50 feet	50 feet	25 feet
	1 inch	$1\frac{1}{4}$ inches	$1\frac{1}{2}$ inches	2 inches	$2\frac{1}{2}$ inches	3 inches	4 inches
150	2.7						
200	4.8						
250	6.9	2.4					
300	9.9	3.4					
400		5.8	2.4				
500		8.9	3.7				
600		12.6	5.2				
700			7.0				
800			9.0	2.1			
1000			13.6	3.2			
1200				4.5			
1500				6.9	2.4		
2000				12.2	4.2		
3000					9.3	3.6	
4000						6.3	
5000						9.6	1.4
6000						13.6	1.7

For Interpolation

To other pressure: $\Delta P_2 = \dfrac{P_1 + P_a}{P_2 + P_a} \times \Delta P_1$

To other volume: $\Delta P_2 = \dfrac{(Q_2)^2}{(Q_1)^2} \times \Delta P_1$

ΔP_1 = Pressure loss from table for a certain volume and pipe diameter at 100 psig (table values are valid for a pressure range of ± 10% of the entrance pressure psia)
ΔP_2 = Pressure loss in question for volume or pressure other than table value
P_1 = 100 psig
P_2 = Pressure other than 100 psig
P_a = Atmospheric pressures
Q_1 = Volume of free air from table
Q_2 = Volume other than table value.

*Assumes lubrication is at tool and air in hose does not have oil from line oiler.

air line must be great enough so that sufficient pressure can be maintained at the heading.

Lighting

A string of light bulbs, evenly spaced, is the usual type of general lighting. One-hundred-watt bulbs are often spaced at 50 feet.

Floodlights are used for lighting work areas, and are permanently mounted in strategic locations on jumbos and other working structures.

Power is usually 110 to 120 volts, single phase, supplied by dry transformers spaced throughout the tunnel. Insulated wires separated about 6 inches are used for distribution. Power for additional lighting can be tapped off these lines by using "pig tails." The wires may be held to the rock, concrete, steel or timber by various methods such as wooden dowels, expansion bolts or concrete inserts.

Drill jumbos sometimes have compressed-air driven floodlights since explosive handling should not be done with electric power on the jumbo.

Communications and Signal Systems

The efficiency and safety of any tunnel project is improved by the use of well-planned communications and signal systems.

Typical communications system terminals on a tunnel project are:

- Tunnel headings
- Tunnel-boring machine operator's console
- Intermediate points along tunnel alignment
- Shaft—top, bottom and landings
- First aid station
- Superintendent's office
- Project engineer's office
- Project manager's office
- Shops—mechanic's, electrician's
- Batch plant
- Compressor house
- Inspector's office.

In addition to those listed above, locations for signal system terminals might include:

- Materials transfer points
- Turnouts
- Passing sections
- Blind curves in alignment.

One satisfactory means of controlling access to rail haulage passing sections utilizes a system of colored lights, visible to locomotive operators. Pull cords, hanging from the tunnel crown within reach of the locomotive operator, are used to regulate the lights on both sides of the passing section. Distance from the signal lights to the passing section must be greater than the stopping distance of the train.

A closed-circuit television system can be effectively used to monitor conditions and operations at many locations on a tunnel project, including materials transfer points, shaft landings, tunnel alignment blind spots, and the construction yard.

Telephones in the tunnel or at noisy locations should be equipped with signal devices which can be seen as well as heard when the telephone rings.

Water Supply

Water line diameter is determined by the volume and pressure requirements and the length of line. Tables 11-25 and 11-26 give approximate friction losses for pipe and hose. Friction losses for pipe fittings are tabulated in most piping handbooks. Light-gauge, welded steel pipe with bolted couplings is commonly used in tunnels. Tees and valves should be spaced at about 1000-foot centers.

Drainage and Dewatering

A certain amount of groundwater entering a tunnel being driven upgrade will flow to the shaft or portal by gravity. It should be confined to a ditch located on one side or it can flow on the invert of machine-excavated tunnels. In order to maintain their capacity, such waterways must be kept open and free from sediment.

When the water inflow exceeds the capacity of the open channels, when the tunnel is driven downgrade or when the flowing water would have a deleterious effect on the tunnel, pumping through pipelines is necessary. The water is collected in sumps spaced as necessary along the tunnel and pumped through a discharge

Table 11-25. Water friction in 100 feet of light-gauge, spiral-welded steel pipe.

SIZE		4-INCH I.D.		5-INCH I.D.		6-INCH I.D.		8-INCH I.D.		10-INCH I.D.		12-INCH I.D.		
DISCHARGE														DISCHARGE
GALLONS/ MINUTE	CUBIC FEET/ SECOND	VELOCITY FEET/ SECOND	LOSS FL	VELOCITY FEET/ SECOND	LOSS FL	VELOCITY FEET/ SECOND	LOSS FL	VELOCITY FEET/ SECOND	LOSS FL	VELOCITY FEET/ SECOND	LOSS FL	VELOCITY FEET/ SECOND	LOSS FL	GALLONS/ MINUTE
1	0.002228													1
5	0.01114													5
10	0.02228													10
15	0.03342	0.38281	0.02166											15
20	0.04456	0.51042	0.03519	0.01228										20
25	0.05570	0.63802	0.05176	0.40838	0.01801	0.28360	0.00759							25
30	0.06684	0.76563	0.07236	0.49002	0.02496	0.34032	0.01042	0.19146	0.00262					30
40	0.08912	1.0208	0.12135	0.65337	0.04135	0.45376	0.01739	0.25527	0.00435					40
50	0.11140	1.2760	0.18050	0.81671	0.06139	0.56720	0.02576	0.31910	0.00647	0.20425	0.00223			50
60	0.13368	1.5312	0.25118	0.98005	0.08518	0.68065	0.03576	0.38292	0.00901	0.24510	0.00311			60
70	0.15596	1.7864	0.33160	1.1434	0.11204	0.79409	0.04699	0.44674	0.01168	0.28595	0.00406			70
80	0.17824	2.0462	0.42713	1.3067	0.14315	0.90753	0.05959	0.51056	0.01493	0.32680	0.00515			80
90	0.20052	2.2969	0.52835	1.4700	0.17715	1.0209	0.07411	0.57438	0.01821	0.36765	0.00637	0.25530	0.00263	90
100	0.22280	2.5521	0.64019	1.6334	0.21374	1.1344	0.08871	0.63821	0.02219	0.40850	0.00768	0.28637	0.00318	100
125	0.27850	3.1901	0.97129	2.0417	0.32451	1.4180	0.13363	0.79776	0.03156	0.51063	0.01132	0.35459	0.00470	125
150	0.33420	3.8281	1.3448	2.4501	0.44738	1.7016	0.18613	0.95731	0.04567	0.61725	0.01611	0.42551	0.00655	150
175	0.38900	4.4662	1.7932	2.8585	0.59373	1.9852	0.24478	1.1168	0.06042	0.71488	0.02066	0.49643	0.00861	175
200	0.44560	5.1042	2.3422	3.2668	0.75949	2.2688	0.31330	1.2764	0.07664	0.81701	0.02624	0.56735	0.01099	200
225	0.50130	5.7422	2.8721	3.6752	0.94120	2.5524	0.38845	1.4359	0.09508	0.91913	0.03226	0.63827	0.01353	225
250	0.55700	6.3802	3.4890	4.0835	1.1494	2.8360	0.46956	1.5944	0.11502	1.0212	0.03886	0.70918	0.01624	250
275	0.61270	7.0183	4.1759	4.4919	1.3684	3.1196	0.56214	1.7550	0.13630	1.1233	0.04631	0.78010	0.01937	275
300	0.66840	7.6563	4.9150	4.9002	1.6105	3.4032	0.65821	1.9146	0.16051	1.2255	0.05428	0.85102	0.02249	300
325	0.72410	8.2943	5.7364	5.3086	1.8797	3.6868	0.76401	2.0741	0.18535	1.3726	0.06632	0.92194	0.02600	325
350	0.77980	8.9324	6.6157	5.7170	2.1556	3.9704	0.87631	2.2337	0.21266	1.4297	0.07122	0.99286	0.02969	350
375	0.83550	9.5704	7.5520	6.1253	2.4606	4.2540	0.99470	2.3932	0.24145	1.5318	0.08132	1.0637	0.03373	375
400	0.89120	10.208	8.5430	6.5337	2.7837	4.5376	1.1189	2.5528	0.27168	1.6340	0.09154	1.1347	0.03778	400
425	0.94690	10.846	9.6440	6.9420	3.1065	4.8212	1.2559	2.7123	0.30327	1.7361	0.10277	1.2056	0.04220	425
450	1.0026	11.484	10.689	7.3504	3.4627	5.1048	1.4000	2.8719	0.33809	1.8382	0.11333	1.2765	0.04680	450
475	1.0583	12.122	11.841	7.7587	3.8357	5.3884	1.5508	3.0315	0.37244	1.9404	0.12488	1.3474	0.05186	475
500	1.1140	12.760	13.044	8.1671	4.2253	5.6720	1.6984	3.1910	0.41028	2.0425	0.13681	1.4183	0.05684	500
550	1.2254	14.036	15.695	8.9838	5.0526	6.2392	2.0430	3.5010	0.49068	2.2467	0.16365	1.5602	0.06765	550
600	1.3368	15.312	18.566	9.8005	5.9413	6.8065	2.4026	3.8298	0.57713	2.5410	0.19253	1.7020	0.07916	600
650	1.4482	16.588	21.662	10.617	6.9304	7.3737	2.7860	4.1483	0.66933	2.6552	0.22199	1.8438	0.09184	650
700	1.5596	17.864	24.974	11.434	7.9892	7.9409	3.2114	4.4674	0.87877	2.8595	0.25444	1.9857	0.10469	700
750	1.6710	19.140	28.494	12.250	9.1142	8.5081	3.6867	4.7865	0.88045	3.0637	0.29033	2.1275	0.11877	750

Table 11-25. Continued.

SIZE		4-INCH I.D.		5-INCH I.D.		6-INCH I.D.		8-INCH I.D.		10-INCH I.D.		12-INCH I.D.		DISCHARGE
DISCHARGE														
GALLONS/ MINUTE	CUBIC FEET/ SECOND	VELOCITY FEET/ SECOND	LOSS FL	VELOCITY FEET/ SECOND	LOSS FL	VELOCITY FEET/ SECOND	LOSS FL	VELOCITY FEET/ SECOND	LOSS FL	VELOCITY FEET/ SECOND	LOSS FL	VELOCITY FEET/ SECOND	LOSS FL	GALLONS/ MINUTE
800				13.067	10.307	9.0753	4.1436	5.1056	0.99572	3.2680	0.32833	2.2694	0.13434	800
850				13.884	11.564	9.6425	4.6198	5.4247	1.1103	3.4723	0.36617	2.4112	0.15076	850
900				14.700	12.883	10.209	5.1461	5.7438	1.2371	3.6765	0.40799	2.5530	0.16799	900
950				15.517	14.355	10.776	5.6977	6.0630	1.3699	3.8808	0.44899	2.6949	0.18494	950
1,000				16.334	15.807	11.344	6.2740	6.3821	1.5084	4.0850	0.49439	2.9367	0.20253	1,000
1,100						12.478	7.5432	7.0203	1.8022	4.4935	0.59068	3.1204	0.24190	1,100
1,200						13.613	8.9200	7.6585	2.1311	4.9020	0.69401	3.4041	0.28427	1,200
1,300						14.747	10.400	8.2967	2.4851	5.3105	0.80399	3.6877	0.32940	1,300
1,400						15.881	11.983	8.9349	2.8448	5.7190	0.92634	3.9714	0.37957	1,400
1,500						17.016	13.757	9.5731	3.2444	6.1275	1.0564	4.2551	0.43012	1,500
1,600								10.201	3.6668	6.5360	1.1940	4.5388	0.48300	1,600
1,800								11.487	4.6099	7.3531	1.5011	5.1061	0.50321	1,800
2,000								12.764	5.5777	8.1701	1.8283	5.6735	0.73742	2,000
2,500								15.955	8.6566	10.212	2.7980	7.0918	1.1245	2,500
3,000								19.146	12.294	12.255	3.9737	8.5102	1.5885	3,000
3,500										14.297	5.3321	9.9286	2.1276	3,500
4,000										16.340	6.8655	11.347	2.7588	4,000
4,500										18.382	8.6257	12.765	3.4662	4,500
5,000										20.425	10.572	14.183	4.2165	5,000
6,000										24.501	15.112	17.020	6.6974	6,000

Note: 1 foot of water = 0.4335 psi.

Table 11-26. Water friction in 100 feet of smooth bore hose.* (*Courtesy Gorman Rupp Co.*)

FLOW (GALLONS/ MINUTE)	VELOCITY (FEET/ SECOND)	FRICTION HEAD (FEET)	VELOCITY (FEET/ SECOND)	FRICTION HEAD (FEET)	VELOCITY (FEET/ SECOND)	FRICTION HEAD (FEET)	VELOCITY (FEET/ SECOND)	FRICTION HEAD (FEET)	VELOCITY (FEET/ SECOND)	FRICTION HEAD (FEET)	VELOCITY (FEET/ SECOND)	FRICTION HEAD (FEET)	VELOCITY (FEET/ SECOND)	FRICTION HEAD (FEET)	VELOCITY (FEET/ SECOND)	FRICTION HEAD (FEET)	FLOW (GALLONS/ MINUTE)
	$\frac{1}{2}$ INCH		$\frac{3}{4}$ INCHES		1 INCH		$1\frac{1}{4}$ INCH		$1\frac{1}{2}$ INCHES		2 INCHES		$2\frac{1}{2}$ INCHES		3 INCHES		
1.5	1.6	2.3	1.1	0.97													1.5
2.5	2.6	6.0	1.8	2.5													2.5
0.5	5.2	21.4	3.6	8.9	2.0	2.2	1.3	0.74	0.9	0.3							5
10	10.5	76.8	7.3	31.8	4.1	7.8	2.6	2.64	1.8	1.0	1.0	0.2					10
15			10.9	68.5	6.1	16.8	3.9	5.7	2.7	2.3	1.5	0.5					15
20					8.2	28.8	5.2	9.6	3.6	3.9	2.0	0.9	1.3	0.32			20
25					10.2	43.2	6.5	14.7	4.5	6.0	2.5	1.4	1.6	0.51			25
30					12.2	61.2	7.8	20.7	5.4	8.5	3.1	2.0	2.0	0.70	1.4	0.3	30
35					14.3	80.5	9.1	27.6	6.4	11.2	3.6	2.7	2.3	0.93	1.6	0.4	35
40							10.4	35.0	7.3	14.3	4.1	3.5	2.6	1.2	1.8	0.5	40
45							11.7	43.0	8.2	17.7	4.6	4.3	2.9	1.5	2.0	0.6	45
50							13.1	52.7	9.1	21.8	5.1	5.2	3.3	1.8	2.3	0.7	50
60							15.7	73.5	10.9	30.2	6.1	7.3	3.9	2.5	2.7	1.0	60
70									12.7	40.4	7.1	9.8	4.6	3.3	3.2	1.3	70
80									14.5	52.0	8.2	12.6	5.2	4.3	3.6	1.7	80
90									16.3	64.2	9.2	15.7	5.9	5.3	4.1	2.1	90
100									18.1	77.4	10.2	18.9	6.5	6.5	4.5	2.6	100
125											12.8	28.6	8.2	9.8	5.7	4.0	125
150											15.3	40.7	9.8	13.8	6.8	5.6	150
175											17.9	53.4	11.4	18.1	7.9	7.4	175
200											20.4	68.5	13.1	23.4	9.1	9.6	200
225													14.7	29.0	10.2	11.9	225
250													16.3	35.0	11.3	14.8	250
275													18.0	42.0	12.5	17.2	275
300													19.6	49.0	13.6	20.3	300
325															14.7	23.5	325

NOTE: 1 foot of water = 0.4335 psi.

*For various flows and hose sizes, table gives velocity of water and feet of head lost in friction in 100 feet of smooth bore hose. Sizes of hose shown are actual inside diameters.

pipe by means of centrifugal pumps. The sumps should be large enough to handle expected surges and to cope with possible pump stoppages. They should be deep enough so that the required frequency of cleaning is reasonable. If the water contains abrasive sediment, a settling compartment in the sump will save wear in the pumps and pipelines.

Water in the heading is collected in low spots or temporary sumps and pumped back to the first permanent sump by means of portable submersible pumps, either compressed air or electric powered (see Figs. 11-19 and 11-20). The discharge pipe is usually hung on the tunnel sidewall.

Since the quantity of groundwater flows cannot be predicted with any accuracy, and since the flows usually diminish with time, the dewatering system should have a maximum of flexibility. Automatic float switches should be provided at main sumps. The pump characteristics should be compatible with a broad range of pressures and flows.

Consideration must be given to the disposal of the tunnel drainage. Local environmental regulations may dictate the use of settling ponds or other forms of treatment.

CONCRETE PLANT

General

The concrete plant consists of storage facilities, batch plant, mixer, transport equipment, placing equipment, forms and ancillary equipment. In urban areas, it is frequently possible to purchase transit-mixed concrete delivered to the job site, thus eliminating the need for storage and batch-mix facilities. However, the dependability of the supply should be confirmed.

Batching and Mixing

The batch plant should be located so that ample storage of aggregates can be provided, depending on the nature and dependability of local deliveries. Stockpiling and reclaiming systems will depend on the volumes to be stored and the available storage areas. Protection against winter cold and summer heat may be necessary.

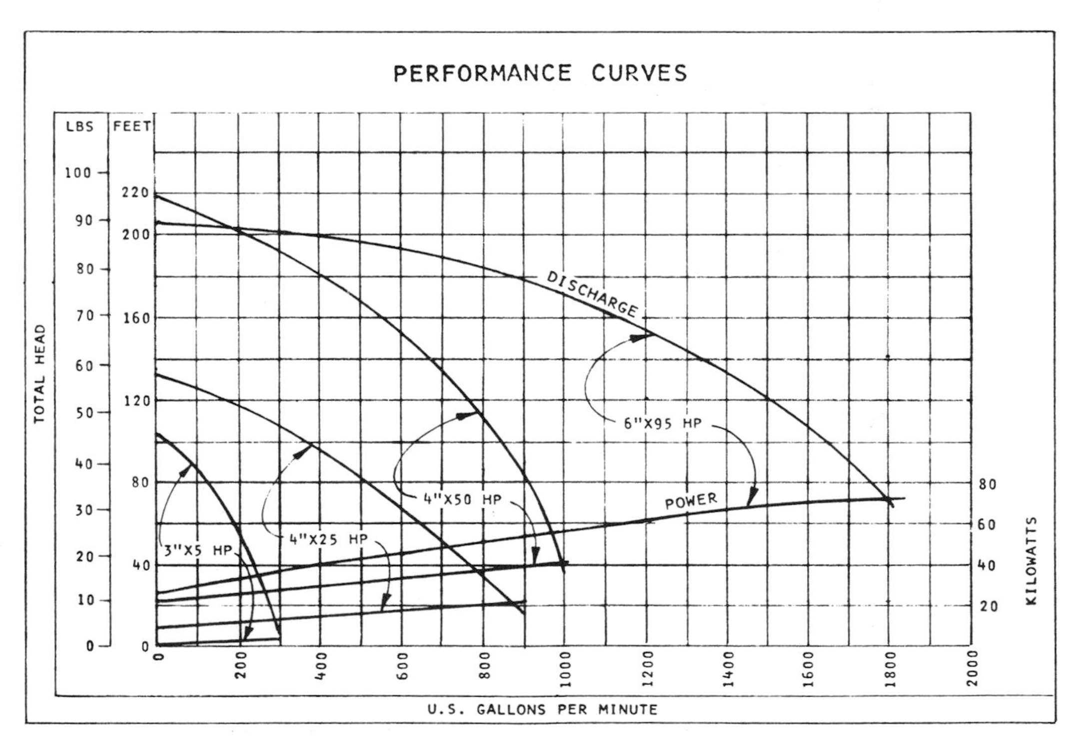

Fig. 11-19. Typical submersible electric motor-driven pump.

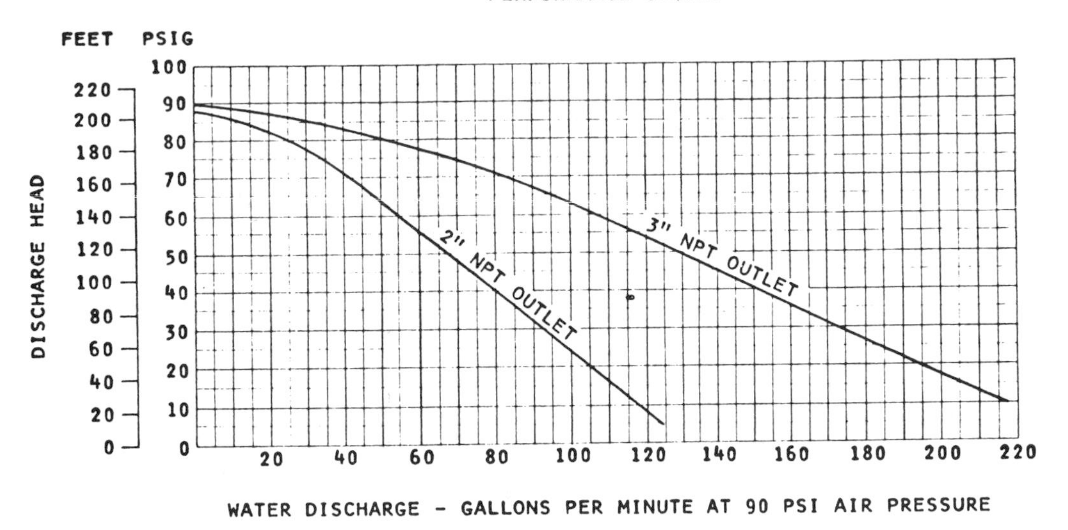

Fig. 11-20. Typical submersible double diaphragm air pump.

Cement is invariably handled in bulk, so the batch plant will include cement storage facilities of adequate capacity. Bins for handling the required sizes of aggregates are also needed. Batching of cement and aggregate is by weight and that of water by either weight or volume. See Fig. 11-21 for an example of a typical mobile batch-mix plant. Automatic batching is available and sometimes required.

Mixing is usually done at the batch plant unless the transport time would be excessive. Tilting mixers are most commonly used, although horizontal axis and turbine mixers are also satisfactory. When mixing in the tunnel, the horizontal axis type is usually indicated due to space limitations. The mixing plant should have ample capacity for the concrete placing schedule.

Transport

Mixed concrete is transported into the tunnel in transit mixers, agitators or pneumatic placers. These can be mounted on either pneumatic-tired truck chassis or railroad cars. When the haul is extremely long, dry batches for mixing in the tunnel are normally transported in compartmented rail cars, each compartment holding one batch with the cement being isolated from the aggregates. Special agitators for use with rail cars discharge successively through each other, so that switching in the tunnel is unnecessary. They are powered by compressed air or electric motors which must be connected at the placing site.

The Flocrete Placer, by virtue of its quick coupling devices for compressed air and discharge, is designed to be used for transportation as well as for placing the concrete. When mounted on rail chassis, a train of these placers can be quickly discharged by successively connecting each one to an articulated discharge-compressed air station and, when empty, uncoupling it from the train and moving it on through the station. When the tunnel track is smooth, the Flocrete Placers can be used for hauls up to about 30 minutes in duration. Hauls of 7500 feet are common and hauls of as much as 15,000 feet have been reported.

Concrete can also be transported through the tunnel by pipeline, using a positive displacement pump. Under ideal conditions, a distance of 1000 feet can be achieved. Greater distances are possible when using two or more individual setups in tandem. Transporting through pipelines by means of compressed air supplied by a pneumatic placer has been reported in some mines, but is not commonly used in tunnels.

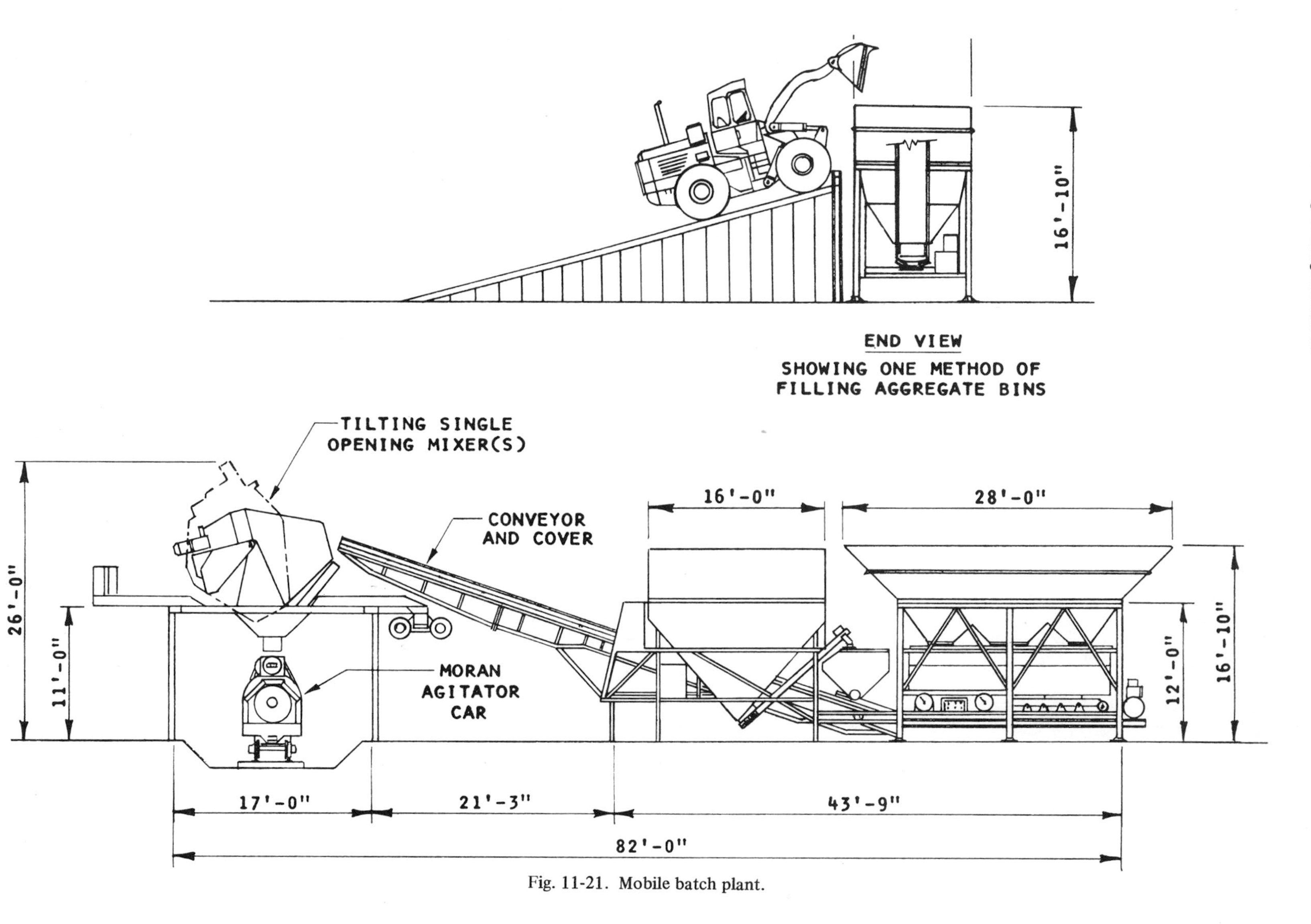

Fig. 11-21. Mobile batch plant.

Mixed concrete can be dropped through vertical pipes into the tunnel, either at a working shaft, or at other locations. It is not necessary to keep the pipe full and there seems to be no practical limit to the distance dropped, friction and the displaced air apparently limiting the velocity. The pipe must, of course, discharge into a terminal which can absorb the energy of the falling material and deliver it without segregation into the conveyance in the tunnel.

Placing

Various methods are used for placing the concrete in its final position in the tunnel. Simple low pours, such as a sidewall footing, can usually be made by merely dumping the concrete out of its transport vehicle, using a chute if necessary. For higher pours into open forms, the concrete can be elevated by means of a belt conveyor. Invert concrete must usually be placed on a prepared or cleaned invert and consequently must be conveyed from the end of the track or roadway to the point of deposit. This can sometimes be accomplished with a belt conveyor. For fast-moving systems, an invert bridge may be used. This is a long, movable platform supported on concrete curbs, on dowels in the invert, or on the invert itself. It provides space underneath for roadbed removal, invert clean-up and placing and finishing the invert. It also supports the concrete conveyance, which may be a belt conveyor or the transport vehicles for bringing the concrete into the tunnel, as well as screeding and finishing equipment.

Concrete behind closed forms, such as the arch, or the entire circumference of a small circular tunnel placed monolithically, is placed through a pipeline. See Fig. 11-22 for typical arrangement of arch concrete placing equipment. Two different types of placers are available, positive displacement pumps and pneumatic placers.

Several makes of positive displacement concrete pumps are available. Basically, they are piston pumps designed to accept concrete from an integral agitator or remixer and force it through a pipeline. Various sizes and capacities are available, usually requiring discharge pipe of from 4 to 8 inches in diameter. Horizontal distances of 1000 feet and vertical lifts of 100 feet or more can be handled. For maximum limits, a smooth, workable mix is necessary. Maximum aggregate size should be less than one-quarter of the pipe diameter, and the aggregate should not be harsh, well-graded, rounded particles being best. Slump should be about 4 inches and the sand and cement content should be relatively high. Air entraining admixtures are helpful in preventing segregation and consequent blockages at bends or other constructions. Compressed air boosters on the line assist in handling marginal cases and, at the slickline going into the form, assist in filling the crown.

The low discharge velocity of the concrete pump permits its use as a conveyance for concrete to all other areas of the tunnel if desired. It can even be used to pump the concrete vertically through ports in the crown of an arch form, although this system will ordinarily not fill the crown as completely as will the slickline method. Figure 11-23 shows the relationship between pumping rate, pumping distance and line pressure for positive displacement pumps.

Pneumatic placers are available in several makes. Typically, they consist of a pressure chamber charged with a given volume of concrete. The concrete is discharged by means of compressed air and then recharged for another cycle. These units are of small dimension compared to concrete pumps of equal capacity. Although some specification writers prohibit their use, when properly installed and operated, they can safely produce an acceptable product. Although the discharge is at high velocity, when equipped with suitable energy dissipaters, they have been used to place concrete into open formwork.

Forms

Most of the concrete placed in a tunnel must be confined and shaped by the means of forms. While the design of concrete forms is beyond the scope of this chapter, it should be noted that the moving of the forms is a material handling problem and that the forms may frequently be used to facilitate other material handling

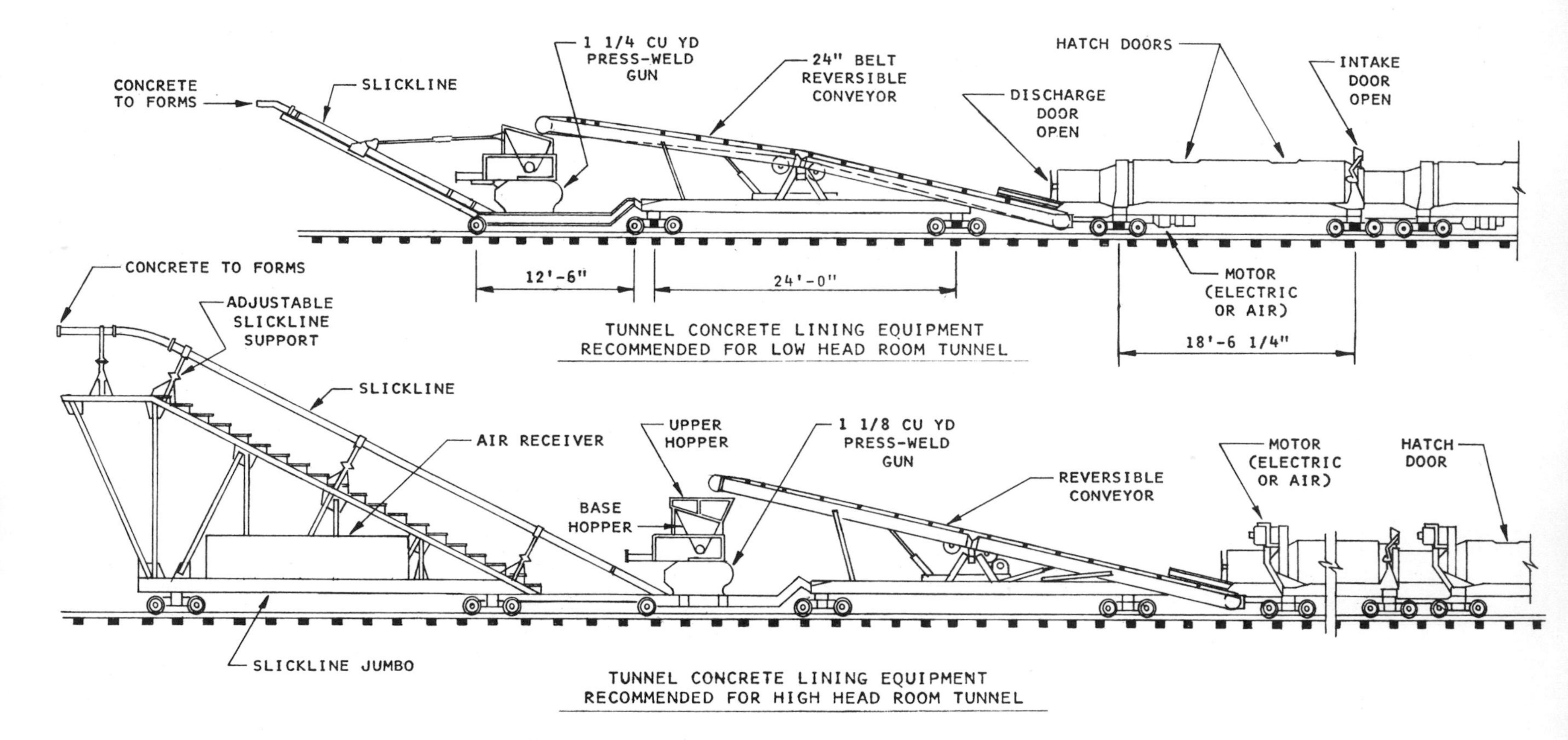

Fig. 11-22. General arrangement of arch concrete placing equipment.

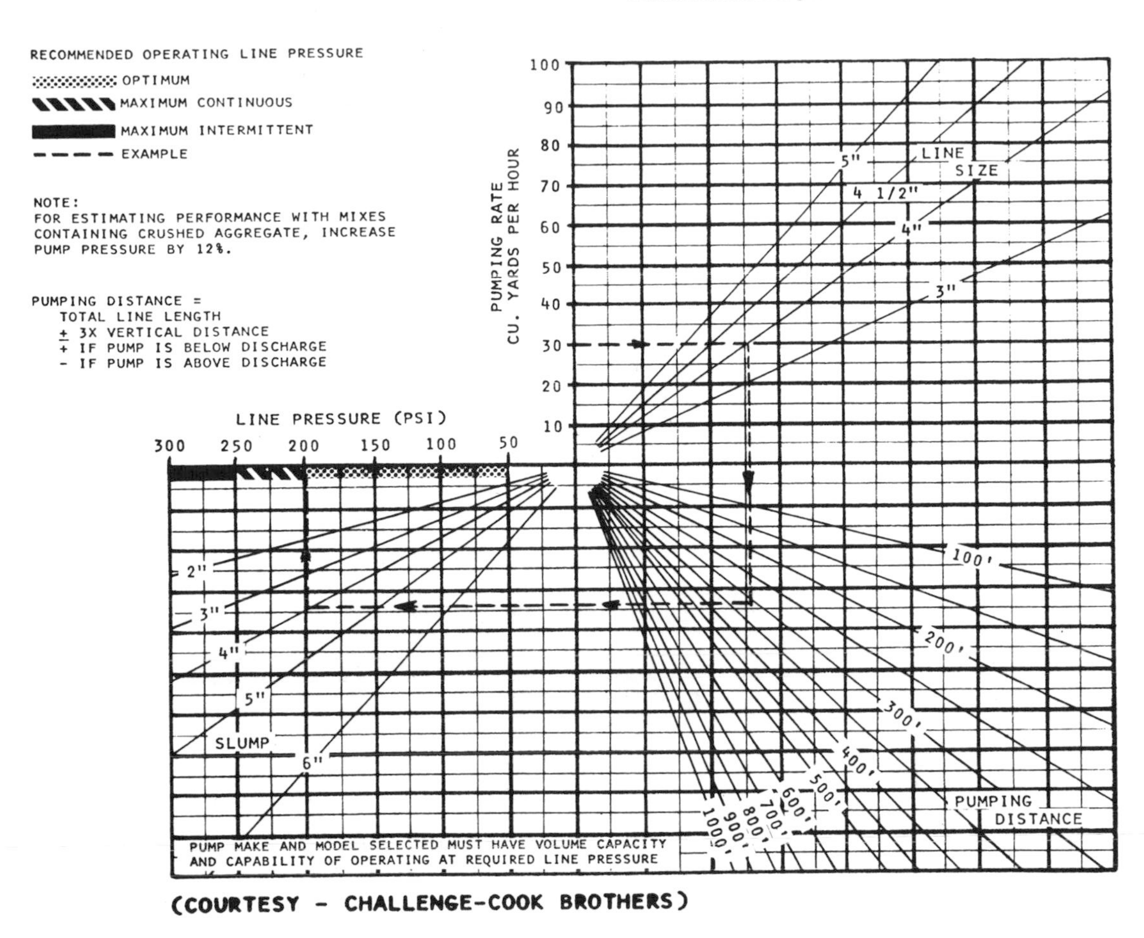

Fig. 11-23. Concrete pump performance estimator.

problems in the tunnel. For example, invert forms may be used as support for an invert bridge or for a finishing screed and platform.

Because of the many reuses which can usually be made, most tunnel forms are fabricated from steel. In circular tunnels, the forms can be designed so that the full circle is concreted monolithically or the invert may be placed separately from the arch and sides. In either case, the forms are usually stripped and moved by means of a traveler. For continuous placing, the forms may be designed telescopic; that is, the last form is retracted, moved through the entire line of forms, and reset at front. See Fig. 11-24 for examples of telescopic and non-telescopic forms.

A well-designed set of forms will include provisions for bulkheading, access doors for observation and vibration, provisions for mounting and operating form vibrators, provisions for bracing against uplift and horizontal movement, provisions for alignment and provisions for utilities such as electric power.

SURFACE PLANT

General

Surface plant can be classified into the following categories.

1. Access and yards.
2. Operational facilities.
3. Shops.
4. Warehousing and storage.
5. Sanitary and medical facilities.
6. Offices.
7. Camp facilities.

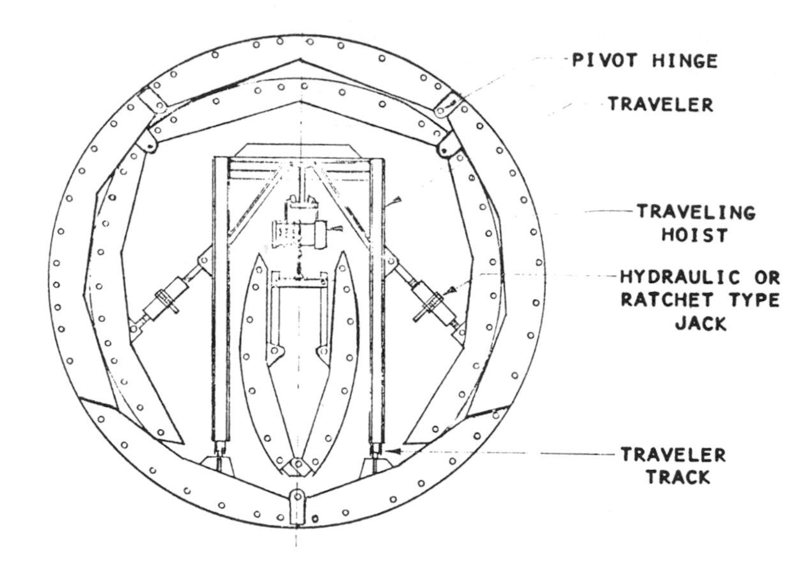

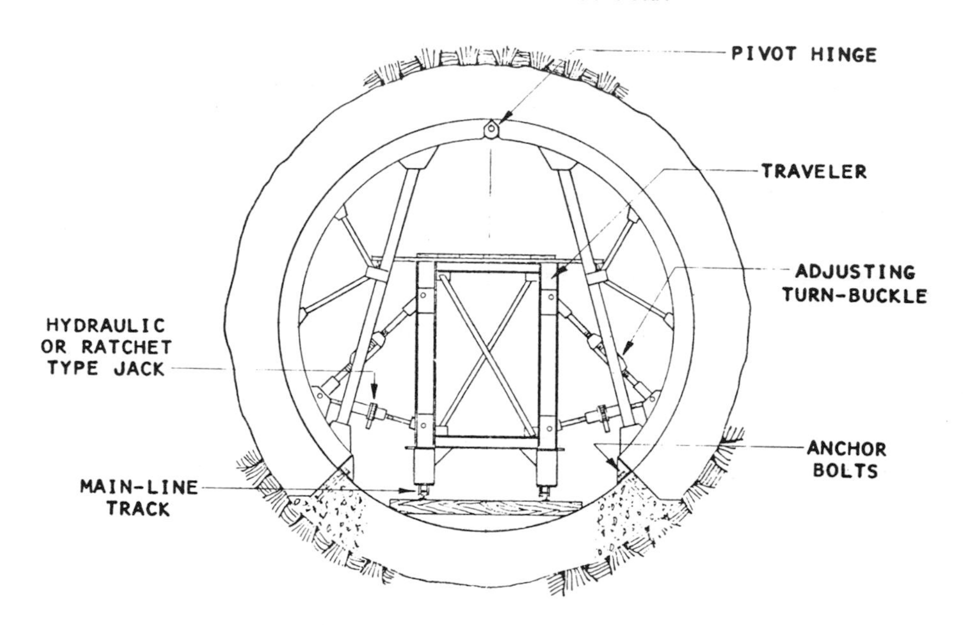

Fig. 11-24. Forms for placing concrete lining.

8. Utilities.
9. Environmental requirements.

The extent to which these items will be required depends on the nature of the job and its location, with large projects in remote locations requiring the most extensive facilities. Type of construction, weatherproofing, etc., will be influenced by the climate and other local conditions. For many applications, buildings can be composed of single or multiple trailers.

ACCESS AND YARDS

Access to the work may vary from a simple driveway from a city street to miles of roadway through mountainous terrain. The quality of the access must be adequate to assure that workmen, supplies and equipment can expeditiously reach the site. For large projects involving much traffic, roads should be paved. Railroad access may be indicated in some cases for equipment and materials delivery.

When access to the work site is difficult, time can usually be gained if the owner constructs access roads during the design and bidding period for the tunnel.

Yards should be arranged for good communication between the various facilities. Special attention should be given to arrangement and location of storage and parking areas. If rail haulage is to be used in the tunnel, the railroad system must be extended to the yard for servicing of warehouses, powder magazines, repair shops and storage yards.

In urban areas and rugged mountainous terrain, there may not be sufficient space for an ideal yard layout. In such cases, it is necessary to make the best use of whatever is available.

Operational Facilities

These include batch-mix plants, air compressor stations and power sub-stations. Their layout should take local conditions into account. For example, the mode of delivery of cement and concrete aggregates will influence the location and layout of the batch-mix plant and associated storage facilities. In cold climates, a compressor house located adjacent to the repair shop will provide supplemental heating. In a hot climate, this would be undesirable.

Shops

Every job needs a repair shop for the repair and maintenance of equipment. Required facilities would depend on the job, but might include metal working machinery, welders, bit grinders, drill steel reconditioning equipment, battery shop, electrical shop, tire shop, etc. Depending on the types of equipment to be serviced, pits, hoists and cranes may be necessary.

Other shops may be needed for specialized job requirements. These might include a woodworking shop, a reinforcing steel cutting and bending shop and a steel fabrication shop.

Warehousing and Storage

Enclosed storage is needed for most repair parts, job supplies and various materials and small tools. The warehouses should be designed for easy handling and good identification and retrieval of stored items.

Large or heavy items can frequently be stored in the open. Again, provision for storing, handling and retrieval must be provided.

If the job requires explosives, a magazine for high explosives and a separate magazine for detonators is necessary. Their type of construction and location is dictated by safety regulations. Where safety regulations do not prohibit, a make-up house for preparing primers is also provided.

Sanitary and Medical Facilities

In addition to the usual sanitary facilities, a change house must be provided for the workmen. This will include lockers for the street clothes and racks or suspended hooks for storing and drying the work clothes between shifts. Showers, wash basins and toilet facilities must be provided, as well as adequate heat and ventilation. The change house should be located conveniently to the tunnel access.

Minimum medical facilities would comprise at least one first aid station with supplies for treating minor cuts, bruises and burns, and handling emergency cases. In remote areas, or where other circumstances warrant, a job hospital may be indicated. Compressed air tunnel jobs require access to a medical lock, together with adequate examination facilities.

Offices

Office space should be provided for project administrative, engineering and clerical staffs. The extent and type of facilities will depend mainly on the size and nature of the job. Even in urban areas, combinations of trailers are frequently found most satisfactory.

Camp Facilities

In remote areas, a construction camp may be necessary. Its extent will vary with the duration and size of the job, as well as with its location. Quarters may include houses, cottages and dormitories. A suitable mess hall with adequate kitchen facilities will be needed, as well as a commissary. Entertainment and recreational facilities are a must, their nature depending on the particular situation. The need for other services, such as laundry, post office, banking and automobile service, will depend on the proximity to established communities.

Utilities

The yard and camp area must, of course, be provided with necessary utilities. Electric power, water supply and sewage disposal are practically always needed. Other possibilities include natural or bottled gas, refuse pickup, telephone service and radio communication.

Environmental Requirements

In remote areas, the location and final landscaping of muck disposal areas must be given adequate consideration. Water discharged from the tunnel may require treatment before disposal. Clean-up and landscaping of yard and camp areas is usually required, and replacement or protection of trees is demanded in our National Forests.

In urban areas, many more restrictions are found. Yard areas may have to be guarded from public view by suitable fencing. Compressors, fans and other equipment must be muffled and muck bins, batch plants, etc., insulated to minimize noise. Dust and other atmospheric pollutants must be controlled. Consideration must be given to traffic regulations and the flow of traffic near the project site.

12 Shotcrete

EDWARD E. MASON

Principal
Dolmage, Mason & Stewart
(Retired)

and

ROBIN MASON

Principal Engineer
A.A. Mathews Division of CRS Group Engineers

DEVELOPMENT OF SHOTCRETE FOR TUNNELS AND SHAFTS

Shotcrete has been defined as concrete or mortar conveyed through a hose and pneumatically projected at high velocity onto a surface. The term "shotcrete" will be used in this chapter only with reference to accelerated-set coarse aggregate sprayed concrete. In 1909, the term "gunite" and designs for a "cement gun" were patented by the Allentown Cement Company, thereby giving air-placed mortar the name by which it was known until lately.

Introduction of sprayed mortar or gunite underground probably took place at the Brucetown Experimental Mine of the Pittsburgh Bureau of Mines in 1914. Subsequently, it found use as a protection of rock surfaces underground from atmospheric deterioration, and on occasion, presumably, as a temporary support measure. It failed to develop into an acceptable support material, however, due to a tendency to spall from minor rock pressures. It can be applied only in relatively thin layers made thinner on average by the irregularities of the excavation surface. Thus, adhesion problems between layers are multiplied and further aggravated by excessive shrinkage, caused by the high cement content. However, occasional successful applications are reported, such as its use in conjunction with rock bolts in a tunnel in the U.S. in 1952, and in a second tunnel reported by Keifer.[1] No rock criteria are reported for either.

European development largely paralleled that of this continent, although the European literature suggests more rigid quality control. Following World War II, emphasis was placed on underground development as an economic necessity in the multitude of hydroelectric and related civil engineering projects in the Alpine countries (Austria, Switzerland and Northern Italy) and in Sweden. In 1952, sprayed mortar was used successfully as the sole support and lining of pressure and non-pressure tunnels in the Swiss Maggia hydroelectric development.

The next few years saw the development of sprayed concrete or "shotcrete." Equipment capable of placing 1 inch of aggregate was developed, thus facilitating the mixing of aggregates and cement without preprocessing for reducing moisture, and bringing the cement content into manageable proportions. An accelerating-set and hardening admixture also was developed, making it possible to place shotcrete on wet surfaces and against heavy water flows.

Its development occured largely with experiences gained in the following projects. (Cross-sectional areas are given in square feet, followed by lengths.)

- 1953-1954, Prutz-Imst, Austria: 270-380 square feet, 89,000 feet.[2]
- 1955-1958, Schwarzach (Los Birql), Austria: 485 square feet, 15,000 feet.[2]
- 1957-1960, Serra Ripoli, Italy: 2-1180 square feet, 1392 feet.[3]
- 1956-1962, Monastero Pressure Tunnel, Italy: 640 square feet, 25,000 feet.[4]
- 1960-1962, La Planicia, Venezuela: 2-570 square feet, 2200 feet.[5]
- 1962-1963, Kaunertal Pumped Storage Project, Austria: About 43 miles of tunnels and inclined shafts, 105-215 square feet.[5]
- 1967, Milano Subway, Italy: 360 square feet, 660 feet (personal files).

In the first two examples, machines were developed for placing coarse aggregate shotcrete, and shotcrete's function in limiting loosening in both chemically and structurally unstable rocks was demonstrated. Its effectiveness was shown further in unconsolidated heterogeneous slide material and in soft wet ground at Serra Ripoli, Monastero, and on occasion at La Planicia. In the last instance, also, with loosening halted or prevented, the excavation remained stable for 12 months, while a conventionally supported twin suffered local failures from progressive loading. At Kaunertal, effectiveness of shotcrete in conjunction with grouted rock bolts was demonstrated in very heavy ground, wet plastic mylonitized sericite schists, where conventional steel supports combined with steel forepoling had failed.

In 1967, a section of tunnel was driven through unconsolidated gravels for the Milano Subway, incurring less surface settlement than occured in an adjacent shield-driven tunnel. Steel arches and wire mesh were used as supplementary reinforcement in some but not all of these projects.

In the meantime, parallel development of shotcrete practice was occurring in the Scandinavian countries. First major projects reported in Sweden were at the Holjes[6] (1958-1960) and the Lossens (1959-1969) hydroelectric projects, and in Norway, at the Tokke (1963) hydroelectric project. Rock conditions, particularly in Sweden, are more favorable than those in the Alpine regions. Nevertheless, there exists a wide variety of geologic conditions. Development in Sweden emphasized the use of shotcrete without reinforcement, wire mesh or other conventional tunnel support elements. Swedish experience also has been exported around the world, to India, Africa, Hong Kong, Mexico and South America. The 1960-1962 experience in Venezuela was exported to Chile and Peru by a spray concrete equipment manufacturer. By 1965-1966, Japan also appears to have joined in the development, although little has been reported in English of their early experiences.

This continent lagged, probably due to ample supplies of alternative economical support materials. Experience with gunite underground also had left a general suspicion among engineers and contractors alike as to the integrity of the method, and much of this suspicion remains. Early examples of shotcrete usage in North America were as follows.

- 1967-1968, Canadian National Railways Tunnel: 515 square feet, 9360 feet.[7]
- 1968, Tahachapi No. 1 Tunnel: 710 square feet, horseshoe.[8]
- 1968, Balboa Tunnel: 200 square feet, 3800 feet.[9]
- 1968, Luck Friday Shaft: Hecla Mining Co.[10]

In summary, the effectiveness of shotcrete as a

preventive of rock loosening has been demonstrated in a variety of geologic conditions. Its use is advantageous in soft ground, particularly where non-uniform cross-sections occur.

SHOTCRETE METHOD

Two processes of placing shotcrete exist, the wet-mix and dry-mix processes.

Wet-mix Process

The wet-mix process consists of mixing measured quantities of aggregate, cement and water, and introducing the resulting mix into a vessel for discharge pneumatically or mechanically through a hose to final delivery from a nozzle. It has the advantage of rigidly controlling the water/cement (w/c) ratio of the product. Existing equipment can handle maximum aggregate size to minus $\frac{3}{4}$ inch. Further, successful methods have been devised to introduce quick acting accelerators to the delivery hose. Pumping the very low slump concrete is commonly a problem, so a slightly higher than desirable water content is used. By use of accelerators, such concrete can be made to adhere overhead, but ultimate strength usually suffers. However, the method has been found convenient for use with less skilled operators, in particular in the limited size access workings of mines, the major percentage of which generally are dry.

Dry-mix Process

Dry-mix shotcrete consists of a mixture of damp aggregate and cement fed into a placing machine, fed at a uniform rate into an airstream to travel through a hose to the nozzle. The water of hydration is added at the nozzle before discharge to the surface. Water is manually controlled. Powdered accelerators are added to the dry-mix as it is fed into the placer. If liquid, the accelerator is mixed with the feed water before it goes to the nozzle. The dry-mix process is currently the prevalent method of applying coarse aggregate shotcrete in underground structures.

Mixes

Shotcrete quality depends on the quality of materials employed, mixing of ingredients, cement content, w/c ratio and degree of compaction.

Aggregates. These should conform to ASTM standards. They should be well graded, without gaps between or excesses in any size fraction. Thus, optimum compaction can be gained, benefiting density, impermeability and compressive strength, and minimizing rebound. Aggregate having splintery shapes or containing flat, elongated particles tends to decrease compaction. There is also a sharp increase of surface area of angular aggregate in comparison with rounded aggregate of the same sieve sizes, resulting in a cement-poor mix. Hence, quality suffers measurably.

The self-sorting action of shotcrete application results from the rebounding of mainly coarse particles. Evidence indicates that poorly graded aggregates may be capable of producing quality shotcrete, although the quantity of rebound may be greater than acceptable.

Current shotcrete specifications are tending to become more performance oriented than in the past. For example, specifications for Washington Metro projects no longer require shotcrete aggregates to comply with certain gradations, but only that the final product must meet specified strength. The suggested gradation limits shown in Table 12-1 and plotted

Table 12-1

U.S. STANDARD SIEVE SIZES	PERCENT PASSING BY WEIGHT
$\frac{3}{4}$ inch	100
$\frac{1}{2}$ inch	75–95
$\frac{3}{8}$ inch	65–87
No. 4	50–70
No. 8	35–55
No. 16	20–40
No. 30	10–30
No. 50	5–20
No. 100	2–10
No. 200	0–6

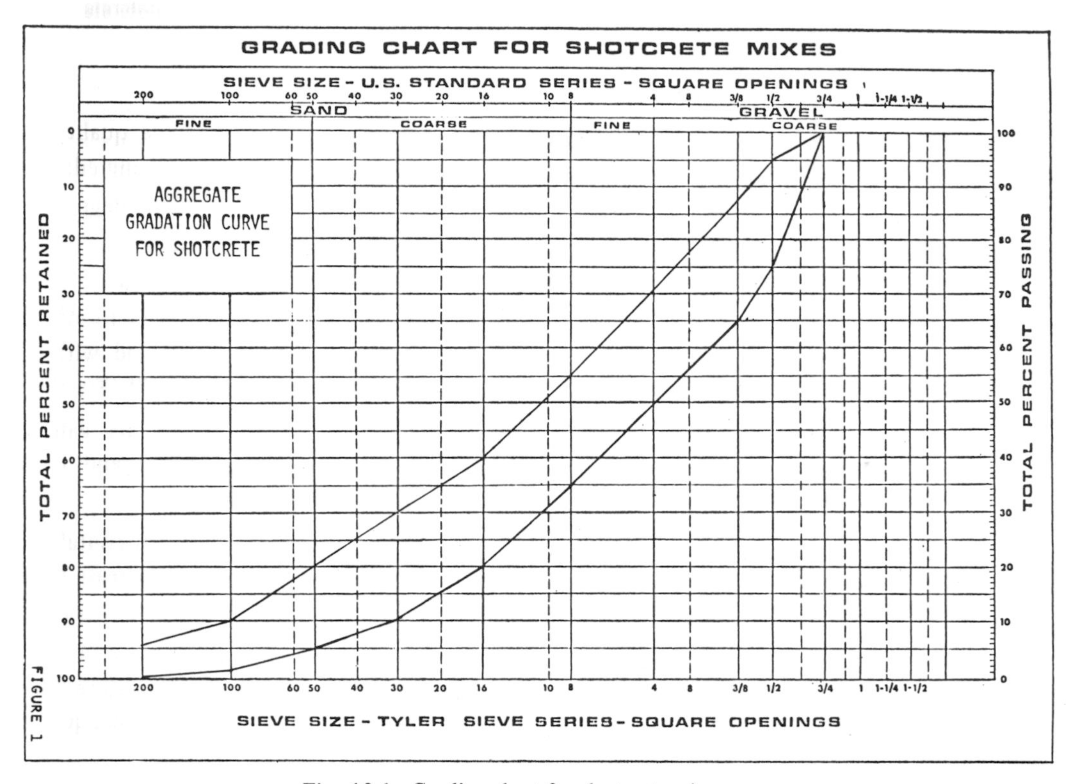

Fig. 12-1. Grading chart for shotcrete mixes.

graphically in Fig. 12-1 would be acceptable for most civil works.

Cement. Portland cement Types I, II, and III are equally acceptable. Selection of cement type often depends on compatibility with available accelerators. Optimum setting times are obtained with the most compatible cement-accelerator combinations. Compatibility of a cement of particular manufacture and a particular accelerator requires testing. Where shotcrete may be exposed to sulfate attack, a satisfactory sulfate-resistant cement (such as Type II) is required. Type III cement, a high early strength cement, may not require the use of accelerators.

Cement content of a mix is selected to satisfy the required specified strength of in situ shotcrete. Rebound raises the cement content of in situ shotcrete.

Cement content is also controlled by the strength requirements in relation to the maximum aggregate size. Unnecessarily high strength requirements will require excessive cement content, resulting in excessive shrinkage and cracking. For example, optimum cement content for the minus $\frac{3}{4}$-inch aggregate mix in the Vancouver Tunnel was 650 lb/cubic yard; at 775 lb/cubic yard, significant shrinkage cracking occurred. For minus $\frac{3}{8}$-inch mix, minimum cement content required was 700 lb/cubic yard.

Water. Water for mixing should be reasonably clean and free from suspended silt and organic matter, and from alkali and other dissolved mineral salts. Water which complies with applicable concrete specifications is acceptable.

In general, concrete strength rises as the w/c ratio is lowered (Fig. 12-2). However, Zynda[11] has shown that an optimum w/c ratio exists in the air-placing of mortar, below which the product suffers. The optimum w/c ratio occurs at the point of maximum density. To obtain maximum density, sufficient water must be added to overcome bulking of the material. At maximum density, the mass becomes plastic and will begin to flow or sag. "The internal

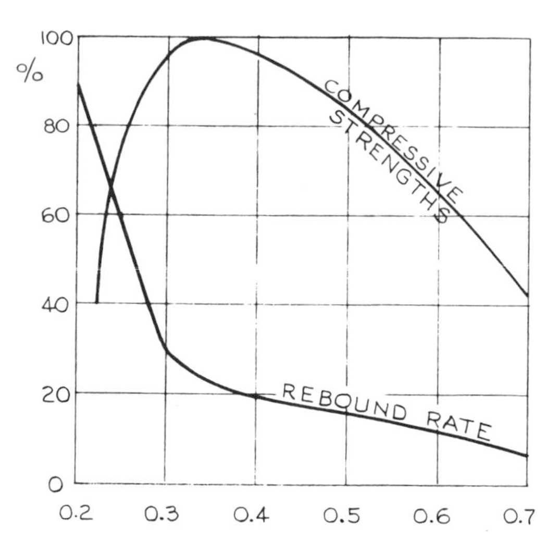

Fig. 12-2. Effect of w/c values on compression strength and rebound rate.

forces causing bulking must also increase rebound as the same forces resist compaction." Thus, the target is to place the materials at the "wettest stable consistency or point of incipient sag." The actual sag point is beyond the visual point of impending sag. The latter should guide the operator in manipulating the water content used. Correct water content becomes evident as a damp sheen forms momentarily on the surface of the newly placed shotcrete. Parker[12] reports an average in situ w/c ratio of 0.31 for a number of shotcrete test panels examined.

Accelerators. A conspicuous feature of tunneling with shotcrete has been its capability to perform satisfactorily under damp to wet ground conditions and to control heavy water flows. This has been due to the employment of inorganic accelerating set admixtures developed in Alpine Europe. Recently, organic accelerators have also been developed. A good example occured in the Mexico City Drainage tunnels.

A water flow of 900 gallons/minute was brought under control in the course of routine shotcrete support work and the tunnel wall was shotcreted. To permit shotcreting, the water flow was diverted through 1-inch and 2-inch diameter pipe nipples embedded in shotcrete. Commonly, these streams are piped through plastic hoses to drainage. For permanent support in colder climates, hoses may be embedded in shotcrete of sufficient thickness to prevent freezing. On occasion, the streams may be plugged to reestablish the water table.

The use of accelerators makes it possible to attain early set and hardening of shotcrete, thus facilitating its placement in the normal underground conditions of moderate to low temperatures.

Through the use of accelerators, the thickness of individual layers that can be placed is increased. Their quick-setting qualities permit early blasting against a new application (2 hours at Vancouver). They reduce rebound, too. Inorganic accelerators are very caustic, however.

Accelerator dosages are normally from 2 to 6% of the cement content by weight. Dosages are kept as low as possible for two reasons: cost, and, the ultimate strength reduction which accompanies the use of inorganic accelerators. In the field, two levels of accelerator usage should be considered, a lower level for normal usage and a higher level for primary coats of minimum thickness in difficult conditions of water or loose ground. The former should not produce a strength reduction of more than 30% from that obtainable with unaccelerated shotcrete, whereas in the latter usage, ultimate strength loss is unimportant. The ground has been temporarily stabilized, or is being drained, and structural thicknesses of normal strength shotcrete will follow.

Principal active accelerator ingredients in inorganic accelerators are sodium aluminate and sodium hydroxide with sodium carbonate and calcium hydroxide as buffers. Ingredient proportions are varied to meet variables in proportioning of the four principal Portland cement compounds. A certain type of accelerator may not be equally compatible with the cement of different manufacturers. Thus, cements and accelerator should be tested to permit selection of the best combination.

An effective liquid accelerator developed by Stabilator AB and Sika has been marketed for some time. Dosage is metered by a pump that feeds into the water line directed to the nozzle. This liquid accelerator also has an inorganic base that provides a strength loss of between 20 to 30% in 28 days. Lately an organic accelerator is being marketed which claims no strength loss in 28 days. It has shown much promise in detailed tests sponsored by the Urban Mass Transit Administration at the Atlanta Test Cavern, particularly when liquefied and fed directly into the water line rather than as dry powder into the cement-aggregate mix. Compared to the inorganic accelerator tested, the organic product produced faster setting time, lower 10-hour and higher 28-day strengths. Organic accelerators are not caustic, and are therefore safer to use.

To test accelerator-cement compatibility, the procedure outlined by Mahar[13] is more appropriate for shotcrete accelerators than the commonly specified ASTM C266. However, as the setting time varies with the w/c ratio, the w/c ratio must be specified. A w/c ratio of 0.35 is appropriate. Specified minimum setting times of three minutes initial set and twelve minutes final set have been successfully utilized on a number of projects. Due to the difficulty in achieving good reproducibility, and the difference in mixing action between cement patty and shotcrete ingredients, it is inadvisable to utilize the test results for other than verification of component compatibility.

Strength

Shotcrete has basically the same strength characteristics as concrete, and compression tests are the most common measure of overall quality and strength. Shotcrete strength is difficult to describe quantitatively due to a lack of standardization of testing and reporting procedures. While shotcrete is most effectively sampled by cutting either cubes or cores from test panels or the finished work, lack of standardization of size, configuration, test procedures and conversion factors to normal concrete sampling have resulted in confusion. Another difficulty arises in discussing European practice, which emphasizes the use of test panels to test shotcrete quality. Such a practice lends itself to "improvement" of the strengths by the shotcrete crew, and resulting high values may not be representative.

Shotcrete, even more than concrete, is a construction material of variable quality. A number of factors which influence shotcrete quality, as indicated by its compressive strength, are given below.

1. Shotcrete is essentially a low-slump concrete which has been consolidated upon impact. However, in order to remain in place without sloughing, shotcrete must set and harden rapidly. The required rate of hardening depends on the orientation of the surface on which it is placed, the thickness placed, the water content (slump) of the shotcrete, and the presence of water flows on the placement surface.
2. In order to produce the necessary rapid set and hardening, accelerators are frequently used in underground work. While accelerators produce the required short-term benefits, in the long term, conventional inorganic accelerators produce a strength loss usually between 20 and 30% at 28 days, compared with unaccelerated shotcrete.
3. The 28 days strength loss due to use of inorganic accelerators depends on several factors, such as the admixture dosage and admixture-cement compatibility. The required admixture dosage in turn depends on the reactivity of the cement due to its temperature, age, accelerator reactivity, the extent of prehydration which occurs when cement is added to moist aggregates, the thoroughness of mixing and field conditions.
4. When spraying shotcrete directly overhead, higher accelerator dosages are normally used than when spraying vertical surfaces. Higher accelerator content decreases the setting time and prevents loss of adhesion and sloughing which might occur if less accelerator were used. For spraying vertical surfaces, accelerator dosages can be minimized since the risk of sloughing is lower, and accelerator material costs are important.
5. The quality of shotcrete is affected by the quality of materials and each operation in preparation and placement. Variation in shotcrete quality can result from variations in the following: moisture present in the aggregates; mixing of cement and aggregates; blending of dry-mix and accelerator;

water quantity; mixing of water and dry-mix at the nozzle; materials velocity and subsequent impact compaction; and angle of application to the receiving surface. For overhead spraying, the nozzleman suffers considerable discomfort from rebound, which tends to adversely affect quality of workmanship. Material velocity is frequently decreased by lowering air pressure or increasing nozzle distance to the placement surface to relieve discomfort from rebound.

These factors explain the large divergence of strength values reported by writers using idealized test panels results, and field personnel who are involved with in situ testing. On the other hand, it has been demonstrated on the Washington Metro that shotcrete with an in situ strength equivalent to 5000 psi concrete can be produced in large scale, when required by design.

Development of compressive strength with time has been documented by Sallstrom.[14] He described tests made with nine combinations of one powdered and two liquid accelerators with a 1:4 mix of cement with minus $\frac{5}{8}$-inch sandy gravels. This mix was sprayed horizontally onto a panel attached to a wall. Two methods of modified penetration tests were correlated with the standard compressive strengths, and each sample was tested three ways. A compressive strength of 40 psi was found almost immediately on completion of spraying onto the form, and 155 psi in 30 minutes for the accelerated mixes. Relative times for these values for mixes without accelerator were 7 and 12 hours, respectively. However, it also was found that after three days, the shotcrete without accelerator rose to higher values than the accelerated material. Three to four of these tests suffered a 15% decrease from the unaccelerated values in 28 days, the remainder of a 30 to 43% decrease. Sallstrom also notes a 28-day strength of 5680 psi for unaccelerated shotcrete.

Common practice is to specify shotcrete compressive strengths at 8 hours, 3-day and 28-day ages by testing 3-inch diameter cores or, alternatively, 3-inch cubes. The intent of the procedure is to determine the early strength, when loosening pressures and ground movements must be impeded and the 28-day strength as a final or design value. Three-day tests are used mainly to give early warning of deficiency. The specified 8-hour, 3-day and 28-day strengths are generally about 800 psi, 2200 psi and 4000 psi, for cores with a length to diameter ratio of 2. However, different criteria are applied for each phase of quality control testing: preconstruction testing and performance testing during construction.

Flexural strength, or the ability to withstand beam loading, may be of particular importance to shotcrete's supporting function, particularly in blocky ground. Various relationships between flexural strength (or modulus of rupture) and compressive strength have been determined; for example, by Parker[12] and Seabrook.[15] The ratio of flexural to compressive strength appears to vary from 0.30 to 0.18 at one-day age and 0.18 to 0.15 at 28 days. For quality control purposes, it is more convenient to test compressive samples.

Another important property of support shotcrete is its bond to the ground surface. While testing methods are still under development, tests indicate that bond or pull-out strengths may be approximately one-quarter the compressive strength.[16] For practical purposes, the bond strength must be maximized in the field by careful surface preparation and spraying of the first shotcrete layer.

Quality Control Specifications

Quality control specifications were primitive for the Canadian National Railways Vancouver Tunnel in British Columbia. However, quality control was maintained by testing 3-inch cubes, cut from test panels during preconstruction testing, and from shotcrete slabs removed from the tunnel arch during construction. Since the cube strengths were substantially above specified compressive strength of 4000 psi, and far above measured in situ shotcrete stresses, shotcrete quality was deemed satisfactory and the initial support lining was approved to serve as the final lining as well.[7]

Difficulties in achieving specified strengths at the New Melones Tunnel in California, Hunter

Tunnel in Colorado and Nanticoke Cooling Water Tunnel in Ontario indicated the need for preconstruction testing of materials, equipment and skills. For the Washington Metropolitan Area Transit Authority (WMATA), detailed quality control specifications were prepared and first used on Sections A-4a, A-4b and C-5. Preconstruction testing was specified to ensure adequate quality of product before construction, and 3-inch diameter cores were tested for quality control during construction. Statistical requirements for data interpretation, from ASTM C94, were introduced.

WMATA Standard Specifications, published in 1973, offered the alternative of testing 3-inch cores or cubes for preconstruction testing. In addition, the concept of coefficient of variation and its effect upon required overdesign factors was introduced. Due to the statistical variation, core strengths exceeding the design strength, f′c, were required.

The U.S. Bureau of Reclamation followed a course parallel to WMATA, deriving their own quality control specifications from experience gained in a number of tunnels.[17] In addition, the Bureau performed laboratory and field studies on shotcrete,[16] which enlarged their knowledge of shotcrete properties. Bureau quality control practice is to utilize either 4-inch cubes or $2\frac{1}{8}$-inch cores from test panels for preconstruction testing, and $2\frac{1}{8}$-inch cores from in-place shotcrete for testing during construction.

The American Concrete Institute (ACI) also recommends preconstruction testing in their latest standard specification, ACI 506.2-77, by testing 3-inch diameter cores or 3-inch cubes. Testing during construction may be of cores taken from in-place shotcrete, or from test panels, as determined necessary by the engineer. Recommended preconstruction testing includes submittal of cube or core samples from one test panel per mix. Average core strengths for preconstruction testing must equal design strength f′c. Both the Bureau of Reclamation and ACI specifications contain factors to permit conversion of cube results to equivalent core values.

The Peachtree Station contract, section CN120 of the Atlanta Metro (MARTA), contains a number of innovations which were also included in revised WMATA Standard Specifications for Section A-11b, in 1977. The procedure outlined for preconstruction testing includes a test for cement and aggregate quality, utilizing standard concrete cylinders, and a statistically meaningful series of core testing from test panels. The required core strength for samples taken during construction is less than f′c, in accordance with ACI Standard 318-71, updated for shotcrete in ACI 506.2-77.

For WMATA A-11b, which specifies in situ strength, f′c, of 5000 psi, preconstruction testing of cores must achieve an average compressive strength of 5540 psi at 28-day age. This test average includes correction factors for comparing cores to cast cylinders, for L/D ratio of cores, for sample size, and for an assumed coefficient of variation.

Testing of shotcrete during construction for A-11b is in accordance with ACI 506.2-77, which requires three 3-inch diameter cores for each sample of the structure in question. The average strength of these cores, when corrected for L/D, must equal or exceed 85% of required strength, f′c (5000 psi), with no single core less than 75%. By adjusting these values for L/D equal to one, a required average strength of 3 inches long, 3-inch diameter cores is 4670 psi with 4120 psi minimum strength.

Shotcrete is commonly used either for initial support and final lining or as initial support only, with a final lining placed inside of the shotcrete layer. In the former case, some rational structural design method is frequently used, unless the shotcrete has merely a protective rather than structural role. If a structural design is utilized, the designer must use some portion of the design shotcrete strength, f′c, as an allowable stress. He therefore must be assured that field shotcrete strengths will not fall below assumed ultimate stress, within the limits of statistical probability. For permanent structures in public use, such as subway stations for WMATA, this assurance is essential.

On the other hand, the U.S. contractual climate favors the use of shotcrete for initial support only, to be installed at the contractor's option, and at his discretion. If such is the case, it may be counterproductive to the owner's interests to insist upon complying with stringent preconstruction testing and quality control specifications. A prudent contractor might decide to avoid such inconvenience and utilize more expensive steel rib supports. In any case, the contractor is responsible for the safety of his work, so the quality of shotcrete used for initial support may not be of critical concern to the owner.

In other cases, shotcrete may be used as a protective and structural final lining, for which a rational structural calculation is difficult to justify. In shotcrete-lined water, utility and hydroelectric diversion tunnels, for example, high compressive strengths may be unimportant. In such cases, overall quality of workmanship may be of greater importance, to minimize construction defects where local failure could occur. Quality control specifications should therefore reflect the need for product uniformity.

SHOTCRETE APPLICATION

Proportioning and Mixing

Moisture content of the mixed coarse and fine aggregates before mixing with cement should not range outside 3 to 6%. Irregularities of moisture content in the feed result in irregularities of feed through the nozzle, adversely affecting shotcrete quality. The natural tendency of granular material to drain should be utilized to control aggregate moisture.

Moisture is largely contained in the sand fraction, where drainage is slower than in the stone fraction. Both will drain given a base from which drainage water can escape. In the 200-inch annual rainfall at the Alto Anchicaya Project, Colombia, sands dug from the river were dumped into a hopper and drained, then stockpiled for 24 hours. The resulting mix with $\frac{3}{8}$-inch gravels was less than 6% moisture. In Mexico it was found impossible to maintain the moisture content within desirable limits in a premixed sand and $\frac{5}{8}$-inch gravel mix during the rainy season. While plugging of drop pipes, hoppers, shotcrete machine and hoses was frequent, the final product was not adversely affected since volumetric batcher-mixers at the heading were utilized and prehydration of cement therefore was minimized.

If dry, the sands should be wetted and thoroughly mixed to an approximate 8% moisture content before mixing with the stone fraction, and moisture should be checked before addition of cement. Overly damp aggregate-cement mixes block delivery lines and increase hydration rates to beyond an acceptable level. Too dry a mix presents wetting problems at the nozzle, with increased dust and decreased compaction. Wetting devices installed on conveyors leading to the shotcrete machine have been successfully utilized.

Transportation and mixing units were developed to mix aggregates and cement in the tunnel for immediate delivery to the placer machines. Proportioning and mixing is done by graduated screws feeding the components at relative volumes to deliver a predetermined mix of cement and pre-mixed aggregates, (Stabilator AB of Sweden trixers); or with adjustable belt feeds delivering concrete mixes from separate bins of cement, sand and coarse aggregates (National Concrete Machinery of Lancaster, Pennsylvania concrete mobiles). Eimco and Conspray also manufacture mixing and transportation equipment for shotcrete.

These units have their obvious advantages, including the provision of ideal accelerator-cement-water reaction conditions. However, they are more sensitive to irregularities of moisture content altering bulking ratios of the aggregates. There is less opportunity also to alleviate segregation from vertical drops, and there is variable wear and tear between aggregate and cement screws. A close scrutiny of the product must be maintained with screen and cement content analyses.

Powdered accelerator should not be added to the dry-mix until it is entering the placer. A mechanical feeder is preferable to manual dosification. The screw feeder type provides positive measure, whereas the vibratory type tends to clog, since accelerators are hygroscopic. Liquid accelerator should be mixed with the mix water before it is directed to the nozzle. Care must be taken to ensure that accelerator feed is geared to the material delivery rate so as not to produce surges of high accelerator content.

Placing Equipment

Two types have been developed for spraying dry-mix:

1. *Dual pressure chamber* design, with intervening air lock, in which dry-mix is introduced in batches into the upper chamber. Then it is delivered under pressure and gravity to the lower chamber when the air lock is opened. From the lower chamber, measured charges under pressure are fed by a spocket wheel into the delivery hose. In the meantime, the upper chamber is being recharged. Feed into the delivery hose is continuous.
2. *Rotary design,* in which dry-mix is fed continuously into the hopper of the machine, where it falls into a series of nine or more holes in a rotating cylinder or bowl. Much as in a revolver, the charge in each hole is activated by air pressure as it passes under cover, and is forced by an airstream into the delivery hose.

Both types of machines can be equally satisfactory. The dual chamber type has the disadvantage of demanding constant labor to perform at capacity. Several rotary types are manufactured, and currently are tending to replace the dual chamber pressure vessel placers. Performance of rotary types is largely a matter of drive power employed. Placers vary in maintenance requirements and in dust-making. If lightly constructed and operated at high rotor velocities, they may require early replacement. The pressure chamber types have less maintenance problems, but their production is generally less.

Ratings for the shotcrete placing machines are from 7 to 11 cubic yards/hour through the nozzle. Maximum delivery distances vary widely with different machines, but can exceed 900 feet horizontally and 300 feet vertically for the better designs. In extreme delivery distances, steel pipe can be substituted for hoses in straight sections, to lessen wall friction.

Conveying

Dry mix is conveyed to the placer machine by the most suitable means. With rubber-tired tunneling equipment, dry-mix bins and placers may be mounted together on a truck, with a hydraulic boom providing a movable platform for the nozzlemen. With tracked equipment, the bins and placers may be mounted on the drill jumbo, as in the Vancouver Tunnel. There, bins were supplied by dry-mix cars which were elevated by a cherry-picker onto the jumbo and dumped into the bins.

Alternatively, the bins or dry-mix cars and the placers may be separately mounted, and worked as a unit on a switch behind the jumbo. An Italian contractor in Milano used dry-mix cars with bottom gates feeding a belt conveyor to the placer. In the Mexican project, screw mixers (trixers) fed directly into the placers, both located on a switch behind the drill jumbo.

Deliveries of dry-mix to headings with access through vertical shafts can be made by dumping the dry-mix down an open pipe. Regardless of delivery method, each batch should await mixing until called for, and be dumped, transported and sprayed before allowable standing time has been exceeded. At the Vancouver tunnel, the mix was allowed to stand two hours, providing hydration had not raised its temperature 10° above the temperature at the time of the mix. Such empirical limits must be reviewed for each project, as cement and accelerator reactivity vary greatly. With portable mixers such as trixers and concrete mobiles, cement is added when the mix is being processed into the placer. The cement and aggregates thus may stand indefinitely or be stored in bins.

Delivery pipes 12 to 20 inches in diameter have been used. They should be installed without bends and hence should be drilled into tunnels outside the shaft perimeter, and grouted to seal out water. The large diameter was found to be necessary in Mexico, because of over-damp deliveries of mixed aggregate. Mixed aggregate cakes and bridges more easily than dry-mix, where included cement apparently provides lubrication. Baffles should be placed at the discharge end to reduce segregation. Height of drop in the Vancouver Tunnel was 120 feet. In the Mexico project, drops ranged from 160 to 730 feet. Discharge at the Orange-Fish Tunnel in South Africa was through a bend in the pipe into Moran cars with drops to 1100 feet. The bend increased segregation radically.

Spraying

Of first importance in constancy of air, water and materials flow into the placer machine and through the nozzle. While minor discrepancies may receive adjustment by a skilled nozzleman, good shotcrete cannot be made when the spray beam varies in composition, or is sporadic.

Air and water must be maintained at constant pressures, and excessive moisture removed from the air with traps if necessary. Air pressure should generally be 55 psi at the intake of a rotary placer, and increased 5 psi for each 50 feet of hose added to the first 100 feet. For the pressure chamber types of placer, air pressure commences at 22 psi. A gauge should be used to monitor air pressures, and the pressures controlled by cracking the placer inlet valves as required. Water pressure should be maintained at 10 to 15 psi above inlet air pressures, to obtain adequate wetting at the nozzle. Placement of the water ring up to 10 feet from the hose outlet has resulted in improved mixing and final product.

The nozzleman should always be able to take a spraying position which is normal to, and at a distance of between 3 to 4 feet from, the rock surface (Fig. 12-3). Thus, staging or an equal substitute generally is necessary. These may take the form of hydraulically operated platforms, or strategically placed platforms on a drill jumbo.

Unsupported shafts or raises can be shotcreted from the muckpile as they are being drawn down. If the muck is overdrawn, staging will be necessary to obtain proper spraying positions.

Surface Preparation

Proper shotcrete to rock bond is probably the most important requirement of shotcrete placement. To obtain optimum bond, the surfaces must be cleaned of dust and loose foreign matter, and left damp for spraying. Cleaning with air and water from the nozzle is inadequate. To properly remove blast deposits and the like, a blowpipe should be used, attached to line pressures of air and water discharging through a squeezed $\frac{1}{2}$-inch nipple. Air and water pressures can be regulated at the valves.

Rebound losses vary with the properties of the mix, its delivery and the convenience of

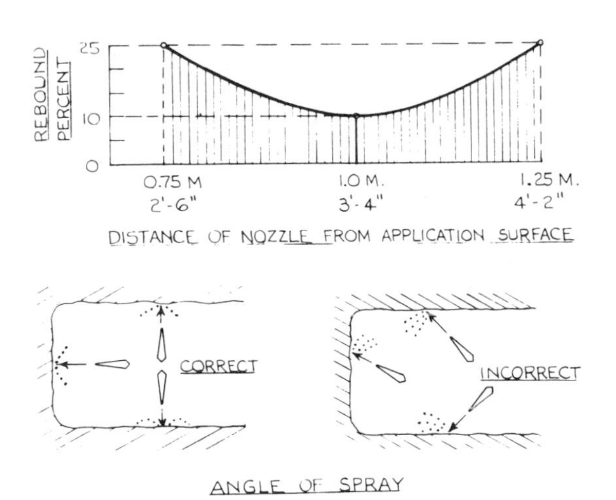

Fig. 12-3. Rebound, distance of nozzle and angle of spray.

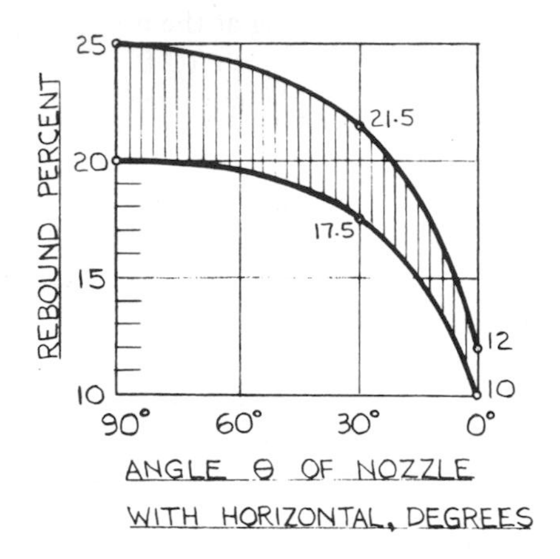

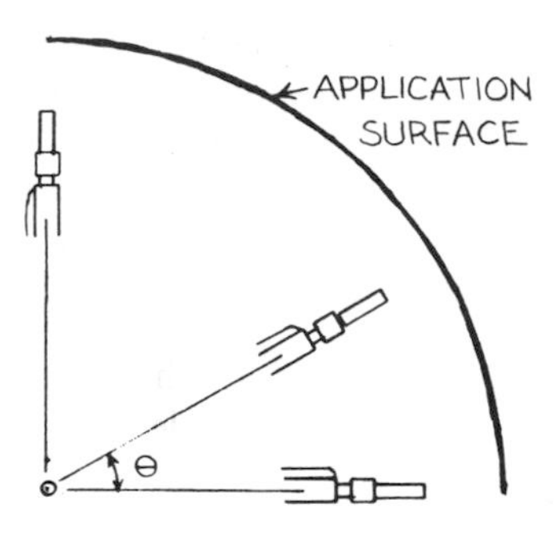

Fig. 12-4. Effect of gunning direction on rebound.

placement (Fig. 12-4). Damp surfaces and water flows increase rebound losses.

Rebound is increased also by gap grading; segregation in the feed; excessive or insufficient discharge velocities; insufficient or pulsating water pressures; irregular material or accelerator deliveries into the placer machine; and improper placer machine operation. Rebound also varies with the hardness of the surface being sprayed. Therefore, it is high for the initial thin layer of shotcrete, but for subsequent layers the sprayed materials are embedded in the plastic initial layer. It follows that rebound is higher for thin coatings than for a thicker layer placed in one operation.

When the nozzleman sprays downwards, the rebound falls back on the sprayed surfaces and is buried as spraying continues. The invert, therefore, is better poured. Rotary placers can be adjusted to serve this purpose, or a cement-enriched mix with a high w/c ratio can be sprayed.

Sequence of Operations

The primary function of shotcrete is to halt loosening of freshly exposed and blast-shaken rock surfaces. Thus, shotcreting should commence as soon as possible after the round is blasted or excavated, whether it is to seal loosening joints and fractures or the minute fissures that form as the ground relaxes. Shotcrete should be applied within two or four hours when in questionable ground. However, rock structures exist with more extended standing times and others are largely self-supporting. In such cases, shotcrete or gunite may serve as a permanent overlay in a largely self-supporting excavation where minor loosening is possible; for example, in underground powerhouse caverns supported by rock bolts.

The tunnel arch requires the first attention. Spraying is sometimes started from the muckpile. In tunnels of 20 feet in height or more, it is possible to extend a movable platform from the drill jumbo over the muckpile for spraying the arch while mucking. The loading equipment would have to suit this situation. Alternatively, mucking and spraying may be done side by side if the width of the tunnel permits. Shotcreting should start from near the face to provide the arch there with the earliest set, and should be capable of withstanding a succeeding blast within two hours of the end of the mucking cycle. The balance of the specified thickness can be completed on subsequent rounds during drilling.

Walls are usually shotcreted in one pass during drilling. In some instances the walls may need earlier attention, as, for example, in the Pastora Interceptor (Mexico City Drainage Tunnels). Here the soft ground broke down under heavy water flows, since the depressed water table was below springline. In this instance, a thin spray coat was necessary to hold the roof temporarily. Except in small tunnels, spraying can be integrated without delaying excavation procedures.

Field Quality Control

A nozzleman's work includes degrees of discomfort that tend to influence his performance. It is more difficult and unpleasant to spray upward into an arch than horizontally. Blind areas also may exist that are difficult to spray from fixed platforms. Both situations tend to produce deficiencies in thickness. In more questionable ground, constant inspection is mandatory, to ensure that spraying is done properly and to necessary thicknesses.

Gauge pins are commonly placed at 3- to 5-foot centers to estimate thicknesses applied. They do not, however, assure the thicknesses implied. At the Vancouver tunnel, percussion test holes of $2\frac{1}{2}$-inch diameter were drilled, three each round, at random spots, to actually check distances to rock.

Mechanized Spraying

The discomfort of the nozzleman's work underground, at times under a shower of rebound, and the necessity of spraying loosening ground as early as possible after blasting, motivated the mounting of the nozzle at some distance ahead of the operator. A hydraulic boom with a rotary head similar to percussion drill mountings on a jumbo is used. The boom is extendable and freely movable in any direction forward of the operator. The nozzle is mounted normal to the boom with additional hydraulics controlling its direction within the arc, and in some cases an automatic back and forward movement in about a 20-degree arc.

Stabilator AB of Sweden has developed probably the most sophisticated type of this equipment, and has patented it under the name of Robot. Mounted on track or rubber-tired vehicles, such equipment permits shotcreting to be commenced immediately after ventilating the blast, and without preparatory scaling. Unless the tunnel is wide, the machine holds up mucking. On this continent, similar nozzle booms have been developed in mines, and are being manufactured by Eimco and Conspray.

Automatic spraying has been employed in at least two instances, both following machine-bored tunnels. At the Heitersberg Tunnel (35 feet in diameter) near Zurich, Switzerland, an automatic system was designed by Stabilator AB to follow a Robbins mole. Two nozzles were mounted that automatically sprayed a complete arc from a precast invert. The operator was limited to adjusting water feed and rate of traverse, and moving the unit forward at the end of each arc. The unit was 70 feet back from the face, however, which, in the less competent lacustrine deposits, necessitated pattern bolting immediately behind the mole. A small manual robot was also mounted close to the cutterhead to permit early support application if necessary.

At the Stillwater Tunnel in Arizona (16 feet in diameter), a nozzle was mounted experimentally on a Joy hydraulic drill boom with rotary head, following a mole. The unit was advanced at a predetermined pace, the nozzle simultaneously moving through an arc rather like a windshield wiper. The experiment has not been repeated, however.

Automatic spraying does not appear to lend itself to the irregular outlines resulting from drilling and blasting.

SHOTCRETE USAGE

Support of loosening pressure has been the main concern of much of drill and blast tunneling. A thin layer of shotcrete is commonly employed to seal an excavation against loosening. Shotcrete also is being used over a broad range of underground conditions and for a variety of purposes.

- Special functions: control of groundwater; local repairs and rehabilitation of caving or caved ground.
- Swedish practice: initial support against loosening, with or without rock bolts; as a final support.
- New Austrian Tunneling Method (NATM): structurally integrated with rock bolts and light steel ribs, to stabilize weak or soft ground and soil.
- WMATA: structurally integrated with rock

bolts and steel ribs in reasonably good ground; as final support capable of meeting strict standards of public use and safety.

Special Functions

An instance of shotcrete use in caved ground was the stabilization of overstressed steel ribs against a massive cave in squeezing ground and subsequent remining through the cave, at the Tide Lake Tunnel for Granduc Mine in British Columbia. Lateral deformation in the shotcrete embedding the steel ribs reached 9 inches within weeks, without seriously cracking the shotcrete. Overstressed liner plate in squeezing ground at the Carley V. Porter Tunnel in California was similarly stabilized.

For the Mexico City Drainage Tunnels, shotcrete was used extensively in conjunction with consolidation grouting to control groundwater flows. In addition, shotcrete equipment and crews were utilized as "firemen" to try to limit, stop or advance through cave-ins in unstable ground, often in headings where shotcrete was not routinely used.[18]

Swedish Practice

One of the largest sources of tunneling experience with shotcrete is that assembled in Sweden and Norway. The geology of the Scandinavian Precambrian shield undoubtedly has influenced the direction of its development, particularly in its avoidance of auxiliary steel members. The rocks, gneisses and intrusives generally are harder and more competent than in Alpine conditions. Alberts[19] explains the Swedish practice as follows: "Shotcrete is forced into open joints, fissures, seams and irregularities in the rock surfaces, and in this way serves the same bonding function as mortar in a stone wall."

Basically, the function is to increase the shear strength of the shear planes in the rock and prevent loosening. Recent research, both in Norway and in the U.S., indicates that the shotcrete-rock bond is of greatest importance to resist block movement.[20,21]

As described by Cecil,[22] an initial 2 to 4 inches of shotcrete is applied as a general rule. A second 3 to 4 inches is added only if the first layer shows signs of deterioration.

Shotcrete application in Swedish practice is usually of non-uniform thickness and sometimes discontinuous. Reinforcement of the shotcrete application is a field decision of supervisory staff that have a wealth of experience. Operators also are well grounded in underground conditions, and skilled in their work with particular concern for potential loosening. An excellent analysis of loosening mechanisms and their reaction to various forms of support, including shotcrete, is given in Ward.[23]

In North America, the U.S. Bureau of Reclamation also has utilized shotcrete as a sealing and semi-structural lining in a number of gravity water supply tunnels.[17] Generally, not more than 3 inches thick shotcrete liners are applied in portions of tunnel which would ravel if unprotected, while lengths of tunnel in good quality rock are left unsupported. Where unfavorable ground conditions prevail, however, conventional steel ribs and lagging, followed by cast-in-place concrete, are utilized.

In the Stockholm Subway, shotcrete provides initial support and final lining for cavern stations in excellent quality rock. The lining is also utilized as the architectural finish, utilizing colors and even abstract sculpture and pictures.

New Austrian Tunneling Method (NATM)

The name was attached to differentiate it from the Austrian Method used when timber was the sole supporting means. In these earlier times, several national names were applied to particular methods of conducting multi-heading excavation of a face too large to be supported in full face, such as Austrian or Belgian. The NATM was first developed by Rabcewicz, with early contributions by Muller and Packer, all Austrians.

The NATM is a flexible method of tunnel excavation and support which is adaptable to varying ground conditions from hard rock to soil. Support of the opening is provided by constructing a "carrying arch" of the in situ ground, with shotcrete preventing loosening and rock bolts reinforcing the structure.

Performance of the carrying arch is monitored with instrumentation, and modifications are made as required. In highly stressed ground, the objective is to construct a system to minimize loosening, which contributes to higher final loads, but permits load-relieving plastic deformation to occur, thereby reducing the tectonic load on the system. In soft ground under typical urban conditions, the method is adapted to minimize all deformations, thereby controlling surface subsidence.

For large openings in soft or overstressed ground, a typical support system utilized in the NATM is comprised of these elements:

- Welded wire mesh
- 4 to 12 inches of shotcrete
- Lightweight (10 to 25 lb/foot) steel ribs
- Grouted rock bolts.

While all of these elements are deemed necessary, the largest structural contribution is from the rock bolt-induced carrying arch.

Proponents of the NATM have continued to expand the application of this composite system. In the 35-foot span Tauern Tunnel, a section of 1000 feet presented a problem in unexpected ground conditions.[24] Excavation was in a zone of crushed graphitic phyllite schists. Designed support was 6 inches of shotcrete reinforced with 46-lb/foot steel ribs to distribute pressure on the liner, and 20-foot rock bolts spaced at 30 inches. In most of the length, maximum displacements reached 10 inches, which could be tolerated and repaired. A small section reached 30 inches at one point. Voids were initiated behind the shotcrete with partial destruction of the liner.

The problem was resolved by lengthening anchors, some thickening of shotcrete, and provision of 6-inch wide longitudinal contraction joints in the shotcrete liner. On occasion, these open slot contraction joints closed completely in two to three days. Damages were repaired and voids grouted in a matter of six to eight weeks and routine advance was resumed.

In a double-track section of the Munich Subway,[25] NATM replaced a conventional rigid final concrete support design. The work was done downtown and subsidence was the critical factor. The ground was dense, micaceous sand overlying hard to very hard clays at springline of the excavation. Maximum span was 40 feet, and maximum height 31 feet.

Initial stages of excavation were two vertically elliptical sections at the outside walls, from invert to above the springline of the arch. With these stabilized, the arch was completed. Construction of the inner remnants followed, which included inside walls between the trackways and the invert curvature. Stabilization of the outside sections provided pillar-like support for the large roof section, the most sensitive section of the work. Outer ring shotcrete was 10 to 11 inches, reinforced with light steel ribs. System anchoring was with fully grouted rock bolting.

Final subsidence was $\frac{3}{4}$ inches, 50% resulting from excavation and 50% from lowering the groundwater. Savings in construction costs, including thinner final concrete liner, amounted to 30% of the rigid concrete liner design.

For initial support, the NATM has been found most useful in two applications: large tunnels in tectonically stressed rocks, often at great depth, and shallow soft ground tunnels, frequently of varying cross-section (such as subway stations and crossover structures). Final lining of these works is generally of reinforced cast-in-place concrete.

Shotcrete used for initial support and final lining with or without rock bolts in good quality rock is also referred to as the NATM in Austria, but in application is no different from the Swedish Method.

WMATA Rock Cavern Stations

For the Washington Metropolitan Area Transit Authority (WMATA) subway work, shotcrete has played a key role in initial support and final structural lining in nine rock cavern stations. These caverns are approximately 60 feet wide by 45 to 56 feet high, located in hard metamorphic rocks with varying degrees of weathering, jointing and shear zone occurrence.

The caverns are supported with initial shotcrete, rock bolts and heavy steel ribs embedded in a final shotcrete lining. Free-standing, nonstructural architectural linings are installed inside of the structural support. Apart from station and crossover structures, shotcrete is used on WMATA projects predominantly for initial support.

Cavern construction generally utilized a pilot tunnel located in the center of the crown, from which prebolting of the future crown was performed for Dupont Circle Station.[26] The main function of the pilot tunnel, driven before the station is contracted, is to provide detailed geotechnical information for bidders and subsequent contract administration. Main cavern excavation followed in five steps at Dupont Circle and three steps at Rosslyn Station.[27] The entire top heading was excavated in one step at Rosslyn Station, utilizing drilled-in spiling bars cantilevered ahead of the face from steel ribs.

SHOTCRETE SUPPORT DESIGN

Shotcrete design, as with all components of underground support, is imprecise. For normal conditions, that is ignoring cases where shotcrete is used for control of groundwater and caving ground, design may described for three distinct cases:

1. Initial and final support against loosening pressures are best dealt with by "rules-of-thumb" or experience factors.
2. Initial support in ground which reacts plastically, either highly altered rock at great depth, or in shallow soft ground tunnels, may be dimensioned with procedures developed for NATM.
3. Initial support and final lining of large openings in jointed rock, where use of a composite shotcrete-steel rib beams analysis may be appropriate.

Loosening Pressure Design

For initial support and final lining against loosening pressures, a number of design procedures have been proposed; for example, by Sutcliffe[28] Heuer,[29] the U.S. Bureau of Reclamation[17] and Cecil.[22] Basically, these analyses attempt to rationalize shotcrete behavior by crude modeling or simply experience rules. Both methods are useful, as they are mainly based on successful experience and practical limitations. For example, Heuer proposes a hoop stress analysis which produces a required shotcrete thickness. An additional 2 to 4 inches of thickness is then added, to bring the results into line with experience and practical application, and to account for irregularities in excavation.

Other design analyses have been proposed, such as peripheral shear or diagonal tension of sliding blocks.[13]

While all of these procedures are useful, their use should be restricted to confirmation of design based upon experience and a thorough study of the anticipated ground behavior. For permanent linings particularly, attention to reported previous design failures (for example, by Selmer-Olsen) describes a number of failures attributed to swelling clays contained in fault zones, which were affected by post-construction changes in moisture. He also reports failures due to frost action.

NATM Design

The NATM design theory and calculations have been described and updated by several writers.[24,31] The theoretical model is the formation of a plastic zone or "carrying arch," which, if not properly supported and reinforced, will proceed to progressive failure, first of the walls and finally of the arch. Failure occurs by shearing the upper and lower sidewalls. Such a failure mode has been documented in overstressed ground conditions.[18,24,32]

In the NATM design procedure outlined by Golser,[24] the calculations for dimensioning of shotcrete, rock bolt and steel rib components are used for an initial design for various classifications of anticipated ground conditions. Rock mechanical properties and experience from previous projects are utilized in the initial design procedures.

During construction, the ground conditions encountered are classified by contractor's and owner's representatives and the initial design for that classification utilized. Extensive monitoring of ground and shotcrete pressures, convergence of tunnel walls and arch and carrying arch deformation are performed, and this data utilized to confirm or change the classification, or modify the design within the classification. For large tunnels in overstressed ground, rock bolting provides a much larger contribution to support than the shotcrete lining, so that an increase or decrease of rock bolting is the most frequent change within a classification.

More sophisticated analysis of the NATM has been utilized for several projects. Rabcewicz[33] describes a finite element analysis of the as-built Tarbela Dam gate chamber to have been less than successful. On the other hand, Laabmyr[25] reports some success with using analytical methods for a section of the Munich Subway. It is doubtful if such analyses are used as more than theoretical verification of design and procedures resulting from experience.

Underground Rock Cavern Design

A number of underground powerhouse caverns have been constructed with rock bolt support, shotcrete being applied as a protective covering long after excavation is completed (for example, at the Mica Powerhouse[34]). For station caverns constructed in rock for the Washington Metro, however, a composite of shotcrete and steel rib initial support and final lining has been utilized. Rock bolts, also utilized, have not been considered as part of the permanent support structure.

Due to the lack of confining pressure from low overlying cover, groundwater and unfavorable jointing and shear surface orientations, steel rib-reinforced shotcrete linings are employed. Design procedures of these structures vary with the designer's confidence in the supporting function of shotcrete, but generally follow reinforced concrete design. The structure is designed to support theoretical loads caused by rock blocks of assumed dimensions. Rock block dimensions are related to the known joint and foliation orientations. Monsees[35] describes the design procedure for Bethesda and Medical Center Stations, where elastic beam elements and boundary elements were used to model the composite lining and interacting rock. Moments and thrusts were calculated using the SAP-IV computer program and total stresses compared to allowable stresses and those measured in the previous Dupont Circle rock cavern station. For normal conditions of tunnel design, such design methods may not be practical.

SUPPLEMENTARY SUPPORT

Rock Bolts

Rock bolts are used to secure locally loosened blocks of rock, or in systematic fashion at predetermined spacing to restrain loosening and plastic deformation in the arch of a tunnel, alone or in conjunction with shotcrete in both forms. In the latter instance, bolting should follow placement of the first 2 to 3 inches of shotcrete, to gain full advantage of the high early strength obtainable in shotcrete, and to increase the standing time of the arch.

Expansion anchors are frequently used in good rock, for initial support. In the softer rocks, where anchorage is difficult to obtain with an expansion anchor, some form of bolt grouted with either cement mortar or polyester resin is necessary.

Used in conjunction with shotcrete, bolts are most conveniently installed from the back of the drill jumbo. It is possible to devise an effective support arrangement in the stronger rocks with tension bolting in conjunction with a less than structural thickness of shotcrete. For larger cross-sections, grouted radial bolting may be necessary in addition to structural shotcrete. For the high tectonic pressure tunnels driven with the NATM, such as Tauern and Arlberg, the measured effect of bolting is much greater than that of the reinforced shotcrete liner.

While post-tensioned expansion bolting has reached a certain sophistication in large underground structures such as powerhouse complexes and domed nuclear chambers, untensioned grouted bolting is becoming more commonly used in tunnels and in mine stoping. Many of these stopes have spans as wide as are found in powerhouses. Certain of the larger Canadian and South African mining companies have opted for the untensioned grouted bolt, consisting of driving standard rebar steel into a freshly grouted hole. Much of Swedish practice is such.

In North America, the use of polyester resin encapsulated rock bolts has become prevalent. Although more costly than cement mortar rock bolts, they are more convenient to install and harden to design strength in minutes. The quick hardening attribute of polyester resin rock bolts makes them particularly appropriate for use with quick-setting shotcrete under unfavorable conditions.

In Austria, quick-setting mortar rock bolts are used which contain a glass vial of accelerator at the end of the holes, thus permitting rapid bolt tensioning if required. For the NATM, an entire range of cement mortar rock bolts is employed, including pointed hollow tubes with holes in the tube walls. These anchors are driven into alluvium and gravels with a pneumatic hammer and then grouted under pressure.

Wire Mesh Reinforcement

The use of wire mesh reinforcement is obviously based on an educated distrust of unreinforced concrete. Placed in structural thickness and properly bonded, shotcrete flexes with the rock in its readjustments and appears capable of withstanding the stresses measured to date. Local failures tend to form slowly and at an early stage, and can be repaired without catastrophic consequences.

The general use of steel wire mesh has disadvantages. It may have an unfavorable structural effect as it ties the whole shotcrete structure together, thus hindering it from reacting freely to localized ground rearrangements. Thus, on at least one occasion in the Mexico City Drainage Tunnels, a major tunnel collapse was initiated where only a local failure was present. Installation of mesh over irregular rock surfaces requires large quantities of shotcrete for cover. Rebound pockets form points of weakness behind the mesh. Due to the unavoidable vibration of the mesh as it is being sprayed, bond between rock and shotcrete suffers.

For the NATM initial support application, wire mesh is generally incorporated. However, in much of this work, light steel ribs are also used.

In portions of the Mexico City Drainage Tunnels, wire mesh reinforced shotcrete ribs, 2 feet wide and 1 foot thick, were used to provide additional rigidity. This practice, which originated in Sweden, was not a cost-effective one on the Mexico project.

In recent years, the use of steel fiber in shotcrete has been tested in a number of applications. For a rock slope stabilization, Kaden[36] reports that all contractors bidding alternative designs of wire mesh shotcrete and fiber shotcrete chose the latter, due to the inherent saving in total shotcrete volume.

Use of steel fiber shotcrete underground has been limited. Testing of fiber shotcrete shows slightly increased flexural and tensile strengths over conventional shotcrete, with no appreciable change in compressive strength. Fiber does not give the same level of tensile strength as wire mesh and shotcrete, however.

Steel Ribs

Ribs constitute the last line of defense in reinforcing a shotcrete liner when its shear strength is considered unequal to presumed loadings. Heavy steel ribs were utilized in the Mexico City Drainage Tunnels for advancing through a caved zone of volcanic rock metamorphosed to clay.[18] Shotcrete support of wall plate drifts, and embedment of the ribs, ensured an uneventful construction of this difficult section.

In the NATM, steel ribs are lightweight, U-shaped in cross-section, and used only for reinforcement. Ribs are bolted together, sometimes with sliding joints where contraction slots are left in the shotcrete initial liner, and fixed to the ground with rock bolts. No attempt is made to support the rib base, nor is a base plate provided. In zones of overbreak, the ribs are sometimes deformed into the overbreak area, so as to reduce quantities of embedment shotcrete.

As previously described, steel ribs provide substantial capacity to the support systems for Washington Metro stations. The technical necessity of such strong supports in some cases is blurred by contractual complications, where designer and construction supervisor are two separate parties.

Spiling

Steel spiling bars and sheets have been utilized in conjunction with shotcrete in several reported cases. In Mexico, steel spiling bars used with the aforementioned mesh-reinforced shotcrete arches resulted in a catastrophic failure of the lining following a blast. Approximately a 65-foot length of tunnel lining, tied together with spiling bars, collapsed while the overlying soft rock maintained itself essentially intact.[18] In similar ground, spiling bars used with, and welded to, steel ribs were successful, as were conventional shotcrete and radially-placed grouted rock bolts.

Steel spiling sheets were used in some sections of the Tauern tunnel, where a landslide debris slope was penetrated. In Munich's Metro Line 81, spiling sheets were used in uncemented sandy gravel. Sheets were driven ahead of the advancing top heading, using fairly heavy steel H ribs as a template. All ground surfaces, including face and invert, were covered with at least an inch of shotcrete.

REFERENCES

1. Keifer, O., "Multiple Layer Shotcrete Lining," ACI Publication, SP-14, 1966.
2. Rotter, E. "Spritzbeton und seine praktische Anwending, im Untertagebau," Symposium at Lesben, Austria, 1960.

3. Zanon, A., "Excavation of Super Highway Tunnels in Very Difficult Formations," translated by B.E. Hartmann, Terrametrics, Golden, Colorado.
4. Quadrio, Curzio P., "New Methods for the Construction of Tunnels," Il *Monitore Tecnico* 1.
5. Rabcewicz, L.V., "The New Austrian Tunneling Method," *Water Power*, 1964.
6. Karlsson, L. and Fryk, J.O., "Lining for the Intake Water Tunnel of the Holjes Hydroelectric Plant in Sweden," *Hoch-und Tiefbau* 50, 1963.
7. Mason, E.E., "The function of Shotcrete in Support and Lining of the Vancouver R.R. Tunnel," *Tunnel and Shaft Conference*, University of Minnesota, 1968.
8. Cecil, O.S., "Site Inspection at Tehachapi Mountians Crossing of the California Aqueduct, Tunnel No. 1, North Portal."
9. Blanck, J.A., "The Balboa Outlet Tunnel," Monorock Publication, 1969.
10. Miner, G.M. and Hendricks R.S. "Shotcreting at Hecla Mining," *Rapid Excavation Symposium*, Sacramento, California, 1969.
11. Zynda, S.G., "Properties of Sand-mix Shotcrete," ACI Publication SP-14, 1966.
12. Parker, H.W., Fernandez-Delgado, G. and Lorig, L.J., "Field-oriented Investigation of Conventional and Experimental Shotcrete for Tunnels," DOT FRA OR&D 76-06, 1975.
13. Mahar, J.W., Parker, H.W. and Wueller, W.W., "Shotcrete Practice in Underground Construction," DOT FRA OR&D 75-90, 1975.
14. Sallstrom, S., "Improving Initial Compressive Strength of Shotcrete with Accelerating Agents." *International Symposium on Large Permanent Underground Openings*, Oslo, 1969.
15. Seabrook, P.T., "Properties of Shotcrete on Construction Projects," ASCE and ACI Publication SP-54, 1977.
16. Rutenbeck, T., "Shotcrete Strength Testing–Comparing Results of Various Specimens," ASCE and ACI Publication SP-54, 1977.
17. U.S. Bureau of Reclamation, Engineering and Research Center, "State-of-the-art Review on Shotcrete," U.S. Army Engineer Waterways Experiment Station, Report S-76-4, 1976.
18. Mason, R.E., "The Use of Shotcrete for Tunneling through Different Ground," SME-AIME Fall Meeting, 1976.
19. Alberts, C., "Instant Shotcrete Support in Rock Tunnels," *Tunnels and Tunneling.*
20. Holmgren, J., "Thin Shotcrete Layers Subjected to Punch Loads," ASCE and ACI Publication SP-54, 1977.
21. Fernandez Delgado, F., Mahar, J.W. and Parker, H.W., "Structural Behavior of Thin Shotcrete Liners Obtained from Large Scale Tests," ASCE and ACI Publication SP-54, 1977.
22. Cecil, OS., "Shotcrete Support in Rock Tunnels in Scandinavia," ASCE, January, 1970.
23. Ward, W.H., Coats, D.J. and Tedd, P., "Performance of Tunneling Support Systems in the Four Fathom Mudstone," *Tunneling '76 Proceedings*, Institution of Mining and Metallurgy, London, 1976.
24. Golser, J., "The New Austrian Tunneling Method (NATM): Theoretical Background–Practical Experience," ASCE and ACI Publication SP-54, 1977.
25. Laabmyr, F., "Soft Ground Tunnel for the Munich Metro," ASCE and ACI Publication SP-54, 1977.
26. Brierley, G.S. and Cording, E.J., "The Behavior During Construction of the Dupont Circle Subway Station Lining," ASCE and ACI Publication SP-54, 1977.
27. Bock, C.G., "Rosslyn Station, Virginia: Geology, Excavation and Support of a Large, Near Surface Hard Rock Chamber," *AIME Proceedings, Rapid Excavation and Tunneling Conference*, 1974.
28. Sutcliffe, H. and McClure C.R., "Large Aggregate Shotcrete Challenges Steel Ribs as Tunnel Support," *ASCE*, November 2, 1969.
29. Heuer, R.E., "Selection/Design of Shotcrete for Temporary Support," ASCE and ACI Publication SP-45, 1974.
30. Selmer-Olsen, R., "Examples of the Behavior of Shotcrete Linings Underground," ASCE and ACI Publication SP-54, 1977.
31. Rabcewicz, L.v. and Golser, J., " Principals of Dimensioning the Support Systems for the New Austrian Tunneling Method," *Water Power*, March 1973.
32. Sperry, P.E. and Heuer, R.E., "Excavation and Support of Navajo Tunnel No. 3," *Rapid Excavation and Tunneling Conference Proceedings*, 1972.
33. Rabcewicz, L.v and Golser, J., "Application of The NATM to the Underground Works at Tarbela," Water Power, September 1974.
34. Imrie, A.S. and Campbell, D.D., "Engineering Geology of the Mica Underground Powerplant," *AIME Proceedings, Rapid Excavation and Tunneling Conference*, 1976.
35. Monsees, J.E., "Station Design for the Washington Metro System," ASCE and ACI Publication SP-54, 1977.
36. Kaden, R.A., "Fiber-reinforced Shotcrete, Ririe Dam and Little Goose (CPRR) Relocation," ASCE and ACI Publication SP-54, 1977.

13
Sunken Tube Tunnels

JOHN O. BICKEL

Associated Consultant
Parsons, Brinckerhoff, Quade & Douglas

and

DONALD N. TANNER

Assistant Vice President
Senior Professional Associate
Parsons, Brinckerhoff, Quade & Douglas

GENERAL DESCRIPTION

Under favorable conditions, the sunken tube method is the most economical construction for any type of underwater tunnel crossing. Tunnel sections in convenient lengths, usually 300 to 400 feet, are placed into a predredged trench, joined, connected and protected by backfilling the excavation. The sections may be fabricated in shipyards on shipways, in dry docks, or in temporary construction basins serving as dry docks, depending on the type of construction and available facilities. A prerequisite for this method is a soil with adequate cohesion which permits dredging of the trench with reasonable side slopes which will remain stable for a sufficient length of time to place the tubes and backfill. If necessary, part of the trench may be excavated in rock, but where the tunnel is completely in rock, construction by conventional tunneling in free air or by tunnel-boring machines may be more economical.

The top of the tunnel should preferably be at least 5 feet below the original bottom to allow for an adequate protective backfill. Where this is not practical due to grade limitations and bottom configuration, the tunnel may project partly above the bottom and be protected by a backfill extending about 100 feet on each side of the structure and confined within dykes. The fill must be protected against erosion by currents with a rock blanket, protective rock dykes or other means. There have been cases where ships in confined channels have used anchors as turning pivots in wharping operations. This practice, although contrary to navigation rules, may require deeper backfill over a tunnel in such a location or special protection by rock cover, concrete slabs or other means to protect the tunnel against damage.

Tides and currents must be evaluated to establish conditions to be met during dredging and tube sinking operations. If there are shellfish areas near the tunnel site, their location must be determined and precautions taken during construction to prevent damage from silting due to dredging or backfill spillage.

Dredging and backfilling operations should be executed in such a manner as to limit, to the maximum degree possible, disturbance in the natural ecological balance at the construction site. In addition to securing Corps of Engineers and Coast Guard Permits for construction, other governmental agencies having jurisdiction over environmental protection, natural resources or local conditions must be consulted. Approval of these agencies should be obtained in the preliminary design stage.

STEEL SHELL TUBES

Configurations

The basic element of this type of tube is a steel shell which forms a watertight membrane and, in combination with a reinforced concrete interior lining, provides the necessary structural

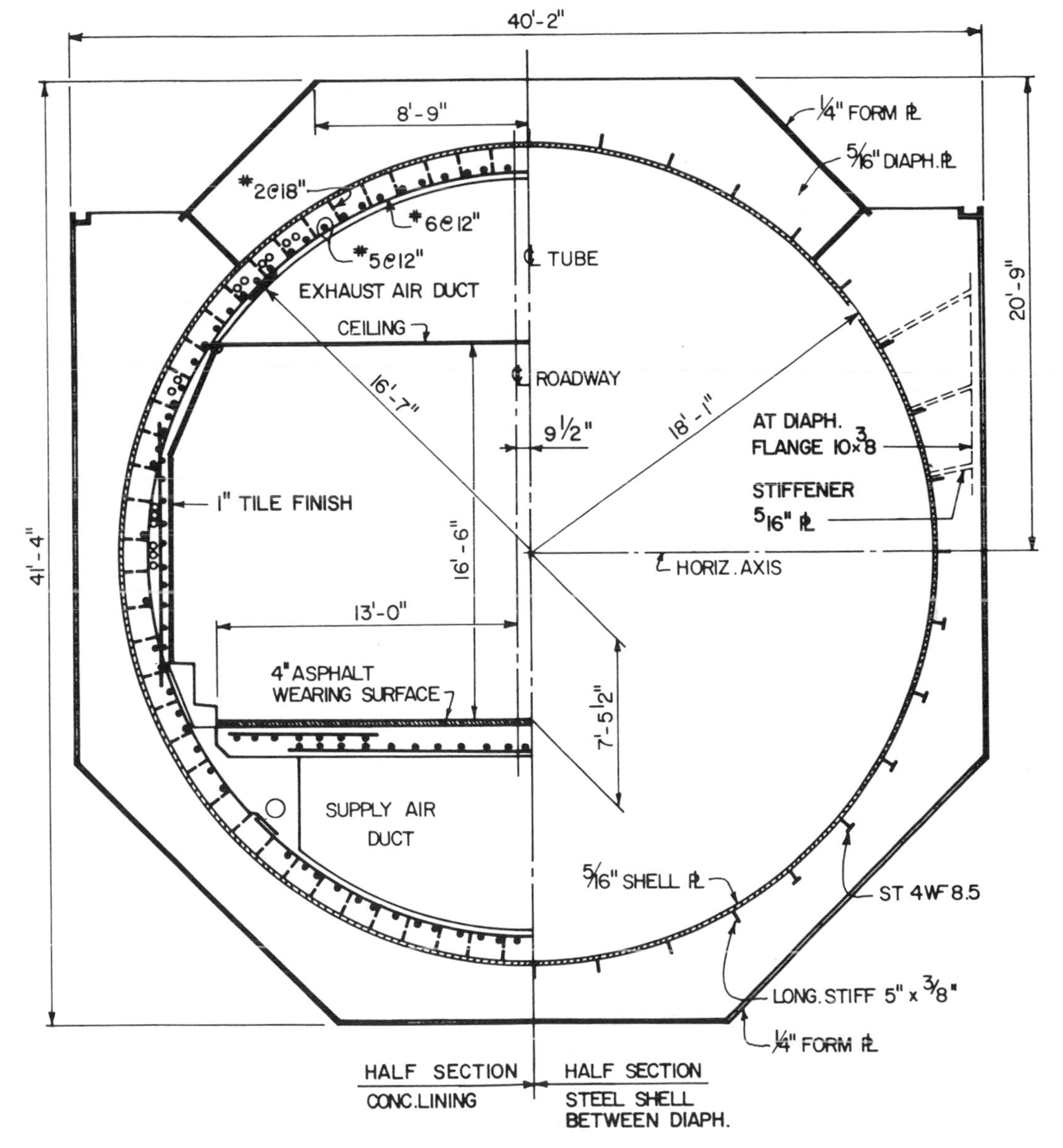

Fig. 13-1. Second Hampton Roads Tunnel.

strength for the finished tunnel. The shell may have a circular cross-section or any other configuration to suit the purpose of the tunnel. Figure 13-1 shows the cross-section of a two-lane tunnel on an interstate highway. Its circular steel shell has a diameter of 36 feet, 2 inches, made of $\frac{5}{16}$-inch welded steel plate. It is stiffened by external diaphragms spaced 14 feet, 10 inches apart and external longitudinal stiffening ribs. The interior is lined with a minimum thickness of reinforced concrete. An exterior concrete envelope of 2-foot minimum thickness, confined by $\frac{1}{4}$-inch steel-form plates attached to the shell, protects the shell against corrosion and acts as ballast against buoyancy. The space below the roadway slab forms a fresh air supply duct; the segment above the ceiling is an exhaust duct.

Figure 13-2 shows a twin circular shell tube for a four-lane highway tunnel. The two shells are connected with transverse steel diaphragms spaced 17 feet, 6 inches on centers with horizontal stiffening trusses between diaphragms. The interior reinforced concrete lining has a minimum thickness of 1 foot, 6 inches. Exterior keel concrete, 2 feet thick (minimum), and concrete filling the space between the two shells provides weight to overcome buoyancy and, in addition, to protect the shell plate against corrosion. Where not covered by the exterior concrete, the shells are protected against corrosion by $2\frac{1}{2}$ inches of gunite reinforced with wire mesh.

Figure 13-3 shows a tube for two rapid transit tracks, separated by a gallery providing service access and an emergency ventilation exhaust duct for fire protection. The shell plate is $\frac{3}{8}$ inches thick. It is stiffened by interior transverse steel ribs spaced 6 feet on centers and two longitudinal vertical interior trusses, encased in the reinforced concrete walls of the gallery. The interior lining of reinforced concrete has a minimum thickness of 2 feet, 3 inches. The exterior shell is protected against corrosion by an impressed current cathodic protection system. Ballast pockets 2 feet, 6 inches deep on top of the tube are filled with gravel to provide adequate weight to overcome buoyancy.

Figure 13-4 shows a tube with four bores arranged in pairs vertically for rapid transit. The $\frac{3}{8}$-inch shell plate is supported by interior transverse stiffeners spaced at a maximum distance of 4 feet, $4\frac{1}{2}$ inches longitudinally. The shell is stiffened by a vertical longitudinal truss

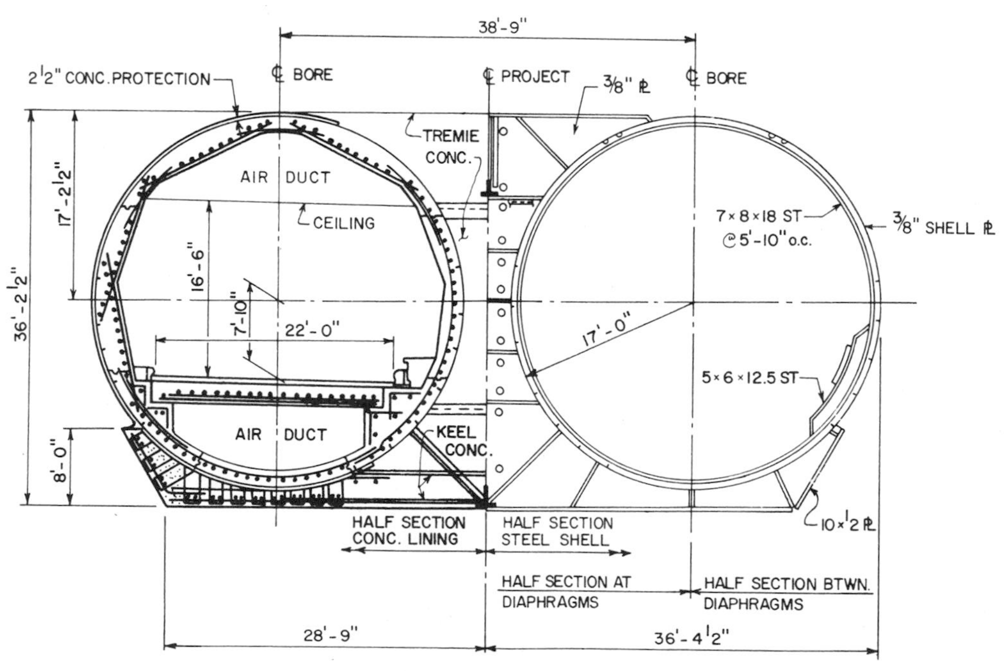

Fig. 13-2. Hong Kong Tunnel.

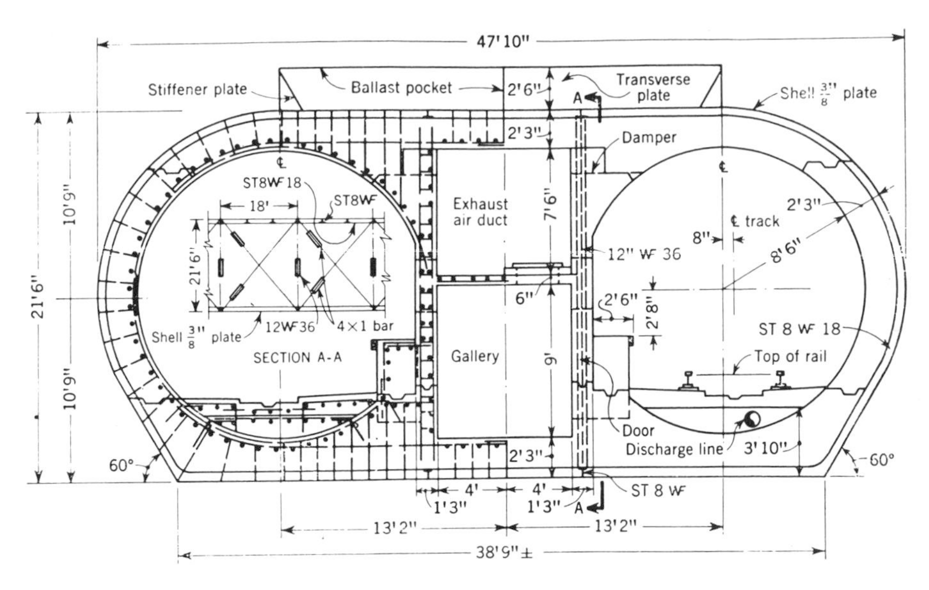

Fig. 13-3. Transbay Tube (San Francisco Bay Area Rapid Transit).

between bores and horizontal pipe struts. The interior lining is reinforced concrete and the shell is cathodically protected against corrosion. Ballast pockets are filled with concrete and rock.

Figure 13-5 shows a rectangular two-lane vehicular tunnel cross-section with side air ducts. The section consists of an exterior steel shell plate supported by transverse structural steel stiffeners spaced at 2 feet, 6 inches on centers with an interior prestressed concrete lining. The steel shell is designed to permit off-site fabrication of the steel shell, and, after launching, to place the interior concrete lining while afloat. After concreting, post-tensioning tendons made up of high-strength strands are pulled through the previously placed ducts. The tendons are then partially tensioned. The section is then towed into position over the trench and lowered into position by placing ballast on top of the tube section. After the section is placed and joined to the previously placed unit, the post-tensioning force is increased in stages as the tunnel is backfilled. The tendon anchorages are inside of the cross-section and therefore can be tensioned from the interior of the tunnel.

The technique of incrementally applied prestressed force to counteract successive backfilling loading stages is defined as staged post tensioning and has been successfully applied in North America for high-rise buildings and foundations. It is necessary to use staged post-tensioning in those cases where the dead load at the time of initial tensioning is not sufficient to counteract the force of the final prestress force.

An alternative technique to staged post-tensioning would be to add temporary tendons inside the tunnel cross-section. This system was used on the La Fontaine (Boucherville) Tunnel in Montreal. The tendons spanned between the roof and invert slabs and were tensioned simultaneously with the other tendons and the temporary tendons were released as the tunnel was backfilled.

This type of cross-section permits utilization of a steel shell, which provides a watertight membrane and permits outfitting while afloat with the advantages of prestressing, which is structurally more suitable for the rectangular configuration with long spans.

Materials and Fabrication of Steel Shells

Steel tubes can be constructed on shipbuilding ways of adequate size. These may be remote from the project site, since the prefabricated steel tubes can be towed considerable distances.

Material for the shell plates, stiffening members and end bulkheads are usually ASTM designation A36 steel. All joints are welded by

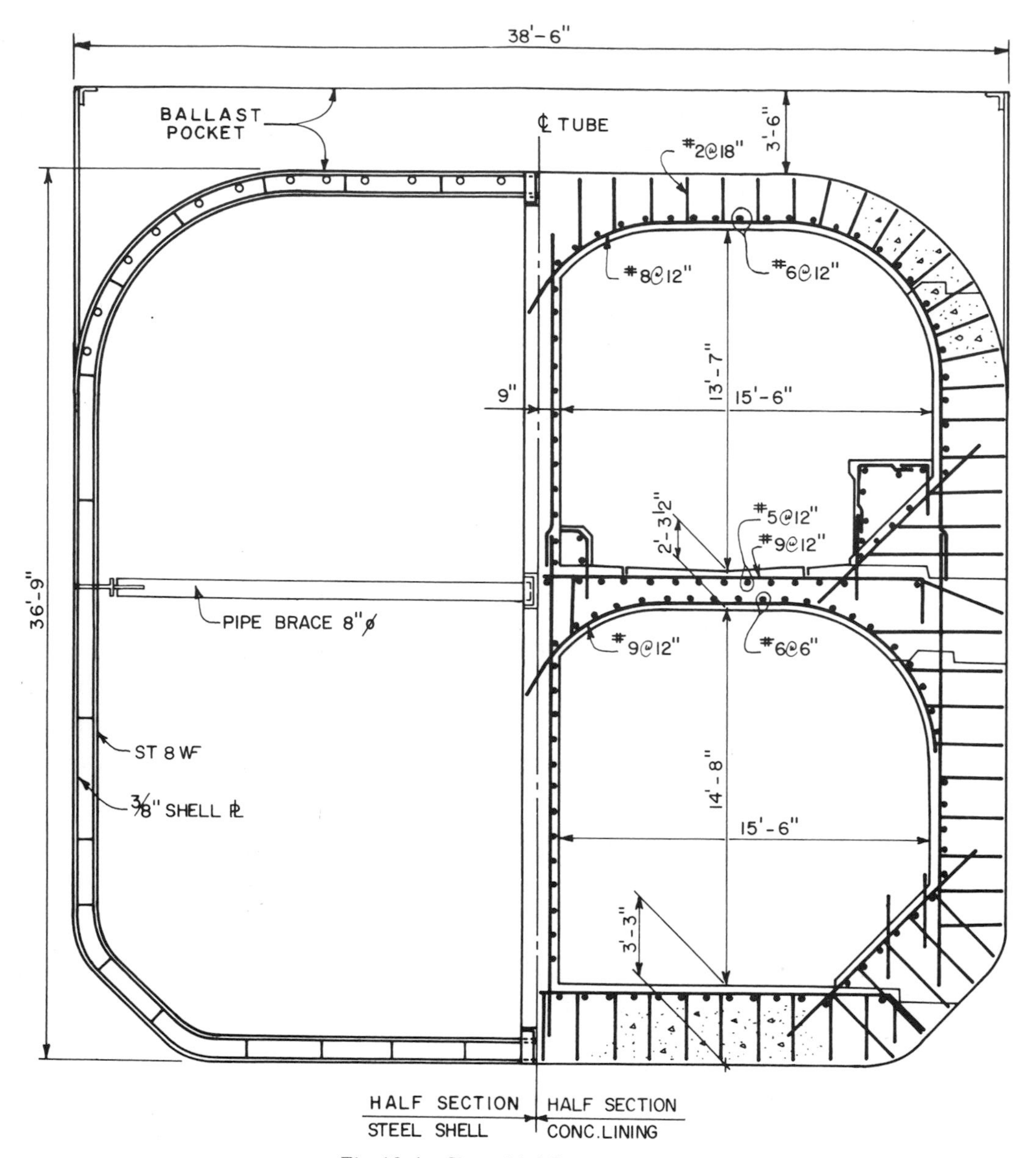

Fig. 13-4. Sixty-third Street Tunnel.

the metal arc-welding process. Joints in shell plates are full penetration butt welds. Where practical, welds are made by the automatic machine process.

Welds (Fig. 13-6) are visually inspected through out fabrication and a certain percentage are tested by radiographic, ultrasonic, dye-penetration and magnetic particle methods. Ten percent of butt welds in the end bulkheads and in the end sections of the steel shells, and 1% of the remaining butt welds in the shell plates are tested radiographically. About 1% of all fillet welds are tested by any of the other (most suitable) methods.

All shell plate welds are tested for watertightness before launching by coating the outside of the shell plate with soap solution and jetting compressed air at 40 psi against the inside, the nozzle being held not more than 3 inches from the surface. Welds in end bulkheads are similarly tested. All leaks are repaired and retested. Tubes are fabricated in convenient subassemblies

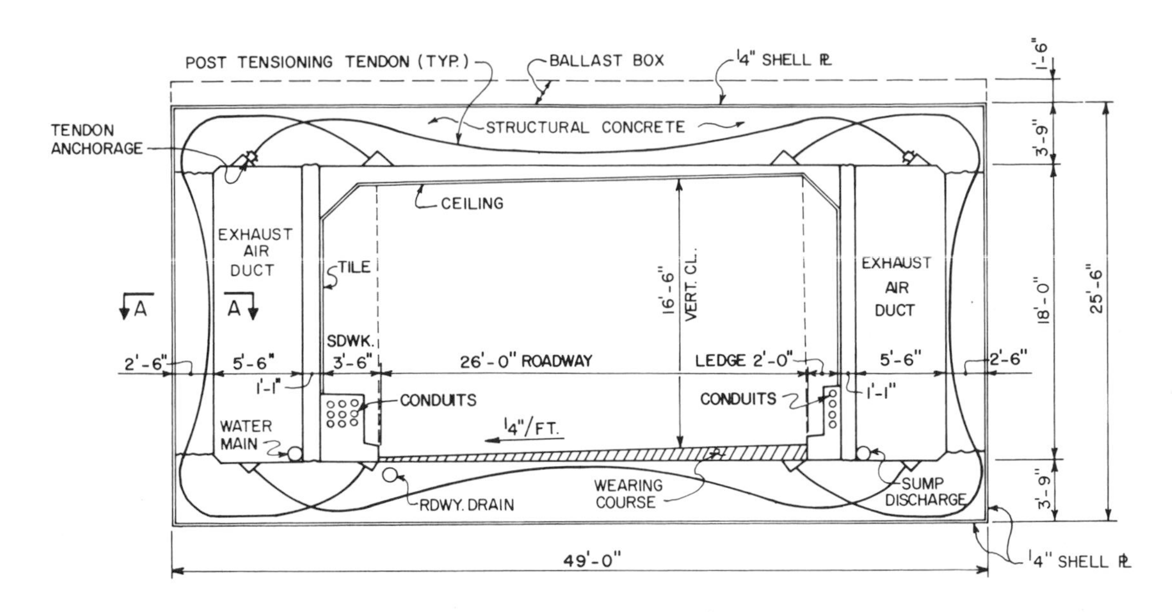

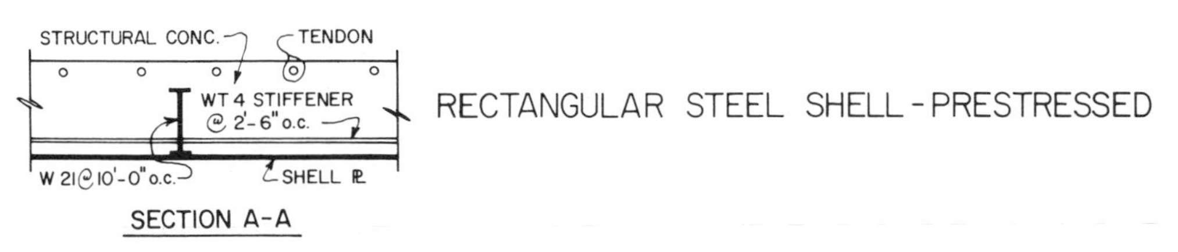

Fig. 13-5. Rectangular steel shell–prestressed.

I. TO DETERMINE THE MAXIMUM SIZE OF DEFECT PERMITTED IN ANY JOINT OR WELD THROAT THICKNESS: PROJECT (A) HORIZONTALY TO (B)

II. TO DETERMINE THE MINIMUM CLEARANCE ALLOWED BETWEEN EDGES OF DEFECTS OF ANY SIZE: PROJECT (B) VERTICALLY TO (C)

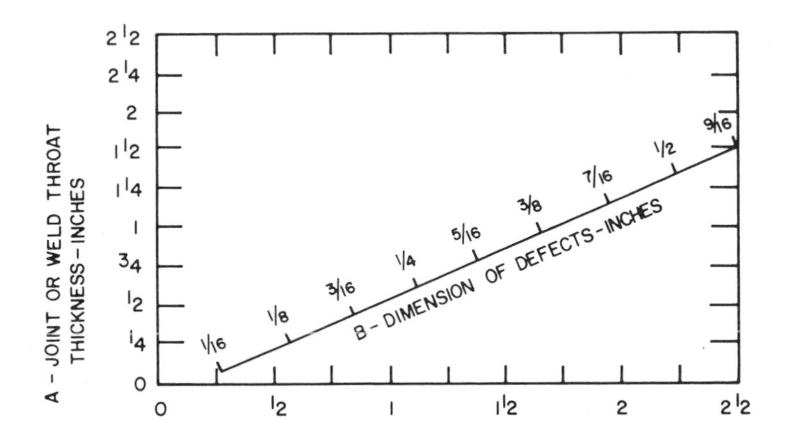

C - MINIMUM CLEARANCE ALLOWED BETWEEN EDGES OF POROSITY OR FUSION DEFECTS - INCHES (LARGER OF ADJACENT DEFECTS GOVERNS)

Fig. 13-6. Modified weld quality requirements (limitation of porosity and fusion defects).

and fitted together on the shipways. The maximum tolerance of the assembled shell measured from the theoretical longitudinal axis is $\pm \frac{1}{2}$ inches in radius, and $\pm 1\frac{1}{4}$ inches in overall length.

During fabrication, tubes are supported on the ways by blocking under diaphragms or stiffener rings. Before launching, the ends of the tubes are closed with watertight bulkheads. It is advantageous to place the interior reinforcing steel before the bulkheads are attached.

Launching of Tubes

Structurally, the tubes are particularly suitable for side launching, but may also be end-launched.

Side Launching. The tubes are supported on launching sleds by a series of parallel ways at right angles to the tube, located under diaphragms designed to transfer the weight of the tube to the sliding ways. The actual launching may be free or restrained, depending upon the declivity of the ways, depth and width of waterway at the launching site. After completion of fabrication of the tube, its weight is transferred from the fixed building ways to the launching cradles or sliding ways which rest on a coating of grease applied to the groundways. Either the groundways or the sliding ways should be equipped with guides to prevent side slipping during launching. The tube is held in position until launch time by a number of triggers, usually four, which are cut simultaneously at launching. Jacks are provided to start the tube, if necessary.

Declivity of the launching ways ranges from 0.5 to 1.8 inches/foot in order to ensure free sliding.

Restrained Launching. The speed during launching is controlled by chains or other suitable drag devices. These control the run-out distance of the tube in the water.

Controlled Side Launching. The tube supported on the sleds is gradually allowed to slide down the ways at a suitable speed, controlled by synchronized winches until it floats off the sleds.

End Launching. The tube slides on two ways parallel to its longitudinal axis. A single trigger holds the tube, after its weight is transferred to the launching ways, until released for the launch. At the inboard end of the sliding ways, the forepoppet is of special construction to support approximately one-half the weight of the tube when the outboard end of the tube is lifted off the ways by buoyancy.

Launching Stresses. These stresses have to be analyzed. Those for side launching are nominal, consisting of hydrostatic pressure on the shell and form plates when the tube enters the water. End launching causes bending moments when the outboard end of the tube floats while the inboard end is still supported on the ways. Buckling stresses in the top of the shell may be critical. In some cases, temporary strengthening by longitudinal steel beams welded to the top of the shell may be required.

Concrete Lining

After launching, the tube is towed to an outfitting pier for placing the interior concrete lining. Structural concrete has a 3000 psi compressive strength. This includes the interior lining, roadway slab, keel and cap concrete, as shown in Fig. 13-7.

Usually the keel concrete is placed prior to launching in order to provide stability and to reinforce the shell and form plates. If the keel concrete is placed after launching, the shell and form plates will probably require additional stiffening. Exterior concrete acting as ballast and protection (Figs. 13-2 and 13-7) has 2200 psi compressive strength. If tremie concrete is used in joints between tubes, it should have a 3000 psi strength. Retarding admixtures to reduce water content and shrinkage are recommended. Before starting concreting, trial mixes are made to determine unit weight of concrete to conform to the design weight used in buoyancy calculations.

Reinforcing Steel. The interior concrete lining is reinforced with a layer of steel at the inside face, as required by design loads and temperature for minimum code requirements. Steel rein-

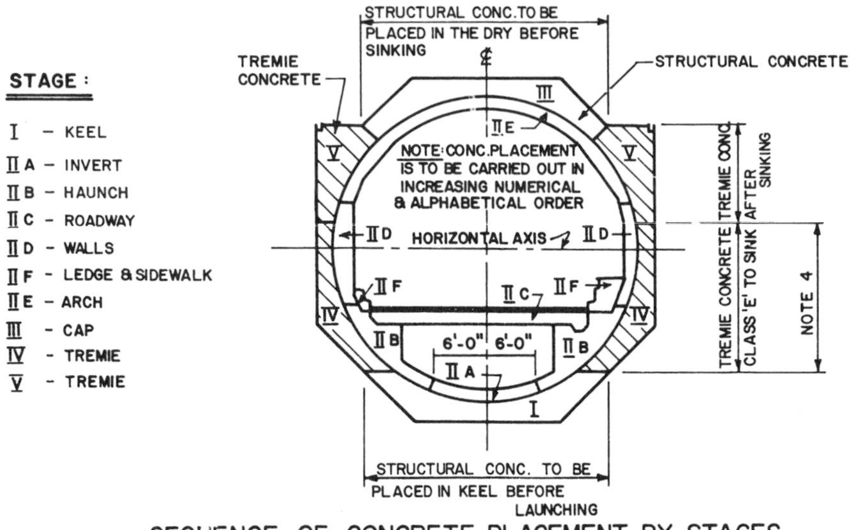

SEQUENCE OF CONCRETE PLACEMENT BY STAGES

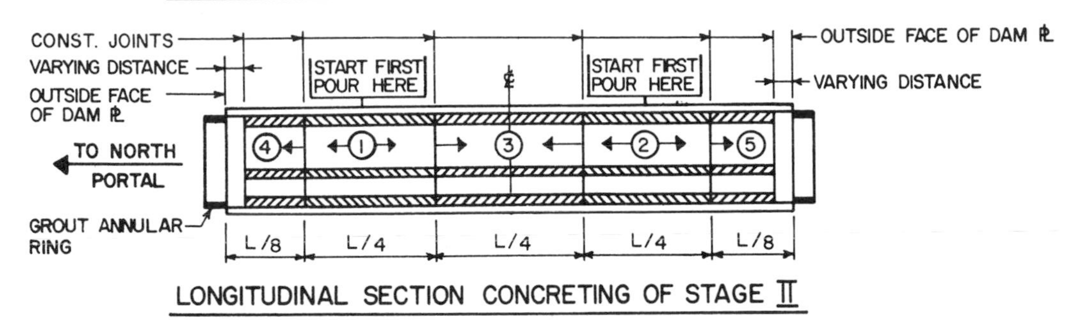

LONGITUDINAL SECTION CONCRETING OF STAGE II

NOTES:

1. EACH SEGMENT OF STAGE II CONCRETE SHOWN IN "SEQUENCE OF CONCRETE PLACEMENT BY STAGES" IS TO BE PLACED IN THE FOLLOWING ORDER: POUR (1) & (2) TO THE CONSTRUCTION JOINTS, POUR (3) TO THE CENTER OF THE TUBE BETWEEN JOINTS, POUR (4) & (5) TOWARDS THE ENDS OF THE TUBE AS SHOWN IN "LONG. SECTION CONCRETING OF STAGE II."
2. STAGE II F SHALL BE MADE AFTER THE TUBES HAVE BEEN SUNK AND OPENED.
3. CARE IS TO BE TAKEN WHILE POURING ALL TUBES TO MAINTAIN THEM IN A LEVEL POSITION.
4. DIMENSIONS TO BE DETERMINED BY THE CONTRACTOR AND APPROVED BY THE ENGINEER.

Fig. 13-7. Concrete pouring sequence.

forcing bars are 60,000 psi yield point ASTM Designation A15, new billet steel. Bars with welded splices or welded to structural steel shall not exceed a 0.30% carbon content. Ring and longitudinal bars are spaced 12 inches maximum. Ties, 18 inches on centers, are not less than $\frac{1}{4}$-inch diameter welded directly to steel shell, or upset bars are threaded into Nelson stud sockets welded to shell. Additional reinforcing is provided at niches and openings in the concrete lining. Length of lap splices is usually 24 times the bar diameter.

Placing of Concrete

Concrete is usually supplied by a plant at the outfitting site for large projects, or may be supplied by transit mix for smaller projects. Concrete is placed into movable steel forms by pump lines extended through access hatches in the top of the shell. Hatch openings about 6 feet, 6 inches by 4 feet are provided at approximately 75-foot centers for form handling. Small openings (12 by 12 inches) may be provided in top of the steel shell to facilitate placing of the arch concrete. Pouring sequence is controlled to prevent excessive hydrostatic pressure on the shell at successive increases in immersion and to maintain the tube in level condition for trim and list. Placing of concrete starts at the invert and proceeds in increments on both the haunch, sidewalls and arch pours. The roadway slab in highway tunnels is also placed at this stage. Length of pours is determined by the above considerations and convenience in the length of forms. Placing should start approximately at quarter points of the tube and proceed in both directions simultaneously. Construction joints are provided with keys. Figure 13-7 shows a typical pouring schedule and sequence.

Buoyancy and stability of tube sections for progressive stages of concreting are checked from the time of launching to towing to the sinking site.

Keel concrete (Fig. 13-7) is usually placed before launching and the structural cap concrete is placed in the dry at outfitting pier.

Shell plates above the keel concrete pour are sounded for voids, which must be grouted at pressures not to exceed 10 psi. Grout plugs are provided and sealed by welding. After all interior concreting is completed, all access hatches and concreting openings in the shell are closed by steel plates welded watertight. Voids between the shell plate and the interior concrete lining are grouted.

Construction Hatches

The end tubes are equipped with construction hatches and steel stacks extending above the water to provide access to the interior for finishing work prior to completion of the adjacent cut-and-cover sections. Hatches are usually circular, with a diameter of 10 to 12 feet. They are closed with welded steel covers after the end bulkheads have been removed and access is available from the cut-and-cover section. The stacks are cut off at least 3 feet below the finish

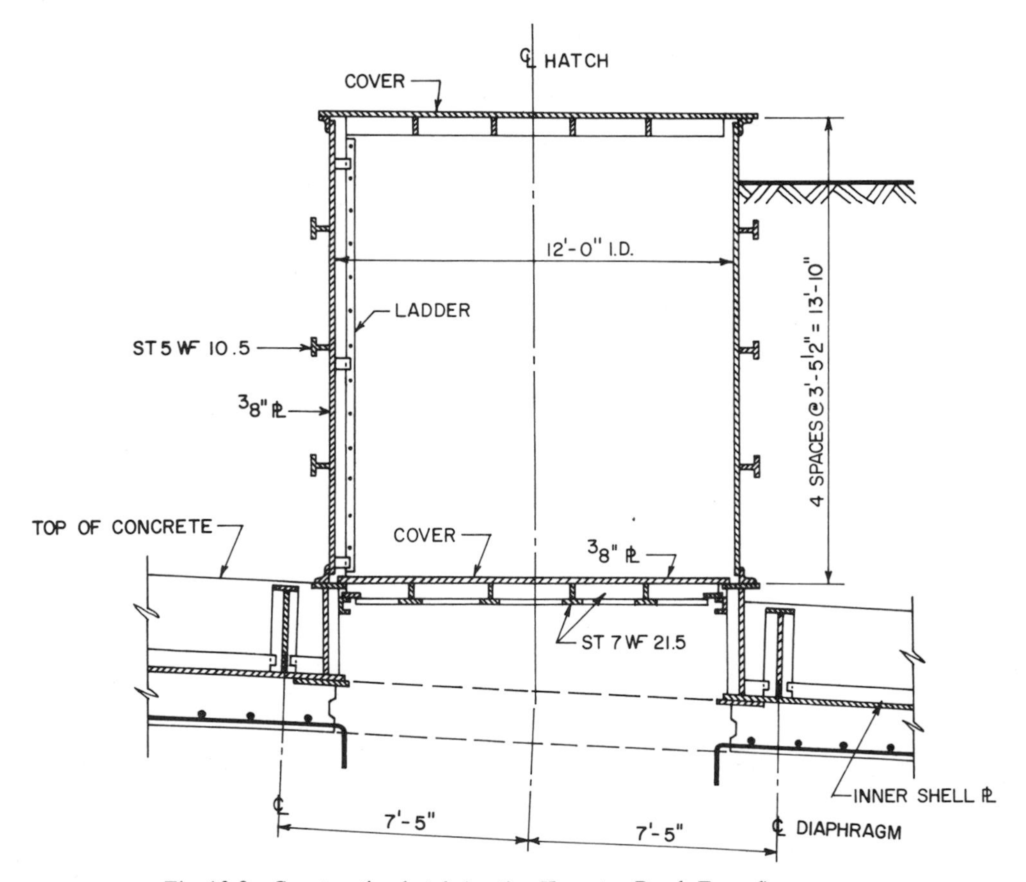

Fig. 13-8. Construction hatch (section Hampton Roads Tunnel).

line of the backfill placed over the tube. During launching and towing, the top of the stack is closed with a temporary cover. Figure 13-8 shows a cross-section of a construction hatch.

Tremie Concrete

Exterior protective and ballast concrete (Fig. 13-7) is placed by tremie while the tube is floating at the outfitting pier. Pockets are filled in stages to reduce pressure on form plates. The concreting sequence is controlled to keep the tube in longitudinal and transverse balance. Sufficient concrete is placed at the outfitting pier to reduce freeboard of the tube to about 2 feet for towing. The remainder is placed after the tube is suspended from the placing barge.

Ballast

To provide weight for sinking the tubes into place, they are ballasted. In a section, such as Fig. 13-1, the exterior concrete is sufficient for this. In sections such as Fig. 13-3, gravel ballast is placed into pockets on top of the section. Water or sand temporarily placed in the interior of the tubes have also been used. Water has to be confined in tanks or in bulkhead compartments to prevent unbalancing in case the tube takes a longitudinal tilt, either accidentally or to conform to final grade. Exterior watertight ballast pockets, filled with water, on top of the tube have also been used, as have pontoons. Interior ballast is removed only after sufficient backfill and/or interior finish concrete or other permanent weight has been added to overcome buoyancy.

CONCRETE TUBES

Configuration

Rectangular reinforced concrete box sections are usually used for tunnels with four or more traffic lanes, particularly where concrete is more economical than steel. Figure 13-9 shows a four-lane highway tunnel with two two-lane compartments and ventilation ducts on both sides.

A design for a highway tunnel with two three-lane roadways; compartments for pedestrians and bicycles; and ventilation ducts is shown on Fig. 13-10.

Construction of Tubes

The sections are constructed in dry docks or in temporary dry basins located near the site, which are flooded for floating the tubes. Unless a dry dock or outfitting basin is available for construction in the dry, these facilities must be constructed with a resultant increased costs and time of construction. The Hong Kong tunnel was originally designed as a four-lane, rectangular, prestressed section, but an alternate tender was made for the twin steel shell section shown in Fig. 13-2. The steel section was constructed, at a saving of several million dollars and one-and-a-half years of construction time. The base and roof slabs and sidewalls are of massive reinforced concrete to provide adequate weight to overcome buoyancy. Partitions are reinforced concrete as determined by structural and functional requirements. Concrete has a compressive strength of 3000 psi, and reinforcing steel a 60,000 psi yield point corresponding to ASTM Designation A-15. Since the tubes are solidly supported in the construction basin, the concreting sequence is not critical. Usually, alternate longitudinal sections varying from 20 to 50 feet are poured, and then the intermediate sections are placed in order to minimize shrinkage. The base slab is poured over a waterproofing membrane and walls and top slab are placed by means of movable steel forms to provide smooth and accurate interior surfaces.

The concrete tube may be heavy enough for sinking, in which case it has to be supported by pontoons for floating into position. If the completed section is buoyant, it has to be ballasted for sinking. Temporary ballast may be placed internally, using water, sand, ballast blocks, or permanent ballast consisting of gravel or lean concrete may be placed on top of the section. In some instances, the 5-foot minimum cover over the top of the tubes has been used as ballast. This is not recommended since tidal erosion may cause loss of cover (ballast).

End bulkheads may be steel, timber with waterproofing or reinforced concrete. Selection of end bulkhead materials is a matter of individ-

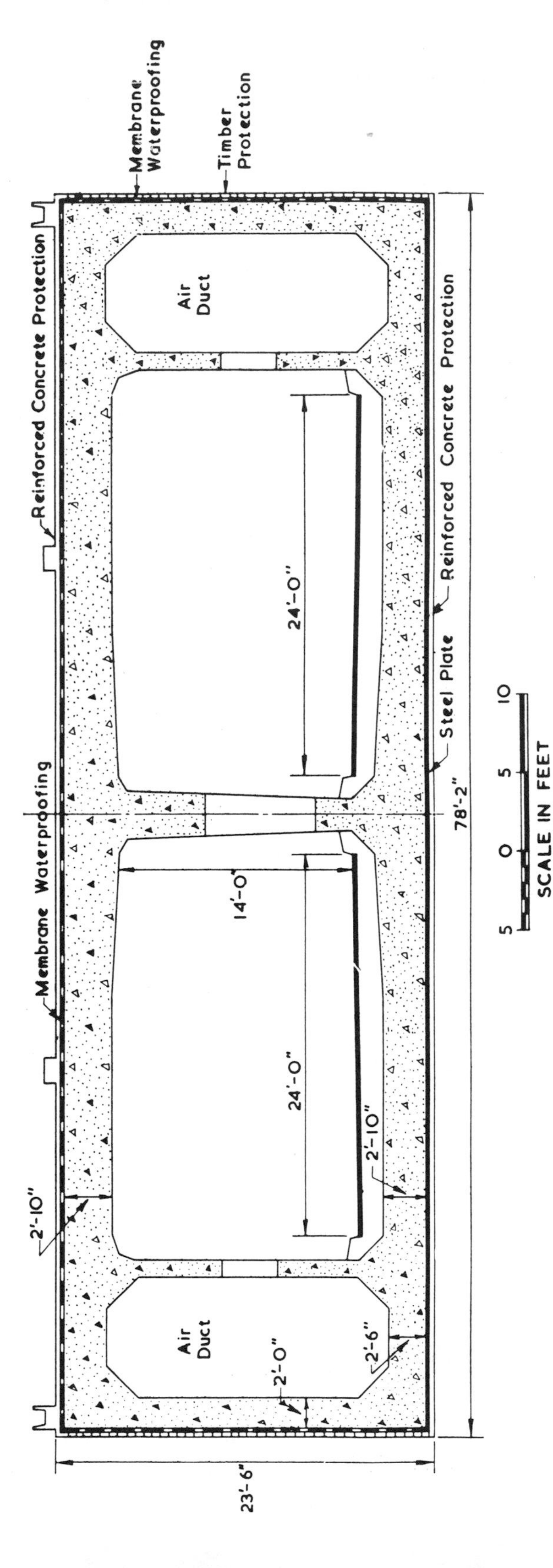

Fig. 13-9. Deas Island Tunnel (four-lane prestressed box section).

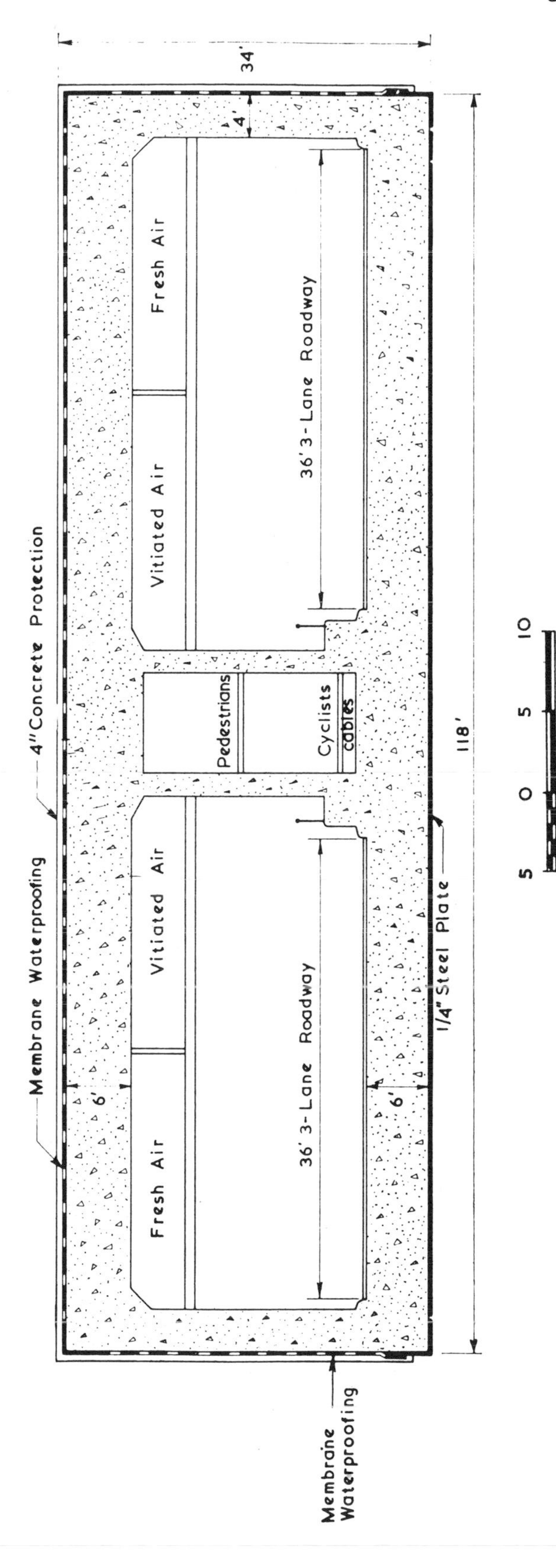

Fig. 13-10. Scheldt River Tunnel (six-lane prestressed box section).

ual preference and economy. Usually, steel bulkheads are easier to install and remove.

Prestressing

Since concrete thickness is largely determined by the weight required to prevent uplift, prestressing is economical only for very wide sections. There was a slight advantage for prestressing in the six-lane design shown in Fig. 13-10.

Obtaining watertightness by prestressing alone, without membrane, is controversial. Due to the large concrete cross-section, achieving adequate compressive stress for this purpose requires very substantial tensioning loads. Figure 13-11 shows a four-lane tunnel, prestressed to a residual compressive stress of about 200 to 300 psi, which is assumed to be watertight.

Waterproofing

Steel membranes have been applied to a number of concrete box tunnels as a waterproof enclosure, made of $\frac{1}{4}$-inch structural steel plate. All joints are welded and anchors are welded to the skin to bond it to the concrete. With temporary stiffening, the plates may serve as a form for the concrete. In most cases, a steel plate has been used to waterproof the base slab and also a portion of the sidewalls.

Multi-ply membranes of fabric and coal-tar layers can be applied to the top of the box sections, at least four plies being used. To protect the membrane against damage during handling and placing of tubes, it is covered with a thin layer of poured concrete, bricks, concrete or asphalt or wood planks. Due to the difficulty of attaching such a membrane, and for its protection, it is not practical to use it on the sidewalls and the bottom.

Plastic membranes made of synthetic neoprene or vinyl type rubbers have been developed for waterproofing concrete structures. Sheets of this material, about $\frac{1}{8}$-inch thick, are attached to the sidewalls with special adhesives. Due to the difficulty of protecting the membrane on vertical sides against damage, the use of this method on sunken tube sections has been limited so far.

Epoxy coatings, mostly of coal-tar epoxy to waterproof the sides and tops of tunnel sections, have had a few applications. Great care is needed in preparing an epoxy mixture in order to retain adequate elasticity and adhesion and in its application to the concrete surface. Protecting the coating on vertical sides against mechanical damage is difficult. The coating may be brushed or sprayed on a completely dry surface. Further developments in this field should be investigated.

FIBERGLASS TUBES

General

Recently reinforced fiberglass pipes ranging from 4 to 12 feet in diameter have been placed

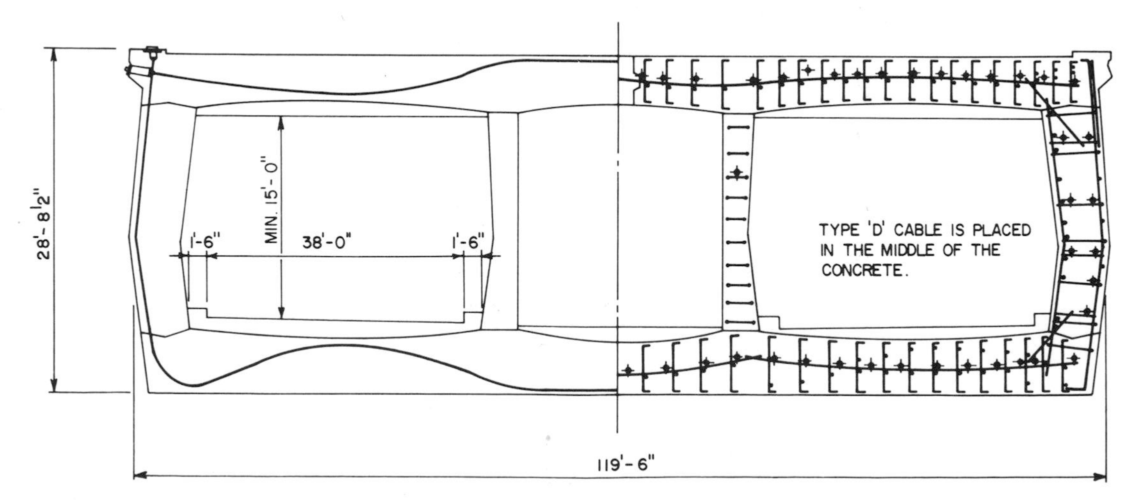

Fig. 13-11. Four-lane prestressed concrete section of La Fontaine (Boucherville) Tunnel.

underwater for lengths of up to 4000 feet. These pipes are used for circulating water cooling lines for power plants. A special report for the Califronia Undersea Aqueduct[1] determined it would be engineeringly feasible to construct 599 miles of floating conduit and 122 miles of partially buried conduit consisting of 34-foot diameter fiberglass pipe in water depths of 60 to 600 feet.

Material

Rib-wall reinforced fiberglass pipe consists of polyester resins, epoxy or vinyl ester with continuous filament winding. Seamless pipe is manufactured in 60-foot lengths and can be fiberglass welded into longer lengths in the shop or field. Material properties are: 106 to 120 pcf density, 3.95×10^6 psi hoop modulus, 5530 psi hoop stress and 9000 psi compressive strength. Fiberglass will not support combustion; nor will it corrode or decay.

Joints and Inserts

Bell and spigot, various mechanical metal and fiberglass-welded joints are used to join sections of pipe together. Metal inserts are installed by overlaying successive layers of resin impregnated glass fabric around the insert.

Fiberglass Tunnel

The design and construction of a vehicular or rail tunnel consisting of a fiberglass shell in lieu of a steel shell is feasible. The fiberglass shell would be designed to withstand the forces of launching and outfitting while afloat during the placing of the interior concrete lining. Shear connectors embedded in the fiberglass shell would connect the shell and the interior concrete lining which would be designed to support the in situ backfill and water loads. The fiberglass shell would provide a waterproof membrane that could withstand abrasion during construction and would eliminate the need for cathodic protection.

WEIGHT CONTROL OF TUBES

Final Weight

Final weight of the completed tube, including main structure, interior finish and backfill, has to be controlled to counteract buoyancy forces. Thickness of concrete is more often determined by the weight required rather than by structural strength.

Weight of Tubes At Various Stages

This has to be controlled in accordance with construction procedures. Tubes built in dry docks are solidly supported by keel blocks and the only criteria are the loading on the supports and on the dock floor. These do not impose any serious restrictions on the construction sequence. When concrete is placed while the tube is floating, as in the steel shell type, the sequence of pouring concrete is governed by several factors: keeping the tube on an even keel; increments of immersion producing hydrostatic pressure on the steel plate; and longitudinal bending moments from unequal loading, as discussed elsewhere in this chapter.

Factors of Safety Against Buoyancy

During various stages, these should not be less than the following.

After placing of the tube in the trench, and before backfill is placed or interior finish including sidewalks, ledges and roadway pavement completed, a minimum factor of safety of 1.10 is required. This may be achieved by adding temporary or permanent ballast. The specific gravity of the water in the bottom of the trench and the possible accumulation of silt from adjacent dredging operations or sludge from other sources must be checked.

After dewatering and removal of the end bulkhead, and prior to completing the interior concrete lining and backfill at the joint, the structure should have a minimum factor of safety of 1.02.

The completed structure should have a minimum factor of safety of about 1.2, not including backfill.

With backfill in place, the factor of safety against uplift of the tunnel is usually at least 1.5 or more.

Checking of weight is done by computation of weight of structural and reinforcing steel and accessories and continuous checking of unit weight of concrete samples. In determining the volume of the concrete the volumes of embedded structural steel, reinforcing steel, conduits and pipes more than 2 in. in diameter, and all boxed out niches should be deducted. The weights of available aggregates are determined and the mix is designed for weight as well as for strength. If local aggregates have insufficient weight, it is usually more economical to use more distant sources of supply of heavier material instead of increasing the concrete volume of the tubes.

PREPARATION OF TRENCH

Dredging Sections

The trench must be deep enough to allow for a foundation course below the tunnel, the height of the tubes and for a minimum of 5 feet of protective backfill. Under certain conditions, a shallower trench may be used for short distances, with the tunnel projecting partly above the natural bottom of the channel. Figures 13-12 and 13-13 show typical sections of the dredged trench. Figure 13-14 shows a configuration of a dredged trench with the tube projecting above the natural bottom and protective dykes. The side slopes depend upon the soil characteristics and may vary from near vertical in rock to 1:4 in soft material. Generally, a slope of 1:1.5 is feasible. In unusually deep trench sections, an intermediate berm may be required.

Payment for dredging may be based on fixed quantities determined by a side slope selected on the basis of borings and slope stability studies, say 1:1.5 in reasonably firm soil, giving the contractor the option to steepen the slope to, say, 1:1 at his own risk, or flatten it to 1:2, with a unit price payment on overdredging if actual soil conditions require this.

Rough Dredging

This is usually done, at the contractor's choice, as a continuous operation. Hydraulic dredges are used where suitable for depth and material and where disposal areas are within economical pumping distance. For greater depths or barge transportation of spoil, clam shell dredges are used. For harder materials, dipper dredges are used.

Fine Dredging

Fine dredging to final dimensions is done two or three tube lengths ahead, to keep the time interval between it and the placing of the tube to a minimum.

Removal of Silt

Just prior to the placing of the foundation course, the trench is checked for accumulation of silt or sloughing of the dredged slopes. Any such accumulation is removed by clam shell, suction dredge or air lift.

PLACING OF TUBES

Several methods have been used to prepare the foundation supporting the tubes. Since the weight of the tubes including backfill is not

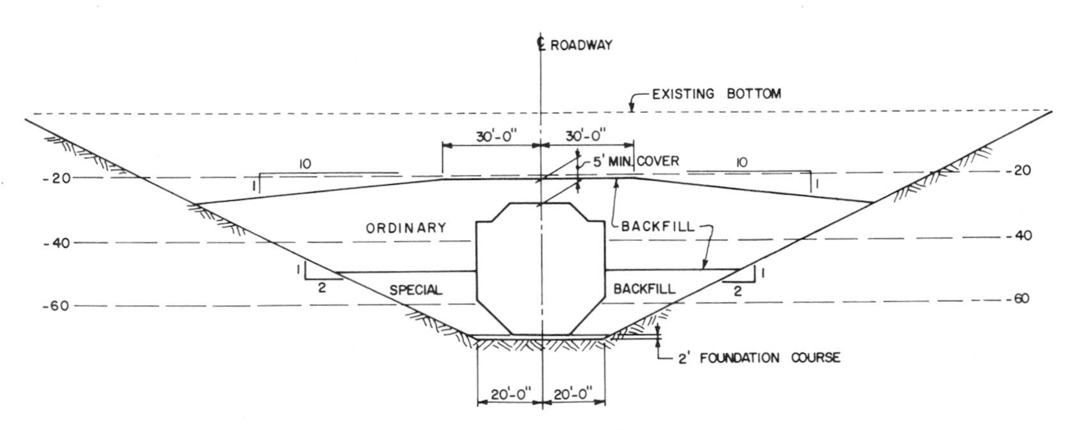

Fig. 13-12. Typical section, dredged trench.

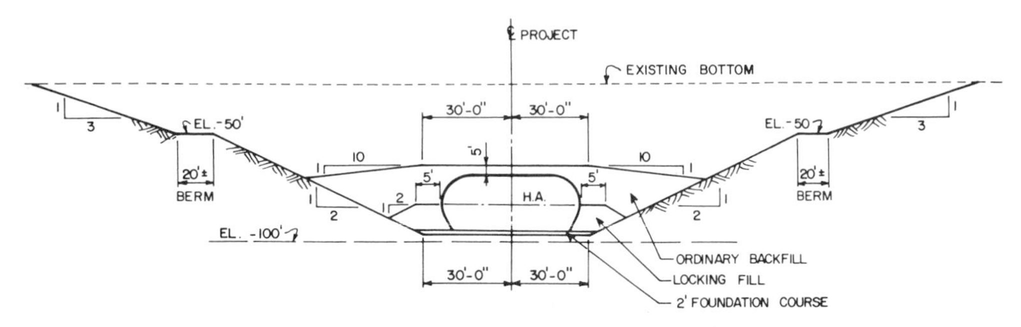

Fig. 13-13. Cross-section, dredged trench with berms.

much greater than that of the displaced soil, accuracy of alignment is the most important function, bearing capacity being a second consideration.

Screeded Foundation Course

The finished dredging is carried to a minimum of 3 feet below the bottom of the tube. A cover of coarse sand or well-graded gravel is placed in the trench and leveled to accurate grade. The gradation of the material may vary from $\frac{1}{4}$ to $1\frac{1}{2}$ inches in light currents, or may go up to 6-inch stones where heavy currents occur. The leveling is done by dragging a heavy screed, made of a grid of steel beams, over the surface in successive passes, adding material as needed. The screed is suspended from winches on a carriage rolling on tracks supported on an assembly of two steel barges yoked together. The tracks are adjustable to parallel the grade. The rig is anchored in correct position. Adjustments are made in the screed suspension to compensate for tide level changes. The foundation for a 300-foot tube can be screeded in two or three set-ups. For the 58 tubes of the San Francisco to Oakland rapid transit tunnel, a special screed rig was built consisting of a steel truss of 240-foot length, supported by floatation tanks which could be partly ballasted with water to reduce buoyancy, permitting anchoring the assembly against tide level changes, thus eliminating adjustments during screeding.

Accuracy of screeding usually allows a tolerance of $\pm 1\frac{1}{2}$ inches in elevation for a 300-foot tube for a highway tunnel. The tubes for the San Francisco Transbay tunnel (Fig. 13-15) were set within a tolerance of $\pm 1\frac{1}{2}$ inch of theoretical grade.

Jetted Foundation Course

A number of tunnels have been built where the sand foundation was jetted under the tubes after they were set to proper grade on temporary supports in the trench. The Danish construction firm, Christiani and Nielsen, developed a patented system of jetting a mixture of sand and water under the tube and pumping off excess water. The rolling current deposits the sand

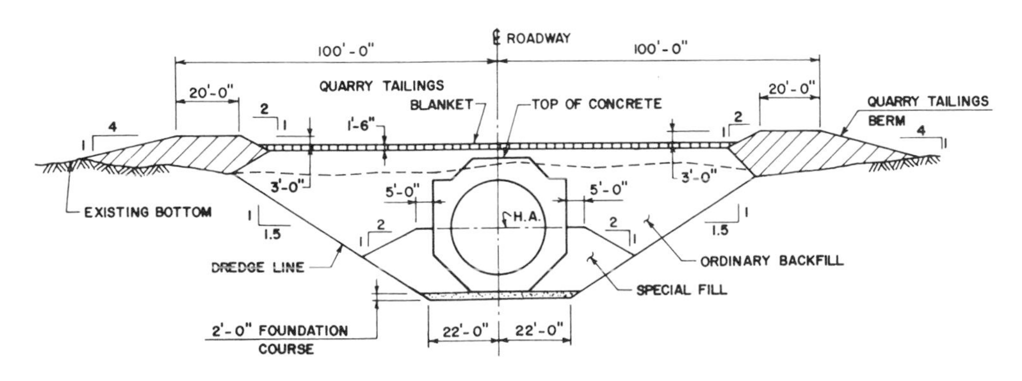

Fig. 13-14. Cross-section, dredged trench—protective berms.

Fig. 13-15. Screeding rig—Transbay Tube.

in a firm layer. The jet pipes, sandhoppers and pumps are mounted on a steel framework, traveling on tracks on top of the tube. During this operation, the tube has to be held securely in position and protected against uplift until the locking fill can be placed.

Pile Foundations

In unusual circumstances, where the soil under the tunnel is too weak to support the tunnel and backfill, and cannot be economically excavated and replaced with firm material, the tubes may have to be supported directly on pile bents. These are driven and cut off under water to exact grade.

A weak layer of soil may be strengthened by driving a series of compaction piles. These are driven so that their tops are 1 or 2 feet below grade and are covered with a screeded foundation course.

Placing Procedure

The tubes are moved from the construction basin or outfitting site with a freeboard of about 12 to 18 inches. Heavy concrete box sections may have less freeboard, and to ensure their buoyancy, they may be supported by pontoons or barges which then form part of the lay barge (Fig. 13-16). The methods of placing vary somewhat according to the type of foundation support and type of joints between tubes. In tidal waters, the actual placing is scheduled during slack tide. In rivers with constant flow or where slack tides are extremely short, the effect of the current on the tube while it is lowered must be analyzed. This is particularly important for wide rectangular sections where water pressure may upset the equilibrium of the tube while it is suspended on wire ropes with a relatively small positive weight. The specific gravity of the water in the bottom of the trench is checked to gauge the adequacy of the ballast.

The lay barge consists of two or more steel barges, sometimes car floats. These are placed parallel, with sufficient space between them to clear the tube, and connected with transverse girder bridges which carry four lowering winches. The assembly is further stabilized with diagonal wire ropes. The lay barge is held in position by wire ropes extended from winches to heavy anchor blocks placed on the bottom at distances of several hundred feet. The anchoring devices must have an adequate factor of safety to resist the maximum current pressure and wind forces on the barge and tube. The capacity of the lowering winches and wire ropes is determined by positive weight of the tube during placing. This may vary between 100 and 400 tons, depending on the size of the tube and the contractor's choice of sinking equipment. Load-limiting devices may have to be installed on the lowering winches to prevent overload on the wire ropes if there is a danger of sudden surges in water level to which the tube cannot respond quickly enough. By imposing speed restrictions on passing vessels during lowering, this danger can be minimized.

In some cases, two whirly cranes mounted on barges have been used to support the tubes during placing. This limits the lowering weight to about 100 tons, but it permits more precise

Fig. 13-16. Lay barge–model, Transbay Tube.

adjustment of the position of the tube without moving the barges.

Ballasting of the tube takes place after it is suspended from the lay barge and positioned over the trench. The type of ballast depends on the construction of the tube. For the Hampton Roads Tunnel (Fig. 13-1), a number of the outside tremie pockets were filled with concrete. The remainder of the pockets were filled as soon as the tube had been lowered into position and connected to the previously placed tube. The Transbay Tube (Fig. 13-3) was ballasted by filling the pockets on top of the tube with gravel. The Deas Island Tunnel (Fig. 13-9) used water ballast in tanks inside the tubes.

Placing on Screeded Foundation. After ballasting, the tube is slowly lowered until it nearly touches the foundation. Its end is kept clear of the end of the previously placed tube. After a final alignment check, it is then moved against the other tube as required to make the connection at the joint and set down on the foundation. If the weight of the tube is light, as when handled by cranes, additional ballast must be placed as soon as the connection is made–either by filling all tremie pockets with concrete or by temporary heavy concrete blocks to secure the tube against uplift.

Placing on Jetted Foundation. This follows the same procedure, except that the tube is temporarily supported on the bottom of the trench. In early projects, two low-capacity pile bents driven to grade provided this support. The pile bents were designed to support the positive lowering weight of the tube, but not the completed tunnel and backfill. Recently, four temporary foundation blocks were placed beneath the tube. These were attached to steel shafts penetrating into the interior of the tube, where they were connected to hydraulic jacks. After lowering of the tube, its vertical position is adjusted by jacking against these blocks bearing against the bottom of the trench. This requires access to the inside of the tube through stacks extending above water. The stacks remain in place until the foundation course is in place

Fig. 13-17. Survey tower (Second Hampton Roads Tunnel).

underneath the tube and must be protected against damage from ships during this time. As soon as the connection to the previously placed tube is made, the foundation course is jetted into the space between the underside of the tube and the bottom of the trench.

Alignment Control

In order to check the horizontal and vertical position of the tube during placing operations, temporary survey towers (Fig. 13-17) are mounted on the tube near the ends. These project above the water and carry survey targets.

To control transverse level (roll) of tubes and to ensure coincidence of the target location with that of the tube below, a vertical plummet was installed in the survey towers of the Transbay Tube in San Francisco in relatively deep water.

On the first tube placed at each end, two survey towers are mounted, one near each end to correctly place the tube. The tower at the outboard end is left in place until the next tube is placed. The position of the inboard end of each following tube is governed by the tube already in place so that a survey tower on these tubes is needed only at the outboard end. After the towers have served their purpose, they are disconnected by divers and reused.

Instruments for checking the alignment are mounted on each shore. On long tunnels or where horizontal curves occur, additional instruments are set on pile-supported platforms in the water. Laser beams are not sufficiently accurate for placing controls, due to their spread over long distances. Theodolites are used. Where more elaborate survey towers with vertical plummets are used, instruments can be mounted on them, sighting on targets located on the shores.

A final check of the position is made after the tube has been placed and connected to the previously placed section.

JOINTS BETWEEEN TUBES

Tremie Joints

This joint (Fig. 13-18) has been used in a number of steel shell tubes and is still applied when relatively few sections are required. A circular steel collar plate, of the same diameter as the tube shell, is welded to the outside of the end bulkhead and projects 4 feet. To the lower half of this collar on the outboard end of the tube, a 2-foot wide hood plate is welded with a filler, half of its width projecting beyond the end of the collars. A similar hood plate is welded to the upper half of the collar at the inboard end of the tube. The fillers provide an annular clearance of 1 inch between hood plates and collar. When the following tube is lowered, its inboard collar plate fits into the lower hood of the previous tube. On the horizontal diameters, where upper and lower hood plates meet, each carries a heavy welded bracket with a hole for a 5-inch diameter steel pin. The hole in the upper bracket is round; the one in the

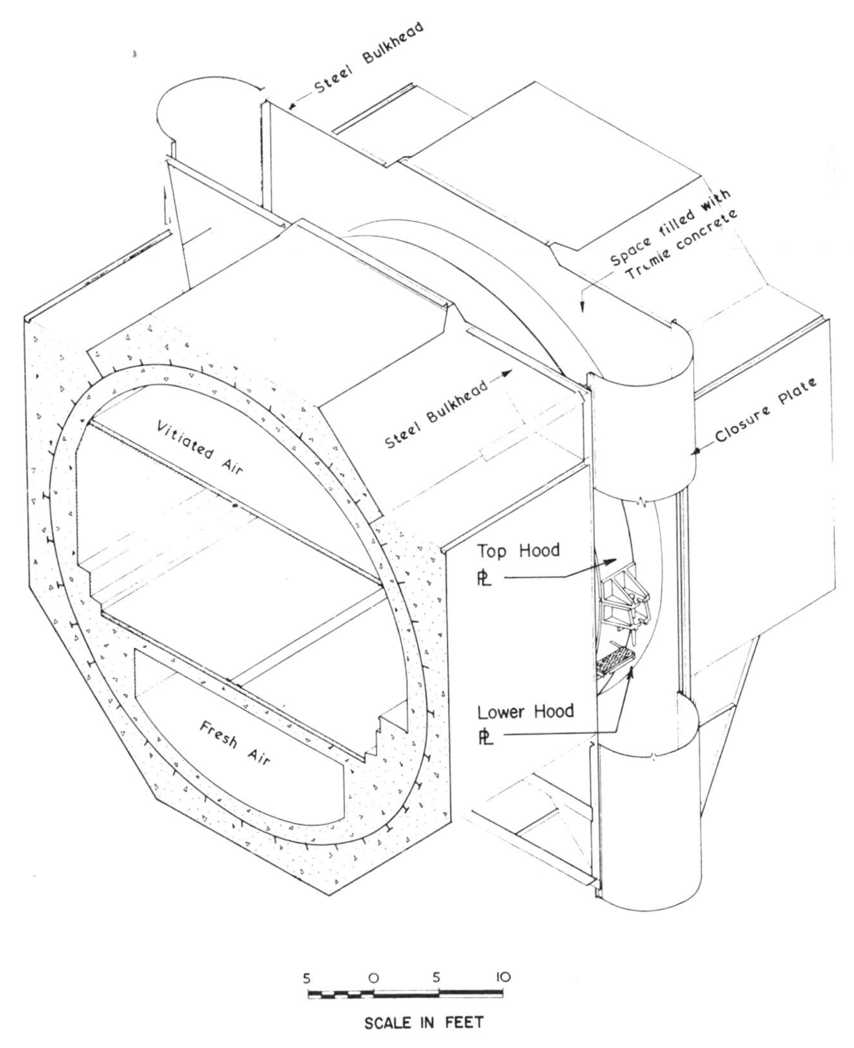

Fig. 13-18. Tremie joint.

lower is oblong to provide about $\pm 1\frac{1}{2}$ inch tolerance. After the tubes are pulled together by four steamboat ratchets, operated by divers, to match the brackets, the divers insert the tapered end steel pins, the steamboat ratchets on the brackets are tightened and the pins are secured in the brackets by wedges driven through slots in the pins (Fig. 13-19).

Curved closure plates are inserted in guides attached to the vertical edges of the square dam plates. Either sheet pile sections or steel angle guides are used. Sometimes, a series of sheet piles in a curved configuration have been used instead of closure plates. The piles are driven into the bottom of the trench, the foundation course having been omitted at the joints, so that an enclosed space is provided by them and the end dam plates, which is filled with tremie concrete. To prevent concrete from flowing inside of the joint, the annular space between hood plates and collars is caulked by divers. Care must be taken to ensure that the tremie concrete flows across the bottom of the joint, and that at the top is level with the top of the dam plates. This provides an adequate seal of the joint with only minor leaks, if any. To de-water the joint, valves are provided in the end bulkhead of the previous tube. The joint is then entered through the watertight doors mounted in the end bulkheads. Before opening the bulk-head doors, the air in the joint should be tested for explosive gases due to decomposition of organic matter in the water trapped in the joint.

Liner plates (Fig. 13-20) are then welded to T-sections on the interior of the collar plates to form a continuous watertight steel membrane for the tunnel. The space between these liner plates and the collar plates is filled with grout under 10 psi pressure. Then the interior portions and the dam plates are removed, and the interior concrete lining is placed, thus completing the joint.

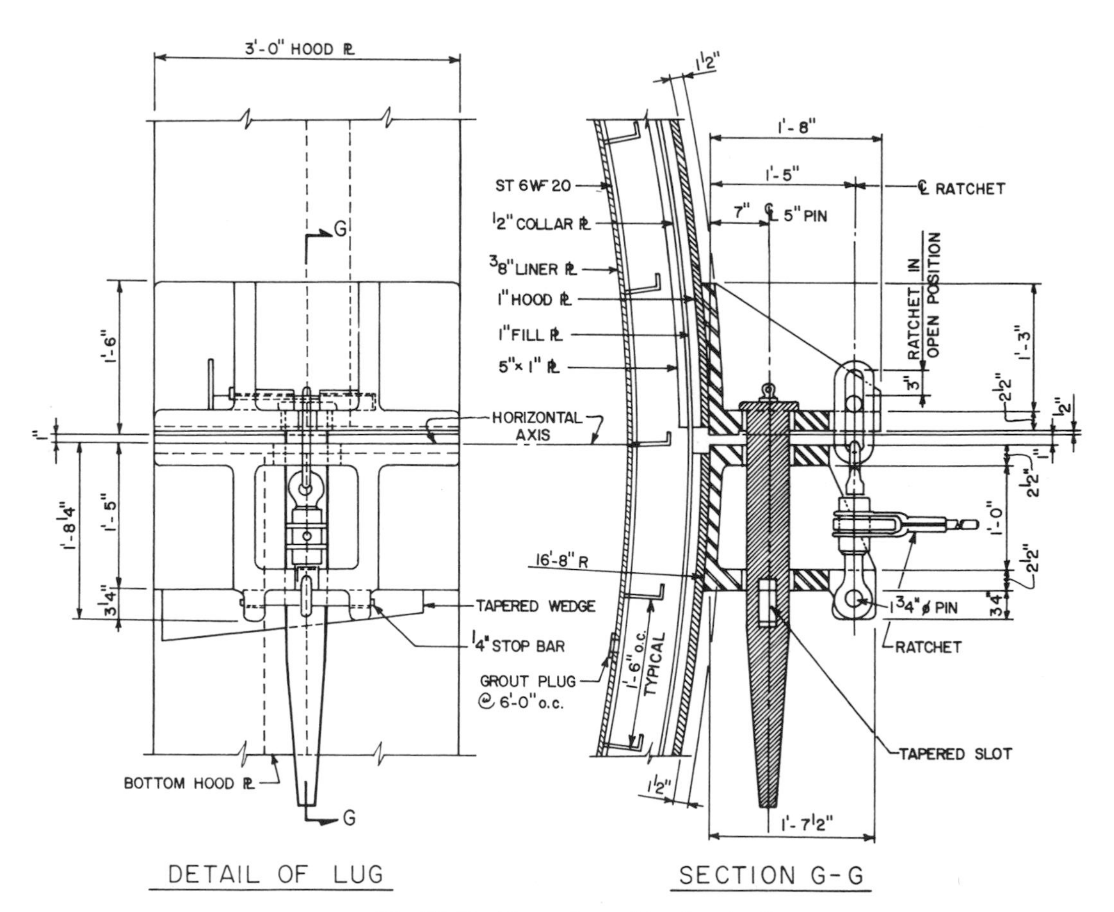

Fig. 13-19. Pin, wedge and bracket detail.

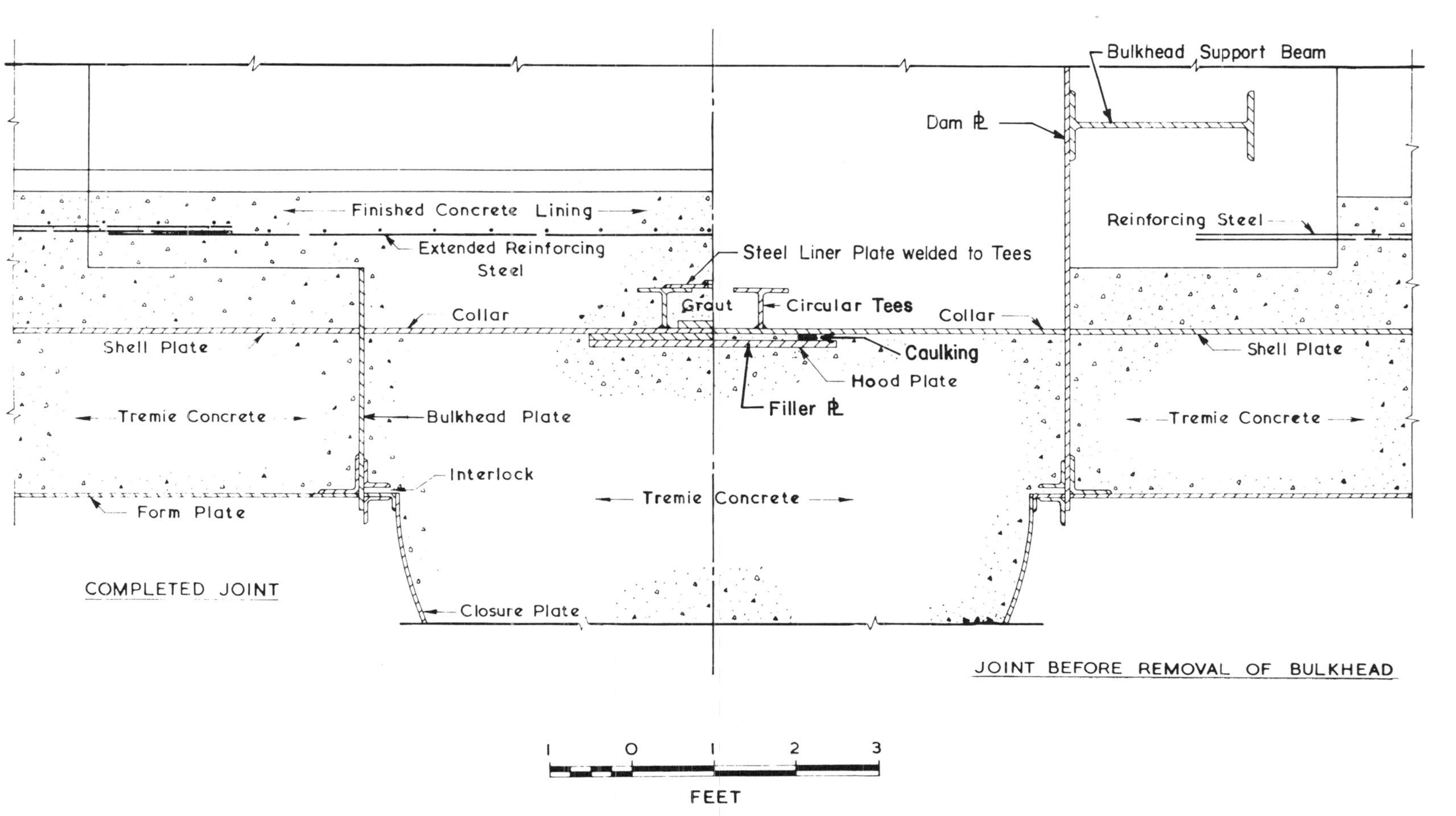

Fig. 13-20. Detail–liner plate.

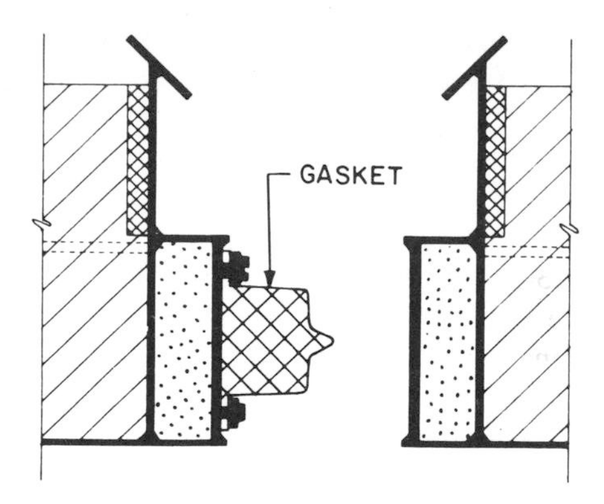

Fig. 13-21. Rubber joint with single gasket (Deas Island).

Rubber Compression Joint

In this type of joint, the initial seal of the joint is provided by the compression of rubber gaskets attached to the face of one tube and bearing against a smooth surface on the joining tube. While various shapes of gaskets have been used, the principal is the same. The tube being placed carries a continuous gasket on the periphery of the face. After lowering, the tube is moved close to the end of the tube in place so that hydraulic coupling jacks, extending from one of the tubes, can be engaged into mating parts on the other tube. These jacks pull the tube into contact and give an initial compression to the gasket. This seals the joint sufficiently so that it can be drained from the inside, which brings into action the entire hydrostatic pressure on the far end of the tube, compressing the gasket to its limit, temporarily sealing the joint. After dewatering, the joint can then be entered through the doors in the end bulkheads and the permanent connection made.

A single gasket of large dimensions was used on the Deas Island Tunnel and on similar concrete box tunnels in Europe, as shown in Fig. 13-21. The tip of the gasket provides the initial seal. Under final pressure, the body of the gasket is compressed to about half of its height and carries the entire load.

A double ring gasket of smaller cross-section was used in the Transbay Tube of BART in San Francisco (Fig. 13-22). The gaskets are attached to a continuous bracket type extension of the tube structure. The cantilever lip on the outer gasket is deflected by the pull of the jacks and provides the initial seal. Upon dewatering of the joint, the gaskets are compressed to one-half

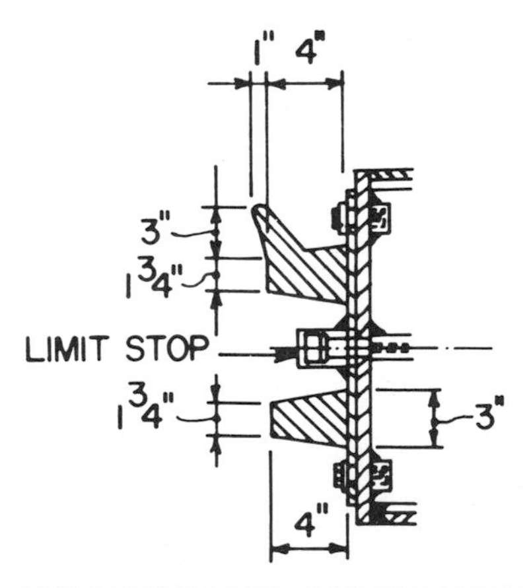

Fig. 13-22. Rubber joint with double gasket (Transbay Tube).

of their height. A steel bar welded to the face provides a definite (limit) stop to the movement which avoids any variation in the final spacing of the joint as shown in Fig. 13-23. Valved drain lines bleed the water between the gaskets to the interior during compression. The gaskets made the joint (Fig. 13-24) watertight, permitting the welding of the interior closure plate and completion of the concrete lining.

Coupling Jacks. Large single coupling jacks (Fig. 13-25) mounted in the center of the bulkhead in each bore have been used in Deas Island and other similar tunnels. They must be controlled from the inside of the tube, and a diver has to enter the open joint to check that they are properly engaged. Once pulling has started, they are no longer accessible in case trouble develops.

Multiple, externally mounted jacks—similar to large automatic railroad couplers—were used in the San Francisco Transbay Tube. Two of these were mounted on each side of the tube: one near the top, the other near the bottom. They are at all times accessible to divers for checking. Should one malfunction, the other three are sufficient to close the joint. After the gasket is compressed, the coupler bars are wedged in position and the hydraulic jacks are removed for reuse. The hydraulic control hose lines were extended to a control panel on the lay barge. The system worked exceptionally well for all 58 tubes.

Closure Joint

In long tunnels with many tubes, it is usually most expeditious to start placing the tubes from both ends, with the closure joint (Fig. 13-26) somewhere out in the waterway. In the 58-tube

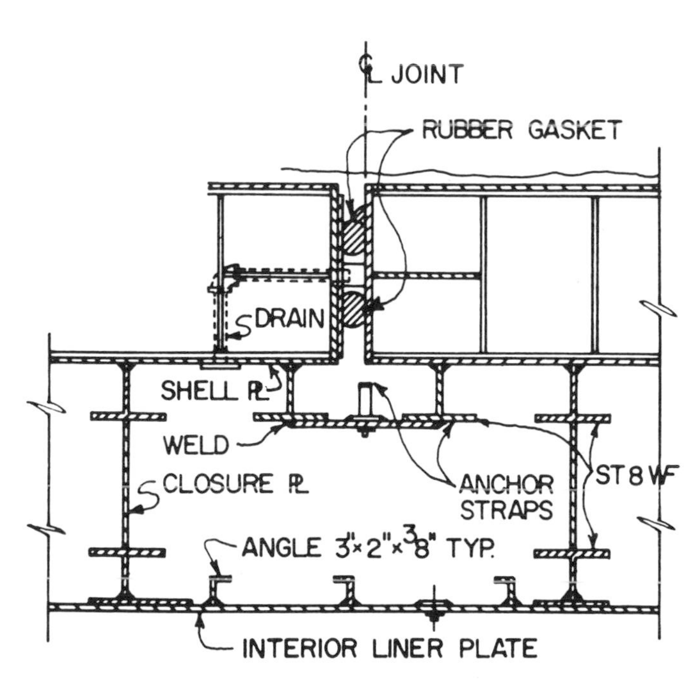

Fig. 13-23. Rubber joint with double gasket (Transbay Tube).

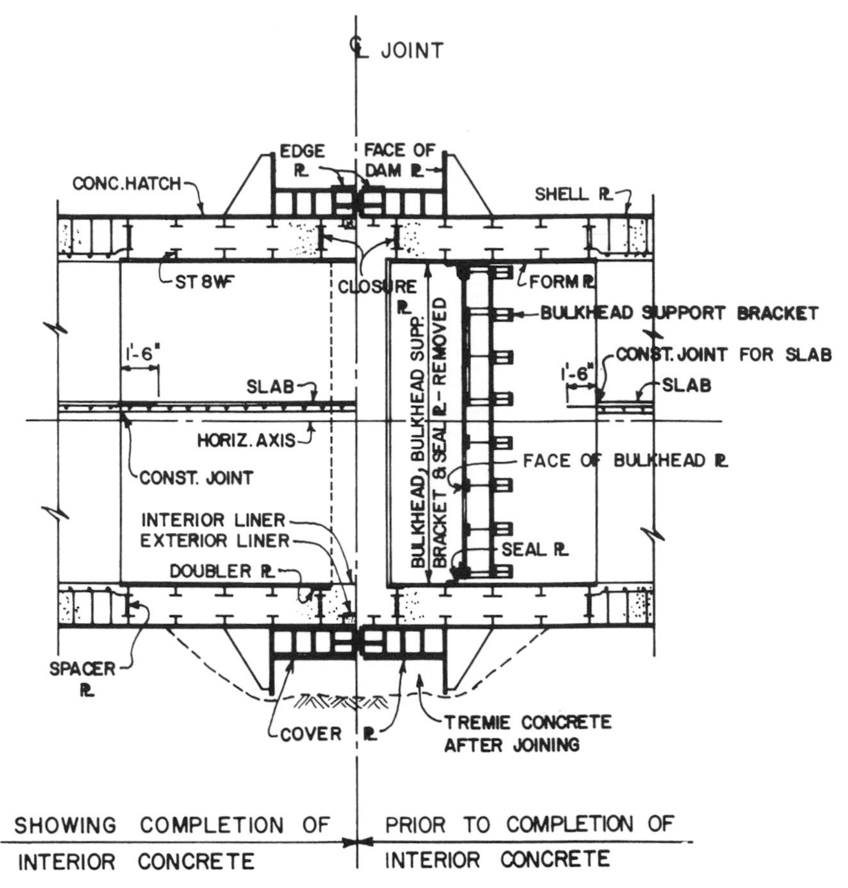

Fig. 13-24. Typical joint cross-section (Transbay Tube).

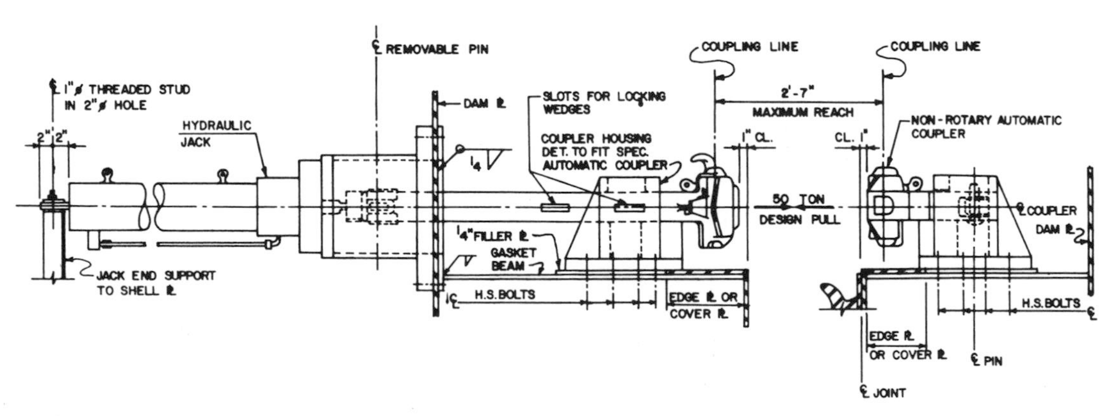

Fig. 13-25. Coupling jacks (Transbay Tube).

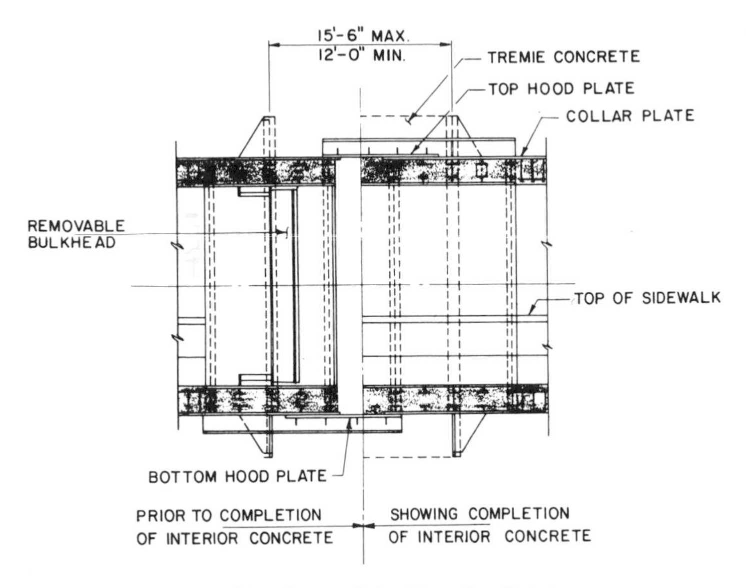

Fig. 13-26. Closure joint (Transbay Tube).

Transbay Tunnel, Tube No. 37 was the last tube placed. To provide for fabrication tolerances, which may be cumulative, and for adequate clearance in placing the last section into the space, the final joint must allow for adjustment. Rubber gaskets cannot be used because there is no compression available. A tremie joint with extra long hood plates was used in the Transbay Tube, providing a 24-inch allowance for clearance. It was sealed in the same way as described for the regular tremie joint. Due to very tight fabrication control and limit stops in the joints, the actual tolerance required was less.

Special Joint for Seismic Movements

The San Francisco Transbay Tube for BART is located in Zone 3 seismic area (Fig. 13-27). Although not crossing any active faults, the tunnel structure, including the joints, is capable of elastically absorbing deflections from seismic waves originating anywhere in the area's major faults. Where the ends of the submerged tube section join the massive ventilation buildings, particularly at the San Francisco side, differential movements in any direction may occur between tube and building, amounting to as much as 4 inches. A patented composite telescoping and sliding joint allows for these motions. For transverse sliding or rotation displacement, two neoprene gasket rings are attached to the tube end, held under compression against a Teflon-covered steel plate on the building side by a series of short steel wire ropes, which deflect under movement. A telescoping joint of similar construction permits longitudinal motion. Provisions are made for retightening of the ropes by threaded sockets and nuts if needed. Flexible neoprene covers attached to the outside of the joints protect the outer gaskets and Teflon areas from the backfill.

BACKFILL

Locking Fill

In order to securely lock the tubes in position after they are connected, a special fill is placed in the trench to about half the height of the tube. This is a well graded material which compacts easily on placing. It may be sand or coarser material, depending on available sources. Material used has been well graded, usually ranging in size from $1\frac{1}{2}$ inches to Number 100 mesh size, but sizes up to 8 inches have been used in strong currents.

Sand may be placed through tremie pipes or by clam shell bucket, which is opened only when reaching the bottom. This is also used for coarser material.

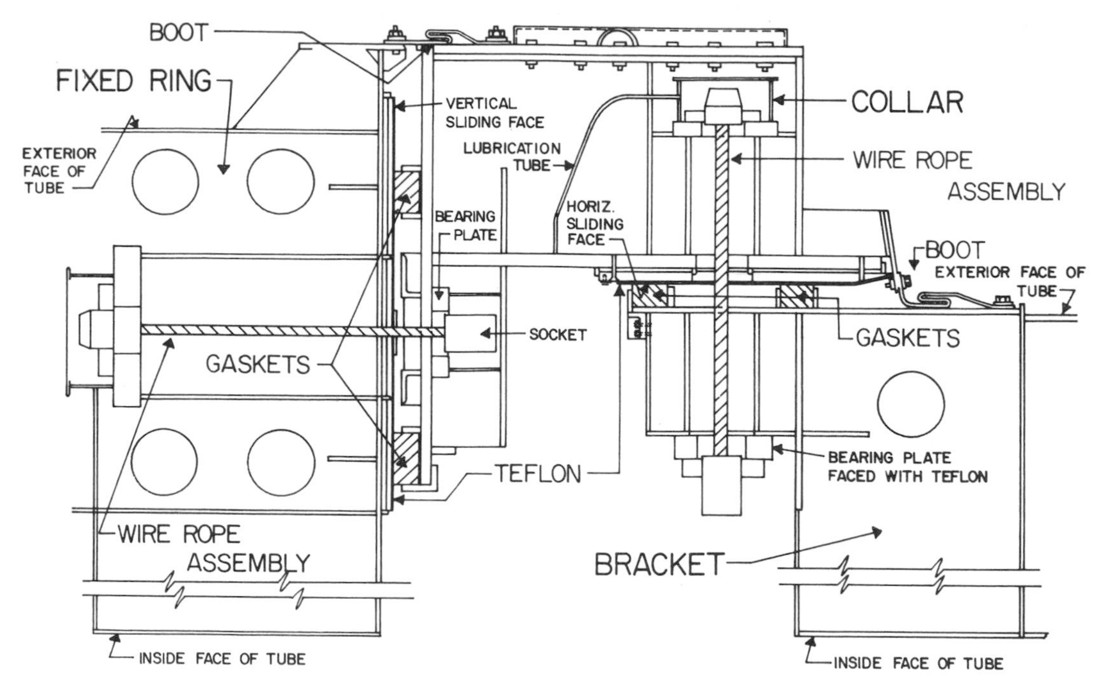

Fig. 13-27. Seismic joint (Transbay Tube).

Ordinary Backfill

Ordinary backfill to fill the trench to a depth of at least 5 feet above the tube may be any reasonably firm material available. In long tunnels, soil excavated from the trench may be used as backfill over tubes which have been placed in other parts. Backfill should be free from clay balls, chemically inert, and material passing the 200 mesh sieve should not exceed 20% by weight.

Protection Against Scouring

Where part of the tunnel projects above the original bottom, or where strong currents prevail, the backfill is protected with a rock blanket 2 or 3 feet thick. The size of the stone depends upon erosive action of current and usually varies between 1 inch and 100 lb. The width of the blanket should be at least 100 feet on each side of centerline and confined within dykes.

DESIGN OF TUBES

Selection of Cross-section

For a railroad, transit or vehicular tunnel, selection of cross-section is dependent on vertical and horizontal clearances, number of lanes or tracks, type of ventilation system and required air duct areas. Typical cross-section configurations are circular, octagonal, arch and rectangular. The number of bores is dependent on the number of tracks or lanes. Transit and railroad tunnels are usually ventilated by the piston action of the train, and except for fire protection do not require air ducts. Vehicular tunnels over 500 feet in length usually require ventilation. A full transverse ventilation system requires an exhaust and supply air duct, and a semi-transverse system requires either a supply or exhaust duct. Ideally, the exhaust duct is located above the roadway and the supply duct beneath it. The configuration for a single bore is best suited to a circular or octagonal shape tunnel which structurally is the most efficient. The arch shape is suited to single-bore tunnels with a semi-transverse ventilation system. Rectangular shapes for single or multiple bores have center or side duct locations which reduce the depth of the structure and dredging. For a single bore with a semi-transverse supply ventilation system, location of the supply duct above the roadway has been used but it requires reversible fans for emergency exhaust during a fire.

Loading Condition

For the design of the tubes, the loading condition consists of dead load, water pressure, earth pressure and superimposed live load. The latter load can consist of temporary loads such as a sunken ship, increased load due to dewatering during construction or permanent loads due to surcharge. For a circular cross-section, the loading conditions are as follows.

Loading Condition	General Equation	
Uniformly Distributed Load – Top W = #/ft. R = ft. 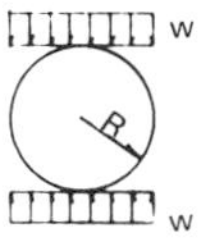 w w Variable	$M\theta = wR^2\left[0.25 - \frac{\sin^2\theta}{2}\right]$ $N\theta = -wR \sin^2\theta$ $V\theta = -wR \sin\theta \cos\theta$	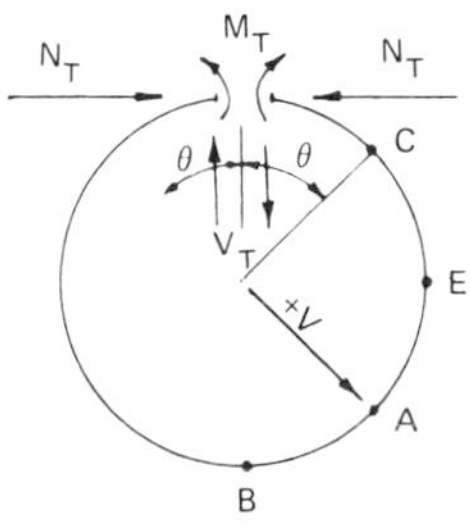 +M = Compression on outside of ring + Tension - Compression FIGURE 28 Loading Conditions and Moments Thrust and Shears
Uniformly Distributed Load – Side 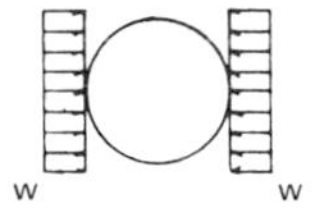 w w Variable	$M\theta = \frac{wR^2}{2}\left[\frac{1}{2} - \cos^2\theta\right]$ $N\theta = -wR\left[\cos^2\theta\right]$ $V\theta = +wR \sin\theta \cos\theta$	
Wedge Shape Horizontal Load 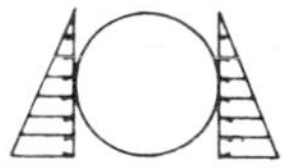 2 wR 2 wR	$M\theta = wR^3\left[0.250 - 0.125\cos\theta - \frac{\cos^2\theta}{2} + \frac{\cos^3\theta}{6}\right]$ $N\theta = -wR^2 \cos\theta\left[0.125 - \frac{\cos^2\theta}{2} + \cos\theta\right]$ $V\theta = wR^2 \sin\theta\left[0.125 + \cos\theta - \frac{\cos^2\theta}{2}\right]$	
Quadrant Loading P P wR 0.2145 wR $P = wR^2\ (1 - \pi/4)$ $P = 0.2145\ wR^2$	$0 \leqslant \theta \leqslant \pi/2$ "θ" in Radians $M\theta = -wR^3\left[0.327 - \theta\frac{\sin\theta}{2} + \frac{\sin^2\theta}{2} - 0.521\cos\theta + \frac{\cos^3\theta}{6}\right]$ $N\theta = -wR^2\left[\sin^2\theta - \theta\frac{\sin\theta}{2} - \frac{\sin^2\theta\cos\theta}{2} - \frac{\cos\theta}{48}\right]$ $V\theta = -wR^2\left[\cos\theta\sin\theta - \theta\frac{\cos\theta}{2} - \frac{\sin\theta\cos^2\theta}{2} + \frac{\sin\theta}{48}\right]$ $\pi/2 \leqslant \theta \leqslant \pi$ "θ" in Radians $M\theta = -wR^3\left[0.434 - 0.0208\cos\theta - \frac{\cos^2\theta}{2} - \frac{\pi}{8}\sin^2\theta\right]$ $N\theta = -wR^2\left[0.214\sin^2\theta - 0.0208\cos\theta\right]$ $V\theta = -wR^2\left[0.0208\sin\theta + 0.214\sin\theta\cos\theta\right]$	
Buoyancy Forces 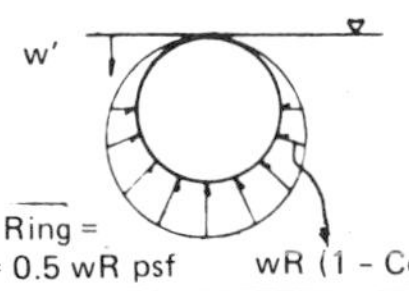 w' Wt. Ring = w' = 0.5 wR psf wR (1 - Cos θ)	$M\theta = 0$ $N\theta = -wR^2\left[1 - \frac{\cos\theta}{2}\right]$ $V\theta = 0$	
Water Surcharge 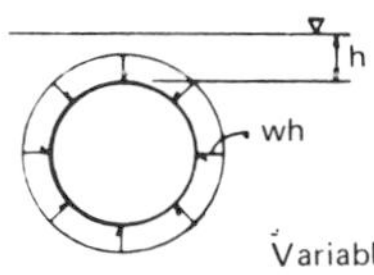 h wh Variable	$M\theta = 0$ $N\theta = -whR$ $V\theta = 0$	

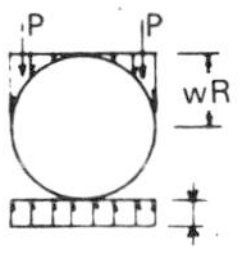

Fig. 13-28. Loading conditions and moments, thrusts and shears.

1. Uniformly distributed load–top.
2. Uniformly distributed load–side.
3. Wedge-shaped horizontal load–side.
4. Quadrant loading–top.
5. Buoyancy forces. (If the weight of the shell of the tube is uniformly distributed around the circumference and is just sufficient to cause full submergence, there will be no bending in the shell. This can be proven by combining the moments and thrusts of Cases XVI and XVIII of the article by Paris.[2] For Case XVIII, the loads will be reversed from those shown by Paris. The assumption that the weight is uniformly distributed is not strictly correct, since some of the weight is concentrated near the bottom, in the roadway slab, the sidewalks and ledge, etc.; also part of the tremie concrete is not placed until after the tube rests on the bottom of the trench. Analysis has shown that these variations have a relatively slight effect on the stresses in the shell of the tube.
6. Water surcharge. When the tube is completely submerged, the additional uniform water pressure caused by the depth of water above the top of the tube causes a uniform compression around the ring, but no bending moments. The loading conditions are shown diagrammatically in Fig. 13-28. For rectangular cross-sections, quadrant loading does not exist unless the top of the tunnel section is sloped at its extremities.

Design Loads

Design loads for stress analysis are based on boring information and the specific gravity of water at the site. Usual values are:

Submerged earth:	70 pcf
Moist earth in air:	120 pcf
Water:	64 pcf
Lateral coefficient of earth pressure:	0.27

For stability against buoyancy, assuming 3000 psi concrete with $1\frac{1}{2}$-inch maximum size aggregate, the unit weight of concrete is usually 145 pcf; structural and reinforcing steel, 490 pcf; and water, 64 pcf.

For conditions of flotation, the usual weight of concrete is assumed to be 147 pcf and steel and water weights are the same as for stability against buoyancy.

Factors of safety against buoyancy were discussed earlier in this chapter. Stability calculations should be based on the net volume of concrete. Weight of all conduit, pipe, reinforcing steel and structural steel should be included. The calculations should be based on the full length of tube. The critical condition for stability against flotation occurs after the tube is on the bottom and connected to the previously placed tube. After the joint has been dewatered and the end bulkhead has been removed, and prior to completing the interior concrete at the joint, with no backfill in place, the factor of safety should not be less than 1.02. If this factor of safety cannot be maintained, temporary ballast, consisting of sinking blocks, should be provided.

Water level is usually assumed to mean sea level. Where the fill over the tube extends above water level, maximum moments occur in the tubes with minimum water level. Where the surface of the fill is below water level, the moments are not affected by the elevation of the water surface, but the direct thrusts in the shell of the tube will increase with higher water levels. Under these conditions, the maximum compressive stress in the shell of the tube will be somewhat higher under high water conditions; but the relative increase in the compressive stress will be small, since a greater portion of this stress results from bending moments.

Live loads (surcharge) depend on the future use of the land above the tube. Usually, 500 psf is assumed unless definite loads are known. In industrial areas, a surcharge of 750 psf is not unusual. The effect of these loads are evaluated at normal working stress. Temporary loads, such as a sunken ship (1000 psf), have been considered as a surcharge with a 25% increase in allowable stress. Provision for surcharge for sunken ships is not usually required.

Design Stresses and Codes

Design practice for tubes in the U.S. is usually governed by the requirements of the AISC Specification for Design, Fabrication and Erection of Structural Steel for Buildings and the ACI Standard 318—Building Code Requirements for Reinforced Concrete Design. These codes are used in lieu of AASHTO Standard Specifications for Highway Bridges or AREA Specifications for Railway Bridges, since the loads on a tube are essentially non-repetitive and the theory of working stress is utilized rather than ultimate strength or plastic design theory. Welding requirements are in accordance with AWS Structural Welding Code for Bridges. In addition, where appropriate, the requirements of the ABS Rules for Building and Classing Steel Vessels and the ASME Boiler and Pressure Vessel Code Section IX Welding Qualifications are utilized.

Materials for the tube usually consist of structural steel meeting the requirements of ASTM designation A36, reinforcing steel 60,000 psi meeting the requirements of ASTM designation A15 new billet steel, structural concrete (3000 psi), tremie concrete (2200 psi) and prestressed concrete (5000 psi) meeting the requirements of ACI 318.

Allowable stresses should be in accordance with the referenced codes. Usual design practice permits a 25% increase in allowable stress for temporary loading conditions. Stresses during outfitting, while temporary in nature, are usually designed for normal stress. Design of the terminal tube bulkheads is for dewatered moist soil at increased working stress.

Structural Systems

Structural systems for trench tunnels consist of a structural steel shell with a reinforced concrete lining acting compositely, reinforced concrete and prestressed concrete. Generally, octagonal, circular and arch configurations utilize the composite section, whereas rectangular sections re usually reinforced or prestressed concrete.

Octagonal composite sections have a circular steel shell backed by external longitudinal stiffeners and ring diaphragms, lined with reinforced concrete on the inside for structural strength. Outside the circular steel shell are octagonal-shaped form plates, attached to the shell by the diaphragms and radial angle struts. The space between the octagonal steel form plates and shell plate is filled with concrete, a portion placed under water to furnish corrosion protection to the shell and diaphragms and weight to overcome buoyancy, and the remainder placed in the dry to provide structural strength in the in situ condition (see Fig. 13-1).

Circular composite sections have a circular steel shell stiffened by internal transverse structural steel tees. The shell and tees are designed to act as a ring girder during construction and, in the final in situ condition, to act compositely with the reinforced concrete lining. The steel shell usually requires protection against corrosion. On the Transbay Tube (Fig. 13-3) and Sixty-third Street (Fig. 13-4) Tunnels, a cathodic protection system was used, whereas for the Hong Kong Tunnel (Fig. 13-2), a coating of gunite was used to provide corrosion protection.

Arch and rectangular composite sections with steel shells and reinforced concrete linings have structural systems similar to the circular composite section.

Reinforced concrete and prestressed concrete sections usually have a rectangular rigid frame structural system and require construction in the dry with an exterior waterproofing system.

Design Analysis—Steel Octagonal Shell Tubes

Three distinct design stages require investigation:

Stage I—Fabrication and launching.
Stage II—Outfitting.
Stage III—Final condition.

Stage I. Fabrication and launching stresses are a function of the type of structural system, the method of fabrication, the length of tube modules, the method of erection on the ways, blocking between ways for support of the modules, way spacing and declevity and the method launching. During these assembly stages, each module and tube section should be self-supporting or should be locally reinforced for local buckling, lifting and launching stresses. In addition, transverse deflections of the modules should be checked, and, if required, spiders should be installed to maintain the diameter of the shell plate in order to permit proper fit-up for the circumferential butt welds between the tube modules.

After pouring the keel concrete, the weight of the steel shell and keel concrete are transferred from the temporary blocking to the launching sleds (or fore and aft poppets) prior to launching. For end launching, the fore and aft poppet loads may require reinforcement of the shell and diaphragms within the poppet lengths. Similarly, for side launching, the steel, shell and diaphragms may require local strengthening over each sliding way. In evaluating these stresses, the keel concrete should be considered to be a longitudinal beam acting with the shell plate and diaphragms. End launching may require temporary longitudinal reinforcing of the top of the shell. Stresses in the shell and form plates should be investigated for water pressure at launching. Fabrication and launching stresses are usually the responsibility of the contractor, since they are dependent on his fabrication methods and plant.

Stage II. During outfitting while floating, except for the end bulkheads, the weight of the tube is uniformly distributed longitudinally and is supported by the buoyancy forces, thereby causing no longitudinal bending moments. The end bulkheads, however, impose concentrated loads which are balanced by buoyancy forces that are uniformly distributed along the length of the tube with resulting "hogging" moments.

Additional longitudinal moments are imposed on the tube during concreting operations, since the weight of the fresh concrete cannot be uniformly distributed along the length of the tube. These moments are resisted by the steel shell acting as a beam.

In addition to the longitudinal moments, the steel shell and stiffening diaphragms are subjected to circumferential bending moments resulting from the exterior water pressure. Usually, the form plates are not made watertight in order to minimize the pressure acting on them and to permit the shell plate to resist all the water pressure.

A typical sequence of concrete placement by stages is shown in Fig. 13-7. The length of pour for each stage is a function of convenience in form placing operations, the quantity of concrete that can be conveniently placed in a single shift and the longitudinal moments and shears that the shell can withstand. In so far as possible, pours should be symmetrical about the transverse and longitudinal centerlines of the tubes in order to maintain trim of the tube and reduce the water pressure on the shell plate. Experience has shown the stages and sequence indicated on Fig. 13-7 usually result in the most favorable design and construction conditions.

Moments, shears and head of water acting on the shell plate and diaphragms due to the unbalanced weight of the bulkheads and the fresh concrete are calculated for each stage and length of pour. Usually, the critical condition occurs during the placing of the haunch pour for sequence 5. At this stage, the shell and diaphragms are fixed at the top of the keel concrete and the head of water against the shell extends from the top of the keel to approximately the horizontal axis. The pressure of the fresh concrete against the inside of the shell is usually neglected. The resulting circumferential moments are resisted by the shell plate and the diaphragms, with longitudinal distribution of loads by the shell plate and longitudinal stiffeners.

The critical unit compressive stress for buckling of a curved panel under uniform compression is given in Roark and Young.[3] (Table 35-Case 13)

$$\sigma = \frac{1}{6} \cdot \frac{E}{1-\nu^2} \cdot \left[\sqrt{12(1-\nu^2)\left(\frac{t}{R}\right)^2 + \left(\frac{\pi t}{b}\right)^4} + \left(\frac{\pi t}{b}\right)^2\right]$$

where

- σ = critical unit compressive stress for buckling (psi)
- $E = 30 \times 10^6$ psi
- ν = Poisson's ratio (0.3 for steel)
- t = thickness of shell plate (inches)
- R = radius of the centerline of the shell plate (inches)
- $b = \frac{2\pi R}{n}$ = peripheral spacing between stiffeners (inches)
- n = number of stiffeners—usually equally spaced around the circumference.

The moment capacity of the shell is determined by taking the critical buckling stress, dividing by a factor of safety–usually 2–and multiplying the result by the section modulus of the shell.

Within the elastic range, the local buckling strength of the shell, loading in torsion or transverse (beam) shear, is given by formulas set forth in Brockenbrough and Johnston.[4] (page 91-Formula 4.27)

$$\sigma_{crs} = 0.632\, E\left(\frac{t}{R}\right)^{5/4} \times \left(\frac{R}{L}\right)^{1/2}$$

where

σ_{crs} = local buckling strength in torsion (psi)
$E = 30 \times 10^6$ psi
t = thickness of shell plate (inches)
R = centerline radius of the shell (inches)
L = length in inches between diaphragms.

The above formula is valid when (page 91 Formula 4.27)

$$10\left(\frac{t}{R}\right)^{1/2} < \frac{L}{R} < 3\left(\frac{R}{t}\right)^{1/2}$$

The upper limit for shear buckling, σ_{crs}, shall not exceed $f_y/\sqrt{3}$ (where f_y = unit stress at yield point). For transverse shear, the buckling stress may be considered, conservatively, as 1.3 times the buckling stress in torsion, but not greater than $f_y/\sqrt{3}$.

The shear capacity of the tube is

$$V = \pi R\, \frac{\sigma_{crs}}{2} \times 1.3.$$

The water pressure on the shell plate is assumed to be uniformly distributed between longitudinal stiffeners, and, therefore, since the load is acting radially against the plate, the stress induced will be axial compression only, with no bending moment. The load on the shell is $p = \dfrac{wh}{144}$ and the thrust in the plate equals $T = pR$ and produces a stress $f_s = pR/t$. This stress is checked against the minimum pressure that could cause buckling of the arch. From Roark and Young,[3] (Table 35 Case 21) the approximate buckling pressure is

$$q' = \frac{Et^3\left(\frac{\pi^2}{\alpha^2} - 1\right)}{12R^3(1 - \gamma^2)}$$

where

q' = critical buckling stress (psi)
$E = 30 \times 10^6$ psi
α = angular spacing of the stiffeners (radians)
R = centerline radius of shell (inches)
γ = Poisson's ratio (0.3).

The longitudinal stiffeners maintain the shape of the shell plate and transfer by beam action between diaphragms the reactions of the plate elements considered above.

The shell plate and the longitudinal stiffeners cannot be considered as separate structural elements, but rather as components of a cylindrical shell spanning between diaphragms. The shell plate alone is capable of resisting water pressure as both a ring structure and as a cylindrical beam. Therefore, as an approximation, the longitudinal stiffeners at the portion of the shell where the exterior pressure is applied is considered as a beam on an elastic foundation. The load on such a beam is proportional to its deflection; a parabolic deflection was assumed. Also, the shell plate was assumed to support the full load at the midpoint between diaphragms. Thus, the net load on the longitudinal stiffener beams is the uniform load minus the parabolic load supported by the shell plate. The stiffener beams (Fig. 13-29) are considered as simply supported at the diaphragms, where

L = span of stiffener beam between diaphragms (inches)
p = load on the plate (psi) $= \dfrac{64h}{144}$
w = maximum linear load (pounds per inch =
b = spacing (inches) between stiffeners = $\frac{\pi}{}R \times \dfrac{n}{180}$, where n = number of stiffeners.

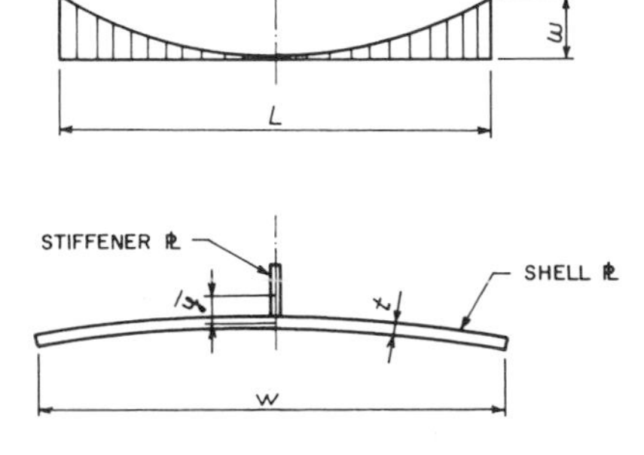

Fig. 13-29. Stiffener load.

The bending moment $= wL^2/48$. The effective width of the stiffener beam is

$$b = \frac{6000t}{(f_y)^{1/2}},$$

as recommended in Brockenbrough and Johnston.[4] (Page 102, Table 4.3)

The section modulus of the stiffener beam is calculated, and the stress is

$$f_s = \frac{wL^2/48}{S_{min}}.$$

The diaphragm is designed as a ring structure, fixed at the top of the keel concrete, resisting the external water pressure on the length of tube between diaphragms. Calculation of bending moments is made per lineal foot of tube (Fig 13-30).

The critical moments, thrusts and shears are resisted by the diaphragm (see Fig. 13-31) which consists of an exterior flange, a web and a shell plate. The form plates, since they are not protected against corrosion, are neglected. The effective width of the shell plate (l) acting with the web and flange is based on an article by Schorer.[5]

$$l = \sqrt{1.56Rt}.$$

R = the centerline radius of the shell plate in inches and t = the thickness of the shell plate. The section properties are calculated and the stresses are determined from

$$f_{max} = \frac{T}{A} \pm \frac{M_c}{I}.$$

Stage III. In its final condition, the precast tunnel consists of a cylindrical structural tube which is substantially circular in shape on the inside, and of octagonal shape on the outside. In order to simplify the calculations, it is assumed that the exterior is also circular. Since the tremie concrete is not considered as part of the structural section, the radius to the center of gravity and the moment of inertia of the section shows considerable variation around the circumference. It is believed, however, that a sufficiently accurate analysis of moments in the tube can be made by assuming the tube to be replaced by a cylinder consisting of the interior reinforced concrete and the structural steel shell with a centerline radius equal to that of the steel shell plate.

Bending moments and direct stresses are computed by means of the formulas and loading conditions tabulated in Figs. 13-28 and 13-30. These formulas were presented in an article by Paris[2] and in *Formulas for Stress and Strain* by Roark and Young.[3] Throughout this analysis, the following conventions apply: positive moment causes compression on the exterior of the tube; tension is considered positive; compression is negative.

Moments, thrusts and shears are calculated for each loading condition and are then combined to determine the final moments, thrusts and shears for design. Usually, the final moments, thrusts and shears show that at all points the thrusts cause compressive stresses at the top and bottom, and moments cause compressive stress on the outside, while at the horizontal axis, the moments cause compressive stress on the inside. It is desirable to have a design section such that most of the compressive stress is resisted by the concrete and all of the tensile stress is resisted by the steel.

Usually, the design of the steel diaphragms show that the outside flange at the top of the tube, which is in compression, requires an excessive area due to the combination of bending stress with direct compression. To minimize this condition, the cap concrete is placed in the dry so that the diaphragm is encased in structural concrete. To protect the exterior flange, the concrete extends 2 inches above the diaphragm.

At the horizontal axis, the interior concrete shell is in compression. Experience has shown that a reasonable flange area for the diaphragm can be provided to carry the tensile stress. While the tremie concrete is not included in the structural design, it provides lateral support for the diaphragms.

Because of this design section, the location of the center of gravity (CG) of the structural shell varies considerably around the circumference of the tube. It coincides approximately with the geometrical axis at the top and bottom, but is located between the geometrical axis and the interior face of the concrete at the horizontal axis. Taking the tube as a whole, the CG

Loading Condition	Top	45^{0}	Horizontal Axis	135^{0}	Bottom
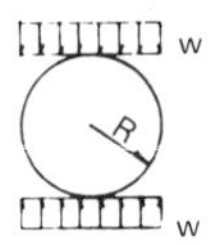 Uniformly Distributed Load — Top $W = \frac{\#}{ft.}$ w R = ft. R w Variable	$M_T = +0.25\ wR^2$ $N_T = 0$ $V_T = 0$	$M_c = 0$ $N_C = -0.5\ wR$ $V_C = -0.5\ wR$	$M_E = -0.25\ wR^2$ $N_E = -wR$ $V_E = 0$	$M_A = 0$ $N_A = -0.5\ wR$ $V_A = +0.5\ wR$	$M_B = +0.25\ wR^2$ $N_B = 0$ $V_B = 0$
Uniformly Distributed Load — Side 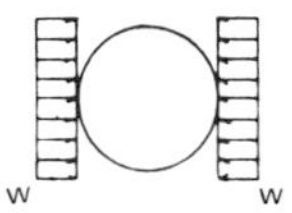 w w Variable	$M_T = -0.25\ wR^2$ $N_T = -1.0\ wR$ $V_T = 0$	$M_C = 0$ $N_C = -0.5\ wR$ $V_C = +0.5\ wR$	$M_E = +0.25\ wR^2$ $N_E = 0$ $V_E = 0$	$M_A = 0$ $N_A = -0.5\ wR$ $V_A = -0.5\ wR$	$M_B = -0.25\ wR^2$ $N_B = -1.0\ wR$ $V_B = 0$
Wedge Shape Horizontal Load 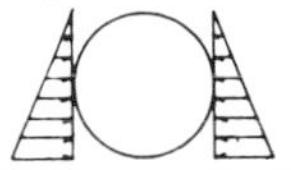 2 wR 2 wR	$M_T = -0.208\ wR^3$ $N_T = -0.625\ wR^2$ $V_T = 0$	$M_C = -0.0295\ wR^3$ $N_C = -0.412\ wR^2$ $V_C = +0.413\ wR^2$	$M_E = +0.25\ wR^3$ $N_E = 0$ $V_E = +0.125\ wR^2$	$M_A = +0.0295\ wR^3$ $N_A = -0.59\ wR^2$ $V_A = -0.589\ wR^2$	$M_B = -0.292\ wR^3$ $N_B = -1.375\ wR^2$ $V_B = 0$
Quadrant Loading 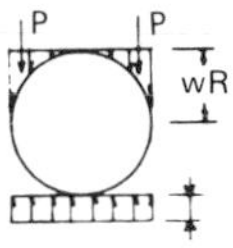 P P wR $P = wR^2\ (1 - \pi/4)$ 0.2145 wR $P = 0.2145\ wR^2$	$M_T = +0.027\ wR^3$ $N_T = +0.0208\ wR^2$ $V_T = 0$	$M_C = +0.009\ wR^3$ $N_C = -0.03\ wR^2$ $V_C = -0.0608\ wR^2$	$M_E = -0.0419\ wR^3$ $N_E = -0.215\ wR^2$ $V_E = -0.208\ wR^2$	$M_A = -0.003\ wR^3$ $N_A = -0.122\ wR^2$ $V_A = +0.092\ wR^2$	$M_B = +0.0447\ wR^3$ $N_B = -0.0208\ wR^2$ $V_B = 0$
Buoyancy Forces 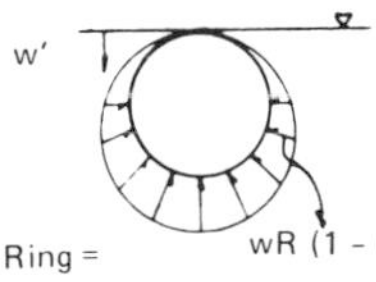 w' Wt. Ring = wR (1 - Cos θ) w' = 0.5 wR psf	$M_T = 0$ $N_T = -0.5\ wR^2$ $V_T = 0$	$M_C = 0$ $N_C = -0.647\ wR^2$ $V_C = 0$	$M_E = 0$ $N_E = -1.0\ wR^2$ $V_E = 0$	$M_A = 0$ $N_A = -1.35\ wR^2$ $V_A = 0$	$M_B = 0$ $N_B = -1.5\ wR^2$ $V_B = 0$
Water Surcharge 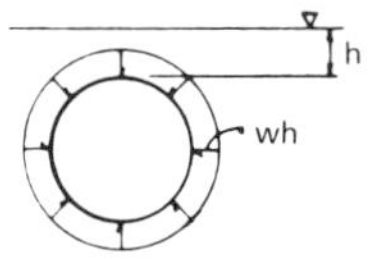 h wh Variable	$M_T = 0$ $N_T = -whR$ $V_T = 0$	$M_C = 0$ $N_C = -whR$ $V_C = 0$	$M_E = 0$ $N_E = -whR$ $V_E = 0$	$M_A = 0$ $N_A = -whR$ $V_A = 0$	$M_B = 0$ $N_B = -whR$ $V_B = 0$

Fig. 13-30. Equations for moments, thrusts, and shears.

axis of the shell is more in the form of an ellipse than a circle.

In computing the bending moments and thrusts at various points around the shell, it was assumed that the shell itself was relatively thin and was circular in shape. With this assumption, a relatively small change in radius of the cylinder has little affect on the computed moments and thrusts.

In applying the results of the moment calculations to the actual structure, it is necessary to take into account the variation in location of

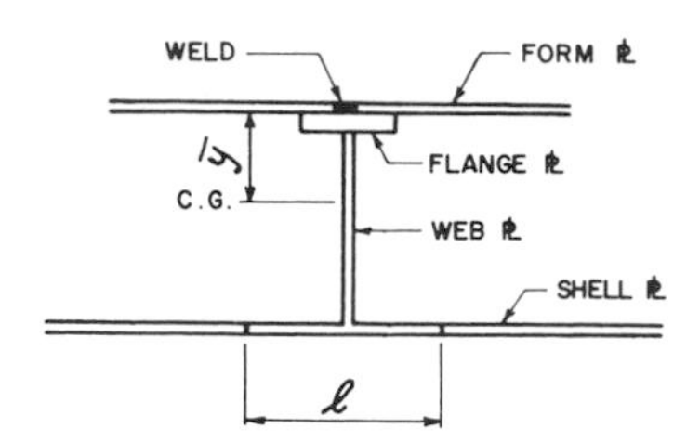

Fig. 13-31. Diaphragm cross-section.

the CG of the composite steel and concrete section, and the fact that the locus of the CG is not a circle. One method is to assume the locus is a circle with radius equal to the mean of the CG radius at the top and bottom and that at the horizontal axis. In this case, the thrust at the horizontal axis is applied outside the CG of the structural section, and the resulting moment about the CG, as the center of moments, is less than the moment computed from the Paris[2] formulas. However, due to the approximation in the application of these formulas, and in the desire for a conservative design, the stresses are computed as follows.

1. At the top and bottom of the tube, the thrust (N) is applied at the geometrical axis of the section.
2. At the horizontal axis, the thrust is applied at the CG of the combined steel and structural concrete section (gross section).

Calculation of Stresses in Concrete and Steel. Stresses are computed for the length of section between diaphragms by the "moment of inertia" method as follows.

1. Assume a design section.
2. Assume the location of the neutral axis and compute the net area and the net moment of inertia.
3. Compute the tensile and compressive stresses by the standard formula:

$$f = \frac{N}{A} + \frac{M_c}{I}.$$

4. Compute the location of the neutral axis from the values of the stresses obtained above. If this is more than one inch from the location assumed, repeat the calculations, using a new location for the neutral axis.
5. When the correct location of the neutral axis is found, a final check of the stresses is made by a balance of forces: summation of the forces $= N$; summation of moments $= M + Ne_1$, where M and N are computed by the Paris[2] formulas, and the moments are computed about the neutral axis, and e_1 = distance between line of action of N and the neutral axis.

Note that the line of action N is at the CG of the gross structural section or the geometrical axis.

Calculation of Shearing Stresses. The transverse shear at any point in the shell of the tube is equal to the derivative (rate of change) of the bending moment at that point.

Since the bending moments caused by uniform vertical loads are equal and of opposite sign to the moments produced by equal horizontal loads, the uniform earth loads on this shell may be simplified by considering only a vertical load equal to (1 - 0.27) or 0.73 times the applied uniform vertical load.

The shear caused by a load above the top of the tube may be calculated for a net uniform unit load of

$$q = 0.73 \times (\text{total vertical unit load}).$$

For this loading, the moment M at an angle from the top is

$$M\theta = qR^2\left(0.25 - \frac{\sin^2\theta}{2}\right)$$

$$V\theta = \frac{dm}{Rd\theta} = -\,qR\sin\theta\cos\theta = -\,qR\frac{\sin 2\theta}{2}.$$

The maximum shear, V_m, will occur at 45°:

$$V_m = \frac{-qr}{2}.$$

For this loading condition,

$$M_t = \text{moment at the top}$$

$$M_t = \frac{qr^2}{4}.$$

Hence,

$$V_m = \frac{-2M_t}{R}.$$

The variation of the shear with the angle can be indicated as follows:

θ	V/V_m - sin 2θ
15°	0.50
22.5°	0.707
30°	0.866
45°	1.000

Assuming that the shears from the combined triangular and quadrant earth loads vary in the same way as the uniform earth loads, the maximum value of shear from all the loads may be approximated as:

$$V_m = 2M_t/R.$$

Note that for the triangular and the quadrant earth loads, a slightly higher value of V_m could be obtained by using the moment at the bottom rather than the moment at the top, but the effect on the final stresses is negligible.

Unit Shearing Stresses. The values of V obtained above are the total shear in the shell of the tube at various points around the circumference. The unit shearing stresses are computed by using the standard formula:

$$v = \frac{VQ}{I}$$

where

v = the shear per linear inch along the circumference of the tube at the point checked

V = total shear on the section

Q = static moment about the CG of the portion of the net section outside of the point where the unit shear is being checked

I = moment of inertia of the net working structural section about its CG.

The calculations of unit shears are required to check the stress in the welds between the flange and shell plates and the diaphragm web plates, the unit shear stress in the concrete, the bond stress of the reinforcing steel and the bond stress between the concrete and the shell plate.

Determination of Unit Shears. The unit shears are determined at the following sections, where it is usually assumed the tube is located in the maximum depth of water with a minimum of 20 feet of submerged backfill.

1. At the 45-degree axis, using the net structural section (without the exterior concrete) and the actual shear.
2. At the top, using the top working section and the external shear at 15 degrees from the vertical.
3. At the horizontal axis, using the working section at that axis and the exterior shear at 15 degrees from the horizontal.
4. Usually, the results obtained above indicate that no further check of unit shears is warranted.

The unit shears are determined at the terminal tubes where the depth of water is minimum and the fill is partially above water:

1. At $22\frac{1}{2}$ degrees from the horizontal, using the heavy flange (same as the horizontal axis), the actual net structural section and the actual external shear.
2. At $22\frac{1}{2}$ degrees from the horizontal, using the light flange (same as the 45-degree axis), the actual net structural section and the actual external shear.
3. At $22\frac{1}{2}$ degrees from the vertical.

In order to find the unit shears for 1, 2 and 3, it is necessary to find the moments and thrusts at these sections in order to determine the net structural working section. Expressions for the moments and thrusts at any section for each of the loading conditions are given in Fig. 13-30.

Design Analysis—Steel Shell, Circular Tubes

As previously described, these circular composite sections have an exterior steel shell, usually $\frac{3}{8}$ inch in thickness, stiffened by internal structural steel tees. These transverse members act with the effective width of the shell plate to form stiffening rings that resist the applied loads during launching and outfitting. In addition, the rings support the shell plate, longitudinally.

At points where the tube is supported on the launching ways, larger tees are provided that are designed to carry the weight of the tube between supports.

The loading and design conditions are similar to those for the octagonal tube. However, since the shape is circular rather than elliptical, the design assumptions more closely approach the theory.

During fabrication, the shell is analyzed for ring and longitudinal stresses due to its own weight. The analysis is based on the theory developed by Schorer.[5]

From Schorer,[5]

$$fr = \frac{rw}{t}$$

where

fr = maximum unit ring stress
r = outside radius of shell plate (inches)
w = weight of shell plus stiffeners (lb/inch^2)
t = thickness of shell (inches).

Also:

$$fe = \frac{wL^2}{2tD}$$

where

fe = maximum longitudinal unit stress as a simple beam
L = distance between ring girders (inches)
D = outside diameter of shell (inches).

The ring girder carries the weight of the tube between supports, plus any additional weight (such as reinforcing steel for the concrete lining). The ring girder stresses depend on the method of support. Once the contractor has established his method of support on the shipways, the ring is analyzed using formulas from Roark and Young.[3] The width of shell plate to be used as a flange of the ring girder is determined by the formula $B = 1.56\sqrt{rt} + C$ given by Schorer.[5]

After launching, during outfitting, the tube is analyzed for longitudinal bending moments due to non-uniformly distributed loads. To evaluate the effect of water pressure on the shell, the plate is assumed fixed at the ends of the span at the adjacent stiffeners. The plate stresses are determined using the theory given in Timoshenko.[6] The actual stress in the plate will be less than calculated due to the additional stiffening resulting from circumferential curvature.

The ring stiffeners will be subjected to water pressure, which is also dependent on the draft. The effect of the pressure and the weight of the tube can be analyzed by computer using the Stress Program considering the ring made of the tee and the width of the plate (as defined before, based on Schorer[5]. The load applied corresponds to a width equal to the spacing of the stiffeners.

During the outfitting process, longitudinal moments and shears will be developed on different sections of the tube. These moments and shears must be below the capacity of the tube. To investigate the magnitude and location of these moments, as well as draft and trim of the tube during outfitting, computer programs are usually used.

To determine the bending capacity of the tube, the buckling stress of the shell is determined.[4]

A factor of 1.3 is used to correct for buckling due to bending, and a factor of safety of 1.8 is used to determine a working stress. Multiplying this working stress by the tube section modulus, the moment capacity is obtained.

To determine the shear capacity, U.S.S Steel Design Manual[4] should be consulted. A factor of 1.3 is used to convert torsional shear buckling into transverse shear buckling, and using a factor of safety of 2, the shear capacity is found by this formula:

$$V = s\pi Rt$$

where

s = shear allowable stress (psi)
t = thickness of plate (inches).

Design of Rectangular Tubes

Loading conditions similar to those shown in Figs. 13-28 and 13-30 would apply to a rectangular box type tunnel section. The design would be based on well known methods of rigid frame analysis, for which computer programs are also available.

Design Analysis by Computer

In the past, steel shell tunnels have been analyzed using the Paris[2] formulas to determine moments, thrusts and shears acting on the diaphragms of the steel shell and the composite in situ structure.

Current tube design techniques utilize computer technology to analyze the forces and stresses during fabrication, outfitting and in the in situ structure.

A frame analysis program, either STRESS or STRUDL, is utilized to analyze the composite tube or the steel diaphragms. The cross-section of the structure is subdivided into a series of segments which are joined at nodes as shown on Fig. 13-32. For the composite structure, a more accurate analysis would locate the member nodes at the center of gravity of the composite steel and concrete. Usually, this refinement is not required. For this section shown in Fig. 13-33, the tremie concrete is not considered in the analysis.

Based on the gross section properties of the structural concrete, the area, moment of inertia and shear area of each segment is computed. The section properties are those of the average section between nodes. This is a simplifying assumption, since the effects of the steel in the section is neglected in computing the section properties. A further refinement could be introduced by using the cracked section properties of the composite structure, but this would re-

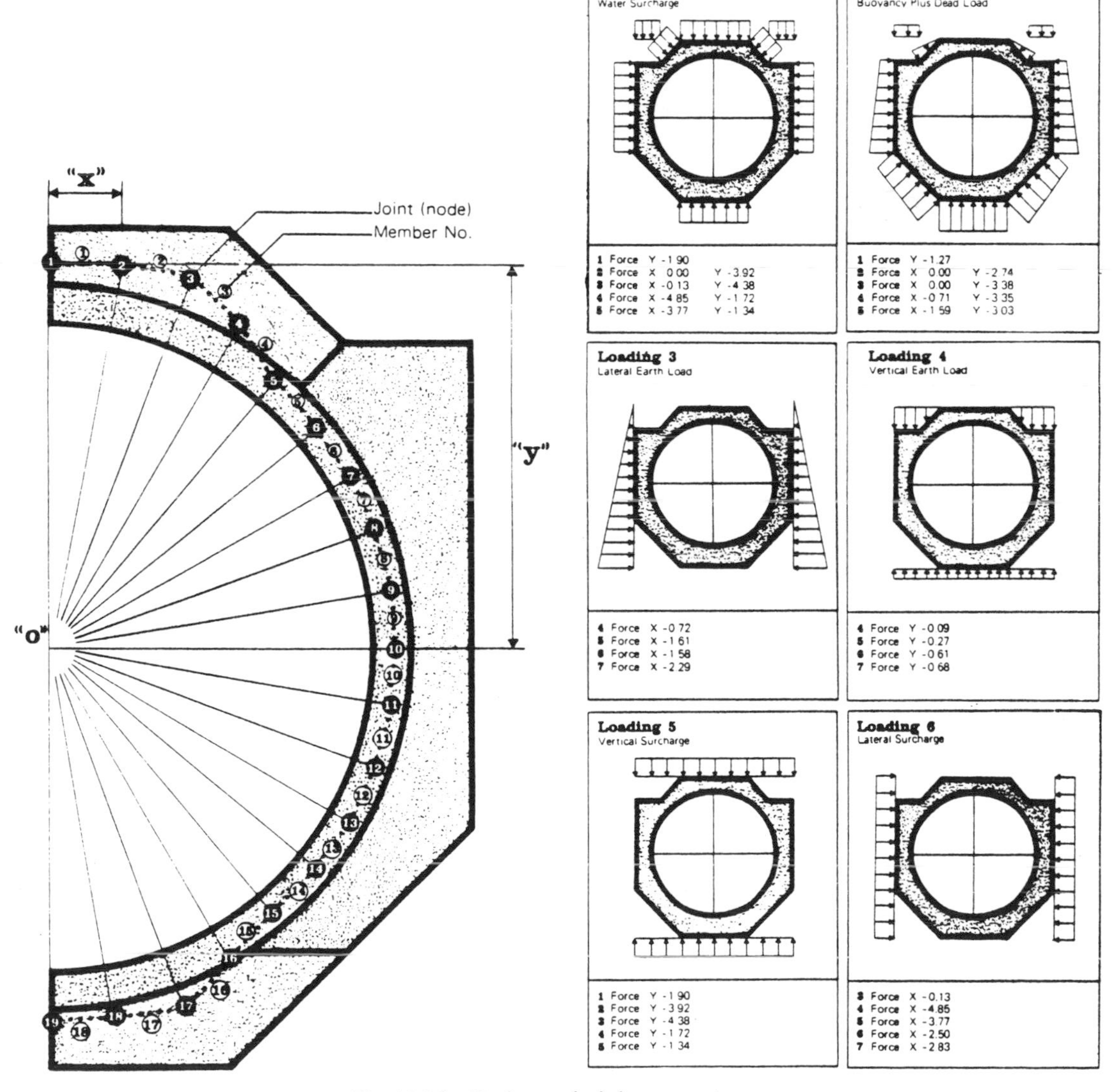

Fig. 13-32. Design analysis by computer.

COMPUTER INPUT

```
STRUCTURE TUNNEL
NUMBER OF JOINTS 19
NUMBER OF SUPPORTS 2
NUMBER OF MEMBERS 18
NUMBER OF LOADINGS 34
TYPE PLANE FRAME
JOINT COORDINATES
1 X  0.00 Y 19.50 SUPPORT
2 X  3.42 Y 19.37
3 X  6.91 Y 18.98
4 X  9.88 Y 17.11
5 X 11.86 Y 14.13

LOADING COMBINATIONS
LOADING 11A STA 129 + 89 BACKFILL TO TOP OF SECTION
COMBINE 1 3.14 2 1.00 3 0.46 4 1.00 5 0.00 6 0.00
LOADING 11B STA 129 + 89 GROUND AT +7. KA
COMBINE 1 3.14 2 1.00 3 0.27 4 1.00 5 4.06 6 1.10
LOADING 11C STA 129 + 89 1 KSF SURCHARGE, KA
COMBINE 1 3.14 2 1.00 3 0.27 4 1.00 5 5.06 6 1.37
LOADING 11D STA 129 + 89 SURCHARGE + FLOOD, KA
COMBINE 1 3.68 2 1.00 3 0.46 4 1.00 5 5.06 6 2.33
```

COMPUTER OUTPUT

MEMBER FORCES FOR MEMBER 1 STA 129 + 89

LOADING	JOINT	AXIAL FORCE	SHEAR FORCE	MOMENT
23	1	92.098	- 3.741	185.36
23	2	- 92.098	3.741	-198.16
24	1	110.738	-10.748	-272.32
24	2	-110.738	10.748	235.53
25	1	116.662	-12.442	-376.12
25	2	-116.662	12.442	333.54
26	1	153.365	-12.056	-149.19
26	2	-153.365	12.056	107.93

COMPUTER INPUT

```
ANALYSIS OF PRECAST TUBE SECTION
SECTION IS TRANSFORMED TO CONCRETE
+ STRESS = TENSION
+ MOMENT = TENSION ON INSIDE
+ AXIAL FORCE = TENSION

CASE 1   TOP

DISTANCE BETWEEN DIAPHRAGMS, L = 60 INCHES
DEPTH OF WEB, W1 = 34.00 INCHES
WIDTH OF FLANGE, W2 = 12.00 INCHES
THICKNESS OF FLANGE, T2 = 1.000 INCHES
THICKNESS OF EXTERIOR CONCRETE, T3 = 36.00 INCHES
THICKNESS OF SHELL PLATE, T4 = 0.3125 INCHES
RE-BARS-NUMBER 9 AT 12.00 INCHES, 2.00 INCHES COVER
THICKNESS OF INTERIOR CONCRETE, T5 = 18.00 INCHES
N = 9.30
```

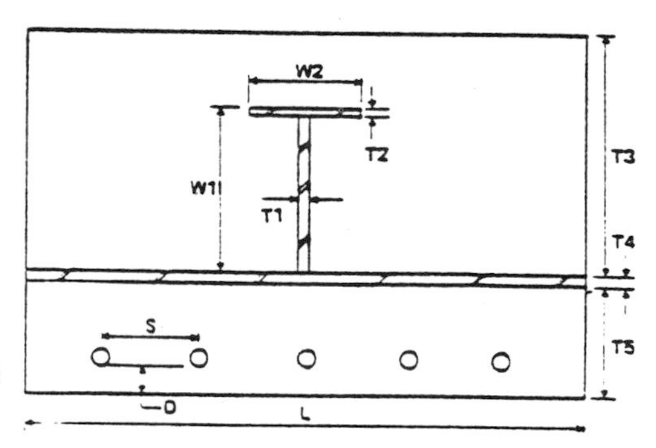

Fig. 13-33. Computer input and output.

quire many cycles of computation and would not greatly increase the accuracy of the results.

The frame is analyzed for the loading conditions shown in Fig. 13-28. These loadings are computed for each section. The loadings are then analyzed and combined to produce output which gives maximum and minimum thrust, shear and moment for each beam at each node. While this can be easily stated, it is, in fact, a complicated task which requires judgment to ensure the proper combinations have been considered which will result in the maximum tension, compression and horizontal shear in the tube. Figure 13-33 shows an example of the computer input and output.

While the above discussion is referenced to an octagonal tube, the same technique is valid for any cross-section.

A further refinement is to consider the uniformly distributed base pressure to be replaced by a series of springs which join the frame at the beam nodes. These springs represent the elastic modulus of the foundation soil. This analysis considers the tube to act as a beam on an elastic foundation and results in a more accurate analysis of the forces in the tube. This is particularly true when analyzing a tube with a wide, flat base. It also has the advantage of simplifying some of the computer input since the output automatically gives the base pressure.

After completion of the analysis for the tube, the next step is to check the tube for stresses and to adjust the reinforcing steel, diaphragm plates or possible concrete thickness to ensure that all stresses are at or below the permissible levels. This is done with a computer program, which, given the size and spacing of reinforcing steel plates and concrete thickness, as well as axial force and moment, will compute the stresses in the section. The analysis is based on working stress design for a cracked section under combined moment and axial force.

The program uses a transformed-area method of analysis. The steel in the section is transformed to an equivalent area of concrete. Since all tension concrete is neglected in computing the moment of inertia of the section, the location of the neutral axis is determined by trial and error. The section may or may not be cracked, depending upon the magnitude and sense of the applied axial force (see Fig. 13-33).

The axial force is assumed to act through the centroid of the gross section (i.e., tension concrete included), whereas the moment of inertia is computed about the centroid of the net section. Thus, in a cracked section, an eccentricity $e = Yn = Yg$ results, inducing a moment equal to Ne. In the program, this moment is added to the original applied moment.

Following this analysis, the horizontal shear between the shell plate and concrete is computed using VQ/It. For the exterior portions of the concrete, this shear can be resisted by the longitudinal stiffeners which act as shear lugs. For the interior portions of the concrete, the horizontal shear is resisted by the ties which support the reinforcing bars as well as by bond. The bond stress computed by this method can be large, and it may exceed the bond values found in codes for composite construction. However, experience has shown this type of construction to function well, and since the shell is a closed system, some bond slip would not be of great consequence, as there is no place for the steel plate or concrete to go. If extremely high bond stress is found, shear studs can be welded to the interior steel shell plate.

Another computer program has been developed to check the moments and shears in a tube during outfitting. This program determines the depth of the tube section in the water after each concrete pour. The moments and shears in the tubes are used to check the shell and form plates for shear and bending stress, while the water depth is used to compute the hydrostatic pressure on the shell plate as well as the forces in the diaphragm.

An initial loading consisting of the uniform weight of the tube plus the concentrated weight of the end bulkheads is followed by any number of loadings, each consisting of one or more sets of uniform loads which represent the concrete pours. It is not necessary to place the loadings symmetrically about the center of the tube, although the unsymmetrical loadings will cause the tube to list with an uneven draft. The uneven draft will increase the water head on one end of the tub, which may cause maximum pressure on the shell or form plate. The concrete unit weight for each set of pours may be different.

A shear and moment diagram is produced for

each loading applied to the tube. The results are generated at each end of every pour and at all maximum and minimum points.

REFERENCES

1. "Special Report, California Undersea Aqueduct," U.S. Department of the Interior, January 1975.
2. Paris, James M., "Stress Coefficients for Large Horizontal Pipes," *Engineering News Record* 87, *No. 19,* November 10, 1921.
3. Roark, Raymond J. and Young, Warren C., *Formulas for Stress and Strain,* Fith Edition, McGraw-hill, New York.
4. Brockenbrough, R.L. and Johnston, B.G., *U.S.S. Steel Design Manual,* 1974.
5. Schorer, Herman, "Design of Large Pipe Lines," *Trans. ASCE* 98, 1933.
6. Timoshenko, S., *Theory of Plates and Shells,* First Edition, Article 3, p. 15, Fig. 9.

14

Cut-and-Cover Construction

D.J. MORTON
PRINCIPAL ENGINEER

Chief Design Engineer (Retired)
Subway Construction Branch
Toronto Transit Commission

Shallow depth tunnels, such as utility tunnels, sewers and rapid transit tunnels, are often built with the cut-and-cover method. For depths of 35 to 45 feet, this is usually cheaper and more practical than underground tunneling, and depths of up to 60 feet are quite common in rapid transit cuts. This chapter will discuss the design and construction of the larger cast-in-place concrete structures used for utility corridors or transportation tunnels for pedestrian, vehicular or rapid transit traffic. The tunnel is designed as a rigid frame box structure, and due to the limited space available for a construction site in urban areas, it is usually constructed within a neat line excavation using braced supporting walls. Where construction space permits, such as in open areas beyond urban development, it is usually more economical to utilize the open slope method of construction. Where the tunnel alignment is beneath a city street, the cut-and-cover construction will cause interference with traffic and other urban activities. This disruption is lessened through the use of decking over the excavation immediately following removal of the first lift of excavation. This deck is left in place with construction proceeding below it until the stage is reached for final backfilling and surface restoration.

TUNNEL DESIGN

Loading

The cut-and-cover structure must be designed to have sufficient structural capacity to resist safely and effectively all loads and influences that may be expected over the life of the structure. This will include the long-term development of water and earth pressures and the influences of subsequent excavations alongside the completed structure.

The structure must be proportioned for the following loads where they exist.

Dead load (D)
Vertical earth load (V_1)
Surcharge (V_2)
Horizontal earth pressure (H)
Buoyancy (B)
Live load (L)
Vehicle loading (L)
  Vertical wheel loads

Centrifugal force
Longitudinal braking and tractive force
Rolling force
Earthquake forces (E)
Contraction or expansion forces due to temperature or moisture changes, shrinkage or creep (T)

Dead loads to be considered in the design of the structural member consist of:

1. The weight of the member itself;
2. The weight of all materials of construction incorporated into the structure and to be permanently supported by the member;
3. The weight of permanent equipment; and
4. The forces due to prestressing.

Vertical earth loads are calculated by multiplying the depth of cover by the assumed unit weight of the compacted fill, assumed as 120 pcf for dry fill, 130 pcf for moist fill and 70 pcf plus hydrostatic pressure for submerged fill below the groundwater table. The American Concrete Institute requires that these loads be multiplied by an appropriate load factor to determine actual design loads. These loads should be considered as dead load when applying load factors.

Shallow structures of the size and width of subways or vehicular tunnels constructed near the surface may impose undue restrictions upon future loadings over the tunnel if the structure is designed only for the actual earth cover. For example, structures that pass under city streets must be designed to permit special vehicle loadings in excess of normal axle loads such as trucks moving transformers, boilers, vaults or large structural members. It is often desirable in developing air rights over tunnels in off-street areas that they be capable of directly supporting moderate building loads without distress. For these reasons, the tunnel should be designed for a minimum vertical load equivalent to 8 feet of cover irrespective of the actual cover.

Surcharge loading may result from adjacent building foundation loading, surface traffic loading or other surface live loading. Except for surcharge arising from adjacent building foundations, this should be considered to be live load when applying load factors. The minimum surcharge loading should be a uniform vertical load applied at the surface of 200 psf.

Horizontal earth pressure is dependent upon the nature of the original ground through which the cut is being taken and will require investigation by a soils engineer who can make specific recommendations for loadings. Table 14-1 is a guide of how different soils may affect the horizontal earth pressures on the structure. These loads should be treated as dead load when applying load factors to arrive at design loads.

It is important that the short-term and long-term changes in horizontal earth pressure be considered. The horizontal earth pressure (H) recommended by the soils engineer will be the fully developed long-term loading. During the life of the tunnel, there may be substantial changes to this loading. Immediately following construction, the actual short-term earth pres-

Table 14-1.

SOIL TYPE	N VALUE BLOWS/FOOT	CHARACTERISTICS	LATERAL PRESSURE
Dense sand	Greater than 30	Difficult to drive a 2 × 4 stake with a sledge hammer.	Low
Loose to medium sand	Less than 30		Moderate
Hard clay or silt	Greater than 30	Can be indented by thumb nail.	Moderate
Medium to stiff clay	Less than 30	Can be indented by thumb; plastic enough to be molded.	Medium
Soft clay	Less than 5		High

sures may be considerably less than H and, in the event of an excavation parallel and adjacent to the tunnel, unbalanced lateral pressures may occur with a pressure equal to H applied to one side and a lesser pressure applied to the other. Under this condition, the tunnel may or may not be restrained against horizontal sway, depending upon the bracing used in the adjacent excavation. For these reasons, the tunnel should be designed for both short-term and long-term values of H, and the tunnel members should be proportioned for unbalanced lateral pressures. There are differing opinions on whether the structure should be assumed restrained against horizontal translation or proportioned for stresses resulting from side-sway caused by the unbalanced pressures. This requirement may depend on a number of local factors; however, it is recommended that the tunnel be proportioned for side-sway if it is a single-story structure, and, in the case of two or more stories, side-sway should be considered in the upper story only with the lower stories assumed to be braced against horizontal translation. This provides a competent factor of safety against future mishap resulting from adjacent construction, and it will be found in proportioning the structural members of the tunnel that they are not unduly increased in size through consideration of this loading.

The short-term or reduced loading resulting from adjacent construction is determined by multiplying H by a reduction factor. In design of the Toronto subway, the value of this reduction factor is taken as 0.5.

Buoyancy and the effects of lateral hydrostatic pressure must be considered whenever the presence of groundwater is indicated. The possibility of future major changes in the groundwater elevation should also be considered.

Live loads may include pedestrian loading, vehicular or train loading and moving equipment loads. In a tunnel one story high, the live loading caused by pedestrians, vehicles or moving equipment within the tunnel need not be considered, because the loads through the invert slab will be transmitted directly into the ground. These loads will effect the proportioning of the tunnel members only if the tunnel is two or more stories in height.

Earthquake forces will normally not affect the proportioning of members forming part of a buried structure. Under most soil conditions, the soil will be found to be so stiff relative to the structure that no structural damage can occur as a result of earthquake deformations. Where earthquake forces are a factor, the structure should be designed for maximum ductility and toughness and the design checked against earthquake requirements. Where it is shown that these forces can be neglected without affecting the structural safety of the tunnel, they need not be considered in the calculations.

Special provisions for contraction or expansion forces usually need not be allowed where normal construction practice of locating an expansion joint every 40 to 65 feet along the length of the tunnel is followed.

Load Combinations

In designing the tunnel and its structural members, all of the loads listed herein should be considered to act in the following combinations and the structure proportioned for the combination that produces the most unfavorable effects:

$D + V_1 + V_2 + L + H$ (on both sides of structure)

$D + V_1 + V_2 + L + 0.5H$ (on both sides of structure)

$D + V_1 + V_2 + L + H$ (on one side) $+ 0.5H$ (on other side).

In addition, where indicated by the water table, the tunnel should be checked against buoyancy and lateral hydrostatic pressure. It is also important that the tunnel be checked for earthquake forces when located in a seismic zone requiring consideration of seismic forces. If the tunnel is being constructed with unconventional spacing of construction joints and expansion joints, it may be necessary to make special provisions in proportioning the structural members for forces resulting from contraction or expansion forces.

Frame Analysis

In the analysis of the structural frame of the tunnel, the loading combinations of the previous section are applied and the shears and

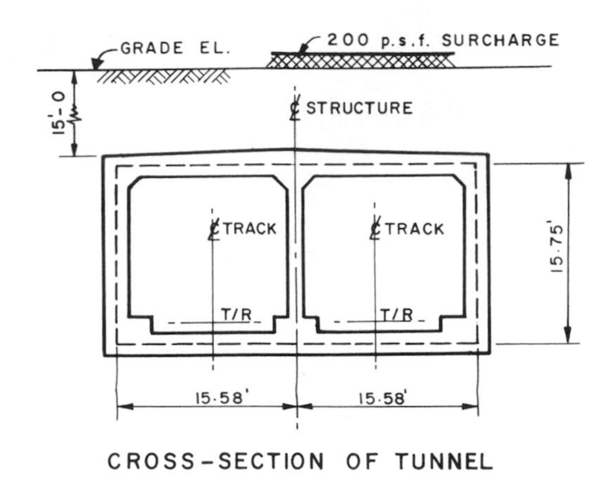

CROSS-SECTION OF TUNNEL

LOADING

VERTICAL ROOF LOADING

WEIGHT OF FILL = 120 p.c.f.

EARTH COVER	15 X 120 =	1800
SURCHARGE		200
WEIGHT OF SLAB	2 X150 =	300
	TOTAL	2300 p.s.f.

VERTICAL INVERT REACTION

FROM ROOF		2300
WEIGHT OF WALL		310
	TOTAL	2610 p.s.f.

HORIZONTAL LOADING (full pressure)

FOR SIMPLICITY OF CALCULATIONS THE HORIZONTAL DESIGN-PRESSURE IS ASSUMED TO BE UNIFORM AND EQUAL TO THE CALCULATED EARTH PRESSURE AT MID-HEIGHT OF THE WALL.

$H = K_A \gamma H$

$= \cdot 3 (120 \times 23 \cdot 9) + \cdot 3 (200)$

$= 920$ p.s.f.

LOADING COMBINATIONS

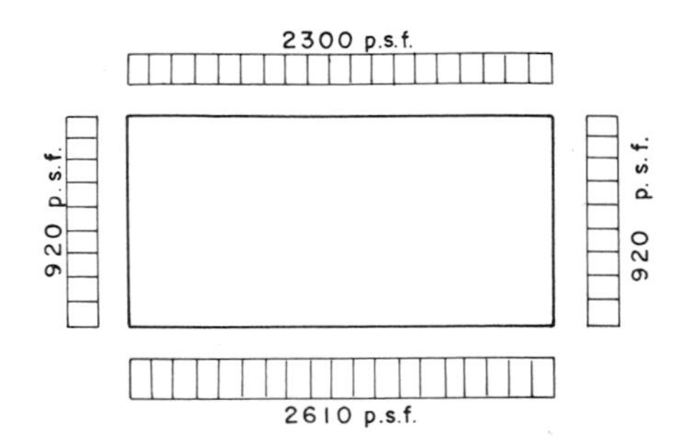

CASE I. FULL VERTICAL AND FULL HORIZONTAL LOADS.

Fig. 14-1. Design calculations for cut-and-cover box structure.

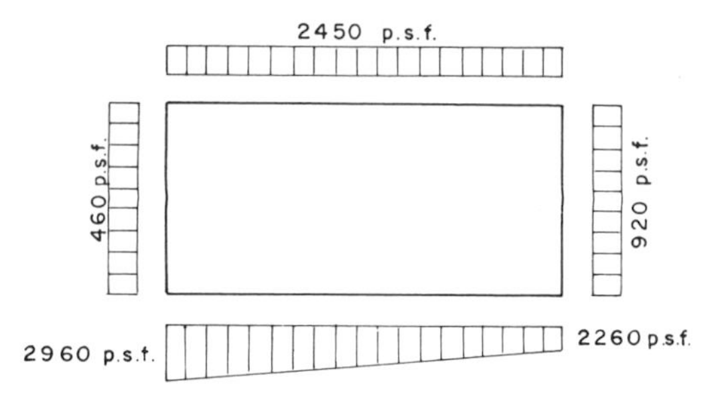

CASE 2. FULL VERTICAL LOAD,
ONE-HALF HORIZONTAL LOAD ON LEFT SIDE,
FULL HORIZONTAL LOAD ON RIGHT SIDE.

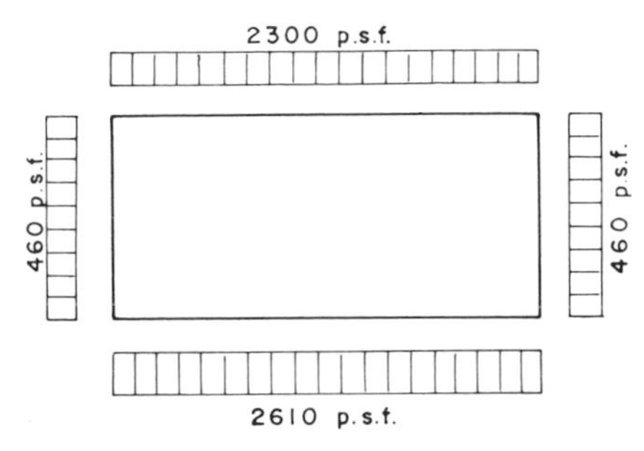

CASE 3. FULL VERTICAL LOAD,
ONE-HALF HORIZONTAL LOAD ON BOTH SIDES.

MOMENT DIAGRAMS

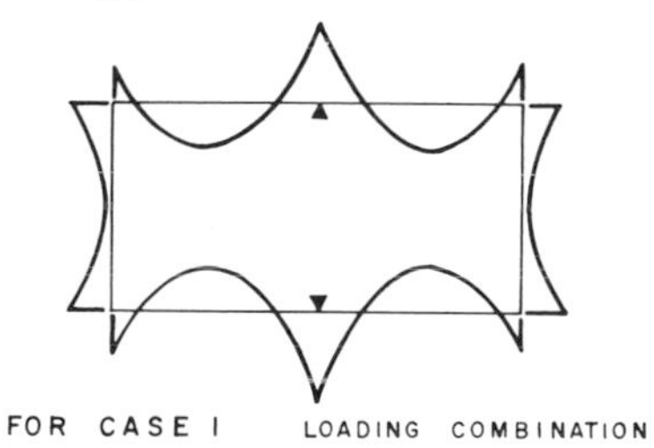

FOR CASE I LOADING COMBINATION

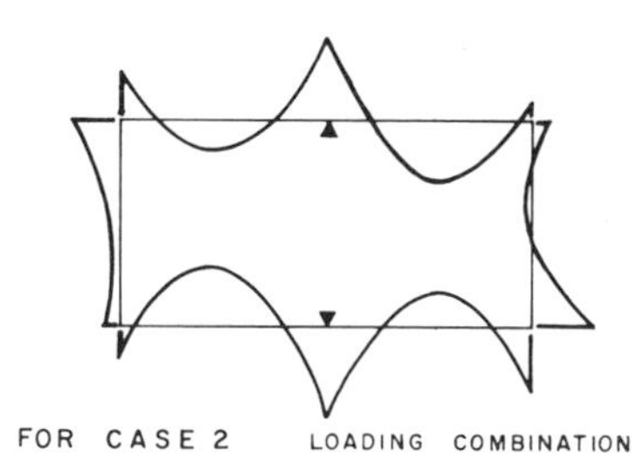

FOR CASE 2 LOADING COMBINATION

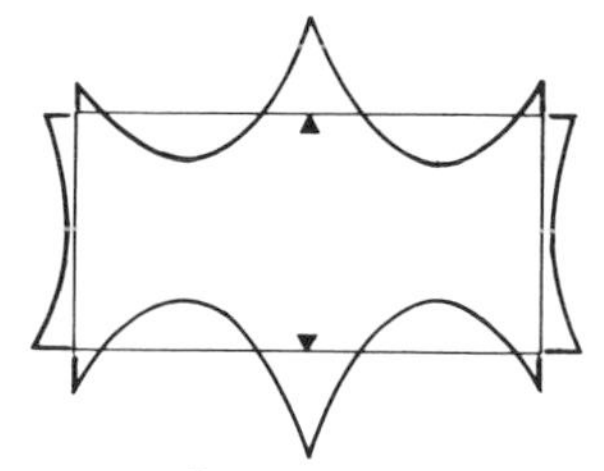

FOR CASE 3 LOADING COMBINATION

Fig. 14-1. (*Continued*).

moments for each element of the frame determined through rigid frame analysis based on the theory of elastic frames. This may be done using the moment-distribution methods or, more probably, using a computer program whence shears and moments can be determined. The usual practice is to assume that vertical load reactions are uniformly distributed over the bottom of the slab. This assumption results in maximum slab moments and therefore is conservative. The moment and shear diagrams are plotted, and from these diagrams the concrete members and reinforcing steel may be proportioned to provide a safe and serviceable structure. (Fig. 14-1)

Joints and Waterproofing

It is necessary, in the case of most tunnels (and especially those used for pedestrian, vehicular or rapid transit traffic), that the structure be made watertight. Where the tunnel is located below the water table, problems with leaks and water infiltration are unavoidable and will require post-construction remedial work, such as grouting and caulking to seal off the leaks. However, leaks can be held to a minimum through good design and quality construction.

Many specifications require complete waterproofing of the tunnel structure after the concrete is placed. This is unnecessary and has been misleading because few of the waterproofing membranes used in the past have had a service life equal to the required service life of the tunnel, and the membrane cannot be replaced. A watertight tunnel is best assured by specifying a 4000 psi strength concrete with a low W/C ratio and carefully placing it to ensure a dense, uniform, high-quality concrete. Experience indicates that the outside walls and slabs should have a minimum thickness of 2 feet. If expansion joints are located every 40 to 65 feet along the tunnel to prevent possible shrinkage or thermal cracks, these walls and slabs can be expected to be watertight.

Leaks

When leaks occur, they are to be expected at the expansion joints. These joints are necessary to provide for shrinkage and thermal movement that may occur in the tunnel structure. The temperature variations in an underground tunnel will normally be much less than experienced by above-ground structures; therefore, the joint width may be much less. Experience in Toronto has indicated that $\frac{1}{4}$-inch wide joints spaced 40 to 65 feet along the underground tunnel are adequate. The joints should be formed with a preformed joint filler and provided with a polyvinyl plastic water stop. The design of the water stop and the concrete placing methods used when installing the water stop are of utmost importance in assuring a watertight joint. A wide, flexible water stop will often become damaged or displaced during concrete placing, and therefore the water stop should be narrow but of sufficient width to secure against the possibility of pulling out of the structure. It should have sufficient thickness to be rigid and resistant to abuse during construction. The concrete should be carefully placed and vibrated around the water stop with constant supervision provided for this most important step in securing a dry tunnel. If this water stop is designed so that it can be correctly installed and the concrete embedding it properly placed, there will be very little need for post-construction remedial work to seal points of water infiltration (see Fig. 14-2).

Bentonite, in the form of panels, tubes and powder, is sometimes used as additional protection at the joints. Bentonite is a decomposed volcanic ash that swells when wet to fill cracks and crevices through which water might infiltrate. In areas where waterproofing is of partic-

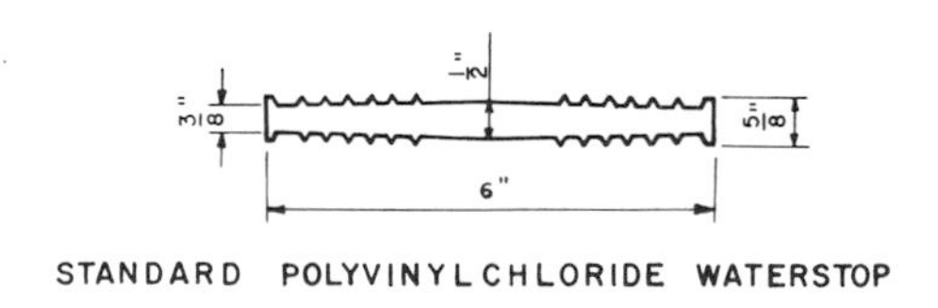

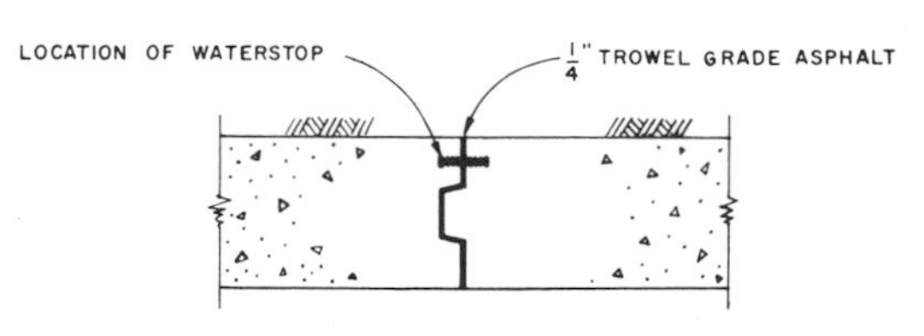

Fig. 14-2. Typical water stop used in Toronto Subway.

ular importance, this further protection may be considered advisable.

If leaks do occur and remedial measures must be taken, the usual method is to pressure-grout either the defective concrete or the soil in the vicinity of the leak from within the tunnel. A neat cement grout mixed to form a thick fluid grout is usually preferred, although proprietory chemical grouts may be used under conditions in which the cement grout is unsuccessful. Initial efforts normally involve pressure grouting the defective concrete by drilling a number of grout-holes into the defective area. Grout is pumped into the grout-holes to seal the cracked or porous concrete. If the leak cannot be arrested in this manner, grout-holes may then be drilled through the concrete wall and the grout injected into the surrounding ground under pressure. In fine-grain soils, such as fine sand or coarse silt, it may be possible to get more effective penetration of the grout using a proprietory chemical grout or a cement grout with a small percentage of bentonite added.

SHEETING METHODS

Neat line excavation requires that sheeting be installed and braced to support the vertical faces of the excavation and to prevent detrimental settlement of the ground, utilities and buildings at the side of the cut. The design and construction of the sheeting will depend upon the ground conditions and the importance of preventing surface settlement. Surface settlement is caused by movement of the soil surrounding the excavation as a result of bottom heave, inward movement of the sheeting or loss of ground through the sheeting. The volume of settlement will be approximately equal to the volume of lost ground. To evaluate the need for underpinning of adjacent installations, the engineer must be able to predict the probable amount of settlement and the effect upon the installations. The amount of movement will depend upon the soil type, the size of the excavation, the construction methods and the quality of workmanship. The accuracy of predictions for the amount of this movement based on soil tests and theory is very uncertain and experience and observational data are needed as a guide to judgment.

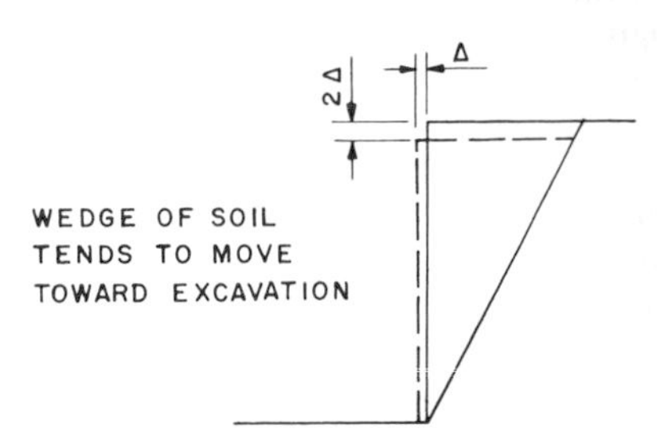

Fig. 14-3. Settlement due to wedge failure.

Cuts in Sand

Settlement resulting from excavation in sand will occur due to inward movement of the sheeting or loss of soil through the sheeting.

In the event of movement involving a failure wedge (as shown in Fig. 14-3), settlement will occur within an area extending beyond the edge of the cut a distance equal to $\frac{H}{2}$. If the surface settlement is equal in volume to the loss of ground resulting from this inward movement, then it is evident that the surface settlement resulting from this type of wedge failure will be approximately double the inward movement. A number of observations during construction of the Toronto subway in sand and silty soils have tended to confirm this analysis.

To prevent settlement caused by cuts in sand, therefore, requires that inward movement of the sheeting be prevented and care be taken to avoid loss of ground through the sheeting.

Inward movement may occur as a result of:

1. Elastic strain of the bracing members of the sheeting;
2. Inelastic strain or failure of the bracing and sheeting; or
3. Lateral yield of the sheeting and soil below the bottom of the excavation.

In a 40-foot wide cut, elastic strains in the bracing can be in the order of $\frac{1}{4}$ to $\frac{1}{2}$ inch. This could permit inward movement of the sheeting and wedge failure resulting in possible settlement of as much as 1 inch. A considerable reduction in this type of settlement can be achieved through prestressing of the struts to about 50% of the calculated design load.

Settlement through inelastic strain or failure of the sheeting system may result from improper design. Inadequate design loads or inattention to design detail may be frequent causes. The struts are usually attached to the walers or soldier piles using clip angles. The designer may assume that the load is transferred into the strut through the end-bearing and the clip angles used only for positioning of the strut. If, in fact, the strut is cut too short for the field positions of the sheeting and the load transfer from sheeting to strut made dependent upon the clip angles, inelastic deformation of the clip angles may be expected. Inward movement and surface settlement will be the result.

Inward movement of sheeting in sand or sandy subsurface materials resulting from lateral yield of the sheeting below the bottom of the excavation does not appear to be a significant cause of settlement. Care should be taken that sufficient penetration of the sheeting below the bottom of the excavation is provided. The designer must ensure that the mobilized passive pressures on the soil are less than the pressures that will cause yielding of the soil.

Open cuts employing soldier pile and lagging sheeting involve installation of the soldier piles prior to the start of excavation with timber sheeting installed as the excavation progresses. Ground may be lost as a result of sand running into the excavation before installation of the lagging or as a result of groundwater bleeding through the lagging and carrying soil fines with it. Settlement resulting from these causes is dependent upon the workmanship used in the construction, and the amount cannot be readily predicted. If voids occurring behind the lagging are immediately filled with clay or a lean soil-cement packing and the lagging is wedged tightly against the line of excavation, the amount of settlement will normally be within tolerable limits. However, where heavy building loads occur within a distance beyond the cut equal to the depth of excavation, a more elaborate construction method should be considered. The building foundations may be underpinned or a continuous cut-off wall with prestressed bracing used. The cut-off wall may consist of sheet piling, a slurry trench wall or a drilled wall. This type of wall has been used successfully in a number of excavations to prevent loss of ground through the sheeting and is usually more economical than underpinning.

Cuts in Stiff Clays

Settlement resulting from excavations in stiff clays will occur due to much the same reasons as in sand; except due to the cohesive strength of the soil there is less probability of loss of ground from runs into the cut before installation of the lagging. There is very little information available with which to predict the area over which settlement may occur; however, in stiff clay it is suggested that installations or buildings falling within a distance from the edge of the cut equal to the depth of excavation be considered within the zone of possible settlement. Struts or tiebacks used to brace the sheeting should be prestressed where it is considered that inward movement may cause unacceptable surface settlement or damage to adjacent installations. Continuous cut-off walls should be used in lieu of underpinning to support heavy building loads.

Cuts in Soft Clays

When a cut is excavated in soft clay, the clay located at the sides of the cut acts like a surcharge and yielding of the clay near the bottom of the excavation occurs. The clay yields laterally toward the cut near the bottom of the excavation and the bottom of the cut rises. Lateral movements associated with cuts in soft clay are substantially greater than with sand or stiff clay deposits. As a result of these lateral movements, settlement occurs at the ground surface. The extent of the settlement zone beyond the edge of the cut and the amount of settlement will be dependent upon the type of soil, dimensions of the excavation and the construction procedures. Figure 14-4 is a summary of settlement records adjacent to open cuts in soft clays in Oslo, Norway. From this graph, it is evident that settlement may occur for distances up to

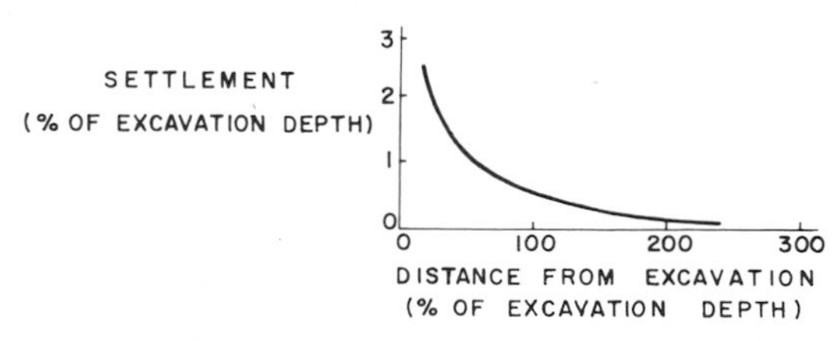

Fig 14-4. Graphical illustration of settlement records in soft clay adjacent to open cuts in Oslo, Norway.

three times the depth of the excavation from the edge of the cut, although most extensive settlement will occur within a distance of H beyond the cut. When the depth of excavation exceeds the critical height, H_c, base failure will occur with substantial surface settlement and damage. To ensure a reasonable factor of safety against this type of failure, it is usually recommended that depths greater than about $5\frac{c}{\gamma}$ not be attempted without deep penetration of the sheeting into the soil below the excavation, preferably with penetration into a firm stratum (where c should be taken as the undrained shear strength of the soil below the base of the excavation and γ as the bulk density of the retained soil above the base of the excavation).

To minimize lateral movements and loss of ground in soft clays, it would be necessary to construct the supporting walls or sheeting before the excavation commences.

This can be done using sheet piling, slurry trench walls or drilled walls. Depending on the nature of the ground and the effects of settlement, a further precaution is to install bracing before excavation proceeds by digging narrow trenches across the cut into which the struts can be installed and prestressed. Precautions against bottom heave can be taken where necessary by extending the sheeting walls the required depth below the bottom of the excavation. This type of construction will add considerable cost to the tunnel project and is used only where subsoil conditions demand it. This is particularly true where it is found necessary to install the struts in trenches prior to ground excavation. The presence of the struts interferes with the general excavation between the sheeting walls and precludes efficient use of heavy construction excavators.

The results of alternative methods, such as leaving high berms on each side of the cut until the bracing is installed, and prestressing, should be considered. This involves two stages of excavation with a center trench removed initially, followed by removal of the side berms after installation of bracing. This allows the use of heavy construction excavating equipment and, if adequate berms are left in place, this will normally reduce inward movement of the sheeting and surface settlement to within tolerable limits.

Earth Pressures

Earth pressure envelopes used in the design of braced sheeting have been developed from data made available through a series of measurements taken on braced excavations. A majority of the measurements available were taken on strutted excavations for subway cuts in the decade of the 1940's in Berlin, Chicago and New York

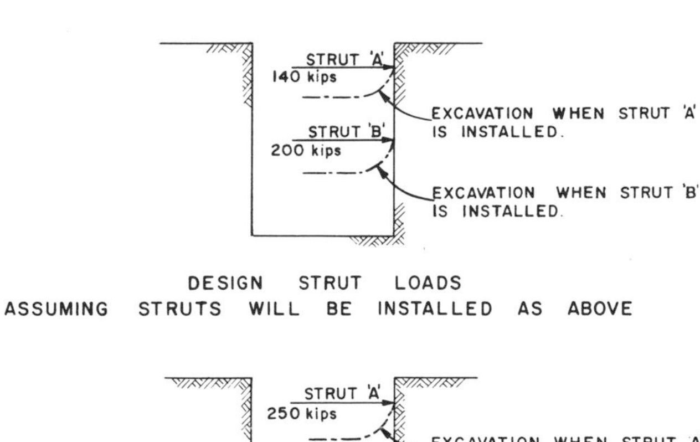

Fig. 14-5. Effect of excavation procedure on strut loading.

and, in the past decade, in Oslo, San Francisco and Toronto. The pressure envelopes do not necessarily indicate actual pressure distribution on the sheeting but rather represent design pressure values that, if used, can be expected to provide a safe and serviceable structure. The pressure envelopes assume that the installation of struts and bracing will follow as soon as practicable after excavation is taken below the level of the strut. The method of construction will be an important factor in determining the actual distribution of pressure on the sheeting. For example, if the excavation in Fig. 14-5 is taken 10 or 20 feet below the planned strut level without leaving side berms before the strut is installed, the sheeting will yield inward toward the excavation. This will cause the load to transfer to the upper struts, resulting in overloading and possible failure.

The pressure distribution envelopes will be valid only where the method of construction is in accordance with recommended procedures and for cuts from 25 to 50 feet in depth. For depths above 50 feet, the pressure envelopes should be used with caution until more instrumentation data are available to verify the design criteria.

Recent subway excavations in Toronto have involved cuts of up to 65 feet in depth in dense sand and silt and in mixed faces of dense sand and hard clay. The sheeting was designed in accordance with the recommended pressure envelopes and, although the strutting was not instrumented, the satisfactory performance of the sheeting would seem to validate the recommendations for the depths of cut in these particular soil types.

The sheeting or soldier piles will be subject to unpredictable amounts of movement at the support points provided by the struts. For this reason, the elastic method of design for sheeting or soldier piles should not be considered valid and these members should be designed by plastic design methods taking into account the continuity of the soldier piles or the sheeting at the strut support points.

Deflections of sheeting or soldier piles may be controlled to some extent by preloading of struts, particularly where movement of soil has to be kept to a minimum to prevent settlement of adjacent sensitive structures.

Design Pressure Envelopes. Recommended design pressure envelopes for use in calculating working loads are shown in Fig. 14-6. Despite care to ensure uniform construction practices, there will still be a considerable scatter of actual loading in the struts. Loads calculated from these pressure envelopes take into account this scatter and can be used under normal circumstances to provide a safe structure by directly applying an appropriate load factor or safety factor to the calculations.

There is no clear distinction possible for the

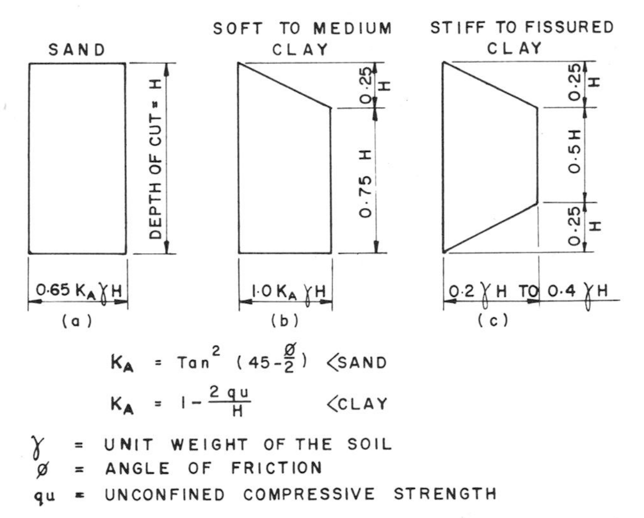

Fig. 14-6. Apparent pressure diagrams suggested by Terzaghi and Peck for computing strut loads in braced clay.

behavior of cuts in clay on the basis of whether the clay can be described as soft or stiff. The behavior of the cut depends upon the dimensions of the excavation as well as the nature of the clay. Peck recommends that sheeting for all braced cuts in clay be designed on the basis of the maximum pressure calculated from Fig. 14-6. Sheeting for braced subway cuts in stiff to hard clay have been satisfactorily designed in Toronto using the lower values of pressure indicated in Fig. 14-6. Instrumentation of two of the cuts having a depth of about 40 feet indicated a wide range of scatter in the individual strut loadings but provided evidence that would confirm the validity of the design assumptions.

Arching in Soils. Excavations in sand and sandy silt materials braced with soldier piles and timber lagging will experience horizontal arching action, transferring pressure from the lagging to the soldier piles. This arching effect, if taken into account, allows the designer considerable reduction in the thickness of lagging. This is a valid design consideration; however, due to a lack of theoretical data, it is necessary to establish empirical rules through observation and experience. In soils exhibiting arching action, and where this pressure transfer has been taken into account, it is recommended that the thickness of timber lagging be not less than $\frac{1}{24}$ of the span between soldier piles to limit deflections and lateral movement of the lagging.

Common Types of Sheeting

The sheeting usually follows the physical dimensions of the tunnel, allowing just enough space to permit the construction. Bracing may be strutting or tie-backs, as shown in Fig. 14-7, although strutting is more common for the excavation widths associated with tunnels. Tie-backs are advantageous where struts would interfere with construction operations within the excavation. By judiciously locating the struts vertically, it is usually possible in tunnel construction to avoid any significant interference by the struts.

Soldier Piles and Lagging. A common type of bracing used in deep excavations is soldier piles and lagging (Fig. 14-8). Steel soldier piles are placed prior to the start of the excavation by driving or drilling at a spacing of 8 to 10 feet. Horizontal lagging is placed between the soldier piles as the excavation progresses. The minimum thickness of lagging should be $\frac{1}{24}$ of the span between soldier piles. The soldier pile should be designed as continuous members using plastic design techniques with support points at the strut levels. A bottom support point with partial fixity may be assumed at a point approximately 1 foot below the bottom of excavation, depending on the stiffness of the ground.

Sheet Piling. Another type of sheeting used in

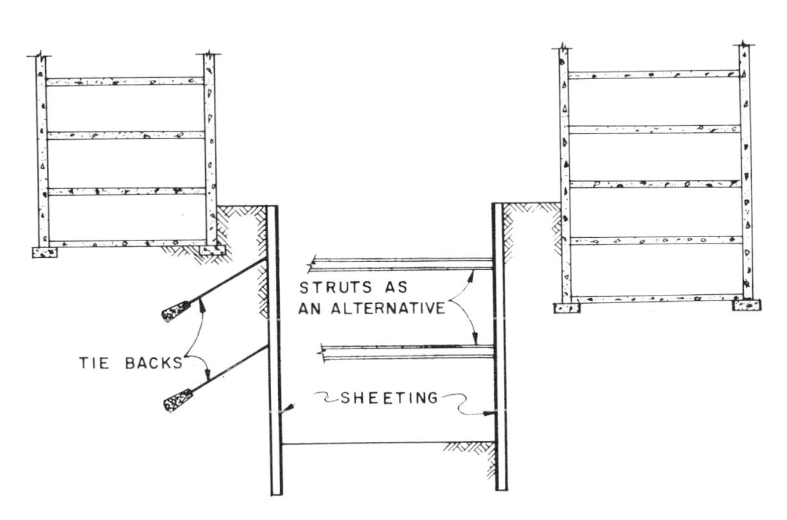

Fig. 14-7. Sheeting for cut-and-cover excavation showing alternative methods of bracing.

Fig. 14-8. Soldier piles and lagging used in construction of Toronto Subway.

soft grounds is sheet piling. Sheet piling is used where there is a danger of bottom heave in soft clay soils or, in the case of sand, to reduce the danger of settlement caused by soil runs into the excavation before placing the lagging. Sheet piling may also be used to avoid dewatering where there is a risk that lowering the water table may cause subsidence in the surrounding ground. Settlement of this type is usually not a problem in dense sand or stiff and hard clay strata; however, in loose sand or soft clay, substantial settlement can occur as a result of lowering of the water table. Sheet piling may be used to avoid this and consequent settlement. It should be noted that sheet piling is not satisfactory in hard clay, very dense sand or rock-bearing strata due to the difficulties of driving it into place. The sheet piling is normally designed as continuous sheeting with supports at the lines of struts similar to soldier piles.

Continuous Concrete Walls. Recent construction techniques involving construction of continuous concrete shoring walls have been used successfully to reduce settlement adjacent to an open excavation. The continuous concrete wall may be used under the same conditions as sheet piling and has the advantages of greater stiffness and causing less noise and vibration during construction. The noise of driving sheet piling may be unacceptable in the area of construction and the vibration may cause settlement as serious as that which it was being installed to prevent. This may be especially true in loose sand deposits. Continuous concrete walls are also frequently used in cut-off wall construction to avoid the cost of underpinning adjacent buildings. This appears to provide satisfactory results provided the struts are prestressed and installed as soon as practicable following excavation.

There are three methods of constructing continuous concrete sheeting:

1. Slurry trench walls
2. Soldier pile tremie concrete walls
3. Drilled walls.

Fig. 14-9. Slurry trench wall used in New York during construction of the World Trade Center.

Slurry trench walls (Fig. 14-9) are constructed by machine excavating a trench along the neat line of the tunnel, which is filled with a bentonite slurry to prevent caving of the sides of the trench. The excavation is carried out using a clamshell bucket or a multibit drill excavator. Reinforcing cages are dropped into the trenches and concrete cast by the tremie method, displacing the slurry. Excavation between the walls then progresses in the same way as with sheet piling.

Soldier pile tremie concrete walls were used in San Francisco on excavation for the BART subway. Soldier piles spaced at approximately twice the wall thickness were placed in predrilled slurry-filled holes and wedged into position. The soil between the piles is then removed by a clam bucket type of excavator continuing to use slurry to hold the walls of the trench. Concrete is then placed between the soldier piles by tremie method displacing the slurry. The result is a continuous concrete wall reinforced with structural steel soldier piles.

Drilled walls used in subway construction in Toronto (Fig. 14-10) are a variation of the soldier pile tremie concrete walls used in San Francisco. The end result again is a continuous concrete wall reinforced with soldier piles (Fig. 14-11). Holes approximately 30 inches in diameter are drilled at about 5-foot, 6-inch centers along the line of the excavation, using a slurry to hold the sides of the hole, and 21-inch wide flange soldier piles are lowered into the predrilled holes and concrete tremied into place. The remaining soil between the soldier piles is then removed using the 30-inch drill, and the space between piles is filled with concrete. This does not provide a watertight wall and is not satisfactory as a means of groundwater control. However, where the wall is being installed to prevent loss of ground or inward movement of the sheeting during excavation, it has proven satisfactory.

A similar method has been used by drilling holes, filled with slurry at spacing less than their diameter, and filling them with concrete. The spaces between these concrete piles are drilled with the same diameter tool cutting into the

Fig. 14-10. Drilled concrete wall used in Toronto during subway construction.

adjacent concrete piles. Filling these spaces with concrete, displaying the slurry, results in an unreinforced, tight concrete wall.

Upside-down Construction. Instead of excavating to the full depth using temporary bracing, permanent steel may be installed as the excavation progresses, especially in stations with several levels. This has been called "upside-down construction." Its advantage is a saving in the amount of temporary steel. Where it is used with sheet pile or soldier pile walls, great care must be used when pouring the concrete walls, in submerged soil, to prevent leakage along the struts where they are embedded in concrete. Permanent steel must also be available to keep progress with the excavation.

DECKING

Decking is used to minimize surface disruption to normal activities during the construction period. Through the use of decking, the duration of interference can be limited to a short time at the beginning of construction while the surface and initial lift of excavation is removed, and again for a short duration at the completion of construction for deck removal, final backfilling and surface restoration. Decking may be required for maintenance of vehicular traffic over large cuts running beneath, and parallel with, a vehicular right of way or where any cut crosses a street or road.

Design

Materials used in the design and construction may consist of steel plating, wooden timbers and concrete slabs. Steel plates are most commonly used for small cuts of 3 to 4 feet in width. In larger excavation, such as vehicular underpasses or subway tunnels, timber decking is most common in North America. In Europe and Asia, more use has been made of concrete slab and steel plates. These differences in practice reflect the differing economies (where there is an available source of timber for decking, this will be the most economical and flexible material). Timber is lightweight for handling and is easily cut to fit job conditions using power saws.

Specifications prepared by the American Asso-

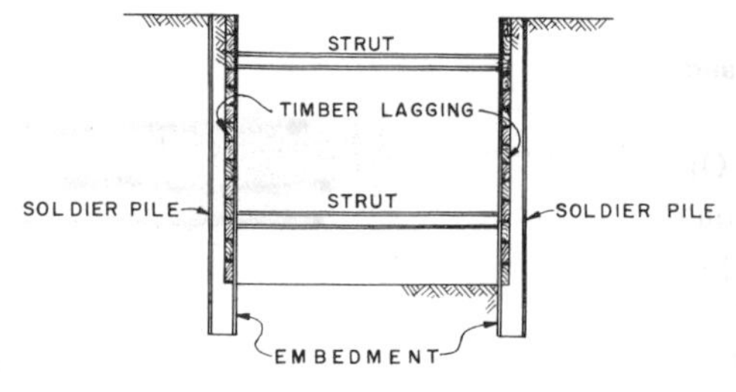

SECTION THROUGH EXCAVATION

LOADS

HORIZONTAL EARTH PRESSURE = $0.65\, K_A (\gamma H + S)$

γ = UNIT WEIGHT OF EARTH = 120 p.c.f.

H = HEIGHT OF CUT = 32 FT.

S = SURCHARGE = 200 p.s.f.

K_A = ACTIVE EARTH PRESSURE COEFFICIENT = 0.26

DESIGN EARTH PRESSURE = 700 p.s.f.

LOADING CONDITIONS FOR PILES AND STRUTS

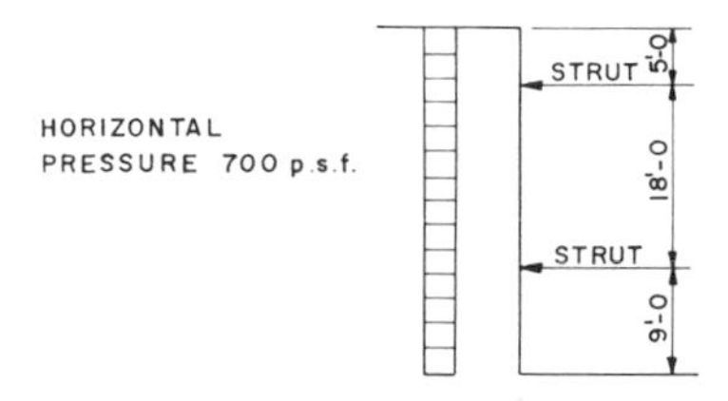

CASE I. EXCAVATION TO FULL DEPTH

LOADING CONDITIONS FOR PILES AND STRUTS

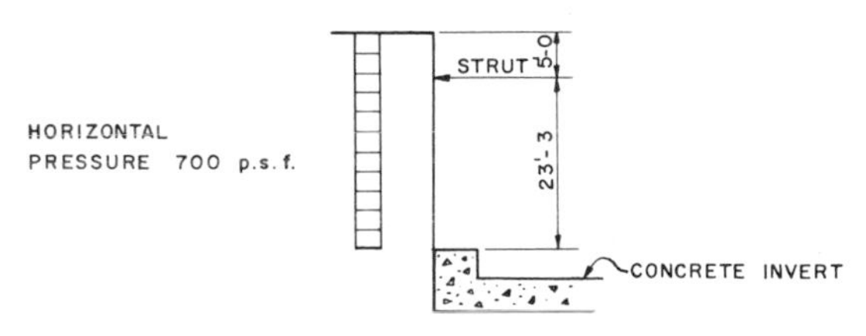

CASE 2. INVERT PLACED AND LOWER STRUT REMOVED FOR CONSTRUCTION OF TUNNEL WALL

METHOD OF DESIGN

1. SOLDIER PILES – DESIGN USING PLASTIC DESIGN METHODS WITH PARTIAL FIXITY AT THE BASE – MAX. RECOMMENDED SPACING 10'-0.

2. STRUTS – DESIGN FOR CALCULATED END REACTION FROM SOLDIER PILES PLUS A VERTICAL LOAD EQUAL TO WEIGHT OF THE STRUT PLUS VERTICAL CONSTRUCTION LOADING.

3. TIMBER LAGGING – DESIGN FOR 700 p.s.f. OR REDUCED LOADING IF HORIZONTAL ARCHING IS TAKEN INTO ACCOUNT – MINIMUM THICKNESS OF LAGGING TO BE 1/24th OF THE CLEAR SPAN BETWEEN SOLDIER PILE FLANGES.

Fig. 14-11. Design calculations for soldier piles and lagging.

ciation of State Highway Officials are generally used in the design of permanent highway structures and bridges. These specifications establish the minimum design live loads due to vehicle traffic, taking into account the occasional overload resulting from infrequent heavy truck loads. Decking is a temporary structure for temporary maintenance of vehicular traffic and it is unreasonable to design for the same loading as used for permanent highway bridges. The engineer should determine the maximum vehicle loadings permitted by normal highway licensing of vehicles in the local jurisdiction, and during the period of construction, traffic over the decking should be limited to these normal vehicle loads. Overloading from special heavy trucks carrying loads such as transformers or boilers should be prohibited by posting of signs establishing load limits or control of vehicle permits. With assurance of cooperation and enforcement by local authorities, the engineer should proceed with design on the basis of this loading. The design stresses and member sizes should be selected to limit live load deflection to not more than $\frac{1}{500}$ of the deck span.

The decking is usually designed with structural steel beams or girders supported on soldier piles and spanning across the width of excavation. The deck beams may rest directly onto the soldier piles or be carried by a cap beam spanning between the piles. A cap beam is used whenever irregular pile spacing is necessary due to the presence of utilities or boulders. Decking formed from timber, steel plate or concrete slab is installed over the deck beams to provide the traffic surface. The decking may be 10 or 12 feet in length, spanning from deck beam to deck beam. The decking is frequently not fixed to the deck beams, although a positive means must be provided to prevent it from moving off the support due to vibration from the traffic. This will be some form of stop fixed to the decking to be flush with the edge of the deck beam. The decking should be surfaced with an anti-skid compound, such as a trap rock seeded epoxy coating, if it is to be used by vehicular traffic.

Prior to deck installation, the soldier piles on each side of the structure are installed at 8- to 10-foot centers. During this construction opera-

Fig. 14-12. Decking during installation for subway construction in Toronto.

tion, it is usually necessary to effect one traffic lane at a time. Following installation of soldier piles, traffic is usually diverted around the cut area, and the initial lift of excavation of about 12 feet is removed using machine excavation (except near utilities, where hand excavation is necessary). The deck beams are then installed and decking is placed. In wide cuts, where it is impractical to divert traffic around the full width of the cut at one time, split decking may be used. With this technique, the deck is installed one-half width at a time and a mid-span splice is made in the deck beams. An example of this construction method is shown in Fig. 14-12.

EXCAVATION AND GROUNDWATER CONTROL

Excavation Methods and Equipment

In excavating a cut for a tunnel, one of three methods is usually selected. For shallow cuts, the excavation is usually made from the surface, using backhoe equipment and dumping the excavated material directly into trucks or into a stockpile along the side of the excavation. This is an efficient means of excavation for cuts up to approximately 20 feet in depth.

For deep cuts, the excavation may be carried out using trucks with haul ramps from the surface to the bottom of the excavation or using clamshell hoists to lift the excavated material from the bottom of the cut and onto waiting trucks. If haul ramps are used, the grade should not exceed 15%. In general, where space will permit a haul ramp, trucks hauling directly from the bottom of the excavation will be the most efficient means of removing the excavation material. When clamshell hoists are used, the excavated material may be pushed to within reach of the clamshell to provide for more efficient operation. Excavation by clamshell is slow and inefficient and will be employed only on deep excavations or where a confined work site makes it impractical to use other methods. Clamshells are frequently used in later stages of excavation, when it has been necessary to remove the haul ramp. To permit operation of trucks and tractors within the excavation, consideration must be given to properly locating bracing and struts for the sheeting. The lower strut must be at least 12 feet above the bottom of the excavation to permit equipment operation beneath it.

Groundwater Control

Improper control of groundwater is often a cause for settlement and damage to adjacent installations. There are several ways of dewatering excavations and it is important that the method selected be suitable for the proposed excavation.

For small excavations, ditches may be used to collect water into a sump, from where it may be pumped. This is the most economical method. However, the groundwater seeping into the cut may carry soil fines with it, causing loss of ground and settlement. To reduce this loss of fines, straw is sometimes packed behind the sheeting. This can be reasonably effective, but generous use of straw can in itself create voids leading to subsidence. Burlap may be used as an alternative to straw, with generally better results.

In larger excavations in permeable soil, the water table may be lowered using wellpoints or deep wells. This method of groundwater control can often be effective in sand or coarse silt deposits, but cannot be used in fine silt or clay soils, because of their low permeability. Gravity-fed wellpoints are usually considered effective in subsoils in which 90% or more of the grain sizes are larger than about 0.05 millimeters; however, this appears to vary, depending upon actual site conditions. In borderline soil conditions requiring groundwater control, a test well should be installed to determine the effectiveness of the proposed system. Each stage of wellpoints will effectively lower water to a depth of approximately 15 feet. Where it is necessary to lower the water table more than 15 feet, a multiple stage of wellpoints or an eductor system may be installed. An alternative method of dewatering would be to use deep wells with submersible pumps.

The spacing of wells will depend upon the permeability of the soil and the desired rate of drawdown of the water table. Typical spacing for wellpoints or eductors may be 5- to 20-foot centers. Deep wells may be spaced at 20- to 50-foot centers for effective groundwater control.

In fine-grained soils, such as silts or clayey silts, wellpoints will generally be unsuccessful in controlling groundwater. The excavation may be carried out using drainage ditches and sumps, with care taken to avoid loss of ground through the sheeting or more elaborate construction methods used. These may include a cut-off wall of sheet piling or concrete; or soil stabilization by freezing. If a cut-off wall is used, care must be taken in determining the depth of penetration required below the bottom of the excavation. The cut-off wall must be carried below the depth of the excavation a sufficient depth to prevent failure through subsurface erosion, causing piping or a blow-up of the bottom of the excavation. This type of piping failure occurs due to the differential in hydrostatic head on each side of the sheeting. Water seepage under the toe of the sheeting may cause springs to form in the bottom of the excavation, carrying fine grains of soil with it. This may eventually undermine the sheeting, causing failure and flooding of the excavation. Furthermore, this type of seepage will disturb the bottom of the excavation, making it unsuitable as a subgrade to support the tunnel foundation. To evaluate the effect of the upward seepage pressures, a flow net may be drawn. For the construction of this diagram and its use, refer to a standard textbook on soil mechanics. Designing to prevent subsurface erosion is complicated by the non-homogeneous nature of soils. Empirical rules developed from actual observations, supplemented by local experience, are the normal basis of determining the required depth of penetration (see also Section 5).

CONCRETE

Mix Design

The objectives of concrete mix design are production of a concrete mix having the properties required for the specific application with respect to:

- Strength
- Impermeability
- Durability
- Placeability
- Resistance to segregation and bleeding.

The strength, impermeability and durability of the concrete is greatly dependent upon the quality of the cement paste. The quality of the cement paste is dependent upon the ratio of water to cement and the extent of curing. As the paste is thinned with water, its quality is reduced, resulting in concrete with reduced strength, impermeability and durability. For reference, the reader is directed to the American Concrete Institute Manual of Concrete Practice.

Placeability and resistance to segregation and bleeding are affected by the aggregates selected and the cement content. The use of large aggregates and lower cement contents result in reduced placeability and increased probability of segregation during placing. In cut-and-cover construction involving long, vertical drops with the aid of elephant trunks into heavily reinforced walls and slabs, these become important properties of the concrete.

The minimum cement content and maximum aggregate sizes may be influenced by local materials and local conditions and it may be necessary to experiment with trial batches to determine placeability and resistance to segregation under actual placing conditions.

In cut-and-cover construction for the Toronto subway, vertical drops of up to 60 feet were involved. The concrete used had a maximum aggregate size of $\frac{3}{4}$ inches with a minimum cement content of 525 lb/yard^3 of concrete. The W/C ratio specified was 0.59 with a maximum slump of 3 inches. The concrete was placed with vertical elephant trunks suspended within 6 feet of the bottom of the form. The hardened concrete was dense, impermeable and essentially free of segregation.

In thick walls or slabs, a high cement content can cause excessive temperature rise during curing, resulting from heat liberated during hydration of the cement. High curing temperatures may cause excessive surface cracking in the concrete due to a high temperature differential between the ambient conditions and the center of the concrete member. To control this temperature rise, it may be necessary to use larger size coarse aggregate and a pozzolanic mixture such as fly ash. Both measures will permit the use of less cement and a subsequent reduction in heat of hydration.

Concrete having the best qualities of strength, impermeability, durability, placeability and resistance to segregation should be designed to have the following properties.

1. Water/cement ratio selected to produce the required strength, impermeability and durability.
2. Sufficient cement content to ensure placeability without segregation.
3. Maximum slump of 3 inches to control segregation.

A number of admixture agents are available that may be added to concrete. The two most used admixtures are air-entraining and water-reducing admixtures added to improve the durability and placeability of the concrete.

Water-reducing agents with air-entrainment increase the plasticity of concrete having a given water/cement ratio and improve placeability. They are useful when placing concrete by means of a pump or when using a tremie.

Mixing and Placing.

Thorough mixing of the concrete ingredients is essential for uniformity of strength, durability, impermeability and workability of concrete. Equipment and methods used for measuring and mixing concrete should be carefully checked for their ability to produce a consistent and uniform product. Two methods used for mixing are:

1. Plant-mixed concrete, in which the concrete is completely mixed in a stationary mixer and delivered to the job site by truck; and
2. Truck-mixed concrete, in which the concrete is completely mixed in a truck mixer while en route to the job site.

Whichever method is used, the capacity of the equipment should be checked to ensure that the equipment and facilities have sufficient capacity to ensure a continuous supply of concrete to the work site at the intended rate of concrete placement.

The mixing methods and equipment should conform with ASTMC94 or CSA A23.1 to ensure proper quality of delivered concrete.

In transporting and placing concrete, it is of utmost importance that methods and procedures be followed to prevent segregation of the coarse aggregate from the cement past or bleeding of the water to the surface. Care must be taken at points of discharge, such as the ends of chutes, conveyor belts and hoppers, where a change in direction of the flowing concrete can cause segregation. The flow of the concrete into the form should be controlled at all times using a placing method that ensures that the concrete drops vertically into the center of the form or whatever container receives it. If the concrete is permitted to strike against the sides of the forms or reinforcing bars as it drops to the bottom of the form, segregation will invariably occur. Chutes and elephant trunks should be located so that the concrete can be delivered to points within 5 to 6 feet vertically or horizontally of its final location. In narrow wall forms, it is usually necessary to permit the concrete to drop vertically for the full height of the forms. Where this drop exceeds 6 feet, it is important that a grout cushion of 3-inch thickness be spread at the bottom of the form to cushion the impact of the falling concrete and control segregation. Care should always be taken to ensure that conveying equipment be kept free from hardened concrete and foreign materials.

The equipment most commonly used on the construction site for moving the concrete from the delivery trucks to the forms are crane buckets, conveyor belts, pumps and elephant trunks.

Conveyor belts can be used to transport concrete to forms provided precautions are taken to prevent segregation at the end of the belt and loss of mortar on the return belt. When using long conveyor belts, it may also be necessary to protect the concrete from loss of slump through evaporation. Segregation results largely from failure to control the flow of concrete as it leaves the end of the belt, with the larger aggregate being thrown farther than the fine aggregate and cement paste. This can be corrected by the use of suitable hoppers and drop chutes.

Pumped concrete can be readily moved in straight pipe for distances up to 1000 feet using pipes having a 6- to 8-inch diameter. Pumping capacities vary from 15 to 65 yards^3/hour. It is important that the pipe be of steel or rubber, as it has been found that aluminum pipe can have a detrimental effect upon the quality of the concrete. Although low-slump concrete can be pumped, higher slumps (over 3 inches) are commonly used to increase the efficiency of pumping. For this reason, an air-entraining and water-reducing agent should be used with this method of concrete placing.

Forms. Concrete should be placed in forms that are rigid and designed so that they will not shift or bulge under the weight or pressure of the plastic concrete or construction loads imposed upon them. Forms should be oiled to facilitate release from the set concrete and to prevent absorption of moisture into the forms from the concrete; or, if they are not oiled, they should be thoroughly wetted just prior to placing concrete.

Where concrete is being placed in contact with set concrete, the surface of the set concrete should be properly prepared to ensure bond. Laitance or soft mortar should be removed and the surface saturated with water for 4 hours prior to concreting, and a 1- to 3-inch layer of grout should be spread over the surface of the hardened concrete immediately before placing the new concrete. The grout should be a mix similar to the concrete mix but with the coarse aggregate omitted.

The concrete should be placed as nearly as possible in its final position. It should not be allowed to pile up and then be moved horizontally by vibration or tamping. The concrete should be placed in horizontal layers at such a rate that each layer is placed while the previous layer is still soft and the two layers can be vibrated together.

Vibrators. The most common method of consolidating concrete is by use of vibration. Vibrators usually make it possible to place a stiffer concrete mix and, when used properly, are effective in ensuring a dense uniform concrete. When using a vibrator, skilled workmanship is important. Insufficient vibration will cause honeycombing and under-consolidation; excessive vibration will cause segregation.

The Tremie Method. This is usually used to place concrete under water. This involves use of a straight pipe with a receiving hopper attached at the top. The pipe is long enough to reach the bottom of the wall pour and is usually made up of a number of sections which can be individually removed to permit adjustment to the length of tremie. Concrete is placed into the hopper at the top and flows through the tremie to the bottom of the wall. The lower end of the tremie is kept submerged in the fresh concrete to maintain a seal and to force the concrete to flow into position under pressure. Concrete placed by the tremie method must be quite fluid, having a slump in the order of 6 to 7 inches. A high cement content or water-reducing agent will be required to ensure the required properties of strength, durability and impermeability.

Cold Weather Concreting

When ambient air temperatures fall below 40°F additional precautions should be taken to ensure a consistent concrete quality. The mixing water and aggregates for the concrete should be heated so that the concrete mix has a temperature between 60° and 80°F at the time of placing and the site should be enclosed and heated to ensure that surfaces against which the new concrete is to be placed are at above 40°F. If heat is supplied within enclosures, the heaters should be arranged to give a uniform distribution of heat and the humidity within the enclosure maintained at 95%. The temperature of the set concrete should be maintained at 70°F for 3 days plus 40°F for an additional 4 days. If high early strength concrete is used, this may be reduced to 50°F for 3 days plus 40°F for an additional 4 days. It is important in withdrawing heat that the rate of change in the concrete temperature be limited to not more than 20°F per day until the ambient temperature is reached.

Hot Weather Concreting

In hot weather, the temperature of the concrete should be at or below 80°F at the time of placing and the concrete protected through shading from direct sunlight and cured by sprinkling with cool water or application of wet burlap for at least 3 days after placing. The most effective method for controlling the temperature of the concrete at the time of delivery at the site is through the use of crushed ice as a substitute for mixing water. Stock piles of aggregate should also be shaded from direct sunlight.

WATERPROOFING

Although the exercise of strict concrete controls will reduce the tendency of shrinkage cracks to form, their presence cannot be entirely eliminated. Structures located in permeable soils and below the level of the groundwater table will be subject to leakage. For underground structures in general, and vehicular tunnels in particular, water infiltration (leakage) would be unacceptable, since it would result in unsightly streaking of wall and ceiling finishes. Further, in structures subject to freezing, thawing cycles such leakage could prove hazardous.

In consideration of the foregoing, engineers consider it both prudent and cost-effective to prescribe some form of waterproof protection for these underground structures. In developing this subject further, the discussion will be confined to the external waterproofing of structures, integral and internal waterproofing will not be discussed, since their reliability is uncertain.

Waterproofing Materials

Waterproofing material falls into three categories:

- Fabric or membranes
- Asphalts or tars
- Protective materials.

Fabric. One of the essential ingredients in the waterproofing of a structure is the fabric which forms the base to which the bitumen is poured or mopped on (or both). Earlier, it was found that the most effective way to build a substantial barrier to the entrance of groundwaters through base slabs, walls or roofs of structure was to employ a membrane fabric. The inpregnated felts (normally applied in the roofing trades) tend to be too friable and fragile for application to underground structures. Hence, cotton with a close mesh, impregnated with bitumen or, more recently, glass fiber material similarly impregnated, are the basic fabrics used in building up a membrane waterproofing. The individual layers are referred to as plies and generally vary between one and four. Plies in excess of four have been used on rare occasion but are very difficult to build up and hold. Currently, four-ply waterproofing would be considered the maximum number of plies, after which, if more protection is required, brick-in-mastic should be specified.

Prior to laying the first ply, an asphalt or tar primer is applied to that part of the concrete structure for initial sealing and compatibility with the build-up of the membranes. Each ply is mopped on with a hot bithumastic coating with overlapping of the seams of the individual sheets. When a section of the structure has received the last ply, a cover coating of hot bithumastic is mopped on using a heavy-duty cotton mop.

For base slabs, a 6-inch ground slab of lean concrete is placed on the subsail and the membranes are applied on top of this, as described above.

In hot weather, the membranes applied to vertical walls exposed to the sun should be protected and partly backfilled as quickly as possible.

Asphalts or Tars. The impermeable medium is actually the asphalt or tar applied. Because of its availability and ease of handling, asphalt is the preferred product to be used in the waterproofing of structures. Tar is in less use today as a waterproofing product. Whether asphalts or tars, the requirements are that they be heated to a requisite temperature and applied hot to the fabric plies. Generally speaking, the product is liberally applied and mopped on to ensure that the fabric is totally covered with product—and no "holidays" present.

Protective Materials. Built-up membrane waterproofing after application to an underground structure must be protected from all damages that might result prior to the backfilling and final grading of the structure. Unprotected waterproofing is subject to abrasion, cutting and other damage from workmen, materials and equipment working on the project. For base slabs and roof, very quick to apply and relatively inexpensive is a thin layer of regular or lightweight concrete. For vertical wall structures, however, a more easily applied inpregnated protection board (felt-urethane) is mopped over the membrane waterproofing and backfilled with soils to furnish required support.

Alternative Protection

Where waterproofing protection beyond that furnished by a four-ply membrane is required, a brick-in-mastic should be considered. The technique of laying-up this excellent form of waterproofing protection was developed for the subway systems in this country and abroad. The cost of labor and materials in those earlier years made brick-in-mastic a very cost-effective method of waterproofing.

The general construction technique in the building of a brick-in-mastic protection is as follows.

For horizontal surfaces like base slabs and roofs, a single layer of bricks, nominally 8 by 4 by $3\frac{1}{2}$ inches, separated by $\frac{1}{4}$-inch joints, is laid down on a 6-inch ground slab of lean concrete, after which the hot bitumen fills the $\frac{1}{4}$-inch joints between brick. Then the whole surface is mopped with hot bitumen. It is usual to asphalt-prime the horizontal surface and apply one ply of membrane waterproofing before initially laying the brick.

For vertical wall surfaces, the nominal thickness of the brick-in-mastic protection course is $8\frac{3}{4}$ inches by bricks 2 by 4 inches wide plus $3\frac{1}{4}$-inch joints. To form a trough in which to place this protection requires an outside form or wall, usually a thin concrete or supported plank. The protection is raised in approximately 3-foot lifts (the depth a workman can reach an arms length to place the lowest course of brick). As the protective wall is completed, it is backfilled and supported, in order that the next 3-foot trough height can be laid.

An alternative of brick-in-mastic for vertical surface is the use of an all liquid mastic—commonly referred to as Waylite Mastic, (a mixture of lightweight concrete and hot bithumastic). The protective thickness can thus be reduced from $8\frac{3}{4}$ inches to one of from 3 to 4 inches. Further, if the exterior form can be properly constructed and sealed against the escape of the hot liquid, then a significant height of a wall can be poured in a single lift—something to be preferred over a maximum lift of 3 feet for brick-in-mastic.

Other Forms of Waterproofing Protection

Plastic Membranes. Synthetic membranes such as neoprene and polyvinyl sheets have been used for waterproofing. They are attached to sidewalls with special adhesives. This application requires great care and so far not much performance experience is available. Continuous developments in the search for an economical, reliable material of this type are going on and should be investigated.

Epoxy Coatings. Coal tar epoxy combinations have been used for waterproofing. A coating is sprayed or painted on the outside or inside surface, which must be completely dry to obtain good adhesion. Here too few performance records are available, but the latest developments should be investigated.

Bentonite Panels. A form of waterproofing protection for vertical surfaces in current use is that of bentonite panels. Present subway construction in this country, where stations and line sections are being constructed by the cut-and-cover method, allows the bentonite sheets to be hung or secured directly to the soldier beams and lagging. The final structure wall can then be poured directly against this—thus providing an impervious barrier and producing a dry interior of subway section after the structure is backfilled.

15
Subway Construction

D.J. MORTON, PRINCIPAL ENGINEER

Chief Design Engineer (Retired)
Subway Construction Branch
Toronto Transit Commission

The shape and structure of a city evolves around its transportation system. The subway is the highest volume transportation system available, and therefore can have the greatest influence upon a city's structure. A subway system will encourage the development of high density population and work centers within the station influence zones, and encourage a city form characterized by an intense central business district and a dense residential population in station areas. These centers develop because a larger number of people are able to travel to common points of destination when traveling by subway than by any other mode of transportation. Our urban areas are constantly undergoing change and growth, and the extent and nature of these changes in the future will be dominantly influenced by the choice of transportation facilities. Subways can and will be a major factor in reshaping urban areas, increasing population densities and revitalizing central business districts.

The transportation and land use objectives of the urban area must first be defined and the subway then used as a tool in achieving these objectives. The relationships between residential or work centers and the spaces between these centers will be influenced by the choice of subway route location and station siting. In planning a route location, the designer must consider:

1. Operating costs and integration with surface feeder systems;
2. Effects on the community with respect to redevelopment and disruption of existing community structure; and
3. Construction costs.

The final selection of subway route and construction method must be made on the basis of providing the best overall transportation system at least cost to the community.

SELECTION OF ROUTE AND METHOD OF CONSTRUCTION

The designer planning a subway project is often faced with selecting between an on-street or off-street route alignment and between tunneling or cut-and-cover construction. The costs to be considered are the construction costs, right-of-way costs, and construction nuisance costs such as traffic disruptions or interference with normal commercial and other urban activities.

An on-street alignment reduces the cost of acquiring right-of-way and the need for demolition to existing buildings. However, on a heavily traveled street this construction will cause considerable disruption to normal vehicular traffic flow during the construction and a temporary loss of access or reduction in access to private business or residential properties. This situation is relieved somewhat through the use of tunneling rather than cut-and-cover construction, but even when tunneling is selected as a construction method, it is often limited in soft ground to the running tunnels between stations being constructed by cut-and-cover methods. In selecting an on-street alignment the designer must therefore consider the costs and difficulties of maintaining vehicular traffic and access to private properties.

Vehicular traffic can be maintained through the installation of timber decking at the surface, supported on steel deck beams spanning the width of the cut. The steel deck beams are usually supported directly onto soldier piles used in the shoring system or on cap beams spanning from soldier pile to soldier pile. With the installation of decking, disruptions to vehicular traffic and access to private properties are reduced to short construction periods during installation and removal of the deck and during placing of materials through the deck from the surface. The design of the decking system is covered in more detail in Chapter 14.

An on-street alignment using cut-and-cover construction also affects a major disruption to the existing street surface and utilities. The street surfaces must be replaced and the utilities maintained and replaced during construction. During excavation and construction of the subway structure the utilities may be suspended directly from the deck beams or diverted around the construction site. Water mains, sewers, telephone and hydro ducts can usually be supported directly from deck beams; however it is preferable, where possible, to divert high risk utilities, such as high pressure gas mains, around the excavation. Where gas mains must be supported within the excavations, they are usually supported on a separate system to isolate the mains from traffic vibrations. During restoration the utilities may be supported directly on the backfill over the subway or on a post-and-beam structure bearing directly onto the subway roof slab. If a post-and-beam support is used, the posts tend to interfere with the general backfilling operation, reducing the overall quality of backfilling and increasing the construction costs. Therefore where thorough inspection is provided to assure proper compaction of the backfill, support of the utilities directly on the backfill material should be favored.

The exposure of utilities during the subway construction often affords the utility companies an opportunity to improve or update their buried plant. If the utility companies express an interest in this, an agreement must be prepared to establish the basis for sharing in the restoration costs. A similar agreement may be necessary with the roads authority to permit the inclusion of street improvements during surface restoration.

The disruptions to traffic, private property access and utilities are much more serious on a primary traffic artery than on a secondary city street. The demand corridor for a subway usually parallels a primary traffic artery that is already congested with traffic, and it is often the traffic congestion that has been the most evident reason for considering a subway. By selecting an alignment within the street allowance of a secondary street paralleling the primary traffic artery, it may be possible to achieve minimum property acquisition costs with acceptable inconvenience and disruption. The practicality of this consideration depends upon the distance between the preferred alignment, within the primary street, and a suitable secondary street. Because of the discontinuity of secondary streets, this solution will generally result in a combination of on-street and off-street construction.

An off-street alignment involves construction on a private right-of-way usually paralleling a primary traffic artery that represents the traffic demand corridor. Disruption to surface traffic occurs only at cross streets which may be decked over to maintain traffic during the construction period. This is an effective means of reducing construction nuisance but involves high costs of right-of-way acquisition and building demolition. The right-of-way acquisition costs are offset through use of the restored right-of-way for redevelopment, parking lots and park lands. Where the right-of-way is used for redevelopment, the air rights over the structure may be sold or leased on a long term basis. Terms of the sale or lease should define the easement requirements of the subway structure and the maximum loading permitted and require that all building plans for construction over the subway be submitted to the subway operating authority for approval. To facilitate preparation of these agreements the contract drawings used for construction should indicate the design loadings used in the design of the subway structure.

The suitability of an off-street alignment is usually dependent on the value of buildings that will be demolished. Lesser valued buildings can usually be purchased and demolished with air rights made available for resale after construction. More expensive buildings may be underpinned or tunneled underneath using conventional mining methods. Where the subway is constructed beneath existing buildings, right-of-way costs are reduced to the cost of acquiring a construction easement and a permanent right-of-way easement for the subway structure. In assessing the value of these easements it is necessary to consider the value that the private property owner will place on the loss of future foundation capacity and flexibility for redevelopment.

Buildings to be demolished within the right-of-way are usually removed before construction in a separate contract and the subway construction contractor provided with a cleared right-of-way except for buildings to be tunneled under or underpinned. Buildings requiring support during the construction are usually enumerated in the tender documents. Where the building support can be adequately provided through the provision of a cut-off wall or a simple underpinning wall, the design is usually made the responsibility of the contractor. However, where the subway passes directly under a building, the underpinning requirements can become fairly complex and the examination of the building, search of building records and design of the underpinning should be carried out by the designer with the underpinning requirements fully detailed on the contract drawings. Where the building foundations or underpinning members for the building bear directly on the subway structure, care should be taken to effectively isolate the building from structurally borne vibrations originating within the subway.

Tunnel construction is frequently considered for the running structure between stations to reduce disruptions to the surface and the cost of acquiring the right-of-way. For shallow subway alignment in soft ground, tunneling will often be more expensive than cut-and-cover construction; however, it permits construction of the subway with very little public inconvenience except at the mucking shafts. The cost of tunneling is most dependent on the soil conditions encountered. If running sand, soft clay or ground water is encountered, the increased construction costs and risk of surface or building settlement will often favor other means of construction despite the advantages of reduced right-of-way and construction nuisance. Subway stations are seldom constructed in soft ground using tunneling methods, due to the shallow depths of cover involved, the large face diameter and the increased construction costs.

Cut-and-cover construction usually involves a twin box structure constructed in a trench excavation cut to the neat lines of the structure and using braced sheeting to support the sides of the cut. After completion of the structure the cut is backfilled and the surface restored. Decking is used to minimize interruptions to pedestrian and vehicular traffic where required. Since the amount of surface disruption, traffic interference and cost of right-of-way is directly affected by the width of cut, consideration should be given to possible means of reducing the width of the box structure required. One possible method is the use of over-and-under construction in lieu of the more conventional twin box structure. This involves locating the tracks one above the other instead of side by side as in twin box structures. Through this method of construction it is possible to reduce the width of excavation by approximately 15 feet between station areas and involves considerably less interference to traffic and urban facilities. This structure is most suited to line stations, due to difficulties arising at terminal stations in providing turnback and crossover facilities. Cut-and-cover construction will often represent the least cost construction method in the shortest construction period for a shallow subway alignment; however, it is necessary to consider these cost advantages against the increased right-of-way costs, community disruptions and construction nuisances.

Table 15-1 compares the advantages and disadvantages of on-street and off-street alignments employing cut-and-cover and tunnel construction. From this table we see that an off-street cut-and-cover alignment will often result in the least overall project costs and shortest construction period, while tunneling will offer the least construction nuisance and traffic interference.

ROUTE SURVEYS AND PRECONSTRUCTION INSPECTIONS

Topographical Surveys

Initial topographical surveys are usually carried out with the aid of aerial photographs which can be enlarged to be used for conceptual planning and in preparing topographical maps showing contours, building locations and sidewalk and above-grade utility positions. The aerial photographs, when enlarged to a scale of 1 inch to 1000 feet, may be used for conceptual planning by overlaying the proposed route locations over the photograph. This type of photograph may also be used effectively during presentation meetings as an aid to indicate the proposed route and its relationship with existing topography and buildings. Survey plans of the topography may be prepared to a scale of 1 inch to 200 feet and may be used by the designer to relate preliminary vertical and horizontal alignment proposals to existing topography and buildings. These survey plans are also used in calculations for excavation and fill quantities for the project cost estimates.

In the preparation of detailed contract drawings it is normally necessary to supplement this aerial photography with on-the-ground surveys using survey crews to accurately locate street lines, curbs, building lines, property lines and critical elevations. On-the-ground surveys will also be required to locate and establish survey monuments for a horizontal control line and bench marks for the vertical control. This control line usually consists of monuments spaced 1000 to 3000 feet apart and surveyed to an overall accuracy of 1 in 20,000. Physical features of the site are referenced to this line for field measurements and subsequent plotting in the office. The final alignment is referenced to the control line and located on-site using chainages and offsets related to the control monuments. During construction these control line monuments must be protected from damage or loss. Sufficient measurements should be taken to allow reestablishment in the event that a monument does become damaged during construction.

The subway transportation facility will encourage new development or redevelopment and it is often necessary for the operating

Table 15-1. Unweighted comparison criteria for alignment.

RATINGS	BEST PERFORMANCE SCORE 4	SCORE 3	SCORE 2	WORST PERFORMANCE SCORE 1
	ON-STREET		OFF-STREET	
CRITERIA	TUNNELING	CUT-AND-COVER	TUNNELING	CUT-AND-COVER
Right-of-way costs	4	4	3	2
Construction costs	2	3	2	4
Construction time	2	3	2	4
Construction nuisance	4	2	4	3
Construction interference with traffic	4	2	4	3

authority to examine building plans for proposed construction over or adjacent to the subway. These plans may be for the sale or lease of air rights over the subway or for the imminent construction of a planned building or facility. In either case, the purpose of the examination will be the protection of the subway structure and operation.

The examination of redevelopment plans is facilitated if the proposed structure can be referenced to the original control monuments used in the construction of the subway or to the reestablished monuments in those cases where the original monuments cannot be retained. This long-term requirement for a means of accurately locating the completed subway structure should be kept in mind when locating control monuments for initial construction. Monuments should be placed in locations least likely to be disrupted in the near future and should be tied to a sufficient number of local and relatively permanent features to allow reestablishment if lost in the future. All control monuments and ties to them should be shown on contract drawings used for the construction to provide the operating authority with a permanent record and a permanent means of locating the subway at the surface.

Subsoil Stratigraphy

A preliminary subsoil exploration program typically involves locating bore-holes at a spacing of approximately 300 to 500 feet along the proposed alignment to a depth of a few feet below planned subgrade for the subway. Records are kept during the boring program indicating the type of subsoil encountered and the location of groundwater tables. The relative density of the ground is determined by performing standard penetration test and samples are taken of the soil for laboratory testing and analysis. Typical laboratory tests performed are unit weight, grain size analysis and water content, including liquid and plastic limits, unconfined compression tests and consolidation tests.

Where groundwater is encountered, it is common to install observation pipes or piezometers to enable the engineer to take long-term readings of the groundwater to establish the permanent groundwater table and seasonal fluctuations to it. This record can be of importance in the design of the structure in determining the type and need for waterproofing, as well as being essential information to the contractors in tendering the proposed construction. In station areas and areas where construction difficulties are anticipated due to the nature of the ground, it will be necessary to supplement this preliminary subsoil program with additional bore-holes. The subsoil conditions determined through the subsoil exploration program may be plotted on profile plans drawn to a scale of 1 inch to 200 feet horizontally and 1 inch to 20 feet vertically. These drawings, with the information from the laboratory tests, can then be bound into a Soils Report, which is made available to the designers during the design of the structure and the contractors during the tendering and construction of the proposed structures.

Utility Surveys

In built-up urban areas the subway designer must establish the location of all existing utilities before he is able to determine the final vertical or horizontal alignments. This will involve reviewing the plans of all utility companies in the area and checking with the utility companies to determine the possible existence of as-constructed information that may supplement other available plans of existing utility plant and facilities. A thorough search of existing building and utility records is required to determine the presence and location of existing and abandoned underground plant such as pipes, ducts, tanks, foundations and vaults. An on-site survey must be carried out that will locate and identify all manholes, catch basins, utility poles, valves, hydrants, etc. that may be useful in checking the actual locations against the plans. Inverts of sewers should also be checked at manholes to compare actual elevations with those shown on available utility plans.

Future plans for utilities should also be discussed with the individual utility companies so that, where necessary, the subway alignment can be established to make provision for them. It is usually possible to relocate these proposed utilities to avoid the subway but in some instances the alignment for the subway may have to be selected to avoid conflict with the proposed utility structure. In selecting the final subway alignment, care must be taken to ensure that all existing and proposed utilities can be successfully redirected over, around or under the subway structure. This design activity can often be one of the longest in the preparation of contract documents, due to the scarcity of accurate information, and should be scheduled early into the design activity network.

Building Inspection

Most subway construction involves excavation in built-up urban areas close to existing building structures and the possibility of settlement or other forms of damage to these existing structures is a necessary risk involved in the construction. Under these conditions, subject to the requirements of the insuring authority, it is good practice to carry out a thorough inspection on these existing building structures and record the conditions of them prior to construction. These records serve as protection for both the owners of the buildings and the contractors carrying out the construction, as they will substantiate any legitimate claim for damages on the part of the property owner, or when used by the contractor, the record provides a basis for refuting any claim considered not to be legitimate.

The building report should indicate all structural and superficial damages present in buildings adjacent to or within the right-of-way of the subway construction and have attached to it records of building elevations and photographs to provide an overall and complete account of the building condition immediately prior to start of construction. The report should then be signed by the owner, the contractor and the authority responsible for the subway construction. This type of procedure, when agreed to by all parties concerned, will substantially reduce the time and work involved in evaluating subsequent claims for building damage as a result of the construction works. Irrespective of the form taken, it is important that building inspections be carried out on existing buildings and their preconstruction condition recorded. The extent to which buildings should be inspected beyond the neat lines of the cut is dependent to some extent on the ground conditions. A common approach is to carry out building inspections on all buildings having footings that fall within the zone of influence of the subway construction. The zone of influence is defined by assuming a plane 45 degrees to the horizontal drawn through the invert at the neat line of the excavation. A larger zone of influence may be considered in adverse ground conditions or where considered advisable by the engineer.

ALIGNMENT AND TRACK DESIGN

Detailed planning of the geometry for the horizontal and vertical alignment will involve preparation of roll drawings showing plan and profile drawn at a scale of 1 inch equals 200 feet horizontally and 1 inch equals 20 feet vertically. On these drawings the topography, subsoil conditions and critical utility locations should be plotted and an alignment for the subway prepared which minimizes disruptions to existing utilities and traffic, and minimizes the cost of

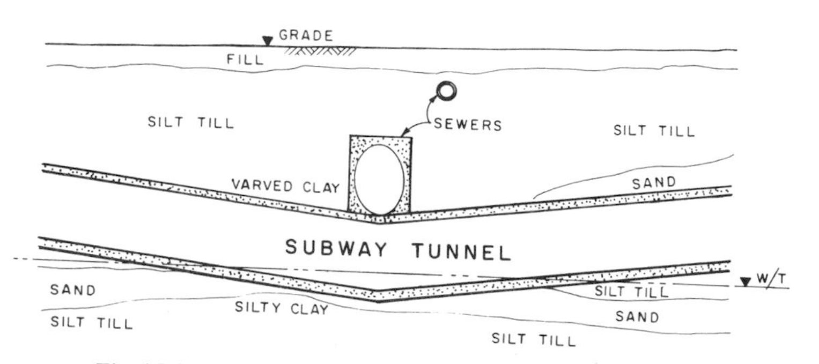

Fig. 15-1. Plan and profile drawing for alignment planning.

acquiring the right-of-way and of constructing the subway structure. (Fig. 15-1).

In developing horizontal and vertical alignment criteria, consideration is given to the operating characteristics of the equipment and minimum levels of passenger comfort and safety. In subway construction the operating speeds and the type of equipment using the track can normally be defined within a very limited range, therefore it is possible to design the track with superelevation to provide a reasonably balanced lateral acceleration and a high level of passenger comfort without introducing speed restrictions or special suspension systems on the vehicles.

Horizontal Alignment

The horizontal alignment consists of tangents and circular curves joined by spiral transition curves. The length of the spiral transition curve and the amount of superelevation is related to the design speed of the subway vehicle at that location in the structure. To determine the amount of required superelevation, the designer should prepare a speed-distance curve for the proposed alignment.

Figures 15-2 and 15-3 indicate the acceleration and deceleration speed-distance characteristics of the vehicles in use on the Toronto subway and the San Francisco subway. As can be seen from these two diagrams, there can be a wide variation in performance of the transit vehicles and it is necessary at the onset of the design program to establish an assumed performance criteria. In the Toronto subway the maximum permitted vehicle speed is 55 mph, whereas the speed limit is 80 mph on the San Francisco transit system. From this vehicle performance

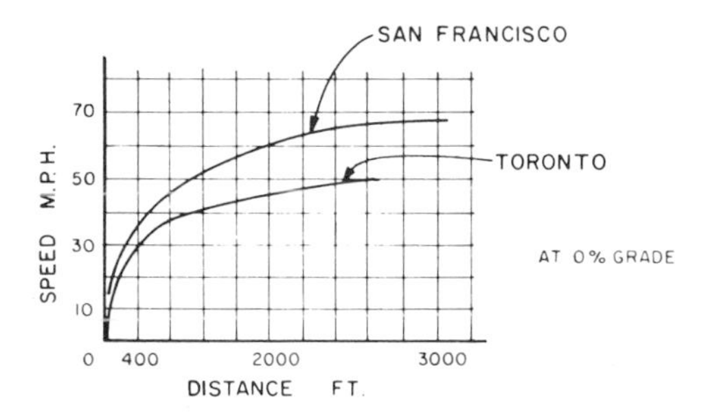

Fig. 15-2. Speed distance curves for normal acceleration, Toronto and San Francisco.

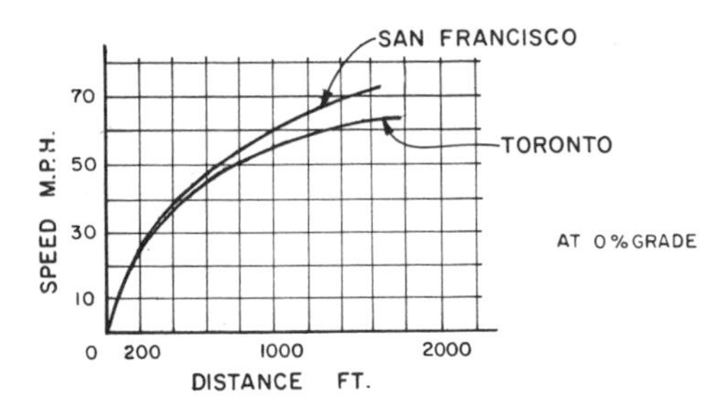

Fig. 15-3. Speed distance curves for normal deceleration, Toronto and San Francisco.

criteria and from knowing the maximum permitted speed within the system, the designer is able to plot the vehicle speed at any point between station stops. From this information he is able to calculate the required amount of superelevation to limit unbalanced lateral acceleration to an accepted limit and to determine the rates of horizontal curvature required to avoid speed restrictions between station stops.

The desirable minimum length of tangent between curves in horizontal alignment is normally 500 feet, and the absolute minimum length of tangent is 75 feet. Although transit vehicles are able to operate on smaller radii, the desirable minimum radius of curvature for mainline running track should be established as 1000 feet. Often the minimum radius will be larger than this to meet the criteria for maximum unbalanced lateral acceleration without requiring that limits be placed on the speed of operation.

Wherever possible, horizontal curvature should be avoided throughout the length of the platform areas in stations and, due to the outswing of the train vehicle, the tangent through these stations should be extended at least 75 feet beyond the end of the station platforms. Curvature occurring within the station area or immediately beyond the end of platforms will require that the platform be cut back to provide for the outswing and inswing of the transit vehicle. This will result in a variable width of gap between the vehicle and the platform and introduce a condition that could increase the risk of mishaps to passengers while boarding and leaving trains.

The unbalanced lateral force experienced by passengers on the San Francisco, Toronto and Washington subways has been limited by placing limitations on the amount of unbalanced super-

	MAX. OPERATING SPEED	MAX. PERMITTED SUPERELEVATION	MAX. PERMITTED UNBALANCED SUPERELEVATION	MAX. PERMITTED LATERAL ACC. FORCE AT MAX. OPERATING SPEED
San Francisco	80 m.p.h.	6 inches	$2\frac{3}{4}$ inches	0.04 g
Toronto	55 m.p.h.	4 inches	$2\frac{1}{2}$ inches	0.04 g
Washington	75 m.p.h.	4 inches	$4\frac{1}{2}$ inches	0.08 g

Fig. 15.4 Design Criteria For Passenger Comfort on Horizontal Curves

elevation, where unbalanced superelevation is defined as the difference between the equilibrium superelevation and the actual superelevation. Figure 15-4 indicates the criteria used in San Francisco, Toronto and Washington for maximum permitted superelevation and maximum permitted unbalanced superelevation. It is normal practice to design for balanced superelevation and to design on the basis of unbalanced superelevation only where the equilibrium superelevation exceeds the maximum permitted.

Vertical Alignment

In tunnel construction the vertical alignment is often established on the basis of ground conditions. Construction costs are minimized by locating the tunnels in the most competent ground for mining operations and by avoiding, wherever possible, construction in water-bearing soils. Although deeper stations require more stairs and escalators, and structure for these facilities, these costs are usually more than offset if the alignment avoids the added cost of working in soft or running ground or water-bearing soils which require extensive shoring or compressed air facilities. A disadvantage of deeper stations is the problem of vertical transportation and the increased travel time required by the passengers. The alignment should be established to avoid disruption to major utilities. In soft ground tunneling to reduce the risks of settlement at ground surface, the minimum depth of cover over the tunnels is usually in the order of one tunnel diameter, depending upon the nature of the ground at and above the crown of the tunnels.

In cut-and-cover construction, costs are usually minimized by adopting a profile that is as shallow as possible but which avoids major utility costs. Shallow stations also improve vertical transportation from street level and reduce the overall user travel time. Typically a minimum cover of 6 to 8 feet above the roof of the proposed subway structure is established to allow shallow utilities to pass over the top of the structure. The deeper utilities are then diverted around or under the subway structure.

Parabolic vertical curves are used to connect the tangent grade in the vertical profile. The normal length of the vertical curve is determined on the basis of passenger comfort and safety. The formula in use by authorities in Toronto and Washington is that the length of the vertical curve shall equal: $(G_1 - G_2) \times 100$, where $(G_1 - G_2)$ is the algebraic difference between the grades being connected by the vertical curve in percentage. The minimum length of any vertical curve is usually limited to 200 feet.

Criteria for maximum grades normally reflects the characteristics of the vehicles to be operated on the track. This is usually limited to between $3\frac{1}{2}$ and 4%. A minimum grade is required to facilitate drainage of seepage water within the structure and this is usually between 0.3 and 0.35%. It is good practice to limit the vertical grades within the station areas to the minimum grades required for drainage purposes. Vertical grades greater than this increase the risk of a train rolling out of position during boarding and unboarding activities and thus endangering passengers. Although there is no specific criteria for the minimum length of constant profile between vertical curves, most operating authorities will want some limitation placed on this constant profile to avoid a roller coaster effect. In San Francisco and Washington the specified minimum length of constant profile is 100 feet, whereas in Toronto it is 500 feet. However, a number of vertical curves in succession connected by minimum length tangents would not be desirable.

It is sometimes advantageous to consider a vertical grade that ascends into the station and descends out of the station. This achieves a construction in which the stations are as close as possible to grade to optimize vertical transportation conditions and the running structure is at a greater depth for safer tunneling conditions or increased utility clearances. This type of profile also utilizes gravity to reduce power requirements during train acceleration. The advantages of this type of profile are dependent on the station spacing, method of construction and ground conditions.

Track Design

Trackwork design practices of subway systems vary within the many existing systems. The type of operation, the vehicles, operating speeds and safety are the most important design criteria to be considered.

Generally North American systems use the standard railway gauge of 4 feet, $8\frac{1}{2}$ inches, with Toronto using 4 feet, $10\frac{7}{8}$ inches and San Francisco using 5 feet, 6 inches. Rail varies from 85

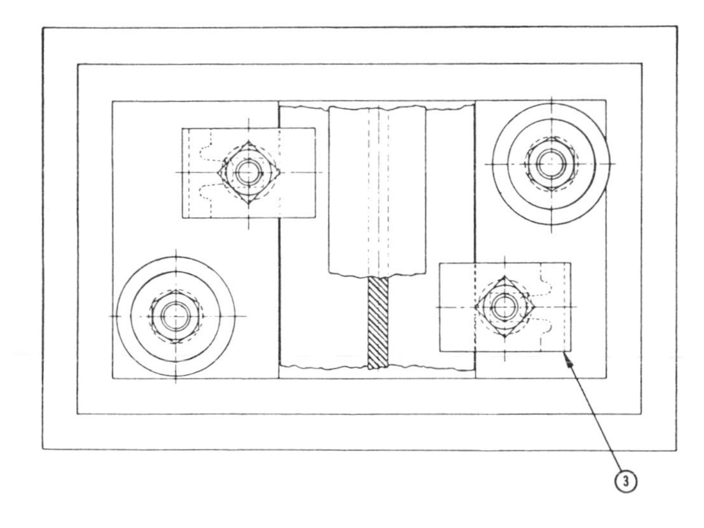

Track Fastening Assembly

1. 100 LB. ARA-A RAIL STD ROLLED
2. STEEL TIE PLATE
3. COMPRESSION RAIL CLIP
4. RUBBER PAD
5. ¾ INCH DIA. HEAT TREATED BOLT
6. CAST STEEL WASHER
7. FIBRE INSULATING SLEEVE
8. FIBRE INSULATING WASHERS
9. DOUBLE COIL LOCK WASHER
10. STEEL WASHER
11. 7/8 INCH DIA. ANCHOR BOLT
12. ANCHORS

GROUT PAD

Fig. 15-5. Toronto track fastening detail.

lb/yard to 119 lb/yard with the majority of systems using rail of 100 lb/yard. Rail cant in all but one system is 1:40 (Cleveland uses 1:20).

Rail installation in open cut is usually carried out with wood ties and ballast. Limited use has been made of precast concrete ties in lieu of wooden ties and of direct fixation of the rail to concrete slabs on grade. In structure a variety of rail fastenings are being used, tie and ballast, short ties embedded in concrete, tie and ballast paved with asphalt and direct fixation to concrete inverts (Fig. 15-5). All stress the important elements of electrical insulation and ample holding power. All steel-wheeled systems use the running rails as electrical conductors, usually for both the signal system and the traction power system. Spacing of ties and direct fixation anchors vary with each system, and often vary within a system, from $22\frac{1}{2}$ inches to 36 inches with the average being about 24 inches for continuous welded rail.

The continuous welded rail is utilized by all systems. Welding of the rail is being done in both the shop and in the field. Shop welding is either electric flash butt welds or oxyacetylene welds. Thermit welding, both pre-heat and self-heat methods are being used in the field.

All subway systems use the A.R.E.A. standards in the design of special trackwork for switches and crossovers. However, modification to this standard is carried out by most authorities to satisfy their own safety practices and design requirements. All special work is anchored to wood ties and ballast and all rail in a crossover is jointed.

The location and number of crossovers will affect the design of the structure and are usually determined by the operational procedures of the operating authority. Crossovers are required at all terminals for turnback operation and can be located ahead of, or behind, a station. Emergency crossovers are located at intervals as deemed necessary by the authority to provide emergency operation and to permit a short turn service in the event of a failure in the system. This facility precludes the possibility of total loss of the system due to one failure.

Storage tracks located within a system again are provided at the discretion of the operating authority but are installed to permit temporary storage of a bad order train that would otherwise tie up an entire system. Storage tracks can be used as emergency crossover by installing the necessary switching at both ends. Provision of crossovers and/or temporary train storage track for emergency operation is dependent on the operating requirements of the system and the station spacing. However, it is recommended that 3 to 5 miles be considered the maximum track distance between emergency facilities. Where storage tracks are provided, the track grade should be established at the minimum grade required for drainage wherever there is the possibility of a stored train rolling out of position and onto mainline track.

SUBWAY CLEARANCES

Clearance Envelope

The vehicle clearance envelope is established on the basis of the dynamic outline of the vehicle and normally is drawn 2 inches clear of this dynamic profile. All fixed equipment, structure and manways are located clear of this vehicle clearance envelope.

The dynamic outline of the vehicle requires giving consideration to the static dimensions of the vehicle, vertical and horizontal oscillation

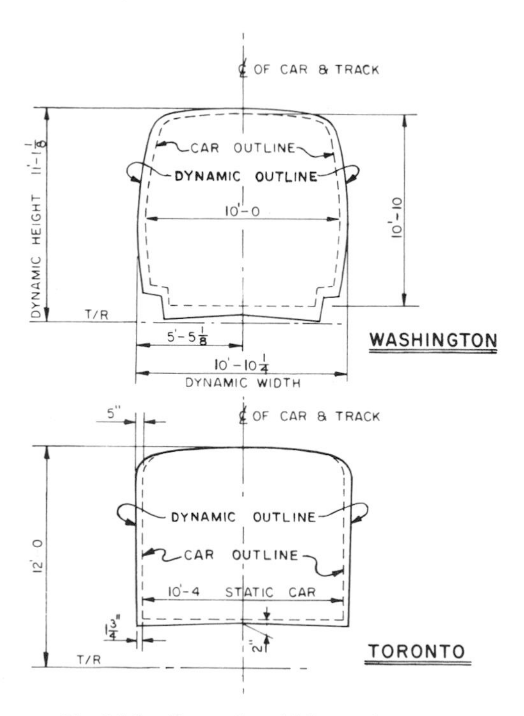

Fig. 15-6. Dynamic vehicle envelopes.

due to wheel and rail wear, and vertical and horizontal movements within the car suspension system. Figure 15-6 indicates the dynamic outline for Toronto and Washington subway cars operating on tangent.

This envelope must be increased for operation on horizontal curves as a result of superelevation and the inswing and outswing of the vehicles. Typical allowances for superelevation and curvature are shown on Fig. 15-7. In circular tunnel construction, these allowances are provided by selecting a tunnel diameter that exceeds

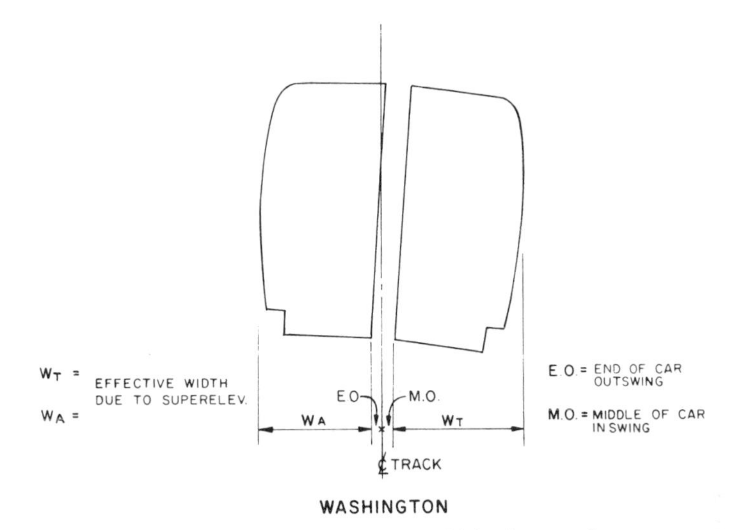

Fig. 15-7(a). Additional dynamic vehicle clearance for curves.

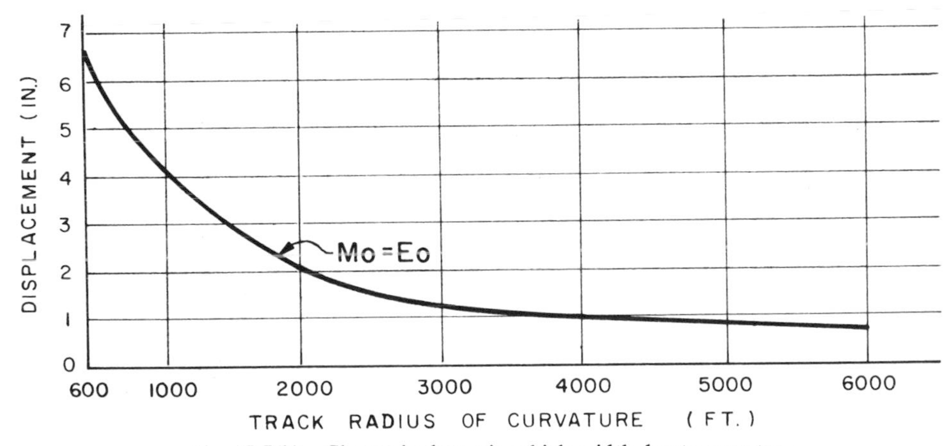

Fig. 15-7(b). Change in dynamic vehicle width due to curvature.

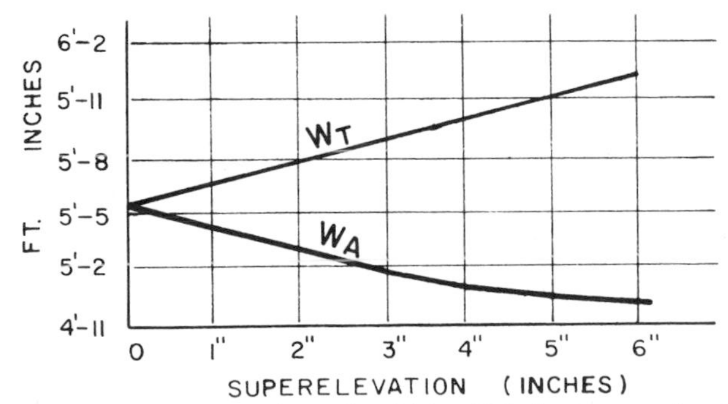

Fig. 15-7(c). Change in dynamic vehicle width due to superelevation.

minimum clearances on tangent to provide the minimum clearance allowances on specified curvature and superelevation. On small radius curves and maximum superelevation, where it would be impractical to provide for full vehicle outswing clearance plus full equipment clearance, it is sometimes possible to allow vehicle encroachment within the assigned equipment space by relocating the fixed equipment to avoid conflict. This practice is occasionally necessary in circular tunnels. However, where used, this practice places a permanent limitation on the installation of fixed equipment within the restricted zone. In cut-and-cover construction, full allowance for clearances is more easily attained by widening the structure and vehicle encroachment within the manway, or equipment spaces should not be permitted.

Clearance to Structure

Within the tunnels it is necessary to provide adequate allowances beyond the vehicle clearance envelope for manways, equipment installation and construction tolerances.

A continuous manway having a minimum width of 2 feet is recommended throughout all tunnels. Wherever separate tunnels enclosing each trainway are used, cross passages not more than 1000 feet apart should be provided to provide access between subway lines for maintenance personnel and a means of egress from an endangered trackway for evacuating passengers.

Equipment space provided between the clearance envelope and the fixed structure must be adequate to provide for lighting, power, signal, communications and supervisory cables and equipment that will be installed in the operating subway. Typical cable and equipment installed in a cut-and-cover subway tunnel is shown in Fig. 15-8. The type of equipment and space required for it must be determined before a final selection is made for the tunnel cross-section. Fig 15-9 shows a typical clearance diagram.

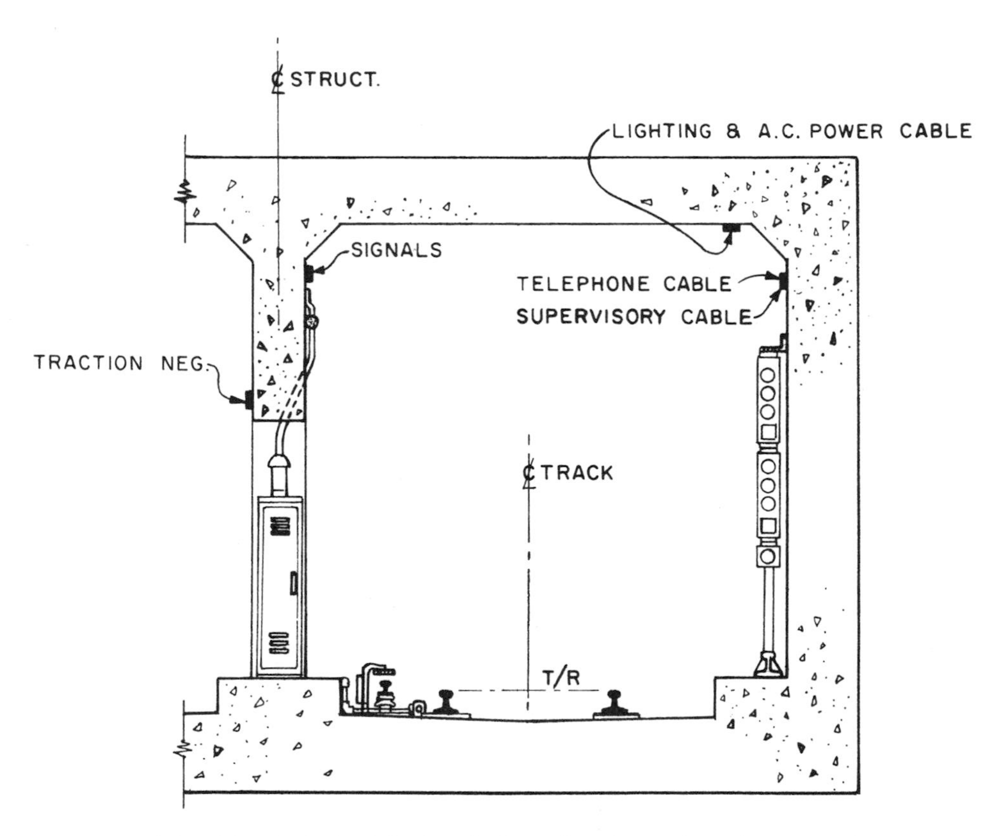

Fig. 15-8. Typical installation of cable and equipment in a subway tunnel.

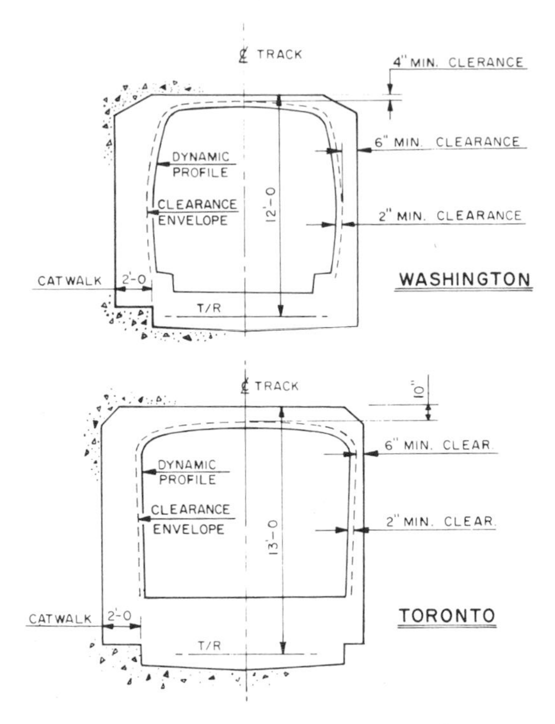

Fig. 15-9. Vehicle to structure clearance diagram.

SUBWAY STRUCTURES

Tunnel Structures

Subway constructed by mining methods in soft ground may be lined by either flexible or rigid tunnel lining.

Flexible Tunnel Lining. This lining is most commonly used employing a segmental liner of cast iron, steel or precast concrete.

The flexible tunnel liner is designed to serve both as a temporary and permanent ground support and must be designed for the following considerations:

1. The liner must be able to provide immediate bearing capacity against external loads from earth pressure.
2. Because the liner is to form the permanent tunnel structure, it must be able to resist the long-term influences and the development of external earth pressures and water pressures without detrimental deformations or leakage. This will include the influence of subsequent tunneling or deep excavations alongside the completed liner.
3. Where water is present in the ground, the lining must form a watertight structure. This requires that the joints be manufactured to permit caulking and that the liner segment be a watertight membrane.
4. The liner must resist high axial stresses from the propulsion jacks during advancement of the shield. (For further information see Section 6: Shield Tunneling)

Types of Primary Tunnel Liners. The oldest type, used for over 100 years, consists of cast iron segments. More recently liners of fabricated steel or reinforced concrete precast segments have been used. These are illustrated in Figs. 15-10, 15-11 and 15-12. In the dense London clay, boltless, hinge joint concrete liners have been used (Fig. 15-13). For further information on preliminary tunnel liners, see Chapter 6.

Rigid Tunnel Lining. This usually consists of a flexible primary liner followed by a rigid cast-in-place concrete liner. In Washington subway construction, the temporary primary liner was constructed using steel ribs and timber lagging. The tunnels were shield driven using a digger shield and a permanent reinforced concrete liner was later constructed to complete the tunnel. (See Fig. 15-14)

Concrete Secondary Liner. This is sometimes later constructed to provide structural adequacy to resist long-term earth pressure loadings and a watertight subway structure. The secondary lining is designed as a rigid structure and the members proportioned through elastic analysis to limit working stresses to specified design values.

Tunnel Shield. A tunnel shield is usually employed in the construction of subway tunnels in soft ground, to provide protection and facilitate excavation and liner erection (see Chapter 6).

Groundwater Control. In water-bearing ground, groundwater control may be through the use of

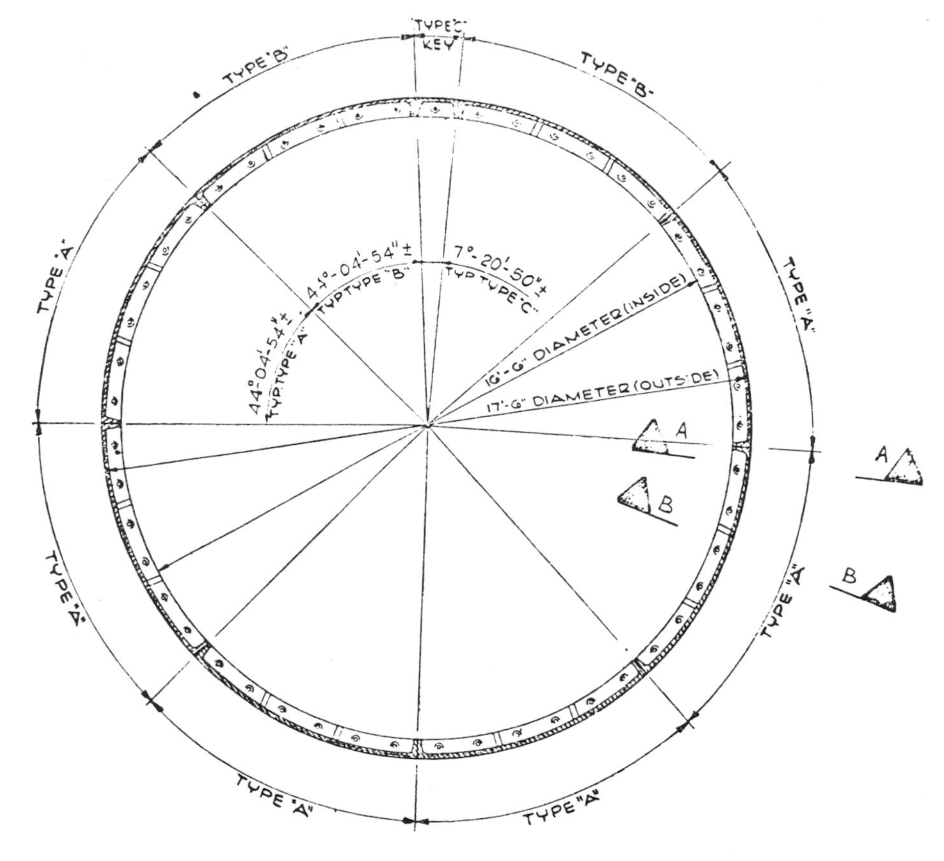

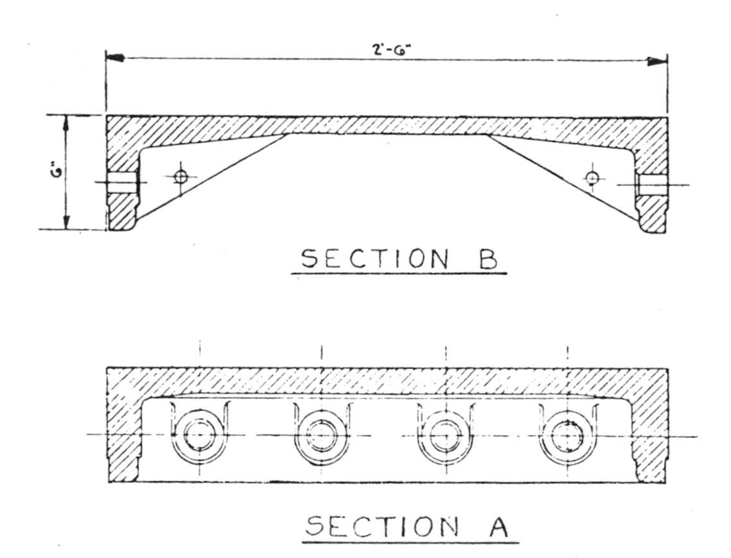

Fig. 15-10. Washington cast iron tunnel lining.

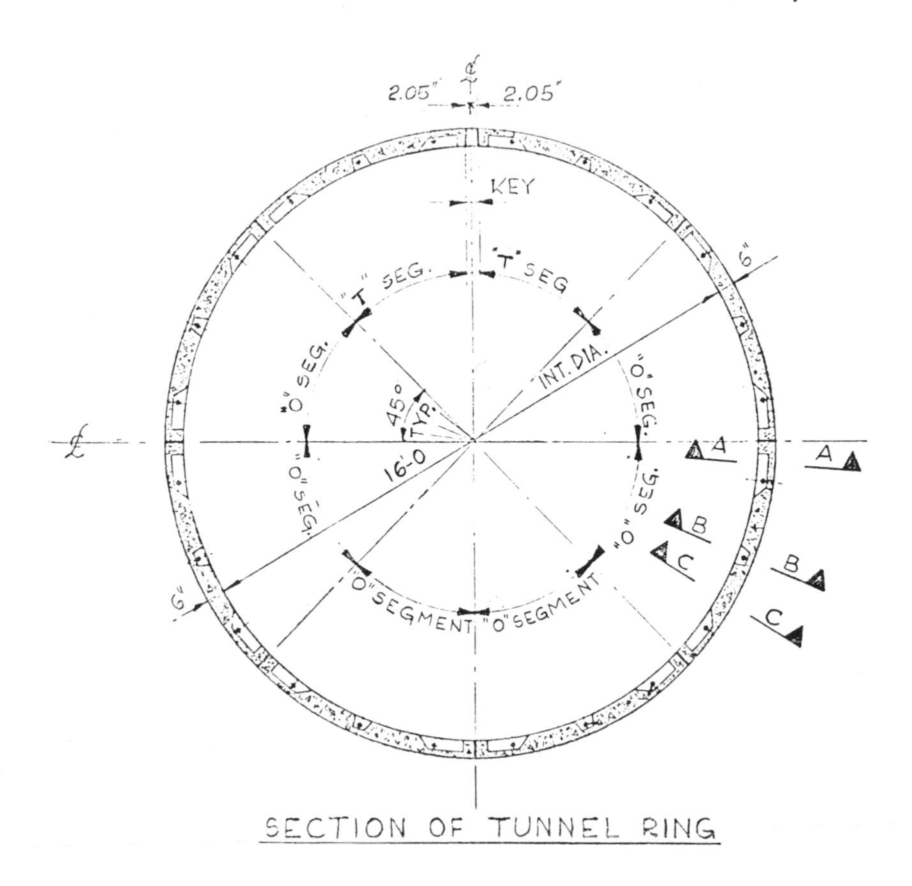

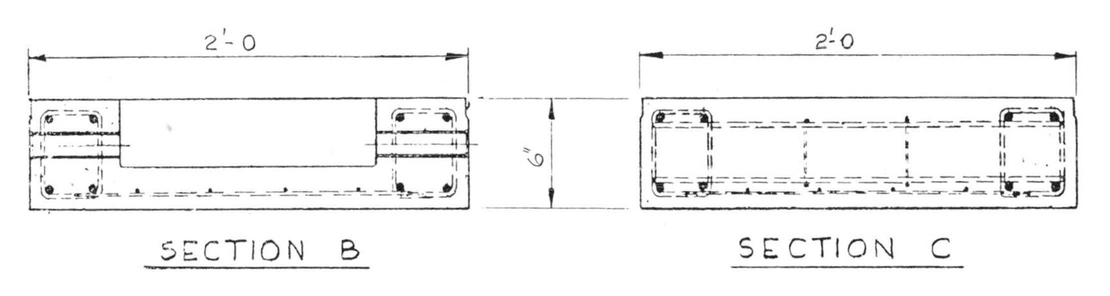

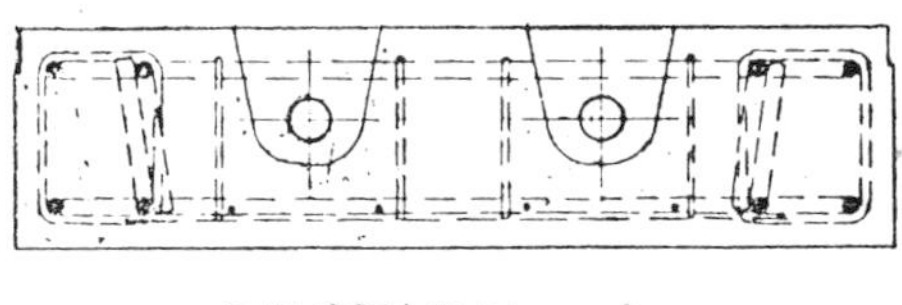

Fig. 15-11. Toronto precast concrete tunnel lining.

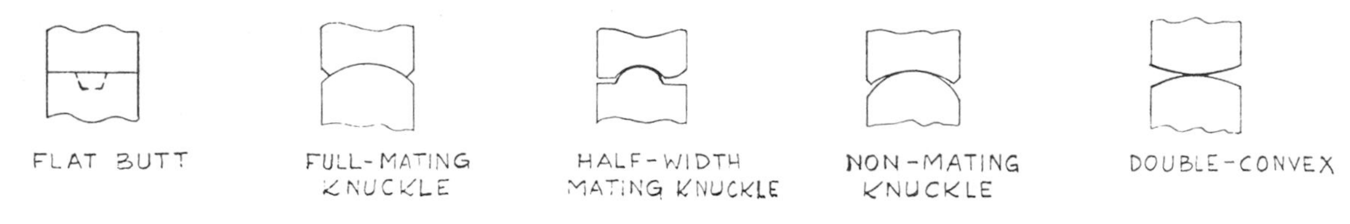

Fig. 15-12. Types of longitudinal joints.

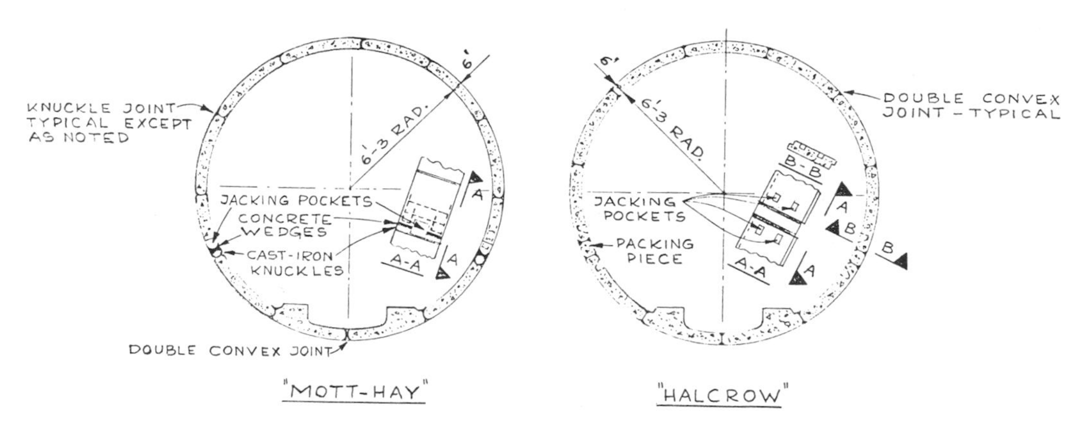

Fig. 15-13. London precast concrete tunnel lining.

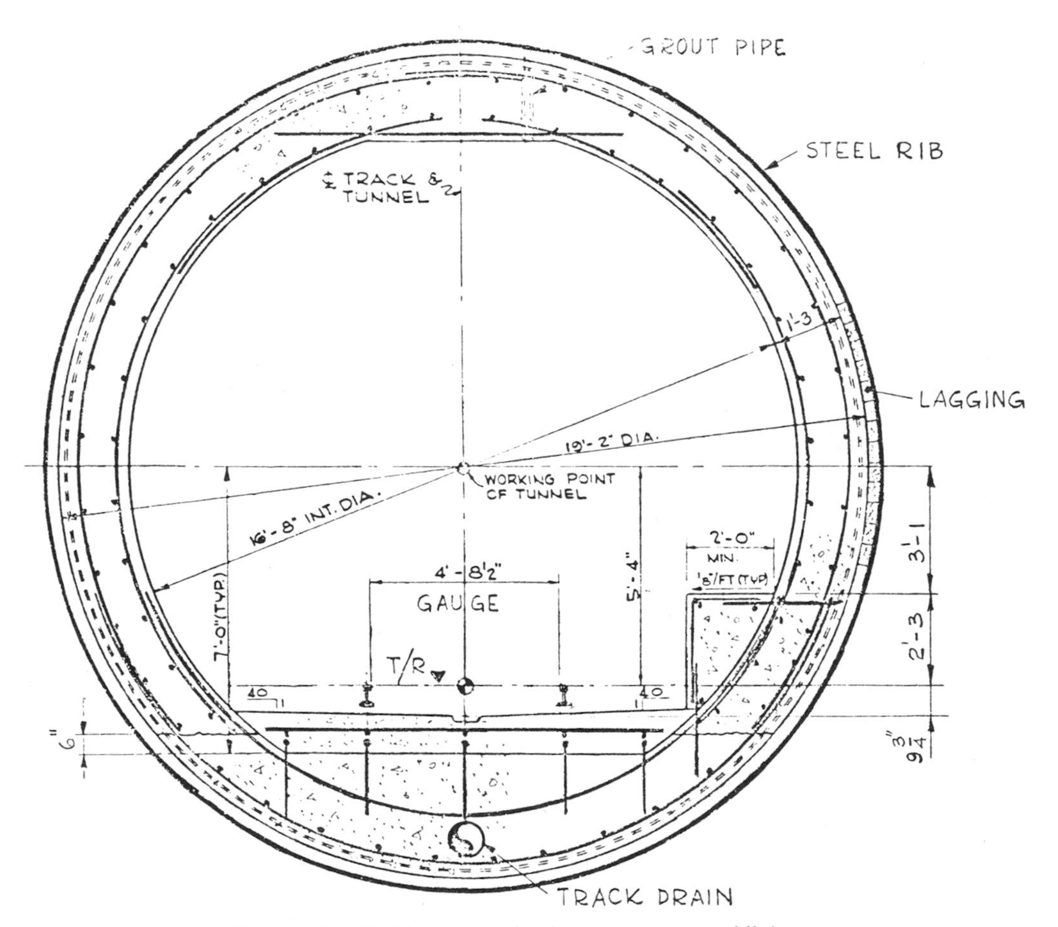

Fig. 15-14. Washington cast-in-place concrete tunnel lining.

dewatering techniques. The shallow profile frequently selected for subway tunnels requires special consideration of air losses at the face where compressed air is used. This may involve pregrouting of coarse grained soils or use of clay blankets and limits on the minimum size of compressed air plant to be used. The minimum compressed air plant capacity specified in soft ground tunneling in Toronto when working two 17-foot diameter faces is 7500 feet^3/minute. Due to the shallow depth and the increased risk of loss of air at the face, dewatering through the use of deep wells or eductors may be considered as an alternative means of groundwater control. In selecting a method of construction the designer and contractor must consider the relative costs of installing and operating the plant facilities for groundwater control and, in the case of compressed air conditions, the increased labor costs and reduced productivity.

Unit Price Contracts. Such contracts for tunneling (in which the risk of increased construction costs due to unknown conditions is borne by the owner instead of the contractor) are being used to an increasing degree in recognition of the uncertainties and risks involved. This leads to more competitive bidding by the contractors and removes the need for a contingency allowance in the construction bid. A typical tunnel contract will allow payment on a footage basis for the excavation, supply and erection of tunnel liners, with additional unit prices tendered for the supply of groundwater control facilities such as air plant or dewatering equipment. The contract should also provide for extra payment for the excavation, supply and erection of tunnel liners for work carried out in compressed air. These extra payments will vary with the amount of compressed air used. Provisions should also be made for extra payment for removal or tunneling through unforeseen obstructions, such as old pilings, foundations or other hidden obstructions. Due to the lead time required in the supply of tunnel liners, some authorities pre-tender for the supply of the tunnel liners and have them delivered to a storage yard from where they are supplied to the tunnel contractor as required. This can result in an overall reduction in construction time and a possible reduction in the cost of the liners through bulk purchasing if a number of tunneling contracts are involved.

Cut-and-cover Subway Structures.

In cut-and-cover construction, the subway tracks are usually enclosed in a double box structure with a supporting center wall or beam with columns. The track centers are located as close together as possible while providing adequate clearances for equipment, manways and structure to permit a minimum width of excavation. In a typical box section each trainway will have a clear width of about 14 feet depending upon the width of vehicle and clearances to be provided for equipment installations and manways. (See Fig. 15-15)

Exterior walls and slabs should be formed from a dense, uniform, carefully cast, high-quality concrete with a recommended minimum strength of 4000 psi. The minimum wall and slab thickness is commonly specified to be 2 feet, and this concrete is carefully placed to ensure uniform density and provide a watertight structure. Waterproofing membrane is not commonly used except for roof slabs of station areas where water infiltration would damage station finish materials. In station roof areas existing practice varies from application of butyl sheet membrane, which was specified in San Francisco, to a brush-on coat of asphalt dampproofing used in Toronto. For waterproofing, where considered necessary, see Chapter 14.

Expansion or construction joints are located at regular intervals to control shrinkage and thermal stresses and to minimize cracking. Actual construction practice varies with different authorities. In San Francisco construction joints were provided at 35- to 50-foot spacings to control shrinkage stresses and minimize cracking. All construction joints were bonded and provided with non-metallic waterstops. It was considered that this construction would adequately allow for shrinkage and that, due to the relatively uniform internal temperatures in the subway structure, thermal stresses need not be considered. This construction was considered to reduce the probability of water infil-

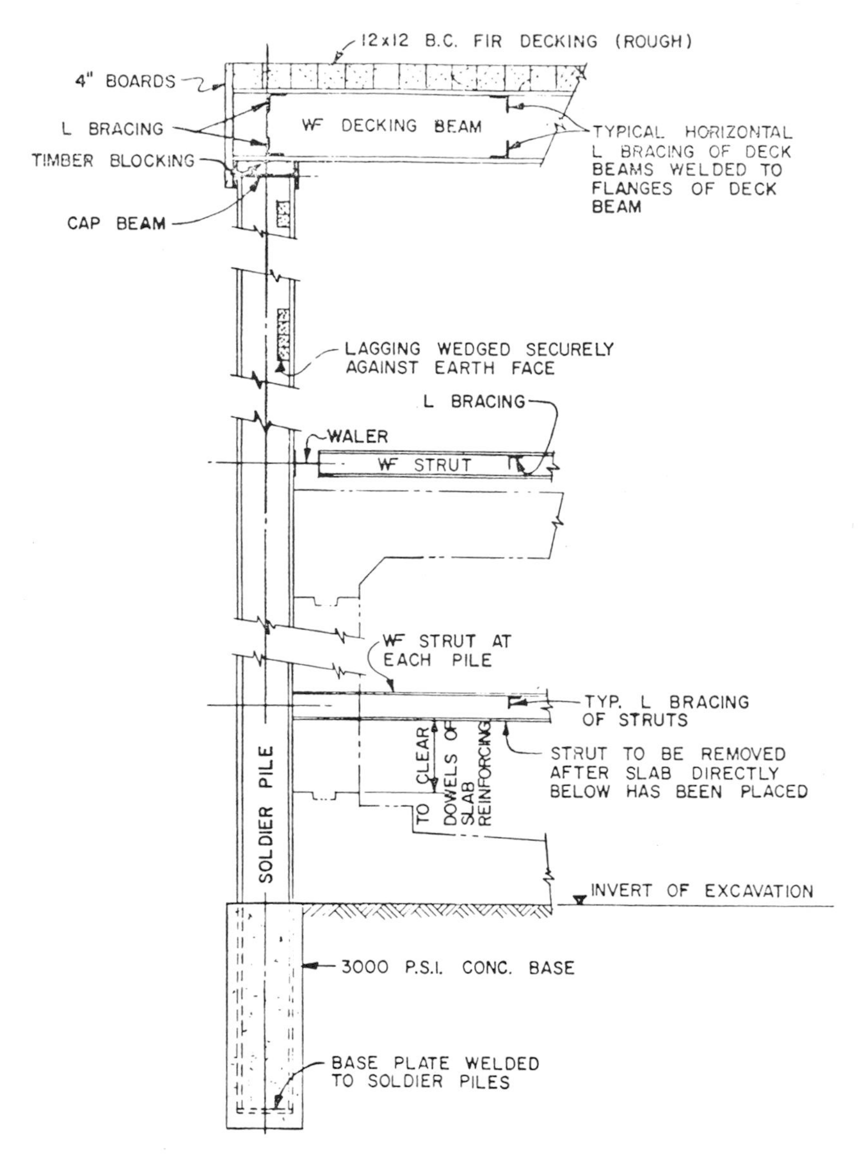

Fig. 15-15. Typical decking arrangement.

tration through elimination of expansion or contraction joints. Designers in Washington provided waterstopped unbonded construction joints at a maximum spacing of 50 feet to accommodate shrinkage and thermal strain. Further waterproofing protection was provided through installation of a bentonite-filled tube embedded into the concrete at the exterior face of the joint. In Toronto, $\frac{1}{4}$-inch expansion joints are provided at 40- to 65-foot spacing to allow both contraction and expansion of the structure due to shrinkage and thermal stresses. The joints are provided with plastic waterstops to prevent groundwater infiltration. Further protection at station areas is provided through installation of a 2-inch deep layer of bentonite powder across the joint for the full width of the structure. (See Fig. 15-16)

The waterstops must be designed to provide a long service life and to ensure a proper installation under difficult construction conditions. Concrete slabs and walls in subway construction may vary from 2 to 5 feet in thickness with the concrete placed by elephant trunk or buckets

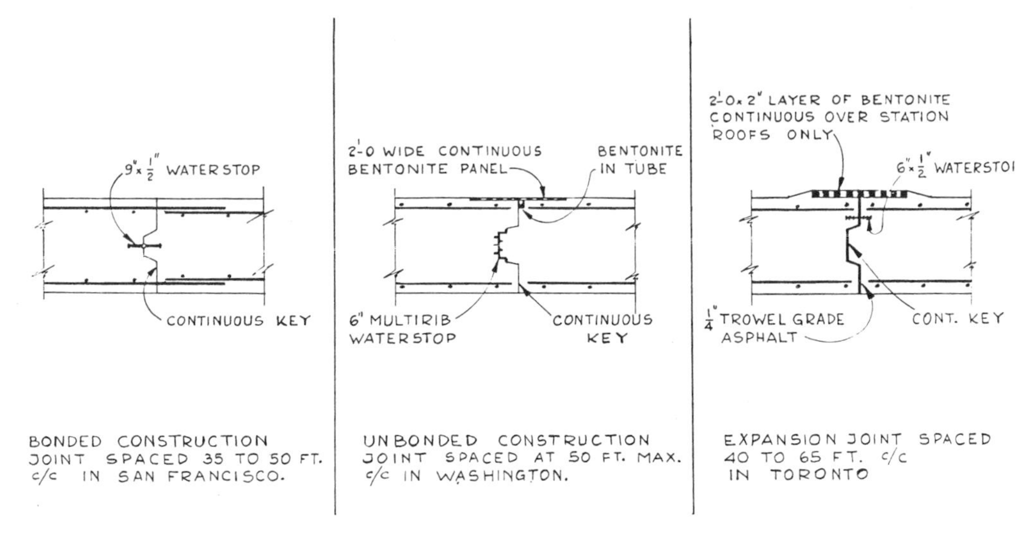

Fig. 15-16. Joints used in subway construction.

up to 2 yards in size. Under these conditions a wide flexible waterstop is easily displaced during construction. Careful attention to the selection of the waterstop and the casting of the concrete is required to ensure a continuously welded waterstop that is embedded in both concrete pours on each side of the joint.

The design of the subway structure must make allowance for:

1. *Dead load*
   Weight of the completed structure
   Weight of the supported backfill
   Horizontal loads form lateral earth pressures
   Horizontal loads from water pressure.
2. *Live loads*
   Subway vehicles
   Subway equipment
   General occupancy usage
   Surcharges from surface traffic loading.
3. *Earthquake forces.*

The weight of the supported backfill may be taken as 130 to 135 lb/foot^3 and the loading calculated by determining the depth of fill from the top of the subway structure to the existing or proposed grade. A uniform surcharge of 200 lb/foot^2 is usually added to this. For covers less than 8 feet over running structure between stations, it is common to design for a minimum loading equivalent to 8 feet of backfill. This design policy adds very little to the cost of the subway and allows for unexpected future surcharge loading and a degree of flexibility in developing future air rights over the subway. Eight feet of backfill is the equivalent loading of an 8- to 10-story building and, after removal, it is possible to support such a building directly onto the subways.

Horizontal earth pressures should be based on recommendations made by a soils engineer. There is a wide range of practice with regard to the application of unbalanced lateral earth pressure-loading. In Toronto the structure is designed for three separate loading combinations:

1. Full vertical load and full horizontal load.
2. Full vertical load, full horizontal load on one side and half horizontal load on the other side.
3. Full vertical load with half horizontal load on both sides.

In San Francisco the subway was designed for full vertical and full horizontal load without allowance for unbalanced horizontal loadings, although consideration was given to horizontal earthquake loading. For further discussion of lateral earth pressure-loading refer to Chapter 14.

Subway vehicle loads are dependent on the

passenger and maintenance vehicles which will be operated within the system. Standard vehicle loadings used in San Francisco, Toronto and Washington are shown in Figure 17. In addition, a loading allowance must be made for dynamic impact, braking and tractive forces and a lateral rolling force resulting from rocking of the vehicle suspension system. The impact loading allowance varies with the type of structure, but for general design purposes may be assumed to be 30% of the static vehicle loading. Braking and tractive forces are horizontal loads that must be applied in the longitudinal direction of the structure at the center of gravity of the transit vehicle. This force may be assumed for general design purposes to act 6 feet above the top of rail and be equal to 25% of the static vehicle loading. The lateral rolling forces are applied downward on one rail and upward on the other and are usually assumed to be equal to 10% of the static vehicle loading.

Except in soft clays the soils surrounding subway structures are found to be sufficiently stiff compared to the structure to preclude any structural overstressing resulting from earthquake design forces. In severe earthquake zones the structure should be designed on the basis of static and live load and checked for seismic requirements.

Twin box construction is the most common type of structure in cut-and-cover subway construction. This section usually involves a double box structure with the trainways located adjacent to each other and is economical and rapidly constructed when use is made of reusable traveling forms. These are easily moved after each concrete pour and are usually made up in three sections, invert forms, wall forms and roof forms. The wall forms and roof forms are designed to travel on wheels or runners which rest upon the invert concrete. (Fig. 15-18)

A variation of the conventional double box structure is the over-and-under box section in which the trainways are located one above the other. This is an economical section for construction requiring a greater depth of excavation but reducing the width of surface disruption and cost of right-of-way.

Slurry wall construction is sometimes employed to avoid costly underpinning where the alignment is near existing buildings. When the walls have been completed, the earth between

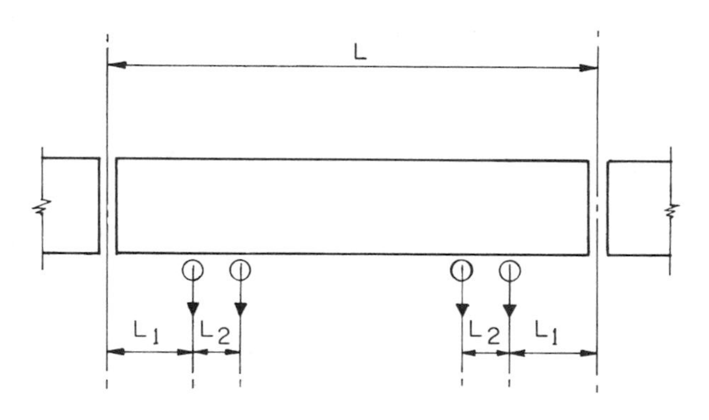

	SAN FRANCISCO	WASHINGTON	TORONTO
L	70'-0	75'-0	75-0
L_1	5'-9	8'-3	8'-0
L_2	8'-0	6'-6	5'-0
TOTAL VEHICLE WEIGHT	50 TONS	60 TONS	70 TONS
AXLE LOAD	25 KIPS	30 KIPS	35 KIPS

Fig. 15-17. Subway vehicle loadings.

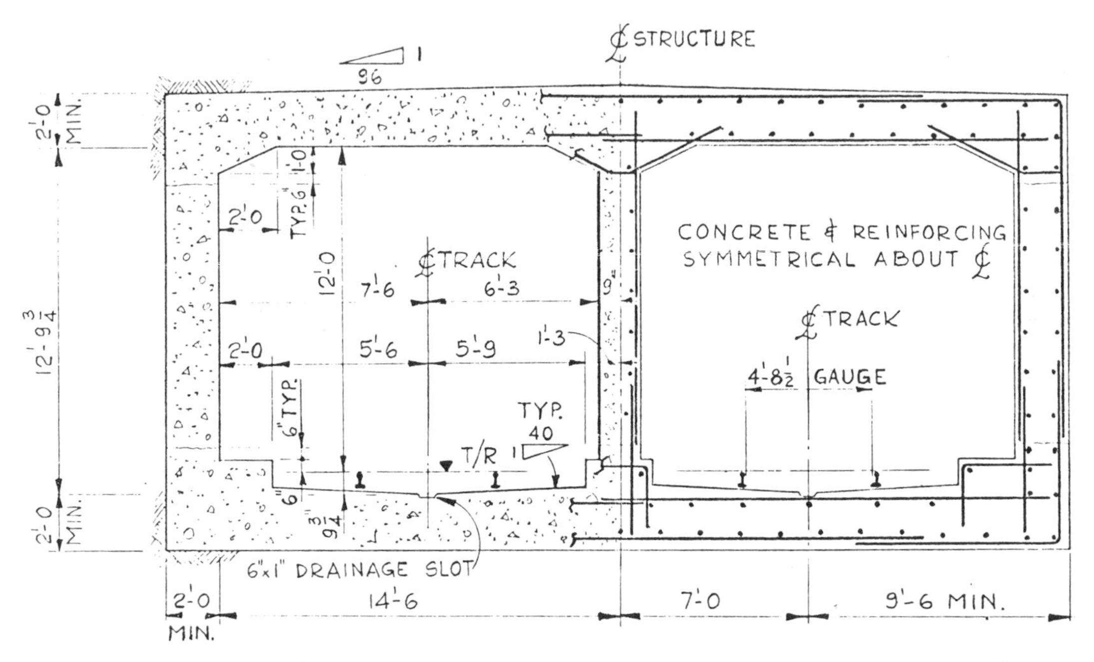

Fig. 15-18. Standard double box section in Washington.

them is excavated to the underside of the roof slab. Using the ground as a form the concrete for the roof slab is then placed. The excavation above the roof may then be backfilled and utilities and the surface restored.

The next stage of construction involves removal of the earth between the subway walls by mining from the ends of the structure until the level of the subway floor is reached. The concrete subway floor is constructed and the rectangular box becomes the subway structure. An experimental section was constructed in Toronto employing this technique, but, due to costs, it was not continued. Nevertheless, this remains a useful tool in subway construction and warrants consideration where expensive underpinning would be required using conventional cut-and-cover constuction. (See Fig. 15-19)

Slurry wall construction may also be used to reduce the duration of surface disruption in heavily traveled traffic arteries. The street surface may be restored while construction continues beneath the subway roof slab to complete the subway structure.

Soldier pile and tremie concrete wall construction was used extensively in San Francisco for the cut-and-cover construction (see Chapter 14). The subway roof slab was cast to bear directly onto this wall and the interior of the tremied concrete wall was finished with a reinforced concrete wall having a minimum thickness of one foot. Through this method of construction, the soldier pile and tremie concrete wall served both as a temporary shoring wall and as part of the permanent structure.

This method of construction offers advantages similar to slurry wall construction in that it permits construction of a rigid cut-off wall in the vicinity of existing buildings, where extensive underpinning may otherwise be required. It also minimizes the duration of construction period by permitting an early restoration of the street surface while construction of the subway continues below the underside of the roof slab. (See Fig. 15-20)

SUBWAY STATIONS

Station structures include the trainway for trains, boarding and off-boarding platforms, stairs and passageways, concourse areas for fare collection, bus transfer platforms and service rooms. If the running structure is circular tun-

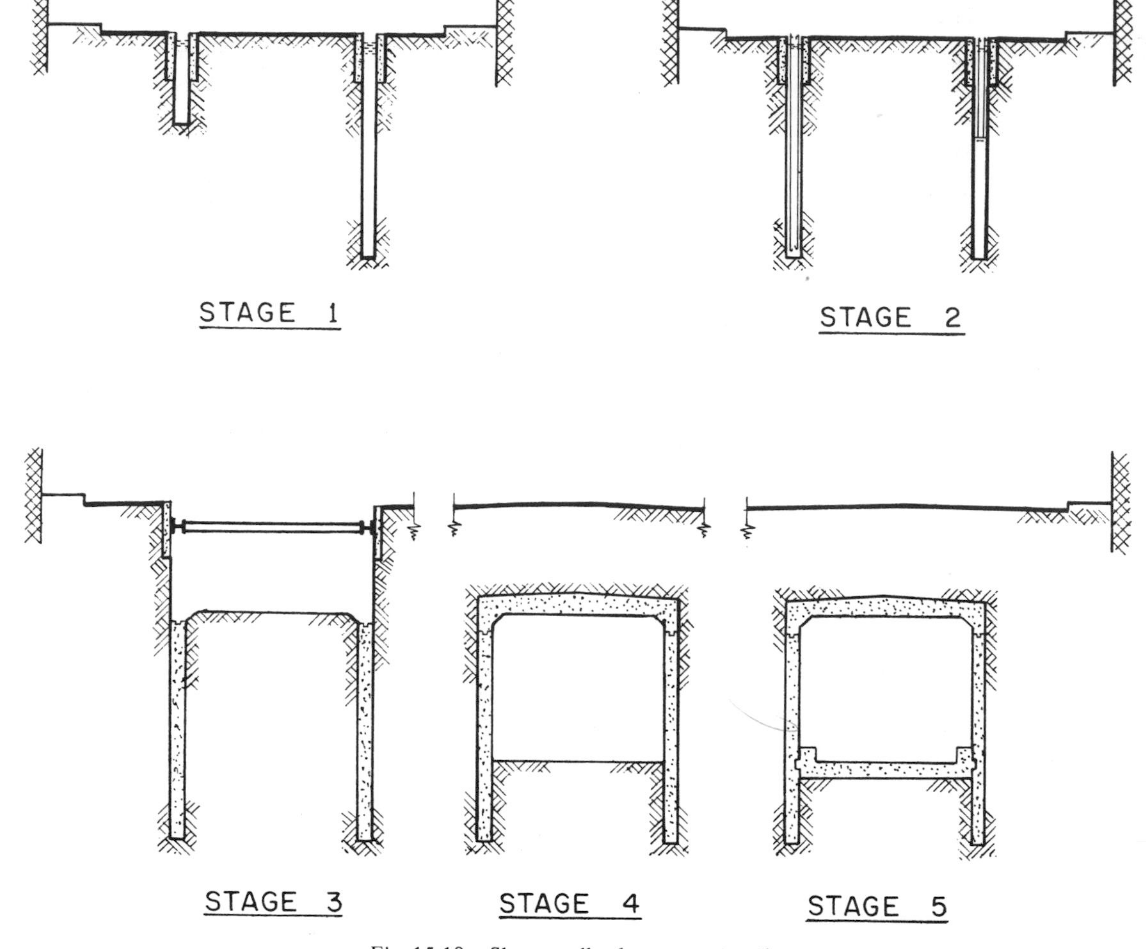

Fig. 15-19. Slurry wall subway construction.

nel, the stations may be constructed using a large diameter tunnel or center platform cut-and-cover structure. On subway lines in which the running structure is constructed in double box structure, the stations are usually of the side platform type, except at terminals where center platforms are normally used.

Tunnel Stations

In deep stations, mining methods have often been used in the station construction using either flexible segmental lining or rigid cast-in-place concrete lining.

The stations are usually split into two levels with an upper level housing the control area concourse and the lower level housing the tracks and platforms. The upper level can usually be located under shallow cover and constructed by cut-and-cover methods. The lower level and the inclined tunnel for stairs and escalators is then constructed by mining methods.

Normally a tunnel shield is employed and, although the mining methods are similar to that used for line structures, there are notable differences in construction staging due to the size of tunnel face and shape of the station.

In station construction in Toronto segmental cast iron liners were used and the joints caulked for watertightness. As a precautionary measure to protect the station finish and to direct any leaks away from the finish, the segments were lined on the inside with a heavy vinyl sheet. The stations consist of two 24-foot diameter tunnels at 48-foot, 3 inch centers which enclose the tracks and train platforms. A shorter 18-

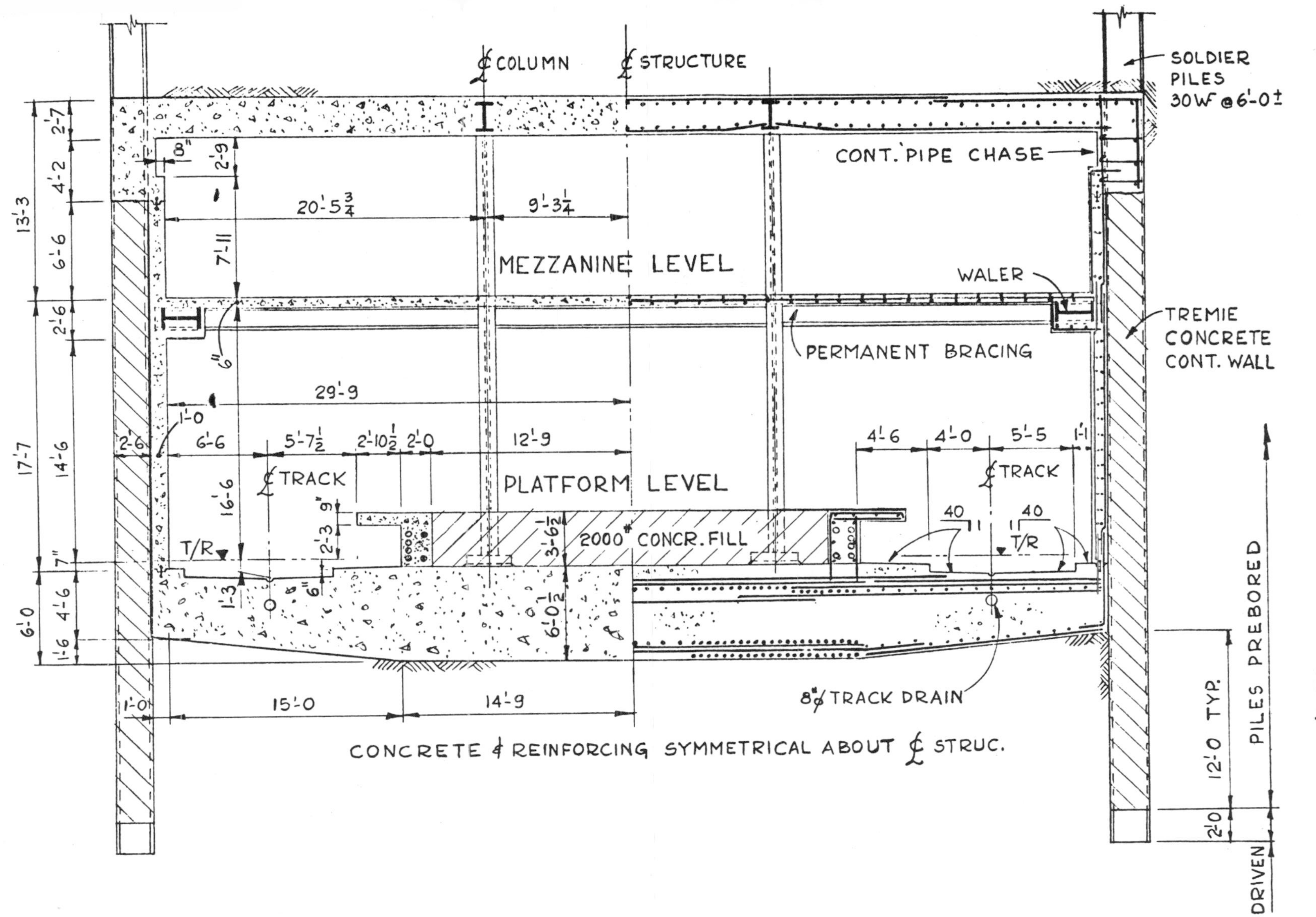

Fig. 15-20. Soldier pile and tremie concrete station structure in San Francisco.

foot diameter tunnel was constructed between the two large tunnels to house connecting passageways, escalator pits and service rooms.

In the construction of the stations the 16-foot diameter running tunnels were initially driven through the station areas. The running tunnel served as a pilot tunnel and was later enlarged to 24 feet. The 24-foot diameter tunnel was driven by shield with the 16-foot diameter liners dismantled as the shield moved forward. The 18-foot diameter service tunnel was constructed by hand, working through headings from the 24-foot tunnels. The inclined stair and escalator tunnels were then hand-constructed using mining methods to the underside of the control area concourse. This concourse was constructed by cut-and-cover methods after completion of the tunnels. (See Fig. 21 and 22)

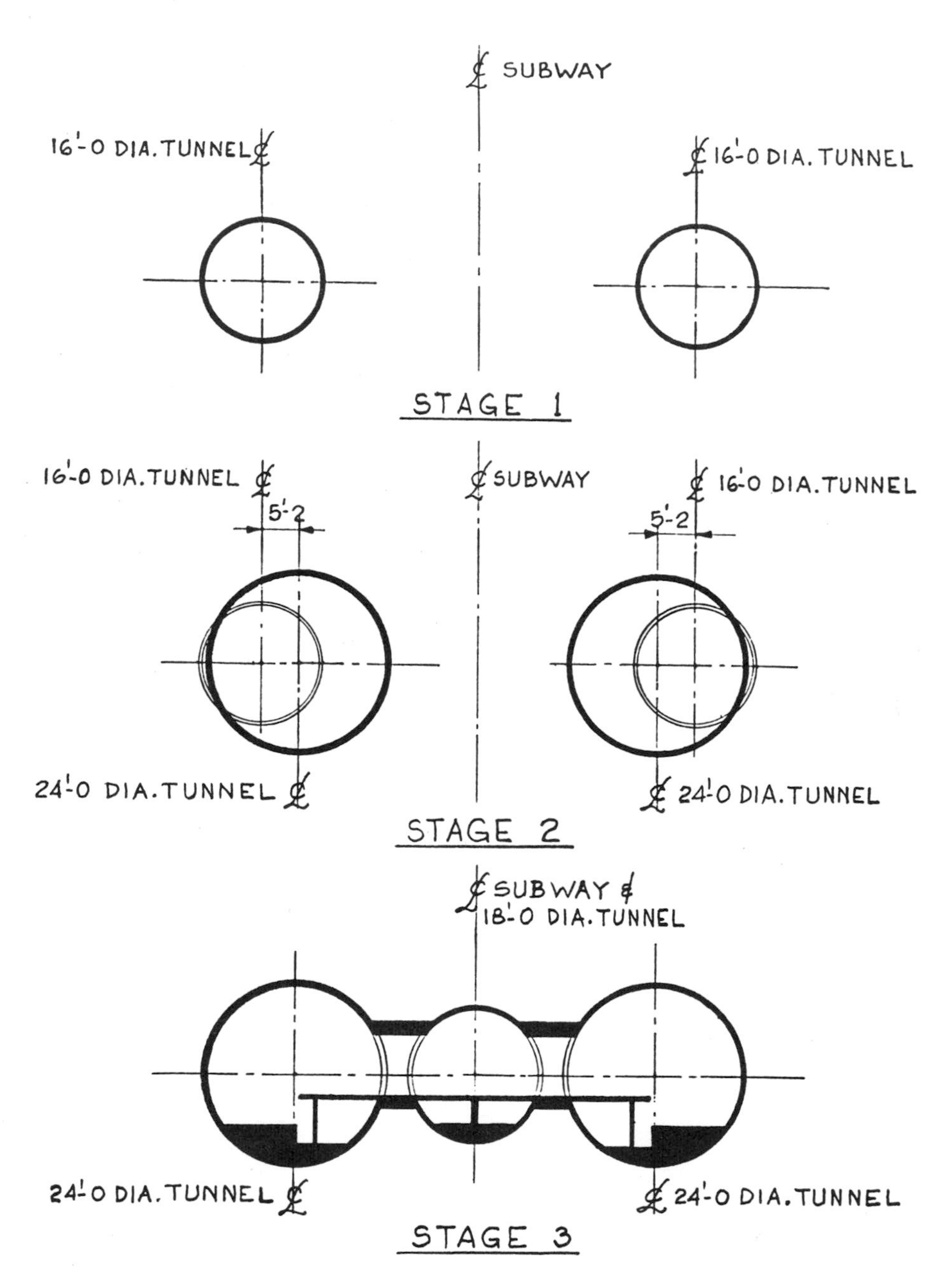

Fig. 15-21. Stages of construction for Toronto subway station.

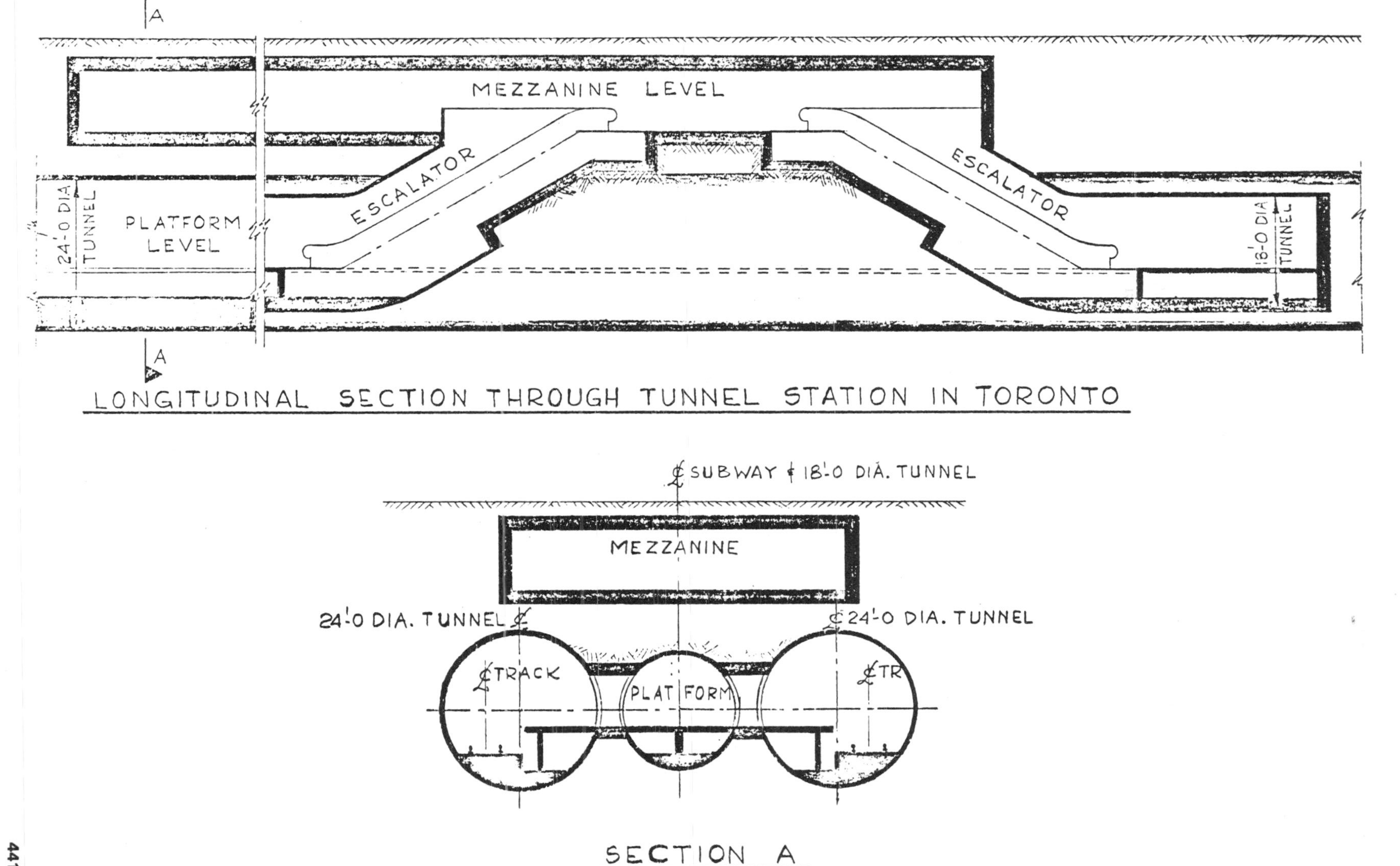

Fig. 15-22. Sections through deep tunnel station in Toronto.

Cut-and-cover Stations

Side platform stations are formed by widening the structure within the station area by about 20 feet to permit a 12-foot wide platform on each side of the two trainways. Stairs and escalators are located at the sides of the platform or off the end leading to an upper level housing the control area concourse. This type of station is commonly used when the running structure is cut-and-cover as it permits the closest possible track center spacing and the narrowest possible structure whenever the tracks are located side by side. (See Fig. 15-23)

The over-and-under station design is a variation of the side platform design adopted were it has been advantageous to reduce the overall width of excavation. The line structure is widened about 12 feet in the station area to provide platforms with stairs and passageways leading to a fare collection concourse. Figure 15-24 indicates a typical over-and-under station with the concourse at the side of the structure. This type of construction avoids the necessity of a mezzanine level over the subway to cross over the trains to gain access to the platforms. (See Fig. 15-24)

Center platform stations are usually constructed at terminals and at line station locations where the running structure has been constructed using separate circular tunnels for each track. At terminals the center platform is preferred because, in many subway systems, trains enter terminal stations on either of the two tracks and passengers are able to board a train on either track from the same platform. In tunnel construction involving a separate tunnel for each track, the center platform station is adopted to provide a safe distance between the two parallel tunnels during mining operations. Stairs and escalators leading to the fare collection concourse may be located either at the end of the platform or within the platform area. If they are located within the platform, a minimum platform width of 30 feet should be adopted to avoid constricted passenger movement around the stair and escalator areas. If a 30-foot platform width is adopted, a width of over 20 feet of undisturbed ground will be left between separate tunnels. This separation ensures that each face can be mined independent of the other, except in the most adverse ground conditions. (See Fig. 15-25)

Platform Dimensions and Exits

The length of train platform is dependent on the length of trains proposed for operation and may be from 20 to 50 feet longer than the longest proposal train. If center platform stations are being considered, the width of platform may vary from 30 to 35 feet. If the platform is too narrow it may constrict patron movements and cause conflict between oppos-

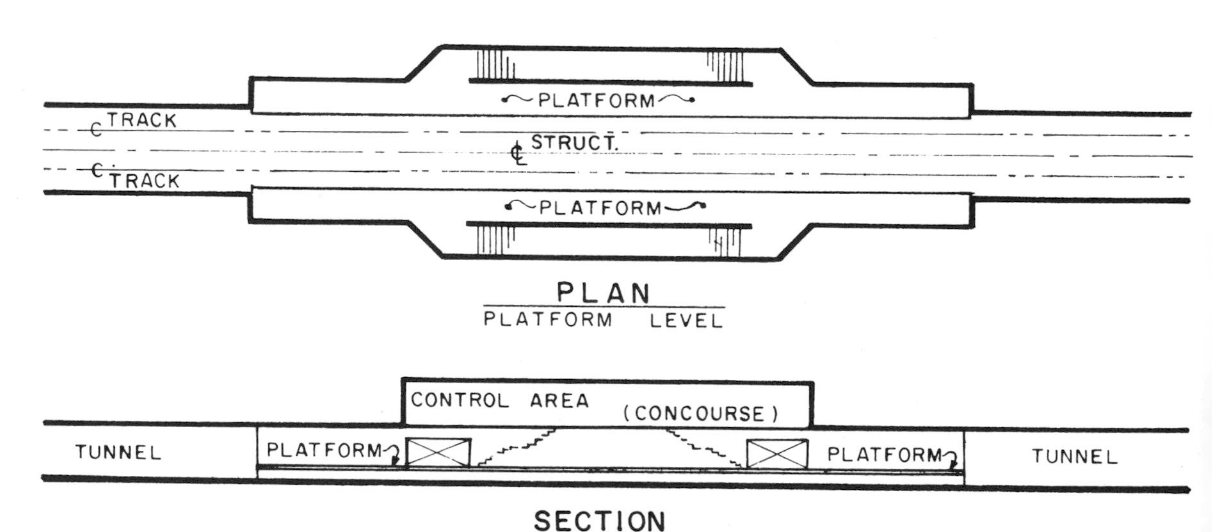

Fig. 15-23. Side platform—twin tunnel station.

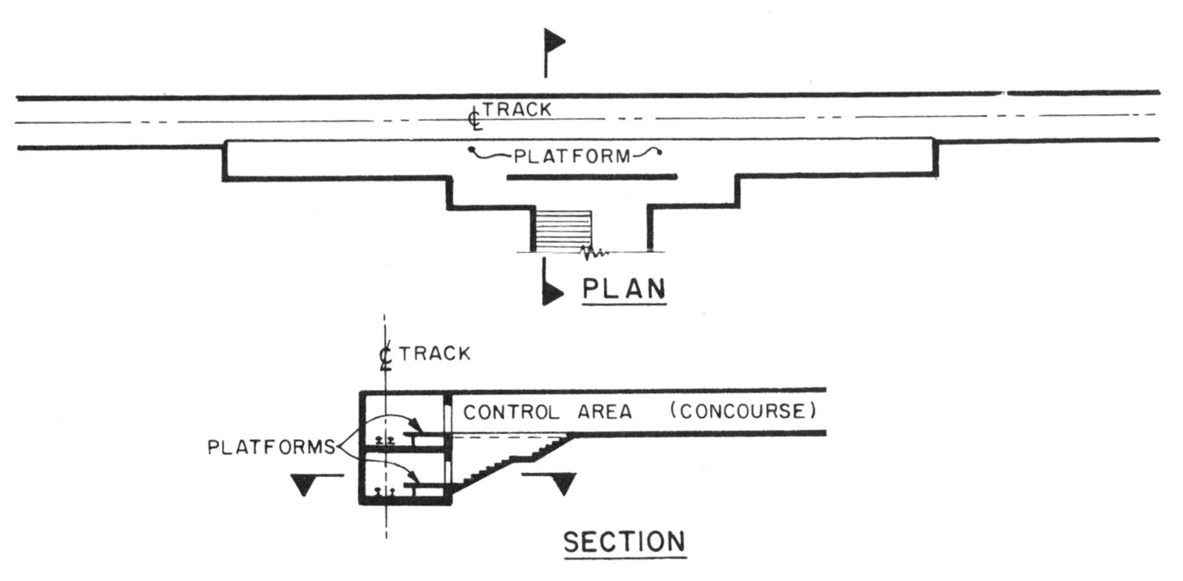

Fig. 15-24. Side platform—over and under tunnel station.

ing traffic directions. This may be a substantial risk to patrons standing near the edge of the platform. A minimum distance of 8 feet is recommended between the edge of platform and the nearest barrier or wall running parallel to the platform edge. If the stairs from the platform are located within the platform area, then a minimum platform width of about 30 feet should be established. This permits a space of 14 feet in which the designer can locate the stairs or escalators and enclosing walls or structure. On side platforms the width may vary from 12 to 15 feet with 12 feet recommended as the minimum safe width.

At least two separate passageways should be provided leading from each train platform to the control area concourse. The passageways should connect to the train platform so that the distance from the end of the platform to the nearest passageway is less than 200 feet. This is the recommended upper limit for the dead-end distance on a subway station platform. Two separate exits should be provided from each platform, but they are usually permitted to

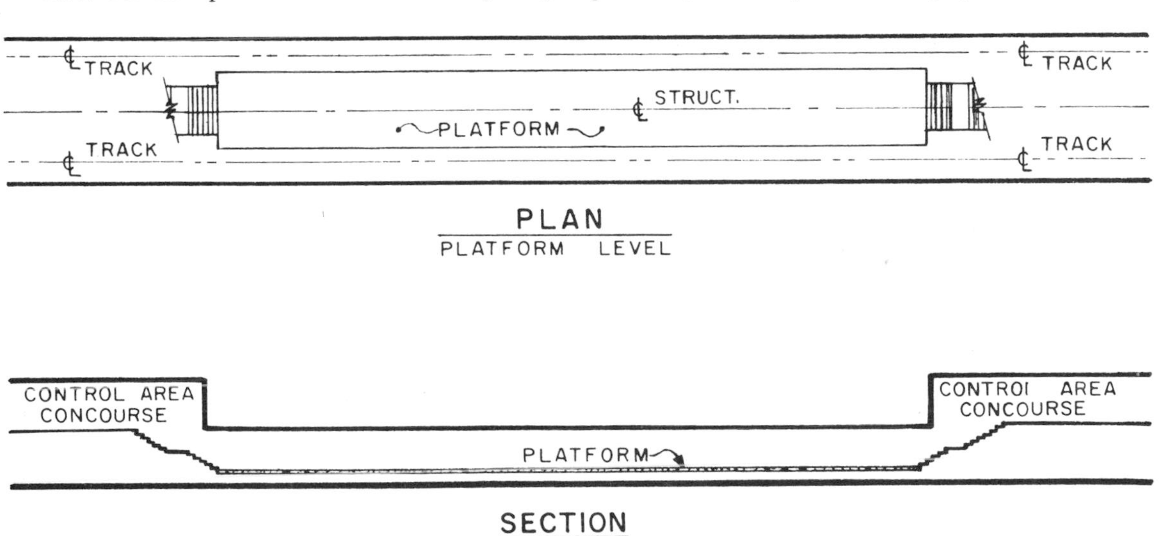

Fig. 15-25. Center platform station.

connect to a single control area concourse to allow grouping of fare collection equipment. Under these conditions the control area concourse will represent the only means of egress from platform by the patrons. This is an acceptable condition where, through the use of noncombustible construction materials in the stairs and passageways, the hazards of a fire in these areas are minimized and smoke control facilities are provided which will prevent migration of smoke from any source into the control area or path of egress. This will require remotely controlled reversible fans and dampers. Where this arrangement of passageways and fare collection equipment is used, care must be taken to ensure that all areas containing a fire load, such as equipment rooms, storage rooms or concessions, are effectively separated from the passageways and concourse by use of a fire separation which will provide a barrier against the spread of fire and smoke into the path of egress.

DRAINAGE

Most tunnels will encounter problems with groundwater infiltration. The extent of the problem will depend on the ground conditions, the type of structure, the quality of construction and the effort made to seal off leaks.

In tunnel construction attempts to seal leaks usually involve caulking of joints from the inside of the tunnel and grouting through grout holes to attempt a seal from the outside. Leaks around bolt holes are sealed using plastic grommets over the bolts.

Leaks are best prevented in cut-and-cover construction through careful attention to the detailing and construction of joints and the use of a uniform, dense, high-quality concrete. Where leaks do occur, attempts to seal these through grouting are usually made.

Despite care during construction and remedial efforts taken after construction, some water infiltration will occur and provision must be made to control it. In addition to infiltration through leaks, drainage water will result from rainfall at open portals, shafts, and stairs and from washwater used in periodic cleaning of the subway structure (floor, walls, roof, etc.). Another source of drainage water could be from an emergency such as street flooding, broken water main or fire-fighting.

A longitudinal drain should be provided along each trainway which leads to a sump at the low point. This may be continuously open trench sized to carry the drainage flow to the sump or a shallower gutter leading to catch basins connected to a drain pipe just below the invert slab. It has been found that a drain pipe of 8-inch diameter at a minimum slope of 1% or an open trench with an area of about 60 square inches sloped at a minimum of 0.3% is adequate under most circumstances. Each sump should be provided with two 500-gallon/minute submersible discharge pumps connected to a municipal sewer.

In addition to track drainage it is necessary that floor drains be adequately provided to prevent seepage water from spilling over large floor areas or collecting at low points and causing offensive odors. Floor drains should be installed adjacent to all outside walls in station areas near points of probable water infiltration (e.g., expansion joints) and at low points in pits. At outside walls, floors should be provided with terrazzo bases, in which small gutters can be tooled to lead seepage water to floor drain locations.

The exterior walls in below-grade structures should have a cavity between the structural wall and the finish wall to avoid moisture damage to the station finish in the event of groundwater infiltration. The base for the wall finish may be sloped with drainage holes provided at the low points to lead the water from behind the finish wall to floor drains.

16
Safety Provisions

ROBERT J. JENNY

President
Jenny Engineering Corporation
South Orange, New Jersey

GENERAL SAFETY RULES

At one time tunnel accidents claimed one life for each one-half mile of tunnel constructed. However, increased concern over construction safety led to improvements in the miner's working conditions, with a subsequent reduction in the frequency of deaths and disabling accidents. This section on safety provisions was prepared while bearing in mind the fact that accidents are not inevitable, and that accident prevention can be profitable by recognizing and eliminating major hazards.

The major causes of accidents are 1) uncontrolled contact between men and materials or equipment: 2) failure of temporary structures; 3) inherent constructional hazards such as the use of explosives; and 4) unsafe practices of individual workers or carelessness. Minimizing these hazards can reduce other indirect accident costs not covered by insurance, such as lost time of injured employee; lost time of co-workers due to general work stoppage when an accident occurs; lost time of supervisory staff to assist the injured, investigate and prepare a report; and the cost of damaged equipment or property. Adherence to safety regulations in the sensitive tunnel construction industry can pay off in increased profits especially well.

Compliance With Official Regulations

The safety rules presented herein are not the sole criteria for safety procedures or requirements. The U.S. Bureau of Mines, state and local codes, as well as the Federal Safety and Health Regulations for Construction issued under the Construction Safety Act of 1969, and the Occupational Safety and Health Act of 1970, including all current amendments (called Federal Regulations), must be complied with when applicable. The more stringent requirements of applicable codes will govern. Some of these requirements have been presented herein to bring the reader up to date on current practice.

Safety Engineer and Safety Program

An incentive safety program designed by a competent safety engineer is recommended in the early stages of construction planning. All employees should be motivated to actively support the program.

All employees should be instructed on the recognition of hazards, the observance of precautions and the use of protective and emergency equipment in enclosed underground spaces. When the employees are underground, an accurate record of their location should be kept on the surface.

On tunnel operations with 25 or more employees underground, at least two rescue crews should be trained in rescue procedures and the use of oxygen breathing apparatus, as well as the use of fire-fighting equipment. At least one crew should be trained in routine safety operations. A safety inspection of the job site, including all materials and equipment, should be made at least once per shift.

In New Jersey, the construction code requires a full-time safety director on projects with more than 25 employees.

Emergency Measures

Tunnel evacuation plans and procedures should be developed and made known. These plans should incorporate a separate communications system independent of the tunnel power supply and provisions for emergency hoisting in shafts.

Protective Clothing

Tunnel personnel should wear protective head gear, footwear and any other special garments that applicable codes require. Safety glasses, rubber gloves, goggles or face shields should be made available for use when required, and should be worn when handling corrosive, toxic, or injurious substances. Working molten metal (such as welding) requires fire-resistant clothing and face shields. Moving machinery necessitates the wearing of snug-fitting clothes, and sharp, rough and splintery material handling requires protective gloves. In New York the state's Industrial Code Rule governing tunneling operations (called New York State Regulations) also requires waterproof clothing in wet areas and safety belts for shaft workers where the drop is 10 feet or more.

General Precautions

Each specific working area in tunnel construction has unique hazards which should be mitigated.

Miscellaneous Areas. Safe means of access and egress should be available in all working areas and all ladders and stairways should comply with applicable code requirements. Subsidence areas presenting safety hazards should be fenced and posted. In New Jersey, the Construction Safety Code requires the surface working site to be walled or fenced to a height of 8 feet.

Tunnel. The roof, face and walls of rock work areas should be tested frequently and loose rock scaled down or supported. Rock bolts should be tested frequently for proper torque with a calibrated torque wrench. All steel or timber sets should be installed to prevent movement into the tunnel and lateral bracing should be provided between sets to further stabilize the support. Damaged or dislodged tunnel supports must be repaired or replaced immediately. Proper designed shields, forepoling and other devices should be provided as required for soft ground tunneling. Suitable provisions for breasting the face should also be incorporated (see Chapter 5).

Walls, ladders, timbers, blocking, wedges and supports should be inspected for loosening following blasting operations. Where found unsafe, corrections must be made immediately.

Safety belts should be worn by crews on skips and platforms in shafts when the clear distance between the skip or cage and the sides of the shaft is greater than 1 foot unless guard rails or cages are provided. The New York State Regulations provide much more extensive and detailed precautions to be observed in shafts and hoisting operations.

Caissons. In caisson work in compressed air a protective bulkhead should be erected in the working chamber when the chamber is less than 11 feet in height and the caissons are at any time suspended or hung so that the bottom of the excavation is more than 9 feet below the deck of the working chamber while work is in progress. Shafts must be made tight and hydrostatic or air pressure tested. The test pressure must be stamped on the outside shell about 12 inches from each flange. Accurate and accessible gauges should be located in the lock and on either side of each bulkhead. Caissons greater than 10 feet in diameter or width should be provided with a man lock and shaft exclusively for employees.

LOCALIZED OPERATIONAL HAZARDS

All equipment to be used during a shift should be inspected before use.

Blasting

Blasting operations must comply with Subpart U of the Federal Regulations, and with state or local codes governing explosives and blasting procedures. Underground transportation and handling of explosives is described later in this Chapter. The New Jersey Department of Labor and Industry Safety Regulations No. 23 governing the use of explosives is an excellent guide for engineers and contractors and is highly recommended for review whenever the use of explosives is contemplated.

Drilling

Before drilling the roof must be scaled down by an experienced miner and the area inspected and made safe from all potential hazards. The men working below jumbo decks should be warned of the possibility of residual explosives. Lifter holes should not be drilled through blasted rock or water and air lines buried in the invert should be identified by signs. No one should be allowed on a drill boom while the drill is in operation and no one except the operator should be allowed on a moving jumbo. Jumbos should have storage receptacles for drill steel, a mechanical heavy materials lifter, stair access to decks for at least two persons and removable guard rails on all sides and at the back of platforms if the deck is over 10 feet high. When a jumbo is being moved, equipment should be secured and the booms should be in a safe position. Drills on columns should be anchored firmly before drilling and retightened frequently. Scaling bars should always be sharp and in good condition.

Hauling

Powered mobile equipment should be provided with adequate brakes, audible warning devices and lights at each end. Visible or audible warning should be given before equipment is started and moved. Cabs should have clean windows constructed of safety glass.

Adequate backstops or brakes should be installed on inclined conveyor drive units. No one should ride on power-driven chains, belts or bucket conveyors, in dippers, shovel buckets, forks or clamshells, in the beds of dump trucks or on haulage equipment.

Electrically powered mobile equipment should not be left unattended unless the master switch is in the off position, all operating controls are in the neutral position and the brakes are set. Parked railcars should be blocked securely. Means should be provided to prevent overtravel and overturning at dumping locations and at all track dead ends. Rocker-bottom or bottom dump cars should have positive-locking devices. Supplies, materials and tools other than hand-tools should not accompany employees on man-trip cars. Equipment which is being hauled should be protected against sliding or spillage.

Hoisting

Hoisting machines should be worm-geared or powered both ways, and if the power is stopped, the load should not move. Power hoist controls should have a non-locking switch or control and a device to deactivate the power should be installed ahead of the operating control. Hoist machines with cast metal parts should not be used. All anchorages of hoists should be inspected at the beginning of each shift, and every hoist should be annually tested to at least twice the maximum load.

An enclosed covered metal cage designed with a safety factor of 4 should be used to raise and lower personnel in the shaft. The cage must be load-tested prior to use and the exterior should be free of projections or sharp corners. Only closed shackles should be used in the cage rigging and a positive locking device should be installed on the cage to prevent the door from opening accidentally while in operation.

Maximum rates of speed for transporting persons should be established and adhered to, and signal codes should be employed in the operation of the hoist.

The New York State Regulations provide the following factors of safety for hoisting rope for passenger hoists.

Car Speed (feet/minute)	Factor of Safety
50	7.5
100	7.9
200	8.6
300	9.2
400	9.8
500	10.3
600	10.7
700	11.1
800	11.3
900	11.4
1000	11.6
1200	11.8
1500	11.9

FIRST AID STATION

Equipment

A weatherproof first aid kit should be provided containing materials recommended by the consulting physician which conform to Red Cross standards, with individual sealed packages for each item. The contents of the first aid kit should be checked by the employer before being released for use and at least weekly to ensure that expended items are replaced. Equipment for prompt transportation of an injured person to a physician or hospital, or communications for ambulance service, should be provided. The New York State Regulations also require that blankets and one stretcher per 100 workers underground be made available.

Attendent(s)

Sufficient competent personnel (with at least one person currently certified in first aid training by the U.S. Bureau of Mines or the American Red Cross) should be available either on or near the work site to perform first aid treatment or any rescue work that may be required in the tunnel.

Medical Service

Provisions should be made to ensure the availability of medical personnel for continual consultation and of prompt medical attention in case of serious injury. The telephone numbers of the doctors, hospitals and ambulances should be conspicuously posted.

FIRE HAZARDS

Limitations on Combustible Materials

The carrying of matches, lighters or other flame-producing smoking materials should be prohibited in all underground operations where fire or explosion hazards exist. Gasoline or liquefied petroleum gases should not be taken underground. Paper, combustible rubbish or scrap wood should not be allowed to accumulate. Only a one day's supply of diesel fuel should be stored underground, and oil, grease or fuel should be well sealed and kept a safe distance from sensitive areas. Only approved fire-resistant hydraulic fluids should be used in hydraulically operated equipment. Air that has passed through underground oil or fuel storage areas must not be used to ventilate working areas. When compressed gas cylinders are being moved to a new location underground, the safety caps for protecting the cylinder valves should be secured in place.

Limitations on Burning and Welding

Noncombustible barriers should be installed below welding or burning operations in or over a shaft or raise. During, and for thirty minutes after welding or flame cutting underground, a person with a fire extinguisher should be ready to extinguish any fire.

Fire-fighting Equipment

Fire extinguishers should be provided at the head and tail pulleys of underground belt conveyors, at 300-foot intervals along the belt line, and wherever combustible materials are stored. These extinguishers must be suitable for extinguishing fires of wood, oil, grease and electrical equipment.

Smoke Masks and Inhalators

Bureau of Mines approved self-rescuers (in good condition) should be available near the advancing heading on the haulage equipment and in areas where employees could be trapped by smoke or gas.

Electrical Equipment

Electrical cables should not be taken or used underground unless they have an armored non-combustible casting or jacket continuously throughout their length. Powerlines should be well separated or insulated from water lines, telephone lines and air lines. Oil-filled transformers should not be used underground unless they are in a fire-resistant enclosure to contain the contents in the event of a rupture.

Caution must be exercised in relying on some local codes governing electrical equipment; for example, the New York State Regulations are far more comprehensive with regard to electrical equipment safety in tunnels than many other codes.

VENTILATION DURING CONSTRUCTION

Detection of Noxious and Explosive Gases

Testing for the presence of gases should be conducted frequently. The allowable quantities below were taken from "Threshold Limit Values," Safety Regulation No. 3, by the State of New Jersey Department of Labor and Industry.

Carbon monoxide	50 ppm maximum allowed
Carbon dioxide	5000 ppm maximum allowed
Methane	1% maximum allowed
Nitrogen dioxide	5 ppm maximum allowed
Nitric Oxide	25 ppm maximum allowed
Hydrogen Sulfide	10 ppm maximum allowed

The presence of other flammable or toxic gases, dusts, mists and fumes should be determined and if 1.5% or higher concentrations are detected, employees should be evacuated and power cut off to the affected area until concentrations are reduced to 1% or less. A record of all tests should be kept on file. Field oxygen analyzers may be used to test for oxygen deficiency below a minimum value of 20%.

Air Quality Maintenance

Tunnels should be provided with mechanically induced reversible-flow primary ventilation for all work areas. Ventilation doors should be self-closing and remain closed regardless of direction of flow. When primary ventilation has been discontinued for any reason, employees should be evacuated and qualified personnel should examine the tunnel for gas and other hazards before activating power or readmitting employees to the work areas.

The supply of fresh air should not be less than 200 cubic feet per minute per employee and the velocity should be at least 30 feet per minute where conditions can produce harmful dusts or gases. Respirators should not be used in place of environmental controls except in welding, blasting and lead-burning operations. Internal combustion engines other than mobile diesel should not be used underground. After blasting, smoke and fumes should be exhausted through the vent line to the outside air.

HANDLING AND STORAGE OF EXPLOSIVES

Handling and Transportation Underground

The blaster should be fully qualified to handle and use explosives safely as required.

Additional explosives or blasting agents should not be taken underground beyond the amount required for one blast. Explosives and blasting agents should be hoisted, lowered or conveyed in a powder car. The hoist operator should be notified before explosives are transported in a shaft and personnel, materials, supplies, detonators or equipment should not be transported in the same conveyor with the explosives. Explosives should not be transported in a locomotive during man-haul-trip, or in an unmarked vehicle. A physical separation of 24 inches should divide the compartments of detonators and explosives in a vehicle, and no one, except the operator, his helper and the powderman should be permitted to ride in a vehicle or train transporting explosives and blasting agents.

Trucks used for the transportation of explosives underground should have the electrical system checked weekly for electrical hazards. A written record of such inspections should be kept on file.

Storage of Explosives and Detonators

Explosives must be stored in the types of facilities required by the Internal Revenue Service Regulations, 26 CFR 181, Commerce in Explosives.

Permanent storage of explosives or blasting agents should not be permitted in any underground operation until at least two modes of exit have been provided. However, New Jersey and New York State Regulations, for example, prohibit storage of explosives in tunnels where men are employed. Where permanent storage is allowed, magazines should be at least 300 feet from any shaft or active area.

Smoking and open flames should not be permitted within 50 feet of explosives and detonator storage magazines. Blasting caps, detonating primers and primed cartridges should not be stored in the same magazine with other explosives or blasting agents. Nor should detonator magazines be located closer than 50 feet to a magazine containing explosives.

INACTIVE HEADINGS

Access to unattended underground openings should be restricted by gates. Unused chutes, manways or other openings should be tightly covered, bulkheaded or fenced off and posted.

Short Duration

Should tunneling operations be interrupted, the heading should be securely supported. Hydraulic pressure or collapsible rams or struts should not be used for securing the faces of inactive headings. If a shield invert is below the water table, watchmen should be on duty at the heading at all times when excavation is suspended. However, closed circuit television may be used in some cases to monitor headings in lieu of maintaining the watchmen underground.

Long Duration

If a tunnel is inactive for a relatively long period of time, it is recommended that a bulkhead with a valve be installed at the face. The valve should be opened to check for water pressure or noxious fumes before the bulkhead is removed.

COMPRESSED AIR WORK

Medical Regulations

A licensed physician qualified and experienced in treating decompression illness and willing to enter a pressurized chamber should be available whenever work is in progress. In addition, a fully equipped first aid station and a vehicle equipped with one litter should also be available.

Prospective compressed air workers should be examined by the physician to determine if they are physically qualified for the work. If a worker is ill or injured, or has been absent for ten days, he should be examined before returning to work under compressed air. In addition, compressed air workers should be re-examined at least annually. These examinations should be kept on record along with any decompression illnesses reported by the physician. The records should be available for inspection and a copy should be sent to the Bureau of Labor Standards following a death, accident, injury or decompression illness.

Under the new Federal Regulations, a medical lock must be maintained whenever air pressure in the working chamber is increased above the atmospheric pressure. The lock must conform to the specifications in the Regulations.

Identification badges should be furnished to the compressed air workers indicating the worker's name, the nature of his job, the address of the medical lock, the phone number of the physician on the project and instructions that, in case of unknown or doubtful cause of illness, the wearer should be rushed to the medical lock. The badge should be worn on and off the job at all times.

Records and Communication

There should be one person present representing the employer who is knowledgeable about and responsible for complying with compressed air regulations. Every employee should be instructed in the rules concerning safety. The time of decompression should be posted on each lock as outlined in the decompression tables in this section. Also, signal codes should be posted at work place entrances. Communications should be maintained at all times among the following locations: working face, work chamber side of the man lock door, the man lock, lock attendant's station, compressor plant, first aid station, emergency lock and special decompression chamber.

For each shift, a record of each employee's time under air pressure and his decompression time should be kept by an employee remaining outside the lock near the entrance, and a copy should be submitted to the physician after each shift.

Compression and Decompression

Every employee going under air pressure for the first time should be instructed on how to avoid excessive discomfort.

Compression procedure:

First minute: Up to 3 psig maximum, hold to determine if any discomfort is experienced.

Second minute: Raise uniformly at a maximum rate of 10 psi/minute.

When employees signal discomfort, hold the existing pressure for 5 minutes. If the discomfort has not ceased after 5 minutes, reduce pressure gradually until it has ceased. If the discomfort persists, release employee from lock. No one should be subjected to a pressure greater than 50 psi. Decompression to normal atmospheric conditions must be in accordance with decompression tables[1] (see Table 16-1).

Table 16-1 shows the total decompression time for gauge pressures from 0 to 50 psi and working periods from $\frac{1}{2}$ to 8 hours and over.

Decompression proceeds by two or more stages, with a maximum of four for a working chamber pressure of 40 psi or over.

Stage 1 consists of a reduction in ambient pressure ranging from 10 to a maximum of 16 psi, but in no instance will the pressure be reduced below 4 psi at the end of stage 1. This reduction in pressure in stage 1 will always take place at a rate not greater than 5 psi per minute.

Further reduction in pressure will take place during stage 2 and subsequent stages as required at a slower rate, but in no event at a rate greater than 1 psi per minute.

If repetitive exposure to compressed air is required (more than once in 24 hours), the physician should establish and be responsible for compression and decompression procedures. The physician is also responsible for decanting methods, if these methods are required. In decanting, no longer than 5 minutes should elapse in atmospheric pressure before recompression.

Man Locks and Muck Locks

Decompression of employees from a compressed air atmosphere must always take place, except in emergency. Except when the air pressure is below 12 psig and there is no danger of rapid flooding, each bulkhead in tunnels of 14 feet or more in diameter, or an equivalent area, should have at least two locks—one a man lock, the other, a materials lock. If only a combination man and materials lock is required, the lock should be able to hold an entire heading shift. A lock attendant, responsible to the physician, should be at the controls of the man

Table 16-1. Total Decompression Time (Minutes).

WORK PRESSURE PSIG	WORKING PERIOD (HOURS)										
	$\frac{1}{2}$	1	$1\frac{1}{2}$	2	3	4	5	6	7	8	OVER 8
0–12	3	3	3	3	3	3	3	3	3	3	3
14	6	6	6	6	6	6	6	6	16	16	33
16	7	7	7	7	7	7	17	33	48	48	62
18	7	7	7	8	11	17	48	63	63	73	87
20	7	7	8	15	15	43	63	73	83	103	113
22	9	9	16	24	38	68	93	103	113	128	133
24	11	12	23	27	52	92	117	122	127	137	151
26	13	14	29	34	69	104	126	141	142	142	163
28	15	23	31	41	98	127	143	153	153	165	183
30	17	28	38	62	105	143	165	168	178	188	204
32	19	35	43	85	126	163	178	193	203	213	226
34	21	39	58	98	151	178	195	218	223	233	248
36	24	44	63	113	170	198	223	233	243	253	273
38	28	49	73	128	178	203	223	238	253	263	278
40	31	49	84	143	183	213	233	248	258	278	288
42	37	56	102	144	189	215	245	260	263	268	293
44	43	64	118	154	199	234	254	264	269	269	293
46	44	74	139	171	214	244	269	274	289	299	318
48	51	89	144	189	229	269	299	309	319	319	
50	58	94	164	209	249	279	309	329			

lock whenever men are in the working chamber or in the lock.

If the air pressure is 12 psig or above, decompression must be regulated by automatic controls, supplemented by manual controls to allow the lock attendant to override the automatic controls if required. Manual controls for an emergency must also be provided inside the man lock. The man lock must contain the following equipment: a clock and a continuous recording pressure gauge outside the lock, a pressure gauge, a clock and a thermometer inside the lock. In addition, 4-inch minimum diameter observation ports should be installed so the lock occupants can be observed from the chamber and free air side. Ventilation should be provided and the temperature should be at least 70°F in the lock. The lock must contain 30 cubic feet of air space per occupant, and have 5 feet clear headroom minimum at the center. Also, each bulkhead should have a pressure gauge on both faces.

When locks are not in use and employees are in the working chamber, lock doors should be kept open to the working chamber. Also, if the working force is disabled, provisions should be made for rescue parties to enter the tunnel quickly.

A special decompression chamber to accommodate the entire force of employees being decompressed at the end of a shift should be provided whenever the time of decompression exceeds 75 minutes. This chamber is commonly known as the "Luxury Lock."

Special Decompression Chamber

The headroom in the special decompression chamber should be at least 7 feet. For each person, there should be 50 cubic feet of airspace, 4 square feet of walking area and 3 square feet of seating space exclusive of lavatory space. The rated capacity of the chamber will be based on the stated minimum space per employee and should be posted. The capacity should not be exceeded, except in case of emergency. Each special decompression chamber should be equipped with the following: clocks, pressure gauges, valves to control the supply and discharge of air, an oral communication system among the occupants, attendant and compressor plant, and an observation port at the entrance.

Seating space, not less than 18 by 24 inches wide, should be provided per occupant, and normal sitting posture permitted. Proper and adequate toilet and washing facilities, in a screened or enclosed recess should also be provided. Fresh, pure drinking water should be available. Community drinking vessels should be prohibited.

Unless the special decompression chamber is serving as the man lock to atmospheric pressure, the chamber should be adjacent to the man lock on the atmospheric pressure side of the bulkhead.

Compressor Plant

At all times a thoroughly experienced, competent and reliable person should be on duty at the air control valves as a gauge tender, regulating the pressure in the working areas. During tunneling operations, one gauge tender only should regulate the pressure in two headings, provided the gauge and controls are all in one location.

The low air compressor plant capacity should permit the work to be done safely and provide a margin to meet emergencies and repairs. Low air compressor units should have at least two independent sources of power supply. The compressors should be of sufficient capacity to maintain the necessary pressure in the working chamber even during periods of breakdown, repair or emergency.

Switching from one independent source of power supply to the other should be done periodically to ensure the workability of the apparatus in an emergency. Duplicate low-pressure air feedlines and regulating valves should be provided between the source of air supply and a point beyond the locks, with one of the lines extending to within 100 feet of the working face. All high- and low-pressure air supply lines should be equipped with check valves. Low-pressure air will be regulated automatically but manual valves should be provided for emergency.

The air intakes should be located at a place where fumes, exhaust gases and other air contaminants will be at a minimum. Gauges indicating the pressure in the working chamber should be installed in the compressor building, the lock attendant's station and the employer's field office.

Bulkheads and Safety Screens

Intermediate bulkheads with locks, or intermediate safety screens or both, are required where there is a danger of rapid flooding. The New York Regulations limit the length between work face and bulkhead to 1000 feet if the possibility of rapid flooding exists.

In tunnels 16 feet or more in diameter, where there is a danger of rapid flooding, hanging walkways should be provided from the face to the man locks as high in the tunnel as practicable, with at least 6 feet of head room. Walkways should be constructed of noncombustible material. Standard railings should be securely installed throughout the length of all walkways on open sides. Where walkways are ramped under safety screens, the walkway surface should be skid-proofed by cleats or by equivalent means. Bulkheads used to restrain compressed air should be tested to prove their ability to resist the highest air pressure expected to be used.

Ventilation and Air Quality

The working chamber should be well ventilated. The air in the work areas should be analyzed at least once per shift and a record of analyses kept. Test results must fall within the threshold limit values set forth by the applicable regulations, otherwise immediate corrective action should be taken. During the entire decompression period, forced ventilation of fresh air must be provided.

Whenever heat-producing equipment is used, a positive means of removing the head build-up at the heading should be provided. The temperature of all working chambers should be maintained at temperatures not in excess of 85°F.

Sanitation

Clean, heated, lighted and well-ventilated dressing rooms and drying rooms should be provided for all employees engaged in compressed air work. Such rooms should contain suitable benches and lockers. Bathing accommodations (showers at the ratio of 1 to 10 employees per shift), equipped with running hot and cold water and with suitable toilet accommodations (1 toilet for each 15 employees per shift) should also be provided. All parts of caissons and other working compartments should be kept in a sanitary condition.

Fire Prevention and Protection

Proper fire-fighting equipment must be available for use at all times. While welding or flame-cutting is being done, a firewatch with extinquisher should stand by. Shafts and caissons containing flammable material of any kind should be provided with a fire hose arranged so that all points of the shaft or caisson are within reach of the hose stream.

Tunnels should be provided with a 2-inch minimum diameter water line extending into the working chamber and to within 100 feet of the working face. The line should have hose outlets with 100 feet of fire hose attached and maintained as follows: one at the working face and one immediately inside of the bulkhead. In addition, hose outlets should be provided at 200-foot intervals throughout the length of the tunnel, and 100 feet of fire hose should be attached to the outlet nearest to the location of flammable material or any area where flame is being used.

Fire hose should be at least $1\frac{1}{2}$ inches in nominal diameter and the water pressure and supply should at all times be adequate for efficient operation of the type of nozzle used. The powerhouse, compressor house and all buildings housing ventilating equipment should have at least one hose connection in the water line. A fire hose should be maintained within reach of wood structures over or near shafts. The compressor building should be constructed of noncombustible material.

In addition to the fire hose protection required on every floor of every building used in connection with compressed air work, there should be at least one approved fire extinguisher of the proper type for the hazard involved. At least two approved fire extinguishers should be provided in the working chamber as follows: one at the working face and one immediately inside the bulkhead (pressure side). Extinguishers in the working chamber must use water as the primary extinguishing agent and may not use any extinguishing agent which could be harmful to the employees in the working chamber. Highly combustible materials should not be used or stored in the working chamber.

Man locks should be equipped with a manual fire extinguishing system that can be activated inside the man lock and also by the outside lock attendant. In addition, a fire hose and portable fire extinguisher should be provided inside and outside the man lock. The portable fire extinguisher should be the dry chemical type. Equipment, fixtures and furniture in man locks and special decompression chambers should be constructed of noncombustible materials. Bedding and like materials must be chemically treated to be fire-resistant. Head frames should be constructed of structural steel or open framework fireproofed timber. Temporary surface structures within 100 feet of the shaft, caisson or any tunnel opening should be built of fire-resistant materials.

Oil, gasoline or other combustible material should not be stored within 100 feet of any shaft, caisson or tunnel opening. However, oil may be stored in suitable tanks in fireproof buildings if the buildings are at least 500 feet from any tunnel-connected building or opening. Positive means should be taken to prevent leaking flammable liquids from flowing into the tunnel-connected openings or buildings. The handling, storage and use of explosives must comply with all applicable regulations in connection with compressed-air work.

Electricity

All lighting in compressed-air chambers should be by electric methods exclusively. Two independent electric lighting systems, with independent sources of supply, should be used. The minimum intensity of light on any walkway, ladder, stairway, or working level should not be less than 10 foot-candles, and in all work areas the lighting should at all times enable employees to see clearly.

All electrical equipment and wiring for light and power circuits must comply with the requirements of the National Electrical Code for use in damp, hazardous, high temperature, and compressed-air environments. External parts of lighting fixtures and all other electrical equipment, when within 8 feet of the floor should be constructed of noncombustible, nonabsorptive, insulating materials, except that metal may be used if it is effectively grounded. Portable lamps should be equipped with noncombustible, insulating sockets, approved handles, basket guards and approved cords. The use of worn or defective portable and pendant conductors should be prohibited.

DECOMPRESSION TABLES

The Federal Regulations provide the following explanation and tables concerning decompression times.

Explanation

The decompression tables are computed for working chamber pressures from 0 to 14 psi, and from 14 to 50 psi gauge inclusive by 2 psi increments, and for exposure times for each pressure extending from one-half to over 8 hours in Table 16-1 (for exposure times varying from $1\frac{1}{2}$ hours to over 8 hours see the Bureau of Reclamation publication[1]). Decompressions will be conducted by two or more stages with a maximum of four stages, the latter for a working chamber pressure of 40 psi gauge or over.

Stage 1 consists of a reduction in ambient pressure ranging from 10 to a maximum of 16 psi, but in no instance will the pressure be reduced below 4 psi at the end of stage 1. This reduction in pressure in stage 1 will always take place at a rate not greater than 5 psi per minute.

Further reduction in pressure will take place during stage 2 and susequent stages as required at a slower rate, but in no event at a rate greater than 1 psi per minute.

The decompression table, Table 16-1, indicates the total decompression time in minutes for various combinations of working chamber pressure and exposure time.

Another decompression table[1] indicates various combinations of working chamber pressure and exposure time for the following:

1. The number of decompression stages required.
2. The reduction in pressure and the terminal pressure for each required stage.
3. The time in minutes through which the reduction in pressure is to be accomplished for each required stage.
4. The pressure reduction rate in minutes per pound for each required stage.

Important note: The pressure reduction in each stage is accomplished at a uniform rate. Do not interpolate between values shown in the decompression tables. Use the next higher value of working chamber pressure of exposure time should the actual working chamber pressure of the actual exposure time, respectively, fall between those for which calculated values are shown in the body of the tables.

REFERENCES

1. Occupational Safety and Health Administration; Safety and Health Standards.
2. New York State Industrial Code.
3. Safety Regulations by the State of New Jersey.
4. Internal Revenue Service Regulations, 26 CFR 181 Commerce in Explosives.

17

Tunnel Finish

DONALD N. TANNER

Assistant Vice President and Senior Professional Associate
Parsons, Brinckerhoff, Quade & Douglas

The interior finish of a tunnel is dependent upon its functional use and the requirements for structural support. All tunnels except those mined through sound, competent rock require a structural lining. This lining can consist of rock reinforced by rock bolts and wire mesh, shotcrete, segmental steel or precast concrete liners, a steel envelope, reinforced concrete or combinations thereof. With the exception of highway tunnels, the interior surface of the structural lining constitutes the tunnel finish.

Railroad, rapid transit and highway tunnels usually require ventilation, and therefore the interior surfaces of these tunnels should be smooth in order to reduce air turbulence and consequent frictional losses.

All highway tunnels of considerable length must be ventilated in order to dilute to acceptable levels the pollutants contained in vehicle emissions. This ventilation is provided by introducing fresh air into the tunnel and exhausting the polluted air. Where required, the air is transported in ducts located exterior to the roadway clearance envelope. For circular, arch and horseshoe-shaped tunnels, these ducts are ideally located above the roadway clearance envelope and separated therefrom by a suspended ceiling. This ceiling is part of the tunnel finish.

Effective tunnel lighting requires that the finish surfaces within a highway tunnel provide sufficient reflectivity consistent with established lighting criteria (see Chapter 20). Finish materials which can satisfy these requirements are necessary for the tunnel sidewalls and the ceiling.

In addition to finished sidewalls and ceiling, highway tunnels require sidewalks, roadway pavement and miscellaneous items of tunnel finish. These items are utility niche frames and doors, sidewalk railings and police booths.

Tunnel finishes are subject to vibrations, fires, explosions, dampness and the attrition of frequent washing with detergents, brushes and high-pressure water jets. The presence of moisture and engine exhaust products in the tunnel atmosphere—especially exhausts from diesel-powered trucks—creates a corrosive environment as well as darkening of the finish surfaces, which detracts from their light-reflection qualities.

Frequent, high-quality maintenance cannot always be depended on. Therefore, selection of materials with impervious surfaces that resist soiling will help to compensate for maintenance deficiencies. Joints between finish materials should be minimized, and filled with durable materials.

Criteria for the selection of materials for tunnel finish are corrosion-resistance, fire-resistance, noise-attenuation, durability, cleanability and reflectivity. Further, these criteria must be applied within the framework of public safety, acceptable aesthetic standards, and economical cost.

CAST-IN-PLACE CONCRETE CEILING

This type of suspended ceiling construction (Fig. 17-1) was used in the Holland Tunnel, the first subaqueous motor vehicle tunnel, which was constructed during the years 1920 to 1927. To date, approximately 90% of tunnel ceilings have been built by this method.

Description. The ceiling is a reinforced concrete slab, with a thickness of 4 to 6 inches, which spans transversely between the tunnel sidewalls and interior supports. These interior supports vary from 1 to 3 (or more) in number and consist of composite concrete and structural steel beams or structural steel stringers which span longitudinally between ceiling hangers. Ceiling hangers vary in longitudinal spacing from 4 to 12 feet on centers and transfer the ceiling load to the tunnel structural lining. Transverse expansion joints are provided in the ceiling slab and vary in spacing from 20 to 50 feet.

Geometry and Drainage. The ceiling soffit is located 13 to 16 feet, 6 inches above the tunnel roadway, in accordance with AAHSTO Standards for Federal Interstate Defence Highway clearances. The longitudinal profile of the ceiling is parallel to the roadway grade, and the transverse slope of the ceiling is usually parallel to the pitch of the roadway surface in order to provide transverse drainage of water from condensation or leakage above the ceiling surface, which is removed by sidewall drains, located on the upgrade side of each expansion joint, and discharged into the roadway drainage system. These drains are usually $1\frac{1}{2}$ to 2 inches in diameter and are of hard copper with suitable radius bends to permit cleaning.

Design Loads. Design loads are the dead load of slab and ceiling finish and a 20 psf live load which provides for maintenance equipment and personnel. Where the ceiling forms the floor of an exhaust air duct, suction loads must be considered. The total suction pressure in the exhaust air duct can be determined from Chapter 19 and usually does not exceed 50 psf (10 inches of water).

Design Stresses. These stresses are in accordance with ACI Building Code Requirements for Reinforced Concrete. The use of this code is preferable to that of the AAHSTO Code for Highway Bridges since the live loads are essentially static rather than repetitive, as provided for in AAHSTO. Reinforcing bars are deformed and in accordance with the requirements of ASTM A305. Design strength of the concrete is usually 3000 psi using Type II Portland cement with a maximum aggregate size of $\frac{3}{4}$ inch. Type III cement (high early strength)

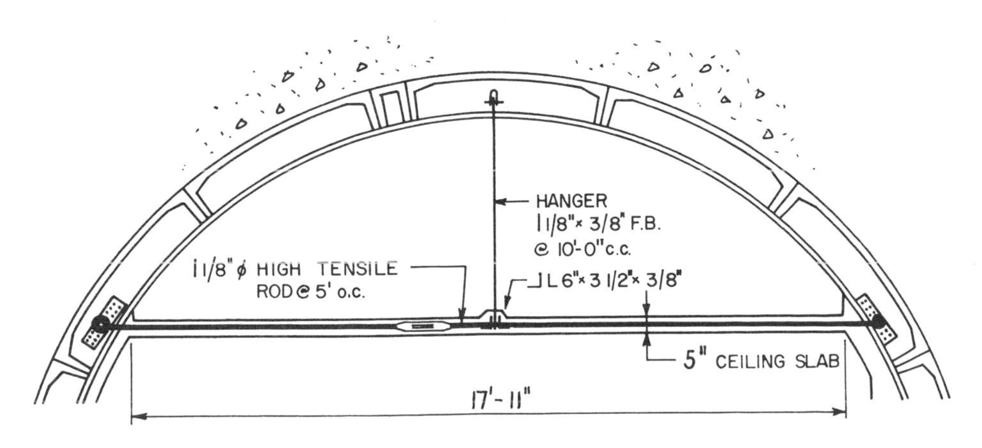

Fig. 17-1. Typical cast-in-place concrete ceiling, Holland Tunnel.

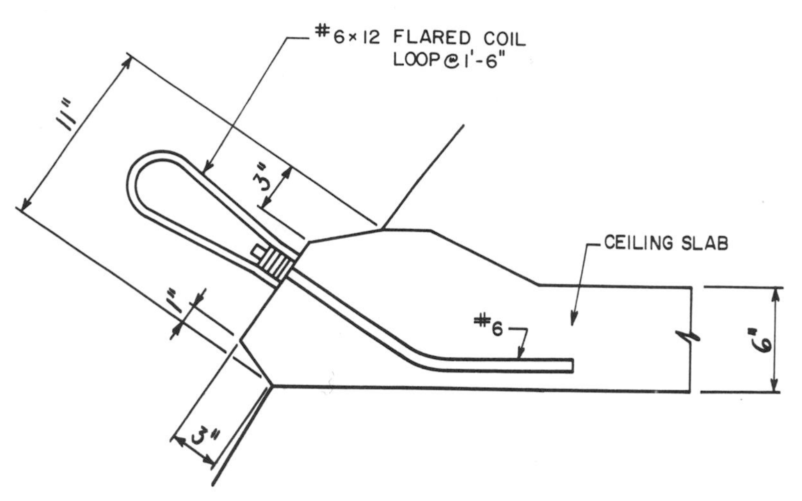

Fig. 17-2. Cast-in-place concrete ceiling, sidewall ceiling support, key with loop anchor and dowel, Wilson Tunnel, second bore.

may be used where accelerated form stripping is desired.

Sidewall Ceiling Support. A typical sidewall support for the ceiling, shown in Fig. 17-2, utilizes a loop anchor insert which permits a threaded dowel to be screwed in after stripping the structural lining forms. A shallow key within the structural lining without a haunch is shown. Addition of a haunch requires more complicated forming but is preferable.

Figure 17-3 illustrates a deep non-doweled key within the structural lining without a haunch. This type of support requires the main reinforcing steel to be interrupted and spliced with larger diameter reinforcing bars located clear of the key.

Another type of sidewall ceiling support is shown in Fig. 17-4. This support consists of a haunch that does not utilize a key or dowel. This detail has a non-bonded support with a joint filler which permits deflection of struc-

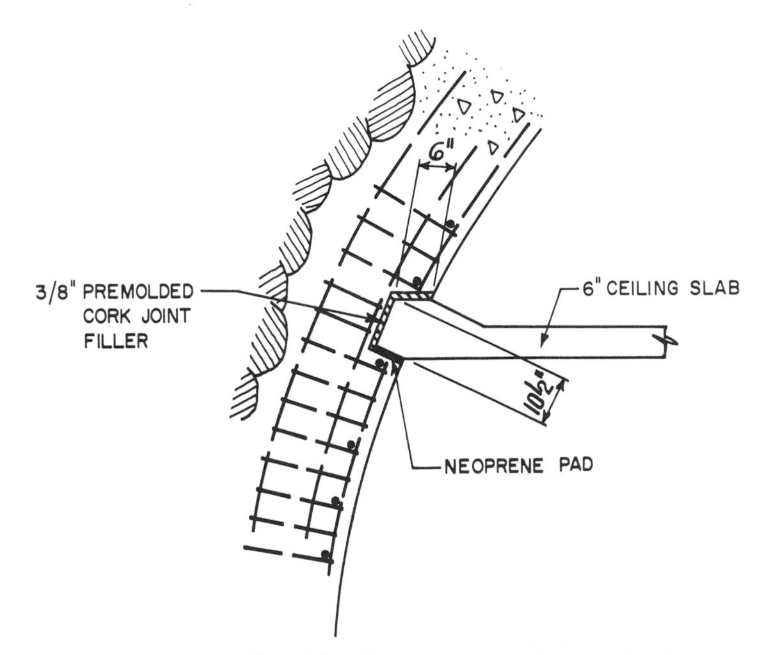

Fig. 17-3. Cast-in-place concrete ceiling, sidewall ceiling support, deep non-dowled key, Big Walker Tunnel.

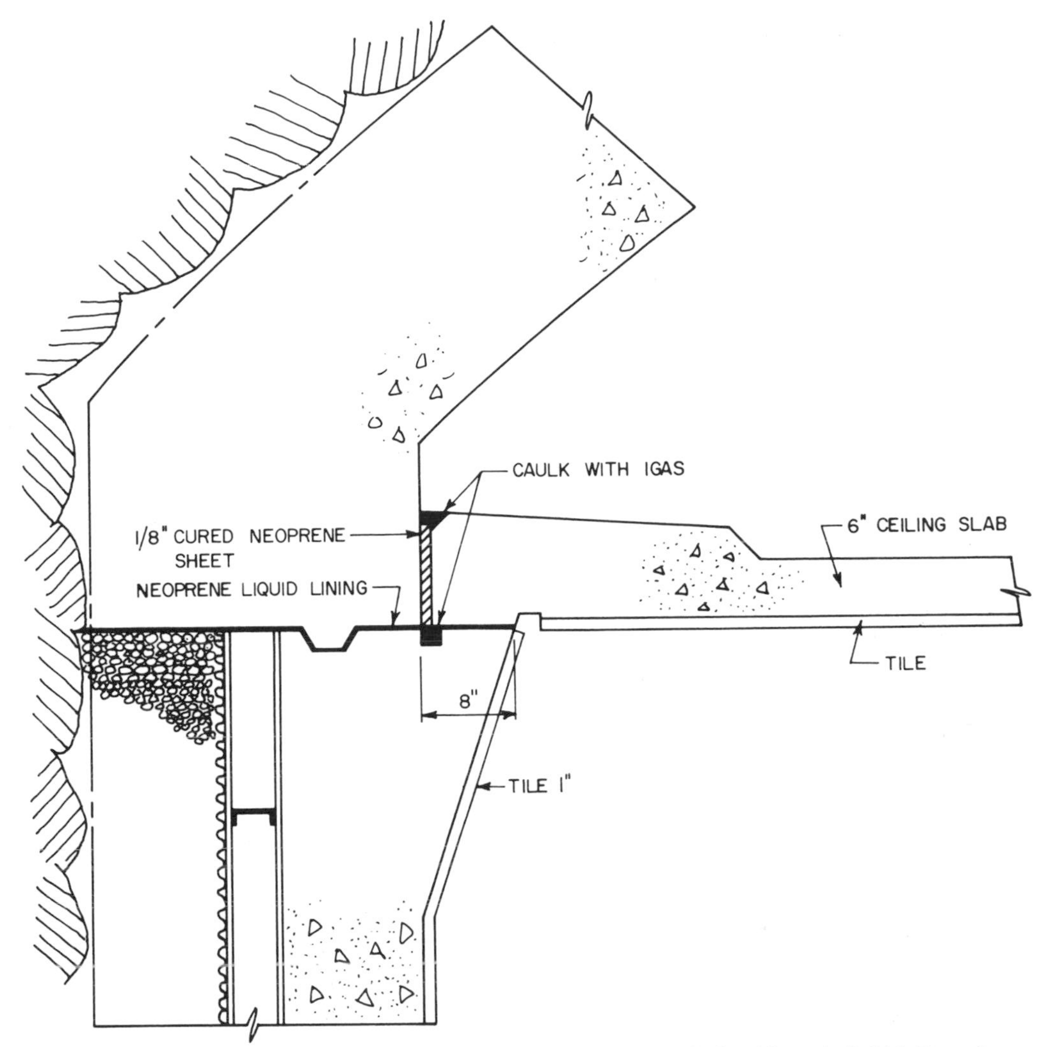

Fig. 17-4. Cast-in-place concrete ceiling, sidewall ceiling support, isolated haunch, Lehigh Tunnel.

tural lining without transfer of load to the ceiling.

Deflection of the structural lining caused by variations in earth, rock or water pressure can result in axial loads (tension or compression) being transmitted to the ceiling after it has been installed. Consideration should be given to this possibility in the design of the ceiling where keyed and doweled-type supports are used.

A typical sidewall ceiling support used for shield tunnels is shown in Fig. 17-1. The tension tie rod is connected to the primary lining during the excavation and then the interior concrete lining is placed with a box-out as shown. The box-out is filled in when the ceiling concrete is placed.

Ceiling Hanger Assemblies. The dead loads of the cast-in-place ceiling are usually greater than the suction loads, and therefore the hanger assembly is in tension. Typical details are shown in Fig. 17-5. The hanger assembly consists of a top insert, anchored in or to the structural lining, an adjustable rod or flat bar hanger and a bottom connection to the interior longitudinal support. Where the interior ceiling support is a reinforced concrete beam, a concrete insert is used for the bottom connection to the ceiling slab. Where the interior support is a composite

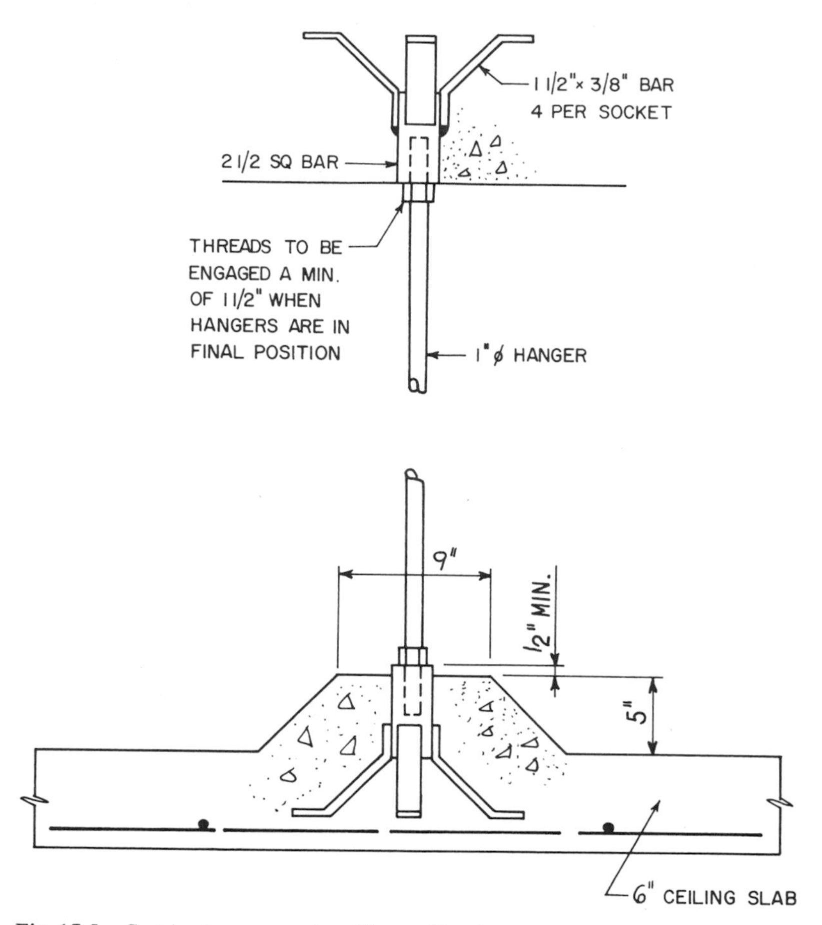

Fig. 17-5. Cast-in-place concrete ceiling, ceiling hanger assembly, Vista Ridge Tunnel.

concrete and steel beam, the hanger is connected to the longitudinal structural steel stringer, which consists of a T or double angles.

The hanger assembly is fabricated from structural steel meeting the requirements of ASTM A36 or stainless steel conforming to ASTM A276 Type 304. Protection of the hangers against corrosive exhaust gases and dampness has ranged from painting with red lead with shop and field finish paint to coal tar enamel coatings. Recently corrosion-resistant finishes, such as epoxies, ceramic and metal compounds and metallic zinc paint, have been used. In several ceiling installations, bare stainless steel hangers have provided excellent service.

Ceiling Hanger Adjustments. Vertical adjustment of the hanger assembly is provided to compensate for construction and fabrication tolerances of the hanger and tunnel lining in order that the ceiling to roadway clearance can be maintained. Since the structural lining of the tunnel is usually constructed on chords rather than following the vertical curve, varying length hangers are required throughout the length of the vertical curve. In order to minimize the number of variable length hangers, an adjustable length hanger is desirable. Further, in order to prevent bending in the hanger, the top insert should be set plumb, or a hinged connection should be provided at the top of the hanger. These adjustments are accommodated for by threaded top and bottom hanger connections, by field-drilled holes, by slotted holes or combinations thereof.

Ceiling Finishes. The soffit of the cast-in-place concrete ceiling is adaptable to varied types of finishes such as unfinished concrete, painted concrete, ceramic or glass tile and veneers of ceramic or metallic materials. (For selection of the type and color of the finish, see p. 473).

Installation. Forms for placing the cast-in-place concrete ceiling are usually steel, faced with plywood and supported by a traveler that is laterally and vertically adjustable. The traveler is designed to permit passage of construction equipment and is mounted on pneumatic tires or steel flanged wheels on rails.

The form length varies from 150 to 175 feet, depending upon the spacing of construction and expansion joints and the length of pour desired. The construction sequence is usually threefold: one form section supports the previously placed concrete; another supports the concrete being placed; and the final supports the reinforcing steel in preparation for the section to be placed.

Travelers have been designed to permit lowering and bypassing the self-supporting forms and, in other cases, each section of form has been supported by its own traveler. Choice of methods is a question of economy and individual preference. The number of three-phase form units is dependent upon the desired construction schedule and form stripping time. Most specifications require a minimum strength of 1500 psi in the concrete, which is usually determined by the breaking strength of cylinders cured in the tunnel.

Where a ceramic tile ceiling is used as a finish, a layer of gummed paper, sticky side up, is placed on the ceiling form. This is dampened and tiles are placed face side down on the paper. The joints between tiles are then filled with fine sand, or, in some cases, a lean cement-sand grout. Next, a hat-shaped wire mesh, 2 inches in depth, is set in place on top of the tile. Then a sand-cement grout is placed to a depth of 1 inch. After the grout has set, the reinforcing steel and structural steel stringers are installed. Then the ceiling concrete is placed using pneumatic pumping equipment. The average cycle of advance is one week. When an unfinished concrete ceiling, glass tile or a veneered finish is used, the gummed paper and plywood are omitted. Inserts for attaching the glass tile or veneer are fastened to the forms and the ceiling concrete is placed directly in the form. Exhaust and/or supply air port castings are also installed prior to placing concrete.

Advantages and Disadvantages. The cast-in-place concrete ceiling has many advantages. It is suitable for all types of ceiling finishes and requires a minimum number of fabricated metal parts for mechanical adjustment to compensate for construction and fabrication tolerances. Horizontal curves can be accommodated by an adjustable form shoe which will compensate for variations in width. This adjustable shoe also eliminates chipping of concrete where sidewall tights occur.

The disadvantages are few, but are of major consequence. Repair or replacement of the ceiling and/or finish is a slow and expensive operation that requires special equipment and usually necessitates closing the tunnel to traffic. Care must be taken in the design and placement of the concrete mix in order to minimize shrinkage and subsequent cracking of the ceiling. Infiltration of drainage water through these cracks damages the finish and can cause structural disintegration of the concrete slab. In addition, tile or veneered finishes are hazardous, since during a fire these finish materials can explode off the ceiling, endangering fire-fighters with falling debris and making treacherous footing.

METAL PANEL CEILING

This type of suspended ceiling construction was first used in the Downtown Elizabeth River Tunnel, which was built during the years 1950 to 1952. Support details are shown in Fig. 17-6.

Description. This type of ceiling consists of a cold-formed steel or extruded aluminum metal panel filled with concrete. The panels vary from 2 to 4 inches in thickness with a length of 6 to 13 feet and a width of 1 to 2 feet, 6 inches. The soffit and sides of the panel are finished with a white or light-colored porcelain enamel. The panels span transversely between longitudinal structural steel stringers, which are supported at the tunnel sidewalls by brackets and interior hangers suspended from the tunnel structural lining. Ceiling hangers are located at 12-foot, 7-inch centers longitudinally. The stringer is continuous over the hanger supports

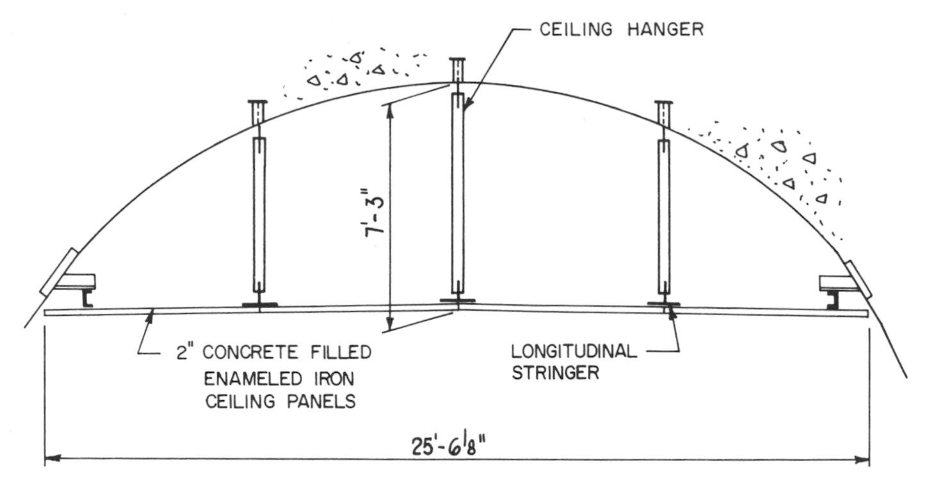

Fig. 17-6. Metal panel ceiling, typical cross-section, Midtown Elizabeth River Tunnel.

and has cantilever end spans which vary from 1 to 3 feet. Transverse expansion joints are located at 25 feet, 2 inches on centers.

Geometry and Drainage and Design Loads. Vertical clearances, drainage and design loads are similar to those used for the cast-in-place type ceiling.

Design Stresses. These are also similar to those used for the cast-in-place ceiling, except that the design of the metal panels conforms to the Light Gage Cold Formed Steel Design Manual published by the AISI. Structural steel design is in accordance with the requirements of the AISC Manual of Steel Construction. Design strength of the concrete is 3000 psi with a maximum aggregate size of $\frac{3}{4}$ inch. Where extruded aluminum panels were used, a lightweight (110 pcf) 3000 psi concrete with expanded shale aggregate was used for panel concrete. Structural steel stringers and brackets conform to the requirements for ASTM A36 steel.

Panel Material. Material for steel panels is 14-gauge commercial grade enameling iron. Extruded aluminum panels fabricated from Alloy 6063-T42[3] have been used in one tunnel. After installation of the ceiling, spalling of the porcelain enamel finish occurred. This spalling may have been due to the use of an unsuitable alloy or to improper enameling techniques. Studies by the aluminum industry indicate that a controlled 6061 Alloy might eliminate the spalling. Aluminum panels for tunnel ceilings have not been used in any other tunnel in the U.S. Typical cross-sections of steel and aluminum panels are shown on Fig. 17-7.

Sidewall Ceiling Support. A typical sidewall support for the metal panel ceiling, shown in Fig. 17-8, consists of a steel bracket which permits a 2-inch vertical and a $1\frac{1}{2}$-inch lateral adjustment of the ceiling. These brackets are located on the same centers as the hanger inserts.

Ceiling Hanger Assemblies. Fig. 17-9 illustrates a typical tension hanger which consists of a stainless steel rod with hinged top and bottom connections. Where the dead load of the ceiling approaches the suction load, the hanger assembly is designed as a pipe column, as shown in Fig. 17-10. Both types of hangers provide installation adjustment, but the column type hanger is fixed at the bottom by welding after the ceiling is installed.

The ceiling hanger assembly for the Second Hampton Roads Tunnel is similar to that shown in Fig. 17-10. However, in lieu of the top insert, a percussion-drilled hole 6 inches deep by $1\frac{3}{8}$ inches in diameter was provided. A polyester resin cartridge was placed in the hole and the upset bolt was spun into the structural lining. Maximum time of rotation was 10 seconds. The bolt was held in position for 40 seconds until the gel had set. After 15 minutes, each

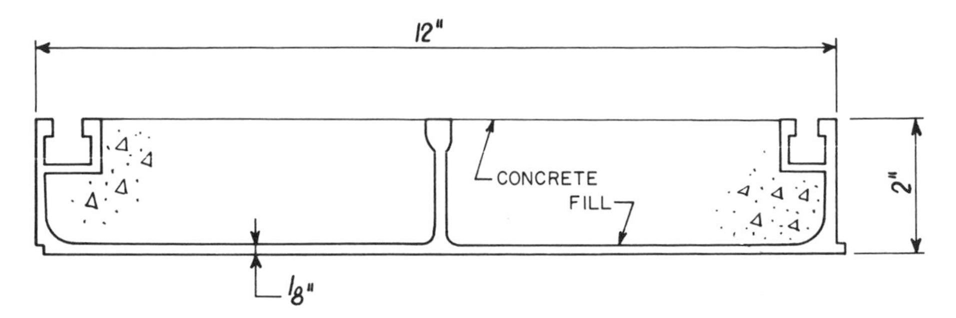

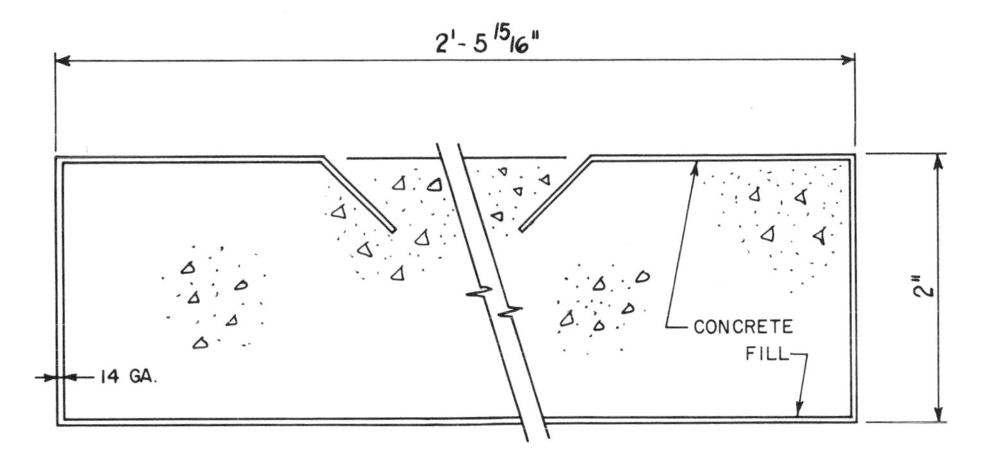

Fig. 17-7. Typical metal ceiling panels.

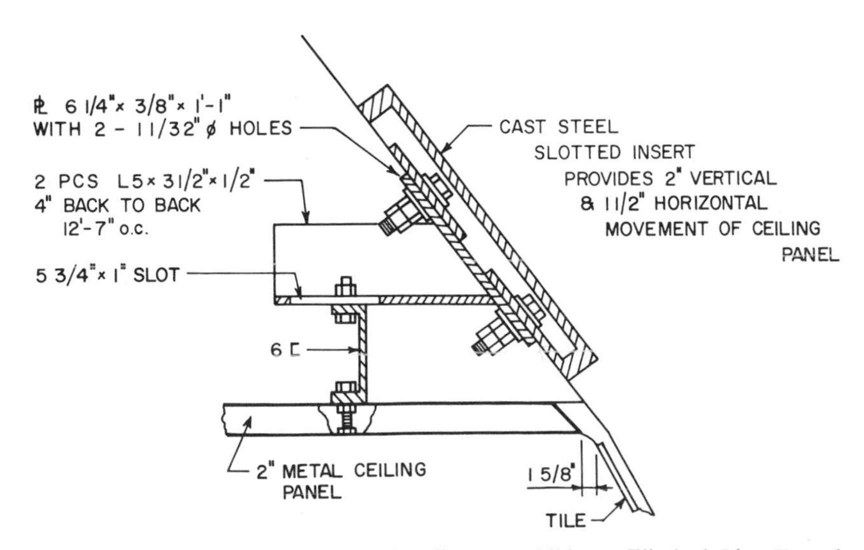

Fig. 17-8. Metal panel ceiling, typical sidewall support, Midtown Elizabeth River Tunnel.

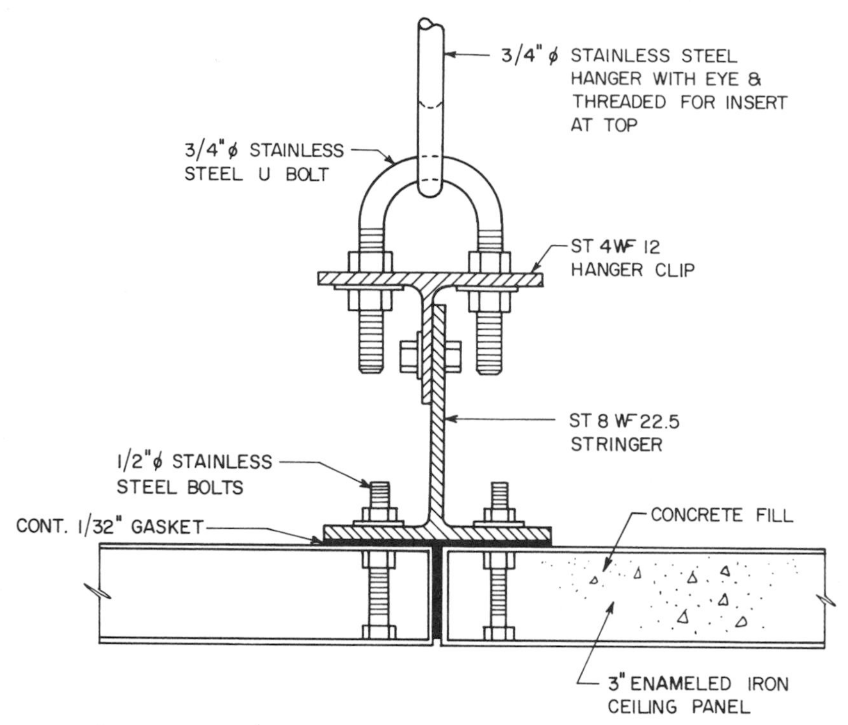

Fig. 17-9. Metal panel ceiling, tension rod ceiling hanger assembly, Downtown Elizabeth River Tunnel.

bolt was tested to twice the design load. This method of installation permits compensation for construction tolerances.

Ceiling Joints. At the center of the ceiling, adjustment of the panels is provided for as shown in Fig. 17-11. This adjustment eliminates fabrication of special panels where sidewall tights occur. Joints between panels are gasketed to provide airtight and watertight connections.

Panel Connections. Panels are supported at each corner by stainless steel bolts, which are connected to the stringers. Adjacent panels are connected longitudinally by four stainless steel bolts. Stainless steel conforms to ASTM A276 Type 304 and gasket material is a white, fire-resistant (self-extinguishing) neoprene. These gaskets are talc-extruded to the shapes shown in Fig. 17-12.

Gaskets are cemented to the panels during erection of the ceiling and are prevented from falling through the joints by their T and L shapes. Alignment of joints between panels is maintained by slotted holes in the supporting stringers. The bulb shape of the transverse gaskets ensures a watertight joint while providing for longitudinal adjustment of joints that is necessary for erection tolerances and maintenance of the expansion joint width.

All structural steel brackets, stringers and hangers are coated with corrosion-resistant finishes similar to those used for the cast-in-place type ceilings. Metal panels are temporarily protected against shipping and installation damage by covering with adhesive paper or a plastic film that is removed after installation of the ceiling.

Installation. Installation of the metal panel type ceiling is a three-phase operation. The initial operation consists of the installation of the hangers and erection of the stringers to line and grade. Panels are assembled on a traveler similar in design to those used for the cast-in-place type ceiling. The traveler is usually 25 feet in length and is adjustable in height. Concrete is usually hoisted to the traveler platform for placing in the panels. Pneumatic placement can be used,

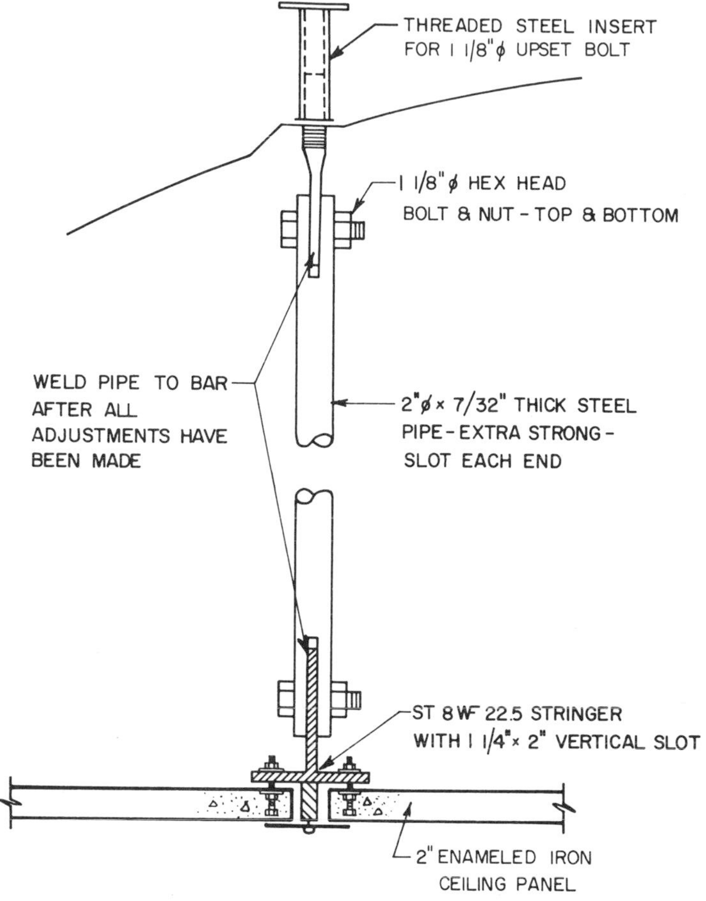

Fig. 17-10. Metal panel ceiling, pipe hanger ceiling assembly, Midtown Elizabeth River Tunnel.

but has not been considered economical in the past. On the recently completed Second Hampton Roads Tunnel, concrete was placed in the panels off-site. After curing, the panels were shipped to the site and lifted to the traveler using a cherry picker with a vacuum type lifting device. The final stage is the raising of the traveler platform to the level of the stringers and attachment of the panels to them. Since the metal panels have been designed as forms, the panels are usually fastened to the stringers as soon as the concrete fill has set (usually the next morning). The number of traveler units depends upon the desired construction schedule. After final adjustments have been made, all bolts are tightened and welded connections are made.

Advantages and Disadvantages. The major advantages of the suspended metal panel type ceiling are maximum off-site fabrication, ease of replacement, factory control of fabrication and finishes and speed of erection. If the ceiling is damaged during tunnel operation, the metal panels can be replaced within a short time by unbolting the damaged panel and inserting a new panel in the ceiling. Extra panels, complete with concrete fill, are stored at the completed tunnel for this purpose. In addition, during a fire, there are no tiles or veneered finishes to explode off the soffit due to the creation of superheated steam created by free water in the concrete. The concrete fill in the metal pans is contained and can only explode upwards into the air duct. Mechanical adjustments of all parts of the ceiling permit closer tolerances and result in high aesthetic standards of finish. Also, the high corrosion-resistance of the porcelainized metal panels minimizes the effects of leakage in the structural lining.

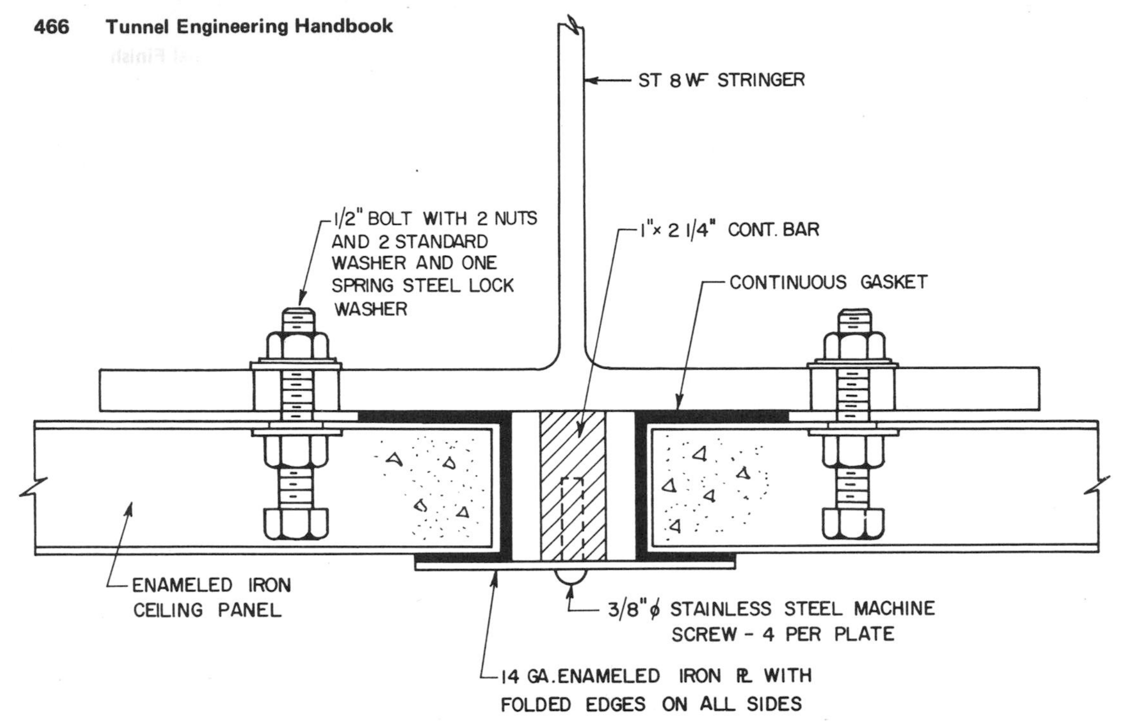

Fig. 17-11. Metal panel ceiling, center panel adjustment, Midtown Elizabeth River Tunnel.

The principal disadvantage of the metal panel type ceiling is lack of familiarity with it.

PRECAST CONCRETE PANEL CEILING

This type of ceiling construction was first used for the Hong Kong Tunnel, which was opened to traffic in 1972. The ceiling is similar in concept and design to the suspended metal panel type, except that a precast concrete panel is used in lieu of the metal panel.

VENEERED CEILINGS

This type of construction consists of direct attachment of the ceiling finish to the structural lining. The veneer thickness is usually 1 to 2 inches.

Tile Veneer. The most common example of a veneered ceiling is ceramic tile. The tile can be installed prior to placing the structural lining, in a similar manner as described for the tile finish

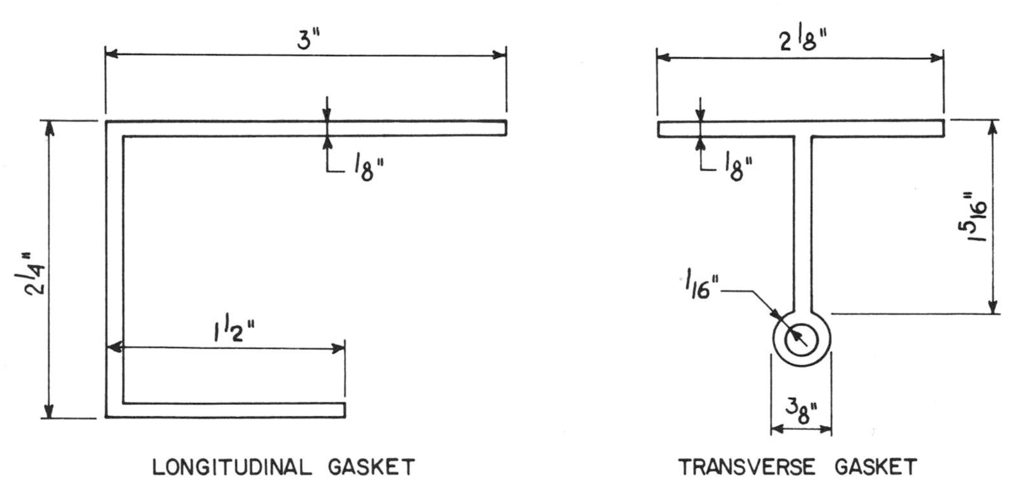

Fig. 17-12. Metal panel ceiling, ceiling gaskets, Midtown Elizabeth River Tunnel.

for the cast-in-place ceiling. In many cases, the tile has been installed after placing the structural lining. The former method of bonding the tile to the structural lining is usually used in cut-and-cover tunnels, while the latter method is used for all other types of tunnels where a suspended ceiling is not required.

Metal Panels. Another type of veneered ceiling consists of the attachment of porcelain enamel metal panels to the structural lining. This was first used on the cut-and-cover section of the Downtown Elizabeth River Tunnel. The $1\frac{1}{2}$-inch thick, 2-foot, 6-inch by 7-foot, 6-inch panels were attached by T bolts after placing the structural cut and cover roof slab. Typical details are shown in Fig. 17-13. Lightweight vermiculite concrete fill was used to prevent "tin canning" and to minimize noise due to "drumming." Recently, a similar type of porcelain enamel metal pan veneer ceiling was installed in the Tyne Tunnel.

Ceramic Glass Veneer. A test section of Pyram panels was installed in the Brooklyn Battery Tunnel in 1969. These aluminum-framed, ceramic glass panesl were manufactured by Corning Glass Works. The 3-foot by 8-foot panels were attached $1\frac{1}{2}$ inches below the existing cast-in-place concrete ceiling, which had a glass tile finish that had been installed during construction in 1950. Leakage through the structural cast iron and concrete lining and low temperatures caused large areas of the tile to fall off. The Pyram panels were selected for the repair of the existing ceiling since they had a minimum thickness, a fail-safe material and a durable and washable fire-resistant finish, and since large size panels could be rapidly installed. The test installation has made an improvement in the appearance of the ceiling, but it is too soon to evaluate the performance of this installation.

Advantages and Disadvantages. Veneered type ceilings have advantages similar to the various

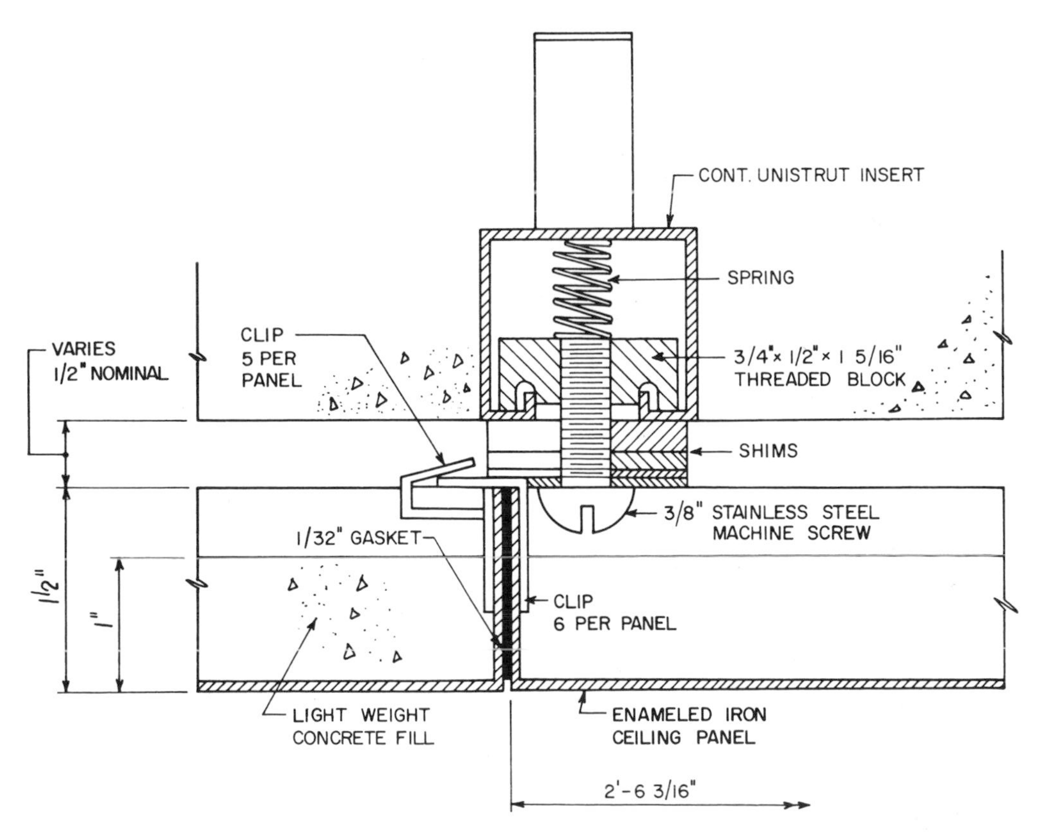

Fig. 17-13. Veneered ceilings, metal panels, Downtown Elizabeth River Tunnel.

types of suspended ceilings. Their principal disadvantage is that the veneer is susceptible to separation from the structural lining due to the formation of ice caused by low temperatures and leakage. It is also a hazard in the event of a tunnel fire.

TUNNEL SIDEWALL FINISHES

With the exception of highway tunnels, the interior surface of the structural lining constitutes the tunnel sidewall finish. Paint or an epoxy coating applied directly to the structural lining, ceramic tile veneer and furred-out walls with various types of finish have been used for highway tunnel sidewalls.

Paint and Epoxy Coating. These finishes are applied directly to the structural lining. They have the lowest initial cost and require no space allowance, thus resulting in the minimum interior tunnel dimensions. (The advantages and disadvantages of these finishes are discussed on p. 473.)

Ceramic Tile Veneer. This veneer is, and has been, the most prevalent finish for tunnel sidewalls. The thickness of the finish is usually 1 inch and consists of a mortar scratch coat applied to the structural lining, roughened by $\frac{1}{4}$- inch steel wires welded to the sidewall forms at approximately 2-inch centers. When the forms are stripped, the edges of the grooves are slightly fractured and provide an additional bonding surface for the scratch coat. The scratch coat also serves as a leveling course to compensate for forming tolerances required for placing the structural sidewalls. After the scratch coat has cured, the ceramic tiles are cemented in place using a sand-cement mortar. The joints between tiles are filled with a float coat consisting of white or gray neat cement.

The average rate of advance is 2000 square feet of wall per 8-hour shift. The tile work starts after the tunnel has been holed through and the sidewalks have been placed, and is usually concurrent with electrical and mechanical finish work in the tunnel.

Porcelain Enameled Steel Tile. This material was used for sidewall finish in the Detroit-Windsor Tunnel in 1928. This installation is probably the first use of porcelain finish in a tunnel. After 49 years of continuous service, these porcelain-enameled steel tiles, while requiring major renovation, are an example of material durability.

Metal Panels. The IJ Tunnel in Amsterdam, completed in 1968, has a metal panel sidewall finish. These panels are a greenish-gray procelain-enameled steel which can be easily removed and replaced in the event of damage. The wall panels are slightly inclined from the vertical in manner such that sound is reflected upwards, where much of it is absorbed by an acoustic ceiling. This ceiling consists of perforated aluminum panels enclosing rock wool batts. Both wall and ceiling panels are isolated from the structural lining by special attachments which incorporate blocks of neoprene.[1] (The efficiency of this sound-reducing system is discussed on p. 473.

Furred-out Walls. Recently, furred-out wall construction has been used for the Mersey 2 and Tyne Tunnels in England. The finish sidewalls consist of coated steel panels supported by a structural steel framework which is attached to the structural lining.

The wall construction for the Tyne Tunnel is 20-gauge porcelain-enameled steel panels approximately 12 feet wide, bent to follow the contour of the tunnel. On the reverse side of the panel, a $\frac{1}{2}$-inch thick asbestos insulation board is bonded with a neoprene base adhesive and the exposed face of the board is sealed with a chlorinated rubber. The asbestos board increases the rigidity of the panel and acts as a sound deadener. The panels were attached to the structural steel frame with Z clips at the side flanges. Vertical joints between panels were enclosed with a full height stainless steel cover strip.

The steel sidewall panels for the Mersey 2 Tunnel are similar to those used for the Tyne Tunnel. The steel panels were finished with a field applied epoxy paint.

A furred-out porcelainized enameled iron wall panel system is shown in Fig. 17-14.

The panels are 8 feet, 6 inches long by 2 feet,

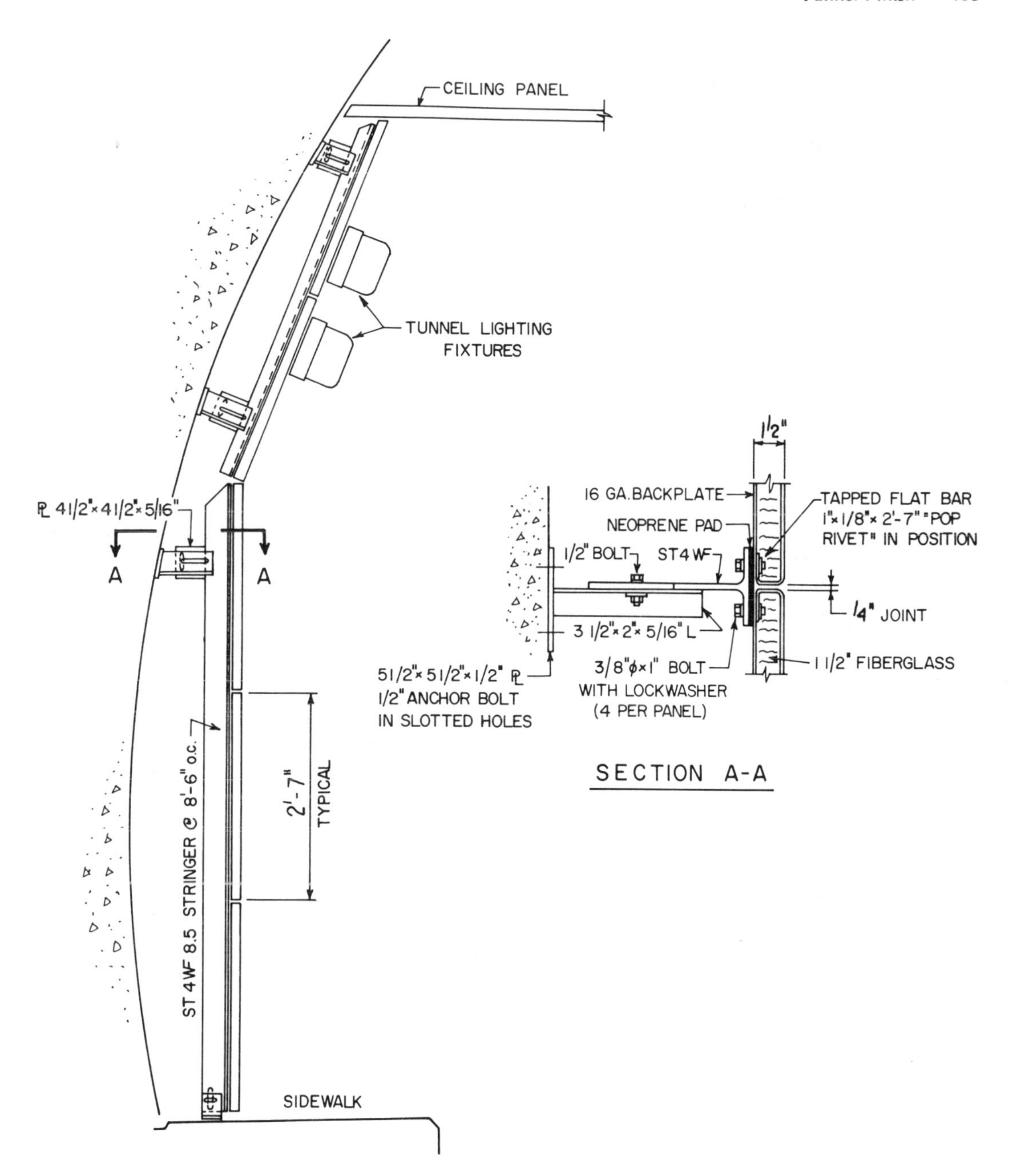

Fig. 17-14. Furred-out metal wall panel.

7 inches high and $1\frac{1}{2}$ inches thick. The panel is 14 gauge with a 16-gauge back cover plate pop riveted into place after porcelainizing. To provide rigidity, the panels are filled with fiberglass, which is bonded to the panel after porcelainizing. The support system consists of a vertical structural steel tees spaced 8 feet, 6 inches on centers, to which the panels are attached. The tees are attached to the concrete tunnel lining with adjustable clips which are expansion-bolted into place, thus providing for construction tolerances. All structural steel members are coated with corrosion-resistant materials. All hardware is stainless steel.

Other types of furred-out walls have been considered, such as precast concrete panels and structural glazed tile. Furred-out walls require a structural support system which should be designed for lateral loads due to air pressure created by moving traffic and ventilation. The finish wall and the structural support system usually require a space of 4 to 6 inches inside the structural tunnel lining. This air space between walls permits installation of branch ducts for distribution of ventilation air from the air ducts to the tunnel roadway and provides free drainage of seepage water through the structural lining without damage to the interior finished wall. In addition, utility conduits can be surface-mounted on the wall of the tunnel structural lining. Since the furred-out wall panels can be readily removed, access for repair of leaks and conduits and replacement of damaged wall panels can be easily accomplished. This type of wall construction permits compensation for construction tolerances required for alignment and placing of the concrete lining and is ideally suited to unlined rock tunnels.

SIDEWALKS

Sidewalks in the tunnel provide a means for emergency exit for motorists in the event of fire or accident, access for maintenance personnel and a pathway for traffic-control police. In addition, utility conduits are enclosed within the sidewalk and the sidewalk protects the structural lining of the tunnel from vehicular damage.

Figure 17-15 shows a sidewalk, level with the top of the curb and ranging in width from 2 feet, 6 inches to 4 feet and, on the opposite side of the tunnel, frequently called the ledge side, a safety barrier. This type of construction is prevalent in rock tunnels where intermittent vehicular police patrol is provided.

For subaqueous vehicular tunnels, a raised sidewalk and a ledge are usually provided, as shown in Fig. 17-16. The type of construction shown is for a cast-in-place concrete sidewalk and ledge. Many tunnels have been built using an architectural terra cotta tile for facing of the servicewalks in lieu of the construction shown. In this type of tunnel, continuous police supervision is usually provided. The raised sidewalk is ideally suited to a circular or arch tunnel configuration and has the advantage of greater room for utility conduits and pull boxes. Also, it keeps the tunnel patrolman in clear view of the traffic, while protecting him from the traffic in event of accident, and alleviates somewhat the suction air blast from passing vehicles.

Sidewalk patrol cars have been utilized in several tunnels to provide motorized police surveillance of traffic. These vehicles have gasoline engines or battery-operated electric motors and travel on the sidewalk at speeds of up to 30 mph. The car is guided by a continuous rail anchored to the sidewalk. A typical guide rail detail is shown in Fig. 17-17. The vertical leg of the angle provides lateral guidance for a pair

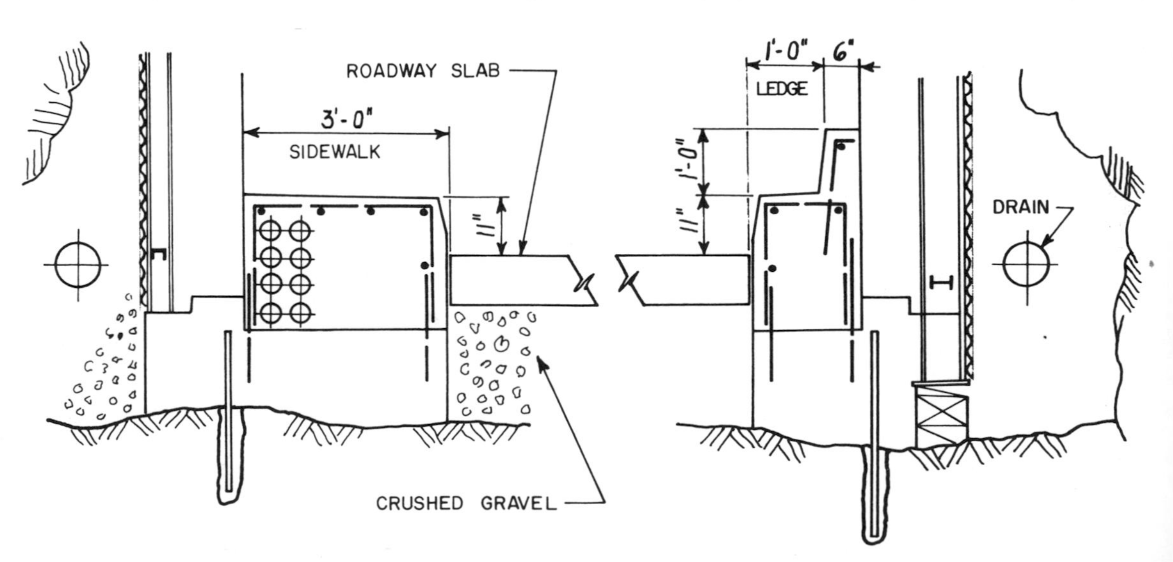

Fig. 17-15. Rock tunnel sidewalk, Lehigh Tunnel.

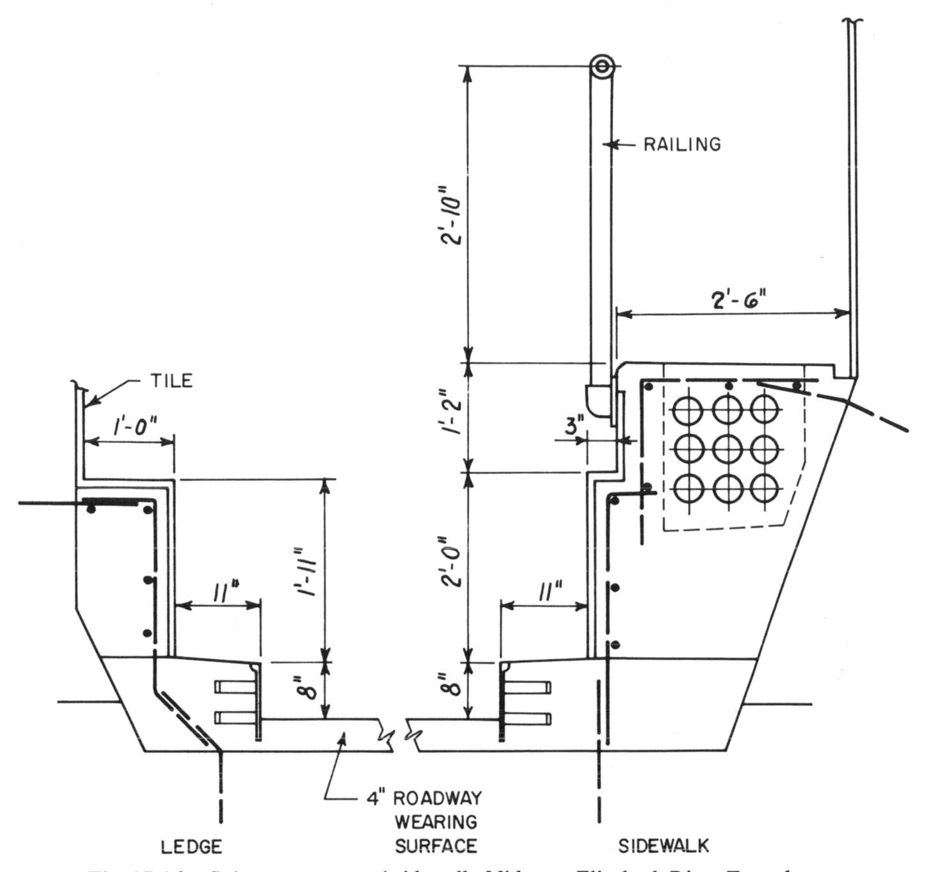

Fig. 17-16. Subaqueous tunnel sidewalk, Midtown Elizabeth River Tunnel.

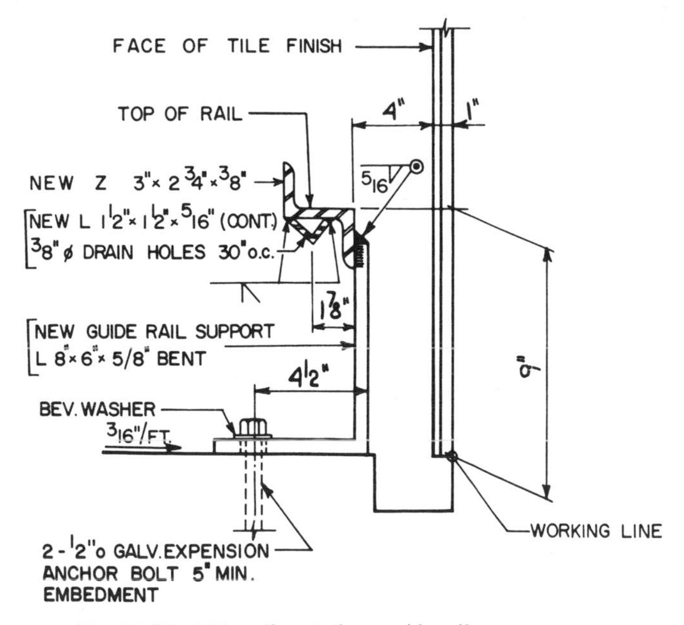

Fig. 17-17. Sidewalk patrol car guide rail.

of wheels mounted horizontally on the car. Longitudinal guidance is provided by a vertical wheel, mounted on the car, which runs along the vee of the inverted angle.

ROADWAY DESIGN

Vehicular tunnel roadways in the U.S. should be designed in accordance with the requirements of AASHTO for a HS20 with a check for military highway loading. While there are no published geometric standards for lane widths in tunnels, recent practice has been to provide a 12-foot width for traveled lanes with a 1-foot shoulder on each side. Where traffic density is high and design speeds of 60 mph are provided, 8- to 10-foot shoulders are preferred. Most tunnels built prior to 1950 have traveled lane widths ranging from 10 to 12 feet without shoulders. Design speeds, stopping sight distances, horizontal and vertical curvature and superelevation are discussed in Chapter 2.

Structural roadway slabs are usually reinforced concrete with a 4-inch wearing surface. In circular tunnels that have air ducts beneath the roadway, these slabs span transversely to support the roadway loads. In rock tunnels and rectangular sections, the roadway slab is uniformly supported by a crushed gravel subgrade or by the base slab of the rectangular section. In most older tunnels with air ducts beneath the roadway, the roadway slab consisted of structural steel beams embedded in concrete, spaced from 30 to 60 inches on centers.

Roadway wearing surfaces have consisted of granite paving blocks, hard burned brick, asphaltic concrete and concrete. In recent years, brick and paving blocks have not been used since they have a higher initial cost, give a noisier ride and require more maintenance than asphalt or concrete pavement. The relative merits of asphalt versus concrete wearing surfaces are controversial. Except in areas where ice and snow are prevalent, the asphalt wearing surface is generally the preferred surface.

Selection of the roadway surface requires evaluation of initial costs, maintenance and replacement costs, traffic density, ride comfort, quietness and performance records for safety. These factors must be evaluated for each tunnel.

MISCELLANEOUS FINISHES

Miscellaneous finishes consist of utility niche frames and doors, sidewalk railings and police booths.

Niche Frames and Doors. These are required for niches containing electrical control equipment for traffic, power, communications and fire-fighting equipment. Where possible, this equipment is mounted exposed within a tiled niche. Generally, power conduits are embedded in the concrete lining on the ledge side of the roadway and all control and communication equipment is located on the sidewalk side of the tunnel. This separation ensures electrical isolation and permits ease of maintenance without traffic interruption. Where required, equipment is recessed in a niche and provided with a framed door. Door and frame materials have been bronze, ebony asbestos, aluminum, stainless steel and painted carbon steel. Selection of material should be based on initial, maintenance and replacement costs. Current practice is to use anodized aluminum.

Sidewalk Railings. These railings are usually provided where raised sidewalks are used (Fig. 17-16). Most sidewalk railings are of pipe construction, iron, aluminum or porcelainized aluminum or steel, and have been painted.

Police Booths. In subaqueous vehicular tunnels, where sidewalk police patrol is required for traffic control, police booths should be provided. These booths are usually located at intervals such that the entire tunnel is under surveillance of such tunnel police. The booth is located adjacent to a control center for communications. It also provides a heated and air conditioned shelter and seat for the policeman to protect himself from air blast, noise and temperature during his tour of duty.

The booth consists of an aluminum frame with glass panels to give maximum visibility with pass-through doors on each end of the booth.

Recently, with the installation of closed circuit television and motorized sidewalk patrol vehicles, the use of foot patrolmen in tunnels has been considerably reduced without impairing surveillance (see Chapter 25).

TUNNEL FINISHES AND MATERIALS

The interior finish of a highway tunnel is, to the public, generally the measure of its success. Regardless of the design and construction innovations and economics, acceptance is measured by ride comfort, convenience, aesthetics and safety. However, in addition to these factors, the interior finishes have to satisfy visibility, lighting criteria, fire-resistance, noise attenuation, minimal maintenance and durability.

Visibility. Visibility within a tunnel is enhanced by the use of light-colored finishes on the walls and ceiling. However, this concept is not universal. Many designers, particularly in Europe, are of the opinion that the function of the ceiling as a reflective surface is not important. A number of tunnels with black or dark-colored ceilings have been recently built in Europe. These ceilings require minimum washing and therein may lie the motivation for the selection of a dark finish. In a two-lane tunnel, it can be postulated that light reflected from the sidewalls may be sufficient for adequate visibility in the tunnel. If it is assumed that the driver is guided by the lighted sidewalls, which are comfortably within his span of vision at eye level, and that they help him see cars ahead and also allow him to anticipate changes in roadway curvature, a reflective light-colored ceiling may not be necessary. However, conceding the validity of this assumption, a dark ceiling may be oppressive and have a detrimental effect on some occupants of vehicles, particularly those suffering from claustrophia.

In wide tunnels accommodating three or more lanes of traffic, a light-colored reflective ceiling assumes more significance.

In addition to the possible depressing feeling of a dark ceiling, there is another objection to its use. The ceiling and roadway form two parallel planes which appear to converge in the distance. As a result, the ceiling and roadway, assuming both were dark in color, would not present a backdrop where objects can be seen in silhouette. A well-illuminated, light-colored ceiling provides a backdrop that facilitates discernment of objects on the roadway and changes in roadway curvature. Further, the increased reflectivity will reduce capital and operating costs for tunnel lighting.

Paints and Epoxy Finishes. The designers of early highway tunnels for many installations used unfinished concrete ceilings and, in some cases, unfinished concrete sidewalls. These surfaces soon became darkened and provided poor visibility. Following this initial trend, sidewalls were tiled with white, at least to the springline, to increase visibility and, soon after, ceilings began to be tiled with white. Thus, prior to 1950, it was almost an axiom that a highway tunnel required a light-colored tiled ceiling and walls.

Increased costs of labor and material and new paint technology in the 1950's and 1960's led to experimentation with paints and epoxy finishes that were applied directly to the concrete structural lining. Under laboratory conditions, these surface finishes appeared to be durable, light-reflective and economical. Field application of these finishes required extensive preparation of the structural lining, including removal of fins, air bubbles, steam-cleaning and sandblasting. Most of these surface coatings were required to be applied under conditions of controlled humidity and dry, clean surfaces.

Although of low initial cost, most of these surfaces have had a very short life (five years) and, depending on field technology, considerably less. At this date, they have not yet established a satisfactory performance record for durability, low maintenance and aesthetic acceptance.

Pyrok Surface Treatments. These treatments are proprietary vermiculite materials used for sound-absorption, fire protection, anti-condensation, thermal insulation and interior tunnel finish. These materials have been used extensively during the past several years in England for tunnels and subways.

The principle of the Pyrok system for vehicular tunnels is to introduce a permeable membrane between the concrete tunnel lining and the Ceramicoat finish. This membrane is formed with a type of Pyrok which is cellular in structure and approximately $\frac{1}{2}$ inch in thickness. The Pyrok is sprayed or trowled on the concrete forming vertical panels approximately 3 by 5 feet with joints between panels. After the permeable layer of Pyrok is finished smooth, a finish of Ceramicoat is sprayed on the panels. In theory, if seepage through the concrete lining occurs, the moisture will move laterally by capillary attraction towards the joints and thence down the wall, without leaving unsightly stains on the finish. An advantage of this system is that construction irregularities, fins and air bubbles are concealed by the applied Pyrok finish.

Ceramicoat finish can be high-gloss, satin-matte or complete matte and is available in all colors. The finish is similar to tile, has excellent reflective qualities and can be readily washed with detergents and mechanical brushes and gives a permanent low-maintenance surface. One of the salient features of this finish is its anti-graffite properties. A special chemical additive enables defacing of the finish caused by felt pens, and aerosol paints to be quickly and easily wiped off using a special cleaning fluid.

While Ceramicoat can be directly applied to the concrete lining, it is preferable to remove fins and other surface imperfections in order to achieve a high class finish. Direct application of the Ceramicoat finish has the disadvantage that cracks and seepages in the concrete lining can mar the finish and result in an unsightly surface. Another disadvantage of this finish is its incompatibility with silicone form oils. Further, the surface hardness of the finish and its durability in comparison with ceramic tile has, as yet, not been demonstrated. However, the in-place capital costs of Pyrok finishes are comparable to paints and epoxy finishes and both are considerably more economical than ceramic tile.

Ceramic Tile. Traditionally, a light-colored ceramic tile has been the accepted and proven finish for walls and ceilings in highway tunnels.

The usual tile size is $4\frac{1}{4}$ by $4\frac{1}{4}$ inches, but 6 by 6, 6 by 12 and other sizes have been used. Generally, the manufacturing tolerances for size and warpage versus the labor costs for a reduced number of tile units have favored the $4\frac{1}{4}$ by $4\frac{1}{4}$-inch size tile.

Ceramic tile for tunnels should be vitreous glazed and of uniform thickness and should not vary from the nominal size by more than $\frac{3}{64}$ of an inch. It should have a cushioned edge, backs should be free from glaze, warpage should not exceed 0.02 inch in a length of $4\frac{1}{4}$ inches from a plane surface, wedging or crooked edges exceeding 0.70 of 1%, and from other imperfections such as pressing cracks, dents, swelling and chipping. These tiles are special for tunnel use and their specifications for absorption, crazing, thermal shock, weather, glaze hardness and bond can be found in most tunnel tile specifications.

The hard, durable color and finish of ceramic tile is not conducive to noise attenuation. However, their finish and color fastness, with lack of craze, has proven them to be the most universally accepted and proven tunnel finish material. They have a successful record of maintenance abuse, having withstood erosion and loss of glaze and color from mechanical abrasions and chemical reactions due to washing and scrubbing. They are highly fire-resistant, light-reflective and economical. Further, they are available in all colors and sizes. Since they are special materials, replacements (10% of total tiled surface) are usually provided for at the time of manufacture and stored at the tunnel site. Despite the ease of replacement, tile replacement is difficult to conceal, and the tiles are subject to displacement due to vibration and temperature changes. The joints between tiles become dirty, as well.

Porcelain Enamel. The application of porcelain enamel, a combination of glass and inorganic color oxides, was first used as a tunnel finish material on the Detroit-Windsor Tunnel for metal wall tile in 1928 and for a metal panel ceiling on the Downtown Elizabeth River Tunnel in 1950.

The porcelain enamel finish for tunnel usage should meet the requirements for acid resistance "Class A" of the Porcelain Enamel Institute, Inc., current edition for "Test for Acid Resistance of Porcelain Enamel."

Enamel coatings are applied on the front, sides and back of the metal panels. Total coating thickness varies between 0.003 and 0.006 inch. The soffit side of the panel, including the outside flanges, is clad with a ground and finish coat of porcelain enamel. The back of the panel and the inside of the flanges receive a ground coat.

The procelain enamel finish is usually medium gloss so that its light-reflecting quality will not produce undue glare, nor, on the other hand, absorb enough light to reduce illumination to the extent of creating unsafe driving conditions. Several tunnels have used a "ripple" surface finish. This type of finish is not recommended since it tends to increase specular light-reflection. Also, the textured surface harbors dirt and, hence, requires increased washing.

The porcelain finish must be factory applied, and requires reasonable care and protection during shipping and installation of the panels. Over 25 years of low-maintenance experience have clearly established the durability, colorfastness and abrasion- and impact-resistance of this finish under all traffic conditions. It is highly fire-resistant and is available in all colors. It is ideally suited for ceiling and curtain wall finish. Unfortunately, like ceramic tile, its hard durable surface is not conducive to noise attenuation.

FIRE-RESISTANCE

All materials used for interior tunnel finish should be non-flammable and not emit toxic gases or smoke during and after installation. Particular attention should be given to coatings, gaskets and cements and other finish materials to ensure that they do not create fire, loss of structural support or environmental or operational hazards during construction or operation of the tunnel.

Minor fires within highway tunnels are not uncommon. For example, it has been estimated that prior to 1949 over 50 such fires occurred annually in the Holland Tunnel. Damage to the unfinished concrete ceiling and tile walls due to these minor fires was inconsequential.

Shortly after the Holland Tunnel fire in 1949, the concept of a suspended metal panel ceiling with a concrete fill was initiated. The purpose of this ceiling was to limit structural damage in the event of a major fire and explosion and to permit ease and speed of erection of a new ceiling in order to provide ventilation for resumption of traffic. This innovative ceiling was installed on the Downtown Elizabeth River Tunnel in 1952. The fire-resistance of a porcelain-enameled metal panel ceiling with concrete fill has not been subject to a major fire; however, many minor conflagrations have resulted in only minor damage to the porcelain finish and the structural integrity of the concrete-filled metal panels.

Design of the concrete-filled metal panels was based on fire tests.[3] These tests indicated that for a minor fire (1200°F for 6 minutes) no damage to the porcelain enamel finish or the metal panel occurred. Minor damage, consisting of slight permanent deflection of the panel, occurred for a 1200°F, 31-minute fire. Further, tests for a 1575° to 1600°F fire showed that the panel conducted the heat and dissipated it over a large area such that the metal did not approach its melting point. However, permanent distortions under this condition would require panel replacement.

NOISE-ATTENUATION

Two types of noise are associated with vehicular tunnels: those within the tunnel, which are due to reverberation of sound off the interior surfaces of a highway tunnel and are generated principally by vehicular traffic, and, secondly, the noise produced at the ventilation buildings due to exhaust air and ventilation fans. The former affect the motorist and tunnel patrolman, while the latter have an impact on the general environment in the vicinity of the ventilation structures and portals.

Noise-attenuation at the ventilation structures is a determinant problem. The factors to be considered are the general level of noise in the area of the ventilation buildings and the required, if necessary, attenuation. Attenuation of fan and exhaust noise can be provided by installing noise absorbing baffles in the evase stacks to within reasonable levels, but requires increased fan horsepower (see Chapter 19).

Reduction of noise within the tunnel has not been resolved. Acoustical ceilings, tilting of sidewalls and various other noise-attenuating methods have been used at Velsen and other European Tunnels and the Hong Kong Tunnel.[4] At best, the sound reduction has been 10%.

Fortunately, with increasing traffic, most vehicles are equipped with air-conditioning. Hence, the noise in the tunnel is isolated from the passenger by closed vehicle windows.

However, non-air-conditioned cars and tunnel police are still subject to air blast and other tunnel noises. Tunnel finishes that are light-reflective and durable and meet all other requirements as set forth herein do not, by their reflective nature, attenuate noise.

To date, no satisfactory tunnel finish has been found that will attenuate noise while providing a satisfactory level of maintenance. The current state of the art of noise-attenuation is based on absorbing noise within a cellular material. This material can consist of glass wool, asbestos compounds or synthetic materials, which in combination with corrosion-resistant metals fabricated in egg crate or perforated sandwiches, will absorb noise and provide a washable and maintenance-free finish. Eventually, however, with the residue from exhaust gases and dust in the air, the cellular structure becomes clogged and the noise-attenuation becomes less effective. Use of baffle type surfaces without cellular sound-absorbing material is less effective than the sandwich construction and still leaves the problem of cleaning the baffles.

Considerable study and increased technology are required in this area.

REFERENCES

1. "Tunnel Panelling," *Tunnels and Tunnelling,* p. 368, September 1971.
2. "The Holland Tunnel Chemical Fire, New Jersey–New York," Report by National Board of Fire Underwriters, May 13, 1949.
3. Ricker, R.W. and Manley, C.R., "Aluminum Panels Pass Fire Tests," *Ceramic Industry,* June 1957.
4. Permanent International Association of Road Congresses, *XIII Congress,* p. 29, Tokyo 1967.

18
Service Buildings

JOHN O. BICKEL

Associated Consultant
Parsons, Brinckerhoff, Quade & Douglas

BUILDING REQUIREMENTS FOR HIGHWAY TUNNELS

Ventilation equipment, electrical switchgear and other mechanical equipment required for the operation of a roadway tunnel have to be housed in buildings. Space has to be provided for housing service vehicles; emergency trucks; shops for electrical and mechanical repairs; and storage of material and equipment needed for operation and maintenance of the tunnel. Facilities for personnel engaged in the operation have to be provided.

VENTILATION BUILDINGS FOR SUBAQUEOUS TUNNELS

Depending on the length of the tunnel, one, two or more ventilation buildings are required. They are usually located directly over the tunnel for the most efficient arrangement of the ventilation ducts. They house the fans and appurtenances, fire pumps, the emergency power plant, the switchgear and controls, the workshop for electrical and mechanical repairs, toilet facilities and a personnel elevator. The main control room is usually located in one of the ventilation buildings.

Foundations. The foundations of the ventilation buildings are incorporated into the cut-and-cover part of a sunken tube tunnel. In shield-driven tunnels, the ventilation structures are usually placed over, and form part of, the shafts used for starting the shields (Fig. 18-1).

Superstructure. The superstructure of the ventilation building may be steel frame or concrete construction. Walls in many of the early tunnels were brick. In more recent tunnels, precast concrete or other preformed, architectural panels have been used. The architectural appearance should harmonize with the surroundings and may be subject to approval by a local review board. In some cases, there may be restrictions for the height of a building which could have an influence on the type of ventilation fans selected. Figure 18-2 shows a recent building.

Fan Room. The size of the fan room is determined by the number and type of ventilation fans. Vane axial fans require less floor area than centrifugal fans of the same capacity. Adequate space must be provided for removal of fans, or parts of them, for repairs and main-

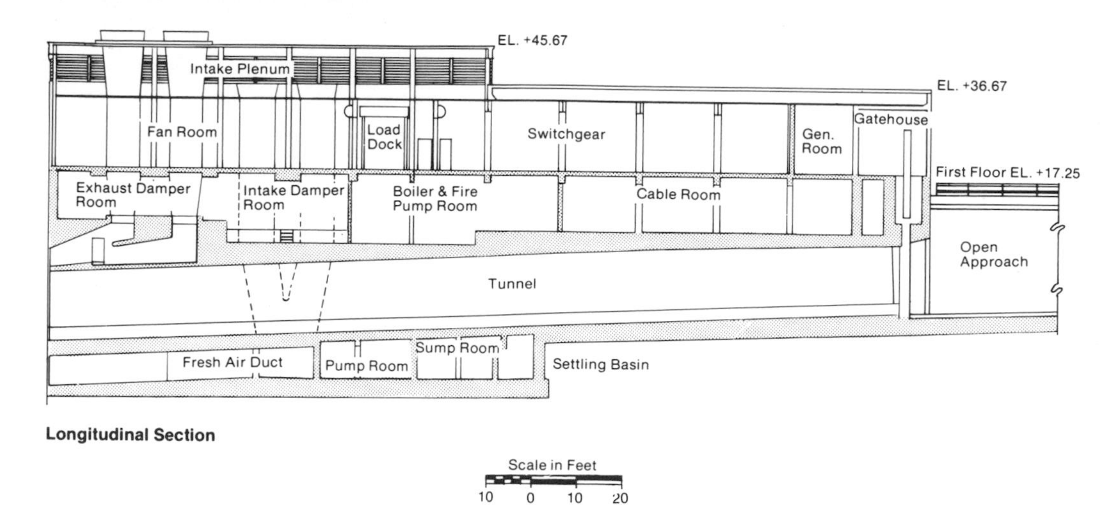

Fig. 18-1. Substructure, ventilation building, Second Hampton Roads Tunnel.

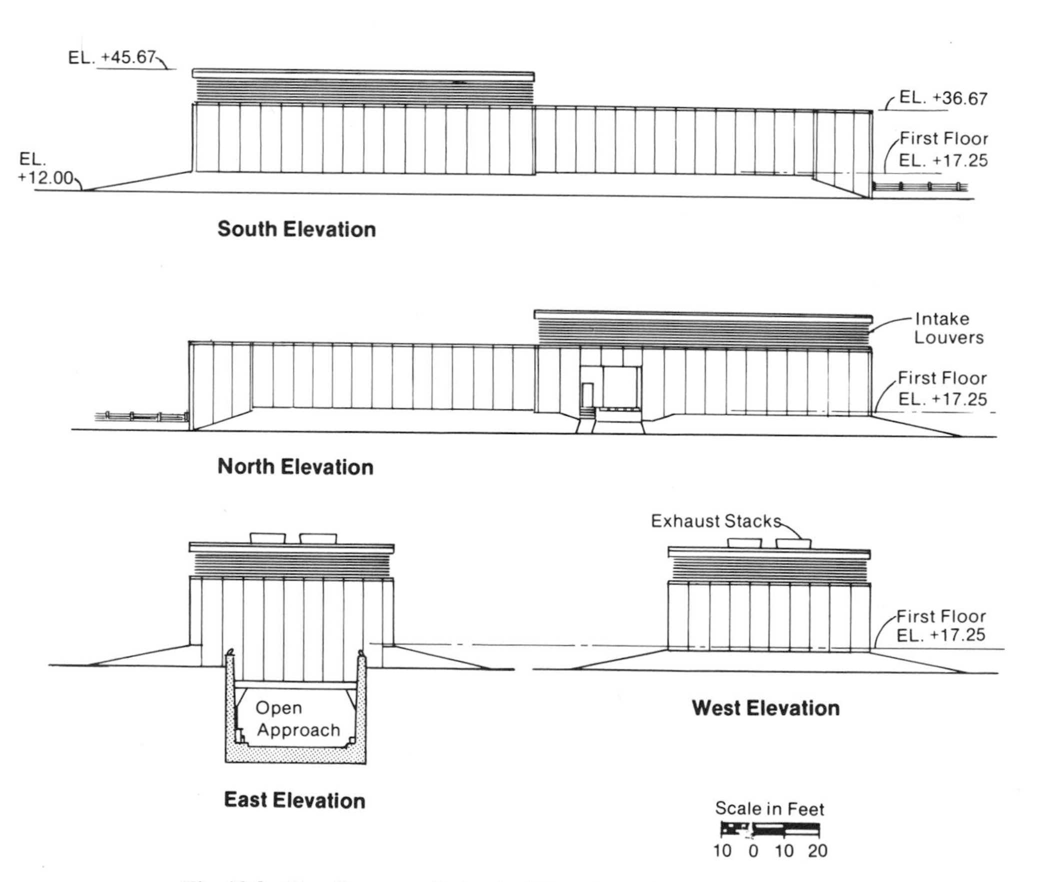

Fig. 18-2. Elevations, ventilation building, Second Hampton Roads Tunnel.

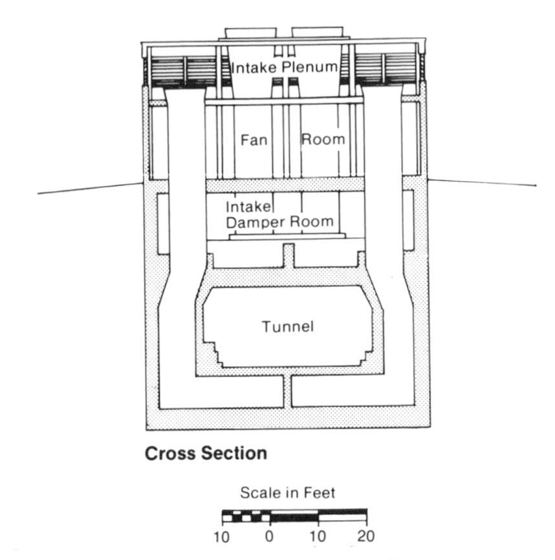

Fig. 18-3. Fan room vane axial flow fans.

tenance. Trolley beams and hoists are required to move equipment, capable of handling the heaviest piece to be handled. Figure 18-3 shows a plan and section of the Hampton Roads Tunnel's ventilation building with vane axial fans. Louvered openings are provided in the walls for taking in air for the supply fans. They are equipped with weatherproof louvers of corrosion-resistant metal, and screens to keep out birds. In very exposed locations, rain baffles may be installed behind the intake louvers. The stacks for the exhaust fans are extended through the roof of the fan room and protected with bird screens. Acoustical baffles may be required for air intake and exhaust openings for noise abatement if fan sounds are objectionable in the surroundings. Drainage inlets are needed for the area of the supply fans for water, which may blow through the intake louvres. Figures 18-3 and 18-4 show fan room plans for vane axial and centrifugal fans.

Switchgear Room. This room contains the transformers and switchgear for electric power to all motors and their control. Space is provided in front and back of the equipment for maintenance. The room must be adequately ventilated to dissipate the heat given off by the transformers. Figure 18-5 shows a typical switch-gear room.

Control Room. This room provides space for the central control board, the operator's desk and all indicating and monitoring devices for ventilation, traffic surveillances and control and communication systems. A battery room for the control system is located next to the control room.

Miscellaneous Rooms. Other rooms are provided for the fire pumps and valves, an air compressor, an air tank and an emergency generator (if used). The room for the latter requires ventilation to engine-cooling and combustion air.

Workshop. The workshop contains work benches, equipment and tools for maintenance and repair of electrical and mechanical equipment. Space is also provided for spare parts

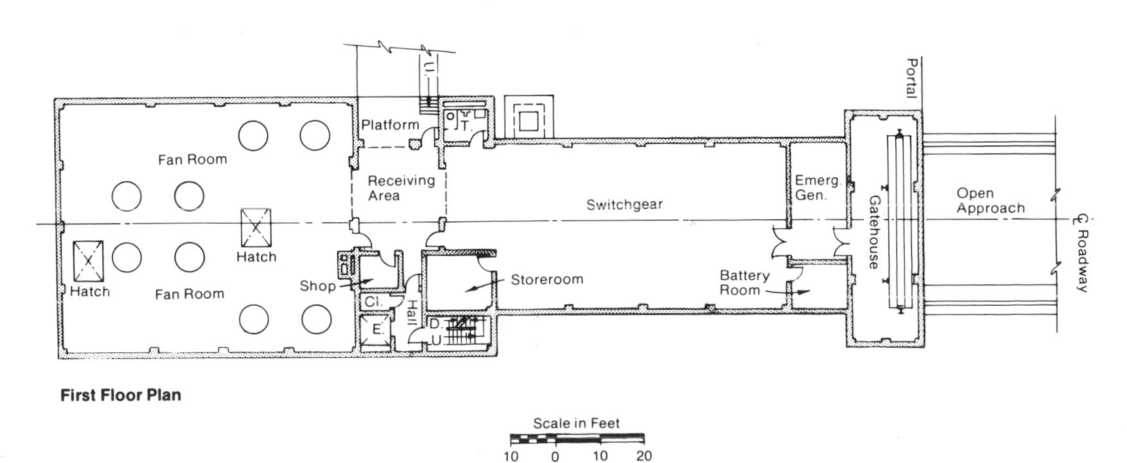

Fig. 18-4. Plan—fan and switchgear rooms, vane axial fans, Second Hamption Roads Tunnel.

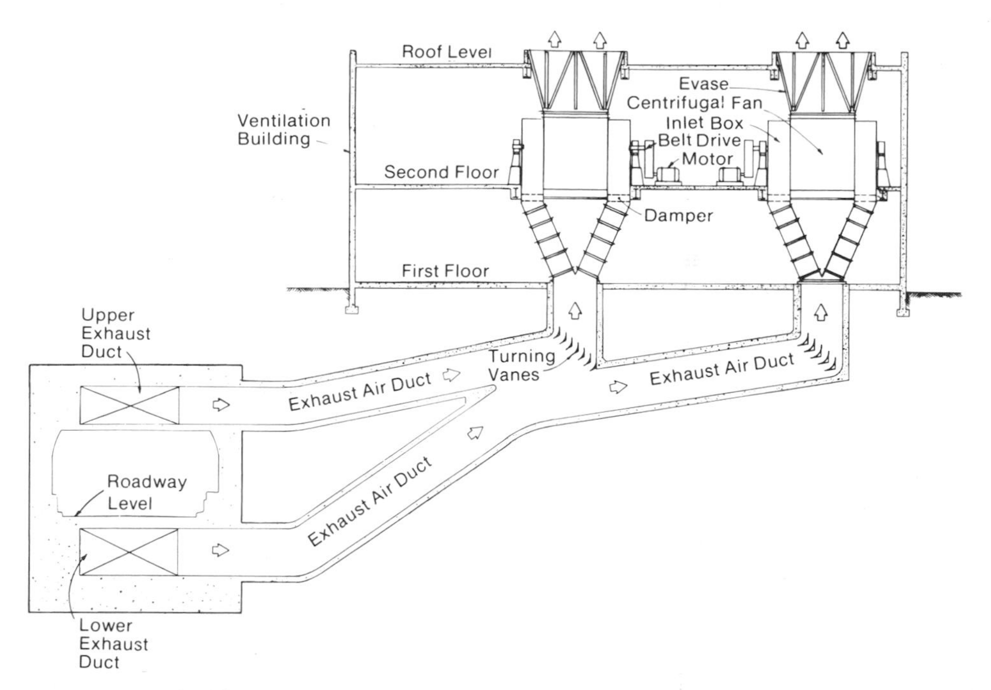

Fig. 18-5. Fan room, centrifugal fans, Downtown Elizabeth River Tunnel.

kept on hand. These include elements for control relays, relays, contacts for the circuit breaker, bulbs for illumination and indicating lamps and valves and gaskets for pumps and pipe systems. Fluorescent lamps for tunnel lighting may be stored in incidental areas of the building.

Heating Plant. Switchgear and control rooms and all work areas and washrooms are heated, usually by steam from a plant located in the basement of the building. Unit heaters are mounted in the switchgear room, radiators or convectors in other spaces. Air conditioning in the control room may be required in hot climates.

Elevator. An elevator for passengers provides access to all levels of the building from the roadway to the top. It usually has space for four people and has to comply with the safety laws of the area.

VENTILATION BUILDINGS FOR MOUNTAIN TUNNELS

For mountain tunnels the ventilation buildings are incorporated in the portal structures. Centrifugal fans are more often used than vane axial fans. The building has to provide space for similar equipment, as described above (Fig. 18-6). In some very long European highway tunnels, such as the new 14-kilometer St. Golthard tunnel in Switzerland, three intermediate ventilation stations with inclined shafts from the surface are provided in addition to the portal ventilation buildings.

GARAGE AND SERVICE BUILDINGS

Buildings are required to house emergency trucks, tunnel-washing trucks and other service vehicles, space for tunnel guards and the emergency crew and a shop for vehicle maintenance.

Emergency Truck Garages. These are located near the upper ends of the open approaches at both ends of a subaqueous tunnel with quick access to the roadway. They have toilet facilities for the stand-by crews of the crash truck. A lighter tow-car is also stationed there (Fig. 18-7). In the larger twin tunnels, such as the Lincoln Tunnels in New York, the crash truck stations are built between the adjacent tunnel portals.

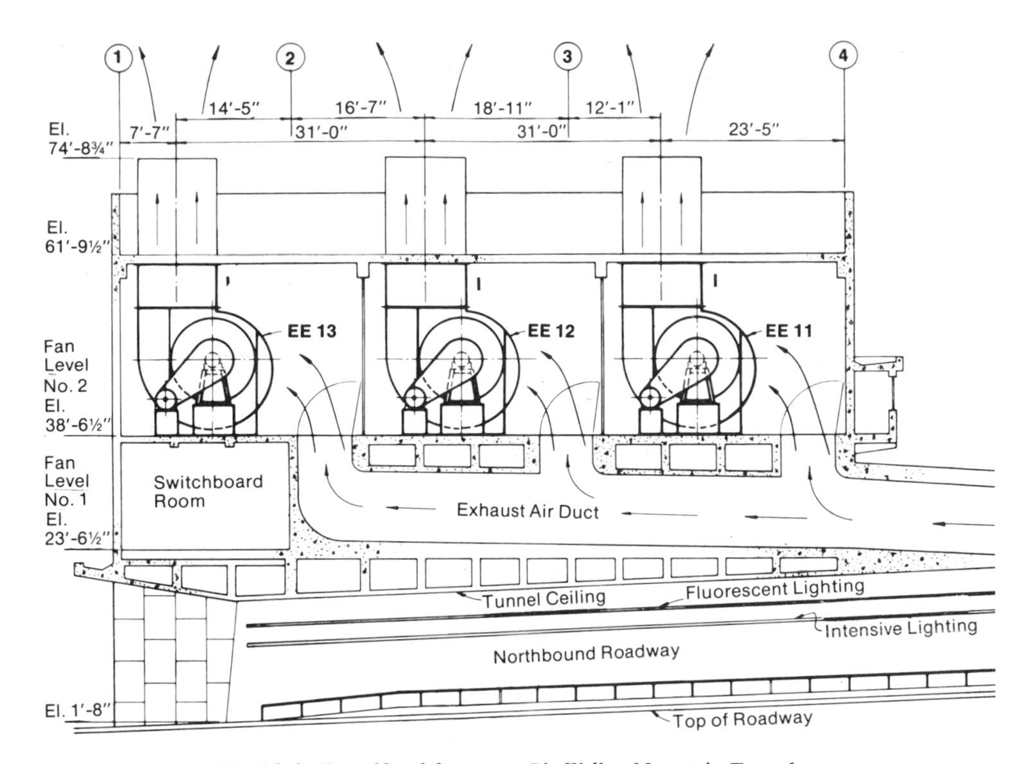

Fig. 18-6. Centrifugal fan room, Big Walker Mountain Tunnel.

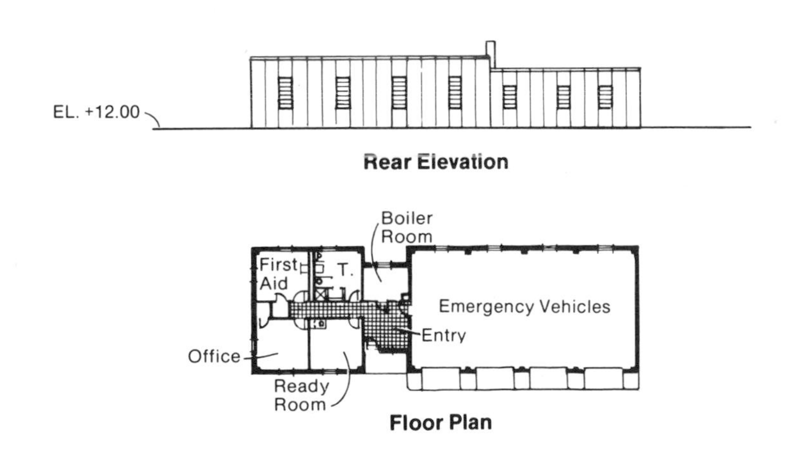

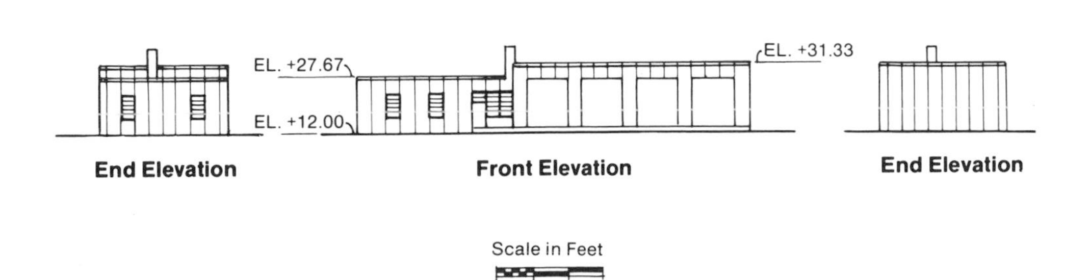

Fig. 18-7. Crash truck garage, Second Hampton Roads Tunnel.

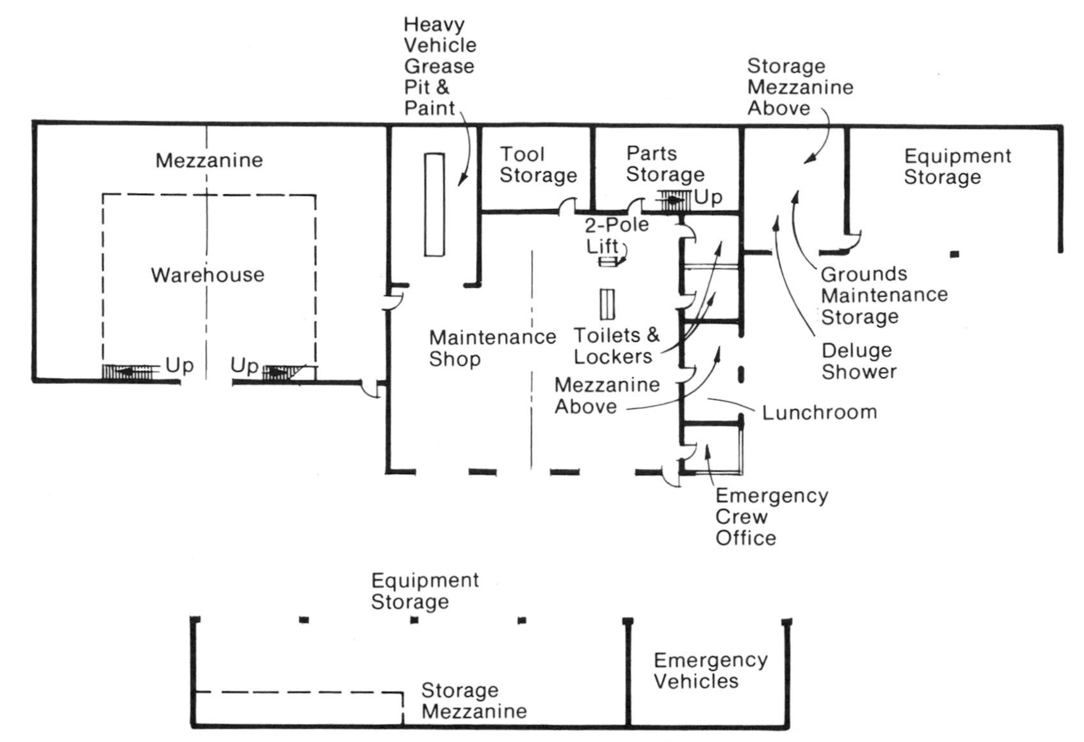

Fig. 18-8. Service vehicle garage, and maintenance shops, Downtown River Tunnel.

Service Vehicle Garage. This garage may be combined with one of the crash truck buildings. It provides space for the washing truck, the platform truck for maintenance of tunnel lights and several service cars for patrol and maintenance personnel (see Fig. 18-8).

Maintenance Shop. This is for routine maintenance of all service vehicles and may be in a separate building or combined with the garage. It contains work benches with the necessary machine tools, storage for tools and replacement parts, a grease pit, burning and welding equipment, hot and cold water, a steam supply and an air compressor. In cold climates, a closed storage for sand and salt for roadways should be included. Fuel pumps for vehicle supply are located at the maintenance shop.

19
Tunnel Ventilation

ARTHUR G. BENDELIUS

Vice President
Senior Professional Associate
Parsons, Brinckerhoff, Quade & Douglas

INTRODUCTION

Whereas tunnels themselves date back to early civilizations, the ventilation of tunnels has taken on greater significance only within the past 100 years, as increasing quantities of combustion products and heat have become more troublesome. Ventilation of a tunnel is required whether it is of the vehicular, rail or rapid transit type.

The products of combustion generated in a vehicular tunnel can cause discomfort and illness to vehicular passengers, and therefore ventilation is required to dilute the contaminants. The chief purpose of a vehicular tunnel ventilation system is to provide a good, respirable atmosphere within the tunnel which will not have ill effects on the health of the tunnel occupants. The visibility within the tunnel will also be aided by the dilution effect of the ventilation air.

A railroad tunnel presents a somewhat different problem to the ventilation engineer. Besides combustion products, the heat generated by the locomotives within the tunnel must also be removed to permit proper engine cooling. An evaluation of the natural ventilation in a rapid transit tunnel (subway) must be made to determine if the heat emitted from the train is adequately being removed from the tunnel. Mechanical ventilation (or, possibly, cooling) is required if the natural ventilation does not adequately remove the heat. All tunnels must also be provided with a means of emergency ventilation.

GLOSSARY

Air Horsepower. The theoretical horsepower required to drive a fan assuming no losses in the fan or an efficiency of 100%.

Bidirectional Traffic. Traffic in a two-lane tunnel with each lane handling traffic in a different direction.

Consist. A group of rail vehicles making up a train.

Mixed Traffic. Traffic stream containing both trucks and buses along with passenger cars.

Threshold Limit Value-Time Weighted Average (TLV-TWA). The values for airborne toxic materials which are to be used as guides in the control of health hazards and represent time-wieghted concentrations to which nearly all workers may be exposed 8 hours/day over extended periods of time without adverse effects.[1]

Traffic Density. The number of vehicles occupying a unit length of the traffic lanes of a roadway at any given instant, usually expressed in vehicles/mile.

Traffic Volume. The number of vehicles that pass over a given section of a lane or a roadway during a time period, usually one hour.

Unidirectional Traffic. All traffic in a tunnel traveling in the same direction.

Vehicle Headway. The interval, in time, between individual vehicles measured from head to head as they pass a given point, usually in seconds.

Vehicle Spacing. The interval, in distance, from head to head of successive vehicles, usually in feet.

VEHICLE EMISSIONS

The exhaust emissions from internal combustion engined vehicles, both spark-ignition and compression-ignition type, are of concern to the tunnel ventilation engineer. Each of these engine types generates exhaust gases having different characteristics.

The majority of the passenger cars on the road today are powered by spark-ignited engines fueled by gasoline. Trucks, on the other hand, have both spark-ignited and compression-ignited engines. The compression-ignited engine is most prevalent in large buses; however, some small buses do have spark-ignited engines.

The *spark-ignition or Otto cycle engine* utilizes a volatile liquid or gas fuel. Gasoline is the predominant fuel in road vehicle engines. However, some LPG-fueled and propane-fueled vehicles have appeared on the roads. The major constituents of the exhaust from a spark-ignited engine are carbon monoxide, carbon dioxide, sulfur dioxide, oxides of nitrogen and unburned hydrocarbons. Table 19-1 lists these exhaust gas components and their relative percentage by volume.

Table 19-1. Typical composition of spark-ignited engine exhaust.[2]

Carbon monoxide	3.0000%
Carbon dioxide	13.2000%
Oxides of nitrogen	0.0600%
Sulfur dioxide	0.0060%
Aldehyde	0.0040%
Formaldehyde	0.0007%

Adapted from Reference 2

The *compression-ignition or Diesel cycle engine* utilizes a liquid fuel with a low volatility. The fuels for these engines range from kerosene to crude oil but are usually diesel oil. Nitrogen dioxide, carbon monoxide, carbon dioxide and sulfur dioxide are the major components of diesel engine exhaust, as shown in Table 19-2.

Types of Vehicle Emissions

Carbon Monoxide. Carbon monoxide (CO) is an odorless, toxic gas which is present in the exhaust of both spark- and compression- ignited engines. In the human body, carbon monoxide combines with the blood hemoglobin much like oxygen to form carboxyhemoglobin (COHb). Carbon monoxide has an affinity for blood hemoglobin 300 times that of oxygen; therefore, inhaled carbon monoxide interferes with the capacity of the blood to transport and release oxygen to the tissues. This means that carbon monoxide is absorbed in preference to oxygen. Figure 19-1 presents the relationships among the level of carboxyhemoglobin in blood,

Table. 19-2. Typical composition of compression-ignited engine exhaust.[2]

Carbon monoxide (maximum)	0.100%
Carbon monoxide (average)	0.020%
Carbon dioxide	9.000%
Oxides of nitrogen	0.040%
Sulfur Dioxide	0.020%
Aldehyde	0.002%
Formaldehyde	0.001%

Adapted form Reference 2

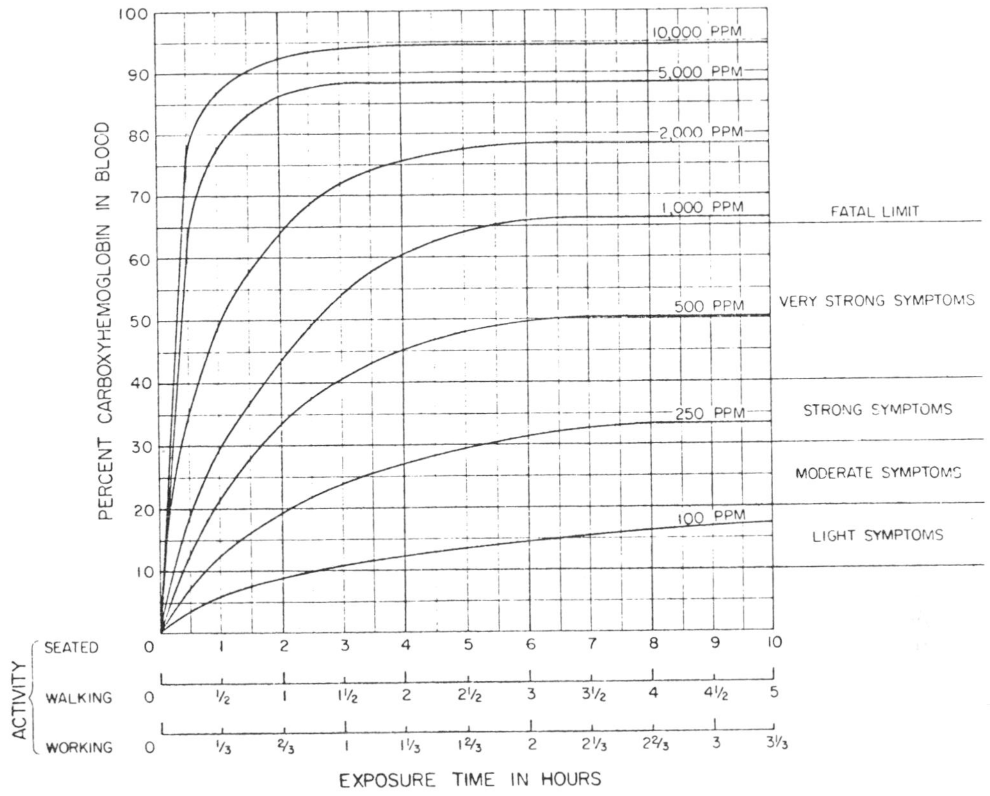

Fig. 19-1. Relation of carboxyhemoglobin in blood and carbon monoxide content of air, exposure time and activity.[2] (Adapted from Ref. 2)

carbon monoxide content of inhaled air, exposure time, level of activity, and the resultant human reaction. As noted in Fig. 19-1, the fatal carboxyhemoglobin concentration is reached at approximately 65%. At about 10% carboxyhemoglobin, the first toxic effects become evident. Carbon monoxide poisoning is a reversible process; the carboxyhemoglobin level in the blood decreases as exposure to carbon monoxide free air is continued, as can be seen in Fig. 19-2. Characteristics of carbon monoxide are shown in Table 19-3.

Carbon Dioxide. Carbon dioxide (CO_2) is one of the most important products of combustion, but is toxic only in concentration levels which are well above those found in vehicular tunnels, as noted in Table 19-4 and 19-5.

Oxides of Nitrogen. Of the many oxides of nitrogen, only two are of concern as air contaminants produced by internal combustion engines: nitric oxide (NO) and nitrogen dioxide (NO_2). The toxic effects of these two are similar except that NO_2 is five times more toxic than NO.

Nitric oxide is a colorless, odorless gas formed during high-temperature combustion. The amount of nitric oxide produced increases with an increase in flame temperature. Nitric oxide is slightly soluble in water and has a great affinity for the blood hemoglobin (forming methemoglobin). This then produces a shortage of blood oxygen.

Nitrogen dioxide is a reddish-orange-brown gas which is almost insoluble. It has a characteristic pungent odor and is both irritating and toxic. Approximately 95% of the nitrogen dioxide inhaled remains in the body. Concentrations of 100 to 150 ppm are dangerous for exposures of 30 to 60 minutes. Nitrogen dioxide unites with water in the lungs to form nitrous acid and nitric acid. The nitric acid formed in the lungs destroys the alveoli and thus reduces the ability of the lungs to trans-

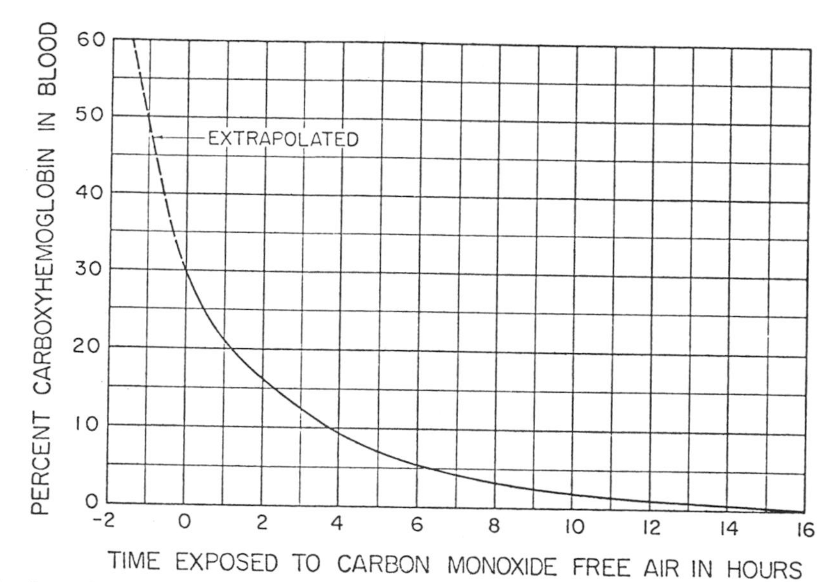

Fig. 19-2. Reduction of carboxyhemoglobin in blood when breathing carbon monoxide-free air and walking slowly.[2] (Adapted from Ref. 2)

port oxygen. Nitrous acid is a potent mutagen. Nitrogen dioxide also combines with hydrocarbons and sunlight to form smog, causing reduction of visibility by absorption of light. Figure 19-3 shows the transmittance of light at various nitrogen dioxide concentrations.

Table 19-3. Properties of carbon monoxide.[3]

Approximate molecular weight	28
Specific gravity relative to air	0.968
Density at 70°F	0.0724 lb/foot^3
Density at 32°F	0.0780 lb/foot^3
Gas constant, *R*	55.19 foot-lb/lb/°R

Adapted from Reference 3

Lead. Lead is a cumulative poison which affects the nervous, blood and reproductive systems, ultimately leading to lead poisoning and death. Although lead is found in air, water and food, most of the lead in the atmosphere is emitted by motor vehicles burning leaded fuels. Lead compounds such as tetra-

Table 19-4. Dilution of spark-ignited engine exhaust gas.[4,2]

	EXHAUST GAS COMPOSITION (PPM)	LEVEL AFTER DILUTION* (PPM)	THRESHOLD LIMIT VALUE TIME WEIGHTED AVERAGE
Carbon monoxide	30,000	200.00	50
Carbon dioxide	132,000	880.00	5,000
Nitrogen dioxide }			5
Nitric oxide }	600	4.00	25
Sulfur dioxide	60	0.40	5
Aldehyde	40	0.27	NA
Formaldehyde	7	0.02	2

*Diluted to maintain 200 ppm of carbon monoxide using a dilution ratio of 150 to 1.
Adapted from Reference 4, 2

Table 19-5. Dilution of compression-ignited engine exhaust gas.[4,2]

	EXHAUST GAS COMPOSITION (PPM)	LEVEL AFTER DILUTION* (PPM)	THRESHOLD LIMIT VALUE TIME WEIGHTED AVERAGE
Carbon monoxide	1000	6.70	50
Carbon dioxide	90,000	600.00	5000
Nitrogen dioxide }			5
Nitric oxide }	400	2.70	25
Sulfur dioxide	200	1.30	5
Aldehyde	20	0.13	NA
Formaldehyde	11	0.07	2

*Diluted to maintain 200 ppm of carbon monoxide using a dilution ratio of 150 to 1.
Adapted from Reference 4, 2

ethyl lead were used as a gasoline additive beginning in 1924 to increase the octane rating of fuel (approximately 2.7 grams/gallon of lead are added). Of this total lead added to the gasoline, approximately 70% is emitted from the exhaust, 30% falling out and 40% becoming airborne. Most of this remains in suspension for extended periods of time.

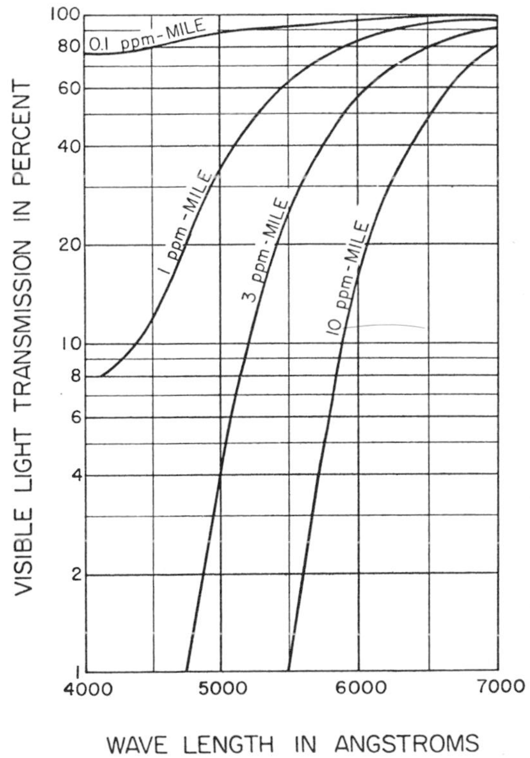

Fig. 19-3. Transmittance of visible light at several concentrations of nitrogen dioxide and viewing distances.[5]

Hydrocarbons. These are the most complex of all of the internal combustion engine emissions, although they only comprise a small portion of the total vehicle emissions. The major portion of the hydrocarbon emissions from vehicles are methane, ethane, propane, ethylene, acetylene, pentane and hexane. Hydrocarbons are known to aid in the formation of photochemical smog, but by themselves are not in large enough quantities in the tunnel environment to exceed the threshold limit value.

Sulfur dioxide. Sulfur dioxide (SO_2) is the most prevalent oxide of sulfur found in vehicle exhaust gas. Sulfur dioxide, a non-flammable, non-explosive, colorless gas, oxidizes in the atmosphere to form sulfuric acid and then reacts with other pollutants to form toxic sulfates. These sulfates affect the respiratory system. At concentrations greater than 3 ppm, sulfur dioxide has a pungent irritating odor. Sulfur dioxide is not a major component of vehicular exhaust gas and, therefore, appears in the tunnel environment in quantities considerably lower than the accepted level at which there is a perceptible response.

Aldehydes. These organic compounds are present in the internal combustion engine exhaust gas. All organic aldehydes are irritants either to the skin or mucous membranes or both. The irritant nature of these compounds provides sufficient warning to preclude serious health effects.

Odor. This is undoubtedly the most unpleasant part of vehicle exhaust. The most offensive odor in vehicle emissions comes from the diesel engine. There is, however, little knowledge on this subject since the methods of odor identification are extremely subjective and thus prone to wide variation. Pollutants such as aldehydes, nitrogen dioxide and sulfur dioxide have all been identified as possible contributors to odor.

Particulates. Most of the particulates in vehicle exhaust are those produced by incomplete combustion of hydrocarbon fuels. They tend to stay in suspension indefinitely and are within the respirable size. A further danger is the absorption of gases such as sulfur dioxide and oxides of nitrogen by these particulates, thus carrying these corrosive gases farther into the lungs. The particulates are the cause of smoke and haze present in many tunnels. The smoke is usually most dense when a significant number of diesel engined vehicles are present in the traffic stream.

Other gases are included in the vehicular exhaust stream, such as nitrogen, hydrogen and water vapor. However, these are not present in large enough quantities to be harmful.

Vehicle Emission Rates

History. Early work was done in this area by the U.S. Bureau of Mines around 1920.[6] This early research was conducted, and a series of tests performed, to enable the governing commissions to establish design criteria for the Holland Tunnel. Results of these tests have been used as a basis for design of most of the tunnels in the U.S.

These tests were, of course, conducted with vintage 1920 cars. Improvements in the automobile industry within the last 50 years make these data somewhat out of date. More recent work has been done by the Colorado Department of Highways in the mid-1960's, relating to the Straight Creek Tunnel.[7] However, this research and associated testing dealt with vehicle emissions at elevations of 5500 to 10,500 feet. The majority of tunnels built in the U.S. are at elevations well below 5500 feet.

A thorough evaluation has been prepared for tunnels located at or above 7000 feet by the Institute for Highway Construction of the Swiss Institute of Technology and published as Bulletin No. 10, "Report of Committee on Tunnel Ventilation to Swiss Department of Highways."[2] This work and the more recent tests conducted in Switzerland[8] are applicable to all types and sizes of vehicles, since they are based on the percent of carbon monoxide emitted and the vehicle weight.

An emission rate for each of the exhaust pollutants mentioned above can be computed in varying degrees of accuracy for both the spark-ignited and the compression-ignited engines. However, it can be seen in Tables 19-4 and 19-5 that if the level of carbon monoxide is maintained at or below 200 ppm (a 150 to 1 dilution ratio), all other constituents of vehicle exhausts will be well within the threshold limit value-time weighted averages TLV-TWA for each material. The TLV-TWA is defined by the American Conference of Governmental Industrial Hygienists as "the values for airborne toxic materials which are to be used as guides in the control of health hazards and represent time weighted concentrations to which nearly all workers may be exposed 8 hours per day over extended periods of time ("day after day") without adverse effects."[4]

Emission Rates for Moving Vehicles. The method of computing the emission rate of carbon monoxide will be considered for tunnel design. But the same approach could be used for all vehicle-emitted pollutants. For the spark-ignited engine, an approximate emission rate of 0.88 $feet^3$/ton/mile is used.[9] Future reductions in pollutant emissions by emission control or more efficient engines will require reevaluation of this factor. Such an evaluation should not be attempted until the technological advances providing the reductions have been proven.

For the diesel engine vehicle, a carbon monoxide emission rate of 0.026 $feet^3$/ton/mile should be used.[10]

Vehicle Emission Rates during Idle. When traffic is halted, the vehicle engine idles, thus producing up to 10% carbon monoxide in the exhaust gases. In this mode of operation, the engine consumes a low amount of fuel; thus, the total carbon monoxide emitted may not be much more than that for normal traffic.

Based on the gasoline-fueled engine of passenger cars consuming approximately 0.24 gallons/hour and having a 5% carbon monoxide exhaust content, the emission rate for idling passenger cars is 10.6 $feet^3$/car/hour.[9]

The emission rates outlined above for both moving and idle vehicles relate only to gasoline-fueled, spark-ignited engines. The emissions from compression-ignited engines are not as simple to determine. A compression-ignited engine should be operated with 20 to 40% excess air in order to reduce the smoke produced to an acceptable level. This, however, reduces the maximum power capability of the engine. Some amount of smoke will have to be accepted to enable the diesel-engined vehicle to attain high power and good acceleration. This problem is intensified when these engines are not properly maintained, as they are often deliberately reset to gain additional power.

Because of the numerous unknowns inherent in the operation of compression-ignited engines, in the past it has been the practice to assume all compression-ignited engines to be gasoline-fueled, spark-ignited engines for the purpose of computing the ventilation required. Recent tests in Switzerland have resulted in a method of determining the effect of diesel engine exhaust on the tunnel atmosphere (visibility), since the most hazardous component of diesel exhaust gases in normal conditions, is not the toxic gases but the smoke emitted. This smoke creates the haze found in most vehicular tunnels. It is composed mostly of carbon particles. The Swiss tests resulted in the establishment of a relationship between visibility and the concentration of diesel smoke. Figure 19-4 shows the test results from reduction of visibility.

An acceptable level of smoke must be established to ensure safety and comfort. From a safety standpoint, the stopping distance of a vehicle should be an excellent guide. For a vehicle traveling at 37 mph and having a stopping distance of 300 feet, from Fig. 19-4 it can be seen that this is between 2 and 4 Dmg/m^3, depending on the level of lighting in the tunnel.

Although the smoke emission rate for a supercharged diesel is lower than that for a naturally aspirated engine, an average value of 150 Dmg/ton/mile has been recommended for the present.[11]

Elevation Effect. At higher elevation, the air fuel mixture in spark-ignition engines becomes richer due to the reduced air pressure. The resultant increase in gasoline consumption for both partial and full load generates a considerable increase in carbon monoxide production at these higher elevations. If the carburetor is adjusted for sea level and the car operates at a higher elevation, the emission rate is even further increased.

For the compression-ignition engine, the reduced air pressure at the elevated location results in less excess combustion air and, therefore, creates an increase in smoke production. Smoke from diesel engines at higher elevations and steep grades can become intolerable within the tunnel environment, especially if the engines are not in proper adjustment and good condition.

Based on these facts, an adjustment to the vehicle emission rate is required at elevated locations. For elevations up to approximately 1300 feet, no correction is required. The correction factors shown in Fig. 19-5 should be applied for elevations from 1300 feet to 7000 feet.

There is an extremely limited amount of data available on the emission rates of vehicles at elevations above 7000 feet. The Colorado Department of Highways report prepared for the Straight Creek Tunnel[7] (elevation, 11,090 feet at center) contains test data for elevations from approximately 5500 to 11,350 feet (Figs. 19-6 and 19-7 show the results of these tests and can be utilized to approximate the emission rates at elevations beyond 7000 feet). A caution should be noted that the data available for these elevated conditions are limited. If possible, some site testing should be considered prior to initiation of design for a tunnel in such a location.

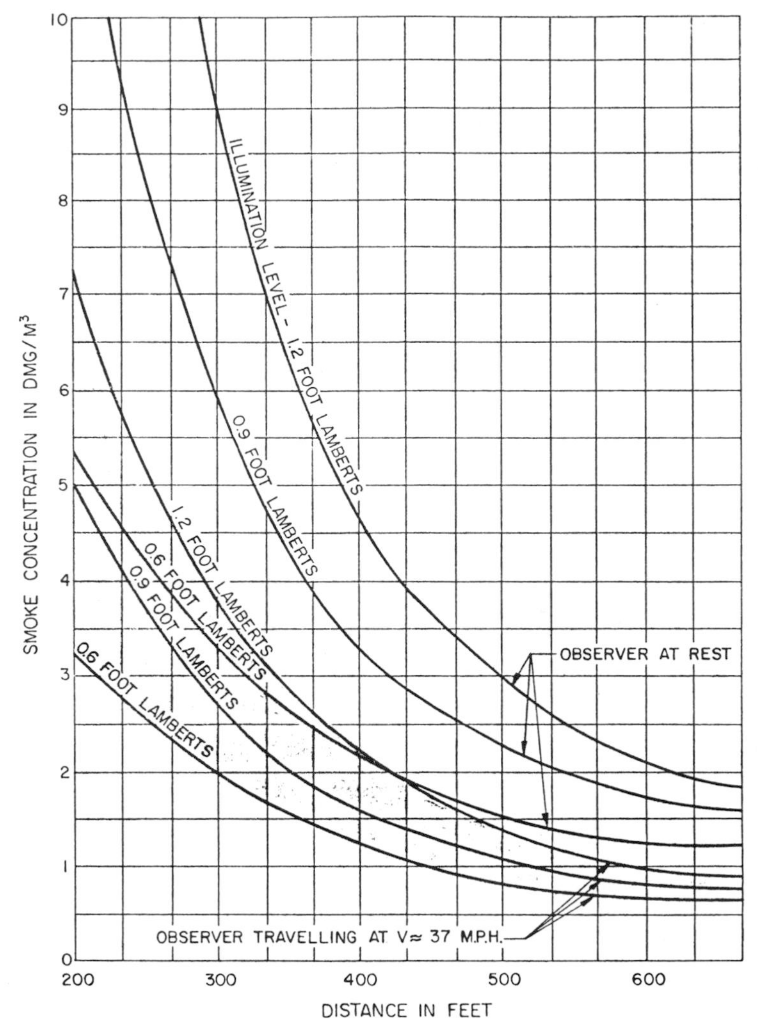

Fig. 19-4. Reduction of visibility created by smoke from diesel engines.[8] (Adapted from Ref. 8)

The effect of increased elevation is also severely felt by the human body and its ability to react to a carbon monoxide laden environment.

Grade and Speed Effect. The grade of the tunnel roadway and the vehicle speed have a tremendous effect on the emission rates.

On an upgrade, the increased fuel consumption results in a sizable increase in pollutant emission. The reduced fuel consumption on a downgrade produces a lower pollutant emission rate than on a level roadway. However, on extremely steep downgrades, the effect of a driver utilizing the vehicle's engine in lower gear ratios for braking will create a slightly increased pollutant emission rate.

Figures 19-8 and 19-9 are the results of tests conducted to evaluate the effect of grade and speed on vehicle emissions.[8] Figure 19-8 shows the factor for gasoline-engined passenger cars for grades of −6 to +6% and vehicle speeds of 3 to 37 mph. When using these figures for mixed traffic in one lane on an upgrade roadway, the truck and bus maximum speed will be a restricting factor on the car speed.

Figure 19-9 shows the effect of graded speed on the smoke generated by diesel-engined trucks and buses. The maximum vehicle speed shown

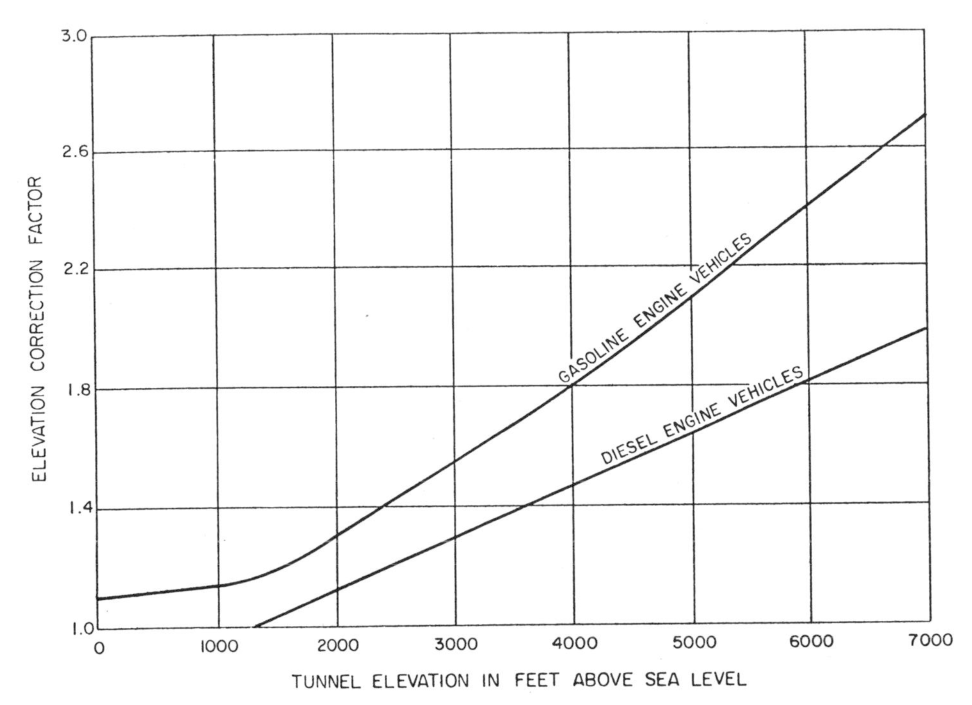

Fig. 19-5. Elevation correction factor for vehicle emissions.[12] (Adapted)

in this figure is based on an average vehicle with 11 hp/ton.

For vehicle speeds beyond those shown in these two figures, special studies may be required to determine the effect on vehicle emissions.

TRAFFIC

The tunnel ventilation engineer must be aware of the traffic characteristics of each tunnel for which he designs a ventilation system. The vehicle density, the traffic volume and the traf-

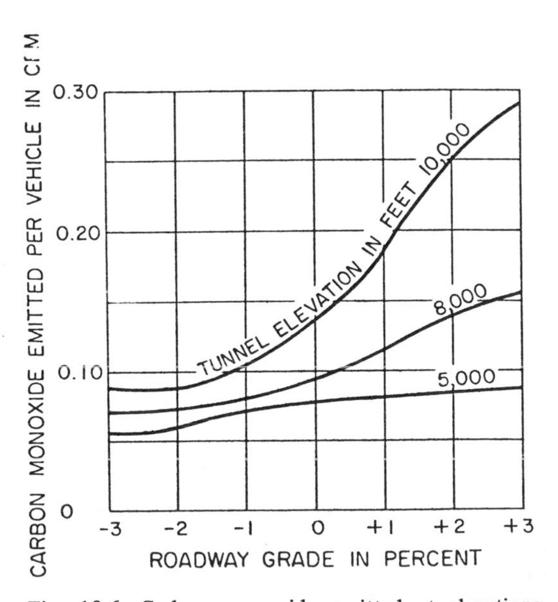

Fig. 19-6. Carbon monoxide emitted at elevations above 5000 feet by vehicles traveling at 40 mph.[7] (Adapted)

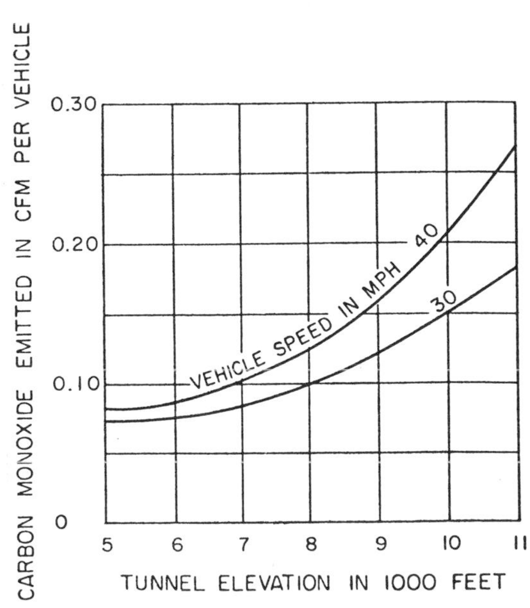

Fig. 19-7. Carbon monoxide emitted at elevations above 5000 feet on +1.68% grade.[7] (Adapted)

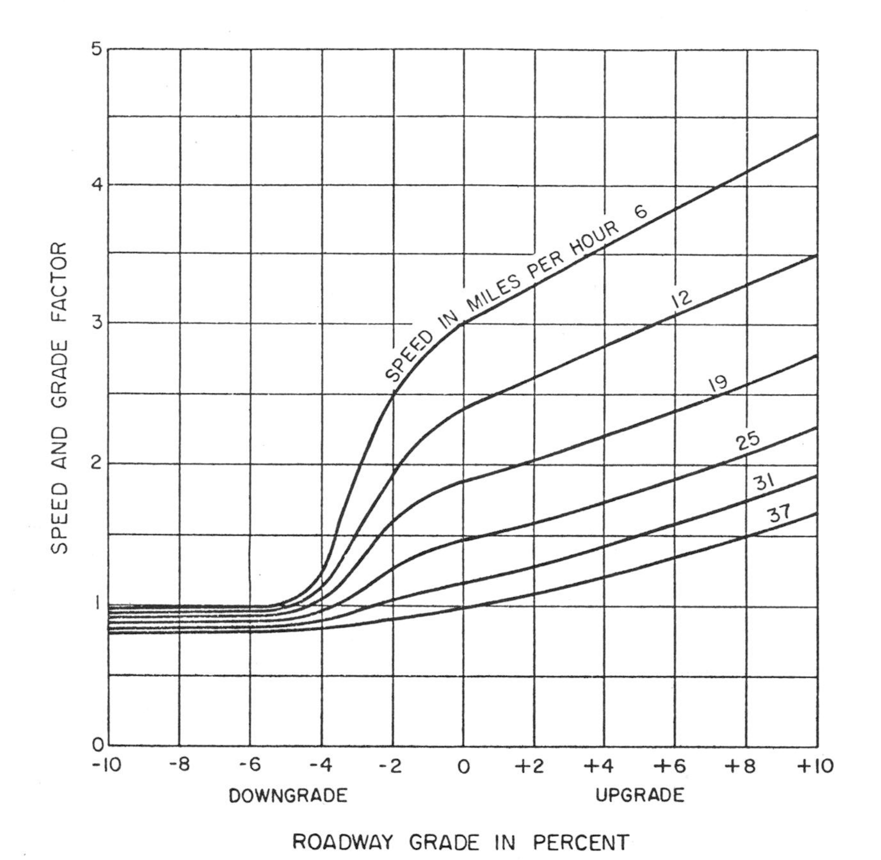

Fig. 19-8. Speed and grade factors for spark-ignited engine vehicles.[12] (Adapted)

fic composition have a direct relationship to the amount of carbon monoxide emitted within the tunnel. With this in mind, an evaluation is required to determine the proper design traffic volume. The maximum roadway capacity may not be the appropriate peak design values due to factors such as tunnel location and lane width.

Traffic Volume

A tunnel located in or near a highly populated urban area will undoubtedly utilize the maximum tunnel roadway capacity; however, a mountain tunnel located in a rural area will most likely not be utilized to full roadway capacity. The urban tunnel will be confronted with normal day to day congested traffic conditions, which also must be considered.

The maximum capacity of a standard 12-foot lane in a well-lighted, well-ventilated tunnel is approximately 2000 vehicles/hour. This is based on two lanes moving in the same direction. This maximum volume will be reduced by opposing traffic, grades, reduced land widths and vehicle mix. The effect of reduced lane width and lateral clearances (tunnel width) on free flowing traffic can be noted in Figs. 19-10 and 19-11. The number of trucks and buses in the traffic stream will, because of the varied vehicle length, cause a reduction in the traffic density (see Fig. 19-12 for a given vehicle speed).

In most instances, the traffic engineer will furnish a traffic analysis from which the proper tunnel design vehicle flow can be determined. There are times, however, when such an analysis is not available, such as during the early stages of a study, or when the traffic conditions to be studied occur at vehicle speeds other than the roadway design speed. In these cases, the ventilation engineer must be able to make a proper judgment and select a traffic volume.

The traffic volume can be computed utilizing the following relationships:

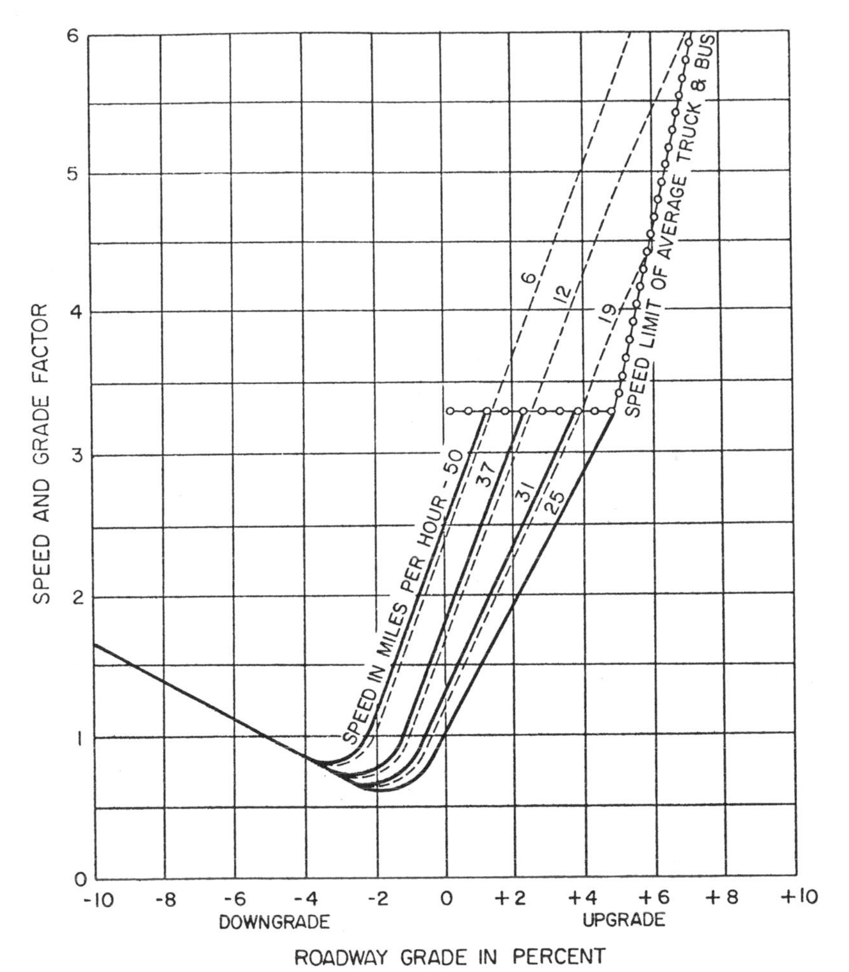

Fig. 19-9. Speed and grade factors for compression-ignited engine vehicles.[13] (Adapted)

$$TRV = \frac{3600}{AHY}$$

where

TRV = traffic volume (vehicles/hour)
AHY = Average headway (seconds).

$$TRV = \frac{TRS \times 5280}{VES}$$

where

TRS = traffic (vehicle) speed (miles/hour, mph)
VES = average vehicle spacing (feet, measured front to front).

An excellent set of guidelines to the average headway are the recommendations contained within the Highway Capacity Manual published by the National Research Council.[15] This manual states that "minimum headways vary from $\frac{1}{2}$ second to 2 seconds depending on the driver and traffic conditions." The average has been found to be $1\frac{1}{2}$ seconds. Since this minimum relates to an open roadway and a tunnel is restricted, a minimum headway for a tunnel of 1.8 seconds is more appropriate. This is compared to other methods used in the past and plotted in Fig. 19-13. Applying these values, the average vehicle spacing can be computed as follows.

$$VES = AHY(1.467\ TRS) + VEL$$

where

VEL = average vehicle length (feet).

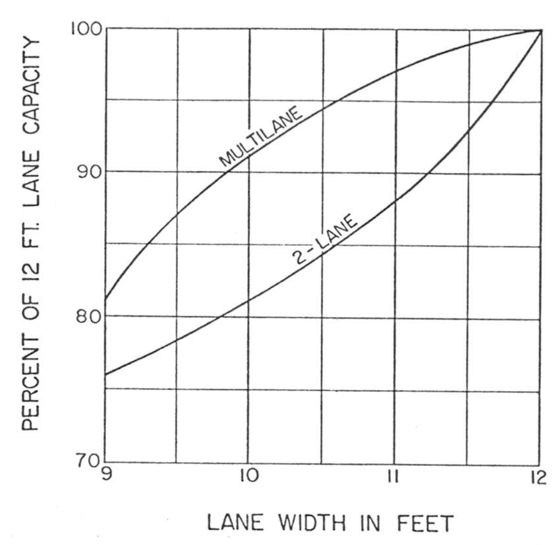

Fig. 19-10. Effect of lane width on lane capacity.[14] (Adapted)

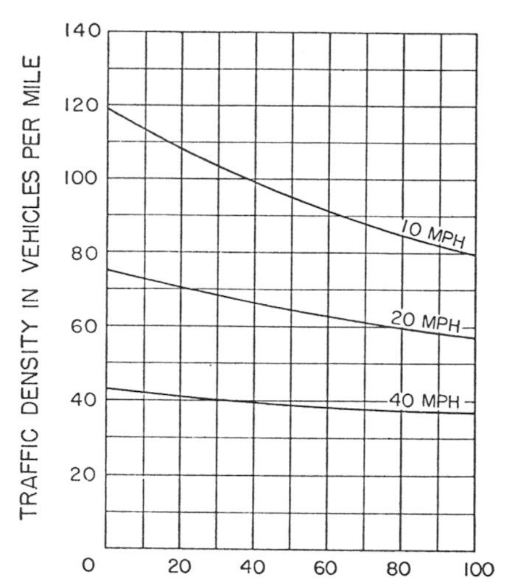

Fig. 19-12. Effect of trucks in traffic mixture on traffic density (based on 18-foot average car length and 40-foot average truck length.)

The traffic density can be computed as follows.

$$TRD = 1/VES$$

where

TRD = traffic density (vehicles/foot of roadway).

The preceding equations provide the necessary traffic data to evaluate the contaminants emitted within the tunnels. The curve plotted in Fig. 19-14 provides a guide to the maximum traffic volume at selected average vehicle speeds.

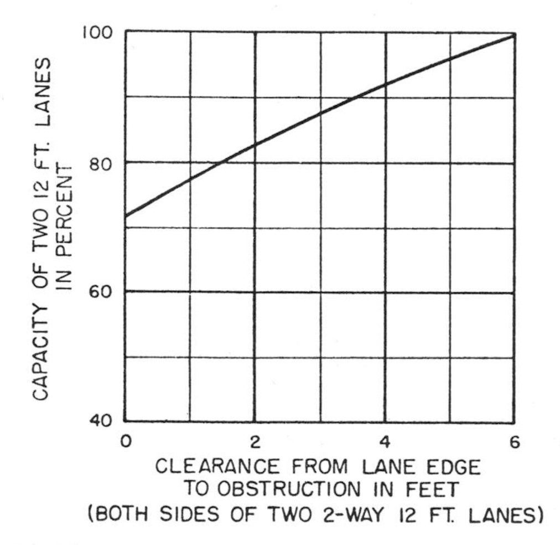

Fig. 19-11. Effect of lateral clearance on lane capacity.[14] (Adapted)

Traffic Composition

The percent of trucks and buses in any stream of traffic is another factor to be considered in the determination of the traffic capacity of a tunnel roadway. The presence of trucks and/or buses in a lane of traffic will decrease the density. The presence of an upgrade roadway within the tunnel will restrict the speed of the traffic stream. Figure 19-15 indicates the maximum truck and bus speed on upgrades. The number of trucks and buses in a traffic stream also bears heavily on the pollutant emissions, especially the oxides of nitrogen and smoke emitted by diesels.

An urban tunnel, such as one of those located in New York City, might experience a truck/bus percentage of 15, whereas the rural tunnel may see only 4 or 5% trucks/buses. The location of the tunnel should be the major factor in determining the truck/bus percentage in the traffic stream.

PERMISSIBLE CONTAMINANT LIMITS

Limits for the various contaminants present in the tunnel environment are required to protect the safety and health of both the motorist passing through and the attendant patrolling a tunnel. Carbon monoxide has been shown to

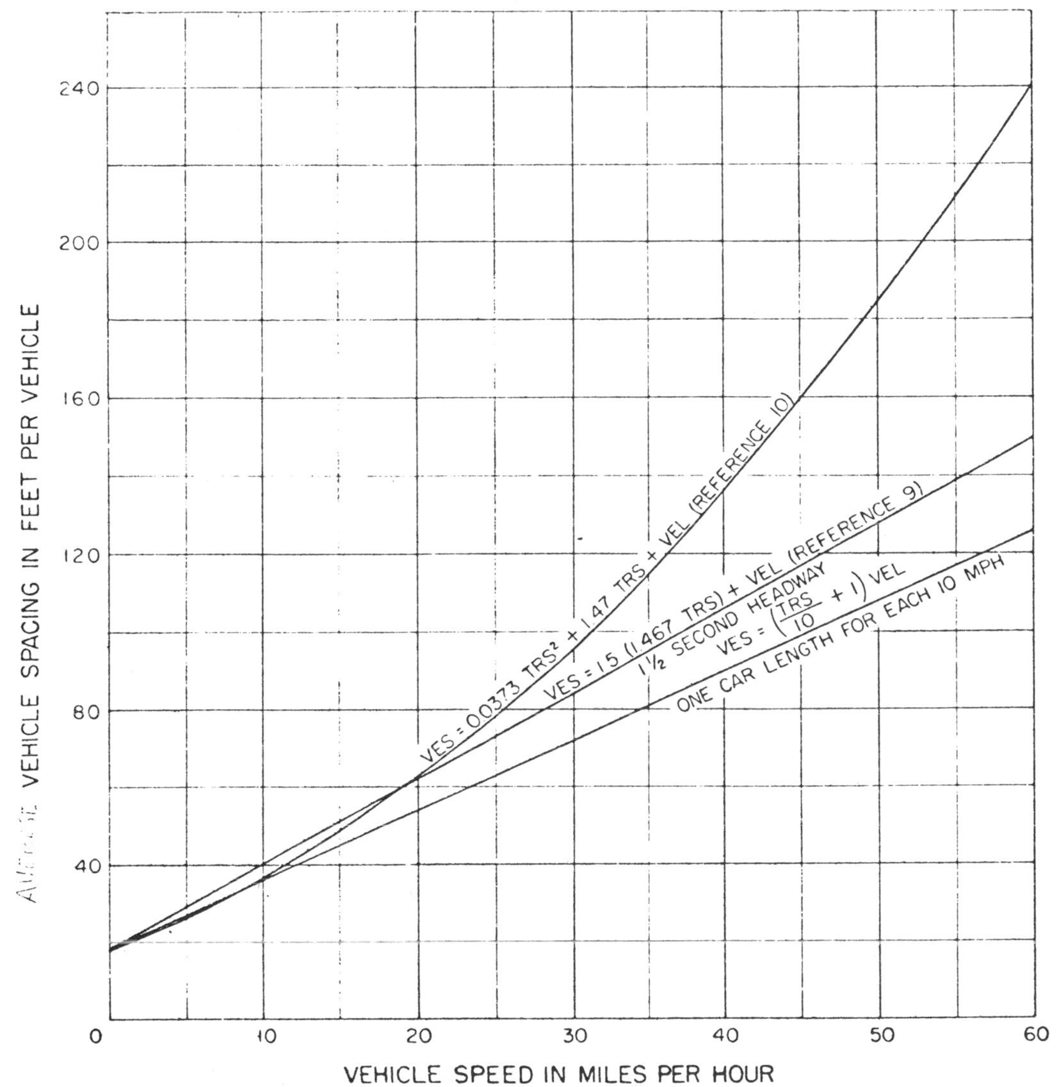

Fig. 19-13. Comparison of several methods of evaluating vehicle spacing.

be the key contaminant relating to health and safety within the tunnel. Consideration should also be given to the contaminants which create visual obscurity within the tunnel.

The first limits established in the U.S. were based on recommendations of the Bureau of Mines and were contained in a report regarding the ventilation of vehicular tunnels.[6] These recommendations were to provide sufficient air to prevent the carbon monoxide concentration from exceeding 4 parts in 10,000 (400 ppm) at the vehicle driver position. It was felt that exposure to this level of carbon monoxide for a 45-minute period provided an adequate factor of safety.

Years later, although the maximum limit on carbon monoxide concentration within the tunnel remained at 400 ppm, the actual value used for most tunnel designs was 250 ppm.[17] This meant that under most normal traffic conditions, the carbon monoxide concentration was less than 250 ppm and, when an emergency situation occurred, a maximum of 400 ppm was permitted. When the diesel-engined truck and bus became prevalent on the roads and in the tunnel traffic stream, it was found necessary

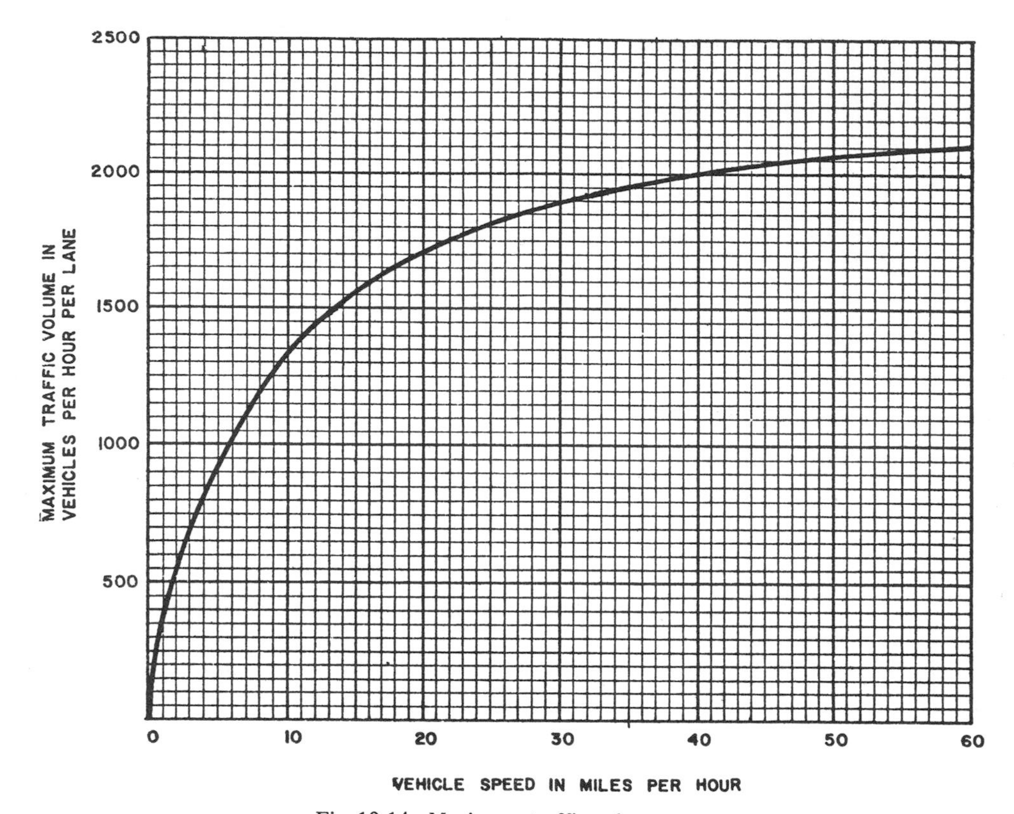

Fig. 19-14. Maximum traffic volume.

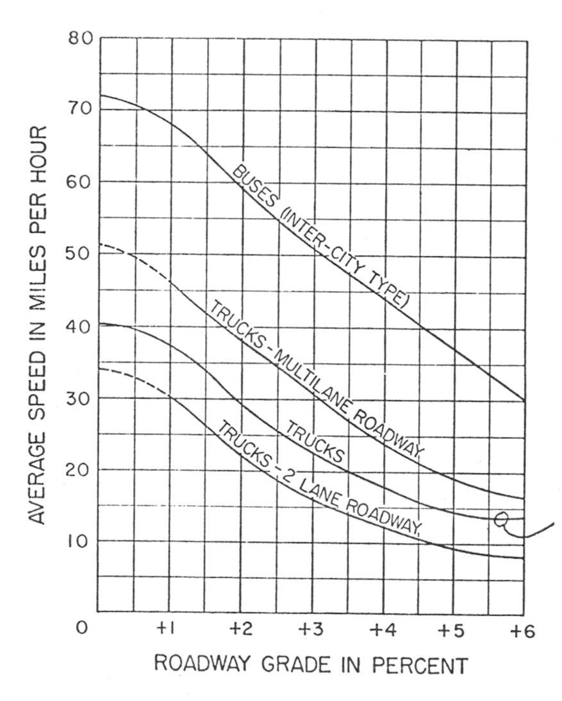

Fig. 19-15. Speed restriction of trucks and buses on upgrade.[14] (Adapted)

to establish the design limit between 200 and 250 ppm in order to maintain adequate visibility within the tunnel.[2] More recent design limits call for a carbon monoxide level of 150 to 250 ppm with an allowable emergency upper limit of 400 ppm.

The limits mentioned above were established for tunnels at or near sea level. If the tunnel is located at a much higher elevation, the carbon monoxide limits must be adjusted accordingly. In the study regarding the Straight Creek Tunnel in Colorado,[7] which is at 11,000 feet above sea level, the recommendation was made to establish a limit of 150 ppm of carbon monoxide for a maximum of 1-hour exposure and 100 ppm as a maximum average for any 1 hour. This tunnel was designed as an unmanned tunnel. Table 19-6 indicates the equivalent simple exposure (no carbon monoxide) for several carbon monoxide concentrations at 11,000-feet elevation.

The proposed new standards for the environment within tunnels are being generated by

Table 19-6. Equivalent exposure to carbon monoxide at 11,000 feet.[7]

EXPOSURE AT 11,000 FEET		SIMPLE EXPOSURE EQUIVALENT
TO (PPM)	FOR (MINUTES)	(NO CARBON MONOXIDE) AT ELEVATION (FEET)
	15	11,250
	30	11,500
50	60	11,750
	120	12,250
	180	12,750
	15	11,500
	30	11,800
100	60	12,500
	120	13,500
	180	14,250
	15	12,000
	30	12,700
200	60	13,750
	120	15,200
	180	16,200

state departments of labor and public health agencies, along with the Federal Highway Administration. In order to fully understand these proposed standards, manned and unmanned tunnels must be defined. A manned tunnel is one in which an attendant is on foot patrol within the tunnel environment, usually on the sidewalk. An unmanned tunnel, on the other hand, does not have any persons working within the tunnel who are not within a vehicle, with the possible exception of an occasional maintenance crew.

The manned tunnel, since it is the workday environment for a person, would fall under the jurisdiction of the U.S. Department of Labor's Occupational Safety and Health Act (OSHA). These standards establish requirements "to assure safe and healthful working conditions for working men and women."[18] OSHA has established the Threshold Limit Value (TLV) as adopted by the American Conference of Governmental and Industrial Hygienists (ACGIH) as the contaminant limits for the working environment. Excursion beyond these limits can be tolerated if (as defined by ACGIH) these excursions are compensated by equivalent excursions below the level during the working day.[19] These permissable excursions can then be applied to the tunnel environment. Table 19-7 illustrates the application of these time-weighted averages to the tunnel environment.

The normal exposure of a motorist in a tunnel is limited to 5 to 15 minutes/day. The motorist is the only "inhabitant" of an unmanned tunnel. Therefore, contaminant levels commensurate with the short-term exposure of the motorist must be selected. An attempt to set these limits have been made by the Federal Highway Administration in their report on tunnel ventilation.[20] Standards such as these are subject to question, since the data available relating to short-term exposure to carbon monoxide are extremely limited. There are other factors that must be considered in

Table 19-7. Permissable Excursions as applied to the tunnel environment.[4, 19]

CONTAMINANT	THRESHOLD LIMIT VALUE (PPM)	PERMISSABLE EXCURSION LIMIT (PPM)	SHORT-TERM EXPOSURE LIMIT
Carbon monoxide	50	75.0	400.00
Nitric oxide	25	37.5	35.0
Nitrogen dioxide	5	5.0	5.0

the tunnel environment, such as the visibility as it affects the traveling motorist. It was the question of visibility which helped establish the accepted standard of 150 to 250 ppm in an unmanned tunnel. The visibility within a tunnel is also a difficult item to evaluate, since the haze-producing contaminants are difficult to identify and quantify.

There is another area, vehicle emissions, which is presently being affected by legislation and will continue to be in the foreseeable future. The first vehicle emission values were established by the tests conducted prior to the Holland Tunnel design.[6] These tests showed carbon monoxide concentrations of from 5 to 9% by volume in the exhaust gas stream. These data formed the basis of design for many of today's vehicular tunnels designed in the U.S. The carbon monoxide concentrations in automotive exhaust gases have fallen off sharply through the years. In 1960, a set of data was published in Switzerland presenting a predicted range of 0.5 to 2.5% carbon monoxide in the automotive exhaust gas stream and 2.5% to 3.5% for the gasoline-engined truck exhaust.[2] These data have been refined further as a result of extensive tests conducted in Switzerland.[8]

Within the U.S. there is a growing trend toward reduction of the emissions from motor vehicles. California state standards required reduction to 1.5% carbon monoxide. Since then, the U.S. Environmental Protection Agency has established standards for exhaust emissions. The establishment of legal emission limits does not assure compliance of all vehicles on the road; therefore, some amount of judgment must be employed when using these legal limits for tunnel design.

In 1975, the U.S. EPA issued a supplement to their Guidelines for Preview of Environmental Impact Statements for Highway Projects. In this supplement the EPA issued the following.

For the users of highway tunnels at or near sea level, it has been determined that an adequate margin of safety would exist if the concentration of CO does not exceed 125 (ppm) and the exposure time does not exceed one hour.

Since the issuance of the EPA supplement, 125 ppm has been widely adopted in the U.S. as the design criteria for tunnels located at or below an altitude of 5000 ft.

The continuing reduction of the contaminant emission of vehicle exhaust provides a question regarding the future design of tunnel ventilation systems. As the emission rates decline, the required air quantities should also decline. However, the trend toward reduced environmental limits could counteract that trend; thus, the required quantities for ventilation of a tunnel could remain unchanged.

VEHICULAR TUNNEL VENTILATION SYSTEMS

All vehicular tunnels require ventilation. This can be provided by natural means, traffic-induced piston effects or mechanical equipment. Ventilation is required to limit the concentrations of obnoxious or dangerous contaminants to acceptable levels. The ventilation system selected should be the most economical solution with regard to both construction and operating costs, which will meet the specified criteria. Naturally, ventilated and traffic-induced systems are considered adequate for tunnels of relatively short length and low traffic volume (or density). The longer and more heavily traveled tunnels should be provided with mechanical ventilation systems.

Natural Ventilation

Naturally ventilated tunnels rely chiefly on meteorological conditions to maintain satisfactory environmental conditions within the tunnel. The piston effect of the traffic provides additional airflow when the traffic is moving. The chief meteorological condition affecting the tunnel environs is the pressure differential between two portals of a tunnel created by differences in elevation, differences in ambient temperatures or wind. Unfortunately, none of these factors can be relied upon for continued, consistent results. A sudden change in wind direction or velocity can rapidly negate all of these natural effects, including the piston effect. The sum total of all pressures must be of sufficient magnitude to overcome the tunnel resistance which is influenced by tunnel length, coefficient of friction, hydraulic radius and air density.

Airflow through a naturally ventilated tunnel can be portal to portal, as in Fig. 19-16(a), or portal to shaft, as in Fig. 19-16(b). The portal to portal flow type functions best with unidirectional traffic, which produces a consistent, positive airflow. As can be seen in Fig. 19-16, the air velocity within the roadway is uniform and the contaminant concentration increases to

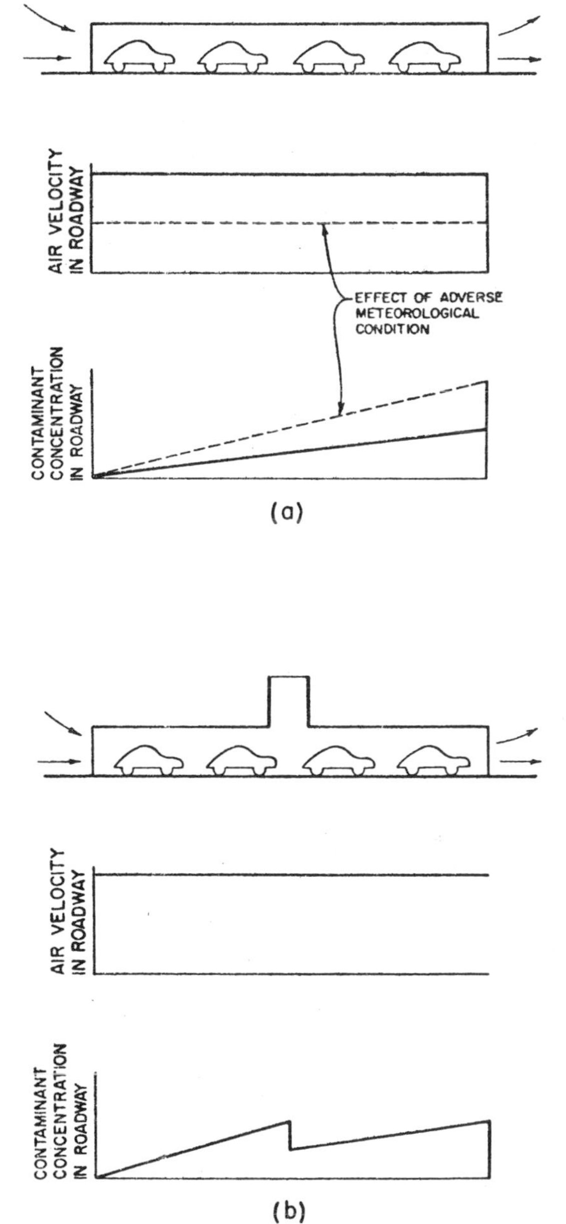

Fig. 19-16. Natural ventilation systems.[2] (Adapted)

a maximum at the exit portal. If adverse meteorological conditions occur, the velocity is reduced and the carbon monoxide concentration is increased, as shown by the dashed line in Fig. 19-16. If bidirectional traffic is introduced into such a tunnel, further reductions in airflow result.

The naturally ventilated tunnel with an intermediate shaft, seen in Fig. 19-16(b), is best suited for bidirectional traffic. However, the airflow through such a shafted tunnel is also at the mercy of the elements. The added benefit of the "stack effect" of the shaft is dependent on air and rock temperatures, wind and shaft height. The addition of more than one shaft to a naturally ventilated tunnel will be more of a disadvantage than an advantage since a pocket of contaminated air can be trapped between the shafts, thus causing high contaminant levels.

Because of the numerous uncertainties outlined above, it is rare that a tunnel of greater than 1000 feet in length is ventilated by natural means. There are exceptions to this, such as the Via Mala Tunnel in Switzerland and the Tenda Pass Tunnel between France and Italy. Via Mala, which is 2050 feet long, is located in an area where the traffic flow is extremely low. The 9000-foot long Tenda Pass Tunnel has a large difference in portal elevations, which creates a large pressure differential and thus adequate airflow. It also is located in an area having favorable wind conditions. Both of these tunnels have recently been outfitted with booster fans. Table 19-8 lists some of the naturally ventilated tunnels of the world.

It has been found necessary to install an emergency mechanical ventilation system in most naturally ventilated urban tunnels over 500 feet in length. This is required to purge smoke and hot gases generated during an emergency and to remove stagnated polluted gases during severe adverse meteorological conditions.

The reliance on natural ventilation for all tunnels over 500 feet should be carefully and thoroughly evaluated, specifically the effect of adverse meteorological and operating conditions. This is particularly true for a tunnel with an anticipated heavy or congested traffic flow. If the natural mode of ventilation is not adequate, a mechanical system with fans must be consid-

Table 19-8. Representative tunnels—natural ventilation.

TUNNEL	TYPE	LOCATION	COMPLETION DATE	LENGTH (FEET)	CONFIGURATION (TUBES/LANES)	PEAK TRAFFIC VOLUME (VEHICLES/HOUR)
Aragnouet Bielsa	Mountain	France/Spain	1972	9882	1/2	NA
East River Road	Mountain	United States	1936	5627	2/4	100
Lioran	Mountain	France	1847	4633	1/2	NA
FDR Drive	Urban	United States	1941	2412	2/6	5800
Big Oak Flat	Mountain	United States	1938	2152	1/2	NA
Maroggia	Mountain	Switzerland	1967	2067	2/4	NA
Blatt	Mountain	Switzerland	1969	1641	2/4	NA
Lake Washington	Urban	United States	1941	1559	2/4	5280
Mesaverde	Mountain	United States	1965	1470	1/2	100
Lake Way	Urban	United States	1950	1463	2/4	2500
Kaltwasser	Mountain	Switzerland	1966	1312	1/2	NA
Wilcox	Mountain	United States	1931	1312	1/2	NA
Ebenrain	Mountain	Switzerland	1969	1263	2/4	NA
Arch Cape	Mountain	United States	1937	1230	1/2	262
Colonial Parkway	Urban	United States	1942	1198	1/2	NA
Susten	Mountain	Switzerland	1945	1050	1/2	NA
Waldo Marin	Urban	United States	1955	1001	2/6	4400
Beaucatcher	Mountain	United States	1929	1001	1/2	1755
Nuuanu Pali	Mountain	United States	1959	1001	2/4	2775
Axenstrasse	Mountain	Switzerland	1939	945	1/2	1600
Washington Highway 12	Mountain	United States	1932	801	1/2	100
Sunset	Urban	United States	1941	744	1/2	770
Fredhall	Urban	Sweden	1967	656	2/6	NA

ered. There are several types of mechanical ventilation systems; the most appropriate for tunnels are outlined below.

Longitudinal Ventilation

A longitudinal ventilation system is defined as any system where the air is introduced to or removed from the tunnel roadway at a limited number of points, thus creating a longitudinal flow of air within the roadway (Fig. 19-17). The injection type longitudinal system has been used frequently in rail tunnels. However, it has also found application in vehicular tunnels. Air is injected into the tunnel roadway at one end of the tunnel, where it mixes with the air brought in by the piston effect of the incoming traffic

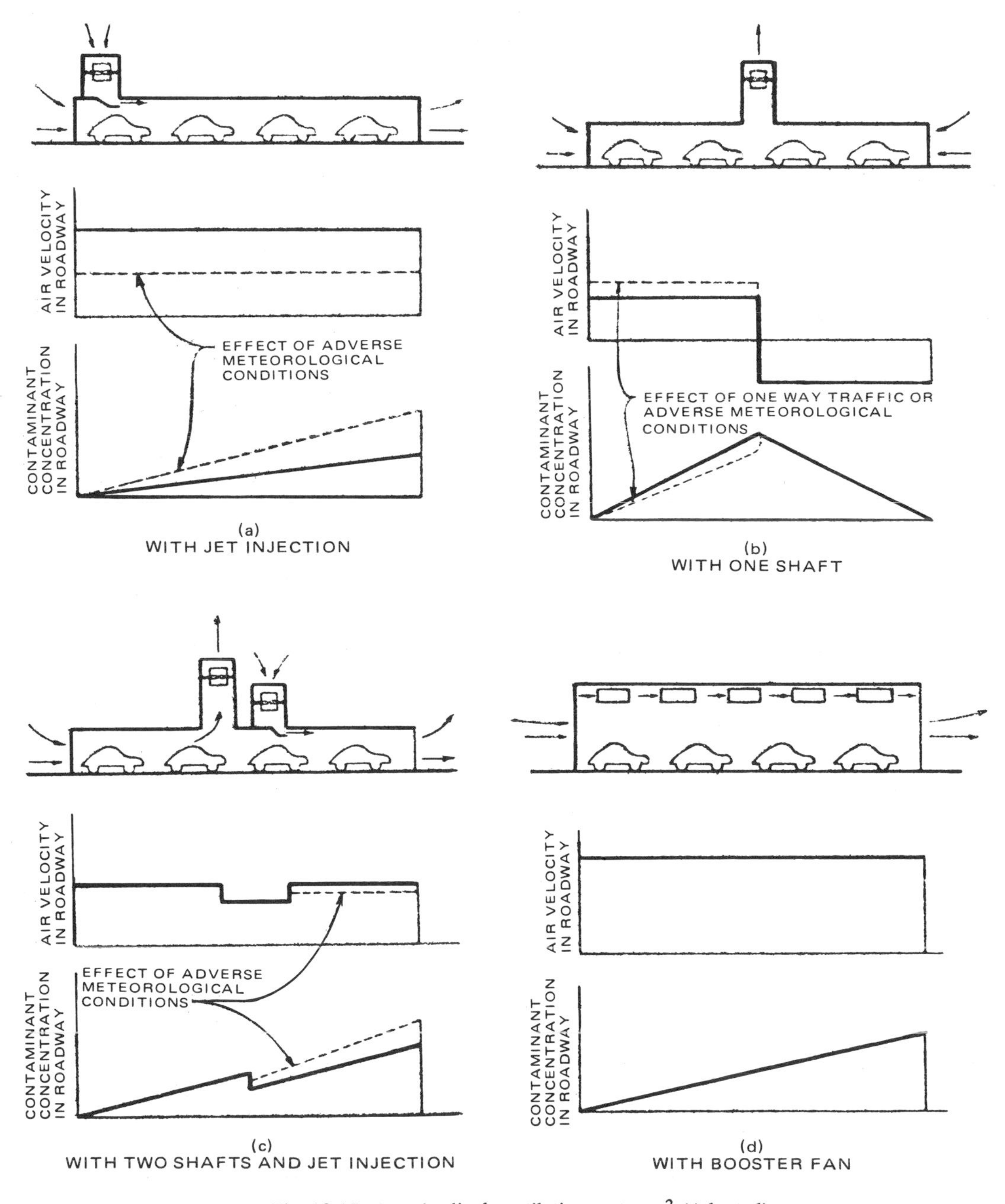

Fig. 19-17. Longitudinal ventilation systems.[2] (Adapted)

(Fig. 19-17a). This system is most effective in a unidirectional tunnel. The air velocity within the roadway is uniform throughout the tunnel, and the concentration of contaminants increases from zero at the entering portal to a maximum at the exiting portal. Adverse external atmospheric conditions can reduce the effectiveness of this system. The contaminant level increases at the exit portal as the flow of air decreases or the tunnel length increases.

The longitudinal system with a shaft is similar to the naturally ventilated system with a shaft except that it provides a positive stack effect (fan induced) Bidirectional traffic in a tunnel ventilated in this manner will show a peak contaminant concentration at the shaft location. This system is applicable only for bidirectional tunnels since for unidirectional tunnels, the contaminant levels become unbalanced.

The longitudinal system with two shafts located near the center of the tunnel, one for exhausting and one for supplying (Fig. 19-17c), will provide a reduction of contaminant concentration in the second half. A portion of the air flowing in the roadway is replaced in the interaction taking place at the shafts. Adverse wind conditions can cause a reduction of flow and a rise in contaminant concentration in the second half of the tunnel along with short circuitry of the fan flows.

Booster Fans. In a growing number of tunnels, a longitudinal ventilation system is achieved with booster fans mounted at the tunnel ceiling. Fig. 19-17(d). Such a system eliminates the need for space to house ventilation fans in a building; however, it may require a tunnel of greater height or width to accommodate the booster fans.

The standard longitudinal ventilation systems (excluding the booster fan system), having either supply or exhaust at a limited number of locations within the tunnel, are the most economical systems since they require the least number of fans, place the least operating burden on these fans and do not require separate air ducts. However, as the length of the tunnel increases, the disadvantages of these systems become pronounced, such as the excessive air velocities in the roadway and smoke being drawn the entire length of the roadway during an emergency. To alleviate these problems, a uniform distribution of air should be considered. Table 19-9 lists some of the longitudinally ventilated tunnels of the world.

Semi-Transverse Ventilation

Uniform distribution or collection of air throughout the length of a tunnel is the chief characteristic of a semi-transverse ventilation system. The supply air version of this system (Fig. 19-18a) produces a uniform level of carbon monoxide throughout the tunnel because the air and the vehicle exhaust gases enter the roadway area at the same rate. In a tunnel with unidirectional traffic, an additional airflow will be created within the roadway area. This system, due to the fan-induced flow, will not be adversely affected by atmospheric conditions. The air will be transported in a duct fitted with supply flues. The best location for the introduction of air to the roadway is at the exhaust pipe level of the vehicles to permit immediate dilution of the exhaust gases. In order to accomplish the air distribution described above, an adequate pressure differential must be generated between the duct and the roadway to counteract the piston effect and atmospheric winds.

During a fire within the tunnel, the air supplied will provide dilution of the smoke. To aid in fire-fighting efforts and in emergency egress, the fresh air should enter the tunnel through the portals to create a respirable environment for these activities. Therefore, for these reasons, the fans in a supply semi-transverse system should be reversible and a ceiling supply should be considered although the use of ceiling supply will not provide for the optimum distribution of contaminents within the roadway. With a ceiling supply system and reversible fans, the smoke will be drawn upward. It has been demonstrated that the environmental conditions within the roadway are affected by the location of the supply inlets (Fig. 19-19).

The exhaust semi-transverse system in a unidirectional tunnel, as shown in Fig. 19-18(b), will produce a maximum contaminant concentration at the exiting portal. In a bidirectional

Table 19-9. Representative tunnels with longitudinal ventilation systems.

TUNNEL	TYPE	LOCATION	COMPLETION DATE	LENGTH (FEET)	CONFIGURATION (TUBES/LANES)	FAN TYPE
Tende	Mountain	France/Italy	1882	10450	1/2	Axial boosters
Musko	Subaqueous	Sweden	1964	9548	1/2	NA
Liberty	Mountain	United States	1924	5905	2/4	Centrifugal
Hallstatt	Urban	Austria	1966	3950	2/2	Axial
Mobile River	Subaqueous	United States	1972	3380	1/2	Centrifugal
Bärenburg	Mountain	Switzerland	1970	3274	1/2	Axial
Rofla	Mountain	Switzerland	1970	3265	1/2	Axial
Limfjord	Subaqueous	Denmark	1969	3100	2/6	Axial boosters
Kiesberg	Mountain	Germany	1970	2953	2/4	Axial
Utsunoya	Mountain	Japan	1958	2770	1/2	Axial
Benelux	Subaqueous	Netherlands	1967	2608	2/4	Axial boosters
Viamala	Mountain	Switzerland	1967	2435	1/2	Axial
Welland Canal	Subaqueous	Canada	1968	2362	2/4	NA
Kennedy	Subaqueous	Belgium	1969	2264	2/6	Axial
Rheinallee	Urban	Germany	1967	2133	2/4	Axial boosters
Rendsburg	Subaqueous	Germany	1961	2100	2/4	NA
Heinenoord	Subaqueous	Netherlands	1969	2015	2/6	Axial boosters
Coen	Subaqueous	Netherlands	1966	1925	2/4	Axial boosters*
Bruser	Mountain	Switzerland	1972	1854	1/2	Axial
Wallring	Urban	Germany	1966	1805	2/4	Axial boosters
Saint-Etienne	Urban	France	1968	1470	2/4	Axial
Flughafen	Urban	Germany	1965	1378	1/2	Axial boosters
Salzburg	Airport	Austria	1960	1224	2/4	Axial boosters
West Rock	Mountain	United States	1949	1200	2/4	Axial
New River	Subaqueous	United States	1960	863	2/4	Axial

*With supplementary semi-transverse system.

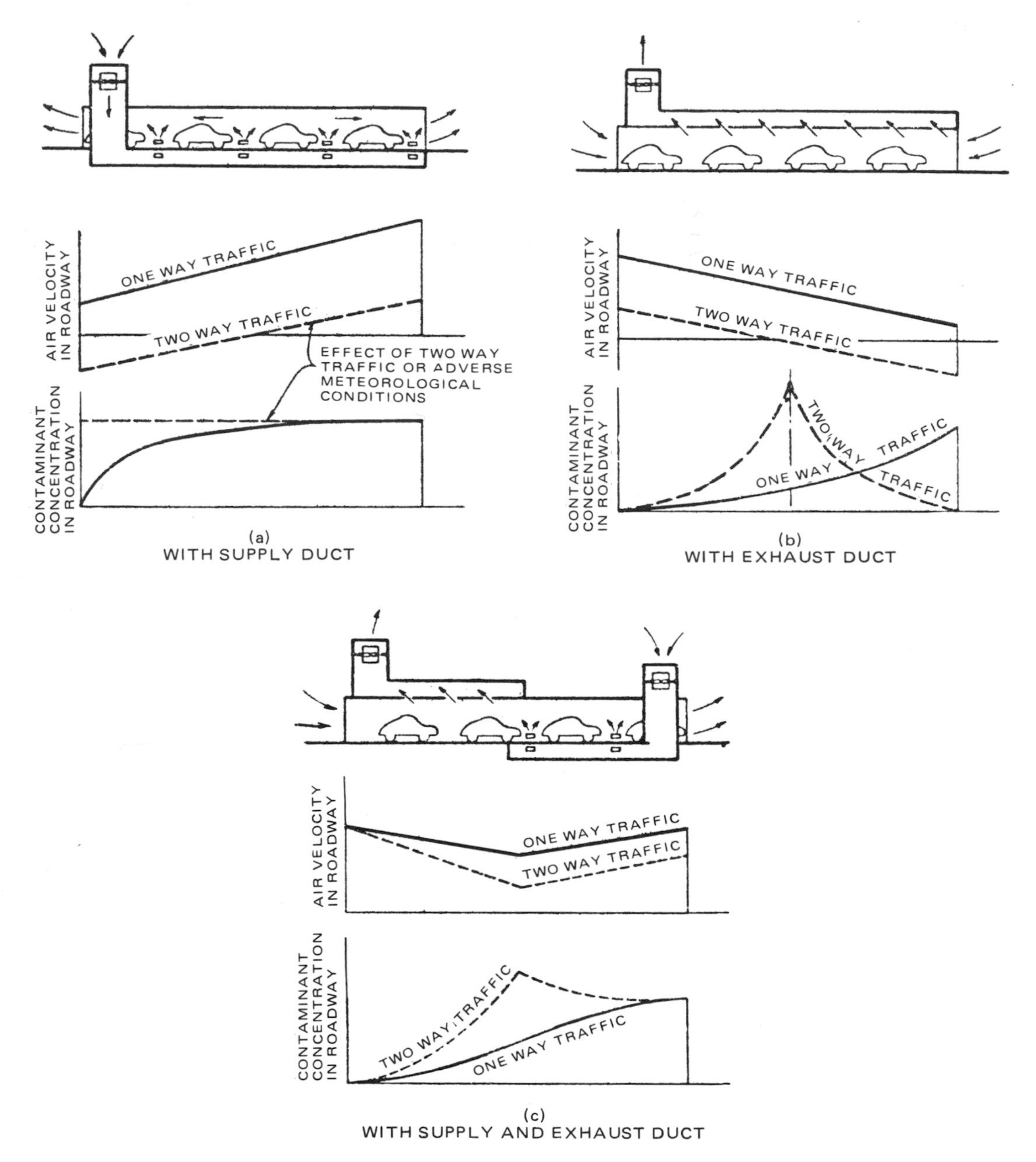

Fig. 19-18. Semi-transverse ventilation systems. (Adapted)

tunnel, a zone of zero fresh air is created near the center of the tunnel, which, of course, produces the maximum level of contaminants.

A combination supply and exhaust system, as shown in Fig. 19-18(c), has been used. Such a system is applicable only in a unidirectional tunnel where the air entering the traffic stream is exhausted in the first half, and air is supplied in the second half to be exhausted through the exit portal.

The semi-transverse supply system is the only one not affected by adverse meteorological conditions or opposing traffic. Semi-transverse systems are used in tunnels up to approximately 3000 feet, at which point the tunnel air velocities near the portals become excessive.

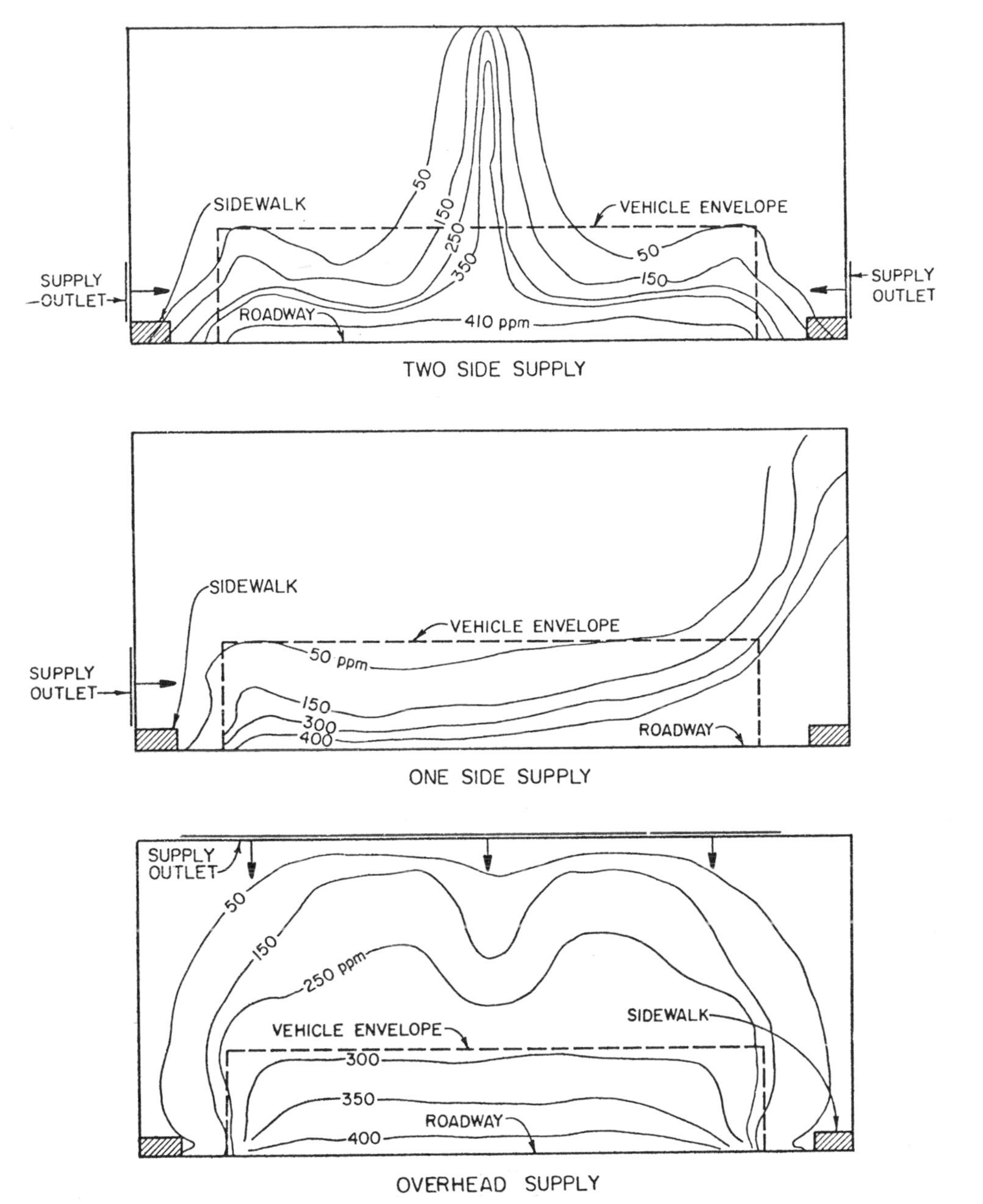

Fig. 19-19. Contour maps of carbon monoxide levels in tunnel with semi-transverse supply ventilation system.[21]

For a listing of tunnels having semi-transverse ventilation systems, see Table 19-10.

Full Transverse Ventilation

For larger tunnels, the full transverse system is used. A full exhaust duct is added to a supply type semi-transverse system which achieves a uniform distribution of supply air and a uniform collection of vitiated air (Fig. 19-20). This system was developed for the Holland Tunnel in 1924. With this system, a uniform pressure will occur throughout the roadway and no longitudinal airflow will occur except that generated by the traffic piston effect, which will tend to reduce contaminant levels. An ade-

Table 19-10. Representative tunnels with semi-transverse ventilation systems.

TUNNEL	TYPE	LOCATION	COMPLETION DATE	LENGTH (FEET)	CONFIGURATION (TUBES/LANES)	FAN TYPE
Mersey I	Subaqueous	Great Britain	1953	10,614	1/4	Centrifugal
Guadarrama I	Mountain	Spain	1963	9407	1/2	Axial
Mersey II	Subaqueous	Great Britain	1971	8200	1/2	Axial
Cross Harbour	Subaqueous	Hong Kong	1972	6080	2/4	Axial
Bankhead	Subaqueous	United States	1941	5390	1/2	NA
Dartford	Subaqueous	Great Britain	1962	4692	1/2	Axial
Glion	Mountain	Switzerland	1971	4374	2/4	Axial
Squirrel Hill	Mountain	United States	1953	4225	2/4	Centrifugal
Baregg	Mountain	Switzerland	1969	3840	2/4	Axial
Mosi	Mountain	Switzerland	1965	3747	1/2	Axial
Fort Pitt	Mountain	United States	1960	3603	1/2	Centrifugal
Baytown	Subaqueous	United States	1953	3000	1/2	Centrifugal
Louvre	Urban	France	1968	2825	1/2	Axial
Kajiwara	Mountain	Japan	1963	2670	2/4	Axial
Kasumigaseki	Urban	Japan	1964	2660	2/4	Axial
La Havane	Subaqueous	Cuba	1958	2410	2/4	Axial
Kambara	Mountain	Japan	1968	2343	2/4	Axial
Deas Island	Subaqueous	Canada	1958	2165	2/4	Axial
London Airport	Airport	Great Britain	1955	2060	2/4	NA
Vieux-Port	Subaqueous	France	1967	1942	2/4	Centrifugal
Lac Superieur	Subaqueous	France	1971	1883	2/4	Axial
Parc Des Princes	Urban	France	1971	1870	2/4	Axial
Lowry Hill	Urban	United States	1971	1496	3/6	Axial
Etoile	Urban	France	1970	1247	1/2	Axial
Haneda	Subaqueous	Japan	1964	984	2/4	Axial

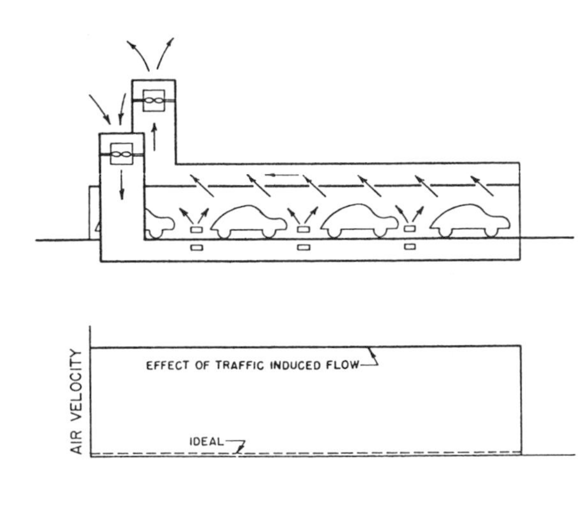

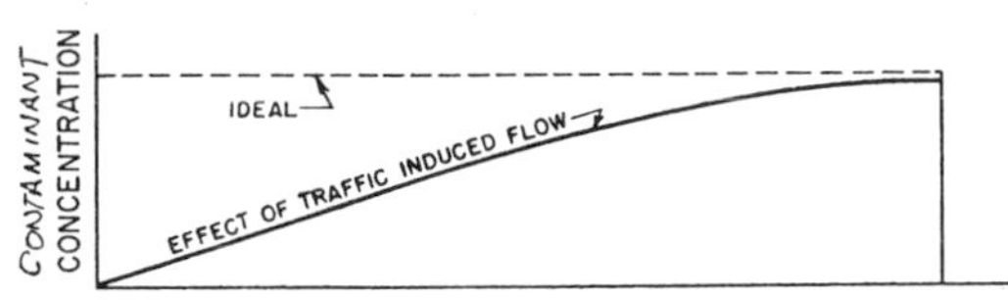

Fig. 19-20. Transverse ventilation systems.[2] (Adapted)

quate pressure differential between the ducts and the roadway is required to assure proper distribution of air under all ventilation conditions.

The desirable location of the supply air inlets from the standpoint of rapid dilution of exhaust gases is at the level of the vehicle emission, with the exhaust outlets in the ceiling. This arrangement was established by the full-scale tests conducted by the U.S. Bureau of Mines. The air distribution can be one-sided or two-sided. Recent studies have demonstrated the variations in the contaminant profiles within the roadway depending on the method of distribution (Fig. 19-21).

There are many variations and combinations of these systems. A combined system for a unidirectional tunnel, approximately 1300 feet long, is shown in Fig. 19-22. Section 3 utilizes a full transverse system because of the upgrade roadway; section 2 utilizes a semi-transverse supply with a longitudinal exhaust; and the remainder of the tunnel (section 1) is a semi-transverse supply system. Such a system would not be recommended for a long tunnel.

Table 19-11 contains a summary list of the world's vehicular tunnels with transverse ventilation systems.

Typical tunnel cross-section configurations are shown in Fig. 19-23.

VENTILATION EQUIPMENT AND FACILITIES

Once a determination is made that mechanical ventilation is required and the air quantity has been computed, the necessary equipment, such as fans, motors, dampers and sound attenuators, and the facilities to house them, must be considered.

Fans

A fan creates air movement by aerodynamic action. It is a constant volume device since it delivers the same gas volume irrespective of the gas density.

There are three basic types of fans predominant in tunnel ventilation systems identified as axial, centrifugal and propeller. The decision to apply each type is based on the fan flow, pressure requirements and space arrangements.

Axial Flow Fan. The flow of air through this fan is virtually parallel to the impeller shaft (Fig. 19-24). The radial component of velocity is nearly zero. The axial fan impeller with airfoil blades rotates in a cylindrical housing. There are two types of axial fans, tube axial and vane axial. The tube axial fan is usually used in systems requiring development of pressures up to $2\frac{1}{2}$ inches water gauge. The vane axial fan is a tube axial fan with guide vanes on one or both sides of the impeller to correct the rotary motion imparted by the impeller. These vanes improve the pressure characteristics and operating efficiency, thus making it possible to use the vane axial fan for applications requiring up to 10 inches of pressure, higher for special design fans.

Axial fans exhibit performance characteristics, as represented by the curves shown in Fig. 19-25. The characteristic dip in the axial pressure curve at 40% to 50% of wide open volume is caused by aerodynamic stall, during which the blades cease to function in the normal manner, although flow continues. The maximum efficiency occurs nearer to free delivery than for the centrifugal fan. The optimum operating

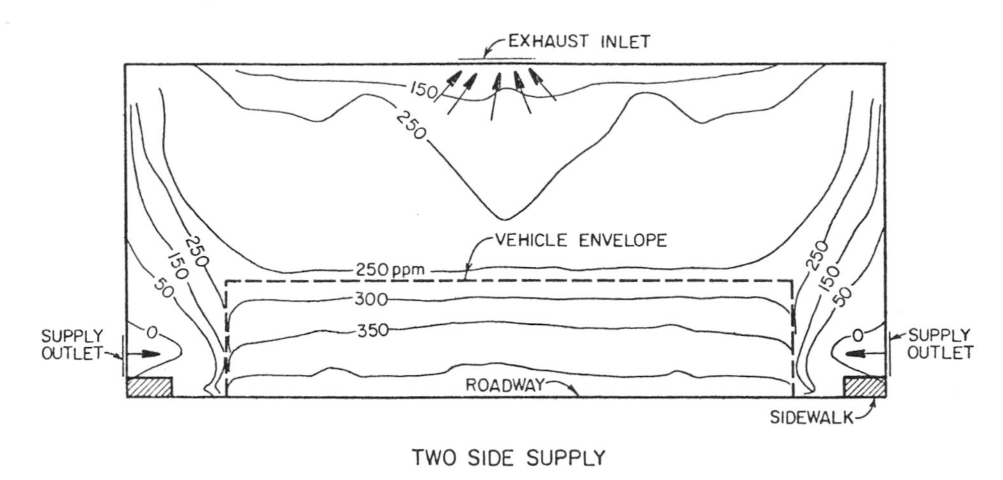

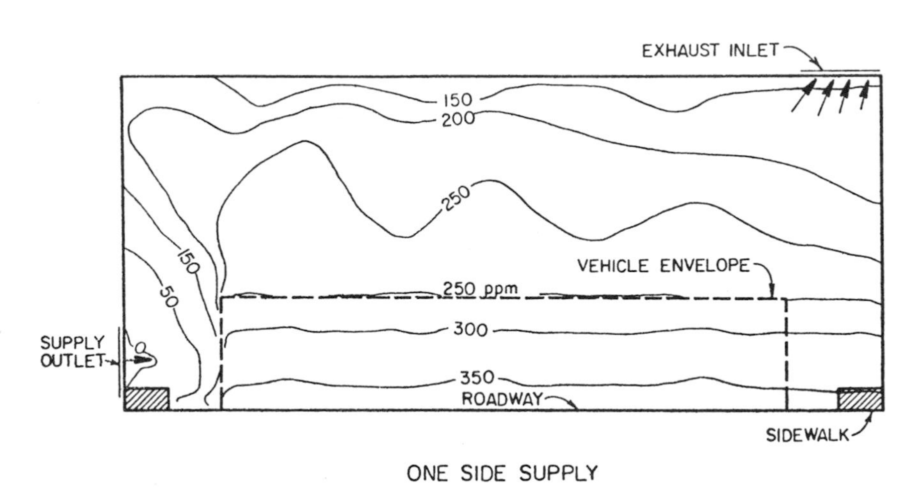

Fig. 19-21. Contour maps of carbon monoxide levels in tunnel with transverse ventilation system.[21]

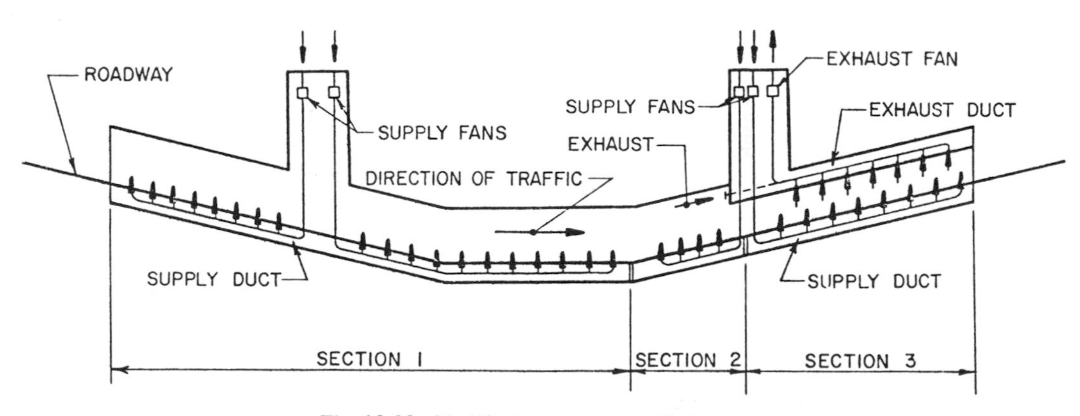

Fig. 19-22. Modified transverse ventilation system.

Table 19-11. Representative tunnels with transverse ventilation systems.

TUNNEL	TYPE	LOCATION	COMPLETION DATE	LENGTH (FEET)	CONFIGURATION (TUBES/LANES)	FAN TYPE
Mont Blanc	Mountain	France-Italy	1965	38,060	1/2	Centrifugal
San Bernadino	Mountain	Switzerland	1967	21,640	1/2	Axial
Chiyoda	Urban	Japan	1964	11,810	1/2	Axial
Kanmon	Subaqueous/mountain	Japan	1958	11,360	1/2	Axial
Belchen	Mountain	Switzerland	1970	10,435	2/4	Axial
Brooklyn Battery	Subaqueous	United States	1950	9120	2/4	Centrifugal
Holland	Subaqueous	United States	1927	8530	2/4	Centrifugal
Lincoln	Subaqueous	United States	1957	8216	3/6	Centrifugal
Eisenhower	Mountain	United States	1973	8148	1/2	Centrifugal
Grand St. Bernard	Mountain	Switzerland-Italy	1964	8100	1/2	Axial
Hernandarias	Subaqueous	Argentina	1969	7869	1/2	Axial
Baltimore Harbor	Subaqueous	United States	1957	7710	2/4	Centrifugal
Hampton Roads I	Subaqueous	United States	1957	7480	1/2	Axial
Hampton Roads II	Subaqueous	United States	1976	7300	1/2	Axial
Queens Midtown	Subaqueous	United States	1940	6400	2/4	Centrifugal
Croix-Rousse	Mountain	France	1952	5840	1/4	Axial
Imalso	Subaqueous	Belgium	1931	5805	1/2	Centrifugal
Thimble Shoal	Subaqueous	United States	1961	5740	1/2	Centrifugal
Tyne	Subaqueous	Great Britain	1967	5476	1/2	Axial
Baltimore Channel	Subaqueous	United States	1961	5450	1/2	Centrifugal
Detroit Windsor	Subaqueous	Unites States-Canada	1930	5125	1/2	Centrifugal
Callahan	Subaqueous	United States	1961	5085	1/2	Centrifugal
Big Walker	Mountain	United States	1972	4229	2/4	Centrifugal
Elizabeth River II	Subaqueous	United States	1962	4190	1/2	Axial
Maas	Subaqueous	Netherlands	1941	3527	2/4	Axial
IJ	Subaqueous	Netherlands	1968	3405	2/4	Axial
I-95 Mall	Urban	United States	1974	3400	2/8	Centrifugal

Table 19-11. Continued

TUNNEL	TYPE	LOCATION	COMPLETION DATE	LENGTH (FEET)	CONFIGURATION (TUBES/LANES)	FAN TYPE
Caldecott	Mountain	United States	1965	3350	3/6	Axial
Wagenburg	Mountain	Germany	1968	2707	1/2	Axial
Velsen	Subaqueous	Netherlands	1957	2520	2/4	Axial
Clyde	Subaqueous	Great Britain	1964	2500	2/4	Centrifugal (axial boosters)
179 Street	Urban	United States	1961	2400	1/2	Centrifugal
Sumner	Subaqueous	United States	1932	2362	1/2	Centrifugal
Piazza Fiume	Urban	Italy	1966	2346	1/2	Axial
Battery Street	Urban	United States	1954	2200	2/4	Axial
Durnstein	Mountain	Austria	1959	1550	1/2	Axial
Avenue Louise	Urban	Belgium	1957	1220	1/4	Axial

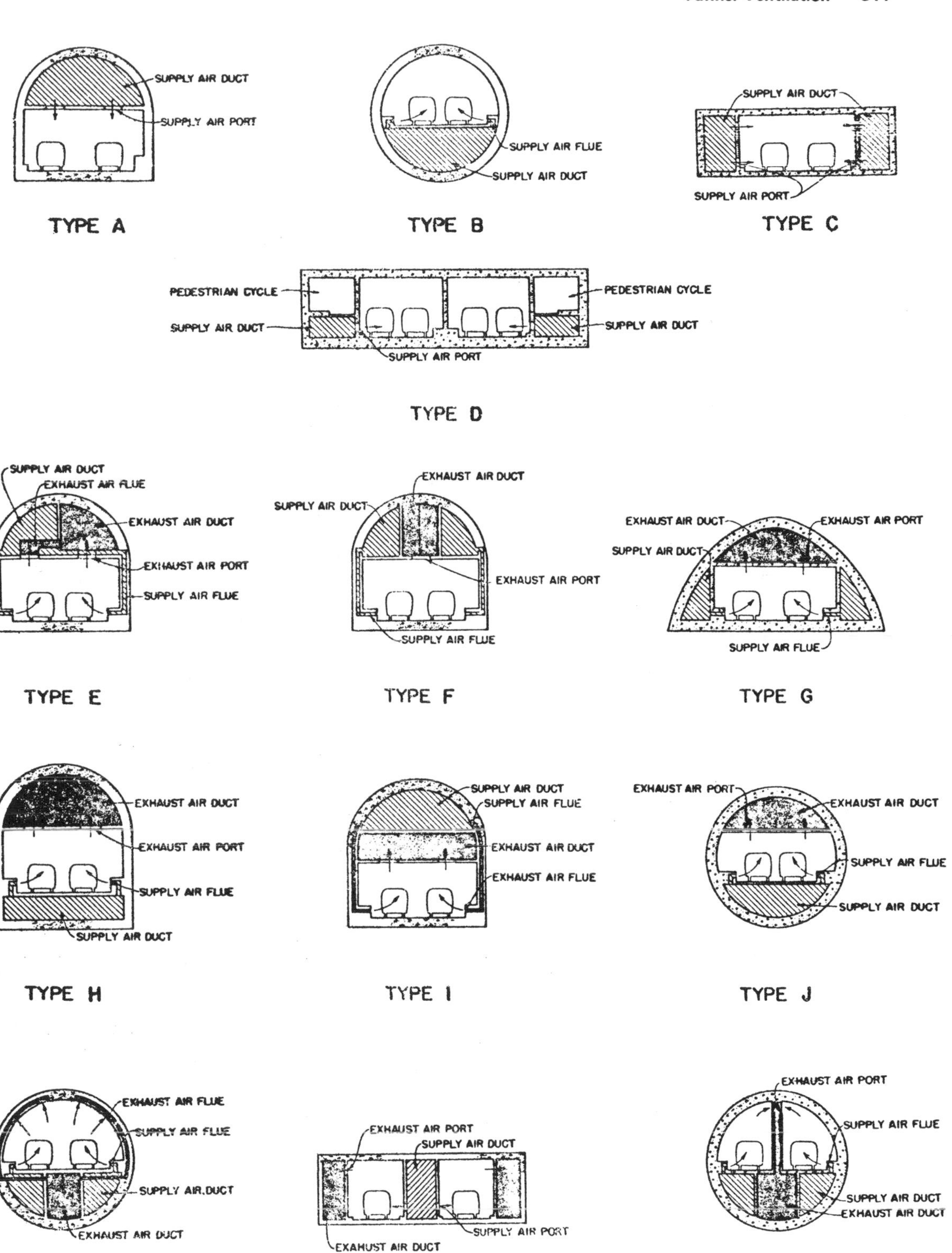

Fig. 19-23. Typical tunnel cross-sections with ventilation configurations.

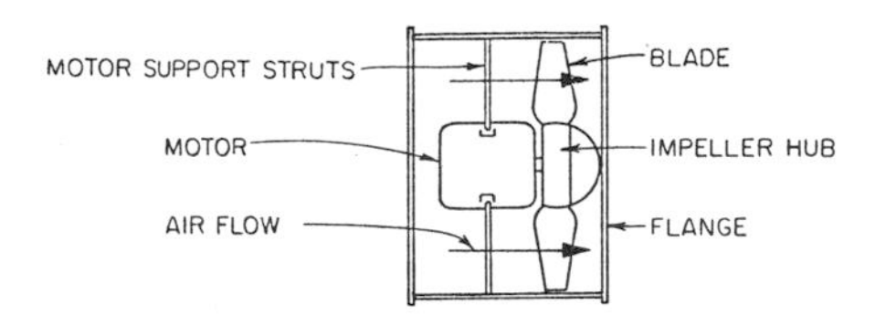

TUBE AXIAL FAN

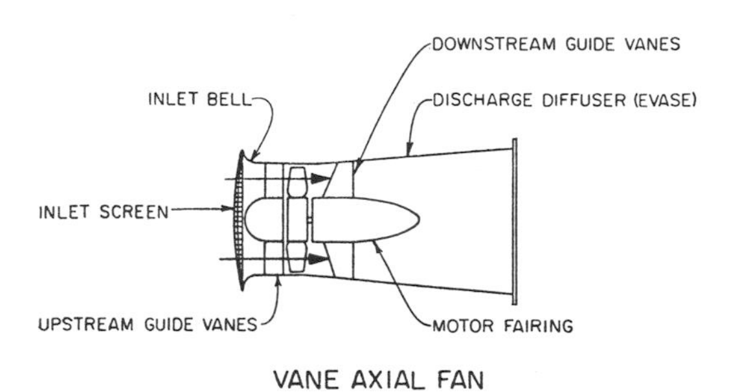

VANE AXIAL FAN

Fig. 19-24. Axial flow fans.

range is from 60 to 80% of wide open air volume. The shape of the horsepower curve is significant, since it has a non-overloading characteristic within the normal selection range. As can be seen in Fig. 19-25, the horsepower characteristic falls off as flow increases above the maximum efficiency point. The total efficiencies of axial flow fans can range between 70% and 80%.[22] The level of sound generated by an axial fan is lowest near the maximum efficiency point.

The pressure developed by an axial fan is influenced by the hub diameter, tip clearance and staging. These characteristics are established by the manufacturer during the design phase. A small hub to housing diameter ratio results in low pressure, whereas a large hub to housing diameter ratio results in high pressure. The maximum pressure is obtained with minimum tip clearance, which is usually between 0.1 and 0.2% of impeller diameter. This effect is shown in Fig. 19-26. To obtain high pressure characteristics, strict manufacturing tolerance must be adhered to, to maintain minimum tip clearance. Staging involves the installation of two or more impellers in the same fan housing. This has the effect of fans operating in series; that is, doubling the developed pressure at the same flow volume. While both tube and vane axial fans can be staged, the tube axial must use contrarotating impellers to eliminate air rotation and increase efficiency. Contra-rotating tube axial fans in series could produce triple the developed pressure of a single stage unit at the same flow rate.[22]

The volume delivered by a fan at a fixed speed can be varied by changing the blade pitch angle. Axial fans can be supplied with factory set, adjustable or controllable blade pitch. The fan with fixed pitch is limited to one flow rate unless speed is changed, whereas the air volume delivered by the adjustable and controllable

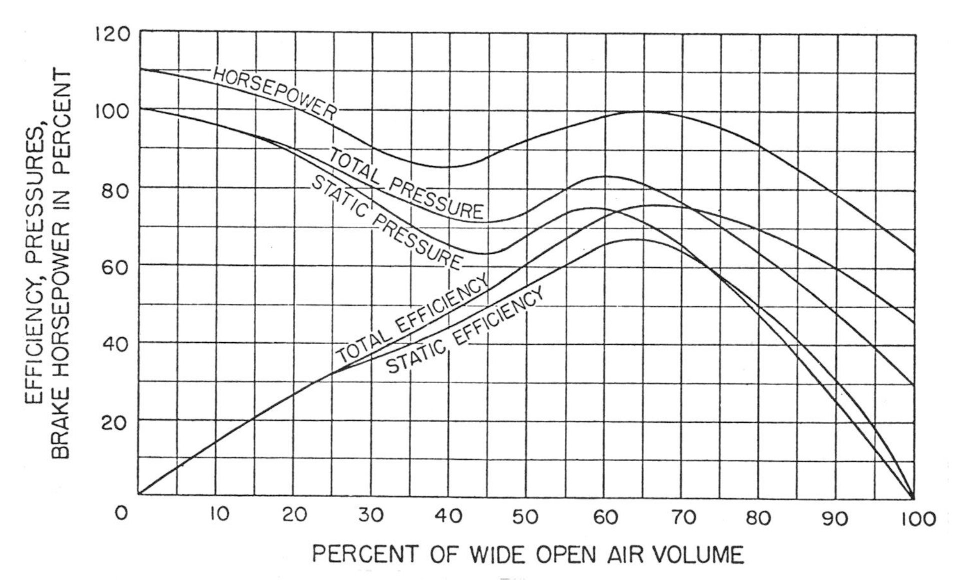

Fig. 19-25. Axial fan performances characteristics.

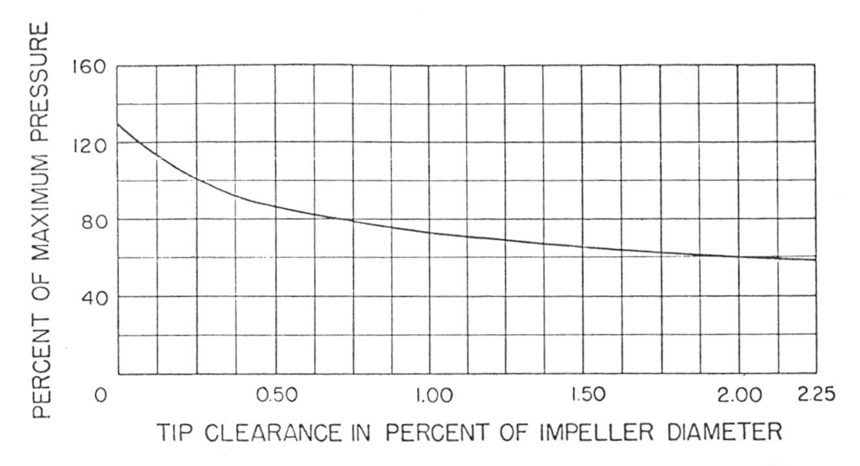

Fig. 19-26. Effect of tip clearance on developed pressure in axial fans.[22]

type can be varied by changing the blade pitch. The adjustable type requires that the fan be stopped to permit the blade pitch adjustment to be made. The ability to adjust the blade pitch while the fan is in operation is the feature of the controllable type. There are only a few rare cases where the controllable pitch feature can be justified in Vehicular tunnel design, therefore careful consideration should be given before controllable pitch fans are used for tunnel ventilation, since the controllable pitch mechanism is an added maintenance item. The necessary diversity of airflow in a tunnel can be achieved by the use of multiple two-speed fans, making the use of controllable pitch fans of doubtful justification. The cost of controllable pitch can only be justified on fans with large horsepower motors.

The effect of blade pitch change can be noted in Fig. 19-27, where at low blade angles the pressure increase is smooth with a drop in volume. However, at the higher blade angles, a point is reached where the fan can no longer develop the pressure required to deliver the air volume, as demonstrated by the accentuated stall points.

Inlet bells shown on the vane axial fan in Fig. 19-24 should be added to an axial fan with a free air intake to reduce the inlet losses and obtain efficient operation.

A discharge diffuser, as shown in Figure 19-24, is added to an axial fan to convert additional energy.

Axial fans may be driven by any number of methods such as belt, chain, direct connected or floating shaft (see *drives*). The most reliable

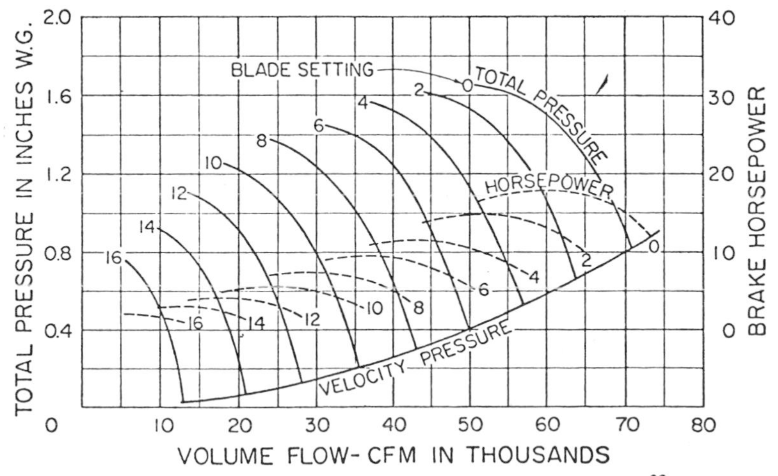

Fig. 19-27. Effect of blade pitch change on axial fan performance.[23] (Adapted)

method is the direct connected with the impeller mounted directly on the motor shaft. In this case, the motor bearings are the fan bearings; therefore, there are no separate bearings or drive mechanisms to maintain.

Centrifugal Fans. These consist of a wheel rotating within a scroll-shaped housing or casing, as shown in Fig. 19-28. The wheel has a series of blades on the periphery and the casing has an inlet on the wheel axis with a discharge outlet at 90 degrees to the inlet. The centrifugal force created by rotating the air trapped between the blades and the kinetic energy imparted to the air by its velocity when it leaves the wheel are the major producers of pressure in a centrifugal fan. The scroll-shaped casing converts high-velocity pressure at the blade tip to static pressure.

The air enters the centrifugal fan parallel to the wheel shaft and is discharged at a 90° angle tangential to the wheel. This turn reduces somewhat the efficiency of centrifugal fans due to the shock losses created. The efficiency of centrifugal fans will range from 45 to 85%.

Centrifugal fans can be obtained with either a single or double inlet. The single-inlet fan, called single-width, single-inlet (SWSI), has a single-width wheel with one casing inlet on the side of fan opposite from the drive. The double-width, double-inlet fan (DWDI) has a double-width wheel with casing inlets on both sides. The DWDI fan will deliver twice the volume of air that a SWSI fan of the same wheel diameter and rotational speed can deliver.

Centrifugal fans are available with four basic blade configurations: radial, forward curved, backward curved and airfoil. Each has its own performance characteristics, as outlined below.

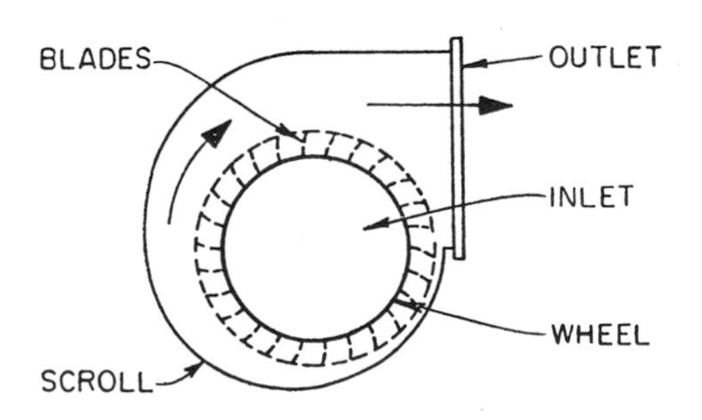

Fig. 19-28. Centrifugal fan.

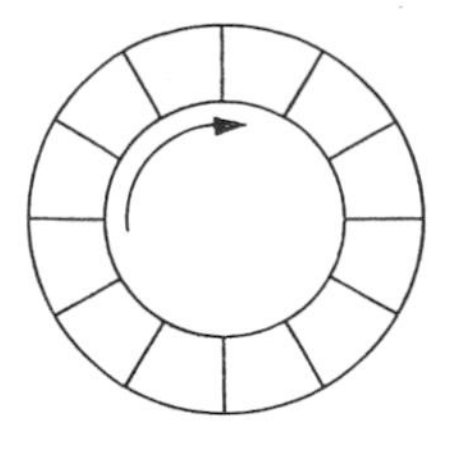

STRAIGHT RADIAL

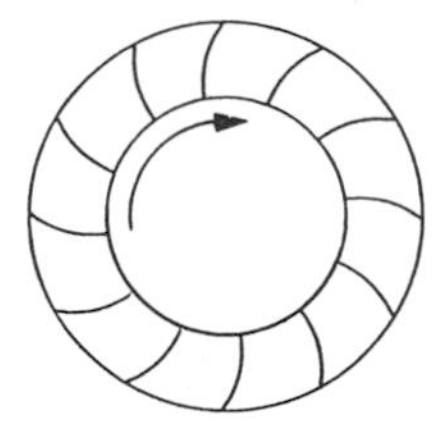

FORWARD CURVED

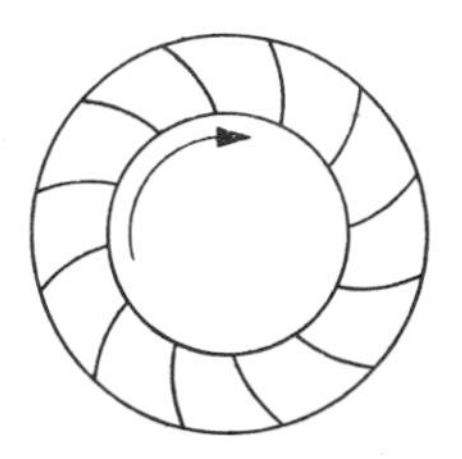

BACKWARD CURVED

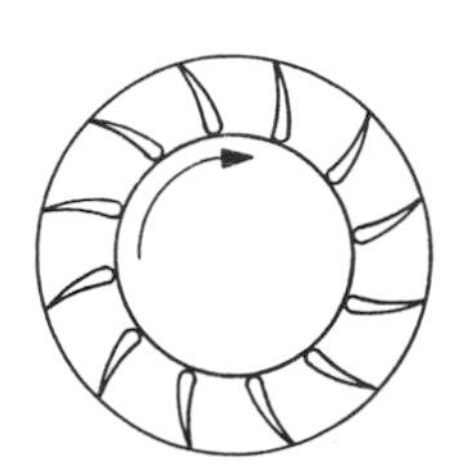

AIR FOIL

Fig. 19-29. Centrifugal fan blade configurations.

The *radial blade* or *paddle wheel* fan is the oldest and simplest form of centrifugal fan. Consisting of a series of flat plates mounted radially, as shown in Fig. 19-29, the radial blade fans usually have low efficiency and are therefore bulky. They are applied only in situations having moderate pressure requirements and are therefore not appropriate for tunnel applications. The typical performance curves shown in Fig. 19-30 indicates that the horsepower rises to a maximum at wide open volume, thus requiring special attention to motor and drive selection to prevent overloading.

The *forward curved (FC) blade* has a scoop effect on the air. The leaving air velocity is higher than for any other blade type; thus, the forward curve blade moves more air than any other centrifugal fan of the same size and speed. As can be noted from Fig. 19-30, the horsepower rise toward a maximum is as severe as that for the radial blade; therefore, the fan motor and drive are again affected. The FC fan is not used in tunnel ventilation systems.

The *backward curved (BC) blade* provides the family of centrifugal fans with its highest efficiencies. By curving the blades backward, the airflow through the blades is improved by the

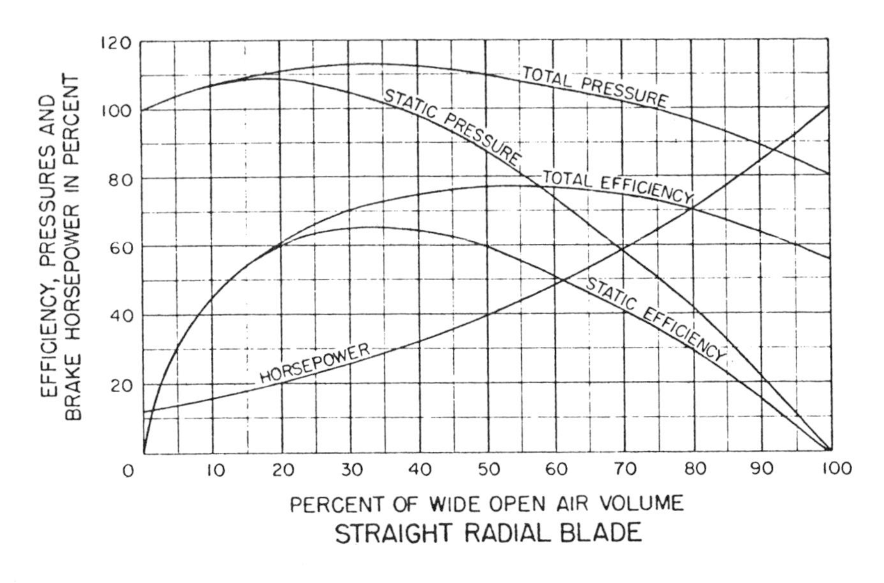

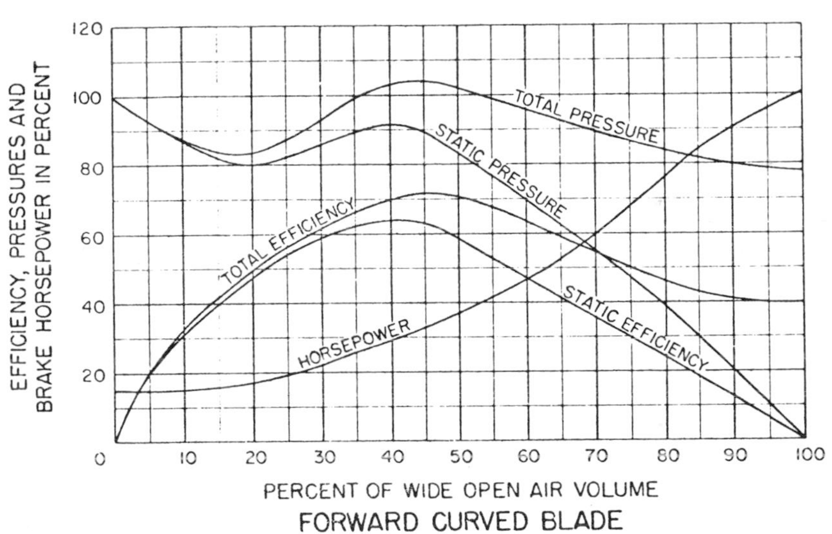

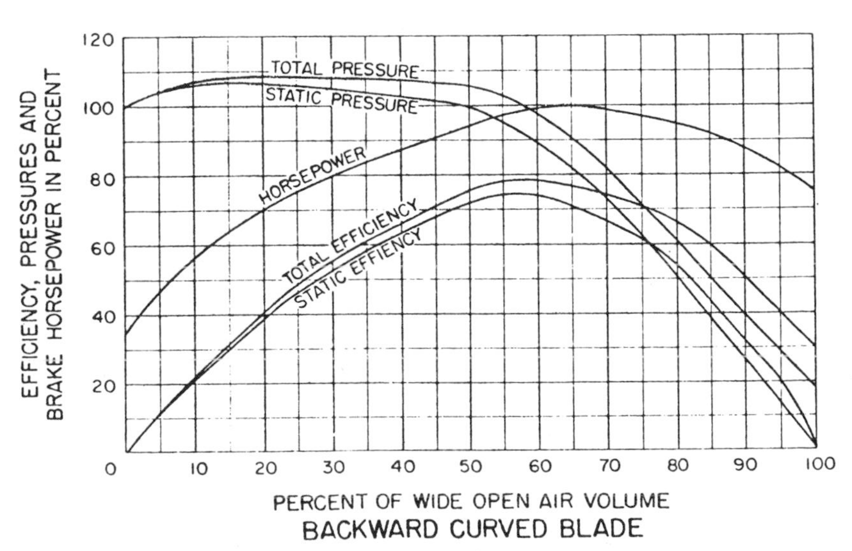

Fig. 19-30. Typical centrifugal fan performance characteristics.

reduction of shock and eddy losses. The BC blade develops more pressure by centrifugal force and less by velocity conversion. Higher tip speeds are used in BC fans, so the wheels must be of sturdier construction than other types. The maximum horsepower occurs within the normal operating range, as seen in Fig. 19-30.

The *airfoil bladed* centrifugal fan is a refinement of the backward curved type. The airfoil shape of the blade produces an increase in efficiency and a decrease in generated sound.

The backward curved and airfoil bladed fans are the ones of greatest interest to the tunnel ventilation engineer. The high efficiency of these types provides the least space requirements, along with the smallest horsepower requirements. These factors are extremely important when the size of tunnel ventilation systems is considered.

Propeller Fans. A simple form of an axial fan is the propeller fan. The propeller fan has had a limited use in tunnels due to its severe pressure limitation. The maximum pressure for most propeller fans is from $\frac{1}{2}$ to 1 inch water gauge. The singular advantage of propeller fans is the large volume of air they can handle at low operating costs along with low capital costs.

The propeller fan blade is often fabricated of sheet metal and set at an angle to the hub. The shape of the blade affects capacity and sound characteristics.

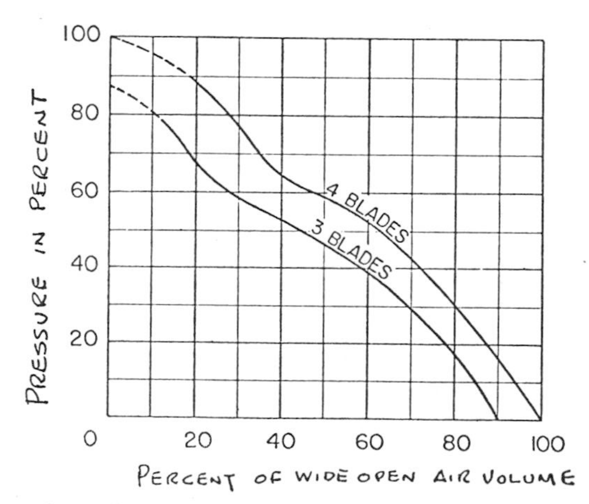

Fig. 19-31. Typical propeller fan performance characteristics.

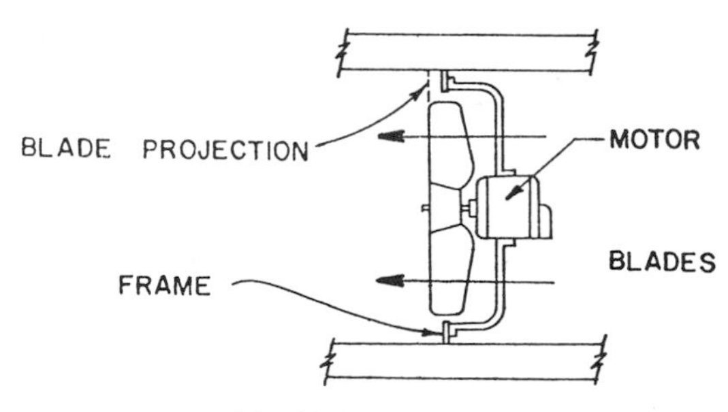

Fig. 19-32. Propeller fan.

These fans can be either direct-driven or belt-driven; however, if a direct drive is used, there is no way to adjust volume, unless the blade pitch angle is adjustable.

The horsepower of a propeller fan increases rapidly with increasing pressure, as shown on the typical propeller fans' pressure volume curves. Pressure volume curves for a typical propeller fan are shown in Fig. 19-31. This can cause motor burn-outs if the motor is not generously sized.

The air enters the propeller fan from all directions and leaves axially. As the pressure increases, recirculation of air occurs, thus reducing the net airflow. Therefore, steps must be taken to prevent this loss of airflow. In a free flow condition, the discharge edge of the blades are mounted flush with the frame. When the propeller fan is installed in ductwork, the discharge edge of the blades is extended beyond the frame, thus permitting centrifugal discharge of air and reduction of recirculation (Fig. 19-32). This will enable the fan to develop maximum pressure. The duct must be at least 25% larger than the impeller diameter to achieve this maximum.

Fan Characteristics

Flow reversibility is frequently required in a tunnel ventilation system. An axial fan or a propeller fan can be reversed electrically by reversing the rotation of the motor. Electrically reversing a vane axial fan will result in anywhere from 40% to 60% flow in reverse direction. There are axial fans available which approach 100% reversibility. These are being used on rapid transit systems as emergency ventilation fans. Some amount of efficiency is

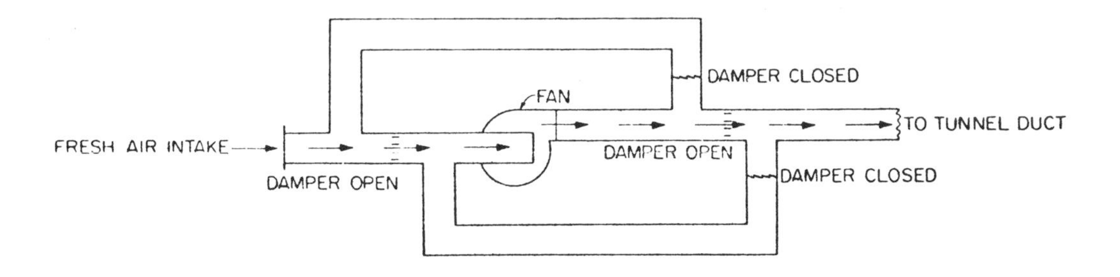

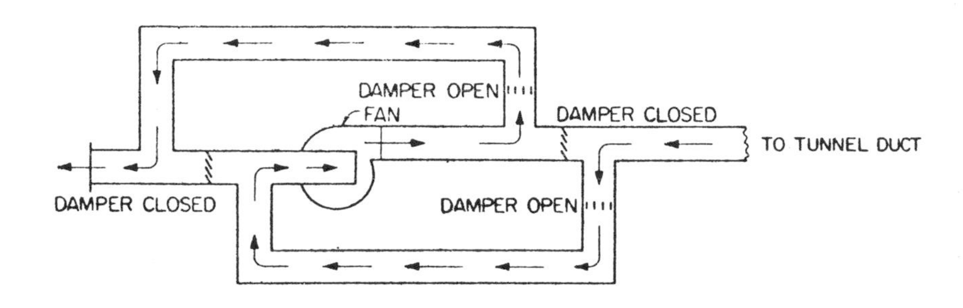

Fig. 19-33. Duct runaround system for centrifugal fans.

sacrificed in both flow directions to obtain this degree of reversibility. Care should be exercised in using a fan of this type for continuous operation because of the low efficiency.

The direction of airflow in a centrifugal fan cannot be reversed by reverse rotation. In order to use a centrifugal fan in a reversing situation, a duct runaround system must be constructed, as shown in Fig. 19-33. As is noted, this requires a number of additional dampers along with ductwork which can be both space-consuming and costly.

Parallel operation of two or more fans is possible and must be evaluated. Two identical fans operating in parallel will deliver twice the volume as one fan at the same pressure. Figure

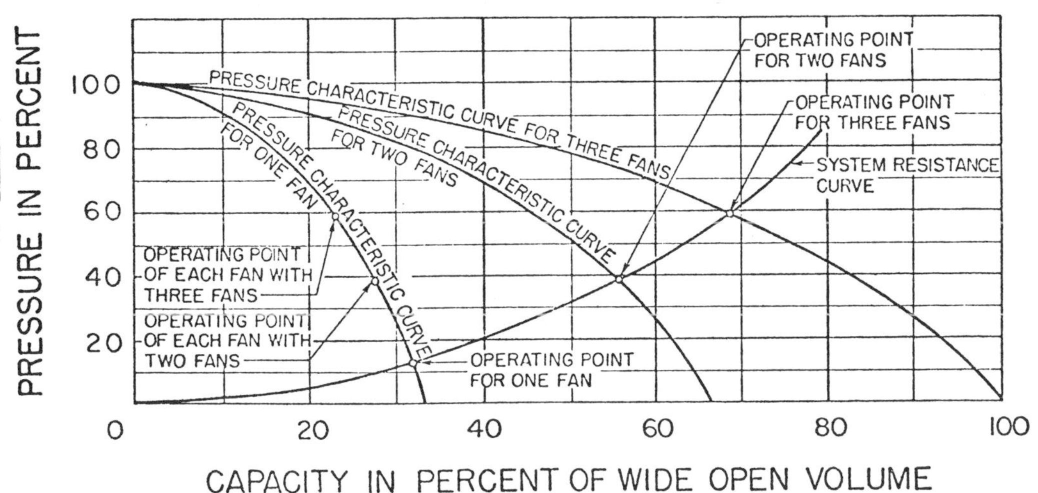

Fig. 19-34. Parallel fan operation.

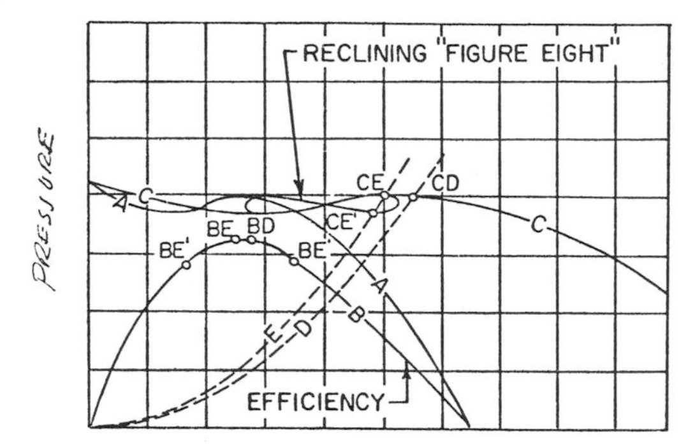

Fig. 19-35. Parallel operation of forward curved centrifugal fan.[24]

19-34 illustrates the effect of parallel operation of three fans on the performance curve. When the fan curve has a more complicated shape, such as a dip (as in Fig. 19-35), the relationships are not as simple. The dip creates an extremely complex combined curve. Care must be taken to select the fans so that none of the possible system resistance curves lie on the reclining "figure eight." System curve E would not provide a stable operation, whereas D would. As seen on Fig. 19-34, two fans will not deliver twice as much air as one fan alone when both are operating in the same system.

Two or more fans can be operated in *series* with the outlet of one fan connected to the inlet of the next fan. This arrangement for two fans will produce the same volume as one fan at an appropriate multiple of the total pressure (Fig. 19-36). For an evaluation of a series operation, only the total pressures can be used, since the static pressures are not additive.

Volume Control. The volume of air delivered by a fan can be controlled in a number of ways, both mechanically and electrically. In tunnel ventilation applications, the primary concern is reduction of airflow, and in turn reduction of horsepower during periods of light traffic to reduce energy costs.

Throttling dampers in the duct system at the fan discharge is the simplest, yet the most costly method. Although this method does reduce the flow, it accomplishes this reduction by increasing the system resistance, but not necessarily reducing the horsepower; in fact, it may increase. This method also poses a problem if

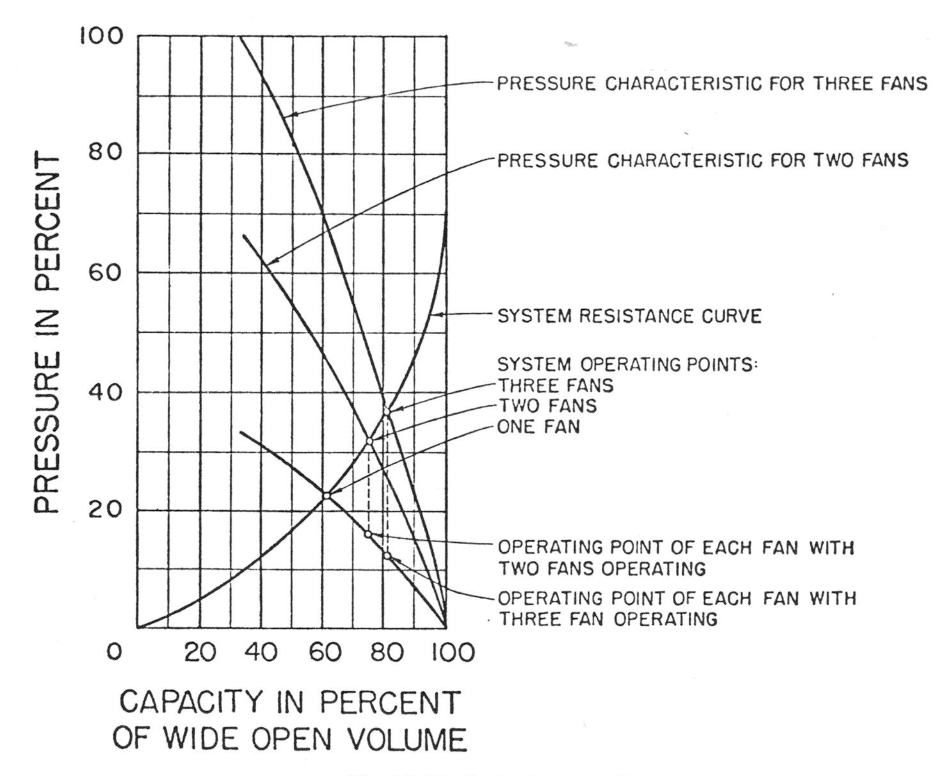

Fig. 19-36. Series fan operation.

the pressure increase places the fan near the operational stall point of the fan. Throttling dampers should not be used in a tunnel ventilation system.

The *bypass method* utilizes relief of air to atmosphere, thus reducing the flow into the system, but the fan continues to handle the same volume. Some small pressure reduction is usually achieved due to the reduced flow in the system; however, any power reduction will be extremely small; therefore, the bypass method is not appropriate for tunnel ventilation systems.

Speed regulation is the ideal method of fan volume control. This can be accomplished by some form of variable speed coupling, by gear box, by drive or by electrical means, by a multiple speed motor. The effect of speed reduction on fan performance is shown in Fig. 19-37. This method is ideal, since as the flow is reduced, the horsepower is also reduced, thus cutting power consumption.

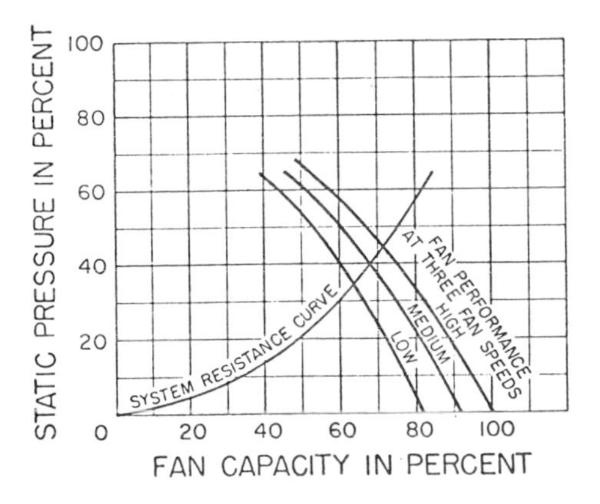

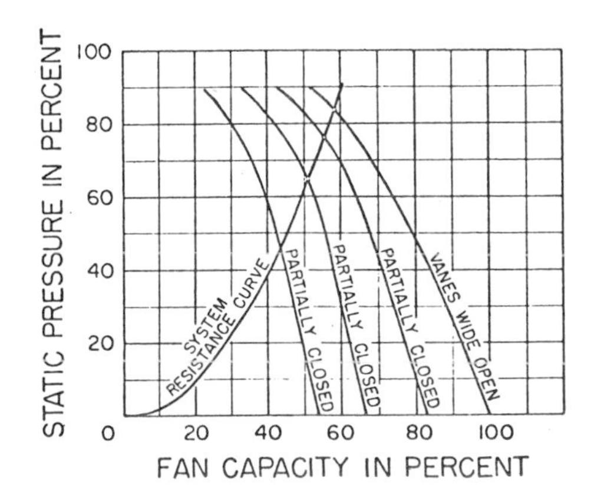

Fig. 19-37. Effect of volume control on fan performance.

Inlet vane dampers can be used for control of air volume delivered by a centrifugal fan. As the vanes turn, the relative direction of the airflow entering the fan wheel is altered, thus varying the amount of work done by the blades. The effect of this is to modify the fan performance curve, as shown in Fig. 19-37, instead of throttling the airflow. This results in a power reduction.

Blade pitch control can be used to change the volume of air passing through an axial fan. Such a system has the effect of changing the fan performance curves, as shown in Fig. 19-27. A lower power consumption is achieved with this method, although the fan may actually be operating at a point of lower efficiency. A pitch control system of this type will permit pitch change while the fan is in operation. This approach has also been applied to multi-stage axial fans (*see axial flow fan* for limitations).

Air volume in a multi-stage axial fan can be controlled by proper operation of the stages. By shutting down one or more stages, a variety of flow rates can be achieved.

Fan Sound. The sound generated by a fan is a function of fan design, airflow, pressure tip speed and fan efficiency. The sound generated within a fan by air turbulence is transmitted partly through the fan outlet and partly through the fan inlet. Bearing, drives and vibration harmonics are also sources of fan sound.

The only valid basis for evaluation and comparison of fan sound is the actual sound power levels generated by the fan. These data should be obtained from the fan manufacturer and, if possible, based on actual sound tests. The fan outlet velocity has no bearing on the fan-generated sound; thus, any selection based on outlet velocity alone is meaningless. To minimize the sound generated by a fan, the following items should be considered.

1. Select the fan near its point of peak efficiency.

2. Design the air system for smooth flow and low resistance.
3. Provide good fan inlet conditions.

In order to properly evaluate the effect of fan selection on fan sound, it is necessary to consider the sound generated across the entire sound spectrum. Fig. 19-38 shows the relative differences in sound generated by axial fans and centrifugal fans across the entire sound spectrum. The vane axial fan generates very little sound at low frequency but has a significant sound peak at the blade passing frequency (blade passing frequency equals number of blades X rpm) and secondary peaks at multiples of this frequency. The centrifugal fan, on the other hand, exhibits high sound characteristics in the low frequency ranges with less significant peaks at the blade passing frequencies.

Vibration. Vibration is induced in fans by unbalanced centrifugal forces and by aerodynamic forces. If a fan is rigidly mounted to the structure, the full force of this vibration is transmitted to the structure. When considering rigid mounting for a fan, the fan must be accurately balanced to minimize the unbalanced forces. To reduce the transmission of vibration to the structure, some form of anti-vibration mounting must be considered. These devices do not suppress the vibrations at the fan; in fact, they may permit an increase of vibration. They do, however, reduce transmission of these vibrations to the base structure. Transmissibility is the ratio of the force transmitted to the structure to the force generated by the vibration. The amplitude of any vibrations can be reduced by adding mass such as an inertia block under the fan.

The most widely used materials for vibration isolation of fans are steel and rubber. Steel springs are used for all fans having rotational speeds below 700 rpm and they may be used for fans of all speeds. Rubber in shear isolators are used for fans with rotational speeds above 700 rpm. Steel springs exhibit nearly perfect elasticity, and therefore require some form of restraint in a fan isolation application. The rubber used in vibration isolation devices is likely to decompose at temperatures above 150°F or in the presence of grease and oil.

The amount of deflection of the isolator must be chosen carefully to give proper natural frequency. This depends on the disturbing frequency and the allowable transmissibility.

Fan Selection. A fan can be selected only after the total airflow and pressure requirements have been determined. If at all possible, in

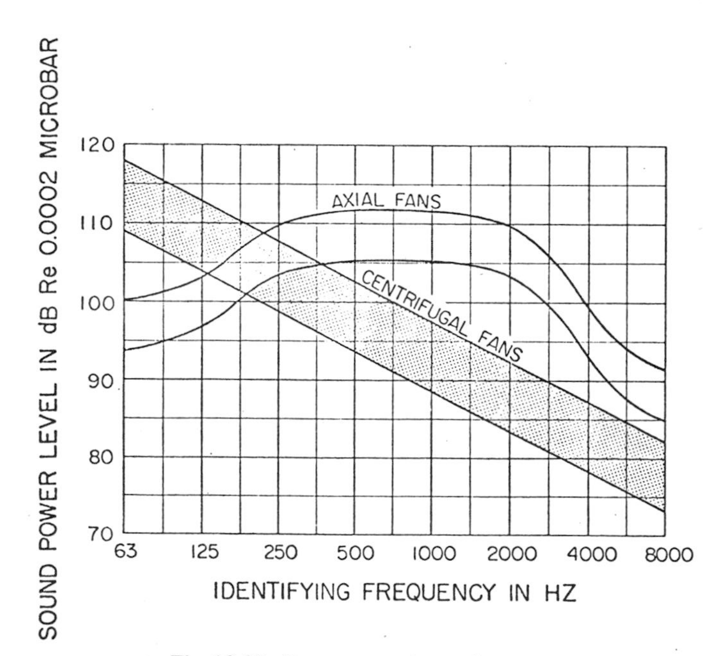

Fig. 19-38. Fan-generated sound.

order to reduce costs and reduce possible manufacturing delays, standard manufactured fans should be selected for a tunnel ventilation system in lieu of those requiring special designs. There will often be several fans, all of which will meet the system performance requirements; therefore, capital cost, operating costs and maintenance costs must be taken into account prior to the engineer making a final fan selection. There are several factors which must be evaluated before proceeding with the fan evaluation:

- Air density
- Possible combinations of fans to be operated
- Inlet and outlet conditions
- Type of drive
- Allowable sound levels.

A *system resistance curve* can be plotted for each ventilation duct system. The majority of tunnel ventilation systems will be in the completely turbulent range, thus exhibiting a $P = CQ^2$ relationship. This results in a curve similar to that shown in Fig. 19-39. When a fan performance curve is superimposed on such a system resistance curve, the point of operation can be determined readily, as noted on Fig. 19-39. These curves form the best method of selecting a fan for a tunnel ventilation system, since the full fan curve is visible and the possibility of unstable operation can be detected and avoided. Although the use of total pressure in the fan selection process is the correct method, occasionally the fan diameter is not known and the

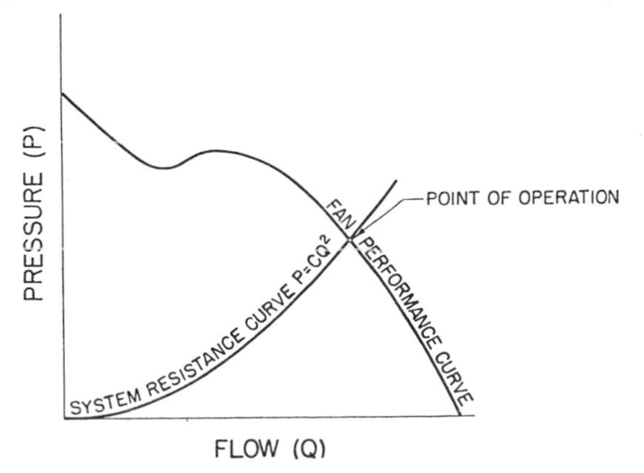

Fig. 19-39. Typical system resistance curve.

static pressure is used instead of the total pressure. This approach is acceptable provided the curves and relationships are properly identified.

Some fan manufacturers present their fan performance data in what are called "multi-rating tables." These tables are relatively simple to use; however, they do not provide the overview of a fan's performance characteristics as does a performance curve. Therefore, it is recommended that before the final fan selection is made, a performance curve be obtained for each tunnel ventilation fan at each of its operating speeds and conditions.

In most tunnels, there will be more than one fan operating on a system, usually in parallel. All combinations of fans operating in parallel at all speeds must be evaluated to assure that all design operating points are well within the stable operating range of the fan (Fig. 19-34 illustrates a set of fan performance curves prepared for three fans in parallel). The major problem with this type of fan operation is the possible instability created when the fan curve displays a noticeable dip, as appears in Fig. 19-35. The fans must be selected near the maximum efficiency point and must remain within the stable portions of the curve for all points of operation. If there is a question regarding the accuracy of the fan duty calculations, a margin should be allowed for additional pressure rather than for additional air volume.

Fan Laws. Fans exhibit certain properties which are defined by the fan laws. These laws relate the performance variables for any homologous series of fans. The performance variables include fan size, rotational speed, air density, capacity, pressure, horsepower and efficiency. To properly use the fan laws, the dimensions of the fans considered must be proportional to the dimensions of the rated fan.

Air Density. Fan performance curves and rating tables developed within the U.S. are based on standard air which has a temperature of 70°F, a barometric pressure of 29.29 inches of mercury and a density of 0.075 pcf. In Great Britain, it is 0.764 pcf, 60°F and 30 inches of mercury.

When the density of the inlet air to a fan

changes, there is a proportional change in power and pressure developed by the fan. The fan, being a constant volume device, delivers the same total air volume regardless of the density. From the fan laws,[24]

1. Pressure developed varies directly with density; and
2. Power consumed varies directly with density.

To correct the pressure and power for density variation, the values computed for standard air conditions must be multiplied by the ratio of the densities as follows.

$$PRES = PRES_o \times \left(\frac{DEN}{DEN_o}\right)$$

where

$PRES$ = pressure at modified density
$PRES_o$ = pressure at original density
DEN = modified density
DEN_o = original density

$$BHP = BHP_o \times \left(\frac{DEN}{DEN_o}\right)$$

where

BHP = brake horsepower at modified density
BHP_o = brake horsepower at original density

The density ratio can be computed as follows.

$$\frac{DEN}{DEN_o} = \left(\frac{460 + TEM_o}{460 + TEM}\right) \times \left(\frac{BAR}{BAR_o}\right)$$

where

TEM_o = original air temperature
TEM = modified air temperature
BAR_o = original air barometric pressure
BAR = modified air barometric pressure

For U.S. standard conditions:

$$\frac{DEN}{DEN_o} = \left(\frac{530}{460 + TEM}\right) \times \left(\frac{BAR}{29.92}\right)$$

$$\frac{DEN}{DEN_o} = \left(\frac{530}{29.92}\right) \times \left(\frac{BAR}{460 + TEM}\right).$$

A tabulation of specific gravities of air is presented in Table 19-12.

$$DEN = DEN_o \times SGB \times SGT$$

where

SGB = specific gravity based on barometer
SGT = specific gravity based on temperature.

The system resistance should be computed and the fan selected assuming standard air conditions, since most available fan data are based on this standard. After the fan is selected, the necessary adjustments should be made. If the

Table 19-12. Relative specific gravity of air at various temperatures and elevations.[25]

TEMPERATURE (°F)	SGT*	ELEVATION (FEET ABOVE SEA LEVEL)	BAROMETER (INCHES MERCURY)	SGB**
0	1.152	0	29.92	1.000
20	1.104	500	29.38	0.982
40	1.060	1000	28.86	0.964
50	1.039	1500	28.33	0.947
60	1.019	2000	27.82	0.930
70	1.000	2500	27.32	0.913
80	0.982	3000	26.82	0.896
100	0.946	4000	25.84	0.864
120	0.914	5000	24.90	0.832
160	0.855	6000	23.98	0.801
200	0.803	7000	23.09	0.772
300	0.697	8000	22.22	0.743
400	0.616	9000	21.39	0.715
500	0.552	10000	20.58	0.688

*SGT = specific gravity (temperature) of standard air at 29.92 inches mercury.
**SGB = specific gravity (barometer) of standard air at 70°F.

fan data are available and presented for the required non-standard conditions, the system resistance must then be computed based on the actual density.

The effect of humidity on fan selection is usually neglected in tunnel applications. Saturated air, however, is slightly lighter than dry air, the difference in density being only approximately 2.0% at 60°F and 4.3% at 120°F.

Fan Construction. The fan selected for tunnel ventilation systems must be designed to withstand the environment encountered. This includes air quality, pressure and temperature. Most tunnel fan housings are constructed of carbon steel with an applied protective coating such as paint. The centrifugal fan wheel is usually also fabricated of steel, whereas aluminum alloys are being used for axial fan impellers.

The pressure developed by a fan will determine the amount of structural reinforcement required. The designations adopted by the Air Moving and Conditioning Association (AMCA), as shown in Fig. 19-40, are used to establish the class of fan construction required.

The supply air systems should be evaluated at the maximum density conditions, the lowest ambient temperature air to enter the fans, to assure sufficient motor capacity. In a tunnel, temperature consideration is also made for the exhaust portion of the ventilation system, since during an emergency such as a fire, high temperatures could be reached at the fan. The usual temperature requirement for these systems ranges from 250° to 300°F. It has been recommended to install a deluge water system on the inlet side of tunnel exhaust fans.

To facilitate installation and maintenance, the housings of most large tunnel fans must be split; that is, they must be of bolted, gasketed construction, thus permitting removal of a portion of the housing without disturbing the remainder of the fan. A typical example of split housing for an axial fan is shown in Fig. 19-41. In the case of the horizontal axial fan, the removal of the top portion of the housing is usu-

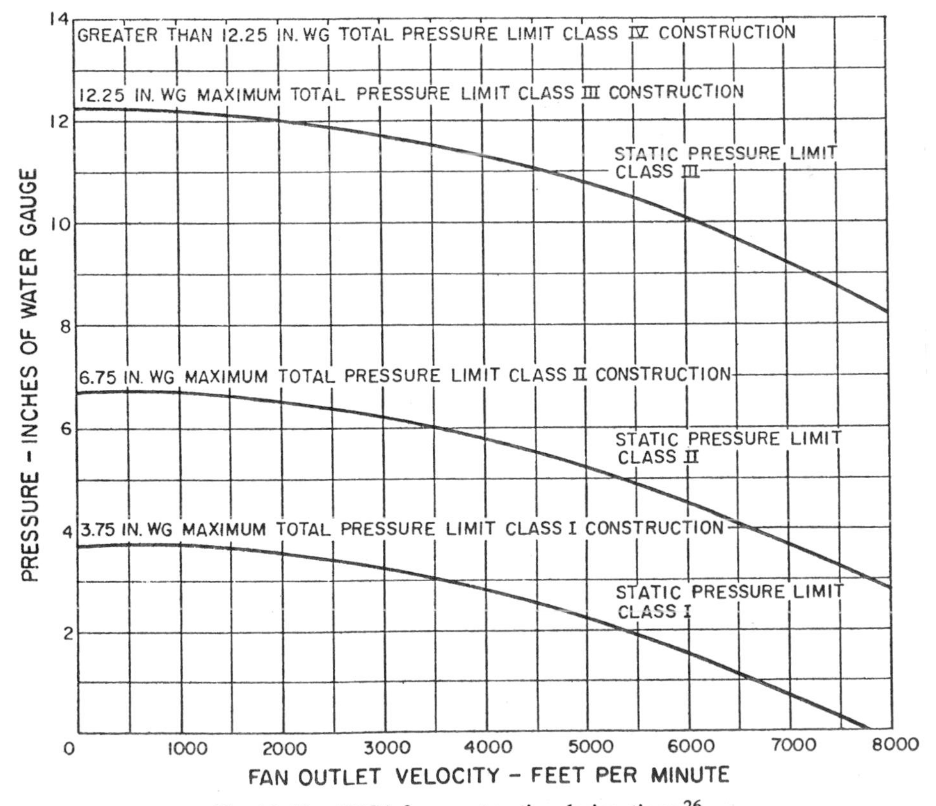

Fig. 19-40. AMCA fan construction designations.[26]

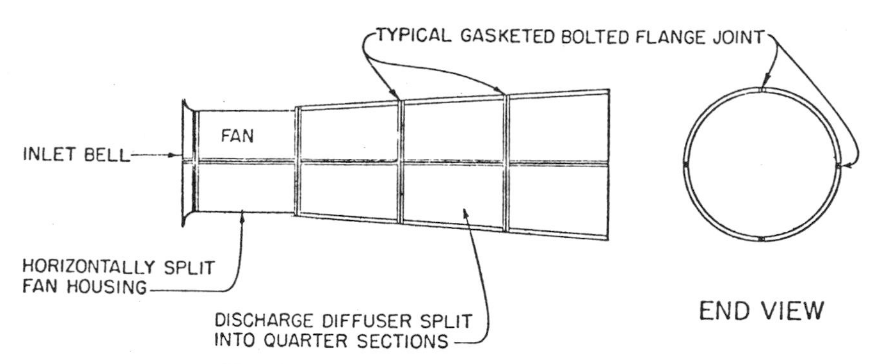

Fig. 19-41. Typical horizontal axial fan housing.

ally required to gain access to the impeller and the motor of a direct driven fan. Removal of the centrifugal fan wheel and shaft will require removal of the top portion of the centrifugal fan housing.

Motors

Most tunnel ventilation fans are driven by electric motors. The selection of a motor for a fan is based on the full load horsepower requirements, the fan speed and the starting characteristics. The type of drive to be used must also be considered.

The torque required to accelerate the air and the rotating mass and to overcome friction are considered retarding torques for the motor. During the starting phase, the motor torque must exceed these retarding torques if the fan is to be accelerated. Fan torque varies as the square of the speed; thus, the torque required is zero at a stop fan condition and increases with increasing speed. Therefore, all motor torque is available for fan starting and a percentage is available for accelerating at other speeds (Fig. 19-42).

Most tunnel fan motors are of the polyphase AC, either squirrel cage or wound rotor, non-synchronous-speed induction type. Both of these types experience little slip at rated speed. The synchronous speed of a motor is a function of the alternating current frequency and of the number of poles in the motor (NSYN = 120F/N poles).

Squirrel cage motors are constant speed machines which, with the addition of special windings, can be made to operate at multiple speeds. Class B insulated motors are usually used for fans. It is suitable for continuous operation at slight overloads, and the starting current is normal (approximately six times the full load current).

Wound rotor or slip ring motors of the continuous rated type are used where adjustable speed motors are required. Wound rotor motors are not suitable for tunnel ventilation systems, as they require secondary resistance and their speed control is not positive under light load conditions.

Motor enclosures are either open or enclosed. Examples of open type motors are drip-proof and splash-proof. The open motor should not be used if the ambient air at the motor contains any harmful material. The totally enclosed motor (TE) prevents free interchange of air between the inside and the outside of the motor. However, it is not to be considered airtight. The totally enclosed fan-cooled motor (TEFC) has a built-in fan which cools the motor by directing air over the motor. Totally enclosed non-ventilated motors (TENV) have no internal

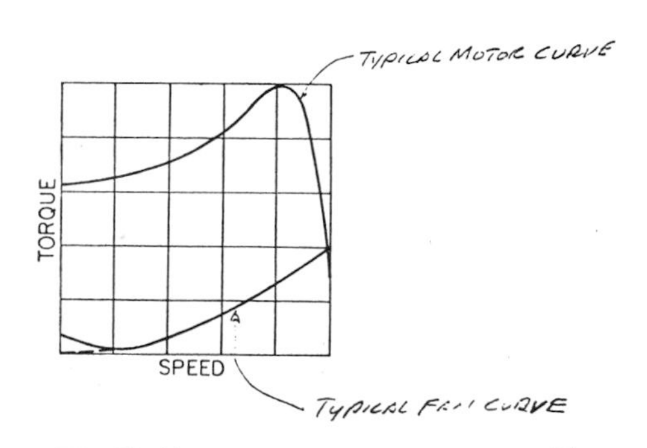

Fig. 19-42. Typical speed-torque curves.[24]

fan. Therefore, by comparison, a larger frame size is required. If air is flowing over the motor, the frame size can be reduced. A totally enclosed air over motor (TEAO) is rerated to account for the cooling effect of the air flowing over the fan. This is the type of motor used for direct connected axial fans. The explosion-proof and the dust ignition-proof type motors are special designs not frequently used in tunnel applications.

Motor winding insulation is classified based on the maximum temperature for which it is designed. These temperatures are shown in Fig. 19-43. The temperature of the insulation must not exceed the limit, as shown in this figure. The maximum temperature may be considered to be the sum of the ambient temperature and the temperature rise above the ambient at the hottest accessible spot in the motor. The standard ambient condition for motor design is 40°C (104°F). If the ambient temperature is greater than 40°C, the temperature rise must be reduced, as seen in Fig. 19-43. At elevations higher than sea level, the allowable temperature rise must be reduced to compensate for the reduction in air density, which reduces the cooling effect. In tunnel ventilation systems, the maximum gas temperature a motor will be exposed to must be evaluated. The most critical situation is a direct connected air over motor, usually on an axial fan operating in an exhaust mode where the gas temperatures of the fan could exceed 300°F during a severe fire within the tunnel.

The National Electric Manufacturers Association (NEMA) has adopted standards regarding all phases of motor design and performance, and therefore should be consulted for greater in-depth information.

By NEMA standards, a motor should be capable of operating successfully under a plus or minus 10% voltage and frequency variation provided that the frequency variation does not exceed plus or minus 5%.

Motor starters are either full voltage, reduced voltage or split winding. Starters will also provide overload protection in addition to being able to energize and de-energize the motor circuits. The full voltage or across-the-line starter is the simplest and least costly type; however, if limitations are imposed on starting currents, then a reduced voltage starter will have to be used.

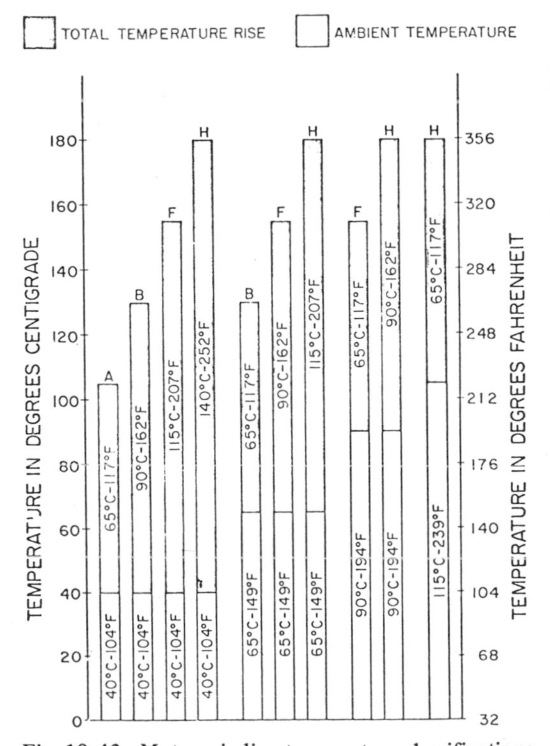

INSULATION CLASS	MAXIMUM ALLOWABLE TEMPERATURE	
A	105°C	221°F
B	130°C	266°F
F	155°C	311°F
H	180°C	356°F

Fig. 19-43. Motor winding temperature classifications.

Fan Drives

Generally, fan drives may be the direct or indirect type. The direct connected drive means that the fan wheel or impeller is rotating at the same speed as the motor, and the torque is transmitted directly through a fixed shaft. This method is used mostly in axial and propeller fans where the impeller is mounted directly on the motor shaft. This arrangement places the motor bearings in the role of fan bearings. They must be designed to withstand the weight and thrust of the fan impeller. The use of a floating shaft is also considered to be a direct connected drive; however, in this circumstance, a separate set of bearings is required for the fan impeller. The indirect drives, which allow greater flexibility in motor location and fan speed, include both variable and non-variable types. Belts, chains and gears are forms of indirect, non-variable speed devices. This does not mean that these do not allow fan speed adjustment by modification of sheaves or gears. The V-belt drive is probably the most widely used type of fan drive; however, the chain-driven fan was frequently used in the past for tunnel ventilation systems. The chain drive produces less slip than the belt drive, however, it has more parts and requires lubrication and more maintenance in general. With the advent of efficient V-belt drive, the use of chain drive is minimal.

Variable speed drives, either the magnetic or fluid type, are used where a fan speed variation is required, usually with a single speed motor. These devices also eliminate transmission of mechanical vibrations through the connector. A variable pitch sheave used with a V-belt drive will permit a wide range of speed variation.

Sound Attenuation

Prior to evaluation of the amount of sound attenuation required, a set of design goals must be established. The fan sound will be transmitted both through the discharge and the inlet. The portion of the sound that passes through to the tunnel roadway is usually not critical, since some degree of attenuation is achieved by the ductwork, and the ambient sound level within the tunnel due to the movement of traffic will be high.

The sound transmitted to the neighborhood surrounding the fan structure deserves the most critical look. An evaluation of the existing community sound surrounding the fan installation is necessary. The most accurate method of determining this is to survey the local conditions by having a set of sound readings taken. These tests must be related to the appropriate time of day when conditions are most severe, such as when the generated sound is highest and the surrounding sound is lowest. The effect of the fan sound on occupants of nearby buildings must be evaluated, especially during off-peak traffic hours. Before establishing the sound design goals, the local anti-noise ordinances must be reviewed to assure that these goals will be within any limits that are established by these ordinances.

Attenuation. Once the fan sound level is determined and the design goals established, the amount of noise must be removed by some form of attenuation. All natural and built-in attenuation should be explored before considering the installation of a sound attenuator device. Distance is an excellent natural form of attenuation. Barrier walls erected between the source and the receiver are also extremely good attenuators. Other factors, such as the sound absorption of the fan room, ducts, plenums and other enclosures, are also to be considered.

If, after the evaluation of all of this natural or built-in attenuation, there remains an amount of noise yet to be removed, noise attenuating devices must be considered. These devices usually consist of a series of acoustical baffles designed for smooth air passage. The acoustical material is usually contained behind perforated metal walls. These can be weatherproofed to protect the acoustic fill. They are available in two shapes, cylindrical and rectangular. The cylindrical type is designed to be mounted directly on an axial fan inlet or outlet, while the rectangular model can be built in a modular manner, as shown in Fig. 19-44, to be connected by duct to a fan or mounted architecturally.

Selection of these units should be based on the unit manufacturer's noise reduction data, usually presented as dynamic insertion loss (DIL).

Dampers

A damper is a device used to control the flow of air in a ventilating system duct. The control of airflow is accomplished by varying the resistance to flow created by the damper much as a valve does in a water system.

There are two general categories of dampers, those having sliding blades and those having rotating blades. The rotating type can be furnished either with a single or with multiple blades. The application of dampers in a tunnel ventilation system, in the majority of instances, is in a fan shut-off operation. This requires

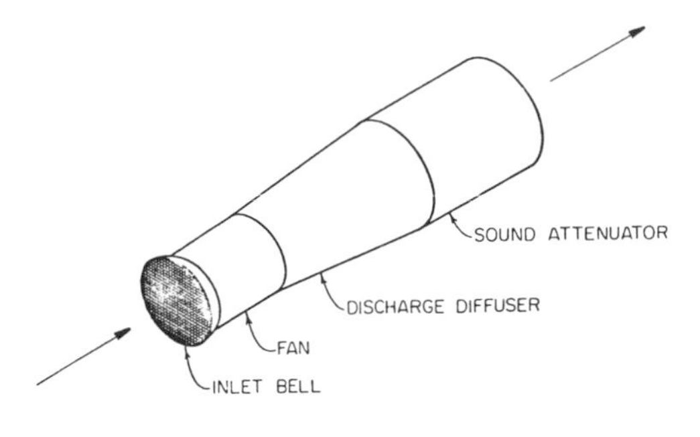

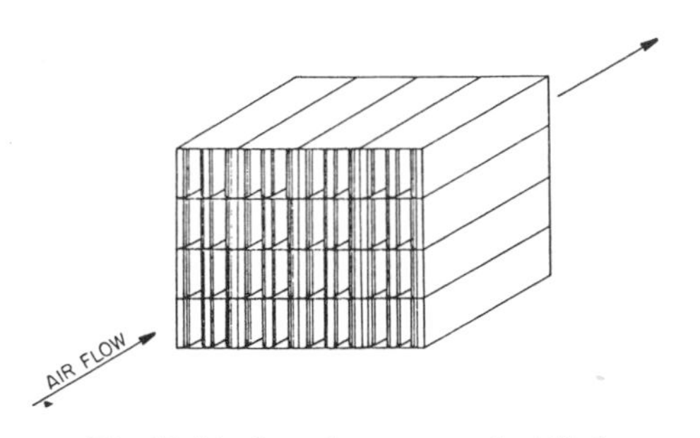

Fig. 19-44. Sound attenuator installations.

only two positions of operation, full open and full closed. There are some instances, however, where dampers are used to control airflow or pressure.

The sliding blade or guillotine-type damper is the simplest form appropriate for fan shut-off operation. There are a minimum number of moving parts to this damper, thus reducing the maintenance cost. There is also no damper resistance loss when the damper is in its full open position. One severe shortcoming, however, is the added space required to store the blade when the damper is in its full open position.

The simplest rotary blade damper is the single blade type, which can either have a rectangular or a round shape. This damper is also highly suitable for a fan shut-off operation due to its low resistance when in the full open position. There are, however, restrictions on the size of the blade from a structural design standpoint. Also, the space required on both sides of the damper for the open damper blade is often a problem.

The multiple blade damper is the most widely used of the rotary blade types. The blades in the multi-blade damper can be arranged to rotate either in parallel or in opposed action (Fig. 19-45). The opposed blade action damper is most appropriate for airflow control because of its flow pressure characteristics, while the parallel blade action damper is used chiefly for a fan shut-off application due to its excellent low leakage characteristics.

The inlet vane dampers used for volume control on centrifugal fans contain a series of radial pie-shaped blades which rotate on a series of shafts set in a ring frame.

Dampers used in tunnel ventilation systems should be constructed to withstand the maximum pressure and temperatures anticipated. Fan shut-off dampers should be designed to withstand the maximum shut-off pressure of the fan, with a factor of safety added to allow for shock loadings due to sudden closure of the damper. The blades of all rotary type dampers should have an aerodynamic shape to reduce resistance losses. The frames of all dampers should be of structural steel.

The damper operator should be of the type to provide a power drive in both the closing and the opening mode. A minimum allowance of 50% of required capacity should be added to allow for deterioration of the damper mechanism.

Leakage characteristics of a damper are most important in a fan shut-off application. The air leakage through a closed damper must be spec-

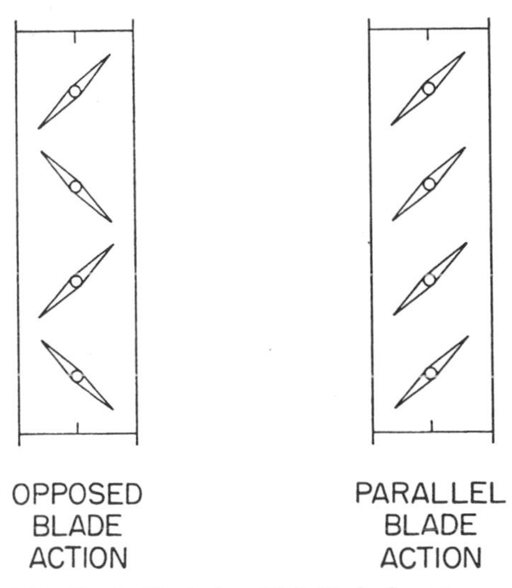

Fig. 19-45. Typical multiple blade damper.

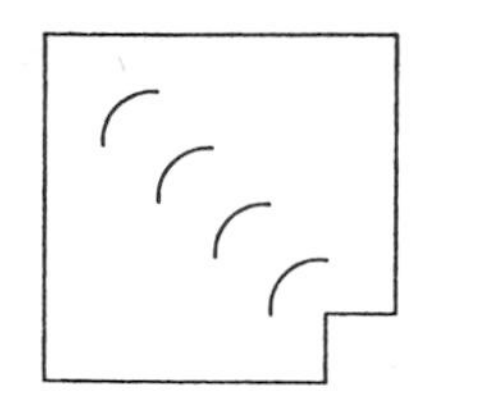

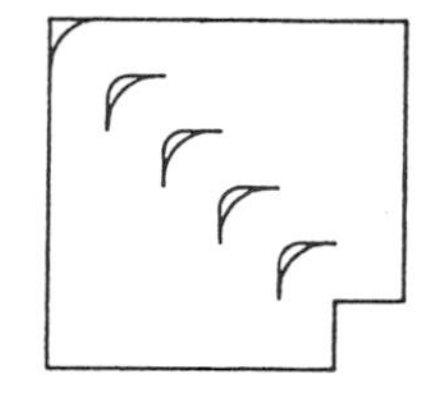

Fig. 19-46. Square duct elbow turns.

ified for a particular pressure drop across the damper. The allowable leakage must be that amount of air which will be acceptable to the ventilation system without severe deterioration of the ventilating effect. This leakage of airflow must not be of a magnitude to permit windmilling of the idle fan or fans.

The pressure drop characteristics across a full or partially opened damper should be included in the computation of the system resistance.

Turning Vanes

Turning vanes should be installed in all square duct elbows to reduce the resistance loss. The type of vane employed will depend upon the duct velocity, as the loss is directly related to the velocity pressure, as shown in Fig. 19-46. For example, the 25% velocity pressure saving shown for the double thickness vanes over the single thickness vanes at 4000 fpm becomes $\frac{1}{4}$-inch water gauge reduction in system loss. For the ranges of horsepowers considered in tunnel ventilation systems, this difference can be significant in the amount of power saved.

Turning vanes are designed to maintain a reasonable aspect ratio of the space between the vanes. Based on a curve ratio of 0.7 and an aspect ratio of 5 to 1 for the flow passages between the vanes, the following relationships hold.

$$NV = \frac{5W}{D} - 1$$

where

NV = number of vanes
W = width of duct (feet)
D = depth of duct (feet) (Fig. 19-47).

$$P = \frac{W}{NV + 1}$$

where

P = spacing between vanes (feet) (Fig. 19-48).

$$R = 3.33P$$

where

R = radius of inner vane surface (feet).

$$r = 2.33P$$

where

r = radius of outer vane surface (feet) (Fig. 19-48).

Turning vanes are usually constructed of metal having corrosion-resistance properties.

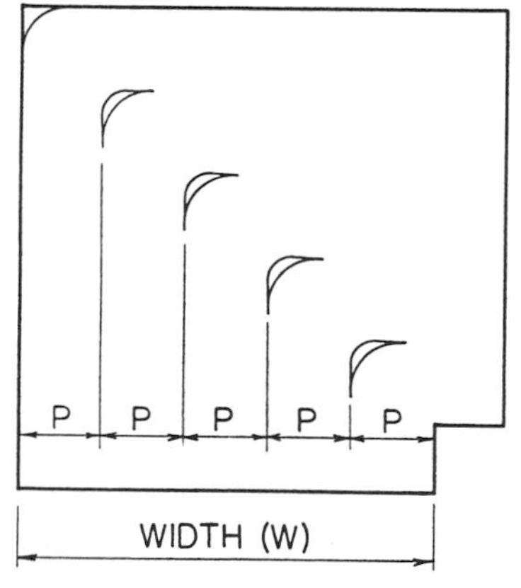

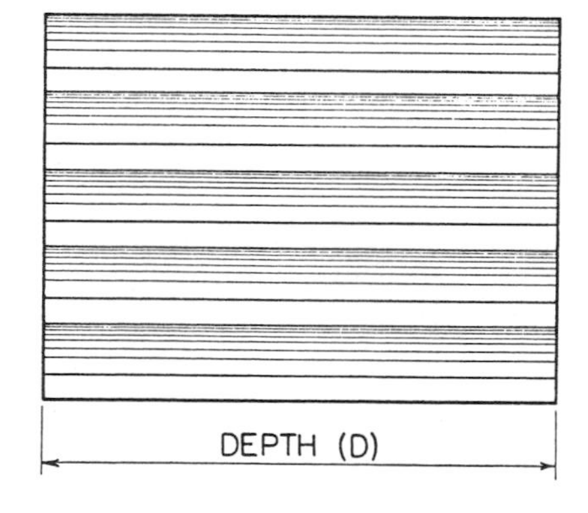

Fig. 19-47. Vaned elbow.

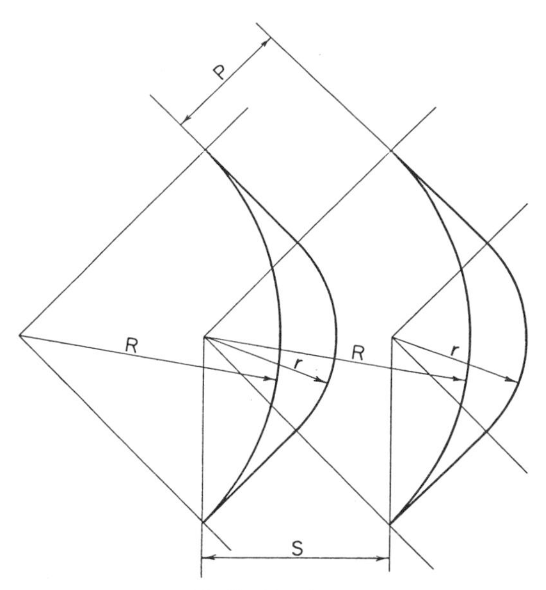

Fig. 19-48. Turning vane.

Ducts

The ducts which transport the air within the tunnel are usually part of the tunnel construction. In most tunnels, these ducts are a constant cross-sectional area throughout. The surfaces of these ducts must be relatively smooth to assure minimal air friction losses. Streamlining of obstructions, such as hangers, especially in areas of high velocity, is recommended.

The maximum air velocity in these ducts has, in most instances, been held below 6000 feet/minute. There are, however, cases where 7000 feet/minute velocities have been used.

The various duct connections from the fans to the tunnel ducts are usually either of concrete (as a part of the structure) or of sheet steel. The steel ducts must be reinforced against the maximum pressure conditions anticipated. The interior surface must be smooth, with all bracing located on the exterior. General practice has been to use $\frac{1}{4}$-inch steel plate for most of these applications.

Access must be provided for all ducts and near all equipment.

Air measuring ports are required at strategic locations to permit balancing and testing of the tunnel ventilation system. These ports should be large enough to permit entrance of a proper sized Pitot tube. They should be fitted with a screw-type plug or cap to permit closure during system operation.

The ventilation air must be transported between the ducts and the roadway. Various types of flues and ports are used to accomplish this task. Supply air flues designed to carry the supply air from the duct to the roadway are usually constructed of steel. They may also be constructed of any material which will withstand forces, such as pouring of the concrete or other construction activities. The interior surfaces of these flues must be smooth to minimize air friction losses. All of the flues are usually made of the same size, and regulation of airflow through each flue is accomplished by use of a simple plate-type damper. A typical flue and damper arrangement is shown in Fig. 19-49.

Ports are simply slots or openings in a wall or ceiling to permit passage of air. They are sometimes sloped in a direction of airflow as shown on Fig. 19-50. Ports can be used for either supply or exhaust air. To assure even air distribution along the length of the duct, a plate-type damper is often used or the spacing of the ports is varied.

Ventilation Buildings

Ventilation buildings are required to house the tunnel ventilation fans, electrical equipment, controls and auxiliary equipment and facilities. Relatively short tunnels will most likely have only one ventilation building, whereas larger tunnels will have two, three or more buildings. Each building will usually serve one or more ventilation zones.

The general arrangement of the ventilation building will depend to a great extent on the type of fans and ventilation system employed, and on the type and length of tunnel. A subaqueous tunnel will by nature require a vertical building arrangement to take advantage of the space above the tunnel and below grade, as shown in Fig. 19-51. A cut-and-cover tunnel in an urban area utilizing an underground ventilation building could appear, as in Fig. 19-52. In a tunnel such as this, often the space available for the ventilation structure is limited. A mountain tunnel utilizing centrifugal fans is shown in Fig. 19-53.

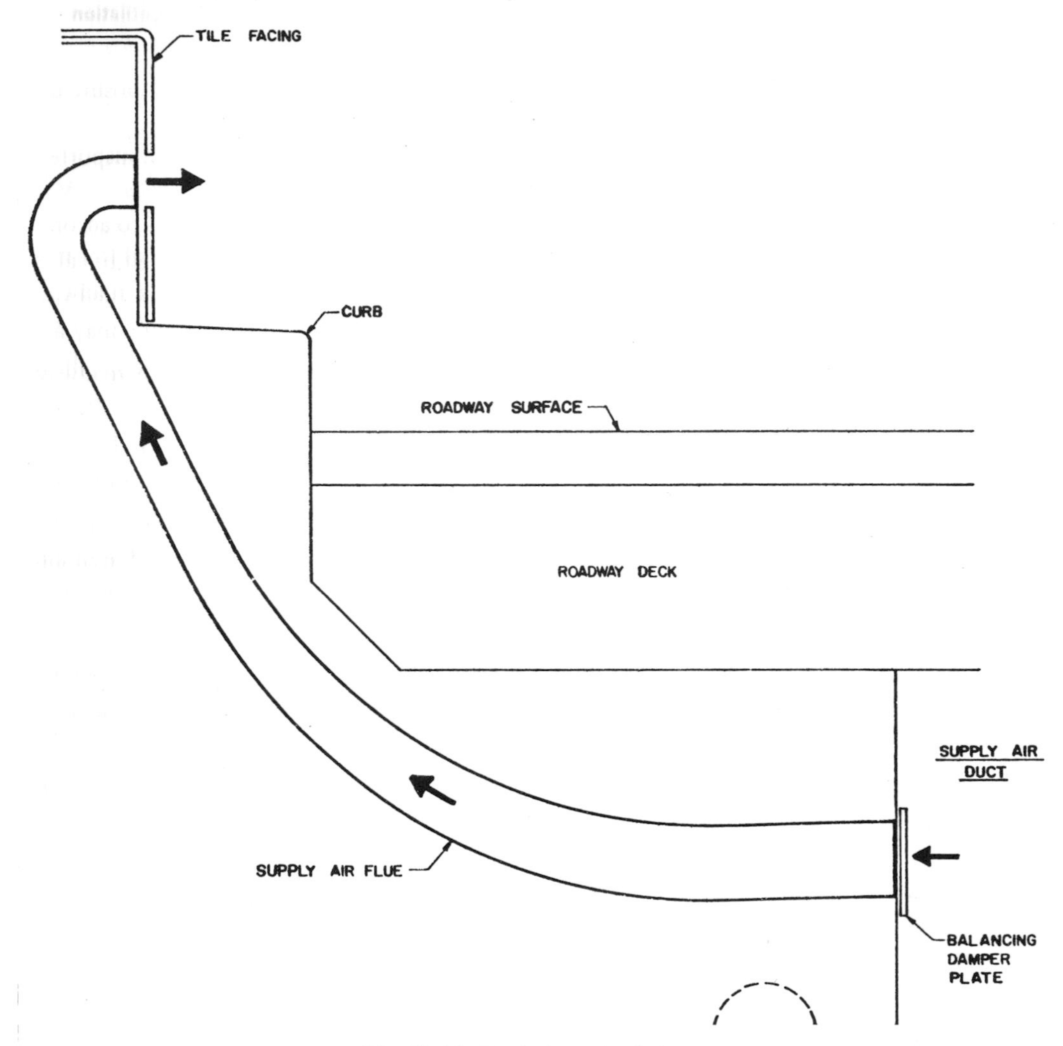

Fig. 19-49. Typical supply air flue.

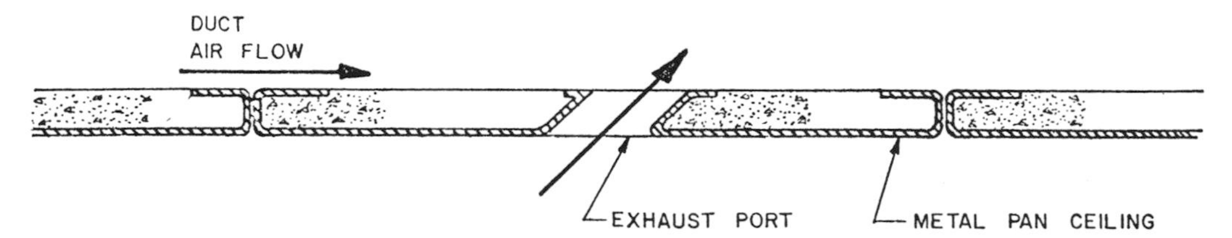

Fig. 19-50. Typical ceiling exhaust air port.

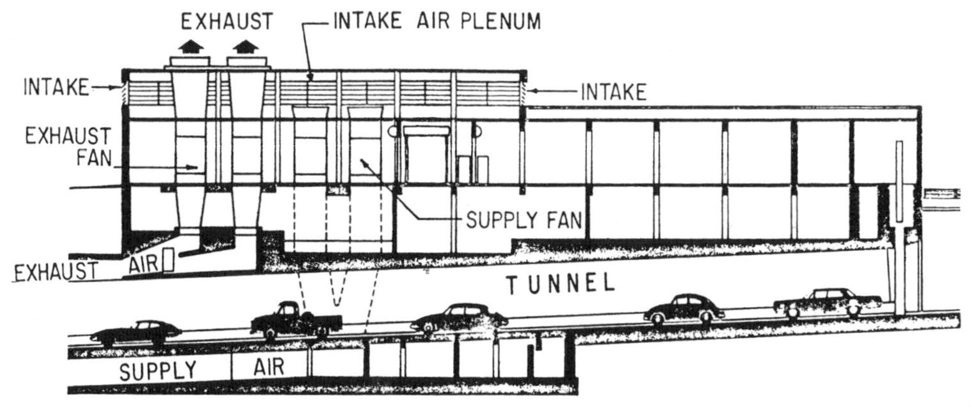

Fig. 19-51. Typical ventilation building for a subaqueous tunnel.

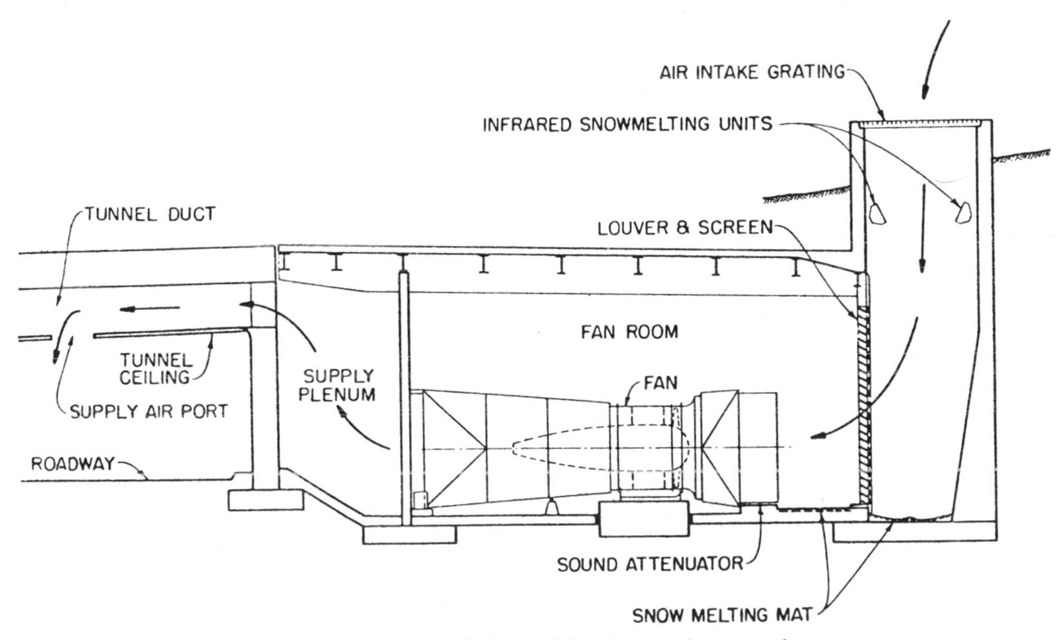

Fig. 19-52. Typical ventilation building for an urban tunnel.

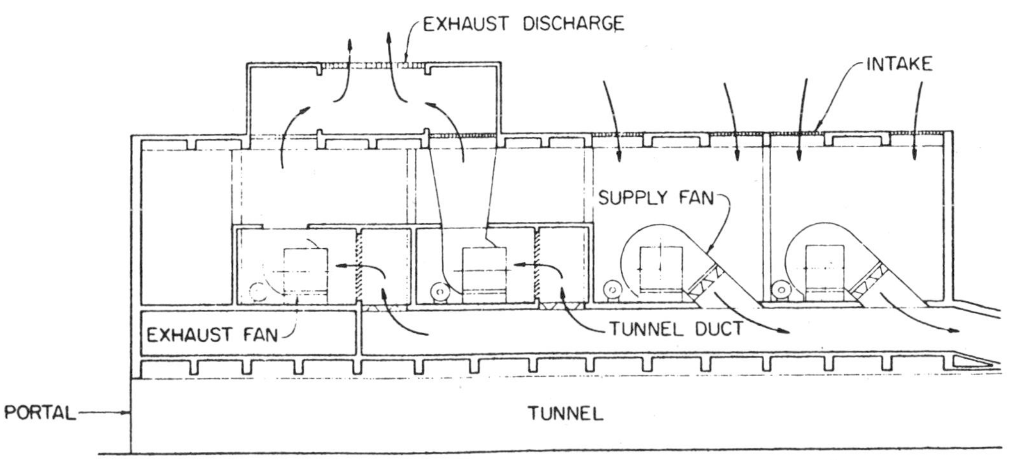

Fig. 19-53. Typical ventilation building for a mountain tunnel.

During the process of arranging the ventilation building layout, consideration must be given to the space required for the airflow to and from the fans and for the proper service and removal of equipment. When considering maintenance, both routine service and removal of equipment is important. Enough space must be allowed around each fan for service activities and for removal. Along with accessibility, provision must also be made to lift the equipment during service activities. On a centrifugal fan, the impeller, shaft, housing and motor will be the items most likely requiring lifting. The impeller, motor and housing on a horizontal axial fan will most likely be moved. On a vertical mounted axial fan, a rail system can be provided which will permit ready removal of the entire fan for service.

Testing

Factory tests should be conducted on all specially designed and built major operating components of the tunnel ventilation system. These tests would normally cover fans, motors and dampers to verify the predicted performance of this equipment. They will assure that there will be minimal delays when the equipment is permanently installed. It is also preferable to test the fan along with its associated motor and shut-off damper, if possible. The fan tests should be conducted in accordance with the standard AMCA test procedure. It may be necessary, however, under some circumstances, to adjust the set-up outlined by AMCA if the fan is too large for the particular test facility. Sound tests, as outlined in the AMCA standards, should be conducted along with the airflow tests mentioned above.

Mechanical testing of the fan wheel or impeller should be considered, especially if the fan rotational speed is high, as in the axial fan. Overspeed testing in a vacuum pit is one method of mechanical testing the impeller.

Dampers should be tested for leakage, especially if the fan operating pressures are high. The dampers should be tested against the fan shut-off head, too. If the damper is being used to modulate pressure and flow, a more complicated testing would be required, where flow and pressure for all intermediate positions of the damper could be demonstrated.

Site Testing. Once the ventilation system has been installed, the airflow pressure and sound should be measured and recorded. This procedure will assure compliance with design and will aid in the balancing process. The shop test usually does not mimic the actual site conditions, while site testing will evaluate the actual flow conditions.

AIR VOLUME

The volume of air necessary to satisfy the dilution requirements of carbon monoxide and the other contaminants outlined earlier should be computed based on the average carbon monoxide emission rate of the vehicles.

Evaluation

The average carbon monoxide emitted by a lane of moving traffic can be computed as follows.

$$COV = COK \times FGS \times \frac{WGT}{2000} \times TRV \times \frac{TUL}{5280} \times FSF \times FEL$$

where

COV = average carbon monoxide emitted per lane of traffic (feet^3/hour)
COK = average carbon monoxide emitted for moving traffic (feet^3/ton/mile)
FGS = speed and grade factor (Fig. 19-8)
WGT = average vehicle weight (lb)
TRV = traffic volume (vehicles/hour)
TUL = length of tunnel (feet)
FSF = reserve factor
FEL = elevation correction factor (Fig. 19-5).

The average smoke produced by a lane of moving diesel-engined vehicles can be computed as follows.

$$DSC = DSE \times FGS \times \frac{WGT}{2000} \times TRV \times \frac{TUL}{5280} \times FSF \times FEL$$

where

DSC = average diesel smoke concentration per lane (Dmg/hour)
DSE = average diesel smoke emitted from moving diesel-engined vehicle (Dmg/ton/mile)
FGS = speed and grade factor (Fig. 19-9).

For idle traffic, the following can be used to compute the average carbon monoxide emitted.

$$COV = COH \times TRD \times \frac{TUL}{5280} \times FSF \times FEL$$

where

COH = average carbon monoxide emitted during idle (feet^3/vehicle/hour)
TRD = traffic density (vehicles/mile).

Emission values must be computed for each type of vehicle in the traffic mixture; thus, the total carbon monoxide emission per lane is:

$$COVL_i = COVP + COVT$$

where

$COVL_i$ = carbon monoxide emitted per lane
COVP = carbon monoxide emitted by passenger cars
COVT = carbon monoxide emitted by trucks and buses.

Using this value of $COVL_i$, the total carbon monoxide emitted per bore can then be computed as follows.

$$TCOV = COVL_1 + COVL_2 + COVL_3$$

where

TCOV = total carbon monoxide emitted per bore or tube (feet^3/hour).

At this point, the carbon monoxide values should be corrected for non-standard atmospheric conditions, as noted earlier. The temperature experienced at the tunnel location has a minor effect on the amount of carbon mon-

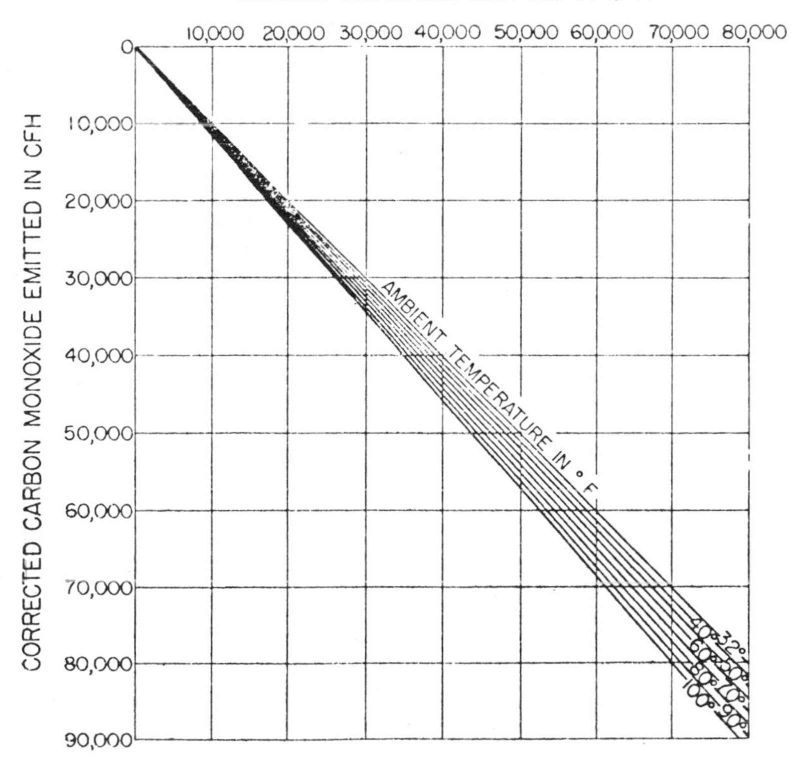

Fig. 19-54. Temperature correction chart.

oxide emitted. This correction takes the following form.

$$\text{TCOVC} = \text{TCOV} \times \frac{\text{TEMA}}{492}$$

where

TCOVC = corrected average carbon monoxide emission per bore (feet^3/hour)
TEMA = average maximum temperature at tunnel location (°R) (°R = °F + 460)

A quick correction can be obtained from the chart in Fig. 19-54. Using these corrected emission rates, the total air quantity requirements can be computed as follows.

$$\text{QAR} = \frac{\text{TCOVC} \times 10^6}{\text{PPM} \times 60}$$

where

QAR = final air quantity required (feet^3/minute)
PPM = allowable level of carbon monoxide (ppm).

The final air quantity can be obtained from the chart in Fig. 19-55.

At this point, it should be determined if the air quantity desired for maintenance of a proper level of carbon monoxide in the tunnel will be sufficient to maintain a level of safe visibility. Since the smoke emitted from the diesel-engined vehicle will create a visibility problem in the tunnel, this is the factor which is used to evaluate visibility (as follows).

$$\text{QAR} = \frac{\text{TDSC} \times 0.5885}{\text{DSA}}$$

where

TDSC = total diesel smoke emitted per bore or tube (Dmg/hour)
DSA = allowable concentration of diesel smoke (Dmg/m^3).

Adjustment of Air Volume

Most tunnel ventilation systems are divided into ventilation zones to aid in emergency smoke purge operations. In many cases, the zones are arranged based on the location of ventilation buildings. This is clearly shown on the diagram of the Holland Tunnel ventilation system in Fig. 19-56.

One of the requirements established by the Federal Highway Administration (FHWA) regarding tunnel ventilation is that, in any single ventilation zone, a capacity equal to approximately 85% of the total capacity is required when one fan is out of operation. This can be achieved by providing a spare or standby fan or by installing a minimum of four fans in each ventilation zone. With this configuration, and if the fans are selected properly, the three fans

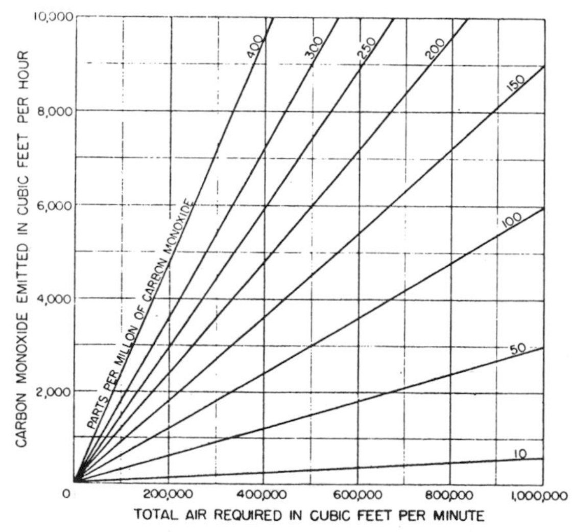

Fig. 19-55. Final airflow rate chart.

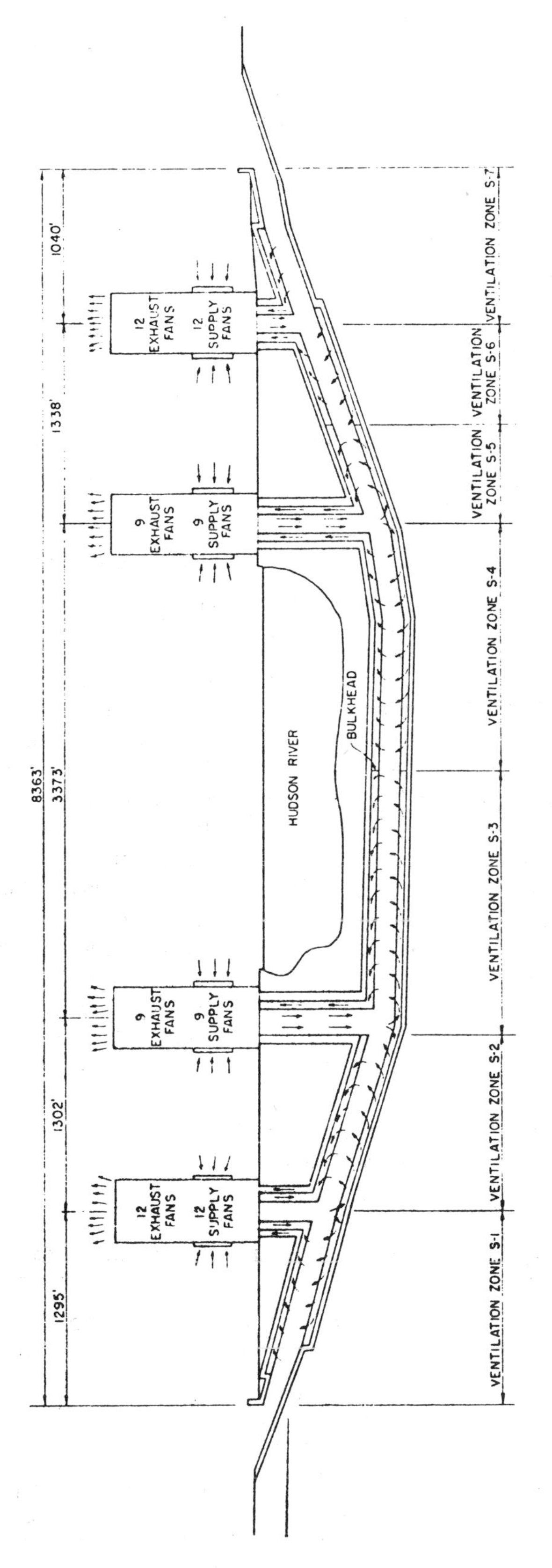

Fig. 19-56. Ventilation diagram of the Holland Tunnel.

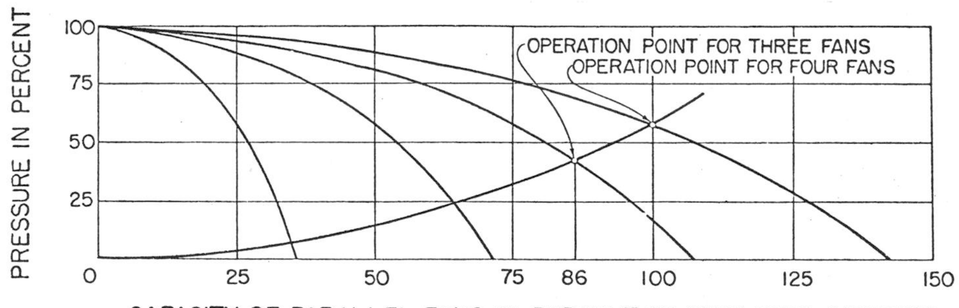

Fig. 19-57. Operation of four fans in parallel in one ventilation zone.

when operating in parallel will provide approximately 85% of the total four-fan capacity. This can be seen on the fan and system curves shown on Fig. 19-57. Using this approach, the cost of a standby fan is eliminated.

The air volume requirements within the tunnel roadway will continuously vary throughout the day as the traffic volume varies. In order to minimize the power of consumption, some form of flow adjustment must be considered. This adjustment must be such as to reduce the fan horsepower along with the fan flow rate. Numerous methods of accomplishing this have been outlined above.

The fan airflow should be varied to maintain a preselected level of carbon monoxide within the tunnel. This variation can be achieved by varying the number of fans operating in the multiple fan system, using multiple speed motors, using controllable pitch fans (axial) or using inlet vane control (centrifugal). Any combination of these methods can be used as is shown in the examples cited in Table 19-13, where both the effects of multiple fan operation and multiple speed motors are shown. The use of multiple fans and two-speed motors is the most suitable method of air volume control.

PRESSURE EVALUATIONS

General

The air pressure losses in the tunnel duct system must be evaluated in order to be able to compute the fan pressure requirements and the power required to drive the fan.

The fan selection should be based on fan total pressure instead of static pressure, since the fan total pressure is equal to the difference in total pressure across the fans, whereas the static pressure is not equal to the difference in static pressure across the fan. The fan total pressure (FTP) has been defined by the Air Moving and Conditioning Association (AMCA)[27] as the algebraic difference between the total pressure at the fan discharge (TP_2) and the total pressure at the fan inlet (TP_1), as shown in Fig. 19-58:

$$FTP = TP_2 - TP_1.$$

The fan velocity pressure (FVP) is defined by

Table 19-13. Typical fan operational modes.

CARBON MONOXIDE LEVEL	TWO FANS PER ZONE, THREE-SPEED MOTORS	FOUR FANS PER ZONE, TWO-SPEED MOTORS
1	1 fan, low speed	2 fans, low speed
2	2 fans, low speed	3 fans, low speed
3	2 fans, medium speed	4 fans, low speed
4	2 fans, high speed	2 fans, high speed
5		3 fans, high speed
6		4 fans, high speed

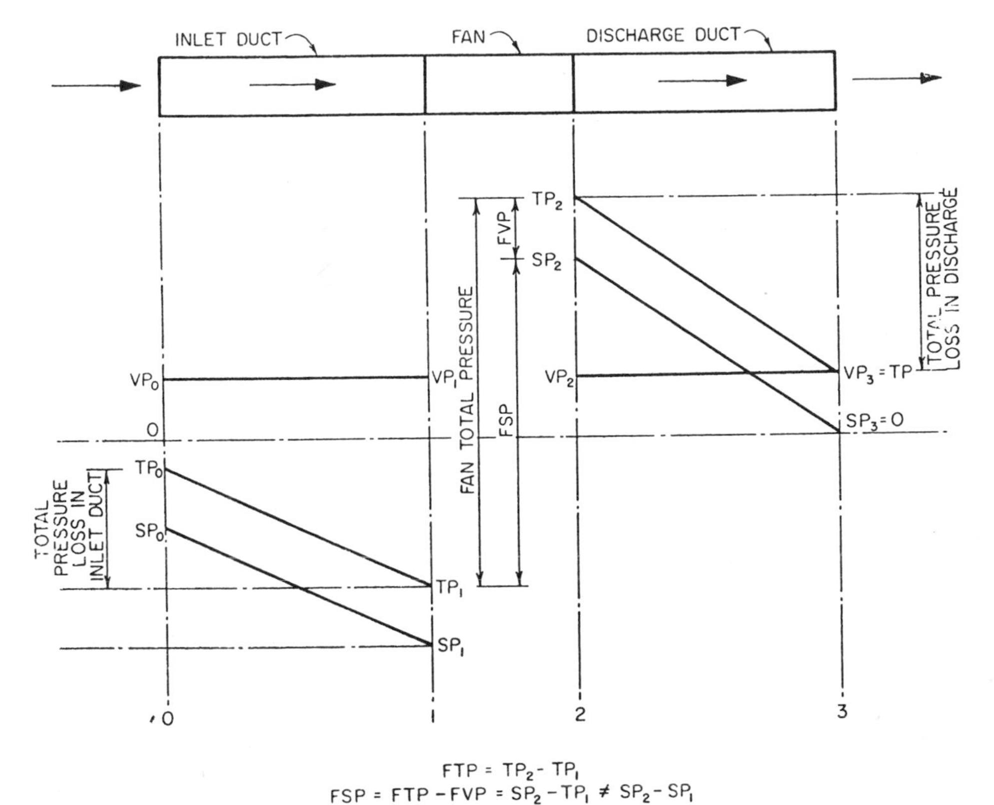

Fig. 19-58. System pressure definitions.

AMCA to be the pressure corresponding to the air velocity and air density at the fan discharge:

$$FVP = VP_2.$$

The fan static pressure (FSP) is equal to the difference between the fan total pressure and the fan velocity pressure:

$$FSP = FTP - FVP.$$

A fan must have the total pressure at the fan discharge (TP_2) equal to the total pressure losses (ΔTP_{23}) in the discharge duct and the exit velocity pressure (VP_3):

$$TP_2 = |\Delta TP_{23}| + VP_3.$$

Likewise, the total pressure at the fan inlet (TP_1) must be equal to the total pressure losses in the inlet duct system and the inlet pressure:

$$TP_1 = |\Delta TP_{01}| + 7P_0$$

However a proper fan selection must deliver the total pressure requirements. This is the fan total pressure (FTP) equal to the sum of the total pressure losses in the inlet system (ΔTP_{01}), the total pressure losses in the discharge system (ΔTP_{23}) and the exit velocity pressure (VP_3).

$$FTP = |\Delta TP_{01}| + |\Delta TP_{23}| + VP_3$$

The loss of pressure in an air duct system is due to the effect of friction and dynamic factors. These factors manifest themselves in the many components of a duct system, such as the main ducts, elbows, transformations, duct entrances and exits, obstructions and equipment, as typified by Fig. 19-59.

Straight Ducts

Straight ducts used in tunnel ventilation systems can be classified into two categories: those which merely transport air and thus have constant area and constant air velocity, and those which uniformly distribute (supply) or uniformly collect (exhaust) air.

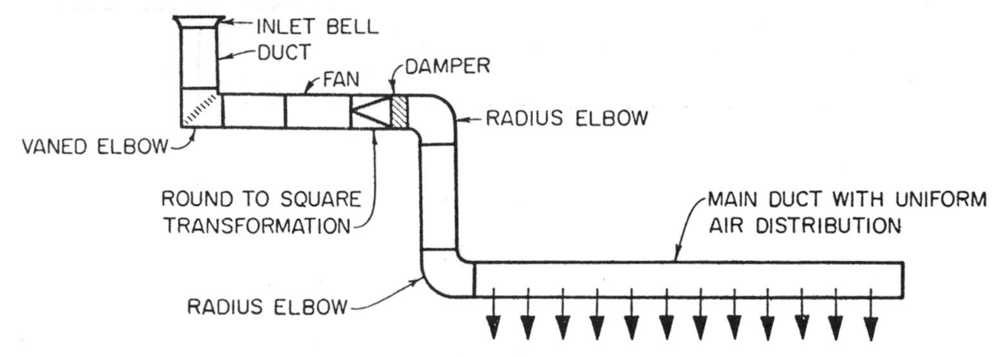

Fig. 19-59. Typical tunnel supply ventilation system.

There have been several methods developed which will predict the pressure losses in a duct of constant cross-sectional area and uniform distribution or collection of air. The two most significant methods are presented below. The first method, presented in 1929, was developed for the Holland Tunnel design[28] and has been in use for many years. The following relationships will give the total losses of pressure at any point in the duct (Fig. 19-60).

Supply Duct

$$P = P_1 + \frac{12y}{D}\left\{\frac{{V_0}^2}{2g}\left[\frac{aLZ^3}{3m} - \tfrac{1}{2}(1-K)Z^2\right] + \frac{bLZ}{2gm^3}\right\}$$

where

P = total pressure loss at any point in duct (inches of water)

P_1 = pressure at bulkhead (inches of water)

y = density of air (pcf)

D = density of water (pcf)

V_0 = velocity of air entering (or exiting) duct (feet/second)

g = acceleration of gravity (feet/second/second)

a = first numerical constant related to coefficient of friction for concrete (0.0035)

b = second numerical constant related to coefficeint of friction for concrete (0.01433)

L = total length of duct (feet)

$Z = \dfrac{L - X}{L}$

X = distance from duct entrance (or exit) to any selected location (feet)

m = hydraulic radius (feet)

K = numerical constant which takes turbulence into account (0.615).

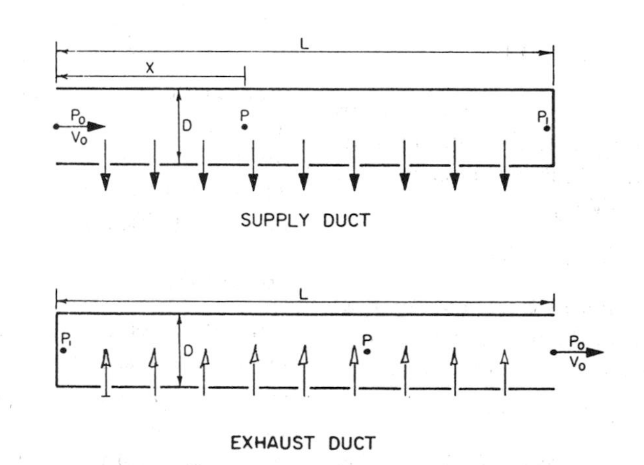

Fig. 19-60. Duct pressure analysis as per Singstad.[28] (Adapted)

Exhaust Duct

$$P = P_1 + \frac{12y}{D}\left\{\frac{V_0^2}{2g}\left[\frac{aL}{(3+C)m}Z^3 + \frac{3}{2+c}Z^2\right] + \frac{bL}{2gm^3(1+C)}Z\right\}$$

where

P = total pressure loss at any point in duct (inches of water)
P_1 = pressure at bulkhead (inches of water)
Y = density of air (pcf)
D = density of water (pcf)
V_0 = velocity of air exiting duct (feet/second)
g = acceleration of gravity (feet/second/second)
a = first numerical constant related to coefficient of friction for concrete (0.0035)
b = second numerical constant related to coefficient of friction for concrete (0.01433)
c = numerical constant relating to turbulence at exhaust ports (0.25)
L = total length of duct (feet)
X = distance from duct exit to any selected location (feet)
$Z = \frac{L - X}{L}$
m = hydraulic radius (feet).

The second method was presented 30 years later, in 1959, in Switzerland[2] (Fig. 19-61).

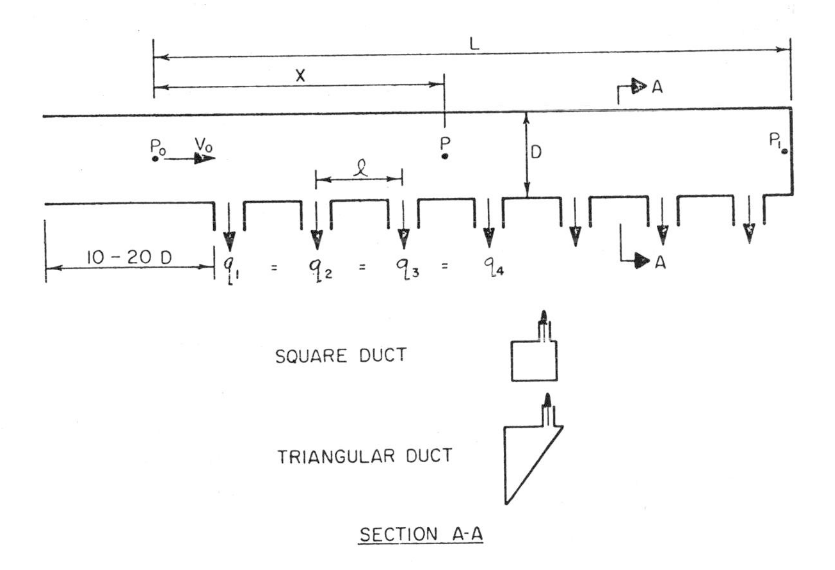

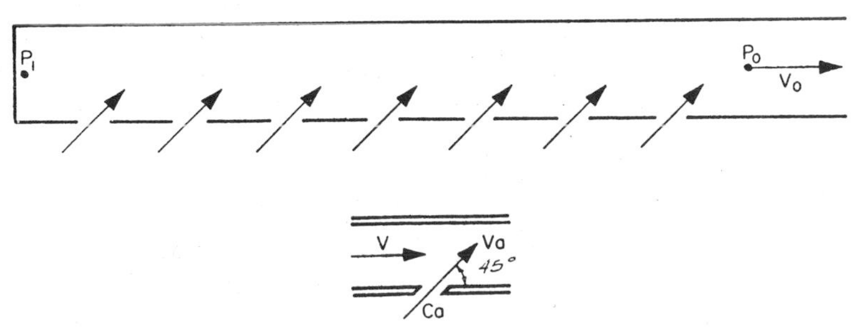

Fig. 19-61. Duct pressure analysis.[2] (Adapted)

Supply Duct

$$P = P_1 + \frac{12}{D}\frac{Y}{2g}V_0^2\left[\left(\frac{X}{L}\right)\left(-\lambda\frac{L}{D}+2R_0\right) + \left(\frac{X}{L}\right)^2\left(\lambda\frac{L}{D}-2R_0+R_1\right) + \left(\frac{X}{L}\right)^3\left(-\frac{1}{3}\lambda\frac{L}{D}+\frac{2}{3}(R_0-R_1)\right)\right]$$

where

P = total pressure loss at any point in duct (inches of water)
P_1 = pressure at bulkhead (inches of water)
Y = density of air (pcf)
g = acceleration of gravity in (feet/second/second) = 32.2
D = Density of water (pound/cf)
V_0 = velocity of air entering duct (feet/second)
X = distance from entrance of duct to selected location (feet)
L = total length of duct (feet)
λ = friction factor
D = hydraulic diameter (feet)
R_o = pressure regain coefficient at point 0 (Figure 19-62).
R_1 = pressure regain coefficient at point 1 ($R_1 = 0.21$ for square ducts; $R_1 = 0.48$ for triangular ducts)

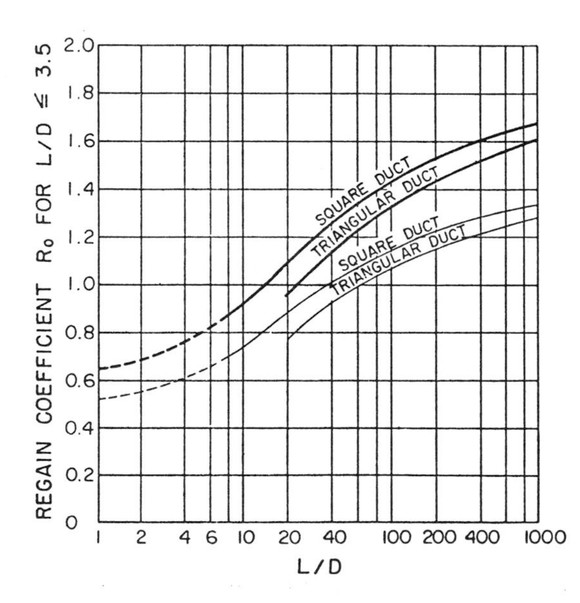

Fig. 19-62. Pressure regain coefficient.[2] (Adapted)

Exhaust Duct

$$P = P_1 + \frac{12}{D}\frac{Y}{2g}V_0^2\left[\left(\frac{X}{L}\right)\left(\lambda\frac{L}{D}+2\bar{K}\right) - \left(\frac{X}{L}\right)^2\left(\lambda\frac{L}{D}+\bar{K}\right) + \left(\frac{X}{L}\right)^3\left(\frac{1}{3}\lambda\frac{L}{D}\right)\right]$$

where

P = total pressure loss at any point in duct (inches of water)
P_1 = pressure at bulkhead (inches of water)
Y = density of air (pcf)
D = density of water
g = accelerationg of gravity (feet/second/second) = 32.2
V_0 = velocity of air exiting duct (feet/second)
X = distance from exit of duct to selected location (feet)
L = total length of duct (feet)
λ = friction factor
D = hydraulic diameter (feet)
K = average pressure conversion coefficient (Fig. 19-63).

The two methods outlined above produce somewhat differing results.

Transport Duct Losses

The pressure losses in a transport duct having constant cross-sectional area and constant velocity are due to friction alone and can be computed using standard air flow equations.[29]

Dynamic Losses

The dynamic losses created by eddying airflow are found in sudden changes in direction of airflow and changes in the magnitude of the air velocity. The general relationships necessary for the evaluation of pressure losses due to changes in flow direction or velocity are known.[29]

Supply Air Flues

The loss of pressure as the air passes through the supply air flue, if the flue area and cross section remain constant, can be computed as follows.[2]

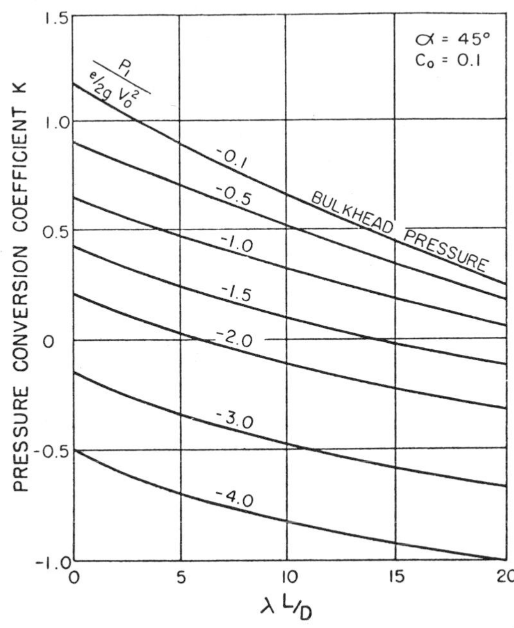

Fig. 19-63. Average pressure conversion coefficient.[2] (Adapted)

$$PLF = \frac{12}{D}\frac{Y}{2g}\left[\left(1 + C_{EF} + \lambda\frac{LF}{DF} + C_{BF} + C_{XF}\right)V_F^2 - (1 - C_E)V^2\right]$$

where

PLF = pressure loss through flue (inches of water)
Y = density of air (pcf)
D = density of water (pcf)
g = acceleration of gravity (feet/second/second)
C_{EF} = loss coefficient due to change in direction of airflow at entrance to flue (Fig. 19-64)
λ = flue friction coefficient
LF = length of flue (feet)
DF = equivalent diameter of flue (feet)
C_{BF} = dynamic loss coefficient relating to change of airflow direction within the flue (Reference 29)
C_{XF} = dynamic loss coefficient relating to flue exit (if there is not exit deflection grille, then C_{XF} = 0)

FLUE ENTRANCE CONFIGURATIONS	C_{EF}	C_E
1	2.00	0.90
2	0.65	1.00
3	0.50	1.00
4	0.28	0.85
5	0.20	0.60

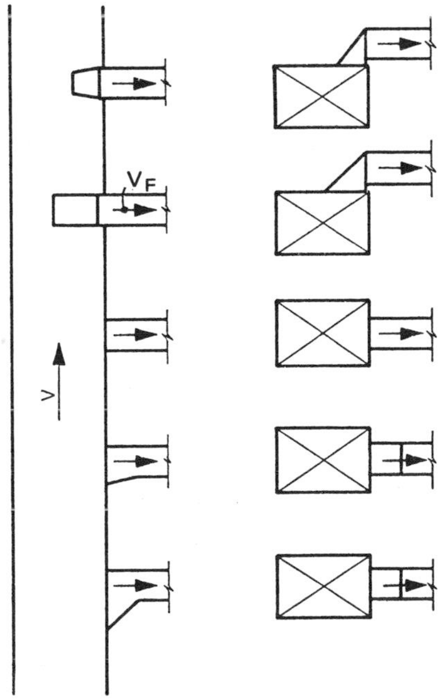

Fig. 19-64. Entrance pressure loss.[2]

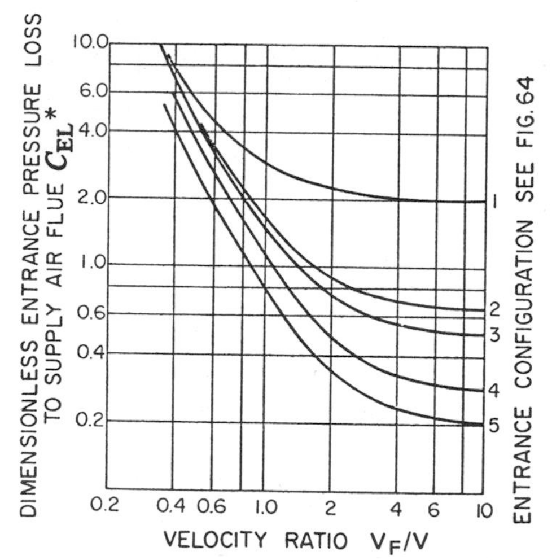

Fig. 19-65. Entrance pressure loss.[2]

V_F = air velocity in flue (fps)

C_E = loss coefficient due to change in direction of airflow at entrance to flue (Fig. 19-64)

V = air velocity in main duct (fps).

Using the values of C_{EL} obtained from Fig. 19-65 the following can be applied.

$$\mathrm{PLF} = \frac{12}{D}\frac{Y}{2g}\left[(C_{EL})V_F^2 + \left(1 + \lambda\frac{LF}{DF} + C_{BF} + C_{XF}\right)V_F^2 - V^2\right]$$

where

C_{EL} = pressure loss coefficient at entrance to flue (Figure 19-65)

$$C_{EL} = C_{EF} + \frac{C_E}{(V_F/V)^2}$$

The overpressure required in the main duct to maintain constant air distribution into the roadway must be at least equal to PLF; where these overpressures are larger than PLF, provisions must be made to adjust the size of the flue opening.

Exhaust Air Ports

The exhaust air ports (Fig. 19-50), which are usually located in the tunnel's ceiling, can be evaluated as square-edged orifices, unless the exhaust port has a branch duct connection, in which case it should be evaluated in a fashion similar to the supply air flues.

Equipment Losses

All equipment located in the ventilation air steam will add resistance to the flow of air.

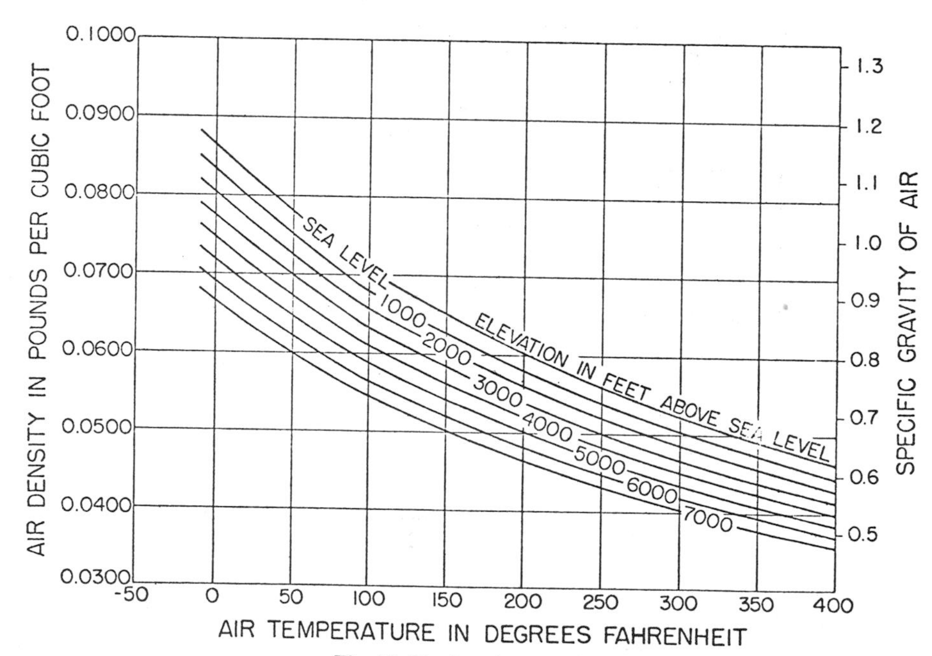

Fig. 19-66. Density correction.

These losses must also be added to the system pressure losses. Louvers and dampers are the most prominent pieces of equipment, other than the fan, to be found in a tunnel ventilation system.

Density Correction

The air density for a tunnel ventilation system should be selected based on the minimum temperature and the maximum barometric pressure anticipated, thus assuring adequate fan power for all meteorological conditions. Figure 19-66 provides the necessary data for various tunnel elevations and air temperatures.

Fan Horsepower

The fan horsepower required to deliver the specified air quantity at specified pressure[30,31] can be found from:

$$AHP = \frac{Q \times FTP}{6356}$$

where

AHP = air horsepower
Q = flow rate of fan (feet^3/minute)
FTP = fan total pressure (inches of water).

$$BHP = \frac{Q \times FTP}{6356\ M_T}$$

where

BHP = brake horsepower
M_T = total efficiency (%).

CONTROL

General

The control of a tunnel ventilation system can be either manual or automatic, with either remote or local operation. A basic operational philosophy is involved regarding whether there will be a human operator continuously present at the tunnel control. A manually operated tunnel will require an operator to be present at all times; whereas a fully automatic system can function without the attendance of an operator. However, a fully automatic system is not completely without some human participation, in that there are a number of system operating conditions which must be monitored in order to prevent serious equipment breakdown. It is still felt by many tunnel designers that, except for relatively short tunnels, the human element must be included in the tunnel operation since only the human can make the required judgment in the event of emergency conditions. The type of control selected for the ventilation system should be consistent with those selected for all other equipment, such as pumps, lights and power.

Manual Control

A manual control system is usually operated from a control room located within the facility which provides a centralized location for all control indication and monitor functions. In many vehicular tunnels, the ventilation system control is combined with the traffic control into one location. All necessary control and monitoring systems must be provided to enable the operator to start, stop or adjust the ventilation system to suit the tunnel traffic and tunnel air quality conditions. The operator responds to the output from some form of monitoring of these conditions.

The most predominant environmental factors to be monitored for fan operation are carbon monoxide concentrations, visibility and traffic flow. Any one or combination of these factors can be used as a control factor. The use of a carbon monoxide analyzer system which produces an indication of the carbon monoxide concentration in ppm, and a visibility measuring system which produces an indication of the percent visibility, along with knowledge of the historical pattern of traffic flow, will allow the operator of a manual system to anticipate the airflow required to suit the tunnel conditions. For a full description of these monitoring systems, see p. 468.

A typical control board for a manual system would include the numerous recorders required to present the output of the monitoring systems, the necessary fan controls and the systems' status indications.

Automatic Control

The operation of fans responding automatically to changes in carbon monoxide concentrations, visibility, traffic flow or the calendar is all possible with a fully automatic tunnel ventilation control system. Any one or a combination of the above factors can be used to control the fans.

A tunnel ventilation system operating automatically from either carbon monoxide concentrations or visibility monitoring devices will have fan operation adjustments made whenever a change in the tunnel environment takes place. There is usually a damping or time delay built into the control system to prevent unnecessarily frequent fan speed adjustments. When the fans are controlled by a system based on traffic flow, it is not the specific tunnel environment that is monitored. The traffic flow information, along with the anticipated carbon monoxide concentration and visibility generated by the traffic, is used to set the fan operation program. This is usually based on historical data gathered during tunnel operation. A calendar-base control system functions along the same lines using the anticipated contaminant levels at each hour of the day as a basis for a control program. This information is gathered during historical experience. The traffic or calendar-base systems must also be provided with either a manual or automatic override based on carbon monoxide or visibility. The most elaborate control system would be one operating on a calendar base with automatic overrides for carbon monoxide, visibility and traffic flow.

Local Control

All control systems, whether automatic or manual, must be provided with local control within sight of the equipment to facilitate safe maintenance.

Indication

In any tunnel ventilation system, there are operational conditions which must be monitored to assure continued operation and to prevent damage to the ventilation equipment. Provision must be made in the control room (or in a remote location, in the case of a fully automatic system) to permit indication of the ventilation system operational status and annunciation of the system operational conditions.

Indication of the operational status of the fans and associated equipment, such as fan on/off, damper open/close and fan speed, can be accomplished by the use of indicating lights on the control board. The operational conditions being monitored can be set to trigger an alarm which signals that a particular operational limit has been exceeded. The following are typical examples of these operational limits: motor winding overtemperature, bearing overtemperature and high carbon monoxide concentration.

A closed circuit television system installed within the tunnel can also be considered as part of the control system. It can be used to evaluate the traffic flow situation and aid in selecting the proper action to be taken with regard to fan operation. For a full description of closed circuit TV systems, see Chapter 22.

ENVIRONMENTAL MONITORING SYSTEMS

This section covers the systems used to monitor the environment within a tunnel and provides signals which are used for fan operation and control. These systems include the carbon monoxide analyzing systems and the visibility monitoring systems. Other systems are used for the control of fans, such as traffic counters and closed circuit television, but these are covered elsewhere in this handbook.

Carbon Monoxide Analyzer

The carbon monoxide analyzer is a device which, by processing a sample of tunnel air, can determine the amount of carbon monoxide present in the sample, thus determining the carbon monoxide concentration within the tunnel environment. The units for measuring carbon monoxide within the tunnel are ppm (parts per million) by volume (100 ppm = 0.01%). There are two types of carbon monoxide analyzers currently in use in vehicular tunnels. They are the catalytic reaction type and the non-dispersive infrared type.

In the *catalytic reaction* type, the air sample

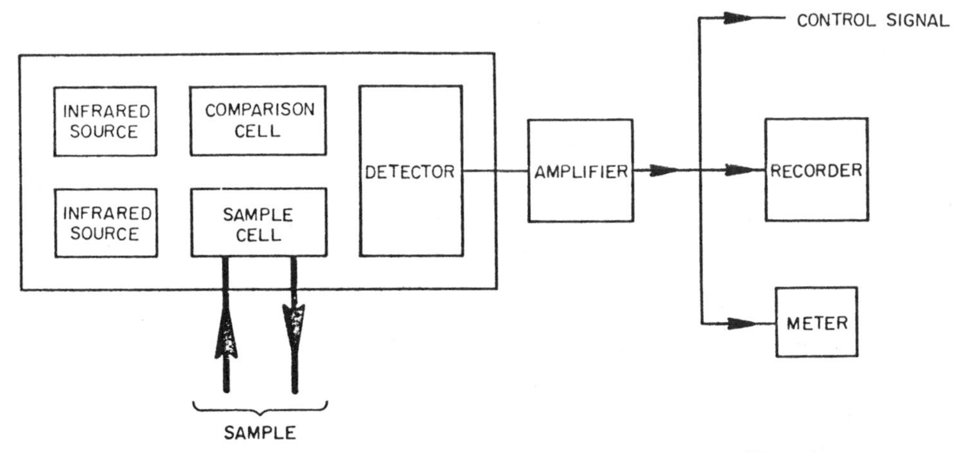

Fig. 19-67. System schematics of a non-dispersive infrared carbon monoxide analyzer system.

is filtered and cleaned as it enters the analyzing unit to remove all airborne solid particles, all hydrocarbons and all moisture. When the air sample passes through the analyzer cell, the carbon monoxide oxidizes as it comes in contact with the catalyst, usually hopcalite. (The heat generated by this reaction is directly proportional to the amount of carbon monoxide in the air sample.) Then, by use of thermalcouples, a signal output is obtained, and after amplification it can be used to record or control.

The *non-dispersive infrared type* of carbon monoxide analyzer utilizes two infrared beams, one passing through a comparison cell and the other through a sample cell (see Fig. 19-67). The sample gas, depending on its makeup, will absorb varying amounts of infrared energy. It is this variation in the infrared beam reaching the detector that provides the measurement of the gas concentration present. The signal from the detector is then amplified and used to record or control.

For the vehicular tunnel, the catalytic type of analyzer appears to be most appropriate. It is less expensive, more rugged and simple to maintain. The infrared system is more expensive and requires more specialized maintenance. The infrared system does have a faster response time characteristic and has lower sensitivity, as shown in Table 19-14, but for the levels of carbon monoxide encountered in a vehicular tunnel, the catalytic reaction type is most satisfactory.

Table 19-14. Comparison of carbon monoxide analyzing systems.[32]

	CATALYTIC	INFRARED
Rise	100 seconds	10 seconds
Initial response	20 seconds	3 seconds
Sensitivity	10 ± 5 ppm	2 ± 2 ppm

There are several methods of arranging the system to transport the sample to the analyzers. The most widely used is one where the air sample is drawn from the sample point in the tunnel to the location of the analyzers, and the analyzer signal then is transmitted to the recording units in the control room. This approach places all the analyzers in one location for ease of maintenance and service. There are, however, situations where the length of sample line becomes excessive, thus creating an extensive delay in response time. It is then that the second method can be employed, placing the analyzers as close to the sample point as possible. In this method, the air sample lines are extremely short and only the signal is transmitted from the local analyzer unit to the recorders on the control board. There are several disadvantages of this latter method, one being the difficulty in servicing and maintaining the equipment in a limited space within an active tunnel.

Equipment

Pump. The vacuum pump is the most critical part of the analyzer system and is the one piece of equipment which usually provides the greatest number of problems. The pump must be large enough and of sufficient capacity to draw the sample from the sample point to the analyzer within a total response time of not greater than two to three minutes. Historically, turbine, positive displacement and diaphragm pumps have been used, with the diaphragm type being the most recent and enjoying the most success.

Sample Point Terminal Filters. These are required at the end of the sample tube to mechanically filter out particulate matter prior to its entering the sample lines. Various types of terminal filters have been used for this purpose. The extreme environmental conditions experienced within the tunnel mandate that this type of equipment be properly designed and properly protected, especially where the sample is taken in the tunnel walls. In this location, the pressurized water sprays often used for tunnel washing can easily reach the filter, thus creating problems with moisture. There are presently several methods being used to preclude this problem. A three-sided cover permitting air to enter in the bottom only and a terminal filter with an extended skirt are but two of these methods.

A recent study has shown that the sidewall location of the terminal sample point may be a poor one. As shown in Figs. 19-21 and 19-22, the profiles drawn from this theoretical study indicate that the lowest concentrations of carbon monoxide may occur at the sidewall sample point. To counter this problem, an additional sample point should be located in the exhaust gas stream. This location will provide a reference, as it will show the average carbon monoxide present in the tunnel air. The sidewall location will provide an indication of the relative gas concentrations throughout the tunnel.

Visibility Detectors

Visibility detectors have been developed which can measure the visibility of the tunnel air by means of the light-scattering principle. There are units available which measure light transmittance directly and others which use a comparison method. The comparison method makes use of a dual cell—one being a calibrated optic, the other being the tunnel air sample. A light beam is alternately passed through each of these two cells. This system, because of the comparison method used, is not affected by the deterioration of the optics exposed to the tunnel air.

There is presently a limited level of experience with this type of equipment in the U.S. and further testing and evaluation should be conducted prior to extensive use of this equipment.

VENTILATION OF RAILROAD TUNNELS

General

This section pertains to railroad tunnels; more specifically, tunnels through which rail vehicles travel propelled by diesel engine locomotive power. (Tunnels in which electrically propelled trains operate are covered later in this chapter.)

The most important environmental factors in rail tunnels are the presence of heat and the products of combustion. Excessive heat will raise the tunnel ambient air temperature to a point at which the locomotive engine can no longer function. The presence of excessive products of combustion will create a hazard to health and a haze within the tunnel. Any criteria established must therefore consider the health of the train occupants, both crew and passengers, and the ability of the locomotive units to function properly within the tunnel.

The diesel locomotive is, in actuality, an electrically-driven vehicle with the diesel engine driving an electric generator supplying power to the electric traction motors. The throttle, which usually has eight positions, controls engine speed. The diesel engines, in 70% of the locomotives in the U.S., are of the two-stroke cycle type. The horsepower of these engines ranges from 700 to 4000.

Hazards of Diesel Exhaust Gases

A general discussion of diesel engine exhaust gas composition was provided earlier. The composition of railway locomotive diesel exhaust does not differ markedly from that of the truck or bus diesel engine. However, in the railroad tunnel, the significant contaminant becomes the oxides of nitrogen and not carbon monoxide, as in the vehicular tunnel. This fact requires that the oxides of nitrogen must be studied to a greater depth. It can be shown that, based on the composition of diesel exhaust gas, if the oxides of nitrogen are maintained within specified acceptable limits, all other exhaust gas contaminants will also be maintained within acceptable limits.

The establishment of criteria for oxides of nitrogen must consider the length of time personnel are exposed to the environment, since the nature of the effect on the human body is time-dependent. The American Conference of Governmental Industrial Hygienists, in their 1980 edition of *Industrial Ventilation*,[4] present a threshold limit value of 5 ppm for nitrogen dioxide and 25 ppm for nitric oxide. These threshold limit values are, as defined earlier, for exposures of 8 hours/day over a working lifetime. Therefore, since the travel time through a railroad tunnel is on the order of 1 hour, levels of 5 ppm for nitrogen dioxide and 37.5 ppm for nitric oxide can be tolerated for these short periods of exposure without adverse effects.[1]

The exhaust gases are emitted from the top of most diesel electric locomotives, as shown in Fig. 19-68. This creates a phenomena which aids the ventilation of any railroad tunnel. The stratification effect created in the crown of the tunnel and in the annular space by the temperature gradients remains stable, and thus a percentage of the exhaust gas contaminants will remain in the crown of the tunnel and not interact with the train (Fig. 19-68). Tests have shown that about only 45% of the emitted exhaust gases descend from the tunnel crown to interact with the air in the spaces at the sides of the locomotive.

The effect of airflow in the annular space has been determined from tests conducted at Cascade Tunnel.[33] These tests show that approximately 50% of the total airflow within the annular space relative to the train is effective for cooling and combustion.

A method for determining the oxides of nitrogen emitted from a diesel engine locomotive is contained in Reference.[34]

Heat

Heat from the engine combustion process is released by the railroad locomotive as it travels through the tunnel. The total heat is composed of heat from the exhaust gas, jacket water, lubricating oil, braking resistors and radiation. The heat from exhaust gases, jacket water, lube oil and braking resistors is emitted from the top of the locomotive, whereas the radiated heat is emitted from all surfaces of the locomotive (Fig. 19-69).

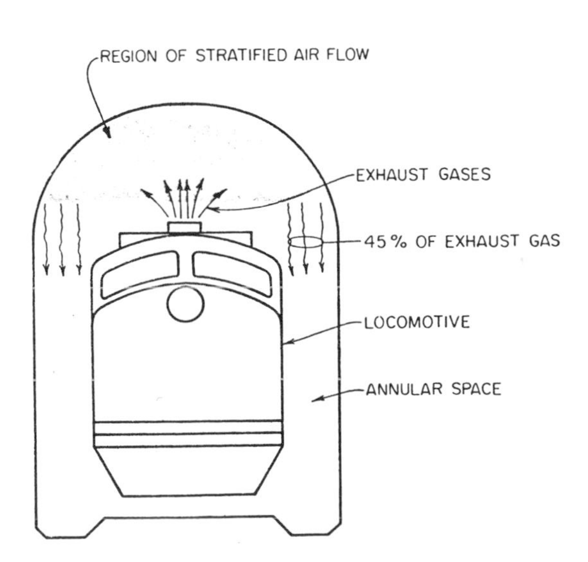

Fig. 19-68. Cross-section of a railroad tunnel showing the effect of stratified airflow.

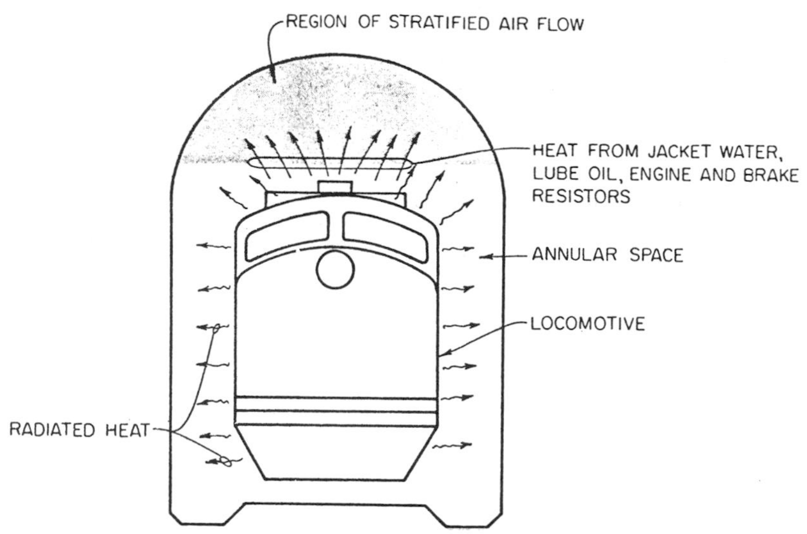

Fig. 19-69. Typical railroad tunnel cross section showing effect of stratified flow and heat rejection.

The final dissipation of this heat is important to the proper operation of the locomotive unit, since air is required for the engine cooling. If the inlet air temperature is raised by this heat to a point above the maximum operating temperature of the engine, the engine will shut down on a high water temperature condition. This is usually evident where a large number of locomotive units are operating in a single consist. As in the case of the air contaminants, only a portion of this heat affects the locomotive inlet air conditions due to the stratified airflow in the crown of the tunnel.

The total amount of the radiated heat, along with the heat from the exhaust gas, jacket water, lube oil and braking resistors, interacts with the inlet air. An expression for the temperature of the inlet air of a specific locomotive unit follows.

$$TIA = \frac{TAA + HER + RH1\,(HEE + JHW + HLO + HBR) \times N}{DEN \times CP \times EAF \times QRT}$$

where

TIA = temperature of inlet air to $N + 1$ locomotive unit (°F)

TAA = temperature of tunnel ambient air ahead of consist (°F)

HER = heat radiated from one locomotive (Btu/minute)

RHI = portion of heat from the exhaust gas, jacket water lube oil and braking resistor which interact with the inlet air.

HEE = heat rejected with engine exhaust gas (Btu/minute)

HJW = heat rejected from jacket water system (Btu/minute)

HLO = heat rejected from lubricating oil sys- (Btu/minute)

HBR = heat rejected from braking resistors (Btu/minute)

DEN = density of air (lb/foot^3)

CP = specific heat of air (Btu/minute/°F).

Values for engine heat rejection in the case of specific design, should be obtained from the engine manufacturer.

The effect that the inlet air temperature has on engine operation can be judged by the design air inlet temperature of most locomotive units of 115°F. This value applies to over-the-road service. However, most locomotive engines can operate at temperatures greater than 115°F for short periods of time. Tests have shown that units have operated with inlet air temperatures exceeding 150°F. The allowable intake temperature for each locomotive type should be obtained from the engine manufacturer when a specific disign is being contemplated.

Ventilation

Ventilation is required in all railroad tunnels to remove the heat generated by the locomotive units and to change the air within the tunnel, thus flushing the tunnel of air contaminants. Ventilation can take the form of natural, piston effect or mechanical ventilation. While the train is in the tunnel, the heat is removed by an adequate flow of air with respect to the train (Fig. 19-70), whereas the air contaminants are best removed when there is a positive airflow out of the tunnel (as shown also in Fig. 19-70(c)).

The three major forms of railroad tunnel ventilation are outlined below.

Piston Effect With an Open-ended Tunnel. In this type of tunnel, shown in Fig. 19-70(a), the only means of ventilation is the piston effect of the train on the air. The air in the annular space flows in the same direction as the train, and there is a net flow of air through the tunnel in the direction of the train. This method is satisfactory for short tunnels since the flushing effect is good, and because the oxides of nitrogen levels and the heat have not had an opportunity to build up.

Piston Effect With a Portal Gate. The addition of an operable portal gate to the end of a tunnel, as shown in Fig. 19-70(b), will greatly enhance the ability of the ventilation system to remove the heat generated by the locomotive units. However, since there is no net flow from the tunnel, the tunnel under these conditions is not flushed of the air contaminants. This fact precludes the use of a portal gate on a railroad tunnel without mechanical ventilation.

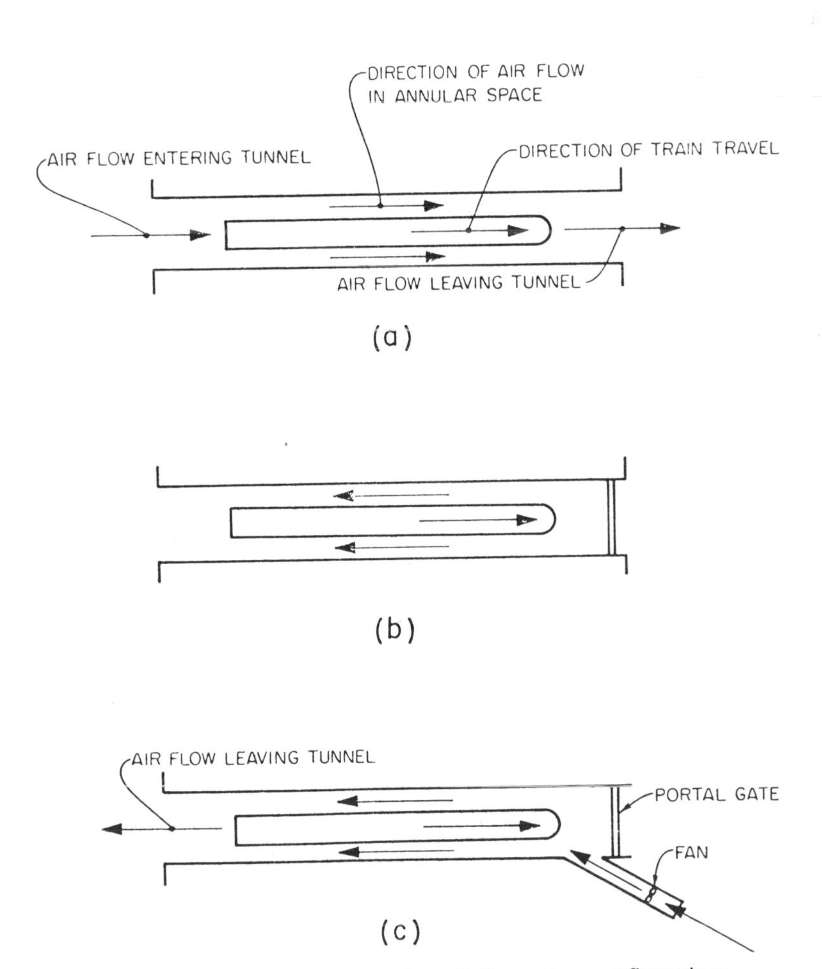

Fig. 19-70. Railroad tunnel ventilation system configurations.

Mechanical Ventilation With Portal Gate. The addition of a mechanical ventilation system, as shown in Fig. 19-70(c), provides the ultimate in the ventilation of a railroad tunnel. The heat, along with the air contaminants, will be removed, since there will be adequate airflow with respect to the train and an adequate net flow through the tunnel.

Pressure

To properly design a railroad tunnel ventilation system, it is necessary to be able to predict the pressures generated in the tunnel for various train speeds as well as the airflow surrounding the train. These airflows must satisfy the cooling requirements of all locomotive units in the consist. The ventilation system components must be designed to withstand the maximum pressures generated by the train and the fans.

The following relationships were developed[22] to predict the piston effect of the train within the tunnel and the pressures in the tunnel. For a definition of the variables in these relationships, see Fig. 19-71.

Tunnel Portal Gate Open

$$U = \frac{\sqrt{\epsilon}}{\sqrt{\epsilon} - 1} V \qquad \text{Eq. 1}$$

where

U = air velocity in unobstructed tunnel (feet/second).

$$\epsilon = \frac{\dfrac{\phi}{(1-\phi)^2}\left[\phi - \dfrac{Lf_t}{4R_t(1-\phi)}\right]}{\sum_{1}^{n} K_i + \dfrac{f_w(L_T - L)}{4R_w}} \qquad \text{Eq. 2}$$

where

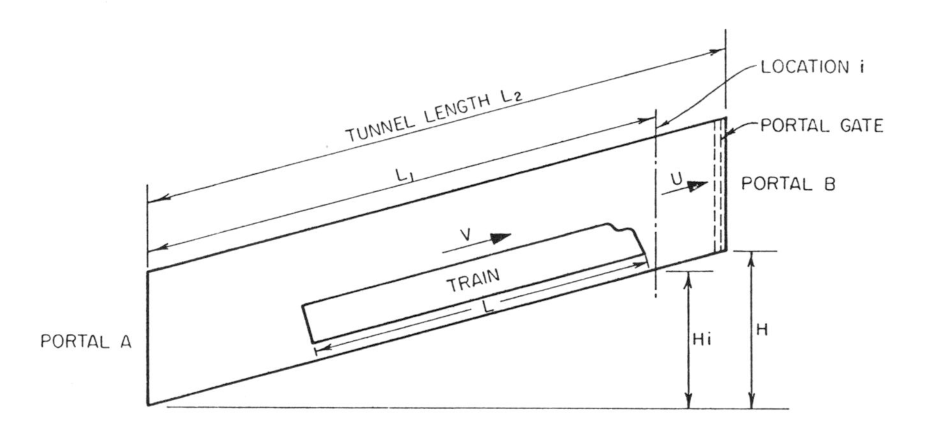

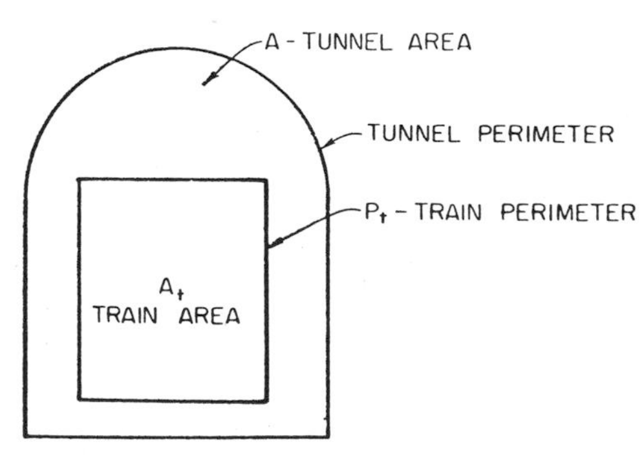

Fig. 19-71. Identification of variables in railroad tunnel.

V = train velocity (feet/second)

$\phi = \dfrac{A_t}{A}$

L = train length (feet)

f_t = friction factor of train surface

f_w = friction factor of tunnel wall

R_t = hydraulic radius of train (feet)

K_i = localized tunnel loss coefficient

L_T = total tunnel length (feet)

R_w = hydraulic radius of unobstructed tunnel (feet)

A = tunnel cross-sectional area (feet^2)

A_t = train cross-sectional area (feet^2)

Tunnel Portal Gate Closed, Fans Not Operating

$$\frac{\Delta P}{\gamma} = \frac{P_A - P_B}{\gamma} - \Delta H + \frac{V^2}{2g}\left[\frac{P_t f_t L}{4A(1-\phi)^3} + \left(\frac{\phi}{1-\phi}\right)^2 + \frac{\phi^2}{(1-\phi)^3}\frac{Lf_w}{4R_w}\right] \qquad \text{Eq. 3}$$

where

ΔP = pressure above portal B pressure at location 1 (lb/foot^2)

P_A = barometric pressure at portal A (lb/foot^2)

P_B = barometric pressure at portal B (lb/foot^2)

H = difference in elevation between portals (feet)

g = acceleration of gravity (feet/second/second)

P_t = train perimeter (feet)

δ = unit weight of air (lb/foot^3).

Tunnel Portal Gate Closed, Fans Operating

$$\frac{\Delta P}{\gamma} = \frac{P_A - P_B}{\gamma} - \Delta H_1 \pm \frac{U^2}{2g}\left[\sum_1^n K_t + \frac{f_w(L_1 - L)}{4R_w}\right] + \frac{(U \pm V)^2}{2g}$$

$$\left[\frac{P_t f_t L}{4A(1-\phi)3} + \frac{\phi}{1-\phi}^2\right] \qquad \text{Eq. 4}$$

$$\pm \frac{(U \pm V\phi)^2}{2g}\left[\frac{Lf_w}{4R_w(1-\phi)^2}\right]$$

where

H_1 = difference in elevation between location 1, where pressure increment is wanted, and portal A (feet) (when location 1 is chosen at the gate, $H_1 = H$)

L_1 = total distance between location 1 and portal A (feet).

By continuity:

$$\text{QRT} = (U - V) \times A. \qquad \text{Eq. 5}$$

"Equations 1 and 2 give the explicit value of the air velocity in the tunnel, in front of and behind the train, for a level tunnel with the gate open. Only a straightforward substitution of numerical values is required. Equation 3 is used to determine the pressure change produced by piston effect in the portion of the tunnel between the train and the gate. That pressure change could be either above or below atmospheric pressure, depending on the direction of the train movement. However, that value must be determined by successive approximations, since the value of U, air velocity in the tunnel, depends on that pressure difference. The computational procedure starts by assuming an estimated value for U and determining the corresponding ΔP, which is also the pressure rise against which the fans are operating. With this computed ΔP and the fan characteristic curve, the total fan discharge can be determined. That fan airflow must equal the tunnel airflow: i.e., U (assumed above) multiplied by the area of the tunnel; otherwise the procedure must be repeated until an equality is attained. For condition 2, fan not operating–gate closed, the value of U equals zero."[35]

In order to complete the above computations, the friction factors for both tunnel and train must be known. The tunnel friction factor which corresponds to the coefficient f in the Darcy-Weisbach equation for friction losses in pipe flow, as developed from test data for a concrete-lined railroad tunnel with ballasted roadbed, is 0.0133. This value was verified by computations taking into account relative roughness and the Reynolds number with excellent correlation.

The airflow in the annular space between the train and the tunnel wall is influenced by the train friction factor. Friction factor values for both freight and passenger trains were developed based on tests conducted at Cascade Tunnel as follows.[33]

Train Type	Friction Factor
Freight	0.143
Passenger	0.065

The makeup of the trains passing through the tunnel vary, as there are many sizes and shapes of railroad cars. A method was developed to determine the cross-sectional area, perimeter and hydraulic radius of an equivalent train. A weighted average of the area and hydraulic radius can be computed by grouping the cars in classes on the basis of the Association of American Railroads' (AAR) mechanical designation list for each car type. This method permits rapid determination of the area and hydraulic radius of an equivalent train having the same length as the train being considered. An example for a 116-car train is shown below.

Determination of Equivalent Car Characteristics

Total cars: 116

Train length: 5560 feet

CLASS	EHR	NUMBER OF CARS	CAR TYPE
I	2.4-2.6	66	E, WC, R5, ES
II	2.6-2.8	6	R, EP, RP
III	2.0-2.4	8	H, C3, H4, GH
IV	Less than 2.0	16	GC, G, C4, C6, F

$$\text{EHR} = \frac{(66 \times 2.6) + (6 \times 2.8) + (8 \times 2.4) + (16 \times 2.0)}{96}$$

$$= 2.49 \text{ feet}$$

Flushing

There is a requirement in all railroad tunnels for a flushing cycle after the train has passed through the tunnel. The ventilation of the tunnel during the train passage is never 100% effective. There will always be air contaminants remaining in the tunnel. The natural ventilation effect created by the pressure differential due to a difference in elevation may be sufficient for a short tunnel to provide the necessary flushing. A fan system will be required to provide this flushing effect for longer tunnels or in cases where the train headway requires a more rapid flush cycle.

Existing Railroad Tunnels

Included in this section is a description of several railroad tunnel ventilation systems.

Cascade Tunnel. The Cascade Tunnel is a 7.79-mile long single track tunnel located near Wenatchee, Washington on the Great Northern Railway (now the Burlington Northern).[36] The tunnel was originally built for electrified train service in 1929. The tunnel is straight and on a uniform grade of 1.57%. Concrete-lined throughout, the tunnel cross-sectional area is 314 square feet. A tunnel ventilation system was installed in the 1950's when diesel-powered locomotives were run through this tunnel for the first time. The ventilation equipment is located at the east portal.

This ventilation system was designed to provide for the removal of heat, smoke and air contaminants from the tunnel. The system includes two axial flow fans, shut-off dampers and a portal gate, along with all auxiliary equipment. A standby generator provides electrical power in the event of power failure. The fans are two-stage reversible axial type direct driven by 800 HP AC induction motors.

The fan shut-off dampers are of the multi-bladed type and are designed to withstand 14.4 inches of pressure.

The portal gate is a steel vertical lift counterweighted type which operates in vertical guideways. The gate is closed to prevent leakage of air from the tunnel when the fans are operating. The gate will open on a counterweight arrangement if the power fails. Gate operation is interlocked with the fans and the track block signals to assure that the gate is in the closed position prior to fan starting and in the open position prior to a train passing the east portal.

An 1100-hp diesel engine-driven generator provides the required emergency power for one fan, a portal gate and selected auxiliary loads.

The fans and gate are controlled remotely.

Only one fan is used to ventilate the tunnel when a train is traveling upgrade (eastbound) through the tunnel. Both fans are operating in parallel to flush the tunnel of combustion gases after the train has left the tunnel.

Flathead Tunnel. This is a 7-mile long rail tunnel located in northwest Montana.[37] It was built as a part of the relocation of the Great Northern Railway (now the Burlington Northern) in 1971 and forms a part of the Libby Dam project. The tunnel is straight, with a uniform grade of +0.46% (with the exception of the 230 feet at the eastern portal, which has a -0.20% grade, as shown in Fig. 19-72). The ventilation system is capable of flushing the tunnel of contaminants and providing cooling for the locomotive units of an eastbound (upgrade) train. The ventilation system, as shown in Fig. 19-73 includes two axial flow fans, shut-off dampers, a relief damper and a portal gate, along with all auxiliary equipment. Two standby diesel engine-driven generators are provided to furnish power to the ventilation system and all auxiliaries during a power failure.

The fans are horizontal, two-stage reversible axial flow fans direct-driven through floating shafts by AC induction motors. These fans are equipped with a blade pitch control system which permits automatic adjustment of the fan airflow to compensate for the wide ambient temperature variation experienced at this location. This system maintains a constant mass flow of air through the fan and limits the brake horsepower to 2000 per fan.

The relief damper, provided to prevent an excessive build-up of pressure in the tunnel, is a four-module multi-bladed type damper, controlled by pressure regulators within the tunnel. The fan shut-off dampers are of the multi-bladed type designed to withstand 60 inches of pressure. The portal gate is of welded structural

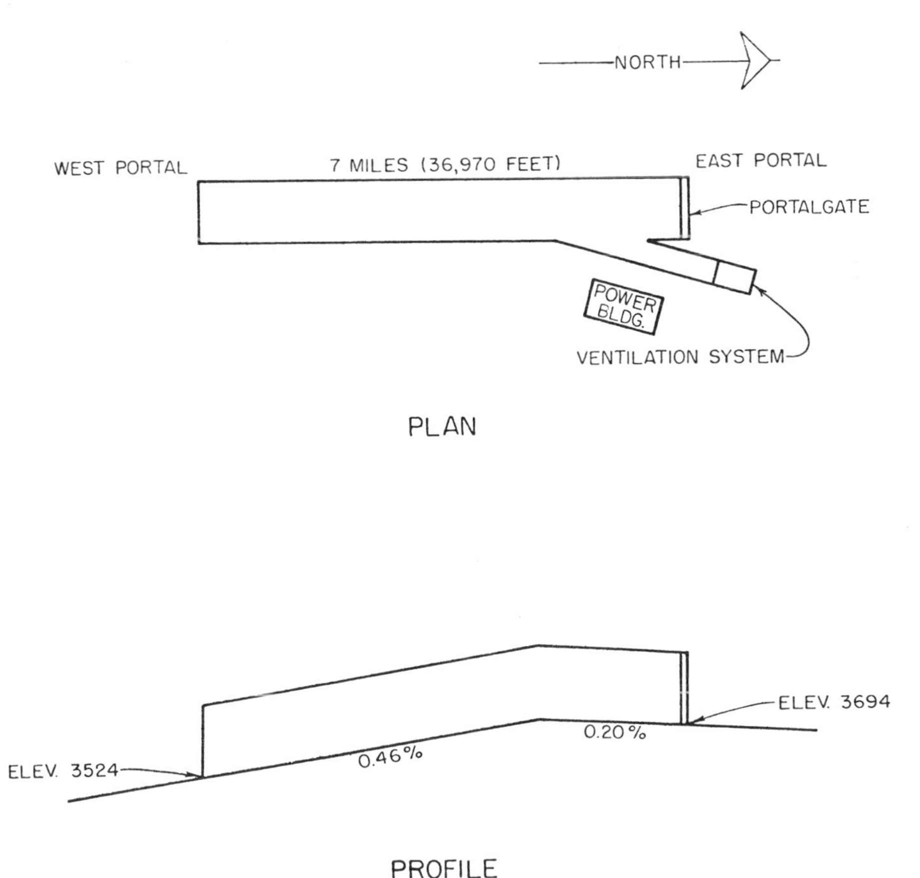

Fig. 19-72. Flathead tunnel.[37]

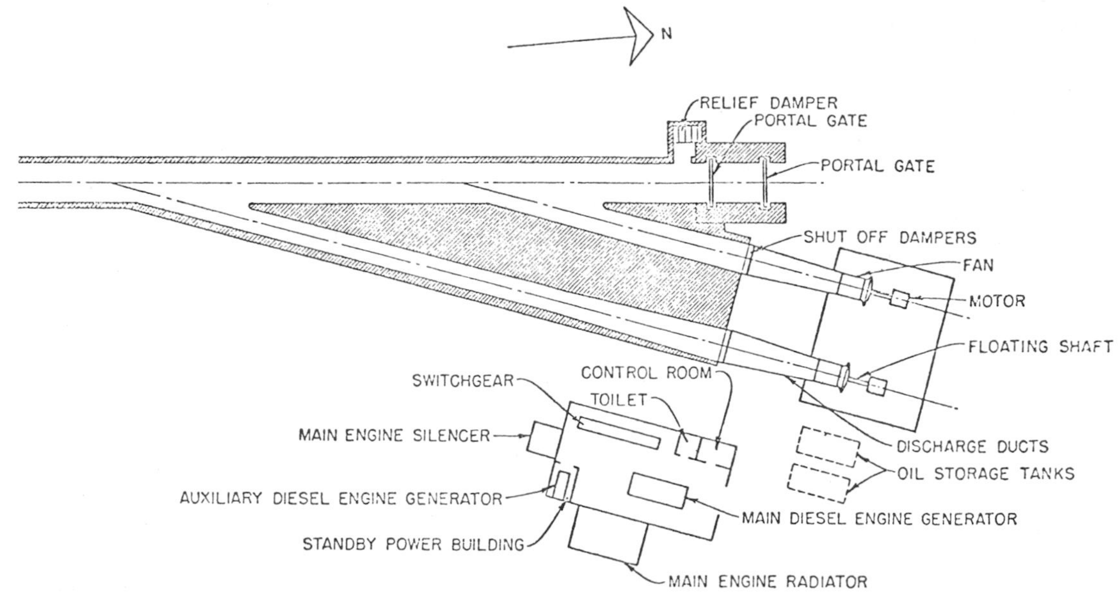

Fig. 19-73. Arrangement of ventilation system at flathead tunnel.[37]

steel construction designed for uniform air pressure of 60 inches of water. It is supported by rollers bearing against steel supports located in the walls. Rubber seals minimize leakage through the gate. The gate has a fail-safe mechanism which will open the gate upon loss of power. The center portion of the gate is built of frangible material which will allow a train to go through the gate, should any of the gate mechanisms fail, without major damage to the train or gate. Two gates are installed, although only one is used at any time.

A 2000-hp diesel engine-driven generator provides emergency power for one fan. A smaller unit provides power for the auxiliary loads.

One fan is utilized to provide cooling for the upgrade trains which require it, while both fans are utilized to flush the tunnel in less than 20 minutes after a train leaves the tunnel.

VENTILATION OF RAPID TRANSIT TUNNELS

The portions of rail rapid transit systems which are located below the surface in subway structures require control of the environment.

Normal Ventilation

During normal train operation, the "piston effect" of the moving train provides air motion within the trainway and stations. In most of the systems built in early years, this air movement has been sufficient to maintain adequate environmental conditions within the subway facilities. However, with the growing emphasis on higher performance transit vehicles, which are capable of attaining speeds of 80 mph, the amount of heat generated often exceeds the ability of the piston effect ventilation to remove it from the subway. Coupled with this is the increasing acceptance of air conditioned environments in the lives of the transit patrons. Thus, a set of environmental criteria must be established, such that many of the newer systems being designed and constructed in the world today will be considering cooling of the station facilities. This occurs on systems such as those in Caracas, Washington, Baltimore and Atlanta.

Emergency Ventilation

The other aspect of subway ventilation is that which is required for emergency operation. The major purpose of such a system is to remove heat and smoke generated during a fire in order to permit the safe egress of passengers and the entrance of fire-fighting personnel. When an emergency situation occurs, the trains are halted and the "piston effect" ventilation ceases. This requires the incorporation of a mechanically induced ventilation system into the subway.

Rapid Transit Tunnels

In rapid transit systems, there are two types of tunnels: the standard subway tunnel, which is usually located between stations and normally constructed beneath surface developments with shafts communicating with the surface, and the long tunnel, usually crossing under a body of water, such as the Transbay Tube in San Francisco, or through a mountain, such as the Berkeley Hills Tunnel, also in California. The ventilation concepts for these two types will be different, since in the long tunnel there is usually limited ability to locate a shaft at any intermediate point, as can be accomplished in the standard subway tunnel.

Subway Environment

The objectives to be sought in controlling the environment within a subway system are, first, to provide a suitable environment for patrons as well as for operating and maintenance personnel; and, second, to minimize the exposure of the equipment to high temperatures, thus prolonging its life. This must be considered for all public and non-public spaces within the stations and the normally occupied space in the trainways. Also to be considered is the control of haze, odor and vapor, and the purging of smoke during a fire.

The Urban Mass Transit Administration of the U.S. Department of Transportation funded a recent research project dealing with the subway environment. The rapid transit properties in the U.S. and Canada sponsored and partially funded this project through the Institute for Rapid Transit and the Transit Development Corporation. This "Subway Environmental Research Project"(SERP) had as its objective preparation of a handbook,[38] along with the required methodologies to permit proper evaluation of the subway environment.

Consideration of the environment within the subway system involves three major components: criteria, analysis and control.

Criteria. Criteria for the subway environment must be based on the reaction of the human patron to the surrounding environment. The factors to be considered in this regard are temperature, humidity, air movement, noise and vibration. The capacity of human beings to endure the environment is a function of age, occupation, health, acclimatization and the natural variation in human beings. The most critical of these is the thermal factor.

Criteria should be established for maximum air velocity, rapid pressure change and air quality, including gas particulate and odor contaminants.

A thorough definition of these criteria and their method of determination is covered in the *Subway Environmental Design Handbook*, recently published by the U.S. Department of Transportation.[38]

Analysis. After defining the criteria for a subway system, an analysis of where the heat is in the system and how the airflow created by piston effect of trains and/or fans serves to dissipate this heat and influence the thermal load analysis is required. Until recently, there have been limited available techniques to permit such an analysis. A series of sophisticated design tools have been developed as part of the SERP which now permit this form of evaluation.

If the heat generated by the equipment and the people in a subway is greater than the capacity of the ground or of ventilation to dissipate it, the temperature will rise. All forms of electrical energy input to the subway system eventually are dissipated as heat. The trains account for approximately 85 to 90% of all the heat generated within the system. Heat generated within line sections by trains from their traction, braking and air conditioning systems will be at a substantially higher rate in subway systems now under construction, or being planned for the future, than that which exists in most systems today. This is due to the higher speed and acceleration requirements for vehicles, which necessitate significantly higher power input and resultant power losses. Of these inputs, the major portion is derived from braking and accelerating. There are presently underway many efforts to seek ways of reducing this heat such as by the use of regenerative braking, fly wheels, etc.

Approximately 50% of the total heat input attributable to train operation can be assigned to braking—or approximately 45% of the total heat input of the subway system. Since braking occurs in the vicinity of the station platform, it is the basic cause of much of the environmental problem.

The heat released into the tunnels or stations is partly transmitted into the surrounding soil and partly carried forward by piston action, or mechanical ventilation (if such is used), and eventually exhausted to the atmosphere through the shafts and access openings in the stations. The soil acts as a heat sink when the air in the trainway is at a higher temperature than the soil's temperature, and as a heat source when the reverse takes place.

The heat sink, where it is effective, is a "natural" cooling mechanism, as is the piston effect of moving trains. Cooling is accomplished by the exchange of hot inside air with cooler outside air.

A reliable estimate of the piston effect ventilation, along with the heat sink, and the impact of both on the thermal conditions within the subway, is required prior to determining what measure must be taken to meet established environmental criteria.

One of the analytical tools developed as a part of the SERP is a computer model—the Subway Environment Simulation (SES) program[49] which is in the public domain and available.

The SES is a user-oriented model; that is, both the required input information and the output produced are tailored for use by design engineers concerned with practical environmental problems. This computer model provides a dynamic simulation of the operation of multiple trains in a subway and permits continuous readings of the air velocity, temperature and humidity throughout the stations, tunnels and ventilation shafts. In addition, the program has been designed to provide readings of the maximum, minimum and average values for air velocities, temperatures, and humidities during any preset time interval. The program also will compute estimates of the station cooling and heating loads necessary to satisfy established environmental criteria, as well as the percentage of time that such environmental criteria are exceeded.

Control. Control relates to the methods used to maintain the desired environment within the subway facility. There are many types of control systems which can be employed in this regard.

Control of the temperature rise in the subway system involves the balancing of the heat from the system against losses of heat to the surrounding heat sink and to the air flowing in the system. When the heat gain from the system is greater than the losses to the sink and to the air, then the temperature in the system will rise. Such a rise in temperature will increase the rate of heat flow to the sink. Additional airflow may be required to control the temperature rise. This increase in airflow would reduce the temperature rise but would also reduce the cooling effect of the heat sink. Thus, under extreme conditions of maximum heat gain and maximum outside air temperature, an imbalance of heat gain and heat losses may result in an overall temperature rise in the system. Under conditions of reduced heat gain and reduced outside air temperature, normal airflow rates resulting from train piston effects may provide the desired control over temperature rises. Frequently, the airflow rate required to provide this control will also be sufficient to replenish the system air with outside air at a desirable rate. If the rise in temperature is not desirable, a system of mechanical ventilation or cooling by refrigeration must be considered.

A knowledge of the available environmental control equipment and its application is required to effect a complete solution for the subway system. In the development phase of the system, the environmental engineer must consider the variety of system concepts appropriate to subway environmental control and the applicability of these systems in the optimization of subway construction and operation. These systems include, in addition to station platform air conditioning, supply and exhaust ventilation systems, trainway or under platform exhaust systems to remove heat, tunnel line section ventilation systems and various other possibilities and combinations. Descriptions of standard environmental control equipment and systems for general cooling applications are found in ASHRAE Handbook.[39,40]

The major contributor to the heat problem in

the subway system is the transit vehicle and its propulsion system. Much greater attention must be addressed to this equipment, since optimization of the cost of the vehicle along may not optimize the cost of building, owning and operating the entire rapid transit facility. Consideration should be given in the future to propulsion systems which provide for rejection of the waste heat out of the enclosed portions of the system and possible reuse of this energy.

Systems Descriptions

The highlights of several of the systems designed within the past ten years are described below.

BART. The Bay Area Rapid Transit (BART) system, in the San Francisco area, was the first new system to be designed in the U.S. in a number of years. Of the total system length of 75 miles, approximately one-third (or 24 miles) is aligned in subway. There are 20 underground stations. San Francisco has an extremely mild climate, and thus it was found that ventilation alone would be adequate to satisfy the patrons' comfort objectives in the subway station's public places. Mechanical ventilation at the rate of six air changes per hour will result in a maximum average ambient station air temperature of 4° above the maximum outdoor ambient. The piston effect ventilation is found to provide approximately three air changes per hour, thus achieving a station average temperature of 10° to 12°F above outside ambient. In the tunnels, emergency ventilation fans were installed. These fans are reversible with approximately identical airflow generated in both directions. A maximum of 15 mph air velocity and a minimum of 6 mph air velocity in the annular space were selected as criteria. The purge time was set at a maximum of 30 minutes.

Fans were selected so that when operated in unison in supply or exhaust modes, these criteria will be met. The criteria established for air velocities were maximum exit velocities at surface gratings of 350 feet/minute in public sidewalks and 1000 feet/minute in street areas other than crosswalks. Gratings were located to minimize disturbance to pedestrians.

Caracas Metro. The Metro in Caracas is a proposed system approximately 12 miles in length (Catia-Petare Line), which is the initial phase of a four-line system.[41]

Approximately 90% of the Catia-Petare Line is underground, with 19 underground stations. The initial part of the system was designed based on system-wide environmental studies. It was found that the ground sink temperature is extremely high, 87°F stabilized. This meant that the use of mechanical cooling to remove the heat in the stations was necessary. The heat sink effect of the ground was negligible, and, because of the limited temperature differential, the ventilation was inadequate to remove the heat.

A system of air cooling by mechanical refrigeration was developed along with underplatform exhaust and tunnel exhaust systems. This permitted consideration of the subway stations and the tunnel as separate environments with different criteria. A higher ambient temperature is to be maintained in the tunnels than in the stations. This required minimizing of the interaction of station environment with tunnel environment. This is achieved by a center exhaust fan in each tunnel segment located midway between stations, thus creating a flow of air from the stations toward the center of the tunnel. The environmental control equipment to be located in each station consists of two built-up supply air cooling systems of 80,000-feet^3/minute capacity each, one located at each end of the station mezzanine. These systems have a capability of 100% recirculation of air, which is used during the evening rush hour peak to take advantage of the 5°F temperature differential between station and outdoor air temperatures. The air is distributed to the station through ducts located over the trainway (Fig. 19-74). The return air is taken directly above the trainway, as shown in this figure. Chilled water is supplied to the station cooling systems from remote off-site refrigeration plants. The chillers and primary chilled water pumps are located in the plant, and secondary booster pumps are located in each station. The condenser water for these plants will be provided by cooling towers located on the roof of each plant.

Also included in each station is an underplatform exhaust system Fig. (19-77) which provides a means of removing some of the under-

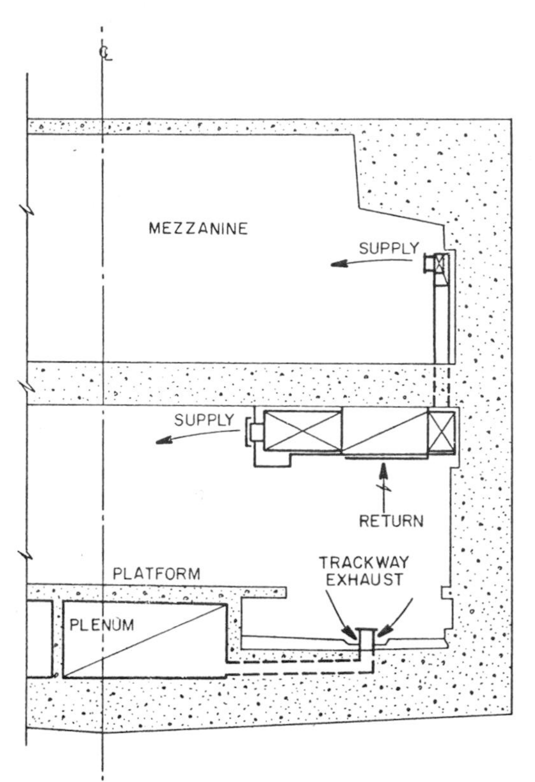

Fig. 19-74. Partial section through station air distribution system (Caracas).[41]

car heat from the train area while it is located in the station. These systems have a capacity of 70,000 feet^3/minute/trackway and are considered at this point to be 40% efficient.

Washington Metro. The new transit system built in Washington, D.C., which is 98 miles long with 47 miles of subway, has air conditioned vehicles and air cooled subway stations (50) along with ventilated tunnels.

A method of envelope cooling was developed[42] to achieve the desired environmental conditions. The chilled air is discharged, as shown in Fig. 19-75, to create an envelope encompassing the platform area. Such an approach requires that the system minimize the interaction of the piston flow air from the tunnels with the station environment to minimize disturbance of the cooling envelope. In order to achieve this, the connection of the tunnel to the station has been modified, as shown in Fig. 19-76. A series of fans have been installed in shafts along the line to provide added ventilation when required.

MARTA. The Metropolitan Atlanta Transit Authority (MARTA) system will be 52 miles

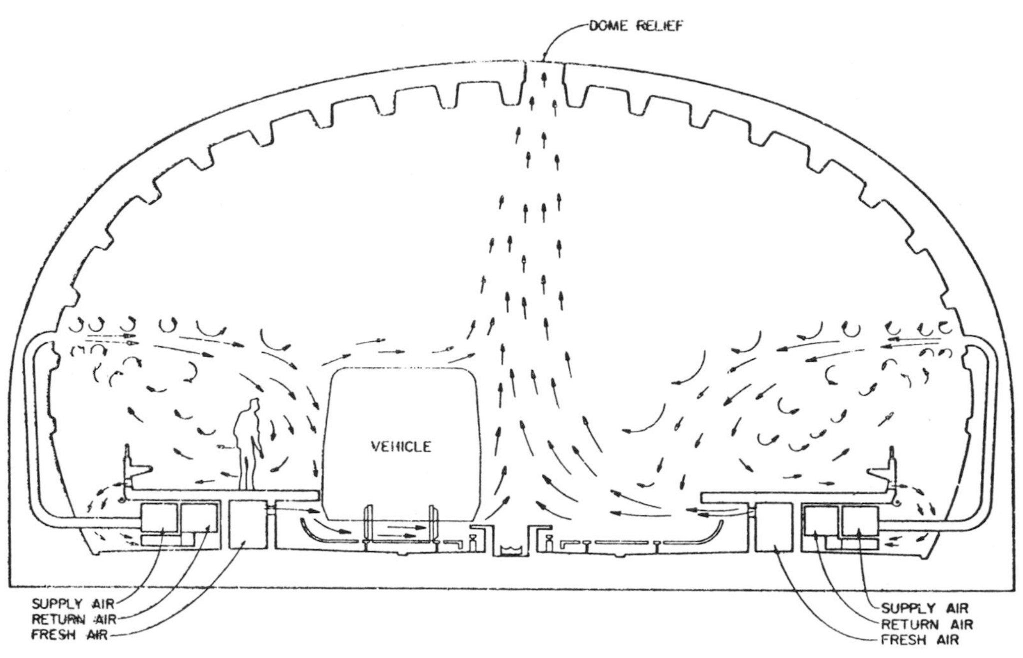

Fig. 19-75. Typical air distribution (WMATA).[42]

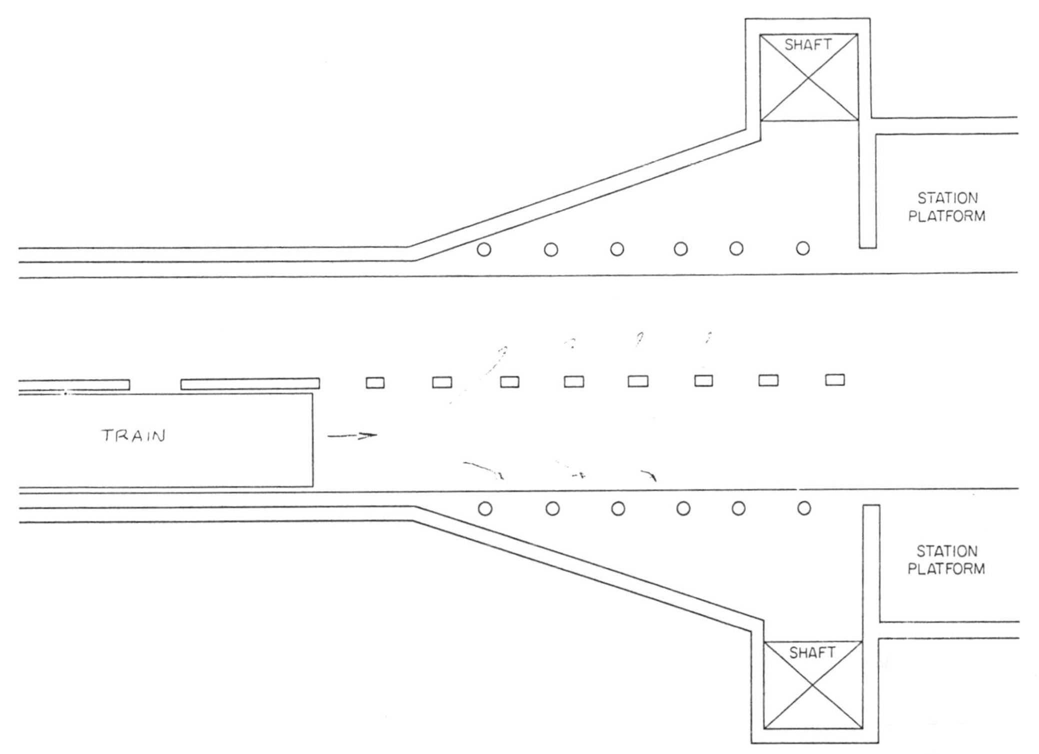

Fig. 19-76. Tunnel connection to station showing expansion chambers (WMATA).[42]

(84 kilometers) in length, including approximately.[43] 10 miles of subway and 42 miles of aerial and at-grade construction. There are 12 distinct subway sections, six of which are contiguous to underground stations, and the remaining are isolated; that is, not connected to stations. Most of the subway construction in Atlanta will be by the cut-and-cover method.

The MARTA underground station environmental control systems will be designed to maintain average platform temperatures of 85°F dry-bulb with 50% relative humidity during the peak evening rush period, thus providing a suitable transitional environment from the street to the transit vehicle. It was determined that a signficant portion of the train heat can be kept out of the station by providing an effective tunnel heat relief exhaust system, which included the consideration of mid-tunnel exhaust fans and a 100-foot break in the center dividing wall located at the station platform, as shown in Fig. 19-77. The air pushed ahead of the train would

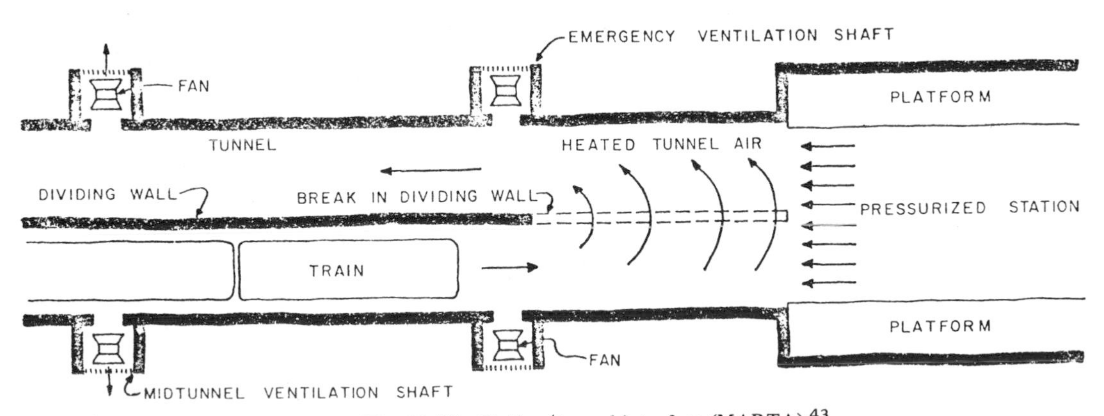

Fig. 19-77. Station/tunnel interface (MARTA).[43]

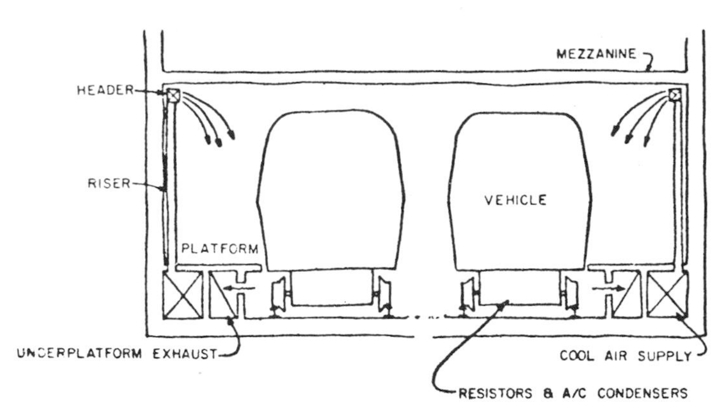

Fig. 19-78. Station distribution (MARTA).[43]

be relieved through the break in the dividing wall into the opposite trainway, thus minimizing the interaction of heated tunnel air with cool station air. The subway stations are to be provided with air cooled by refrigeration, supplied at the platform areas only. This air will be transported by underplatform ducts to vertical risers, to horizontal headers, from which the air will be distributed to the patron areas on the platforms (Fig. 19-78). Mechanical equipment rooms will be located at both ends of the platforms and will house fans, filters and chilled water coils. None of the supply air will be recirculated and will be permitted to exfiltrate through the station portals and serve as make-up air to the tunnel exhaust systems.

The adjoining tunnels will be provided with ventilation fans for both normal and emergency operations. During the peak traffic times of normal operation, mid-tunnel fans having a capacity of 140,000 feet^3/minute can be operated in each trackway near the mid-point of each tunnel to extract tunnel heat. The dividing wall between trackways will be cut back for 100 feet (30.5 meters) from the end of each platform to minimize the flow of heated air into the station (see Fig. 19-77).

For emergency ventilation, specially designed, "100%" reversible fans will be located in each trackway at the point where the break in the dividing wall begins, as shown in Fig. 19-77. The mid-tunnel fans will also be a part of the emergency ventilation system, and will be reversible so operation can be coordinated with the station-end emergency fans.

Chilled water for station air conditioning systems will be generated by electrically-driven chillers located in plants either local to the station or remote and serving several stations.

Rapid Transit Tunnels

The TransBay Tube Tunnel in San Francisco is a 19,000-foot crossing of San Francisco Bay and is a part of the BART system.[44] The tunnel is ventilated during normal operation by the piston effect generated by trains which travel at speeds up to 80 mph through this tunnel. During an emergency, when the purge of the tunnel is required, a mechanical ventilation system can be operated. This system consists of four large centrifugal exhaust fans located in the two ventilation structures (Fig. 19-79). These fans exhaust air through an exhaust duct located in the cross-section, as shown in Fig. 19-80.

A series of automatic dampers are set in the separating wall between the duct and the trainway. In operation, several of these dampers can be opened in either tunnel at or near the site of the heat or smoke generation. This type of operation will maintain a flow of air toward the emergency site and prevent movement of smoke to the remainder of the tunnel. This will permit safe egress of passengers and entrance of fire-fighting personnel toward the fire.

State of the Art

The evaluation of the environment within the subway system has taken great strides in recent years. Since the BART system was the first new system to be designed in approximately 30

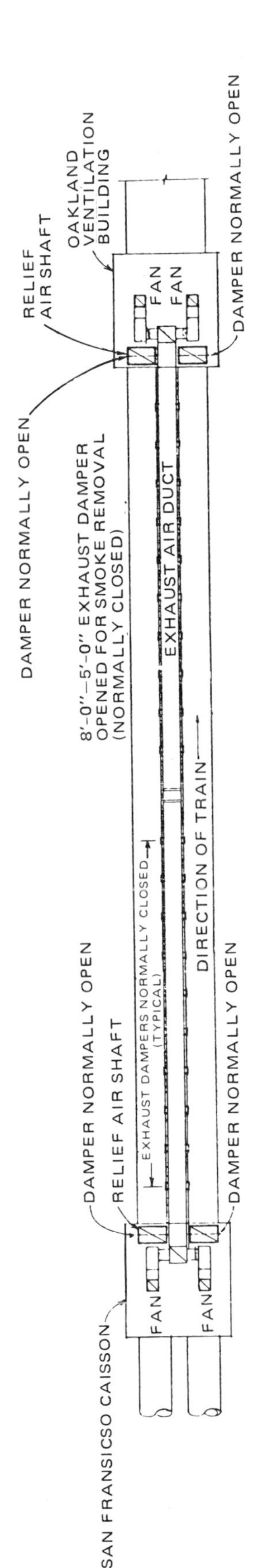

Fig. 19-79. Air flow diagram (Trans Bay Tube).

years, the available technology was limited. Efforts have progressed from the analytical tools which were developed for BART[45] to the sophisticated analytical tools developed as a part of the Subway Environmental Research Project.[38] The published handbook[38] resulting from this project contains the most up-to-date technical data available for evaluation of the subway environment.

VENTILATION DURING CONSTRUCTION

There is a requirement, during the construction of any tunnel, for ventilation. This is true whether the tunnel is constructed by blasting, boring or the placing of prefabricated tubes in a trench. Such temporary ventilation is necessary to provide a suitable, safe working environment for the construction workers. Since, during the construction process, many flammable or toxic airborne contaminants in the form of gases, dust, mists and fumes are released, removal of these contaminants can only be achieved through a form of ventilation. Such a temporary ventilation system is usually separate from the permanent ventilation to be installed in the finished tunnel, since the purposes and the fan capacities are usually not compatible, although some portions of the permanent ventilation system, such as the ducts or shafts, may be utilized for the temporary system.

The criteria for the working environment within a tunnel under construction is documented in the Safety and Health Regulations for Construction as issued by the Bureau of Labor Standards of the U.S. Department of Labor.[45] The threshold limit values (TLV) adopted by the American Conference of Governmental Industrial Hygienists (ACGIH)[19] have been established by the Department of Labor as the legal limits of contaminants for the tunnel environment during construction.[45]

These regulations require that all harmful airborne contaminants, either flammable or toxic, be removed by ventilation if they exist within the tunnel environment in hazardous quantities, as determined by the threshold limit values. The temporary ventilation system must be of the mechanically induced reversible type designed for the minimum flow requirements

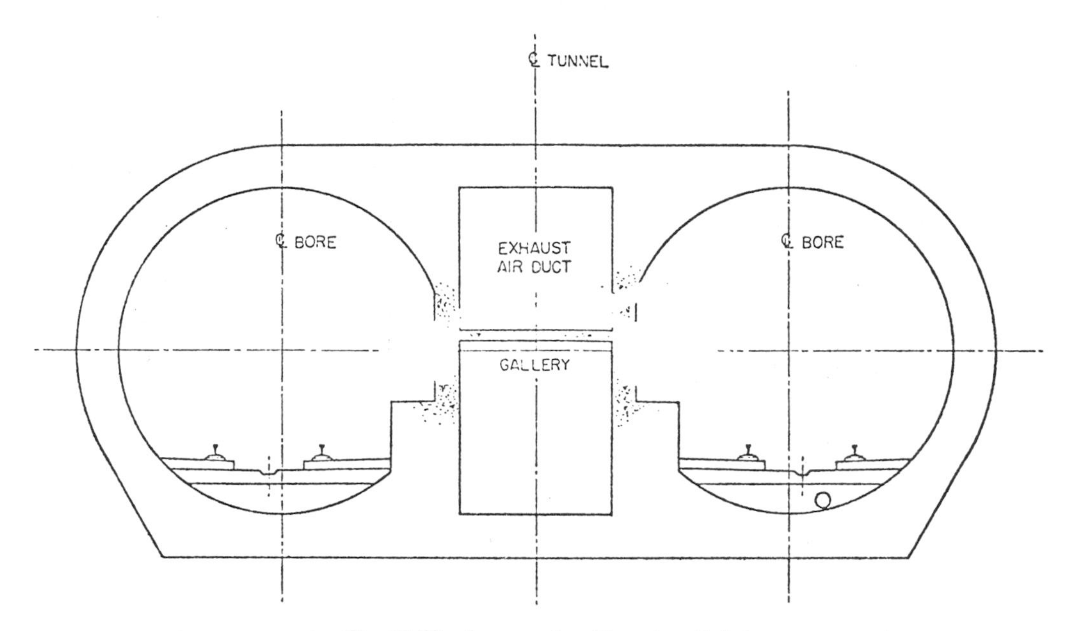

Fig. 19-80. Cross-section (Trans Bay Tube).

outlined in the regulations. There are several of these minimum airflow requirements, such as a minimum of 200 feet^3/minute/worker employed in the tunnel, and a minimum longitudinal velocity of 30 feet/minute where blasting or drilling is taking place.

Instrumentation must be available to permit a quantitative measurement of the levels of all hazardous contaminants and the level of oxygen within the tunnel environment. Oxygen must be maintained at a level of 20% or greater.[45]

The only type of internal combustion engine permitted in a tunnel under construction is the mobile diesel type. The amount of ventilation air required for each piece of mobile equipment powered by diesel can be obtained from the Bureau of Mines Information Circular Number 8363.[47]

Diversity of use of this diesel-powered equipment should be taken into account when computing the total air requirements.

The arrangement of the temporary ventilation system should be such as to minimize the chances of the construction work being halted because of equipment failures. Ventilation plants should be located at both ends of the tunnel. These plants should consist of a minimum of two fans, each of these fans sized to handle tne full air quality. This provides 100% standby capacity.

REFERENCES

1. *Industrial Ventilation, Sixteenth Edition*, American Conference of Governmental Industrial Hygienists, Lansing, Michigan, pp. 13-1, 1980.
2. Stahel, M., Ackeret, J. and Haerter, A., *Die Luftung der Autotunnel*, Mitteilung Inst. für Strassenbau, Eidg. Techn. Hochschule, Zürich, Switzerland, 1961.
3. *Air Quality Criteria for Carbon Monoxide*, AP-62, U.S. Department of Health, Education and Welfare, Public Health Service, Washington, D.C., pp. 2–3, 1970.
4. *Industrial Ventilation*, opcit, pp. 13-1 thru pp. 13-20, 1980.
5. *Air Quality Criteria for Nitrogen Oxides*, AP-84, Environmental Protection Agency, Washington, D.C., pp. 2–6, 1971.
6. Fieldner, A.C. *et al.*, *Ventilation of Vehicular Tunnels*, Report of U.S. Bureau of Mines to New York State Bridge and Tunnel Commission, and New Jersey Interstate Bridge and Tunnel Commission, American Society of Heating and Ventilating Engineers, New York, 1921.
7. "Research Investigations and Studies to Establish the Ventilation Requirements for the Straight Creek Highway Tunnels," Tippetts-Abbett-McCarthy-Stratton for State of Colorado Department of Highways and U.S. Department of Commerce, Bureau of Public Roads, October 1965.
8. Haerter, A., *Fresh Air Requirements for Road Tunnels*, Proceedings: International Symposium on the Aerodynamics and Ventilation of Vehicle Tunnels, BHRA Fluid Engineering Canterbury, England, 1973.

9. Haerter, A., op cit, pp. B1–6.
10. Haerter, A., op cit, pp. B1–8.
11. Haerter, A., op cit, pp. B1–7.
12. Haerter, A., op cit, pp. B1–14.
13. Haerter, A., op cit, pp. B1–15.
14. *Highway Capacity Manual*, Special Report 87, Highway Research Board, Washington, D.C., pp. 89 1965.
15. Highway Capacity Manual, op. cit, pp. 52.
16. *Highway Capacity Manual*, Bureau of Public Roads, U.S. Department of Commerce, Washington, D.C., pp. 3, 1950.
17. Atkinson
18. Occupational Safety and Health Act.
19. Threshold Limit Values for Chemical Substances and Physical Agents in the Workroom Environment for 1980 American Conference of Governmental Industrial Hygienists, Cincinnati, Ohio, pp. 3, 1980.
20. Rogers, S.J., Roelich, F., Jr. and Palladino, C.A., *Tunnel Ventilation and Air Pollution Treatment,* Mine Safety Appliances Research Corp. for Federal Highway Administration, Washington, D.C., 1970.
21. Chen, Ti.C., *Air Quality Distribution in Highway Tunnel Ventilation Flow Studies in Air and Water Pollution*, The American Society of Mechanical Engineers, pp. 9, 10, 11, 1973.
22. Osborne, W.C. and Turner, C.G. (Co-Eds.), *Woods Practical Guide to Fan Engineering*, *Second Edition*, Woods of Colchester, Ltd. Colchester, England, pp. 130, 131, 135, 1964.
23. *Joy Fan Catalogue*, Joy Manufacturing Company, New Philadelphia, Ohio.
24. Jorgensen, R. (Ed.), *Fan Engineering*, *Sixth Edition*, Buffalo Forge Company, pp. 226, 227, 258, 290, 1961.
25. Jorgesen R., op cit., pp. 32.
26. AMCA Standards Handbook, Publication 99, Air Moving and Conditioning Association Inc. Park Ridge, Illinois, Standard 2408-66, 1967.
27. AMCA Test Code for Air Moving Devices, Standard 210-67 Air Moving and Conditioning Association Inc., Part Ridge, Ill. 1967.
28. Singstad, O., *Ventilation of Vehicular Tunnels*, World Engineering Congress, Tokyo, Japan, 1929.
29. *Handbook of Fundamentals*, American Society of Heating, Refrigerating and Air Conditioning Engineers, Inc., New York, 1977.
30. Osborne, W.C., op cit. pp. 138.
31. Jorgesen, R., op cit.
32. Rogers, S.J., op cit., pp. 173.
33. "Libby Dam Project Railroad Tunnel Ventilation Research and Test Program at Cascade Tunnel," Parsons, Brinckerhoff, Quade & Douglas, Inc., August 1966.
34. Berger, L.B. and McGuire, L.M., *Observations on the Use of a Diesel Freight Locomotive through a Railway Tunnel*, U.S. Department of the Interior, Bureau of Mines, June 1946.
35. Aisiks, E.G. and Danziger, N.H., *Ventilation Research Program at Cascade Tunnel*, Great Northern Railway, American Railway Engineering Association, pp. 114, October 1969.
36. Aisiks E.G., op cit., pp. 108.
37. Libby Dam Project, Tunnel Ventilation System Department of the Army Corps of Engineers, Seattle District, Seattle, 1966.
38. Subway Environmental Design Handbook, Volume I Principles and Applications, U.S. Department of Transportation, Second Edition (NTISN^{o} PB-254788), Washington D.C., 1976.
39. *ASHRAE Handbook*, *Equipment*, American Society of Heating, Refrigerating, and Air Conditioning Engineers, Inc., New York, 1979.
40. *ASHRAE Handbook*, *Systems*, American Society of Heating, Refrigerating, and Air Conditioning Engineers, Inc., New York, 1976.
41. Bendelius, A.G. and Metsch, N.W., "*Environmental Control for the Caracas Metro*," *ASHRAE Journal*, 1973.
42. Soloman, I.M., *Subway Station Air Conditioning for the Washington Metropolitan Area Transit Authority*, Paper 73-RT-7, Rail Transportation Division, American Society of Mechanical Engineers, New York, 1973.
43. Bendelius, A.G., *Subway Environmental Control for the MARTA System*, Intersociety Conference on Transportation, Atlanta, Georgia, 1975.
44. McCutchen, W.R., Ventilation Considerations for the Trans Bay Tube, Symposium Proceedings–Ventilation Aspects of Hostile Environments," ASHRAE, New York, 1974.
45. Danziger, N.Y. and Aisiks, E.G., *Analysis and Control of Thermal Environment in Subway Systems*, Presented at the Rail Conference of the American Transit Association/Institute for Rapid
46. "Safety and Health Regulations for Construction," *Federal Register* 36, *No. 75*, *Part II. Section 1518.800* (Tunnels and Shafts): 7393-7394, Washington, D.C., April 17, 1971.
47. *Mobile Diesel-powered Equipment for Non-gassy, Non-coal Mines and Tunnels Approved by the Bureau of Mines 1951–1961*, Bureau of Mines, U.S. Department of the Interior, IC8363, February 1968.
48. ASHRAE Handbook, Applications, Chapter 14, Enclosed Vehicular Facilities, A S of H R and A C Engineers, New York, 1978.
49. Subway Environmental Design Handbook Volume II, Subway Environmental Simulation Computer Program (SES), NTISN^{o} PB-254-789, Part 1–3, 1975.

20
Tunnel Lighting

MICHAEL A. MARSZALOWICZ

Professional Associate
Parsons, Brinckerhoff, Quade & Douglas

LIGHTING OF HIGHWAY TUNNELS

The lighting in tunnels, particularly daytime lighting, poses a number of problems. This is due to the impracticability, both technically and economically, of lighting a tunnel interior as brightly as the open road. On bright sunlit days, a great difference, in the order of 100 to 1, exists between the exterior and interior luminances, and this manifests itself as a severe visual task to the motorist because of the inability of the human eye to adapt to sudden light changes instantaneously. It takes the eye longer to adapt from light to dark than from dark to light. Moreover, because the motorist is in transit, the optical system is not in a steady state. This creates a transition problem, in addition to other problems, such as flicker from the light sources due to spacing, to which the techniques normally employed in the solution of interior lighting problems are not readily applicable.

Lighting of railroad tunnels and rapid transit tunnels, with relatively simple requirements, is discussed in the latter part of this chapter.

Definitions of Terms

Access Zone. That portion of the open approach of the highway immediately preceding the tunnel facade or portal.

Adaptation. The process by which the sensitivity of the retina of the eye adjusts to more or less light than that to which it was adjusted during a period immediately preceding. The resulting sensitivity of the retina is termed the *state of adaptation.* The luminance causing this state of adaptation is termed the *adaptation level.* The time necessary for the complete adjustment is known as the *adaptation time*, and the time-dependent effect itself is the *adaptation.*

Adaptation Point. The point on the road where the adaptation of the eye of a motorist approaching a tunnel begins to be influenced by the presence of the dark tunnel entrance.

Brightness. The luminous flux per unit of projected area and unit solid angle, either leaving a surface at a given point in a given direction or arriving at a given point from a given direction; the luminous intensity of a surface in a given direction per unit of projected area of the surface as viewed from that direction.

Exit Zone. That portion at the end of the tunnel which during daytime appears to be a brilliant "white hole" when a motorist has driven for several minutes in the tunnel interior.

Flicker. The result of periodic luminance changes in the field of vision, due to the spacing of the lighting fixtures.

Footcandle. The incident illumination on a surface 1 foot^2 in area on which is distributed a uniform light output of 1 lumen. It equals 1 lumen/foot^2.

Foot Lambert. The unit of photometric brightness (luminance). A perfectly diffused surface emitting or reflecting light at the rate of 1 lumen/foot^2 would have an equivalent brightness of 1 foot Lambert.

Glare. The sensation produced by brightness within the visual field that is sufficiently more intense than the luminance to which the eyes are adapted to cause discomfort or loss in visual performance and visibility.

Interior Zone. (Sometimes known as the normal day zone.) The length of tunnel between a point just beyond the entrance transition zone, where eye adaptation is no longer a consideration for visual perception, and the exit portal.

Lamp Lumen Depreciation Factor. One of the many factors used in the evaluation of a maintenance factor causing a reduction of lumen output of a lamp during its life cycle.

Louver. A series of baffles used to shield a source from view at certain angles or to absorb unwanted light.

Lumen. The unit of measure of the quantity of light. The amount of light that falls on an area of 1 foot^2, every point of which is 1 foot from a source of 1 candela (candle).

Luminaire. A complete lighting device consisting of a light source together with its direct appurtenances.

Luminaire Ambient Temperature. The lumen output of fluorescent lamps is affected by the ambient temperature in which they operate and must be considered when obtaining the maintenance factor (Fig. 20-1).

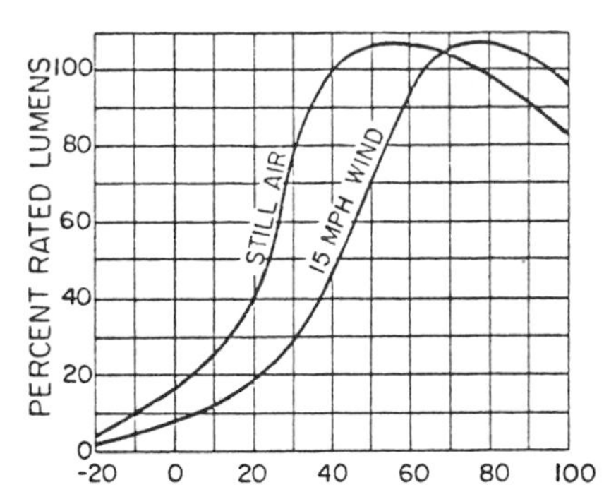

Fig. 20-1. Light output versus ambient temperature.[1]

Luminaire Dirt Depreciation Factor. Another maintenance factor, causing loss of light on the work plane due to the accumulation of dirt on the luminaire.

Luminaire Surface Depreciation. This depreciation of initial luminaire lumen output results from the adverse changes in metal, paint and plastic components. Glass, porcelain and processed aluminum depreciate negligibly and are easily restored to their original reflectance. Baked enamels and other painted surfaces have a permanent depreciation because all paints are porous to some degree. For plastics, acrylic is least susceptible to change, but in certain atmospheres, transmittance may be reduced by usage over a period of 15 to 20 years. For the same usage, polystyrene will have lower transmittance than acrylic and will exhibit a faster depreciation. Because of the complex relationship between the light controlling elements of luminaires using more than one type of material, (such as a lensed troffer), it is difficult to predict losses due to deterioration of materials. In addition, some materials are more or less adversely affected, depending on the atmosphere in which they are installed. At present, no factors are available.

Luminaire Voltage. High or low voltage at the luminaire will affect the output of most lamps. For incandescent units, deviations from rated lamp voltage cause approximately 3% change in lumen output for each 1% change in primary voltage deviation from rated ballast voltage. When regulated output ballasts are used, the lamp lumen output is relatively independent of primary voltage within the design range. Fluoresent luminaire output changes approximately 1% for each $2\frac{1}{2}$% change in primary voltage.

Maintenance Factor. The ratio of the in-service lumens of a lighting system to the initial lumens, which can be determined with reasonable accuracy by the evaulation of various light loss factors which eventually contribute, in varying amounts, to the in-service illumination levels. Some of these light loss factors are luminare ambient temperature, voltage to luminaire, ballast factor, luminaire surface depreciation, room surface dirt depreciation, burn-outs, lamp lumens depreciation and luminaire dirt depreciation.

Matte Surface. A surface from which the reflection is predominantly diffuse, with or without a negligible specular component.

Optical Guidance. The means by which visible aids indicate the course of the road direction. In tunnel application, this aid is derived from the light sources as well as curb delineation and lane marker lines.

Reflectance. The ratio of the flux reflected by a surface or medium to the incident flux. This general definition may be further modified by the use of one or more of these terms: specular, spectral, diffuse.

Room Surface Dirt Depreciation. The accumulation of dirt on room surface reduces the amount of lumens reflected and inter-reflected to the work plane and is one of the many factors to be considered in the evaluation of the design maintenance factor.

Threshold Zone. The first section at the entrance end of a tunnel, where the first decrease in daylighting takes place. This decrease can be accomplished by a reduction in daylighting with the use of screening or by the use of artificial lighting or by a combination of both.

Transition Zone. The section at the entrance end which follows immediately after the threshold zone and contains diminishing light levels until the interior zone levels are reached. It is sufficiently long to provide for adequate eye adaptation time from open road brightnesses to interior tunnel brightnesses (Fig. 20-2).

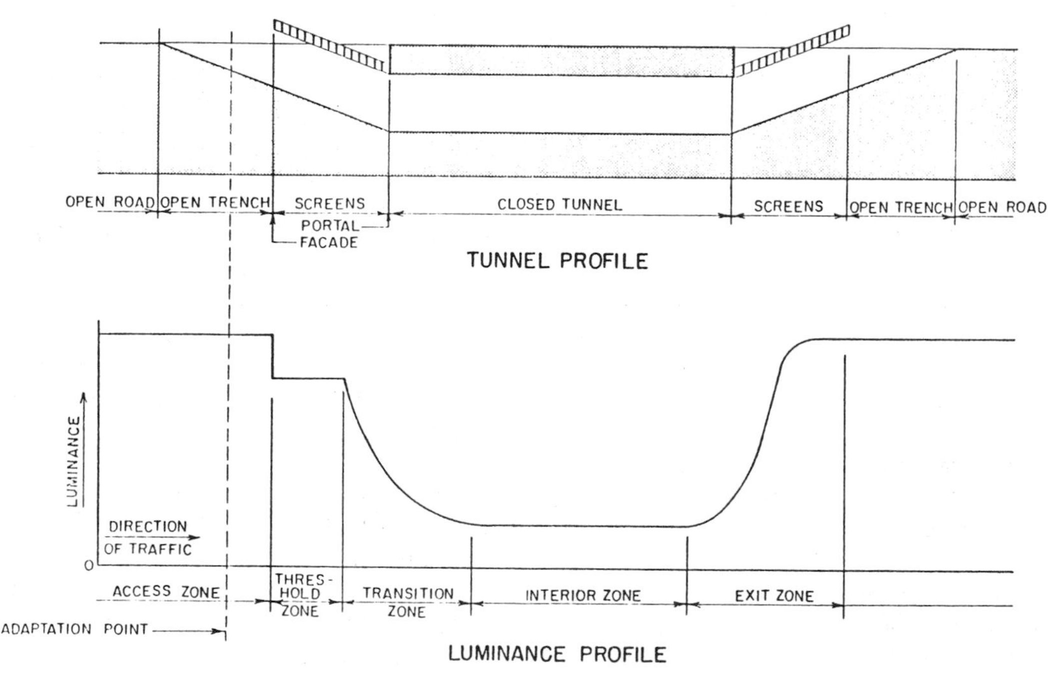

Fig. 20-2. Tunnel lighting nomenclature.[5]

Tunnel Lighting Nomenclature. For the terminology used in discussions on tunnel lighting, see Fig. 20-2.

Tunnel Types. For profiles of the various types of vehicular tunnels, see Fig. 20-3.

Utilization Factor. (Also known as the coefficient of utilization.) The ratio of the lumens that reach the work plane to the total lumen output of the bare lamps. It accounts for the luminaire photometric characteristics, mounting heights, room dimensions and surface reflectances.

PHYSIOLOGICAL CONSIDERATIONS IN TUNNEL LIGHTING DESIGN

The speed with which a motor vehicle can be safely operated depends entirely on the information that the motorist can quickly obtain about his environment. In gathering this information, the most important sense is vision. It is the only means by which the motorist can detect and discern objects and obstacles in the path of the vehicle. Therefore, the amount of information obtained depends on the lighting conditions. Obstacles and objects must be visible at such a distance that the necessary maneuvers can be executed in an effective manner.

Since the ability to see is dependent on the lighting conditions, which are subject to sudden birghtness changes, especially during daytime hours, a great deal depends on the sensitivity of the eye. Before criteria for tunnel lighting design can be established, it is important that the characteristic behavior of the human eye be understood.

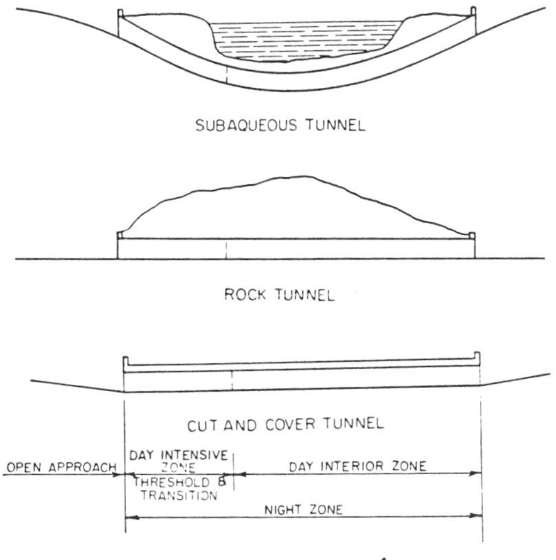

Fig. 20-3. Tunnel types.[4]

The response of any portion of the retina at a given moment depends upon the light that fell on that region during the just preceding period of time and is called *adaptation*. The resulting sensitivity of the part of the retina under consideration is termed the *state of adaptation* of that part. One particular state of adaptation can occur as the result of different luminance distributions in the field of vision. Such luminance distributions are equivalent, and this is termed the *adaptation level.* The time necessary for the adjustment is called *adaption time* and the time-dependent effect itself the *adaptation.* Any change in the adapting illumination produces a change in the photochemical substances toward a new equilibrium level. Since these reactions take a finite time, there will be a time delay before the retina reaches maximum sensitivity for a given adaptation level. Eye adaptation is much faster when the eye's field of view transfers from a dark to a bright environment than when it transfers from a bright to a dark environment.

In the lighting of tunnel entrances, two problems in particular are closely connected with the nature of the adaptation process. The first is that, with a given state of adaptation, only objects with a luminance not far below that adaptation level can be seen. This induction effect is likely to arise when a tunnel entrance has a substantially lower luminance than the open road. As a result of this effect, the tunnel entrance will give the impression of being a black area (commonly referred to as the "black hole"). If the tunnel entrance luminance is below the threshold for that particular state of adaptation on the open road, no details in the interior will be discerned. The second property of the function of the eye which presents a problem is the complexity of the adaptation to changes in the local illumination. Quick changes in the illumination are not always followed immediately by adaptation, as a result of which both the ease of vision and the visual performance decrease. In extreme cases, this decrease can be so strong and so sudden that for a time no observation can be made at all.

Tests made for the purpose of establishing the minimum intensity required at the tunnel entrance reveal the the relation between the interior intensity, L_2, and the outside intensity L_1, should follow the curve shown in Fig. 20-4. This curve is based on being able to see an object approximately 8 inches square with a contrast C against its immediate background in 0.1 seconds in 75% of the cases at the distance of approximately 325 feet. If L_3 is the luminance of the object and its contrast with the immediate surroundings, L_2, is defined by the relation $C = [(L_2 - L_3) \div L_2]100$ and expressed in percent, then the curve gives the values of L_2 relative to L_1 for an object contrast of 20%. For a large range of L_2, this means that L_2 should not be less than $0.10\ L_1$.

Adaptation tests indicate that the visibility of the obstruction does not alone determine the rate at which the entrance lighting can be progressively reduced into the tunnel, but also the uncertainty of vision due to the after-images in the eyes which only disappear gradually. Considering this, the rate of reduction of L_2 against time should not be greater than that shown in Fig. 20-5 in order to provide for 75% of the drivers.

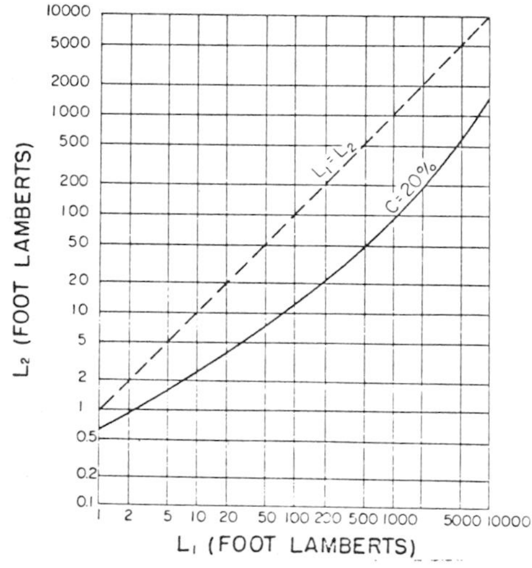

Fig. 20-4. Relationship between L_1 and L_2.[5]

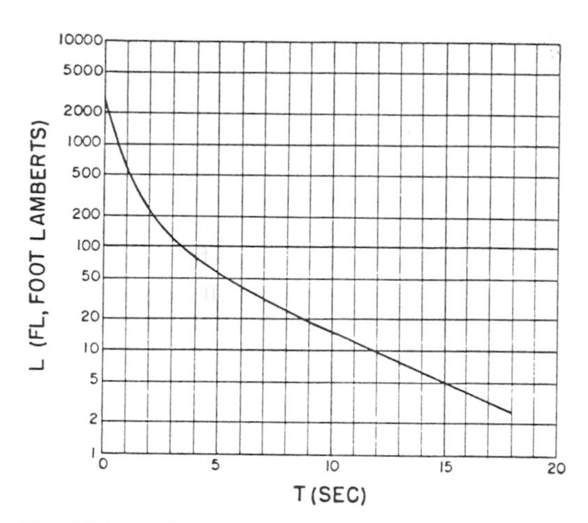

Fig. 20-5. Relationship between luminance and time in tunnel for satisfactory adaptation.[5]

Figures 20-4 and 20-5 are the basis for determining the lighting of tunnel entrance sections. The highest intensity should be extended for at least 325 feet beyond the point at which eye adaptation starts. This point may be directly at the portal, or it may be a considerable distance ahead of it, depending upon the geometry of the tunnel entrance.

Another consideration in tunnel lighting is *flicker*, which results from periodic luminance changes in the field of vision. The impression made on the motorist by these periodic changes is dependent on the repetition frequency. With low frequency–lower than approximately 1 cycle/second(cps)–each turn is preceived as such. On the other hand, if the frequency is higher than a certain value, the periodic character of the lighting cannot be distinguished any longer. This frequency is called *flicker-fusion frequency* and, as a rule, amounts to 50 to 80 cps. Flicker frequencies which lie below the fusion frequency should be avoided. Experiments have shown that the greatest annoyance is experienced generally when the repetition frequency is between 5 and 10 cps and, again, as a rule, the annoyance becomes negligible at frequencies below 2.5 cps and above 13 cps. At a speed of 60 mph (88 feet/second), the spacing of point lighting sources would be 35 feet for the lower frequency and 6.8 feet for 13 cps. At the greater spacing, it would be very difficult to achieve uniformity of luminance on the walls and roadway because tunnels usually have a low height in comparison to their width. With the lesser spacing, the center-to-center distance becomes so close that a line source of light is usually chosen as opposed to a point source of light. Furthermore, the point sources must be dimmed rather than be switched off in order to keep an installation free from annoying flicker when the luminance level in the tunnel must be reduced during the night.

Table 20-1. Summary of recommended day interior maintained illumination levels in footcandles.

AUTHORITY	WALLS	ROADWAY
IES	1.4–2.18	5–10
AASHTO	1.4	5
CIE	1.3–2.1	4.4–7.3

LUMINANCE LEVEL IN THE TUNNEL INTERIOR

Recommendations for tunnel interior lighting levels, as well as methods of measurement, presently vary among authoritative sources. The *Fifth Edition* (1972) of the *IES Handbook* suggests at least a 5-footcandle horizontal maintained illumination level on the roadway. For tunnels up to 2000 feet in length, it may be desirable to use up to 10 footcandles in this zone. AASHTO recommends, for use in most situations, an average maintained illumination on tunnel walls of 5 footcandles. These values are based on a wall reflectance factor of at least 70%. When the reflectance factor is less than 70%, designed footcandles should be increased to compensate for the lower reflectance. This statement suggests that the wall luminance is the parameter which should be kept constant. This authority further qualifies the foregoing illumination levels with, "Since the conditions which influence lighting requirements vary considerably, the need for judgment in applying these guide values is apparent. They are considered suitable for most situations but should not be adopted universally without an evaluation of all factors which may dictate a need for modification." The Commitee Internationale de l'Eclairage (C.I.E.)[3] recommends a luminance in the interior of long tunnels of at least 2 to 5 candelas/square meter (cd/m^2) = 0.88 to 1.46 foot Lamberts (fL) or, with a 20% reflectance, 4.4 to 7.3 footcandles (fc) and preferably 15 to 20 cd/m^2 = 4.5 to 6 fL, 6 to 8 fc. These values are minimum maintained values measured on the road surface. The walls should have at least the same level of luminance.

Based on a reflectance factor of 70% for the tunnel walls and ceiling, and a reflectance of 20% for the roadway, Table 20-1 summarizes these recommended values.

While it appears that there is considerable disparity in these recommendations, there are conditions under which the extremes can be shown to be adequate for daytime interior lighting. Where there is sufficient time for complete adaptation, such as in long tunnels, the interior lighting can approach the criteria for nighttime street lighting, in which case neither dimming nor switching will be required. On the other hand, if only partial adaptation can be effected, then the higher values appear to be the better choice during daytime hours. In many cases, economic factors, as well as the availability of the proper lighting equipment, will play an important role in the final interior lighting level.

EXIT LIGHTING

During the daytime, the exit of the tunnel appears as a bright hole to the motorist. Usually, all obstacles will be discernible by silhouette against the bright exit and thus will be clearly visible. This visibility by silhouette can be further improved by lining the walls with tile or panels having a high reflectance and thus permitting a greater daylight penetration into the tunnel. This effect is shown in Fig. 20-6.

When the exit is not entirely clear but partially screened by a large object such as a truck, then a different visibility requirement is present. In this instance, if a smaller object is following the large object, the smaller object may not be readily visible. Whereas the large object screening the exit would be clearly visible, the following small object would have equivalent brightness to the large one and, therefore, would not be readily discerned by silhouette. This situation is greatly improved by extending the normal

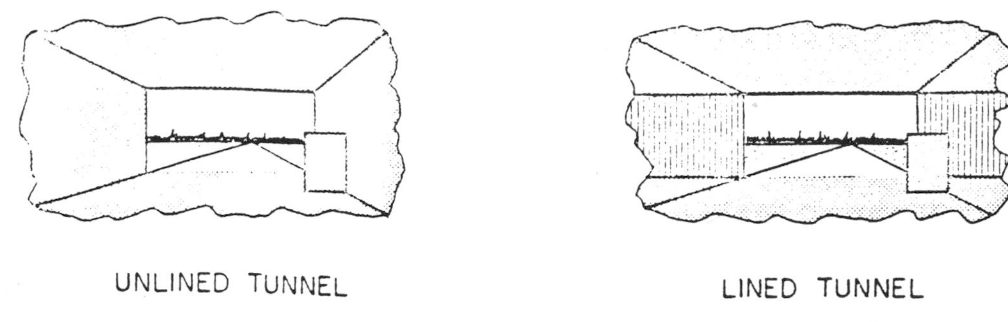

Fig. 20-6. Effect of natural light penetration on walls at tunnel exit.[4]

interior zone daytime lighting system to the exit portal. This effect is shown in Fig. 20-7.

Under certain conditions and during certain periods of the day, sunlight may penetrate directly into the tunnel exit, creating an extremely difficult visual condition. Coupled with the use of glazed walls and glossy paints, the resulting specular reflection may create considerable interreflections, thus preventing the motorist from identifying the tunnel outlines clearly, as well as creating a situation where discomfort and disability glare would be experienced. Under these conditions, discernment of objects by silhouette may be difficult and a hazardous traffic situation may result. Even sun screens will not offer much help, so the situation should be avoided. Reorientation of the tunnel exit or the use of materials with diffuse reflectances (matte or flat) should prove helpful.

ENTRANCE LIGHTING

The most critical section of tunnel lighting is the entrance section which comprises the threshold and transition zones (Fig. 20-2). Recommended illumination and luminance levels at the threshold and transition zones vary somewhat in different countries. Figures 20-8, 20-9, 20-10, 20-11 and 20-12 illustrate these variations. Figure 20-13 shows a composite of the various recommendations for comparison purposes. It is interesting to note that, except in Japan, there is general agreement on the ratio of the first luminance reduction from the expected maximum open road luminance. This reduction is 8 and 10 to 1. This means that the order of magnitude level of luminance just inside the entrance portal should be between 290 and 350 fL for operation during daylight hours. To produce these levels solely by artificial means using fluorescent lamps would require a great number of luminaries on the walls and ceiling and perhaps could not be justified economically. While the use of high intensity point sources would permit reduction in the number of units, the accompanying source glare and the risk of producing annoying flicker might prove to be objectionable and unacceptable.

For these reasons, some of the most recent

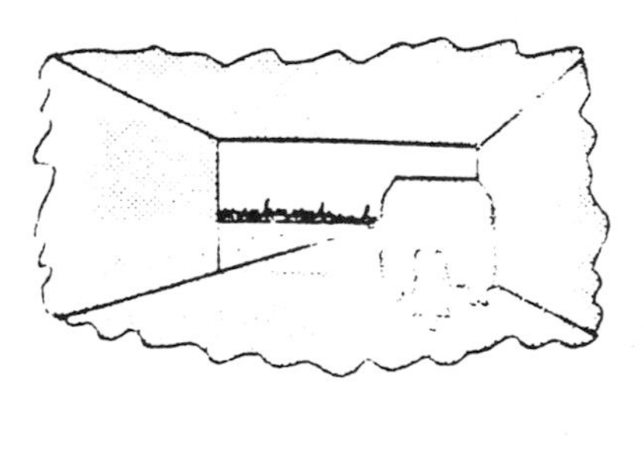

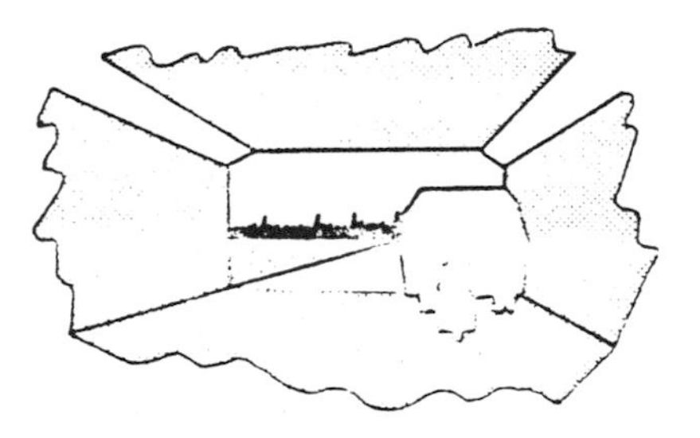

Fig. 20-7. Effect of normal interior zone daylighting at the tunnel exit.[4]

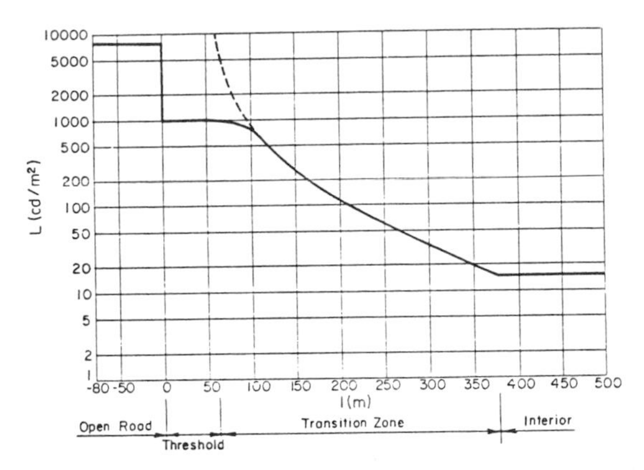

Fig. 20-8. Proposed CIE tunnel entrance daytime luminance levels.[3]

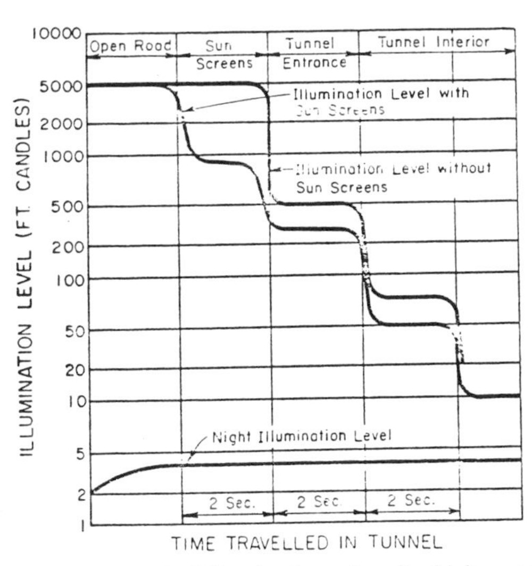

Fig. 20-10. Typical illumination values for high-speed, high-volume tunnel entrances.[6]

tunnel lighting designs in Europe and on this continent have incorporated a grid at the tunnel entrance to screen out natural daylight in the proper amounts to assist in the eye adaptation process. The subject of sun screens will be treated later in this chapter, but as presently designed, they have not proven to be the ultimate answer to the first step in reduction of natural daylight.

Another and very important aspect of the entrance lighting that can be observed from the recommendations shown in Fig. 20-13 is that there is apparently a difference in the assumed transition time required for adaptation from the open road luminance to the tunnel interior luminance. The CIE recommendation is based on the curve of Fig. 20-5, which is the result of laboratory adaptation tests for partial adaptation considered to be sufficient to avoid disturbing negative after-images. The IES and the Canadian Good Roads recommendations, on the other hand, agree on 4 seconds as sufficient adaptation time from open road luminance to tunnel interior luminance. There is no evidence presented in these latter recommendations, however, to substantiate that 4 seconds is sufficient for satisfactory adaptation except that the IES recommendation states that "the time that the motorist requires in passing from the highest to the lowest illumination, should be 4 seconds or more." As a result of these differences in adaptation time, the lighting designer is faced with a perplexing decision which might be difficult to resolve. While the conservative approach is certainly the safest, from the motorist's point of view, it is also the more costly. Conversely, the alternate approach is definitely the less safe approach and, if safety is a prime factor in the design of the highway system, then the former approach to adaptation time appears to be the better choice.

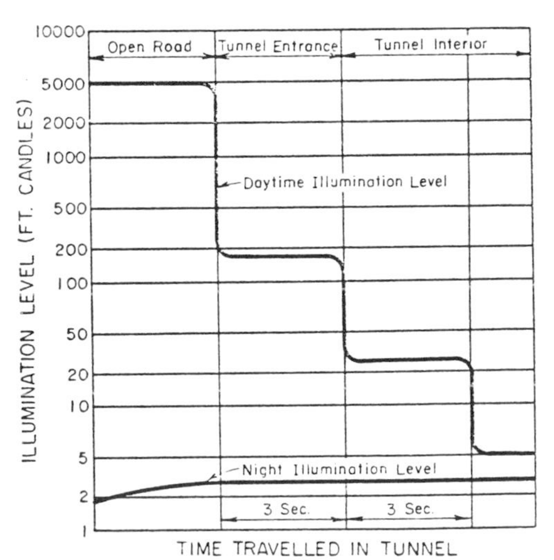

Fig. 20-9. Typical illumination values for low-speed, low-volume tunnel entrance.[6]

Other factors to be considered in the evaluation of the lighting levels, as well as transition times, are tunnel orientation, latitudes, geographical location, approach grades, terrain and adaptation point. An example of favorable

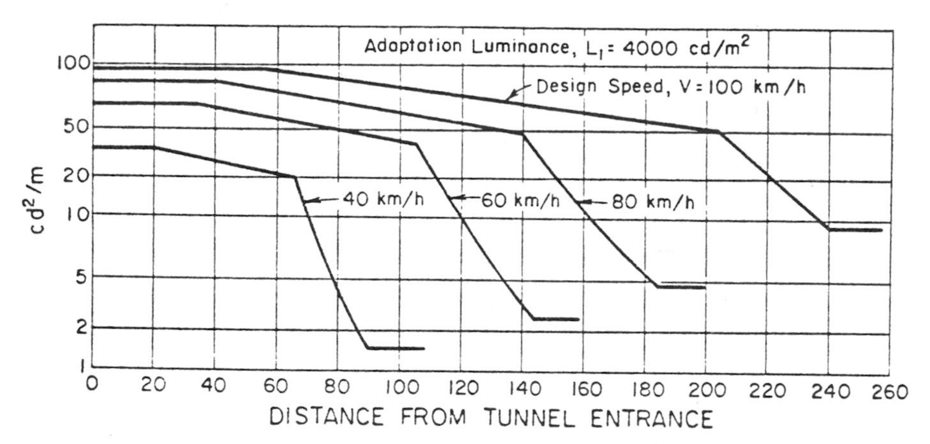

Fig. 20-11. Tunnel entrance lighting curves.[6]

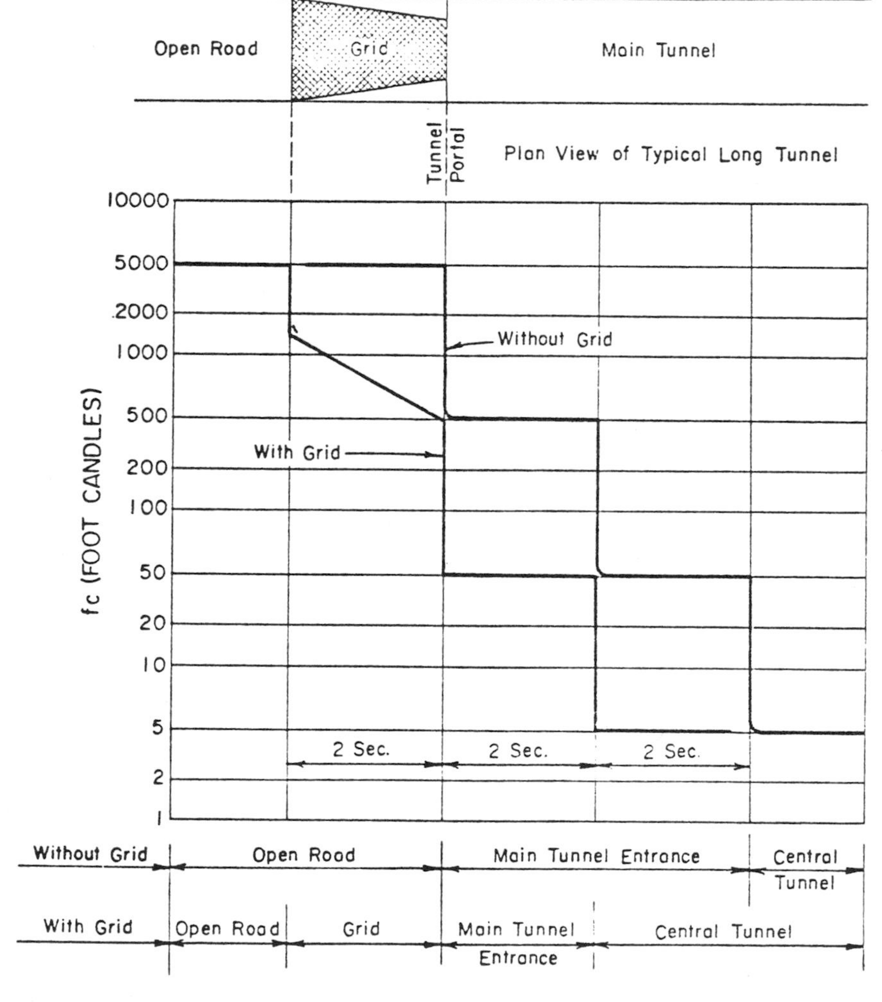

Fig. 20-12. Daytime illumination for typical long tunnel.[11]

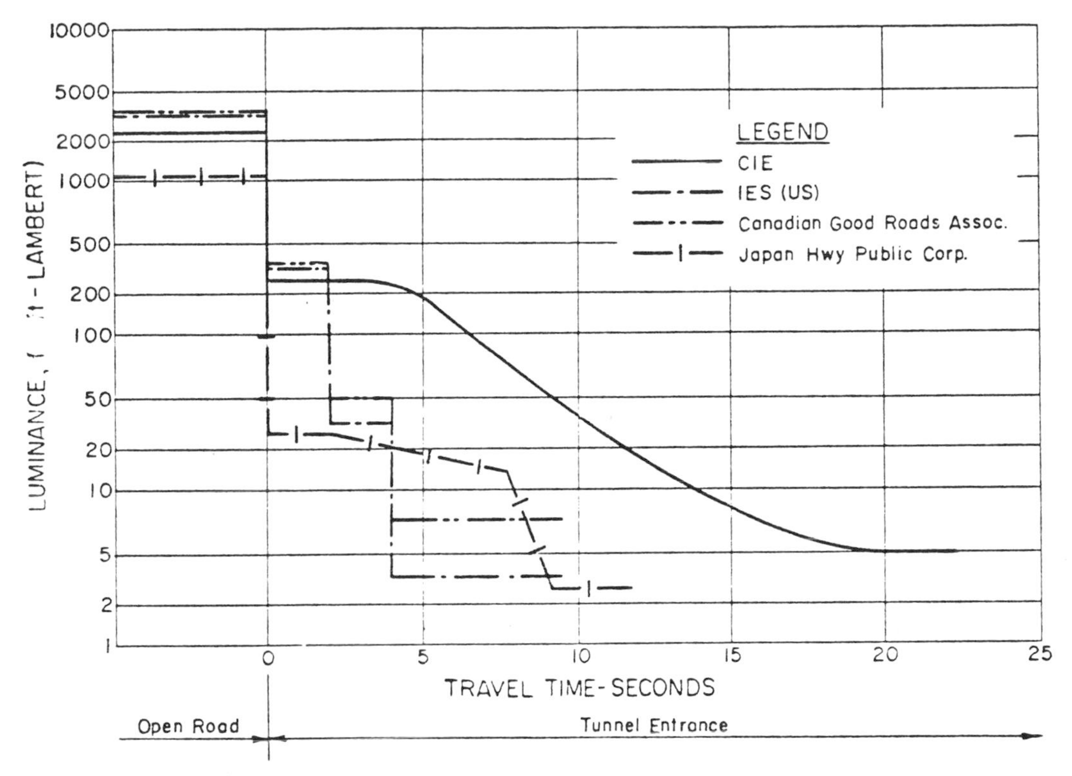

Fig. 20-13. Comparison of recommended tunnel entrance luminaire levels.

conditions where the tunnel lighting problem can be readily solved using conventional equipment is a tunnel located in the higher latitudes where the sun or high sky brightnesses never come into the field of view; the orientation is east-west; the location is through a mountain; and the approach is on an upgrade, which allows the tunnel ceiling (as well as the walls) to act as an appropriate backdrop. An example of unfavorable conditions is a location of a subaqueous tunnel in the lower latitudes having a southeast or southwest orientation, where the sun or its reflection over open water would be directly in the field of view during morning or afternoon rush hours, on a downgrade approach to the entrance portal, where the roadway surface (an eventually soiled and usually poor reflecting surface) becomes the principal backdrop for silhouette discernment of objects or stalled cars. A good example of such a severe visual task is the Hampton Roads Tunnel and Bridge Crossing between Hampton and Norfolk, Virginia.

TUNNEL CLASSIFICATION

Underpasses

From the point of view of adequate supplementary lighting, there is no definite line of demarcation between an underpass and a short tunnel. AASHO defines an underpass as a covered highway whose length is between 75 and 100 feet. Other authorities, such as the IES and the CIE, generally recognize all covered highways as tunnels and do not recognize an underpass as a separate and distinct structure. These authorities indicate that for lighting purposes every artificial or natural covering of a road, irrespective of the length and nature of the covering, is considered to be a tunnel.

Short Tunnels

IES and CIE define a short tunnel as one where, in the absence of traffic, the exit and the area behind the exit can be clearly visible from a point ahead of the entrance portal. For lighting purposes, the length of a short tunnel is usually limited to approximately 150 feet. Some tunnels up to 400 feet long may be classified as short if they are straight, level and have a high width and/or height to length ratio. Thompson and Fansler[4] define a short tunnel as one which requires no more than 5 seconds to travel through it at the posted speed limit or the design speed. For example, if the speed limit is 60 mph, a tunnel no longer than 440 feet is considered a short tunnel.

Long Tunnels

For lighting purposes, IES defines a long tunnel as one which is straight, generally over 150 feet in length, and where the exit brightness takes up too small a part of the driver's field of vision to serve as an effective background for silhouette discernment. CIE defines a long tunnel as one, from the viewpoint of vision, where the exit and the area behind the exit cannot (under the traffic conditions normal for that tunnel) be perceived from a point at some distance in front of the entrance.

LIGHTING OF SHORT TUNNELS

In most cases, a lighting system is not required inside short tunnels for adequate driver visibility. Daylight penetration from each end, plus the silhouette effect of the opposite end brightness, generally will assure satisfactory visibility. On the other hand, tunnels between 75 and 150 feet in length may require supplementary daytime lighting if the daylight is restricted due to roadway depression, tunnel curvature or the proximity of tall builidings in urban areas. Short tunnels appear to the approaching driver as a black frame, as opposed to the black hole which is usually experienced in long tunnels. This effect illustrated in Fig. 20-14, indicates that adaptation to the lower level inside the tunnel will not have an opportunity to occur. After entering the tunnel, the central part of the field of view is taken up by the brightness of the exit so that the brightness of the walls and ceiling has insignificant adaptation stimuli.

Obstructions in the short tunnel can be seen if they are high enough to silhouette against the bright environment of the exit. If the vertically projected dimension of the dark frame across the roadway is less than the height of the smallest object that must be seen, then the object will be silhouetted against the exit brightness and will be visible. Figure 20-15 illustrates this for various tunnel lengths on a straight and level roadway with the driver's eye at a height of 3.75 feet and at a distance of 475 feet (stopping sight distance at 60 mph) from the entrance portal.

Due to daylight penetration from both portals, a short tunnel on a straight and level roadway can be as much as 75 feet long and not require daytime lighting for an obstacle height of 6 inches. Silhouetting can be enhanced by lining the structure wall with a light-colored material to reduce the darkness within the tunnel.

Where local conditions warrant the installation of an artificial lighting system for daytime operation, the level of luminance within the entire length of the tunnel should be at least 0.1 of the expected maximum open approach

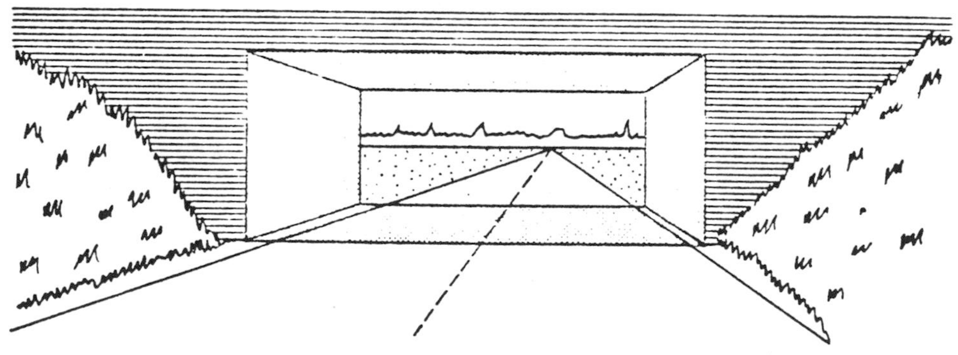

Fig. 20-14. A short tunnel appears as a dark frame.[5]

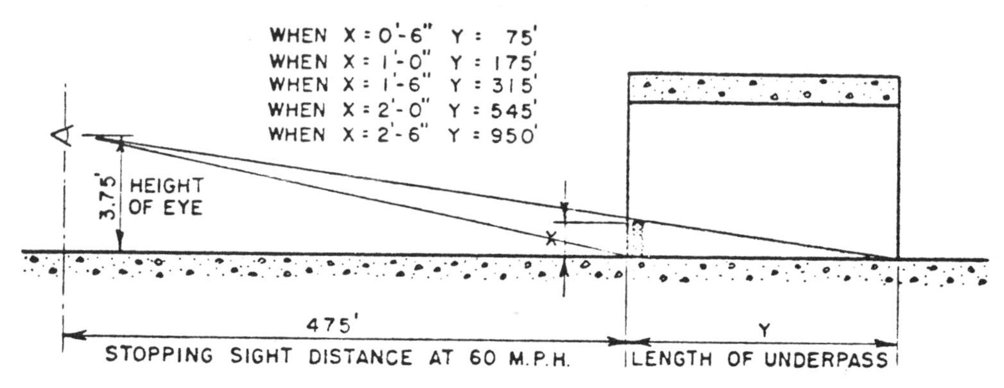

Fig. 20-15. Vertical dimension of dark frame.[5]

luminance. This requirement may mean that a luminance level of 300 to 350 fL is needed for satisfactory visibility. Where it is not practical to achieve these levels, visibility can be improved by providing an opening in the ceiling about midway through the structure to permit daylight penetration, as illustrated in Fig. 20-16. In effect, the opening provides two structures and may satisfy the requirement for lighting during the day. Walls and ceilings lined with a light-colored, matte-finished material will enhance the penetration of natural light into the structure. A design with a funneled-up ceiling at the portals will further permit additional daylight penetration.

LIGHTING OF LONG TUNNELS

Lighting in long tunnels should generally follow the luminance levels illustrated in Fig. 20-8 for satisfactory visibility during daylight hours. The system should be flexible enough to permit its operation at night at a reduced level.

The long tunnel requires two daytime lighting levels, one for the intensive zone (the entrance zone comprising the threshold and transition sections) and another for the normal day zone (interior zone).

The Entrance Zone

The entrance zone is the most critical area, because without sufficient portal brightness, the entrance will appear to the approaching driver as a black hole. The most severe visual task is not when the driver is passing through the plane of the portal shadow, but when he is outside of it and is trying to see within the portal shadow. The point in front of the portal at which must be discerned objects within the tunnel is dictated by the safe stopping sight distance at the posted speed. At 60 mph, this would be 650 feet in front of the portal and, in most cases, will be at a point ahead of the adaptation point. This means that the motorist's eye is at that time adapted to the ambient luminance

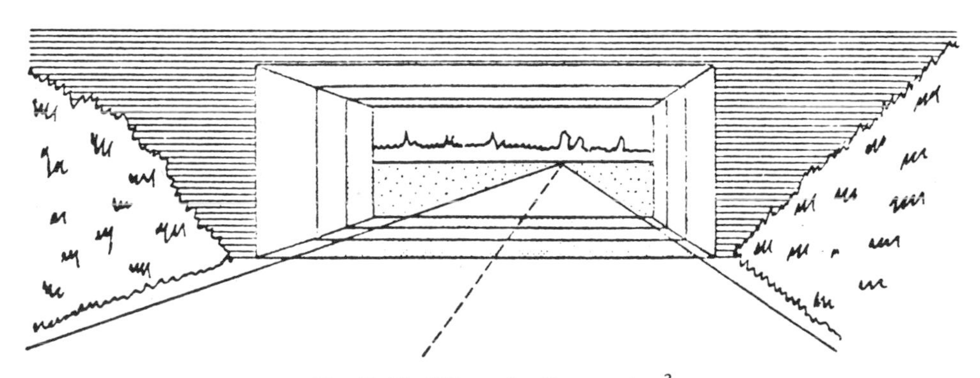

Fig. 20-16. Effect of ceiling opening.[3]

level of the open road. According to the recommendations of IES, CIE and Schreuder,[5] the portal luminance level should be $\frac{1}{10}$, and preferably $\frac{1}{8}$, of the maximum expected open approach luminance. This portal luminance should be displayed on the walls, ceiling and roadway so that these surfaces can serve as an effective background for discernment of objects by contrast and silhouette. The effects of the so-called black hole would then be very drastically reduced or even eliminated. The first reduction in luminance should occur in the threshold zone and should be at a constant level. The length of the threshold zone is dependent on the driving speed and the position of the adaptation point which, in turn, is dependent on the construction and dimensions of the access zone and the tunnel portal facade.

Stopping Sight Distance

It was pointed out earlier that at a distance of 325 feet, a critical object with a contrast of 0.20 against its background (I_2) can be seen in 0.1 second if the object's background luminance is at least $\frac{1}{10}$ of the approach zone luminance. The distance at which the critical object can be seen must correlate with the driving speed and the safe stopping sight distances. If the test distance of 325 feet is to be the safe stopping sight distance on a wet road, the corresponding vehicle speed would be about 42 mph. In a similar manner, the safe stopping sight distance should be used as the distance to the critical object for other speeds in determining the length of the threshold zone. From the geometric relation shown in Figure 20-17 the adaptation point can be calculated from:

$$d_a = (H - h) \cot l$$

where

H = Portal height in feet.
h = Height of driver's eye above pavement (usually taken as 3.75 to 4 ft.).
l = Vertical angle in degrees (Schreuder[5] suggests 7°; Ketvirtis[6] suggests 25°).

Knowing the location ahead of the portal of the adaptation point, Ketvirtis[6] suggests that the

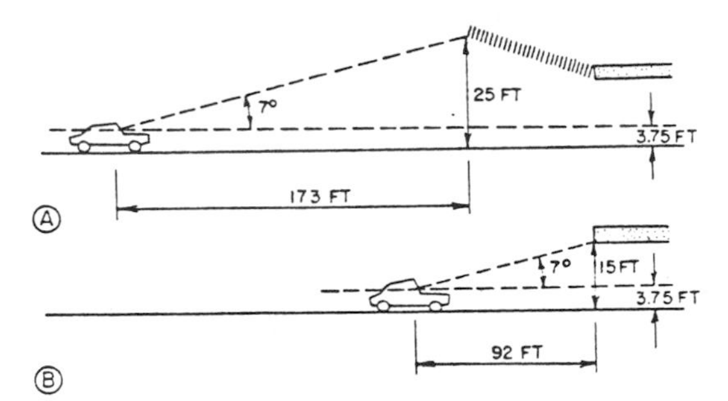

Fig. 20-17. Location of adaptation point.[5]

total length of the supplementary cone (threshold plus transition) can be computed from:

$$L_s = (V_s \times t_a) - d_a$$

where

V_s = Speed (ft/sec).
t_a = Adaptation time (sec., assumed to be reasonable at 8 seconds).
d_a = Adaptation point (ft).

The length of the threshold zone can be calculated from:

$$L_t = \frac{a}{h - a} \times d_s$$

where

a = Hazardous object height (1 ft.).
h = Height of driver's eye position (4 ft.).
d_s = Safe stopping sight distance.

Table 20-2 shows safe stopping sight distances for various design speeds.

Table 20-2. Minimum stopping sight distance.

DESIGN SPEED (MPH)	STOPPING SIGHT DISTANCE	
	WET PAVEMENT*	DRY PAVEMENT**
30	200	158
40	300	236
50	450	327
60	650	434
65	750	489
70	850	554
75	950	622
80	1050	696

*Rounded for design.
**Computed.

Adaptation Point

Figure 20-17 illustrates the method by which the location of the adaptation point can be determined. The vertical angle shown is a value taken as an average for the shielding angle of vehicle windshields. The two profiles demonstrate that, as the portal height increases, the distance of the adaptation point in front of the portal also increases. The raising of the facade, consequently, is a very effective expedient to start the adaptation process sooner and as a result to reduce the length of the threshold zone. Further aids are dark finish of the facade and other neighboring surfaces.

When the motorist has passed the adaptation point, his eyes begin to adapt to the dark area of the tunnel entrance. If the luminance within the transition zone of the tunnel decreases in accordance to the curve of Fig. 20-5, inconvenience caused by after-images is prevented. These luminances must always exist at points equal in distance to the safe stopping sight distance in front of the observer. For this reason, the origin of the curve in Fig. 20-5 is placed at the end of the threshold, and together they form the recommended luminance level at the tunnel entrance. This level is illustrated in Fig. 20-18, showing the recommended luminance profile for design speeds of 40, 50 and 60 mph.

Entrance zone lengths of 1210, 1350 and 1540 feet for design speeds of 40, 50 and 60 mph, respectively, are shown in Fig. 20-18 when the level of luminance in the interior is 5 fL. If the interior zone luminance is more or less than 5 fL, the lengths would be, conversely, more or less.

Haze

At the entrance of tunnels, the presence of haze due to the exhaust emissions of vehicles may create a visual problem, particularly when there is no open-grid construction for screening of daylight through which the fumes can escape. Whenever this haze is illuminated by the sun, the luminance can become very high, and it may prove difficult to see through it because of the veiling effect.

Reference is made to p. 568, where L_1 was defined as the outside luminance, L_2 as the luminance of the immediate background of the critical object and L_3 as the luminance of the object itself. The contrast of the object against its background was defined as:

$$C = \frac{L_2 - L_3}{L_2}.$$

A veiling luminance L_v between the object

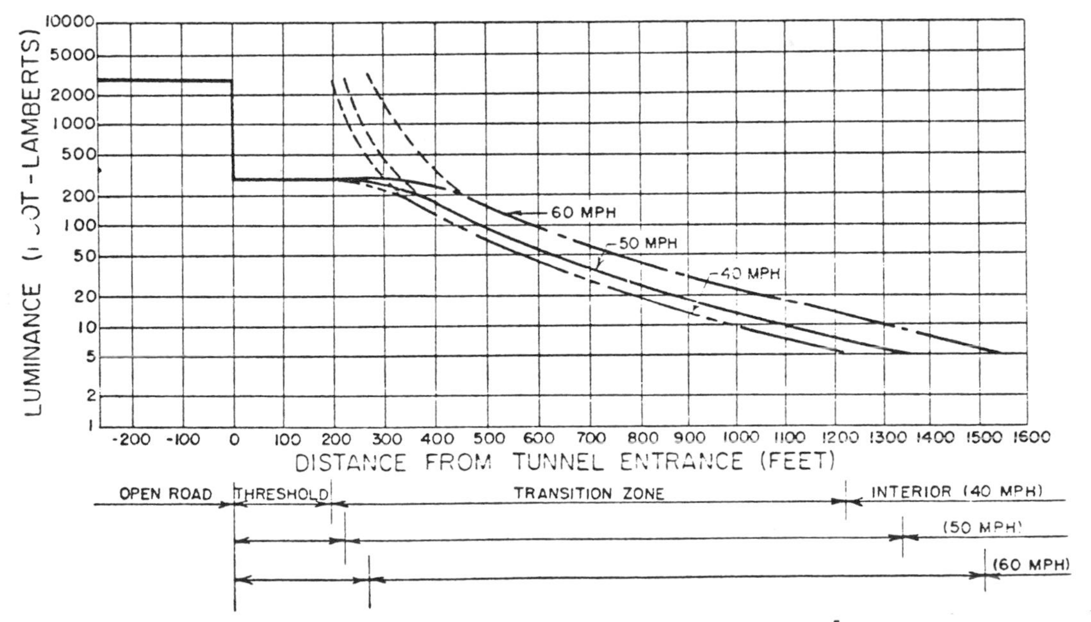

Fig. 20-18. Recommended luminance profile at tunnel entrance.[5]

and the observer increases the luminance of the object to $L_3' = L_3 + L_v$ and that of the tunnel to $L_2' = L_2 + L_v$. The new apparent contrast, C', between the object and the tunnel background becomes:

$$C' = \frac{L_2' - L_3'}{L_2'} = \frac{(L_2 + L_v) - (L_3 + L_v)}{L_2 + L_v}$$
$$= \frac{L_2 - L_3}{L_2 + L_v}$$

or

$$C' = C\frac{L_2}{L_2 + L_v} \text{ and } \frac{C'}{C} = \frac{L_2}{L_2 + L_v}.$$

The presence of the interposing veiling brightness acts on the eye the same way as a source of stray light which, within the eye, produces a superimposed veiling luminance upon the retinal image of the object to be seen. This alters the luminance of the image and its background; hence, the contrast. The ratio of the apparent contrast to the original threshold contrast, shown in the equation above, will always be less than 1 because of the presence of L_v in the denominator. This means that the object, which was previously seen at the threshold level, will not be seen under conditions of a lower contrast with the same background luminance, L_2. To bring the object back to threshold visibility, the background luminance, L_2, would have to be increased. Figure 20-19 shows a curve of the visual performance criterion function indicating the variation of the background luminance with the task contrast. Illustrated on it is an example of the additional background luminance required for a decrease in task contrast to bring the object to threshold visibility.

Another manifestation of a veiling brightness is that produced by the reflection of the top deck of a car's dashboard through the inclined windshield on a bright, sunlit day. This also

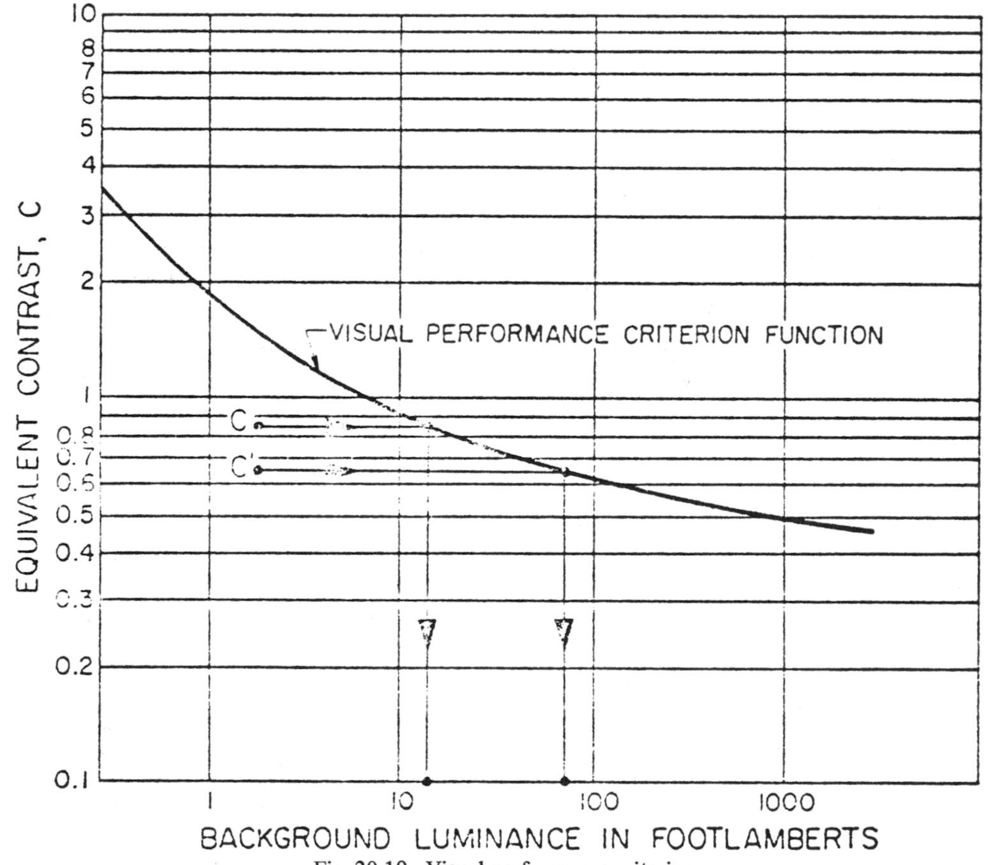

Fig. 20-19. Visual performance criterion.

appears to the motorist as a bright screen between the object and the viewer. A dirty and dusty windshield will also produce a similar effect.

Theoretically, it should be possible to "see through" all of these veiling brightnesses and sources of stray light, whether they occur individually or collectively, by providing sufficient background luminance. However, in practice, overcoming these effects by providing brightnesses using artificial means is neither practical nor economical.

NIGHT LIGHTING

The requirements for lighting at night in short and long tunnels are identical. The lighting must be designed to avoid flicker and glare, and the illumination level should be the same throughout the entire length. The ratio of the roadway lighting inside the tunnel to that of the roadway outside the tunnel should not be greater than 3:1 and preferably should not exceed 2:1. Wall and ceiling illumination in tunnels during the night should be maintained between 0.7 and 2.0 fc. Fluorescent luminaires are recommended for night lighting to provide continuous uniform, low brightness lighting.

At night, the most severe visual task occurs at the exit of the tunnel when the roadway outside the tunnel has no lighting or has a lower level of illumination than inside the tunnel, and the exit appears as a dark hole to the motorist. The common cause for this situation is excessive lighting inside the tunnel. Many tunnels have the same level of illumination for night lighting as for daytime lighting in the interior zone. Inasmuch as the normal daylighting system usually must provide substantially more brightness than is required for night lighting, a single lighting system, without provisions for varying the amount of illumination, will not satisfy both operating requirements. Lighting in excess of the optimum amount is detrimental and may create a hazardous condition at the exit.

In instances when it is not desirable to reduce the night lighting level to a ratio of, at most, 3 to 1 with the approach roadway lighting, then consideration should be given to providing transition lighting. The zone for the transition lighting should be located at each end of the tunnel on the approach roadway and should be about 500 to 600 feet long.

Approach roadways to many tunnels are of the depressed type and these sections of the highway should be provided with roadway lighting. For the comfort of the motorist, the nighttime lighting level in this type of depressed highway should be 50% greater than that required on the open road. If none is provided on the open road, then the lighting level in this section should be 50% of the level within the tunnel.

When sunscreens are applied at the entrance and/or exit of the tunnel, the nighttime lighting system must also be extended under and to the end of the sun screens. These sections of the lighting installation must also operate at dusk and at low levels of outside illumination during the daytime.

Very short tunnels which are straight and on a flat grade, and which are not provided with lighting for daytime operation, can usually be provided with satisfactory nighttime visibility by the proper positioning and mounting height of pole-mounted street lighting luminaires at each end of the tunnel. This principle of obtaining artificial light penetration into the tunnel is illustrated in Fig. 20-20. When higher than usual mounting heights are used for the luminaires adjacent to the tunnel portals, care must be exercised not to exceed the uniformity ratio; otherwise, some tunnel lighting may be necessary.

USE OF DAYLIGHT IN TUNNELS

Natural daylight can be employed in lighting the interior of tunnels in three ways:

1. The use of subdued daylight at the entrance portal of long tunnels by the use of daylight screening.
2. The application of light slits in the roof of short tunnels.
3. The use of the light that falls into the tunnel through the exit.

Sun Screens

Daylight screening as a means for reducing ambient access luminance has been installed

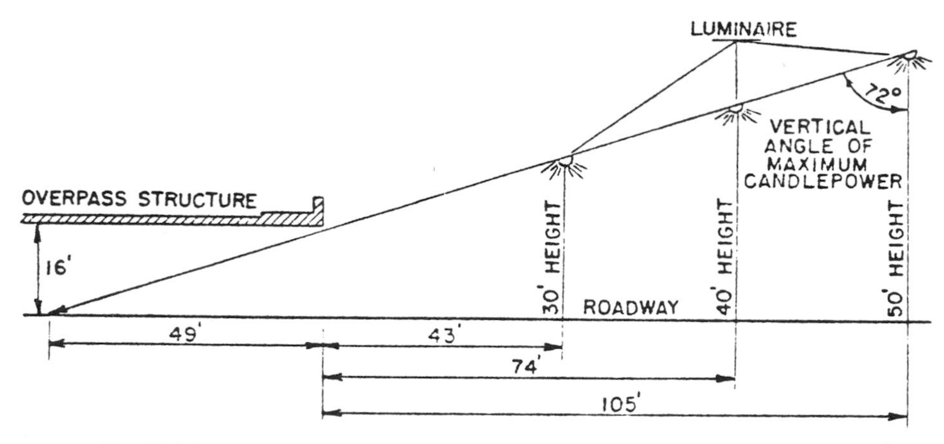

Fig. 20-20. Location of roadway lighting luminaires adjacent to underpass portals.[4]

in many European tunnels and, to a limited extent, in this country (Chesapeake Bay Bridge-Tunnel and Harbor Tunnel, Baltimore). All of these tunnels use a form of louver spanning over the roadway and are constructed with vertical or inclined slats and in various geometric shapes and patterns.

Since louvers must be designed with an upper cutoff angle sufficient to prevent direct sunlight penetration to the roadway, they do not perform with a constant transmission factor when the ambient access illumination level varies from sunrise to sunset and from summer to winter as the sun's altitude gets progressively lower in the sky. The fact that a sun screen has a variable transmission factor was reported by Narisada (1971) in a paper presented at the fiftieth meeting of the Highway Research Board. The paper, titled "Latest Research in Tunnel Lighting in Japan," reported that the transmission factor is not always constant and that the illumination under the louvers tends to saturate at about 280 to 465 fc when the outdoor illumination exceeds approximately 3700 fc. As a result, there will be times, under certain atmospheric conditions, when artificial lighting will be required under the louvers to supplement daylight so that the portal shadow, in effect, will not be transferred to the beginning of the louver.

The important characteristic for a louver to operate successfully is that the transmission factor must remain constant under all conditions of outdoor illumination levels. It appears, however, that additional research is required in this field before a satisfactory solution is reached. For these reasons, sun screens are not recommended as a viable solution to the tunnel entrance visibility problem.

Daylight Slits

These can sometimes be used in tunnels of medium length. An estimate can be made of the effect of such slits by the following formula if we assume that the slit is very long and that it can be considered as a diffuse light source:

$$E = \frac{\pi B}{2} (\sin\gamma_2 - \sin\gamma_1)$$

where E is the illumination at a point P (Fig. 20-21), B the luminance of the slit and γ_1 and

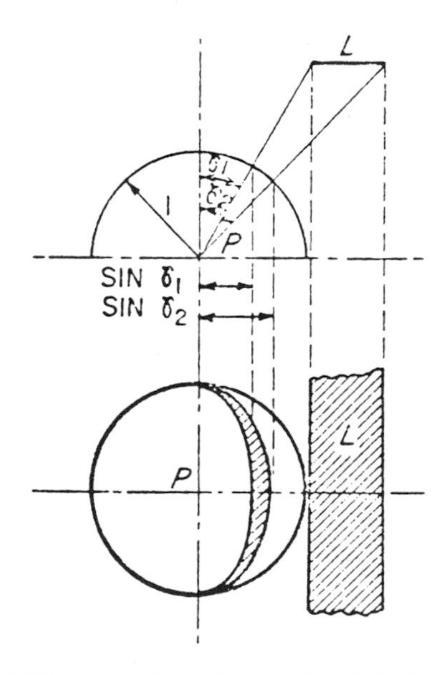

Fig. 20-21. Determining transmission of daylight slits.[5]

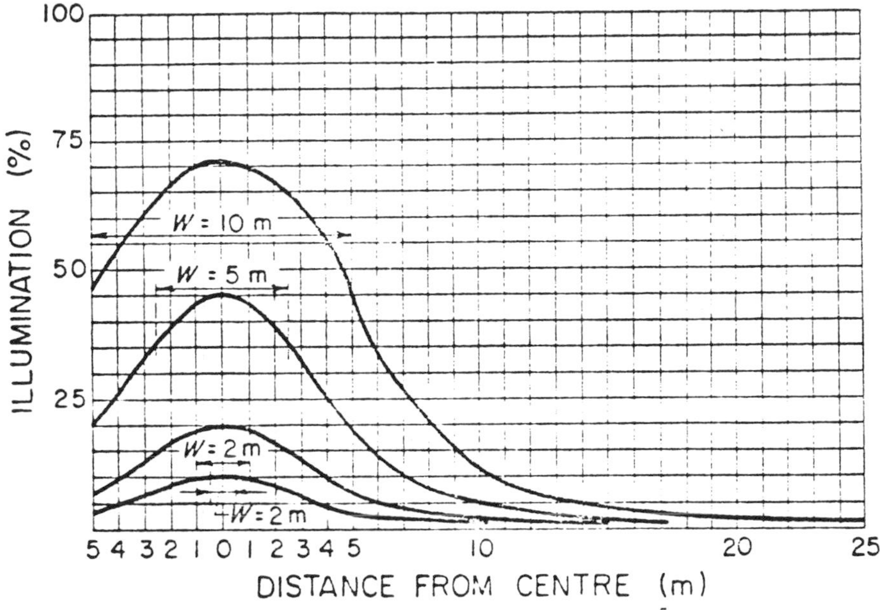

Fig. 20-22. Transmission of daylight slits.[5]

γ_2 are the angles at which the edges of the slit are seen. Figure 20-22 illustrates some of the results obtained from the use of this formula.

Daylight Entering a Tunnel Portal. Figure 20-23 illustrates the amount of horizontal (E_H) and vertical (E_V) illumination penetration of daylight in terms of the percentage of outside illumination. Figure 20-21 is plotted from measurements and from calculated data. It shows that the contribution to tunnel illumination by the penetrating daylight can be neglected after about 20 feet from the portal.

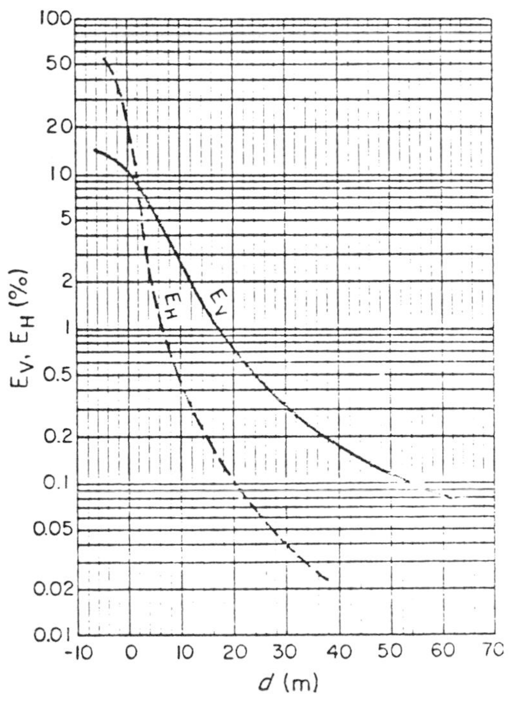

Fig. 20-23. Penetration of daylight.[5]

TUNNEL LINING

Roadway brightness has less effect on the road user's visibility than wall and ceiling brightness, except in wide tunnels. Generally, roadway brightness can be assigned a secondary consideration in the lighting design because, with sufficient wall and ceiling brightness, adequate roadway brightnesses and light reflections from obstructions will automatically follow.

The interior walls and ceiling of the tunnel are important adjuncts of the lighting system, and their brightnesses and uniformity depend on the reflectance quality of the surface. While the ceiling may not play an important part as a background against which objects may be silhouetted, except in tunnels with an upgrade approach, its light color and high reflectance is desirable because of the higher wall and roadway brightnesses that will result. The surfaces should be of a material that will not deteriorate with age and chemical attack, will not readily soil and can be easily cleaned. A light-colored matte (non-specular) finish surface with a reflectance factor of 70% or greater is recommended.

The ceiling should also consist of a durable and reflective surface. Epoxy paint coatings have been applied with mixed success.

The tunnel roadway surface should have as high a reflectance factor as possible and not less than 20%. To enhance reflective qualities, natural baked flint is used in England in conjunction with black asphalt. Some European tunnels have used equivalent artificial materials. Concrete, with the same additives, would perhaps produce almost ideal tunnel pavement material.

TUNNEL LIGHTING LUMINAIRES

The interiors of existing long tunnels are lighted by four basic types of luminaires. The first type, in some of the older tunnels, consisted of incandescent units recessed into the tunnel wall near the ceiling. This provided intermittent flush windows which could be easily cleaned. However, it was also inflexible, was not very effective and produced annoying flicker.

Fluorescent Lamp Luminaires

Fluorescent lamp luminaires for application to tunnel lighting have been developed consisting of two tandem-connected lamps mounted in a protective pyrex glass tube and sealed at the ends to connection devices, providing watertight integrity. They are inserted into wall- or ceiling-mounted, spring-loaded sockets. The lamps are usually located in close proximity to the corner intersection of the wall and ceiling. The in-service record of this type of luminaire is excellent and is still being used in newly constructed tunnels. However, with the trend toward higher interior entrance zone brightnesses, the use of this type of fixture becomes very costly when its installation cost is compared to a multi-lamp reflector type luminaire for equivalent interior brightnesses.

Multi-lamp-enclosed Fluorescent Lamp Luminaire

This lamp luminaire with internal reflector is placed either on the walls or in the angle between the ceiling and the walls and is one way of adding new lighting to an old tunnel. This type of luminaire has been installed in some recently completed tunnels. It offers considerably more flexibility over the pyrex tube fixture but must be watertight, bug-tight and dust-tight for efficient lighting and maintenance. This is illustrated in Fig. 20-24(a). The particular advantage of this fixture is the control of the photometric distribution by the design of the reflector in conjunction with the enclosing diffuser and the higher lumen output per unit length.

Multi-lamp Fluorescent Luminaire With Reflector-deflector

This luminaire, developed by the California Division of Highways, is similar to the internal reflector type except that it has deflector louvers to direct the light to the walls and ceiling, and delivers only indirect light to the roadway. This is illustrated in Fig. 20-24(b).

Supplementary Mercury Vapor Lamps

The George Washington Bridge Lower Level Tunnel in New York and the Thorold Tunnel in Ontario employ a new concept in tunnel lighting which merits attention. Both tunnels use a combination of fluorescent and mercury vapor units in the intensified entrance zone for daytime operation. The George Washington Bridge Tunnel uses continuous fluorescent lighting for night lighting and for the normal day zone. Supplemental daytime entrance zone lighting is accomplished by the use of mercury vapor lamp luminaires. The fluorescent are 6-foot, 350-milliampere slimline lamps wired so that two lamps are connected in series with a choke across a 1000-volt, three-phase supply. The supply voltage can be reduced to 800 volts or raised to 1200 volts to provide variable intensity lighting. In addition, entrance zone lighting, consisting of 400-watt clear mercury liminaires, spaced 5 feet on centers for the first 150 feet, and 10 feet on centers for the next 100 feet, is provided for the eastbound tunnel. These luminaires are recessed into the tunnel ceiling and are close to, and directed toward, the walls to produce 100 fc on the walls in the first zone and 65 fc in the second zone. The

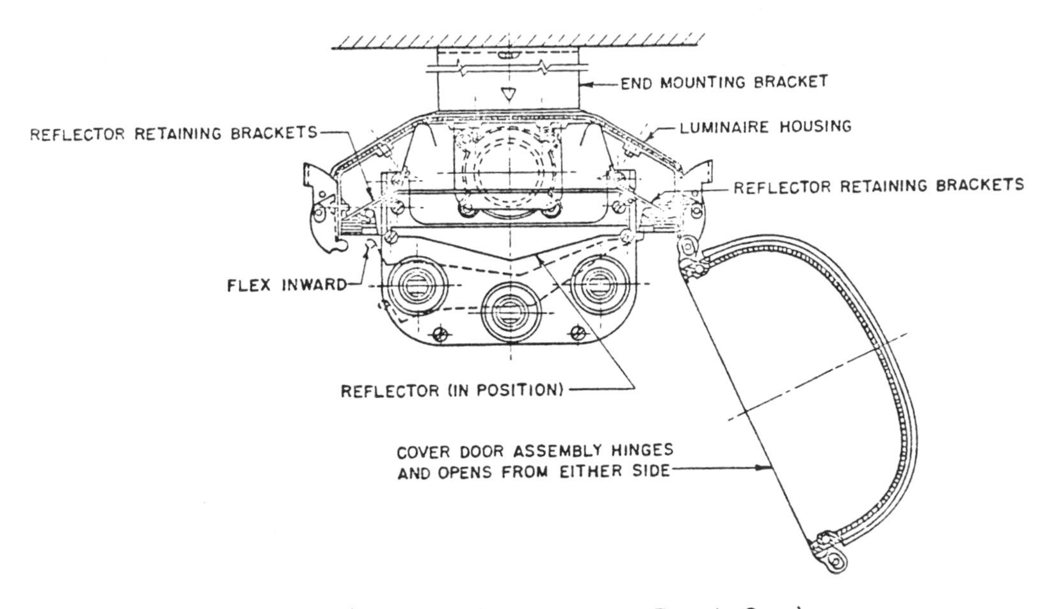

a. (Courtesy of Westinghouse Electric Corp.)

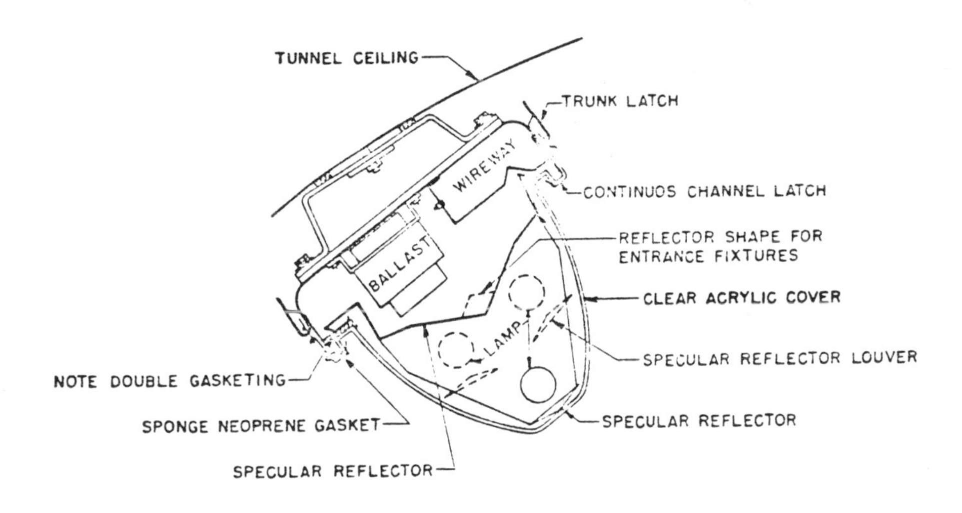

b. (Courtesy of California Division of Highways)

Fig. 20-24. Typical multi-lamp tunnel lighting fixtures.

mixing of fluorescent and mercury lamps, according to Thompson and Fansler,[4] with the yellow tile tunnel lining, produced a pleasing and effective lighting in the daytime.

In the Thorold Tunnel, the main tunnel lighting fixture consists of two T12 72-inch fluorescent lamps mounted in series and enclosed in a toughened glass tube supported at each end by a fixture holder containing the necessary lamp ballast. For the intensified zone, a unit consisting of a cast base, a prismatic glass cover and a 250-watt clear mercury vapor lamp is used. There are two continuous rows of fluorescent units in each of the two traffic tubes, one row on each side of the road. The fixtures are mounted on haunches at the angles between the wall and the roof. The supplementary mercury vapor lamp units in the intensified zone are mounted on the ceiling in such a manner that the maximum candle power is directed onto the side walls. They are located over the curbs so as not to infringe on the roadway clearance height. For the first 200 feet, these units are located on $2\frac{1}{2}$-foot centers and, for the next 150 feet, they are located on 5-foot centers. Ballasts for the mercury units are mounted in the fresh air passage in the center between tunnels. The fluorescent units are operated at 600 milliamperes for daytime conditions and at 300 milliamperes at night. A photocell sensor located on the roof of the service building, through appropriate relays and control contactors, switches the intensified lighting and dims the main lighting.

The Luminous Gallery

A new and rather unique approach to tunnel lighting design is the luminous gallery method. Good quality intensified illumination at the tunnel entrance can be achieved by using high-intensity light sources (mercury vapor, high-pressure sodium or metal halide) if the light beam intensities are dispersed and diffused before permitting them to enter the tunnel space. This can be accomplished by mounting luminaires behind translucent screens (Fig. 20-25). The screen material can be of the diffusing or refractory type. The geometry of this arrangement may consist of a 6-foot by 3-foot

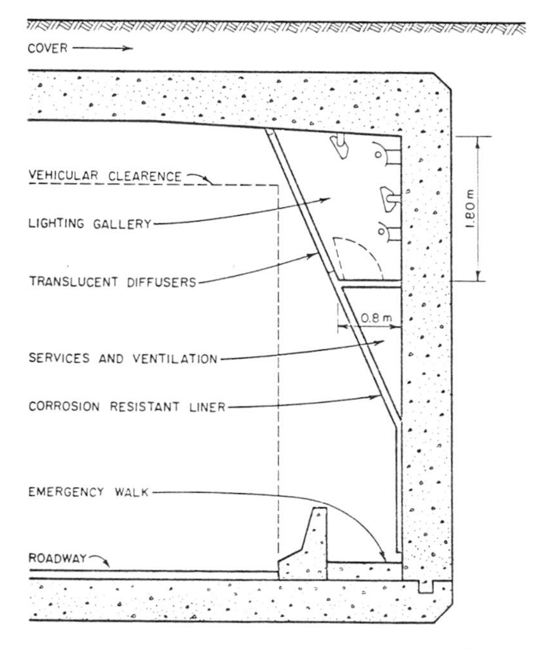

Fig. 20-25. Gallery lighting arrangement.[6]

walk-in gallery fitted in the upper corners of the tunnel. Such a scheme is obviously more applicable to a rectangular tunnel section than to a circular section, where the space above the walkway is restricted. In a tunnel with a rectangular section, a recessed gallery may be preferred, but such a design requires a wider tunnel span and consequently increases the cost of the structure. The advantages of gallery illumination are such that, in addition to increased visual comfort and motorists' guidance, the overall field optics can be successfully coordinated. By applying the method of diffusion or refraction, the light is distributed over a much wider area of gallery facing, reducing source brightness. If the tunnel walls and pavement are treated according to visibility requirements, brightness coordination between major planes forming the visual field can be achieved, and the contrasts can be kept within a reasonable range. Such conditions lead to a reduction in discomfort glare, and increased lighting effectiveness.

Maintenance of the tunnel walls is simplified, and mechanical equipment, such as rotating brushes, may be successfully employed. Lighting equipment can be inspected and lamps changed without the use of vehicles which would block the outside lanes. All work can be performed using electrician's tools only.

While no actual installation of the luminous gallery method exists, full-scale models were constructed in connection with the William R. Allen Expressway development in Toronto, Canada. They were used for the purpose of testing the feasibility of gallery lighting as well as to take various readings and to make observations. Two rows of continuous fluorescent lamps operated at 600 milliamperes and 300 milliamperes and two rows of 400-watt mercury vapor lamp units were spaced on 18-inch centers. Using this type of equipment, it was learned that levels of illumination between 5 and 800 fc were feasible. The glare-control uniformity of the road surface, as well as uniform gallery-face brightness, was achieved without any major difficulties. To change the levels of illumination, switching of the mercury vapor lamps and varying the output of the fluorescent lamps were used. The control of glare was accomplished by diffusing glass or plastic materials. It was also learned that gallery lighting completely eliminates the effects of flicker.

MAINTENANCE

Reliability of the tunnel lighting system is extremely vital and its continuous operation without interruption must be assured. The lighting system depends on many factors that should be recognized and considered at the design stage. Maintenance of the lighting is a most important factor, and the proposed maintenance program must be reflected in the initial design. Figure 20-26 illustrates the effects of a preventive maintenance program.

The amount of maintenance required depends on the location, type and volume of the traffic; the type and capacity of the ventilation system; the tunnel cross-section and shape; the ceiling and wall finish; operating speeds in the tunnel; the type and location of luminaires; grades and alignment within the tunnel; and the electrical system and supply.

A tunnel must have sufficiently high initial illumination to compensate for the many factors that will reduce it while it is in service. These factors fall into two categories: light loss factors not to be recovered and light loss factors to be recovered. In the first category are luminaire ambient temperature, voltage to luminaire, ballast factor and luminaire surface depreciation. In the latter category are room surface dirt depreciation, burn-outs, lamp lumen depreciation and luminaire dirt depreciation. A more detailed evaluation and analysis of these factors and their contribution to determining the overall maintenance factor for the purposes of calculating the in-service illumination during the initial design phase can be obtained in a paper prepared by the Design Practice Committee of the IES, "General Procedure for Calculating Maintained Illumination."

The most important factor, with the exception of the luminaires, is the surface treatment of the walls and ceiling. Reducing the reflection factor of these surfaces may be more detrimental to lighting effectiveness than any other single factor. Vehicular tunnels must be finished with an interior surface that will not deteriorate as time progresses and as chemicals attack it, that will not readily soil and that can be easily cleaned. These attributes are characteristic of a light-colored matte-finished tile or porcelain-enameled panel with an initial reflectance of 70% or higher.

Luminaires should be sealed to prevent the

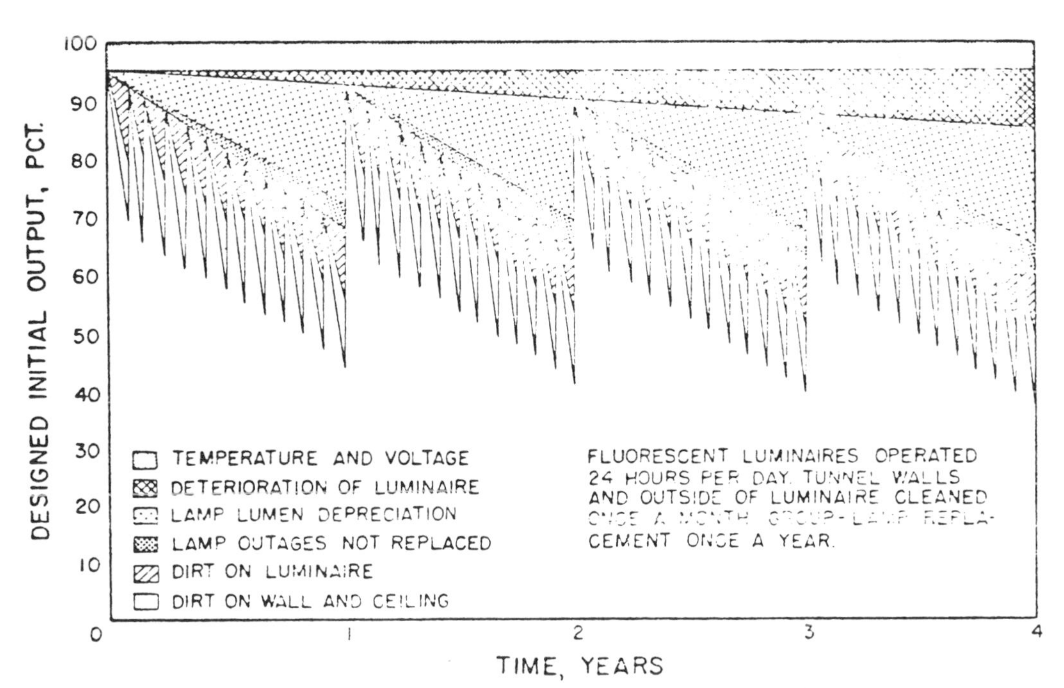

Fig. 20-26. Effect of preventive maintenance program.[4]

entry of dust and water when the tunnel is being cleaned by high-pressure water spray. They should be designed to permit quick and easy internal cleaning. The materials used should be resistant to alkaline deposits, to concentrated exhaust fumes and especially to harsh cleaning solutions that must be used to thoroughly clean the tunnel walls and ceiling.

Lamps should be replaced on a group replacement program, which not only will help to maintain the light output at the desired design level, but will also provide for a balanced maintenance work load. The magnitude of the relamping task will generally dictate that group relamping must be done by sections. To replace lamps on a burn-out basis may be acceptable in certain cases, but is usually false economy from the standpoint of lighting output and equipment maintenance costs. Fluorescent ballasts and lamp holders can be damaged by very old lamps that develop rectification and sputtering. Even though some lamps may operate for years, eventually their light output will be reduced to a point beyond where it is uneconomical to keep them in service.

The inside of the luminaire will require cleaning more often than the frequency of lamp replacement, but the maintenance schedules should still prescribe cleaning each time a lamp is replaced. Some tunnels require a program of cleaning on a weekly basis, whereas others may not require cleaning for several weeks. Locations in the more northern latitudes where freezing conditions may prevail for long periods make cleaning operations difficult and frequently impractical. All of these factors must be recognized in the initial design of the lighting system.

EMERGENCY LIGHTING

Complete interruption of the tunnel lighting even for an instant cannot be tolerated, which requires an uninterruptable power supply (see Chapter 21).

For dual utility power sources, one-half of the tunnel lighting is connected to each supply, so that in case of failure at least one-half of the system remains energized until transfer of the entire load to the remaining source.

For single utility service and standby diesel-generator, one-sixth of the tunnel lighting is connected to an emergency circuit, which in case of power failure is immediately transferred to the emergency battery system until the diesel-generator picks up to carry one-half of the tunnel lighting.

DESIGN COMPUTATIONS

Recent interest in better tunnel lighting has generated the need for taking a closer look at the accuracy of predicting illumination levels. For nearly 50 years, lumen or flux methods for calculation of the illumination level using coefficients of utilization factors have been employed. These original methods used coefficients of utilization which were determined by empirical methods for lighting equipment available at the time. Later developments used mathematical analysis as a method for computing coefficients of utilization data. All of these methods were based on the theory that average illumination is equal to lumens divided by the work area over which they are distributed. The newer mathematical methods of analysis, taking into account the concept of inter-reflection of light, have led to progressively more accurate coefficients of utilization data and have been culminated in the new IES-approved method for calculating illumination, the Zonal Cavity Method. This method improves older systems by providing increased flexibility in lighting calculations as well as greater accuracy, but does not change the basic concept that footcandles are equal to flux over an area.

For the details in the procedures to be followed in the use of the Zonal Cavity Method of illumination computation, refer to the *IES Handbook*.[1] The application of the Zonal Cavity Method to the solution of tunnel lighting problems, however, is of particular interest to the tunnel lighting designer. In this connection, therefore, the techniques involved, in applying this method to tunnels, were developed and prepared by Faucett.[7]

To complement the newly developed accuracy in predicting the initial average illumination levels in tunnels, the designer should also avail himself of the new procedure for evaluating the maintenance factor for the determination of an equally important accurate prediction of the overall average in-service illumination level. A comprehensive procedure for developing a meaningful maintenance factor will be found in the *IES Handbook,*[1] latest edition.

REFERENCES

1. *Illuminating Engineering Society (IES) Lighting Handbook.*
2. American Association of State Highway and Transportation Officials, *Guide for Roadway Lighting.*
3. Commission Internationale de l'Eclairage (CIE), *1973 on International Recommendations for Tunnel Lighting.*
4. Thompson, J.A. and Fansler, B.L., "Criteria for Highway Tunnel Lighting Design," *Public Roads* 35, *No. 4*, October 1968.
5. Schreuder, A. *The Lighting of Vehicular Tunnels*, Phillips Technical Library, Eindhoven, The Netherlands, 1964.
6. Ketvirtis, A., Foundation of Canada Engineering Corps, 1971.
7. Faucett, Robert E., Senior Application Engineer, General Electric Co., "The Zonal-cavity System Applied to Tunnels," Paper presented at the National Technical Conference of IES, August 1969.

21
Power Supply, Distribution and Control

MICHAEL A. MARSZALOWICZ

Professional Associate
Parsons, Brinckerhoff, Quade & Douglas

PRIMARY POWER FOR HIGHWAY TUNNELS

General Requirements

Electric power service to a highway tunnel and its distribution system have to provide maximum reliability for continuous operation. While interruption of power to the ventilation equipment can be tolerated for a short time, any (even momentary) failure of tunnel lighting, especially in subaqueous tunnels, is not acceptable, since a tunnel's sudden plunging into darkness could lead to panic and cause accidents.

Reliability of primary power may be provided by two feeders originating in separate parts of a utility distribution system. If such independent sources are not available or are not economically justified, a single utility feeder with a standby emergency engine-generator supply may be used. This would have to be supplemented by an uninterruptable emergency supply for a minimum of tunnel lighting until the emergency power takes over.

The choice between these two systems should be based upon an evaluation of overall economy and reliability.

While a dual service system offers certain advantages over the single service system, its application to a vehicular tunnel should not be made simply on the basis of the availability of the two services but rather on the overall reliability of the prospective services. In evaluating a dual service system, consideration should be given to the geographic location, the proximity of the services, the method of distribution, the history of scheduled and unscheduled outages (especially with due regard to their frequency and duration) and the distance from the prime generating source.

Geographic location pertains to the assessment of natural disturbances to which the utility distribution system might be subject. Such disturbances would consist of lightning, floods, earthquakes and severe winds.

Utility charges for initial installation of a dual supply and yearly demand charges for the service should be compared with initial investment for emergency generating plants and their operation and maintenance costs.

Single Utility Service

An example of the single utility service distribution system with a standby electric generator

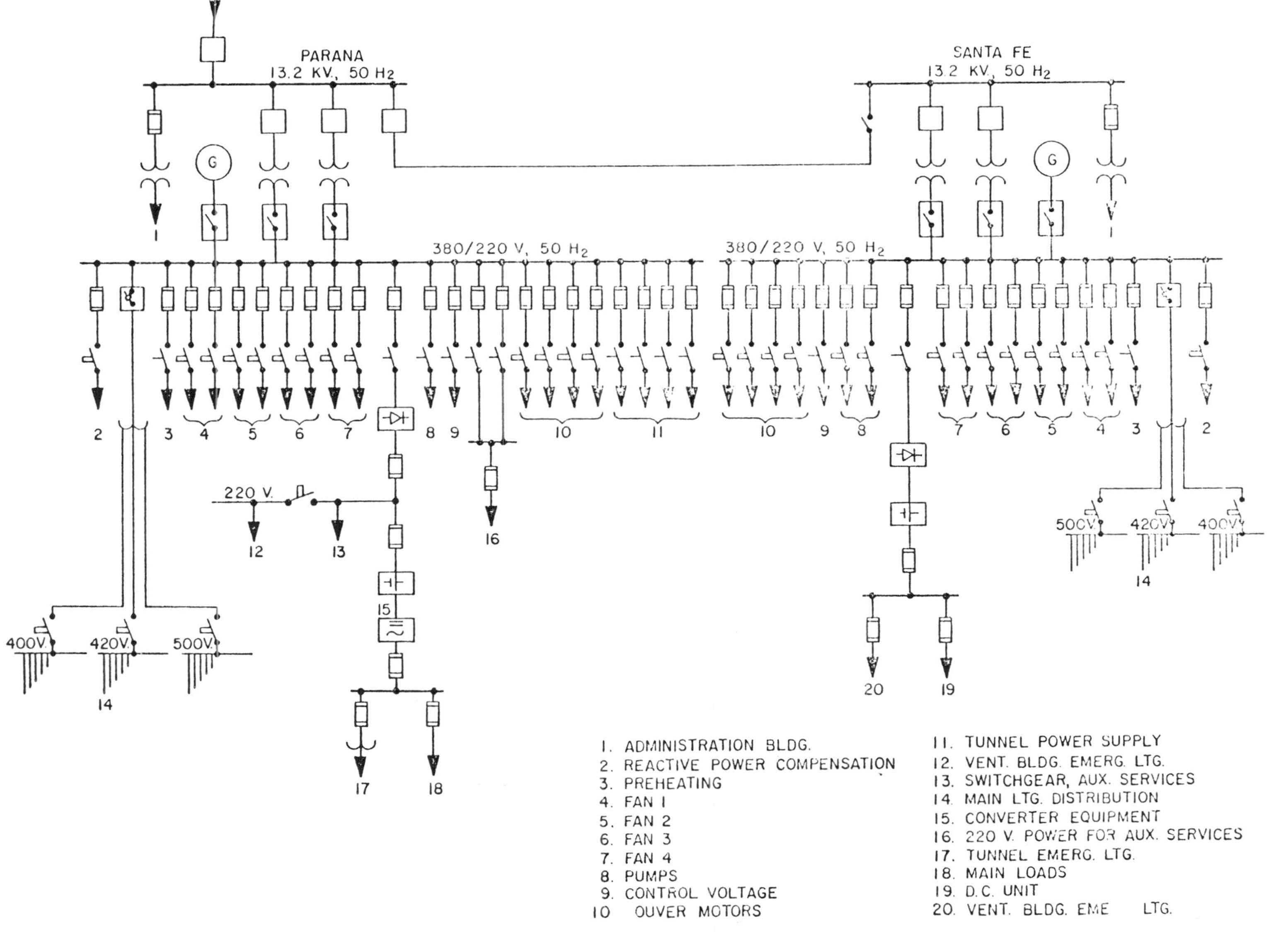

Fig. 21-1. General diagram of electrical equipment, single utility service.

and uninterruptable power supply back-up is shown in Fig. 21-1. This distribution system is employed in the Parana Tunnel in Argentina. Service of Parana through a 13.2-kilovolt, 50-hertz radial line. Energy is supplied to the Santa Fe side by means of a high-voltage tie cable installed in the fresh air duct below the roadway.

This layout permitted the splitting of the power more or less equally, each substation drawing approximately one-half of the total consumed power (which, in this instance amounts to approximately 2.7 megavolt-amperes). During normal operating conditions, a coincidence factor of 0.5 to 0.6 can be assumed.

By means of two transformers of 1250 kilovolt-amperes each, the high voltage is stepped down to the service voltage of 380/220 volts. Both administration buildings at the beginning of the approach roads contain one additional 100-kilovolt-ampere transformer, each of which is connected by cable to the high-voltage system. These transformers provide the local supply, such as illumination of streets, buildings and parking lots.

In order to keep the traffic moving inside the tunnel even during a system breakdown, emergency sets consisting of a diesel engine-driven three-phase synchronous generator with a capacity of 600 kilovolt-amperes at 0.9 PF are provided at the substations on either side of the tunnel. During system undervoltage conditions or failure, undervoltage relays connect these emergency generating sets to the 380-volt buses of the respective low-voltage distribution, thus continuing the power supply for the tunnel lighting which has been partly reduced by phase control, for 50% of the ventilation, for the traffic signal circuits, as well as for the communication equipment.

To bridge the gap between the system failure and the connection of the emergency generating units to their respective service buses, an additional 25-kilovolt-ampere converter unit for instantaneous use was installed. It consists of a 220-volt battery with a capacity of 145 ampere-hours at a 2-hour discharge rate, a DC motor and single-phase AC generator, and ensures the continuous supply of the traffic control and signaling equipment, as well as the loudspeaker and TV monitoring units, and provides for the emergency lighting to be switched on instantaneously, illuminating the tunnel without interruption. The batteries of the converter are continuously charged from the system by a rectifier unit. In addition, the batteries supply the control voltages for switchgear operation.

Dual Utility Service

The origin of the two feeders and their approach to the facility should be as widely separated as possible. They should preferably be on the two sides of the river for a subaqueous tunnel or on opposite ends of a mountain tunnel. In some cases, service from two different utilities may be available. As to methods of distribution, underground distribution is much more reliable, from the standpoint of a non-electrical disturbance, than is an overhead distribution.

The history of scheduled and unscheduled outages can be readily obtained from the utility company. Their frequency and duration will be greatly influential on evaluating service reliability. When outages are frequent and of long duration, then the prospective services cannot be very reliable.

Finally, the closer that the facility is to a prime generating source, the less likelihood there will be for the distribution to encounter unfavorable exposure and the more reliable will be its service.

A most important aspect to establish in considering a dual service and distribution system is the absence of concurrent outages on both services. It should be noted that, in this system, no provisions are made for an instantaneous back-up of the tunnel lighting. Consequently, a careful and thorough investigation should be made into the service reliability.

A dual utility service and distribution system is employed in the Bridge-Tunnel Facility of the Hampton Roads Crossing between the cities of Norfolk and Hampton, Virginia (Fig. 21-2). At the Norfolk shore, a 34.5-kilovolt, three-phase, 60-hertz overhead utility line terminates in a substation which transforms the subtransmission voltage to a service voltage of 13.2 kilovolts. At the Hampton shore, a subtrans-

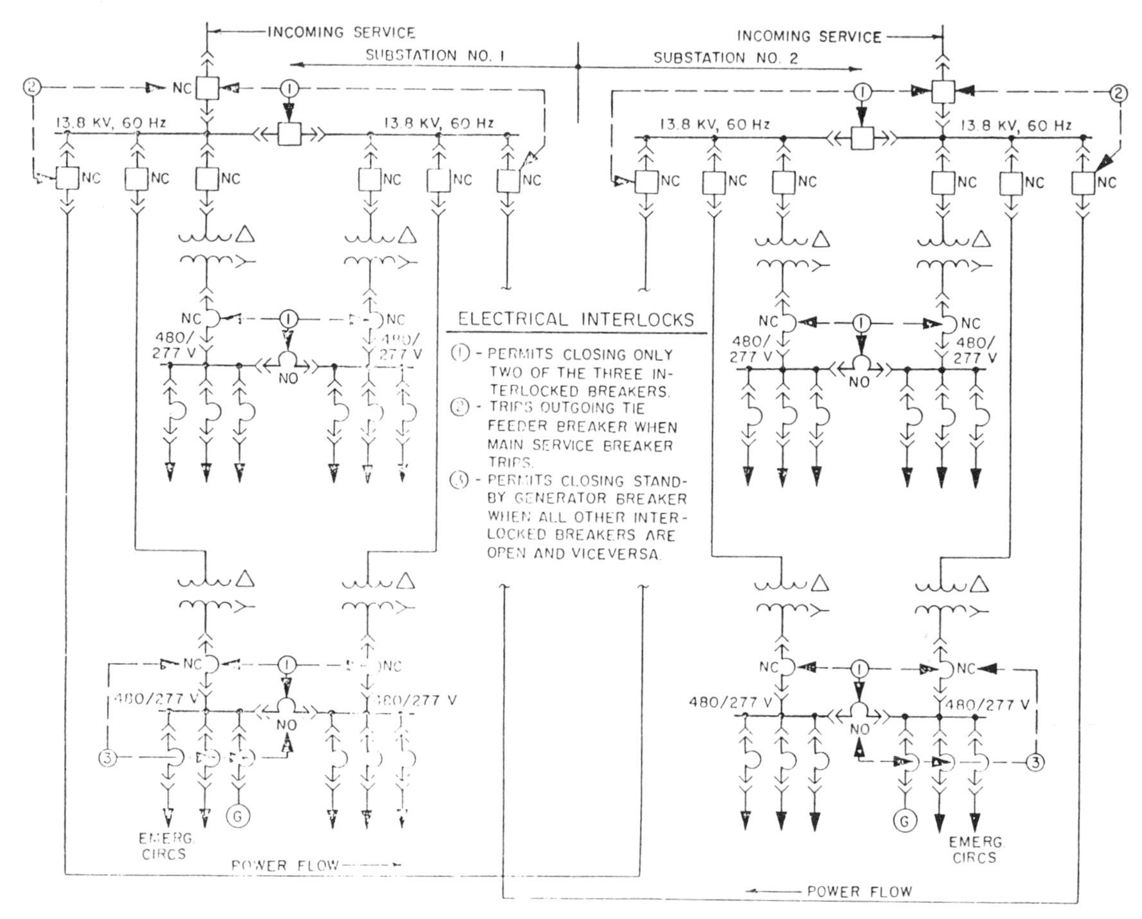

Fig. 21-2. Dual service distribution arrangements.

mission voltage of 23 kilovolts, three phases, and 60 hertz is similarly transformed, to a service voltage of 13.2 kilovolts. Both of these service voltages are extended from the shore-based substations by feeder cables, installed underground and aerially under the bridge deck to man-made islands on each side of the shipping channel. On the islands, the incoming service feeders terminate in free-standing, metal-clad, dead-front switchgear and control centers housed in a structure which also contains the tunnel ventilation fans, drainage pumps and fire pumps.

These feeder cables connect to the supply side of 15-kilovolt main power air circuit breakers, which, in turn, connect to one section of a sectionalized primary bus. The primary bus sections are coupled together through a bus tie circuit breaker. The switchgear is double-ended and contains primary feeder breakers, transformer sections and utilization voltage sections, consisting of a circuit breaker section for radial feeders and a motor control section for vent fans and auxiliaries. The utilization voltage is 480/277 volts, three phases, and 60 hertz. The low-voltage buses are also sectionalized and appropriately inter-tied by bus tie circuit breakers. There are two double-ended substations in each ventilation building. There are also two 13.2-kilovolt tie feeders installed in the tunnel walls to interconnect a substation in each ventilation building.

Normally, the primary and secondary bus tie breakers are open and the whole system operates as two parallel radial systems entirely independent of each other, each being powered by one of the two incoming utility services. Both tie feeders are normally energized and carrying power so that, upon failure of one source, half of the electrical system in the tunnel and each ventilation building remain operational. Each utility source and all major components of the distribution system, such as power transformers, tie feeders and primary and secondary buses, have sufficient rating to carry not only their own normal loads, but also the normal loads of their counterparts. All primary circuit breakers, as well as secondary bus tie breakers, are electrically operated by local control at the breaker, as well as remote controlled from the central control room in one ventilation building. The primary and secondary bus tie circuit breakers are interlocked in a manner which prevents the two utility electric sources from being connected together at any point in the distribution system. All switching of loads from one source to the other source is done in open transition.

The primary bus tie circuit breaker in each substation permits both transformers of a double-ended substation to operate during an electric service anomaly. This arrangement makes use of full transformer inventory under emergency operating conditions. In addition, it permits the isolation of a faulted primary bus section, and the restoration of interrupted loads is quickly effected through the closing of the secondary bus tie breaker. Except for conditions created by an area outage, this arrangement provides adequate assurance against a total loss of power, and the installation of an electric generator as a back-up for the electric service for powering a limited capacity of tunnel lighting and ventilation equipment is not considered to be necessary. On this project, an emergency generator of limited capacity was installed for the purpose of providing standby electric power for the portal tide gates, tunnel drainage pumps, fuel oil transfer pumps and emergency lighting in both ventilation buildings. The generator provides power to this critical equipment during times of severe storms, when both utility services might be inoperative and sea water might be expected to reach the flood level, in which case it would be necessary to close the tide gates and operate the tunnel drainage pumps as the need arises.

EMERGENCY POWER FOR VEHICULAR TUNNELS

Limited capacity emergency power in the form of on-site standby engine-driven electric generators should be provided to supply power to critical loads during a service interruption of the normal source. In order to maintain traffic in the tunnel at reduced capacity, only the essential loads should be sustained by the standby generators. Loads which can be considered as essential are ventilation, tunnel lighting, traffic lights, closed circuit television, fire alarms, a carbon monoxide monitoring system, drainage pumps and fire pumps. Ventilation and tunnel lighting loads should be reduced to a minimum during these emergency operating conditions in order to keep the capacity of the standby plant within reasonable economic means.

Limited Power for Ventilation

Under emergency conditions, power for tunnel ventilation should be limited to supplying one ventilation fan of each type in each ventilation building operating at its lowest design speed. This would provide a ventilating capacity equivalent to that which would normally be required under light traffic densities. If the emergency occurs during peak traffic hours, then the traffic density within the tunnel can be regulated by patrolmen stationed at the entrance portal, or by the traffic control system so that the carbon monoxide can be kept within the prescribed safe concentration level. This permits the facility at least to remain operational.

The engine-generator, besides automatic starting upon failure of primary power, should be equipped for manual starting from the main control board and locally at the unit. The engine should have an automatic speed control, and should be provided with air intake, lubricating oil and fuel oil filters. Adequate cooling and combustion air must be available. The generator should have automatic voltage regulation, and in damp locations it should be equipped with electric heating coils. Instruments or indicating lights on the main control board and at the units must show the proper functioning of all components.

Emergency Power for Tunnel Lighting

For emergency power back-up for the tunnel lighting system in the short period between a power failure and the acceptance of the critical loads by a standby generator, an additional standby power source must be installed to preclude total blackout. This can be accomplished by a static or rotating battery-operated inverter of limited capacity. The instantaneous switching of one-sixth of the tunnel lights to the inverter has been shown to be satisfactory. To keep the physical size of the standby battery to a minimum, its rating should be sufficient only to carry on for the length of time required for the generator to be started and switched on-line. For an extended period of time, the standby engine-generator should have capacity to power at least half of the tunnel lights.

POWER DISTRIBUTION

Several factors affect the selection of system service voltage and system utilization voltage for a tunnel facility, and each varies in its influence, depending upon the voltage class being considered. These factors are as follows.

1. Load magnitude.
2. The distance power is to be transmitted.
3. Utilization device availability as a function of voltage.
4. Safety, codes and standards.

Incoming Service Voltage

For reasonably long tunnels, those 2000-feet long and longer, the medium-voltage levels are most commonly used. These levels consist of nominal voltages comprising 2400, 4160, 6900 and 13,800 volts. These medium-voltage levels may be used as transmission or utilization voltages, or both. Therefore, the power transfer considerations, as well as the utilization voltage considerations, are of prime concern in selecting the facility medium voltage. Since the bulk of the electric connected load consists of electric motors driving ventilation fans, considerable economies may be realized when these motors are operated at the service voltage, since substation transformer capacities can be reduced. When costs are being compared at different motor voltage levels, consideration must be given not to the motor costs alone but also to the cost of adequate starting equipment and that part of the power distribution system which is different for the two utilization voltages being considered.

When all of these factors are combined, a convenient ready reference bar chart, as shown in Fig. 21-3, can be prepared. The heavy vertical lines that extend across two voltage bars indicate the horsepower at which the costs are the same for either voltage. The minimum horsepower indicated in each voltage class is based on the recommended minimum rating that is consistent with good motor design. These horsepower ratings are consistent with the most favorable conditions as to type of motor, rated speed and application. The upper limits of horsepower ratings at a given voltage are based primarily on considerations other than motor design. The economic limit and availability of starting equipment are reached much sooner than the motor design limit and should be used as the main guide in determining the maximum horsepower rating at a given voltage rating. Although Fig. 21-3 shows, in case 3A, that the 6600-volt motor does not become more economical than the 440-volt motor until the 1000-hp rating is exceeded, such a high horsepower rating at 440 volts will likely involve starting problems and will certainly require special starting equipment.

The selection of the medium-voltage level for service and distribution is dependent basically upon the amount of power that is to be transferred. Figure 21-4 shows that, in the majority of the cases, 4160 volts will suffice below 10,000-kilovolt-ampere load capacity and 13,800 volts will suffice above a 20,000-kilovolt-ampere load capacity. For the range between 10,000 and 20,000 kilovolt-amperes, either of the two voltages may prove to be the most economical. These guidelines should not be taken, however, as hard and fast rules, and each circuit should be evaluated on its own. There may be some instances where, even though loads could be less than 10,000 kilovolt-amperes, 13.8 kilovolts could prove to be the more suitable and desirable primary voltage.

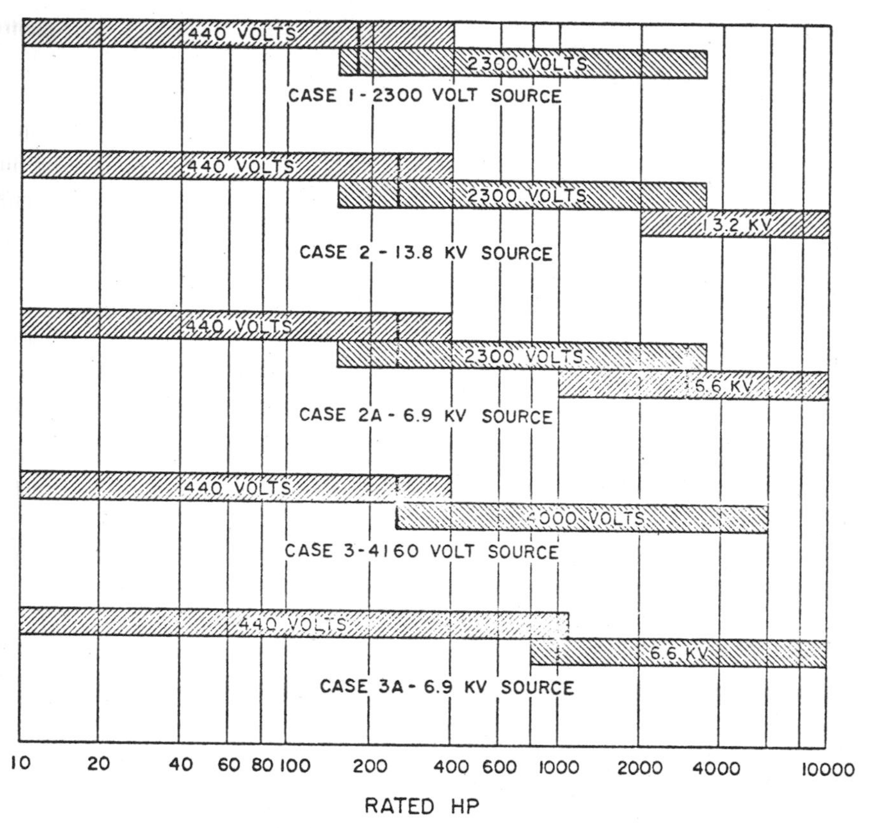

Fig. 21-3. Recommended motor ratings.

Utilization Voltage

Where the service voltage cannot be applied directly to powering large loads such as ventilation fans, then a 480-volt, three-phase, utilization voltage can be applied to most of the loads in a tunnel. This voltage is suitable for application to motor loads between $\frac{1}{2}$ and 400 hp. In addition, the dual voltage of 480/277 volts, three phases, four wires is quite suitable for tunnel lighting where fluorescent or HID lamps are employed.

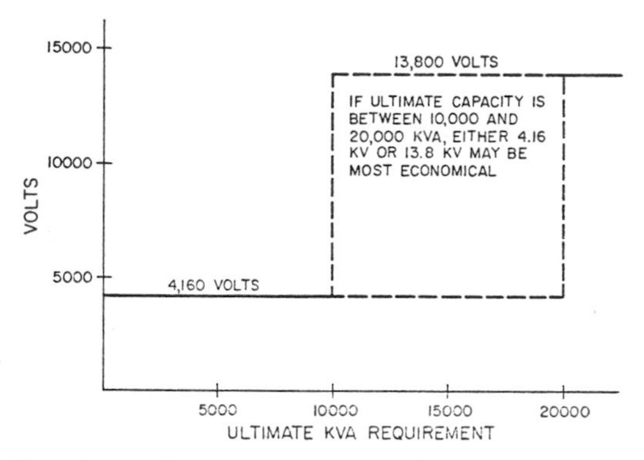

Fig. 21-4. Most economical medium voltage as a function of kilovolt-amperes.

When the low-voltage class is selected as the utilization voltage, there is seldom any reason, from the standpoint of economic considerations, for selecting a 240-volt or 208-volt system over a 480-volt system. This is shown in the graph of Fig. 21-5. It is due to the fact that the 240- and 208-volt systems have higher losses and greater voltage regulation, as a result of the higher voltage drop in the distribution wiring for conveying an equivalent amount of power. Where there are 120-volt loads to be serviced, this can be readily done by installing local wall- or floor-mounted dry type transformers, and this will still result in more economy generally. A rule of thumb on the economics of the higher voltage system is that money will be saved if 33 to 50% of the load can be served directly from the 480-volt system and if feeder lengths average 200 feet or more.

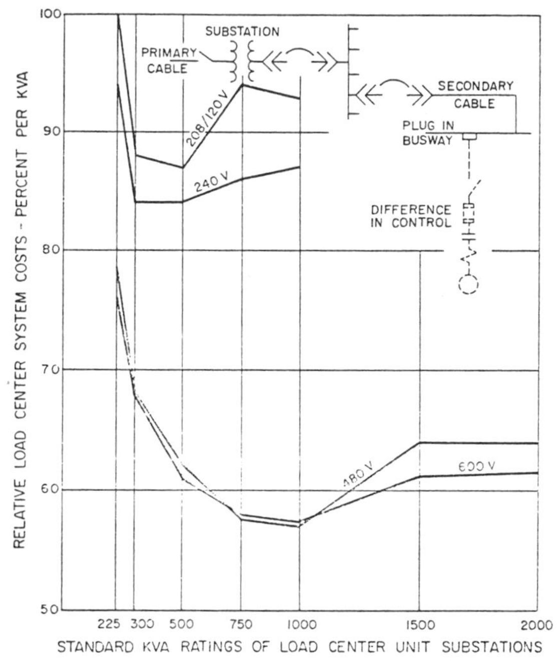

Fig. 21-5. **Load center versus substation size.**

Much higher loads can be transmitted, and generally at a lower cost, with the 480/277-volt system. A voltage drop limit of 1% permits a drop of not more than 2.08 volts in the lower voltage system and a corresponding percentage drop in the higher voltage system permits a drop of 4.8 volts. By combining the reduced current for a particular load with the higher allowable voltage drop, it is possible to supply, with the 480-volt system, twice the load at twice the distance over the same conductor.

Main Transformer Station

A main transformer station for powering the ventilation fans, tunnel lighting and miscellaneous loads is usually located in the ventilation buildings, where the bulk of the load is concentrated.

After the service and utilization voltages have been selected (in some cases, this may be the same voltage and the need for a transformer station may be precluded), the selection of the substation size then becomes a matter of obtaining the most economical balance of the overall components in the distribution scheme, which include primary cable and switchgear, transformers, secondary switchgear and cables.

Overcurrent and short circuit protective practices generally dictate the relative kilovolt-amperes of primary feeder loading along with corresponding ratings of the low-voltage substations. For example, when the primary voltage is 13.8 kilovolts, it is frequently desirable to have a loading of 4000 to 7500 kilovolt-amperes per primary feeder. If a transformer, rated at 500 kilovolt-amperes, were selected, it would require individual primary overcurrent protection for each such transformer, whereas if 750-, 1000-, or 1500-kilovolt-ampere units were selected, then primary voltage feeder loadings of the order of magnitude just mentioned could be obtained without individual primary overcurrent protection.

At the lower primary voltages of the order of 2.4 to 4.16 kilovolts, the most economical primary feeder loadings would be in the order of 1000 to 3000 kilovolt-amperes per feeder. In this case, transformer ratings could be smaller and not require primary overcurrent protection. In any event, substations larger than 2000 kilovolt-amperes with a 480-volt secondary are not recommended generally because of the overall increase in system cost. This is due to the increased cost of switchgear at these larger substation ratings.

In a tunnel facility, it is desirable to connect all cirtical loads directly to the secondary switchgear rather than to some point downstream, such as a power distribution panelboard. Because some unit loads may be large, such as a 400-hp fan, and other unit loads may be small, such as a 5-hp tide gate motor, their feeder protective device short circuit capabilities will have a bearing on the selection of transformer ratings and may be used as a guide for sizing the station. As an example, if the 30,000-ampere symmetrical interrupting rating of a circuit protective device is not to be exceeded, then the largest transformer size that can be used at a 480-volt secondary is 1000 kilovolt-amperes. Similarly, a 750-kilovolt-ampere transformer rating is the largest that can be used in order not to exceed a short circuit interrupting rating of 25,000 amperes. This requirement often dictates the selection of two 1000-kilovolt-ampere substations instead of one 2000-kilovolt-ampere unit.

Where short circuit currents of magnitudes exceeding ratings of protective devices are available, the use of current limiting reactors or current limiting busway may be used to good advantage in obtaining the most economical system in certain cases.

Switchgear

The power transformers are usually integrated into the primary and secondary switchgear lineup and the switchgear itself is usually located in a room adjacent to, or very near, the ventilation fan room. Figure 21-6 is an example of the space required for the switchgear and motor control center associated with a subaqueous tunnel approximately 7500 feet long. The arrangement shown makes the most efficient use of the floor space by making the operating aisle common to both switchgear lineups. The common aisle need be only sufficiently wide to permit the withdrawal for test and maintenance of one primary circuit breaker plus about two additional feet to permit passage during inspection.

Although both oil and air types of power circuit breakers are available for application at the primary medium-voltage levels (2.4 through 13.8 kilovolts), the air type is used almost exclusively in indoor installations and is quickly superseding the oil type up to 34.5 kilovolts. In addition, some oil circuit breakers are not rated to close and latch against their momentary rating, otherwise there is relatively little difference between the oil and air types from the overcurrent protection standpoint.

Power transformers integrated in switchgear

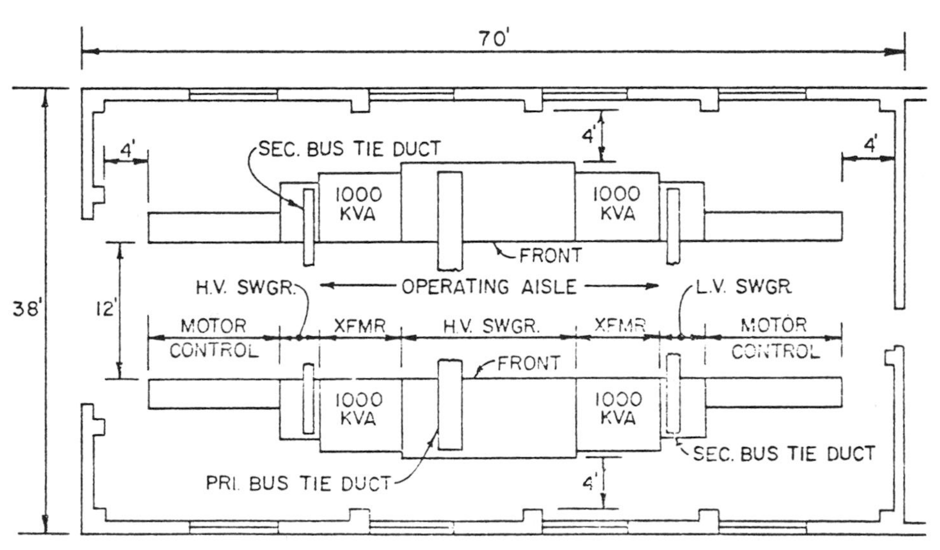

Fig. 21-6. Plan of switchgear room.

lineups are usually of the silicon-liquid-insulated, open dry-type or gas-filled sealed dry-type. The oil type is seldom used because of the requirement for vault type construction when used indoors. Open dry type transformers should not be used in dust-laden or corrosive atmospheres.

Any of the basic low-voltage circuit protective devices may be used in the low-voltage section of the switchgear lineup. Amongst these are power circuit breakers, molded case circuit breakers, circuit breaker and fuse combinations, fused safety and interrupter switch combinations and current limiting fuses. Power circuit breakers lend themselves very well as main circuit breakers, as well as bus tie circuit breakers, because they can be electrically operated both locally and remotely and are readily adaptable to all varieties of interlocking schemes. Molded case circuit breakers can be applied on feeder circuits of lower capacity, can be provided with drawout features and can be adapted for electric operation when applied within their interrupting rating. Fuses can be applied on feeder circuits supplying tunnel lighting where a ground fault of one phase resulting in a single fuse operation is not considered detrimental but rather a desirable feature, since it permits the remaining phases to continue operating. The same operation can be obtained on tunnel lighting feeders by using a single pole instead of three-pole circuit breakers.

Motor control for critical and major units of electric equipment can be readily integrated into the switchgear lineup, although in some cases it may be more advantageous to locate the motor control near the controlled equipment.

CONTROLS

Controls, as discussed herein, will be limited to those associated with the operation and functioning of electric motor-driven devices; switching of the primary and secondary power distribution system under normal and emergency operating conditions; and operation and control of the tunnel and approach lighting systems during daytime and nighttime hours. (For controls associated with traffic, carbon monoxide concentration level communications, TV surveillance and monitoring, fire alarm, haze detectors and traffic volume and count, refer to other chapters of this handbook.)

Motor Controls

For a vehicular tunnel facility, motor controls may be automatic or manual or a combination of both. In unmanned tunnels, controls are usually automatic with overriding features for local manual control when maintenance and testing are required. In a manned tunnel, the controls are predominantly manual and are usually initiated from a central control room. Even in a manned facility, some equipment, such as tunnel drainage pumps, will be automatically operated.

Ventilation Fan Control

Figure 21-7 is a wiring diagram for the control of a two-speed ventilation fan motor associated with the mechanical ventilation of a manned tunnel facility. In this control scheme, ventilation fan motors are started and stopped manually by pushbutton actuation, either remotely, from a main control board, or locally, at the motor. Associated with each fan motor is an auxiliary blower motor and a motorized damper. During a normal fan motor starting sequence, the motor may be started at either high or low speed. Pressing either button causes the damper motor to open the damper and at the same time start the auxiliary blower. Interlocking contacts DL15 and SS prevent the main fan motor from starting until the damper is opened at least 15 degrees (or some other suitable predetermined opening) and air flow is sensed by a sail switch in a downstream duct of the auxiliary blower. The auxiliary blower provides cooling air in a closed duct connection to the main motor to provide positive cooling. Time delay relays TDH and TDL are provided to abort the starting directive in the event of an incomplete starting sequence. A speed-sensing contact (TACH) is connected in the low-speed contactor holding coil (MIL) circuit to prevent this contactor from immediately closing when the low speed is selected while the motor is operating at high speed. The speed-sensing tachometer contact is set at a speed slightly above low speed.

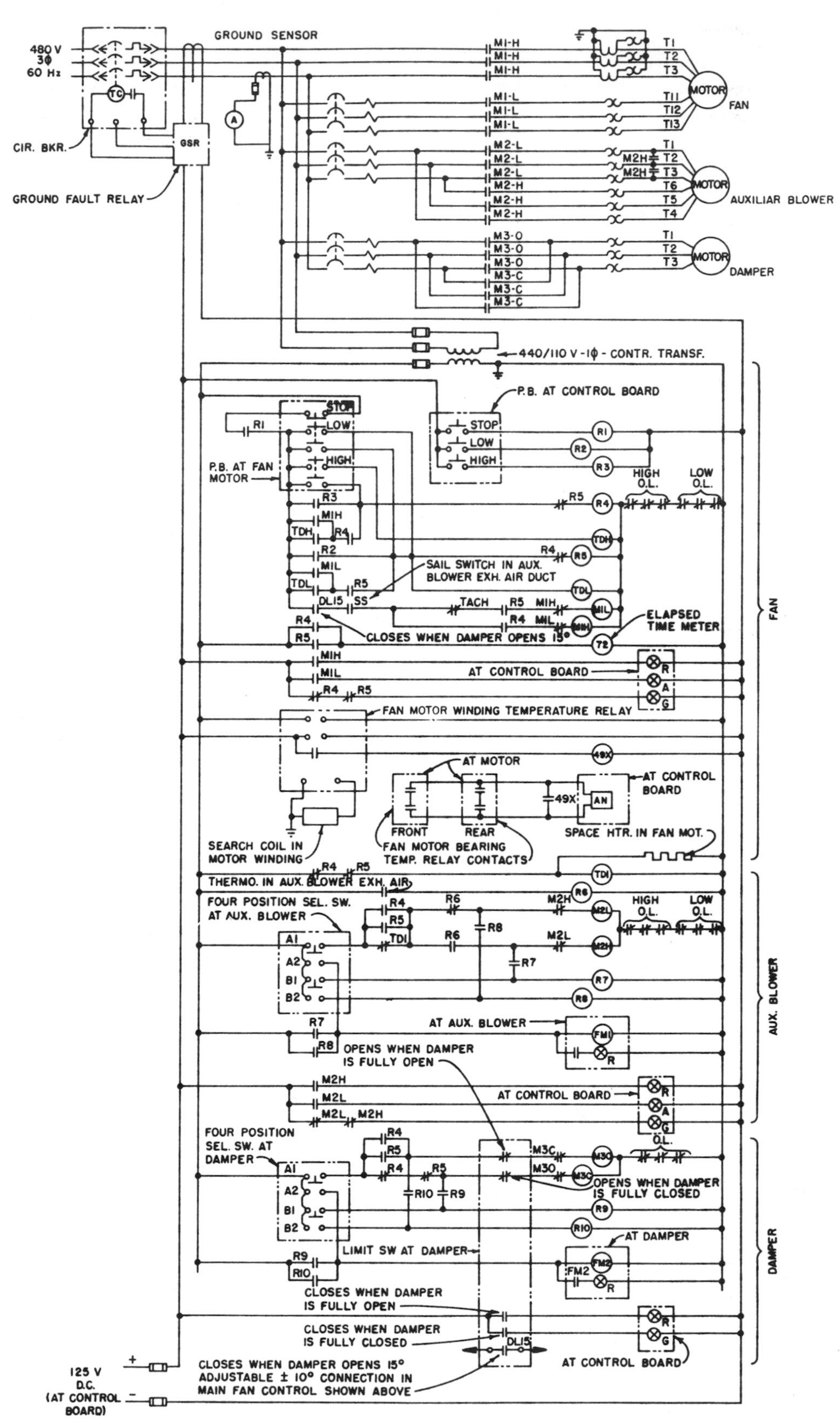

Fig 21-7. Control diagram for a two-speed ventilation fan.

Its function is to prevent induction generator action when speed reduction is selected.

In addition to the conventional motor current overload relays, the motor is equipped with a fan motor winding temperature relay as well as front and rear fan motor bearing temperature relay contacts. These combine to operate an annunciator drop on the main control board when malfunction is sensed.

A space heater is provided within the motor housing to prevent condensation on the motor windings during idle times and adverse ambient temperature and humidity conditions. A thermostat is placed in the main fan motor cooling air duct outlet to automatically select the speed of the auxiliary cooling air blower, depending on the ambient temperature of the cooling air.

Upon shutdown of a main ventilation fan, time delay relay TD1 permits the auxiliary blower to operate for an additional predetermined amount of time as a precautionary measure to preclude the certain rise in winding temperature which will result from the residual heat stored in the motor frame.

Four-position selector switches are located at the auxiliary blower and motorized damper to permit local manual operation in all of the operating modes for maintenance purposes. A flashing red light becomes energized in each position of the selector switch, except the automatic position, to alert the maintenance mechanic to restore this switch to automatic position when the maintenance procedure is complete.

Figure 21-8 is a wiring diagram for the control of a three-speed ventilation fan motor for either automatic or manual control. The normal operating mode is automatic, but provisions are incorporated to permit manual operation through the mode selector switch SSMA.

Automatic Mode. Here the fan motors are controlled in pairs with one of the fans preselected as the lead fan. The lead fan is selected by switch SSIFL. Under expected normal carbon monoxide concentration level, the lead fan is programmed to operate continuously at low speed. With an increase in carbon monoxide concentration level, the second of the paired fans will start and operate at low speed. On further increase in the carbon monoxide level, both fans will operate at medium speed and then finally at high speed. When excessive carbon monoxide is sensed, an alarm will be sounded at a remote monitoring location so that remedial action can be immediately taken.

By means of speed-responsive relays, energized via an adjustment network fed by a tachometer generator on the motor, each fan is caused to operate at low speed or any other speed in response to an initiating signal. During operation, transition from low speed to either medium speed or high speed, or from medium speed to high speed, is accomplished by dropout of the energized contactor and immediately closing of the desired contactor. However, when speed is to be decreased, the energized contactor drops out immediately, but the motor speed is allowed to decrease to slightly above the desired speed, either medium or low, before that contactor closes. Transition or incomplete sequence is indicated by action of an indicating light dimmer.

Control circuits and devices under both modes of control are arranged for fan motor shutdown and lockout by means of the stop buttons and the motor protective devices. Interlocking with the associated damper operator initiates its operation and prevents motor operation with insufficiently opened dampers.

Manual Operation. In this mode, initiating control signals originate at the pushbutton stations located at the control center and also at the motor. The circuit provides under voltage and incomplete control sequence release.

Under automatic operation, initiating control signals are received from relays at the carbon monoxide recorder and analyzer equipment, or other automatic devices, such as traffic density detectors or haze indicators.

Drainage Pump Control.

Drainage pumps are always provided in subaqueous tunnels at the low point in the tunnel, or at some other location of the low point, and at each portal. These pumps lift collected fluids accumulated due to rain at the portals and in mid-tunnel by tunnel-washing operations and spillage. Figure 21-9 is a diagram for the

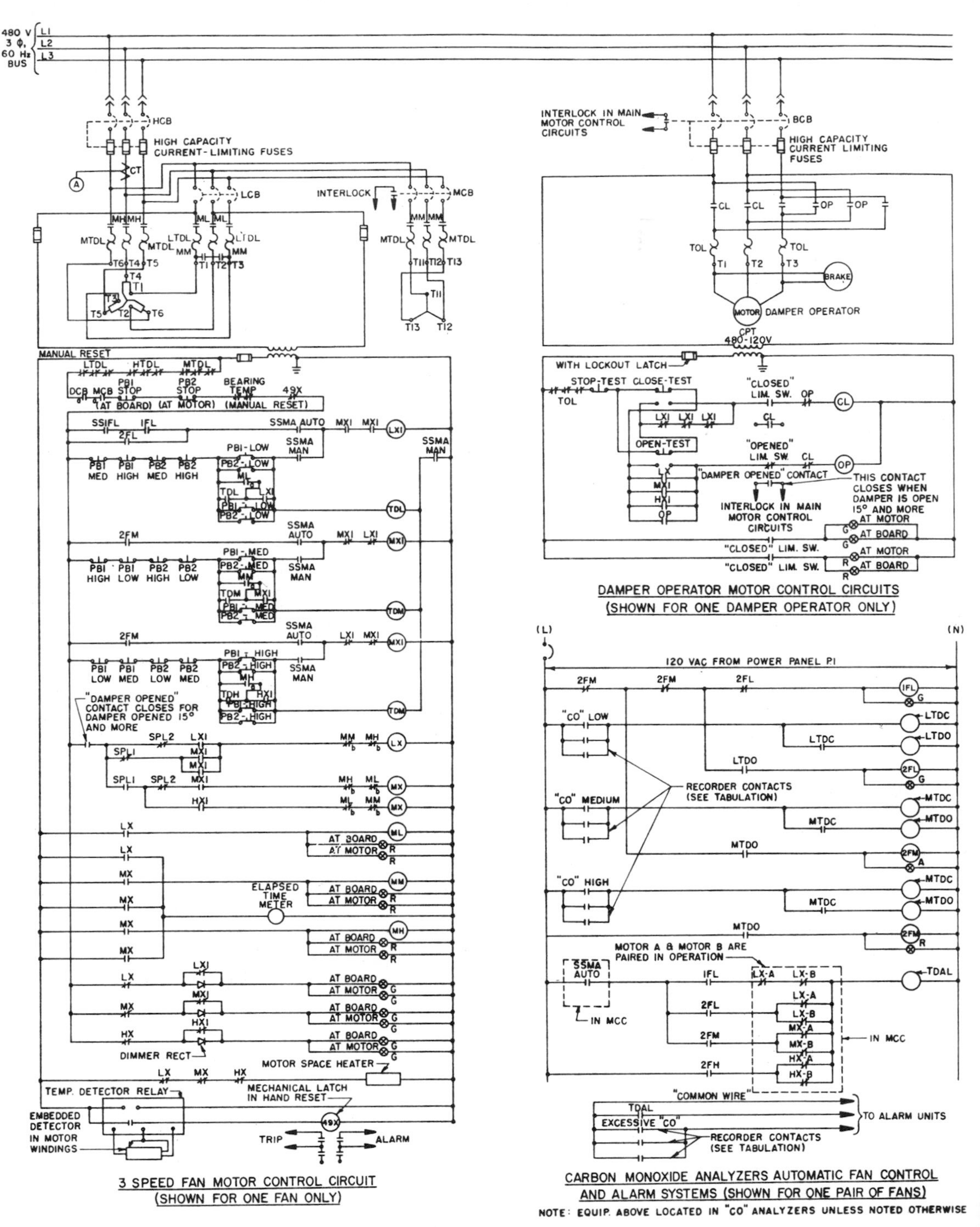

Fig. 21-8. Control for three-speed fan motor for automatic or manual peration.

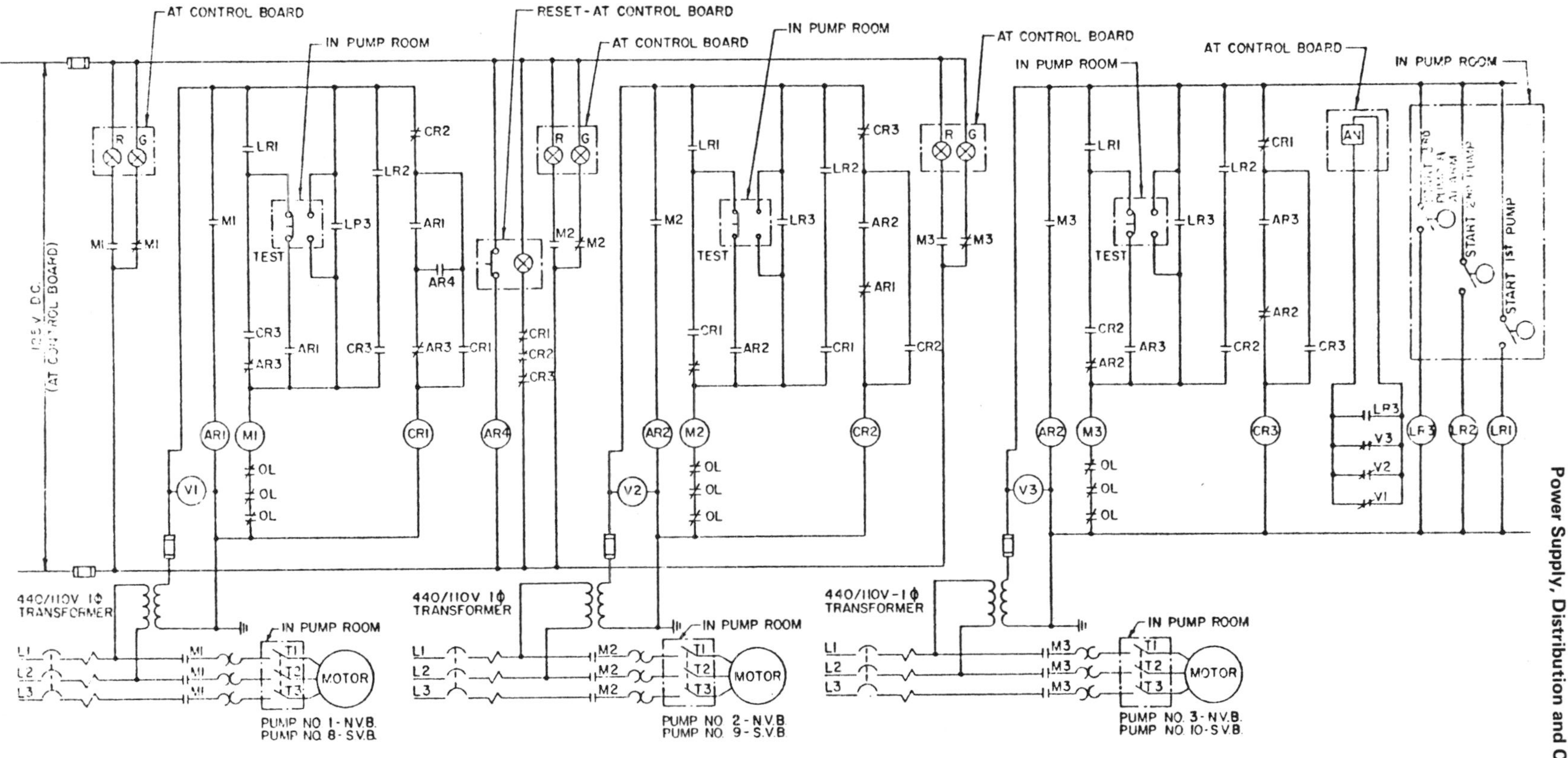

Fig. 21-9. Three-pump control scheme.

control of three drainage pumps in an automatic, electric, alternating sequence. Pump operation is initiated by float-operated pilot switches which respond to low, intermediate and high water levels. The pumps are operated in a 1-2-3, 2-3-1 or 3-1-2 sequence. In any cycle, the lead pump will start first at the preselected low water level. If required, the second pump of a sequence will start and operate when the intermediate water level is reached. If the water continues to rise beyond the capacity of the two operating pumps, the third pump starts and assists the pump-down. When the high water level is reached an alarm is sounded at the control center on the annunciator. A momentary contact button is provided at each pump for maintenance and test purposes. A reset button is provided in the control room to reestablish the automatic, electric, alternating sequence in event of a momentary system power failure. A control voltage monitoring relay is connected in each pump control circuit to annunciate control voltage failure caused by local component failure.

Figure 21-10 is a diagram for the control of two drainage pumps in an automatic, electric, alternating sequence. Water level sensing in this control scheme is also accomplished by means of float-operated pilot switches. At the first preset water level, the first pump will start. If required, the second pump will start when the second water level is reached. When a third water level is reached, an annunciator alarm will sound at the control center. At the pumps, a momentary contact test button is provided for maintenance and test purposes.

Power Distribution System Contol

In tunnel operations, continuity of power to the critical load components is extremely important. Upon the failure of a vital device or a distribution feeder, it is imperative to quickly institute switching maneuvers so that normal operating conditions can be restored as soon as possible. Since critical loads are located in widely separated portions of a tunnel facility, the most effective way to accomplish a rapid switching operation is to provide central control of important switching devices. In the distribution scheme shown in Fig. 21-2, all of the primary circuit breakers as well as the secondary transformer main and bus-tie circuit breakers are remotely controlled from central control. This permits rapid switching maneuvers which would result in the isolation of a faulty transformer, feeder or bus, and also the reinstatement of the affected load by switching to redundant elements of the distribution system.

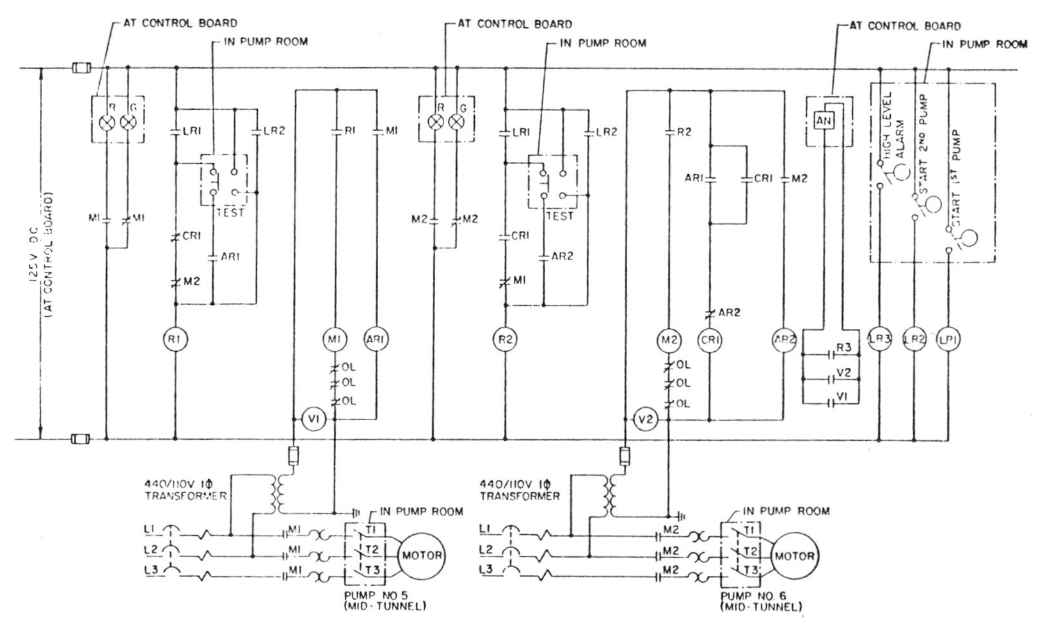

Fig. 21-10. Two-pump control scheme.

Tunnel Lighting Control

Tunnel lighting operates on a 24-hour basis with lighting levels selected during both the night and the day. Control of these lighting levels can be manual or time clock-governed, or can be provided by daylight-sensitive sensors which respond to ambient daytime light levels. In tunnels equipped with single lamp fluorescent fixtures, lighting levels are controlled by switching off the supplementary rows of lamps in the entrance zone at night and by reducing the flux of the interior lighting by lowering the voltage by use of multi-tap transformers or ballasts. Intensity should not be reduced by switching off alternate lamps, leaving dark spots.

Greater flexibility is obtained in multi-lamp fluorescent fixtures by turning off lamps according to changes in daylight intensity. This is controlled by light-sensitive elements mounted outside the tunnel. In one case, three levels of threshold intensity above the interior level of 11 FC were used: 67 FC, 143 FC and 219 FC. The 11 FC are also maintained at night to facilitate traffic surveillance by closed circuit television.

An extremely sophisticated daytime-lighting control is used in the IJ Tunnel is Amsterdam, where the daytime tunnel lighting level is automatically and electronically controlled by 12 light-sensitive sensors from a maximum entrance level of 300 lux to a minimum nighttime level of 50 lux. The adjustments are effected in two different ways: first, by switching lamps on and off, and, secondly, by controlling the energized lamps' luminous flux. Starting from the highest outdoor illumination level, which is reached around mid-day, the extra lighting in the two transition zones is switched off in two steps by so-called "twilight switches" as the outdoor illumination decreases. After the last switching step, the automatic dimming installation comes into operation, and, as the outdoor lighting level continues to decrease gradually, the tunnel illumination level decreases, accordingly, to a minimum nighttime level. At an outdoor illumination level of about 2000 lux (around sunset and sunrise), the lamps in the grid zone are switched on. These lamps are connected to the dimming installation and only operate at low intensity. After sunrise, the cycle described above is reversed.

Central Control Board

In a manned tunnel, a control center is provided, into which all information regarding traffic, air quality, visibility, support equipment status and condition is channeled, observed by experienced operating personnel who take appropriate action according to changing requirements or deviations from normal conditions. For this purpose, a central control board is usually installed on which all of the control, status, read-out and indication displays are located. Some of the devices normally located on the central control consists of the following.

- Carbon monoxide level recorders.
- Haze level recorders.
- Televistion monitors.
- A mimic tunnel diagram incorporating traffic light indications.
- Main ventilation fan control stations.
- Drainage pump status indications.
- Fire alarm.
- Electric distribution system switching controls.
- Status of changeable message traffic signs.
- Equipment overrating alarm annunciation.
- Telephone and radio communications.
- Fog and ice detection indication.

To ensure good supervision, as well as safe and rapid countermeasures in response to incoming signals, care should be taken in the design and equipment arrangement of the control board. Consideration should be given to range as well as angle of sight to the various displays from the operator's normal viewing position. This normal position is usually behind a control desk, which should be situated at the focal point of the display. Additional important considerations are the interior design of the room containing the display. Lighting should be complementary to the visual task. The air temperature and humidity conditions should be conducive to operator comfort, and the color scheme should be pleasing.

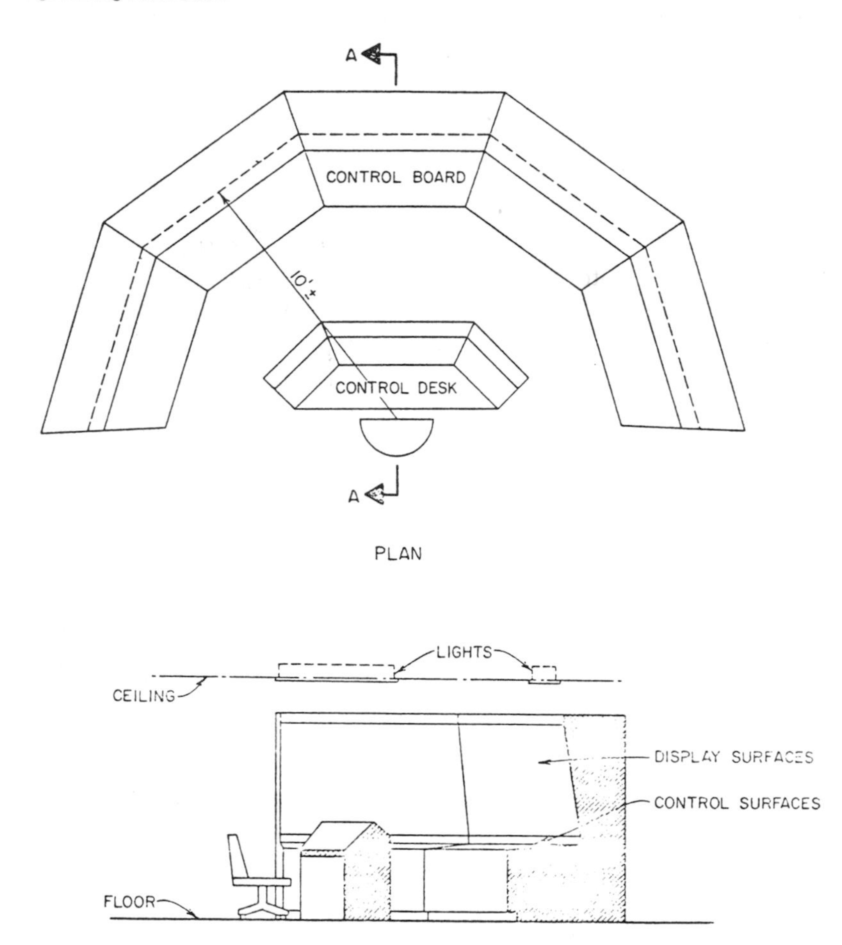

Fig. 21-11. Control board layout.

Figure 21-11 shows a plan and sectional layout of one of many equipment arrangements that would satisfy the foregoing requirements. The display surfaces of the control board should contain all of the registrations of check-back indications which relate to the traffic status and condition in the tunnel and approaches at all times. These indications should be displayed in a manner which relates to the physical location of equipment within the total facility. In this respect, a mimic tunnel and approach diagram, which contains indications of the status of traffic lights, as well as current traffic control instructions posted by all changeable message signs, will be extremely helpful, not only to the operator on duty, but also to the succeeding operator at shift change. Also displayed should be those check-back indications which report traffic conditions, such as flow, density, local or overall congestion, as well as tunnel and approach environmental conditions and other indications listed (also any other visual indication which would assist the operator in quickly responding to changing traffic conditions). The control surfaces should have those devices which permit the operator to respond to any changing operating conditions. Such devices would include pushbuttons and indication of the operating condition of the main ventilating fans, remote control of important switching components in the electrical distribution system and other control functions of similar nature.

INDICATORS AND ALARMS

Sensors, for the automatic detection of apparatus or system abnormal operating conditions, should be installed to initiate the sounding of an alarm in the control center. The alarm should sound early in the abnormality to permit the operator sufficient time to respond with remedial measures before permanent damage occurs. Some of the conditions for which indicators and alarms should be provided can be found among the following.

1. Main ventilation fan motor winding and bearing high temperature.
2. High water level in the drainage pumping system sump.
3. Emergency generator low oil pressure and high water temperature.
4. High carbon monoxide concentration level in the tunnel.

It may also prove desirable to provide protection against malfunction on apparatus which is costly to replace or requires a long period of down-time, thus resulting in the operation of the facility at reduced capacity.

POWER SUPPLY FOR RAILROAD TUNNELS

In tunnels utilizing steam and diesel engines as the prime source for locomotion, electric power is furnished for the purpose of powering force draft fans in order to provide adequate ventilation and cooling of the locomotive and, in addition, to assure a thorough flushing of the tunnel after the train has passed through. Normal power is usually furnished by the local electric utility company at subtransmission voltages between 13.8 and 34.5 kilovolts and transformed down by an on-site utility substation to a utilization voltage of 2.4 or 4.16 kilovolts for supplying large horsepower ventilation fans, and further transformed to 277/480 volts for auxiliary power equipment and lighting.

To forestall the probability of an extended outage of the electric service occurring at a time when a train may be passing through the tunnel, a second and reliable source of power must be provided. In urban and suburban areas, a reliable second source of power can usually be obtained from the utility company at reasonable initial and operating charges. The reliability of this second service is greatly enhanced when it originates from another substation or, better yet, from another power plant, and the routing of the circuit is over a different geographical area. In remote and isolated areas, on the other hand, a second and reliable utility service may not be readily available, in which case the second source of power should be provided by an on-site electric generating unit, automatically started upon the sensing of a voltage failure in the normal power supply. (See Fig. 21-12 for a representative power distribution diagram for a long railroad tunnel with diesel operation requiring ventilation.)

Power supply for traction in electrically-operated railroad tunnels is not within the scope of this handbook. These tunnels do not usually need ventilation.

POWER SUPPLY FOR RAPID TRANSIT TUNNELS

Auxiliary electric power, as distinguished from traction power, is required in a rapid transit tunnel for ventilation, as well as for trackway lighting. Ventilation is required to dissipate smoke and heat generated by a fire on the train or in the line section in front of or behind a train, and the lighting is required to provide for adequate visibility during evacuation of a disabled train and for trackway maintenance purposes.

In addition to ventilation and lighting, auxiliary electric power is also provided for exhaust duct dampers, drainage pumps and fire pumps.

Lighting fixtures are normally located in the tunnel trackway on 40- to 50-foot centers on the walkway side of the tunnel and generally consist of 50-watt incandescent lamps or 40-watt fluorescent lamps. The use of fluorescent lamps is becoming more and more popular because of their higher lumen/watt output rating. Because fluorescent lamps also lend themselves to operation from a 277-volt source, the distribution branch circuits can be rated at 480/277 volts. At these branch circuit voltage ratings, installation economies are realized because larger circuit loads at greater lengths can be effected within the allowable voltage regulation criteria and with the conventional lighting branch circuit wire sizes.

These foregoing principles are exemplified in the recent rapid transit tunnel installation under the San Francisco Bay between Oakland

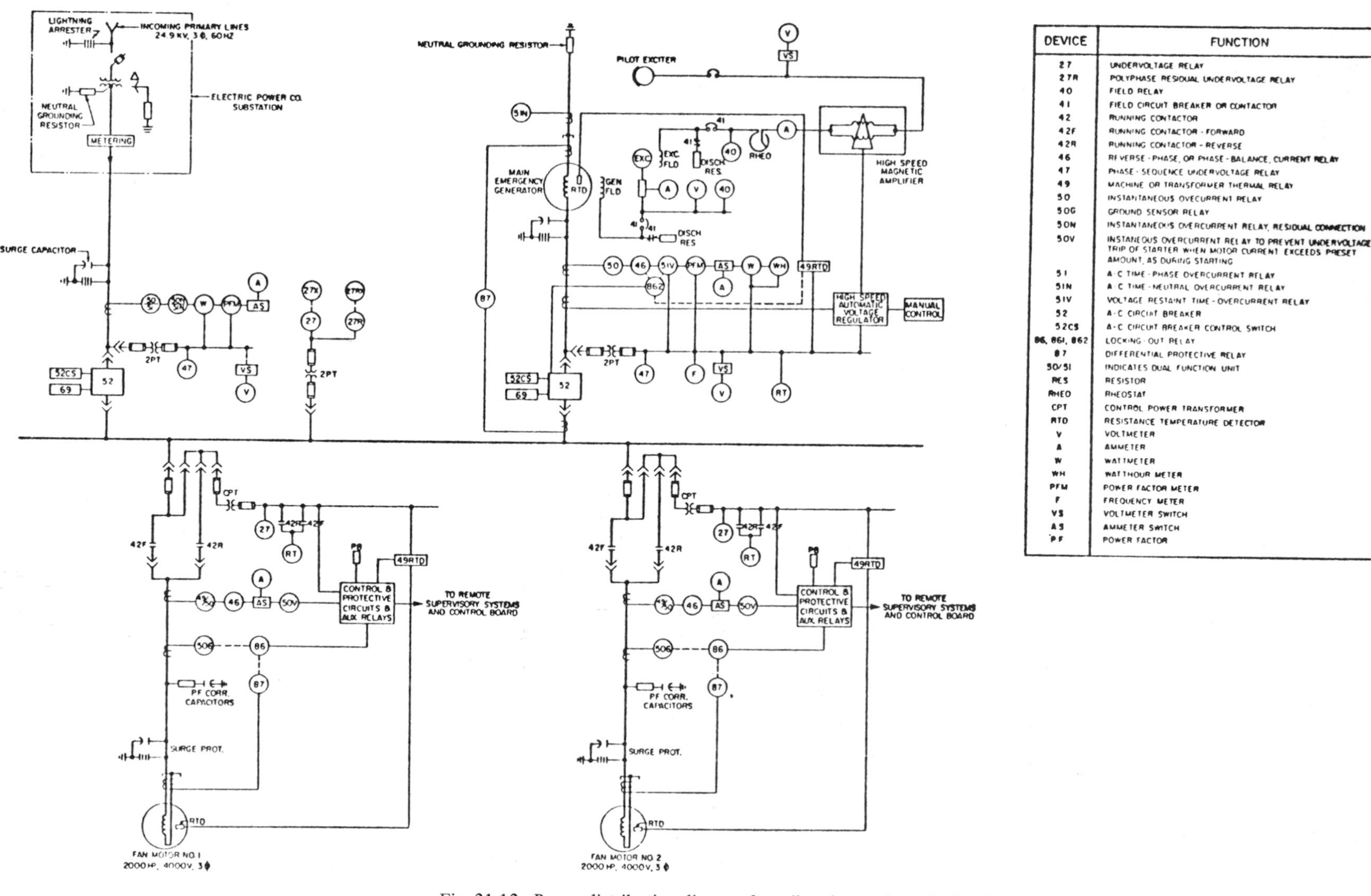

DEVICE	FUNCTION
27	UNDERVOLTAGE RELAY
27R	POLYPHASE RESIDUAL UNDERVOLTAGE RELAY
40	FIELD RELAY
41	FIELD CIRCUIT BREAKER OR CONTACTOR
42	RUNNING CONTACTOR
42F	RUNNING CONTACTOR - FORWARD
42R	RUNNING CONTACTOR - REVERSE
46	REVERSE - PHASE, OR PHASE - BALANCE, CURRENT RELAY
47	PHASE - SEQUENCE UNDERVOLTAGE RELAY
49	MACHINE OR TRANSFORMER THERMAL RELAY
50	INSTANTANEOUS OVECURRENT RELAY
50G	GROUND SENSOR RELAY
50N	INSTANTANEOUS OVERCURRENT RELAY, RESIDUAL CONNECTION
50V	INSTANEOUS OVERCURRENT RELAY TO PREVENT UNDERVOLTAGE TRIP OF STARTER WHEN MOTOR CURRENT EXCEEDS PRESET AMOUNT, AS DURING STARTING
51	A-C TIME - PHASE OVERCURRENT RELAY
51N	A-C TIME - NEUTRAL OVERCURRENT RELAY
51V	VOLTAGE RESTAINT TIME - OVERCURRENT RELAY
52	A-C CIRCUIT BREAKER
52CS	A-C CIRCUIT BREAKER CONTROL SWITCH
86, 861, 862	LOCKING - OUT RELAY
87	DIFFERENTIAL PROTECTIVE RELAY
50/51	INDICATES DUAL FUNCTION UNIT
RES	RESISTOR
RHEO	RHEOSTAT
CPT	CONTROL POWER TRANSFORMER
RTD	RESISTANCE TEMPERATURE DETECTOR
V	VOLTMETER
A	AMMETER
W	WATTMETER
WH	WATTHOUR METER
PFM	POWER FACTOR METER
F	FREQUENCY METER
VS	VOLTMETER SWITCH
AS	AMMETER SWITCH
PF	POWER FACTOR

Fig. 21-12. Power distribution diagram for railroad tunnel ventilation fans.

and San Francisco, California on the Bay Area Rapid Transit system (BART). This tunnel is approximately 20,000 feet long and consists of two trackway tubes separated by a longitudinal pipe gallery and an exhaust ventilation duct. The pipe gallery serves as a convenient space for equipment location and distribution of the lighting circuits, in addition to mid-tunnel pumping equipment. At or near each shore, ventilation buildings are located to house the tunnel exhaust fans. From these points, two 4160-volt feeders are extended from each ventilation building to points within the tunnel near the midpoint. At their source in each vent building, these feeders are each connected to independent service buses. Lighting transformers and panelboards are located at the midpoint of the pipe gallery and at intervals of 5000 feet therefrom towards each shore. The transformers step down the voltage to a 480/227-volt, three-phase, four-wire, 60-hertz utilization voltage with solidly grounded neutral. Number 10 wire branch circuits, operating at 480/227 volts, are extended from these panelboards to the tunnel lighting fixtures. These branch circuits are limited to lengths not exceeding 2500 feet. Two such branch circuits are installed in parallel, in separate conduits, and each connecting to a separate service and each serving alternate lighting fixtures.

Because of the degree of automation incorporated into the operation of BART, the Transbay Tube trackway lighting system is normally de-energized, and is energized only for purposes of evacuating an incapacitatated train or for trackway maintenance purposes. The trainway lighting system is controlled at each ventilation building and by remote control from the central dispatching center. The remote control feature will assure quick energization of the lights when a call comes in from the train attendant via the on-board radio communications link.

22
Signal and Communication Systems

DONALD N. TANNER

Assistant Vice President
and Senior Professional Associate
Parsons, Brinckerhoff, Quade & Douglas

and

MICHAEL A. MARSZALOWICZ

Professional Associate
Parsons, Brinckerhoff, Quade & Douglas

All vehicular tunnels require traffic control and communication systems. Their complexity depends on their location, the traffic volume, the length of the tunnel, the approach highway system and the climatic conditions. The minimum requirements for a single two-lane tunnel with straight approaches are, in addition to carbon monoxide detectors, a means for controlling ventilation, a method of surveillance of traffic (either by guards stationed in the tunnel or, more recently, by closed circuit television monitors), a fire alarm system, emergency telephones, overheight detectors (if ceilings restrict vertical clearance) and signs on the approaches indicating restrictions for use of the tunnel. In short tunnels, such as long underpasses in city streets, fire alarm stations and emergency telephones connected to the nearest city police and fire stations may suffice.

Where traffic requires multiple lanes divided between parallel tunnels and complicated highway approaches, the minimum requirements are expanded and may involve computerized control programs and elaborate signing. Other requirements might include extended traffic controls augmented by vehicle detection loops to indicate traffic flow, expanded communication systems including telephone and radio and, in certain climates, fog and ice detectors.

A control center is usually located in one of the ventilation buildings. The operation may be manual, partly automatic, fully automatic or a combination of these. They must be supervised by a competent operator who has the power to activate, adjust or override any automatic control. Figure 22-1 lists the major control elements for different requirements.

In view of the initial costs and maintenance,

	MINIMUM MANUAL SYSTEM	SOPHISTICATED AUTOMATIC SYSTEM
1. *Equipment status system*		
Power	•	•
Fans	•	•
Dampers	•	•
Drainage pumps	•	•
Fire pumps	•	•
Tunnel lighting	•	•
2. *Climate surveillance system*		
Wind velocity and direction		•
Tide gates		•
3. *Safety surveillance system*		
Fire alarm	•	•
Carbon monoxide monitoring	•	•
Haze detection		•
Fog detection		•
Ice detection		•
Vehicle detection		•
4. *Communications system*		
Sound-powered telephone	•	•
External bell telephone	•	•
Radio		•
Police radio		•
Maintenance radio		•
5. *Traffic control system*		
Traffic lights in tunnel	•	•
Lane-use control signals		•
Variable message signs		•
Closed circuit television monitoring		•
Overheight vehicle detection	•	•
Inspection stations	•	•

Fig. 22-1. Tunnel control system elements.

the extent of automatic controls justified should be carefully analyzed.

TRAFFIC SURVEILLANCE AND CONTROL SYSTEMS

The following systems serve to observe and control traffic through tunnels and on their approaches.

Closed Circuit Television

This system permits the operator at all times to observe traffic in the tunnel and on the open approaches. It has largely replaced the guards formerly located in the tunnel, except in some of the older installations, and even in many of those a television system has been installed. It consists of a series of television cameras, mounted on the ceiling over the roadways or on the sidewalls, which transmit video signals over cable or by microwave to monitors in the main control room. Cameras are spaced so that every part of each traffic lane is observed. Depending on grades and curvature, the spacing may vary from 200 to about 600 feet. They are usually oriented facing oncoming traffic, placed over the centerline of the roadway (if clearance permits) or in the cove over the service walk.

Tunnel Cameras. These cameras are solid state, except for the videcon tube, and are equipped with a fixed focus and are fixed in position. They must operate within the range of the lowest and the highest ambient temperature expected, and light levels from 7 to about 300 footcandles. Scanning rates are those of standard commercial cameras.

Outdoor Cameras. These cameras cover the open approaches and should be mounted at sufficient height, either on masts or on the roof of a ventilation building, and should have adequate tilt, pan and zoom capability to focus on any part of the open approach. They should be equipped with defogging and deicing devices. Automatic iris control adjusts the output with varying brightness and cuts down headlight glare.

All cameras are enclosed in watertight housings to protect them against tunnel-washing operations or rain.

Power Supply. The power supply is by closed loop, DC-maintaining operating performance at specified line voltage and temperature variations.

Transmission. Transmission from cameras to monitors is by double-shielded, coaxial cable for extra protection against pick-up of interference. In long tunnels, transmission from the far end approach camera location to the control room may be by way of a microwave system, which can also transmit tilt, pan and zoom controls.

Monitors. Mounted on the control board, monitors must be arranged for ease of observation by the operator. They should give a clear picture with adequate contrast under all conditions. They are solid state, designed to operate on a closed circuit system continuously. A 14-inch size is adequate.

Two systems are available:

1. Each camera is connected to its own monitor, displaying traffic conditions continuously; and
2. With a sequential system for dual tunnels, cameras are connected to a switching unit in pairs, one from each tunnel, showing alternately on a monitor, each picture displayed for 5 seconds (this reduces the number of monitors by half).

The first system is more reliable. If one monitor or camera fails, traffic surveillance is still possible. If, in the sequential system, a switch fails, one tunnel is blanked out. Also, in case of a tunnel incident, it is desirable to continue uninterrupted observation of the incident, in which instance surveillance of the other tunnel will not be possible.

Vehicle Detection

In order to pass the optimum number of vehicles through a tunnel under any traffic conditions, stoppages inside the tunnel should be avoided. Congestion usually starts on the upgrade part of the roadway, where drivers have a tendency to slow down. Traffic density and speed are detected by a system of induction loops, embedded in the pavement under each lane in the tunnel, under the open approaches and sometimes, for a certain distance, in the approach highway. Pairs of wire loops, separated by 15 to 20 feet, are used for calculating vehicle speed and occupancy. As a vehicle passes over an induction loop, an impulse will be sent to a roadside amplifier, which, through an encoder and transmitter, relays coded information to the computer located in the control center. The most essential loop locations are at the start of the upgrade sections. When preset limits of traffic density are reached, the computer will trigger audible alarms and indicators in the control center. Variable message signs at the entering portal will instruct motorists to pause there, then go, then prepare to stop, stop or resume speed. These instructions will reduce volume entering the tunnel, increase vehicle spacing and thereby ensure an even flow of traffic through the tunnel.

Traffic Lights

Tunnel traffic lights, displaying green, amber and red, are installed for each traffic lane, mounted over the centerline or on the sidewalls. When ceiling-mounted, clearance limitations may restrict the lens size to 4-inches. When sidewall-mounted, lenses up to 6 inches may be used. They should be spaced so that, from any position on the roadway, at least one light is visible within stopping distance. They are enclosed in watertight housing to withstand tunnel-washing operations. Traffic lights are also installed at the portals and at the upper ends of the open approaches. For lanes where traffic may at times be reversed, the lights are double-faced. In straight tunnels, spacing of 600 to 700 feet is adequate, whereas curved tunnels require special attention.

Control of traffic lights is located on the control board in the central control room. Pilot lights on the board duplicate the setting. In some tunnels, auxiliary controls are installed in the tunnel to permit patrolmen to actuate them in emergencies. Where computerized systems are used, the light control is integrated into the computer programs. Control circuits are interlocked so that if one light is turned to red, all lights on the upstream side are automatically turned red, while the signals on the downstream side remain green to permit traffic to clear the tunnel.

Lane-use Signals

Lane-use signals on approach highways to tunnels are special overhead configurations of red, amber and green lights intended to permit or to prohibit the use of specific lanes or to alert to the impending prohibition of a lane's use. They are distinguished from traffic signals by their shape and symbols. Supplementary signs are sometimes required to explain their intent. Lane-use signals are specifically used over reversible lanes but may also be installed for other reasons. The principal uses are as follows:

1. To prohibit use of certain lanes during certain hours.
2. To indicate a lane blocked by accident or breakdown.
3. In preparation of a lane reversal.

The configuration and meaning of lane use signals are listed below.

A steady downward green arrow: Lane open for normal traffic.

A steady amber X: Prepare to vacate lane.

A flashing amber X: Motorist may make a left turn from lane with caution.

A steady red X: Prohibits occupancy of lane and alerts motorist to obey all other traffic controls.

Lane use signals are installed on the approach roads and may be extended as far as the nearest off-ramps. Control of the signals originates in the control room. In computerized control systems they are included in control programs.

Vehicle Message Signs

These signs are installed on the approach highways, entrance and exit ramps, and at the tunnel entrances. Their number and extent depends on the complexity of the highways. They carry messages covering lane-use directions, special restrictions, hazardous road conditions, speed limits and instructions in case of overheight vehicle detection. Signs may be installed in the tunnel if clearances permit.

Overheight Detectors

These detectors, to exclude overheight vehicles from entering the tunnel, should be installed on the approach roads. Their quantity and locations depend on local conditions. A light beam is projected across the lane at the permissible clear height and is targeted on a photoelectric sensor. If the beam is interrupted, the light cell sends a signal to the operating center and to the inspection station located at the head of the open approach. It also actuates message signs directing all trucks to pass through the inspection station. Another detector may be installed farther downstream to detect a vehicle ignoring the instructions, which will automatically turn traffic lights to red to stop all traffic and alert the operator, who will dispatch a patrolman to the location. The detectors are also included in computer programs, if provided. The light source and the photoelectric cell are mounted on posts or overhead frames. Poles must be rigid enough to resist sway due to wind forces so that false beam interruption will not occur.

SAFETY SYSTEMS

Fire Alarm System

Coded fire alarm stations are installed at distances of about 300 feet on the tunnel walls along the service walk. They are recessed in niches and are conspicuously marked. If actuated, they send coded signals to the control room and to the crash truck stations.

Carbon Monoxide and Haze Recorders

These recorders should be installed in the control room to assist the operator in controlling the tunnel ventilation system in order to maintain proper air quality. (For a full description, see Chapter 19, on tunnel ventilation.)

Climatic Indicators

Used to warn against unexpected driving hazards on the approach roads due to weather conditions caused by fog conditions or ice formations, climatic indicators constitute an essential part of the safety system.

Fog Detectors. These detectors measure visibility and, if this drops below a certain predetermined value, they send a signal to the control center, alerting the operator to actuate warning signs and reduce speed limits. Under computerized controls, these signs and speed limits will be automatically activated upon receipt of the signal by the computer. The detectors should be placed in areas which are subject to fog formation.

Ice and Frost Detectors. Detection systems have been developed which, upon anticipation of conditions conducive to ice formation, will send a signal to the control center, where the operator will actuate warning signs and alert maintenance crews to take remedial action. Such action also may be instigated automatically by the computer. The detectors consist of temperature indicators in the pavement of approach highways or overpasses, where such conditions first occur, and instruments measuring ambient temperatures and humidity.

These climatic devices are still subject to considerable improvements to reduce the number of false alarms. A feasibility study on ice and snow detection and warning systems has been prepared by the Midwest Research Institute in Kansas City.

COMMUNICATION SYSTEMS

To ensure safe and efficient operation and traffic control, and to serve the convenience of the motorists, a number of communication systems are used.

Sound-powered Telephones

Telephones permit communication among the tunnel interior, the control room, the crash truck stations, the tunnel police stations and the truck inspection station. Because it is independent of any power source, this system is considered to be the most reliable under all operating conditions. Telephones are installed in watertight boxes in tunnel wall niches adjacent to the fire alarm stations. They are equipped with magnetic ringing devices. A flashing white light may be incorporated into the traffic light box to call guards to the telephone.

External Telephones

Connections are provided in the control room, in all electrical switchboard rooms, in the tunnel police stations, in the maintenance building and at other vital points to the local public telephone system. Under certain conditions, direct lines to city police and fire departments may be desirable. At the tunnel plazas, pay telephones may be installed for the convenience of drivers.

AM Rebroadcast Radio System

This system provides continuous commercial radio receptions to motorists in the tunnel over a reradiating antenna installed the full length of the tunnel under the ceiling or on a wall. It also enables the operator to transmit emergency messages and instructions to drivers by means of a voice-modulating generator covering the entire AM spectrum, regardless of the station to which radios may be tuned. Another system transmits a low-power signal at a fixed frequency set on a quiet spot of the AM band, and motorists are directed, by an illuminated sign, upon entering the tunnel, to tune to this frequency. Fewer motorists would be reached by this method, and tuning to a frequency without a signal is difficult. The first system is, therefore, preferable. Motorists are requested by signs at the entrance to turn on their radios.

Police and Maintenance Radio

Operating on different frequencies of the VHF band, this system allows police and maintenance crews to communicate with the central control, also from inside the tunnel, and with state or city police and maintenance departments. The frequencies are usually under the jurisdiction of the city or state.

CONTROL CENTER

All tunnel operations, including power distribution, mechanical and electrical systems and traffic signals, are controlled from the control center, usually located in one of the ventilation buildings. The control board for power, ventilation, tunnel lighting and mechanical systems is described in Chapter 21.

Display Board

Additional sections of the control board include displays of the status of all traffic control devices, traffic lights, variable message signs, lane-use signals, a vehicle detection system and traffic display monitors. Equipment malfunctioning and abnormal system operating conditions will actuate audible and visual alarms on the display board through an annunciator.

Control Console

A control console with all devices for complete traffic control should be located with a line of sight to the display board from the operator's position. It should be arranged to permit the operator to perform all necessary functions from a seated position. For a dual tunnel computerized system, the following are included: a line printer, a computer CRT output, a CRT input keyboard, two television master monitors, video switches, controls for outside television cameras, an escutcheon control panel, telephones, radio control and a speaker, and a digital clock. Manual control of traffic control components is conducted from the escutcheon panel.

Computer Control

In order to regulate flow of traffic on multilane tunnel systems and to furnish the operator information for quick and precise action, traffic control computer systems have been developed. The one used on the Hampton Roads Crossing illustrates this. Under certain conditions, such as detection of congestion in the tunnel, the computer initiates automatically the preprogrammed actions and warning signals to alert the operator. In other cases, such as incidents, lane closings or lane reversals, the operator manually activates computer programs to control traffic signals and variable message signs. The computer also controls read-outs on the map display board and on the control console escutcheon panel. By switches on the escutcheon panel, the operator can override all actions of the computer and manually change the setting of each sign and signal. All such changes will immediately be shown on the map display board. The computer, through the CRT monitor, provides the operator with information to perform complicated traffic control maneuvers efficiently. He types into the CRT keyboard the appropriate code to initiate the desired maneuver. Changing traffic conditions manually has to be done in proper sequence, allowing adequate time between steps, and will take longer. Messages will be instantly displayed on the board.

Data can be accumulated by the computer for generating reports on operations such as 24-hour volume counts in 15-minute increments, summaries of incidents and changes in road conditions.

MODES OF OPERATION

Modes of operation are illustrated by two facilities: a single two-lane, bidirectional tunnel with manual operation, and a dual tunnel with a complex approach highway system.

Single-lane Tunnels

A one-lane tunnel is equipped with traffic surveillance and controls, sound-powered emergency telephones, fire alarms and overheight detectors. The following modes of operation will apply.

- *Normal traffic*, with both lanes open, traffic signals on green and speed signs at normal permitted speeds. Operator observes traffic on closed circuit television or receives reports from tunnel patrolmen of abnormal conditions.
- *One lane closed* by temporary closure. Operator sets traffic signals upstream of stoppage on red and signals in opposite lane approaching stoppage on amber and dispatches emergency crew to location. If possible, traffic will be directed alternately around the breakdown by patrolmen as signals are changed from red to amber.
- *Both lanes stopped* by serious accident or fire. All upstream signals in both lanes, including open approaches, are set on red, signals on the downstream side remain green to permit traffic to leave the tunnel and to permit emergency vehicles to enter. Traffic signals on approach highways are changed to amber, and signs are set to reduced speed.

Overheight Detectors. These are located on the approach highway at a distance permitting an alerted guard stationed near the upper end of the open approach to divert the overheight vehicle. If an inspection station is provided, all trucks are directed by illuminated sign, actuated by the overheight detector to stop for inspection.

Radio Antenna. An antenna may be installed the full length of the tunnel to permit reception of broadcasts, including emergency instructions over the car radios and communications for patrol vehicles inside the tunnel.

Dual Tunnels

With a total of four or more traffic lanes, dual tunnels require more elaborate control and communication systems. This is illustrated by the Hampton Roads project on I-64, completed in 1976. It consists of dual eastbound and westbound highways and bridges connecting two two-lane tunnels crossing Hampton Roads between Willoughby Split, Norfolk and Hampton, Virginia. There are five zones of control.

Nine modes of operation are provided. Vehicle detection systems, closed circuit television monitoring of traffic and carbon monoxide indicators provide the operator with the necessary information to activate computerized control systems which automatically, in sequence and by zones, activate variable message signs, speed, lane-use signals and traffic lights. Traffic density and overheight detection systems are automatic and computer controlled.

Radio, sound-powered telephones and walkie-talkies permit communication between mobile and foot patrol police and the controller. Together with closed circuit television and vehicle detection, these systems provide flexible surveillance and control of operations.

For identification, EB lanes are even numbered, 2 and 4, and WB lanes are odd numbered, 1 and 3, the lower number indicating the right-hand lane.

Normal Operation. All lanes eastbound (EB) and westbound (WB) are open and operating at posted speeds as follows.

- Zone A—variable EB. Traffic enters at 55 mph and approximately $\frac{2}{3}$ of the way through the zone, speed is reduced to 45 mph.
- Zone A—variable WB. Traffic enters at 45 mph and approximately $\frac{2}{3}$ of the way through zone A, the speed is increased to 55 mph.
- Zone E—variable EB. Traffic exits zone

D at 45 mph and approximately $\frac{3}{4}$ of the way through zone E, speed is increased to 55 mph.

- Zone E–variable WB. Traffic enters zone at 55 mph and approximately $\frac{2}{3}$ of the way through zone E, the speed is reduced to 45 mph.

In Zone A–EB and zone E–WB, trucks, trailers and campers are directed by variable message signs and lane-use signals into lanes 1 and 2, where they remain until exiting zone E–EB and zone B–WB.

One-lane Operation–Either EB or WB. In the event that one lane of traffic must be partially closed for a short time, either EB or WB, lane shifts are actuated by the controller. Variable message signs, including speed reduction signs, lane-use signals and traffic lights, are automatically activated by computer in sequence by zones starting from the most remote zone to the zone to be closed.

After traffic has cleared the zone where the lane has been closed, it is automatically directed by computer back into normal dual-lane operation by lane-use signals, and variable message signs, including increased speed limits.

In zone C, the tunnel, lane shifting is prohibited except under local direction of tunnel police, and only at restricted speeds, which are manually set by the operator.

To reopen a closed lane, the operator, after checking by closed circuit television and communication with policemen at the lane closure, manually actuates variable message signs, lane-use signals, traffic lights and speed controls. These are automatically activated by computer, starting from the closed zone in reverse sequence.

Emergency Operation. In the event of an accident or fire, traffic is stopped in both lanes, either EB or WB. Detection of the emergency is by observation of the closed circuit television monitors in the control room; by notification via radio or sound-powered telephone by patrolmen; or by vehicle detectors.

The operator manually actuates, by zone, variable message signs, including speed controls, lane-use signals and traffic lights, which automatically, by a precomputerized sequential program, stop and/or slow down traffic from the scene of the emergency to the zone most remote from the accident.

After determining the nature and severity of the accident, the operator dispatches the necessary emergency equipment and additional police to the scene. Local traffic control is accomplished by voice communication via radio, walkie-talkie or sound-powered telephone between police at the site and the operator.

To reopen the dual lanes closed, the operator, after checking the closed circuit television and after checking with the police at the site of the emergency, manually actuates, in reverse sequence, variable message signs, lane-use signals and traffic lights, which are automatically activated by computer, starting from the closed zone to the most remote zone from the closed zone.

Single Tunnel Operation. In the event that one tunnel or approach bridge through zones B and D has to be closed for a long duration for maintenance, traffic is diverted into a single lane in the tunnel remaining in service. Then traffic from the opposite tunnel, which is to be closed, is directed into a single lane (either 3 or 4) and diverted at the crossover into the tunnel and approach bridges in service. This results in bidirectional traffic in a single tunnel.

- To close the WB tunnel and approach bridges, the following procedures are followed.

  1. The maintenance crews (or police) are ordered by the operator to remove and relocate the removable guard rail barrier sections at each of the median crossovers on the Willoughby Spit, and on the Hampton side.
  2. The roll-away sequential arrowboards at each crossover median are then placed into proper postion.
  3. Patrolmen at each crossover stop traffic, and portable barricades are placed into

position in preparation for merging two lanes of traffic to one in a preconceived pattern.

4. At the Hampton end (while traffic halted), the portable supporting posts for the vertical panel chanelizers are inserted into the railway sockets.
5. A patrol car is dispatched from the Hampton side to verify the fact that the EB lane 4 and approach bridges are clear of all traffic.
6. When the patrolmen inform the operator that all is in readiness to permit bidirectional traffic in one tunnel, the electronic strobe flashing lights located on top of the chanelizer, in the form of a flashing directional arrow, are energized at the same time that the variable message signs, changes in posted speed limits and lane-use control signals are placed into operation.
7. WB traffic at the Norfolk side is directed into land 3, proceeds through the crossover and travels westbound in lane 4 through the tunnel. After exiting zone B, the WB traffic in lane 4 is directed into the Hampton crossover to normal WB lanes.

- To reopen the WB tunnel and restore normal dual EB and WB operations, the following procedures are followed.

  1. A partol car inspects the WB roadway (lane 3) from Hampton to Willoughby Spit to ensure lane clearance and then takes position in the unused area of the Willoughby Spit crossover.
  2. At the direction of the operator, and after the portable barricades are removed, the signs at Willoughby Spit are then blanked out or activated to indicate normal operation for the WB roadway. The EB signs would still indicate lane 4 operation in the WB direction in order to clear the EB roadway (lane 4) of WB traffic.
  3. After replacement of crossover guard rail sections to close the medians, the patrol car proceeds to the Willoughby Spit crossover, then westerly along the EB roadway in lane 4 to Hampton to ensure the clearance of the lane of all vehicles, portable barricades and vertical panel chanelizers.
  4. The operator is informed that lane 4 is cleared for EB operations. Then signs at the Hampton side (zones A and B) are blanked out or activated to indicate normal dual-lane operations for EB lanes 2 and 4.

Overheight Vehicle Detection. All traffic traveling eastbound on I-64 passes under an overheight vehicle detector. When the detector records an overheight vehicle, an alarm is automatically set off in the Inspection Station. Simultaneously, a sign located approximately 600 feet eastward of the detector is energized. This sign has alternately flashing amber lights with an audible alarm and bears the following legend: "When Flashing, Trucks Stop for Inspection—$\frac{1}{2}$ Mile." Trucks merge to lane 2 and exit at the Inspection Station for overheight vehicles.

Upon inspection, if the vehicle is over 13 feet 6 inches (the official clearance in the tunnels), the driver is directed to proceed eastward in lane 2 to ramp G and exit the project via John Street in Hampton. In order to prevent back-up of trucks in lane 2 waiting for inspection, a height detector at the Inspection Station is used to automatically measure the height of all trucks passing through the Inspection Station. This detector flashes a red light, informing the policeman of an overheight vehicle passing through the height indicator. All legal height vehicles pass through the Inspection Station under the height indicator, at reduced speed, and merge into land 2 with a minimum of delay and increased safety.

An additional overheight detector is located 400 feet west of the Inspection Station to record overheight vehicles entering the project via ramp F. Overheight vehicles automatically set off an alarm located in the Inspection Station. Simultaneously, a sign located at the point where ramp F merges with I-64 eastbound is energized. This has alternately flashing amber lights with an audible alarm and bears the following legend: "When Flashing, Trucks Stop for Inspection—600 Feet."

As a precautionary measure, another overheight detector is located 1100 feet eastward of the Inspection Station. This detector serves as a back-up. In the event that an overheight vehicle does not heed the first alarm and continues past the Inspection Station, the second detector will reactuate the alarms in the Inspection Station and in addition sound an alarm in the North Crash Truck Shelter and in the Central Control Room. Also, after a preset adjustable delay, a flashing sign, located at Station 839 + 25, will be automatically energized. This sign bears the legend: "Prepare to Stop when Flashing." At a further adjustable delay, the lane-use signals at the east end of zone B change from green arrows to red *X*'s bringing both lanes (2 and 4) to a full stop. In progressive sequence, lane-use signals turn from green arrows to red *X*'s westward for lanes 2 and 4, to the westerly limit of zone B. Speeds in zone A (for lanes 2 and 4) progressively decrease for traffic in the EB direction.

Provisions have been made to enable the policeman at the Inspection Station to override the automatic operation and manually actuate the lane-use signals and speed limit signs while at the same time in communication with the operator.

In the interim, for either automatic or manual operation, the policeman at the Hampton Inspection Station notifies, by sound-powered telephone, the operator and the policeman at the North Crash Truck Shelter of the overheight vehicle that has not stopped.

The policeman at the North Crash Truck Shelter, alerted by the alarm, proceeds to the west end of zone B. When all EB traffic is stopped, he manually sets a height indicator, at the west limit of zone B, to verify the overheight vehicle. In the interim, he directs traffic past the red *X*'s until the height indicator flashes a red light, indicating the overheight vehicle. This vehicle is then diverted onto the North Island. After diversion of the overheight vehicle, the policeman at the North Island notifies the operator, who manually activates a computerized program which returns lane-use signals and speed control signs in reverse sequence to normal operation.

The North Crash Truck Shelter policeman notifies the operator to reduce the speed limit in lane 3, zone C, WB to permit entrance of the overheight vehicle to WB traffic in order for it to exit the project. After notification by walkie-talkie from the policeman, the operator restores normal operation speeds in lane 3 in the reverse sequence.

Similar overheight detectors are provided for WB traffic entering zone E. Diversion procedures are similar to those outlined for EB traffic.

Dangerous Cargo Detection. Transportation of dangerous cargo in *Interstate* and Foreign Commerce is regulated by the Code of Federal Regulations, Title 49, Parts 100 to 999, for all persons using public highways.

Section 397.9, Title 49 prohibits the transportation of dangerous cargo within a tunnel. Section 390.3, Title 49 permits state and subdivisions thereof to make local laws relating to safety which will not prevent full compliance with Title 49. The Virginia Toll Revenue Facilities, in which the Hampton Roads Bridge-Tunnel facility is included, has published and issued, "Rules and Regulations Covering the Use of the Hampton Roads Bridge-Tunnel Project." These regulations stipulate the type, quantity and required container for shipment of hazardous materials through the facility. In addition, all tank trucks, tractor-trailers, housecars and campers with gas bottles over 60-lb capacity are prohibited from using the facility.

A series of advance warning signs are located along the east and westbound lanes of I-64 in advance of control zones A and E, to advise drivers of vehicles transporting prohibited materials that they are legally required to leave I-64 at the next exit and seek an alternate route.

Drivers who continue along I-64 and enter control zones A and E are warned by signs to stop for inspection at the Inspection Stations.

Inspection personnel will question the drivers and check manifests, bills of lading and other documents pertaining to the transport of materials or articles. If necessary, the inspector may open sealed compartments and inspect and weigh containers.

After inspection, vehicles determined to be in violation of the project regulations will be directed to leave the facility, under police escort, at the nearest exit ramp.

In the event that tank trucks, tractor-trailers, housecars or campers do not stop for inspection and continue onto the approach bridges (zones B and D), the policeman at the Inspection Station will manually set in motion the same procedures as those stopping and apprehending overheight vehicles prior to their entering zone C (the tunnel).

Fog Detection. If visibility on the roadways is reduced to less than 750 feet at the posted speed, fog sensors provided on the EB roadways automatically activate audible alarms on the central control board and the operator manually activates, in sequence by a precomputerized program, variable message signs indicating fog conditions which will automatically, by zone and in sequence, reduce speed limits on EB and WB roadways. The reduced speed limits, depending upon the degree of visibility, will be selected by the operator.

Ice Detection. If frost, ice or snow creates hazardous driving conditions, sensors provided on the EB roadways automatically activate audible alarms on the central control board and the operator manually activates precomputerized variable message signs indicating ice conditions, which will automatically, by zone and in sequence, reduce speed limits on EB and WB roadways. The reduced speed limits, depending upon temperature, will be selected by the operator.

23
Fire Protection

ARTHUR G. BENDELIUS

Senior Professional Associate
Vice President
Parsons, Brinckerhoff, Quade & Douglas

Fire protection in a tunnel is necessary in order to protect life and property from the damages created by fire. Fire protection includes any system or equipment which aids in the fight against fire, such as water supply, deluge, sprinklers, fire extinguishers and fire alarms.

The vehicular and rail rapid transit tunnels are usually provided with fire mains and fire extinguishers, and, in some instances, with water deluge systems on the exhaust fans. The rail tunnels, because of their extreme length, are usually not provided with fire mains. The installation of sprinkler systems in the roadway area of a vehicular tunnel is rare. Only a limited number of tunnels have such systems because of extensive cost of an installation of this nature.

DESIGN QUANTITIES

The most significant aspect of tunnel fire protection is the water supply, which is usually transported in a fire main located within the tunnel. The minimum flow and pressure requirements must be established prior to implementation of design.

Flow Rate

The National Fire Protection Association (NFPA), in its tentative standard on fire protection for limited access highways, tunnels, bridges and elevated structures,[1] recommends a minimum water flow rate for vehicular tunnels of 1000 gallons/minute at adequate pressure. This will provide sufficient flow for four 250-gallon/minute hose streams (using the standard $2\frac{1}{2}$-inch fire hose). Such a water quantity will also be sufficient for any auxiliary use the system is put to, such as wall-washing.

Pressure

There must be sufficient available pressure at each hose station to permit the use of a proper length of hose and nozzle to create an adequate hose stream. The most appropriate length to consider for design purposes is 100 feet of $2\frac{1}{2}$-inch rubber-lined fire hose. Figure 23-1 indicates the pressure loss for varying flow rates through the standard $2\frac{1}{2}$-inch fire hose. The required pressure at the nozzle to deliver the

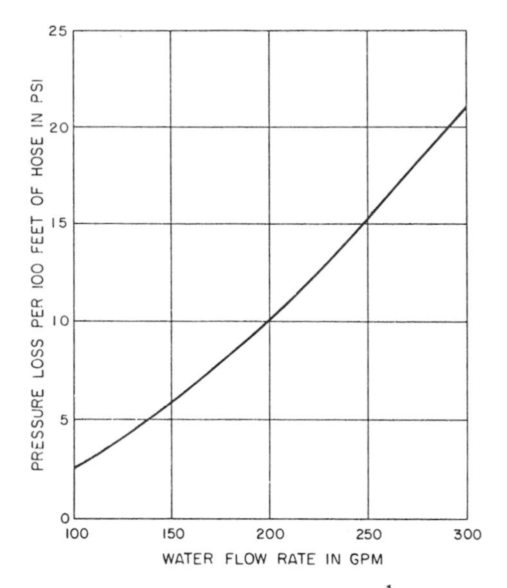

Fig. 23-1. Friction loss per 100 feet of $2\frac{1}{2}$-inch rubber-lined fire hose.[2] adapted.

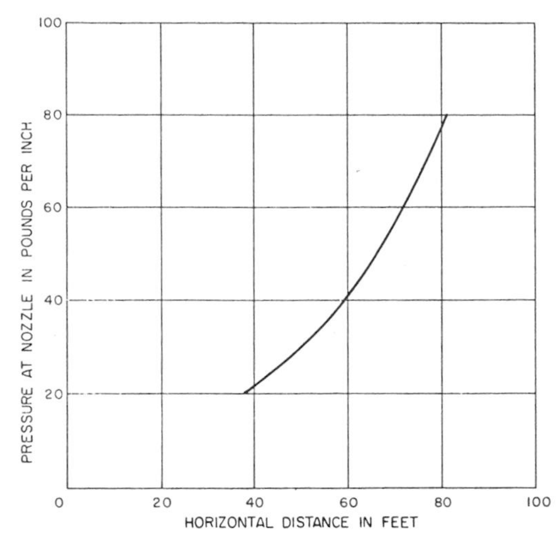

Fig. 23-3. Approximate effective horizontal reach of fire hose streams for $1\frac{1}{8}$-inch nozzle.[3] adapted.

desired quantity of water and create a good water stream must also be determined. Figure 23-2 provides a guide to this pressure. A water flow rate of 250 gallons/minute per hose will require approximately 60 psi pressure at the hose valve when $2\frac{1}{2}$-inch hose with a $1\frac{1}{8}$-inch nozzle is used. This combination will produce an effective horizontal fire stream of 60 to 70 feet (Fig. 23-3). A nozzle pressure of 45 psi will be sufficient for the use of spray nozzles.

In a subaqueous tunnel, the fire hose valve located at the low point of the tunnel will, because of the large static head, have a greater available pressure than the fire hose valve located near the portal. Therefore, the fire hose valve located near the portal will, in most cases, be the most critical from the standpoint of maintaining a minimum pressure requirement. In a mountain tunnel of fixed grade, the available pressure at opposing ends of the tunnel will vary greatly due to the large difference in elevation. This fact must be considered when a water supply system is being designed for a mountain tunnel.

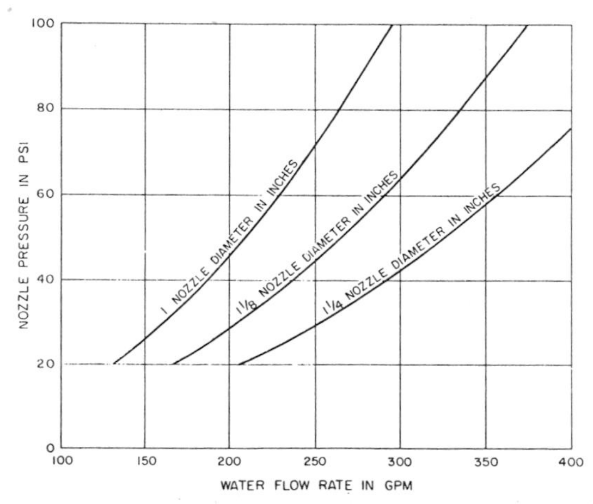

Fig. 23-2. Discharge through solid stream nozzles.[2] adapted

Should the pressure, or, in some cases, the water flow rate, be insufficient to provide the required pressure at the nozzle, booster fire pumps must be considered, as outlined later in this chapter.

WATER SUPPLY

The selection of a source of water for the fire protection system of a tunnel must be made early in the design process. In the case of a tunnel located in an urban area, the choice is relatively simple, since large quantities of water are normally available from a municipal water supply system. A municipal water supply system should be the prime choice as a water source for any tunnel when it is available. This is a valid approach provided that the municipal water lines available are relatively close and have an adequate flow rate and pressure to meet the flow and pressure criteria required at the tunnel. If, however, the municipal water lines are not within a reasonable distance from the tunnel, it may be less costly to provide other means of water supply. This could be a well or reservoir system, along with a storage tank and a booster pump.

In cases when the municipal water supply system is available but does not have sufficient flow rate or pressure at the tunnel site, a booster pump or a storage tank and a booster pump will be necessary to provide the required capacity.

When all the possible avenues to use the municipal water supply system have been exhausted, a well or reservoir system must be considered. In order to minimize the cost of such a system, a storage tank should be used with a well or reservoir supply. The well or reservoir would be used to provide makeup water to the tank, thus reducing the size of the well required and the capacity of the well supply system. The tank should be sized to handle the fire-fighting requirements of the tunnel and the ventilation equipment for a minimum of 30 minutes. Using the recommended flow rate of 1000 gallons/minute, a tank of 30,000-gallon capacity would be required. A thorough economic analysis should be made prior to selecting the water supply system for each tunnel. If the cost of the well and additional pumping required is considerable, all attempts should therefore be made to use a municipal water supply system.

Where possible, it is advisable to have a dual

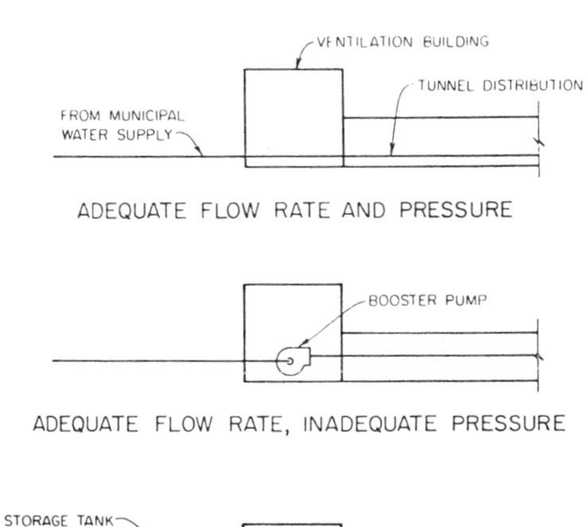

Fig. 23-4. Typical water supply arrangements.

source of water supply for the tunnel fire protection system; that is, one line from each end of the tunnel. This is usually only practical in a subaqueous or urban tunnel, where the municipal water supplies are close at hand. Figure 23-4 shows schematic diagrams of the three basic types of water supply systems utilized in vehicular tunnels.

FIRE MAINS

The supply of water into and through a tunnel for purposes of fire protection is transported in a fire main located in a protected location within the tunnel cross-section similar to those shown in Figs. 23-5 and 23-6.

Tunnels

Consideration must be given to whether the fire main should be wet or dry. The wet fire main is filled with water at all times, whereas the dry type is filled only when water is required in the tunnel.

Wet Fire Main. The wet main is, of course, more advantageous, since the water is available immediately when required during a fire. The major disadvantage occurs in the more northern climates, where freezing weather is prevalent, thus presenting the possibility of having the water in the main freeze.

Freeze protection of the fire main can be accomplished in several ways. Heating the water in the main to keep its temperature above 32° F is the most positive method. However, the cost of such a heating system in a tunnel facility can be considerable. Another approach would be to install the main in concrete within the tunnel structure, thus providing insulation against the freezing temperatures encountered in the roadway or in the air ducts. A third method is to use a bleeding or circulating system, which keeps a continuous movement of water in the pipe, thus minimizing the possibility of freezing. This can be accomplished by a circulating pump, or, if external pressure is available, by a simple bleeding system, such as that shown in Fig. 23-7. This system uses the available municipal water pressure to keep the water flowing.

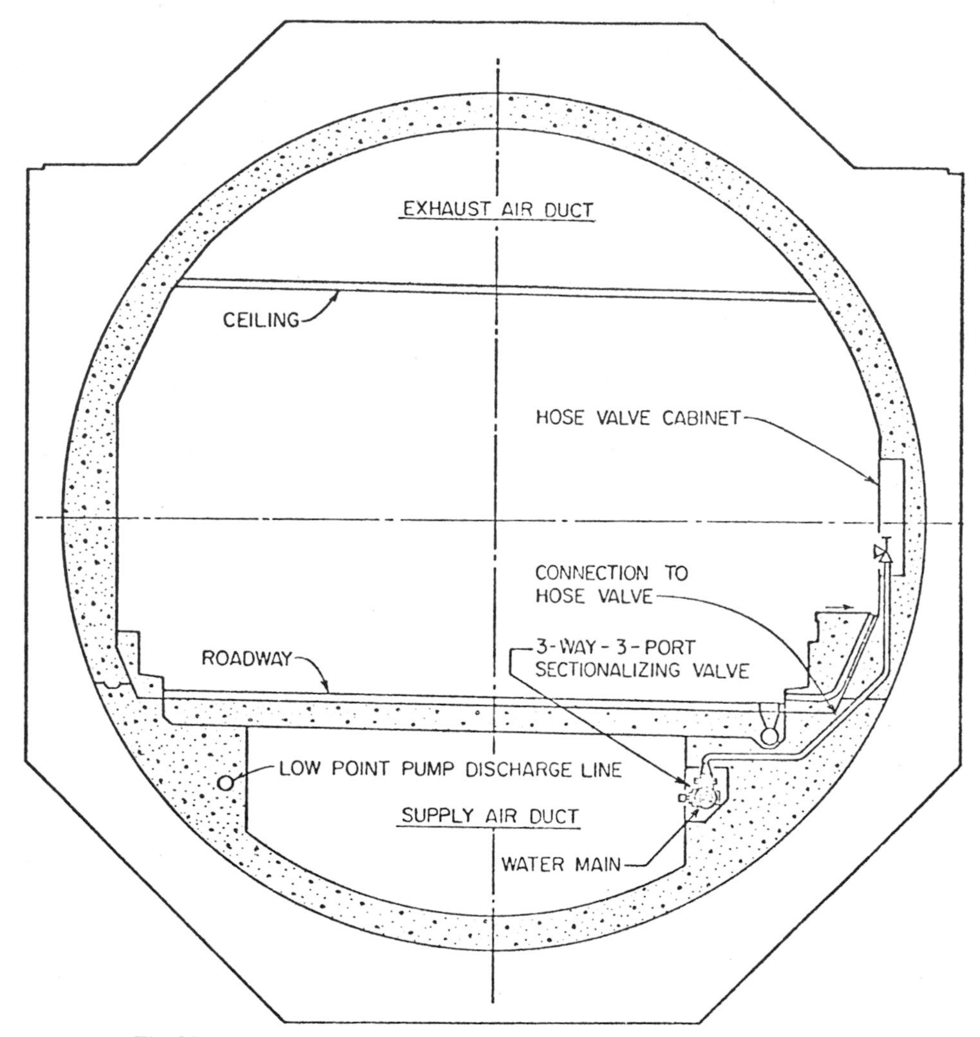

Fig. 23-5. Location of fire mains and hose valves in a subaqueous vehicular tunnel.

Dry Main. The dry main has a serious disadvantage which becomes more significant in the longer tunnels; that is, the length of time required to completely fill the main and thus have water available for fire-fighting operations. As an example, an 8-inch fire main in a 19,000-foot tunnel would require 30 minutes to fill using the normal municipal water pressure. This situation creates a severe handicap during an emergency. Whenever possible, a wet line should be used.

Arrangement. The fire main in a tunnel should be provided with two sources of supply, preferably one at each end of the tunnel. The main should be suitably valved to prevent interconnection of water supplies. A method of valving for this interconnection is shown in Fig. 23-8. The fire main within the tunnel should be arranged and valved in such a way as to permit repair of a damaged section of the system without reducing the fire protection level of the facility. Sectionalizing valves can be used, as shown in Fig. 23-9, to accomplish this.

Open Approach

The water supply system for fire protection within the tunnel should be extended into the open approaches of a subaqueous tunnel where

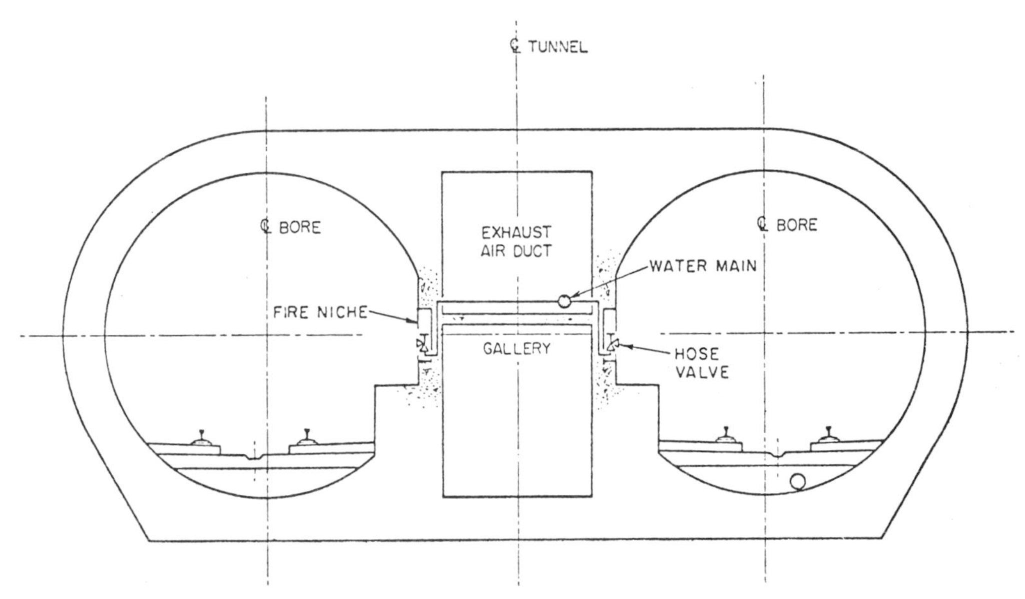

Fig. 23-6. Fire mains and hose valves in a twin tube subaqueous rapid transit tunnel.

there remains a restricted roadway. This will provide fire protection on this portion of the roadway which has limited access.

Buildings

Water supply for the fire protection of the buildings associated with the tunnel should be treated with standpipes according to the rating category defined by local codes, the National Fire Code 4 and the local authorities having jurisdiction.

Size

The fire main should be sized to minimize the pressure drop through the system and to retain sufficient pressure to achieve the required value at the hose valves. This should be verified whether the driving force is created by a booster pump or a municipal water supply.

Material

The piping material used for the fire main is usually steel sufficient to withstand the maximum pressure of the system.

FIRE HOSE STATIONS

Fire Hose Connections in the Tunnel

Fire hose connections are required to be located within the tunnel roadway or trackway area.

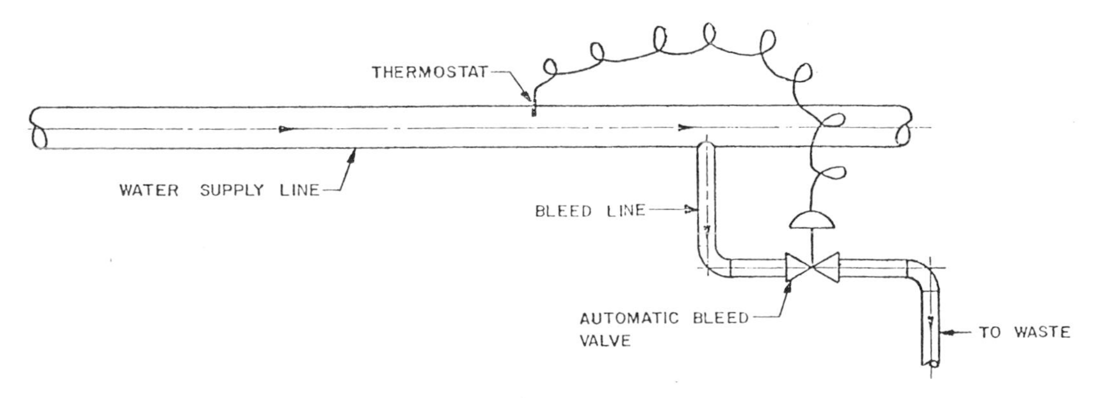

Fig. 23-7. Automatic bleed valve in fire main.

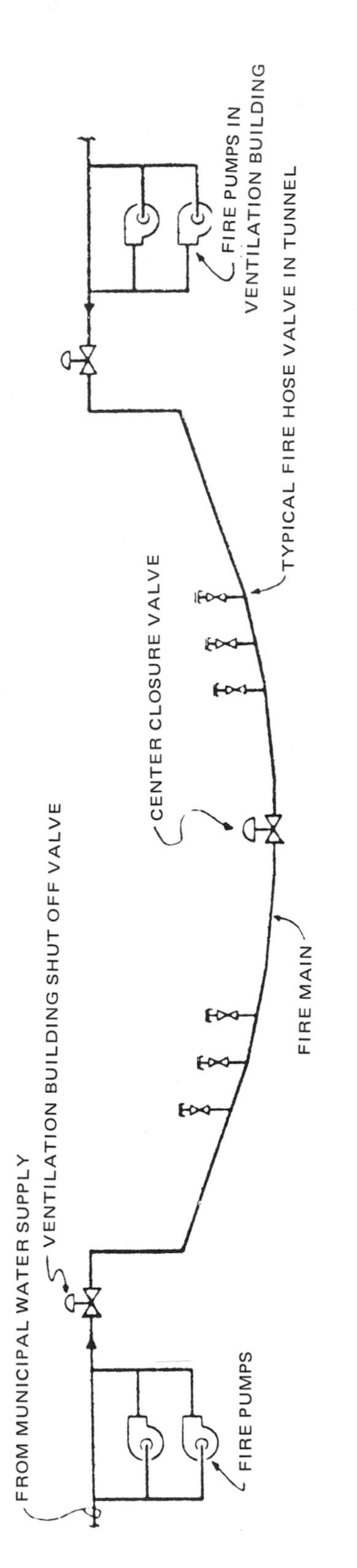

Fig. 23-8. Fire main with booster pumps for dual municipal supply.

These should be conspicuously marked and located at a maximum spacing of 150 feet. In wide vehicular tunnels, spacing of 300 feet on each side, with a staggered arrangement to achieve an equivalent spacing of 150 feet, would be appropriate.

The fire hose stations in each tunnel should consist of a fire hose valve, a fire extinguisher and, possibly, fire hose. In many tunnels, the fire hose has not been installed at the fire hose station, but rather is carried on the emergency vehicle, as noted in Chapter 25. Elimination of the hose from the fire hose station minimizes the amount of hose to be maintained, since unused fire hose is subject to rapid deterioration. For location of the fire hose station in the tunnel cross-section, see Figs. 23-5 and 23-6.

The fire hose valves should be compatible with the standard used by the local fire departments. The valve should be approved for fire service. The majority of fire departments in the U.S. utilize the $2\frac{1}{2}$-inch standard.

The enclosure for the fire hose station should be provided with a door to protect the equipment. This door should be marked to permit ready identification of the station. In a number of tunnels, a system is installed which will provide a signal to the control room when the door of the fire hose station is opened.

Open Approach

The same type of fire hose station used in the tunnel should be installed in the open approach. Another method would be to install hydrants at the surface behind the open approach retaining walls. This method provides for greater access to hose connections with vehicular fire-fighting equipment from the peripheral service roads above, and it does eliminate the depth at which water piping connections must be made.

Buildings

The fire hose stations located within the buildings associated with the tunnel should be installed as required by the National Fire Code[4] and local codes and in accordance with the authorities having jurisdiction.

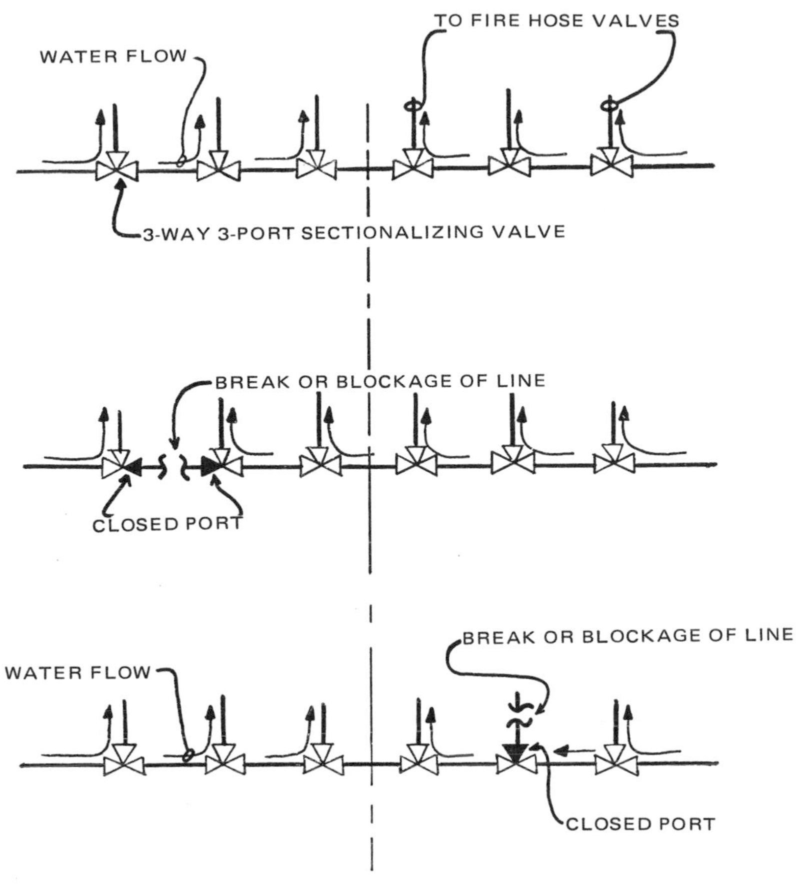

Fig. 23-9. Fire main sectionalizing valve arrangement.

PROTECTION OF EXHAUST FANS

An important part of the fire protection system of any tunnel is the capability to purge the hot gases generated during a fire. This is usually accomplished by the use of fans which are capable of removing the smoke and heat in order to facilitate both the egress of tunnel occupants and the entrance of fire-fighting personnel.

In vehicular tunnels, this purge function is usually accomplished by either the normal exhaust ventilation system, or by a reversible supply system, whereas, in rapid transit systems, a set of fans is installed for emergency ventilation (see Chapter 19).

To provide the assurance of continuing operation of the fans during emergency conditions, consideration must be given to the protection of the critical components such as the power supply, motor, fan controls, bearings and impeller.

The power supply should be safeguarded against fire, flooding and impact, by placing it in protective conduits, possibly imbedded in concrete. The experience in the Montreal Metro fire of December 1971 highlights the need for such protection.[5] As an added precaution, fans located in rapid transit systems should be provided with power from two directions. (For further discussion of power supplies, see Chapter 21).

High-temperature insulation should be provided for all fan motors which could possibly be exposed to hot gases from a fire. In a direct-connected axial fan, this is a must, while, in a centrifugal fan, the motor can be separated from the hot gases. Elimination of the normal thermal protective devices in a motor should be considered in the case of an emergency fan which would be near the intense heat of a potential fire. This step should not be considered for a fan which is used for normal ventilation.

The fan controls and control wiring should be located so as to be excluded from the stream of hot combustion gases and protected from all possible damage, during either normal operation or emergency operation of the system.

The fan itself (namely, the impeller and the bearings) is also susceptible to damage from excessive heat. During the Holland Tunnel fire of 1949,[6] the hot gases heated the fan wheel shaft to a temperature sufficiently high to soften the babbitt bushings of the bearings and thus cause the shutdown of three fans. In the case of axial fans, the required close tolerance between the impeller and the housing is a cause for concern when hot gases pass through the fans, causing serious distortion of the housing and interrupting the fan operation, thus creating a serious condition during a fire.

As protection against this problem of fan damage, deluge systems have been installed in several tunnels, providing the cooling effect of a water spray on the hot gases in the duct or on the critical surfaces.

The tunnels operated by the Port Authority of New York and New Jersey have been provided with a water spray arrangement on the fan equipment. These systems were installed after the 1949 Holland Tunnel fire[6] and consist of two nozzles located so that their spray is directed onto the shaft and bearings, thus cooling the critical portions of these fans. The water is initially supplied from a storage tank, and, if the hot gas flow continues, a pumper connection is utilized to provide the required continued flow rate.

On the Hampton Roads and the Elizabeth River Tunnels in Virginia, a deluge system has been installed to protect the axial fans in the exhaust ventilation system. These deluge systems consist of eight water spray nozzles arranged on each fan, as shown in Fig. 23-10. Each nozzle has a capability of delivering 50 gallons/minute of water at 100 psig nozzle pressure. The water flow is controlled by a deluge valve, which, in turn, is actuated by heat detectors located in the exhaust air duct.

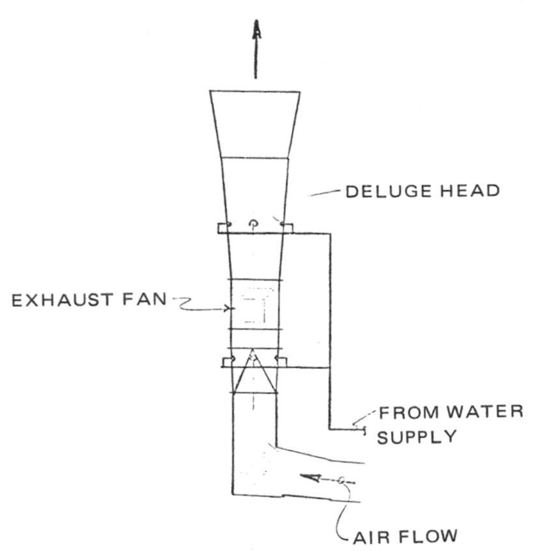

Fig. 23-10. Exhaust fan deluge system arrangement.

There is a limited amount of design or test data available regarding fan deluge systems. Most of these systems in the U.S. were installed after the Holland Tunnel fire, and little, if any, research was done prior to their installation. The report regarding the Montreal Metro fire[5] has suggested the incorporation of such spray systems on fans for rapid transit facilities.

FIRE PUMPS

Fire pumps will be required in the tunnel fire protection system to boost the water supplies available from the municipal system's tanks and reservoirs. The fire pump is, in fact, a booster pump used to boost pressure and flow.

Centrifugal Fire Pump

The standard fire pump is the centrifugal type, which has been specially approved for fire pump service. The major advantage of the centrifugal pump is that the discharge flow is reduced when the head increases, thus preventing a build-up of pressure in the piping system.

The standard fire pump sizes, by rated capacity, are 500, 750, 1500, 2000 and 2500 gallons/minute with pressure ranging from 200 psi for horizontal pumps and 70 to 280 psi for vertical pumps. There is also a series of smaller special service pumps having rated capacities of

200, 300 and 400 gallons/minute, with pressure ratings of 40 to 100 psi. These special pumps have an overload limit of 130%.

Horizontal Type. The horizontal shaft, single-stage, double-suction volute pump is the one most often used in tunnel fire protection systems. In this pump, the suction inlet water flow separates and enters the impeller through the eye on both sides.

Pump Characteristics. A typical set of capacity characteristic curves for a fire pump are shown in Fig. 23-11. These include a plot of total head, brake horsepower and efficiency versus discharge water flow rate and are based on constant speed operation.

The total head of a horizontal centrifugal pump can be defined as the vertical distance between the level of the water supply source and the level of the water equivalent to the discharge pressure. This includes all pipe friction and fitting losses, as shown in Fig. 23-12. For evaluation of these losses, an appropriate hydraulic handbook should be consulted.

Specific speed of a pump relates head, capacity and speed, and can be defined as the rotational speed of a geometrically similar impeller that will deliver 1 gallon/minute at 1 foot of water head and can be computed as follows.

$$NS = \frac{\text{rpm} \times \text{gpm}^{1/2}}{\text{HD}^{3/4}}$$

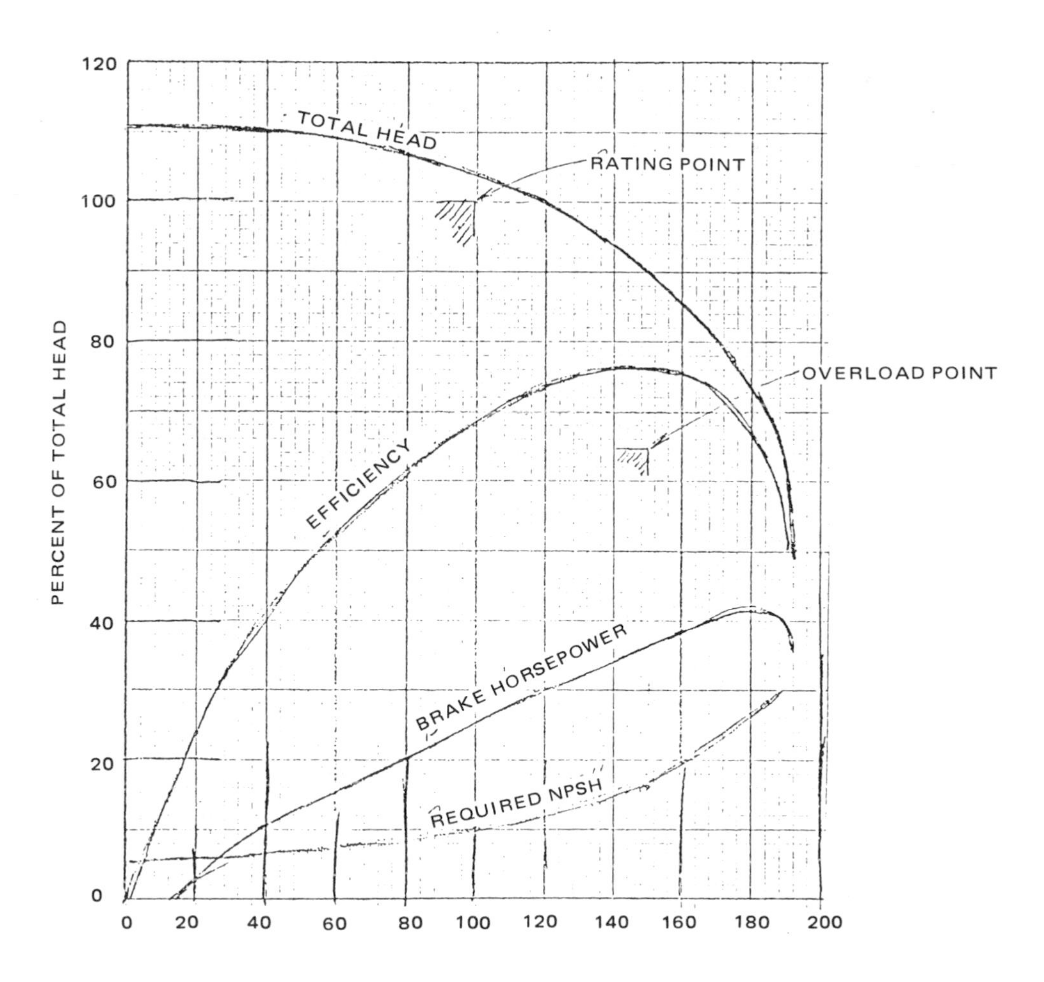

Fig. 23-11. Typical fire pump performance curves.[2]

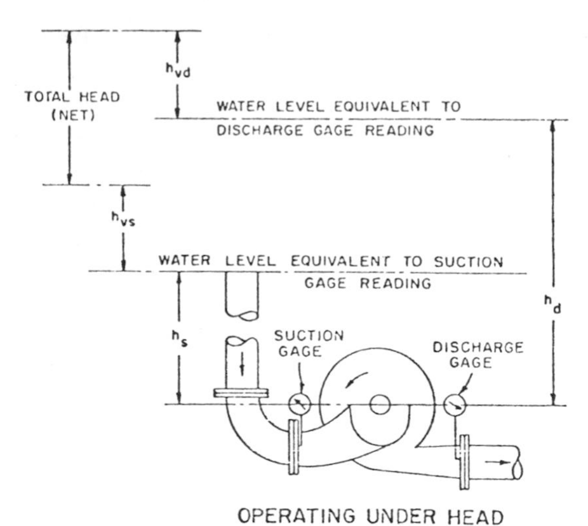

Fig. 23-12. Typical head of horizontal centrifugal fire pumps.[2]

where

NS = specific speed (revolutions/minute)
rpm = rotational speed (revolutions/minute)
gpm = discharge flow rate (gallons/minute)
HD = total head (feet of water).

Specific speed is a useful guide for determining the maximum suction lift or minimum suction head for a pump. A pump of low specific speed will accommodate a larger suction lift than one with a higher specific speed, provided both have equal capacity.

Net positive suction head (NPSH) is the pressure head that causes water to flow into the impeller of pump. The available NPSH in the system must always be equal to or greater than the required NPSH of the pump at the design operating conditions. The required NPSH of a pump can be determined from the pump performance curves, as seen on Fig. 23-11.

The suction lift of a fire pump should not exceed 15 feet, as recommended in the National Fire Code.[4] This should be reduced by 1 foot for each 1000 feet of elevation of the pump location.

Cavitation in a pump can cause severe damage and possibly failure. This effect is created when the pressure at the pump inlet drops below the vapor pressure of the water, thus creating a bubble of gas which, when exposed to the higher pressure regions of the pump, collapses. This collapse, depending on its severity, causes the damage to the pump. Therefore, proper design steps must be taken to prevent cavitation.

Performance Curves. The acceptable shape of a pump head curve is affected by three limiting factors: shut-off, overload and rated capacity, as shown in Fig. 23-11.

- *Shutoff head.* The total head of a horizontal centrifugal pump should not exceed 120% of the rated head at 100% capacity with a pump operating at rated speed and the discharge line blocked.
- *Overload.* The total head should not be less than 65% of the pressure at rated capacity at 150% of rated capacity.
- *Rating.* The capacity head curve should pass above the point of rated capacity and head to allow for normal variation in pump construction and dimension.

All of these factors must be considered when selecting a fire pump.

Pump Selection. Fire pumps must be carefully selected based on the system resistance characteristics curve, the pump performance curve and the factors outlined above, as shown in Fig. 23-13. The pump capacity and pressure must be sufficient to adequately meet the design and code requirements.

The brake horsepower for a fire pump can be estimated by using the following formula.

$$\text{Bhp} = \frac{\text{gpm} \times \text{HD}}{1710 \times \text{EFF}}$$

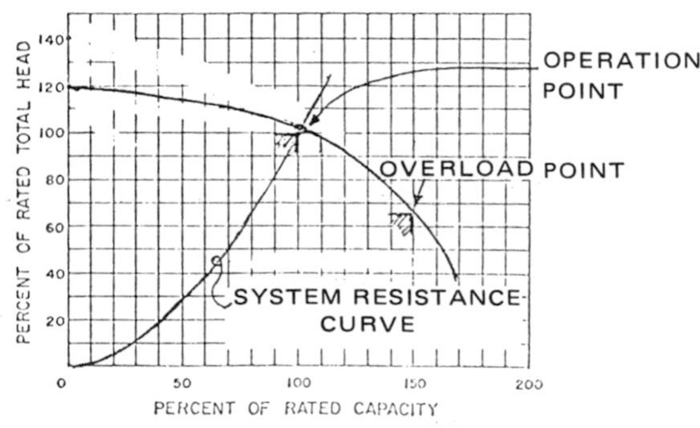

Fig. 23-13. Standard performance and system curves for fire pumps.[2]

where

Bhp = brake horsepower
gpm = discharge flow rate (gallons/minute)
HD = total head (psi)
1710 = conversion factor
EFF = efficiency (percent/100): also water horsepower/input horsepower.

Approvals. The National Fire Code[4] recommends that only approved equipment be used for fire pump installations. The pump, when installed in accordance with the National Fire Code, should perform satisfactorily. The pump manufacturer should be responsible for testing the pump equipment, obtaining equipment approval and providing certified performance curves.

Fire Pump Drives. Most fire pumps are driven by electric motors, since a high degree of reliability is usually developed by the overall tunnel power system, and the chances of power loss is remote. Also, in many of these tunnels, a standby emergency generator driven by an internal combustion engine is available for critical services (such as the fire pumps). However, in the instances where the electrical power reliability is low, an internal combustion engine drive should be considered for the fire pump drive. The fire pump motor is not required to be specially for fire service but it should comply with all NEMA specifications. All electric equipment and wiring in a fire pump installation should comply with the National Electric Code,[7] except where modified by Volume 6 of the National Fire Code.[4]

Motor controllers for fire pumps are specially approved and listed for alternating current fire pump motors at standard voltages up to 600. An approved controller is a completely assembled unit, wired and tested and ready for service. Motor controllers are listed for combined and automatic operation, or for manual operation only.

The internal combustion engine, powered by diesel fuel, gasoline and natural gas, is listed for fire pump service.

Auxiliary Devices. There are a number of auxiliary devices which are required for the proper functioning of an approved fire pump installation.[4]

- A *relief valve* on the pump discharge to remove excess pressure during pump operation. (These are required on all fire pump installations with adjustable speed drives and with shut-off of pressure above 100 psi or above the pressure rating of the system.)
- *Horse valves* for test purposes, usually $2\frac{1}{2}$ inches, approved for fire hose connection.
- An *automatic air release* valve on top of the pump casing, necessary when the pump is to be automatically or remotely controlled.
- *Circulating relief valves,* necessary on all automatic or remotely controlled fire pumps to ensure adequate cooling of the pump when the discharge flow rate is low. (At a pressure set slightly above the rated pressure, this valve begins to circulate water.)

Pump Arrangement. It is recommended that 100% standby pump facilities be installed in the tunnel fire protection system.

Pump Control. Most tunnel fire pump installations are arranged for automatic or remote manual control. The horizontal centrifugal fire pump, which is under automatic control, should be installed with a suction head to avoid priming problems.

Pump Rooms. The fire pump should be located in a dry enclosure, protected from fire, explosion, flood, dirt, corrosion and unauthorized access. The enclosure should be provided with light, heat, ventilation and floor drainage. Figure 23-14 shows a typical tunnel pump room.

Acceptance Tests. Tests should be conducted on each completed tunnel fire pump installation, in accordance with Volume 6 of the National Fire Code,[4] to demonstrate the adequacy of the pump suction and the ability of the pump to deliver the quantity of water in accordance with the performance data.

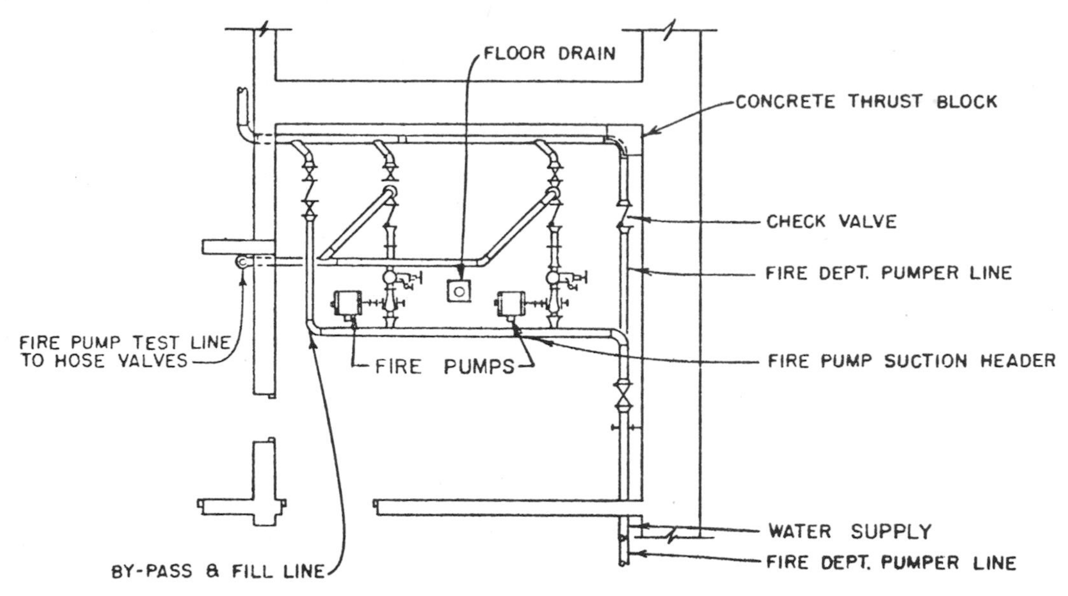

Fig. 23-14. Fire pump room arrangement.

FIRE EXTINGUISHERS

Portable fire extinguishers should be provided within the roadway area of a vehicular tunnel for use in arresting small fires. The extinguisher should be of the multi-purpose, dry chemical type with an ABC rating. The minimum capacity should be one with a 4-A; 40-B:C rating. The maximum size considered convenient for this use should be a 20-lb unit.

The extinguishers should be spaced such that the distance between units is not greater than 150 feet on each wall of the tunnel. It is preferable, where possible, to have fire extinguishers on both sides of a vehicular tunnel roadway. A staggered spacing arrangement would place the extinguisher at a maximum distance of 75 feet from any possible fire.

The extinguisher should be mounted in a well marked flush wall enclosure, preferably with a door. Often the fire extinguisher is located in the same enclosure with the fire hose valve.

The cabinet should be arranged so that when the door is open or the extinuisher removed, an audible alarm is sounded in the control center to alert the operator of the use of this equipment. This arrangement will discourage and signal unauthorized use or theft of fire-fighting equipment.

FIRE ALARM SYSTEM

A complete description of fire alarm systems and the interacting surveillance and communications systems, such as telephone, radio, and closed circuit television, can be found in Chapter 22.

REFERENCES

1. "Tentative Standard on Fire Protection for Limited Access Highways, Tunnels, Bridges and Elevated Structures," NFPA No. 502 PT, *1972 NFPA Technical Committee Reports,* Boston Massachusetts, 1972.
2. *Fire Protection Handbook, Fourteenth Edition*, National Fire Protection Association, Boston, Massachusetts, 1976.
3. Freeman, J.R., *ASCE Transactions* **XII**.
4. National Fire Code, Volume 6, National Fire Protection Association, Boston, Massachusetts, 1980.
5. Donato, G., *Montreal Metro Fire, December 9, 1971,* International Metropolitan Railway Committee, International Union of Public Transport, New York, 1972.
6. *The Holland Tunnel Chemical Fire, May 13, 1949,* Report by The National Board of Fire Underwriters, New York, 1949.
7. National Electric Code, National Fire Protection Association, Boston, Massachusetts, 1980.

24
Drainage Systems

ARTHUR G. BENDELIUS

Vice President
Senior Professional Associate
Parsons, Brinckerhoff, Quade & Douglas

A drainage system is required in all tunnels to remove water from rainfall, tunnel-washing operations, vehicle drippings or fire-fighting operations, or from any combination of these sources. Drainage of a tunnel can be accomplished either by a gravity flow system or by a pumped system. For tunnels with continuous grades, a gravity flow system will suffice, provided that the collected water can be disposed of at the lower end of the tunnel. It will most likely be necessary to utilize a pumped system in a tunnel having a change in grade which creates a low point within the tunnel.

DESIGN QUANTITIES

The drainage system design must be predicated on a proper determination of the anticipated flow rate; that is, the peak discharge rate of the water to be drained. This water can be from rainfall, tunnel-washing operations, fire-fighting operations, vehicle drippings or seepage.

Rainfall

In order for an adequate drainage system to be designed, information must be available regarding the amount of expected rainfall in the areas to be drained. Three key factors in determining the amount of water to be drained from rainfall are intensity, frequency, and time of concentration. Frequency of a storm, expressed in years, is the average number of years between occurrences of a storm of a given or greater intensity. The probability of more severe storms increases as the design period increases. For the normal drainage design, a 10-year storm frequency is used; however, for a tunnel where flooding is a serious concern, such as in a subaqueous tunnel, a 100-year storm frequency should be considered.

Time of Concentration. The time of concentration is defined as the time required for runoff from the most remote point of the drainage area to arrive at the drainage outlet. The most remote point is the point from which the time of flow is greatest, not necessarily the greatest distance. The chart shown in Fig. 24-1 provides estimates of surface flow time.

Intensity. This is the amount of rainfall within a specific period of time, usually given in inches/hour. The rainfall intensity can be determined

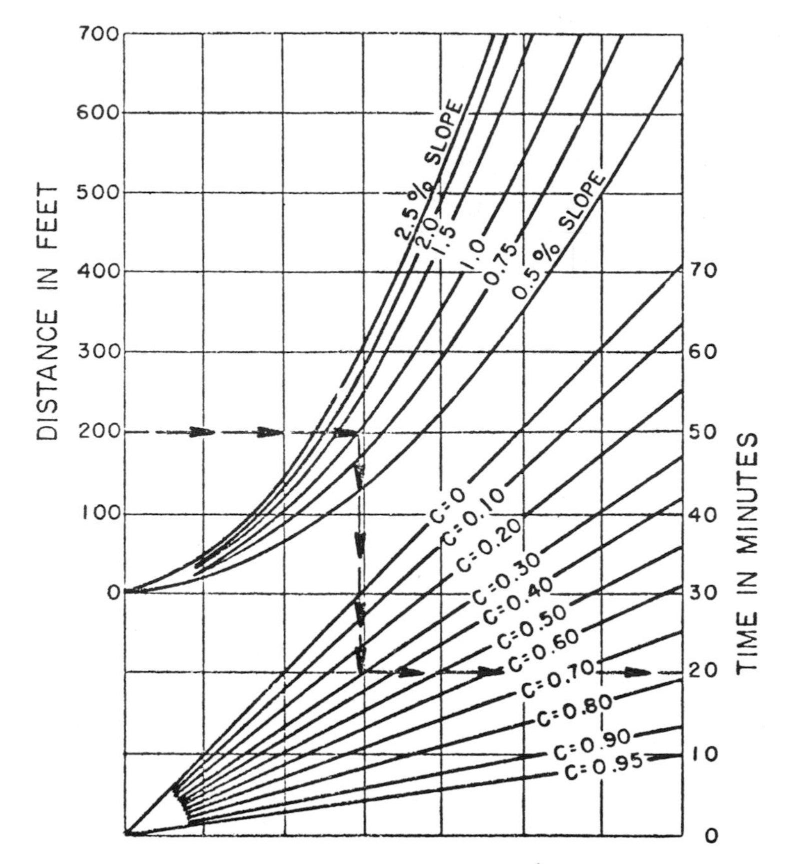

Fig. 24-1. Surface flow time curves.[1]

by using the frequency and the duration and entering the proper rainfall intensity curve. A sample curve is shown in Fig. 24-2. Intensity-duration curves for many areas of the U.S. are published by the U.S. Weather Bureau.[2]

Runoff. The amount of water left from a rainfall after the losses from evaporation, transpiration and infiltration is the runoff. The method commonly used when computing runoff for small areas is the Rational Method,[1] as shown in the following formula.

$$Q = CAI$$

where

Q = peak runoff rate (feet^3/second)
C = runoff coefficient
A = drainage area (acres)
I = rainfall intensity for the time of concentration and storm frequency selected (inches/hour).

The above formula, while not dimensionally correct, is satisfactory since 1 foot^3/second is approximately equal to 1 acre inch/hour.

The *runoff coefficient* is the ratio of the rate of runoff to the rate of rainfall at an average intensity when the entire drainage area is contributing. A list of runoff coefficients is shown in Table 24-1. For most tunnel applications, the value of the runoff coefficient for surface drainage will be about 0.90, since most of the drainage areas will be sloped pavement. Therefore, we can modify the formula for tunnel applications to:

$$Q = \frac{0.90 \times AF \times I}{4.35 \times 10^4}$$

where

AF = drainage area (feet^2).

For open areas, such as vertical vent shafts and fan discharges, the following would apply,

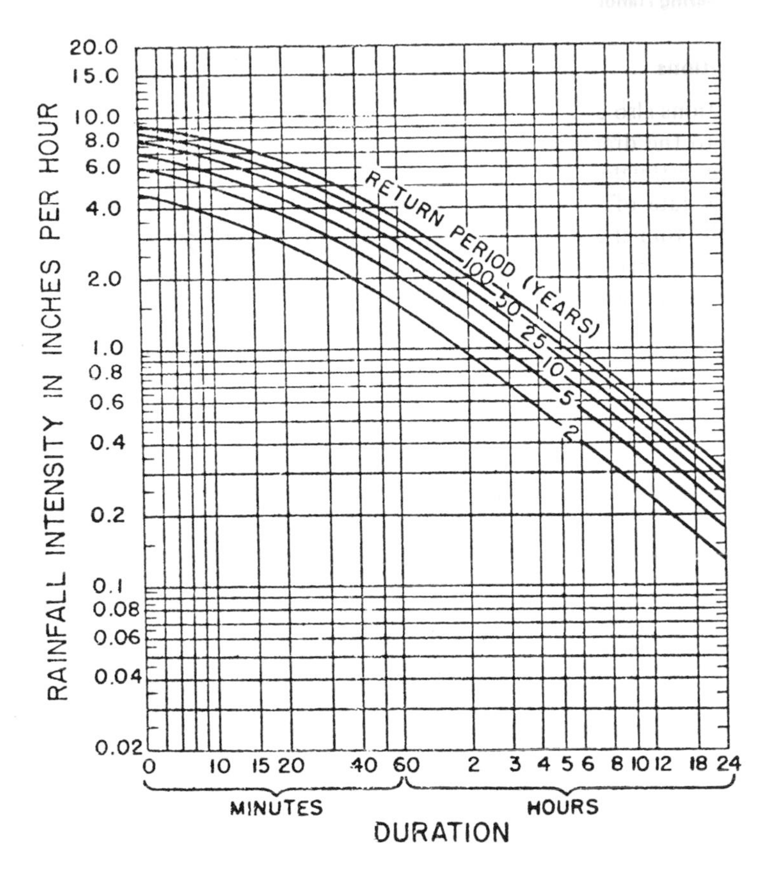

Fig. 24-2. Rainfall intensity curves.[2] adapted from reference 2

since all of the rainfall is assumed to be falling into the opening:

$$Q = \frac{AF \times I}{4.35 \times 10^4}.$$

Tunnel-washing Operations

In most vehicular tunnels, the walls and ceilings are washed periodically, usually in a two-stage procedure. First, a detergent is applied; secondly, the detergent and the dirt are rinsed off with water. In a typical tunnel of 7500 feet in length, this process includes a quantity of water and detergent equal to 1500 gallons and a rinse quantity of 15,000 gallons; the process takes approximately one hour. Thus, an average washwater flow rate for this typical tunnel would be approximately 275 gallons/minute.

Table 24-1. Selected values of runoff coefficients.[1,3]

SURFACE TYPE	RUNOFF COEFFICIENT* (C)
Concrete	0.80–0.95
Asphaltic	0.70–0.95
Roofs	0.75–0.95
Gravel Roadway	0.40–0.60
Grassed surface	0.10–0.35

*For flat slopes, use the lower values; for steep slopes, use the higher values.

Fire-fighting Operations

Fire-fighting operations also contribute a sizable quantity of water to the drainage system when used. This can be estimated by determining the maximum flow rate of water which could be pumped into the tunnel during a fire emergency within the capability of the existing fire protection system. For a determination of these quantities, see Chapter 23.

Vehicle Drippings

Vehicle drippings have been shown to be of minimal consequence and, if the system is designed to handle all the above quantities, the water from vehicle drippings will be adequately handled.

Seepage

Most tunnels through hills and mountains have water seepage problems. Surface water penetrates through fissures and percolates through permeable soils. Concrete liners are not completely watertight, and water may find its way through cracks in the lining.

Cut-and-cover tunnels can be waterproofed and, with good control, the number of leaks in such a tunnel can be minimized. Seepage in underwater tunnels, either the shield-driven or the sunken-tube type, is usually limited, and it can be controlled by caulking joints where leaks do appear in segmented liners.

OPEN APPROACH DRAINAGE

The portions of the tunnel roadway which extend beyond the portals are classed as the tunnel approaches. In the cases where the approach road slopes down into the tunnel and cannot be drained by gravity external to the tunnel, the drainage water must be included in the tunnel drainage system. This is true in a subaqueous tunnel where the open approaches are below the surrounding grade. The open approach drainage system should be designed to minimize the influx of water flow from the approach into the tunnel.

The *quantity of drainage water* on the open approach can be computed from rainfall alone, since this value, in most instances, will be the greatest.

Straight Open Approaches. On straight open approaches without superelevation, transverse interceptors placed approximately 300 feet on centers, with the first one located immediately outside the tunnel portal, are most effective in preventing the runoff from entering the tunnel. The actual interceptor spacing will depend on grade, inlet capacity, and pavement type. An additional interceptor should be placed inside the portal. The interceptors are approximately 18 inches in width, extending from curb to curb, and are covered with heavy cast iron gratings with slots parallel to the centerline of the roadway.

Superelevated Approaches. When the approach is provided with a superelevation, the approach drainage inlets must be placed at regular intervals along the low curb. (For comparison of these two arrangements, see Fig. 24-3.) The drainage water is carried by gravity from the approach inlets into the tunnel drainage system (Fig. 24-6). The gravity line should be a minimum of 8 inches in diameter, with clean-outs located at the required intervals.

TUNNEL DRAINAGE

Roadway Drainage

The roadway drainage system for a tunnel can be either open or closed.

The *open type* consists of a continuous gutter recessed into the curb and has been used in many tunnels (Fig. 24-4). This system, however, may permit propagation of a fire of burning fuel, in the event of a serious accident. The *closed system,* on the other hand, will prevent such propagation, since the drainage liquid enters the inlets located at the curb lines, then passes through a closed gravity flow system to a pump station. For this reason, the closed system should be used in all situations. The drainage inlets should be spaced 50 to 75 feet apart on both sides of a level roadway and on the low side of a superelevated roadway.

The *drainage inlet* design is important, since it

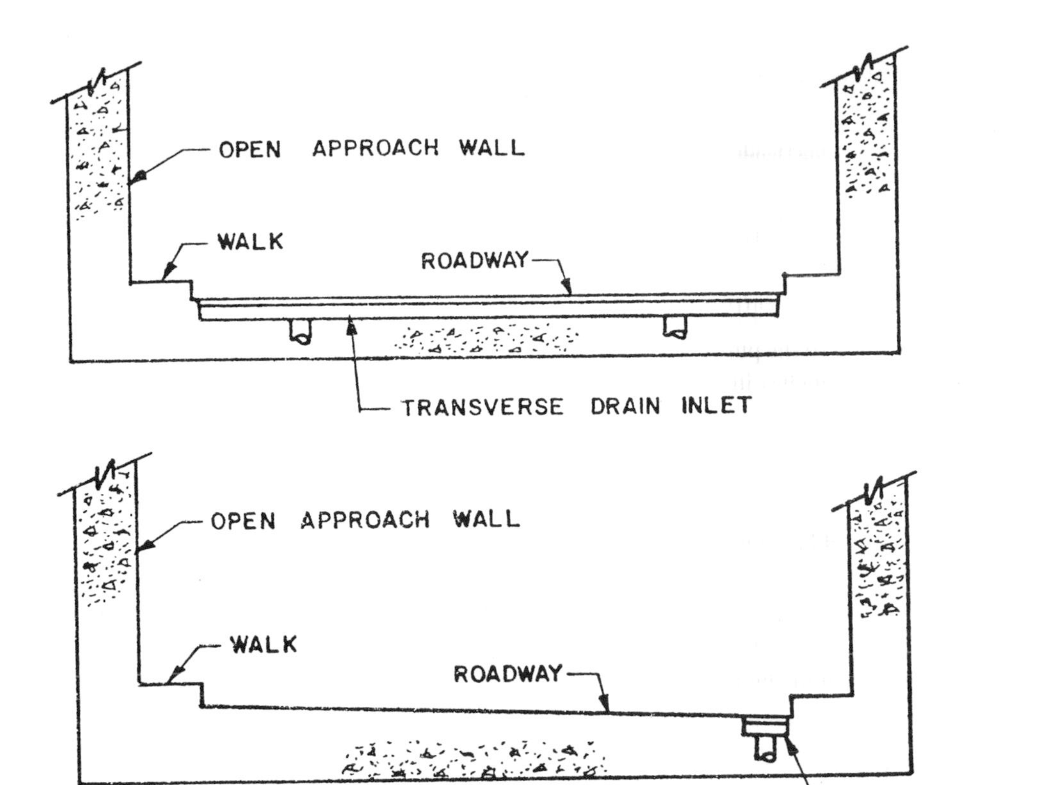

Fig. 24-3. Open approach roadway drainage arrangements.

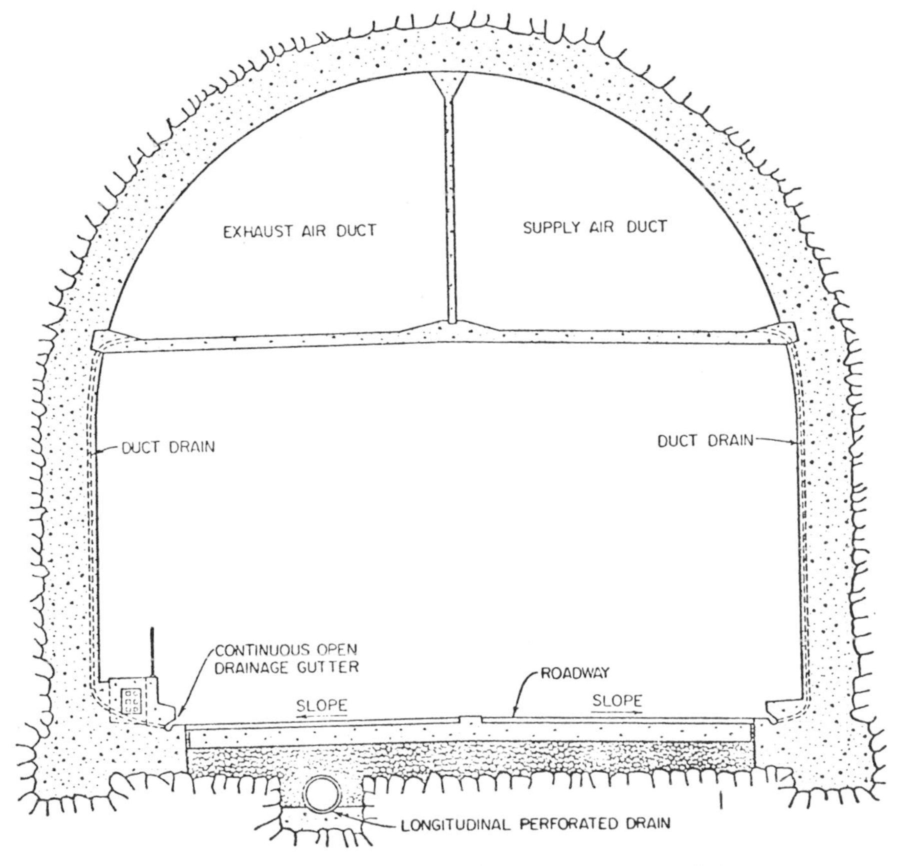

Fig. 24-4. Open gutter drainage system in mountain tunnel.

must remain clear of debris which would prevent influx of water.

The *gravity drain line* carrying the drainage water from the inlets to the pump station should be a minimum of 8 inches in diameter. Clean-outs should be located every 100 feet, with suitable access.

Miscellaneous Tunnel Drainage

There are several areas in the tunnel in which water can collect and should be drained to maintain proper operation of the tunnel. This water may be from wall-washing, seepage or fire-fighting operations. Figure 24-5 shows clearly some of these for a typical subaqueous trench type tunnel cross-section. The diagram includes the overhead air duct, the sidewalk gutter and electrical boxes. All of these can be drained by gravity to the roadway. The size of the line is significant only from the standpoint of possible blockage by debris, since the flow is extremely small. A minimum 2-inch diameter pipe is recommended.

DRAINAGE PUMP STATIONS

Any tunnel from which the drainage water cannot be removed by gravity must be provided with one or more pump stations having the necessary capacity to remove the maximum drainage demand. There are two basic locations for tunnel pump stations: at the low point of a tunnel, in particular a subaqueous type, and at the portals of any tunnel (Fig. 24-6).

Low Point Pump Station

The purpose of the low point pump station is to collect all of the drainage water within the tunnel and pump it out to the portal pump station.

The low point pump station in a subaqueous vehicular tunnel could have a configuration

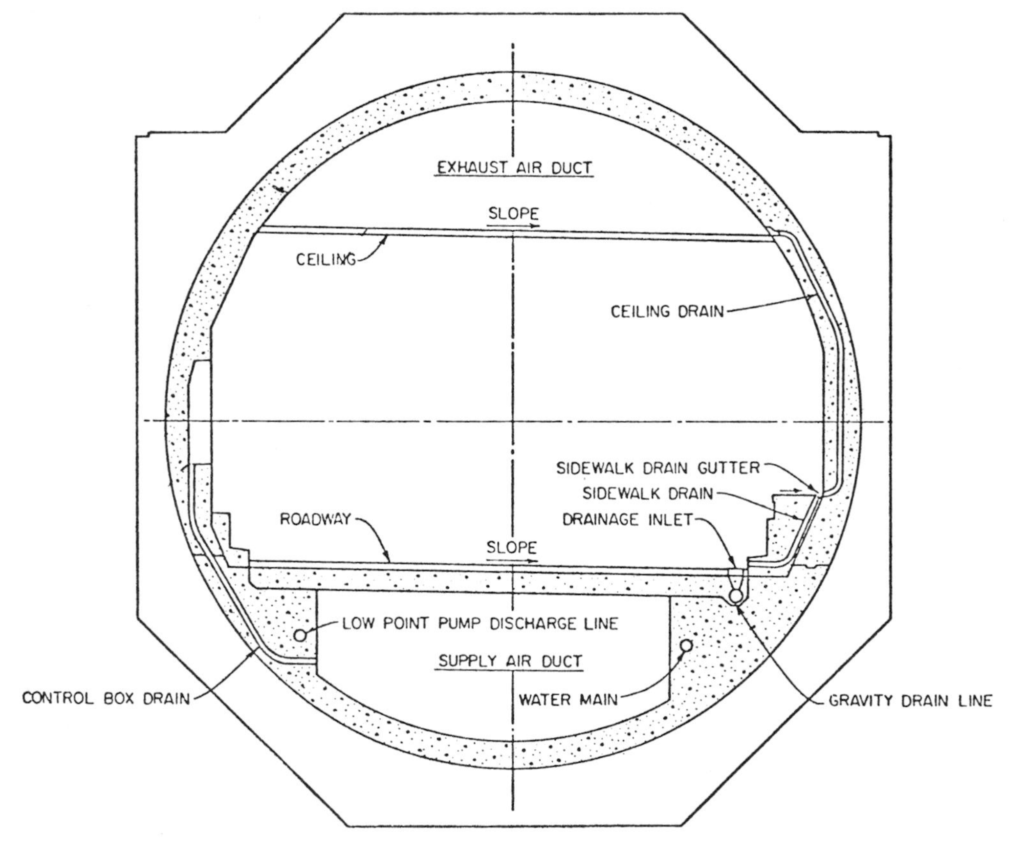

Fig. 24-5. Cross-section of subaqueous tunnel showing miscellaneous drains.

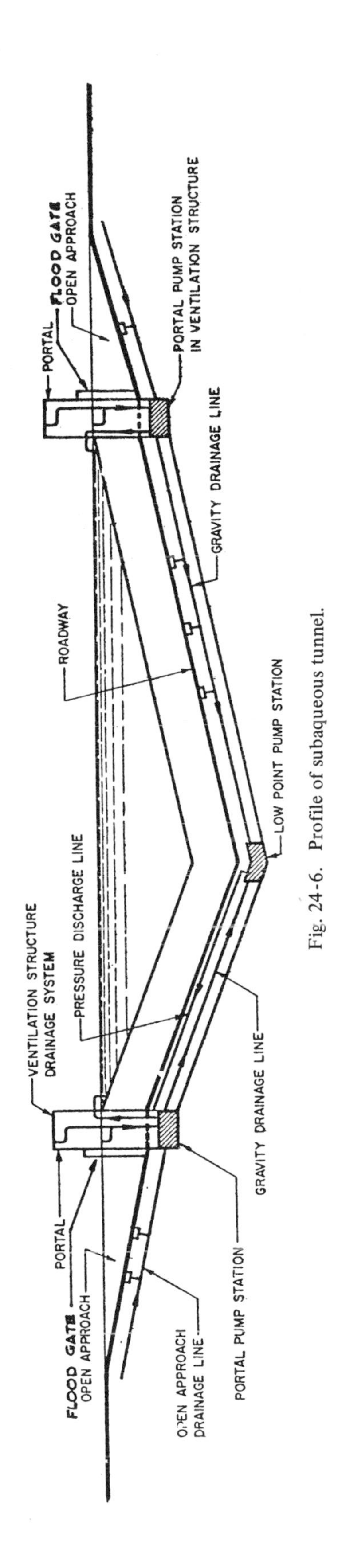

Fig. 24-6. Profile of subaqueous tunnel.

similar to that shown in Fig. 24-7. This pump station is located within the supply air duct.

The pump type most appropriate for the low point pump station is a vertical dry pit or horizontal centrifugal type, because of the usual limited available headroom in such locations.

Portal Pump Station

The portal pump station in a subaqueous tunnel usually collects the water pumped from the tunnel low point and the drainage from the open approach and from the ventilation structures, as shown in Fig. 24-6.

The portal pump station for a mountain type tunnel would be required if the water cannot be drained by gravity.

A typical arrangement for the portal pump station in a subaqueous tunnel is shown in Fig. 24-8. Almost any type of pump can be used in the portal pump station since headroom is usually not a serious problem.

The portal pump station may be constructed by either of two methods. The first is "in place" construction, where the structure at the portal is formed in such a manner to create the necessary settling and holding chambers, and pump room. The pumps and associated piping are then installed in the field. Such an arrangement is shown in Fig. 24-9. The second method employs a prefabricated package pump station similar to that shown in Fig. 24-9. Package pump stations are most appropriate in locations where they can be installed from the surface directly into the structural enclosure. They usually are not suitable for installations within a tunnel, such as at the low point, due to the difficulties arising in transporting and installing them within confined spaces.

The discharge from the tunnel pump station can be taken into a closed system, such as a sewer, or to an open body of water. The point of discharge should be selected to utilize the shortest run of piping and to minimize the disturbance to any open body of water considered.

Pumps

There are three broad categories of pumps available: reciprocating, rotary and centrifugal. However, since only the centrifugal pump is used

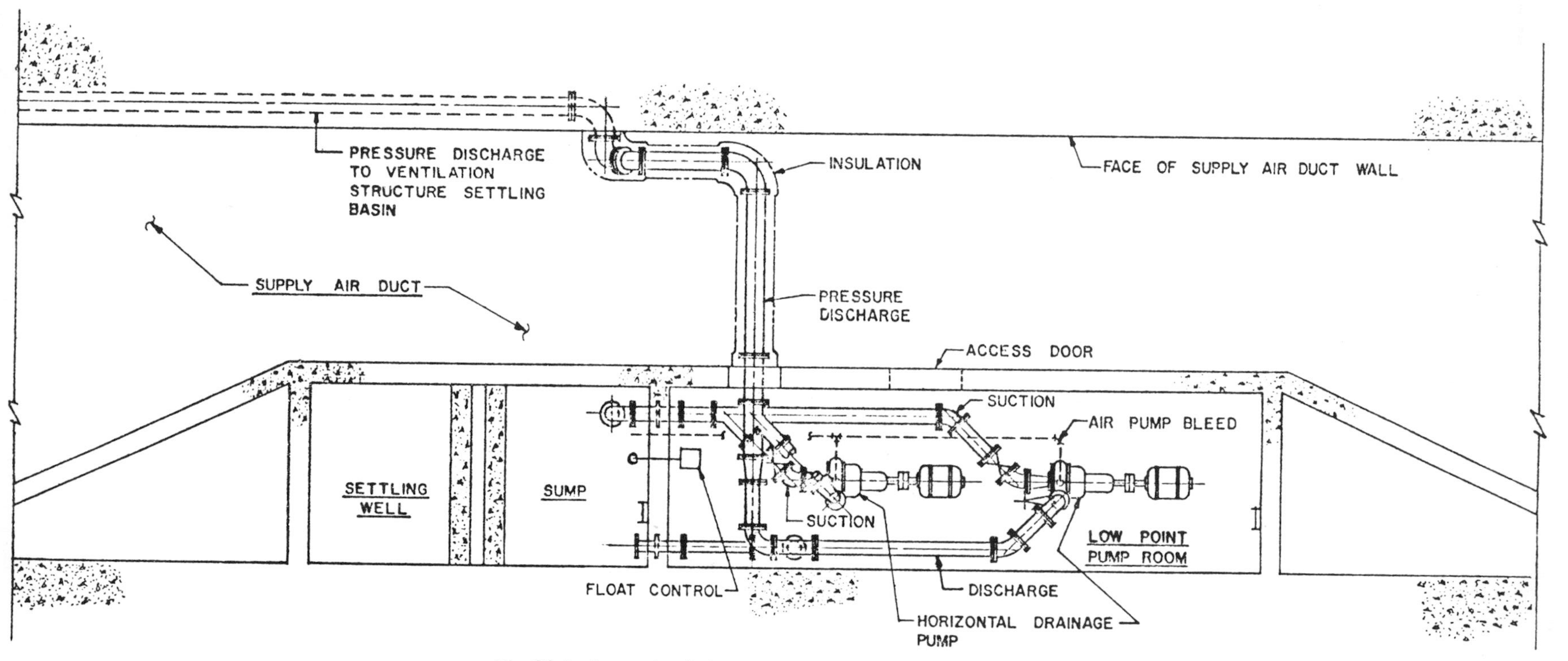

Fig. 24-7. Low point drainage pump station in subaqueous tunnel.

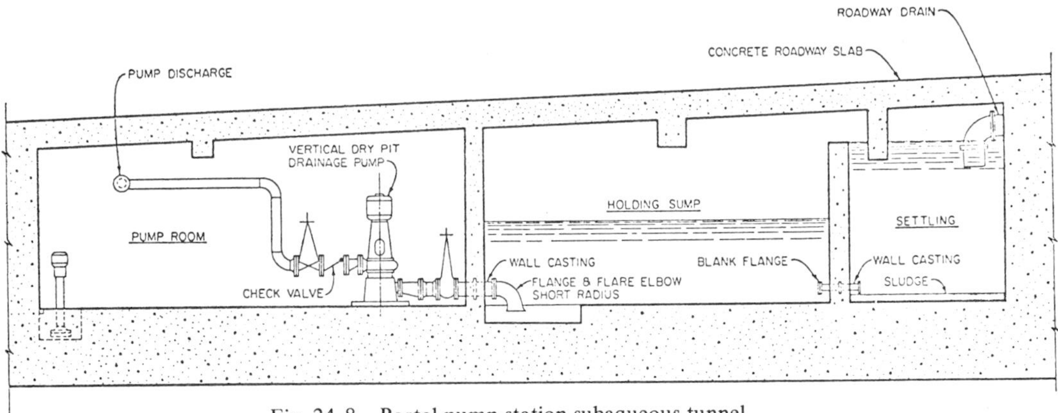

Fig. 24-8. Portal pump station subaqueous tunnel.

in tunnel drainage applications, only this type of pump will be dealt with in this section.

The Centrifugal Pump. This pump can be operated against a completely blocked discharge without overloading the drive, and is the most suitable pump for tunnel drainage service. Centrifugal pumps can be obtained in several arrangements: horizontal, vertical dry pit, submersible and vertical wet sump. Each of these is suitable for a particular situation or situations.

The *horizontal pump* is suitable where the headroom is limited and space in the plan is available or when high flow capacity and pressure are required. This pump, with a horizontally split casing, is easy to maintain, as the entire pump is accessible to mainentance personnel.

The *vertical dry pit pump* can be utilized, much like the horizontal pump, where the vertical space is limited but plan space is available adjacent to the sumps. This pump requires less floor space than does the horizontal type. The entire pump is accessible for ease of maintenance. The vertical dry pit pump is a type most often used in a package pump station.

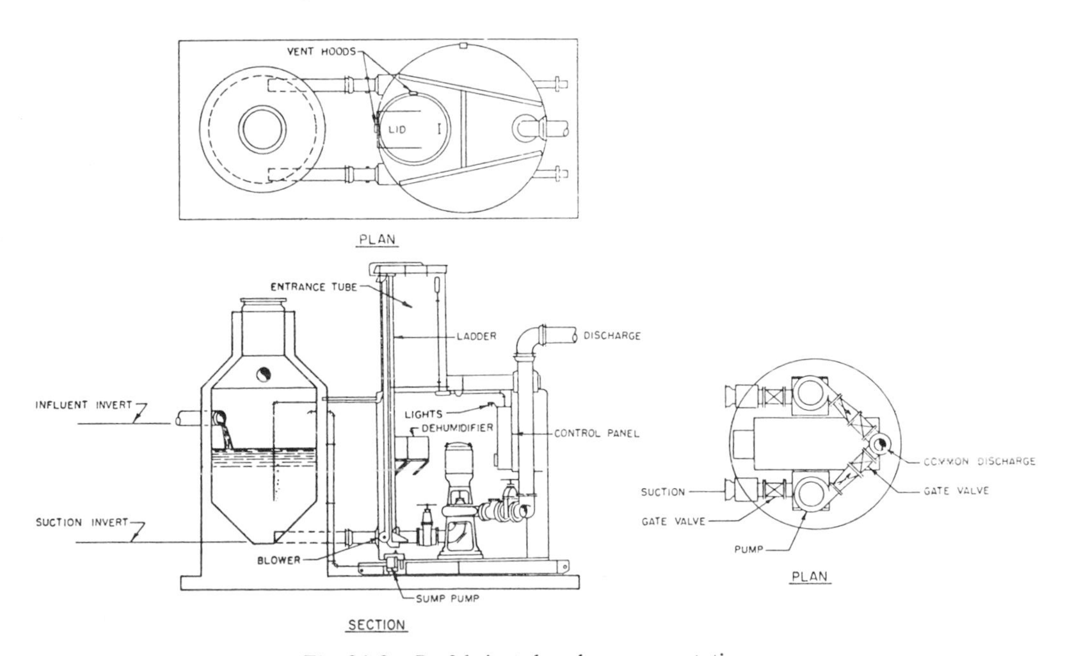

Fig. 24-9. Prefabricated package pump station.

The *vertical sump pump* is appropriate where floor space is limited but vertical space is available above the sumps. The impeller and the bearings of this type of pump are located below the water, thus creating some problems with maintenance. When the pump must be removed, considerable space is required to lift the pump.

The *submersible pump* is installed completely below the surface of the water in the sump. The only connection from the pump to the space above the sump is the discharge pipe, the power cable and the removal cable, thus complicating maintenance in a deep sump. However, in a shallow sump, maintenance is relatively easy. It is not recommended where, in order to service the pump, traffic in the tunnel must be interrupted. This pump requires a minimum of floor area and vertical space.

In all cases, the centrifugal pumps outlined above should not, if possible, be installed in tunnel drainage systems requiring a suction lift, but rather should have a flooded suction. If a suction lift is unavoidable, a reliable priming system must be installed.

Pump Drives. The majority of tunnel drainage pumps are driven by electric motors. However, there are instances where an internal combustion engine drive is appropriate. This would be when emergency pumping during a power failure is absolutely necessary.

Pump Arrangement. In most tunnel drainage systems, more than one pump should be installed to provide an adequate factor of safety in what is a critical system, particularly in a subaqueous tunnel. Other benefits of multiple pump installation include: smaller installations and servicing loads, reduced electrical starting loads and cable sizes, smoother pumping and overall installation economies. The use of two pumps in each pumping station, each having 100% capacity, implies full spare capacity, but an extremely large pumping increment. However, if three pumps are used, each having 50% capacity, the system will have a margin of safety and a smaller pumping increment.

Pump Selection. The performance of a centrifugal pump can be defined by the water flow rate and the total head, and can be graphically displayed in a characteristic curve such as that shown in Fig. 24-10. A pump should be selected to operate at a point where the pump curve intersects the system resistance curve with an attempt to maximize the pump efficiency (Fig. 24-11).

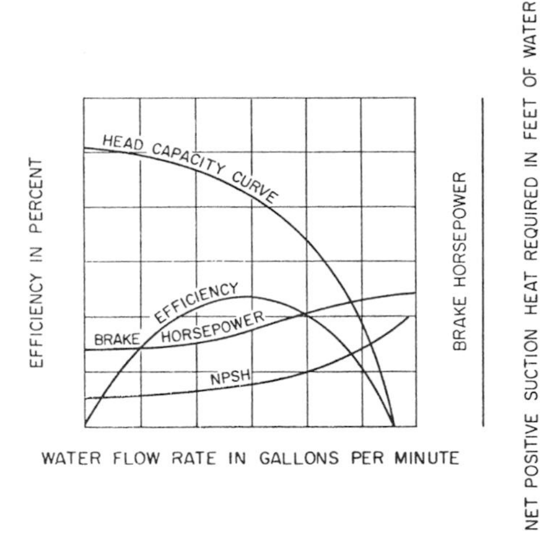

Fig. 24-10. Centrifugal pump performance curves.

In most tunnel drainage systems, the pumps are installed in parallel arrangement so that each pump can operate singly or in parallel. Therefore, the pump selection must take this fact into account. When two pumps operate in parallel, they will not deliver twice the water flow rate of one pump operating on the same system (Fig. 24-12).

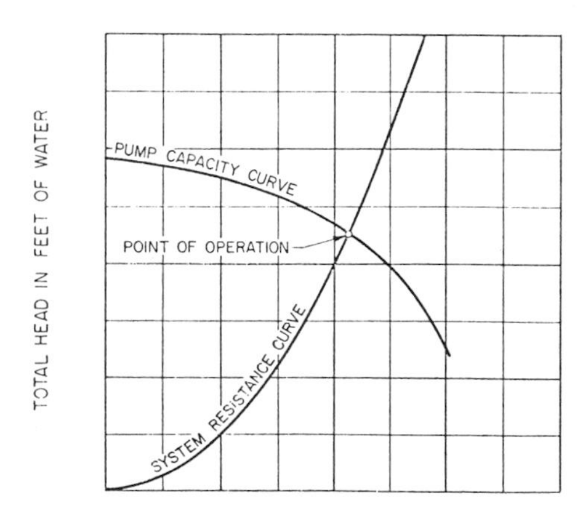

Fig. 24-11. Pump selection curves.

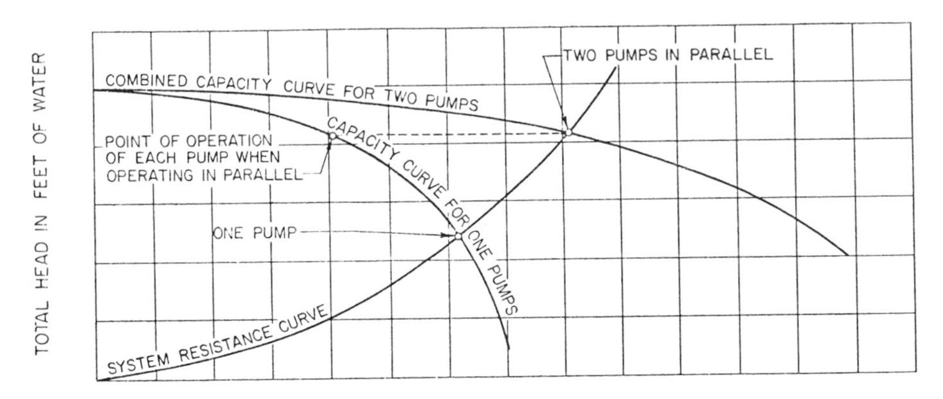

Fig. 24-12. Parallel pump operation performance curves.

It is rare that drainage pumps in a tunnel will be operated in series. However, when two pumps are operated in this mode, they will deliver a selected quantity of water at twice the rated head, as shown in Fig. 24-13.

The *total head,* or pressure, against which a drainage pump must operate is made up of two components: static head, and friction and dynamic losses.

Static head is the maximum total height to which the water must be raised by the pump. In a low point pump station, this would be the vertical distance from the surface of the water in the low point sump to the highest point of the discharge piping near or at the tunnel portal.

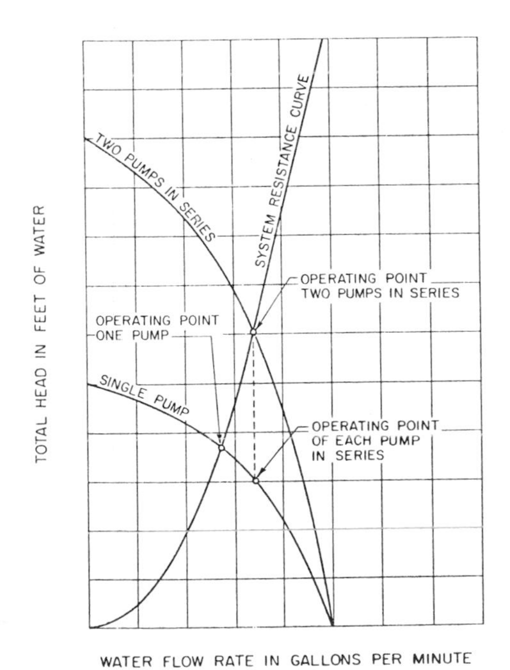

Fig. 24-13. Series pump operation performance curves.

The *friction and dynamic losses* are due to the water velocity in the piping system. These losses can be estimated by referring to a hydraulic data book.

The *required net positive suction head (NPSH)* of the pump must be less than the available NPSH in any water pumping system in order to provide sufficient pressure at the pump suction to prevent flashing and subsequent loss of prime. The required NPSH is shown on the pump characteristic curves (Fig. 24-10).

The *hydraulic horsepower* is the work required to change water from one elevation, pressure and velocity to another elevation, pressure and velocity, and can be computed from the following formula.

$$\text{Hhp} = \frac{(\text{gpm})(\text{TH})}{3960}$$

where

Hhp = hydraulic horsepower
gpm = water flow rate (gallons/minute)
TH = total head (feet of water)
3960 = conversion factor.

Brake horsepower is the work required as input to the pump and is always greater than the hydraulic horsepower due to losses within the pump. These losses are accounted for by use of the pump efficiency as follows.

$$\text{Bhp} = \frac{\text{Hhp}}{\text{EFF}}$$

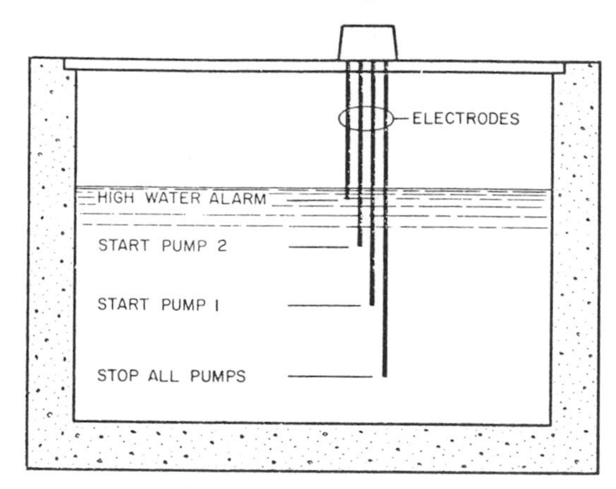

Fig. 24-14. Electrode water level control.

or

$$Bhp = \frac{(gpm)(TH)}{3960(EFF)}$$

where

Bhp = brake horsepower
EFF = pump efficiency (percent/100).

Pump Control. All tunnel drainage pump stations must be provided with automatic pump control systems which will permit automatic starting and stopping of pumps based on the amount of water in the sump. The pump operation should be sequenced and alternated to obtain equal operating time on each pump.

These systems usually consist of some form of water level detection and associated pump controls. There are several types of level detection and control devices suitable for the tunnel drainage system.

The *float type* has probably been used more than any other type in tunnel drainage systems. There are, however, several disadvantages to both the mechanically linked ball float and the chain or tape operated float. A single float is usually used to detect a number of water levels and control points; and, if the mechanical portion of the system is jammed, no additional pumps will start, thus possibly creating a flooding condition. The float type level detector in a tunnel drainage system usually requires a complicated installation.

The *electrical conduction electrode* or probe type of water level detection is easy and inexpensive to install. Despite the fact that the effectiveness of these probes are affected by foreign material in the drainage water, they remain as the most effective in the tunnel environment from both reliability and maintainability standpoints. Separate probes are required for each control point, as shown in Fig. 24-14. An advantage is that the control relays can be located remotely from the probe location.

The *mercury float switch* is a simple and inexpensive water level detection device. However a main disadvantage is its wide detection range and the fact that it is seriously affected by turbulent water. A typical installation is shown in Fig. 24-15.

There are other types of water level detection devices, such as the newly developed sonic type. However, there is at this time limited experience in tunnel drainage systems with this equipment.

The pump control point should be established to allow a minimum pumping time of four minutes for each pump.

Chambers

All pump stations require one or several chambers designed to hold and treat the drainage water. A full evaluation of the drainage water treatment is presented in the next section of this chapter.

The *settling well* or sump, as shown in Figs. 24-7 and 24-8 is the first line of water treatment. It should be sized to provide adequate time for solids to settle out of the water, based on normal inflow rates. The settling well should

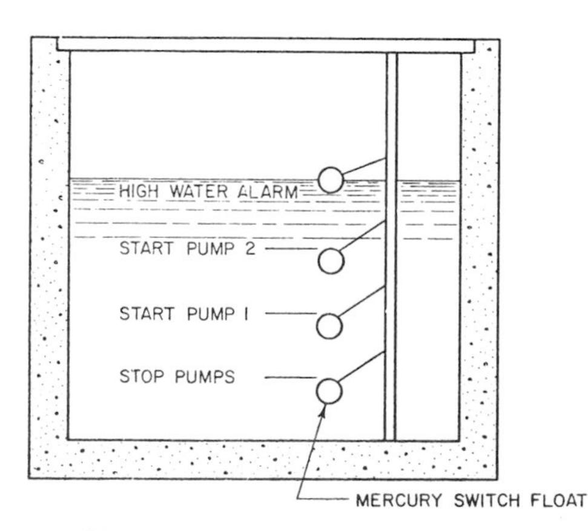

Fig. 24-15. Mercury float switch level control.

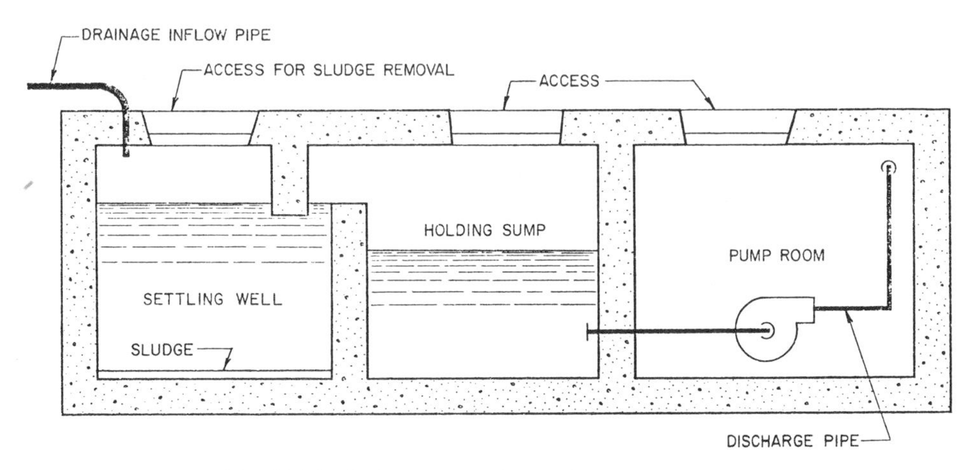

Fig. 24-16. Drainage pump station chamber arrangement.

be equipped with a skimming weir, as shown in Figs. 24-7 and 24-8, to prevent floating materials, from clogging the pumping system or being discharged out of the system.

The *holding or storage sump*, which receives the water from the settling well, should be sized to prevent a rapid cycling of the drainage pumps and to allow a minimum of four minutes of running time for each pump.

A provision must be made in all installations for periodic removal of sludge from the sumps. Access manholes and proper drainage facilities are necessary, as shown in Fig. 24-16.

WATER TREATMENT

The drainage water from a vehicular tunnel contains some contaminants from vehicle drippings and tunnel-washing operations. It is the washing operations which produce the severest problem. The effects of the detergents used in the washing operation on the drainage water quality must be evaluated. The detergent is usually diluted by water to a ratio of 700:1 during the wash-rinse cycle.

It has been shown that, for many tunnel installations, a treatment system, as shown in Fig. 24-17, will be adequate to mitigate the effect of the tunnel operation on water quality. There must be sufficient settling time to permit sedimentation of solids and to permit skimming of floating materials. Also, periodic removal and proper disposal of sludge, which accumulates at the bottom of the settling well, must be included in the operation of the system.

The critical pump station in a subaqueous tunnel from a standpoint of settling is the one at the low point, where all of the washwater is collected and then pumped to one of the portal ventilation structures. The water will enter this sump at the rate to 100 to 200 gallons/minute during washing operations.

The type and location of the drainage discharge must be carefully considered lest the receiving body of water be seriously affected by the discharge. Subsurface discharge should be considered for open bodies of water.

Treatment of drainage from a rail tunnel is normally not necessary, since this type of tunnel is not washed and, therefore, detergents are not introduced into the drainage system.

FLOOD PROTECTION

The only tunnels subjected to serious flooding are subaqueous, either in tidal areas or in flood plains of rivers. Where possible, flooding should be prevented by raising the elevation of the approaches above maximum flood levels. Where this is too expensive or impractical, flood gates must be installed (Fig. 24-6).

It may be possible to raise the approach walls above the flood level and install flood barriers at the upper end of the open approaches. This would be less expensive and would prevent flooding of the tunnel and the open approaches.

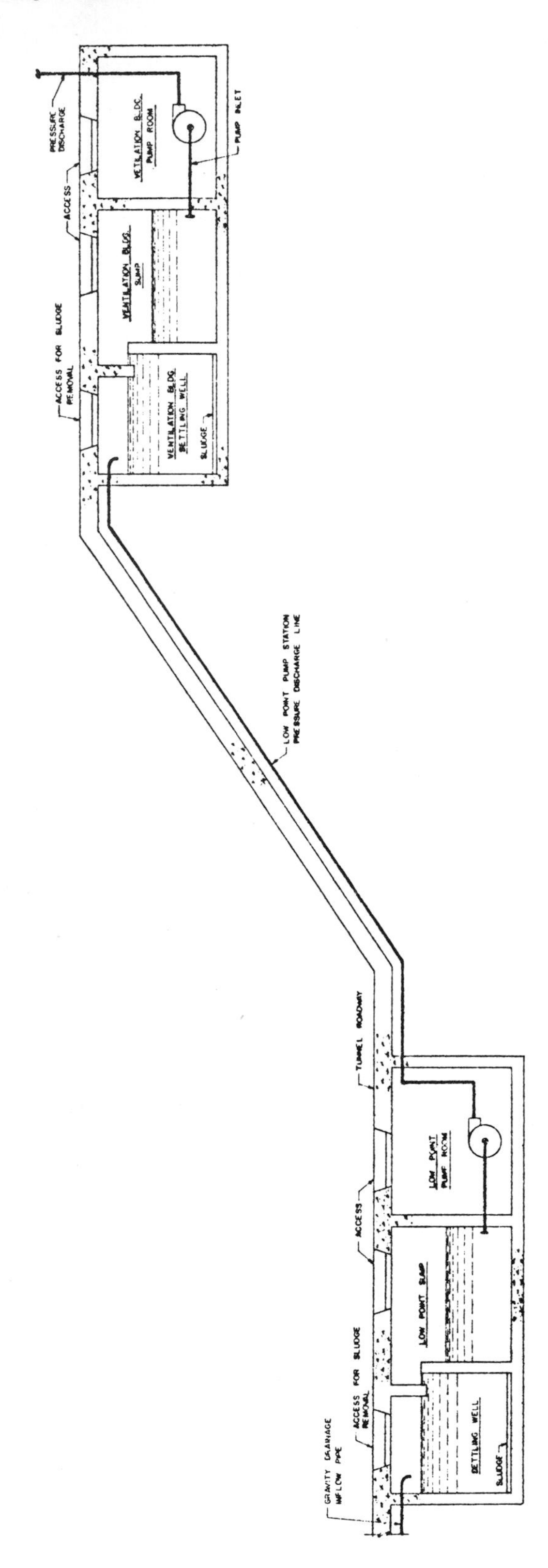

Fig. 24-17. Drainage discharge treatment system by settling basins.

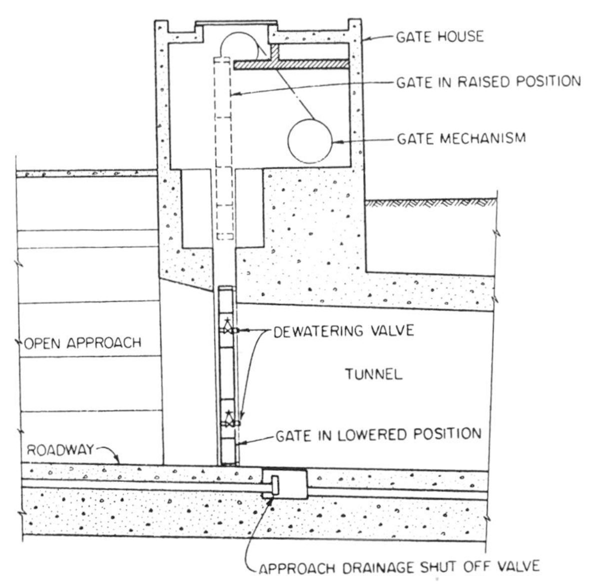

Fig. 24-18. Flood gate installation at portal of subaqueous tunnel.

Flood Gates

These gates provide a closure at the portal to restrain the rising flood waters and permit collection of flood water at the portals. A typical installation appears as in Fig. 24-18. These gates are usually of the slide type, constructed of steel and designed to withstand the hydraulic forces encountered during maximum flood conditions. The gate travels in vertical guideways and seats against the roadway. Seals are required to minimize the leakage into the tunnel. An enclosure is provided above the gate to house the operating machinery and the gate while it is in its raised or stored position (Fig. 24-18).

Maximum *leakage criteria* should be established for each portal gate installation. The criteria should be such that the drainage pumps which are capable of functioning in an emergency can remove the amount of water leaked into the tunnel without major flooding occurring within the tunnel.

Valves are required on the gate (Fig. 24-18) to permit rapid drainage of the water collected in front of the gates prior to raising the gates. This water can be drained and permitted to enter the tunnel drainage system prior to raising the gates. Gate valves of a minimum $2\frac{1}{2}$-inch size with threaded connection for $2\frac{1}{2}$-inch hose are most appropriate for this application.

In subaqueous tunnels where the open approach drainage system is connected to the tunnel roadway drainage system, a means must be provided to isolate these two systems during a flooding condition to prevent the ingress of flood waters through the open interconnected drain pipes. A method of providing such isolation is to place a closure device on the tunnel side of each portal gate. A shear gate type of valve installed as shown in Fig. 24-18, located in a pit, will provide this. The use of an open pit permits inspection of the valve to assure that the pipe leading from the open approach will be clear of debris and can be sealed tightly against the flood waters.

Testing. The flood gate should be tested against a head of water equal to the maximum flood level anticipated. This is required in order to assure that the leakage criteria are not exceeded. Construction of a watertight bulkhead on the open approach side of the gate will permit development of such a head and testing of the gate.

Operation. When flood conditions are imminent, the gate is lowered after the necessary cover plates have been removed. The drainage system isolation valve, along with the dewatering valves on the gate, must be in the closed position at this time.

DRAINAGE OF RAIL TUNNELS

Rail tunnels include both those carrying railroad and those carrying rapid transit trains, with both ballast and concrete roadbeds. A railroad tunnel is usually drained by installing a perforated pipe below the track ballast, as shown in Fig. 24-19. Where there is no ballast used, such as in a rapid transit tunnel, an open channel or drainage trough is often used, as shown in Fig. 24-20. The drainage water is carried through this channel to inlets located at specified intervals. These inlets permit the water to enter the gravity flow line, thus transporting the water either to a low point sump, in the case of a subaqueous tunnel, or to the low portal, in the case of a mountain tunnel.

The water quantity anticipated in the rail tunnel drainage system will consist of water from either fire-fighting operations, vehicle drippings, seepage or rainfall on the approach tracks and on openings to the surface.

Subway

Drainage in the subway portion of a rail rapid transit system is accomplished by use of a center channel to collect and transport water, which is then piped to pump stations located at each low point.

System Descriptions. The *Transbay Tube* is a $4\frac{1}{2}$ mile subaqueous rapid transit crossing of the San Francisco Bay in California. It is a part of the Bay Area Rapid Transit system. This tunnel has two low points, as shown in Fig. 24-21. The drainage system consists of two low point pump stations, four intermediate gallery pump stations located at changes in grades, and two ventilation structure pump stations. The four intermediate gallery pump stations, as shown in Fig. 24-22, are arranged so that two comprise the intermediate pump stations at grade changes and two are included in the low point pump stations. Table 24-2 shows the pump capacity for each of these arrangements.

The drainage water is collected and transported in the open channel located between the rails in the concrete roadbed and then collected either at the intermediate points or at the low

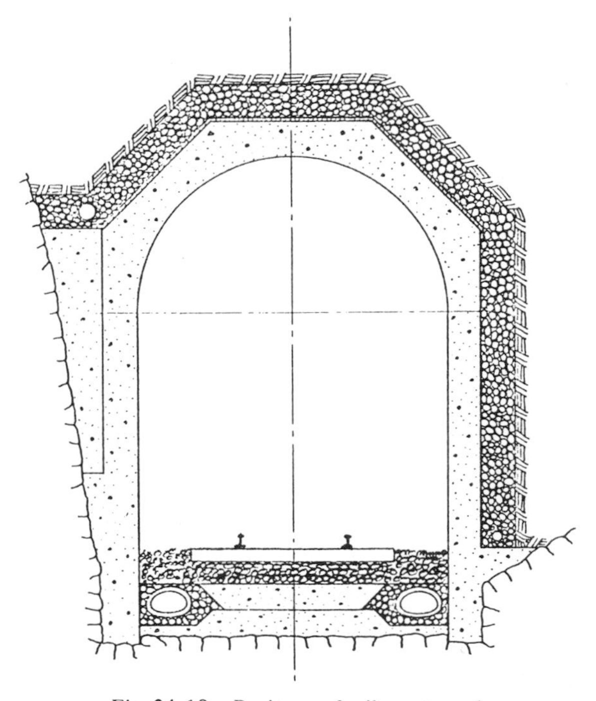

Fig. 24-19. Drainage of railway tunnel.

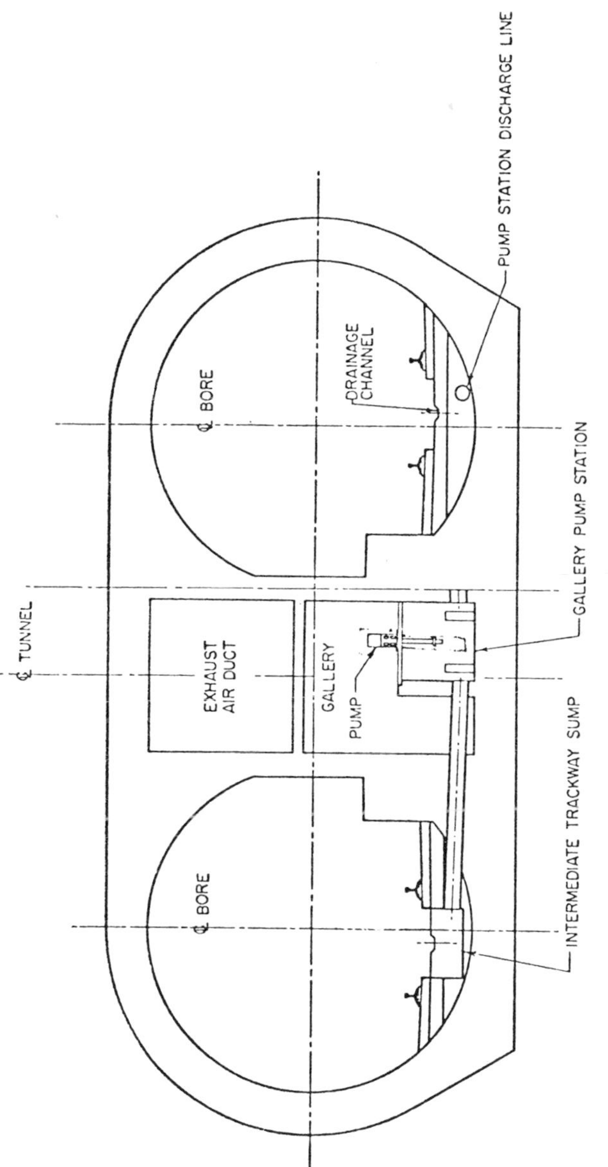

Fig. 24-20. Drainage of subaqueous rapid transit tunnel.

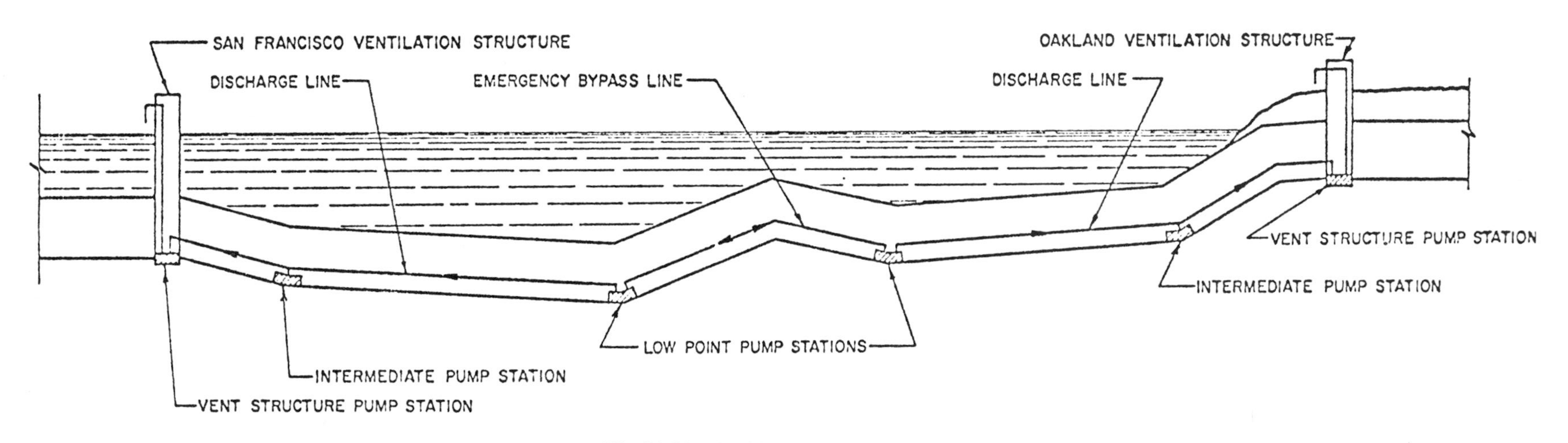

Fig. 24-21. Profile of Transbay Tube.

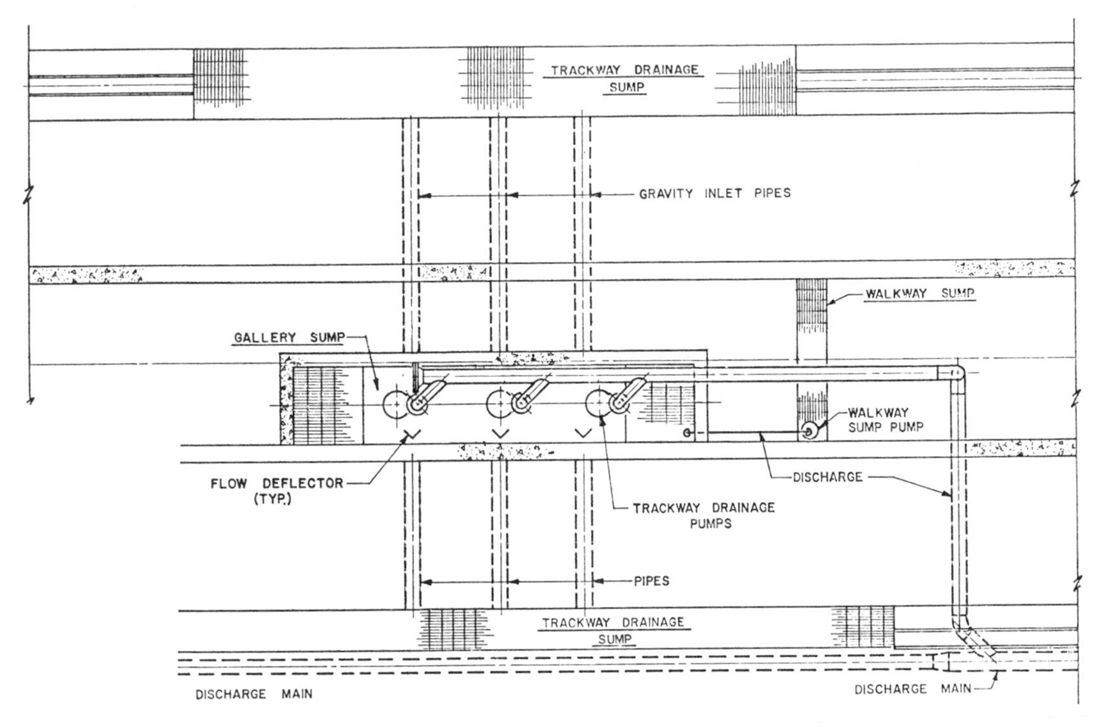

Fig. 24-22. Gallery pump station Transbay Tube.

points. At the intermediate point, the water is collected in a trackway sump and then flows by gravity into the intermediate sumps, from which it is pumped through the main discharge line to the pump station located in the ventilation structures. At the low point, the water is collected in a trackway sump, drained to the gallery sump, then pumped into the main sump of the low point pump station. From the low point sump station, the water is pumped through the main discharge line to the vent structure pump station and then discharged to the surface.

This system is equipped with an emergency bypass arrangement, whereby either low point pump station can pump its effluent to either ventilation structure. This provides a method of removing water, should there be a break in either end of the discharge line. A recirculating arrangement has also been built into this system to provide the means to exercise the pumps with a minimum of water in the sumps.

The *Potomac River crossing,* on the Huntington Route of the Washington Metro, is a 6000-foot subaqueous rail tunnel. It is a double-track, two-bore tunnel. The crossing profile is shown in Fig. 24-23 and has one low point. The water is drained from the trackbed by an imbedded drainage pipe. The water will flow in an open channel to drain inlet sumps (Fig. 24-24), which are spaced approximately 300 feet on centers. Through these inlets, the water enters the imbedded gravity drain line and then

Table 24-2. Transbay Tube Drainage system pump capacities.

	NUMBER OF PUMPS	CAPACITY OF EACH PUMP	PUMP TYPE
Gallery	3	250	Vertical sump
Low point	2	500	Horizontal
Ventilation structure			
San Francisco	2	500	Vertical sump
Oakland	2	500	Vertical dry pit

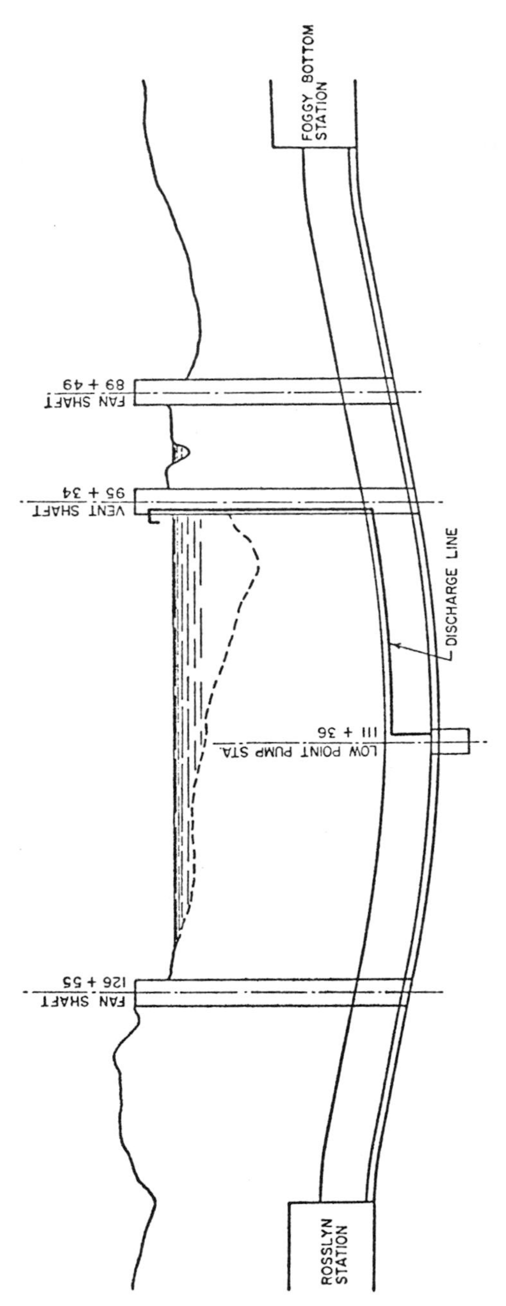

Fig. 24-23. Profile of Potomac River crossing.

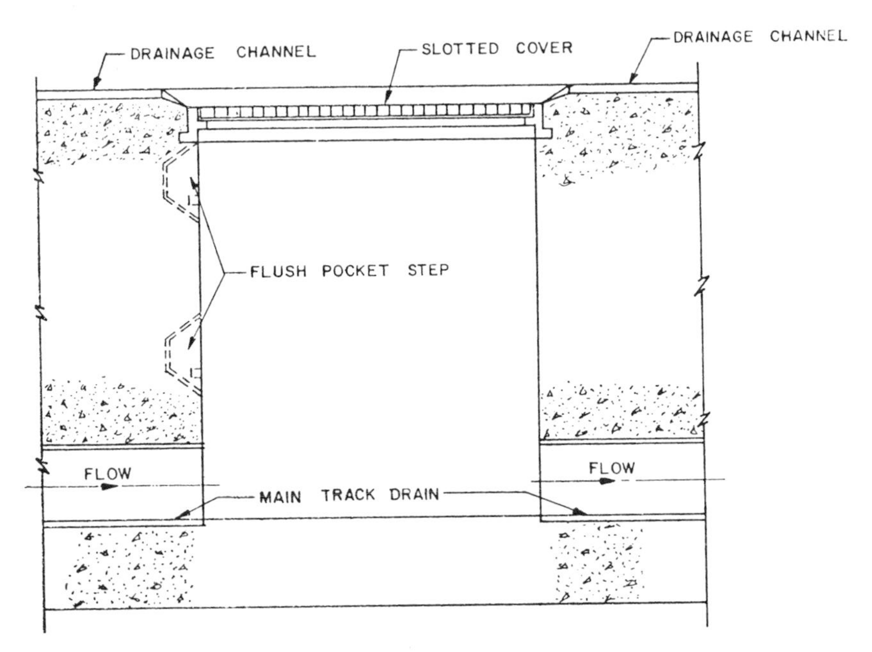

Fig. 24-24. Trackway drainage inlet sump.

flows to the low point pump station. The low point pump station is of a package type construction with two pumps. The water is then pumped to the surface through a pressure discharge line.

DRAINAGE OF SEEPAGE

Almost all mountain tunnels have water infiltration problems. Surface water penetrates through fissures and percolates through permeable soils. Attempts to seal off the rock by grouting with either cement or chemicals usually are not successful. Concrete linings are not completely watertight. Water will find its way through shrinkage cracks in the linings into the interior tunnels. There, it can freeze and cause an unsightly appearance in vehicular tunnels.

If water appears in considerable quantity during tunneling operations, longitudinal drainpipes should be installed behind the sidewalls, with laterals at regular intervals to the main tunnel drain lines.

REFERENCES

1. *Drainage of Asphalt Pavement Structures, Manual Series No. 15, First Edition,* The Asphalt Institute, Maryland, 1966.
2. "Rainfall Intensity Atlas of the United States," Technical Paper No. 25, U.S. Department of Commerce, Weather Bureau, Washington, D.C., 1961.
3. Design and Construction of Sanitary & Storm Sewers, WPCF Manual of Practice No. 9 (ASCE Manual and Report on Engineering Practice No. 37), Water Pollution Control Federation, Washington, 1970.

25
Tunnel Operation and Maintenance

JOHN O. BICKEL

Associated Consultant
Parsons, Brinckerhoff, Quade & Douglas

Operation and maintenance of tunnels falls into two categories: contol of traffic-related functions, and maintenance of the physical plant. The first includes control and supervision of traffic; emergency services; operation of ventilation; and lighting and signals systems. The second is charged with keeping all operational devices in proper repair by scheduled maintenance and tests, cleaning of the tunnel interior and maintenance of roadway pavement and structures.

SUPERVISION OF OPERATION

Automatic Operation. Relatively short tunnels, such as long underpasses of city streets, are generally operated with automatic controls, without direct supervision. Ventilation is controlled by detection devices, as described in Chapter 19. Emergency telephone and fire alarm stations in the tunnel, and alarms for excessive contamination of tunnel air, should be connected to the nearest police or highway maintenance station.

Supervised Operation. Highway tunnels of greater length should be operated under continuous supervision by trained personnel, even if the ventilation is normally under automatic control (see Chapter 19). This applies particularly to all functions involving traffic safety, and is especially important in subaqueous tunnels with their greater pyschological hazard in case of accidents.

TRAFFIC SUPERVISION

Need of Supervision. Maintenance of traffic flow and quick action in case of accidents are essential in all tunnels, with their confined space, to prevent potential panics. This can take various forms: guards stationed in the tunnel, initiated in the early subaqueous tunnels (Holland Tunnel) and still widely used; patrol cars circulating in the tunnel; and observation by closed circuit television, combined with patrol cars or special patrol vehicles.

Tunnel Guards. Guards stationed in the tunnel are best suited for quick action in emergencies and for enforcement of traffic rules, such as maintenance of minimum spacing of cars and keeping in line. Before availability of reliable haze detectors, they also reported impaired visibility in the tunnel to the ventilation control station.

The number of tunnel guards is determined by the length and changes of the roadway grade of the tunnel. There should be enough guards to keep every part of the roadway under observation.

Limits of tunnel duty by general experience should not exceed two hours at one stretch, alternating with two hours outside duty, to avoid excessive exposure to tunnel air and the noise and visual effects of the traffic. Federal and state health codes may have regulations to be observed. Guards should be trained in emergency actions, use of fire extinguishers, actuating alarms, assisting the crew of the emergency trucks and first aid.

TUNNEL PATROL VEHICLES

Since guard duty in tunnels is unpleasant, attempts have been made to substitute moving patrols.

Police Patrol Cars. Some tunnels use regular police cars patrolling continuously, or being on standby for call in conjuction with a television observation system. Their effectiveness in emergencies is open to question. In a severe breakdown or collision in heavy traffic, and particularly in case of fire, their quick access to the site may be blocked by traffic until the offgoing traffic ahead of the accident has cleared the tunnel.

Service Walk Vehicles. These have been developed by the Port Authority of New York, to be used with remote television monitoring and radio communication. They are self-powered, one-man vehicles running on the service walk along guide rails. They are equipped with fire extinguishers, first aid kits and some emergency tools, and are in radio communication with the control room. Normally stationed outside the portals and dispatched by radio, they can reach trouble spots quickly. This method has proven effective and saves personnel.

EMERGENCY TRUCKS

Function. Emergency trucks are required to remove incapacitated vehicles from the tunnel as quickly as possible and to carry fire-fighting equipment. Since the truck enters the tunnel against traffic direction, after vehicles ahead of the accidents have cleared the lane, there is usually one stationed in a service building near each portal. For dual tunnels with one-way traffic, an emergency truck is stationed near each portal.

Capacity. The truck should have the capacity to handle the heaviest trucks using the tunnel. It is equipped with a winch of tons-lifting capacity and has enough traction power to tow such a vehicle up the maximum grade. Its wheel base should be short enough so that it can turn in the tunnel with not more than one backing operation. A trailer truck tractor chassis usually serves as a base.

Equipment. The following equipment is carried.

- 500 feet of fire hose to fit the hydrant outlets in the tunnel, with regular and fog nozzles.
- Fire foam containers.
- Large fire extinguishers.
- Fire axes and tools.
- First aid equipment and litter.
- Heavy dollies to support truck wheels with flat tires so the truck can be towed without lifting.

Crew. A well trained operator should be on standby around the clock, usually in three shifts. Additional emergency personnel must be available at all times, either guards in the tunnel or guards from the outside, riding on the truck.

TOW CARS

Capacity. For removing stalled passenger cars from the tunnel, a smaller tow car is used. It should have enough traction power to tow any such car up the maximum grade. A heavy four-wheel drive jeep or the chassis of a $1\frac{1}{2}$ to 2-ton delivery truck may be used.

Equipment. The tow car should be equipped with a 2-ton hoist, dollies to support wheels with flat tires, a large fire extinguisher, tools and a first aid kit. For winter use, it is equipped with a snowplow attachment.

WASHING TRUCK

Washing of Tunnel. In order to maintain the efficiency of the lighting system and to preserve the general appearance, the tunnel walls and ceiling must be cleaned regularly. Depending upon the type of traffic, this may have to be done weekly or at least once a month. Some dirt may be removed by flushing with water only, but the deposits of soot and oily residue, especially from diesel engines, need application of detergents and mechanical action to break through the film. General practice is to spray detergent solution on the surfaces and after a short interval to scrub with brushes and flush with water.

Detergent Application. The best detergent should be determined by tests. Early washing trucks had two sets of nozzles, the one in front for detergent and one in the back for flushing. To give the detergent more time for action, it is preferable to spray it from a separate small truck, equipped with a tank of about 300-gallon capacity, an engine-driven pump and spray nozzles.

Washing Truck. A tank truck holding about 1500 gallons of water, equipped with a pump providing about 120 psi pressure, powered by a separate gasoline engine, serves as the washing truck.

Scrubbing brushes and spray nozzles are mounted on the truck. The rotary brushes for side walls and ceiling are mounted on swinging arms, with adequate pressure applied by hydraulic cylinders on the walls and on the ceiling, or by counterweights for the latter. High-speed, long-bristle, soft brushes, as for car-washing, may be used, but fairly stiff, long bristles are more effective. A two-lane tunnel is cleaned in two passes during low-traffic periods.

High-pressure water jets have been used instead of brushes, with satisfactory results (Potapsko River Tunnel). Detergent is applied as mentioned and rinsing water from relatively fine nozzles is sprayed on under 500 to 600 psi pressure.

MAINTENANCE AND SERVICE CARS

Street Cleaner. For roadway cleaning, a standard motorized sweeper with a water container, a rotary brush and a storage container is used.

Platform Truck for Lighting Maintenance. This is a light platform truck of convenient height with storage racks for fluorescent tubes and containers for other replacement parts. For maintaining high-mounted approach lights, it may be equipped with an extension ladder.

General Service Cars. These cars are needed for transporting maintenance personnel and tunnel guards and for general utility purposes. Their type and number depends upon the extent of the project. Equipment for ground maintenance consists of lawn mowers, sprinklers and snow removal tools.

ELECTRICAL MAINTENANCE

The electrical system is the heart of tunnel operation and requires first class maintenance.

Power Distribution System. The stystem includes the primary supply, transformers, switchgear, control equipment and distribution wiring.

Switchgear and control equipment must be kept clean and in good operating condition. All switching devices, relays and circuit protection elements must be tested periodically and checked for proper calibration. Their contacts must be inspected and, if necessary, dressed smooth.

Transformers, if liquid cooled, should have samples of the fluid tested once a year by a laboratory for contamination and insulating properties. Air cooled transformers should be cleaned periodically to prevent dust accumulations from interfering with the cooling.

Wiring and cables, where accessible, should be inspected. In particular, all connections must be kept tight. Junction boxes exposed to tunnel-washing operations should be kept watertight and inspected for water accumulation, and gaskets should be replaced, if necessary. Where drains are provided, they should be kept clean.

Tunnel Lighting. A scheduled maintenance and lamp replacement program should be followed. Fixtures should be inspected for water accumulation from washing operations and be kept watertight. Since fluorescent lamps in tunnel service have a very long life, often in excess of 20,000 hours, random replacement may be acceptable, but will result in occasional dark spots.

Local conditions and experience will determine the best method.

Fire Alarm Communication and Signal Systems. Fire alarm boxes should be tested monthly for proper operation. Service telephone stations should be operated regularly. Traffic controls should be observed continuously for proper functioning and all defective lamps in traffic signals and in illuminated signs must be replaced promptly, as must the defective tubes in closed circuit television observation systems. Radio communications between all fixed and mobile stations should be tested regularly.

MECHANICAL MAINTENANCE

Ventilation System. The system must be kept clean. Dust accumulations, especially, must be removed from the exhaust ducts and dampers. Belt and chain drives should be maintained at proper tension, inspected for wear and replaced in time. The lubrication for chain drives should be kept at a proper level. For vane axial fans with built-in drive, the cooling system for the bearings requires special attention and the bearings should be checked for unusual sounds. The damper operation and its interlock with the fan controls should be checked regularly.

Emergency Diesel Generators. Such generators for standby service for drainage pumps and emergency lighting, if installed, should be operated once a month and all controls checked.

Lubrication. Lubrication of all bearings should be done on a regular schedule. This applies to all motors, fans, pumps, generators, and engines.

Drainage Pumps. These should be checked regularly. Multiple pumps should have their sequencing changed at regular intervals, depending on the amount of use, so that they operate in rotation.

Fire Pumps. Fire pumps should be operated monthly, by using the bypass valve, for pressure and discharge.

Vehicle Maintenance. All routine vehicle maintenance should be done on the premises in an adequately equipped shop. On a large project, major vehicle repairs are often also done on the premises. In smaller tunnels, where the more elaborate shop facilities may not be warranted, these may be done outside.

OPERATING PERSONNEL

In general, the following staff for operation and maintenance of tunnel project is required, the actual number being dependent upon the magnitude of the facility.

- General Manager or Superintendent with an office staff.
- Supervisor of operation.
- Supervisor of maintenance.
- Tunnel operators, on a three-shift basis, who should be trained electricians.
- Electricians for electrical maintenance, working in single shifts, but available for emergency service on a 24-hour basis.
- Mechanics for all mechanical maintenance.
- Tunnel guards for three shifts.
- Emergency vehicle operators for three shifts. These operators may also do some vehicle (or other) maintenance around their stations while standing by for calls.
- Maintenance men for tunnel cleaning and ground maintenance.

Index

CHAPTER 4 Tunnel Surveys and Alignment Control

CHAPTER 5 Soft Ground Tunneling

CHAPTER 6 Shield Tunnels

CHAPTER 7 Rock Tunnels

CHAPTER 8 Mixed Face Tunneling

CHAPTER 9 Shafts

CHAPTER 10 Tunnel-Boring Machines

CHAPTER 11 Materials Handling and Construction Plant

CHAPTER 12 Shotcrete

CHAPTER 13 Sunken Tube Tunnels

CHAPTER 14 Cut-and-Cover Construction

CHAPTER 15 Subway Construction

CHAPTER 16 Safety Provisions

CHAPTER 17 Tunnel Finish

CHAPTER 18 Service Buildings

CHAPTER 19 Tunnel Ventilation

CHAPTER 20 Tunnel Lighting

CHAPTER 21 Power Supply, Distribution and Control

CHAPTER 22 Signal and Communication Systems

CHAPTER 24 Drainage System

CHAPTER 25 Tunnel Operation and Maintenance